Handbook of Elemental Speciation: Techniques and Methodology

Editor-in-chief

Rita Cornelis

Ghent University, Belgium

Associate Editors

Joe Caruso

University of Cincinnati, USA

Helen Crews

Central Science Laboratory, UK

Klaus Heumann

Johannes Gutenberg-University Mainz, Germany

WILEY

Telephone (+44) 1243 779777

Email (for orders and customer service enquiries): cs-books@wiley.co.uk
Visit our Home Page on www.wileyeurope.com or www.wiley.com

Other Wiley Editorial Offices

John Wiley & Sons Inc., 111 River Street, Hoboken, NJ 07030, USA

Jossey-Bass, 989 Market Street, San Francisco, CA 94103-1741, USA

Wiley-VCH Verlag GmbH, Boschstr. 12, D-69469 Weinheim, Germany

John Wiley & Sons Australia Ltd, 33 Park Road, Milton, Queensland 4064, Australia

John Wiley & Sons (Asia) Pte Ltd, 2 Clementi Loop #02-01, Jin Xing Distripark, Singapore 129809

John Wiley & Sons Canada Ltd, 22 Worcester Road, Etobicoke, Ontario, Canada M9W 1L1

Wiley also publishes its books in a variety of electronic formats. Some content that appears in print may not be available in electronic books.

Library of Congress Cataloging-in-Publication Data

Handbook of elemental speciation : techniques and methodology / edited by Rita Cornelis ... [et al.].
p. cm.
Includes bibliographical references and index.
ISBN 0-471-49214-0
1. Speciation (Chemistry) – Technique. 2. Speciation (Chemistry) – Methodology. 3. Environmental chemistry – Technique. 4. Environmental chemistry – Methodology. I. Cornelis, Rita.

QD75.3 .H36 2003
544–dc21

2002193376

British Library Cataloguing in Publication Data

A catalogue record for this book is available from the British Library

ISBN 0-471-49214-0

Typeset in 10/12pt Times by Laserwords Private Limited, Chennai, India
Printed and bound in Great Britain by Antony Rowe Ltd, Chippenham, Wiltshire
This book is printed on acid-free paper responsibly manufactured from sustainable forestry in which at least two trees are planted for each one used for paper production.

"'ICES

Handbook of Elemental Speciation: Techniques and Methodology

Contents

List of Contributors

F. R. Abou-Shakra
Micromass UK Ltd, Wythenshawe, UK

K. L. Ackley
University of Cincinnati, Cincinnati, USA

J. I. G. Alonso
University of Oviedo, Oviedo, Spain

T. Berg
Danish Veterinary and Food Administration, Søborg, Denmark

J. Bettmer
Johannes Gutenberg-University, Mainz, Germany

I. Bontidean
Lund University, Lund, Sweden

B. Bouyssiere
Université de Pau et des Pays de l'Adour, Pau, France

P. Brereton
Central Science Laboratory, York, UK

J. Buffle
CABE, Sciences II, Geneva, Switzerland

C. Cámara
Universidad Complutense de Madrid, Madrid, Spain

J. A. Caruso
University of Cincinnati, Cincinnati, USA

H. Chassaigne
EC, Joint Research Center, Institute for Reference Materials and Measurements, Geel, Belgium

C. C. Chéry
Ghent University, Ghent, Belgium

R. Cornelis
Ghent University, Ghent, Belgium

H. Crews
Central Science Laboratory, York, UK

E. Csöregi
Lund University, Lund, Sweden

K. de Cremer
Ghent University, Ghent, Belgium

E. Dabek-Zlotorzynska
Environment Canada, Ottawa, Canada

J. H. Duffus
The Edinburgh Centre for Toxicology and the University of Edinburgh, Edinburgh, UK

A. N. Eaton
Micromass UK Ltd, Wythenshawe, UK

H. Emons
EC, Joint Research Center, Institute of Reference Materials and Measurements, Geel, Belgium

J. R. Encinar
University of Oviedo, Oviedo, Spain

M. E. Foulkes
University of Plymouth, Plymouth, UK

K. G. Heumann
Johannes Gutenberg-University, Mainz, Germany

G. M. Hieftje
Indiana University, Bloomington, USA

J. Hlavay
University of Veszprém, Veszprém, Hungary

R. S. Houk
Iowa State University, Ames, USA

K. Keppel-Jones
Environment Canada, Ottawa, Canada

G. Köllensperger
Universität für Bodenkultur Wien, Vienna, Austria

A. M. Leach
Indiana University, Bloomington, USA

R. Lobinski
Université de Pau et des Pays de l'Adour, Pau, France

D. M. McClenathan
Indiana University, Bloomington, USA

R. K. Marcus
Clemson University, Clemson, USA

R. Macarthur
Central Science Laboratory, York, UK

B. Michalke
GSF National Research Center for Environment and Health, Neuherberg, Germany

R. M. Olivas
Universidad Complutense de Madrid, Madrid, Spain

H. M. Ortner
Technische Universität Darmstadt, Darmstadt, Germany

K. Polyák
University of Veszprém, Veszprém, Hungary

M. Potin-Gautier
Université de Pau et des Pays de l'Adour, Pau, France

P. Quevauviller
European Commission, DG Research, Brussels, Belgium

W. Schuhmann
Ruhr University Bochum, Bochum, Germany

J. Szpunar
Université de Pau et des Pays de l'Adour, Pau, France

R. M. Town
The Queen's University of Belfast, Belfast, Northern Ireland

J. F. Tyson
University of Massachusetts, Amherst, USA

F. Vanhaecke
Ghent University, Ghent, Belgium

E. Welter
HASYLAB at DESY, Hamburg, Germany

C. Zhang
Tsinghua University, Beijing, China

X. Zhang
Tsinghua University, Beijing, China

Preface

The recognition of the fact that, in environmental chemistry, occupational health, nutrition and medicine, the chemical, biological and toxicological properties of an element are critically dependent on the form in which the element occurs in the sample has spurred a rapid development of an area of analytical chemistry referred to as speciation analysis. In contrast to its biological meaning, the term speciation in chemistry refers to the distribution of an element among defined chemical species, i.e. among specific forms of an element defined as to isotopic composition, electronic or oxidation state, and/or complex or molecular structure.

The areas of speciation analysis have been undergoing a continual evolution and development for the last 20 years. The area most frequently referred to is speciation of anthropogenic organometallic compounds and the products of their environmental degradation, such as methylmercury, alkyllead, butyl- and phenyltin compounds, and simple organoarsenic and organoselenium species. The presence of a metal(loid)–carbon covalent bond ensures a reasonable stability of the analyte(s) during sample preparation. The volatility of the species allows the use of gas chromatography with its inherent advantages, such as the high separation efficiency and the absence of the condensed mobile phase, that enable a sensitive (down to the femtogram levels) element-specific detection by atomic spectroscopy. Much effort has been devoted by the European Commission Measurement and Testing Program to raise the standards of accuracy of speciation measurements in terms of appropriate calibration and method validation using certified reference materials.

An insight into endogenous metal species in biological systems has remained for a long time a challenge to the analyst. Indeed, millions of years of evolution have resulted in a great variety of biological ligands with different functions and a significant coordinating potential for trace elements. They include small organic acids, macrocyclic chelating molecules, and macromolecules, such as proteins, DNA restriction fragments or polysaccharides. The complexity and the usually poor understanding of the system (the majority of trace element species with biological ligands have not yet been discovered!) have been the major obstacles on the way to the identification and characterization of the endogenous metal complexes with biomolecules. Their generally poor volatility in comparison with organometallic species calls for separation techniques with a condensed mobile phase that negatively affects the separation efficiency and the detection limits.

A fundamental tool for speciation analysis has been the combination of a chromatographic separation technique, which ensures that the analytical compound leaves the column unaccompanied by other species of the analyte element, with atomic spectrometry, permitting a sensitive and specific detection of the target element. Recent impressive progress toward lower detection limits in ICP MS, toward higher resolution in separation techniques, especially capillary electrophoresis and electrochromatography, and toward higher sensitivity in electrospray mass spectrometry for molecule-specific detection at trace levels in complex matrices allows new frontiers to be crossed. Analytical techniques allowing direct speciation in solid samples are appearing.

Speciation analysis is a rather complex task and a reference handbook on relevant techniques and methodology has been awaited by all those with an interest in the role and measurement of element species. These expectations are now fulfilled by the *Handbook of Elemental Speciation* of which the first (of the announced two) volume is now appearing. This first volume brings a collection of chapters covering comprehensively different aspects of procedures for speciation analysis at the different levels starting from sample collection and storage, through sample preparation approaches to render the species chromatographable, principles of separation techniques used in speciation analysis, to the element-specific detection. This already very broad coverage of analytical techniques is completed by electrochemical methods, biosensors for metal ions, radioisotope techniques and direct solid speciation techniques. Special concern is given to quality assurance and risk assessment, and speciation-relevant legislation.

Although each chapter is a stand-alone reference, covering a given facet of elemental speciation analysis written by an expert in a given field, the editorial process has ensured the volume is an excellent introductory text and reference handbook for analytical chemists in academia, government laboratories and industry, regulatory managers, biochemists, toxicologists, clinicians, environmental scientists, and students of these disciplines.

Ryszard Lobinski

Pau, France, October 2002

Technical Abbreviations and Acronyms

Abbreviations

AAS	atomic absorption spectrometry
AC	alternating current
AED	atomic emission detection
AES	atomic emission spectrometry
AF	atomic fluorescence
ANOVA	analysis of variance
CE	capillary electrophoresis
CGC	capillary gas chromatography
cpm	counts per minute
CRM	certified reference material
CVAAS	cold vapour atomic absorption spectrometry
CZE	capillary zone electrophoresis
DC	direct current
ES	electrospray
ESI	electrospray ionization
ETAAS	electrothermal atomic absorption spectrometry
ETV	electrothermal vaporization
EXAFS	extended X-ray absorption fine structure spectroscopy
FAAS	flame atomic absorption spectrometry
FID	flame ionization detector
FIR	far-infrared
FPD	flame photometric detector
FT	Fourier transform
FPLC	fast protein liquid chromatography
GC	gas chromatography
GD	glow discharge
GLC	gas–liquid chromatography
GSGD	gas-sampling glow discharge
HGAAS	hydride generation atomic absorption spectrometry
HPLC	high performance liquid chromatography
Hz	Hertz
ICP	inductively coupled plasma
i.d.	internal diameter
IEF	isoelectric focusing
IDMS	isotope dilution mass spectrometry
INAA	instrumental neutron activation analysis
IR	infrared
ISFET	ion-selective field effect transistor
ISE	ion-selective electrode
ITP	isotachophoresis
LA	laser ablation
LC	liquid chromatography
LED	light emitting diode
LOD	limit of detection
LOQ	limit of quantification
MAE	microwave-assisted extractions
MIP	microwave-induced plasma
MS	mass spectrometry
NIR	near-infrared
NMR	nuclear magnetic resonance
o.d.	outer diameter
OES	optical emission spectrometry
PBS	phosphate buffer saline
PIXE	particle/proton-induced X-ray emission
QA	quality assurance
QC	quality control
QF	quartz furnace
REE	rare earth element
RPC	reversed phase chromatography
RSD	relative standard deviation

SE	standard error
SEM	scanning electron microscope
SFC	supercritical fluid chromatography
SFE	supercritical fluid extraction
SFMS	sector-field mass spectrometer
SIMS	secondary ion mass spectrometry
SPME	solid-phase micro-extraction
TD	thermal desorption
TEM	transmission electron microscope
TIMS	thermal ionization mass spectrometry
TOF	time of flight
UV	ultraviolet
UV/VIS	ultraviolet–visible
XAFS	X-ray absorption fine structure
XRD	X-ray diffraction
XRF	X-ray fluorescence

Units

μg	micrograms
ng	nanograms
pg	picograms
fg	femtograms
mL	millilitres
L	litres
cL	centilitres

Symbols

M	molecular mass
M_r	relative molecular mass
r	correlation coefficient
s	standard deviation of sample
σ	population standard deviation

CHAPTER 1
Introduction

R. C. Cornelis
Laboratory for Analytical Chemistry, Ghent University, Belgium

H. M. Crews
Central Science Laboratory, Sand Hutton, York, UK

J. A. Caruso
University of Cincinnati, Ohio, USA

K. G. Heumann
Institut für Anorganische and Analytische Chemie, Mainz, Germany

'Speciation', a word borrowed from the biological sciences, has become a concept in analytical chemistry, expressing the idea that the specific chemical forms of an element should be considered individually. The underlying reason for this is that the characteristics of just one species of an element may have such a radical impact on living systems (even at extremely low levels) that the total element concentration becomes of little value in determining the impact of the trace element. Dramatic examples are the species of tin and mercury, to name just these two. The inorganic forms of these elements are much less toxic or even do not show toxic properties but the alkylated forms are highly toxic. No wonder analytical chemists had to study elemental speciation and devise analytical techniques that produce qualitative and quantitative information on chemical compounds that affect the quality of life.

Before embarking on the definitions of elemental speciation and species, it may be interesting to give a short historical setting of this emerging branch of analytical chemistry. Analytical chemistry began as a science in the early 19th century. A major milestone was the book by Wilhelm Ostwald 'Die Wissenschaftlichen Grundlagen der analytischen Chemie' (Scientific Fundamentals of Analytical Chemistry) in 1894 [1]. A personality who contributed substantially to the development of analytical chemistry and chemical

Handbook of Elemental Speciation: Techniques and Methodology R. Cornelis, H. Crews, J. Caruso and K. Heumann
 ISBN: 0-471-49214-0

analysis was Carl Remigus Fresenius. In 1841 he published a very interesting book on qualitative chemical analysis [2]. It was followed over the next 100 years by a series of standard works on qualitative and quantitative analysis by several generations of the Fresenius family and by the publications by Treadwell [3, 4], Feigl [5] and Kolthoff [6, 7], to name just these few. The interest remained largely focused on inorganic analytical chemistry. The term 'trace elements' dates back to the early 20th century, in recognition of the fact that many elements occurred at such low concentrations that their presence could only just be detected. During the following 60 years all efforts were focused on total trace element concentrations. Scientists developed methods with increasing sensitivity. It was only in the early 1960s that questions were raised concerning the chemical form of the trace elements and that the need for an analytical methodology developed subsequently. This development has been growing exponentially to the point that research on trace element analysis today appears almost exclusively focused on trace element species.

Extensive literature is available on the speciation and fractionation of elements. Newcomers have to absorb a wealth of highly specialised publications and they miss the broader overview to guide them. This handbook aims to provide all the necessary background and analytical information for the study of the speciation of elements.

The objective of this handbook is to present a concise, critical, comprehensive and systematic (but not exhaustive), treatment of all aspects of analytical elemental speciation analysis. The general level of the handbook makes it most useful to the newcomers in the field, while it may be profitably read by the analytical chemist already experienced in speciation analysis.

1 DEFINITION OF ELEMENTAL SPECIATION AND OF FRACTIONATION

The International Union for Pure and Applied Chemistry (IUPAC) has defined elemental speciation in chemistry as follows:

(i) *Chemical species.* Chemical element: specific form of an element defined as to isotopic composition, electronic or oxidation state, and/or complex or molecular structure.
(ii) *Speciation analysis.* Analytical chemistry: analytical activities of identifying and/or measuring the quantities of one or more individual chemical species in a sample.
(iii) *Speciation of an element; speciation.* Distribution of an element amongst defined chemical species in a system.

When elemental speciation is not feasible, the term fractionation is in use, being defined as follows:

(iv) *Fractionation.* Process of classification of an analyte or a group of analytes from a certain sample according to physical (e.g., size, solubility) or chemical (e.g., bonding, reactivity) properties.

As explained in the IUPAC paper [8], it is often not possible to determine the concentrations of the different chemical species that sum up to the total concentration of an element in a given matrix. Often, chemical species present in a given sample are not stable enough to be determined as such. During the procedure, the partitioning of the element among its species may be changed. For example, this can be caused by a change in pH necessitated by the analytical procedure, or by intrinsic properties of measurement methods that affect the equilibrium between species. Also in many cases the large number of individual species (e.g., in metal–humic acid complexes or metal complexes in biological fluids) will make it impossible to determine the exact speciation. The practice is then to identify various classes of the elemental species.

2 PROBLEMS TO BE SOLVED

While the incentive to embark on speciation and fractionation of elements is expanding, it becomes more and more evident that the matter has to be handled with great circumspection. Major questions include: What are the species we want to measure? How should we sample the material and isolate the species without changing its composition? Can we detect very low amounts of the isolated species, which may represent only a minute fraction of the total, already ultra-trace element concentration? How do we calibrate the species, many of these not being available as commercial compounds? How do we validate methods of elemental analysis? All of these questions will be carefully dealt with in the first volume of the handbook. The second and third volumes will extensively address elemental species of specific elements and the analysis of various classes of species.

Advances in instrumentation have been crucial to the development of elemental speciation. There has been a very good trend towards lower and lower detection limits in optical atomic spectrometry and mass spectrometry. This has allowed the barrier between total element and element species to be crossed. While the limit of detection for many polluting species is sufficient for their measurement in a major share of environmental samples, this is not yet the case for human, animal and perhaps plant samples at 'background levels'. The background concentration of elemental species of anthropogenic origin was originally zero. Today they are present, because they have been and continue to be distributed in a manner that affects the life cycle. However, because we cannot measure them in living systems it does not mean that their presence is harmless. At the same time it is also highly plausible that a certain background level of these anthropogenic substances can be tolerated without any adverse effect. In order to assess the impact of low background levels of element species we will have to develop separation and detection techniques that surpass the performance of the existing speciation methodology.

In the mean time research teams are very resourceful in developing computer controlled automated systems for elemental speciation analysis. These are or will become tremendous assets for routine analyses. Preferably systems should be simple, robust, low cost and if at all possible portable, to allow for fieldwork. Although currently limited it can be postulated that once there are more regulatory or economic motives, this technology would develop rapidly. Moreover, sensors will play a major role in rapid detection of elemental species. Simple screening methods will be increasingly popular because they can provide speedy and reliable tests to detect elemental species and give an estimate of their concentration.

Good laboratory practice and method validation are a must to produce precise and accurate results. To this end, it is evident that provision has to be made for elemental species data in more certified reference materials (CRMs) reporting on elemental species. This need will become even more acute once legislation becomes specific and cites elemental species instead of the total concentration of the element and its compounds, as is presently the general rule. This type of legislation may be politically charged, because every species carries a different health risk or benefit. The toxicity may vary by several orders of magnitude among species of the same element. This may lead to some confusing and dangerous conclusions. A product may be legally acceptable on the basis of total concentration but when that total consists of some very toxic species it may constitute a real hazard. The opposite may also be true. This can be exemplified by the occurrence of arsenic in food. Whereas the total arsenic in some fish derivatives, such as gelatine, often exceeds the accepted limit, the product should not be rejected, because the arsenic is mainly present as arsenobetaine, a non-toxic arsenic species, as opposed to the toxic inorganic arsenic species.

3 SPECIATION STRATEGIES

'Strategy' signifies 'a careful plan or method' or 'the art of devising or employing plans or stratagems toward a goal' [9]. Ideally scientists hope to learn everything about the elemental species they study: to start with its composition, its mass, the bio- and environmental cycle, the stability of the species, its transformation, and the interactions with inert or living matter. This list is not exhaustive. The work involved to achieve this goal is, however, challenging, if not impossible to complete. Therefore a choice has to be made to identify the most important issues as elemental speciation studies are pursued. A first group of compounds to be studied very closely are those of anthropogenic origin. Although they fulfil the requirements for which they were synthesised, unless they happen to be synthetic by-products or waste, their long-term effect on the environment, including living systems, has often been ignored or misjudged. One of the most striking examples is the group of organotin compounds. They have surely proven to be the most effective marine anti-fouling agents, fungicides, insecticides, bacteriostats, PVC stabilising agents, etc. However, the designers of these compounds never anticipated what the negative effect would be on the environment. Their disturbing impact on the life cycle of crustaceans constituted the first alarming event. In the mean time these components have become detectable, with concentrations now increasing in fish products and even in vegetables from certain areas. Little thought was given that organotin compounds would be serious endocrine disruptors, or that they would have such a long half-life. Hence they will continue to be a burden on the environment and ultimately on mankind itself into the distant future. The organotin compounds are only a minute part of the total amount of tin (mainly inert tin oxides) to be found in contaminated areas. Determination of the total tin concentration would surely not be appropriate.

In speciation studies, a lot of attention must be paid to the stability. Species stability depends on the matrix and on physical parameters, such as temperature, humidity, UV light, organic matter, etc. Next comes the isolation and purification of the species, the study of the possible transformation through the procedure, their characteristics and interactions. New analytical procedures have to be devised, including appropriate quantification and calibration methodologies.

Besides the suspect elemental species of anthropogenic origin, there is the barely fathomable domain of the species that developed along with life on earth. For many elements nothing is really known, or only a few uncertain facts can be stated. Whereas the total trace element concentration may be static, the species may be highly dynamic. They will change continuously with respect to changes in the surrounding environment, depending on chemical parameters such as pH value or concentration of potential ligands for complex formation, the physiological state of a cell, and state of health of a living entity. Therefore, thermodynamic but also kinetic stability of elemental species in the environment has to be taken into account. Unstable species in the atmosphere are predominant and this steady transformation requires special analytical procedures. Although species in living cells can be stable covalent compounds when the element forms the core of the molecule (such as Co in vitamin B_{12}), most elemental species exhibit very low stability constants with their ligands. These compounds are, however, very active in reaching the target organs. This to say that a reliable speciation strategy will include stability criteria for the species and awareness of possible transformations.

Understanding the fate of the trace elements in the life cycle is of paramount importance. When, through natural or anthropogenic activities, metal ions enter the environment and the living systems, only a small fraction will remain as the free ion. The major share will be complexed with either inorganic or organic ligands. Natural methylation of metal ions under specific conditions is prevalent. The new species can be much more toxic, as is the case with methylated mercury, or less toxic as in the case of arsenic. In the case of mercury, the concentration of ionic mercury in water may be very low (a few $ng\,L^{-1}$) and that of methylmercury only 1 % of total Hg. Unfortunately, this accumulates to $mg\,kg^{-1}$ levels in the top predators of the food chain, with methylmercury making up 90 to 100 % of the total Hg concentration. Metal ions will also be incorporated into large molecular structures such as humic substances. Elucidation of the many, often-labile species will take many more years. Trace element speciation has become important in all fields of life, and concerns industry, academia, government and legislative bodies [10].

It is obvious that it would have been impossible to accomplish the stated aims and objectives of this handbook without the wholehearted cooperation of the distinguished authors who contributed the various chapters. To them we express our sincere appreciation and gratitude.

4 REFERENCES

1. Ostwald, W., *Die Wissenschaftlichen Grundlagen der analytischen Chemie (Scientific Fundamentals of Analytical Chemistry)*, Engelmann, Leipzig, 1894 [citation in Baiulescu, G. E., *Anal Lett.*, **33**, 571 (2000)], 5th revised edition of the book published by Steinkopff, Dresden, 1910.
2. Fresenius, C. R., *Anleitung zur qualitativen chemischen Analyse, 1841, followed by Anleitung zur quantitativen chemischen Analyse oder die Lehre von der Gewichtsbestimmung und Scheidung der in der Pharmacie, den Künsten, Gewerben und der Landwirthschaft häufiger vorkommenden Körper in einfachen und zusammeng. Verbindungen (Introduction to the quantitative chemical analysis or the instruction of the determination of weight and the separation in pharmacy, arts, paints and frequently occurring agricultural compounds in their elementary and correlated compounds)*, 4th edn. Vieweg, Braunschweig, 1859.
3. Treadwell, F. P., *Kurzes Lehrbuch der analytischen Chemie – Qualitative Analyse,* two volumes, *(Concise Handbook of Analytical Chemistry – Qualitative Analysis)*, Deuticke, Leipzig, 1899.
4. Treadwell, F. P., translated by Hall, W. T., *Analytical Chemistry,* Vol. I, *Qualitative Analysis,* Vol. II, *Quantitative Analysis*, John Wiley & Sons, Inc., New York, Chapman & Hall, London, 1932 and 1935 (original German text 1903 and 1904).
5. Feigl, F., translated by Oesper, R. E., *Chemistry of Specific, Selective and Sensitive Reactions*, Academic Press, New York, 1949.
6. Kolthoff, I. M., Elving, P. J. and Sandell, E. B., *Part I. Treatise on Analytical Chemistry, Theory and Practise*, Vols 1–13, Interscience Publications, John Wiley & Sons, Inc., New York, 1959–1976.
7. Kolthoff, I. M. and Elving, P. J., *Part II – Analytical chemistry of the elements, Analytical chemistry of inorganic and organic compounds*, Vols 1–17, Interscience Publications, John Wiley & Sons, Inc., New York, 1961–1980.
8. Templeton, D. M., Ariese, F., Cornelis, R., Danielsson, L. -G., Muntau, H., Van Leeuwen, H. P. and Lobinski, R., *Pure Appl. Chem.*, **72**, 1453 (2000).
9. *Webster's Third New International Dictionary*, 1976.
10. Ebdon, L., Pitts, L., Cornelis, R., Crews, H. and Quevauviller, P., *Trace Element Speciation for Environment, Food and Health*, The Royal Society of Chemistry, Cambridge, UK, 2001.

CHAPTER 2

Sampling: Collection, Storage

2.1 Sampling: Collection, Processing and Storage of Environmental Samples

Hendrik Emons

EC Joint Research Center IRMM, Geel, Belgium

1 INTRODUCTION

Public awareness and scientific understanding of the various compartments, processes and problems in the environment have been significantly improved particularly in the last three decades. To a large extent this can be attributed to the progress of environmental analysis. Analytical chemistry itself has gained from the needs of environmental sciences, technology and legislation. It is now widely accepted that human activities which influence the chemical composition of the environment have to be systematically controlled. Therefore, procedures for the analysis of an increasing number of elements and chemical compounds in air, water, sediment and soil have been developed and this is still going on. But modern environmental observation has to provide more effect-related information about the state of our environment and its changes with time. Therefore, it cannot be based only on investigations of abiotic environmental samples. Rather environmental studies and control need to take much more account of the situation in the biosphere. This includes the transfer of contaminants, mainly of anthropogenic origin, into plants, animals, and finally also into human beings. As a result, biomonitoring plays an increasing role in modern environmental observation programs. For that, selected biological organisms, called bioindicators, are used for the monitoring of pollutants either by observation of phenomenological effects (loss of needles, discoloring of leaves, etc.) or by measurement of chemical compounds taken up by the specimens. The latter approach is based on the chemical analysis of appropriate bioindicators and adds another dimension of complexity

Handbook of Elemental Speciation: Techniques and Methodology R. Cornelis, H. Crews, J. Caruso and K. Heumann
 ISBN: 0-471-49214-0

to the sample matrices encountered in environmental analysis.

In general environmental analysis should contribute answers to the following questions:

- Which pollutants appear where, when and at which concentration?
- How mobile and stable are pollutants in ecosystems?
- Which transformations occur with originally anthropogenic emissions in ecosystems?
- Where are pollutants accumulated or finally deposited and in which chemical form?
- Which short- and long-term effects do they have with respect to mankind and the environment?

All these questions are expanding both the chemical nature and the concentration ranges of analytes to be included when considering environmental studies and control. For evaluating the necessity, toxicity, availability, distribution, transformation and fate of chemicals, the identity and concentration of chemical species [1] rather than those of total elements have to be studied in very complex systems and samples. Speciation analysis adds new requirements to the usual boundary conditions in environmental analysis, in particular with respect to analyte stability and preservation of original chemical equilibria. Such aspects are also partially considered in the so-called 'organic analysis' of compounds outside the POP (persistent organic pollutants) group, but the thermodynamic and kinetic properties of different redox states or chemical complexes are adding new dimensions of lability and reactivity to analytical chemistry.

Sampling is always the first step of the total analytical process (Figure 2.1.1) and its design and implementation has a decisive influence on the final analytical result. For the purpose of this chapter the term 'sampling' will include the collection of specimens from the environment (accompanied in many cases by non-chemical operations for on-site sample preparation), followed by intermediate storage and often by a mechanical processing of the collected material up to samples which are appropriate for the subsequent steps of chemical analysis, namely analyte separation and determination. In the following text, general aspects and specific requirements for sampling, sample handling and sample storage, in the speciation analysis of environmental samples will be discussed.

2 GENERAL ASPECTS OF ENVIRONMENTAL SAMPLING

The second step after the definition of the analytical problem of interest consists of the careful selection and problem-specific design of the sampling procedure. For that one has to consider not only all relevant properties of the analyte of interest, the matrix and the chosen analytical techniques, but also a number of parameters that are necessary for the final evaluation and assessment of the analytical data. Unfortunately, a large number of environmental analyses are still wasted because of insufficient sampling and sample handling strategies.

Obtaining representative samples is of utmost importance. This includes accessing representative sampling sites for the purpose of the study, which

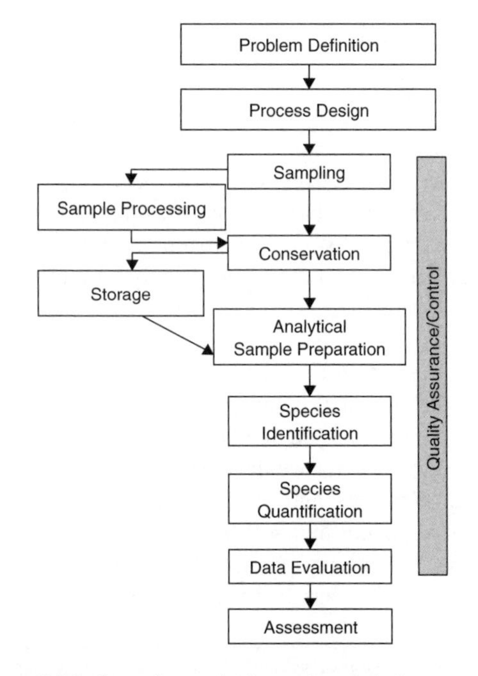

Figure 2.1.1. General steps in the total analytical process.

is often easier to achieve for the control of local emission sources than for the observation of larger 'normally' exposed areas or even background regions. One needs geographical, meteorological and biological data, and information about human activities for the site selection [2]. Sampling within a selected area is commonly planned by taking into account a combination of pre-existing environmental knowledge, statistical approaches (grid sampling) [3] and financial considerations. The heterogeneity of our environment on all scales, from ecosystems via populations and individual specimens down to the molecular level creates challenging demands on sampling concepts. Frequently, one or more screening studies have to precede the actual environmental sampling campaign in order to select the final sampling points.

Secondly, representativeness refers also to the kind of material to be sampled. The selection of representative environmental specimens from the various environmental compartments, ecosystems, etc. depends mainly on the question (groundwater quality, fate of industrial emissions, forest health, etc.), but also on the available knowledge about key indicators of the environmental situation. The frequently described 'flow circles' of elements between the atmosphere, hydrosphere, biosphere, pedosphere and lithosphere are certainly not sufficient for the proper design of sampling procedures for chemical speciation. The physicochemical properties of the target compounds have to be considered in more detail to estimate the transfer, transformation, deposition and accumulation of chemical species in environmental compartments and specimens. In this respect it may be useful to classify 'chemical species' from the point of view of the nature of their primary interactions with the surroundings, namely hydrophilic species with Coulomb forces or hydrogen bonding, and lipophilic species with hydrophobic interactions. In addition, characteristic pathways of species uptake, transport, accumulation and transformation within the biosphere, namely within food chains, should be considered. By taking these factors into account useful sample sets of environmental indicators can be composed which may significantly improve the information content of environmental analysis.

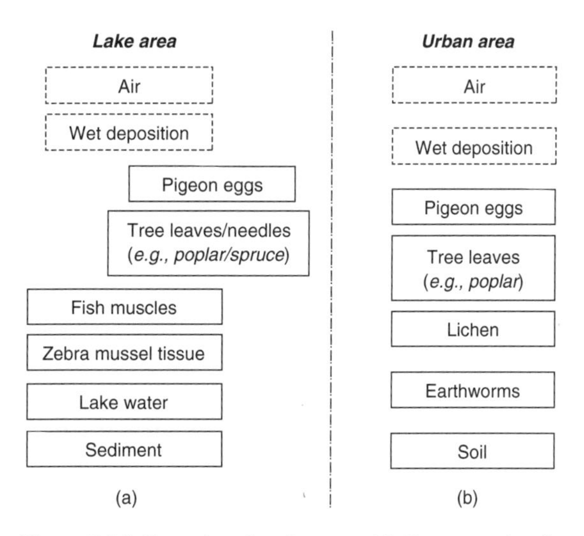

Figure 2.1.2. Examples of environmental indicator sets (modified from [31]): (a) limnic ecosystems: lake area; (b) terrestrial ecosystems: urban area.

Examples for two types of ecosystems are illustrated in Figure 2.1.2.

Another requirement for representativeness consists in the number of specimens which have to be sampled for subsequent analysis. Ideally this parameter should be calculated on the basis of the known variations in the chemical composition of the sampled population which arise from its geochemical/-physical or biochemical/-physical diversity and the variation in imissions in the studied area. An additional uncertainty for biological specimens originates from the natural heterogeneity within organisms and even organs (see also Section 4). All this information is rarely known in detail from screening investigations and therefore, this lack limits the precision (and often even the accuracy) of the environmental information decoded by the chemical analysis of the sampled material.

Representative (and reproducible) environmental sampling has also to consider the variations of climate/weather conditions, seasonal fluctuations of species concentrations in bioindicators and the different exposure time of the samples. An example for the variation of the methylmercury content in mussel tissue is shown in Figure 2.1.3a and can be compared with the corresponding data for total mercury (Figure 2.1.3b). Obviously,

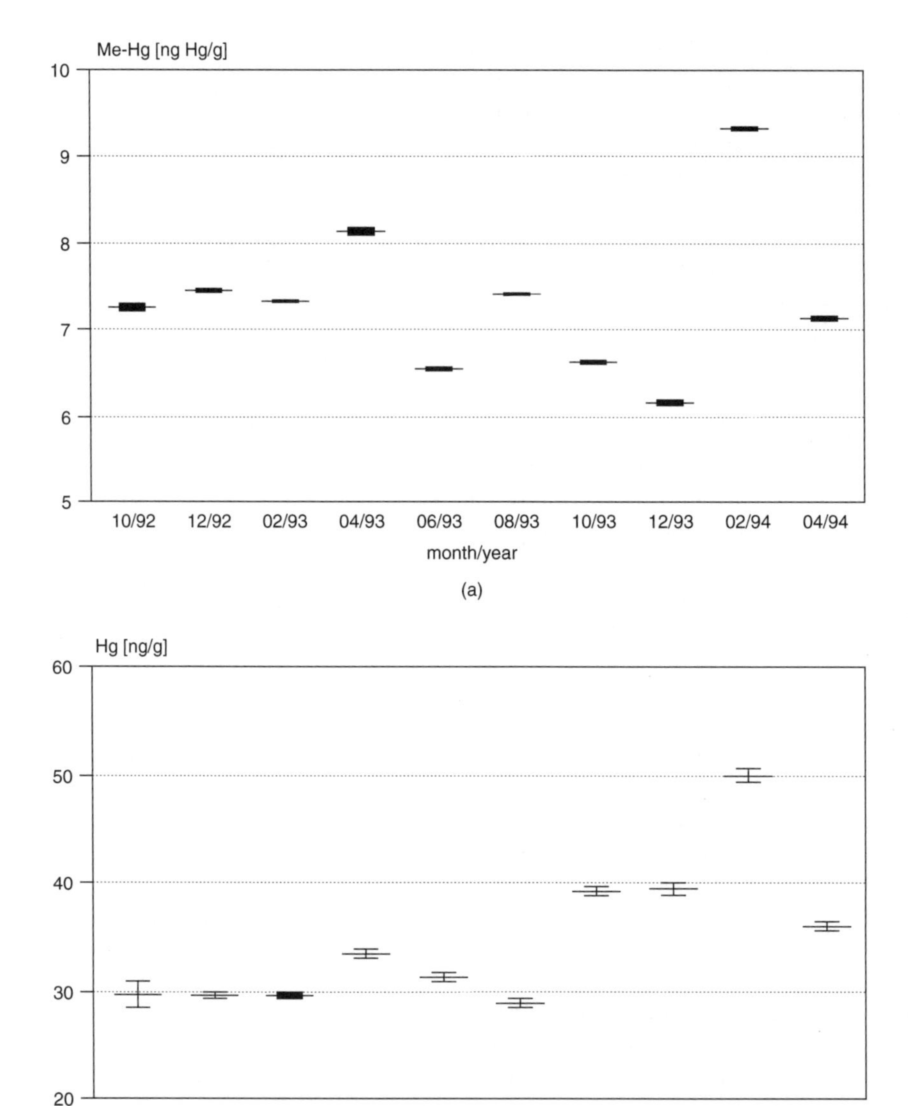

Figure 2.1.3. Seasonal variation of (a) methylmercury and (b) total mercury concentrations related to fresh mass in homogenates of common mussel tissue from the German Wadden Sea between October 1992 and April 1994 (means and standard deviations).

not only it is necessary to document extensively all available parameters (location, time, climate, emission sources, population data such as density, unusual damage etc.) in connection with environmental sampling campaigns, but also standardization of sample selection (and handling/manipulation) is recommended for obtaining useful environmental information from the sampled material. Corresponding standard operating procedures have been developed for various objectives of environmental studies and should provide also the basis for repetitive sampling (see below). But one should always keep in mind that the remaining variation in the representativeness of sampled material makes often a greater contribution to the uncertainty in the final environmental information than any other subsequent analytical step.

In general the sampling procedure should be developed in advance with as much care and detail as possible. Experts with the necessary knowledge in environmental sciences have to be consulted, but the analytical chemist should be also involved from the beginning to avoid or at least reduce any sample manipulations which could influence the final analytical data.

3 SAMPLING FOR SPECIATION ANALYSIS

The key requirement of speciation analysis consists of the preservation of the species information during the whole analytical process (Figure 2.1.1). One could distinguish two principal strategies for achieving this goal: on the one hand, one may keep the chemical species of interest unchanged during all critical steps of their analysis, and on the other hand, the species may be quantitatively transformed at an early stage into suitable derivatives for further separation, accumulation and quantification methods. In practice there is usually a mixture of both extremes, but we will consider primarily the aspects of species-retaining sampling in the following. Therefore, chemical stability and volatility of the analytes of interest are of great importance for the sampling procedures.

In addition, one has to take into account that speciation analysis of environmental samples constitutes, in most cases, trace or even ultratrace analysis. Therefore, both contamination and loss of the analytes have to be avoided along the whole sample pathway from the sampling site to the analytical laboratory. All materials which are in contact with the sample must be checked in advance as a possible source of contamination or adsorption. Sampling devices with stainless steel surfaces which are often used for field sampling (knives, scalpels, drills, etc.) should be avoided in sampling for metal speciation. Careful cleaning of all sampling tools is mandatory. This can be done, for instance, by washing with concentrated or 10 % nitric acid, depending upon the material and the sample matrix, followed by several rinsing steps with distilled water. But the application of washing solutions which activate the adsorption sites at the surface of the tools (such as various concentrated acids for glassware) should be avoided to minimize species losses. The cleaning procedures for the sampling devices as well as the protective measures for the sampling team (wearing of gloves, obeying safety instructions etc.) have to be integrated into the planning and documentation of sampling.

3.1 Air

The collection of samples for 'speciation analysis in the traditional meaning' is mainly focused on the investigation of metal(loid) compounds in the gas phase. Hydrides, methylated and permethylated species of As, Hg, Pb, Se etc. have been sampled from terrestrial sources such as waste deposits or from aquatic ecosystems such as ocean water surfaces. The main techniques include the application of cryotrapping, solid adsorbent cartridges, polymer bags or stainless steel canisters (with coated inner surfaces). The gas phase can be transferred into a pre-evacuated sampling device [4] or sucked with the help of a pump. In most cases the air has to pass through a filter (often 0.45 μm) to remove particles and aerosols up to a predefined size from the sample. The advantage of trapping techniques lies in the integrated accumulation of the analyte and its partial separation from other sample constituents. On the other hand, any sample treatment has to be carefully validated with respect to possible changes of the original species pattern and contamination by used materials. Analyte loss can be caused by photolytic or surface-catalyzed decomposition, hydrolysis, oxidation, adsorption on container walls or absorption. Critical parameters during collection and sample preservation include, in particular, temperature, light intensity, humidity, oxygen content and aerosol concentration. For quantitative analysis one has to adjust for any temperature and pressure variations during the transfer of the original gas phase into the sampling device and during any subsequent operations because of the much larger influence of these parameters on analyte concentrations in the gaseous state in comparison to condensed phases. Recently, the stability of various volatile As, Sb and Sn species was studied as a function of the temperature and a reasonable recovery was obtained for most of the species which were kept at 20 °C in the dark for 24 h [5].

At present the main problems occur in the quantitative sampling of the original air composition under environmental on-site conditions and its preservation. Therefore, the collection of the whole gas at the sampling location into containers, followed by its analysis within a relatively short time (about a day), seems to be the safest way at the moment. The sampling devices can be balloons, cylinders with inlet and outlet valves, canisters or bags. Special consideration should be given to the selected surface. Container walls coated with an inert polymer such as PTFE are recommended. Moreover, contamination by exhaust fumes from the pump has to be avoided. If the species of interest are stable enough, liquid absorbents, adsorbent cartridges or solid-phase microextraction can be applied. All adsorption techniques offer the advantage of sampling larger amounts of air and integrated analyte preconcentration. The same is true for the most widely used sampling approach for volatile species at present, cryotrapping. Volatile metal(loid) species have been sampled from urban air [6], landfill gas [7] or gases from domestic waste deposits [8] with the help of a U-shaped glass trap, filled with a chromatographic packing material (SP-2100 10 % on Supelcoport, 60/80 mesh), and cooled by liquid nitrogen or a mixture of acetone/liquid nitrogen. In most cases an empty cold trap has to be placed in front of the analyte trap for the removal of water vapor at −40 °C. Vacuum filling of stainless steel containers is the official sampling method of the US EPA for volatile compounds in monitoring urban air [9]. The sequential sampling of volatile Hg species using a noble metal trap in series with an activated carbon trap has been reported [10]. Gas from a sewage sludge digester has been collected in inert plastic Tedlar bags allowing sampling volumes of 10 L [5].

An almost unsolved problem is the validation of the sample integrity during all operations. Most of the compounds of interest for speciation studies in the air are not very stable. For instance, some of the volatile metal(loid) species can be transformed by reactions with co-trapped reactive air components such as ozone [11]. Therefore, the development of field-portable instrumentation for sensitive, reliable, fast and efficient on-site speciation analysis will be necessary.

Other targets for the speciation analysis of air are metal compounds in the liquid or solid state, i.e. aerosols and dust particles. The latter samples are collected on various filters (membranes of cellulose or quartz, glass fibers) and further treated as other solid material (see Section 3.4). Aerosols are sampled by using impactors, filters, denuders, electrostatic separators etc. [12]. Their preservation with respect to species integrity is very difficult because of their usually high reactivity and it seems to be more promising to apply *in situ* methods for the direct on-site analysis of aerosol components.

3.2 Water

The sampling technique varies with the properties of the species and the water type of interest (groundwater, freshwater: river or lake, seawater, tap water, wastewater, interstitial water of soil or sediment phases, atmospheric precipitation: rain, snow etc.). It is also influenced by the location (open sea, inner city or somewhere else) and the desired sampling depth below the water surface. Environmental waters are not as chemically homogeneous as water commonly available in the analytical laboratory and the aspects of representativeness (see Section 2) and homogeneity have to be obeyed for sampling.

Volatile analytes can be obtained by an online combination of purging the water with an inert gas such as helium followed by cryogenic trapping as described above. This approach has been used, for instance, for the investigation of methyl and ethyl species of Se, Sn, Hg and Pb in estuarine water [13]. One has to be extremely careful to avoid contamination from the ambient air. Moreover, the formation of biofilms which can act as reaction sites for biomethylation on surfaces of the sampling devices has to be avoided.

Water sampling for speciation analysis should take into account all precautions for trace analysis of such matrices (see for instance [14]). They range from avoiding any contamination from the sampling vessel itself up to the selection of appropriate materials for the sampling and storage devices. Any metal contact has to be avoided, also within the pumping and tubing system. Polycarbonate or polyethylene bottles are recommended for most of the metal species. Mercury species have to be kept in glass bottles [15]. Wet precipitation (rain, snow) is preferably collected with the help of automatic wet-only samplers [16, 17], which can also be used for sampling such specimens for mercury speciation [18].

But several of the routine operations in water analysis cannot be applied to all cases of speciation studies. On the one hand, side filtration (usually with 0.45 μm membrane filters of cellulose or polycarbonate) or centrifugation is necessary for many purposes to remove bacteria and other reactive nondissolved constituents from the water sample. On the other hand, such separation techniques should be checked not only as sources of contamination but also with respect to their influence on original species distributions between the solution phase and the interfaces between particulate matter and water. The latter are prominent adsorption and reaction sites for many species of interest. The frequently recommended acidification of water samples not only stabilizes metal ions in solution by reducing their adsorption on container walls and bacterial activity, but also changes acid–base equilibria and coupled redox and complex formation equilibria, which would prevent the determination of such speciation patterns (see Chapter 5.9). Overall the validation of sampling techniques with respect to species-retaining operations has always to be performed with reference to the target species and general approaches do not exist.

Preservation of original water samples can be a major problem as discussed in Section 5. Therefore, a more promising approach for speciation analysis of dissolved species consists of the application of *in situ* measurements as described in other chapters of this handbook (see for instance Chapter 5.9).

3.3 Biological material

Most of the speciation analysis in biological environmental samples has been directed to the determination of organometallic constituents and redox states of trace elements. For the design of the sample collection one has to take into account that the biosphere varies much more widely in its physical and chemical properties relevant to species distribution and transformation than the abiotic environmental media air, water, sediment and soil. Therefore, only more general aspects will be discussed here. Detailed sampling protocols have to be developed separately for the specific studies depending on the particular requirements of the problem of interest.

Liquid samples from animals or plants (blood, urine, plant juices) have rarely been collected for speciation analysis until now. The main reason seems to be that environmental speciation studies were focused on such specimens which accumulate the compounds of interest and which are easily available in larger amounts for analyzing their ultratrace constituents. In principle, biological fluids from the biosphere could be handled in a comparable manner to the corresponding human samples (see other chapters of this handbook). This commonly includes filtration or centrifugation of the fresh sample material followed by short-term storage at $-4\,^{\circ}C$ in the dark. Alternative techniques are shock-freezing and preservation as a frozen sample or lyophilizing and storage as a dried sample. Naturally, the latter method can only be applied to chemical species which are very stable. Any addition of chemical preservatives, including acidification, should be avoided.

'Solid' biological materials such as tissues have attracted much more attention for speciation analysis in recent years because they are regarded as an important deposit of potentially hazardous compounds. The selection and identification of the biological specimens in the natural environment require appropriate biological and ecological knowledge. Moreover, a scientifically sound interpretation of the analytical data can only be performed if sufficient information about the environmental situation (exposure characteristics, population density and health, etc.) and ecological functions (trophic level, uptake and transport routes for chemical compounds) as well as relevant biological and biometric parameters (age, sex, variations of biological activity with daytime and season, surface area of leaves or needles, mass ratios of individual specimens and sampled organs, etc.) of the studied organism are available. In addition, the sampling strategy has to take into account not only the natural heterogeneity within a biological population, but also that within an individual organism. To fulfill the requirements of representativeness sufficient numbers of individual specimens should be sampled randomly within the selected area and for most studies they have to be mixed and homogenized (see Section 4). It has also been shown that different parts of plants (for instance, algae) [19] accumulate many trace elements to a different extent which has to be considered for the final sample composition. Therefore, detailed sampling procedures and protocols are mandatory. They can be developed for the specific purpose of the investigation on the basis of existing standards for long-term biomonitoring programs such as environmental monitoring and specimen banking [20–22].

During sample collection and further sample manipulations one has to consider that the so-called solid biota are from a chemical point of view very fragile materials with significant water content. Usually the amount of analyte in the chemically complete biological sample (leaf, plant stem, liver, kidney, muscle tissue, egg etc.) is of interest and one has to preserve during sampling the total chemical composition of the specimen as much as possible. Therefore, the removal of the specimens from their natural environment should be carefully planned. Species transformation and loss can already occur during collection at the sampling site. Degradation depends on the chemical nature of the species and may be influenced by biochemical processes such as enzyme activity. This is usually more critical in animal organs than plant samples. A key parameter for chemical reaction rates is the temperature. Consequently, one can diminish species transformations by decreasing the temperature as much and as early as possible. At present, shock-freezing of the desired samples in the gas phase above liquid nitrogen seems to be the safest technique and can be performed immediately at the sampling site. It offers the additional advantage of an inert gas atmosphere for the stored samples. If this approach is not feasible within the specific project a short-term preservation of the biological material at $-20\,^{\circ}C$ is recommended.

The first preparation steps with the samples have to be performed mostly at the sampling site before freezing of the material. Plant samples from natural ecosystems are often modified by adhering material such as dust, soil or sediment particles. A general rule does not exist for separating such abiotic material and one has to decide which of the surface-attached constituents can be considered as an integral part of the sample. Even gentle washing of freshly cut plant organs can partly remove some of the relevant compounds and extensive washing procedures should be avoided in most cases. Marine samples such as algae or mussels can be carefully cleaned of sediment by shaking them in the surrounding water [20, 23].

Another operation of concern is the dissection of target organs which should preferably be performed immediately on the sampling site. Animal organs have to be extracted as fast as possible to minimize species transformations. All dissection tools must be selected with respect to possible contamination of the analytes of interest. Many speciation studies are aimed at metal(loid) compounds and therefore stainless steel knives etc. should be avoided. One can use titanium knives, tools which are coated with titanium nitride or ceramic scalpels. In general all precautions against contamination of the sample should be carefully carried out. Sample containers made from polymeric material such as PTFE, polyethylene or polycarbonate are often adequate. The sampling crew has to wear protective gloves and the contamination is further minimized by working as early as possible on site in mobile laboratories equipped with clean-bench facilities. It has been shown that the dissection of mussels, i.e. the separation of mussel tissue from the shell, can be performed after deep-freezing and cryostorage of the whole mussels in the laboratory [24].

Overall the on-site preparation and preservation of fresh biological material, i.e. samples which still include all liquid components (such as water) belonging originally to the target specimen, is recommended for speciation analysis.

3.4 Sediment and soil

Most of the investigations which have been reported under the heading 'speciation' of sediments, soils or related samples were actually directed to operationally defined fractionation [1] mainly by sequential extraction procedures. This is not within the scope of this chapter on chemical speciation in the sense of analyzing redox states and binding partners for the trace elements of interest.

The most difficult and critical aspect of soil and sediment sampling is representativeness. Environmental specimens are very heterogeneous in their chemical composition and an extensive screening of each sampling site would be necessary for scientifically sound investigations. But usually the number and distances of the lateral and vertical (depth) sampling points are selected on the basis of available geological information by applying statistical models. Corresponding sampling grids and procedures are described in the literature [25–27].

One has to take into account for the adaption of sampling concepts for speciation analysis that larger amounts of such heterogeneous material must be collected in comparison to other environmental specimens for obtaining representative samples. Therefore, appropriately sized sampling tools (shovel, corer, spoon, knife etc.) and containers made from materials such as titanium, ceramics or plastics (e.g. polyethylene) are necessary. Depending on the problem of interest the soil can be collected by initial separation of the mineral layers or as intact depth profiles. The latter approach, which is also very common in the sampling of lake or marine sediments, is achieved by using different corer types (with plastic tubes inside) depending on the soil type, sampling depth and required sample mass. Recommendations for soil sampling can be found, for instance, in [28]. In general, large sample constituents (>2 mm) and larger parts of plants (roots, branches) are manually removed from the collected material.

Sediments can be collected with the help of grab or core samplers. Sediment traps are used in dynamic flow systems such as rivers for sampling at least part of the suspended matter. On-site operations include the decantation of water, the removal of particles larger than 2 mm and sometimes a wet sieving of the sediment. There are different approaches concerning the particle size separation, but 20 or 63 μm are the most commonly used filters. Particulate matter from aquatic ecosystems is collected either by filtration (often 0.45 μm) or by continuous flow centrifugation for processing larger sample volumes.

Special procedures for the species-retaining sampling of soil or sediment have not been described and validated until now. The preparation of harbor and coastal sediments as reference materials for matrix-matched speciation analysis of tributyl tin has been reported [29, 30]. But the application of air drying and the described sample processing (see Section 4) cannot be recommended for all other analytes of interest. Problems also arise from changes in the oxygen concentration during sampling, especially for originally anoxic sediments. Even the definition of the sample composition is difficult for speciation purposes, because the interfaces between solid particle and water, biofilms and pore water can be prominent reaction sites for species transformations and it is almost impossible to preserve their original state during sampling at present. Therefore, the development of both *in situ* methods for speciation analysis (in particular for sediments) and new approaches to species-retaining sampling procedures specifically designed for certain groups of chemical species, will be necessary.

4 SAMPLE PROCESSING

Most of the solid materials collected from the environment have to be prepared by physical operations (grinding, drying, etc.) before any analytical pretreatment in the narrow sense (extraction etc.) can be applied. The influence of such sample manipulations, called sample processing in the following, on the total speciation analysis is sometimes underestimated and should always be carefully controlled by the analyst with respect to contamination, loss or transformation of the analyte and relevant matrix modifications.

Usually the collected material has to be homogenized and divided into chemically authentic aliquots for repetitive analysis. If one is not interested in the 'fractionation' of the sample constituents [1] with respect to their surface attachment to different matrix components or the differentiation between inner- and extracellular species in biological specimens and comparable questions, grinding is performed for the purpose of homogenizing and creating large surface-to-volume ratios for subsequent species extractions.

For biological specimens sample processing with the requirement of minimizing possible changes in their chemical composition can be based on the developments and standards which have been acquired within the frame of environmental biomonitoring and specimen banking programs [22, 31]. The immediate shock-freezing of the sampled material in the gas phase above liquid nitrogen shortly after its removal from the natural environment on the sampling site and the further storage at temperatures below $-130\,^{\circ}C$ provides a raw material with very good mechanical properties for crushing. During grinding the samples should be continuously cooled (preferably with liquid nitrogen) to avoid species transformation or even loss [32] due to the local heating effects from friction. The cryogrinding of a broad range of biological specimens, including fatty or keratin-rich samples (such as liver, hair) and fibrous material (plant shoots etc.), is possible by using a vibrating rod mill (for instance, CryoPalla™) operated under continuous cooling with liquid nitrogen [33]. In many cases a pre-crushing of larger portions of frozen material has to be performed. The grinding at such low temperatures offers not only the advantage of a diminished probability for chemical transformations in the sample, but one can also achieve a fine powder with small particle sizes without sieving. This is shown for two different biological matrices in Figures 2.1.4 and 2.1.5. But

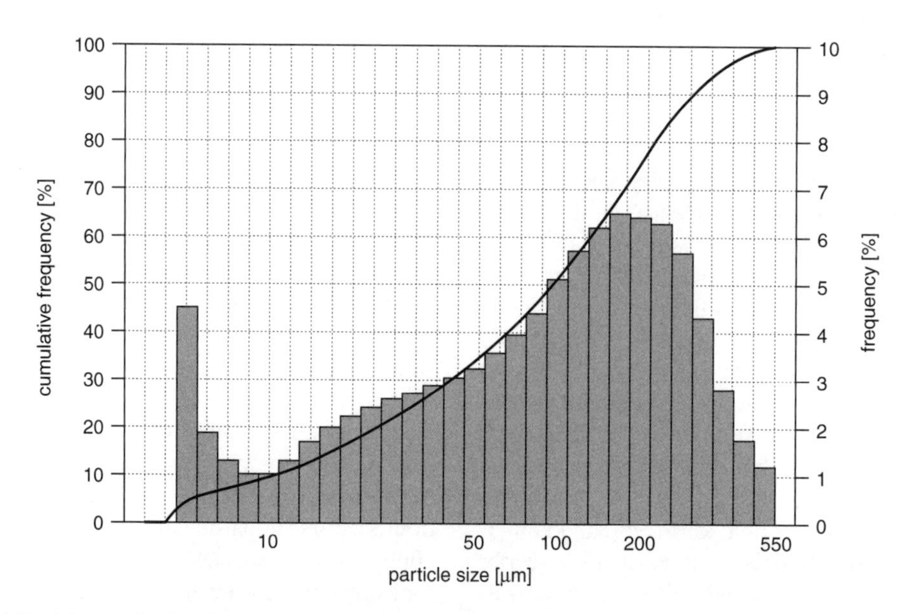

Figure 2.1.4. Particle size distribution after cryogrinding of pine shoots.

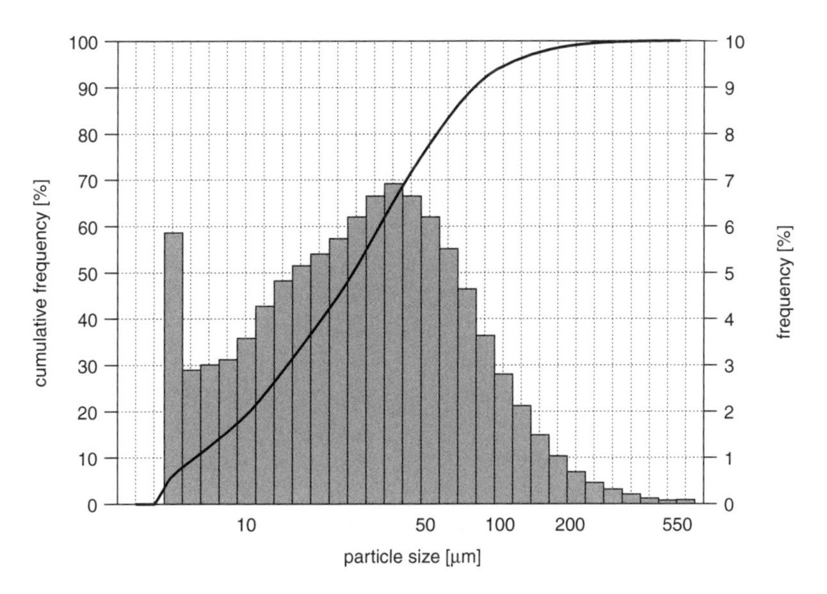

Figure 2.1.5. Particle size distribution after cryogrinding of fish liver (bream).

soil samples can also be finely ground and homogenized by this technique [34].

Such ground material cannot be used for fractionation between groups of species which are differently immobilized at natural occurring surfaces in a sample of interest, but it is well suited for the analysis of the chemical composition of the sample including speciation (see, for instance [35, 36]). Moreover, the finely powdered material is often very homogeneous [37] which simplifies its proper aliquotation into small subsamples. But the natural heterogeneity of analyte distributions in biological specimens can restrict the achievable homogeneity even in such ground samples. This is demonstrated for the case of nickel in fish liver (bream) in Figure 2.1.6. The element (and

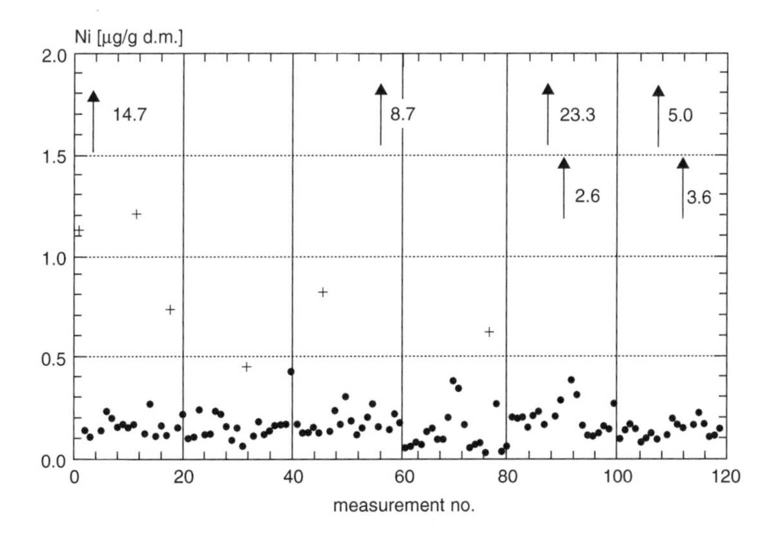

Figure 2.1.6. Nickel concentration related to dry mass in a finely cryoground homogenate of fish liver (bream) determined by solid sampling–atomic absorption spectrometry with sample in-weights of about 680 μg.

consequently also the species) distribution in the powder, which had a particle size histogram almost identical to that shown in Figure 2.1.5, indicated several microparticles with extremely high Ni content. An estimation of the minimum sample mass required for repetitive analyte determinations with an accepted uncertainty of 5 % according to [38] led to a sample in-weight of about 6700 mg which is well above common analytical parameters [37].

Other approaches for cryogenic sample grinding and homogenization include the use of planetary or ball mills [30, 39]. For all techniques the transfer of ultratrace amounts of surface material from the mill into the sample is unavoidable and this contamination has to be considered. The moving parts and the container walls should be made of PTFE or preferably of titanium (which cannot be an analyte afterwards!) in order to minimize the critical contamination.

Many speciation studies have been performed with dried samples and most of the certified reference materials available at present for the quality assurance of species determinations in environmental matrices have been also dried. The BCR procedures of sample processing for CRM 463/464 Tuna Fish (methylmercury content) and CRM 627 Tuna Fish (dimethylarsinic acid and arsenobetaine contents) included mincing of dissected fish muscles, freeze-drying, grinding in a ZrO_2 ball mill, sieving and mixing. Comparable techniques were applied to the preparation of CRM 477 Mussel Tissue (content of butyl tins) [30]. The NIST has prepared a fresh (frozen) SRM 1974a Mussel Tissue (*Mytilus edulis*) certified for its methylmercury content by cryogenically ball milling dissected mussel tissue. This homogenized frozen tissue was blended in an aluminum mixing drum and divided into aliquots. A part of the processed material was prepared as SRM 2974 by subsequent freeze-drying, blending and radiation sterilization (^{60}Co) [40]. Two marine bivalve mollusc tissue materials of NIST (SRM 2977 Mussel Tissue (*Perna perna*); SRM 1566b Oyster Tissue), which have been certified for their methylmercury content, were blended as partially thawed tissues in a cutter mixer, afterwards frozen, freeze-dried, powdered with the help of a jet-mill and radiation sterilized [41]. But one has to be cautious in applying these or other procedures, which have been developed for the preparation of reference materials, to projects directed to the analysis of original species patterns in biological environmental samples. The preservation of original species concentrations of the collected specimens is not necessary for the further use of CRMs and aspects such as cost-efficient processing and long-term storage of large sample batches and the subsequent shipment of stable subsamples all over the world influence significantly the selection of techniques. Therefore, sample manipulations such as sieving are acceptable for CRM production, but should be avoided if one is interested in the speciation for the purpose of environmental monitoring or research. This also holds true with respect to the adaption of processing procedures for soils and sediments. For instance, the BCR candidate reference materials CRM 462 Coastal Sediment (organotin content) was air dried and, after temporary storage and transportation at −20 °C, dried at 55 °C, sieved (1 mm), finely ground using a jet mill with a classifier, sterilized at 120 °C, and homogenized in a mixer [30]. Due to the lack of analytical methods for the direct speciation of organotin compounds in solids one cannot evaluate the species transformations and losses during this sequence of processing steps, but they certainly modify the original sample to a larger extent.

The drying process can lead to sample changes which may significantly influence further speciation analysis. One should keep in mind that drying can remove more than just water molecules from the fresh sample and would consequently alter the general chemical composition of the material. For biological samples prepared for speciation analysis, oven-drying should be avoided. But even gentle freeze-drying can change the species pattern in the original material as was shown in the case of methyl- and butyltin species in fish muscles [42]. One of the most significant limitations of using dried materials for speciation analysis consists of their different extraction properties in comparison to fresh (wet) samples. For example, the extraction yield of arsenic species from fine powders of algae (*Fucus vesiculosus*) and common mussel (*Mytilus edulis*) tissue by using methanol/water decreased from 95.5 and 97 % (wet samples) to 70 and 75.5 %, respectively, by extracting dried aliquots. Obviously the freeze-drying not only removes about 80 % (algae) and 95 % (mussel tissue) of the original sample mass as 'water', but it also reduces significantly the extractability of As species [43].

In general, steps of sample manipulation should be minimized, fresh and nonsieved material should be analyzed if possible and the whole sample processing for speciation analysis has to be evaluated for its influence on changes in the original chemical composition.

5 SAMPLE STORAGE

The preservation of the chemical integrity of the sample has to be the main objective of storage. This concerns first of all the avoidance of loss and contamination of the species of interest, but also of changes in sample mass or matrix properties relevant for speciation analysis. The selection of storage methods for speciation of environmental samples depends on the type of sample material, the storage time, the size and number of samples and the available financial resources. It usually also has to be designed with respect to the target species because generalizations or stability predictions cannot be obtained from studies about the stability of specific species in a matrix under certain conditions due to insufficient knowledge about the chemistry of trace element species in such complex matrices. A major problem for most of the chemical species consists of the lack of analytical methods to determine them directly in a solid sample. Therefore, the different storage approaches for solids cannot be evaluated in a straightforward manner. Commonly long-term studies are performed by using the analytical procedures which are based on the determination of chemical species in the liquid or gaseous state and which are available at the beginning of the stability investigation. Consequently, one has to consider also uncertainties of long-term reproducibility of the analytical sample preparation and of the methods for species separation and quantification in addition to aspects such as homogeneity and contamination control.

Species transformations in closed vessels, i.e. reactions between sample constituents themselves and/or with the inner container surfaces, can be stimulated by temperature, light, microbiological activity or pH changes (in the case of liquids). The key storage parameter which has to be selected carefully is the temperature. In most cases it can be easily adjusted and controlled. The storage of deep-frozen samples seems to be the safest approach for many environmental samples and analytes. But one has to develop and to apply species-retaining rethawing procedures and gentle methods for subsequent rehomogenization of the sample. The container material has to be selected with respect to the storage temperature and possible contamination. Polymer materials such as PTFE or polyethylene are frequently used, but storage at very low temperature such as in the gas phase above liquid nitrogen ($<-150\,^{\circ}C$) has to be carried out in other materials, for instance scintillation glass vials. The container size should be selected in order to minimize the ratio of container surface area to sample volume and therefore species contamination and loss. The samples should be kept in the dark, or at least in appropriately colored or wrapped bottles. One has to take into account that freeze-dried samples are usually hygroscopic and have to be kept in a humidity-controlled environment. Recently, it has been recommended to determine the optimal ranges of water activity and water content of dry biological material, where hygroscopicity is low without danger of product deterioration, and to adjust them for long-term storage [44]. Moreover, it may be necessary to reduce the access of oxygen to the storage medium (always advisable for liquid samples) [45] in order to avoid oxidation reactions. This is simultaneously achieved in the case of liquid nitrogen cooling.

Air samples cannot be stored directly for future speciation analysis. The preservation of original gaseous components in a cold trap is possible for several days in liquid nitrogen [46]. The long-term storage of gaseous species does not seem to be feasible.

Liquid samples are often stored at 4 °C for a limited period. But bacterial activity and exchange reactions or adsorption/desorption at container walls can still be significant. Acidification for diminishing the adsorption of species should be avoided because of the pH dependence of many of the species equilibria in solution. Organometallic species may decompose under the influence of light, microorganisms or suspended particles [47]. Systematic investigations have been performed into the stability of a number of species in aqueous solution (mainly calibrants) within the frame of BCR and SM&T feasibility studies for new reference materials [30, 48]. For instance, it was found that trimethyllead was stable in a BCR candidate reference material 'artificial rainwater' during storage in the dark at 4 °C for 12 months. But about 10 % of this species had decomposed after 3 years [30]. One has to take into account that other solution components which are present in real environmental waters can facilitate species degradation. Therefore, conclusions from stability experiments in pure aqueous standards cannot be generalized for other water types. Species preservation can also be achieved by freezing liquid samples. Using that approach tributyltin was stable in water for 3 months [49]. But one has to check if the liquid–solid–liquid phase transitions change the species pattern of interest in the samples as has been indicated for selenomethionine and Sb (V) [50]. The lack of stabilizing interactions with solid matrix components (such as biological tissues) can induce more easily transformations for many of the dissolved species (ions, metal–ligand complexes) during drastic temperature changes and phase transitions.

Biological solid samples can be preserved with respect to their chemical composition at very low temperatures in the dark for many years [31]. To prevent any microbiological activity the storage temperature should be below −130 °C. This can be easily and efficiently achieved by storing the samples in special cryocontainers in the gas phase above liquid nitrogen. Such methods have been used for more than 20 years in environmental specimen banking. But one has to take into account that freezing and thawing may induce sample changes such as alterations of protein structures or chemical interactions between metal ions and biomacromolecules, complexes, etc. Other approaches developed for the preservation of organometallic species in biological reference materials [30] included the removal of most of the water in the original sample to reduce bacterial activity (but see Section 4). Dry mussel tissue did not show any further transformation of butyltin species during dark storage at −20 °C for 44 months, whereas phenyltins were not stable. The concentrations of methylmercury and two arsenic species (arsenobetaine, dimethylarsinic acid) did not change during storage experiments with dry fish tissues in brown bottles between −20 °C and 40 °C for 12 months and 9 months, respectively [30]. But stability studies of chemical species in biological samples have only been performed until now for a limited number of analytes and matrices. Unfortunately, the relevant thermodynamic and kinetic properties for most of the chemical species in their natural biological microenvironment are not sufficiently known to design a general strategy for the preservation of original speciation patterns. At the moment rapid deep-freezing and species analysis as fast as possible seem to be advisable.

For sediments, soils and particulate matter the preparation techniques for storage which are commonly applied in trace element analysis such as cooling to 4 °C, freeze-drying, oven-drying or air-drying, cannot ensure the preservation of chemical species. It has been shown within feasibility studies for the development of reference materials [30] that the tributyltin content of a harbor sediment which was air-dried, sterilized at 80 °C, ground and sieved (75 μm) did not change during a storage period of 12 months at 20 °C. But this species degraded in a coastal sediment stored under the same conditions [30]. Butyltins were successfully conserved in a sediment after freezing and lyophilization for at least 1 year whereas the transformation of phenyltins could not be avoided [51]. Methylmercury was stabilized in a sediment by γ-radiation which had destroyed bacterial activity [52]. But as discussed above, validated preservation and storage methods for such complex samples in their original chemical composition still have to be developed and deep-freezing of wet samples may be the safest approach at present.

6 FURTHER CHALLENGES

Speciation analysis requires samples with unchanged chemical composition, and appropriate sampling procedures for complex environmental specimens are still in the developing phase at present. New methodical approaches are necessary for species-retaining sample collection, preservation and chemical characterization. This concerns in particular the representative sampling of very heterogeneous matrices such as sediments or soils, the species-oriented sampling of biological material and the conservation of less stable species patterns. Speciation studies in biological organisms have to take into account the natural biological compartmentation, and therefore an increasing number of investigations may even need the sampling of single cells. In the future sampling also has to provide reproducibly study materials with intact physical and chemical structures including undisturbed interfaces for the investigation of species- and matrix-specific interactions in real microenvironments.

Progress with such challenges can only be expected if the procedures for speciation analysis undergo the same rigorous quality assurance and control as was developed for total element determinations mainly in the last 20 years and as is practiced now by the laboratories with an appropriate sense of responsibility worldwide. For that matrix-matched certified reference materials (CRM) are mandatory [43]. The demand for 'fresh' (nondried) biological and abiotic environmental CRMs, which are less manipulated than the presently available ones and which are certified for various species patterns, has to be fulfilled in the future in order to establish speciation analysis as a scientifically sound analytical activity. Quality assurance for the whole speciation procedure from sampling to data evaluation has to be established with the emphasis on controlling the trueness of the results with respect to their information content concerning the original environmental specimen, and not only in regard to the reproducibility of analytical data.

Moreover, fundamental chemical investigations are necessary for the systematic design of complete procedures which allow the preservation of the original species information. They have to be directed to the elucidation of the thermodynamics and kinetics of species states and processes in complex natural systems such as biological cell compartments, to the understanding and quantitative description of metabolic reactions under stress conditions which can take place during sample collection, and to all other processes which influence the result of sampling.

7 REFERENCES

1. Templeton, D. M., Ariese, F., Cornelis, R., Danielsson, L. G., Muntau, H., van Leeuwen, H. P. and Lobinski, R., *Pure Appl. Chem.*, **72**, 1453 (2000).
2. Lewis, R. A., Stein, N. and Lewis, C. W. (Eds), *Environmental Specimen Banking and Monitoring as Related to Banking*, Martinus Nijhoff, Boston, MA, 1984.
3. Guy, P., *Sampling for Analytical Purposes*, John Wiley & Sons, Ltd, Chichester, 1998.
4. Schweigkofler, M. and Niessner, R., *Environ. Sci. Technol.*, **33**, 3680 (1999).
5. Haas, K. and Feldmann, J., *Anal. Chem.*, **72**, 4205 (2000).
6. Pécheyran, C., Lalere, B. and Donard, O. F. X., *Environ. Sci. Technol.*, **34**, 27 (2000).
7. Feldmann, J., Koch, I. and Cullen, W. R., *Analyst*, **123**, 815 (1998).
8. Feldmann, J., Grümping, R. and Hirner, A. V., *Fresenius' J. Anal. Chem.*, **350**, 228 (1994).
9. Evans, G. F., Lumplein, T. A., Smith, D. L. and Somerville, M. C., *J. Air Waste Manag. Assoc.*, **42**, 1319 (1992).
10. Sommar, J., Feng, X. and Lindqvist, O., *Appl. Organomet. Chem.*, **13**, 441 (1999).
11. Szpunar, J., Bouyssiere, B. and Lobinski, R., Sample preparation techniques for elemental speciation studies, in *Elemental Speciation. New Approaches for Trace Metal Analysis*, Caruso, J. A., Sutton, K. L. and Ackley, K. L. (Eds), Chapter 2, Elsevier, Amsterdam, 2000, pp. 7–40.
12. Klockow, D., Kaiser, R. D., Kossowski, J., Larjava, K., Reith, J. and Siemens, V., Metal speciation in flue gases, work place atmospheres and precipitation, in *Metal Speciation in the Environment*, Broekaert, J. A. C., Gücer, S. and Adams, F. (Eds), Springer, Berlin, 1990, pp. 409–433.

13. Amouroux, D., Tessier, E., Pécheyran, C. and Donard, O. F. X., *Anal. Chim. Acta*, **377**, 241 (1998).
14. Helmers, E., Sampling of sea and fresh water for the analysis of trace elements, in *Sampling and Sample Preparation*, Stoeppler, M. (Ed.), Chapter 4, Springer, Berlin, 1997, pp. 26–42.
15. Ahmed, R. and Stoeppler, M., *Analyst*, **111**, 1371 (1986).
16. Klockow, D., *Fresenius' Z. Anal. Chem.*, **326**, 2880 (1985).
17. Grömping, A. H. J., Ostapczuk, P. and Emons, H., *Chemosphere*, **34**, 2227 (1997).
18. Augustin-Castro, B., Untersuchungen zum Quecksilberkreislauf in der Umwelt, Ph.D. Thesis, University of Essen, Berichte des Forschungszentrums Jülich 3694, Jülich (1999).
19. Amer, H., Emons, H. and Ostapczuk, P., *Chemosphere*, **34**, 2123 (1997).
20. Umweltbundesamt (Ed.), *Verfahrensrichtlinien für Probenahme, Transport, Lagerung und Chemische Charakterisierung von Umwelt- und Humanorgan-Proben*, Erich Schmidt Verlag, Berlin, 1996.
21. Giege, B., Odsjö, T., Barikmo, J., Hirvi, J.-P., Petersen, H. and Petersen, A. E., *Manual for the Nordic Countries. Nordic Environmental Specimen Bank – Methods in Use in ESB*, TemaNord 543, ISBN 92-9120-662-8, 1995.
22. Becker, P. R., Wise, S. A., Thorsteinson, L., Koster, B. J. and Rowles, T., *Chemosphere*, **34**, 1889 (1997).
23. Amer, H. A., Ostapczuk, P. and Emons, H., *J. Environ. Monit.*, **1**, 97 (1999).
24. Schladot, J. D. and Backhaus, F., The common mussel as marine bioindicator for the environmental specimen bank of the Federal Republic of Germany, in *Specimen Banking*, Rossbach, M., Schladot, J. D. and Ostapczuk, P. (Eds), Chapter 4.2, Springer, Berlin, 1992, pp. 75–87.
25. Kateman, G., *Chemometrics – Sampling Strategies*, Springer, Berlin, 1987.
26. Guy, P. M., *Sampling of Heterogeneous and Dynamic Material Systems*, Elsevier, Amsterdam, 1992.
27. Einax, J. W., Zwanziger, H. W. and Geiss, S., *Chemometrics in Environmental Analysis*, VCH, Weinheim, 1997.
28. del Castro, P. and Breder, R., Soils and soil solutions, in *Sampling and Sample Preparation*, Stoeppler, M. (Ed.), Chapter 5, Springer, Berlin, 1997, pp. 43–56.
29. Quevauviller, Ph., *Method Performance Studies for Speciation Analysis*, The Royal Society of Chemistry, Cambridge, 1998.
30. Quevauviller, Ph. and Maier, E. A., *Interlaboratory Studies and Certified Reference Materials for Environmental Analysis. The BCR Approach*, Elsevier, Amsterdam, 1999.
31. Emons, H., Schladot, J. D. and Schwuger, M. J., *Chemosphere*, **34**, 1875 (1997).
32. Lambrecht, S., Emons, H., Matschullat, J. and Rossbach, M., in preparation.
33. Koglin, D., Backhaus, F. and Schladot, J. D., *Chemosphere*, **34**, 2041 (1997).
34. Arunachalam, J., Emons, H., Krasnodebska, B. and Mohl, C., *Sci. Total Environ.*, **181**, 147 (1996).
35. Shawky, S. and Emons, H., *Chemosphere*, **36**, 523 (1998).
36. Jakubowski, N., Stuewer, D., Klockow, D., Thomas, C. and Emons, H., *J. Anal. At. Spectrom.*, **16**, 135 (2001).
37. Rossbach, M., Giernich, G. and Emons, H., *J. Environ. Monit.*, **3**, 330 (2001).
38. Pauwels, J. and Vandecasteele, C., *Fresenius' J. Anal. Chem.*, **345**, 121 (1993).
39. Schladot, J. D. and Backhaus, F., Collection, preparation and long-term storage of marine samples, in *Sampling and Sample Preparation*, Stoeppler, M. (Ed.), Chapter 7, Springer, Berlin, 1997, pp. 74–87.
40. Donais, M. K., Saraswati, R., Mackey, E., Demiralp, R., Porter, B., Vangel, M., Levenson, M., Mandic, V., Azemard, S., Horvat, M., May, K., Emons, H. and Wise, S., *Fresenius' J. Anal. Chem.*, **358**, 424 (1997).
41. Tutschku, S., Schantz, M. M., Horvat, M., Logar, M., Akagi, H., Emons, H., Levenson, M. and Wise, S., *Fresenius' J. Anal. Chem.*, **369**, 364 (2001).
42. Shawky, S., Emons, H. and Dürbeck, H. W., *Anal. Commun.*, **33**, 107 (1996).
43. Emons, H., *Fresenius' J. Anal. Chem.*, **370**, 115 (2001).
44. Rückold, S., Grobecker, K. H. and Isengard, H.-D., *Fresenius' J. Anal. Chem.*, **370**, 189 (2001).
45. Gomez-Ariza, J. L., Morales, E., Sanchez-Rodas, D. and Giraldez, I., *Trends Anal. Chem.*, **19**, 200 (2000).
46. Pécheyran, C., Quetel, C. R., Lecuyer, F. M. M. and Donard, O. F. X., *Anal. Chem.*, **70**, 2639 (1998).
47. van Cleuvenbergen, R., Dirkx, W., Quevauviller, P. and Adams, F., *Int. J. Environ. Anal. Chem.*, **47**, 21 (1992).
48. Quevauviller, Ph., de la Calle-Guntinas, M. B., Maier, E. A. and Camara, C., *Mikrochim. Acta*, **118**, 131 (1995).
49. Valkirs, A. O., Seligman, P. F., Olson, G. J., Brinckman, F. E., Matthias, C. L. and Bellama, J. M., *Analyst*, **112**, 17 (1987).
50. Lindemann, T., Prange, A., Dannecker, W. and Neidhart, B., *Fresenius' J. Anal. Chem.*, **368**, 214 (2000).
51. Gomez-Ariza, J. L., Morales, E., Beltran, R., Giraldez, I. and Ruiz-Benitez, M., *Quim. Anal.*, **13**, S76 (1994).
52. Quevauviller, P., Fortunati, G. U., Filippelli, M., Baldi, F., Bianchi, M. and Muntau, H., *Appl. Organomet. Chem.*, **10**, 537 (1996).

2.2 Sampling of Clinical Samples: Collection and Storage

Koen De Cremer

Laboratory for Analytical Chemistry, Ghent University, Belgium

1 INTRODUCTION

Accuracy, quality assurance, quality control, repeatability, reproducibility, true value, validation, uncertainty of the measurement, ... and so many more concepts have become common language in any analytical laboratory. Indeed, these parameters should be carefully considered during an analysis but it is not a given fact that when these criteria are fulfilled the results will be perfect. This will depend strongly on the quality of preanalytical steps such as sampling, sample handling and conservation of the measured species. When analyzing total concentrations of an element, factors such as contamination and conservation are already considered to be very important. When, however, there is a need for speciation between different complexes of an element in a certain matrix, conservation of the species becomes of paramount importance. It is no longer sufficient to bring the total amount of an element into solution before measurement. In addition, throughout the procedure, the sample should contain only the elemental species present in the original matrix and, moreover, in the original ratio. So, before separating and analyzing the sample it is necessary to define the conditions under which the different species to be analyzed remain stable. In practice, this means

Handbook of Elemental Speciation: Techniques and Methodology R. Cornelis, H. Crews, J. Caruso and K. Heumann
 ISBN: 0-471-49214-0

mostly that the conditions during separation can vary only slightly from those occurring in the analyzed matrix. This will be demonstrated using the stability for the vanadium–transferrin complex under different conditions (pH, salt concentration, acetonitrile concentration, ...) that are commonly used during chromatography experiments. In this chapter we will first consider the presampling procedures. Afterwards we will describe the recommended sampling and storage procedures for blood, urine and tissues. Then a more detailed description of contamination hazards and storage conditions for specific trace elements will be given with a discussion of preservation of vanadium species at the end.

2 PRESAMPLING STEPS

2.1 Collection and storage of sample information

Before starting the collection of samples it is important to define *a priori* the exact goal of the experiment. This is the only reliable way to draw up a questionnaire about the donor and the sample to provide all the relevant information needed to evaluate the final results. Such a questionnaire may come in very handy when rather odd (outlying) results are obtained. The length of the questionnaire depends on the kind of element and on the aim of the study. For example, when looking for dose–effect or dose–response relationships, some parameters are of more importance than they might be for the establishment of reference values. In either case, information about the identity of the sample, the time, place and method of sampling, and possible additives, as well as about the person who carried out the sampling, should be gathered. Also a timed record of the solid food and liquid intake may be necessary to interpret certain results. More specific information on the content of such a questionnaire, in particular for measuring reference values, is given in the article of Cornelis *et al.* [1]. In the case of speciation analysis, information on parameters such as pH and conductivity should also be collected at the time of sampling and, maybe even more important, on the addition of chemicals during sampling. If possible, additives should be avoided.

During the set-up of a biological experiment, researchers should also be aware of a fundamental characteristic of all forms of life called biological rhythmicity, which up to now has more often than not been overlooked. It may be interesting to cite here the words by Burns *et al.* [2]: 'daily statistically significant fluctuations occur in all of the normal biological variables studied in the experimental animals and humans. However, many researchers are not aware of the negative impact biological rhythmicity can have on experimental design and/or data interpretation'. One of the most common pitfalls consists of data transfer from the diurnally inactive laboratory animal to the diurnally active human. This can be avoided to a certain extent by reversing the 12 : 12 hours light : dark cycle from the laboratory animals (spread over 10–14 days) so that their active period (dark) coincides with the active period of the researcher during daylight. Other pitfalls concern the frequency of sampling, synchronization of the laboratory animals, shifting of a certain rhythm and plotting data on an 'hours after treatment' basis versus a 'time of day' basis. For experiments dealing with, e.g. enzymes, these thoughts should be certainly borne in mind.

2.2 Cleaning and evaluation of the instruments

Careful selection of the instruments during sampling and storage will reduce greatly the risk of contamination. A collection tube completely free of trace element contaminants does not exist and neither is there a 'standard' collection tube that possesses ideal characteristics for all the trace elements. For example, when doing speciation analysis of aluminum, collection tubes or instruments made of glass cannot be used. Plastics with a cadmium-based softener or zinc-doped stoppers are to be avoided when analyzing for cadmium and zinc, respectively. For each element these instruments should be evaluated separately. While some metals, including cobalt, copper, iron and selenium present no significant risk of contamination, others, such as aluminum, cadmium, chromium, manganese and nickel pose quite a significant risk [3]. They warrant attention with respect not only to the sampling itself but also to the selection of the material to be used. Table 2.2.1 gives an overview of the presence of trace elements in laboratory ware [4]. This table can be used to get a first impression of the possible contamination hazards that might occur for the trace element it is desired to study. Evaluation of the containers or vessels and instruments can be done by different methods, but they all have some drawbacks.

A first, rather elaborate, way is to determine the concentration of the element in the collection tube and instruments. If available, a very suitable technique is NAA (neutron activation analysis). Another option consists of analyzing a certified sample, e.g. blood or plasma, after storage in the container for a length of time (e.g. 5 days at $-20\,^{\circ}$C), typical for the study, and containing a very low concentration of the element. If the certified value is reproduced the contamination can be considered negligible. This implies that the analytical procedure is under control, which again can only be confirmed by the analysis of a blood sample certified for about the same concentrations of the trace element as found in the real samples. This latter includes a major drawback for this method, because there are few elements certified in blood samples at the low concentration interval needed for this kind of study. Another, more common, possibility is to rinse the containers with mild acid (0.03 mol L^{-1} HNO_3 or HCl), EDTA, or with a solution that includes the major ligands (amino acids and peptides) that are present in biological matrices, and to analyze the leaching solution. Table 2.2.2 gives a summary of the leaching of trace elements from laboratory ware in the presence of 6 M HCl or 9 M HNO_3. Table 2.2.3 lists the concentration range of trace elements in ultra pure acids [4]. In a way, rinsing with acid is irrelevant because the pH is too acidic compared to biological matrices. Compilation of a solution with possible biological ligands is also not as straightforward as it should be. To exclude the risk of contamination completely, a very elaborate cleaning procedure can be used as described by Versieck and Cornelis [5]. This procedure consists of washing laboratory ware as follows: wash with distilled water; soak for 2 days in 30 % H_2O_2; rinse with Milli-Q water; boil for 8 h in a 1 : 1 (v/v%) mixture of 65 % nitric acid and 96 % sulfuric acid, both of Suprapur analytical grade; rinse with Milli-Q water; boil twice for 8 h in

Table 2.2.1. Trace elements in laboratory ware. Reprinted from *Talanta*, Vol. 29, Kosta, Contamination as limiting . . ., pp. 985–992, 1982, with permission from Elsevier Science.

Material	Concentration range [mg kg^{-1}]			
	100	10–0.1	0.1–0.01	0.01–0.001
Polyethylene and polypropylene	Na, Zn, Ca, (Al, Ti)[a]	K, Br, Fe, Pb, Cl, Si, Sr	Mn, Al, Sn, Se, I	Cu, Sb, Co, Hg
PVC	Na, Sn[b], Al, Ca	Br, Pb, Sn, Cd, Zn, Mg	As, Sb	–
Teflon	K, Na	Cl, Na, Al, W	Fe, Cu, Mn, Cr, Ni	Cs, Co
Polycarbonate	Cl, Br	Al, Fe	Co, Cr, Cu, Mn, Ni, Pb	–
Glass	Al, K, Mg, Mn, Sr	Fe, Pb, B, Zn, Cu, Rb, Ti, Ga, (Cr, Zn)[c]	Sb, Rb, La, Au, (As, Co)[c]	Sc, Tl, U, Y, In[c]
Silica	–	Cl[d], Fe, K	Br, Ni, Cu, Sb, Cr	Sb, Se, Th, Mo, Cd, Mn, Co, As, Cs, Ag

[a] Al and Ti high in low-pressure polyethylene (used as catalyst).
[b] Heavy metal compounds used alternatively as stabilizers in certain types of PVC.
[c] Not certified in the NBS reference material 617; determined by the cited author.
[d] Chlorine only high in synthetic quartz.

Table 2.2.2. Trace elements leached by hydrochloric acid and nitric acid from plastic containers after 1 week of contact (in ng cm^{-2}). Reprinted from *Talanta*, Vol. 29, Kosta, Contamination as limiting ..., pp. 985–992, 1982, with permission from Elsevier Science.

Material	6 M HCl			9 M HNO_3		
	10	10–1	1	10	10–1	1
Polyethylene (HP)	Na, Pb, Al	Tl, Cr, Zn	Ca, Sn, Cu, K, Mg, Ba, Ni, Cr, Cd, Sr, Se	Ca	Na, Fe, Se, Cu	Cr, Mg, Pb
Polyethylene (LP)	Ca	Zn, Na, Al, Fe, Cu, K, Ba	Sn, Co, Ni, Pb, Mg, Se, Cd, Sr	Na	Zn, Fe, K, Pb, Ni, Sn, Sr	Mg, Ca, Cu, Se, Cr, Cd, Te, Ag, Ba
Polycarbonate	Fe, Ca, Sn, Pb	Na, Cu, Cr, Cd, Ba, Al	Mg, Te, Se, Ni, Sr	–	Al, Na, Fe, Ca, K	Cu, Zn, Ni, Se, Cr, Cd, Pb
Teflon	Fe	Cu, Zn, Cr, Al, Ba, Ca, Na, Pb, Te, K, Mg, Sn, Sr	Cd, Se, Ni	Ca, Fe	Mg, Al, Na, Zn, Pb, Ba, Cu, Ni, K, Sn	Cr, Te, Cd, Sr, Se

Elements are ranked in order of decreasing amounts introduced into the acid.

Table 2.2.3. Trace elements in ultrapure acids. Reprinted from *Talanta*, Vol. 29, Kosta, Contamination as limiting..., pp. 985–992, 1982, with permission from Elsevier Science.

Range ($\mu g\,L^{-1}$)	HCl	HF	HNO_3	H_2SO_4	$HClO_4$
1	Al, Si, S	B, Si, P, S	Si, S, K	Ca, Co, Cu, K, Mg, Na, (Se)	Cr, Fe, Na
1–0.1	Na, Mg, P, Ca, Fe	Na, Al, Ti, Ca, K, Fe, Cu, Zn	Al, Ti, Fe, Na, P, Mg, Ca, B, Cu, Cr	Mn, Ni, Sn, Sr, Tl	Ni, Sn, Br, K, Pb, Tl, Zn
0.1–0.01	B, Ti, V, Cu, Zn, Sn, Ba	Cr, Mn, Co, Zr, Cd, Pb	Zn, Ni, Ba, Pb, Cd	–	Cd, Sr
<0.01	Mn, Co, Cr, Ni, Zr, Cd, Pb	Cd, Ba	Mn, Co, V	–	–

Milli-Q water; rinse with Milli-Q water, and finally steam-clean for 6–8 h with Milli-Q water. The equipment is then dried upside down on a Teflon foil or polypropylene tray at 55 °C in an especially reserved oven. Afterwards, the instruments are stored in an airtight plastic transport container until use. All steps are carried out in clean laboratory conditions (class 100) except boiling for 8 h in the 1 : 1 mixture of concentrated acids. Other high purity items are cleaned in a similar way. When steam cleaning is not possible, e.g., when the items are too small, they are boiled for an additional 6–8 h in Milli-Q water. A shorter method consists of using 0.5 % HNO_3 or 1 % EDTA with a final rinse in distilled water [6]. These procedures are undoubtedly too elaborate and too expensive for routine laboratory work in hospitals and are not necessary for all trace elements. Some authors even consider these procedures not necessary for short periods of storage (24 h). They tested several types of tubes, cleaned and uncleaned. The only significant contamination was an increase of 6 % in the aluminum content of uncleaned heparinized tubes [7]. Besides, when evacuated tube systems (ETS, 'vacutainers') are used for sample collection, and this is the case in the great majority of the hospitals, a washing procedure cannot be applied. Therefore, one of the previous described techniques should be used to evaluate this kind of container for their metal content at regular time intervals. This can be done by rinsing with EDTA or with 10 % HCl [3]. The criterion of acceptability is a maximum contamination level equivalent to 1 % of the normal level in the decontaminating reagent used [3]. If occasionally any other material is to be used, immersion for 24 h in 20 % HCl at 50 °C, followed by three to four rinsings with demineralized water is recommended. Pineau *et al.* [3] describe some regulations for routine work in hospitals which should be effective in eliminating most of the polluting factors: (1) avoid as much as possible, ordinary cloth and cotton clothing, the use of cosmetics that are rich in zinc and aluminum, and metallic articles of jewelry; (2) during the sampling and analysis, the doors and windows of the room should be kept closed, the air conditioning turned off and no smoking should be allowed; (3) no corrosion may be present and do not use any concentrated volatile acid (e.g. HCl) in the work room; (4) wash the floor frequently with distilled water, but not during analysis; (5) keep all reagents, containers and tubes in a dust-free area; (6) keep out all metallic objects.

A rather elegant way to circumvent most problems of contamination (as far as the metal is concerned) is the use of a radiotracer. However, contamination of exogenous ligands still can occur and for ethical reasons the use of radiotracers is not always possible for experiments with human subjects. For nonhuman experiments and when the nuclear facilities are available, the use of a radiotracer should be considered because this offers some major advantages, e.g. for the detection of very low concentrations of the metal species [8, 9].

In conclusion, there is no general solution to exclude the risk of contamination, and the gravity of this problem depends strongly on the element that is under study. To minimize the risk of contamination some recommendations are offered but it is up to the researcher's clear mind to evaluate the potential risk for contamination and to take all the precautions needed.

3 COLLECTION AND STORAGE OF BLOOD

In recent literature, one is aware of the importance of a reliable methodology to obtain a representative blood sample. Therefore, sample collection guidelines are drawn up in different analytical and medicinal fields [1, 10, 11]. For the collection of blood, ETS systems are the most widely used drawing systems, although syringe systems have also been used. A newer sampling technique, i.e. microdialysis, will be discussed simultaneously for blood, urine and tissues in a separate paragraph. The use of syringe systems should be avoided in the case of trace element analysis because they can cause significant contamination, usually originating from the upper part of the plunger made of rubber. In particular, lead and manganese are sensitive to this kind of contamination. In addition, the transfer of blood into a tube for centrifugation represents another potential risk of contamination [3]. The area of skin from where the blood will be collected is cleaned with Milli-Q water and ethanol and allowed to dry by evaporation. When dealing with experimental animals, these will in most cases first be anesthetized with diethyl ether. It is better not to use narcotics that need to be injected into the body, because they can have a potential affinity for the metal under study. When the patient or the animal is continuously connected to a catheter provided with a 'lock' (dialysis treatments) one should aspirate (not infuse) the content of this lock before sampling to avoid contamination from heparin or other constituents that were present in the lock. Collection from sites near a functioning graft, fistula or active intravenous line should be avoided [10]. After putting on gloves (powder free, otherwise can contain e.g. zinc), insertion of the needle into the vein can proceed. The pressure of stasis should be low. This can be accomplished with the patient supine. If the veins are not clearly visible a tourniquet can be applied, but no longer than 1 min in order to avoid hemoconcentration of the blood sample. Needles are color coded by gauge size and the larger the gauge size, the smaller the needle. Sizes 19 through 23 are most commonly used. Selection of a suitable gauge size is important because too small a gauge may damage the blood cells and increase the risk of hemolysis. For the same reason the speed of the blood flow into the ETS tubes should be controlled. If this speed is too high the red blood cells may be damaged when they hit the tube wall, causing hemolysis. The use of a stainless steel needle for the collection of blood is generally not suitable when examining trace metals. Analysis of blood illustrates that the highest contamination occurs in the first 20 ml of blood sampled, especially for iron, chromium, nickel, cobalt and manganese. For the measurement of these elements the alternative is the use of a polypropylene intravenous cannula, mounted on a trocar. Propylene and Teflon catheters also do not induce contamination when analyzing cesium, molybdenum and vanadium if a sufficient quantity of blood is withdrawn [12]. Advocating siliconized needles for trace element determinations can be very misleading, because some types of needles are only siliconized on the outside. If so, the hazard of contamination of the blood sample remains a serious possibility. The potential contamination of the needle can be tested in the same manner as for the laboratory ware, e.g. by comparing the elemental concentrations in blood that has and has not been in contact with the needle [1].

In general, polypropylene or Teflon catheters should preferably be used for venipuncture, and if this is not possible, siliconized needles can be applied. One should bear in mind that these needles can release a little aluminum, chromium and nickel. The use of stainless steel needles in hospitals is considered unrealistic because it would require an unacceptably large quantity of blood (>20 ml) to minimize contamination.

To collect the blood, there is a choice between open systems and the previous mentioned evacuated systems. For open systems it appears that polyethylene tubes provided with stoppers offer the best guarantees for most trace elements. If stoppered polyethylene tubes are not available, polystyrene tubes may be used [3]. In the case of the evacuated systems, tubes produced before 1985 showed potential contamination of chromium, iron, nickel and zinc originating from the rubber stoppers. Tubes equipped with siliconized stoppers (special trace element tubes) show less contamination, e.g., for zinc [13], but it is suggested that, for other difficult elements such as aluminum, cobalt, chromium, nickel and manganese, vacuum tubes, even those especially developed for trace element analysis, should not be used. They can be used, with caution for lead. Vacuum tubes containing a serum separator (e.g. a gel) are generally rejected from trace element analysis [3].

The use of an anticoagulant is very problematic, as most anticoagulants are either polyanions (e.g. heparin) or metal chelators (e.g. EDTA or citrate) and therefore have a high affinity for metals. For speciation research this is disadvantageous in two ways. Firstly, because of their great affinity for most metals, these ligands can bind metal ions originating from the wall of the container or any other exogenous metal ion and thus contaminate the sample to a greater extent. Therefore, this contamination hazard must be evaluated for each particular element under investigation and the blank value must be reported for each batch of anticoagulant. Secondly, metal complexes present in the original sample, but with a lower stability constant, will be destroyed through addition of these anticoagulants. In this way, the results will not reflect the original composition of the sample. Therefore, the use of anticoagulants (and other additives) should be avoided as much as possible in order to preserve the original condition of the sample. So, speciation of metal species in serum samples is to be preferred to speciation in plasma samples. ETS tubes with a red stopper contain no additives (Vacutainer®, Becton Dickinson), while tubes with a green or lavender stopper contain respectively heparin and EDTA. The use of tubes without additives is strongly recommended.

When feasible, in the case of adults, the first milliliters of blood will serve to rinse the needle and will not be used for trace element speciation, but kept apart, e.g. for clinical analyses. The blood is allowed to clot spontaneously and the samples are transported to a clean laboratory. After clotting, the serum is separated from the erythrocytes by centrifugation at 2500 rpm for 20 min. After the first spin, serum is pipetted into other clean tubes and centrifuged for a second time at 2500 rpm for 15 min to remove remaining blood cells. Finally, the serum is pipetted in clean tubes and stored at $<5\,^{\circ}C$ (short periods only) or frozen at $-20\,^{\circ}C$ or less (plastic tubes only) [5]. Pipet tips used to aspirate serum samples should be made of propylene or polycarbonate. These tips have been shown to release cadmium, iron, nickel, chromium, molybdenum, palladium and mercury and should therefore be rinsed with 10 % HCl and demineralized water and tested for absence of contamination [3]. If possible, analysis of the samples should occur on fresh material within a few days after sampling. If not, samples should be deep-frozen until analysis immediately after sampling. Standardization of the clotting and separation procedures and avoidance of hemolysis are important. Whole blood samples that are stored at −20 °C (>24 h) will be hemolyzed [14]. Hemolysis can result in increased concentrations of some elements, e.g. lead, manganese and zinc, in serum due to release of these elements out of the packed cells where these are present in a much higher concentration. Therefore serum is separated from blood cells before freezing and storage. The degree of hemolysis should be assessed by measuring hemoglobin concentrations in serum samples and a criterion for rejection of hemolyzed samples should be put forward.

4 COLLECTION AND STORAGE OF URINE

Guidelines also exist for the collection of urine [1]. Collection of urine poses more problems than taking blood samples because of the many potential sources of contamination, either in the environment (e.g. occupational medicine in a polluted environment) or in the method of collection and the container used [3]. In our laboratory we tried to collect urine of rats directly from the bladder with the use of the ETS system after anesthetizing and before dissecting the animals. The purpose of this technique was to minimize the risk of contamination. However, this technique was not successful because most of the time the bladder turned out to be empty. This was probably due to a reflex of the rats to empty their bladder in stress situations. We solved this problem by using metabolic cages for the separate collection of urine and feces. By doing this, the chance for contamination seriously increases. Dust or other foreign material in the neighborhood can fall into the collection vessels. We partially circumvented this problem by restricting the sampling intervals to periods of 1 h and by the use of a radiotracer. When sampling urine of humans, a more rigorous collection method should be applied. Depending on the purpose and the circumstances of the measurement, sampling can request only one spot sample or the collection period can be extended to 24 h. All time intervals in between can be used. Some authors favor successive and separate collections in order to limit contamination. Afterwards, these samples are mixed together in the laboratory [3]. On the other hand, short collection intervals (without mixing) can give additional information about a possible rhythmicity of an element's behavior. In morning urine, the element concentration is often relatively high. In case it is impossible to collect 24 h samples, some authors suggest to use this morning urine and to correlate the concentrations of the elements to the creatinine elimination. This would compensate for the effects of dilution [15].

Urine should be voided directly into an acid-washed polyethylene container that can be closed with an airtight lid. The subject should be instructed to minimize contamination of the sample by avoiding contact with the inside of the container or lid and to close the container immediately after voiding the urine. In between sample sessions the container should be wrapped into a polyethylene bag. In general it is recommended that measurement of the parameters given in Table 2.2.4 are included. When some samples show values outside the expected range this should be mentioned in the sample report.

After collection of the sample it is advisable to divide the urine into subsamples (different aliquots) after vigorous shaking for a few minutes. The samples should be kept at 4 °C in the refrigerator (short period) or frozen at −20 °C. The stability of the different metal compounds is species dependent (see below). Normally, precipitation of salts and organic compounds occurs resulting in coprecipitation of several trace elements. Often, urine samples are stored in the presence of a preservative (0.03 mol L^{-1} HCl or HNO_3, sulfamic acid,

Table 2.2.4. Recommended parameters to be measured for urine collection.

Parameter	Test/range
Sugar	Negative by strip test
Proteinuria	Negative by strip test
UTI (Urinary tract infection)	Negative for nitrite producing bacteria
Urinary density	1.012–1.030 (reject outlying samples)
Creatinine	7–17 mmol 24 h^{-1}

Triton X-100). This should, if possible, be avoided in speciation analysis because many metal–protein complexes are only stable in a well-defined pH interval (mostly around physiological pH). Addition of an acid will destabilize the metal–protein complex. In one of the following paragraphs it will be clearly demonstrated that it is essential in speciation analysis to maintain as much as possible the physiological conditions during storage and subsequent steps in order to avoid artifacts. However, in the case of aluminum and mercury, it has been shown that, even in frozen samples, the concentration rapidly decreases during the first few days if no additive is present [16]. In these cases, if an additive is necessary, both its influence on the stability of the present species and its purity should be checked.

5 COLLECTION AND STORAGE OF TISSUES

For collection of tissues there is an even greater risk of contamination when compared with the collection of biological fluids such as serum or urine. This is due to the increased amount of handling (cutting, cleaning, homogenization) needed to sample a tissue. A possible solution is microdialysis (see below). Extensive research has been carried out to assess the possible adventitious addition of trace elements to the sample. It was found that contamination of biopsies taken from the liver by means of needle aspiration techniques was much higher compared with biopsies taken with surgical blades. For iron, manganese, copper, zinc, cobalt, chromium and nickel significant additions were detected with the needle aspiration technique, in some cases higher than the natural levels of the elements in the sample [12]. Potential additions from steel scalpels were also examined. It was observed that chromium contamination of frozen muscle tissue samples is about ten times greater than for fresh samples because of the greater friction in cutting. Addition of manganese, antimony and tungsten was also found [17]. It is also important to add that other steps in obtaining and handling biopsies, e.g. taking them with a pair of tweezers, may introduce additional errors. In our laboratory, we sample tissues using a surgical blade or a pair of scissors. After dissection of the animal, organs are rigorously cleaned by removing remnants of fat and connective tissue with a pair of tweezers and a pair of scissors. Depending on the element these instruments can be made out of stainless steel or plastic. Afterwards, the tissues are washed five times with a 0.9 % saline solution at physiological pH (7–7.5). Finally, organs and tissues are stored in plastic vessels with tightly fitting screw caps in a refrigerator at 4 °C for very short periods (hours) or in the deep freezer at −20 °C for longer periods. Tissues spoil at temperatures even slightly above freezing point because chemical reactions (e.g. enzymatic activity) continue to occur within tissues after the death of an animal. This can even happen in a frozen tissue, unless kept extremely cold [18]. Therefore it is very important to perform the analysis as quickly as possible or to

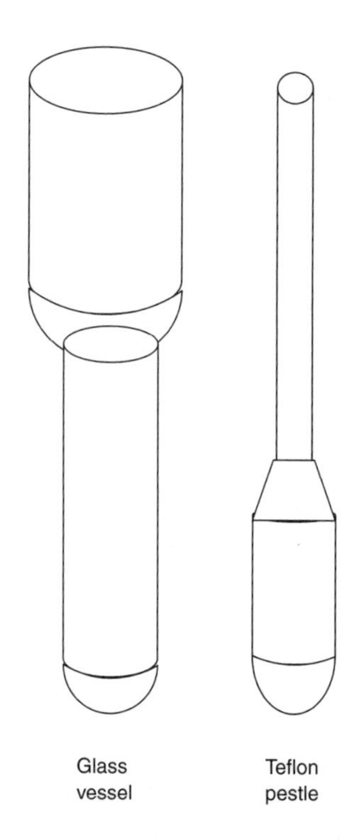

Figure 2.2.1. Evelhjem–Potter homogenizator. Minced tissue is placed in the glass vessel and buffer is added. The tissue is mildly homogenized by moving the Teflon pestle up and down in the glass vessel.

store the tissue in the deepfreezer after sampling. Lyophilization of tissues can be another option, but its influence on the stability has to be checked [19]. At the time of analysis, whole organs (spleen, kidney, testes) or parts of an organ (liver) are first cut into small pieces. In our laboratory these pieces are further homogenized by means of an Evelhjem–Potter homogenizator (Figure 2.2.1). This simple instrument is made of a glass vessel (like a cold-finger) and a Teflon pestle of slightly smaller diameter. Tissue and some buffer are placed in the glass receptacle. By moving the Teflon pestle up and down in the glass vessel the tissue cells are crushed between the pestle and the glass wall. By repeating this handling for a few minutes the tissue is homogenized in a very mild manner. Afterwards you can obtain any cell fraction (membranes, mitochondria, lysosomes, cytosol, ...) by differential centrifugation. Because of the friction with the glass wall during homogenization, possible contamination arising from this glass wall should be checked *a priori* for the element under investigation.

6 MICRODIALYSIS

Microdialysis is a relatively new technique that has been applied extensively in neurosciences to measure neurochemicals *in vivo* [20–22]. Recently the use of this technique has been extended to other biological fields. In theory, microdialysis mimics the passive function of a capillary blood vessel. A small diameter probe (different geometries and sizes available) containing a dialysis membrane is implanted into a tissue (in contact with the extracellular fluid) or a vessel that is continuously perfused with a suitable physiological fluid at a low flow rate with the aid of a pump. Molecular substances with a molecular weight smaller than the cut-off value of the dialysis membrane diffuse along a concentration gradient in or out the probe, depending on the relative concentration on both sides of the membrane. If the concentration of the analyte in the perfusion fluid is lower than that in the tissue itself, the analytes move from the tissue into the probe and consequently they are collected in the perfusate and carried out of the body. This technique offers many advantages compared to classical sampling techniques, i.e. the withdrawal of blood samples or tissue fractions. Substances can be continuously (dynamically) sampled without removing or altering the balance of body fluids. As a function of time, better resolutions are obtained, as sampling intervals are relatively short. Typical flow rates range from 0.5 to 10 $\mu L\,min^{-1}$ with 2 $\mu L\,min^{-1}$ being most commonly used. For a common sampling interval of 5–10 min the sample volume is restricted to 10–20 μL. To obtain optimal results, sampling time and perfusion flow rate should be balanced. Also multiple microdialysis in the same animal is possible, e.g. by placing one probe in a tissue and another in a blood vessel. In this way the passage of a component from the tissue to the vessel or vice versa can be studied. Overall, fewer animals are needed to obtain time-dependent curves. Tissue areas as small as 1 mm^3 can be sampled in either conscious (freely moving) or anesthetized animals. Sample recovery depends on a variety of factors such as sort of tissue, probe design, flow rate, temperature, membrane surface and membrane cut-off value. Probes with a cut-off value ranging from 5 kDa up to 100 kDa are available. However, dialysis efficiency decreases dramatically as the molecular weight of the analyte increases. Because of the use of a semipermeable membrane with a low cut-off value, proteins and degradative enzymes are excluded from the sample. In this way, sample degradation by enzymes is less common and because of the absence of proteins there is no need for a sample clean-up: the sample can be directly injected into analytical instruments for on-line analysis.

The microdialysis probes are manufactured out of biocompatible components. This reduces the risk for contamination of metals in trace element analysis. Most materials are polycarbonate–ether polymer, cellulose (acetate) and polyacrylonitrile for the membranes and fused silica, Teflon or polyethylether ketone for the tubing. The outer diameter of the probes ranges from 250 to 500 μm. The dialysis membranes are usually 1–10 mm long. Procedures for implantation of the probes into the tissues are provided by the manufacturers. In most cases, this involves the insertion of a cannula through which a probe is placed. In general, the implantation technique depends on the durability of the probe and its membrane. Insertion of a probe causes increased blood flow in the surrounding area. Therefore, after insertion an equilibration period should be allowed before starting the sampling. This recovery time depends on the type of tissue, probe and degree of trauma induced. In muscle, no histological changes were observed less than 6 h after insertion. More than 6 h after implantation leukocyte infiltration started and continued for 24 h without morphological changes. After 30 h, scar tissue starts to develop around the probe. This alters the diffusion rate and the probe recovery. It can be concluded that for studies up to 24 h, sampling in some tissues can be done without changes in recovery. For long-term studies, special consideration should be given to scar tissue formation. For sampling tumor tissue, there seem to be fewer problems concerning leukocytes infiltration or tissue changes.

Different probe designs are commercially available. The design is specific for the tissue being studied. In Figure 2.2.2 some designs and their use are depicted. Linear probes are used to sample peripheral tissues (dermal tissue, muscle, adipose tissue, liver and other organs) with minimal damage [23]. Side by side probes and vascular probes are best suited for intravenous sampling. For intracerebral implantations rigid concentric probes are the model of choice. The advantage of loop probes is their larger surface area that makes it possible to enlarge the dialysis surface without lengthening the probe too much. Therefore these are best suited for sampling subcutaneous

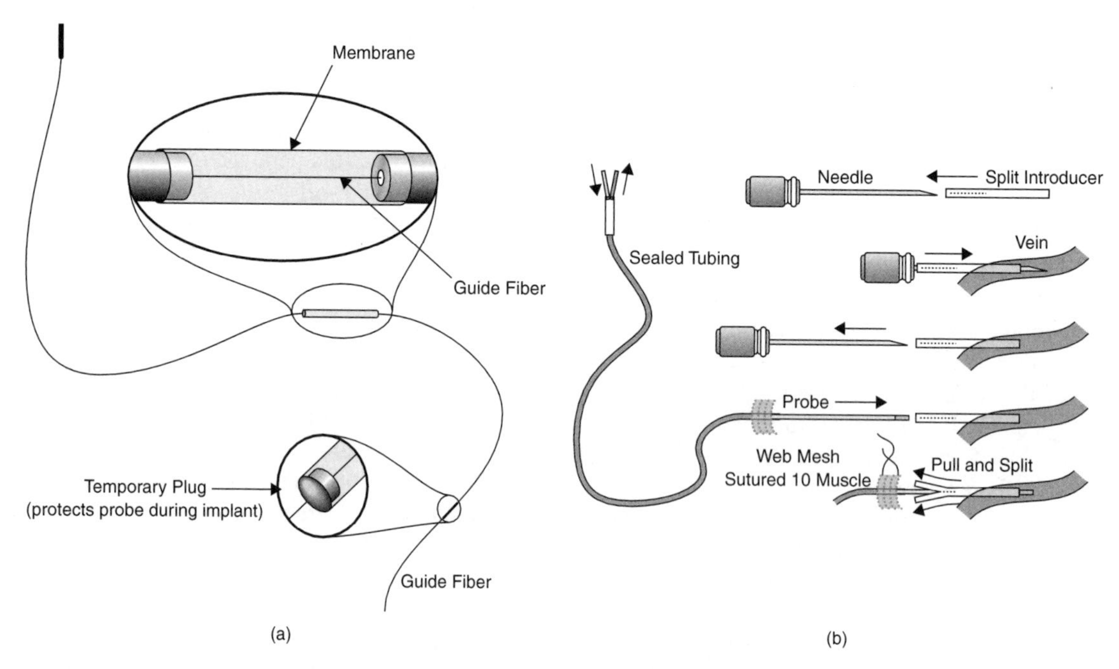

Figure 2.2.2. (a) The linear probe consists of a short length of hollow dialysis fiber attached to narrow-bore inlet and outlet tubes. An aqueous perfusion solution, which closely matches the ionic composition of the surrounding extracellular fluid, is pumped through the probe at a constant flow rate. Low molecular weight analytes diffuse in or out of the probe lumen. Large molecules such as proteins or protein-bound analytes are excluded by the membrane. Molecules entering the lumen are swept away by the perfusion fluid. This dialysate is then collected for analysis. (Reproduced by permission of BAS, Inc.). (b) IV vascular probes were designed for implantation into the rat jugular vein. They are also suitable for other soft tissues. Each probe includes a syringe needle and temporary cannula (split introducer) which aid placement. The thin-walled, plastic introducer slides over the syringe needle, which is then used to pierce the vein. The needle is removed and replaced with an IV probe. A flexible wire mesh on the probe is sutured to the pectoral muscle. The cannula is then pulled out of the vein, leaving the probe behind [23]. (Reproduced by permission of BAS, Inc.)

tissue or the peritoneal cavity. The researcher can choose from these designs or develop their own microdialysis probe fit for purpose. Custom-made probes can also be ordered by some manufacturers.

However, this new sampling method also has some drawbacks. First of all, this sampling technique requires sensitive analytical techniques to detect the low concentrations in the small sample volumes. The technique is also less suited for high molecular weight molecules as they have a slow diffusion rate through the membrane. The membrane itself can exhibit adsorption for some molecules due to residual charges or membrane composition that limit sample recovery. A minimal surgical know-how is needed to implant the probe in a correct and, for the animal (or human), safe manner. Damage of the peripheral tissue has to be avoided as much as possible. Most common effects are direct trauma, circulatory effects (short term) and foreign body effects (longer term). All these facts can influence sample recovery. This brings us to another great disadvantage, i.e. lack of quantitative recovery. Although the amount of analyte does not vary at a particular sampling site (in a limited sampling interval), recovery varies from site to site due to, e.g., differences in tissue volume. Also, within a tissue, there are differences in tissue density. There exist already some calibration procedures and, when used correctly, these yield a good approximation of the analyte concentration. However, there is still a debate going on about the validity of these calibration procedures. Other doubts prevail about changes in tissue permeability induced by inserting the probe, the limited sample area and possible changes in tissue tortuosity at the probe location. If these problems are solved, microdialysis will become a routine sampling method in the clinical research laboratory.

7 SPECIFIC PRECAUTIONS FOR SOME ELEMENTS

In this section the stability and storage of individual trace elements is considered. Gomez-Ariza *et al.* [24], Quevauviller *et al.* [25], Das *et al.* [26] and Cornelis *et al.* [1] have published some excellent reviews during recent years where the reader may find additional information.

7.1 Aluminum

Because 8 % of the earth's crust consists of aluminum, exposure to this element is ubiquitous. This also implies that the risk for contamination is real. Dust on the sample should be avoided at all stages. Therefore sample separation should be done in a clean room (class 100). Samples should be collected in a similar environment and for venipuncture talc-free gloves should be worn. Water and reagents are other sources of aluminum contamination. Blank values for aluminum should be measured in these reagents. All glass and plastic ware must be thoroughly washed with acid or EDTA solutions and then checked for their contributions of aluminum to the sample. From a recent study it is clear that in Teflon containers there is no loss of aluminum for a (relatively short) storage period of 2 h [27]. Also polyethylene containers can be safely used [25]. In the same report it is stated that the stabilization and long-term storage of natural samples at 4 °C was difficult and unachievable in practice without some type of pretreatment, e.g. addition of acid or a complexing agent, and that there is a need for additional investigations for stabilization, e.g. by chemical buffering systems.

Because Al(III) is a 'hard' trivalent metal ion, it binds strongly to oxygen-donor ligands such as citrate and phosphate. Therefore it is important that no additives containing this type of ligand are used during speciation experiments for aluminum. Serum fractionation studies show that most aluminum is protein bound, primarily to the transport protein transferrin. Albumin appears to play no role in serum transport [28]. For the low molecular weight fraction there is little agreement, although some reports indicate that citrate plays a significant role [29]. In a more recent article the speciation of aluminum in various biofluids and tissues is discussed [30].

In serum, aluminum levels of about 1–10 $\mu g\,L^{-1}$ can be expected in healthy persons. In urine, levels fluctuate around 10 $\mu g\,L^{-1}$ for healthy persons [1].

7.2 Antimony

Antimony is a relative toxic element. Trivalent antimony is about ten times more toxic than pentavalent antimony. Organic forms of antimony are less toxic than the inorganic forms. A stability study showed that 25 $\mu g\,L^{-1}$ antimony(III) can be stabilized at 40 °C for 12 months in an aqueous medium, lactic acid 0.1 $mol\,L^{-1}$, or citric acid 0.05 $mol\,L^{-1}$. These solutions were stored in polyethylene bottles. Oxidation of antimony(III) to antimony(V) was likely to occur after 1 month of storage in an ascorbic acid solution (0.06 $mol\,L^{-1}$) even when stored at 4 °C. After 3 months of storage antimony(III) oxidized in aqueous media and already after 1 month in lactic acid at 25 °C. Better results were obtained in citric acid although it did not completely prevent antimony(III) oxidation. Because the total antimony content (50 $\mu g\,L^{-1}$) remained constant at 4 and 25 °C it was concluded that polyethylene is suited to prevent adsorption onto the vessel walls [31]. The use of weak acid media has to be avoided owing to the high instability caused by the hydrolysis of antimony [24, 31, 32]. Another preservation procedure is based on solid-phase extraction using different solid sorbents such as activated carbon or graphite [33].

Antimony is found in serum in concentrations ranging from 0.07 to 0.76 $\mu g\,L^{-1}$ [34].

7.3 Arsenic

The collection of blood and urine for arsenic measurements is also very sensitive to contamination from arsenic in reagents, dust and laboratory ware, so the same precautions should be taken as for aluminum. However, if contamination occurs it will most probably be in the form of inorganic arsenic and not organic arsenic which is metabolized, e.g. in animal tissues. In blood, arsenic is expected to be stable at −20 °C. An extensive study on the stability of common arsenic species such as arsenite [As(III)], arsenate [As(V)], monomethylarsonic acid (MMA), dimethylarsinic acid (DMA) and arsenobetaine in urine shows that low temperature conditions (4 and −20 °C) are suitable for the storage of samples for up to 2 months. For longer periods (4–8 months) the stability of the arsenic species was dependent on the urine matrix. Whereas the arsenic speciation in some urine samples was stable for 8 months at both 4 and −20 °C, other urine samples stored under identical conditions showed substantial changes in the concentration of arsenic(III), arsenic(V), MMA and DMA. The use of additives did not improve the stability of arsenic species in urine [35]. Moreover, the addition of 0.1 $mol\,L^{-1}$ HCl to urine samples produced relative changes in inorganic arsenic(III) and arsenic(V) concentrations. Earlier studies reported that 50 % of arsenite(III) was oxidized to arsenate(V) by dissolved air after 33 days of storage [36]. Therefore, several acids have been proposed for arsenic stabilization in water samples, but as mentioned previously this should be mostly avoided in speciation analysis. A comparison between storage of solutions at room temperature and 4 °C revealed that solutions of organic arsenic species were stable during long-time storage, while solutions of inorganic species were only stable during refrigerated storage [37]. The former BCR (Community Bureau of Reference, Measurements and Testing Programme) has also undertaken projects to evaluate the stability of arsenic compounds [38]. The results showed that pure solutions of arsenic(III), arsenic(V), MMA, DMA, arsenobetaine and arsenocholine were stable for up to 1 year if they were stored in the dark and if the pH values were properly adjusted. No degradation was seen even if the temperature increased to 40 °C. However, degradation was seen when mixtures of arsenic species were prepared, with and without inorganic salts. Another study with a mixture of arsenic species showed oxidation and methylation reactions [39]. After 4 months of storage at 20 °C in the presence of light, all arsenite(III) was converted to arsenate(V). An identical solution stored in the dark at 40 °C showed only slight degradation of arsenic(III) (due to low microbiological activity). Similar experiments with mixtures of arsenic(V), DMA and arsenocholine stored at both 20 and 40 °C revealed the production of MMA after 2 months, while arsenic(III) was formed after 4 months at 20 °C. At 40 °C the formation of arsenic(III) was observed after 2 months and disappeared after 4 months. When the mixture was stored at 4 °C no changes in the fate of the species were observed. In case of the stability of DMA and arsenobetaine in a freeze-dried tunafish, there were no changes seen in stability for 9 months at either 20 or 40 °C [40].

Concentrations in serum of healthy persons vary between 1 and 5 $\mu g\,L^{-1}$, and this level depends on the level of seafood intake. In urine, values for arsenic are around 10 $\mu g\,L^{-1}$ for European citizens. In Japan, however, concentrations can be five times higher [1].

7.4 Cadmium

Cadmium is often used in pigments and as a softener in plastics. For this reason, sample contact with colored stoppers and certain plastics during sampling and processing must be avoided. Glass should also be avoided. In a rainwater sample that was stored at 22 °C in a HDPE bottle, about 40 % of the dissolved cadmium was lost, probably due to adsorption on the walls [27]. It has also been shown that this loss happens during the first 30 min of storage. Storage in a Teflon bottle showed no losses of cadmium. The person who performs the sampling should wear talc-free gloves. No special needle is required because stainless steel needles do not seem to release cadmium. Plastic syringes and test tubes should be cleaned and tested for their ability to release cadmium into the sample. Another important contamination source for cadmium is smoking. Preferably, sample collection and handling should be done by a nonsmoking person.

Speciation of cadmium in liver indicates that cadmium binds to different protein fractions (>400 kDa, 70 kDa and metallothionein, [41]. In the case of a freshly prepared solution cadmium binds to two different isoforms of rabbit metallothionein as is clearly visible in a CE-electropherogram [42]. However, when the solution was allowed to stand for 2 weeks without refrigeration it is obvious that the metallothionein has degraded as is apparent from the lack of well-defined individual components in a CE-electropherogram. For free cadmium, the influence of pH has also been demonstrated [42]. At pH values less than 8, more than 6 mM of free cadmium can be present in solution without significant hydroxide formation. At pH values >8 the signal of free cadmium significantly decreases.

Cadmium concentrations in blood are generally in the range of 0.1–2 $\mu g\,L^{-1}$. Concentrations in urine are usually <1 $\mu g\,L^{-1}$ [1].

7.5 Chromium

For the determination of chromium strict guidelines should be applied in order to obtain reliable data. In contrast to cadmium, no stainless steel needle can be used in the case of chromium. A propylene cannula is compulsory. The first 20 ml of blood cannot be used for chromium analyses. All tubes and plastic ware should be acid washed before sample collection. Unwashed tubes will invariably lead to a too high blank value for chromium. As sweat contains about ten times more chromium than does serum, it is important to avoid contact of the sample with the skin.

Chromium(VI) is much more toxic than chromium(III) and of great concern in public health. The stability of chromium(III) and chromium(VI) in solution was thoroughly investigated as part of a BCR project [25, 43]. Significant losses of chromium(VI) at 20 °C were observed in solutions stored in PTFE containers after the addition of HCO_3^-/H_2CO_3 buffer. The stability of both chromium(III) and chromium(VI) was satisfactory after 228 days of storage in quartz ampoules at 5 °C at concentrations of 40 $\mu g\,L^{-1}$ of chromium(III) and 10 $\mu g\,L^{-1}$ of chromium(VI). In this case the addition of HCO_3^-/H_2CO_3 buffer under CO_2 at pH 6.4 was necessary to avoid chromium(VI) reduction. A promising development shown in this project is the preparation of freeze-dried solutions containing chromium(III) and chromium(VI). The stability of these compounds was demonstrated over a period of 88 days for freeze-dried solutions stored at −20 °C after reconstitution in HCO_3^-/H_2CO_3 buffer under CO_2 at pH 6.4. The stability at 20 °C has also been verified [43]. The redox potential of the chromium(VI)/chromium(III) system depends on the pH of the medium. Chromium(VI) salts (dichromates) are very strong oxidants in sufficiently acid media. Their oxidizing strength decreases significantly with increasing pH which may explain their long-term stability when solutions are kept at a constant pH of 6.4 [25]. In a recent report, storage of water samples was done in the dark at 4 °C at neutral pH and the analysis was done as fast as possible [44].

Chromium is known to be mainly bound to transferrin and albumin in serum [45]. In urine, serum and liver also a low molecular weight was found. In urine, this low molecular weight complex is the most abundant chromium species [46].

The chromium level in serum is about 0.1–0.2 $\mu g L^{-1}$. The concentration of chromium in urine of nonexposed individuals is below 1 $\mu g L^{-1}$ [1].

7.6 Cobalt

No stainless steel needles can be used for the determination of cobalt values in blood. All the receptacles should be acid washed. Cobalt is stable in blood for many years at −80 °C. If possible, determination of cobalt should happen in a class 100 clean room. Additional contamination can arise from jewelry or dental prostheses. It is of major importance to carefully wash the skin before sampling takes place. Because cobalt is an element with a short biological half-life it is important to know the exact time lapse between the beginning of the exposure and time of sampling. A sample protocol with small time intervals is advised.

Cobalt is an essential nutrient and a component of vitamin B_{12}. Next to vitamin B_{12} there are some cyanocobalamin analogs such as adenosylcobalamin, methylcobalamin and hydroxocobalamin [47]. In urine cobalt is excreted both in the inorganic form and in organic forms [48]. The stability of vitamin B_{12} is highest in the pH range 4.5–5 [42]. When working at pH 9 it is likely that cobalt is removed from the porphyrin structure in order to form $Co(OH)_2$. Oxidation of adenosylcobalamin in solution has also been reported [42]. A solution of 1 $mg L^{-1}$ in a deoxygenated buffer that was allowed to stand at ambient temperature for 10 min showed already the oxidation product in its chromatogram. At 108 min after preparation the majority of the adenosylcobalamin was in the form of the oxidized analog.

It appears that values of urine cobalt are in the range 0.1–1 $\mu g L^{-1}$ with values in serum and blood at the lower end of this range [49].

7.7 Copper

Recovery experiments with copper in aqueous solution have shown that about 80 % of dissolved copper in HDPE containers is lost, probably due to adsorption on the walls of the container. This loss mainly occurs during the first 0.5 h of storage [27]. In polypropylene containers the same effect has been demonstrated. In Teflon containers, the concentration of copper remained constant within the experimental error during the 2 h of measurement.

In serum, copper is bound to ceruloplasmin (160 kDa) and albumin (66 kDa) [26, 50]. In a more recent report, a small fraction of the copper is eluting in the dead volume of the column. It is suggested that this fraction consists of either free positively charged copper ions or weakly bound copper [51]. Copper ions added *in vitro* to a serum sample will result in an increase of the albumin bound fraction [52].

Copper serum concentrations range from 0.8 to 1.4 $mg L^{-1}$. In urine, copper concentrations around 0.2 $mg L^{-1}$ have been measured [1].

7.8 Lead

Sampling of blood for lead measurements can be done using disposable sampling devices and containers. However, the lead blank caused by chemicals and materials must be sufficiently below the lead concentration in serum in order to avoid misleading results. In the case of serum samples it is of paramount importance to avoid hemolysis during blood collection and subsequent serum separation. This is needed because 10 % of the lead concentration in whole blood is situated in serum, while the remaining 90 % is concentrated in the packed cells.

This means that even marginal hemolysis will elevate the lead concentration in serum by a factor of 2 or more. Therefore it is necessary to estimate for all serum samples the degree of hemolysis by measuring the hemoglobin concentration.

In serum, lead seems to elute in the fraction ascribed to ceruloplasmin and ferritin [51, 53, 54]. In a recent article, lead eluted in a fraction with different chains of immunoglobulines and low molecular weight components of serum, but no identification of the binding complex is put forward [51]. Lead peaks were seen only in serum of uremic patients.

Because of the environmental concern regarding organolead complexes, the stability of some alkyllead compounds in blood during storage at different temperatures has been studied [55]. Spiked blood was stored prior to the analysis at room temperature, 4, −20 and −70 °C. The results showed that samples can be stored at 4 °C for 1 week, at −20 °C for 2 months and in a deepfreezer for at least 1 year. To secure the stability of the species during transport, a polystyrene box with dry ice was used, which preserved the species integrity for 2 days. The stabilities of these organolead compounds have also been investigated in an aqueous solution [56]. Solutions containing 500 ng L^{-1} of trimethyllead and triethyllead were stable for 3 months, but UV irradiation produced rapid decomposition of these lead species at concentrations of 1.5 μg L^{-1} lead and 2.9 μg L^{-1} lead, respectively. Dimethyl- and diethyllead species decomposed less rapidly under similar conditions and degradation of trialkyllead solutions in daylight was also observed. Trimethyllead was fairly stable over a period of 12 months. It can be concluded that storage in the darkness of alkyllead and especially of trialkyllead is advisable for preservation of the species [24].

As for copper, lead in aqueous samples also shows a significant loss due to absorption on the wall of polypropylene or polyethylene bottles [27]. In Teflon bottles, no loss of lead occurs.

The lead concentration in blood varies from 165 to 296 μg L^{-1}. The measured concentrations for lead depend on the year of sampling (use of leaded gasoline) and the sampled region [1].

7.9 Lithium

In the case of lithium no special precautions seem to be needed for the collection and storage of samples in relation to possible contamination. A well-documented history of the sample is necessary for evaluation of the results. In the literature, no reports on speciation of lithium in serum or urine could be found. There are some reports on the determination of low concentrations in biological samples and on the effect of lithium administration during the treatment of acute mania [57, 58].

In serum of healthy people, lithium concentrations are around the 1 μg L^{-1} level. In urine, normal excretion is up to 60 μg over a period of 24 h [57].

7.10 Manganese

Manganese is a very difficult element to measure in clinical samples because of the multiple possibilities of contamination. One of the difficulties is obtaining water pure enough for dilutions with very low manganese content. The subsequent problem is to conserve purified water, because contamination from ambient air or material occurs quite rapidly. For example, the absorbance of the blank increased substantially after 2 to 3 h in an auto sampler vessel, even when the auto sampler was covered [59]. As with lead, most of the manganese in blood is concentrated in the packed cells, so hemolysis should be avoid. Contamination originating from dust particles, can be eliminated by working in clean room (class 100) conditions. Stainless steel needles cannot be used since they leach manganese into the blood serum. Therefore the use of a Teflon cannula is recommended. Also all the vials and syringes need to be tested for manganese contamination. As in the case of chromium, sweat contains a lot of manganese and therefore cleaning of the skin is compulsory. Even tap water can contain high amounts of manganese (>1 mg L^{-1}).

In serum, the major fraction of manganese co-elutes with a UV peak of unidentified serum components, comprising different chains of immunoglobulines and low molecular weight complexes.

A small amount also elutes from the column in the dead volume, probably as free (solvated) ions [51]. In size-exclusion experiments, manganese was found in serum fractions corresponding to different molecular weight proteins, but no identification was done [60]. Recently, the first two mononuclear manganese citrate complexes were synthesized in aqueous solution near physiological values [61]. These manganese citrate species can be relevant to manganese speciation in biological media and potentially related to the beneficial as well as toxic effects of manganese on humans.

The manganese concentration in the serum of healthy persons is $0.5\,\mu g\,L^{-1}$. In urine of nonexposed persons the manganese concentration amounts to $1\,\mu g\,L^{-1}$ [1].

7.11 Mercury

For mercury, cleaning of the sample containers and checking of their possible contribution to the mercury level in the sample is compulsory. It is generally stated that water samples for mercury determination should be stored in glass bottles and that polyethylene containers are considered unsuitable. Recently, 300 ml poly(ethylene terephthalate) containers have been recommended for the sampling and storage of potable water [62]. Adding 0.5 ml of 20 % m/v potassium dichromate dissolved in nitric acid prevented loss of mercury. In another report on storage experiments in various containers, it was shown that organomercury species were stable for at least 30 days in all containers, except those made of polyethylene. Metallic mercury was stable in all containers except those made of stainless steel or polyethylene. Mercury(II) was rapidly lost from all containers except those made of aluminum, which rapidly converted mercury(II) to metallic mercury, which was stable [63]. In a recent report, the stability of methylmercury (highly toxic and an accumulator in the food chain) and inorganic mercury retained on yeast–silica gel microcolumns was tested and compared with the stability of these species in solution [64]. The columns were stored for 2 months at −20 °C, 4 °C and room temperature. Methylmercury was found to be stable in the columns over the 2 month period at the three temperatures tested while the concentration of inorganic mercury decreased after 1 week of storage even at −20 °C. Formation of methylmercury and dimethylmercury from mercury(II) in the presence of trimethyllead, an abiotic methyl donor, has been observed. These processes became less significant when humic substances were added. Under these conditions methylmercury was preserved at a level of $1\,ng\,L^{-1}$ at 4 °C in the dark for 33 days [65]. Blood samples for total mercury determination can be stored for a few weeks in the refrigerator. Longer periods require storage in the deepfreezer at −20 °C or below. However, stability of the individual mercury species should be tested for long storage periods. The stability of methylmercury chloride and mercury chloride in aqueous solutions was studied by the BCR at 0 °C and at room temperature to validate analytical methods [25]. No detectable effects of temperature were seen after storage for 3 months in the dark. Significant losses of mercury were observed after 100-fold dilution of the initial solution. Methylmercury and inorganic mercury in fish extract solutions were stable for 5 months at 4 °C when stored in the dark [66]. Most data seem to indicate that darkness is necessary for preservation of mercury species during storage in biological matrices and that there is much less influence from surrounding temperature. For total mercury determination in urine, acidification of the samples is recommended (with nitric or acetic acid) to avoid mercury absorption by the container wall. Hence, in speciation analysis any addition of acid should be avoided unless it is proven that is has no influence on the metal species present in the matrix. In the case of mercury it is also important to avoid bacterial growth in the sample as this may reduce some mercury to volatile elemental mercury. For all these reasons it is important to analyze the samples as soon as possible after collection is done. By extension, this also applies to all other elements.

The mercury concentration in serum of healthy persons is about $0.5\,\mu g\,L^{-1}$. In urine of nonexposed individuals the mercury concentration ranges from 1 to $10\,\mu g\,L^{-1}$. Speciation in urine indicated that inorganic mercury was the major form of mercury excretion. Other results indicated as well the presence of methylmercury and ethylmercury [26, 67].

7.12 Nickel

Contamination is a major problem in studying nickel in body fluids. Similar to chromium and manganese, the concentration of nickel in sweat is several times higher than that in serum. Skin should be washed carefully before collecting the sample. In the case of a smoking person, risk of contamination is even higher. Apparently, the use of a stainless steel needle does not compromise the results when the first 3 ml of blood are discarded. However, all materials that come into contact with the sample must be washed by an acid washing procedure. Sample manipulations should be carried out in a clean room (class 100). The concentration of nickel stored in Teflon containers remained constant over a period of 2 h (duration of experiment) [27].

Expected concentrations for nickel in serum and urine are respectively $<0.3\,\mu g\,L^{-1}$ and $<3\,\mu g\,L^{-1}$. A recent report indicated that workers in a nickel refinery are heavily exposed because of the high nickel concentrations that were found in urine [68]. Further speciation research is needed.

7.13 Selenium

For selenium the sampling procedures are essentially free of contamination problems. Standard equipment for sampling of body fluids can be used. In aqueous solution selenium may be present as selenite(IV), selenate(VI), methylated selenium and other forms of organic selenium such as selenium–cysteine and selenium–methionine, where it occurs bound to proteins [25]. Variations in the concentration of selenium species during storage have been reviewed recently [69, 70]. Adsorption and desorption phenomena were important at low selenium concentrations found in environmental samples. Loss of selenium depends on pH, ionic strength, container material and ratio of container surface area per unit of volume. There was no influence of light on inorganic selenium species. Selenium was released from Teflon, polyethylene and polycarbonate containers by 50 % HNO_3. Less leaching was observed with HCl. In a project of BCR solutions of $10\,\mu g\,L^{-1}$ selenium(IV) and $50\,\mu g\,L^{-1}$ selenium(VI) were kept at pH 2 and 6 both in the dark and exposed to light, at three different temperatures (−20, 20 and 40 °C), in two types of containers (PTFE and polyethylene) [25]. The effect of the chloride anion was also tested for its role in oxidizing selenium(IV) to selenium(VI) in basic solutions. Selenium(IV) and selenium(VI) remained stable for 2 months but longer storage periods (6 months) resulted in a decrease in selenium(IV) concentration. Complete loss of selenium(IV) occurred after 12 months when samples were stored in polyethylene containers at pH 2 in the absence and presence of chloride. Selenium(VI) was stable under those conditions for the 12 months tested. The stability of selenium(IV) increased at pH 6, with 2 months being the maximum storage time without risk of selenium(IV) loss. The presence of the chloride ion decreased the risk of losses in some cases. PTFE containers increased selenium(IV) losses at pH 6, especially at $10\,\mu g\,L^{-1}$. Both species were stable at −20 °C for the 12 months tested and losses of selenium(IV) and selenium(VI) were lower at 40 °C than at room temperature [69]. Most of these findings are confirmed in a recent report [70]. However, it is stated there that stability increases with decreasing temperature. The stability order for storage containers was given as Teflon > polyethylene > polypropylene and for pH values was pH 2 > pH 4 > pH 8. The stability of four volatile organic selenium species was also tested under different storage conditions. In a short-term stability test of studied Se species in urine it was observed that after 5 h of storage about 30 % selenite(IV) and 60 % of SeCys were lost [71]. In a subsequent report, the maximum allowed storage time was 1 week [72].

Selenium concentrations in serum vary between 0.04 and $0.16\,mg\,L^{-1}$. In urine, the selenium concentration is about $100\,\mu g\,L^{-1}$. Both concentrations depend heavily on the selenium intake.

In serum, selenium was found to be distributed among three different fractions with approximate molecular weights of >600, 200 and 90 kDa [73]. Up till now, three selenium-containing proteins have been identified in the literature: selenoprotein P, glutathione peroxidase and albumin [74]. For urine, several methods exist to determine the concentration of trimethylselenonium ion (TMSe), selenite(IV) and total selenium [26, 75].

7.14 Tin

Apart from inorganic tin, several organotin compounds are found in nature: tributyltin (TBT), triphenyltin (TPT), dibutyltin (DBT), monobutyltin (MBT), diphenyltin (DPT) and monophenyltin (MPT) [19]. TPT and butyltin species were found to be stable for 3 months in HCl-acidified water stored in polyethylene bottles at 4 °C in the dark and for at least 20 days in brown glass bottles stored at 25 °C [76]. In filtered seawater samples, acidified to pH < 2, TBT was found to be stable when the water was stored in the dark at 4 °C in Pyrex bottles. The addition of an acid to preserve the butyltin concentrations was unnecessary when polycarbonate bottles were used for storage at 4 °C in the dark. Under these conditions, butyltins were stable for 7 months, but TPT showed a reduction from the first month of storage. Storage in Pyrex bottles and acidification showed no improvement. A better preservation of TPT was obtained in C-18 cartridges stored at room temperature and no changes in concentration were observed during the first 60 days. Under the same conditions, butyltins were stable for 7 months [19]. The stability of organotin compounds was found to be dependent on the pretreatment applied for preservation. A significant reduction of the concentration of butyltin was observed when the sample was air-dried under the action of infrared radiation or oven-dried at 110 °C. Lyophilization or desiccation procedures did not affect the stability of butyltin and phenyltin species. Four different storage procedures were evaluated: (1) at 4 °C, (2) freezing, (3) lyophilization and storage in a refrigerator and (4) drying in a desiccator and storage at 4 °C. The butyltin compounds were stable for at least 1 year for the four storage conditions, but a reduction in the phenyltin content was observed after 3 months, freezing and lyophilization being the most reliable procedures for conservation [77]. Also, to avoid losses or changes of butyltin species in oysters and cockles, lyophilization has been shown to be a reliable procedure [19]. Butyltin species were stable over 150 days. A higher stability was found in wet samples stored at −20 °C in the dark. A reduction of 14 % was observed in the TBT content after 270 days, followed by an increase in DBT concentration. Afterwards, a general decrease in TBT and DBT levels and an increase in MBT levels was observed. After 540 days, a general decrease of all the butyltin species was found. This behavior confirms that TBT degrades by stepwise debutylation to DBT, MBT and inorganic tin [19]. Recently, a higher stability (44 months) of organotin compounds in mussels has been reported for lyophilized samples stored at −20 °C in the dark. Temperature, and especially light, affect the stability of butyltin species in this sample. Significant variations were found in the butyltin content after 3 and 6 months of storage at room temperature in daylight and in the dark, respectively. Phenyltin compounds showed a lower stability: a reduction of 30 % was reported for TPT after 12 months for samples stored at −20 °C [78].

For tin in normal human serum, the following values have been reported: 0.502 ± 0.096 ng mL^{-1} (mean) and 0.400–0.636 ng mL^{-1} (range) [79].

7.15 Zinc

In contrast to selenium, the hazard of contamination by zinc during sample collection is very real. Main sources of contamination are collection vials, including the stopper. Careful acid washing of all glassware and plastic ware, followed by rinsing in pure water is recommended. For short-term storage of rainwater in HDPE and PTFE containers there is no loss of zinc [27]. This is rather surprising because HDPE containers have been reported to contain 200 μg mL^{-1} of zinc as metallic impurities (catalyst). It is suggested that the observed stability of dissolved zinc in a snow sample is the result of fortuitous balancing of the loss of zinc from the snow sample with the gain of zinc resulting from it leaching out of the walls of the container. It may also be due to the absence of any biological transformation or lack of formation of any colloidal or ion-exchangeable species likely to be adsorbed onto the container surface [27]. Similar to lead and manganese, the zinc concentration in packed cells is much higher than that in serum. This means that hemolysis should be assessed by measuring the hemoglobin concentration in serum after separation. Moreover, zinc is an element that shows pronounced diurnal changes, so it is important to pay attention at the time of sampling [80]. In case of zinc contamination, there will be an increase in the albumin bound fraction [52]. It has also been shown that the concentration of protein-bound serum zinc in human blood plasma varies depending on whether the sample was taken from a patient who was standing or lying in a recumbent position [81].

Zinc concentration in serum amounts to 1 $mg\,L^{-1}$. In urine, the zinc concentration varies between 50 and 1000 $\mu g\,day^{-1}$.

In serum, zinc elutes in four different fractions. Some zinc elutes in the dead volume of an anion-exchange Mono Q column, probably as free zinc ions. A part of zinc co-elutes with transferrin and also in two other serum fractions, one between immunoglobulin G and transferrin and the other between transferrin and albumin. It is suggested that the last fraction correspond to the α_2-macroglobulin protein (720 kDa) [26, 51]. The distribution of zinc in the plasma of normal and uremic patients differs from each other [51].

8 INFLUENCE OF pH, SALT MOLARITY AND ACETONITRILE CONCENTRATION ON A SELECTED METAL–PROTEIN COMPLEX, i.e. THE VANADATE(V)–TRANSFERRIN COMPLEX

To demonstrate the instability of some metal–protein complexes, we use an example that we have been studying extensively in our laboratory, i.e. the vanadate(V)–transferrin complex [82]. Vanadium is an element that has been far less studied in speciation analysis in comparison to other elements such as arsenic and selenium. However, in recent years the interest in vanadium has grown because of its potential medicinal use as a drug in diabetes and cancer treatment. At the start of our vanadium speciation project, we examined the stability of the vanadate–transferrin complex. This complex is the principal vanadium species in serum and is one of the strongest known vanadium complexes ($\log K = 6.5\,m^{-1}$). However, in comparison with other metal–transferrin complexes, its stability constant is rather low [$\log K(Fe^{3+}) = 22.7\,m^{-1}$ and $\log K(Al^{3+}) = 12.9\,m^{-1}$]. These constants already give an indication for the use of mild separation techniques and conditions. The goal of these experiments was to outline the limits of the pH and salt molarity and acetonitrile concentrations that could be used during separation techniques without altering the vanadate–transferrin complex. All these experiments were carried out by ultrafiltration. First we examined the influence of pH. The result is shown in Figure 2.2.3. As can be seen there is a strong pH dependency of the vanadate–transferrin complex. Around the physiological pH the percentage of vanadate that is bound to transferrin is high while on going to extreme acid media or extreme basic media the binding is rapidly ruptured. This behavior can be explained by the protonation and deprotonation of some amino acids at the protein binding site. Out of this picture it is clear that addition of an acid for preservation purposes, e.g. to urine, would lead to misleading results. This kind of study was also carried out for the vanadate–albumin complex (figure not shown). Here, addition of an acid leads to a higher binding capacity of albumin for vanadate at lower pH values, opposite to the vanadate–transferrin equilibrium.

Also addition of high amounts of salt (Figure 2.2.4) to the sample induces rupture of the vanadate–transferrin binding. Salts are added to the buffer during chromatographic techniques such as anion-exchange or hydrophobic interaction chromatography. The salts are added to generate a gradient, which governs the elution behavior of the analytes from the column. From

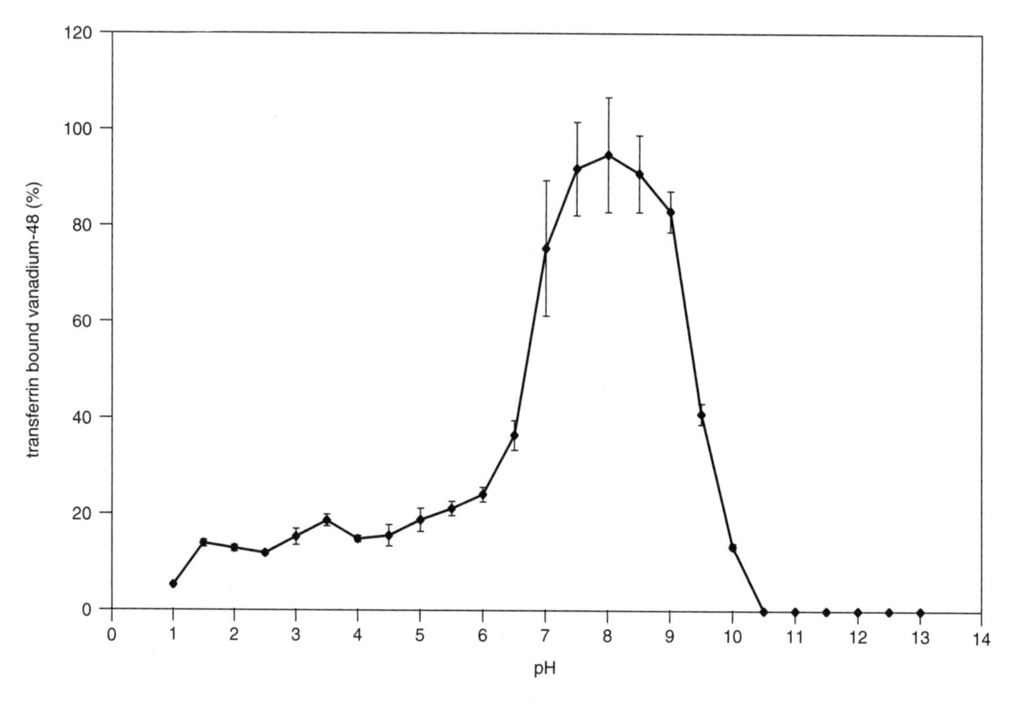

Figure 2.2.3. Stability of vanadium–transferrin complex as a function of pH (mean ± SD). (Reprinted from *Fresenius' Journal of Analytical Chemistry*, Stability of vanadium(V) protein complexes during chromatography, Vol. 363, pp. 519–522, Figures 1–6, 1999, by permission of Springer-Verlag GmbH & Co. KG.)

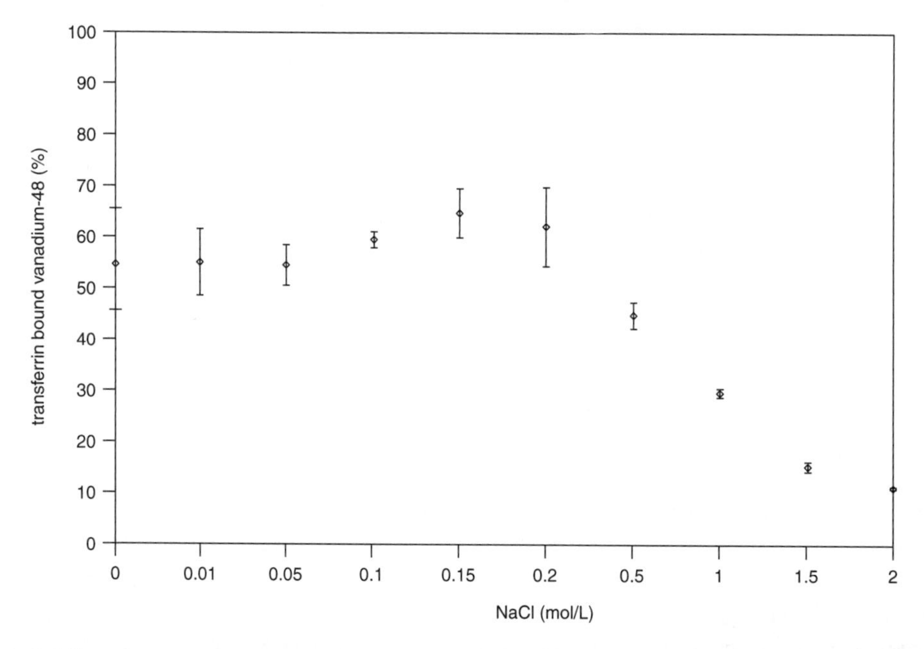

Figure 2.2.4. Stability of vanadium–transferrin complex as a function of NaCl concentration (mean ± SD). (Reprinted from *Fresenius' Journal of Analytical Chemistry*, Stability of vanadium(V) protein complexes during chromatography, Vol. 363, pp. 519–522, Figures 1–6, 1999, by permission of Springer-Verlag GmbH & Co. KG.)

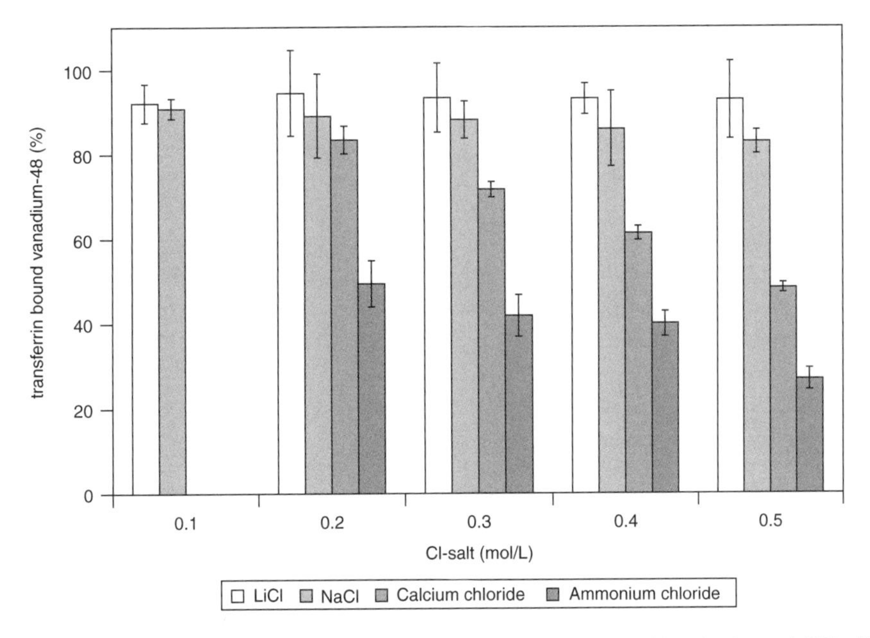

Figure 2.2.5. Stability of vanadium–transferrin complex as a function of different chloride salts (mean ± SD). (Reprinted from *Fresenius' Journal of Analytical Chemistry*, Stability of vanadium(V) protein complexes during chromatography, Vol. 363, pp. 519–522, Figures 1–6, 1999, by permission of Springer-Verlag GmbH & Co. KG.)

Figure 2.2.4 we can conclude that the use of hydrophobic interaction chromatography (high salt concentration at starting point) is not recommended for our kind of study. In anion-exchange chromatography, high salt concentrations are also used, but at the end of the chromatographic run. In this way, limited use of the technique can be considered. We also noticed significant differences between some kinds of salt (Figure 2.2.5). For sodium salts (sodium acetate, sodium bromide, sodium iodide and sodium chloride) no significant differences were found. However, chloride salts (lithium chloride, sodium chloride, calcium chloride and ammonium chloride) exerted different influences on the vanadate–transferrin binding. In case of calcium chloride the reason probably will be the double amount of chloride anions that are present in solution in comparison with the other monochloride salts. For ammonium chloride, this effect originates from a shift in the pH after addition of this salt.

Addition of high amounts of acetonitrile in the buffer also negatively affects the vanadate–transferrin binding (Figure 2.2.6). This figure shows that the use of high acetonitrile concentrations (as in reversed phase chromatography) for separation of vanadate–protein complexes should be avoided.

All these figures show that it is important to investigate the stability of the trace element complexes before embarking on separation procedures. If not, the chromatograms will not reflect the original distribution of the metal species present in your sample, but only yield useless artifacts.

9 CONCLUDING REMARKS

The previous paragraphs indicate that a well-established sampling and storage protocol should be established before starting speciation research. First of all, it is necessary to exclude all known sources of contamination by evaluating and cleaning all the needed instruments, laboratory ware and reagents. Therefore, some evaluation and cleaning methods are suggested in the text. For certain experiments one should also consider the possible influence of the biological time on test subjects. For several trace elements specific precautions

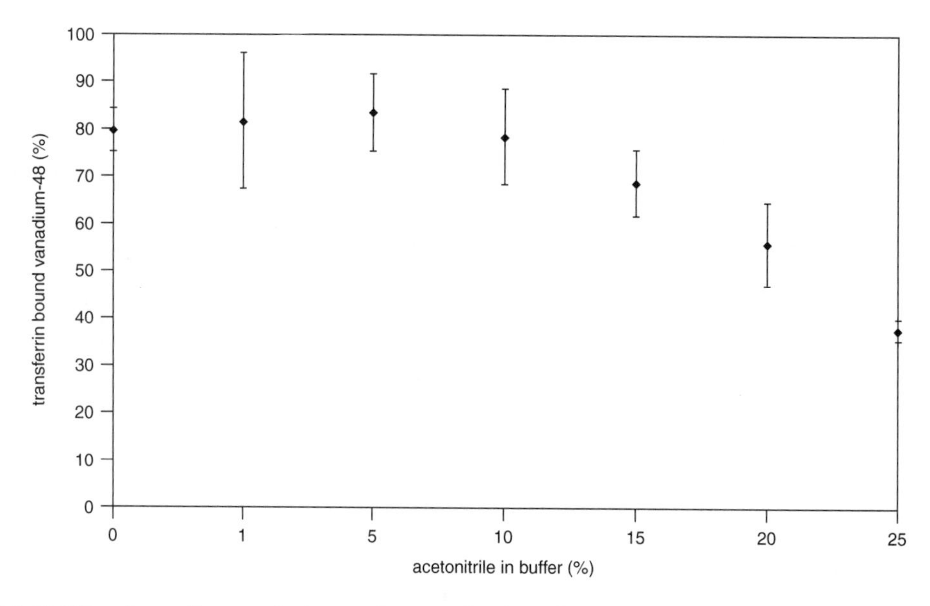

Figure 2.2.6. Stability of vanadium-transferrin complex as a function of acetonitrile concentration (mean ± SD). (Reprinted from *Fresenius' Journal of Analytical Chemistry*, Stability of vanadium(V) protein complexes during chromatography, Vol. 363, pp. 519–522, Figures 1–6, 1999, by permission of Springer-Verlag GmbH & Co. KG.)

are listed. As for the sampling as such, there is a choice between different methods, each with its particular advantages and disadvantages. Apart from the classical sampling methods (evacuated tubes or syringes for fluids, dissection for tissues), a more recent technique is mentioned, i.e. microdialysis. The advantages of this technique are its dynamic and mild characteristics, which are of utmost importance in trace element research. The disadvantages are the difficulties concerning detection and quantification of the analytes.

In the last section of this chapter, it is emphasized that storage and separation conditions for metal species cannot be varied infinitely. The borders of their stability interval under separation conditions should be outlined. This can be done for large molecules by, e.g., ultrafiltration experiments. From the example for the vanadate–transferrin complex it is clear that use of additives (acids, preservatives, anticlotting agents, ...) should be avoided as much as possible. Therefore serum is better suited for speciation purposes than is plasma.

The main goal of this chapter was to give the reader an overview of the most common procedures and pitfalls encountered in speciation research. However, it is left to the reader's judgment to decide to what extent these recommendations should be followed in his/her own research.

10 ACKNOWLEDGEMENTS

KDC is supported by a grant of the Fund for Scientific Research-Flanders (Belgium) (FWO)

11 REFERENCES

1. Cornelis, R., Heinzow, B., Herber, R. F. M., Christensen, J. M., Paulsen, O. M., Sabbioni, E., Templeton, D. M., Thomassen, Y., Vahter, M. and Vesterberg, O., *Pure Appl. Chem.*, **67**, 1575 (1995).
2. Burns, E. R., *Anat. Rec. (New Anat.)*, **261**, 141 (2000).
3. Pineau, A., Guillard, O., Chappuis, P., Arnaud, J. and Zawislak, R., *Crit. Rev. Clin. Lab. Sci.*, **30**, 203 (1993).
4. Kosta, L., *Talanta*, **29**, 985 (1982).
5. Versieck, J. and Cornelis, R., *Trace Elements in Human Plasma or Serum*, CRC Press, Boca Raton, FL, 1989.
6. Alcock, N. W., *At. Spectrom.*, **5**, 78 (1984).
7. Schmitt, Y., *J. Trace Elem. Electrolytes Health Dis.*, **1**, 107 (1987).

8. Cornelis, R., *Analyst*, **117**, 583 (1992).
9. Cornelis, R., *J. Radioanal. Nucl. Chem.*, **244**, 255 (2000).
10. Beto, J. A., Bansal, V. K. and Kahn, S., *Adv. Renal Replace. Ther.*, **6**, 85 (1999).
11. *Procedures for the Collection of Diagnostic Blood Specimens by Venipuncture; Approved Standards* (ed. 4), National Committee on Clinical Laboratory Standards, Wayne, PA, Publication H3-A4, 1998.
12. Versieck, J., *Crit. Rev. Clin. Lab. Sci.*, **22**, 97 (1985).
13. de Haan, K. E. C., de Groot, C. J. and Boxma, H., *Clin. Chim. Acta*, **170**, 111 (1987).
14. Nygren, O. and Nilsson, C., *J. Anal. At. Spectrom.*, **2**, 805 (1987).
15. Neve, J. and Petetz, A., Expression of urinary selenium levels in humans, in *Selenium in Medicine and Biology*, Neve, J. and Favier, A. (Eds), Walter de Gruyter, Berlin, 1988, pp. 189–192.
16. Boiteau, H. L. and Pineau, A., Mercury, in *Quantitative Trace Analysis of Biological Materials*, McKenzie, H. A. and Smythe, L. E. (Eds), Elsevier, Amsterdam, 1988 pp. 553–560.
17. Maletskos, C. J., Albertson, M. D., Fitzsimmons, J. C., Masurekar, M. R. and Tang, C. W., Sampling and sample handling of human tissue for activation analysis, in *Trace Substances in Environmental Health IV*, Hemphill, D. D. (Ed.), University of Missouri, Columbia, MO, 1970, p. 367.
18. Uthe, J. F. and Chou, C. L., *Sci. Total Environ.*, **71**, 67 (1988).
19. Gomez-Ariza, J. L., Giraldez, I., Morales, E., Ariese, F., Cofino, W. and Quevauviller, P., *J. Environ. Monit.*, **1**, 197 (1999).
20. Johansen, M. J., Newman, R. A. and Madden, T., *Pharmacotherapy*, **17**, 464 (1997).
21. Sarre, S., Deleu, D., Van Belle, K., Ebinger, G. and Michotte, Y., *Trends Anal. Chem.*, **12**, 67 (1993).
22. Lunte, C. E., Scott, D. O. and Kissinger, P. T., *Anal. Chem.*, **63**, 773A (1991).
23. BioAnalyticalSystems, http://www.bioanalytical.com.
24. Gomez-Ariza, J. L., Morales, E., Sanchez-Rodas, D. and Giraldez, I., *Trends Anal. Chem.*, **19**, 200 (2000).
25. Quevauviller, P., de la Calle-Guntinas, M. B., Maier, E. A. and Camara, C., *Mikrochim. Acta*, **118**, 131 (1995).
26. Das, A. K., Chakraborty, R., Crevera, M. L. and de la Guardia, M., *Mikrochim. Acta*, **122**, 209 (1996).
27. Sekaly, A. L. R., Chackrabarti, C. L., Back, M. H., Grégoire, D. C., Lu, J. Y. and Schroeder, W. H., *Anal. Chim. Acta*, **402**, 223 (1999).
28. Harris, W. R., Berthon, G., Day, J. P., Exley, C., Flaten, T. P., Forbes, W. F., Kiss, T., Orvig, C. and Zatta, P. F., *J. Toxicol. Environ. Health*, **48**, 543 (1996).
29. Ohman, L. O. and Martin, R. B., *Clin. Chem.*, **40**, 598 (1994).
30. Kiss, T., Jakusch, T., Kilyen, M., Kiss, E. and Lakatos, A., *Polyhedron*, **19**, 2389 (2000).
31. de la Calle-Guntinas, M. B., Madrid, Y. and Camara, C., *Fresenius' J. Anal. Chem.*, **344**, 27 (1992).
32. Smith, A. E., *Analyst*, **98**, 65 (1973).
33. Smichowski, P., Madrid, Y. and Camara, C., *Fresenius' J. Anal. Chem.*, **360**, 623 (1998).
34. Delves, C. E., Sieniawska, C. E., Fell, G. S., Lyon, T. D. B., Dezateux, C., Cullen, A., Variend, S., Bonham, J. R. and Chantler, S. M., *Analyst*, **122**, 1323 (1997).
35. Feldmann, J., Lai, V. W. M., Cullen, W. R., Ma, M. S., Lu, X. F. and Le, X. C., *Clin. Chem.*, **45**, 1988 (1999).
36. Al-Sibaai, A. A. and Fogg, A. G., *Analyst*, **98**, 732 (1973).
37. Jokai, Z., Hegoczki, J. and Fodor, P., *Microchem. J.*, **59**, 117 (1998).
38. Lagarde, F., Asfari, Z., Leroy, M. J. F., Demesmay, C., Olle, M., Lamotte, A., Leperchec, P. and Maier, E. A., *Fresenius' J. Anal. Chem.*, **363**, 12 (1999).
39. Demesmay, C., Olle, M. and Porthault, M., *Fresenius' J. Anal. Chem.*, **348**, 205 (1994).
40. Lagarde, F., Amran, M. B., Leroy, M. J. F., Demesmay, C., Olle, M., Lamotte, A., Munteau, H., Michel, P., Thomas, P., Caroli, S., Larsen, E., Bonner, P., Rauret, G., Foulkes, M., Howard, A., Griepink, B. and Maier, E. A., *Fresenius' J. Anal. Chem.*, **363**, 18 (1999).
41. Wang, J., Dreessen, D., Wiederin, D. R. and Houk, R. S., *Anal. Biochem.*, **288**, 89 (2001).
42. Majidi, V. and Miller-Ihli, N. J., *Analyst*, **123**, 809 (1998).
43. Dyg, S., Cornelis, R., Quevauviller, P., Griepink, B. and Christensen, J. M., *Anal. Chim. Acta*, **286**, 297 (1994).
44. Sacher, F., Raue, B., Klinger, J. and Brauch, H. J., *Int. J. Environm. Anal. Chem.*, **74**, 191 (1999).
45. Cornelis, R., Borguet, F., Dyg, S. and Griepink, B., *Mikrochim. Acta*, **109**, 145 (1992).
46. Feng, W. Y., Qian, Q. F., Ding, W. J. and Chai, Z. F., *J. Radioanal. Nucl. Chem.*, **244**, 321 (2000).
47. Makarov, A. and Szpunar, J., *J. Anal. At. Spectrom.*, **14**, 1323 (1999).
48. Gallorini, M., Edel, J., Pietra, R., Sabbioni, E. and Mosconi, G., *Sci. Total. Environm.*, **150**, 153 (1994).
49. Cornelis, R., Heinzow, B., Herber, R. F. M., Christensen, J. M., Poulsen, O. M., Sabbioni, E., Templeton, D. M., Thomassen, Y., Vahter, M. and Vesterberg, O., *J. Trace Elements Med. Biol.*, **10**, 103 (1996).
50. Gardiner, P. E. and Ottaway, J. M., *Anal. Chim. Acta.*, **124**, 281 (1981).
51. Bayon, M. M., Cabezuelo, A. B. S., Gonzalez, E. B., Alonso, J. I. G. and Sanz-Medel, A., *J. Anal. At. Spectrom.*, **14**, 947 (1999).
52. Cornelis, R., Borguet, F. and De Kimpe, J., *Anal. Chim. Acta.*, **283**, 183 (1993).
53. Gercken, B. and Barnes, R. M., *Anal. Chem.*, **63**, 283 (1991).
54. Owen, L. M. W., Crews, H. M., Hutton, R. C. and Walsh, A., *J. Anal. At. Spectrom.*, **117**, 649 (1992).
55. Nygren, O., *Appl. Organomet. Chem.*, **8**, 601 (1994).
56. Van Cleuvenbergen, R., Dirkx, W., Quevauviller, P. and Adams, F., *Int. J. Environ. Anal. Chem.*, **47**, 21 (1992).
57. Sampson, B., *J. Anal. At. Spectrom.*, **6**, 115 (1991).

58. Keck, P. E., Strakowski, S. M., Hawkins, J. M., Dunayevich, E., Tugrul, K. C., Bennett, J. A. and McElroy, S. L., *Bipolar Disorders*, **3**, 68 (2001).
59. Neve, J. and Leclercq, N., *Clin. Chem.*, **37**, 723 (1991).
60. Bratter, P., Ribas, E. and Schramel, P., *Trace Elem. Anal. Chem. Med. Biol.*, **6**, 1 (1994).
61. Matzapetakis, M., Karligiano, N., Bino, A., Dakanali, M., Raptopoulou, C. P., Tangoulis, V., Terzis, A., Giapintzakis, J. and Salifoglou, A., *Inorg. Chem.*, **39**, 4044 (2000).
62. Copeland, D. D., Facer, M., Newton, R. and Walker, P. J., *Analyst*, **121**, 173 (1996).
63. Bloom, N. S., *Fresenius' J. Anal. Chem.*, **366**, 438 (2000).
64. Perez, M. T., Madrid-Albarran, Y. and Camara, C., *Fresenius' J. Anal. Chem.*, **368**, 471 (2000).
65. Reinholdsson, F., Briche, C., Emteborg, H., Baxter, D. C. and French, W., in *CANAS 95*, Welz, R. (Ed.), Perkin-Elmer, Ueberlingen, 1995.
66. Quevauviller, P., Drabaek, I., Munteau, H. and Griepink, B., *Appl. Organomet. Chem.*, **7**, 413 (1993).
67. Shum, S. C. K., Pang, H. and Houk, R. S., *Anal. Chem.*, **64**, 2444 (1992).
68. Thomassen, Y., Nieboer, E., Ellingsen, D., Hetland, S., Norseth, T., Odland, J. O., Romanova, N., Chernova, S. and Tchachtchine, V. P., *J. Environ. Monit.*, **1**, 15 (1999).
69. Cobo, M. G., Palacios, M. A., Camara, C., Reis, F. and Quevauviller, P., *Anal. Chim. Acta*, **286**, 371 (1994).
70. Gomez-Ariza, J. L., Pozas, J. A., Giraldez, I. and Moralez, E., *Int. J. Environ. Anal. Chem.*, **74**, 215 (1999).
71. Gomez, M. M., Gasparic, T., Palacios, M. A. and Camara, C., *Anal. Chim. Acta*, **374**, 241 (1998).
72. Quijano, M. A., Gutierrez, A. M., Perez-Conde, M. C. and Camara, C., *Talanta*, **50**, 165 (1999).
73. Bratter, P., Gercken, B., Tomiak, A. and Rosick, U., in *Proceedings of the Fifth International Workshop*, Bratter, P. and Schramel, P. (Eds), de Gruyter, Berlin, 1988.
74. Harrison, I., Littlejohn, D. and Fell, G. S., *Analyst*, **2**, 189 (1996).
75. Robberecht, H. J. and Deelstra, H. A., *Talanta*, **31**, 497 (1984).
76. Bergmann, K., Rohr, U. and Neidhart, B., *Fresenius' J. Anal. Chem.*, **349**, 815 (1994).
77. Gomez-Ariza, J. L., Morales, E., Beltran, R., Giraldez, I. and Ruiz-Benitez, M., *Quim. Anal.*, **13**, S76–S79 (1994).
78. Morabito, R., Munteau, H., Cofino, W. and Quevauviller, P., *J. Environ. Monit.*, **1**, 75 (1999).
79. Versieck, J. and Vanballenberghe, L., *Anal. Chem.*, **63**, 1143 (1991).
80. Dawson, J. B., *Fresenius' J. Anal. Chem.*, **324**, 463 (1986).
81. Behne, D., *J. Clin. Chem. Clin. Biochem.*, **19**, 115 (1981).
82. De Cremer, K., De Kimpe, J. and Cornelis, R., *Fresenius' J. Anal. Chem.*, **363**, 519 (1999).

2.3 Food: Sampling with Special Reference to Legislation, Uncertainty and Fitness for Purpose

P. Brereton, Roy Macarthur and H. M. Crews

Central Science Laboratory, Sand Hutton, York, UK

1 INTRODUCTION

An analytical sample typically consists of a few grams of homogenised food. The results of an analysis will tell us how much of a particular species of element that few grams of sample contained at the time it was analysed. However, it is hardly ever the case that it is the objective of a measurement merely to find out the composition of the analytical sample. Typically the objective of a measurement is to determine the level of those species within a food at the time the sample was taken from the bulk. The 'bulk' could be any body of food about which information is required. For example, a batch of food product from a factory; a consignment of food product for import; the agricultural product of a particular region; or the diet of children within a particular country.

If, as is nearly always the case, an estimate of the level of analyte in the bulk is the result required from the 'analysis' then it is important that the analysis (determination of analyte in sample) and sampling (everything that happens to sample prior to analysis), are thought of as a whole 'measurement process' [1]. The quality of the measurement is affected by both analysis and sampling. In fact, it is typical for sampling to make a larger contribution to the uncertainty associated with the result of a chemical measurement than analysis.

This chapter will examine, with special reference to two trace elements (arsenic and molybdenum), the effect of sampling in particular on results produced by measurements. It will discuss the methods that are available (and those under development) for assessing the contribution made by sampling to the uncertainty associated with a

Handbook of Elemental Speciation: Techniques and Methodology R. Cornelis, H. Crews, J. Caruso and K. Heumann
 ISBN: 0-471-49214-0

measurement. It will show how to calculate the effort that should be put into sampling in order to produce samples that are fit for purpose, and for monitoring the quality of the sampling process.

2 SAMPLING TARGETS AND METHODS OF SAMPLING

The choice of plan for taking and preparing food samples is dependent on the matrix, the analyte(s), and the purpose to which results are to be put. The first two factors govern the qualitative features that should be selected for the sampling plan (e.g. sampling tool, storage conditions, sample container). The third factor governs the quantitative performance required of the sampling plan (e.g. maximum acceptable bias, cost of sampling, sample size, number of increments). For speciation measurements in particular, the collection and storage of samples can have a profound effect on the stability of element species.

For example, as part of an ongoing EU project, (G6RD-CT-2001-00473/SEAS; personal communication, Prof. Carmen Camara, University of Madrid, 2002), which is investigating the stability of a variety of elemental species in food matrices, Camara *et al.* have investigated arsenic species in rice. They determined the homogeneity for total arsenic and its species (As(III), As(V), DMA and MMA) in two sets (SEAS rice and NIST SRM 1568a) of rice samples. No problems were found with the homogeneity results both within and between the containers for both samples of rice. They then tested the effect of storage for 2 months of sets of 20 bottles of each rice at room temperature, at 4 °C and at −20 °C in the dark. The stability at these temperatures was evaluated by comparing the results at room temperature (about 20 °C) and 4 °C with those obtained at −20 °C used as reference at the measured time. The stability at −20 °C has been evaluated versus $t = 0$.

The uncertainty U_T was calculated as:

$$U_T = [(\mathrm{CV}_{4\,^\circ\mathrm{C\ or\ }20\,^\circ\mathrm{C}})^2 + (\mathrm{CV}_{-20\,^\circ\mathrm{C}})^2]^{1/2} R_T/100$$

(uncertainty with respect to reference at the same time)

$$U_T = [(\mathrm{CV}_{-20\,^\circ\mathrm{C},\,t})^2 + (\mathrm{CV}_{-20\,^\circ\mathrm{C},\,t=0})^2]^{1/2} R_T/100$$

(uncertainty with respect to reference at $t = 0$)

$$R_T = X_T/X_{\mathrm{reference}}$$

Table 2.3.1 shows the concentration in SEAS rice of each arsenic species at the different times

Table 2.3.1. Stability study of SEAS rice sample along two months of storage at room temperature, 4 °C and −20 °C; where $R_\mathrm{T} = X_\mathrm{T}/X_{-20\,^\circ\mathrm{C}}$ and X_T = mean of 20 replicates at room temperature, 4 °C and −20 °C.

As species	T(°C)	$X \pm$ SD (μg kg^{-1})			R_{T-20}		$R_{T-20,t=0}$	
		$t = 0$	Month 1	Month 2	Month 1	Month 2	Month 1	Month 2
As(III)	20		88.3 ± 4.9	94.9 ± 2.2	1.08	1.02	1.88	2.02
	4		81.9 ± 7.4	92.9 ± 2.3	1.00	1.00	1.75	1.98
	−20	46.99 ±0.75	82.2 ± 7.4	93.2 ± 1.9	1.00		1.74	1.98
DMA	20		29.7 ± 3.6	27.5 ± 2.5	1.14	1.15	1.05	0.97
	4		27.0 ± 1.6	24.9 ± 0.8	1.04	1.04	0.95	0.88
	−20	28.33 ±1.11	26.0 ± 2.3	23.9 ± 0.6	1.00		0.92	0.85
MMA	20		n.d	n.d	0.0	0.0	0.0	0.0
	4		n.d	n.d	0.0	0.0	0.0	0.0
	−20	18.10 ±1.7	n.d	n.d	0.0		0.0	0.0
As(V)	20		10.7 ± 3.9	n.d	0.97	0.0	0.44	0.0
	4		10.5 ± 4.7	n.d	0.95	0.0	0.43	0.0
	−20	24.45 ±1.09	11.0 ± 5.7	n.d	1.00		0.45	0.0

and temperatures. The ratios (R_T) for arsenic species, of the mean values of 20 measurements made at room temperature and 4 °C, versus the mean value of 20 determinations made for samples stored at −20 °C and versus time = 0 are given. For ideal stability, R is 1. Since the mean values of the different arsenic species at different times were not comparable, the values of U_T are not included in Table 2.3.1.

The work indicated no instability in NIST SRM 1568a but found that, for the SEAS rice, MMA and As(V) were not stable under different conditions. No significant differences in the stability at the different temperatures tested were detected and in all cases MMA and As(V) were completely lost by transformation into other As organic species. The sum of the content of the analysed As species in the stability study was constant during the first 3 months, which meant that the total As content remained constant. Also the results obtained after the first month were very similar and no further species interconversion was detected.

Camara *et al.* have postulated that the most likely reason for the instability of the arsenic species in the SEAS sample was the high humidity content (about 18 %) possibly leading to anaerobic activity. In addition the NIST SRM 1568a rice had been irradiated whilst the SEAS had not.

The reader is further directed to Chapter 2.1 where extensive guidance about the collection and storage of biological samples (which includes food matrices) is described. The author gives detailed approaches to sample collection, processing and storage that can be applied to the food sector.

Effective sampling will produce analytical samples that are representative of the bulk. The mean level of the analyte in a large set of representative samples will be equal to the mean level of the analyte in the bulk. Representative samples can also be used to provide information on the spatial (or temporal) variation in the level of the analyte throughout the bulk.

Perfect sampling is not possible. Samples are never wholly representative. The sampling process may be biased in some way towards samples containing particular levels of analyte. Or it may just be the case that the wide variation in analyte concentration throughout a bulk means that a wide range of results is generated by the analysis of the samples and hence it is difficult to gain a precise estimate of the mean concentration.

The sampling process may lead to the contamination of the sample with the species under investigation: for example if stainless steel sampling tools are used. On the other hand, species may be adsorbed onto the surface of an inappropriate container. Otherwise exemplary samples may become unrepresentative of the bulk if they are stored at the 'wrong' temperature; or at the 'wrong' humidity; or if they are exposed to light. Each matrix–analyte combination presents its own challenges and these are intensified when it is crucial that species integrity is maintained. It is not the object of this chapter to provide advice on eliminating the effects of sampling on the species within a sample (good advice is given in Chapter 2.1 that may be applied to food samples), but rather to now provide guidance on how to assess the size of the effect that the sampling process has.

A well-designed sampling protocol will provide information about the extent to which results of analyses are representative, by giving an estimate of the size of the uncertainty associated with the sampling process and will allow an assessment to be made of whether the sampling process is 'in control'. Even if there are good reasons for believing that a particular sampling process has little effect on the composition of a sample, if the size of the effect of the sampling process is not estimated then this belief remains an educated guess.

3 LEGISLATION AND STANDARDS

Recent development of international standards and legislation for sampling food products has been driven by the need for the results of measurements to be internationally comparable. Several international organisations such as the Codex Alimentarius Commission, the European Union and the World Trade Organisation have been responsible for producing legislation and standards relevant to the sampling of foodstuffs.

The creation of 'the single market' within the European Union gave increased the momentum of the harmonisation of national food laws within the Union and there is now a large body of legislation referring to the sampling and analysis of foodstuffs.

3.1 Examples of relevant legislation

'Official Control of Foodstuffs Directive'

Council Directive 85/591/EEC of 20 December 1985 concerning the introduction of Community methods of sampling and analysis for the monitoring of foodstuffs intended for human consumption [2].

Sampling of product types, e.g.
First Commission Directive 87/524/EEC of 6 October 1987 laying down Community methods of sampling for chemical analysis for the monitoring of preserved milk products [3]

Sampling for particular analytes, e.g.
Commission Directive 2001/22/EC of 8 March 2001 laying down the sampling methods and the methods of analysis for the official control of the levels of lead, cadmium, mercury and 3-MCPD (3-monochloropropane-1,2-diol) in foodstuffs [4].

3.2 Codex and WTO

The Codex Alimentarius Commission is the main international body concerned with the setting of food standards. The body is jointly funded by the Food and Agriculture Organisation and the World Health Organisation.

The Codex Committee on Methods of analysis and sampling has produced draft general guidelines on sampling [5]. The guidelines have been produced to ensure that 'fair and valid procedures are used when food is tested for compliance with a particular Codex commodity standard'. The guidelines are addressed to member states in order to enable them to resolve trade disputes.

The World Trade Organisation (WTO) was created in 1995 to provide an organisation capable of updating the implementation of agreements associated with the General Agreement on Tariffs and Trade. One of the Agreements that define the WTO rules has particular relevance to area of food specification and analysis and hence sampling: the Agreement of Sanitary and Phytosanitary Measures (SPS).

The SPS Agreement affirms that members of the WTO should be able to enforce measures to protect human and animal health, and instructs that SPS measures should be harmonised through the application of international standards guidelines and recommendations. The Agreement applies to all SPS measures 'which may, directly or indirectly, affect trade' and defines a sanitary or phytosanitary measure as (amongst other things) 'any measure applied to protect human or animal life or health ... from risks arising from additives, contaminants, toxins ... in foods, beverages or feedstuffs'.

The Agreement defines *harmonisation* as 'The establishment, recognition and application of common sanitary and phytosanitary measures by different members'. As part of the definition it defines 'international standards, guidelines and recommendations' as 'for food safety the standards, guidelines and recommendations established by the Codex Alimentarius Commission relating to food additives, veterinary drug and pesticide residues, contaminants, methods of analysis and sampling, and codes and guidelines of hygienic practice'.

The wide scope of the SPS Agreement (direct or indirect effect on trade) and the specification of the Codex Alimentarius Commission within the Agreement have (for the Members of the WTO) changed the nature of the Codex. Pre-WTO it was a body that made recommendations that could safely be ignored by members who chose to do so; now members of the WTO are effectively obliged to make Codex standards the basis of their national controls.

4 METHODS OF CALCULATING THE UNCERTAINTY ASSOCIATED WITH SAMPLING

The uncertainty associated with sampling has been referred to as 'the uncertainty that dares not speak its name' [6] because its contribution to measurement uncertainty is often neglected. The neglect arises from the lack of standard procedures for the quantitative assessment of sampling quality in the food sector; a perception that gaining information about sampling quality is difficult and expensive; and a lack of communication between samplers and analysts. Nevertheless in the food sector, sampling uncertainty has been shown often to make a larger contribution to measurement uncertainty than the uncertainty associated with analysis (with the possible exception of homogenous liquid samples). Hence no realistic assessment of measurement uncertainty can be made without taking into account the contribution made by sampling.

4.1 Communication

If an analyst is presented with a sample with no information on how that sample was produced then, even if the analytical uncertainty is well characterised, they will be unable to produce a realistic estimate of the measurement uncertainty that applies to the concentration of analyte in the sampling target. Similarly, if customers are presented with the result of an analysis without an assessment of the analytical uncertainty they will be unable to draw appropriate conclusions about even a well-characterised sample taken from a bulk.

4.2 Assessing sampling uncertainty

An essential starting point is to produce a well-characterised sample by following a prescribed (or agreed) sampling protocol. The protocol should have been studied and characterised prior to its use (just as analytical methods should be validated). Methods for characterising the performance of sampling protocols are under development [1, 7–12]. The sample should then by analysed by a well-characterised method, for example a method validated by collaborative trial [13] or by single laboratory validation following recognised protocols [14, 15].

A common statistical approach considers that the target consists of a large set of 'normally distributed' samples, the mean concentration of which is the true concentration of analyte in the sample. If several samples are taken and analysed then the mean result provides an estimate of the mean analyte concentration in the sampling target. The uncertainty associated with this estimate may be calculated by:

$$\text{confidence interval} = \text{mean result} \pm \frac{ts}{\sqrt{n}}$$

where:

t = value for the required confidence level given in t-tables;

s is the standard deviation displayed by the results;

n is the number of samples analysed.

However this approach is only valid if the analytical method has been shown to be unbiased, and the measurements are carried out over several batches of analyses. Analysing the samples in a small number of batches would produce results that do not represent analytical uncertainty, but something closer to analytical repeatability. Although an estimate of the overall measurement uncertainty can be gained, the contributions made by sampling uncertainty and analytical uncertainty cannot be separated. This may lead to problems in identifying the causes of apparently large measurement uncertainties (i.e. quality failures). Also, if a reduction in measurement uncertainty is required, estimates of the contributions made by sampling and analysis are needed to identify the most economical method of achieving the reduction (see Section 5).

4.3 Collaborative trial in sampling

Work in the geochemical and environmental sectors has been carried out to produce methods for the estimation of sampling uncertainty [7, 9, 10]. Methods that rely upon a single sampler taking replicate samples using a single protocol produce uncertainty estimates analogous to analytical repeatability. Methods that employ multiple samplers to take replicate samples using a single protocol produce uncertainty estimates that are analogous to the results of a collaborative trial of an analytical method. They are referred to as collaborative trials in sampling, and should be employed to provide a valid estimate of sampling uncertainty.

The collaborative trial in sampling is carried out by a number of samplers (≥ 8) who, while following the sampling protocol under study, independently use different equipment to take duplicate samples from the sampling target. The duplicate samples must then be analysed in random order under repeatability conditions. Nested ANOVA can then be employed to obtain estimates of: the analytical repeatability standard deviation; the within-sampler repeatability standard deviation (s_1); and the between-sampler standard deviation (s_2) (Figure 2.3.1).

These standard deviations can be combined in the usual way to produce an estimate of the sampling reproducibility standard deviation (sa_R).

$$sa_R = \sqrt{s_1^2 + s_2^2}$$

For unbiased sampling methods the sampling reproducibility standard deviation represents the sampling uncertainty standard deviation, i.e. the contribution made by sampling towards overall measurement uncertainty. Once estimates for sampling repeatability and reproducibility have been established they can be employed in a combined sampling–analytical QA system, and to produce valid estimates for overall measurement uncertainty.

However, like collaborative trials for analytical methods, collaborative trials in sampling are expensive. A further problem is that they are very specific (site, commodity, analyte) and are therefore practicable only for use with bulk commodities.

4.4 Combined sampling–analytical QA

A common element of analytical quality assurance is the duplicate analysis of samples to check that the repeatability standard deviation associated with the results of the analysis of a batch of samples is not larger than the repeatability standard deviation associated with the method of analysis. For a method of analysis with a repeatability standard deviation s_r, the results of two analyses carried out under repeatability conditions should rarely (95 %

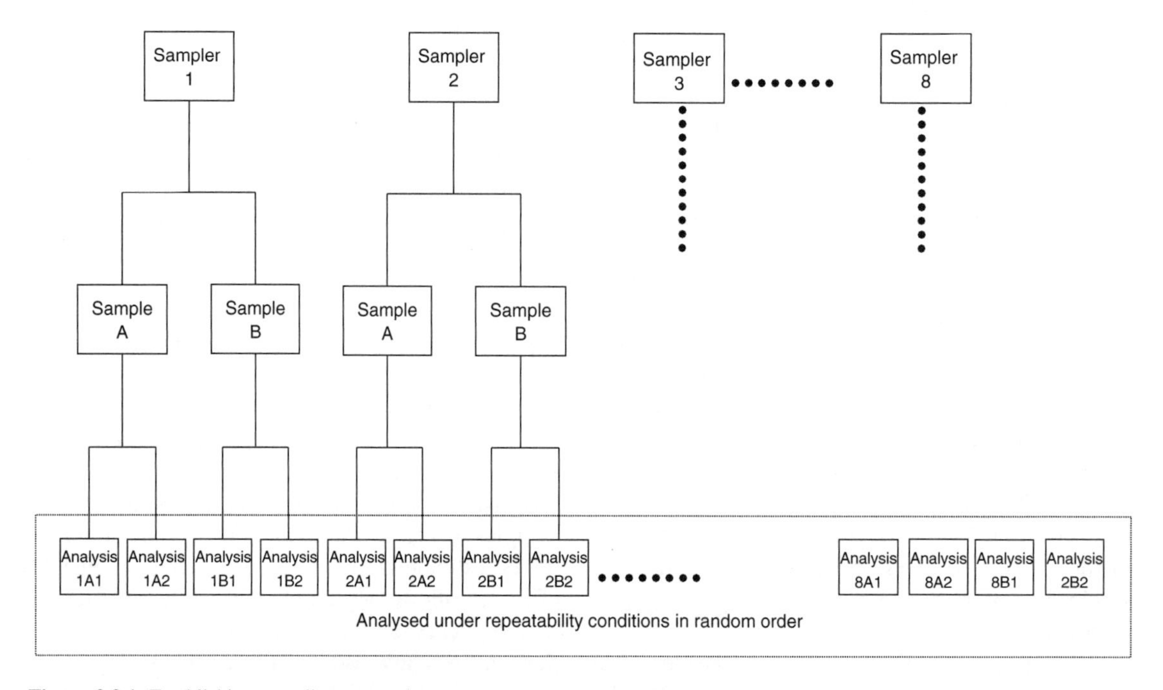

Figure 2.3.1. Establishing sampling uncertainty.

confidence) differ by more than the repeatability limit (r), where the repeatability limit is given by:

$$r = \sqrt{8}s_r$$

This type of quality assurance can be extended to sampling by the analysis of duplicate samples (taken under repeatability conditions). Several protocols can be used to check the quality of sampling, but perhaps the simplest employs the duplicate analysis of duplicate samples (Figure 2.3.2).

If duplicate samples are taken using a protocol assessed as having a repeatability standard deviation of s_1, and each sample is analysed in duplicate under repeatability conditions using a method assessed as having a repeatability standard deviation of s_r, then the mean of the duplicate analysis of each sample should rarely (95 % confidence) be greater than $\sqrt{4s_r^2 + 8s_1^2}$. If the duplicate analyses fall within the repeatability limit (r) of each other (and any additional analytical QA produces satisfactory results) then both sampling and analysis have been shown to be in control. At this stage the sampling reproducibility and standard uncertainty [16] associated with the analysis (u_a) can be combined to give an estimate of the uncertainty associated with the whole measurement (u_c) by:

$$u_c = \sqrt{sa_R^2 + u_a^2}$$

5 FITNESS FOR PURPOSE OF SAMPLING AND ANALYSIS

'Fitness for purpose' of analytical measurements has many definitions. One definition given by Thompson and Fearn [17] is that a method that is fit for purpose results in lower total financial loss than any other possible measurement method. One consequence of this is that a method for making a measurement that is sufficiently accurate is fit for purpose only if there is no other method capable of achieving the same accuracy at a lower cost. As both sampling and analysis contribute to the uncertainty associated with a measurement and the cost of making a measurement, they both need to be looked at to assess a method's fitness for purpose.

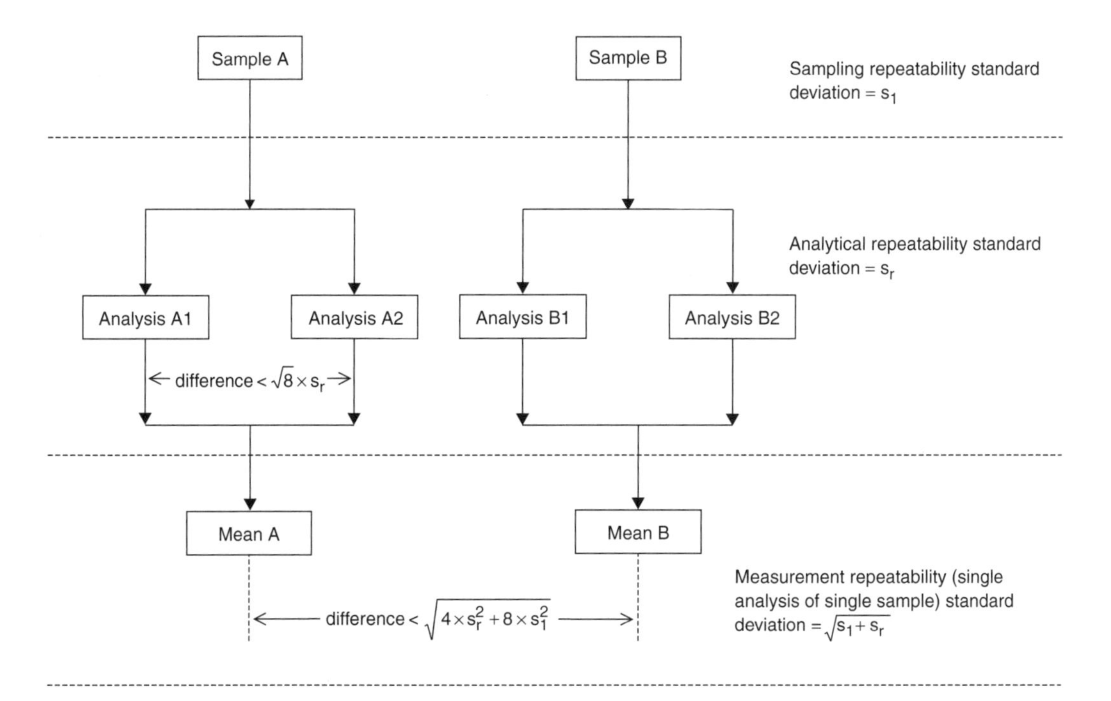

Figure 2.3.2. Combined sampling and analytical quality assurance.

If separate estimates of the uncertainty associated with sampling and the uncertainty associated with analysis (and the cost of these procedures) are available, then it is possible to check whether the proportion of resources devoted to each of the procedures is likely to lead to a measurement method that is fit for purpose.

For example, if a measurement is carried out by taking samples using a procedure with a reproducibility (assessed for example during a collaborative trial in sampling) given by sa_R, and an analytical standard uncertainty given by u_a (assessed for example using methods described in the Eurachem guide [16]); while the cost of producing one sample is £A, and the cost of analysing one sample is £B then the following statements can be made: If the cost of analysis and sampling is proportional to the 'effort' (two samples cost twice as much to analyse as one sample etc.) then

$$Asa_R{}^2 = K_s \quad Bu_a{}^2 = K_a$$

K_s and K_a are constants that reflect the cost at unit variance of sampling and analysis. The cost of producing samples with variance V_s and carrying out analyses with variance V_a is given by

$$\cos t = \frac{K_s}{V_s} + \frac{K_a}{V_a}$$

The variance associated with the measurement $V_m = V_s + V_a$, hence

$$\cos t = K_s/V_s + K_a/(V_m - V_s).$$

At the minimum cost the gradient of this curve is zero. Hence at minimum cost:

$$\frac{K_a}{(V_m - V_s)^2} - \frac{K_s}{V_s^2} = 0$$

Hence at minimum cost the optimum sampling variance (V_s') is given by:

$$V_s' = \frac{V_m(K_s - \sqrt{K_s K_a})}{K_s - K_a}$$

$$\Rightarrow V_s' = \frac{V_m\sqrt{K_s}(\sqrt{K_s} - \sqrt{K_a})}{(\sqrt{K_s} - \sqrt{K_a})(\sqrt{K_s} + \sqrt{K_a})}$$

$$V_s' = \frac{V_m\sqrt{K_s}}{\sqrt{K_s} + \sqrt{K_a}}$$

and the optimum analytical variance is given by

$$V_a' = V_m - V_s'$$

The optimum expenditure on sampling is given by $L_s' = K_s/V_s'$ and the optimum expenditure on analysis is given by $L_a' = K_a/V_a'$

The sum of the optimum expenditures ($L_a' + L_s'$) will be less than the sum of the original expenditures ($L_a + L_s$), while the overall measurement uncertainty will remain unchanged. If the values produced for the optimum expenditures are similar to the current expenditure on sampling and analysis then the measurement is likely to be 'fit for purpose'. If the sum of the optimum expenditures is much less then the current expenditure on sampling and analysis then the apportioning of resources between sampling and analysis should be re-examined.

The data treatments used to provide estimates of the uncertainty associated with sampling; for combined analytical and sampling quality control; and for the assessment of a measurement method fitness for purpose are all based on the assumption that 'normal statistics' describe uncertainties associated with sampling and analysis. This assumption is true (or cannot reasonably be tested) for many applications, although it is a good idea to employ the services of a professional statistician when designing experiments to assess the uncertainty associated with measurements so that this or any other assumptions used can be tested where possible.

6 EXAMPLE: THE MEASUREMENT OF MOLYBDENUM IN WHEAT

6.1 Establishing QA parameters through collaborative trial in sampling

The storage area was approximately 200 ft × 80 ft in total. The grain was stacked against the back wall of the barn and extended forwards towards the access area in the middle span of the barn. The pile of grain covered an area measuring approximately 80 ft^2 and was accessible only on its front sloping edge and top. The pile contained around 800 tonnes and was in the region of 9 ft high. Each section of the stored grain contained a dryer on the top of the pile.

The sampling procedure adopted took general account of a published standard method for the sampling of grain [18]. The standard emphasises the necessity of obtaining a properly representative sample and gives limited details described as 'good practice' which should be followed whenever practicable. It is noted that, as the composition of the lot is seldom uniform, a sufficient number of increments should be taken and carefully mixed to give a bulk (or composite) sample from which the laboratory sample can be obtained. Information is only given for a lot of up to 500 tonnes, in which case increments of 1 kg maximum to produce a composite of 100 kg and in turn a laboratory sample of 5 kg is deemed to be appropriate. Larger or smaller laboratory samples are recognised to be appropriate in some cases.

An 'in-house' sampling scheme was normally followed, in which a sampling spear (about 5 ft in length) containing approximately 100 g of grain was used to take multiple increments to produce a composite sample of around 600 g (or 5 kg if a very heterogeneously distributed analyte is being sampled).

For the purposes of the pilot sampling exercise, in the first instance, a professional sampler followed the normal sampling scheme and produced a composite sample by taking eight sample spear increments at random within each sector. This corresponded to normal practice of taking an eight-increment composite sample per 50 tonnes of grain. Further sampling was then carried out by four nonprofessional samplers. In each case the nonprofessional samplers took six sample increments using the spear, to produce each composite sample. In order to take these six spear increments randomly, use was made of the grid representation given in Figure 2.3.1. Working around the grid a coin was tossed: heads represented a cross in the relevant area of the grid (take a sample); if tails, no sample was taken. Some differences in the ease (and therefore amount collected) and the manner of sampling were apparent among the nonprofessional samplers. In addition to the samples taken by the four nonprofessional samplers one composite sample, again consisting of six incremental spear samples, was also taken to represent the whole of the front sloping edge of the pile. Each of the composite grain samples was blended by hand on receipt and a laboratory sample of approximately one-fifth the size of the composite sample was produced using a sample splitter.

The laboratory samples were analysed by ICPMS in random order under repeatability conditions to determine the concentration of trace elements. The concentration (dry matter basis) of molybdenum found in the samples is shown in Table 2.3.2.

A nested analysis of variance was used to determine the contributions towards the overall variance from between-sampler variation in sampling (sampling reproducibility), within-sampler variation in sampling (sampling repeatability) and analytical repeatability. The results of the analysis of variance are shown in Table 2.3.3.

$$s_r = \sqrt{MS_r} = \sqrt{0.0002617} = 0.016\,\text{mg kg}^{-1}$$

$$s_1 = \sqrt{(MS_1 - MS_r)/2}$$

Table 2.3.2. Results of measurements during collaborative trial in sampling.

Sampler	Sample	Analysis 1 (mg Mo kg^{-1})	Analysis 2 (mg Mo kg^{-1})
1	1	0.385	0.403
	2	0.424	0.435
2	3	0.469	0.446
	4	0.455	0.455
3	5	0.614	0.576
	6	0.494	0.511
4	7	0.531	0.495
	8	0.480	0.506
5	9	0.519	0.496
	10	0.487	0.492

Table 2.3.3. Sources of variation in sampling.

Source of variation	Degrees of freedom	Sum of squares	Mean square estimate
Between-sampler (s_2)	4	0.04301	0.01075
Within-sampler (s_1)	5	0.01055	0.002109
Analytical (s_r)	10	0.002617	0.0002617
Total	19	0.05618	

$$= \sqrt{(0.002109 - 0.0002617)/2}$$
$$= 0.030\,\text{mg kg}^{-1}$$
$$s_2 = \sqrt{(MS_2 - MS_1)/4}$$
$$= \sqrt{(0.0107533 - 0.0021093)/4}$$
$$= 0.046\,\text{mg kg}^{-1}$$

Standard deviation of variation due to sampling:

$$sa_R = \sqrt{s_1{}^2 + s_2{}^2} = \sqrt{0.030^2 + 0.046^2}$$
$$= 0.055\,\text{mg kg}^{-1}$$

The uncertainty associated with the analytical part of the measurement process was studied using methods described in the Eurachem guide to measurement uncertainty [15]. The standard uncertainty associated with the measurement was found to be $0.058\,\text{mg kg}^{-1}$.

Three conclusions can be drawn from this study:

The first conclusion is that both sampling and analysis make a significant contribution towards the measurement uncertainty. Although that statement sounds like it should always be true, if the standard uncertainty (standard deviation) associated with either sampling or (more commonly) analysis is less than 0.3 times the other component, then ignoring its contribution leads to a reduction in the estimated measurement uncertainty of less than 5 %. Therefore analyses carried out for the purposes of quality assurance should produce results that address both the analytical and sampling processes.

The second conclusion is that the standard uncertainty associated with results generated by the measurement of molybdenum in wheat from this supplier is $0.080\,\text{mg kg}^{-1}$ $\left(\sqrt{0.055^2 + 0.058^2}\right)$ or 16 % for samples containing approximately $0.5\,\text{mg kg}^{-1}$ of the metal.

The third conclusion is that the results of the duplicate analysis of a single sample should lie within $0.045\,\text{mg kg}^{-1}$ or $(\sqrt{8} \times 0.016)$ or 9 % of each other, and that the means of the results of duplicate analysis of duplicate samples should lie within $0.09\,\text{mg kg}^{-1}$ $\left(\sqrt{4 \times 0.016^2 + 8 \times 0.030^2}\right)$ or 18 % of each other. These results form the basis of quality assurance tests for the precision of sampling and analysis.

6.2 Fitness for purpose of the measurement of molybdenum in wheat

The cost of taking a sample of wheat for the analysis of molybdenum was £59.62. The cost of analysing a sample was £26.74. These costs were combined with the analytical and sampling uncertainties to provide an estimate of the measurement's fitness for purpose (Tables 2.3.4 and 2.3.5).

$$V_m = 0.058^2 + 0.055^2 = 0.00639$$
$$K_s = 59.62 \times 0.055^2 = 0.180$$
$$K_a = 26.74 \times 0.058^2 = 0.0900$$
$$V_s' = \frac{0.00639 \times \sqrt{0.180}}{\sqrt{0.180} + \sqrt{0.090}} = 0.00374$$
$$V_a' = 0.00639 - 0.00374 = 0.00265$$
$$L_s' = 0.180/0.00374 = £48.16$$
$$L_a' = 0.0900/0.00265 = £34.01$$

Figure 2.3.3 shows how the cost of the measurement varies with the proportion of the total expenditure devoted to sampling. It can be seen that, in this case, the optimum expenditure on sampling and therefore on analysis is not very different from the current expenditure. Therefore the

Table 2.3.4. Costs and uncertainties associated with the measurement of molybdenum in wheat.

	Cost (£)	Standard uncertainty (mg kg^{-1})
Analysis	26.74	0.058
Sampling	59.62	0.055
Measurement	86.36	0.080

Table 2.3.5. Optimum costs and uncertainties associated with the measurement of molybdenum in wheat.

	Optimum cost (£)	Optimum standard uncertainty (mg kg^{-1})
Analysis	34.01	0.051
Sampling	48.16	0.061
Measurement	82.17	0.080

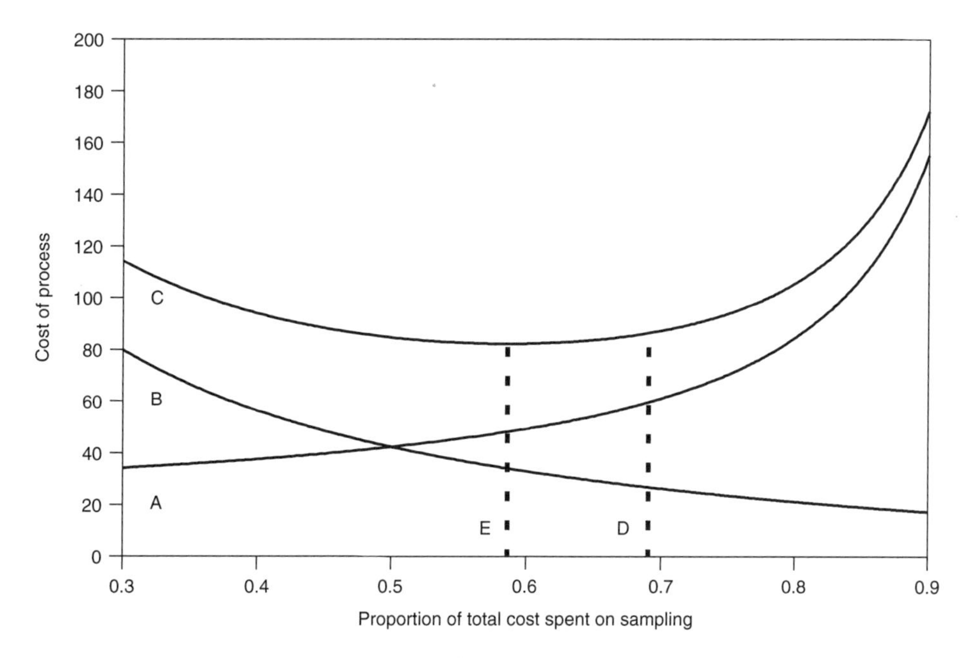

Figure 2.3.3. Variation in cost of measurement with proportion of expenditure devoted to sampling: (A) cost of sampling, (B) cost of analysis, (C) cost of measurement, (D) current division of expenditure between sampling and analysis, (E) optimum division of expenditure.

proportion of expenditure taken up by sampling and by analysis is about right and this aspect of the measurement is 'fit for purpose'.

7 SUMMARY

The aim of a measurement carried out on some sample of food is to gain information about a wider body of food rather than just information about the analytical sample. Thus, both analysis and sampling affect the quality of a measurement.

The need to produce comparable measurements can be met in two ways: through the production of international standards and international legislation on analysis and sampling by bodies such as the codex Alimentarius Commission, the World Trade Organisation and the European Union; and by the use of statistical methods to validate the performance of sampling methods.

Sampling methods can be validated by collaborative trial. However, the process is expensive and will not be suitable for all applications. Methods for combined analytical and sampling quality control are under development and should be used (in combination with standard and/or validated methods of sampling and analysis).

Information gained from the validation of sampling and analytical methods can be used to reduce the cost required to gain a given measurement precision, as well as to provide the information required to validate the performance of a measurement method.

8 ACKNOWLEDGEMENTS

The authors would like to acknowledge the United Kingdom Food Standards Agency who have funded work in this area and been very supportive of the studies.

9 REFERENCES

1. Ramsey, M. H., *J. Anal. At. Spectrom.*, **13**, 82 (1998).
2. Council Directive 85/591/EEC of 20 December 1985 concerning the introduction of Community methods of sampling and analysis for the monitoring of foodstuffs intended for human consumption, *Off. J.*, **L 372**, 0050–0052 (1985).

3. First Commission Directive 87/524/EEC of 6 October 1987 laying down Community methods of sampling for chemical analysis for the monitoring of preserved milk products, *Off. J.*, **L 306**, 0024–0031 (1987).
4. Commission Directive 2001/22/EC of 8 March 2001 laying down the sampling methods and the methods of analysis for the official control of the levels of lead, cadmium, mercury and 3-MCPD in foodstuffs, *Off. J.*, **L 077**, 0014–0021 (2001).
5. Codex Committee on Methods of Analysis and Sampling, Proposed draft general guidelines on sampling, ftp://ftp.fao.org:21/Codex/ccmas23/ma01_03e.pdf, 2001.
6. Thompson, M., *J. Environ. Monit.*, **1**, 19 (1999).
7. Ramsey, M. H., Argyraki, A. and Thompson, M., *Analyst*, **120**, 1353 (1995).
8. Thompson, M. and Ramsey, M. H., *Analyst*, **120**, 261 (1995).
9. Ramsey, M. H., Argyraki, A. and Thompson, M., *Analyst*, **120**, 2309 (1995).
10. Argyraki, A., Ramsey, M. H. and Thompson, M., *Analyst*, **120**, 2799 (1995).
11. Ramsey, M. H., *Analyst*, **122**, 1255 (1997).
12. Thompson, M., *Accreditation Qual. Assur.*, **3**, 117 (1998).
13. Thompson, M. and Wood, R., *J. Assoc. Off. Anal. Chem. Int.*, **76**, 926 (1993).
14. Fajgelj, A. and Ambrus, A. *Principles and Practice of Method Validation*, Royal Society of Chemistry, 2000.
15. Eurachem, *The fitness for purpose of analytical methods: a laboratory guide to method validation and related topics* (1998), http://www.eurachem.ul.pt.
16. Eurachem, Quantifying Uncertainty in Analytical Measurement, 2nd edn (2000), http://www.eurachem.ul.pt.
17. Thompson, M. and Fearn, T., *Analyst*, **121**, 275 (1996).
18. British Standards Institution, Methods for sampling cereals (as grain), BS4510, 1980.

2.4 Sampling: Collection, Storage–Occupational Health

Ewa Dabek-Zlotorzynska and Katherine Keppel-Jones

Environment Canada, Ottawa, Canada

1 INTRODUCTION

The characterization of airborne pollution including metallics is of great importance to many scientific fields, such as epidemiology and toxicology in health research [1–5]. In atmospheric sciences, it is important for modeling atmospheric processes and for environmental control purposes. Occupational health monitoring, on the other hand, relies on contaminant collection and subsequent characterization to evaluate health hazards for humans exposed to pollutants in the workplace. There is wide international agreement that health-related sampling should be carried out based on health-related particle size-selective criteria. Three fractions have been identified: inhalable (particles inhaled during breathing), thoracic (inhaled particles that penetrate beyond the larynx), and respirable (inhaled particles that penetrate to the alveolar region) [6, 7]. In addition, there is ample evidence that the chemical speciation of trace metals associated with particulate matter, and hence their bioavailability, depends on their oxidation state and how the elements are partitioned among the various compounds.

The increasing awareness of problems associated with airborne metallic pollutants has led to a need for defining sampling procedures and adapting existing analytical techniques to identify a wide variety of these compounds. Airborne contaminants can be present in air as particulate matter in the form of liquids or solids; as gaseous material in the form of true gas or vapor; or in a combination of these forms. Most metals in the atmosphere are associated with airborne particulate matter, though some trace atmospheric metals exist also in gaseous form.

In the first part of this chapter, we will focus our attention on sampling techniques used for collecting particulate and volatile pollutants. Total

Handbook of Elemental Speciation: Techniques and Methodology R. Cornelis, H. Crews, J. Caruso and K. Heumann
 ISBN: 0-471-49214-0

and size-fractionating filter collectors such as cascade impactors and other collection devices are addressed. However, the characteristics of these sampling techniques and samplers are discussed only briefly since more detailed descriptions can be found in several monographs [8–11]. An outline of sample handling and storage completes this section. The second part of the chapter will consist of more detailed descriptions of sampling methodologies for specific airborne metal species.

2 COLLECTION OF AIRBORNE PARTICULATE MATTER

Airborne particles are known as 'particulate matter (PM)' or simply 'particles'. These particles are very small solids and/or liquids that are produced by a variety of natural and man-made sources. Airborne particles vary widely in their chemical composition and size. The size of particles may range from 0.005 μm to 100 μm in diameter. The suspended portion (total suspended particulates or TSP, i.e. found floating in air) is generally less than 50 μm.

Suspended particulate matter is not a homogeneous pollutant. In the atmosphere, it typically exhibits a bimodal size distribution with one peak in the range 0.1–2.5 μm, and a second peak in the range 2.5–50 μm. PM_{10} (particulates smaller than 10 μm) include both a coarse fraction ($PM_{10-2.5}$), which is generated mainly mechanically, and a fine fraction ($PM_{2.5}$), which corresponds to secondary PM formed in the atmosphere by chemical reactions. $PM_{2.5}$ are particles of 2.5 μm or less in diameter. The finer particles pose the greatest threat to human health because they can travel deepest into the lungs. Depending on their size distribution, these particulates may be transported up to thousands of kilometers from their source.

A particulate sampling train consists of the following components: air inlet, particulate separator or collecting device, air flow meter, flow rate control valve, and air mover or pump. Of these, the most important component is the particulate separator. The separator may consist of a single element (such as a filter or impinger), or there may be two or more elements in a series (such as a two-stage cyclone or multi-stage impactor) so as to characterize the particulates into different size ranges.

2.1 Filters

Sampling on filters is the most practical method currently available to characterize the particle sizes and chemical composition of airborne particulates. In this case, the composition of bulk airborne particulate samples is performed. Although the individual character of the particulates is lost in this way, their size distribution can still be measured by bulk methods by using sampling devices (e.g. cascade impactors and filters) that collect the particulates according to their size. Airborne particulates acquired by drawing ambient air through filter material remain deposited on the filter, while all gases pass through the filter. The accurate measurement of flow rate and sampling time or sample volume is as important as the measurement of sample quantity.

Filter media of many different types and with many different properties have been designed for or adapted to air sampling requirements. These include fibrous (e.g. glass, quartz), membrane (e.g. cellulose nitrate, polycarbonate, Teflon) and sintered (e.g. silver) filters. Lippmann [12] provides a comprehensive list of commercially available filter types, their sampling efficiencies and manufacturers. No single filter type is suitable for all applications, and thus the choice of filter type for a given application depends greatly on the proposed analysis method for the collected sample. For example, membrane filters have the advantage that they can retain particulates effectively on their surface (good for microscopy), whereas fibrous types have the advantage of providing in-depth particle collection and hence a high load-carrying capacity (good for gravimetric assessment).

Filters are available in a range of dimensions (e.g. 25–100 mm diameter) and pore sizes (e.g. from 0.1 to 10 μm). General criteria which must be considered when selecting filter media are: (1) at least 99 % collecting efficiency for particulates

0.3 μm and greater in size, (2) low hygroscopicity, since a hygroscopicity exceeding 1 mg per piece leads to serious errors in weight concentration measurement, and so an improper estimate of the environmental concentration, and (3) absence of impurities that might interfere with the analysis. As an example, glass fiber or Teflon filters have been found to be unsuitable for the sampling of airborne dust with low platinum contents [13]: only polycarbonate and cellulose nitrate gave blank values as low as 5 pg Pt per total filter.

Additionally, the filters should be mechanically and thermally stable and should not interact chemically with the deposit, even when subjected to strong extraction solvent [8]. The chemical requirements of the filter depend on the nature of the proposed analysis. Spini *et al.* [14] have reported the reduction of Cr(VI) to Cr(III) when cellulose filters were extracted with alkaline solution containing a known amount of Cr(VI). The same results were obtained by an acid dissolution (H_2SO_4) of the filters, which can be explained by cellulose's well-known reducing properties. Therefore cellulose filters must be avoided for chromium speciation in airborne particulates.

Teflon membrane and quartz fiber are the filters most commonly used for particulate chemical analyses, while cellulose fiber filters lend themselves nicely to impregnation for absorbing gaseous precursors, and etched polycarbonate membrane filters are best suited for microscopic analyses.

2.2 Particulate sampling systems

A variety of PM samplers are available for collecting particulates onto a filter. The flow rate through a sampler and filter is usually established according to specifications of the sampling method and is held constant so that an accurate sample and representative volume can be determined. Stationary PM samplers are used to evaluate both outdoor and indoor work environments. Personal breathing zone samplers worn by workers are used to estimate their exposure to workplace pollutants.

2.2.1 Samplers for ambient air particulate matter

The commercial market for PM sampling is largely driven by the need to comply with U.S. ambient air quality standards. Since these standards specify mass concentration within a specific particle size range (PM_{10}) the majority of the reference and equivalent samplers are designed for this purpose.

High-volume samplers are commonly used to collect particulates with aerodynamic diameters less than 10 μm (termed PM_{10}) at flow rates of $1.1-1.4\,m^3\,min^{-1}$. In these systems, PM is collected on a large upwards-facing rectangular filter, which is located inside a large, weatherproof housing. Glass fiber and Hi-Vol samplers have been used for more than 50 years to measure air pollution, and thus procedures for these samplers are well established [8]. One drawback of these units is that frequent inlet cleaning is necessary for accurate size sampling.

For determining particle size distribution, impactors are commonly used. Particles are collected on impaction plates depending on their size: coarser ones on plates in the upper stages, and smaller ones on plates in the lower stages. Two different impactor systems are used for size-fractionated PM collection: virtual impactors and cascade impactors. The physical principles of both types are based on the mass moment of inertia combined with the air resistance of particles, which cause PM to be expelled from a sharply deviated air stream. The resulting aerodynamic diameter (AD) of particles, which is different from the real geometric diameter, therefore defines the size distribution. Impactors are easy to handle, are commercially available and have well-defined cut-off characteristics.

If one is interested in a detailed size distribution measurement of the airborne trace elements, cascade impactors are often used. They allow particles to be sampled and segregated with respect to their size [15]. The cascade impactor consists of a series of nozzles and coated impaction plates, called stages, with each stage collecting progressively smaller particles. Size segregation is dependent on sampler geometry and flow-rate. Cascade impactors can be quite variable in terms of stage numbers and particle 'cut-off' (AD with 50 % collection efficiency) by changing the geometric design and the corresponding flow rate. However, artifacts like particle 'blow-off' or 'bounce-off' from the collection substrate can appear [16].

Virtual impactors, on the other hand, are often constructed for only two or three particle sizes and operate at relatively low flow rates. They differ from cascade impactors in that the impaction plate is replaced by an opening that directs larger particles to one sampling substrate, and the smaller particles to another. This principle is used in the commercially available PM_{10} dichotomous (dichot) sampler, which was the original sampler for inhalable particles. This device uses a virtual impactor to separate the fine ($PM_{2.5}$) and coarse ($PM_{10}-PM_{2.5}$) size fractions at a total flow rate of $16.7\,L\,min^{-1}$. This is achieved by accelerating particles through a nozzle and then drawing 90 % of the flow stream off at right angles. The fine particles follow the right-angle flow stream, while the larger particles continue toward the collection nozzle. Particles are collected on 37 mm filters, which in this unit are generally Teflon or other membrane filters. Separation of the two size fractions minimizes potential interactions between the more acidic fine fractions and the more basic coarse fractions. Relative to cascade impactors, the effects of particle bounce-off and re-entrainment are reduced. However, this method requires that part of the total flow (typically 10 %) be drawn through the virtual surface. Therefore, correction factors must be applied to the coarse channel flow to account for contamination by the fine fraction.

More detailed descriptions of the above sampling devices and other samplers in current use are given in several monographs [9, 10].

2.2.2 *Samplers for particulates in workplace atmospheres*

Sampling and analysis of work atmospheres are simplified by two factors: (1) Industrial hygienists usually know which contaminant or contaminants are present in workroom air from the nature of the process plus a knowledge of raw materials, end products, and wastes. Therefore, identification of workroom contaminants is rarely necessary and, as a rule, only quantification is required. (2) Usually, only a single contaminant of importance is present in the workroom atmosphere and the absence of obvious interfering substances permits great simplification of procedures [17].

One sampling device used is the cyclone sampler. Cyclones have found increasing use in recent years as the first stage in two-stage samplers for respirable mass dust exposure determinations. They use an impeller to impose a circulatory motion upon air entering a cylindrical tube. General principles of centrifugal and gravitational forces are used in the cyclone sampler to separate PM into various size fractions. By design, cyclones used for respirable dust sampling are highly efficient for removal of larger particles (i.e. greater than 10 μm) and are not efficient for particles below about 2 μm. Cyclones are simple to operate, and can be very compact in design. As a result, cyclone samplers have been applied in a number of personal monitoring applications.

Personal exposure monitors (PEMs) are sampling devices worn on the body to estimate an individual's exposure to air pollution. As such, they provide a better representation of what individuals actually breathe than do fixed outdoor monitors or even indoor monitors. PEM design is typically based on cyclone or impactor samplers and cascade impactors have been identified as being particularly useful. Samplers for the inhalable fractions have already been proposed [18] and it is expected that samplers for the thoracic and respirable fractions can be obtained with relatively small modifications to the design and/or mode of operation of existing respirable PM samplers (e.g. cyclones) [19]. In terms of filters, mixed-cellulose ester membrane and polyvinyl chloride (PVC) filters are often used for collecting metallic dusts for chemical analysis.

Among the currently available PEMs, the GSP (Strohlein GmbH, Kaarts, Germany) is one of the most precise with a low sampling bias [20]. The closed-face 37 mm filter cassette (SKC Inc., Eighty Four, PA) which is widely used in the United States, undersamples 50 μm mass median aerodynamic diameter (MMAD) particle distribution by

40 % and 12 μm MMAD particle distribution by 10 %. Other samplers, including the IOM (Institute of Occupational Medicine, Edinburgh, Scotland, UK) sampler (Negretti Automation, Aylesbury, UK, and SKC Inc., Eighty Four, PA), the 'Seven Hole Sampler': (SKC Inc., Eighty Four, PA, and Casella Ltd., London, UK), the CIP10 samplers (Arelco, Fontenay-Sous-Bois, France), the PERSPEC sampler (Lavoro e Ambiente, Bologna, Italy), and the 37 mm open-face filter cassette, do not always perform with the accuracy [20] required by the European Standards Organization standard on the assessments for measurement of airborne particle concentration. Because of the lower flow rates and poorer flow controllers used in these personal devices, they are believed to have poorer precision than ambient samplers [21].

For further information, reviews of workplace measurements of coarse and fine PM are useful [6, 7]. In addition, statistical sampling strategies recommended in developing efficient programs to monitor occupational exposures to airborne concentrations of chemical substances can be found in the manual published by the National Institute for Occupational Safety and Health [17].

3 COLLECTION OF VOLATILE COMPOUNDS

A variety of sampling methods and systems can be employed for volatiles of interest. In general, methods for volatile compounds involve whole air sampling or preconcentration of samples using liquid adsorbents, cryotrapping, adsorbent cartridges or impregnated surfaces [11]. Each sampling method includes the following steps: (1) selection and preparation of sampling media, (2) the actual sampling process, (3) transport and storage of the collected samples. The selection of the optimal sampling method for target compounds (or a class of compounds) depends greatly on the physicochemical nature of these compounds and their expected concentration in air; sample volume must be compatible with the sensitivity of the analysis method, and the expected behavior of targeted compounds during each step of the sampling process must be carefully considered [11].

Sampling of whole air with containers of defined volume such as plastic bags (usually Teflon or Tedlar), cylinders with inlet and outlet valves, or stainless steel canisters is attractive because there is no risk of analyte breakthrough (unlike trapping), and no effect of moisture upon sampling. However, this method has two limitations: (1) the sample volume is limited to a few liters, which, for low compound concentrations encountered in air samples, may not be sufficient for analysis purposes; (2) sample stability during storage is sometimes in doubt due to adsorption on (or desorption from) container walls and chemical reactions between compounds.

Thus, one of the most widely used methods for sampling of gaseous contaminants is preconcentration, either on a suitable solid adsorbent, by cryotrapping or, if the contaminant is reactive, in an absorbing solution contained in a bubbler or impinger or coated on a solid porous support [11]. Porous polymers, such as Tenax-GC, XAD resins, and polyurethane foams (PUF), have found wide application in gas sampling. Other types of sorbents such as various types of charcoal, carbon molecular sieves and other carbon-based sorbents are also widely used, especially for more volatile compounds. All solid adsorbents must be cleaned and tightly sealed prior to use. The cleaning procedure depends on the type of adsorbent.

Cryogenic concentration of volatile compounds is usually performed in an empty tube or a tube filled with glass beads and cooled by liquid nitrogen. Usually, a second cryotrap is needed (cryofocusing) in order to allow narrow bands to enter the gas chromatography columns and thus to enhance resolution. However, the use of liquid nitrogen on-site, carrying a pump as well as the necessary power supply, is not very convenient and the large amounts of liquid nitrogen needed during a sampling trip represent a major hazard during transport. In addition, plugging problems were experienced when sampling atmospheres with high levels of humidity [22].

The application of diffusion denuders in combination with chemical analysis is another suitable method for the determination of a variety of trace gases in the atmosphere [23]. In comparison to techniques using filter methods, the separation of reactive gas and particle phases by diffusion denuder techniques is expected to be less subject to artifacts.

Denuder systems can distinguish between compounds in gaseous and particulate form. In their simplest form, they consist of straight tubes the inside walls of which are coated with a suitable adsorber material acting as a sink for the gas of interest. Gases, due to their large diffusion velocity compared to PM, can reach the walls of the tube and be adsorbed to the coating. After sampling for the desired period of time, the wet coating is washed off and the determination is performed on the liquid extract. Unlike gases, however, PM passes the denuder without touching the walls, as its diffusion is slow. In fact, denuders can be used to trap gaseous components which could interfere with filter measurement. For example, the denuder technique has been used to trap sulfur dioxide, which is responsible for the reduction of sampled Cr(VI) on the filter surface [24].

4 SAMPLE HANDLING AND STORAGE

Adequate contamination control is a prerequisite in all atmospheric trace element research, but is particularly needed when the research is carried out in remote (or even semi-remote) areas. The sampling should be carried out far enough from local sources to avoid contamination, and in atmospheric PM collections it is strongly recommended that a sampling control device be used to monitor wind speed and direction and/or condensation nuclei counts. Furthermore, metal-free PM and deposition collectors should be employed, and they should be thoroughly cleaned with acid prior to use.

As important as contamination control during sampling is the avoidance of contamination during sample handling, storage and chemical analysis. It is therefore strongly advised to have a laminar flow clean bench in the field for all critical sample handling (e.g. loading or unloading of filters). In the laboratory, all critical manipulations should be done at least on a clean bench, but for elemental determination at very low levels the use of a clean room is recommended. To minimize contamination of filters, filter holders have been designed so that filters can be loaded and unloaded in a clean environment rather than in the field. Consideration should be given to the material used to construct the filter holder, particularly when measuring reactive components of particulate matter.

Particles can fall off filters when samples have large deposits and receive rough handling during movement from the field site to the laboratory. Shorter sampling durations and lower flow rates may be required to prevent overloading in very polluted environments, especially those in which fugitive dust is a large contributor. Careful handling during transport will also minimize the loss of particles from the filter handling.

Sample integrity during storage is another important issue in atmospheric trace element research. For particulate matter samples, some elements may undergo changes (e.g. Cr(VI) reduction) as a result of chemical reactions which take place on the collection substrate during sample storage. Such losses may be minimized by storing the samples in closed polypropylene vessels under a pure nitrogen atmosphere [25, 26]. For volatile metal species, sample stability during cryogenic storage has been evaluated and discussed [27].

5 APPLICATIONS

The aim of this section is to focus on various sampling methods used in speciation studies of airborne metals. However, many of these methods are purely operational and do not specifically identify the metal species.

5.1 Lead

Environmental pollution from lead is a problem arising mainly from the use of tetraalkyllead compounds as anti-knock additives in gasoline. Although this use is diminishing, the more stable forms, tri- and dialkyllead, are fairly persistent in the environment. It has been shown that even in remote areas with no direct sources of emission from traffic the concentrations of alkyllead species found are correlated with the introduction and proliferation of motor vehicles. Other sources such as mining, smelting and chemical production only become major contributors in localized areas.

Most reported data concerning lead speciation in atmospheric samples are focused on the determination of tetraalkyllead (R_4Pb) species in the gas phase. However, it should be pointed out that R_4Pb does not account for the total lead present in the gas phase. It has been suggested that R_4Pb decomposes in the atmosphere to produce vapor phase R_3Pb^+ and R_2Pb^{2+} species and inorganic lead PM [28]. A variety of sampling techniques using solid adsorbents operated at ambient temperature or at extremely low temperature ($-40\,^{\circ}C$ or lower) have been reported [28–32]. Due to the practical difficulties incumbent in employing cold-trapping techniques in the field, the use of solid adsorbents operated at ambient temperature is recommended [28, 31]. Among the solid adsorbents (PUF, active charcoal, Amberlites, Chromosorb, Tenax and Poropak), only Tenax and Poropak gave recoveries of more than 90 % for Et_4Pb [32]. The more volatile compounds, such as Me_4Pb and Me_3Pb^+, were adsorbed less on Tenax than on Poropak. Due to the possibilities for R_4Pb decomposition (particularly for tetraethyllead) in the presence of ozone [33], a Teflon tubing pre-filter packed with iron(II) sulfate crystals [31] is included in the sampling train. The consensus opinion also indicates that air samples should be filtered to remove particulate matter, although concern has been expressed that lead alkyls may be adsorbed onto the collected PM [31] leading to low recoveries.

Particulate alkyllead is generally found at $pg\,Pb\,m^{-3}$ levels, some three orders of magnitude below the concentration of vapor phase alkyllead compounds. Ionic alkyl lead species can be trapped in two water-filled gas bubblers connected in series [34] after a pre-filter. This simple procedure is reported to allow reasonable recoveries of ionic species. Both inorganic lead and alkyllead particulate species were effectively removed from gas-phase tetraalkyllead compounds by a highly efficient particle filter.

It should be noted that alkyllead species do not account for all the lead present in the gas phase. The largest fraction is inorganic lead, probably particulates of PbBrCl (not retained on a $0.45\,\mu m$ filter) derived from the combustion of antiknocking additives in gasoline [28].

Particulate lead ambient air samples are often collected using high-volume samplers [35]. Exposure to lead occurs not only through inhalation but also through ingestion of particles, which may be too large to be inhaled, especially by young children. Therefore, the high-volume sampler provides a more complete measure of exposure to airborne lead than the PM_{10} sampler, which by design excludes particulates larger than respirable size.

5.2 Chromium

Cr(III) and Cr(VI) are the most common oxidation states of chromium in the environment. Whereas the Cr(III) species is essential for animals and plants in trace concentrations, Cr(VI) is toxic and carcinogenic [36]. It has been specially classified by the U.S. Environmental Protection Agency (EPA) as a group A inhalation carcinogen [37]. Especially when occupational exposure is likely, inhalation is believed to be a major human exposure pathway for Cr(VI). Occupational exposure to airborne Cr(VI) has been associated with a number of work activities, including metal plating, welding, spray painting, leather tanning, dye and pigment manufacturing, and cleaning of various parts prior to protective painting, especially in the aircraft and automobile industry [38].

Because of the relationship between chromium toxicity and oxidation state, the possibility of changing the valences of chromium due to reduction and/or oxidation must be eliminated. This has led to a need to define appropriate sampling, storage and analytical procedures.

The usual method of workplace monitoring of particulate chromium species is to collect PM samples on filters for subsequent extraction and analysis. Different filter materials have been used during the last two decades, and due to reduction of Cr(VI) on the filters, inconsistent data originating from this problem have been reported [39]. Cellulose nitrate, cellulose acetate, PVC, PTFE and glass fiber filters have been the ones most widely used. Most methods for Cr(VI) occupational monitoring (e.g. NIOSH [40], HSE [41], and OSHA [42]) specify PVC filters for sample collection, since other types of filters are known to cause Cr(VI) reduction [14, 39, 43]. However, as this reduction apparently occurs on a slow time scale, certain filter types such as glass fiber have showed reduction of Cr(VI) to a much lesser extent than the others, so it is recommended as a possible sampling material [14, 39, 44].

Inaccurate estimation of exposure to Cr(VI) is also caused by its instability during sampling, extraction or even sample preparation due to the enrichment of particles on the filter and thus enhanced contact with gaseous species, e.g. SO_2, NO_x, O_3, and/or reaction on the filter with co-collected material, e.g. Fe(II), As(III)-containing components. Due to the pH-dependent electrochemical potential of the reduction of Cr(VI) to Cr(III), losses of Cr(VI) are expected to increase with decreasing pH of the filter surface. The use of the denuder technique to trap sulfur dioxide, which is responsible for the reduction of sampled Cr(VI) on the filter surface, is proposed by Rohling and Neidhart [24]. Losses of Cr(VI) between 16 and 57 % depending on the composition of the PM and the sampling time were found. In addition, if the filters are not treated directly after sampling, storage in closed polypropylene vessels under a pure nitrogen atmosphere is recommended [25, 26]. Under these conditions the chromium species are stable for months [25].

An elegant solution to avoiding problems with Cr(VI) instability is to analyze samples immediately following collection, as recently proposed using field-portable analytical methods [45–47]. In one study, a portable spectrometer and a portable solid phase extraction manifold were transported to the field. An industrial hygiene survey was conducted during sanding and spray painting operations, and air samples were collected and analyzed for Cr(VI) on site. Some area air samples were also collected above an electroplating bath. The results showed this new method to be useful for both environmental and industrial hygiene purposes. Another study developed a simple and readily field-adaptable system for automated continuous measurement of Cr(VI) in airborne particulate matter. The system alternately collected the sample on one of two glass fiber filters. After 15 min of sample collection on one filter, the sampling switched over to the second filter. The freshly sampled filter was washed for 8.5 min and the washings preconcentrated on an anion exchange minicolumn. The washed filter was dried with filtered hot air for the next 6.5 min so that it was ready for sampling at the end of the 15 min cycle [46].

5.3 Mercury

Mercury has long been identified as a potential health and environmental hazard. Unlike most other trace metals, the high volatility of mercury causes it to be present in the vapor form. In the atmosphere, the three main forms of Hg are: elemental Hg vapor (Hg^0), which represents more than 95 % of the global atmospheric mercury burden [48], reactive gas phase Hg (RGM) and particulate phase Hg (TPM). Of these three species, only Hg^0 has been tentatively identified with spectroscopic methods [49] while the other two are operationally defined species, i.e. their chemical and physical structure cannot be exactly identified by experimental methods but are instead characterized by their properties and capability to be collected by different sampling equipment.

RGM is defined as water-soluble mercury species with sufficiently high vapor pressure to exist in the gas phase. The reactive term refers to the capability of stannous chloride to reduce these species in aqueous solution without pretreatment. The most likely candidate for RGM species is $HgCl_2$ and possibly other divalent mercury species. TPM consists of mercury bound or adsorbed to atmospheric particulate matter. Several different components are possible: Hg^0 or RGM adsorbed to the particle surface, or divalent mercury species chemically bound to the particle or integrated into the particle itself. Another species of particular interest is methylmercury (MeHg) due to its high capacity to bioaccumulate in aquatic foodchains and to its high toxicity. Since MeHg is only present at low $pg\,m^{-3}$ concentration levels in ambient air, it is not an important species for the overall atmospheric cycling of Hg.

Sampling and analysis of atmospheric Hg are often done as total gaseous mercury (TGM) which is a fraction operationally defined as species passing through a $0.45\,\mu m$ filter or some other simple filtration device such as quartz wool plugs, and which are collected on gold or other collection material. TGM is mainly composed of elemental Hg vapor with minor fractions of other volatile species such as $HgCl_2$, CH_3HgCl or $(CH_3)_2Hg$.

Gold-coated denuders were developed for removal of Hg vapor from air but have not been applied to air sampling [50]. Potassium chloride-coated tubular denuders followed by silver-coated denuders have been used to collect $HgCl_2$ (RGM) and elemental Hg emissions from incinerators [51] and for gaseous divalent mercury in ambient air [52–54].

For particulate Hg, a variety of different filter methods using Teflon or quartz fiber filters have been applied [55–57]. Recently, a collection device based on small quartz fiber filters mounted in a quartz tube was designed [56]. The mercury collected on the filter can be released thermally, followed by gold trap amalgamation and cold vapor atomic absorbance spectrometry.

An excellent intercomparison for sampling and analysis of atmospheric mercury species was recently reported [58]. Methods for sampling and analysis of TGM, RGM and TPM were used in parallel sampling over a period of 4 days in Tuscany, June 1998. The results for the different methods employed showed that TGM compared well whereas RGM and TPM showed a somewhat higher variability. The relationships for the measurement results of RGM and TPM improved over the sampling period, indicating that activities at the sampling site during set-up and initial sampling affected the results, in particular the TPM results. Additional parallel sampling was performed for two of the TPM methods under more controlled conditions, which yielded more comparable results.

5.4 Platinum group metals

The emission of three metals of the platinum group (PGMs) (Pd, Pt, and Rh) from automobile catalytic converters into the environment is of potential concern for human health. Platinum may affect the health of people through direct contact with platinum in dust, by inhalation of dust, and indirectly through the food chain. A maximum exposure limit over 24 h of $2\,\mu g\,m^{-3}$ of airborne, water-soluble platinum salts has been recommended by the U.S. Occupational Safety and Health Administration. However, there are no data on the effects of exposure to water-insoluble platinum compounds. Environmental concentrations of PGMs have not been shown to directly affect ecosystems or result in direct health risks [59]. However, Pt was bioaccumulated by rats exposed to a model substance which resembled Pt-containing particulates emitted by automobiles [60], and Pt was found to react with DNA [61].

There is not much literature available on the subject of PGMs in air. Several types of filters have been tested in order to avoid high PGM blank values [13, 62]. Some filter materials, e.g. glass fiber or Teflon, showed very high platinum contents and were therefore unsuitable for the sampling of airborne dust with low platinum contents. Only polycarbonate and cellulose nitrate gave blank values as low as 5 pg Pt per total filter. Since knowledge regarding the particle size distribution of Pt metals in airborne particulate matter is critical to allow for a full assessment of their toxic potential and associated risk to human health, there have been several impactor studies done to produce size-fractionation data on this distribution [13, 63].

5.5 Radionuclides

The radioactivity of PM has also attracted much attention. Radioactivity in atmospheric PM has three major sources: emanations of radon and thoron gas, radioactive isotopes from cosmic rays, and anthropogenic radionuclides. Most of these processes produce primary particulates, which later attach to larger particulates by thermal diffusion.

The importance of speciation to characterize the behavior of nuclides in the environment has spawned a very large number of investigations, yet our knowledge is still limited and much more work is needed [64]. This unusual situation arises from the difficulties in assessing radionuclides at very low environmental concentrations and from problems related to the chemical (e.g. sorption, ion exchange) and physical (i.e. radioactive decay) instability of many species. Immediately after their formation by nuclear reactions and/or radioactive decay, radionuclides exist as single, highly kinetically and electronically excited atoms or ions, whose chemical behavior is very difficult to predict and investigate. Very often it is a challenge to collect and analyze samples without changing the identity of the real species. However, an advantage of the determination of radionuclides as compared to nonradioactive trace elements is the practical absence of contamination during sampling and analysis.

Inhaled progeny of radon-222 are the most important source of irradiation of the human respiratory tract and are clearly related to lung cancers, especially among miners of uranium and other minerals [65]. As a natural product of the uranium decay series, ^{222}Rn gas occurs in granitic and similar rocks and thus enters buildings or underground workspaces constructed in these types of geological formations. Its own radioactive decay produces solid radon daughter elements which can attach to aerosols and reach the human respiratory tract in attached or unattached forms [66]. Like other particle-associated contaminants, radon's health effects are heavily dependent on particle size distribution, so size-specific sampling is critical. One option is a modified cascade impactor known as a low pressure Andersen sampler, operated at lower than standard pressures and thus able to select much smaller diameter particles. In a study of collection substrates, stainless steel plates coated with silicone grease provided the best size-selective performance with this sampler [67, 68]. For measurement of the alpha particles from the radon progeny, a ZnS(Ag) scintillation counter was used after waiting 20 min for the decay of ^{218}Po [67], which ensures that the appropriate species are taken into account. Several devices which both collect and analyze air samples have also been developed. The electrical low pressure impactor (ELPI) charges the aerosol particles and then measures the resulting current at each impactor stage to produce size distribution data, though radioactivity can only be measured later after disassembly [69]. The portable RADON-check, however, concentrates the aerosol onto a grid and filter apparatus with in-line radiospectrometry, and can be tuned to the aerosol-size retention function of the bronchi [70].

The radionuclide ^{129}I has both cosmogenic and anthropogenic sources, and with a very long half-life ($T_{1/2} = 15.7 \times 10^6$ year) will be a factor in population dose estimates in spite of not being a radiological hazard at present. An efficient method for sampling iodine is to pump air through a TEDA-activated charcoal filter, which traps >99 % of the main forms of gaseous iodine, namely I_2 and CH_3I. Depending on the analysis method and the data desired, the filter may need to be treated immediately to preserve its isotope ratios [71].

Filter collecting methods using impactors are commonly utilized to characterize particulate radioactivity. Bondietti *et al.*, using impactors to collect particles in a certain size range, found ^{212}Pb and ^{214}Pb mostly on PM with sizes <0.52 μm, with ^{210}Pb on particles of larger radii [72]. Among other elements, ^{90}Sr and ^{137}Cs are found with smaller particles, while ^{95}Zr and ^{144}Ce are often related to larger ones [64].

Tritium is an anthropogenic contaminant of concern for environmental monitoring at nuclear facilities because it is a byproduct of nuclear reactors. Ambient air concentrations of tritium (as HTO) are typically too low for practical measurement using real-time devices; therefore adsorbents must generally be used to collect the atmospheric moisture, which is then analyzed for HTO. Both solid and liquid adsorbents can be used to collect water vapor from ambient air, though the solid types have several advantages. Compared to liquid adsorbent air sampling systems (bubblers or impingers), solid adsorbents are more rugged (e.g. no breakable glass or evaporation problems) and allow for higher sampling rates and lower limits of detection [73].

5.6 Metalloids

Volatile metalloid compounds have been identified in various anthropogenic gases such as domestic waste gas, landfill gas and sewage sludge digester gas [22, 74, 75]. Most of these compounds are thermodynamically unstable and thus undergo chemical transformations. This feature is important for the choice of an appropriate sampling technique. Possible phenomena causing analyte loss are diffusion, oxidation, hydrolysis, photodecomposition, adsorption, and heterogeneous surface-catalyzed breakdown [22].

So far, the method most used for sampling volatile metalloid compounds like volatile metals has been cryotrapping [74–81]. Feldmann and Hirner [75] used chromatographic packing (SP-2100 10 % on Supelcoport −60/80 mesh) in a U-shaped glass tube immersed in liquid nitrogen as the cryogenic liquid, for the sampling of various metalloids in landfill gas [79, 80] and in sewage gas [81]. Glass tubes packed with silanized glass wool have been used as sample collectors for cryofocusing of volatile metal species in urban air [78].

The sampling of volatile metalloid compounds such as methylated, permethylated and hydride species of arsenic, antimony and tin has also been described using Tedlar bags [22]. These bags are very simple to use and a variety of different volumes can be sampled, from 1 to 100 L. In contrast, stainless steel containers have a limited volume of a few liters, unless the sampled air is pressurized. To demonstrate the suitability of Tedlar bags for the sampling of such compounds, a series of stability tests were run using laboratory synthetic and real samples analyzed periodically after increasing periods of storage. The samples were stored in the dark at 20 and 50 °C. Based on the results obtained, storage at 20 °C and analysis carried out by the day after sampling were recommended [22].

5.7 Miscellaneous applications

A number of studies have used size-selective sampling and various pre-treatments of samples to determine trace metal chemical forms or binding forms, and thus obtain speciation information that can help to assess the relative biohazard of trace (and minor) elements associated with airborne particulates [82–94]. These methods include: (1) separation of respirable airborne particles from larger ones; (2) assessment of particle bioavailability through lung absorption by testing their solubility on air filters; and (3) assessment of the probable state of chemical binding in or on the solid particles through sequential extraction procedures. Different leaching solutions were used in single and sequential extraction procedures. Although there is some controversy in the literature over the use of this approach to chemical speciation [95], fractionation ('operational speciation') data from sequential extraction in combination with particle size collection still provide useful information on the source of the elements and on their potential bioavailability.

Such an approach was applied to determine various nickel species present in the airborne dusts of nickel-producing and nickel-using workplaces [91, 92]. Nickel and some of its compounds are identified as potential human carcinogens. A wet chemical procedure is described which apportions the airborne nickel species into four categories: water soluble, 'sulfidic', 'metallic' and 'oxidic' [91]. The suitability of various filter materials for sampling airborne dusts for this fractionation study was also investigated. Sampling of airborne dust is most frequently done with 37 mm membrane filters (personal pumps) or with large 203 × 254 mm sheets of fiber filters (high-volume sampling). If nickel is to be apportioned among the various species in the sample, the filter material must be (a) chemically indifferent toward the sampled dust, (b) chemically resistant to bromine and methanol, and (c) readily wet-ashed to allow complete dissolution of the leached sample residue. Filter media that meet all the criteria are Gelman DM Metricel membranes and quartz fiber filters. Inferior for this purpose of fractionation are cellulose ester membranes (attacked by methanol), PVC and Teflon membranes (hydrophobic and difficult to destroy by acid digestion), and, most importantly, glass fiber filters (surface alkalinity). The natural alkalinity of the glass fiber surface can react with soluble nickel compounds (normal salts) to form basic salts, which are only partly soluble. To avoid undesirable secondary chemical changes in the dust, quartz fiber filters should be used instead of glass fiber filters [91].

In another study the procedure for speciation of As in size-fractionated filter-collected urban particulate matter was described [93]. The samples were collected by Gent-type stacked filter unit and were stored in plastic Petri dishes in a refrigerator at 4 °C until analysis. A mild sequential procedure to differentiate As species between water-extractable, phosphate-extractable and refractory form was utilized.

6 SUMMARY

This chapter has given a brief outline of sampling methodology and samplers used for collecting airborne particulate matter and volatile pollutants for epidemiological and occupational studies. The main motivation for the sampling of ambient airborne particles remains the prediction and prevention of adverse health effects on humans. The evaluation of worker exposure to potentially hazardous agents including metallic contaminants in the workplace is essential to establishing cause–effect relationships between an occupationally related illness and a specific agent(s).

There are a number of samplers that have been developed for epidemiological studies and to monitor individual exposures to health-related pollutants. Of all the airborne particulate matter collection techniques, filter sampling is the most versatile. With appropriate filter media, samples can be collected in almost any form, quantity and state. Sample handling problems are usually minimal, and many analyses can be performed directly on the filter. Application of impactor-based selective sampling to obtain particle-size distribution of airborne particulate matter is of interest for the microcomposition and speciation of respirable submicron particles, which is of particular importance in environmental health. For sampling of gaseous contaminants such as volatile metals and metalloids, cryotrapping and preconcentration on solid adsorbents are widely used methods.

A number of more specific methods have been developed to collect and determine various toxic metal species such as Cr(VI), Hg and Pb compounds or platinum group metals in ambient air and workplace environments. In addition, various approaches (mainly operational) have been proposed to estimate the behavior and fate of atmospherically derived trace metals in order to obtain information on their specific sources and reactions, and thus to assess their potential health effects. These include appropriate selective sampling methods in combination with various sequential leaching procedures.

As links are established between workplace exposure to pollutants and adverse health effects, it becomes more and more important to develop and implement reliable sampling and analysis methods for quantifying these pollutants.

7 REFERENCES

1. Dusseldorp, A., Kruize, H., Brunekreef, B., Hofschreuder, P., de Meerm, G. and van Oudvosrt, A. B., *Am. J. Respir. Crit. Care Med.*, **152**, 1932 (1995).
2. Prichard, R. J., Ghio, A. J., Lehmann, J. R., Winsett, D. W., Park, P., Gilmour, M. I., Drher, K. L. and Costa, D. L., *Inhal. Toxicol.*, **8**, 457 (1996).
3. Samet, J. M., Stonehuerner, J., Reed, W., Devlin, R. B., Dailey, L. A., Kennedy, T. P., Bromberg, P. A. and Ghio, A. J., *Am. J. Physiol. (Lung Cell. Molec. Physiol.)*, **272**, L426 (1997).
4. Smith, K. R. and Aust, A. E., *Chem. Res. Tox.*, **10**, 828 (1997).

5. Chao, C. C., Lund, L. F., Zinn, K. R. and Aust, A. E., *Arch. Biochem. Biophys.*, **314**, 369 (1994).
6. Vincent, J. H., *Analyst*, **119**, 13 (1994).
7. Vincent, J. H., *Analyst*, **119**, 19 (1994).
8. Lodge, J. P. Jr., Methods of air sampling and analysis, Lewis Publishers, Chelsea, MI, 1989.
9. Watson, J. G. and Chow, J. V., Particle and gas measurements on filters, in *Environmental Sampling for Trace Analysis*, Markert, B. (Ed.), VCH, Weinheim, 1994, pp. 125–161.
10. Mark, D., Atmospheric aerosol sampling, in *Atmospheric Particles*, Harrison, R. M. and van Grieken, R. (Eds), John Wiley & Sons Ltd, Chichester, 1998, pp. 29–94.
11. Zielinska, B. and Fujita, E., Organic gas sampling, in *Environmental Sampling for Trace Analysis*, Markert, B. (Ed.), VCH, Weinheim, 1994, pp. 163–184.
12. Lippmann, M., Filters and filter holders, in *Air Sampling Instruments for Evaluation of Atmospheric Contaminants*, 8th Edn, Cohen, B. and Hering, S. V. (Eds), American Conference of Government Industrial Hygienists, Cincinnati OH, 1995.
13. Alt, F., Bambauer, A., Hoppstock, K., Mergler, B. and Tolg, G., *Fresenius' J. Anal. Chem.*, **346**, 693 (1993).
14. Spini, G., Profumo, A., Riolo, C., Beone, G. M. and Zecca, E., *Toxicol. Environ. Chem.*, **41**, 209 (1994).
15. Lodge, J. P. and Chan, T. L., *Cascade Impactor-Sampling and Data Analysis*, American Industrial Hygiene Association, Akron, OH, 1986.
16. Markowski, G. R., *Aerosol Sci. Technol.*, **7**, 143 (1987).
17. Leidel, N. A., Busch, K. A. and Lynch, J. R., *Occupational Exposure Sampling Strategy Manual*, U.S. Department of Health, Education, and Welfare, Cincinnati, OH, 1977.
18. Mark, D. and Vincent, J. H., *Ann. Occup. Hyg.*, **30**, 89 (1986).
19. Vincent, J. H., Tsai, P. J. and Warner, J. S., *Analyst*, **120**, 675 (1995).
20. Kenny, L. C., Aitken, R., Chalmers, C., Fabries, J. F., Gonzales-Fernandez, E., Krombout, H., Liden, G., Mark, D., Riediger, G. and Prodi, V., *Ann. Occup. Hyg.*, **41**, 135 (1997).
21. Wiener, R. W. and Rodes, C. E., Indoor aerosol and aerosol exposure, in *Aerosol Measurement, Principles, Techniques, and Applications*, Willeke, K. and Baron, P. A. (Eds), Van Nostrand Reinhold, New York, 1993, pp. 659–689.
22. Haas, K. and Feldmann, J., *Anal. Chem.*, **72**, 4205 (2000).
23. Slanina, J. and Wyers, G. P., *Fresenius' J. Anal. Chem.*, **350**, 467 (1994).
24. Rohling, O. and Neidhart, B., *Fresenius' J. Anal. Chem.*, **351**, 33 (1995).
25. Dyg, S., Cornelis, R., Griepink, B. and Quevauviller, Ph., *Anal. Chim. Acta*, **286**, 297 (1994).
26. Nusko, R. and Heumann, K. G., *Fresenius' J. Anal. Chem.*, **357**, 1050 (1997).
27. Amouroux, D., Tessier, E., Pécheyran, C. and Donard, O. F. X., *Anal. Chim. Acta*, **377**, 241 (1998).
28. Allen, A. G., Radojevic, M. and Harrison, R. M., *Environ. Sci. Technol.*, **22**, 517 (1988).
29. Radziuk, B., Thomassen, Y., van Loon, J. C. and Chau, Y. K., *Anal. Chim. Acta*, **105**, 255 (1979).
30. de Jonghe, W. R. A., Chakraborti, D. and Adams, F. C., *Anal. Chem.*, **52**, 1974 (1980).
31. Hewitt, C. N. and Harrison, R. M., *Anal. Chim. Acta*, **167**, 277 (1985).
32. Nerin, C., Pons, B., Martinez, M. and Cach, J., *Microchim. Acta*, **112**, 179 (1994).
33. Nielsen, T., Egsgaard, E., Larsen, E. and Schroll, G., *Anal. Chim. Acta*, **124**, 1 (1981).
34. Hewitt, C. N., Harrison, R. M. and Radojevic, M., *Anal. Chim. Acta* **188**, 229 (1986).
35. *Guidance for Siting Ambient Air Monitors Around Stationary Lead Sources*, Off. Air Qual. Plann. Stand., (Technol. Rep.): EPA-454/R-92-009, U.S. Environmental Protection Agency, Washington, DC, 1997.
36. International Agency for Research on Cancer, *Monographs on the Evaluation of the Carcinogenic Risks of Chemicals to Humans: Chromium, Nickel and Welding*, Vol. 49, IARC, Lyon, France, 1990, pp. 257–445.
37. *Health Effects Assessment of Hexavalent Chromium*, EPA 540/1-86-019, U.S. Environmental Protection Agency, Washington, DC, 1984.
38. *Criteria for a Recommended Standard, Occupational Exposure to Chromium (VI)*, National Institute for Occupational Safety and Health, Cincinnati, OH, 1975.
39. van der Wal, J. F., *Ann. Occup. Hyg.*, 29, **377** (1985).
40. Methods no. 7600 and 7604, in *NIOSH Manual of Analytical Methods*, 4th Edn, Eller, P. M. and Cassinelli, M. E. (Eds), National Institute for Occupational Safety and Health, Cincinnati, OH, 1994.
41. Health and Safety Laboratory/HSE, *Methods for Determination of Hazardous Substances*, MDHS Method no. 61, Health and Safety Executive, Sheffield, UK, 1998.
42. Method no. ID-215, in *Hexavalent Chromium in Workplace Atmospheres*, Occupational Safety and Health Administration, Salt Lake City, UT, 1998.
43. Sawarti, K., *Ind. Health*, **24**, 111 (1986).
44. Dyg, S., Anglov, T. and Christensen, J. M., *Anal. Chim. Acta*, **286**, 273 (1994).
45. Wang, J., Ashley, K., Marlow, D., England, E. and Carlton, G., *Anal.Chem.*, **71**, 1027 (1999).
46. Marlow, D., Wang, J., Wise, T. J. and Ashley, K., *Am. Lab.*, **7**, 26 (2000).
47. Samanta, G., Boring, C. B. and Dasgupta, P. K., *Anal. Chem.*, **73**, 2034 (2001).
48. Ebinghaus, R., Hintelmann, H. and Wilken, R. D., *Fresenius' J. Anal. Chem.*, **350**, 21 (1994).
49. Edner, H., Faris, G. W., Sunesson, A. and Svanberg, S., *Appl. Optics*, **28**, 921 (1989).
50. Munthe, J., Xiao, Z., Schroeder, W. H. and Lindqvist, O., *Atmos. Environ.*, **24A**, 2271 (1991).
51. Larjava, K., Latinen, T., Vahlman, T., Artmann, S., Siemens, V., Broekaert, J. A. C. and Klockow, D., *Int. J. Anal. Chem.*, **149**, 73 (1992).

52. Xiao, Z., Sommar, J., Wei, S. and Lindqvist, O., *Fresenius' J. Anal. Chem.*, **358**, 386 (1997).
53. Sommar, J., Feng, X., Gardfelt, K. and Lindqvist, O., *J. Environ. Monitor.*, **1**, 435 (1999).
54. Feng, X., Sommar, J., Gardfelt, K. and Lindqvist, O., *Fresenius' J. Anal. Chem.*, **366**, 423 (2000).
55. Keeler, G., Glinsorn, G. and Pirrone, N., *Water Air Soil Pollut.*, **80**, 159 (1995).
56. Lu, J. Y., Schroeder, W. H., Berg, T., Munthe, J., Schneeberger, D. and Schaedlich, F., *Anal. Chem.*, **70**, 2403, (1998).
57. Berg, T., Bartnicki, J., Munthe, J., Lattila, H., Hrehoruk, J. and Mazur, J., *Atmos. Environ.*, **35**, 2569 (2001).
58. Munthe, J., Wangberg, I., Pirrone, N., Iverfeldt, A., Ferrara, R., Ebinghaus, R., Feng, X., Gardfeldt, K., Keeler, G., Lanzillotta, E., Lindberg, S. E., Lu, J., Mamane, Y., Prestbo, E., Schmolke, S., Schroeder, W. H., Sommar, J., Sprovieri, F., Stevens, R. K., Stratton, W., Tuncel, G. and Urba, A., *Atmos. Environ.*, **35**, 3007 (2001).
59. Rosner, G. and Merget, R., in *Anthropogenic Platinum Group Elements and Their Impact on Man and the Environment*, Alt, F. and Zereini, F. (Eds), Springer, Berlin, 1999, pp 267–281.
60. Artelt, S. and Levsen, K., in *Anthropogenic Platinum Group Elements and Their Impact on Man and the Environment*, Alt, F. and Zereini, F. (Ed.), Springer-Verlag, Berlin, 1999, pp. 217–226.
61. Rosenberg, B., Van Camp, L. and Krigas, T., *Nature*, **205**, 698 (1965).
62. Rauch, S., Lu, M. and Morrison, G. M., *Environ. Sci. Technol.*, **35**, 595 (2001).
63. Zereini, F., Wiseman, C., Alt, F., Messerschmidt, J., Muller, J. and Urban, H., *Environ. Sci. Technol.*, **35**, 1996 (2001).
64. von Gunten, H. R. and Benes, P., *Radiochim. Acta*, **69**, 1 (1995).
65. Yu, K. N., Wong, B. T. Y., Law, J. Y. P., Lau, B. M. F. and Nikezic, D., *Environ. Sci. Technol.*, **35**, 2136 (2001).
66. Park, C., *Progr. Phys. Geog.*, **17**, 473 (1993).
67. Yamasaki, K., Yamada, Y., Miyamoto, K. and Shimo, M., in *Proceedings of IRPA-10*, International Radiological Protection Association, P-1b-42, Hiroshima, 2000.
68. Yamasaki, K., in *Radon and Thoron in the Human Environment*, Katase, A. and Shimo, M. (Eds), World Scientific Singapore, 1998, pp. 73–78.
69. Yamada, Y., Miyamoto, K., Tokonami, S., Shimo, M. and Yamasaki, K., in *Proceedings of IRPA-10*, International Radiological Protection Association, P-1b-05, Hiroshima, 2000.
70. GRIMM Aerosol Technik GmbH and Co. KG, Ainring, Germany. http://www.grimm-aerosol.de.
71. López-Gutiérrez, J. M., García-León, M., Schnabel, C., Schmidt, A., Michel, R., Synal, H.-A. and Suter, M., *Appl. Rad. Isot.*, **51**, 315 (1999).
72. Bondietti, E. A., Papastefanou, C. and Rangarajan, C., Aerodynamic size association of natural radioactivity with ambient aerosols, in *Radon and its Decay Products: Occurrence, Properties, and Health Effects*, Hopke, P. K., (Ed.), American Chemical Society, Washington, DC, 1987, pp. 377–399.
73. Patton, G. W., Cooper, A. T. Jr. and Tinker, M. R., *Health Phys.*, **72**, 397 (1997).
74. Feldmann, J., Grumping, R. and Hirner, A. V., *Fresenius' J. Anal. Chem.*, **350**, 228 (1994).
75. Feldmann, J. and Hirner, A. V., *Int. J. Environ. Sci. Technol.*, **60**, 339 (1995).
76. Pécheyran, C., Quetel, C. R., Lecuyer, F. M. M. and Donard, O. F. X., *Anal. Chem.*, **70**, 2639 (1998).
77. Jiang, S. G., Chakraborti, D. and Adams, F. C., *Anal. Chim. Acta*, **196**, 271 (1987).
78. Pécheyran, C., Lalère, B. and Donard, O. F. X., *Environ. Sci. Technol.*, **34**, 27 (2000).
79. Feldmann, J. and Cullen, W. R., *Environ. Sci. Technol.*, **31**, 2125 (1997).
80. Feldmann, J., Koch, I. and Cullen, W. R., *Analyst* **123**, 815 (1998).
81. Feldmann, J. and Kleimann, J., *Korresp. Abwasser* **44**, 99 (1997).
82. Jervis, R. E., Krishnan, S. S., Ko, M. M., Vela, L. D., Pringle, T. G., Chan, A. C. and Xing, L., *Analyst*, **120**, 651 (1995).
83. Dreetz, C. D. and Lund, W., *Anal. Chim. Acta*, **262**, 299 (1992).
84. Spokes, L. J., Jikells, T. D. and Lim, B., *Geochim. Cosmochim. Acta*, **15**, 3281–87 (1994).
85. Chester, R., Murphy, K. J. T. and Lin, F. J., *Environ. Technol. Lett.*, **10**, 887 (1989).
86. Brunori, C., Balzamo, S. and Morabito, R., *Int. J. Environ. Anal. Chem.*, **75**, 19–31 (1999).
87. Polyák, K., Bódog, I. and Hlavay, J., *Talanta*, **41**, 1151 (1994).
88. Hlavay, J., Polyák, K., Bódog, I., Molnár, A. and Mészáros, E., *Fresenius' J. Anal. Chem.*, **354**, 227 (1995).
89. Bódog, I., Polyák, K., Csikos-Hartyani, Z. and Hlavay, J., *Microchem. J.*, **54**, 320 (1996).
90. Radlein, N. and Heumann, K. G., *Fresenius' J. Anal. Chem.*, **352**, 748 (1995).
91. Zatka, V. J., Warner, J. S. and Maskery, D., *Environ. Sci. Technol.*, **26**, 138 (1992).
92. Füchtjohann, L., Jakubowski, N., Gladtke, D., Barnowski, C., Klockow, D. and Broekaert, J. A., *Fresenius' J. Anal. Chem.*, **366**, 142 (2000).
93. Slejkovec, Z., Salma, I., van Elteren, J. T. and Zemplen-Papp, E., *Fresenius' J. Anal. Chem.*, **366**, 830 (2000).
94. Hoffmann, P., Sinner, T., Dedik, A. N., Karandashev, V. K., Malyshev, A. A., Weber, S. and Ortner, H. M., *Fresenius' J. Anal. Chem.*, **350**, 34 (1994).
95. Bermond, A. P., *Environ. Technol.*, **13**, 1175 (1992).

CHAPTER 3

Sample Preparation

3.1 Sample Treatment for Speciation Analysis in Biological Samples

Riansares Muñoz Olivas and Carmen Cámara

Universidad Complutense de Madrid, Spain

Handbook of Elemental Speciation: Techniques and Methodology R. Cornelis, H. Crews, J. Caruso and K. Heumann
 ISBN: 0-471-49214-0

1 INTRODUCTION: BIOLOGICAL SAMPLES OF INTEREST

There are different approaches in speciation analysis depending on the problem or study that it is being undertaken:

- The most comprehensive one takes as its aim the identification and quantification of all individual forms of the element.
- In some cases, especially for biological samples, some elements are strongly bound inside the tissues and the analyst is not able to obtain complete information about the real species present in the sample. Furthermore, the extraction efficiency of the species detected depends on the sample treatment followed. As an illustrative example, the selenium species determined in the oyster tissue depend very much on the hydrolysis conditions applied [1].
- Some analysis regards only a given species and this is the object of determination. This is the case for many elements with high difference in toxicity among its species (e.g. Cr(VI) and Cr(III), Hg(II) and CH_3Hg^+, etc).
- Often it is necessary only to distinguish between groups of species without specifying the detailed nature of the compounds (e.g. inorganic Hg/organic Hg).
- The term speciation is often used where there is a distribution of an element among different physical fractions: dissolved or particulate (e.g. whole blood fractions: serum, plasma and blood cells).

There are many types of biological samples, and several subsets of this type may be identified. Three general categories can be considered: (1) clinical, including animal and human tissue and body fluids, organs, etc.; (2) vegetables, including tissues of terrestrial and aquatic plants, dissolved organic material and algae; (3) nutritional, food products and related substances.

1.1 Clinical samples

The need for information regarding the mobility, storage, retention and toxicity of species in a biological system is essential. Cornelis *et al.* [2] have classified the elemental species found in clinical samples regarding their toxicity as follows: (i) small organometallic molecules, generally contaminants of food, water or air. Some remain unchanged in the body (e.g. organotin, organolead compounds, methylmercury, ...); (ii) biomarkers of contamination exposure by the toxic species conversion into less toxic ones by the body (e.g. inorganic arsenic species that can be methylated by the organism); (iii) elements with different oxidation states and thus different toxicity (e.g. Cr(III) and Cr(VI); (iv) trace elements that form a metal–ligand complex with a low molecular weight compound, or bound to a protein (e.g. selenium in selenomethionine, etc.); and (v) elements participating in a biomolecule (e.g. Cu in ceruloplasmin). Szpunar has reported another useful classification of metal species in an interesting review concerning bioinorganic speciation [3]. From this classification, six groups of species have been exposed: (a) molecules with a metal–carbon bond, e.g. selenoamino acids and organoarsenic compounds; (b) complexes with macrocyclic chelating agents, including tetrapyrroles, cobalamins and porphyrins; (c) complexes with nucleobases, such is the case of many therapeutic drugs of Pt, Au, Ru, etc.; (d) metal complexes with proteins including enzymes, e.g. metallothioneins which bind metals with sulfur affinity (Cd, Cu, Zn); (e) complexes with polysaccharides or glycoproteins, especially in plants; (f) metallodrugs, such as platinum or ruthenium complexes known in cancer therapy, and gold compounds used as antiarthritic drugs as the most significant examples.

The criteria governing the validity of a clinical sample required for analysis depends on the information needed. Then, definition of sampling and treatment should fit the final purpose of the analysis, e.g. diagnosis of deficiency or toxicity, control of environmental or biological pollution, nutritional surveillance or forensic investigations. Speciation information from these samples can be used with two purposes: (i) to define and evaluate the mechanisms of release of species into the biological matrix and the degree of interaction with the environment; (ii) to know the absorption and distribution mechanisms throughout the organism, their bio-availability, toxicity and excretion pathways [4, 5].

Elemental speciation in this field still represents a great challenge due to the low concentration of the species, their poor stability and the matrix interference. In addition, some aspects have to be solved, such as understanding their role in physiological and pathological processes, the availability and practicability of analytical methods, and finally the ethical considerations for sampling [5].

1.2 Vegetables samples

Most studies related to the analysis of plants have been concerned with the effects of elements on plant growth, element uptake, toxicity, translocation and soil–plant relationship [6]. Many studies with these biological samples come from the ability of some plants to grow in the presence of high heavy metal concentrations. This growth is a result of a variety of tolerance mechanisms, e.g. metal binding at the cell wall, precipitation in vacuoles, and synthesis of metal binding compounds such as proteins, peptides, organic acids and phenol compounds [7]. One of the most abundant mechanisms of metal binding in plants seems to be the synthesis of phytochelatins and small peptides with high cysteine content that have the ability to chelate heavy metals and then to reduce the concentration of free metal ions in the cytosol. A detailed knowledge of the metal-binding characteristics of phytochelatins is necessary to understand the role of these peptides for metal detoxification in plant systems.

1.3 Nutritional samples

Food science in the broadest sense can be extended to include soil chemistry, plant uptake and, at the end of the food chain, the metabolic fate of elemental species when certain foods are consumed by humans or animals. Samples analysed are of a great variety: from the relatively simple, e.g. fruits or vegetables, to more complex such as processed whole meals, diets etc. [8]. A number of incidents have highlighted the need for elemental speciation studies in food, one example being the Minamata Bay incident, in Japan [9], where fish accumulated high amounts of the methylmercury discharged into the bay by the Chisso Corporation factory. This bio-accumulation led to a massive poisoning of local people with fatal consequences.

The chemical form of an element in food determines the mode of absorption in the intestine and the subsequent metabolism processes that may occur. The bio-availability of a species in food may also be measured and is defined as a measure of the proportion of the total amount of a nutrient that is utilised for normal body functions [10].

The amount of a particular species absorbed into the body after ingestion is difficult to measure as the mechanisms of absorption, individual differences in requirements and metabolic control often varies depending on the individual's response [10]. International legislation for food is based on the total element content and not the individual species, but some regulations and guidelines are being introduced [11].

Reports of elemental speciation in all these samples have increased over the last few years. Speciation information may be used to determine the fate of a trace element species in a biological system. In addition, the molecular forms that occur in a biological sample may also be monitored at the cellular level, which is the key to understanding many of the biotransformation processes in most biological systems.

Different steps are generally required for speciation analysis in biological samples: sample collection and handling, storage, evaluation of the species stability until analysis, sample pretreatment, complete species extraction from the matrix, clean-up procedures before separation, preconcentration and derivatisation as sensitivity improvement methods, etc. All these steps, from sampling to analysis mean that the integrity of the species is often a risk and they have been summarised in Figure 3.1.1.

This chapter aims to treat each step separately, discussing in depth the main considerations to take into account depending on the sample type and the information required.

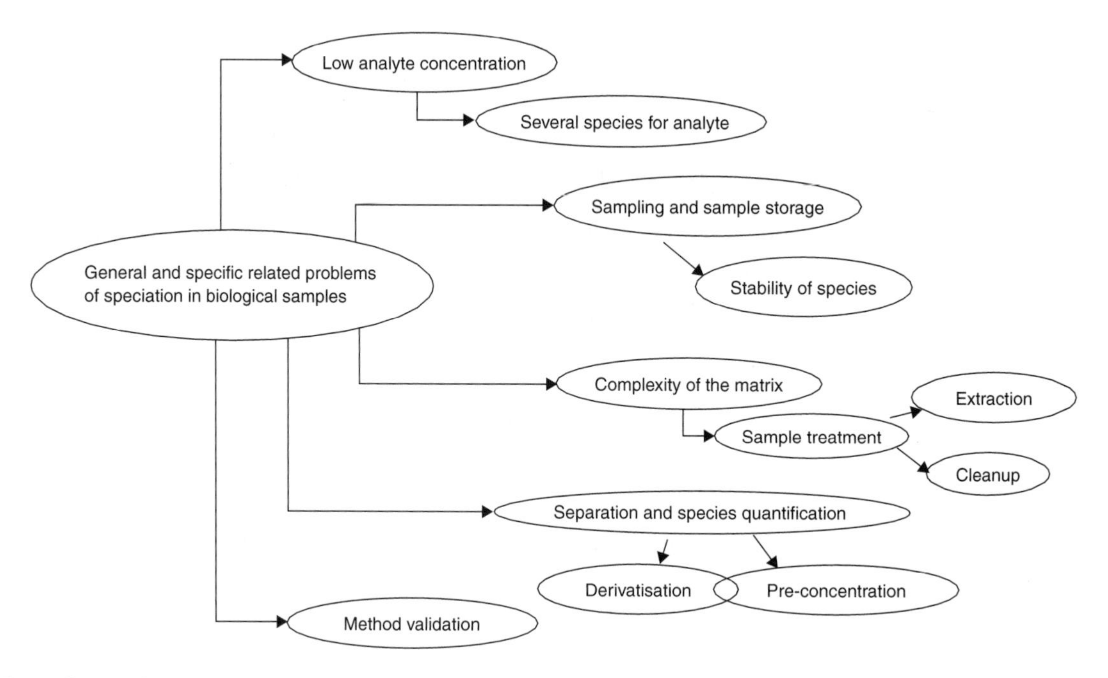

Figure 3.1.1. Main problems of metal speciation in biological samples.

2 ELEMENTS OF INTEREST

It is interesting to have a look at the main chemical species of the elements that have primarily concerned in biospeciation.

2.1 Aluminium

The risk of aluminium in the organism has been associated with several factors; these include a wide variety of uses, e.g. in packing and building materials, paints, industrial exposures, hemodialysis treatment, etc. Additionally, free Al may compete with other metals, causing damages in the activity of some enzymes, and even with DNA, probably by cross-linking the protein chains. Al is suspected of being involved in a number of neurological disorders, e.g. Alzheimer disease, and some liver diseases [12, 13]

Different Al species has been extensively studied in serum and plasma of dialysis patients: their concentration in blood is very high due to the irreversible uptake of Al from dialysate solutions and through the oral intake of Al-containing medicines. The main compounds present in blood are Al–transferrin, Al citrates and Al bicarbonates; they can be deposited in various parts of the body with a very long biological half-life. The species of this element have also been widely studied over the last decades because of some Al–hydroxo complexes that are highly toxic for fish [6, 11, 14].

2.2 Arsenic

Arsenic is an element that raises much concern from both the environmental and human health points of view. The different species of As produce diverse toxicological effects in human, with inorganic forms being more toxic than organic ones. For food control, the objective consists in discerning the toxic inorganic forms from the harmless organo-arsenicals. Although foodstuffs yield total As concentrations below the legal limit, in some cases fish derivatives exceed these values. However, the As present in the latter samples is presumably arsenobetaine and so not toxic. Biomonitoring of occupational exposure to arsenic needs to study different As species (As(III),

As(V), MMA, DMA, As–betaine, As–choline, As–sugars) in a widerange of samples: serum, plasma, urine, tissues, ... It is important to quantify all these species in order to identify the cause of exposure and the metabolism of such species into the body [14, 15].

2.3 Chromium

The difference in toxicity between Cr(III) and Cr(VI) is well known, especially when they are present in living organisms [2]. As a consequence, Cr(III) is an essential trace element; conversely, Cr(VI) is known to be carcinogenic and mutagenic. For instance, if Cr(VI) is inhaled it is able to cross cell membranes and is transported by the blood cells, damaging them. Of particular concern is the regulation of total Cr level in the air (at the workplace) because of the high level of damage caused by this element when inhaled [14]. Differentiation between those forms of chromium is important for assessing food and water safety.

2.4 Copper

Copper is a common contaminant in drinking water, released by the piping for household taps. The corrosion caused by the Cu present in the water results in a layer of CuO_2 with small amounts of other oxides. These species can also be converted into different ones, and even bound to organic matter. We still do not know much about the toxicity of the different compounds that can result from these transformations. In addition, the use of pesticides and fungicides containing this element for many years has led to soil and water contamination. Thus, speciation of Cu is crucial for understanding the mobility and risks to plants and consequently to animals [14]. In contrast, copper is considered an essential element in animal nutrition: in most *in vivo* environments, copper is predominantly associated with macromolecules, but there are other low molecular mass (LMM) species that can traverse the cell membranes and are potentially more toxic [16].

2.5 Mercury

Mercury is probably the element most studied from the toxicological point of view. All forms are considered poisonous; however, the objective will consist in discerning the highly toxic alkyl compounds (mostly methylmercury) from less toxic inorganic Hg forms. Some episodes of health damage by organomercury compounds are well known: in Minamata (Japan), MeHg contamination caused severe brain damage in infants whose mothers ingested contaminated fish during pregnancy. In Iraq, poisoning of human happened because of the consumption of wheat seeds treated with organomercury insecticides. In general, exposure to organic Hg can cause brain damage to a developing foetus since CH_3Hg^+ readily crosses the placenta. It is very important then, from the nutritional and the health point of view, to develop techniques sensitive enough for the monitoring or rapid screening of Hg species in very varied matrices: food (mainly fish tissue), blood, urine, hair, brain tissue, etc. [14, 17].

2.6 Lead

Lead is considered to be a toxic element mainly due to some neurological problems, renal dysfunctions, hypertension and cancer, for which there is evidence for animals but not yet for humans [18]. The most important group of species is the organolead compounds (R_nPbX), which are quite toxic. The more substituted the organic chain the higher the toxicity. Organolead compounds are very labile and easy transformations take place between them. Lead (tetraalkyllead) has been used as an additive in petrol for many years (it is now forbidden in many Western countries) and this is one of the main reasons for concern about lead in biological systems [19, 20].

2.7 Selenium

Selenium has been shown to be essential and toxic, depending on the concentration as well as on the species present. The main essential role identified for selenium has been as a component of the enzyme glutathione peroxidase (GSH-Px), one of the antioxidant defence systems of the body [21]. Keshan disease is the most famous illness caused by a deficit on Se in the human and animal diet, due to the low content in soil and water in that Chinese area [22]. Food supplements containing Se–cysteine have been an object of study in recent years because of their beneficial action in some areas of low selenium content in the soil [14].

The most common selenium species to be considered are the inorganic selenium (Se(IV), Se(VI)), the biomethylated species (DMSe, DMDSe) and the selenium amino acids (Se–cysteine, Se–methionine, Se–cystathionine, ...) and other compounds with higher molecular weight. Selenium speciation has also been undertaken in plants, blood and animal tissues to monitor the occupational exposure to this element as well as problems related to its deficiency.

2.8 Tin

Butyltin and phenyltin compounds are very well known and studied because of their high toxicity for aquatic life. Fortunately, levels of the most toxic compound, tributyltin (TBT) have been regulated in most countries [14]. TBT has become a common contaminant of fish and shellfish. The regulation clearly establishes a daily intake that it is still exceeded in many regions [23]. Nowadays, it is important to measure the levels present in the different tissues and body fluids in order to clarify the metabolism and the transformation or detoxification mechanisms of the organism that has ingested it.

3 SAMPLE PRETREATMENT PROCEDURES

3.1 General points of sampling and storage

The measurement of different trace element species requires a sample preparation and special handling that must be designed for each particular sample. Numerous steps in an analytical procedure increase the risk of losses by adsorption to the containers or due to chemical instability. It enhances the possible contamination from reagents and equipment, as well.

First of all, collection of biological samples needs a number of procedures because the major problem in such work is due to contamination and losses of the trace elements. Some sampling procedures have been presented in Chapter 2.2 of this book.

Concerning the storage of the samples, precautions such as full details on collection, freezing or lyophilisation of samples, addition of preservative or/and additive, etc. must be done by the analyst prior to the analysis as a routine. These procedures have already been treated in Chapter 2.2, of this book.

3.2 Clinical samples

The preliminary steps of pretreatment of clinical samples include, in many cases, the release of species out of the cells, the selection of a particular group of species, desalting procedures, and fractionating techniques.

3.2.1 Release of the species out of the cells

Whole blood is usually centrifuged to obtain serum, which is free of fibrinogen. Otherwise, an anticoagulant can be added to obtain a supernatant or plasma. Serum or plasma is homogenised by vortexing and, at this point, it can be stored at −20 °C or the analyst can proceed with the remaining steps of the analysis [2]. In the case of *tissues*, the chemical speciation begins with the separation of the soluble species from those bound to insoluble compounds. For this purpose, the tissue is subjected to an ultrasonic homogenisation in an isotonic buffer (pH = 7.4). The homogenate is centrifuged and the supernatant, containing the soluble compounds, is subjected to different fractionating techniques that will be discussed later. Speciation of elements bound to insoluble compounds that remain in the precipitate cannot be pursued any further [2]. The distribution of trace elements in tissues can also be approached at the subcellular level, among cytosol, mitochondria and nucleus. Pretreatment of *urine samples* can be limited to a cleaning step by passing it through a C18 cartridge and subsequent filtration and appropriate dilution with water or deproteination using ultrafiltration [24]. However, at room temperature, bacterial action can rapidly cause changes in the sample such as pH and formation of precipitates or change/interconversion of species. Therefore, the pH is important to note when collecting urine samples.

3.2.2 *Selection of the group of species*

The first decision to make for elemental speciation in a clinical sample is whether low molecular mass (LMM) or high molecular mass (HMM) compounds are going to be investigated. The separation of both groups is generally done by ultrafiltration yielding solutions free of protein [25]. The separation is characterised by the cut-off value of the membrane, this being the maximum molar mass of the proteins able to pass through the pores. The ultrafiltrate is collected in the space beyond the membrane and contains all components (with a molecular mass below the cut-off of the membrane) at the same concentration as in the original sample. The most common problem encountered with ultrafiltration is the changes in concentration of the LMM species because the filter can be fouled during the filtration process.

3.2.3 *Desalting*

Desalting is necessary when the ionic strength of the solution does not accord with the conditions required for chromatographic separations. Gel filtration chromatography, with a fractionating range of 1–5 kDa, can be used for this purpose.

3.2.4 *Fractionating techniques*

First of all, the separation of *LMM trace element species* can be performed by ion exchange chromatography (Dowex resins) as they are charged species. Before the chromatographic step, it is necessary to remove the HMM compounds by ultrafiltration to prevent clogging of the column. The pH of this ultrafiltrate must be carefully adjusted to that of the mobile phase before injection. Concerning the *trace element species bound to proteins*, they are usually separated by fast protein liquid chromatography (FPLC), which covers ion exchange, reversed phase, size exclusion and affinity chromatography. The choice of a buffer as chromatographic eluent can be determinant for the type of species analysed: if it contains salts with a strong affinity for the protein it may provoke some replacement reaction [2].

3.3 Nutritional samples

Fresh foods may have to be prepared by using stainless steel knives and/or plastic choppers, depending on the species of interest. Domestic blenders, coffee grinders and food processors fitted with stainless steel, Ti or ceramic blades are very effective for homogenising food samples. Some special considerations must be taken depending on the nature of the sample [8]:

- Liquid foods: milk, wine, fruit juice, etc. only need to be shaken before subsampling; sparkling beverages should be degassed in an ultrasonic bath or by bubbling with any inert gas; those wines which form a precipitate (some red wines) need to be decanted before subsampling.
- Meat: inedible parts (skin, hair, etc.) should be removed and discarded unless a particular contamination study on these materials is required. Meat is homogenised using a blender or equivalent.
- Fish: before subsampling it is necessary to proceed as for meat, avoiding useless parts, e.g. head, fins and larger bones. A food blender is the common way to homogenise samples.
- Vegetables: gross surface contamination should be removed from both root and leafy vegetables by washing with pure water. Analysis of dietary intakes may require peeling root vegetables or discarding the outer leaves of vegetables.

- Fruit: the outer skin or peel may be a useful indicator of surface contamination, and it can be of interest to analyse this separately from the edible portions.
- Whole meals or diets: when a mixture of ingredients is to be analysed there are two options: (i) representative samples of the relevant food can be subsampled; (ii) portions of the whole diet can be combined, homogenised and stored.

4 CLEAN-UP

Biological samples have a fairly complex matrix due to the presence of thousands of compounds of very different nature: hydrocarbons, polysaccharides, lipids, amino acids, glycerides, etc. Therefore, clean-up procedures are necessary in order to remove those compounds or groups of compounds useless for the purposes of analysis. Some of the most common clean-up methods are presented here:

4.1 Elimination of lipids

Lipid decomposition is very fast in biological materials, and it is convenient to extract them completely from the sample. The method of extraction and purification of lipids must involve a mild treatment to minimise their oxidative decomposition, which creates secondary compounds that could affect the stability of the species. One of the fastest and more efficient methods uses a mixture of chloroform and methanol in such proportions that a miscible system is formed with the water present in the tissue. Dilution with chloroform and water separates the homogenate into two layers, the chloroform layer containing all the lipids, and the methanolic layer containing all the nonlipids. Because of the high differences in humidity and/or lipid amount present in the sample, it is sometimes necessary to dilute with distilled water. A purified lipid extract can be obtained by isolating the chloroform layer, if needed. The entire procedure can be carried out in 10 min; it is efficient and free from deleterious manipulation [26].

4.2 Clean-up of extracts

It is also important to consider the clean-up needed in many cases after the extraction step. Some extracts are 'very dirty samples' as a result of a nonspecific extraction method. One of the most common procedures applied is by passing the extract through an alumina column to retain the high saline and other ionic compounds [27]. The removal of the organic matrix from samples with a high organic content (like urine) can be achieved by using a simple treatment consisting of solid phase extraction with C18 cartridges [28]. In addition, this clean-up procedure has been found to give a high stability for some compounds (selenium species) in this complex matrix.

5 EVALUATION OF SEVERAL EXTRACTION PROCEDURES

Sample extraction procedures are often perceived as bottlenecks in analytical methods. In the last few years, various attempts have been made to replace classical extraction techniques in order to reduce the volume of extraction solvents required and to shorten the sample preparation time. Several of these methods have been reported in the literature, using aqueous or organic solvents to solubilise the different compounds; hydrolysis procedures (acid, alkaline and enzymatic hydrolysis); solid phase extraction; and other enhanced techniques, such as supercritical fluid extraction, accelerated solvent extraction, etc.

5.1 Aqueous extraction

Different extraction procedures have been applied for Se speciation in samples like yeast, allium vegetables and other plants. The first approach was a simple hot water extract; the second one was an acid hydrolysis with 0.1 M HCl in a methanolic medium [29]. The extraction needs several hours and the recovery and the selenium species profile is not complete. In the case of arsenic, an ultrasound-assisted extraction with MeOH + HCl (50 % +

10 %) has also been tested in environmental and biological matrices with similar results to those for selenium [30].

5.2 Simple solvent extraction

Most organic solvent extraction methods serve only to isolate species of interest from the major matrix in order to eliminate interference with the subsequent steps of the analytical method, that is, derivatisation, separation and detection. Samples are commonly first digested with basic or acid reagents to completely solubilise the sample, then neutralised prior to extraction with the appropriate solvent. Finally, the species are back-extracted from the organic solvent into an aqueous phase if required by the analytical separation or detection technique. As an illustrative example, some solvent extraction procedures developed for mercury speciation in biological samples include solubilisation by using KOH/MeOH or TMAH/MeOH (methanolic tetramethylammonium hydroxide); extraction by adding dichloromethane, hexane, CH_2Cl_2 [31, 32] or dithizone chelating reagent and chloroform [33]. The organic compounds are back-extracted into the aqueous phase with sodium thiosulfate solution buffered with ammonium acetate or with potassium bromide/copper sulfate mixture to favour the process. In all these cases, recoveries of CH_3Hg^+ higher than 80 % are obtained when applied to different fish tissues and to sea plants.

5.3 Enzymatic extraction

Nonspecific enzymes are capable of breaking down a wide range of proteins into their amino acid components. In this way, enzymes can be employed for digesting biological fluids, mainly blood [34]. Depending on the enzyme and the digestion conditions applied, this method can be applied for elemental speciation. This approach has been widely used for tin and selenium speciation in different biological materials [1, 35]. The main enzymes used with extraction purposes include protease, lipase, trypsin, pepsin, pronase, or mixtures of them. Some authors have found that the addition of additives like methanol or citric acid helps the solubilisation of some compounds (TBT in the case of tin) by complexation, and prevent bacterial growth during the incubation period, sometimes 48 h [36].

Generally, the use of enzymatic hydrolysis processes has shown better results in the release of the species of interest from solid biological samples [29, 35–37]. A study performed by Moreno *et al.* [1] has evaluated the efficiency in both aqueous soluble and solid fractions. The complexity of such a process is clearly evidenced in Figure 3.1.2. Apart from this complexity, the use of enzymatic hydrolysis for extraction in biological samples is very convenient and widely applied.

5.4 Solid-phase extraction (SPE)

SPE was developed by Zhang and coworkers for the extraction of pure organic compounds from aqueous samples [38, 39]. In general, the analyte from a relatively large volume of solution is selectively retained by a solid reagent phase (by a variety of mechanisms) and then released into a relatively small volume of eluent [30]. Actually, it is being used for the extraction and preconcentration of organometallic compounds, after a derivatisation into a volatile and apolar form, which easily evaporates from the aqueous solution into the headspace of a gas chromatograph (GC). Then, it is adsorbed by the apolar phase of the coated silica fibre of a SPME (solid phase micro-extraction) device, establishing equilibrium of the analyte among the three phases: the aqueous phase, the headspace and the fibre coating. After 10 min of exposure time, the fibre is inserted in the GC injection port and the compounds that have been collected are thermally desorbed for subsequent analysis [40, 41]. The accuracy of this approach for biological samples has been tested by means of analysis of two reference materials of fish tissue (NIES-11) and dogfish muscle (NRC DORM-2). Another application of SPE as an extraction/preconcentration method has been the retention of inorganic lead complexed

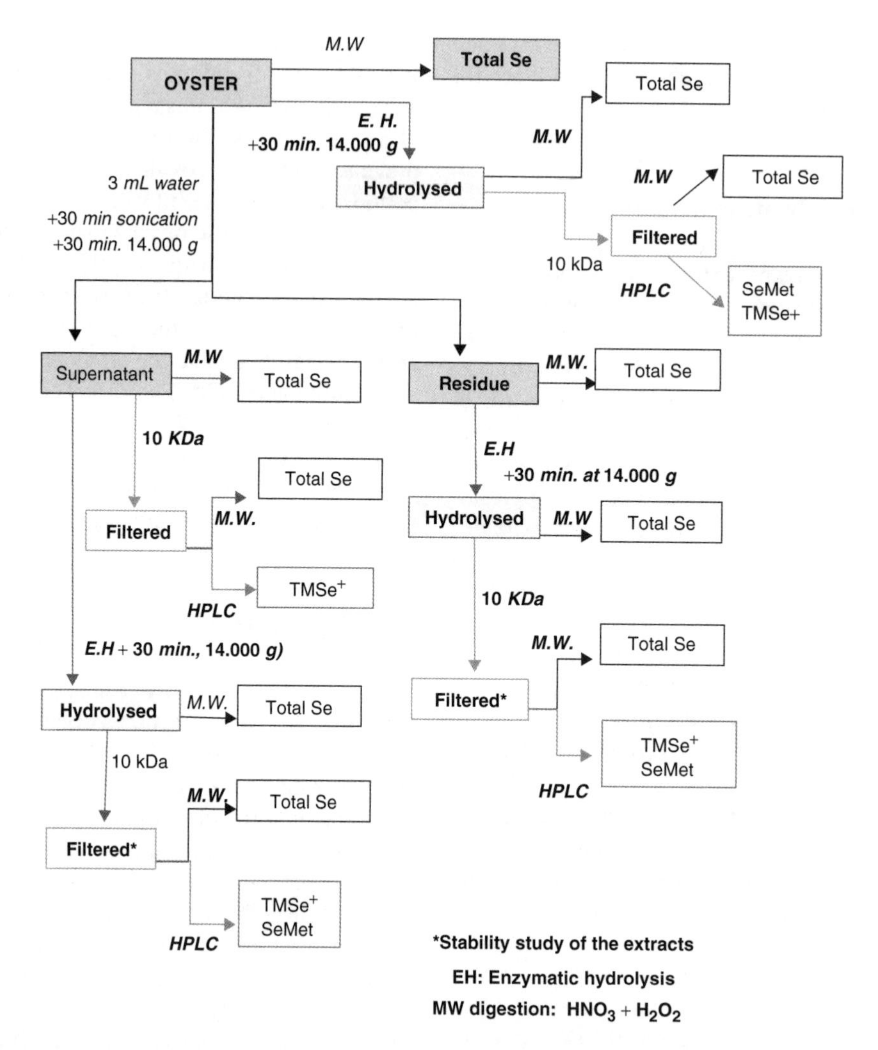

Figure 3.1.2. Scheme of an enzymatic hydrolysis extraction procedure for selenium species in oyster tissue.

with DDC in a column of C18 silica, eluted with acetonitrile, or in a Chromosorb 102 column eluted with methanol. This method was applied to fruit juice [42].

5.5 Supercritical fluid extraction (SFE)

The properties of a supercritical fluid make this extraction technique among the most interesting and promising to obtain extraction recoveries close to 100 %. The properties of supercritical fluids that are attractive from an extraction point of view include: (a) considerably great diffusion coefficients leading to efficient and rapid extractions; (b) low viscosity and absence of surface tension that facilitate pumping in the extraction process; and (c) density close to liquids enabling the greater interactions on a molecular level necessary for the solubilisation. Temperature and pressure changes, near the supercritical point, can affect the solubility by a factor of as much as 100, or even 1000. Moreover, the use of fluids with low critical temperature values (CO_2, N_2O, ...) allows extractions under thermally mild conditions, protecting labile compounds [43].

The main disadvantage is the fairly limited ability to study ionic or highly polar compounds [44]. This extraction technique has been extended to supercritical fluid chromatography coupled to ICP/MS, showing a high resolving power and 100 % sample introduction efficiency. However, there are two main disadvantages: the necessity of using a heated transfer line to make possible the coupling, and the adverse effect to sensitivity caused by the introduction of CO_2 into the plasma [45].

5.6 Accelerated solvent extraction (ASE)

ASE is a new technique that uses organic solvents at high pressures and temperatures above the boiling point. A solid sample is enclosed in a sample cartridge that is filled with an extraction fluid and used to statically extract the sample under elevated temperature (50–200 °C) and pressure (500–3000 psi) conditions for short periods of time (5–10 min). Compressed gas is used to purge the sample extract from the cell into a collection vessel [46, 47]. Two main reasons are responsible for the enhancement of ASE: (i) higher solubility of the analytes and better mass transfer between the two phases, and (ii) disruption of surface equilibrium due to the high temperatures and pressures applied (break-up of the strong solute–matrix interactions, decrease of the solvent viscosity allowing better penetration of matrix particles, etc).

This technique usually requires the use of freeze-dried samples. In spite of the fact that it has been widely used for extraction of organic compounds, the use for metallic species determination is very scarce and only a few papers have been published. Gallagher *et al.* [47, 48] have proposed this technique for the extraction of organoarsenicals in seaweed. After extraction of arsenic compounds and solvent evaporation, the residue was re-dissolved and prepared for HPLC by passing it through a C18 cartridge for sample cleanup. It is important to consider the great influence of the matrix nature on the recovery of this technique as well as the particle size that depends on the number of cycles applied.

Examples of different extraction and other sample pretreatment strategies are summarised in Table 3.1.1.

The difficulties associated with species extraction are illustrated in a recent intercomparison exercise for selenium speciation in a white clover certified reference material (BCR CRM 402) [49, 50]. When comparing the results reported from this study, the extraction efficiencies differ considerably (from 15 to 75 %) and there is even no consensus in terms of the species detected.

6 DERIVATISATION TECHNIQUES TO GENERATE VOLATILE SPECIES

As gas chromatography is often the method of choice in separation of species, the compounds must exhibit high volatility and thermal stability. The number of nonpolar organometallic species is fairly limited (tetraalkylleads, methylselenium, organomercury and naturally metalloporphyrin compounds). Then, derivatisation reactions must be used to transform the polar species into nonpolar ones [19].

6.1 Hydride generation (HG)

The common method for derivatisation is the formation of volatile hydrides. This is applicable to compounds of several elements as As, Sb, Bi, Ge, Pb, Se, Te, and Sn [4, 8, 51]. The inorganic forms of these elements react with sodium borohydride with formation of simple hydrides, but not all species react equally fast and in the same solution conditions. This behaviour can in some cases be used for the separation of the different species. Some of the alkyl derivatives also form volatile hydrides under similar conditions. The advantage of hydride generation is connected with their preconcentration in a trap of liquid nitrogen; they are released by slowly raising the temperature. This gives a chance of simple separation, with subsequent determination of individual species [52]. One of the biggest disadvantages of hydride generation is the interference caused in the liquid phase. Interfering species can appear during the hydride

Table 3.1.1. Decomposition and pretreatment methods for speciation in biological samples. Reproduced by permission of The Royal Society of Chemistry.

Sample type	Compound determined	Sample treatment	Separation detection	Reference
Urine	Me_3Pb^+; Et_3Pb^+	Adjust pH to 10 with NaOH + EDTA to complex metal ions and remove precipitate by centrifugation. Pass the sample though silica gel and desorb Me_3Pb^+ and Et_3Pb^+ using acetate buffer (pH = 3) in 10 % methanol. Add borate buffer (pH = 7).	HPLC	70
	Selenium species	Clean-up: C18 cartridges, previously conditioned with MeOH + H_2O. Dilution into a phosphate buffer (pH = 6).	HPLC-ICP/MS Stationary phase: Spherisorb ODS column Mobile phase: phosphate buffer	28
	Selenite and selenate	Volunteers supplemented with Se-Met for several weeks. Sample dilution 1 + 1 with water.	HPLC-ICP/MS Anion-exchange (Dionex Ionpac AG11-HC) column. Elution with 25 mM NaOH and 2 % MeOH	71
Human serum	Magnesium–protein complexes	Pretreatment: sample dilution two fold with Tris-HCl buffer and saturated with CO_2 (pH = 7.0–7.4). Clean-up with Chelex-100 resins.	IEC-ETV-AAS/IEC-DAD Anion exchange column. Elution with gradient Tris-HCl buffer (pH = 7.4)/0.2M NaCl	72
Plasma, urine, animal tissues	Se, Zn, Cu, and Fe species	Blood centrifuged at 1200 *g* (2 °C) Extraction: 50 mM Tris-HCl (pH = 7.4). Homogenised in N_2 atmosphere. Centrifuged.	SE-HPLC-ICP/MS Mobile phase: 50 mM Tris HCl buffer (pH = 7.4)	73
Blood	Me_4Pb	Hemolyse by freezing at −20 °C (24 h) Extract in *n*-heptane (ultrasonic bath). Centrifuge and withdraw organic layer.	GC	74
	Me_3Pb^+	Add ammonium citrate (pH = 9) + EDTA and DDTC. Extract twice with pentane. Separate and combine organic layers. Derivatise after evaporation with 2 M C_4H_9MgCl in THF. Add pentane and 1 M H_2SO_4. Separate organic phases and dry with anhydrous Na_2SO_4.	GC	74
Plants	Phytochelatins	EDTA in water cleanup + filtration Tris-HCl buffer (pH = 8.6) + mercaptoethanol (to prevent oxidation of thiol groups) Disruption of cells by sonication; centrifugation. Cytosol extraction by filtration (0.22 μm)	HPLC-AAS; HPLC-ICP/MS Reversed phase column Mobile phase: ammonium acetate (pH = 7)	7, 75
	SeMet, Se-MetCysteine; Se-Cysthationine	Hot water extraction. Enzymatic hydrolysis with 10 % protease (shaking 24 h at room temp.). Centrifugation and filtration.	HPLC	76
Yeast, White clover	Se-Cyst, Se-Met, Inorganic Se	0.25 g + water/HCl 0.01 M at 37 °C overnight + MeOH + $CHCl_3$ + H_2O shaken (5 h) + HCl 0.01 M + pepsin/pronase + lipase stirred overnight. Filtration and centrifugation.	HPLC-ETAAS PRP-X100 column Mobile phase: nickel acetate + nickel sulphate	29

		Ultrasonication (2 × 30 min.) in $MeOH/H_2O$. Supernatant filtration (0.45 μm). C18 cartridges and evaporation.	Microbore IEC	50
Yeast	Selenium species	Extraction: • Hot water (90 °C) • 10 % MeOH + 0.2 M HCl • 30 mM Tris-HCl (pH = 7) + 0.1 mM PMSF • 4 % Driselase in 30 mM Tris-HCl + PMSF • 30 mM Tris-HCl (pH = 7) + 4 % SDS. • 20 mM Ammonium phosphate (pH = 4) + 0.15 M NaCl + 0.1 M PMSF + 1 mM EDTA + 5 % SDS • Ammonium phosphate (pH = 7.5) + pronase + lipase incubated 16 h at 37 °C • 25 % TMAH at 60 °C for 4 h.	HPLC-ICP/MS • SE Superdex-200 HR column (1300 kDa limit). Eluent: Tris-HCl buffer (pH = 7) • PRP-X100 column. Eluent: Ammonium phosphate buffer • Inertsil ODS-2 column. Eluent: 0.1 % TFA + 2 % MeOH	77
Oyster tissue	As-Bet, DMA, As-Sugar, As-Residual	Extraction: focused microwave-assisted extraction with $MeOH/H_2O$ (1 : 1) Centrifugation; Evaporation to dryness; Dilution with water up to 10 ml Clean-up: 600 mg C18 snap-cartridges and filtered (0.22 μm) nylon filter	HPLC Anion-exchange column Phosphate mobile phase (pH = 6)	To be published
	TBT, DBT, MBT, TPhT, DPhT, MPhT	Extraction: TMAH 20 % (60 min), addition of HCl (pH = 7); Addition of sodium Ac-/HAc buffer (pH = 4); Extraction with hexane; Derivatisation with $NaBEt_4$	GC-AED	
	TBT, DBT, MBT	Extraction: methanol/HCl/tropolone; Sonication; liquid–liquid partitioning with methylene chloride; back-extraction with iso-octane. Grignard derivatisation (penthylation). Clean-up: silica column	GC-MS	
Oyster tissue	Se-Met, Se-$Cyst_2$, $TMSe^+$, Inorg. Se	Fractionation: 0.2 g + Water. Centrifuged extraction: supernatant and solid fractions + 20 mg protease + Tris-HCl 0.1 M (pH = 7.5). Incubation (24 h at 37 °C). Ultrafiltration through 10 kDa filter	HPLC-ICP/MS • Anion-exchange PRP-X100 column. Mobile phase: ammonium citrate (pH = 4.8) • Cation-exchange PRP-X200 column. Mobile phase: Pyridine formate (pH = 2.8)	1
Fish tissue	TBT, TPhT	Extraction (CO_2 SFE system): 0.15 g + MeOH at 100 °C and 6000 psi. Total time: 15 min.	LC-ICP/MS Mobile phase: MeOH/H2O/Ac- (94 : 5 : 1) at pH = 6. Ion pairing: pentanesulfonate 4 mM.	78
	Organotins	Extraction (focused microwave): • 0.2 g + AcH + $NaBEt_4$. Time: 3 min. • 0.2 g + TMAH + AcH (pH = 5) + isooctane + $NaBEt_4$. Time: 2 min. Clean-up: alumina column	MC-GC/MIP-AES	27
Shellfish tissue	MBT, DBT; TBT	Extraction: 0.2 g freeze-dried + buffer (pH = 7.5) + 10 mg lypase + 10 mg protease. Incubation at 37 °C (4 h).	HG-CT-QF-AAS	35

(continued overleaf)

Table 3.1.1. *(continued)*

Sample type	Compound determined	Sample treatment	Separation detection	Reference
	Arsenite and arsenate	Samples freeze dried and homogenised. Total As: microwave-assisted distillation (0.5 g + KI + ascorbic acid + HCl Extraction: 0.5 g + HCl overnight. Add a reducing agent + chloroform (twice). Back-extract in 1 M HCl. Centrifugation and filtration.	MW-HG-AAS	79
Marine organisms	Arsenic species	Extraction: Water/MeOH (1 : 1) sonicated (x4); filtered; purification with diethyl ether phenol	HPLC-HG-QFAAS HPLC-ICP/OES	80
		• Extraction: methanol/chloroform (1 : 1) • Extraction: methanol/water (1 : 1) Clean-up: C18 cartridges	LC-UV-HG-ICP/OES	81
			LC-UV-HG-ICP/MS Anion-exchange column. Mobile phase: phosphate buffers gradient (pH = 6)	82
Liver	Inorganic selenium	Extraction: 50 mM Tris-HCl buffer (pH = 7) + 0.25 mM glucose at 25 °C under N_2 atmosphere. Centrifugation at 105.000 *g* at 4 °C (1 h).	HPLC	83
Kidney	Se-Cyst, Se-Met, Inorganic Se	Extraction: HCl 0.01 M + pronase + lipase stirred overnight. Filtration and centrifugation.	HPLC-ETAAS PRP-X100 column Mobile phase: nickel acetate + nickel sulphate	29
Human milk	Se-Cyst, Se-Met, Se-Cystamine	Clean-up: removal of fat. Protein precipitation by centrifugation at 25.840 *g* at 8 °C (30 min)	CZE	84
Food supplement	Se-Met	Extraction: enzymolysis with protease K in $NaHCO_3$ (pH = 7) at 37 °C in darkness (20 h). Filtration with nylon filters (45 mm).	HPLC	85
Cereals, Multivitamins tablets	Soluble/insoluble copper, zinc and iron species	• Stomach extraction: enzymolysis with pepsin; incubation for 4 h at 37 °C; pH adjust to 2.5 with HCl • Intestine extraction: enzymolysis with pepsin/pancreatin/; incubation for 8 h at 37 °C; pH adjust to 7.4 with $NaHCO_3$ Centrifugation and filtration through 0.45 μm pore size filters	FI-ICP/MS	86

formation or during their decomposition just before the detection step. In the first case, this is due to an inhibition of the hydride formation or a transformation into other compounds due to interactions with other metals present in solution; in the second case, it can come from other hydride interactions and/or analyte losses [51].

One possibility for hydride generation procedures is the use of an immobilised borohydride reagent. It has been described for arsenic and selenium [53, 54] in drinking water, where the derivatising reagent was immobilised on an anion-exchange resin and the hydride was generated on passage of an acidified sample solution. The optimum tetrahydroborate concentration (0.05 %) was considerably lower than typical values.

6.2 Cold vapour

This is similar to hydride generation. The difference is the volatile product resultant from the reduction reaction: the volatile metal. It is only applicable to Hg and Cd. It has been widely used for MeHg determination in several biological samples due to the ability of this species to react with the reductant borohydride [55, 56].

6.3 Ethylation

Another procedure of derivatisation is ethylation by using sodium tetraethylborate, applied for Sn, Se, Hg, Pb, ... [32, 51, 57, 58]. Ethylation can be accomplished in an aqueous medium; then, derivatisation and extraction frequently occur in the same step. In addition, ethylation permits the simultaneous derivatisation of organo -Sn, -Hg, and -Pb compounds. Nevertheless, liquid–liquid extraction remains necessary and organic solvents are required. Sometimes, there is no discrimination among the different species of an element. This is the case of Pb: Pb^{2+}, Et_3Pb^+, or Et_2Pb^{2+} all react with the reagent $NaEt_4B$:

$$R_3PbX + NaBEt_4 \longleftrightarrow R_3PbEt + BEt_3 + NaX$$

$$R_2PbX_2 + NaBEt_4 \longleftrightarrow R_2PbEt_2 + BEt_3 + NaX$$

6.4 Grignard reactions

Derivatisation using Grignard reagents (alkyl or arylmagnesium chlorides) has a more universal character. This has been applied for Sn, Hg, and Pb speciation [59]. For example:

$$R_3PbX + R'MgX \longleftrightarrow R_3PbR' + MgX_2$$

The main drawbacks of Grignard reagents are their atmospheric instability and their ability to be hydrolysed in presence of water forming $Mg(OH)_2$:

$$R_3PbX + H_2O \longleftrightarrow R'H + \tfrac{1}{2}Mg(OH)_2 + \tfrac{1}{2}MgX_2$$

To avoid this problem, these reagents need to be stabilised in ether and stored under an inert atmosphere. The water remaining is removed after extraction with a complexing agent. Then, sample preparation is laborious and time-consuming, and removal of the excess of the derivatising reagent by water/acid addition is necessary to avoid high blank values.

6.5 Other methods

A very recent approach for derivatisation of amino acids has been accomplished by esterification of the carboxylic acid group using propan-2-ol, followed by the acylation of the amino group with trifluoroacetic acid. The derivatives are then extracted into chloroform and analysed by GC-MS [60]. This new method has been applied to selenomethionine.

There are several other potentially applicable procedures for speciation analysis, based on the formation of volatile chelates, such as trifluoroacetylacetonates, dithiocarbamates [61]; volatile oxides and halides [68]; and carboniles. However, none of them has been reported for speciation analysis in biological materials.

7 PRECONCENTRATION OF THE SPECIES

The speciation analysis of trace elements often needs a preconcentration step due to the very low concentration of the compounds of interest in biological systems and the distribution of these low total levels among several species. For this purpose, metallic species can be trapped or retained in solid adsorbents.

7.1 Amalgam formation

The best and most extensively used amalgam is that formed for mercury preconcentration with gold.

7.2 Cold trap (CT)

The analytes of interest are (after derivatisation) purged from the aqueous sample solution using an inert gas. The gas stream is then dried, and the species are cryotrapped in a fused silica capillary column. Finally, sample introduction into the chromatograph or the detector is accomplished by slowly heating the trap, and the species are released depending on their volatilisation points [63]. A variation of this technique is the capillary cold trap (CCT), in which the trap is reduced to a capillary column. This new design allows the preconcentration and separation of the analytes in the same step. This technique has been applied to Sn, Pb and Hg speciation [51, 64]. Based on this principle, an automatic speciation analyser has been designed and employed for mercury speciation in a fish reference material with excellent results [51, 65].

7.3 High temperature trap

This consists of an electrothermal vaporisation system to raise the temperature during the collection and the evaporation of the analytes. It is a very sensitive and precise method, and it has been used for arsenic determination in biological samples [66].

7.4 Active charcoal retention

This is a preconcentration technique mostly employed for trapping volatile chelates due to the nonpolar nature of such compounds [67].

8 SEPARATION AND IDENTIFICATION STEPS

The hyphenation between chromatography and spectrometric detectors is the best approach for biochemical speciation analysis. There are various factors to be overcome for a successful analysis [20]:

- The first requisite for a hyphenated technique is that the species injected on a chromatographic column leaves the column unchanged, and that no artefact species are generated on the column. The metal–ligand bound should be much stronger than the interaction of the metal or the ligand with the stationary phase.
- Secondly, the mobile phase must have a similar pH and composition to those of the sample matrix. Mobile or stationary phases competing with or displacing ligands from the analyte metal complex should be avoided.
- The third point is that the chromatographic technique should guarantee that the signal corresponds to one particular species. Since biological samples are very complex mixtures containing thousands of compounds, the use of successive separation techniques implying different mechanisms may be required. Generally, metal complexes in biological macromolecules are separated by ultrafiltration using filters (with molecular cut-offs of 500, 5000 and 30 000 Da). Sometimes, a further purification of the compounds is necessary, using different techniques, such as hydrophobicity or electrical charge, affinity chromatography, etc. [2].
- Finally, the simplest method to identify the species of interest is by comparing their retention times with that of a known standard. This approach is fairly successful for organometallic species, but has not the same applicability for biochemical compounds: standards are not available for the majority of the species present. Also, there is a probability of several species having the same retention time. Therefore, analytical techniques offering a characterisation of the metal complex or at least of the ligand are necessary, such as nuclear magnetic resonance (NMR), electrospray-mass spectrometry

(ESI-MS), and liquid liquid mass spectrometry (LC-LC-MS) [20].

The possibilities for an on-line separation technique include gas chromatography, different types of HPLC and electrophoresis. All separation techniques can be coupled to specific detectors (atomic or molecular mass spectrometry) [3]. The reader should refer to Chapter 4 of this book for more information about separation techniques and hyphenated detectors.

9 ACCURACY OF THE DIFFERENT PREPARATION STEPS: NEED FOR ADEQUATE CRMs

Metal speciation in biological samples is much more complex that in any other common sample and so the risk of error is significantly higher. As it is shown in Figure 3.1.1 the specific problems related to speciation in biological samples arise from the low concentration of the different species in a sample, their stability, the inherent difficulties of sample treatment due to the matrix complexity, separation and quantification, among others.

9.1 Sources of error

Sources of error that are likely to occur in speciation analysis have been extensively reported over the last few years [68]. The main errors occurring during sample preparation can be summarised in the following points:

- *Sample extraction:* extraction of chemical species from biological samples is a specially difficult task and the way to perform it depends on the kind of information needed. For instance, to find out whether a metal is bound to a protein of certain molecular mass or to determine the oxidation state of an analyte requires a completely different sample treatment.

  A good assessment of quality assurance is the verification of the recovery, which can be applied by different ways. One of the most widely used is the mass balance in respect of the total atom (amount of analyte found after complete digestion) with respect to the sum of the species found, expressed as the amount of analyte as shown in the example of Figure 3.1.2. However, this is not always feasible because frequently not all the existing species in a sample are correctly identified and/or quantified in the chromatograms (unknown peaks).

  Spiking a sample with known amounts of the species of interest, leaving them to equilibrate and determining the species after extraction constitutes a second way of testing recovery. However, results on recovery based on spiking are not always conclusive, especially in the case of biological samples because it is very difficult to ensure that the spike is bound in the same way as the analyte in the sample. For instance, if an animal tissue is spiked with Se methionine, this aminoacid is part of a protein in the sample and it is obtained as Se-Met in solution after the sample has been hydrolysed.

  A quantitative recovery of the spike does not necessary reflect a quantitative recovery of the analyte from the sample. However, if a quantitative recovery of the spike is not obtained, it implies that the extraction procedure is not adequate since it will certainly not be applicable to naturally bound compounds.

  Recovery must be tested using the standard addition method. It is recommended to add the spike compounds to previously wetted material, allowing the mixture to stand for long enough (24 h) to reach a good equilibration. Freeze-dried materials have to be dried again before analysis.
- *Species stability or transformation:* is an important aspect to be considered during the extraction process. Some authors [68] have reported artificial formation of methyl mercury during treatment of samples containing high concentration of Hg(II). These findings are still the subject of controversy, but this effect seems to be insignificant for biological samples, in which a large proportion of the organic mercury species occurs in the form of methyl mercury. Anyhow, transformation of species must always be checked carefully. For instance, partial transformation of Sb(III) to Sb(V) takes place when

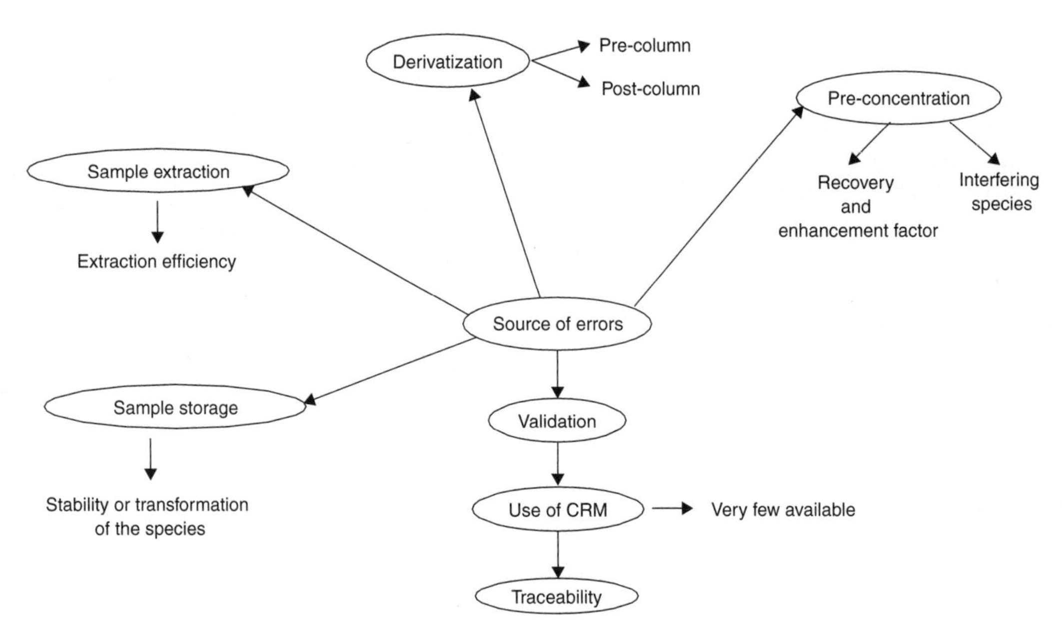

Figure 3.1.3. Main sources of error during sample pretreatment.

solutions are exposed to air. Therefore, each analysis step should be handled with special care depending on the lability of the target species. This risk of error can be evaluated by spiking the sample with those species that can suffer transformation during the sample treatment or the derivatisation process and then analysing the extracts. This problem has been extensively studied for mercury speciation [68]. Only in those cases in which it is established that species transformation does not occur during the extraction procedure can the recovery be calculated by spiking the extracts. However, the recovery assessment can be often overestimated and this risk should be considered [62]. Taking into account the main drawbacks of sample extraction: low efficiency and species transformation, it is important to find a good compromise situation between an acceptable recovery and preservation of species.

- *Derivatisation:* the errors involved are the derivatisation yield (often matrix dependent), which is difficult to determine due to a lack of appropriate high purity calibrates; the high number of steps before or after derivatisation (preconcentration, clean-up), which increases the uncertainty of the analysis. As some of the reaction mechanisms are still not well understood it is understandable that control of all these factors is not an easy task and that the risk of errors increases with increasing the number of steps involved in the analytical process.
- *Preconcentration:* the main problems encountered in this step are related to the retention of some interfering species together with the analytes of interest. A second factor to take into account is the recovery of the species preconcentrated. Ideally, it should be quantitative to assure good accuracy. Finally, the sensitivity enhancement factor is a third point of relevance during this step.

Figure 3.1.3 shows the potential errors that can be encountered during sample preparation and handling.

9.2 Relevance of CRMs

The performance of the sample preparation applied can be validated in respect of accurate and precise species analysis by using extraction solutions

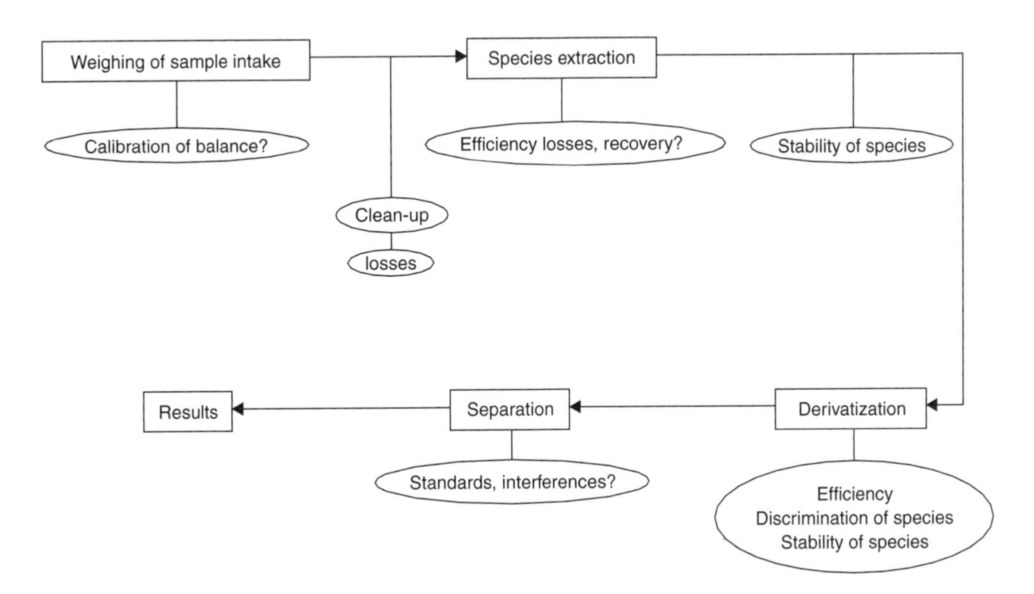

Figure 3.1.4. Different steps involved in a method of validation for speciation purposes.

of matrix CRMs. A key issue in comparing worldwide analytical results is their traceability, in other words when the results are achieved by an unbroken chain of calibrations connecting the measurements process to the fundamental units. As has been previously commented, the chain for speciation analysis in biological samples is much more complex than for total determination since the number of analytical steps involved is much higher.

The use of suitable CRM is very convenient for validation of the different sample preparation steps. When new procedures are applied to a CRM and the results are in good agreement with the certified values (within their uncertainty) it can be assumed that the analytical results obtained are accurate: the method is then validated. However, it is well established that the use of CRMs for method validation requires not only using a CRM of similar sample matrix but also of similar concentrations to those of species of interest to be determined. These facts are extremely important for biological materials in which the extraction efficiency of the species is very much dependent on the matrix complexity or on the type of compounds in which the species are bound in the sample.

The main problem in validating an analytical method by using CRMs is the actual lack of biological CRMs available in the market for speciation purposes, even though an important effort to improve the state of the art in this field is being carried out. This subject is discussed in Chapter 7.2 of this book.

It is important to consider that CRMs allow the user to link results with those of internationally recognised standards, enabling the accuracy of the results to be verified at any desired time. There is an increasing demand for certified materials for chemical species. Nowadays the EU is making a great effort to offer new biological reference materials for speciation purposes and it is soon expected to provide certified oyster tissue for Sn, As and Hg species, rice for As and Se and yeast for Se-Met among others [69].

Figure 3.1.4 summarises the steps involved during an analytical method validation.

10 TRENDS AND PERSPECTIVES

- Improvement in the state of the art for sample treatment, which still is the most limiting step of speciation in biological samples.

- Synthesis of pure standards for the different species to identify unknown species. Pure compounds such as arsenosugars or organic antimony species (except the trimethyl compound) are not available to the scientific community.
- Improvement of knowledge about species stability in original samples and in the extracts will help in maintaining species integrity during the different steps involved in speciation methodology.
- Concerning extraction procedures, the perspectives are focused on increasing the extraction efficiency of the species while the time of the procedure is reduced. The use of microwave radiation as an extraction technique is one of the directions that analytical chemists will have to address in the future. Speciation methods have to be simplified, and the different steps involved (matrix disintegration, analyte extraction, derivatisation and sometimes preconcentration) can in general be better controlled and enhanced under a microwave field, which will be a key future issue in speciation analysis [6]. The latest trend in microwave design is towards a miniaturisation of magnetrons. Research is also focused on reducing solvent consumption. Microwave-assisted procedures will allow one to perform simultaneous extraction into a solvent. Its use will also improve the efficiency of the SPME process. This new approach has been extensively developed and applied to determine organic compounds such as pesticides and up to now very few methods for speciation purposes have been proposed.
- Development of new simple and compact instrumentation to be implemented in laboratories to be used for routine speciation purposes. Instrumentation to be used for derivatisation, preconcentration and species separation having an easy coupling to different detectors. Some instruments have been developed based on the use of capillaries and multicapillaries for cold trapping in the preconcentration and speciation of mercury.
- Need of appropriate certified biological reference materials, which are less manipulated than those currently available and which are certified for species patterns must be fulfilled in the future to enable speciation analysis as a scientifically sound analytical activity. Unchanged biological material as CRMs and appropriate pure standard compounds are indispensable for the further development of an accurate sample treatment with speciation purposes [69].

11 REFERENCES

1. Moreno, P., Quijano, M. A., Gutiérrez, A. M., Pérez-Conde, M. C. and Cámara, C., *J. Anal. At. Spectrom.*, **16**, 1044 (2001).
2. Cornelis, R., De Kimpe, J. and Zhang, X., *Spectrochim. Acta B*, **53**, 187 (1998).
3. Szpunar, J., *Analyst*, **125**, 963–988 (2000).
4. Muñoz-Olivas, R. and Cámara, C., In *Trace Element Speciation for the Environment, Food and Health*, Ebdon, L., Cornelis, R., Crews, H., Donard, O., Quevauviller, P. and Pitts, L. (Eds) Royal Society of Chemistry, Cambridge, UK, 2002, p. 331.
5. Apostoli, P., *Fresenius' J. Anal. Chem.*, **363**, 499 (1999).
6. Sutton, K. L. and Heitkemper, D. T., In *Comprehensive Analytical Chemistry*, Vol. XXXIII, *Elemental Speciation. New Approaches for Trace Element Analysis.* Caruso, J. A., Sutton, K. L., Ackley, K. L. (Eds) Elsevier, Amsterdam, 2000, Chapter 14, pp. 501–530.
7. Leopold, I. and Günther, D., *Fresenius' J. Anal. Chem.*, **359**, 364 (1997).
8. Robb, P., Crews, H. M. and Baxter, M. J., In *Inductively Coupled Plasma Spectrometry and its Applications*, Hill, S. J. (Ed.) Sheffield Academic Press, Sheffield, UK, 1999.
9. Tucher, A., In *Mad Cats and Dead Men at Minamata*, Earth Island Publishing, 1972, Chapter 1.
10. Fairweather-Tait, S. J., *Fresenius' J. Anal. Chem.*, **363**, 536 (1999).
11. Cornelis, R., Cámara, C., Ebdon, L., Pitts, L., Welz, B., Morabito, R., Donard, O., Crews, H., Larsen, E. H., Neidhart, B., Ariese, F., Rosenberg, E., Mathé, D., Morrison, G. M., Cordier, G., Adams, F., Van Doren, P., Marshall, J., Stojanik, B., Ekvall, A. and Quevauviller, P., *Fresenius' J. Anal. Chem.*, **363**, 435 (1999).
12. Massey, R. C. and Taylor, D., In *Aluminium in Food and the Environment*, Royal Society of Chemistry, London, 1988.
13. Viola, R. E., Morrison, J. F. and Cleland, W. W., *Biochem.*, **19**, 313 (1980).
14. Cornelis, R., Crews, H., Donard, O. F. X., Ebdon, L. and Quevauviller, Ph., *Fresenius' J. Anal. Chem.*, **370**, 120 (2001).
15. Gómez-Ariza, Morales, J. L., Sánchez-Rodas, E. and Giráldez, I., *Trends Anal. Chem.*, **19**, 200 (2000).

16. Stephen Reid, R. and Attaelmannan, M. A., *J. Inorg. Biochem.*, **69**, 59 (1998).
17. Marsh, D. O., Myers, G. J. and Clarkson, T. W., *Clin. Toxicol.*, **10**, 1311 (1981).
18. World Health Organisation. In *Elements in Human Nutrition and Health*, WHO, Geneva, 1996.
19. Szpunar-Lobinska, J., Witte, C., Lobinski, R. and Adams, F. C., *Fresenius' J. Anal. Chem.*, **351**, 351 (1995).
20. Lobinski, R. and Szpunar, J., *Anal. Chim. Acta*, **400**, 321 (1999).
21. Simonoff, M. and Simonoff, G., In *Le Sélénium et la Vie*, Masson, Paris, 1991.
22. Gu, B. Q., *Chin. Chim. Med.*, **96**, 251 (1983).
23. Belfroid, A. C., Puperhart, M. and Ariese, F., *Mar. Pollut. Bull.*, **40**, 226 (2000).
24. Larsen, E. H., Pritzl, G. and Hansen, S. H., *J. Anal. At. Spectrom.*, **8**, 557 (1993).
25. Rehberg, P. B., *Acta Physio. Scand.*, **5**, 305 (1943).
26. Bligh, E. G. and Dyer, W. J., *Can. J. Biochem. Physiol.*, **37**, 911 (1959).
27. Rodriguez-Pereiro, I., Schmitt, V. O. and Lobinski, R., *Anal. Chem.*, **69**, 4799 (1997).
28. Quijano, M. A., Gutiérrez, A. M., Pérez-Conde, M. C. and Cámara, C., *Talanta*, **50** (1), 165 (1999).
29. Potin-Gautier, M., Gilon, N., Astruc, M., De Gregori, I. and Pinochet, H., *INT. j. Environ. Anal. Chem.*, **67**, 15 (1997).
30. Tyson, J. F., *J. Anal. At. Spectrom.*, **14**, 169 (1999).
31. Ebdon, L., Foulkes, M. E., Le Roux, S. and Muñoz-Olivas, R., *The Analyst*, **127**, 1108 (2002).
32. Liang, L., Horvat, M., Cernichiari, E., Gelein, B. and Balogh, S., *Talanta*, **43**, 1883 (1996).
33. Hintelmann, H. and Wilken, R. D., *Appl. Organomet. Chem.*, **8**, 533 (1994).
34. Abou-Shakra, F. R., Rayman, M. P., Ward, N. I., Hotton, V. and Bastian, G., *J. Anal. At. Spectrom.*, **12**, 429 (1997).
35. Pannier, F., Astruc, A. and Astruc, M., *Anal. Chim. Acta*, **327**, 287 (1996).
36. Ceulemans, M., Witte, C., Lobinski, R. and Adams, F. C., *Appl. Organomet. Chem.*, **8**, 451 (1994).
37. Gilon, N., Astruc, A., Astruc, M. and Potin-Gautier, M., *Appl. Organomet. Chem.*, **9**, 623 (1995).
38. Zhang, Z. and Pawliszyn, J., *Anal. Chem.*, **65**, 1843 (1993).
39. Zhang, Z., Yang, M. J. and Pawliszyn, J., *Anal. Chem.*, **66**, 844A (1994).
40. Moens, L., De Smaele, T., Dams, R., Van Den Broeck, P. and Sandra, P., *Anal. Chem.*, **69**, 1604 (1997).
41. De Smaele, T., Moens, L., Sandra, P. and Dams, R., *Mikrochim. Acta*, **130**, 241 (1999).
42. Kradtap, S., MS Thesis, University of Massachussets, Amherst, MA, 1996.
43. Lee, M. L. and Markides, K. E., *Chromatography Conferences*, Provo, Ut, 1990.
44. Zoorob, G. K., McKiernan, J. W. and Caruso, J. A., *Mikrochim. Acta*, **128**, 145 (1998).
45. Carey, J. M., Vela, N. P. and Caruso, J. A., *J. Anal. At. Spectrom.*, **7**, 1173 (1992).
46. Richter, B. E., Jones, B. A., Ezzell, J. L. and Porter, N. L., *Anal. Chem.*, **68**, 1033 (1996).
47. Gallagher, P. A., Wei, X., Shoemaker, J. A., Brocknoff, C. A. and Creed, J. T., *J. Anal. At. Spectrom.*, **14**, 1829 (1999).
48. Gallagher, P. A., Shoemaker, J. A., Wei, X., Brocknoff, C. A. and Creed, J. T., *Fresenius' J. Anal. Chem.*, **369**, 71 (2001).
49. A. Pedersen, G. and Larsen, E. H., *Fresenius' J. Anal. Chem.*, **358**, 591 (1997).
50. Emteborg, H., Bordin, G. and Rodriguez, A. R., *Analyst*, **123**, 245 (1998).
51. Dietz, C., Doctoral Thesis, 2001, University Complutense of Madrid, p. 72.
52. Lobinski, R., Rodriguez-Pereiro, I., Chassaigne, H., Wasik, A. and Szpunar, J., *J. Anal. At. Spectrom.*, **13**, 859 (1998).
53. Tesfalidet, S. and Irgum, K., *Anal. Chem.*, **61**, 2079 (1989).
54. Tesfalidet, S. and Irgum, K., *Fresenius' J. Anal. Chem.*, **341**, 532 (1991).
55. Puk, R. and Weber, J. H., *Anal. Chim. Acta*, **292**, 175 (1994).
56. Tseng, C. M., de Diego, A., Martin, F. M., Amouroux, D. and Donard, O. F. X., *J. Anal. At. Spectrom.*, **12**, 743 (1997).
57. Rapsomanikis, S., Donard, O. F. X. and Weber, J. H., *Anal. Chem.*, **58**, 35 (1986).
58. Ashby, J., Clark, S. and Craig, P. J., *J. Anal. At. Spectrom.*, **3**, 735 (1988).
59. Lobinski, R., *Appl. Spectrosc.*, **51**, 260A (1997).
60. Vázquez-Peláez, M., Montes-Bayón, M., García-Alonso, J. I. and Sanz-Medel, A., *J. Anal. At. Spectrom.*, **15**, 1217 (2000).
61. Rigin, V. I., *Fresenius' Z. Anal. Chem.*, **335**, 15 (1989).
62. Quevauviller, Ph., In *Method Performance Studies for Speciation Analysis* 1988, The Royal Society of Chemistry, Cambridge, UK.
63. Skogerboe, R. K., Dick, D. L., Pavlica, D. A. and Lichte, F. E., *Anal. Chem.*, **47**, 568 (1975).
64. Ceulemans, M. and Adams, F. C., *J. Anal. At. Spectrom.*, **11**, 201 (1996).
65. Dietz, C., Madrid, Y., Cámara, C. and Quevauviller, Ph., *Anal. Chem.*, **72**, 4178 (2000).
66. Shickling, C., Yang, J. and Broekaert, J. A., *J. Anal. At. Spectrom.*, **11**, 739 (1996).
67. de Peña, Y. P., Gallego, M. and Valcárcel, A., *J. Anal. At. Spectrom.*, **9**, 691 (1994).
68. Falter, R., *Chemosphere*, **39**, 1075 (1999).
69. Emons, H., *Fresenius' J. Anal. Chem*, **370**, 115 (2001).
70. Blaszkewicz, M., Baumhoer, G. and Neidhart, B., *Fresenius' J. Anal. Chem.*, **325**, 129 (1986).
71. Gammelgaard, B. and Jons, O., *J. Anal. At. Spectrom.*, **15**, 945 (2000).

72. Godlevska-Zylkiewicz, B., Lesniewska, B. and Hulanicki, A., *Anal. Chim. Acta*, **358**, 185 (1998).
73. Suzuki, K. T., Itoh, M. and Ohmichi, M., *J. Chromatogr. B*, **666**, 13 (1995).
74. Nygren, O. and Nilson, C. A., *J. Anal. At. Spectrom.*, **2**, 805 (1987).
75. Grill, E., Winnacker, E. L. and Zeuk, M. H., *Methods Enzymol.*, **205**, 333 (1991).
76. Kotrebai, M., Birringuer, M., Tyson, J. F., Block, E. and Uden, P. C., *Analyst*, **125**, 71 (2000).
77. Casiot, C., Szpunar, J., Lobinski, R. and Potin-Gautier, M., *J. Anal. At. Spectrom.*, **14**, 645 (1999).
78. Kumar, U. T., Vela, N. P., Dorsey, J. G. and Caruso, J. A., *J. Chromatogr. A*, **655**, 340 (1993).
79. Muñoz, O., Vélez, D., Cervera, M. L. and Montoro, R., *J. Anal. At. Spectrom.*, **14**, 1607 (1999).
80. Heumann, K. G., Rottmann, L. and Vogl, J., *J. Anal. At. Spectrom.*, **9**, 1351 (1994).
81. Alberti, J., Rubio, R. and Rauret, G., *Fresenius' J. Anal. Chem.*, **351**, 420 (1995).
82. Dagnac, T., Padro, A., Rubio, R. and Rauret, G., *Talanta*, **48**, 763 (1999).
83. Kobayashi, M., Ogua, Y. and Suzuki, K. T., *J. Chromatogr. B*, **760**, 73 (2001).
84. Michalke, B., *Fresenius' J. Anal. Chem.*, **351**, 670 (1995).
85. B'Hymer, C. and Caruso, J. A., *J. Anal. At. Spectrom.*, **15**, 1531 (2000).
86. Dundar, M. S. and Haswell, S. J., *Analyst*, **120**, 2085 (1995).

3.2 Sample Preparation Techniques for Elemental Speciation Studies

Brice Bouyssiere, Joanna Szpunar, Martine Potin-Gautier and Ryszard Lobinski

Université de Pau et des Pays de l'Adour, Pau, France

1 INTRODUCTION

Hyphenated (coupled, hybrid) techniques based on the combination of a separation technique with atomic spectrometry have become standard in speciation analysis because of their ability to discriminate between the different forms of an element [1–3]. These techniques usually show an outstanding performance for standard solutions but often fail when applied to a real-world sample. The successful speciation analysis of environmental and biological samples critically depends on the sample preparation, whose objective is to isolate analyte species from the matrix,

Handbook of Elemental Speciation: Techniques and Methodology R. Cornelis, H. Crews, J. Caruso and K. Heumann
 ISBN: 0-471-49214-0

convert them into chromatographable species and to preconcentrate them to match the detection limits of the chromatographic detection technique. Wet-chemical sample preparation methods, almost forgotten in total trace element analysis with the advent of ICP AES and ICP MS [4], have a unique role to play in speciation analysis. Indeed, the low analyte concentrations involved (often below $0.1\ \mu g\ kg^{-1}$ or $0.1\ \mu g\ L^{-1}$), the fragility of the analyte compounds and the need for preservation of the organometallic moiety throughout the procedure, and the strict requirements posed by the hyphenated system in terms of the sample volume, polarity and matrix acceptable by the instrumental setup make sample preparation in speciation analysis essential. Another distinct trend is automation which implies the development of faster, possibly single-step, and efficient sample preparation procedures.

The procedure for sample preparation depends on the analytical technique to be used and on the sample type to be analyzed. The polarity (volatility) of the analyte determines the chromatographic technique to be chosen for the separation of the species prior to detection. The separation technique, in turn, sets the requirements for the analyte solution resulting from the sample preparation procedure. Gas chromatographic techniques would require a microliter volume of a nonpolar solvent free of suspended particles containing volatile thermally stable species. An aqueous solution of thermodynamically stable and kinetically inert species (organometallic or metal complexes) free of suspended particles will be required for liquid chromatography-based techniques. In terms of analytes, three major areas can be distinguished: speciation of redox forms, speciation of organometallics (containing a carbon–metal or carbon–metalloid bond), and speciation of metal complexes. Figure 3.2.1 summarizes the principal analytes that are subject of speciation analysis. Gas chromatography is the preferred separation

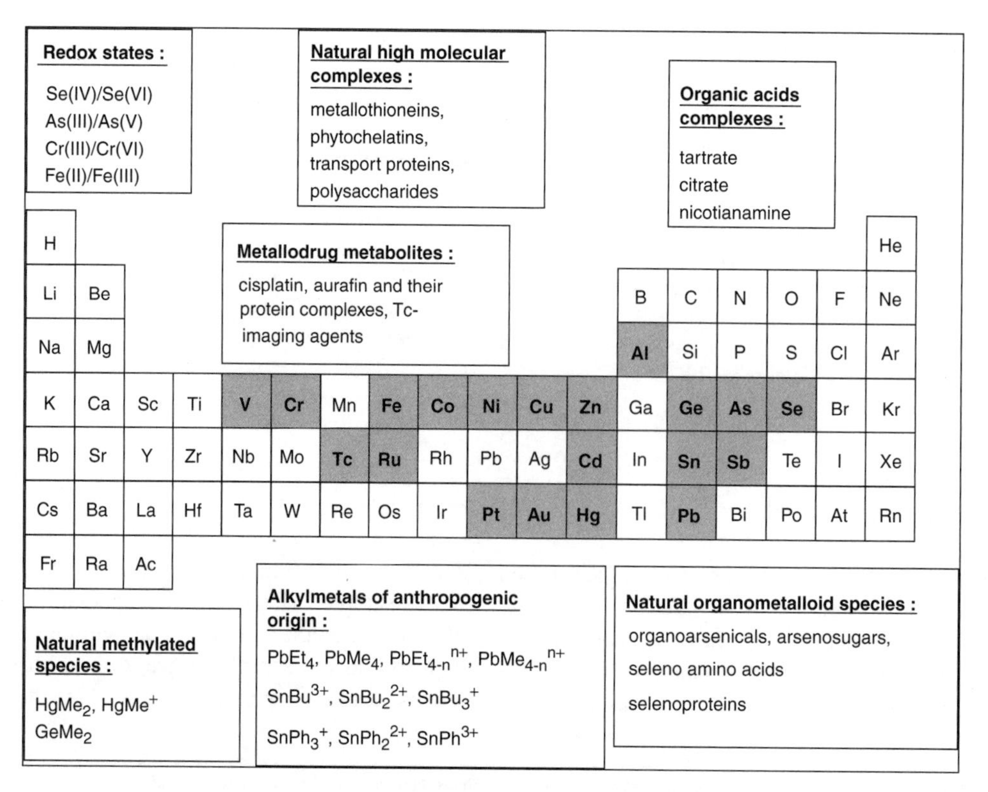

Figure 3.2.1. Species and fields of interest in speciation analysis.

Table 3.2.1. Glossary of the most common steps in sample preparation for speciation analyses.

Filtration	is used to separate particles in gases and aqueous samples. Usually a 0.45 μm filter is used but a 0.2 μm filter is necessary prior to HPLC.
Solubilization	is applied to bring biological materials into solution. It can be achieved by alkaline, acid or enzymatic hydrolysis. Following the solubilization, an aqueous solution of species is obtained but the matrix is not eliminated.
Leaching	or solid–liquid extraction is applied to extract analyte species from solid samples (soil, sediment or biological tissues). Leaching is the most popular method for speciation of metal complexes since leaching agents are pH neutral. Leaching is also the preferred methods for the analysis of soil and sediment samples that cannot be solubilized without the destruction of analytes.
Preconcentration	is necessary to increase the concentration of the analyzed species in the solution introduced on a chromatographic column in comparison with that present in an analyzed sample. The techniques used include cryofocusing, gas solid extraction (trapping) (for analytes already in the gas–phase), solid-phase microextraction (SPME) for analytes in the gas or liquid phase, and liquid–solid extraction (sorption) for analytes in water. Preconcentration can be achieved by solvent extraction and evaporation of the leachate or extract.
Cleanup	is the removal of the matrix components (fats, proteins, high boiling point hydrocarbons) that, if co-introduced on a chromatographic column, would lead to destruction of the column or degradation of its separation properties. Cleanup is usually realized by low resolution chromatographic separation with a mechanism different from that employed for the analytical separation (e.g. passing through a C_{18} column of polar species to be separated by anion-exchange)
Derivatization	is the process of the controlled conversion of species originally present in a sample into forms with improved chromatographic yield or separation coefficient. The most popular is derivatization of ionic or highly polar species into nonpolar species that can be readily separated by GC (e.g. Grignard derivatization).

technique for alkylelement species whereas liquid chromatography is predominantly used in all the other cases.

The choice of sample preparation procedure is also determined by the matrix and the preconcentration factor that needs to be achieved to eliminate the discrepancy between the concentration of the analyte in the sample and the detection limits of the analytical setup. A sample preparation procedure for speciation analysis usually requires a number of steps (Table 3.2.1). They include filtration, preconcentration of analytes, solubilization or leaching in the case of solid samples, cleanup prior to chromatography, and sometimes derivatization of analyte species in order to improve their chromatographic behavior. It is essential that all these steps are carried out with a maximum (quantitative) efficiency and that the original species are not degraded. The number of steps necessary and their duration affects the duration and tediousness of an analytical procedure and should be kept minimal. Only some energy-related samples (shale oil, petroleum, gasoline) containing low polar analytes can sometimes be analyzed directly, usually after dilution.

This chapter discusses sample preparation methods including the recovery of analytes from different matrices and the preparation of their solution ready to be injected onto a gas or liquid chromatographic column. Emphasis is put on procedures that require a minimum of time and a small number of operations having in view the automation of the sample preparation process. Note that before carrying out any sample preparation procedure the issue of stability of chemical species in the sample to be analyzed needs to be critically addressed [5].

2 SAMPLE PREPARATION FOR ORGANOMETALLICS

Because of their easy conversion to volatile and thermostable species, organometallic species are usually determined by hyphenated techniques based on gas chromatography (GC). GC requires an analyte species be presented as a nonpolar thermally stable compound in a nonpolar solvent or in a narrow (cryofocused) band. The sample preparation procedure should therefore include either a step of transfer of analytes from the aqueous solution (water sample, leachate or solubilizate) into an organic solvent (solvent extraction), or a step of formation of a narrow analyte band on a chromatographic sorbent by purge and trap or solid-phase microextraction. The derivatization step to yield a thermally stable species may either follow the solvent extraction step (typically the case of Grignard derivatization) or precede it (usually derivatization with $NaBH_4$ or $NaBEt_4$). An additional preconcentration step, by sorption or evaporation, is necessary where ultra-trace amounts of analytes are determined and/or poorly sensitive detection techniques are used. Flow chart of sample preparation for speciation of organometallic species is shown in Figure 3.2.2.

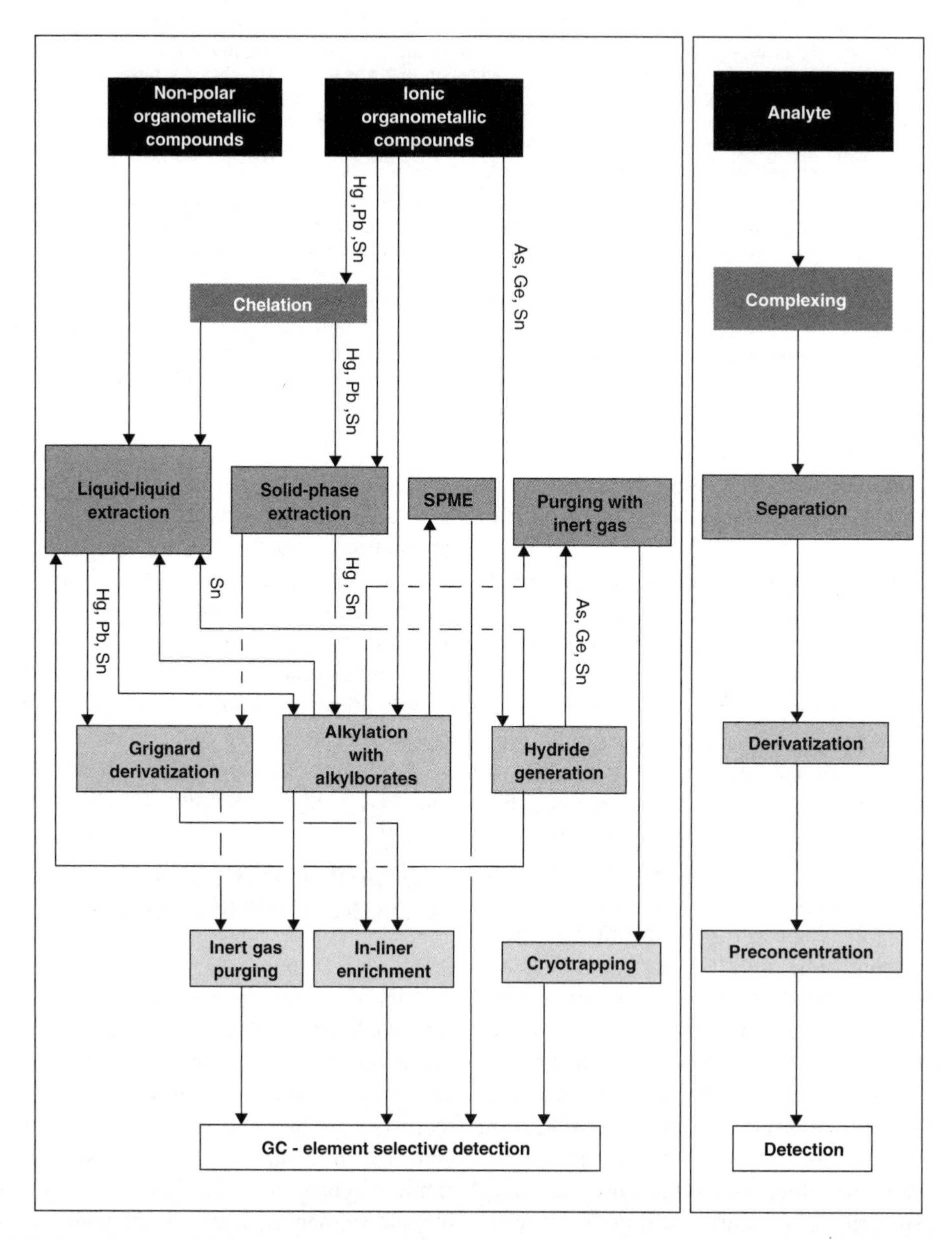

Figure 3.2.2. Flow chart of the sample preparation procedure for speciation of organometallic species.

2.1 Separation of analytes from aqueous matrices

The choice of procedure depends on the matrix. A number of possibilities applicable to aqueous samples include solid-phase extraction (sorption), solid-phase microextraction (SPME), liquid–gas extraction (purge and trap) and classical solvent (liquid–liquid) extraction. The separation of analytes from water is usually combined with their preconcentration, and often with their derivatization. The procedures are classified according to the order of extraction and derivatization steps. When analytes are present at the ng L^{-1} level in relatively clean water samples, *in situ* derivatization followed by an extraction technique with a high preconcentration factor (purge and trap, SPE or SPME) are preferred. In the case of dirty matrices with higher concentrations of analytes, the latter are usually extracted as nonpolar complexes (with DDTC, dithizone or tropolone) which are later derivatized, usually by means of a Grignard reaction.

2.1.1 Sorption (solid-phase extraction)

Solid-phase extraction (SPE), which is becoming increasingly popular for sample preparation in organic analysis, has found application in speciation analysis for organometallics [6–11]. The analytes are extracted by sorption, eluted with a small amount of an organic solvent and derivatized. Its advantages over liquid–liquid extraction include a higher enrichment factor, lower solvent consumption and risk of contamination, and the ease of application to field sampling and automation. Solid-phase extraction is particularly suitable for the analyte preconcentration prior to HPLC; by using an appropriate eluent the analytes can be separated on the SPE cartridge itself or on a connected HPLC column [12, 13].

Only filtered samples can be analyzed, which may constitute a considerable drawback. SPE cartridges in a variety of configurations and sizes have been used but microcolumns proved to be the best choice with respect to low dead volumes and low consumption of the solvent [6]. C_{18} extraction disks, reported to be well suited for the trace enrichment of organics in environmental waters, did not show satisfactory retention of native di- and monobutyl(-phenyl) tin species [6].

Sorbents with immobilized chelating reagents, such as dithiocarbamate [7, 8], dithizone [12] or diphenylthiocarbazone [14], have widely been used. A sampling and storage technique for speciation analysis of lead and mercury in seawater based on the sorption on a column packed with dithiocarbamate resin has been proposed [11]. C_{18} minicolumns modified with sodium diethyldithiocarbamate have been used for the field sampling and preconcentration of mercury species in river water [13].

An interesting curiosity is the use of biomaterials, e.g. dried yeast (*Saccharomyces cerevisiae*) cells, for enrichment of methylmercury from river water [9]. Another is the analytical potential of fullerene, which was investigated as adsorbent for organic and organometallic compounds from aqueous solutions [15].

2.1.2 Solid-phase microextraction

Solid-phase microextraction (SPME) is a preconcentration technique based on the sorption of analytes present in a liquid phase or, more often, in a headspace gaseous phase, on a microfibre coated with a chromatographic sorbent [16, 17]. The fiber is incorporated in a microsyringe. The coating is exposed to the sample in order to extract the analytes onto the coating. The fiber may be immersed in the solution or be placed in the headspace above the sample (Figure 3.2.3(a)). Once equilibrium is reached, the fiber is withdrawn into the needle for protection against contamination (Figure 3.2.3(b)). It is then transferred either to a GC injector where the analytes are thermally desorbed or to an HPLC interface where they are solubilized in the mobile phase used as solvent (Figure 3.2.3(c)).

The absorptive-type polydimethylsiloxane (PDMS) coatings are the most popular in speciation analysis. This nonpolar phase extracts organometallic species such as methyl- and ethylmercury compounds [18–20], alkylead species [21–23], butyl- and phenyltins [24, 25] and methylarsenic compounds [26, 27]. It is commercially available in

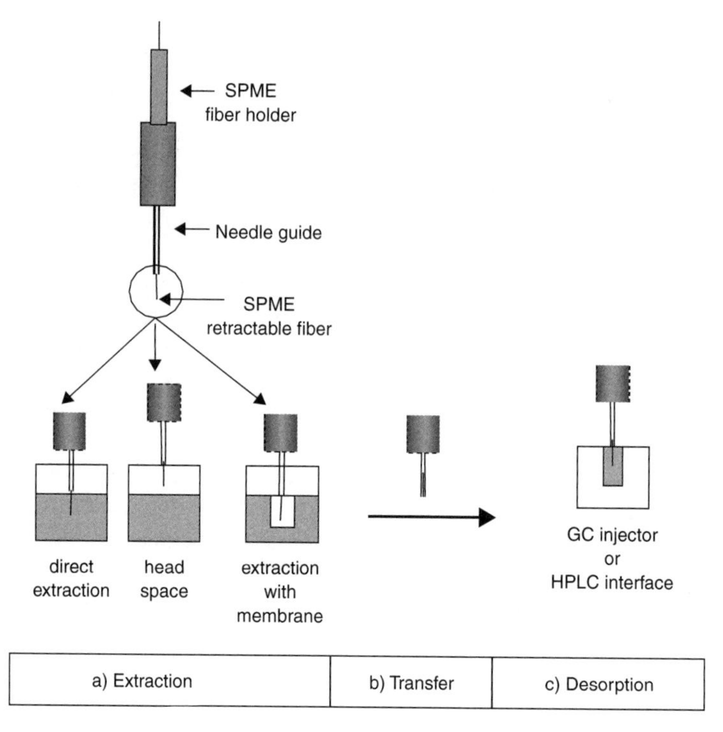

Figure 3.2.3. Principle of solid-phase microextraction.

different thickness and so it can be applied to compounds with different volatilities and distribution constants between the coated and liquid phases. PDMS-coated fibers are very rugged and can withstand a temperature of GC injector of up to 300 °C. An adsorptive carboxen coating seems to be more adapted to the analysis of volatile species such as hydrides of metals and metalloids [28].

SPME is a method that is emerging as an important analytical tool for elemental speciation in environmental samples. This solvent-free technique offers numerous advantages such as simplicity, the use of a small amount of liquid phase, low cost and the possibility of an on-line analytical procedure. Interference problems induced by the complexity of environmental matrices often noted in liquid–liquid extraction are limited by using SPME.

A method based on SPME on a fiber coated (100 μm) with poly(dimethylsiloxane) was developed for the determination of tetraethyllead in water [22]. Speciation of alkyllead and inorganic lead by derivatization with deuterium-labelled sodium tetraethylborate and SPME-GC/MS has been studied [23]. A water sample has also been subjected to headspace SPME of ethylated organometallic species onto PDMS-coated fused silica fibers for sensitive simultaneous determination of organomercury, -lead and -tin compounds [21]. Methylmercury in river water samples was derivatized *in situ* using $NaBEt_4$, and a silica fiber coated with poly(dimethyl siloxane) was placed into the solution. Once equilibrium was reached, the fiber was thermally desorbed into a GC injector [29]. Methyltin chlorides and tetramethyltin were adsorbed from acetone solution onto methylsilicone-coated fibres and transferred to a headspace vial for equilibration with $NaBH_4$ solution [30]. A number of applications concerning organoarsenic species have been published [26, 27].

The high preconcentration factors, often combined with a derivatization step offered by this new tool, facilitate the chromatographic analysis of species at trace and ultratrace levels as is often required in speciation analysis. Very low detection limits can be reached. For example, quantification of butyl- and phenyltins at concentrations lower than 0.1 ng Sn L^{-1} is now possible by combined SPME GC-pulsed FPD or SPME GC-ICP-MS [24, 25, 31].

Applications of SPME to studies of elemental speciation in different environmental and biological matrices have been comprehensively reviewed by Mester *et al.* [32]. Selected applications of SPME to analysis of organometallic species in environmental matrices are presented in Table 3.2.2.

2.1.3 *Solvent extraction*

Solvent (liquid–liquid) extraction can be directly applied to nonfiltered samples with complex matrices and allows the direct transfer of analytes into a nonpolar organic solvent that can be analyzed by GC. Nonpolar species, e.g. tetraalkyllead, are quantitatively extracted from water saturated with NaCl into a volume of hexane 20 times smaller [43]. The same applies to some 'ionic species' with the marked covalent character (e.g. triphenyltin, tributyltin, triethyllead, methylmercury) but the use of water-immiscible solvents with polar character (toluene, ethylacetate) was advised. The properties of organometallic species

Table 3.2.2. Applications of SPME to environmental organometallic compounds analyses.

Compound	Matrix	Mode[a]	Fiber type	Analytical technique[b]	Reference
Tin compounds					
Methyl-	Seawater	H	PDMS, 100 μm	Et-GC-FPD	[30]
Butyl-	Sediment	H	PDMS, 100 μm	Et-GC-ICP-MS	[21]
Butyl-	Urine	H	PDMS, 100 μm	Et-GC-MS-MS	[33]
Butyl-	Sediment, sludge	D	PDMS-7, 100 μm	Et-GC-AES	[34]
Butyl-	Fresh, sea and waste waters	H	Silica fiber Pretreated by HF	Hy-GC-QSIL-FPD	[35]
Butyl-	Sediment	H	PDMS, 100 μm	Et-GC-(HC)-GD-OES	[42]
Phenyl-, hexyl-	Mussel, potato	H	PDMS, 100 μm	Et-GC-ICP-MS	[31]
Lead compounds					
Methyl-	Sediment	H	PDMS, 100 μm	Et-GC-ICP-MS	[21]
Methyl-	Urine	H	PDMS, 100 μm	Et-GC-MS-MS	[33]
Methyl-, ethyl-	Sediment	H	PDMS, 100 μm	Et-GC-(HC)-AES	[42]
Ethyl-, Pb^{2+}	Blood, urine	H	PDMS, 100 μm PDMS/DVB	Et-GC-FID	[36]
Ethyl-	Waters, gasoline	H	PDMS, 100 μm	nd-TD-QF-AAS	[37]
Mercury compounds					
Methyl-	Sediment	H	PDMS, 100 μm	Et-GC-ICP-MS	[21]
Methyl-, Hg^{2+}	Urine	H	PDMS, 100 μm	Et-GC-MS-MS	[33]
Methyl-, ethyl-, phenyl-	Soil	D	PDMS, 100 μm	nd-GC-MIP-AES	[20]
Methyl-, Hg^{2+}	Waters, fish	H,D	PDMS, 100 μm	Et-GC-MS, GC-FAS	[19]
Methyl-	Gas condensate	H	PDMS, 100 μm	nd-GC-MIP-AES	[18]
Methyl-	Soil	H	PDMS, 100 μm	Et-GC-MS	[38]
Methyl-	Fish	H, D	PDMS/DVB, 65 μm	nd-TD/ICP/MS	[28]
Selenium compounds					
Se(IV)	Drinking water	H, D	PDMS, 100 μm	Et-GC-MS	[39]
Arsenic compounds					
Methyl-, Phenyl-	Water, soil	D	PA, 65 μm	d-GC-MS	[40]
Methyl-	Urine	D	PDMS, 100 μm	d-GC-MS	[26]
Methyl-, organo-	Drinking water		In-tube SPME	nd-HPLC-ESI-MS	[41]

[a]H, headspace; D, immersion.
[b]Et, ethylation; Hy, hydridation; nd, non derivatized; d, derivatization.

with a smaller number of organic substituents (e.g. monobutyltin) do not allow their quantitative extraction by any organic solvent. In order to transfer such compounds into an organic phase, the formation of extractable chelate complexes or nonpolar covalent compounds in the aqueous solution prior to extraction is necessary.

Diethyldithiocarbamate (DDTC) is the most often used reagent to form the chelate complexes with organometallic species. The extraction is followed by a derivatization step, usually by a Grignard reaction. Extraction of the complexes of ionic organolead species with DDTC from pH 6–9 into hexane [44–49], organotin [50–52] and organomercury [53, 54] was found to give quantitative recovery. Dithiocarbamates are not as light sensitive as dithizonates which makes the handling easier and the procedure more reliable. The high selectivity of the hexane–tetramethyldithiocarbamate extraction system for ionic alkylleads over Pb^{2+} facilitates greatly the determination of these analytes in matrices containing high levels of Pb^{2+} [55]. Inorganic interferents can be efficiently masked with EDTA [45–47]. For organotin, tropolone has extensively been used [52, 56–60] in addition to DDTC.

An alternative to chelate extraction is the formation of extractable nonpolar species in the aqueous phase by hydride generation [61, 62], ethylation with $NaBEt_4$ [6, 63–66], propylation with $NaBPr_4$ [67], butylation with NH_4BBu_4 [48, 49], or arylation with $NaBPh_4$ [65, 68, 69]. The resultant species should have a boiling point exceeding by 20 °C that of the solvent to be used for their extraction in order to enable the subsequent gas chromatographic separation. Also, it is important that the nucleophilic substituent be different from organic substituents of the species to be analyzed. For example, ethylation will fail for Et_2Pb^{2+} and Et_3Pb^+ since it would lead to the identical species (Et_4Pb) that is also the product of ethylation of Pb^{2+}.

Preconcentration factors in solvent extraction are generally low (typically 1:50 up to 1:250). The methods are rapid, work well for less volatile analytes and are relatively robust in terms of coping with interferences.

2.1.4 *Steam distillation*

Steam distillation has been evaluated as a technique for the separation of methylmercury from natural water samples [70–73]. It was found to give higher and more reproducible recoveries than other extraction techniques but under some conditions it could be responsible for artifactual methylation of Hg^{2+} [71]. The technique is rather slow; the addition of ammonium pyrrolidine dithiocarbamate (APDC) was found to improve recovery (up to 85 % for seawater) and to eliminate the codistillation of inorganic mercury. The method was found to be free of artifacts and to be comparable to nitrogen-assisted distillation with the added advantage of increased samples throughput [72].

2.1.5 *Liquid–gas extraction (purge and trap)*

This technique consists of bubbling an aqueous solution with an inert gas (nitrogen or helium) to extract the nonpolar volatile species into the gas phase. Some species can be extracted directly (e.g. tetraalkyllead, tetramethyltin, dimethylmercury) but others need to be converted into volatile hydrides or ethyl derivatives. Purge and trap methods usually follow hydride generation, which is the oldest but still the most popular method for volatilization of As, Ge, Sn, Sb and of their methyl species [74, 75]. Purging of less volatile tributyl- or triphenyltin hydrides is negatively affected by condensation problems which may lead to nonquantitative recoveries of the analytes. In general, hydride generation is prone to matrix interferences and hydrides are relatively reactive and tend to decompose when subject to harsh instrumental conditions. A rapidly developing alternative for the production of volatile derivatives is ethylation using $NaBEt_4$. It was developed for ionic mercury [76–78], lead [77, 79], tin [77] and selenium [80] species and enjoys a continuously increasing number of applications. Ethyl derivatives of butyl- and phenyltins are not sufficiently volatile to be efficiently purged. Furthermore, this technique is not selective for ethylmetal species.

The purged species are subject to the removal of moisture as in the case of analyses of gases, are cryotrapped and released onto a packed or capillary column. Capillary traps are becoming increasingly popular because of the narrow band of the analyte, compact size, and compatibility with capillary and multicapillary GC columns [77, 81–83]. This method offers high preconcentration factors and allows the introduction of fairly clean samples onto a GC column. Interferences, especially with hydride generation, and the risk of condensation with less volatile analytes are the major shortcomings.

2.2 Separation of alkylelement species from solid samples

Solid samples of interest for speciation analysis of organometallic species can be divided in two major categories: sediment and soils samples and biological materials. Organometallic compounds are apparently not involved in mineralogical processes in sediment and soil and bind onto the surface. Therefore, the complete dissolution of the sample prior to analysis is not considered necessary. In contrast, these compounds may be incorporated in tissues of a living organism. Hence, solubilization of a biological material prior to separation of the analytes is mandatory even if in particular cases some success can be obtained by leaching. Once the analyte species are brought in solution, the latter is treated similarly to water samples as described above.

2.2.1 Leaching methods

The basic approach to releasing organometallic compounds from the sediment involves acid leaching (HCl, HBr, acetic acid) into an aqueous or methanolic medium by sonication, stirring, shaking or Soxhlet extraction with an organic solvent [50, 84–91]. In order to increase the extraction yield the addition of a complexing agent (tropolone, DDTC) is mandatory. Extraction recoveries of organometallic compounds in environmental matrices have been critically discussed by Quevauviller and Morabito [92].

A modification of the leaching procedures is distillation of methylmercury. A sediment, soil or biological tissue sample is suspended in an acidic solution and the mixture is distilled at elevated temperature (ca. 180 °C) with nitrogen [93–97]. The distillate is collected in an ice-cold container. A complexation reagent, e.g. pyrrolidinedithiocarbamate, may be added to improve the extraction recovery. Artifact formation of methylmercury during procedures for its extraction from environmental samples has been studied [98].

A number of leaching methods have been developed for the recovery of butyl- and phenyltin compounds from biological tissues. The extraction efficiencies of 12 of these methods have been critically compared [99]. Acidic conditions, together with the use of tropolone and a polar organic solvent were found to enhance the extraction efficiency, especially for mono- and disubstituted compounds.

For soil and sediments, the need for a cumbersome sample preparation step is the basic weakness of the whole analytical procedure. Indeed, the majority of procedures reported have not only been extremely time-consuming, taking from 1 h to 2 days, but have also usually been inefficient in terms of analyte recovery and, in general, unreliable. As shown by Chau *et al.* [100], only three out of ten sample preparation methods described in the literature for the analysis of sediments were able to recover more than 90 % of Bu_3Sn^+, whereas none of them was able to recover monobutyltin ($BuSn^{3+}$) in a nonerratic and reproducible manner. In general, the more polar the species to be extracted the lower is its recovery, the longer the leaching procedure necessary and the higher the demand for accelerated (supercritical fluid extraction, accelerated solvent extraction or microwave-assisted leaching) are going to be. The classical, supercritical fluid and microwave-assisted techniques have been compared for the extraction of methylmercury from aquatic sediments [101].

2.2.2 Solubilization of biological samples prior to speciation analysis

A suitable digestion (hydrolysis) procedure should allow for the complete destruction of the matrix while the organometallic moiety remains unchanged. Three principal approaches have been developed: (1) acid hydrolysis with HCl [102], (2) alkaline hydrolysis with methanolic NaOH [103, 104] or with tetramethylammonium hydroxide (TMAH) [105, 106], and (3) enzymatic hydrolysis [107–109].

The advantage of the total solubilization method is the 100 % transfer of the species of interest into an aqueous phase that is later subject to extraction and derivatization procedures. The metal–carbon bond seems to resist the degradation even in relatively harsh conditions. However, the resulting solutions are rich in dissolved organic matter and in salts formed during the neutralization of the solution which may negatively affect extraction and derivatization efficiencies.

2.2.3 Supercritical fluid extraction

Substantial progress towards faster and potentially automated speciation analysis of sediments was offered by supercritical fluid extraction (SFE) [89, 110–115]. Equipment cost, however, is high, the extraction step still takes 10–50 min and the recoveries of many species are far from being quantitative.

Analytical strategies involving SFE from environmental matrices in tin, mercury, lead and arsenic speciation studies have been critically discussed [116]. The method was found to be successful for all the analytes if derivatization was performed on the aqueous phase before extraction. Carbon dioxide modified with acetic acid was found to be the most suitable for the extraction of organotin compounds from sediment and biota [116].

2.2.4 Microwave-assisted processes

Microwaves are high frequency (2.45 GHz) electromagnetic waves which are strongly absorbed by polar molecules (e.g. water or mineral acids) and which interact only weakly with nonpolar solvents. Absorption results in dielectric heating; the heat appears in the core of the target sample. The well-known advantages of microwave heating such as absence of inertia, rapidity of heating, efficiency and ease of automation have made it widely used for accelerated extraction of polar compounds into a nonpolar or weakly polar solvent. Many laboratory microwave ovens are commercially available.

The preservation of the organometallic moiety is the prerequisite of a successful leaching/digestion procedure prior to speciation analysis. It can be achieved by a careful optimization of the conditions of the microwave attack [117]. In contrast to common high temperature and pressure acid attack procedures, a focused low power microwave field is preferred for extraction of organometallic species from the matrix. The carbon–metal bonds remain intact.

Microwave-assisted leaching has been shown not only to reduce the time necessary for leaching of organometallic compounds from soil and sediment samples but also to increase the recovery of compounds that are the most difficult to extract. The time necessary for the quantitative extraction of organotin [64, 118–122], organolead [123] and methylmercury [78, 124] was reduced to 2.5–3 min. The values obtained for the extraction recoveries of monobutyltin from CRMs and candidate CRMs by these techniques are among the highest reported in the literature.

The microwave field can also accelerate leaching of analyte compounds from biomaterials [122, 125]. When a suitable reagent is used (e.g. tetramethylammonium hydroxide) the biological tissue can be solubilized within 2–3 min instead of 1–4 h [64, 120, 124]. An even more attractive alternative is the integration of the solubilization, derivatization and extraction steps into a one-step procedure. Hydrolysis with acetic acid carried out in a low-power focused microwave field in the presence of $NaBEt_4$ and nonane has been shown to shorten the sample preparation time for the CGC-MIP AED determination of organotin compounds in biological materials to 3 min [126]. The issue of one-step microwave-assisted extraction–derivatization procedures has been approached chemometrically for methylmercury in biological tissues [127].

2.3 Derivatization techniques for gas chromatography of organometallic species

A number of native organometallic compounds are volatile enough to be separated by GC. They include tetraalkyllead species ($Me_nEt_{4-n}Pb$), methylselenium compounds (e.g. Me_2Se, Me_2Se_2), some organomercury compounds ($MeHg^+$, Me_2Hg) as well as naturally occurring metalloporphyrins. As indicated above they can either be readily purged with an inert gas or extracted into a nonpolar solvent and subsequently chromatographed by thermal desorption, packed column or capillary GC.

The majority of organometallic species exist in quasi-ionic polar forms which have relatively high boiling points and often poor thermal stability. To be amenable to GC separation they must be converted to nonpolar, volatile and thermally stable species. The derivative chosen needs to retain the structure of the element–carbon bonds to ensure that the identity of the original moiety remains conserved. The most common derivatization methods include: (1) conversion of inorganic and small organometallic ions into volatile covalent compounds (hydrides, fully ethylated species) in aqueous media; (2) conversion of larger alkylmetal cations: e.g. $R_nPb^{(4-n)+}$ with Grignard reagents to saturated nonpolar species, and (3) conversion of ionic species to fairly volatile chelates (e.g. dithiocarbamate, trifluoroacetone) or other compounds. The three methods are fairly versatile in terms of organometallic species to be derivatized and the choice depends on the concentration of interest, the matrix and the sample throughput required. Frequently, the derivatives are concentrated by cryotrapping or extraction into an organic solvent prior to injection onto a GC column.

Derivatization (chemical modification) techniques for GC, HPLC and CZE in speciation analysis have been reviewed [128].

2.3.1 Derivatization by hydride generation

Several elements (Hg, Ge, Sn, Pb, Se, Te, Sb, As, Bi, and Cd) can be transformed into volatile hydrides, forming the basis of their determination [74, 75]. The usefulness of this procedure for speciation analysis, however, is severely restricted either by the thermodynamic inability of some species to form hydrides, or by considerable kinetic limitations to hydride formation. Nevertheless, the technique is still essential for some classes of compounds. The chemical reaction of hydride generation, presented on the example of $MeHg^+$ determination is the following:

$$MeHg^+ + NaBH_4 \longrightarrow MeHgH + NaBH_3{}^+$$

Selenium, As, and Sb readily form hydrides only in their lower oxidation states; the higher states need to be reduced beforehand. Thus, all inorganic species of these elements form eventually the same hydride (SeH_2, AsH_3, and SbH_3, respectively) precluding simultaneous chromatographic speciation. Methyl- and dimethylarsonic (or stibonic) acid can be discriminated in one GC run upon hydride generation producing volatile $MeAsH_2$ and Me_2AsH (or $MeSbH_2$ and Me_2SbH, respectively). Trialkyllead species form stable hydrides whereas dialkylleads are nonreactive. Mercury(II) and methylmercury, as well as germanium and methylgermanium species [74] can be transformed to gas chromatographable hydrides. Hydride generation has enjoyed the largest interest for organotin speciation analysis because of its capability for the simultaneous determination of ionic methyl and butyl species in one chromatographic run [84, 108]. Hydride generation with $NaBH_4$ is prone to interferences with transition metals which affect the reaction rate and analytical precision [75].

2.3.2 Derivatization with tetraalkyl(aryl)borates

The vulnerability of hydride generation to interferences in real samples and the restricted versatility can, to a certain degree, be overcome by replacing $NaBH_4$ by alkylborates. The most common derivatization procedures rely on ethylation with sodium tetraethylborate ($NaBEt_4$) which is water soluble and fairly stable in aqueous media. The use of $NaBEt_4$ in speciation analysis has been reviewed [63].

The reaction of derivatization with tetraethylborate, presented for the example of $MeHg^+$ is the following:

$$MeHg^+ + NaBEt_4 \longrightarrow MeHgEt + NaBEt_3{}^+$$

Methyl-, butyl- and phenyltin compounds react readily with $NaBEt_4$ to form thermally stable species that can be analyzed by GC. Whereas methyltins can be purged upon derivatization, other species need to be extracted because of their poorer volatility. All alkyllead species also readily react but only methyllead species can be unambiguously discriminated as the derivatization of ethyl- and inorganic lead will lead to the formation of the same product, namely $PbEt_4$. Methylmercury and inorganic mercury can be determined in one run upon purge-and-trap preconcentration. Derivatization of various selenium species has been demonstrated but showed poor potential for simultaneous analysis. Nevertheless, Se(IV) can be determined selectively and free of interferences by its reaction with $NaBEt_4$ [80].

In situ phenylation using sodium tetraphenyl borate has been studied with some applications [65]. Other reagents, such as sodium tetrapropylborate for organotin or tetramethylammonium tetrabutylborate for organolead [48, 49]. Some efforts concerning multielement and multispecies derivatization using $NaBEt_4$ have been reported [77].

2.3.3 *Derivatization with Grignard reagents*

As mentioned above, hydride generation or ethylation with alkylborates fail for some species or in the cases of very complex matrices. An alternative is derivatization with Grignard reagents, which is fairly versatile but requires an aqueous-free medium for the reaction to be carried out. In practice, it is applicable to extracts containing complexes of an organometallic compound with dithizone, dithiocarbamates or tropolone. Ionic organometallic compounds are transformed to nonpolar species according to the equation (presented on the example of $MeHg^+$).

Grignard derivation of $MeHg^+$:

$$MeHg^+ + RMgX \longrightarrow MeHgR + Mg^{2+} + X^-$$

R can be methyl, ethyl, propyl, buthyl or phenyl and X can be Cl, Br or F.

Whereas Grignard derivatization still remains the primary method for lead speciation analysis [44–46, 129], its position has been gradually eroded for organotin speciation in favor of the less cumbersome and time-consuming derivatization with $NaBEt_4$ [50]. Other applications of Grignard reagents, e.g. for the derivatization of organomercury, have been limited to one research group [53, 54]. An interesting curiosity is the direct Grignard pentylation, demonstrated recently for an organotin-contaminated lard sample [130].

Grignard reagents proposed for derivatization in speciation analysis by GC with plasma source spectrometric detection have included methyl-, ethyl-, propyl-, butyl- and pentylmagnesium chlorides or bromides. Lower-alkyl magnesium salts are generally preferred due to the smaller molecular mass and, hence, the higher volatility of the resulting species, which makes the GC separation faster with less column carryover problems associated with derivatized inorganic forms (which are often present in large excess). In addition, the baseline is more stable and less Grignard reagent-related artefacts occur. Conversely, the low volatility of species derivatized by pentylmagnesium chloride may facilitate concentration by evaporation and, hence, a more efficient enrichment is achieved.

The unreacted Grignard reagent needs to be destroyed prior to the injection of the derivatized extract onto a column, which is achieved by shaking the organic phase with dilute H_2SO_4. As a final step of the procedure, the organic phase is dried over, for example, anhydrous Na_2SO_4, and injected onto the GC column.

2.3.4 *Other derivatization techniques*

The formation of volatile acetonates, trifluoroacetonates and dithiocarbamates is a popular derivatization technique for inorganic GC [131]. Kinetic restrictions or the thermodynamic inability of many species to react, and the small differences in retention times for the derivatized species of the same element make chelating agents of limited importance as derivatization reagents for speciation analysis.

Selenoaminoacids have been derivatized with isopropylchloroformate and bis(*p*-methoxyphenyl) selenoxide [132], with pyridine and ethyl chloroformate [133] or silylated with bis(trimethylsilyl)-acetamide [134]. Selenomethionine forms volatile methyl-seleno-cyanide with CNBr [135, 136]. Selective determination of selenium(IV) and selenium(VI) has been achieved using GC with flame photometric detection (GC-FPD) and GC-MS after derivatization of selenium(IV) with 4,5-dichloro-1,2-phenylenediamine [137].

Methods for the conversion of arsenic compounds to volatile and stable derivatives are based on the reaction of monomethylarsonic acid and dimethylarsenic acid with thioglycolic acid methyl ester (TGM) [138, 139].

2.4 Preconcentration and cleanup

Some extraction techniques, such as sorption (solid-phase extraction) or purge-and-trap, offer a high intrinsic preconcentration factor. Solvent extraction methods have a common disadvantage of yielding a large volume of extract (usually about 1–5 mL). It oversizes considerably the amount which can be introduced onto the capillary column. The discrepancy between the concentration of an analyte species in a sample and the detector's sensitivity is often increased by a dilution factor during the analysis of solid samples. Therefore, an additional preconcentration step (e.g. by evaporation) sometimes needs to be carried out on the leachate or extract containing the analytes.

Purging the extracts with nitrogen or helium in precalibrated tubes, Kuderna–Danish evaporation and rotary evaporation have been the methods of choice. Losses may occur in the preconcentration of the derivatized species, especially for more volatile Me_3Pb^+ species. Better recoveries are obtained when the solution of the organometallic chelates, which are less volatile than tetraalkyl species, is preconcentrated prior to derivatization. Then, however, a minimum volume of 250 μL is required for easy handling during the derivatization step and removal of the unreacted Grignard reagent, inducing a dilution factor of 1 : 250 for capillary GC analysis.

The increasing sensitivity of element selective detectors for GC (sub-picogram absolute detection limits have become a standard) and the wider availability of large volume injection techniques have recently contributed to the elimination of the need for off-line enrichment.

The evaporative preconcentration and large volume injection lead to a co-preconcentration of other substances present in a sample and may be detrimental to the chromatographic column in routine analysis. Therefore a cleanup step of the extract is advised.

2.5 Automation of sample preparation and GC sample introduction

Recent advances in *in situ* derivatization, capillary cryotrapping and multicapillary GC have allowed the development of an automated sample introduction device for an atomic spectrometer for speciation analysis. The scheme of such an accessory for time-resolved introduction of element species into an atomization/excitation/ionization source is shown in Figure 3.2.4. Organometallic compounds (alkyllead, butyltin, methylmercury) and some ions, e.g. Pb^{2+}, Hg^{2+}, Se(IV), As(III), are volatilized *in situ* by means of a suitable derivatization reaction (hydride generation or ethylation with $NaBEt_4$). The derivatives formed are purged from the vessel and pass through a water scrubber (Nafion dryer) to a wide bore (0.53 mm) capillary trap where they are cryofocused at $-100\,^{\circ}C$. Then the trap is heated to release the species on a multicapillary capillary column. The analysis cycle takes less then 5 min. A number of applications have recently been described [81, 82, 140, 141]. A similar design with AAS detection has been successfully used by Dietz *et al.* [83].

The system developed shows two major advantages over the commercial purge-and-trap systems and those described in the literature. It is a free-standing accessory including the separation step so no gas chromatograph is required because of the application of a multicapillary column. The second one consists of using a 30 cm Nafion tube dryer

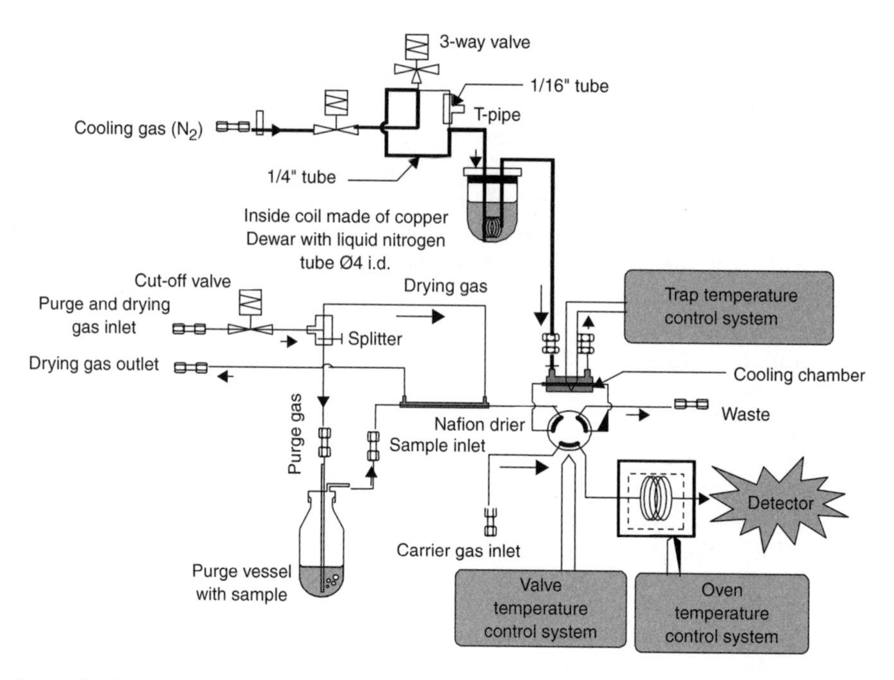

Figure 3.2.4. Device for the time-resolved introduction of organometallic species into an atomic spectrometer.

which makes an external chiller redundant. The result is a compact accessory allowing for species-selective analysis of liquid and solid (with optional microwave cavity) samples.

2.6 Organic samples (gas condensates, shale oils, gasoline)

This category includes various samples of which the most often analyzed are (1) fuels, for alkyl-lead [142, 143] and manganese carbonyl [144, 145] compounds, (2) crude oils, for metallopor-phyrins [146–148], and natural gas condensates for mercury species [18, 149–151]. The analytes are volatile and thermally stable so that samples can usually be injected directly onto a GC column. Gasoline is often diluted ten times to avoid interference with hydrocarbons and the resulting saturation of the plasma.

The detection limits for metals in oils and gas condensates are high in comparison with those for other samples. This is due to the hydrocarbon matrix which gives rise to background interference when it enters into the plasma at the same time as a species of interest. Moreover, carbon compounds can overload the plasma discharge, which has limited thermal energy, and hence reduce the excitation ability. The determination of metalloporphyrins in crude oils is further complicated by their low concentration levels, wide range of molecular weights, the complicated isomerism and the complexity of the crude oil matrix [150, 151]. Hence, a sample pretreatment step is often necessary.

In order to counteract the degradation of GC performance due to the accumulation of nonvolatile residues from the crude oil, removal of the asphaltene and pigment materials has been recommended. This also enables reduction of the maximum elution temperature for a sample, thus protecting the column, reducing the bleed and stabilizing the detector [146]. Other benefits of a sample pretreatment include group isolation of metalloporphyrins and nonporphyrins for GC and preconcentration of metalloporphyrins for trace analysis [146].

An elegant solution to avoid the interference from hydrocarbons was proposed for speciation analysis of mercury in gas condensates [18]. Mercury-containing peaks are collected from the column eluate by an amalgamation trap. They are subsequently released into the plasma as Hg^0 in a flow of helium. Methylmercury and labile ionic mercury can be extracted as complexes with cysteine into the aqueous phase to eliminate the matrix and then re-extracted into hexane prior to GC [149]. A direct approach to speciation analysis of mercury in gas condensates was recently proposed [150, 151].

3 SAMPLE PREPARATION FOR ORGANOMETALLOID SPECIES: ARSENIC AND SELENIUM

Arsenic and selenium are two metalloids with a very rich biochemistry. As a consequence, a number of species containing As–C and Se–C bonds are naturally formed by living organisms and require species-specific analytical methods. With the exception of simple alkylselenium species (Me_2Se, Me_2Se_2, MeEtSe) and methylarsinic and methylarsonic acids these compounds cannot be chromatographed in the gas phase and are usually analysed by HPLC-ICP MS, which requires a custom designed sample preparation. Regardless of the sample preparation the sample to be introduced on the column should be in aqueous solution passing through a 0.2 μm filter.

3.1 Organoarsenic compounds

The species of interest include As(III), As(V), monoarsinic and dimethylarsonic acids, arsenobetaine, arsenocholine and a number of arsenosugars in biological materials. An overview of these techniques can be found in two recent extensive reports [152, 153]. A freezing procedure has commonly been used to preserve a biosample; arsenic speciation in fresh and defrosted samples is carried out [154]. Arsenobetaine in sample extracts that were stored at 4 °C for 9 months decomposed to trimethylarsine oxide and two other unidentified arsenic species [154, 155]. Urine samples can be injected on a chromatographic column directly (filtered) or after dilution with acidified acid [156–159]. Urine samples were collected in polycarbonate bottles and filtered through a 0.45 μm filter [160]. The storage procedure implies the use of acid-washed PE bottles at −10 °C [161] but storage at 20 °C [160] has also been reported. For solid materials the recovery of organic arsenic from the matrix and separation from the matrix are prerequisites for a chromatographic analysis.

For foodstuff samples, defatting, e.g. by leaching with acetone, is the first step [162]. It is necessary to avoid generating an emulsion with the fat, which would make the subsequent clean-up more difficult [163]. In addition, the efficiency of the subsequent methanol extraction step is apparently higher for defatted samples than for nondefatted ones [162]. The uncleaned samples generate problems on the level of ICP MS cones and electrospray ion source.

Extraction of arsenobetaine, arsenocholine and arsenoribosides has usually been performed using methanol [162], methanol–chloroform–water or methanol–water. A comparison study of these methods is available [164]. A methanol–water mixture is recommended for the dry tissues whereas fresh samples can be efficiently leached with pure methanol. With CRM 422 (dried cod muscle certified for the total As content only, but widely used in speciation studies) a precipitate of fatty aspect during CH_3OH–H_2O extraction has been observed [164]. Recoveries of 90 % for fish and 80 % for mussels have been achieved. No degradation of arsenobetaine to more toxic species was observed when an enzymic (trypsin) digestion procedure was applied to the fish [165, 166]. The methanolic extraction is typically repeated 2–3 times followed by preconcentration of the extract by evaporation of methanol using a rotavaporator.

For fatty samples it is recommended to remove the lipids by shaking with diethylether or petroleum [167]. Since some samples of seafood products are prepared in oil and generally tend to have a high salt content an additional cleanup step, e.g. on a strong cation exchanger [162], is required to eliminate the remains of liposoluble compounds not extracted with acetone. The clean-up also avoids the need for periodically reversing and flushing the chromatography column in order to overcome pressure buildup due to accumulation of material on the column.

There has been a surge of interest recently in the use of microwave-assisted procedures for the recovery of organoarsenic compounds from biological tissues [163, 168, 169]. A recent alternative is accelerated solvent extraction (ASE). It is based on performing static extractions at elevated temperatures and pressures. The pressure can be programmed without the use of elevated solvent temperatures that could lead to decomposition of thermally unstable compounds. The optimization of extraction of organoarsenic compounds from seaweed by ASE has been discussed [170, 171].

3.2 Organoselenium compounds

Since selenoaminoacids are soluble in water, leaching with hot water has been judged sufficient to recover selenium species not incorporated into larger molecules. The sample is homogenized with water, sonicated or heated, and ultracentrifuged. The typical recovery of selenium extracted in this way is ca. 10–15 % [172–176] but it may reach 100 % in the case of some selenized yeast. Selenocysteine and some other selenoaminoacids are highly susceptible to oxidative degradation, because the selenol group has a significantly lower oxidation potential than its sulfur counterpart. The carboxymethyl derivative has been synthesized (by addition of iodoacetic acid) to stabilize selenocysteine and prevent its degradation [177].

The low yields of the aqueous leaching procedure for some species and samples have led some workers to use more aggressive leaching media. A trade-off is always necessary between the recovery of selenium from a solid matrix and the preservation of the original selenium species. As shown by Casiot *et al.* [174] the addition of SDS to a leaching mixture increases the yield of Se by releasing selenoaminoacids bound in selenoproteins. The recovery of selenoaminoacids can be increased to above 95 % by degrading the species originally present with a mixture of proteolytic enzymes [175]. An example of the effect of sample preparation procedure on the recovery and speciation for a garlic sample is shown in Figure 3.2.5. Similar results have been reported elsewhere for yeast [174].

Care is advised in the interpretation of literature data since the results depend on the sample

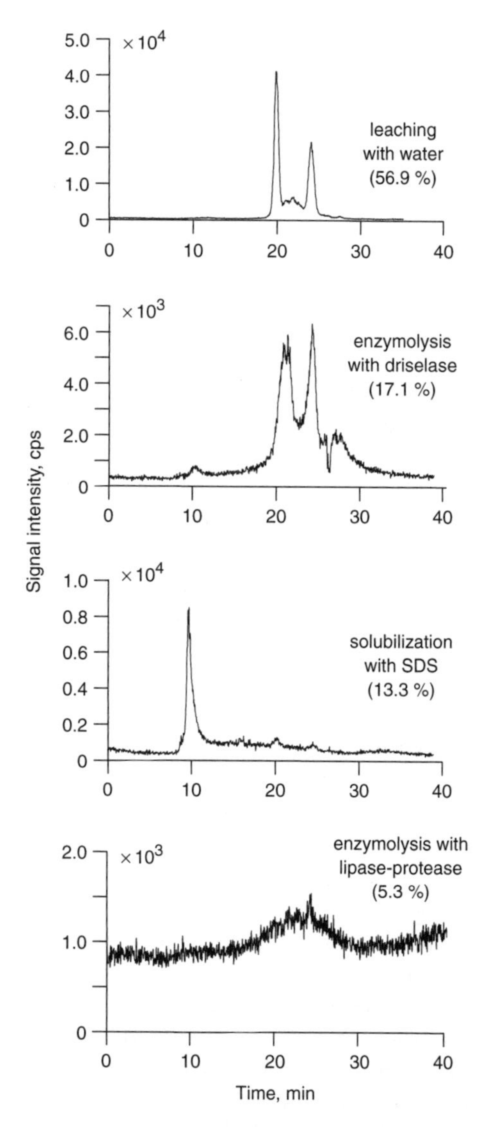

Figure 3.2.5. Effect of the sample preparation procedure on speciation analysis for selenium in garlic.

preparation procedure. This applies in particular to the frequently used statement 'the majority of Se is present as selenomethionine', describing the result of a procedure involving an enzymic digestion. Actually, selenomethionine usually constitutes a part of a larger stable selenoprotein that was destroyed during the sample preparation procedure.

Free selenoaminoacids can be separated by ultrafiltration (in breast milk) [178] or dialysis (algal extract) [179]. A recent paper discusses the use of crown ethers to eliminate salts from urine samples prior to HPLC or CZE of organoselenium compounds [180].

In mammals, speciation of selenium is mostly concerned with the determination of the different Se-containing proteins [181–188]. The most important seem to be selenoprotein P, a major protein which is sometimes used as a biochemical marker of selenium status [183], selenoenzymes such as several glutathione peroxidases and type 1 idothyronine de-iodinase [183, 186, 188] and albumin [182, 186, 187]. More than 25 Se-containing proteins or protein subunits have been detected in rat tissues labelled *in vivo* with ^{75}Se [188]. The sample preparation procedures include equilibration with a physiological buffer and ultracentrifugation.

4 SAMPLE PREPARATION IN SPECIATION OF COORDINATION COMPLEXES OF METALS WITH BIOLIGANDS

The previously discussed cases concerned organometallic and organometalloid species in which the carbon–metal(metalloid) bond conferred species stability in terms of acid–base conditions in the system. An increasingly important class of metallocompounds in speciation and fractionation studies includes coordination complexes. Most of these species are sensitive to pH changes and other coordination ligands present in the system, which requires particular attention during their recovery from samples prior to liquid chromatography or capillary zone electrophoresis. The major bioligands include proteins and small anions with specific functions, such as citrate, ATP, porphyrins or cobalamins. The greatest interest is attached to essential elements, which include some transition metals such as iron, copper and zinc, and toxic metals such as Al, Cr, Pb, Cd and Hg. Metals can be an integral part of metalloproteins and metalloenzymes, e.g. ferritin (Fe, Cu, Zn), β-amylase (Cu), alcohol dehydrogenase (Cd, Zn) and carbonic anhydrase (Cu, Zn), or bind less firmly to transport proteins (albumin, transferrin).

4.1 Biological fluids

The most common body fluids include blood (subdivided by centrifugation into serum and packed cells, and when an anti-coagulant is added the blood is separated into plasma and red cells (erythrocytes)), amniotic fluid, breast milk and urine.

Sampling, sample preservation and preparation prior to chromatography are particularly critical in clinical chemistry because of the low concentrations involved (risk of contamination) the thermodynamical instability of some species, and the complexity of the matrix [189–191].

Sample preparation of serum prior to HPLC includes filtration of sample on a 0.45 μm or 0.2 μm filter [192]. Erythrocytes were subjected to three freeze–thaw cycles to lyse the cells [193, 194], followed by a tenfold dilution with a buffer and centrifugation at 18 000 g to remove fragments of membranes, etc. An alternative procedure recommended by Cornelis *et al.* [190] was based on mixing one part of packed cells with one part of toluene and 40 parts of ice-cold water, followed by centrifugation and 0.45 μm filtration of the lysate.

Blood cells need to be lysed to free their content prior to chromatographic separation. The supernatant was further diluted. When low molecular weight compounds are of interest, ultrafiltration on a 10 kDa filter is carried out [195–197]. The filtrate is protein free and can be analyzed, e.g. for metallodrug metabolites or porphyrins. Breast milk should be centrifuged to remove fat; precipitation of casein with 1 M acetate is optional [198]. Dialysis and purification by size-exclusion chromatography is required if separation by RP HPLC is to be undertaken [198].

Amniotic fluid is a urine-like fluid inhaled and swallowed by the human fetus. Some heavy metals, e.g. Pb can cross the placenta and end up in amniotic fluid. Metal-binding ligands are important in amniotic fluid because of the potential of being transporters to the neurological system [199]. The amniotic fluid sample was centrifuged and the supernatant was stored frozen at −20 °C prior to analysis.

4.2 Plant and animal tissues

Liver and kidney have been the most widely studied organs because of their crucial function in the metabolism of metals.

Washing cells in a Tris-HCl buffer (pH 8) containing 1 M EDTA to remove metal ions reversibly bound to the cell wall has been recommended [177]. In the majority of studies size-exclusion preparative chromatography [200–204] has been preferred to heat treatment [205–207] for the isolation of the metallothionein (MT) fraction from the tissue cytosol. Guidelines for the preparation of biological samples prior to quantification of MTs were discussed with particular attention given to the care necessary to avoid oxidation [208].

Soluble extracts of tissues and cultured cells are prepared by homogenizing tissue samples in an appropriate buffer [209]. Neutral buffers are necessary for extracting MTs since Zn starts to dissociate from the protein at pH 5. Cd and Cu are removed at lower pH values. A 10–50 mM Tris-HCl buffer at pH 7.4–9 is the most common choice. For cytosols containing Cd-induced MTs dilution factors up to 10 have been used whereas for those with natural MT levels equal amounts of tissue and buffer have been found suitable.

Metallothioneins are prone to oxidation during isolation due to their high cystein content. During oxidation disulfide bridges are formed and the MTs either copolymerize or combine with other proteins to move into the high molecular weight fraction. Since MTs may be oxidized by, e.g. oxygen, Cu(I) or heme components, the homogenization of tissues and subsequent isolation of MTs should be normally performed in deoxygenated buffers and/or in the presence of a thiolic reducing agent [210]. β-Mercaptoethanol is added as antioxidant which additionally prevents formation of dimeric forms of MT [211]. Other components added during homogenization include 0.02 % NaN_3 added as an antibacterial agent and phenylmethane-sulfonylfluoride which is added as a protease inhibitor. The homogenization step is followed by centrifugation. The use of an refrigerated ultracentrifuge (100 000 g) is strongly recommended.

As a result two fractions, a soluble one (cell supernatant, cytosol) and a particulate one (cell membranes and organelles), are obtained. Only the supernatant is analyzed for MTs. It is recommended that it is stored at −20 °C under nitrogen prior to analysis [211]. The extraction efficiency by sucrose-Tris at pH 8 was 25–30 % [208]. The percentage of metals in cytosols is typically 50–85 % and 30–57 % in the case of Zn and Cd, respectively [212, 213].

A heat treatment (at 60 °C for 15 min) of the cytosol extracts (especially of concentrated ones) is recommended to separate the high molecular weight fraction, which coagulates from the supernatant (containing MTs which are heat stable). Such a treatment reduces the protein load on the HPLC column not only improving the separation of MT isoforms but also prolonging the column lifetime. Enzymolysis in simulated gastric and gastrointestinal juice has been proposed for meat samples [205].

Filtration of the cytosol (0.22 μm filter) before introducing onto the chromatographic column is strongly advised. A guard column should be inserted to protect the analytical column particularly from effects of lipids, which otherwise degrade the separation [205]. Any organic species that adhere to the column can also bind inorganic species giving rise to anomalous peaks in subsequent runs [205]. A new guard column was used for each injection to prevent adsorption by ligands with a high affinity for cadmium that would otherwise interfere with subsequent injections [205]. An extensive column cleanup was necessary [205].

To avoid contamination of the analytical column by trace elements, buffers should be cleaned by Chelex-100 [205].

Higher molecular weight compounds, referred to as protein complexes with Cd or Ni, were extracted into warm H_2O. The extracts were centrifuged and the proteins were fractionated by successive filtration through membranes with molecular weight cut-offs of 500, 5000 and 30 000 Da [214–216] prior to off-line GF AAS. Guenther and coworkers [213, 217, 218] used low pressure semi-preparative chromatography to separate water-soluble species of Cd and Zn. The use of concentrated surfactants and dithiothreitol are important for the solubilization of high molecular weight proteins and metalloenzymes [219].

Water-soluble polysaccharide species with higher molecular weights can be readily decomposed by enzymic hydrolysis with a mixture of pectinase and hemicellulase to release the dRG-II complex [220]. The same mixture was found to be efficient for extracting the dRG-II–metal complexes from water-insoluble residue of vegetables owing to the destruction of the pectic structure [220].

Some attention has been paid to the analysis of enzymic digests of foodstuffs in the quest of molecular information to contribute to the knowledge on the bioavailability of some elements. Enzymolysis in simulated gastric and gastrointestinal juice has been proposed for meat [221]. The soluble fraction of the stomach and upper intestinal contents of a guinea pig on different diets were investigated for the species of Al, Cu, Zn, Mn, Sr and Rb. The effect of citrate on each of these elements has also been assessed [221]. An enzymic digest of bovine thyroglobulin has been analyzed for iodine species [222].

5 CONCLUSIONS–TRENDS AND PERSPECTIVES

Novel derivatization reagents and instrumental techniques have recently rendered the sample preparation prior to speciation of organometallic compounds by hyphenated techniques faster and easier to automate. The ultimate goal of the development of an automated speciation analyzer seems to be closer than ever. The critical issue remains the certainty that the species arriving at the detector from the chromatographic column is the one originally present in the sample. It seems to be possible to control the stability of species for which standards can be synthesized, such as anthropogenic organometallic pollutants. A lot, however, still remains to be done to control transformations of metal coordination complexes with bioligands during the sample preparation, especially because many of these species remain in complex chemical equilibria in a sample, which are destroyed once a chemical or even a physical (dilution) operation is carried out on the sample. Particular care is advised regarding the species stability during sampling and storage. The recent development of isotope dilution analysis now gives a possibility of improving the accuracy during the quantification and the means of unambiguous determination of the species with a spike of specific compounds with a different isotopic ratio.

6 REFERENCES

1. Lobinski, R., *Appl. Spectrosc.*, **51**, 260A (1997).
2. Quevauviller, P., *Method Performance Studies for Speciation Analysis*, Royal Society of Chemistry, Cambridge, 1998.
3. Caruso, J. A., Sutton, K. L. and Ackley, K. L., *Elemental Speciation. New Approaches for Trace Element Analysis*, Elsevier, Amsterdam, 2000.
4. Lobinski, R. and Marczenko, Z., *Spectrochemical Trace Analysis for Metals and Metalloids*, Elsevier, Amsterdam, 1996.
5. Gomez-Ariza, J. L., Morales, E., Giraldez, I., Sanchez-Rodas, D. and Velasco, A., *J. Chromatogr. A*, **932**, 211 (2001).
6. Szpunar Lobinska, J., Ceulemans, M., Lobinski, R. and Adams, F. C., *Anal. Chim. Acta.*, **278**, 99 (1993).
7. Emteborg, H., Baxter, D. C., Sharp, M. and Frech, W., *Analyst*, **69** (1995).
8. Lansens, P. and Baeyens, W., *Anal. Chim. Acta.*, **228**, 93 (1990).
9. Perez Corona, T., Madrid Albarran, Y., Camara, C. and Beceiro, E., *Spectrochim. Acta, Part B.*, **53B**, 321 (1998).
10. Jian, W. and McLeod, C. W., *Talanta*, **39**, 1537 (1992).
11. Johansson, M., Emteborg, H., Glad, B., Reinholdsson, F. and Baxter, D. C., *Fresenius' J. Anal. Chem.*, **351**, 461 (1995).

12. Sanchez, D. M., Martin, R., Morante, R., Marin, J. and Munuera, M. L., *Talanta*, **52**, 671 (2000).
13. Martinez Blanco, R., Tagle Villanueva, M., Sanchez Uria, J. E. and Sanz-Medel, A., *Anal. Chim. Acta*, **419**, 137 (2000).
14. Salih, B., *Spectrochim. Acta B*, **55**, 1115 (2000).
15. Ballesteros, E., Gallego, M. and Valcarcel, M., *J. Chromatogr. A*, **869**, 101 (2000).
16. Eisert, R. and Pawliszyn, J., *Crit. Rev. Anal. Chem.*, **27**, 103 (1997).
17. Pawliszyn, J. B., *Solid-Phase Microextraction: Theory and Practice* Wiley VCH, Weinheim, Germany (1997).
18. Snell, J. P., Frech, W. and Thomassen, Y., *Analyst*, 1055 (1996).
19. Cai, Y., Monsalud, S., Furton, K. G., Jaffe, R. and Jones, R. D., *Appl. Organomet. Chem.*, **12**, 565 (1998).
20. Mothes, S. and Wenrich, R., *J. High Resol. Chromatogr.*, **22**, 181 (1999).
21. Moens, L., de Smaele, T., Dams, R., van den Broek, P. and Sandra, P., *Anal. Chem.*, **69**, 1604 (1997).
22. Gorecki, T. and Pawliszyn, J., *Anal. Chem.*, **68**, 3008 (1996).
23. Yu, X. and Pawliszyn, J., *Anal. Chem.*, **72**, 1788 (2000).
24. Aguerre, S., Bancon-Montigny, C., Lespes, G. and Potin-Gautier, M., *Analyst*, **125**, 236 (2000).
25. Aguerre, S., Lespes, G., Desauziers, V. and Potin-Gautier, M., *J. Anal. At. Spectrom.*, **16**, 263 (2001).
26. Mester, Z., Vitanyi, G., Morabito, R. and Fodor, P., *J. Chromatogr. A*, **832**, 183 (1999).
27. Mester, Z. and Pawliszyn, J., *J. Chromatogr.*, **873**, 129 (2000).
28. Mester, Z., Lam, J., Sturgeon, R. and Pawliszyn, J., *J. Anal. At. Spectrom.*, **15**, 837 (2000).
29. Cai, Y. and Bayona, J. M., *J. Chromatogr. A*, **696**, 113 (1995).
30. Morcillo, Y., Cai, Y. and Bayona, J. M., *J. High. Resolut. Chromatogr.*, **18**, 767 (1995).
31. Vercauteren, J., De Meester, A., De Smaele, T., Vanhaecke, F., Moens, L., Dams, R. and Sandra, P., *J. Anal. At. Spectrom.*, **15**, 651 (2000).
32. Mester, Z., Sturgeon, R. and Pawliszyn, J., *Spectrochim. Acta*, **B56**, 233 (2001).
33. Dunnemann, L., Hajimiragha, H. and Begerow, J., *Fresenius' J. Anal. Chem.*, **363**, 466 (1999).
34. Tutschku, S., Mothes, S. and Wennrich, R., *Fresenius' J. Anal. Chem.*, **354**, 587 (1996).
35. Jiang, G., Liu, J. and Yang, K., *Anal. Chim. Acta*, **421**, 67 (2000).
36. Yu, X., Yuan, H., Gorecki, T. and Pawliszyn, J., *Anal. Chem.*, **71**, 2998 (1999).
37. Fragueiro, M., Alava-Moreno, F., Lavilla, I. and Bendicho, C., *J. Anal. At. Spectrom.*, **15**, 705 (2000).
38. Beichert, A., Padberg, S. and Wenclawiak, B. W., *Appl. Organomet. Chem.*, **14**, 493 (2000).
39. Guidotti, M., *J. AOC Intern.*, **83**, 1082 (2000).
40. Szostek, B. and Alstadt, J. H., *J. Chromatogr. A*, **807**, 253 (1998).
41. Wu, J., Mester, Z. and Pawlliszin, J., *Anal. Chim. Acta*, **424**, 211 (2000).
42. Orellana-Velado, N., Pereiro, R. and Sanz-Medel, A., *J. Anal. At. Spectrom.*, **16**, 376 (2001).
43. Radojevic, M., Allen, A., Rapsomanikis, S. and Harrison, R. M., *Anal. Chem.*, **58**, 658 (1986).
44. Lobinski, R., Szpunar Lobinska, J., Adams, F. C., Teissedre, P. L. and Cabanis, J. C., *J. Assoc. Off. Anal. Chem. Int.*, **76**, 1262 (1993).
45. Lobinski, R., Boutron, C. F., Candelone, J. P., Hong, S., Szpunar Lobinska, J. and Adams, F. C., *Anal. Chem.*, **65**, 2510 (1993).
46. Lobinski, R. and Adams, F. C., *Anal. Chim. Acta*, **262**, 285 (1992).
47. Heisterkamp, M., de Smaele, T., Candelone, J. P., Moens, L., Dams, R. and Adams, F. C., *J. Anal. At. Spectrom.*, **12**, 1077 (1997).
48. Heisterkamp, M. and Adams, F. C., *Fresenius' J. Anal. Chem.*, **362**, 489 (1998).
49. Bergmann, K. and Neidhart, B., *Fresenius' J. Anal. Chem.*, **356**, 57 (1996).
50. Lobinski, R., Dirkx, W. M. R., Ceulemans, M. and Adams, F. C., *Anal. Chem.*, **64**, 159 (1992).
51. Ceulemans, M. and Adams, F. C., *Anal. Chim. Acta*, **317**, 161 (1995).
52. Bergmann, K., Roehr, U. and Neidhart, B., *Fresenius' J. Anal. Chem.*, **349**, 815 (1994).
53. Bulska, E., Baxter, D. C. and Frech, W., *Anal. Chim. Acta.*, **12**, 545 (1991).
54. Bulska, E., Emteborg, H., Baxter, D. C., Frech, W., Ellingsen, D. and Thomassen, Y., *Analyst*, **117**, 657 (1992).
55. Blais, J. S. and Marshall, W. D., *J. Environ. Qual.*, **15**, 255 (1986).
56. Gomez Ariza, J. L., Morales, E. and Ruiz Benitez, M., *Analyst*, 641 (1992).
57. Harino, H., Fukushima, M. and Tanaka, M., *Anal. Chim. Acta*, **264**, 91 (1992).
58. Forsyth, D. S., Weber, D. and Cleroux, C., *Food Addit. Contam.*, **9**, 161 (1992).
59. Forsyth, D. S., Weber, D. and Dalglish, K., *J. AOAC Int.*, **75**, 964 (1992).
60. Gremm, T. J. and Frimmel, F. H., *Water Res. 1992*, **26**, 1163 (1992).
61. Waldock, M. J., *Mikrochim Acta.*, **109**, 23 (1992).
62. Dowling, T. M. and Uden, P. C., *J. Chromatogr.*, **644**, 153 (1993).
63. Rapsomanikis, S., *Analyst*, 1429 (1994).
64. Szpunar, J., Schmitt, V. O., Lobinski, R. and Monod, J. L., *J. Anal. At. Spectrom.*, **11**, 193 (1996).

65. Minganti, V., Capelli, R. and De Pellegrini, R., *Fresenius' J. Anal. Chem.*, **351**, 471 (1995).
66. Ceulemans, M., Lobinski, R., Dirkx, W. M. R. and Adams, F. C., *Fresenius' J. Anal. Chem.*, **347**, 256 (1993).
67. Heisterkamp, M. and Adams, F. C., *J. Anal. At. Spectrom.*, **14**, 1307 (1999).
68. Hu, G. L. and Wang, X. R., *Anal. Lett.*, **31**, 1445 (1998).
69. Cai, Y., Monsalud, S., Jaffé, R. and Jones, R. D., *J. Chromatogr. A*, **876**, 147 (2000).
70. Bloom, N., *Can. J. Fish Aquat. Sci.*, **46**, 1131 (1989).
71. Horvat, M., Bloom, N. S. and Liang, L., *Anal Chim Acta.*, **281**, 135 (1993).
72. Bowles, K. C. and Apte, S. C., *Anal Chem.*, **70**, 395 (1998).
73. Bowles, K. C. and Apte, S. C., *Anal. Chim. Acta*, **419**, 145 (2000).
74. Grüter, U. M., Hitzke, M., Kresimon, J. and Hirner, A. V., *J. Chromatogr. A*, **938**, 225 (2001).
75. Dedina, J. and Tsalev, D. L., Hydride-Generation Atomic-Absorption Spectrometry, John Wiley, Chichester, 1995.
76. Rapsomanikis, S., Donard, O. F. X. and Weber, J. H., *Anal. Chem.*, **58**, 35 (1986).
77. Ceulemans, M. and Adams, F. C., *J. Anal. At. Spectrom.*, **11**, 201 (1996).
78. Tseng, C. M., de Diego, A., Pinaly, H., Amouroux, D. and Donard, O. F. X., *J. Anal. At. Spectrom.*, **13**, 755 (1998).
79. Craig, P. J., Dewick, R. J. and van Elteren, J. T., *Fresenius' J. Anal. Chem.*, **351**, 467 (1995).
80. de la Calle Guntinas, M. B., Lobinski, R. and Adams, F. C., *J. Anal. At. Spectrom.*, **10**, 111 (1995).
81. Wasik, A., Pereiro, I. R., Dietz, C., Szpunar, J. and Lobinski, R., *Anal. Commun.*, **35**, 331 (1998).
82. Wasik, A., Rodriguez Pereiro, I. and Lobinski, R., *Spectrochim. Acta, Part B*, **53B**, 867 (1998).
83. Dietz, C., Madrid, Y., Camara, C. and Quevauviller, P., *Anal. Chem.*, **72**, 4178 (2000).
84. Segovia Garcia, E., Garcia Alonso, J. I. and Sanz Medel, A., *J. Mass Spectrom.*, **32**, 542 (1997).
85. Pannier, F., Astruc, A. and Astruc, M., *Anal. Chim. Acta.*, **287**, 17 (1994).
86. Tutschku, S., Mothes, S. and Dittrich, K., *J. Chromatogr. A*, **683**, 269 (1994).
87. Liu, Y., Lopez Avila, V., Alcaraz, M. and Beckert, W. F., *J. High Resolut. Chromatogr.*, **17**, 527 (1994).
88. Cai, Y., Rapsomanikis, S. and Andreae, M. O., *Anal. Chim. Acta.*, **274**, 243 (1993).
89. Cai, Y., Alzaga, R. and Bayona, J. M., *Anal. Chem.*, **66**, 1161 (1994).
90. Cai, Y., Rapsomanikis, S. and Andreae, M. O., *J. Anal. At. Spectrom.*, **8**, 119 (1993).
91. Chau, Y. K., Yang, F. and Maguire, R. J., *Anal. Chim. Acta.*, **320**, 165 (1996).
92. Quevauviller, P. and Morabito, R., *Trends Anal. Chem.*, **19**, 86 (2000).
93. Hintelmann, H., Evans, R. D. and Villeneuve, J. Y., *J. Anal. At. Spectrom.*, **10**, 619 (1995).
94. Falter, R. and Ilgen, G., *Fresenius' J. Anal. Chem.*, **358**, 407 (1997).
95. Falter, R. and Ilgen, G., *Fresenius' J. Anal. Chem.*, **358**, 401 (1997).
96. Eiden, R., Falter, R., Augustin Castro, B. and Schoeler, H. F., *Fresenius' J. Anal. Chem.*, **357**, 439 (1997).
97. Padberg, S., Burow, M. and Stoeppler, M., *Fresenius' J. Anal. Chem.*, **346**, 686 (1993).
98. Bloom, N. S., Colman, J. A. and Barber, L., *Fresenius' J. Anal. Chem.*, **358**, 371 (1997).
99. Pellegrino, C., Massanisso, P. and Morabito, R., *Trends Anal. Chem.*, **19**, 97 (2000).
100. Zhang, S., Chau, Y. K. and Chau, Q. S. Y., *Appl. Organomet. Chem.*, **5**, 431 (1991).
101. Lorenzo, R. A., Vazquez, M. J., Carro, A. M. and Cela, R., *Trends Anal. Chem.*, **18**, 410 (1999).
102. Hardy, S. and Jones, P., *J. Chromatogr. A*, **791**, 333 (1997).
103. Fischer, R., Rapsomanikis, S. and Andreae, M. O., *Anal. Chem.*, **65**, 763 (1993).
104. Rapsomanikis, S. and Craig, P. J., *Anal. Chim. Acta.*, **248**, 563 (1991).
105. Ceulemans, M., Witte, C., Lobinski, R. and Adams, F. C., *Appl. Organomet. Chem.*, **8**, 451 (1994).
106. Jimenez, M. S. and Sturgeon, R. E., *J. Anal. At. Spectrom.*, **12**, 597 (1997).
107. Forsyth, D. S. and Iyengar, J. R., *Appl. Organomet. Chem.*, **3**, 211 (1989).
108. Pannier, F., Astruc, A. and Astruc, M., *Anal. Chim. Acta.*, **327**, 287 (1996).
109. Forsyth, D. S. and Iyengar, J. R., *J. Assoc. Off. Anal. Chem.*, **72**, 997 (1989).
110. Wai, C. M., Lin, Y., Brauer, R., Wang, S. and Beckert, W. F., *Talanta*, **40**, 1325 (1993).
111. Vela, N. P. and Caruso, J. A., *J. Anal. At. Spectrom.*, **11**, 1129 (1996).
112. Sun, Y. C., Mierzwa, J., Chung, Y. T. and Yang, M. H., *Anal. Commun.*, **34**, 333 (1997).
113. Liu, Y., Lopez Avila, V., Alcaraz, M. and Beckert, W. F., *J. High Resolut. Chromatogr.*, **16**, 106 (1993).
114. Holak, W., *J. Assoc. Off. Anal. Chem. Int.*, **78**, 1124 (1995).
115. Emteborg, H., Bjorklund, E., Odman, F., Karlsson, L., Mathiasson, L., Frech, W. and Baxter, D. C., *Analyst*, **19** (1996).
116. Bayona, J. M., *Trends Anal. Chem.*, **19**, 107 (2000).
117. Szpunar, J., Schmitt, V. O., Donard, O. F. X. and Lobinski, R., *Trends Anal Chem.*, **15**, 181 (1996).

118. Donard, O. F. X., Lalere, B., Martin, F. and Lobinski, R., *Anal. Chem.*, **67**, 4250 (1995).
119. Lalere, B., Szpunar, J., Budzinski, H., Garrigues, P. and Donard, O. F. X., *Analyst*, 2665 (1995).
120. Szpunar, J., Ceulemans, M., Schmitt, V. O., Adams, F. C. and Lobinski, R., *Anal. Chim. Acta*, **332**, 225 (1996).
121. Schmitt, V. O., Martin, F. M., Szpunar, J., Lobinski, R. and Donard, O. F. X., *Spectra Anal.*, **25**, 14 (1996).
122. Rodriguez, I., Santamarina, M., Bollain, M. H., Mejuto, M. C. and Cela, R., *J. Chromatogr, A.*, **774**, 379 (1997).
123. Rodriguez Pereiro, I., Schmitt, V. O. and Lobinski, R., *Anal. Chem.*, **69**, 4799 (1997).
124. Tseng, C. M., de Diego, A., Martin, F. M. and Donard, O. F. X., *J. Anal. At. Spectrom.*, **12**, 629 (1997).
125. Shawky, S., Emons, H. and Duerbeck, H. W., *Anal. Commun.*, **33**, 107 (1996).
126. Rodriguez Pereiro, I., Schmitt, V. O., Szpunar, J., Donard, O. F. X. and Lobinski, R., *Anal. Chem.*, **68**, 4135 (1996).
127. Abuin, M., Carro, A. M. and Lorenzo, R. A., *J. Chromatogr. A*, **889**, 185 (2000).
128. Liu, W. P. and Lee, H. K., *J. Chromatogr.*, **834**, 45 (1999).
129. Lobinski, R., Dirkx, W. M. R., Szpunar Lobinska, J. and Adams, F. C., *Anal. Chim. Acta.*, **286**, 381 (1994).
130. Jiang, G. B. and Zhou, Q. F., *J. Chromatogr. A*, **886**, 197 (2000).
131. Schwedt, G., *Chromatographic Methods in Inorganic Analysis*, Hüthing, Heidelberg, 1981.
132. Kataoka, H., Miyanaga, Y. and Makita, M., *J. Chromatogr. A*, **659**, 481 (1994).
133. Cai, X. J., Block, E., Uden, P. C., Zhang, X., Quimby, B. D. and Sullivan, J. J., *J. Agric. Food Chem.*, **43**, 1754 (1995).
134. Yasumoto, K., Suzuki, T. and Yoshida, M., *J. Agric. Food Chem.*, **36**, 463 (1988).
135. Ouyang, Z., Wu, J. and Xie, L., *Anal. Biochem.*, **178**, 77 (1989).
136. Zheng, O. and Wu, J., *Biomed. Chromatogr.*, **2**, 258 (1988).
137. Scerbo, R., de la Calle Guntinas, M. B., Brunori, C. and Morabito, R., *Quim. Anal*, **87** (1997).
138. Claussen, F. A., *J. Chromatogr. Sci.*, **35**, 568 (1997).
139. Schoene, K., Steinhanses, J., Bruckert, H. J. and Koenig, A., *J. Chromatogr.*, **605**, 257 (1992).
140. Slaets, S., Adams, F., Rodriguez, I. and Lobinski, R., *J. Anal. At. Spectrom.*, **14**, 851 (1999).
141. Slaets, S. and Adams, F. C., *Anal. Chim. Acta*, **414**, 141 (2000).
142. Kim, A. W., Foulkes, M. E., Ebdon, L., Hill, S. J., Patience, R. L., Barwise, A. G. and Rowland, S. J., *J. Anal. At. Spectrom.*, **7**, 1147 (1992).
143. Rodriguez Pereiro, I. and Lobinski, R., *J. Anal. At. Spectrom.*, **12**, 1381 (1997).
144. Aue, W. A., Millier, B. and Sun, X. Y., *Anal. Chem.*, **62**, 2453 (1990).
145. Ombaba, J. M. and Barry, E. F., *J. Chromatogr. A*, **678**, 319 (1994).
146. Zeng, Y. and Uden, P. C., *J. High Resolut. Chromatogr.*, **17**, 223 (1994).
147. Quimby, B. D., Dryden, P. C. and Sullivan, J. J., *J. High. Resolut. Chromatogr.*, **14**, 110 (1991).
148. Ebdon, L., Evans, E. H., Pretorius, W. G. and Rowland, S. J., *J. Anal. At. Spectrom.*, **9**, 939 (1994).
149. Tao, H., Murakami, T., Tominaga, M. and Miyazaki, A., *J. Anal. At. Spectrom.*, **13**, 1085 (1998).
150. Frech, W., Baxter, D. C., Bakke, B., Snell, J. and Thomassen, Y., *Anal. Commun.*, **33**, 7H (1996).
151. Snell, J., Qian, J., Johansson, M., Smit, K. and Frech, W., *Analyst*, **123**, 905 (1998).
152. Burguera, M. and Burguera, J. L., *Talanta*, **44**, 1581 (1997).
153. Guerin, T., Astruc, A. and Astruc, M., *Talanta*, **50**, 1 (1999).
154. Le, S. X. C., Cullen, W. R. and Reimer, K. J., *Environ. Sci. Technol.*, **28**, 1598 (1994).
155. Muerer, A. J. L., Abildtrup, A., Poulsen, O. M. and Christensen, J. M., *Analyst*, **117**, 677 (1992).
156. Larsen, E. H., Pritzl, G. and Hansen, S. H., *J. Anal. At. Spectrom.*, **8**, 557 (1993).
157. Ma, M. and Le, X. C., *Clin. Chem.*, **44**, 539 (1998).
158. Tsalev, D. L., Sperling, M. and Welz, B., *Analyst*, **123**, 1703 (1998).
159. Zhang, X. R., Cornelis, R., De Kimpe, J., Mees, L. and Lameire, N., *Clin. Chem.*, **43**, 406 (1997).
160. Chana, B. S. and Smith, N. J., *Anal. Chim. Acta*, **197**, 177 (1987).
161. Heitkemper, D., Creed, J., Caruso, J. and Fricke, F. L., *J. Anal. At. Spectrom.*, **4**, 279 (1989).
162. Ybanez, N., Velez, D., Tejedor, W. and Montoro, R., *J. Anal. At. Spectrom.*, **10**, 459 (1995).
163. Dagnac, T., Padro, A., Rubio, R. and Rauret, G., *Talanta.*, **48**, 763 (1999).
164. Alberti, J., Rubio, R. and Rauret, G., *Fresenius' J. Anal. Chem.*, **351**, 420 (1995).
165. Branch, S., Ebdon, L. and O'Neill, P., *J. Anal. At. Spectrom.*, **9**, 33 (1994).
166. Lamble, K. J. and Hill, S. J., *Anal. Chim. Acta*, **334**, 261 (1996).
167. Shiomi, K., Sugiyama, Y., Shimakura, K. and Nagashima, Y., *Appl. Organometal. Chem.*, **9**, 105 (1995).
168. Ackley, K. L., B'Hymer, C., Sutton, K. L. and Caruso, J. A., *J. Anal. At. Spectrom.*, **14**, 845 (1999).
169. Helgesen, H. and Larsen, E. H., *Analyst*, **123**, 791 (1998).

170. McKiernan, J. W., Creed, J. T., Brockhoff, C. A., Caruso, J. A. and Lorenzana, R. M., *J. Anal. At. Spectrom.*, **14**, 607 (1999).
171. Gallagher, P. A., Shoemaker, J. A., Wei, X. N., Brockhoff-Schwegel, C. A. and Creed, J. T., *Fresenius' J. Anal. Chem.*, **369**, 71 (2001).
172. Bird, S. M., Ge, H., Uden, P. C., Tyson, J. F., Block, E. and Denoyer, E., *J. Chromatogr. A*, **789**, 349 (1997).
173. Bird, S. M., Uden, P. C., Tyson, J. F., Block, E. and Denoyer, E., *J. Anal. At. Spectrom.*, **12**, 785 (1997).
174. Casiot, C., Szpunar, J., Lobinski, R. and Potin Gautier, M., *J. Anal. At. Spectrom.*, **14**, 645 (1999).
175. Gilon, N., Astruc, A., Astruc, M. and Potin-Gautier, M., *Appl. Organomet. Chem.*, **9**, 623 (1995).
176. Zheng, J., Goessler, W. and Kosmus, W., *Trace Elements Electrol.*, **15**, 70 (1998).
177. Takatera, K., Osaki, N., Yamaguchi, H. and Watanabe, T., *Anal. Sci.*, **10**, 567 (1994).
178. Michalke, B. and Schramel, P., *J. Chromatogr. A*, **807**, 71 (1998).
179. Fan, T. M., Lane, A. N., Martens, D. and Higashi, R. M., *Analyst*, **123**, 875 (1998).
180. Gammelgaard, B., Joens, O. and Bendahl, L., *J. Anal. At. Spectrom.*, **16**, 339 (2001).
181. Thomson, C. D., *Analyst*, **123**, 827 (1998).
182. Plecko, T., Nordmann, S., Rükgauer, M. and Kruse-Jarres, J. D., *Fresenius' J. Anal. Chem.*, **363**, 517 (1999).
183. Persson Moschos, M., Huang, W., Srikumar, T. S., Akesson, B. and Lindeberg, S., *Analyst*, **120**, 833 (1995).
184. Koyama, H., Kasanuma, Y., Kim, C., Ejima, A., Watanabe, C., Nakatsuka, H. and Satoh, H., *Tohoku J. Exp. Med.*, **178**, 17 (1996).
185. Koyama, H., Omura, K., Ejima, A., Kasanuma, Y., Watanabe, C. and Satoh, H., *Anal. Biochem.*, **267**, 84 (1999).
186. Harrison, I., Littlejohn, D. and Fell, G. S., *Analyst*, **121**, 189 (1996).
187. Deagen, J. T., Butler, J. A., Zachara, B. A. and Whanger, P. D., *Anal. Biochem.*, **208**, 176 (1993).
188. Behne, D., Weiss Nowak, C., Kalcklosch, M., Westphal, C., Gessner, H. and Kyriakopoulos, A., *Analyst*, **120**, 823 (1995).
189. Behne, D., *Analyst*, **117**, 555 (1992).
190. Cornelis, R., De Kimpe, J. and Zhang, X., *Spectrochim. Acta, Part B*, **53B**, 187 (1998).
191. Sanz Medel, A., *Spectrochim. Acta, Part B*, **53B**, 197 (1998).
192. Van Landeghem, G. F., D'Haese, P. C., Lamberts, L. V. and De Broe, M. E., *Anal. Chem.*, **66**, 216 (1994).
193. Bergdahl, I. A., Schuetz, A. and Grubb, A., *J. Anal. At. Spectrom.*, **11**, 735 (1996).
194. Gercken, B. and Barnes, R. M., *Anal. Chem.*, **63**, 283 (1991).
195. Matz, S. G., Elder, R. C. and Tepperman, K., *J. Anal. At. Spectrom.*, **4**, 767 (1989).
196. Poon, G. K., Raynaud, F. I., Mistry, P., Odell, D. E., Kelland, L. R., Harrap, K. R., Barnard, C. F. J. and Murrer, B. A., *J. Chromatogr. A*, **712**, 61 (1995).
197. De Waal, W. A. J., Maessen, F. J. M. J. and Kraak, J. C., *J. Chromatogr.*, **407**, 253 (1987).
198. Makino, Y. and Nishimura, S., *J. Chromatogr., Biomed Appl.*, **117**, 346 (1992).
199. Hall, G. S., Zhu, E. G. and Martin, E. G., *Anal. Commun.*, **26**, 93 (1999).
200. Ang, S. G. and Wong, V. W. T., *J. Liq. Chromatogr.*, **14**, 2647 (1991).
201. Ang, S. G. and Wong, V. W. T., *J. Chromatogr.*, **599**, 21 (1992).
202. Apostolova, M., Bontchev, P. R., Nachev, C. and Sirakova, I., *J. Chromatogr., Biomed. Appl. 29 Oct 1993*, **131**, 191 (1993).
203. Van Beek, H. and Baars, A. J., *J. Anal. At. Spectrosc.*, **11**, 70 (1990).
204. Steinebach, O. M. and Wolterbeek, H. T., *J. Chromatogr., Biomed Appl.*, **130**, 199 (1993).
205. Crews, H. M., Dean, J. R., Ebdon, L. and Massey, R. C., *Analyst*, **114**, 895 (1989).
206. Mazzucotelli, A. and Rivaro, P., *Microchem. J.*, **51**, 231 (1995).
207. Mazzucotelli, A., Viarengo, A., Canesi, L., Ponzano, E. and Rivaro, P., *Analyst*, **116**, 605 (1991).
208. Suzuki, K. T. and Sato, M., *Biomed. Res. Trace. Elem.*, **6**, 51 (1995).
209. Polec, K., Peréz-Calvo, M., García-Arribas, O., Szpunar, J., Ribas-Ozonas, B. and Lobinski, R., *J. Anal. At. Spectrom.*, **15**, 1363 (2000).
210. High, K. A., Methven, B. A., McLaren, J. W., Siu, K. W. M., Wang, J., Klaverkamp, J. F. and Blais, J. S., *Fresenius' J. Anal. Chem.*, **351**, 393 (1995).
211. Van Beek, H. and Baars, A. J., *J. Chromatogr.*, **442**, 345 (1988).
212. Jayawickreme, C. K. and Chatt, A., *J. Radioanal. Nucl. Chem.*, **124**, 257 (1988).
213. Guenther, K. and Waldner, H., *Anal. Chim. Acta.*, **259**, 165 (1992).
214. Lange Hesse, K., Dunemann, L. and Schwedt, G., *Fresenius' J. Anal. Chem.*, **339**, 240 (1991).
215. Lange Hesse, K., *Fresenius' J. Anal. Chem.*, **350**, 68 (1994).
216. Lange Hesse, K., Dunemann, L. and Schwedt, G., *Fresenius' J. Anal. Chem.*, **349**, 460 (1994).

217. Guenther, K. and Von Bohlen, A., *Spectrochim. Acta., Part B*, **46B**, 1413 (1991).
218. Guenther, K., von Bohlen, A. and Strompen, C., *Anal. Chim. Acta.*, **309**, 327 (1995).
219. Leopold, I. and Fricke, B., *Anal. Biochem.*, **252**, 277 (1997).
220. Szpunar, J., Pellerin, P., Makarov, A., Doco, T., Williams, P. and Lobinski, R., *J. Anal. At. Spectrom.*, **14**, 639 (1999).
221. Owen, L. M. W., Crews, H. M., Massey, R. C. and Bishop, N. J., *Analyst*, **120**, 705 (1995).
222. Takatera, K. and Watanabe, T., *Anal. Chem.*, **65**, 759 (1993).

3.3 Sample Preparation – Fractionation (Sediments, Soils, Aerosols, and Fly Ashes)

József Hlavay and Klára Polyák

University of Veszprém, Hungary

1 INTRODUCTION

Increasingly strict environmental regulations require the development of new methods for analysis and ask for simple and meaningful tools to obtain information on toxic fractions of different mobility and bioavailability in the solid phases. Objectives of monitoring are to assess pollution effects on man and his environment, to identify possible sources and to establish relationships between pollutant concentrations [1–6]. Thus, it is necessary to investigate and understand the mechanisms of transport of trace elements and their complexes to understand their chemical cycles in nature. An assessment of the impact of an element cannot be based exclusively on its total concentration. It is often not possible to determine the concentrations of the different chemical species that summarize the total concentration of an element in a given matrix. Concerning natural systems, the mobility, transport and partitioning of trace elements are dependent on the chemical form of the elements. Major variations of these characteristics are found in time and space due to the dissipation and flux of energy and materials involved in the biogeochemical processes, which determine the speciation of the elements.

Chemical species present in a given sample are often not stable enough to be determined individually. During the measurement process the

Handbook of Elemental Speciation: Techniques and Methodology R. Cornelis, H. Crews, J. Caruso and K. Heumann
 ISBN: 0-471-49214-0

partitioning of the element among its species may be changed. This behavior can be caused by, for example, a change in pH necessitated by the analytical procedure. Chemical extraction is usually employed to assess operationally defined metal fractions, which can be related to chemical species, as well as to potentially mobile, bioavailable or ecotoxic phases of a sample. It has been generally accepted that the ecological effects of metals, e.g. their bioavailability, ecotoxicology and risk of groundwater contamination are related to such mobile fractions rather than the total concentration. The use of selective extraction methods to distinguish analytes that are immobilized in different phases of solids is also of particular interest in geochemistry for location of mineral deposits.

Atmospheric speciation of metals has, until recently, received rather little attention because of the real difficulties in measuring even the total concentration of most metals in aerosol samples. Fractionation of trace metals in the atmosphere is somewhat different from that applied to speciation in aqueous media. This comes mainly from two considerations: (i) the mechanism of interaction of the biosphere and the atmosphere, (ii) the mechanism of transport in the atmosphere. Several atmospheric parameters are controlled by aerosol particles, and human health as well as the life of aquatic and terrestrial ecosystems are also affected by the toxic metal content of particles. The total concentration of the elements in the atmospheric aerosol particles can indicate the sources of the pollutant, while chemical fractionation can be used to assess the different defined species present in the aerosol sample.

Fractionation is usually performed by a sequence of selective chemical extraction techniques, including the successive removal or dissolution of these phases and their associated metals. Fractionation has been defined as follows: the process of classification of an analyte or a group of analytes from a certain sample according to physical (e.g., size, solubility) or chemical (e.g., bonding, reactivity) properties [7]. Fractionation by size is carried out by the separation of samples into different particle size fractions usually during sampling. The concept of chemical leaching is based on the idea that a particular chemical solvent is either selective for a particular phase or selective in its dissolution process. Although a differentiated analysis is advantageous over investigations of bulk chemistry of solids, verification studies indicate that there are many problems associated with operational fractionation by partial dissolution techniques. Selectivity for a specific phase or binding form cannot be expected for most of these procedures. There is no general agreement on the solutions preferred for the various components in solids to be extracted, due mostly to the matrix effect involved in the heterogeneous chemical processes [8]. All factors have to be critically considered when an extractant for a specific investigation is chosen. Important factors are the aim of the study, the type of solid material (sediment, aerosol, fly ash, sewage sludge, harbor mud, street dust, etc.) and the elements of interest. Partial dissolution techniques should include reagents that are selective to only one of the various components significant in trace metal binding. Whatever extraction procedure is selected, the validity of selective extraction results primarily depends on the sample collection and preservation prior to analysis. There is a huge amount of literature on specific research areas, in which appropriate leaching protocols can be found for a given problem. Reviews exist on sediments and combustion wastes [9], trace metal speciation in soils [10], and aerosols [11] in the excellent book of Ure and Davidson [12]. The book gives many experimental details as a basis of evaluation of experimental conditions, so one can reliably find even validated leaching methods for any special problem.

In many cases it is impossible to determine the large numbers of individual species. In practice various classes of species of an element can be identified and the sum of its concentrations in each fraction can be determined [13]. Such fractionation is based on different properties of the chemical species, such as size, solubility, affinity, charge, and hydrophobicity. Fractionation may simply involve an actual physical separation, e.g., filtration. Although a direct determination of the speciation of an element is often not possible, the available methods can still be applied to get valuable information. An evaluation of the environmental impact of an element can sometimes be done without the determination of the exact species and, notwithstanding, fractions that are only operationally defined. It is also desirable to measure the total concentration of the element in order to verify the mass balance.

The sequential leaching of metals bound to the specific substrates have been worked out, e.g. for sediments, on the design and study of extraction schemes aiming to investigate the changes in the yearly life cycle of lakes or rivers [12]. Typical substrates are [7]:

- carbonates of calcium, magnesium and iron, which dissolve upon a decrease in pH;
- iron and manganese compounds present in the sediments and changing their adsorption capacities drastically according to the redox conditions (presence/absence of O_2), producing either FeS or FeO(OH) liberating coprecipitated or adsorbed metals at every change;
- organic matter present in sediments undergoing slow degradation, thus releasing the incorporated metals;
- silicates and other refractory minerals, which might contain a high metal concentration, but will not, under any environmental conditions, release them to the aquatic environment.

Measurement of such functionally and operationally defined metal fractions will allow some forecasting of metal release from sediments under certain conditions and represents, therefore, a valuable tool in natural water management. Harmonization of methods should be done because the wide variety of procedures has led to lack of comparability of results even for single aquatic systems. In this paper we present some typical fractionation patterns for a variety of environmentally relevant trace elements in sediments, soils, aerosols and fly ashes as a result of a selective literature survey and our own experiences. Among the solid environmental samples these four types of material were chosen due to their importance in environmental pollution and their effect on human health.

2 SUBSAMPLING, STORAGE, AND PREPARATION OF SEDIMENT SAMPLES

The sampling of environmental pollutants is discussed in detail in Chapter 2.1. In the environmental sampling the act of sample removal from its natural environment can disturb stable or metastable equilibria. Sampling uncertainty may contain systematic and random components arising from the sampling procedure. If the test portion is not representative of the original material, it will not be possible to relate the analytical result to the original material, no matter how good the analytical method is nor how carefully the analysis is performed. Sampling errors cannot be controlled by the use of standards or reference materials. The same applies to the subsampling. Because of the heterogeneity and complex nature of solids, care should be taken during sampling, subsampling, preparation, and analysis to minimize changes in speciation due to changes in the environmental conditions of the system.

For example, appropriate comparability among oxide sediment samples collected at different times and places from a given aquatic system and between different systems can be obtained most easily by analyzing the fine-grained fraction of sediment. Some investigations have pointed to a relation between specific surface, grain size fraction, and the speciation of trace elements in sediments. Suspended particulate matter sampling is mainly carried out by filtration. Such samples are of limited utility for studies of the speciation of elements in solids. In recent years, suspended sediment recovery by continuous-flow centrifugation has commonly been used to obtain sufficient sample for speciation, up to a few grams to carry out all the analysis: particle size distribution, identification of mineralogical phases, total element content, and fractionation by sequential extraction. Etcheber and Jouanneau [14] have provided a comparative study of suspended particle matter separation by filtration, continuous-flow centrifugation, and shallow water sediment traps. Although particles were separated by density rather than size, the continuous-flow centrifugation technique was preferred due to its speed and high recovery rate.

Sample preparation is one of the most important steps prior to analysis. The oxidized sediment layer controls the exchange of trace elements between sediment and overlying water in many aquatic environments. The underlying anoxic layer provides an efficient natural immobilization process for elements. Significant secondary release of particulate metal pollutants can be obtained from the accumulated metals as a result of processes such as [9]:

- desorption from different substrates due to formation of soluble organic and inorganic complexes;
- post-depositional redistribution of trace metals by oxidation and decomposition of organic materials due to microbiological activities;
- alteration of the solid–solution partitioning by early diagenetic effects such as changing the surface chemistry of oxyhydroxide minerals;
- dissolution of metal precipitates with reduced forms (metal sulfides) generally more insoluble than the oxidized form (surface complexes).

Transformations of metal forms during early diagenesis have also been successfully studied by sequential leaching. However, many of these studies did not consider that sample preservation techniques in trace element speciation studies of oxic sediments and sludges are different from those which should be used for anoxic samples [15]. Air and/or oven drying caused major changes in sediment equilibrium by converting fractions relevant to trace element binding into highly unstable and reactive forms [16]. Drying of sediments has also been reported to reduce the quantity of Fe extracted by techniques which remove amorphous iron oxides (hydroxylamine, CH_3COOH, pyrophosphate), suggesting an increase in the oxide crystallinity [17].

In practice, it is usually impossible to relate data obtained from dried sediments to those that existed originally in the field. Such data may even be of limited value in comparing the bioavailable concentrations of trace metals in samples collected within the same environment. Bartlett and James [16] found that manganese extractability changed as a function of storage time. Sieving and mixing in order to obtain a representative sample for bioavailability analysis may lead to precise but inaccurate results. Wet storage of oxidized sediments is inadequate because of microbially induced changes from oxidizing to reducing conditions in the stored sediments. Extractability of the metal with the most insoluble sulfide (Cu) has been reported to decline rapidly during wet storage [17]. Freezing is usually a suitable method to minimize microbial activity. Freezing has been found to enhance the water solubility of metals in the order of Mn (8–17 %) > Cu (7–15 %) > Zn (6–12 %) > Fe (3–7 %) [17]. Storage of anoxic sediments in a freezer was found to cause changes in the fractionation pattern of various metals studied. It has been found that a double wall sealing concept, i.e. an inner plastic vial with the frozen sediment contained under argon in an outer glass vial, is suitable. However, it seems to be impossible to totally avoid changes in the *in situ* chemical speciation of trace elements found in nature, so samples should be extracted immediately to avoid changes in speciation during storage [16].

3 SUBSAMPLING, STORAGE, AND PREPARATION OF SOIL SAMPLES

Seasonal [18, 19] and spatial variability [20, 21] are known to influence significantly the results of sequential extraction schemes in soils. Soil management practices (fertilizing, liming, sludge application) may cause significant seasonal changes in mobile fractions, but also natural seasonal variation of extractable metals in extensively used forest soils or undisturbed ecosystems may occur as well [22, 23]. Seasonal variation of extractable metals is an inherent process that is at least as significant as spatial variability [24]. Soil properties may vary considerably on a micro-scale of about 1–100 mm. Thus, metal solubility and extractability may be affected either directly by micro-inhomogeneity of the total metal contents or by simultaneous variation in soil properties (pH, cation exchange capacity (CEC), organic matter, mineral composition and soil texture) [23].

It has been concluded that the mobility of metals may frequently be underestimated when assessed by chemical extraction of disturbed, homogenized, and sieved soil samples of well-aggregated acidic soils, particularly when anthropogenically polluted.

Sample preparation generally involves the following steps: (1) drying or rewetting, (2) homogenizing and sieving, (3) storage and, occasionally, (4) grinding. Usually, soil samples are air dried prior to extraction. Although changes in the extractability of some elements have already been reported, this problem has only recently received more attention [25, 26]. Air drying prior to extraction is a standard procedure but leads to an increased extractability of Fe and Mn, whereas other metals are more or less unaffected [25, 26]. Several authors have identified possible mechanisms of these changes in metal extractability upon air drying. Drying of samples prior to the determination of mobile metal fractions usually results in unrealistically large amounts of extractable Mn, Fe, Cu, and Zn, and underestimation of Ca, Mg, K, and probably Co, Ni, and V. The changes in extractability upon air drying are related to soil properties, i.e. pH and organic matter content, and to the initial soil moisture conditions.

Although homogenizing and sieving are essential steps in performing representative and repeatable soil analysis, these procedures suffer from some drawbacks. The effects of structure disturbing soil sampling are obviously reinforced, thus creating new surfaces for reactions with metals in the solute phase, giving rise to adverse readsorption or desorption processes during metal extraction [27]. Homogenization of soil material from different horizons may result in erroneous changes in pH and carbonate content of the fine earth. As a conclusion, sample storage seems to be generally less critical to the analysis of extractable metal fractions than does air drying, but it is likely to enhance the effects of air drying in the case of redox-sensitive elements. Occasionally, soil samples are ground prior to extraction. This procedure causes physical breakdown of soil microaggregates, thus potentially altering the extractability of metals from soil samples [28]. The exposure of fresh surfaces may, depending on soil properties, increase the extractability of some metals but potentially may also cause readsorption of metals during the batch process.

4 SUBSAMPLING, STORAGE, AND PREPARATION OF AEROSOL SAMPLES

Metals are transported in the atmosphere primarily on aerosols which can be removed by wet and dry deposition process. The deposition, transport, and inhalation processes are controlled predominantly by the size of the atmospheric aerosols. Thus, the primary type of fractionation is the *aerosol size distribution*. Once deposited, however, chemical speciation in terms of both the dissolved/particulate distribution of the metals in precipitation and the inorganic or organic complexes which the metal may form, plays an important role in controlling the environmental impact of atmospherically deposited metals.

The atmosphere is an important vector of global metal transport between regions, and globally, from land to sea, and from sea to land. Direct atmospheric deposition makes only a minor contribution to the total metal contents of the lithosphere because of the large reservoir of these metals in soils and rocks. The impact of the atmospheric inputs on lake biogeochemistry however, is strongly dependent on both the physical and chemical forms in which metals enter the natural water system. The transport processes are important both spatially and temporally and occur via several atmospheric activities.

Aerosol particles are partly emitted into the atmosphere from sources on the surface (primary particles), while others come into being in the air by gas–particle conversion (secondary particles). These particle generally have sizes (diameters) $<1\,\mu m$ and are called *fine particles*. In contrast, surface dispersion creates particles of diameter $>1\,\mu m$, termed *coarse particles*. Particles due to combustion (fly ashes) can be found in both size intervals. The particles of different origin are of different chemical composition and also of different chemical forms and physical state.

Particle size exercises a strong control on residence time in the atmosphere and on particle dispersion. Large particles are rapidly removed near the source by gravitational sedimentation (typical resident time in the atmosphere of hours to a few days), while small particles have a considerably longer residence time in the atmosphere and are much more efficiently transported. Close to the source the composition of the aerosol will be strongly related to the parent material.

The chemical composition of the particulate emissions is less well known than the chemical composition of the particles in ambient air. Inorganic minerals are mainly emitted from processes, particularly from cement, iron and steel industry, and as fly ash from coal combustion. Organic compounds and soot are mostly released from small combustion sources, mobile sources, and from processes associated with petroleum extraction and refining. The ratio of organic to elementary carbon in these emissions is rather variable, from <1 in the case of emissions from diesel engines, up to 5–10 for low-calorific fuels such as lignite, peat and firewood. The anthropogenic emissions can be divided in three main areas such as (i) incomplete combustion, formation of soot and associated organic compounds (COC) (including small-scale residential combustion, both solid and liquid fuels, and internal combustion engines), (ii) fly ash and particles from the fuel's content of inorganic mineral matter, (iii) industrial processes.

Particles can be generated from *natural* and *anthropogenic* sources. In general, it is estimated that the annual total amount of particles from these sources is about 3000 million tonnes and 400 million tonnes, respectively [29]. Particles from natural sources are overwhelmingly coarse particles, from wind erosion, sea-spray formation and similar processes. Anthropogenic emissions, in contrast, contribute about 60 % to the total fine particle mass in the atmosphere. Behind these estimates lie large uncertainties in terms of source assessment, speciation, and characterization of the atmospheric particles, not to mention the different lifetimes of particles. A rough estimate of their contribution to particulate aerosol mass on a global scale is given in Table 3.3.1.

These emission and formation figures are rather uncertain. It should also be taken into account that the residence times of particles from the respective sources are very variable. The difficulty in assessing the emissions of sea-salt particles from the sea surface, and of soil and desert dust from wind erosion, is partly due to the rapid sedimentation and deposition of these particles. Coarse particles generally have short residence times, typically of the

Table 3.3.1. Estimated contributions to the global atmospheric particulate mass (Tg = terragram) [29].

Source	Annual emission or production ($Tg\,year^{-1}$)	
	Range	Best estimate
Natural		
Wind erosion	1000–3000	1500
Sea salt	1000–10 000	1300
Volcanoes	4–10 000	30
Biological primary particles	26–80	50
Forest fires	3–150	20
Inorganic secondary particles[a]	100–260	180
Organic secondary particles[b]	40–200	60
Anthropogenic		
Direct emissions	50–160	120
Inorganic secondary particles[a]	260–460	330
Organic secondary particles[c]	5–25	10

[a]Oxidation of sulfur dioxide, reduced sulfur compounds and nitrogen dioxide, uptake of ammonia.
[b]Mainly photochemical formation of particulate matter from isoprene and monoterpenes.
[c]Photochemical formation of particulate matter from anthropogenic emissions of VOCs.

order of few days. Fine particles, such as the inorganic and organic secondary particles, have atmospheric residence times of 1–2 weeks.

The frequently used method of relating an element in atmospheric aerosols to its source is to calculate enrichment factors (EF) by employing an indicator element. For crustal aerosols Al is normally used as the indicator element. The enrichment factor can be calculated using the formula as

$$EF_{\text{crust}} = \frac{(C_{(i)}/C_{(\text{Al})})_{\text{aerosol}}}{(C_{(i)}/C_{(\text{Al})})_{\text{soil}}}$$

Those elements that have EF values between 1 and 10 are usually crustal in origin, while elements with EF values in the range of 10–5000 are generally emitted by anthropogenic sources [30].

Aerosol sampling is usually carried out either by collection devices (impactors) or by fibrous or membrane filters. The operation principle of impactors is based on the fact that particles have much larger inertia than gas molecules, which makes their separation in a fluid in motion possible. Since aerosol particles may have different forms and density, impactor data are generally given for *aerodynamic particle diameter*. Usually cascade impactors are used which collect the particles in different size ranges. An important characteristic of a given impactor stage is the particle diameter (cut-off diameter) where the collection efficiency is equal to 50 %. The main characteristics of a Berner impactor [31], widely used in recent aerosol studies in Europe and in the USA, are summarized in Table 3.3.2. The flow rate is 1.9 $m^3\,h^{-1}$ at 20 °C with an exhaust pressure of 150 hPa. The impactor consists of eight stages, while a prestage (No. 9) excludes the sampling of particles with a diameter of 16 μm.

Table 3.3.2. Main characteristics of the Berner-type cascade impactor [31].

Stage no.	Cut-off diameter (μm)	Slit diameter (mm)	Number of slits
9	16	15.9	1
8	8.0	5.0	8
7	4.0	2.7	13
6	2.0	1.2	36
5	1.0	0.70	53
4	0.50	0.60	30
3	0.25	0.42	31
2	0.125	0.30	63
1	0.0625	0.25	128

Aerosol sampling by filtration is based on the passage of air by a pump through a filter substrate placed in a suitable filter holder. *Fibrous filters* consist of mats of fibers made generally of glass, quartz, or cellulose. Membrane filters contain small pores of controlled size, and they are usually composed of thin films of polymeric materials [32]. By using any kind of filters, in the absence of electric forces, larger and smaller particles are captured from the air, pumped through the filter material, by impaction and diffusion, respectively. The characteristics of different types of fibrous and membrane filters widely used for sampling aerosol particles for subsequent chemical analysis are summarized in Table 3.3.3.

The surface reactivity is important for chemical analysis, since some filters react with atmospheric trace gases resulting in sampling artifacts. The filters require pretreatment before sampling which is usually done by acid wash to remove alkaline sites. Teflon, quartz, and Nuclepore filters have

Table 3.3.3. Properties of filters used for particulate sampling with a face velocity of 10 cm s^{-1} [32]. Note that the efficiencies refer to particles with diameters above 0.03 μm.

Filter	Composition	Density ($mg\,cm^{-2}$)	Surface reactivity	Efficiency (%)
Teflon	Polytetrafluoroethene	0.5	Neutral	99
Whatman 41	Cellulose fiber	8.7	Neutral	58
Whatman GF/C	Glass fiber	5.2	Basic (pH 9)	99
Gelman Quartz	Quartz fiber	6.5	pH 7	98
Nuclepore	Polycarbonate	0.8	Neutral	93
Millipore	Cellulose acetate/nitrate	5.0	Neutral	99

been found to give the best substrates for chemical analysis. Teflon and Nuclepore filters are used for inorganic substances, while quartz filters are commonly applied for sampling of organic species. However, in the latter case it should be taken into account that quartz filters can adsorb organic vapors. This can be checked by using a quartz backup filter in the air stream to correct the concentration in the aerosol phase determined on the first filter. It is suggested to use a second filter after a Teflon filter which does not adsorb organic vapors [33]. Consequently, before sampling, it is necessary to desorb volatile organics from the quartz filters at high temperature. Aerosol sampling is usually carried out by filters with a diameter of several centimeters. So the sampling rate is typically some $m^3 day^{-1}$ which gives the amount of aerosols sufficient for the majority of chemical analysis. However, in some measurement programs a large mass of aerosols is needed and high volume of aerosols are sampled. The flow rate in high volume sampling is about $60 m^3 h^{-1}$ with filters of diameter around 25 cm or larger. Glass fiber filters can fulfill this requirement.

5 SUBSAMPLING, STORAGE, AND PREPARATION OF FLY ASH SAMPLES

Fly ash is the fine particulate waste material that remains after incomplete combustion. It is produced in massive quantities, mainly by fossil fuel-based power plants and waste incinerators. Fly ash contains high levels of potentially toxic chemicals, but these are often strongly bound to the particulate matrix and only a fraction poses an immediate threat. Various tests can be used to determine the amounts that will leach from the fly ash matrix (for instance with rainwater) and thus become available for transport to soils, rivers, or groundwater and eventual uptake by organisms. For environmental risk assessment the potentially hazardous fraction, i.e. the fraction that can be leached from the matrix under typical environmental conditions, needs to be determined [34]. Submicrometer fly ash particles may be emitted from power plant smoke stacks in spite of electrostatic filters, and thus enter the environment. Another reason for studying leaching properties from fly ash is to evaluate its potential for re-use in construction materials [35].

Depending on the aim of the measurement, leaching tests can be subdivided into relatively quick compliance tests to verify whether a material meets certain legal criteria (e.g., the European leaching test) [36] and more extensive material characterizations (e.g., pH-stat tests at a range of pH values) [37] or sequential leaching tests [38]. The outcome of such measurements often depends rather critically on the experimental conditions and over the years several measurement protocols have been developed by standardization agencies in various countries. Unfortunately, the results of leaching tests carried out in different countries cannot be compared as long as the test methods have not been harmonized at the international level. For the interpretation of leaching test results it is necessary to understand the leaching processes in terms of the physico-chemical properties of the element species and the matrix. The results can be used to link metal leaching behavior to different types of binding. Particle size of fly ashes is an important parameter since it controls the specific surface area of the combustion material which, evidently, is directly proportional to the amount of toxic metals adsorbed. Hence separation by sieving should be done before leaching procedures. Due to the heterogeneous nature of different fly ash samples, a semi-homogeneous portion can be obtained by separating and using the close particle size fractions. In some cases an agate ball is placed into the bottle and, before removing a test portion from the sample, a strong shaking of the bottle for some minutes is advised. Storage of fly ash sample is usually carried out in a closed bottle.

6 SEQUENTIAL EXTRACTION TECHNIQUES

Sequential extractions have been applied using a series of extractants with increasing extraction capacity, and several schemes have been developed to determine species in solid samples. Although initially thought to distinguish some well-defined chemical forms of trace metals [39], they instead address operationally defined fractions [40]. The selectivity of many extractants is weak or not sufficiently understood, and it is questionable whether specific trace metal compounds actually exist and can be selectively removed from multicomponent systems. Due to varying extraction conditions, similar procedures may extract a significantly different amount of metals. Concentration, operational pH, liquid/solid (L/S) ratio, and duration of the extraction process affect considerably the selectivity of extractants. The conventional approach of equilibration during a single extraction step is the shaking or stirring of the solid phase–extractant mixture. There is no general agreement on the solutions preferred for the extraction of various components in solid samples, due mostly to the matrix effects involved in the heterogeneous chemical processes [8]. The aim of the study, the type of the solid materials, and the elements of interest determine the most appropriate leaching solutions. Partial dissolution techniques should include reagents that are sensitive to only one of the various components significant in trace metal binding.

In sequential multiple extraction techniques, chemical extractants of various types are applied successively to the sample, each follow-up treatment being more drastic in chemical action or different in nature from the previous one. Selectivity for a specific phase or binding form cannot be expected for most of these procedures. In practice, some major factors may influence the success in selective leaching of components, such as:

- the chemical properties of the leaching solutions chosen;
- experimental parameters;
- the sequence of the individual steps;
- specific matrix effects such as cross-contamination and readsorption;
- heterogeneity, as well as physical associations (e.g. coatings) of the various solid fractions.

All of these factors have to be critically considered when an extractant for specific investigation is chosen. Fractions of sequential extraction schemes can be:

(i) *Mobile, exchangeable elements:* this fraction includes the water-soluble and easily exchangeable (unspecifically adsorbed) metals and easily soluble metallo-organic complexes. Most of the recommended protocols seek to first displace the exchangeable portion of metals as a separate entity. Chemicals used for this fraction fall commonly in one of the following groups [27]:

- water or highly diluted salt solutions (ionic strength <0.01 M, e.g. $MgCl_2$);
- neutral salt solutions without pH buffer capacity (e.g. $CaCl_2$, $NaNO_3$);
- salt solutions with pronounced pH buffer capacity (e.g. NH_4OAc).

(ii) *Elements bound to carbonates:* to dissolve trace elements bound on carbonates buffer solutions (e.g. HOAc/NaOAc; pH $= 5$) are commonly used. Zeien and Brümmer [40] have proposed to dissolve carbonates by adding equivalent amounts of diluted HCl to $1\,mol\,L^{-1}$ NH_4OAc/HOAc buffer, addressing specifically adsorbed and surface occluded trace element fractions of soils with $>5\,\%$ m/m carbonates.

(iii) *Elements bound to easily reducible fractions (bound to Fe/Mn oxides):* $NH_2OH \cdot HCl$ at pH 2 is generally used but procedures differ in minor operational details such as S/L ratios, treatment time, interstep washing procedure.

(iv) *Elements bound to easily extractable organics:* NaOCl or $Na_4P_2O_7$ are used mostly.

(v) *Elements bound to moderately reducible oxides:* $NH_2OH \cdot HCl$/HOAc or NH_4O_x/HO_x mixtures are mainly applied.

(vi) *Elements bound to oxide and sulfide fractions:* H_2O_2/NH_4OAc is used most frequently.

(vii) *Elements bound to silicates (residual fraction):* this fraction mainly contains crystalline bound trace metals and is most commonly dissolved with concentrated acids and special digestion procedures, i.e. strong acid mixtures are applied ($HF/HClO_4/HNO_3$) to leach all remaining metals.

As can be seen, a wide range of extraction procedures is readily available for different metals and variations of the extraction conditions are utilized due to varying solid sample composition. The following parameters have to be considered when designing an adequate extraction procedure:

Extractants: chemical and physical interferences both in extraction and analysis steps, respectively.

Extraction steps: selectivity, readsorption and redistribution processes. As extractants the following chemicals are commonly used:

- for cation changing conditions (e.g. NH_4Cl);
- for anion changing conditions (e.g. $NaNO_3$);
- acids (acidity increases with each extraction step from unbuffered salt solutions (NH_4NO_3) to strong acids ($HNO_3/HClO_4$)).

If the single extractants for the different steps are chosen with respect to their ion exchange capacity or reduction/oxidation capacity, each step has to be designed individually following special considerations [23].

Concentration of the chemicals: the efficiency of an extractant to dissolve or desorb trace metals from solid samples will usually be increased with increasing concentration or ionic strength. Thermodynamic laws predict the efficiency of an extractant to dissolve or desorb trace metals from solid samples.

Extraction pH: extractants with a large buffering capacity or extractants without buffer capacity can be used.

Liquid/solid (L/S) ratio and extraction capacity: the relative amount of extractant added to the solid samples has various influences on the results. If, over a sufficiently wide L/S ratio, the capacity of the extractant to dissolve a metal fraction exceeds its total amount present in the solid sample, then the metal concentration in the extract ($mg\,L^{-1}$ extract) will decrease with an increase in L/S ratio. However, the total amount ($mg\,kg^{-1}$) extracted will be constant with increasing L/S ratio. Nevertheless, as sediments, aerosols, fly ashes, and soils are multiphase/multicomponent systems, dissolution of other compounds due to the nonselectivity of the extractant may confuse this behavior.

Extraction time and batch processes: the effect of extraction time is related to the kinetics of the reactions between solid sample and leachant. Extractions may be predominantly based on either desorption or dissolution reactions. For desorption of metal cations from heterogeneous soil systems Sparks [41] has identified four rate-determining steps, namely (i) diffusion of the cations in the (free) bulk solution, (ii) film diffusion, (iii) particle diffusion, and (iv) the desorption reaction. Accordingly, the rates of most ion exchange reactions are film and/or particle diffusion controlled. Vigorous mixing, stirring, or shaking significantly influences these processes. Film diffusion usually predominates with small particles, while particle diffusion is usually rate limiting for large particles. For mineral dissolution, essentially three rate-controlling steps have been identified, namely (i) transport of solute away from the dissolved minerals (transport controlled kinetics), (ii) surface reaction-controlled kinetics where ions are detached from the surface of minerals, and (iii) a combination of both. Mechanical actions, e.g. stirring or shaking, increase the rate of transport-controlled reactions, while they do not affect surface-controlled reactions. Shaking and other batch processes may enhance the dissolution of readily soluble salts effectively, but are unlikely to affect the dissolution rate of less soluble minerals.

Extraction temperature: within the normal range of extraction temperatures (20–25 °C or room temperature), the effect of temperature on extractability of the elements is usually small, but it has to be considered for interpretation of small differences [42].

Finally, the whole procedure has to be optimized with regard to selectivity, simplicity, and reproducibility.

7 STANDARDIZED SEQUENTIAL EXTRACTION PROCEDURE PROPOSED BY BCR

Sequential extraction schemes have been developed during the past 30 years for the determination of chemical binding forms of trace metals in sediment and soil samples. The lack of uniformity of these schemes, however, did not allow the results so far to be compared worldwide or the procedures to be validated. Indeed, the results obtained by sequential extraction are *operationally defined*, i.e. the 'forms' of metals are defined by the procedure used for their determination. Therefore, the significance of the analytical results is related to the extraction scheme used. Another problem was the lack of suitable reference materials. Thus, standardization of leaching and extraction schemes was required with the preparation of sediment and soil reference materials which were certified for their contents of extractable trace element with single and sequential extraction procedures [43]. The Community Bureau of Reference (BCR, now Standards, Measurements, and Testing Programme) has recently launched a programme to harmonize sequential extraction schemes for the determination of extractable trace metals in sediments [44]. BCR has proposed a standardized three-stage extraction procedure (BCR EUR 14763 EN), which was originally developed for the analysis of heavy metals in sediments [45]. This procedure is currently used and evaluated also as an extraction method for soils [46]. So far, the BCR procedure has been successfully applied to a variety of sludge [47], sediment [48], and soil [46] samples. The BCR scheme was recently used to certify the extractable trace element contents of a certified reference material (CRM 601, IRMM). Although this procedure offers a tool for obtaining comparable data, poor reproducibility and problems with lack of selectivity have still been reported [46, 49]. Various research groups have used this technique and found partially discrepancies when applying the scheme. The same extraction scheme has also been used for the determination of extractable elements in soils [50].

8 SEQUENTIAL EXTRACTION SCHEMES APPLIED TO SEDIMENT SAMPLES

Main mineralogical components of sediments, which are important in controlling their metal concentrations, are hydrous oxides of iron and manganese, organic matter, and clay. The degree of interaction between sediment samples and extractant solutions can be altered by changes in experimental parameters such as reagent concentration, final suspension pH, L/S ratios, temperature, contact time, and intensity of mixing. Recently, researchers have tended to use similar extraction protocols, mostly by adapting or modifying the scheme of Tessier *et al.* [51]. Salomons and Förstner [52] have used sequential extraction techniques to determine the chemical associations of heavy metals with specific sedimentary phases. They distinguished five major mechanisms for metal accumulation on sedimentary particles: (1) adsorptive bonding to fine-grained substances, (2) precipitation of discrete metal compounds, (3) coprecipitation of metals with hydrous iron oxides and manganese oxides and carbonates, (4) association with organic compounds, and (5) incorporation in crystalline material. It was pointed out that the standard extraction method should be relatively simple, in order to make routine analysis of large numbers of sediments possible.

Tessier *et al.* [53] have collected sediment samples from streambeds in an undisturbed watershed in eastern Quebec (Gaspé Peninsula). The sediment samples were separated into eight distinct particle size classes in the size range from 850 μm to <1 μm by wet sieving, gravity sedimentation, or centrifugation. Each sediment subsample was then subjected to a sequential extraction procedure designed to partition the particulate heavy metals into five fractions. It was one of the first fractionation process by particle size and chemical bonding for investigation of sediments. A simultaneous sediment extraction procedure for low carbonate sediments, which partitions sediment-bound trace metals (Fe, Mn, Zn, Cu, and Cd) into easily reducible (associated with manganese oxides), reducible (associated iron oxides) and alkaline extracted (bound to organic) metal was investigated. This method was compared to the sequential extraction procedure based on the work of Tessier *et al.* [51]. Both methods showed good agreement for the partitioning of Zn and Cd among the easily reducible, reducible, and organic components of the sediment. The two methods also showed the same general distribution of Mn, Fe, and Cu among the three sediment components, although concentrations of metals recovered by the two methods differed; less Mn and Fe, and more Cu were recovered from sediments by sequential extraction over the simultaneous procedure. The lower recovery of Mn was, in part, attributed to the loss of this metal in the 'in between' reagent rinses required in the sequential extraction procedure. Greater recovery of Cu by the sequential extraction method might be due to the pretreatment of the sediment with strong reducing agents prior to the step used for liberating organically bound metals. Advantages of simultaneous extraction over sequential technique included rapid sample processing time (i.e. the treatment of 40 samples per day versus 40 samples in 3 days), and minimal sample manipulation.

Sediments are the ultimate sinks for pollutants. Before these sediments become part of the sedimentary record (deeply buried), they are able to influence the composition of surface waters. The sediments can be divided in two sections: an oxic surface layer and an anoxic sediment. In anoxic systems when sulphide is present, Zn, Cd, and Cu are likely to be present as sulfides. Remobilization of the deposited sediments is possible when the overlying surface water changes (pH and complexing agent). In addition, changes in the surface water composition may enhance or prevent the removal of dissolved trace metals by particulates and subsequent removal by sedimentation. Remobilization also occurs when sediments are brought from an anoxic to an oxic environment as takes place during dredging and disposal on land. Salomons *et al.* [54] have reviewed the processes affecting trace metals in deposited sediments. The sediment–water system could be divided in three parts: the oxic layer, the anoxic layer and the oxic–anoxic interface. Available data showed that trace metals such as Cu, Zn, and Cd occurred as sulfides in marine and estuarine anoxic sediments. Calculations showed that organic complexation was unlikely and the dominant species were sulfide and bisulfide complexes. Cr and As were probably present as adsorbed species on the sediments. Changes from an anoxic to an oxic environment, as occur during dredging and land disposal of contaminated sediments, might cause a mobilization of some trace metals. The chemical forms of many elements in the sediments of St. Gilla Lagoon (Sardinia, Italy) have been evaluated [55]. Five fractions were separated from sediments by sequential chemical extraction. The metals in each fraction were determined by the total reflection X-ray fluorescence (TXRF) technique. Both principal and trace element distributions in the sequential phases were discussed in terms of pollution sources, metal transport, and deposition/redeposition in air-dried sediments. The use of a sequential extraction procedure could be an effective method for comparative studies between natural and contaminated areas, as well as between areas subjected to different chemical stresses. The results showed that in the examined area the lithogeneous fraction was the most relevant for total metal content. However, under oxidizing conditions among the 'mobile' fractions, the reducible fraction proved to be the most important sink for Zn and Pb, the oxidizable fraction was only relevant for Cu at almost natural level.

Availability of heavy metals depends greatly on the properties of particle's surface, on the kind of strength of the bond and external conditions such as pH, Eh, salinity and concentration of organic and inorganic complexation agents. Most particle surfaces have an electrical charge, in many cases a negative one. In solutions, an equivalent number of ions of opposite charge will gather around the particle, thus creating an electric double layer. The surface charge is strongly affected by pH and the composition of surface. Especially hydrous oxides of Fe, Al, Si, Mn and organic surfaces (e.g. functional amino and carboxyl groups) participate in the H^+ transfer. Lattice defects of clay minerals and the adsorption of ions also contribute to surface charges. The sorption process can be physical or chemical adsorption as well as sorption by ion exchange. Physical adsorption on the external surface of a particle is based on the van der Waals forces or relatively weak ion–dipole or dipole–dipole interactions (about $1\,\text{kcal}\,M^{-1}$).

Two sequential extraction schemes (a modified Tessier procedure [51] with five steps and the three-step protocol developed by BCR [45]) were applied to four sediment samples with different heavy metal contents [56]. The results obtained for partitioning of Cd, Cr, Cu, Ni, Pb, and Zn showed that the metal distributions with the two procedures were significantly different. With the three-step protocol significant amounts of all the heavy metals were extracted with the oxidizing reagent, whereas with the modified Tessier procedure the nonresidual metals were distributed between the second, third, and fourth fractions (CH_3COOH-acetate buffer ($pH = 5$), reducing, and oxidizing reagents, respectively). The residual fraction obtained applying the three-step procedure was in general higher than that obtained using the five-step procedure, except for Cd. The three-step sequential extraction protocol [45] has been evaluated with regard to total recovery, reproducibility, selectivity of extractants, and extent of phase exchanges or redistribution of metals during the extraction [57]. Model sediments of known composition were prepared consisting of humic acid and natural minerals such as kaolin, quartzite, and ochre. It was shown that the chemistry of a metal could be a more important parameter than its actual phase location in the sediment in determining its response to the extractants.

The reproducibility of Tessier's extractions [51] and the total content of Cd, Cr, Cu, Fe, Mn, Pb, Zn, and Ca in river sediment have been evaluated [58]. The accuracy of the dissolution procedures was estimated using a reference material, BCR 145. None of the methods applied proved optimal for all the metals determined. The concentrations of metals extracted by the various reagents were characterized by good reproducibility on species bonded to the carbonates, to iron and manganese oxides, and in the residual fraction; precision was lower in the other cases. The sequential procedure also showed a satisfactory mass balance.

Sequential extraction procedures offer the advantage of simulating, to a certain extent, the various natural environmental conditions. Recently, investigations on bottom sediments of Lake Balaton, Hungary, rivers in its catchment area, and harbors were extensively carried out in spring, summer and fall [59]. A modified sequential extraction procedure of BCR [45] was applied as a four-step sequential leaching for determination of the distribution of seven elements [59, 60]. The fractions were (1) exchangeable metals and metals bound to carbonate, (2) metals bound to Fe/Mn oxides, (3) metals bound to organic matter and sulfide, and (4) acid-soluble metals. The critically examined three-step sequential extraction method [45] should be used to compare the data produced by different laboratories, but the strong acid-soluble fraction of elements could add more valuable information. The four-step method allowed a deeper understanding of the association of elements with the compounds of sediments. The sequential leaching protocol was sufficiently repeatable and reproducible for application in fractionation studies. The amounts of elements removed correlated well with those determined by pseudo-total, acid digestion of sediment. It has been found that elements concentrated mainly in the acid-soluble fraction indicating no serious pollution in the lake sediment. Results were compared to the sediment quality values (SQVs) and sediment background values (SBVs) [61]. SQVs were summarized for 22 metals and metalloids from different area in the USA, Canada, The Netherlands, Norway, Australia, New Zealand, and China. Globally, SQVs for metals and metalloids vary over several orders of magnitude. Regional SQVs can be useful as an initial step in a sediment hazard/risk assessment. The SQVs are numerical, based on total dry weight concentrations of sediments collected from more than 50 different sampling points all over the world (North America, Asia, Europe, and Australia). The SBVs for freshwaters include background values summarized for lakes and streams and the global shale average values. Data clearly showed that the average concentrations of elements have been found to be less than the SQVs and other background data for soils. This means that the sediment is not polluted and, after removing from the bottom, its disposal on the soil is feasible.

9 SEQUENTIAL EXTRACTION SCHEMES APPLIED TO SOIL SAMPLES

To assess metal mobility of trace elements in soils on different time scales and sampling sites, a wide range of extraction schemes have been employed [26, 27]. These methods vary with respect to the extraction conditions: chemical nature and concentration of leaching solutions, solution/soil ratio, operational pH, and extraction time [27]. If more than one leaching solution is used, differences occur due to variation in the extraction sequence. The most critical steps are sampling, sample preparation and the selectivity and accuracy of the leaching procedure [26]. As for total metal concentrations, spatial heterogeneity, as well as seasonal variation of extractable metal fractions may bias the results. The use of correlation coefficients for choosing extractants for assessment of plant availability of elements needs consideration. For extraction of the exchangeable fraction, almost all possible combinations of major cations with either Cl^-, NO_3^- or acetate have been used, with concentrations ranging between 0.05 and $1\,mol\,L^{-1}$, and pH in the neutral range. The solution/soil ratios were varied from 4 : 1 to 100 : 1, the extraction times between 30 min and 24 h. In ideal systems, the relative exchangeability of trace metals is determined by the affinity of the exchanging cation for the solid phase. This affinity increased with increasing valency and decreasing radius of the hydrated cation. Although heterogeneous soil systems may deviate from this ideal behavior, the selectivity of soils for cations was frequently observed to increase according to Na < K < Mg < Ca. Consequently, under comparable conditions, e.g. concentration, extraction time, soil/solution ratio, the efficiency of cations to exchange trace metals increased according to Li < Na < K < Mg < Ca < Ba (< La). Therefore, salts of Ca and Ba were regarded as most effective and selective agents in extracting exchangeable trace metals. Unfortunately, both cations may cause serious background problems (interferences) during determination of Pb and other trace metals. Usually, this can be resolved only by dilution of the extracts by >1 : 10 prior to measurement decreasing the detection limit by the same ratio. For that reason, the use of easily volatilizable salts, i.e. $MgNO_3$ or NH_4NO_3 has been proposed [40, 51]. Compared to $0.1\,mol\,L^{-1}$ solutions of $CaCl_2$ or $BaCl_2$, $1\,mol\,L^{-1}$ NH_4NO_3 in 2.5 : 1 ratio was found to extract about equal amounts of Al, Fe, Mn, Ni, Pb, V, and Zn, and Cd was less efficiently extracted by $1\,mol\,L^{-1}$ NH_4NO_3. This can be explained by a more effective extraction with $CaCl_2$ through the formation of chlorocomplexes with Cd and Cu. Other anions frequently used are either acetate or nitrate. At equal concentrations, the complexing ability increases in the order nitrate < chloride < acetate. The selectivity for extraction of the unspecifically sorbed (exchangeable) fraction should decrease in the same order.

The specifically sorbed fraction is explicitly addressed by only a few methods. In addition to other differences, the wide range of cations used suggests that most methods do not address any 'specifically sorbed' fraction, and, probably even do not extract the same operationally defined solubility group. To extract specifically sorbed trace metals, $Pb(NO_3)_2$ seems to be most adequate, due to its low pK (7.7) and large atomic radius, and it is effective in displacing other trace metals, i.e. Cd (p$K = 10.1$), Ni (p$K = 9.9$), Co (p$K = 9.7$), Zn (p$K = 9.0$) and Cu (p$K = 7.7$), with smaller atomic radius than Pb. $Pb(NO_3)_2$ was found to extract less metal than acetic acid, probably indicating that the latter was more specific [63]. Unfortunately, those trace metals that are constituents of the extractants cannot be determined. Therefore, Zeien *et al.* [40]. proposed $1\,mol\,L^{-1}$ NH_4Ac and $1\,mol\,L^{-1}$ NH_4NO_3 in sequence to extract an operationally defined fraction under optimized analytical conditions by using only one cation (NH_4^+) and decreasing the pH throughout the extraction sequence.

Among the extractants most frequently used to dissolve trace metals bound to carbonates are acids such as HCl and acetic acid (pH = 3–3.5), buffer solutions of HAc/NaAc (pH = 5) and the buffering complexing agent Na_2EDTA at pH = 4.6. Any of these extractants seems to have some potential to extract carbonates from soils, but is probably neither effective for quantitative dissolution (i.e. CH_3COOH buffers), nor selective (cold diluted acids, i.e. HCl), or both, i.e. Na_2EDTA [63].

For extraction of organically bound trace metals, various approaches have been used, e.g. their release by oxidation or dissolution of the organic matter, or through addition of competing, e.g. complexing or chelating, ligands. Among the oxidizing extractants, H_2O_2, either purely or combined with HNO_3 or NH_4Ac, extracted more trace metals from soils than NaOCl [63], i.e. from Fe/Mn oxides. $K_4P_2O_7$ and $Na_4P_2O_7$ have been reported to dissolve organic matter by dispersion and to efficiently complex the released metals [64]. Accordingly, the variation in extraction parameters with concentrations between 0.1 and $1\,mol\,L^{-1}$, solution/soil ratios between 10 : 1 and 100 : 1, and extraction times from 1 h to 24 h indicate that results obtained by different procedures are hardly comparable and are likely to extract non-organically bound trace metals to a varying extent. $K_4P_2O_7$ was found to extract more metals when used before Mn oxide extraction with $NH_2OH \cdot HCl$, while the latter extractant has little effect on the organically bound fraction.

As an alternative to pyrophosphate salts, some procedures employ NaOH [39] or NaOH/EDTA mixture to extract organically bound trace metals by dissolution of organic matter. The selectivity of these methods is considered low, and the extracted metals may precipitate as hydroxides [65]. As an alternative to destruction of the organic ligands, organically bound trace metals may be extracted by competing synthetic chelates, e.g. EDTA or DTPA [46]. In sequential extractions, EDTA or its ammonium salt [51] has been less frequently used than the advantages would suggest. As NH_4EDTA, adjusted with NH_4OH to pH 4.6, has been reported to dissolve considerable amounts of amorphous sesquioxides, it may be less selective than some pyrophosphate methods. Nevertheless, it should be considered as an alternative to extractants with alkaline pH, e.g. $Na_4P_2O_7$, $K_4P_2O_7$, NaOH, or NaOCl. Thus, NH_4ETDA (pH 4.6) can be fitted in a sequence of extractants with decreasing pH that is thought to increase the selectivity by minimizing adverse effects on each subsequent extraction step, i.e. readsorption or precipitation of trace metal compounds [51]. Moreover, the procedure is non-destructive to organic matter and organo-mineral associations, thus creating no new surfaces that may cause adsorption of trace metals during subsequent extraction steps as discussed by Beckett [27]. The dissolution of amorphous sesquioxides is probably limited by choosing a reasonable extraction sequence, extracting organically bound trace metals after removal of the most labile oxide fraction, e.g. the Mn oxides [62], and by a comparatively short extraction time of 90 min, as proposed by Zeien and Brümmer [40]. Accordingly, good correlations were found between organic carbon and NH_4ETDA-extractable metal fractions, although there was an evidence that EDTA extractants could dissolve trace metals from amorphous sesquioxides. Among sesquioxides, the Mn oxides are most susceptible to changes in pE and pH. Therefore, trace metals bound to Mn oxides, i.e. Pb, may be readily mobilized upon changed environmental conditions, e.g. flooding. For that reason, this environmentally significant fraction is separated prior to Fe and Al oxides by most sequential extraction procedures. Essentially, Mn oxides are extracted by reducing agents, e.g. $NH_2OH \cdot HCl$ or hydroquinone, either pure or mixed with NH_4Ac, HAc, or diluted HNO_3.

With higher concentrations (i.e. $0.25\,mol\,L^{-1}$) and higher temperatures during extraction (i.e. 50 °C–100 °C), $NH_2OH \cdot HCl$ extracted considerable amounts of trace metals from sesquioxides with a wide range of crystallinities [27]. As intended by Tessier [51]. $0.04\,mol\,L^{-1}$ $NH_2OH \cdot HCl$ in 25 % HAc (pH = 2 at 85 °C for 5 h) actually should extract most of the sesquioxides, including the crystalline fractions. Zeien and Brümmer [40] proposed $0.1\,mol\,L^{-1}$ $NH_2OH \cdot HCl$ in $1\,mol\,L^{-1}$ NH_4Ac (pH 6, 30 min at 20 °C). Nevertheless, this procedure seemed to be comparably selective, hence it dissolved on an average 37 % (0.12 %–73.9 %) of total Mn, but only 0.02 %–2.9 % of total Fe from a variety of soils. A negative correlation between

Fe and Mn extracted by 0.1 mol L $NH_2OH \cdot HCl$/1 mol L^{-1} NH_4Ac (pH = 6) indicated that only for low levels of Mn oxides present in the soil, this reagent dissolve some Fe oxides up to 2.9 % of total Fe. Since $NH_2OH \cdot HCl$ had little effect on the organically bound metal fraction, it should be applied prior to extractants like $K_4P_2O_7$, $Na_4P_2O_7$, NH_4EDTA [63]. Trace metals bound to Fe and Al oxides were extracted either in one step or were partitioned in two fractions, referred to as amorphous and crystalline Fe oxides. Essentially, trace metals bound to amorphous Fe oxides were removed by various modifications of acid oxalate solution in the dark [27]. To extract either the total amount of Fe oxides, or the crystalline fraction subsequent to removal of the amorphous Fe oxides, acid oxalate solutions were frequently employed either under diffuse illumination or UV radiation at 20 °C–100 °C and solution/soil ratios between 10 : 1 and 50 : 1 for 0.5 to 3 h. The concentrations of the $(NH_4)_2C_4O_4 \cdot H_2O/H_2C_2O_4$ reagents were either 0.175 mol L^{-1}/0.1 mol L^{-1} or 0.2 mol L^{-1}/0.2 mol L^{-1}, occasionally used along with 0.1 mol L^{-1} ascorbic acid. This variety of conditions and the pronounced effects of varied illumination and temperature on Fe extractability suggest that hardly two procedures extract equal amounts of trace metals from soils [27]. Despite differences in the extraction parameters, most procedures may fairly selectively remove the crystalline Fe oxides when employed subsequent to extractions of Mn oxides, amorphous Fe oxides and organic and carbonate fractions [27]. Uncertainties remain whether different extraction conditions may result in dissolution of varied amounts of trace elements from clay minerals.

Several commonly used extraction procedures and the referred fractions are available in the literature. The procedures contain in general the extraction steps as described previously. Slight or significant modifications of these most frequently used procedures are widely reported. Most extraction procedures addressed a wide range of heavy metals but some extraction schemes were developed for specific elements or groups of elements.

10 SEQUENTIAL EXTRACTION SCHEMES APPLIED TO AEROSOL SAMPLES

The information on aerosol chemical speciation indicates the mobility of the elements once the aerosol is mixed directly into natural waters or during scavenging of the aerosol by wet deposition. During mixing of an aerosol with aqueous solutions it is the anthropogenic metals which are preferentially released, with these having potentially the most harmful impact on the biological community. Atmospheric aerosols have important roles in the biogeochemistry and transportation of trace elements in the air. Direct atmospheric deposition makes only a minor contribution to the total metal contents of the lithosphere because of the large reservoir of these metals in soils and rocks. The aerosol particles influence the solar radiation transfer, cloud–aerosol interactions, and control the optical, electrical, and radioactive properties of the atmosphere. Aerosols sampled within the urban environment exhibit a greater solubility than aerosols with a crustal origin and this should be kept in mind when interpreting the results of sequential leaching.

Atmospheric removal occurs by dry deposition of aerosol particles to water, soil, buildings, or plants, or by wet deposition of aerosol particles and gases in rain, fog, hail, and snow. Wet deposition is a very important removal process for those elements associated with small particles and which are predominantly anthropogenic in origin. Approximately 80 % of the atmospheric removal of elements such as Pb, Cd, Cu, Ni, and Zn to the ocean takes place by wet deposition, whereas 40 % of that process for Fe and Al occurs by dry deposition [11]. This is mostly size dependent, so size fractionation is an important control on removal processes. Wet deposition provides a mechanism by which the metals in aerosol particles can be solubilized and the pH of rainwater is a major control on metal solubility in precipitation. Rainwater pH is governed by a balance between the concentration of acid and neutralizing species present in solution. Usually the metal solubility increases as pH decreases. Before aerosol particles are removed by precipitation they are cycled within the atmosphere through clouds and subjected to repeated wetting and drying cycles before removal.

The atmospheric speciation of metals has several difficulties. Measurement of the total concentration of most metals in atmospheric samples is hampered by the lack of proper quality assurance programs. Speciation of trace metals in the atmosphere is different from those occurring in the hydrosphere. Therefore, forms of an element are determined by the mechanism of interaction of the biosphere and the atmosphere, and the mechanism of transport in the atmosphere. For estimation of the elemental budget in the atmosphere dry and wet deposition rates have to be calculated. The dry deposition rate can be calculated from the results of elemental contents of atmospheric aerosols. The deposition velocity of the aerosol particles can be determined with following formula [66]:

$$v = \frac{\sum_{i=1}^{n} c_i v_i}{\sum_{i=1}^{n} c_i} \quad (1)$$

where v is the dry deposition velocity (cm s^{-1}), c_i is the concentration of the ith element (ng m^{-3}), and v_i is the deposition velocity of the ith particle. The dry deposition rate (D_d, mg m^{-2} $year^{-1}$) is calculated with the following formula:

$$D_d = c_i v\, 0.315 \quad (2)$$

where D_d is the dry deposition rate (mg m^{-2} $year^{-1}$), c_i is the concentration of the ith element (ng m^{-3}), v is the dry deposition velocity (cm s^{-1}), and 0.315 is a calculation factor.

Chester *et al.* [67] suggested a sequential leaching scheme for the characterization of the sources and environmental mobility of trace metals in the marine aerosols. The distribution of elements can be reliably determined in three fractions as *environmentally mobile, bound to carbonates and oxides* (Fe, Mn oxides), and *bound to organic matter and silicates* (environmentally immobile) fractions. Aerosols sampled within the urban environment usually exhibit a greater solubility than aerosols with a crustal origin. Particular attention has to be paid to distinguish between environmentally mobile and environmentally immobile fractions because these represent the two extreme modes by which the metals are bound to the solid matrices. The interaction of trace metals in the aerosols with the other receiving spheres (hydrosphere, lithosphere, biosphere) depends greatly on the solubility of metals under environmental conditions. Chester *et al.* [68] have demonstrated that aerosol speciation data can be related to the extent of the solubility of an element in an aqueous medium, with metals in the environmentally mobile form being most soluble, and can provide a framework for assessing the reactivity of the elements once they have been deposited at the surface. Lum *et al.* [69] provided data on the chemical speciation of a number of elements in an aerosol dominated by pollutants and mainly generated at high temperature by applying a sequential leaching scheme to samples of an Urban Particulate Matter Standard Reference Material (SRM 1648). For characterization of the crustal aerosols a five-stage sequential leaching technique was used for soil-sized aerosols [68]. Comparing the data derived from the two studies reveals that the speciation signatures for some elements differ considerably; aerosols collected in a polluted city generally contained more environmentally mobile fractions containing elements such as Pb, Cr, Zn, Cu, and Cd. The crustal aerosol samples consisted of a higher portion in the stable fractions for the all elements studied. The environmentally mobile/bound to silicates fractions can be interpreted with particle surface/particle matrix associations formed as a result of high temperature anthropogenic processes and low temperature crustal weathering processes.

The environmental mobility indicates the exchange of metals adsorbed or condensed on the surface of aerosols. For the separation of this fraction of the total metal content of an aerosol, the metals have to be displaced from the substrate by reversing the binding mechanisms without affecting metals held in other associations. The most commonly used reagents are solutions of either 1 mol L^{-1} NH_4OAc or $MgCl_2$. The pH needs to be sufficiently high that protons neither compete nor react with other phases in the aerosols, but not so high that hydroxides precipitate. A slightly more acidic solution such as $NH_2OH \cdot HCl$ and HOAc can be used to decompose carbonates liberating incorporated trace metals, and dissolving any free heavy metal carbonates. This reducing chemical agent can release heavy metals associated with Fe/Mn oxides. After reduction of Mn(III) and Mn(IV) to Mn(II), and Fe(III) to Fe(II), the species produced are soluble, and metal ions bound to the oxy-compounds are liberated into solution. Finally, the organic and silicate fractions can be decomposed using a combination of HNO_3 and HF. Since SiF_4 is volatile it is lost from the solution, and the metal ions are released from the lattice.

Recently, fractionation by size and chemical bonding on aerosols has been studied for the first time [70]. The sequential leaching technique has been applied to filter-collected aerosols in eight particle size ranges for determination of the distribution patterns of elements. Particular attention was paid to distinguish between the fine and coarse particle size fractions, and the environmentally mobile and environmentally immobile portions. Among several elements Pb showed a unimodal distribution at a maximum of $<1\,\mu m$, indicating unambiguously the single anthropogenic source (traffic) (Figure 3.3.1). The pollution at this site was around $29\,ng\,m^{-3}$, this value being one or two magnitudes lower, due to less traffic and long-range transportation, than others found in different major cities [11]. Emissions from vehicle exhausts dominated the lead contribution to the atmosphere, although smelting operations also contributed to this atmospheric lead load, emitting both PbO and Pb^0. The relative significance of lead sources in the atmosphere has recently changed worldwide as a result of the considerable decline in the use of leaded vehicle fuels. The actual lead compounds in a particular aerosol will depend on the other constituents in the atmosphere and the age of the aerosol. The predominant inorganic aerosol-phase lead species has been identified as $PbSO_4 \cdot (NH_4)_2SO_4$ by XRD [70] and it is suggested that this species arises from transformations of the primary emitted aerosol compounds during atmospheric transport. The distribution of Pb among the three fractions was rather even, the environmentally mobile fraction being about 40 %.

The average geometric mean concentration of Cd was found to be $<1\,ng\,m^{-3}$ and unimodal distribution pattern was obtained (Figure 3.3.2) [70]. The concentration of Cd in aerosols depends considerably on the location, pollution sources, time, meteorological conditions, etc. and ranges from 1 to $300\,ng\,m^{-3}$ in major cities. Cadmium was mostly associated with the environmentally mobile fractions (50 %); smaller amounts of Cd compound were found in the fractions bound to carbonates

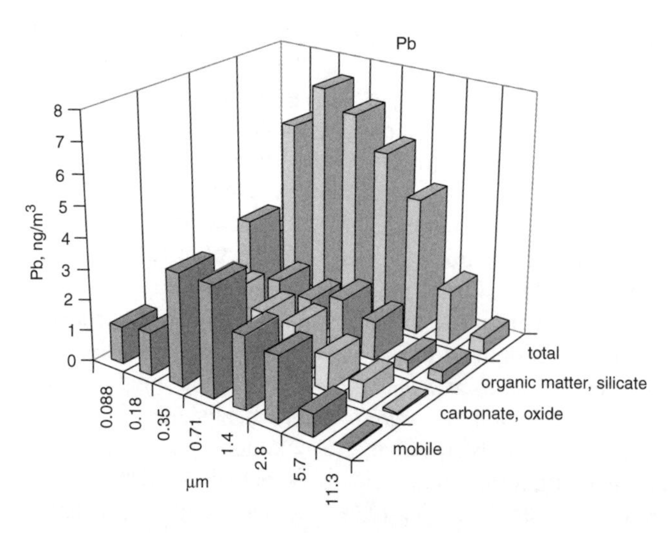

Figure 3.3.1. Distribution of Pb in three fractions as a function of particle size [70].

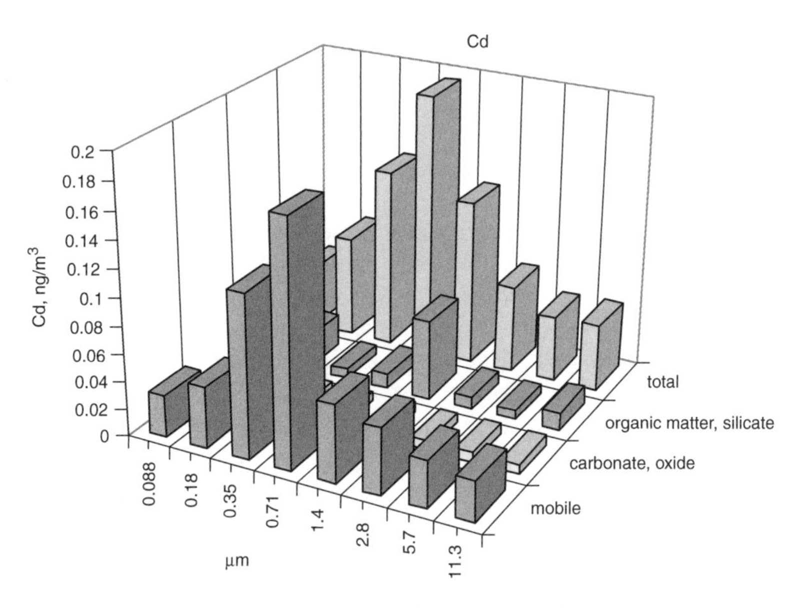

Figure 3.3.2. Distribution of Cd in three fractions as a function of particle size [70].

or oxides (33 %) and bound to organic matter or silicate (17 %). Lum *et al.* [69] have reported that Cd in an urban aerosol was found to be almost completely in exchangeable form. The data of the sequential leaching procedures showed somewhat different patterns, due to the different experimental conditions (leaching steps, reagents, extraction time, temperature, etc.), and the origin of samples. Nevertheless, Cd was mainly found in environmentally mobile form in fly ashes collected at municipal incineration sites, and upon disposal the toxic metal would be transported into the receiving media. High enrichment factor (EF = 4665) values indicate that Cd was emitted from anthropogenic sources [70]. Cadmium liberated during combustion processes has been shown to occur in elemental and oxide forms, whereas emissions from refuse incineration were predominantly as $CdCl_2$. Different metals were strongly fractionated between different aerosol size fractions and this had important implications for all aspects of atmospheric transport from public health to global metal cycling.

There are several projects in which fractionation by particle size has been applied and only the total concentrations of elements are determined. Recently, three size fractions of particulate matter (PM), fine particles ($PM_{2.5}$), (particle size <2.5 µm), coarse particles ($PM_{2.5-10}$) and PM_{10}, were measured at eight sampling sites in four large Chinese cities during 1995 and 1996 [71]. Annual means of PM_{10} concentrations, of which 52–75 % were $PM_{2.5}$, ranged from 68 to 273 $\mu g\,m^{-3}$. Within each city, the urban site had higher annual means of all measured PM size fractions. It was clearly demonstrated that the elements were enriched more in fine particles than in coarse ones. An air quality monitoring program in the Czech Republic has provided data for the concentrations of aerosol and gas-phase pollutants [72]. Fine particulate matter ($PM_{2.5}$) was composed mainly of organic carbon and sulfate with smaller amounts of trace metals. Coarse particle mass concentrations were typically between 10 and 30 % of $PM_{2.5}$ concentrations. The ambient monitoring and the source characterization data were used in receptor modeling calculations.

For evaluation of the atmospheric budget and the environmental effects of trace metals on the biosphere, the calculation of the dry and wet depositions is of vast importance. The relative significance of the two depositional processes varies between locations and is primarily a function of the rainfall intensity in that area. Wet deposition is a very important removal procedure for those elements associated with small particles and which are predominantly anthropogenic in origin. Wet deposition rates are based upon the concentration of trace metals in precipitation samples collected by a wet-only sampler. Using the methods developed for fractionation by chemical bonding, i.e. the three-stage sequential leaching protocols, the dry deposition of aerosols can further be divided and a more reliable estimation can be performed.

In a recent survey the atmospheric budget was calculated for Lake Balaton, Hungary, using a monitoring system of 3 years [65]. It has been reported recently [66] that the ratio D_{dry}/D_{wet} is significant, in particular, for Pb and Zn and usually higher for the others (V, Cr, Ni, Cu, As). In the former study [65] it has been found that, for elements such as Fe, Al, and Cu, the ratio of the two depositions is opposite, i.e. dry deposition plays a more important role in the pollution of the environment (Table 3.3.4.). On the other hand, ratios of D_{dry}/D_{wet} clearly indicated that elements such Mn, As, Cd, Pb, and mainly Zn, were deposited in much higher amounts by wet deposition than by dry deposition.

The environmentally mobile fractions of total dry deposition of elements were also calculated and compared to the total wet deposition. It was obvious that the contribution of mobile fractions of the dry deposition to pollution was, except for Cu, minor. Copper compounds were mainly removed from the atmosphere by dry deposition and one third of the amount of Cu was environmentally mobile. However, if mobile fractions were compared to the total soluble deposition ($D_{drymobile} + D_{wet}$), less than 50 % of Cu came only from dry deposition and much less (4–27 %) from all other elements. This means that in the budget of soluble compounds wet deposition plays the more important role. In the case of elements such as Al, Fe, and Cu, dry deposition has been found to be the major source of pollution, but the mobile parts of dry deposition played a minor role compared to the total soluble deposition.

Furthermore, the soluble fractions of depositions ($D_{drymobile} + D_{wet}$) were compared to the total depositions ($D_{dry} + D_{wet}$). The water quality of the lake has been influenced by the soluble part of the atmospheric depositions, and it has been found that 85–94 % of toxic elements (Pb, Cd, Ni, Zn, and As) were dissolved in the water. The other portions of elements were stable compounds formed under natural environmental conditions, and after precipitation they settle to the bottom of the lake. So, metal compounds, sooner or later, become the part of the bottom sediment, since the fate of dissolved metals depends greatly on the physical and chemical conditions of the bulk water (pH, complex forming capacity, adsorption on clays, quartz, organic matter, biological activities, etc.). In long-term studies the concentrations of the same elements of the lake water have been found to be very low, even lower than the standards permit for drinking waters [65]. Sediments

Table 3.3.4. Total dry (D_{dry}, kg year^{-1}) and wet (D_{wet}, kg year^{-1}) deposition, sum of the dry and wet deposition ($D_{dry} + D_{wet}$, kg year^{-1}), mobile fraction of the total dry deposition ($D_{drymobile}$, kg year^{-1}), sum of the mobile fraction of the dry and wet deposition ($D_{drymobile}/D_{wet}$ kg year^{-1}), ratio of mobile fraction of dry and wet deposition ($D_{drymobile}/D_{wet}$, %) of elements in samples collected around the Lake Balaton [65].

Element	Total D_{dry} (kg year^{-1})	Total D_{wet} (kg year^{-1})	D_{dry} + D_{wet} (kg year^{-1})	Total $D_{drymobil}$ (kg year^{-1})	$D_{drymobil}$ + D_{wet} (kg year^{-1})	(D_{drymob} × D_{wet}) 100 (%)	[D_{drymob} (D_{drymob} + D_{wet})] ×100 (%)	[(D_{drymob} + D_{wet}) (D_{dry} + D_{wet})] ×100 (%)
Al	7980	5670	13650	2131	7801	37.6	27.3	57.1
Fe	19740	13482	33222	4284	17766	31.8	24.1	53.5
Mn	6960	17634	24594	1134	18768	6.4	6.0	76.3
Pb	618	1374	1992	309	1683	22.5	18.3	84.5
Cd	12	54	66	5.4	59.4	10.0	9.1	90.0
Cu	900	336	1236	300	636	89.3	47.2	51.5
Ni	156	516	672	54	570	10.5	9.4	84.8
Zn	618	5604	6222	214	5818	3.8	3.7	93.5
As	138	582	720	39	621	6.7	6.2	86.2

are recognized as a sink and reservoir for metals, metalloids, and other contaminants. Therefore, the accumulation of metals in a lake can be followed by the analysis of bottom sediments.

11 SEQUENTIAL EXTRACTION SCHEMES APPLIED TO FLY ASH SAMPLES

In recent years it has been recognized that the potential toxicity of fly ashes is much more related to the leachable fraction of contaminants, since the total toxic amount of fly ash is not extractable under natural environmental conditions. A lot of leaching tests have been developed worldwide, each characterized by different aims (research or regulatory controls) and different experimental parameters (pH, leachant, stirring device, time of extraction, etc.) [9]. Research groups have been working on the design and study of extraction protocols aiming at the sequential solubilization of metals bound to the specific substrates making up fly ash and known to undergo changes in the yearly life cycle after deposition. Measurement of such functionally defined metal fractions will allow some forecasting of metal release from fly ashes under certain conditions and constitutes, therefore, a valuable tool in disposal management. Harmonization and optimization of methods are necessary because the wide variety of procedures can lead to incomparability of results even for single aquatic systems. It is also desirable to measure the total concentration of the element in order to verify the mass balance. Selective leaching of elements from power plant ashes by water and ammonium acetate and various acids is frequently carried out to determine the nature of the elements and potential mobility during storage [73]. There are some limitations to this procedure, e.g. minerals encased by organic matter or by other minerals may not come into contact with the solvent and, therefore, will not be leached. Elemental interaction (i.e. precipitates, colloids, alkali halides or other water-soluble minerals) may have a relatively large effect, especially in water leaching. The water-soluble minerals are mostly alkali halides. Elements soluble in ammonium acetate are ion-exchangeable in nature, and are mostly associated with organic matter in coal, such as salts of organic acids, and with clay minerals. The elements removed by HCl are associated with acid-soluble minerals such as carbonates, metasulfides and oxides, and also with organic matter functional groups such as carboxyl groups. Hydrofluoric acid is the principal solvent of silicates, while disulfide minerals are mostly removed by nitric acid. A sequential extraction scheme was applied to the determination of binding forms of trace Cd in coal fly ash reference material, NBS 1633a [74]. The sum of the Cd present in the individual fractions showed a good agreement with the certified value of total Cd content. The extraction efficiency was $>85\,\%$. The extractable Cd provides information concerning the binding form of the element.

The chemical fractionation of As, Cr, and Ni in milled coal, bottom ash, and ash collected by electrostatic precipitator from a coal-fired power plant was determined by a sequential leaching procedure [73]. Deionized water, NH_4OAc, and HCl were used as extracting agents and the leachate was analyzed by ICP-AES. Arsenic in the milled coal was mostly associated with organic matter, and 67 % of this arsenic was removed by NH_4OAc. This element was totally removed from milled coal after extraction with HCl. Both Ni and Cr in this sample were extracted by HCl, indicating that water could mobilize Ni and Cr in an acidic environment. The chromium was leached by water from fly ash as a result of the high pH of the water, which was induced during the leaching. Ammonium acetate removed Ni from bottom ash through an ion exchange process. Austin and Newland [75] leached fly ashes from a power plant and a municipal incinerator for 3 h with $0.1\,mol\,L^{-1}$ HCl. They found that Cd was rapidly removed from the ash particles in the initial 5 min of leaching. A combined physical and chemical approach has been proposed to quantify the relative concentrations of elements in the aluminosilicate matrix and in the nonmatrix or surface material of coal fly.

Using several fly ash samples from different sources, it was found that the difference in the

Table 3.3.5. Comparison of results for CW6 sample ($n = 3$, $t(95\,\%) = 4.303$ [76]).

Fractions	Cu		Zn		Pb		Cd		V	
	(mg kg^{-1})	(%)	(mg kg^{-1})	(%)	(mg kg^{-1})	(%)	(mg kg^{-1})	(%)	(mg kg^{-1})	(%)
1	1199 ± 59	59.6	12 046 ± 422	45.9	2355 ± 874	24.8	307 ± 10	74.2	1.1 ± 0.1	5.5
2	302 ± 2	15.0	7603 ± 1417	29.0	2295 ± 707	24.1	90 ± 3	21.7	2.4 ± 0.4	11.9
3	30 ± 13	1.5	1757 ± 522	6.7	418 ± 147	4.4	9 ± 6	2.2	2.7 ± 0.2	13.4
4	311 ± 5	15.5	2482 ± 758	9.4	1473 ± 80	15.5	5 ± 0.5	1.2	4.0 ± 0.1	19.8
5	170 ± 15	8.4	2379 ± 288	9.0	2975 ± 442	31.2	3 ± 1.2	0.7	10.0 ± 0.8	49.5
Σ1–5	2012 ± 70	100	26 267 ± 2732	100	9516 ± 704	100	414 ± 29	100	20.2 ± 1.1	100
Total	1911 ± 379		26 936 ± 4520		8345 ± 2922		428 ± 82		23.2 ± 16.7	

partition depends considerably on the characteristics of the raw material and the operational conditions, i.e. combustion temperature, furnace, etc. So, each type of fly ash should be investigated separately and this fact has to be taken into account when a standard reference material is being produced. A five-stage sequential leaching procedure has been developed and applied to fly ash samples collected at different emission sources [38]. The solvent leaching experiments together with solid-phase examinations carried out by X-ray powder diffraction provided information on the possible environmental impact of particle-associated pollutants. The particulate elements were partitioned into five fractions: (1) *exchangeable elements*, (2) *elements bound to carbonates*, (3) *elements bound to Fe/Mn oxides*, (4) *elements bound to sulphide compounds*, and (5) *elements bound to silicates, residual fraction*. This procedure was optimized for analysis of a fly ash candidate reference material [76]. The total concentration of elements of a candidate fly ash reference material was determined in six laboratories and analytical procedures for the quantitative leaching of inorganic contaminant were harmonized. The average concentration was calculated and compared to the sum of the five fractions determined by the optimized procedure. Data are shown in Table 3.3.5.

Results of the total elemental analysis and the sequential leaching method were compared and good agreement was found. The leachability of the metals proved to be different, so various distribution patterns have been achieved. Copper was found mostly in the environmentally mobile fraction (60 %), while zinc was concentrated in the *exchangeable elements* and *elements bound to carbonates* fractions, (75 %). The sum of the fractions of sequential leaching can also give reliable information on the amounts of elements in fly ash samples. Half of the amount of Pb was identified in environmentally mobile fractions (*exchangeable elements and elements bound to carbonates*), and about 30 % was concentrated in a stable fraction. Results for Cd indicated a great environmental concern, as almost the total amount was accumulated into the *exchangeable elements* and *elements bound to carbonates fractions*, and if the fly ash were disposed of Cd compounds could be transported into the receiving media. Agreement between the two methods (analysis of total concentration and the sequential leaching) is excellent, so the optimised sequential leaching method can reliably be used. Certified reference materials play a key role in the determination of performance characteristics since they stand for the validation of reliable methods. They also permit the comparability of results in different laboratories. If certified reference materials are lacking, the accuracy of any sequential leaching procedure developed can be controlled by an independent method.

12 PRINCIPLES AND APPLICATION OF FIELD-FLOW FRACTIONATION (FFF), SEPARATION OF SUSPENDED PARTICLES INTO SPECIFIC SIZE FRACTIONS

12.1 Normal mode of FFF

An effective way of fractionating is to use field-flow fractionation (FFF). The primary separation in FFF is effected by the action of an external field [77]. The force induced by the field moves molecules toward the wall of a channel and since the molecules cannot penetrate the wall an exponential concentration distribution is built up. If the force affecting the molecules is constant, the thickness of the resulting concentration distribution (layer thickness) depends solely on the diffusion coefficient of the molecule. Slowly diffusing large molecules lie closer to the wall than smaller molecules having higher diffusion coefficients. The separation is amplified by the parabolic flow profile of the liquid pumped through the channel (Figure 3.3.3).

As can be seen the highest flow velocity takes place in the middle of the channel, while closer to the wall the flow is slowed down by the frictional drag. Because of the flow profile, large molecules residing close to the wall are also in slower streamlines than the smaller ones and will therefore be more retained than small molecules. This separation mechanism is called the *normal mode of FFF*.

If the sample particles have a radius of the same magnitude as their layer thickness, the particles will be in the contact with the wall. No exponential concentration distribution will exist and the velocity of the particle is determined only by its protrusion into the fast streamlines. The center of a large particle is in faster streamlines than that of a smaller one, when the particles are in contact with the wall. The elution order is now reversed compared with the normal mode of FFF, i.e. smaller particles elute after larger ones. This is the *steric mode of FFF*. Particles smaller than 1 μm usually elute in normal mode and >1 μm in steric mode.

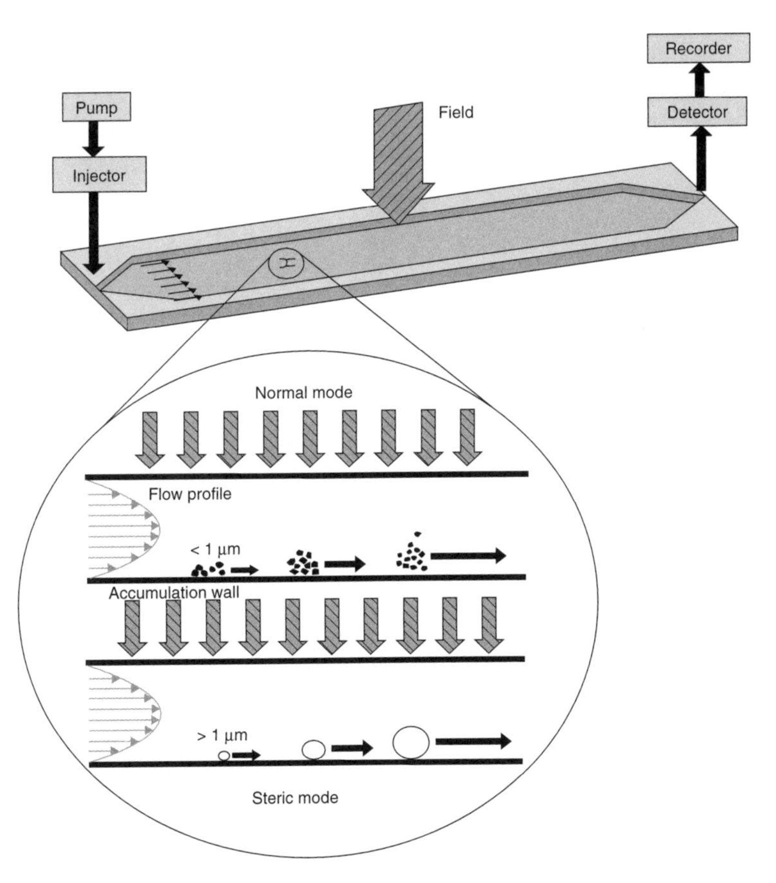

Figure 3.3.3. Principle of field-flow fractionation [77]. Reproduced from ref. 77 by permission of Matti Jussila.

12.2 Thermal FFF (ThFFF)

The effect by which molecules are moved towards the channel wall in ThFFF is thermal diffusion. The upper wall of a thermal FFF channel is heated by electrical cartridge heaters and the lower wall is cooled by water circulation (Figure 3.3.4). The temperature gradient inside the channel induces thermal diffusion of macromolecules usually toward the cooled wall. The thermal diffusion coefficient is relatively independent of molar mass but depends on the interactions between the macromolecule and the carrier liquid. When any exponential concentration builds up against the wall, thermal diffusion is counteracted by ordinary diffusion. The high molar mass selectivity of ThFFF is due to the strong effect of molar mass on the ordinary diffusion coefficient. ThFFF is usually used in normal mode only. The most common application of ThFFF is the molar mass analysis of polymers soluble in organic solvents. The use of water as a carrier is limited by the poor thermal diffusion of polymers in aqueous solutions.

12.3 Flow FFF

The wall elements of a flow FFF channel are made of semipermeable ceramic frit. This allows an additional flow to penetrate the channel perpendicularly to the main flow (Figure 3.3.5). All molecules inside the channel are moved toward the wall by this cross flow. To prevent the sample molecules from being flushed out of the channel, the wall is covered by an ultrafiltration membrane having a cut-off well below the molecular weight of the analyte. The low molar mass carrier can still penetrate the wall elements freely. Because all the sample molecules are affected uniformly by the cross flow, the retention in normal mode is determined only by the diffusion coefficient (D) of the sample. In normal mode operation, D for the sample can be evaluated from its retention time. Using the Stokes–Einstein relationship, D can be converted to the diameter and similarly the fractogram can be turned into a particle size distribution. In steric mode straight calculation of particle size however, is not possible and calibration using standards of known particle size is required. Flow FFF

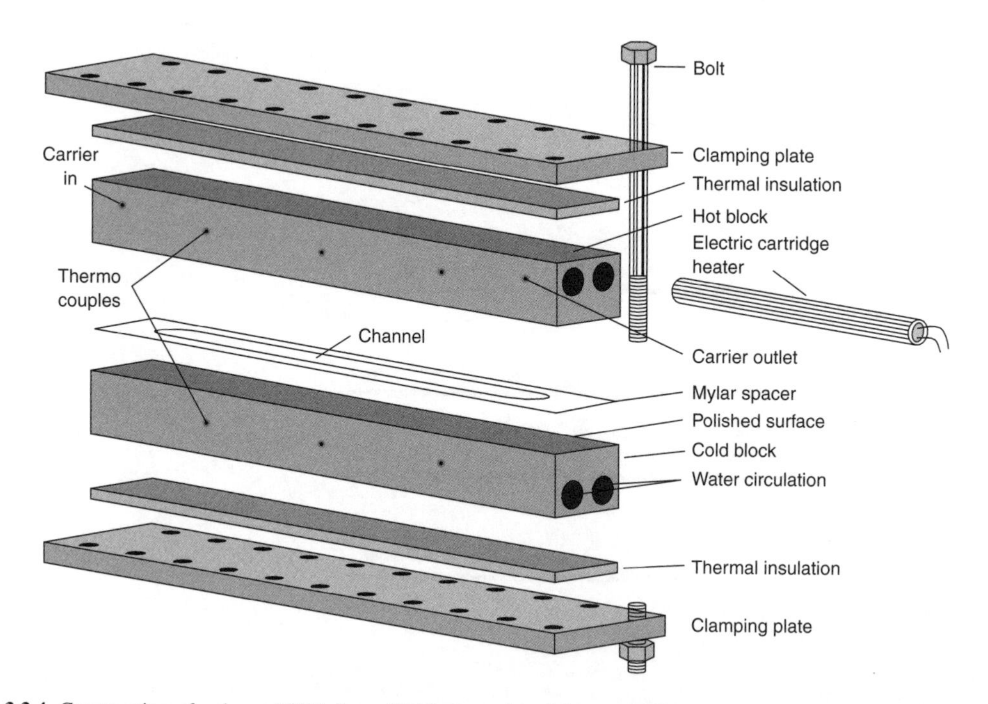

Figure 3.3.4. Construction of a thermal FFF channel [77]. Reproduced from ref. 77 by permission of Matti Jussila.

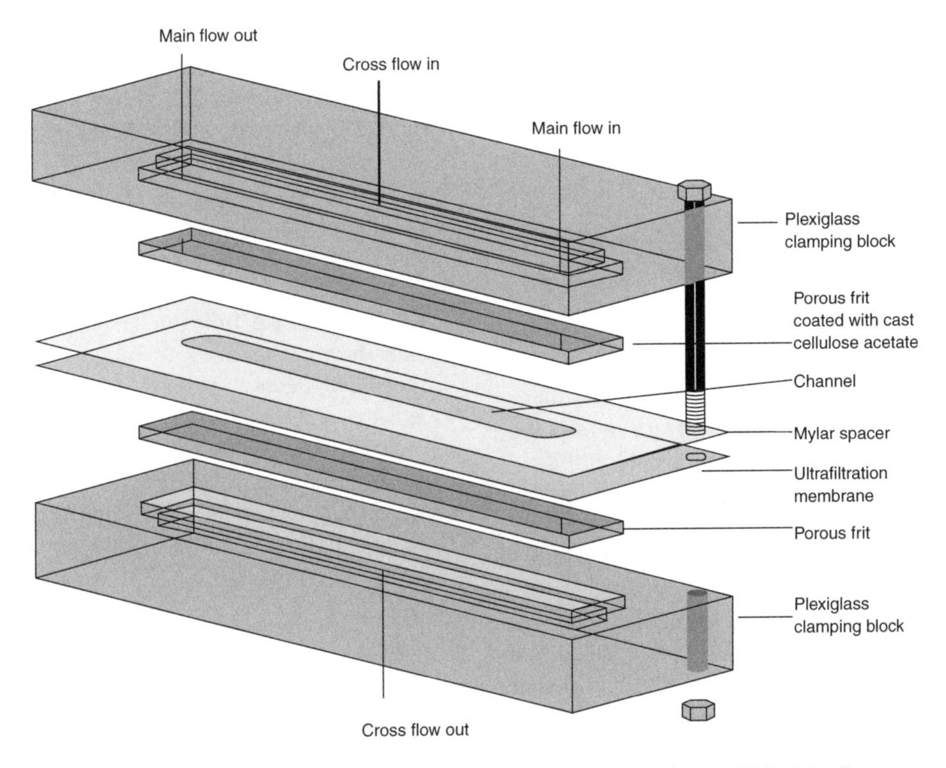

Figure 3.3.5. Construction of an FFF channel [77]. Reproduced from ref. 77 by permission of Matti Jussila.

is applicable to most particulate sample materials suspended in water. The primary developer of this technique is Calvin Giddings and FFF has been proven successful in the analysis of pharmaceuticals, biotechnology products, soils, and foods, among others. A summary of the new developments has been published recently in a book entitled *Field-Flow Fractionation Handbook* [78].

ICP-MS was used for the quantitative measurement of trace elements in specific, submicrometer size-fraction particulates, separated by sedimentation FFF [79]. Fractions were collected from the eluent of the FFF centrifuge and nebulized into an argon ICP mass spectrometer. Measured ion currents were used to quantify the major, minor, and trace element composition of size-separated (Se) colloidal ($<1\ \mu m$) particulates. This approach proved to be ideal for studying the chemistry of clays and other colloidal material. The combination of these two techniques (SeFFF-ICP-MS) has the potential of providing a unique and important tool for the measurement of trace element contamination associated with colloidal suspended matter in environmental water. Using on-channel preconcentration, the flow FFF coupled on-line with ICP-MS has been applied to the study of element distributions in colloids for 28 elements in natural water [80]. The technique was highly flexible and applicable in various particle size ranges. Furthermore, it has a relatively simple theoretical background which gives the possibility to compare theoretically obtained sizes with sizes or molecular weights obtained from calibrations. With optimized conditions, detection limits and reproducibility were sufficient for metal speciation in natural freshwater samples. Sedimentation FFF coupled on line with ICP-MS has opened new possibilities for studies of trace metal adsorption onto natural colloids [81]. Major elements Al, Si, Fe, and Mn were determined simultaneously together with trace elements Cs, Cd, Cu, Pb, Zn, and La. The ability to relate metal uptake to the size and composition of colloids is expected to lead to new insights into uptake processes and into the transport and fate of trace metals in aquatic systems. The FFF techniques need further refinement but can be of great help in studies of element transport in rivers, estuaries, and aquifers.

13 DISCUSSION

Despite their limitations, sequential extraction schemes can provide a valuable tool to distinguish between trace metal fractions of different particle size and solubility. These fractions are empirically related to mobility classes in different solid matrices. The speciation of metals governs their availability to the biota and their potential to contaminate the environment. Available forms of metals are not necessarily associated with one particular chemical species or a specific solid sample component. Speciation of trace elements may vary with time, depending on the solid-phase components that are present, pH, and the number and accessibility of adsorption sites. Soluble and exchangeable forms of metals will decrease with time if there are other solid components present that can adsorb the metal more strongly and have free sites that are accessible (e.g. hydrous oxide, organic matter).

The present state of knowledge on solid matter fractionation of trace elements is still somewhat unsatisfactory because the appropriate techniques are only operationally defined and associated with conceptional and practical problems. With respect to estimating bioavailable element concentrations, one such conceptional problem is the effect of competition between binding sites on the solid substrate and selective mechanisms of metal translocation by the different organisms involved. This situation cannot yet be improved by more sophisticated analytical approaches to fractionation.

On the other hand, the usefulness of a differentiated approach, even if only operationally defined, to the interactive processes between water–biota and solid phases has been clearly proven. The possible environmental implications, e.g. of land disposal of waste material, of acid precipitation, of redox changes in subsoil, and of ingestion of polluted urban dust, can be qualitatively estimated, particularly when the physicochemical conditions of the interacting compartments of the environment are taken into consideration. The method of sequential chemical extraction is the least sophisticated and most convenient technique available for a fractionation assessment. However, we must be certain that we fully understand what is happening during extraction to minimize the risk of producing artifacts and choose standard procedures to ensure that results are comparable. The primary importance of proper sampling protocols has been emphasized, since the sampling error can cause erroneous results even using highly sophisticated analytical methods and instruments. The number of fractionation steps required depends on the purposes of the study. The BCR protocols give a simple guide for most of the solid samples and the results can be compared between different laboratories. Geoscientists and environmental engineers extensively use results of chemical fractionation analysis and scientists have the responsibility to show the pitfalls and limitations of sequential extraction procedures developed. Declaration of the uncertainty of results is a must and greatly improves the quality of these activities.

14 ACKNOWLEDGEMENTS

The authors are indebted to the OTKA T 029 250, AKP 2000-30 2,5 and the Balaton Secretariat of the Prime Minister's Office for their financial support.

15 REFERENCES

1. Cabral, A. R. and Lefebvre, G., *Water Air Soil Pollut.*, **102**, 329 (1998).
2. Pichtel, J., Sawyerr, H. T. and Czarnowska, K., *Environ. Pollut.*, **98**, 169 (1997).
3. Hewitt, C. N. and Harrison, R. M., Monitoring, in *Understanding our Environment*, Hester R. E. (Ed.), The Royal Society of Chemistry, London, 1986, Chapter 1, pp. 1–10.
4. Mehra, A., Cordes, K. B., Chopra, S. and Fountain, D., *Chem. Spec. Bioav.*, **11**, 57 (1999).

5. Barona, A., Aranguit, I. and Elias, A., *Chemosphere*, **39**, 1911 (1999).
6. Schalscha, E. and Ahumada, I., *Water Sci Technol.*, **37**, 251 (1998).
7. Templeton, D., Ariese, F., Cornelis, R., Muntau, H., van Leeuwen, H. P., Danielsson, L.-G. and Lobinski, R., *Pure Appl. Chem.* **72**, 1453 (2000).
8. Martin, J. M., Nirel, P. and Thomas, A. J., *Mar. Chem.*, **22**, 313 (1987)
9. Kersten, M. and Förstner, U., Speciation of trace metals in sediments and combustion waste, in *Chemical Speciation in the Environment*, Ure, A. M. and Davidson, C. M. (Eds), Blackie, London, 1995, Chapter 9, pp. 234–275.
10. Ritchie, G. S. P and Sposito, G., Speciation in soils, in *Chemical Speciation in the Environment*, Ure, A. M. and Davidson, C. M. (Eds), Blackie, London, 1995, Chapter 8, pp. 201–233.
11. Spokes, L. J. and Jickells, T. D., Speciation of metals in the atmosphere, in *Chemical Speciation in the Environment*, Ure, A. M. and Davidson, C. M. (Eds), Blackie, London, 1995, Chapter 6, pp. 137–168.
12. Ure, A. M. and Davidson, C. M., *Chemical Speciation in the Environment*, Blackie, London, 1995.
13. Buffle, J., Wilkinson, K. J., Tercier, M. L. and Parthasarathy, N., in *Reviews on Analytical Chemistry, Euroanalysis IX*, Palmisano, F., Sabbatini, L., Zambonin, P. G. (Eds), Societá Chimica Italiana, 1997, pp. 67–82.
14. Etcheber, H. and Jouanneau, J. M., *Estuar. Coast. Mar. Sci.*, **11**, 701 (1980).
15. Kersten, M. and Förstner, U., *Mar. Chem.*, **22**, 299 (1987).
16. Bartlett, R. and James., B. *Soil Sci. Soc. Am. J.* **85**, 721 (1980).
17. Thomson, E. A., Luoma, S. N., Cain, D. J. and Johansson, C., *Water Air Soil Pollut.*, **14**, 215 (1980).
18. Cuesta, P. A., McDowell, L. R., Kunkle, W. E., Bullock, F., Drew, A., Wilkinson, N. S. and Martin, F. G., *Commun. Soil Sci. Plant Anal.*, **24**, 335 (1993).
19. Linehan, D. J., Sinclair, A. H. and Mitchell, M. C., *J. Soil Sci.*, **40**, 103 (1989).
20. Beckett, P. H. T. and Webster, R., *Soils Fert.*, **34**, 1 (1971).
21. Webster, R., *Adv. Soil Sci.*, **3**, 1 (1985).
22. Vaughn, C. E., Center, D. M. and Jones, M. B., *Soil Sci.*, **141**, 43 (1986).
23. Wenzel, W. W., Brandstetter, A., Pollak, M. A., Mentler, A. and Blum, W. E. H., Seasonal changes of organic matter, pH, nitrogen and some metals in forest topsoils in Austria: A case study of two soils with and without a litter layer, in *Environmental Impacts of Soil Component Interactions, Part II, Toxic Metals, Other Inorganics and Microbial Activities*, Huang, P. M., Berthelin, J., Bollag, J. M., McGill, W. B. and Page, A. L. (Eds), CRC Press, Boca Raton, FL, 1995, Chapter 8, pp. 85–95.
24. Hammer, R. D., O'Brien, R. G. and Lewis, R. J., *Soil. Sci. Soc. Am. J.*, **51**, 1320 (1987).
25. Jones, D. L. and Edwards, A. C., *Commun. Soil Sci. Plant Anal.* **24**, 171 (1993).
26. Wenzel, W. W. and Blum, W. E. H., Assessment of metal mobility in soil-methodological problems, in *Metal Speciation and Contamination of Soil*, Allen, H. E. and Huang C. P., (Eds), Lewis Baco Raton, FL, 1994, Chapter 9.
27. Beckett, P. H. T., *Adv. Soil. Sci.*, **9**, 143 (1989).
28. Gilliam, F. S. and Richter, D. D., *J. Soil Sci.* **39**, 209 (1988).
29. Hinds, W. C., *Aerosol Technology. Properties, Behavior, and Measurement of Airborne Particles,*, John Wiley & Sons, Inc., New York, 1999.
30. Chester, R., Lin, F. J. and Murphy, K. J. T., *Environ. Technol. Lett.*, **10**, 887 (1989).
31. Berner, A., Design principles of the AERAS low pressure impactor, in *Aerosols*, Liu, B. Y. H., Piu, D. Y. H. and Fissan, H. J. (Eds) Elsevier, Amsterdam, 1984.
32. Waldman, J. M., Munger, J. W. and Jacob, D. J., Measurement methods for atmospheric acidity and acid deposition, in *Atmospheric Acidity, Sources, Consequences and Abatement*, Radojevic, M. and Hamson R. M. (Eds), Elsevier, London, 1992, pp. 205–243.
33. McDow, S. R. and Huntzicker, J. J., *Atmos. Environ.*, **24A**, 2563 (1990).
34. Reardon, E. J., Czank, C. A., Warren, C. J., Dayal, R. and Johnston, H. M., *Waste Manage. Res.*, **13**, 435 (1995).
35. Hamilton, K. L., Nelson, W. G. and Curley, J. L., *Environ. Toxicol. Chem.*, **12**, 1919 (1993).
36. Compliance test for granular waste material, CEN Protocol EN 12457–part C, 1996.
37. Characterization of waste – Leaching behavior tests – Influence of pH under steady state conditions – part 1 pH-static test, draft CEN/TC 292/WG6.
38. Bódog, I., Csikós-Hartyányi, Zs. and Hlavay, J., *Microchem. J.*, **54**, 320 (1996).
39. Sposito, G., Lund, L. J. and Chang, A. C., *Soil. Sci. Soc. Am. J.*, **46**, 260 (1982).
40. Zeien, H. and Brümmer, G. W., *Mitteilgn Dtsch. Bodenkundl. Gesellsch*, **59**, 505 (1989).
41. Sparks, D. L., *Kinetics of Soil Chemical Processes*, Academic Press, London, 1989.
42. McLaren, R. G., Lawson, D. M. and Swift, R. S., *J. Soil Sci.*, **37**, 223 (1986).
43. Quevauviller, P., *Analyst*, **123**, 1675 (1998).
44. Quevauviller, P., *Trends Anal. Chem.*, **17**, 289 (1998).
45. Ure, A. M., Quevauviller, P., Muntau, H. and Griepink, B., *Int. J. Environ. Anal. Chem.*, **51**, 135 (1993).
46. Davidson, C. M., Duncan, A. L., Littlejohn, D., Ure, A. M. and Garden, L. M., *Anal. Chim. Acta*, **363**, 45 (1998).
47. Pérez-Cid, B., Lavilla, I. and Bendicho, C., *Analyst*, **121**, 1479 (1996).
48. Mester, Z., Cremisini, C., Ghiara, E. and Morabito, R., *Anal. Chim. Acta*, **359**, 133 (1998).

49. Rauret, G., Sanchez, J. F., Sahuquille, A., Rubio, R., Davidson, C. M., Ure, A. M. and Quevauviller, P., *J. Environ. Monitor.*, **1**, 57 (1999).
50. Davidson, C. M., Ferreira, P. C. S and Ure, A. M., *Fresenius' J. Anal. Chem.*, **363**, 446 (1999).
51. Tessier, A., Campbell, P. G. C. and Bisson, M., *Anal. Chem.*, **51**, 844 (1979).
52. Salomons, W. and Förstner, U., *Environ. Technol. Lett.*, **1**, 506, (1980).
53. Tessier, A., Campbell, P. G. C. and Bisson, M., *J. Geochem. Explor.*, **16**, 77 (1982).
54. Salomons, W., de Rooij, N. M., Kerdijk, H. and Bril, J., *Hydrobiologia*, **149**, 13 (1987).
55. Battiston, G. A., Gerbasi, R., Degetto, S. and Sbrignadello, G., *Spectrochim. Acta*, **458B**, 217 (1993).
56. Lopez-Sanchez, J. F., Sahuquillo, A., Fiedler, H. D., Rubio, R., Rauret, G., Muntau, H., Marin, P., Valladon, B. M., Polve, M. and Monaco, A., *Anal. Chim. Acta*, **342**, 91 (1997).
57. Coetzee, P. P., Gouws, K., Plüddemann, S., Yacoby, M., Howell, S. and den Drijver, L., *Water SA*, **21**, 51 (1995).
58. Accomasso, G. M., Zelano, V., Daniele, P. G., Gastaldi, D., Ginepro, M. and Ostacoli, G., *Spectochim. Acta*, **49a**, 1205 (1993).
59. Weisz, M., Polyák, K. and Hlavay, J., *Microchem. J.*, **67**, 207 (2000).
60. Polyák, K. and Hlavay, J., *Fresenius' J. Anal. Chem.* **363**, 587 (1999).
61. Chapman, P. M., Wang, F., Adams, W. J. and Green, A., *Environ. Sci. Technol.* **33**, 3937 (1999).
62. Miller, W. P., Martens, D. C. and Zelazny, L. W., *Soil Sci. Soc. Am. J.* **50**, 598 (1986).
63. Pickering, W. F., *CRC Crit. Rev. in Anal. Chem.*, **11**(81), 233 (1981).
64. Bascomb, C. L., *J. Soil Sci.*, **19**, 251 (1968).
65. Hlavay, J., Polyák, K. and Weisz, M., *J. Environ. Monitor.*, **3**, 74 (2001).
66. Molnár, Á, Mészáros, E., Polyák, K., Borbély-Kiss, I., Koltay, E., Szabó, G. and Horváth, Z., *Atmos. Environ.*, **29**, 1821 (1995).
67. Chester, R., Lin, F. J. and Murphy, K. J. T, *Environ. Technol. Lett.*, **10**, 887 (1989).
68. Chester, R., Murphy, K. J. T, Towner, J. and Thomas, A., *Chem. Geol.*, **54**, 1 (1986).
69. Lum, K. R., Betteridge, J. S. and Macdonald, R. R., *Environ. Technol. Lett.*, **3**, 57 (1982).
70. Hlavay, J., Polyák, K., Molnár, Á and Mészáros, E., *Analyst*, **123**, 859 (1998).
71. Wei, F., Teng, E., Wu, G., Hu, W., Wilson, W. E., Chapman, R. S., Pau, J. C. and Zhang, J., *Environ. Sci. Technol.* **33**, 4188 (1999).
72. Pinto, J. P., Stevens, R. K., Willis, R. D., Kellogg, R., Mamane, Y., Novak, J., Santroch, J., Benes, I., Lenicek, J. and Bures, V., *Environ. Sci. Technol.* **32**, 843 (1998).
73. Goodarzi, F. and Huggins, F. E., *J. Environ. Monitor.*, **3**, 1 (2001).
74. Petit, M. D. and Rucandio, M. I., *Anal. Chim. Acta*, **401**, 283 (1999).
75. Austin, D. E. and Newland, L. W., *Chemosphere*, **14**, 41 (1985).
76. Polyák, K. and Hlavay, J., *Fresenius' J. Anal. Chem.*, **371**, 838, (2001).
77. Field-Flow Fractionation, http://www.analytical.chemistry.helsinki.fi/research/instruments/fff
78. Schimpf, M., Caldwell, K. and Giddings, J. C., *Field-Flow Fractionation Handbook*, John Wiley & Sons, Inc., New York, 2000.
79. Taylor, H. E., Garbarino, J. R., Murphy, D. M. and Beckett, R., *Anal. Chem.*, **64**, 2036 (1992).
80. Hassellöv, M., Lyvén, B., Haraldsson, C. and Sirinawin, W., *Anal. Chem.*, **71**, 3497 (1999).
81. Hassellöv, M., Lyvén, B. and Beckett, R., *Environ. Sci. Technol.*, **33**, 4528 (1999).

CHAPTER 4

Separation Techniques

4.1 Liquid Chromatography

Kathryn L. Ackley and Joseph A. Caruso

University of Cincinnati, Ohio, USA

1 INTRODUCTION

Separation techniques are key components in elemental speciation analyses. Many sensitive element-specific detectors exist that can rapidly provide total element information, but it is not until a separation technique is coupled with an element-specific detector that the various forms of a particular element in a sample can be determined. A variety of separation techniques have been employed in speciation analyses including gas chromatography (GC), liquid chromatography (LC), capillary electrophoresis (CE), and supercritical fluid chromatography (SFC). LC, however, has emerged as one of the most popular separation techniques for elemental speciation analysis.

Liquid chromatographic separations are carried out by introducing sample onto a chromatographic column filled with a solid stationary phase while a liquid mobile phase is continuously pumped through the column. Early LC stationary phases were made up of a liquid stationary phase coated onto a solid support, but modern LC stationary

Handbook of Elemental Speciation: Techniques and Methodology R. Cornelis, H. Crews, J. Caruso and K. Heumann
 ISBN: 0-471-49214-0

phases are usually comprised of chemically modified silica or polymers. The analytes in the sample interact with both the stationary and the mobile phases while passing through the column, and the extent to which each of the analytes interacts with these phases determines the length of time each analyte resides in the column. Thus, separation is achieved when each of the analytes interacts with the phases to a different extent, and exits the column at different times.

LC has a number of advantages that make it an attractive choice for speciation analyses. LC, unlike GC, is capable of separating non-volatile compounds as well as those that decompose at elevated temperatures. LC is an extremely versatile technique since both the stationary phase and the mobile phase may be altered to achieve the desired separation, and an enormous variety of stationary phases are commercially available. Furthermore, separations can be further enhanced by the addition of additives to the mobile phase. Usually, minimum sample preparation is required, and LC systems are readily interfaced to element-specific detectors such as inductively coupled plasma-mass spectrometers (ICP-MS).

Liquid chromatography is the general term given to chromatographic separations with a liquid mobile phase, but the majority of LC separations performed currently are part of a subset of LC known as high performance liquid chromatography (HPLC). HPLC columns differ from earlier LC columns because the stationary phase particles have smaller diameters. Typical diameters for particles in commercially available HPLC columns are 3–5 μm. The mobile phase is pumped at high pressure through the stationary phase. The increased pressure is a result of pumping a viscous liquid through a column packed with particles of small diameter. Thus, HPLC is often erroneously referred to as 'high pressure liquid chromatography'. The LC work described in this chapter will deal with HPLC separations unless otherwise noted. HPLC separations can be further subdivided by the general type of stationary phase that is used in the separation. The focus of this chapter is not to offer a comprehensive discussion of the theory of LC but rather to provide the reader with an overview of the role of LC in elemental speciation analyses.

1.1 Variables affecting a liquid chromatography separation

A variety of variables may be adjusted to optimize an LC separation. The nature of the chemical species to be separated must first be considered. The interactions between an analyte and the stationary and mobile phases are based upon dipole forces, electrostatic interactions, and dispersion forces. Thus, knowledge of the analyte's structure is useful in predicting how it will behave during a given separation. Generally, compounds that are quite different in polarity or chemical structure will be much easier to separate than compounds that are more similar. (e.g. dibutyl tin chloride and tetraphenyl tin would be easier to separate than dibutyl tin chloride and tributyl tin chloride). The selection of the stationary phase to be used in an LC separation is usually based upon the nature of the analytes that are to be separated. For example, a mixture of anionic species would typically be separated using a column with an anion exchange stationary phase. The nature of the species to be separated may be altered prior to the separation by derivitization to facilitate separation and/or detection [1, 2].

The mobile phase composition is a critical variable when optimizing a separation. The mobile phase may consist of one component such as methanol, or it may be a mixture of solvents and/or aqueous solutions such as buffers. The sample must be soluble in the mobile phase to prevent the precipitation of sample within the column. Buffers may be added to the mobile phase to control pH, and other solutes such as chiral additives, ion pair reagents, and surfactants may be added to enhance a separation. The composition of the mobile phase may also be altered during the separation. Isocratic separations are those in which the mobile phase composition is held constant throughout the separation, but when gradient elution is used, the composition of the mobile phase is changed during the course of the separation. For example, a separation may begin

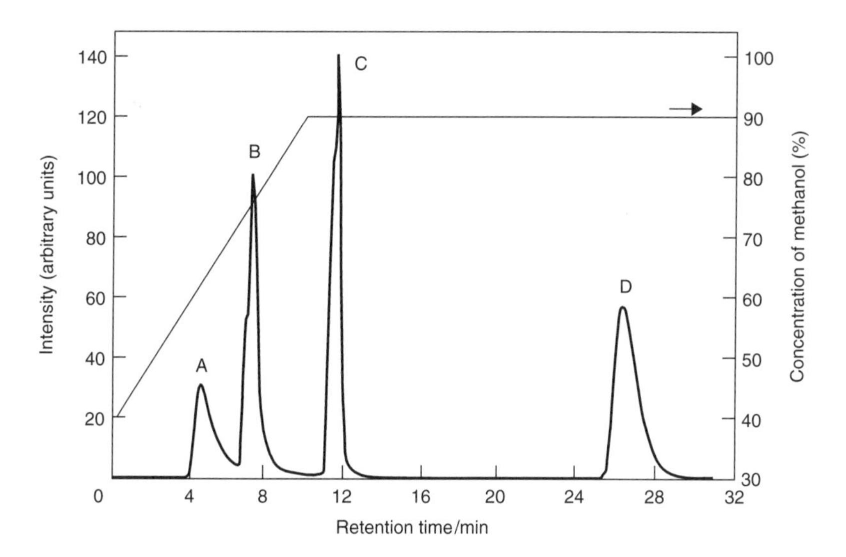

Figure 4.1.1. Separation of 5 ppm of: A, Pb(II); B, triethyllead chloride; and C, triphenyllead chloride; and $100\,mg\,L^{-1}$ of D, tetraethyllead. A Nucleosil C18 column was used with a mobile phase of 8 mM sodium pentane sulfonate at pH 3. Gradient elution was used. The methanol concentration was 40–90 % over 10 min, held at 90 % methanol for 20 min. ICP-MS detection was used [12]. (Reproduced by permission of the Royal Society of Chemistry.)

with a mobile phase consisting of 100 % water. Over the course of the separation, the mobile phase composition may uniformly change over a period of 15 min from 100 % aqueous to 50 % aqueous and 50 % methanol. Gradient separations are often utilized to decrease the time required to perform a separation or to improve resolution and peak shapes. Figure 4.1.1 shows a chromatogram obtained during a gradient separation.

The flow rate of the mobile phase is also a significant variable in the LC separation, for it has a major effect upon the time required to complete a separation. Increasing the flow rate will typically decrease the separation time, but it will increase the pressure inside the column, as more liquid is being forced through the spaces in the stationary phase. One must also consider the method of detection when adjusting the flow rate. Frequently HPLC systems are interfaced to element-specific detectors to perform speciation analyses, and the ability of the detector to accommodate the increased effluent flow rate must be considered.

Temperature may also impact an LC separation. Fluctuations in temperature can cause changes in an analyte's retention time. Thus, HPLC columns are frequently housed in water jackets or ovens to keep the column temperature constant. Generally, increases in temperature cause a decrease in analyte retention time because the mobile phase viscosity is reduced and the rates of diffusion are increased.

1.2 Characterizing chromatographic separations

A basic understanding of the way chromatographic separations are characterized is helpful when studying LC separations utilized in speciation analyses, and a cursory explanation of several significant equations used to characterize LC separations is presented here.

The term 'retention time' refers to the amount of time it takes for an analyte to pass through the column and reach the detector. An unknown compound's retention time in a chromatographic system may aid in the compound's identification. For example, if a plant extract is being analyzed to determine the selenoamino acids present in the sample, and a peak is observed that has the same retention time as selenomethionine, the analyst knows that the sample either contains selenomethionine or another species with the same retention time as selenomethionine. The unknown peak, however, could not be attributed to any of the other selenoamino acids that have a different retention time.

The capacity factor, usually denoted k', is a dimensionless parameter that represents the normalized retention time for an analyte to take into account nonchromatographic contributions to the retention time such as the distance from the column to the detector. The capacity factor is calculated using equation (4.1.1)

$$k' = \frac{(t_r - t_o)}{t_o} \qquad (4.1.1)$$

where t_r is the analyte's retention time and t_o is the time it takes for an analyte unretained by the stationary phase to pass through the column and reach the detector.

The selectivity of a separation, denoted α, refers to how well the chromatographic system is able to distinguish between two different analytes. The selectivity is the ratio of the two analytes' capacity factors.

$$\alpha = \frac{k'_1}{k'_2} \qquad (4.1.2)$$

The resolution, R_s, refers to the efficiency of a separation. The resolution is calculated by dividing the distance between two chromatographic peaks by the average of their widths at the base.

$$R_s = \frac{\Delta t_r}{\frac{1}{2}(w_{t1} + w_{t2})} \qquad (4.1.3)$$

Analyte peaks in LC separations have Gaussian peak shapes caused by longitudinal diffusion, dispersive effects, and the fact that the partitioning of the analyte between the stationary phase and the mobile phase is not an instantaneous process. A mathematical expression for the symmetry of a chromatographic peak is the peak asymmetry factor, A_s. The peak asymmetry factor is determined by drawing a vertical line from the tallest point of the peak to the base of the peak. At 10 % of the peak's height, the width of the peak on the left side of the line is measured along with the width of the peak on the right side. The ratio of the peak 'half-widths' is called the peak asymmetry factor. 'Fronting' occurs when the front portion of the peak is much wider than the back portion of the peak. 'Tailing' occurs when the back portion of the peak is much wider than the front portion of the peak. Figure 4.1.2 provides examples of these types of peaks. Ideally, peaks obtained in LC would have a peak asymmetry factor of 1 and would be as narrow as possible.

Two related terms that are frequently used in LC are plate height, H, and plate count, N. These terms relate to a column's chromatographic efficiency and were derived from theoretical work that treated chromatographic columns as if they were similar to distillation columns [3]. These

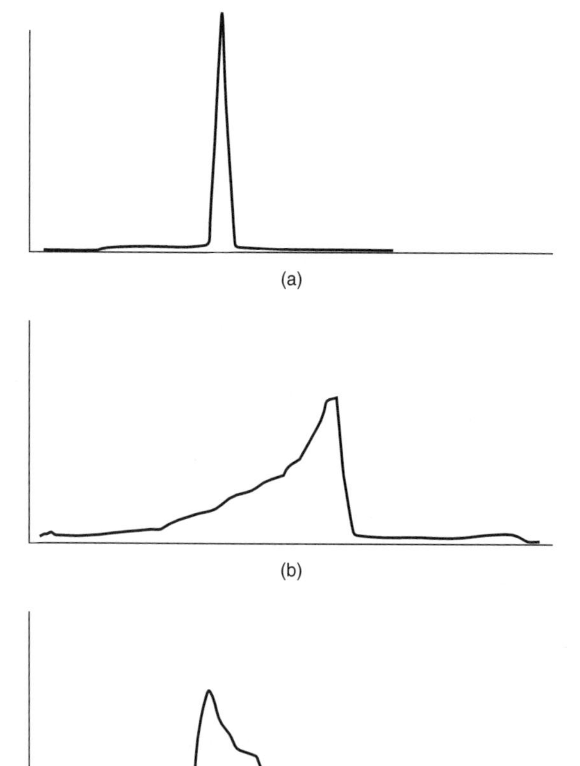

Figure 4.1.2. (a) Ideal chromatographic peak. (b) Chromatographic peak exhibiting 'fronting'. (c) Chromatographic peak exhibiting 'tailing'.

terms are theoretical in nature, and there are no plates actually inside of a chromatographic column. The plate height is determined by dividing the length of the stationary phase, L, by the plate count.

$$H = \frac{L}{N} \tag{4.1.4}$$

Plate count is frequently used to assess column performance. There are multiple ways to calculate N. One method is to determine the width of a peak at half its maximum height, $W_{1/2}$, and substitute that value into equation (4.1.5).

$$N = \frac{5.54\ t_r^2}{W_{1/2}^2} \tag{4.1.5}$$

Regardless of the method used to calculate N, for this discussion it is sufficient to state that columns with a large N value are more efficient than those with smaller N values.

2 LIQUID CHROMATOGRAPHIC STATIONARY PHASES

2.1 Normal phase chromatography (NPC)

Normal phase chromatographic systems are comprised of a polar stationary phase and a nonpolar mobile phase such as hexane. When chromatographic systems consisting of a nonpolar stationary phase and a polar mobile phase were later developed, they were given the name 'reversed phase' since they were the reverse of the established 'normal phase' systems. Unmodified silica or alumina were frequently used in early NPC work, but the peak shapes obtained with these stationary phases were often broad with basic compounds exhibiting tailing and retention times that were difficult to reproduce. These problems were alleviated with the development of bonded stationary phases that have a polar functional group (cyano, diol, etc.) chemically bonded to the silica. Analytes separate in NPC as they are reversibly adsorbed by the polar functional groups of the stationary phase. One of the principal advantages of this technique is that it allows analytes that are insoluble in polar solvents to be separated.

NPC has seen limited use in the area of elemental speciation. The major limitation of this technique is the nonpolar mobile phase, which is incompatible with most elemental detectors used in speciation analyses. Furthermore, many analytes that are separated using NPC may also be separated using reversed phase chromatography, which is compatible with elemental detectors. Nonetheless, examples of the use of NPC for elemental speciation can be found in the literature. Xu and Lesage used an aminopropyl column to separate vanadyl and nickel petroporphyrins [4]. The separation mechanisms were determined to be hydrogen bonding as well as Van der Waals interactions between the petroporphyrins and the amino groups in the stationary phase. Hexane, toluene, and dichloromethane were components in the mobile phase and a fluorescence detector was used. NPC was used to separate organotin pesticides [5]. A column with a cyanopropyl-bonded silica stationary phase was used for the separation. Separated analytes were subjected to UV photoconversion, post-column complexation, and fluorescence detection. The method was used to measure triphenyltin acetate in water adjacent to a potato field sprayed by the pesticide.

2.2 Reversed phase chromatography (RPC)

RPC is one of the most widely used LC techniques in elemental speciation. It is used to separate nonpolar and/or slightly polar species. The polar mobile phases used are typically aqueous or a mixture of water and an organic modifier such as methanol or acetonitrile. The stationary phase is most commonly silica that has been modified through silanization. The silanol–OH groups present on the silica surface are replaced by alkyl chains creating a nonpolar stationary phase suitable for reversed phase separations. Chains with 18, 8, or 2 carbon atoms are the most common. Separation is based on the hydrophobicities of the species with compounds that are the most hydrophobic eluting the latest. Separations may be manipulated by changing a variety of variables including the stationary phase functional group, pH, ionic strength, organic modifier(s) used in the mobile phase, and the gradient program. The pH of the eluent is an important parameter in RPC. In many instances, the pH of the mobile phase dictates if a compound is protonated or deprotonated. This affects the charge of the analyte and its retention on the column. Traditionally, silica-based stationary phases cannot tolerate eluent pHs below 2 or above 7 since cleavage or hydrolysis of the stationary phase may occur. Recently, column manufacturers have produced silica-based reversed phase stationary phases that can tolerate eluent pHs of 2 to 10. Reversed phase stationary phases may also be made of polymeric materials, but silica-based columns are still the most commonly used.

RPC has been widely used in elemental speciation to separate organometallic species. It has been utilized in the speciation of metal porphyrins [6], in speciation studies of platinum in chemotherapy drugs [7], in the determination of tellurium compounds in wastewater [8], and the separation of organotin compounds [9, 10]. Zoorob and Caruso [11] utilized a column with an octadecyl stationary phase and a mobile phase consisting of 80 % water and 20 % methanol to separate the chromium species in azo dye, see Figure 4.1.3. One of the dyes investigated was found to contain uncomplexed and potentially bioavailable Cr(III). Many more examples of RPC used for elemental speciation can be found in the literature. However, these selected examples illustrate the breadth of analytical problems to which this technique can be applied.

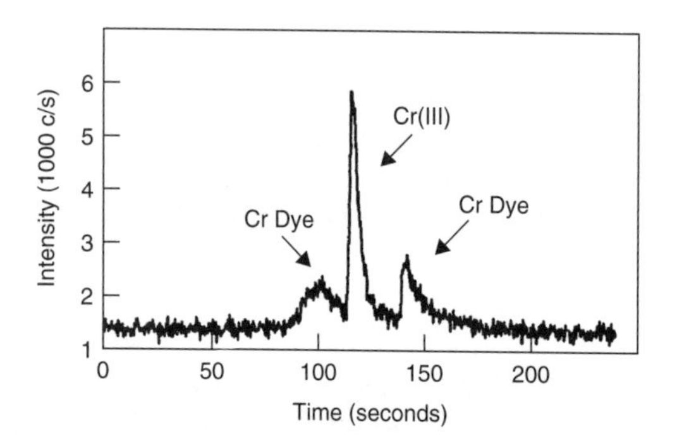

Figure 4.1.3. Reversed phase separation of Acid Blue 193, a commonly used chromium azo dye. The mobile phase composition was water–methanol (80 : 20, v/v). ICP-MS detection was used. [11] (Reprinted from *Journal of Chromatography*, Vol. 773, Zoorob and Caruso, Speciation of chromium dyes by high performance liquid . . ., pp. 157–162, 1997, with permission from Elsevier Science.)

2.3 Reversed phase ion pair chromatography (IPC)

Reversed phase ion pair chromatography is similar to RPC in that RPC stationary phases are used, but an ion pair reagent is added to the mobile phase. An ion pair reagent is a salt with a cation or anion having a polar head group and a nonpolar tail. Examples of ion pair reagents include sodium alkyl sulfonates, tetraalkyl ammonium salts, and triethylalkyl ammonium salts. The concentration of ion pair reagent in the mobile phase is typically between 0.001 and 0.005 M. IPC's popularity results from its ability to simultaneously separate anions, cations, and noncharged species.

The separation mechanism involved in IPC is not totally understood. One widely held theory is that an ionic analyte is electrostatically attracted to the charged end of the ion pair reagent, and an 'ion pair' is formed. The charge neutralization coupled with the nonpolar tail of the ion pair reagent causes the charged analyte to be retained by the nonpolar stationary phase. Another theory is that the hydrophobic portion of the ion pair reagent adsorbs to the stationary phase with the charged portion of the ion pair reagent exposed to passing analytes. Thus, a pseudo ion exchange stationary phase is formed causing charged species that would not have otherwise been retained on the column to interact with the stationary phase. More than likely, both of these mechanisms occur during an IPC separation.

IPC has been used to speciate Pb(II), triethyllead chloride, triphenyllead chloride, and tetraethyllead chloride [12]. Sodium pentane sulfonate was the ion pair reagent. At an ion pair concentration of 2 mM, the inorganic lead and the triethyllead peaks overlapped significantly. At an ion pair concentration of 8 mM, the inorganic lead and triethyl lead were nearly totally resolved (see Figure 4.1.1). This separation was coupled with ICP-MS detection for the analysis of a leaded fuel standard reference material. This separation illustrates how IPC can successfully be used to separate inorganic and organometallic species that are neutral or charged.

Another example of the successful application of IPC is the work of Jiang and Houk [13]. They developed separations for inorganic phosphates, adenosine phosphates, inorganic sulfates, and amino acids. Separations were performed with reversed phase stationary phases, tetraalkylammonium salts as ion pairing reagents, and small amounts of organic modifier (less than 5 %).

2.4 Micellar chromatography

Micellar chromatography is similar to IPC. A surfactant is added to the mobile phase at a concentration above the critical micelle concentration (CMC) so that micelles are formed in the mobile phase. The CMC is characteristic of each surfactant. Below this concentration, surfactant molecules tend to adsorb at surfaces, as in IPC, to minimize solvophobic interactions. At surfactant concentrations greater than the CMC, the surfactant molecules come together to form micelles that have the hydrophobic portion of the surfactant oriented inward and the polar, hydrophilic portion oriented outward towards the polar mobile phase. Figure 4.1.4 shows the structure of a micelle. Micelles can be anionic, cationic, uncharged, or zwitterionic. Analytes may partition inside of the stationary phase, mobile phase, and micelles, making the separation of charged, neutral, hydrophobic, and hydrophilic species possible. Jiménez and Marina [14] have written a review of retention modeling in micellar liquid chromatography. Hydrophobic analytes that may not be soluble in the polar mobile phases used in RPC may be soluble in micellar chromatography eluents since they may partition into the interior of the hydrophobic micelle.

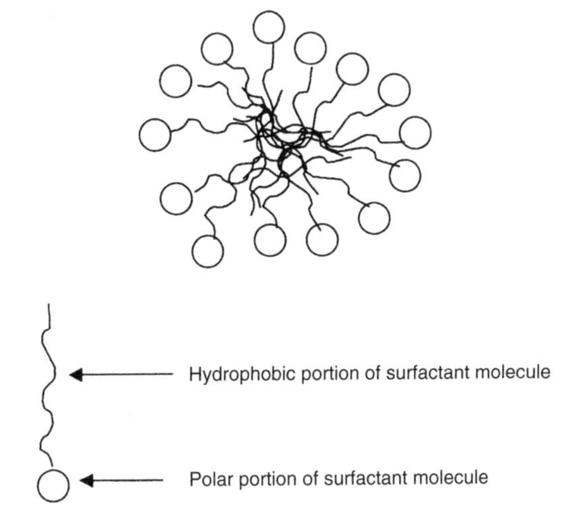

Figure 4.1.4. Structure of the micelle.

Ding *et al.* [15] utilized micellar chromatography for the speciation of dimethyl arsenic acid (DMA), monomethyl arsonic acid (MMA), As(III), and As(V) in urine. Micellar chromatography was selected because proteins found in the urine sample were dissolved by the micelles, so they eluted in the void volume. This was advantageous because clinical samples typically need to be deproteinized to avoid proteins that are insoluble in RPC mobile phases from precipitating inside the LC column. Figure 4.1.5 shows a chromatogram of arsenic species present in urine. Another advantage of this separation was that the chloride ions in the sample were separated from the arsenic species. An ICP-MS was used as the chromatographic detector, and chloride ions interfere with the determination of arsenic by ICP-MS. By separating the chloride ions chromatographically, the interference can be eliminated.

Micellar chromatography has been successfully used to separate alkyltin compounds using a 0.1 M sodium dodecyl sulfate (SDS) micellar mobile phase and a C-18 stationary phase [16]. Three surfactants were studied–SDS which is negatively charged, dodecyltrimethylammonium bromide which is positively charged, and polyoxyethylene(23)dodecanol which is nonionic. SDS was found to separate the analytes. This was not surprising since the organotin species investigated were cationic and would have more electrostatic interactions with an anionic surfactant.

2.5 Ion exchange chromatography (IEC)

IEC is used to separate free ions and easily ionizable species. IEC is a commonly used technique in speciation analyses because metallic species of

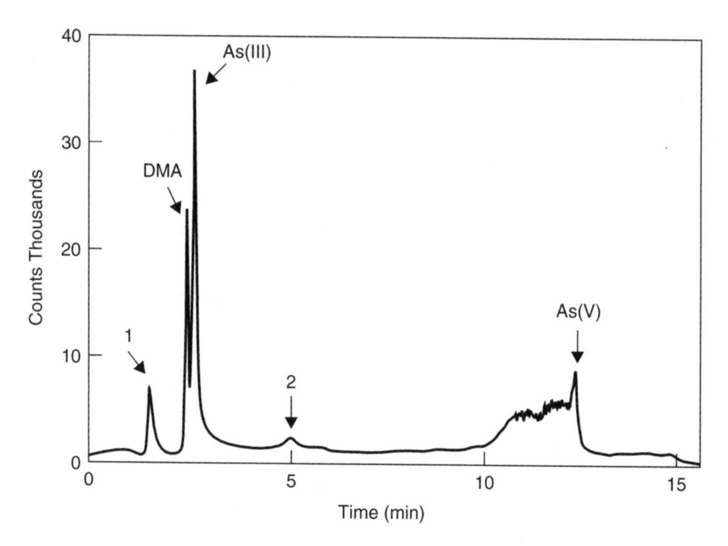

Figure 4.1.5. Chromatogram of arsenic species in urine. Peaks 1 and 2 are different forms of chloride that were visible since ICP-MS detection was used. DMA = dimethylarsinic acid. A Hamilton PRP-1 column was used. This is a reversed phase polymeric column. The mobile phase consisted of 0.05 M cetyltrimethylammonium bromide (CTAB) and 10 % methanol. The pH was 10.2, and the column temperature was 40 °C [15]. (Reprinted from *Journal of Chromatography*, Vol. 694, Ding *et al.*, Arsenic speciation by micellar chromatography, pp. 425–431, 1995, with permission from Elsevier Science.)

interest frequently occur in the ionized form. The stationary phase typically consists of an ionic functional group such as a quaternary ammonium group or a sulfonate group bonded to a substrate such as a polystyrene–divinylbenzene polymer or silica. IEC is frequently subdivided into cation exchange chromatography and anion exchange chromatography depending upon the functional groups present in the stationary phase. The ionic sites on the stationary phase have the opposite charge of the analytes to be separated. Counter ions in the mobile phase maintain electrical neutrality within the chromatographic column. Analytes having a charge that is opposite of that of the charge bearing functional group will interact with the stationary phase electrostatically. The retention time of an analyte increases with increasing electrostatic force. Mobile phases used in IEC separations are typically aqueous solutions of inorganic salts.

IEC is frequently utilized to separate arsenic species. Wang *et al.* [17] used an anion exchange column to separate arsenite and arsenate in coal fly ash extracts. ICP-MS detection was used, and chloride ions, which hinder the detection of arsenic by ICP-MS, were separated from the analytes of interest, thus eliminating the interference. Anion exchange columns have been used to separate arsenic species in other matrices such as soil [18] and fish extracts [19]. Just a few of the many other applications of anion exchange chromatographic columns include the separation of bromate and bromide in drinking water [20], the separation of inorganic and organic antimony species [21], and the separation of Cr(III) and Cr(VI) species [22].

Cation exchange chromatography is also used in elemental speciation. Suyani *et al.* [23] used cation exchange chromatography to separate trimethyltin chloride, tributyltin chloride, and triphenyltin acetate. In some instances, an anion exchange column and a cation exchange column are connected in series to allow for the determination of both anionic and cationic species. Teräsahde *et al.* [24] successfully utilized this technique to separate six arsenic species. Gradient elution was used and the ionic strength of the mobile phase increased and the pH decreased over the course of the separation. The three mobile phase components were dilute nitric acid, water, and carbonate buffer.

IEC has the advantage of using aqueous buffers, which makes this technique compatible with element selective detectors such as an ICP mass spectrometer. IEC is also utilized in speciation analyses for the purpose of sample cleanup prior to analysis and for sample preconcentration.

2.6 Size exclusion chromatography (SEC)

SEC is used to separate macromolecules (MW > 2000 Da) such as proteins, synthetic polymers, and biopolymers according to their size. The separation mechanism is not based on chemical interactions as in other types of LC but rather on the ability of an analyte to penetrate the pores of the stationary phase. Small molecules will penetrate the pores of the stationary phase greater than larger molecules, so large molecules are eluted before small molecules. The mobile phase does not play a large role in the separation, and it is usually selected based on its ability to solubilize the analytes being separated. SEC is often subdivided into gel permeation chromatography and gel filtration chromatography. Gel filtration chromatography refers to the separation of water-soluble macromolecules, and gel permeation chromatography refers to the separation of macromolecules that are soluble in organic solvent. The SEC system must be calibrated with molecules of known molecular weight that have similar physical properties to those of the analytes of interest.

SEC is used in the field of elemental speciation to study such things as metalloproteins [25, 26] and metabolites of metal-containing drugs [27]. It may also be used to separate species of interest from interfering low molecular weight components of the sample matrix. SEC was used to separate protein-bound copper in serum from sodium and phosphate ions that interfered with the determination of copper by ICP-MS [28]. Crews *et al.* [29] used SEC to investigate the cadmium-containing proteins in pig kidney following cooking and *in vitro* gastrointestinal digestion. They found that most of the soluble cadmium in the pig kidney was associated with a metallothionein-like protein. Klueppel *et al.* [30] used SEC with ICP-MS detection to study platinum metabolites in plants. Grass was cultivated with a platinum-containing solution, and the extracted platinum species having a molecular mass less than 10 kDa were separated using SEC. Not all of the metabolites could be resolved using SEC, but the inorganic platinum complexes were separated from the organoplatinum species.

2.7 Chiral liquid chromatography

Chiral molecules have stereoisomers that are nonsuperimposable mirror images of each other (see Figure 4.1.6). These types of stereoisomers are referred to as enantiomers. The most common type of chiral molecule has a tetrahedral carbon atom with four different groups attached to it. However, molecules can be chiral even if they do not have an asymmetric carbon so long as they are nonsuperimposable mirror images of one another. Chiral separations can be very difficult to achieve since most of the physical properties of enantiomers are the same. Yet performing chiral separations is of particular importance when studying pharmaceuticals since biological systems tend to be chiral systems with one enantiomer of a drug having a different biological effect than its chiral counterpart *in vivo*.

Several strategies have been employed to carry out chiral separations including the use of chiral stationary phases, chiral derivatizing agents to form diasteriomers, and the use of chiral mobile phase additives. A drawback to the use of chiral mobile phase additives for LC is that the mobile phase flow rate associated with LC necessitates large amounts of chiral additive to equilibrate the column and perform the separation. Thus, the use of chiral additives to separate enantiomers is primarily used in capillary electrophoresis and thin layer chromatography. Derivatization of chiral

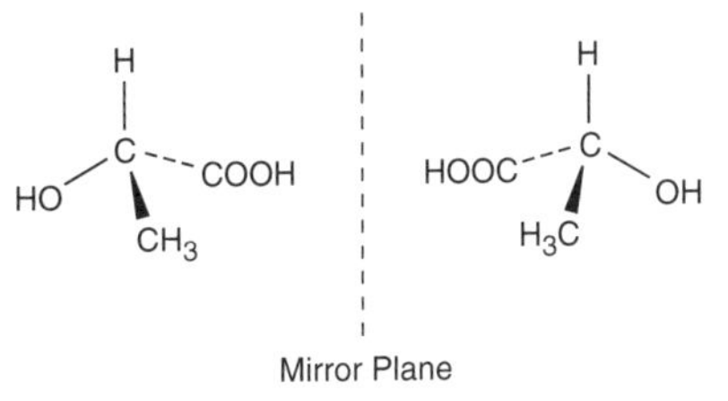

Figure 4.1.6. Two isomers of a chiral molecule.

molecules with a chiral derivatizing agent of known optical purity may be performed prior to separation by LC. The advantage of this technique is that after derivatization, the separation may be performed with commercially available reversed phase LC columns, which are much less expensive than LC columns with chiral stationary phases. However, finding a suitable derivatizing agent can be difficult, and method validation to confirm that the enantiomeric ratio obtained after derivatization is the same as the enantiomeric ratio in the sample is problematic.

The third alternative is the use of LC columns that have chiral selectors immobilized on the surface of the stationary phase. Commonly used chiral stationary phases include cellulosic and amylosic, macrocyclic antibiotic, chiral crown ether, ligand exchange, cyclodextrin, protein, and Pirkle phases [31]. Separation occurs when the chiral analytes form transient diastereomeric complexes with the chiral selectors on the stationary phase. LC columns with chiral stationary phases are expensive and not all enantiomers can be separated with the stationary phases available. Thus, the number of papers appearing in the literature discussing the use of chiral LC for elemental speciation is limited. As the number of stationary phases available grows, the number of papers on this topic is expected to increase as well.

Chiral chromatography has primarily been used in the field of elemental speciation to separate chiral selenoamino acids. Méndez *et al.* [32] separated selenomethionine enantiomers using a ß-cyclodextrin column. *O*-Phthalaldehyde and 2,3-naphthalenedicarboxaldehyde are fluoroionogenic derivatizing reagents that were used to derivatize the selenomethionine enantiomers prior to separation. Fluorimetric and on-line hydride generation ICP-MS detection were used to detect the separated enantiomers. Chiral selenoamino acids have also been separated using a chiral crown ether stationary phase [33, 34]. A 0.10 M perchlorate mobile phase was used, and no precolumn derivatization was necessary. Figure 4.1.7 shows the separation of the selenoamino acids. This separation was used to determine the selenium species present in nutritional supplements [33] and selenium-enriched samples [34]. A teicoplanin-based chiral stationary phase was also used to separate selenomethionine enantiomers in selenized yeast [35]. Teicoplanin is a macrocyclic antibiotic used to achieve chiral separations. Vancomycin is

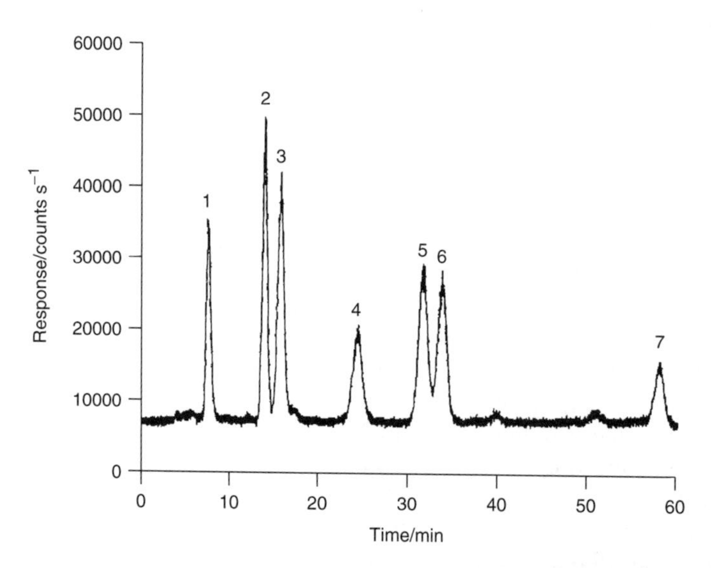

Figure 4.1.7. Separation of selenoamino acids using a chiral crown ether column. The mobile phase consisted of 0.1 M $HClO_4$ at pH 1. The flow rate was 0.5 mL min^{-1} for 35 min, then it was increased to 1.0 mL min^{-1}. Peaks: 1 = L-selenocystine; 2 = L-selenomethionine; 3 = *meso*-selenocystine; 4 = D-selenocystine; 5 = D-selenomethionine; 6 = L-selenoethionine; and 7 = D-selenoethionine [33]. (Reproduced by permission of the Royal Society of Chemistry.)

another macrocyclic antibiotic frequently used for the same purpose.

2.8 Micro liquid chromatography

Standard LC columns usually have internal diameters of 4.6 mm. However, analysts are increasingly finding the use of columns with smaller internal diameters to be advantageous. While no single convention exists for what defines a micro LC column, guidelines have been suggested in a review article by Vissers *et al.* [36]. Chromatography, performed with columns having internal diameters of 0.5–1.0 mm, is referred to as micro LC. Capillary LC generally describes separation with columns having internal diameters of 100–500 μm, and nanoscale LC refers to columns with internal diameters of 10–100 μm. Throughout this chapter, the term micro LC will collectively refer to micro LC, capillary LC, and nanoscale LC.

Performing separations with micro scale LC columns has the advantage of reduced solvent consumption resulting from much smaller flow rates. This may also be advantageous when interfacing micro LC columns with atomic spectrometers that do not tolerate mobile phases with high concentrations of salts or organic modifiers since smaller mobile phase flow rates are used. Because of the reduced column volume, less sample is injected onto the column. The reduced sample requirements make micro LC attractive for analyses when limited amounts of sample are available.

Micro LC columns are frequently made by packing fused silica capillaries. Reversed phase, ion exchange, and ion pairing separations can all be carried out using micro LC columns. The small flow rate and high pressure drop per unit length of small diameter columns pose challenges when selecting/developing a solvent delivery system [37]. Reciprocating and syringe pump systems capable of delivering mobile phases at flow rates on the order of 50–150 $\mu L\,min^{-1}$ are commercially available [36]. For lower flow rates, split-flow techniques may be utilized.

Care must be taken when interfacing micro LC columns to the selected detector to minimize band broadening. Analytes separated by the column exit the column in a narrow band of the mobile phase. Analyte band broadening occurs as the separated analyte continues to diffuse in the mobile phase adjacent to the analyte band. Band broadening causes wider peaks with decreased peak height resulting in a decrease in resolution and an increase in detection limits. To help reduce this problem, tubing connecting the column to the detector should have an internal diameter at least as small as the chromatographic column. Frequently, direct injection nebulization is employed to interface micro LC columns to detectors such as inductively coupled plasma-mass spectrometers to minimize the band broadening that occurs when the column effluent is nebulized into a spray chamber [38–40]. The use of micro LC columns coupled to atomic spectrometers for trace elemental speciation has been reviewed by Garraud *et al.* [41].

Pergantis *et al.* [42] used micro LC for the speciation of arsenic animal feed additives. Three reversed phase columns were investigated, a conventional column with a 4.6 mm i.d., a microbore column with a 1 mm id, and a micro column with a 0.32 mm i.d. (The nomenclature used in ref. 42 was used to refer to the columns rather than the nomenclature described earlier in this section of the chapter.) Graphite furnace-atomic absorption spectroscopy (GF-AAS) was one of the detection methods used. Fractions were collected at 30 s intervals, and the fractions were analyzed by GF-AAS. When GF-AAS was used for the detection, the limit of detection for the micro LC column was three orders of magnitude lower than for the conventional column. The lower limit of detection resulted from all of the micro LC fraction (5 μL) being injected into the graphite furnace. Only 2 % of the conventional LC fraction (500 μL) could be injected into the graphite furnace. However, because the total sample volume for each LC fraction was used in a single GF-AAS analysis, duplicate measurements could not be made. Thus, the analysts found the microbore column offered the best compromise. They also found the flow rates with the micro LC column were easily accommodated by continuous flow-liquid secondary ion–mass spectrometry and direct liquid introduction-mass spectrometry. This research illustrates how micro LC separations can be more advantageous than separations with conventional columns for some applications.

Other applications of micro LC include the speciation of organolead compounds [38], lead and mercury compounds [39], and the speciation of chromium [40].

3 SELECTING A MOBILE PHASE SUITABLE FOR LIQUID CHROMATOGRAPHY WITH ELEMENT-SPECIFIC DETECTION

When selecting a mobile phase, one must consider the type of detector being utilized. Element-specific detection methods such as ICP-MS, FAA, and ICP-AES are coupled with LC to perform elemental speciation analyses, and one must select a mobile phase that will not only achieve the desired separation but will also be compatible with the detector being used.

Liquid sample is aspirated into the flame, in the case of FAA, or the plasma, in the case of ICP-MS and ICP-AES, by a nebulizer. Mobile phases with a high salt content (>0.2 % total dissolved solid) can clog the nebulizer, and should be avoided [43]. Organic solvents such as methanol and acetonitrile are frequently used in LC especially in RPC. However, special precautions must be taken when introducing organic solvent into an ICP. Organic solvents cause plasma instability and may cause high reflective powers or extinguish the plasma all together. Soot may form on the sampler and skimmer cones of the ICP-MS. Even small amounts of organic solvent may have negative effects on the detector. Olesik and Moore [44] reported that both atom and ion emission signals were depressed when organic solvents (<2 % v/v) were present with ICP-AES. Despite this, the use of organic solvents cannot totally be eliminated in LC separations since they are often necessary to achieve a desired separation. Devices such as water-cooled spray chambers and Peltier coolers may be used to minimize the amount of solvent reaching the plasma. Furthermore, plasmas may be operated at increased RF powers to prevent the plasma from being extinguished.

Additional consideration must be made when using gradient elution. Gradient elution causes plasma or flame conditions to change during the course of the separation as the composition of the mobile phase pumped to the plasma/flame changes. This effect is dependent upon the nature of the gradient and the solvents utilized. However, analysts should be aware that gradient separations may cause changes in the ionization/atomization source.

4 DETECTORS USED FOR ELEMENTAL SPECIATION WITH LIQUID CHROMATOGRAPHY

Element-specific detectors are the most commonly used detectors for LC separations when performing speciation analyses. UV and diode array detectors, which are typically used to detect separated organic compounds, are not particularly useful in elemental speciation analyses because the analytes of interest frequently do not absorb UV light. Also, many compounds from the sample matrix may absorb in the UV causing interferences.

Sensitive, element-specific detectors are needed for elemental analysis. The sample concentration of analytes of interest in biological and environmental samples is often in the $\mu g\,L^{-1}$ or $ng\,L^{-1}$ range, so detectors sensitive enough to detect low levels of analyte are desirable. Methods of detection commonly used for elemental speciation analyses include flame atomic absorption spectrometry (FAAS), graphite furnace atomic absorption spectrometry (GF-AAS), inductively coupled plasma-atomic emission spectrometry (ICP-AES), and inductively coupled plasma-mass spectrometry (ICP-MS).

FAAS is one of the most widely used methods for the analysis of single elements in samples. As methods were developed to perform elemental speciation, FAAS was employed for element-specific detection. FAAS instruments have the advantage of being inexpensive, relative to other detectors such as mass spectrometers, and they are available in most laboratories. Also, FAAS units require liquid sample be continuously nebulized into the flame making it compatible with LC systems. However, the chief drawback of FAAS is that it is not sensitive enough for most elemental speciation applications. To improve sensitivity of FAAS detection, work has been done to improve the nebulization process.

GF-AAS is more sensitive than FAAS because the flame atomizer is replaced by an electrothermal atomizer which is often referred to as a graphite furnace. A few microliters of sample are placed in the atomizer. The sample is first evaporated at low temperature and then ashed at higher temperature in the atomizer. Then, the temperature is rapidly increased to 2000–3000 °C, causing rapid atomization of the sample. The atomic absorption is measured in the region above the heated surface [45]. Unlike FAAS, sample is not continuously aspirated into the graphite furnace making it difficult to interface this technique with LC. Typically when GF-AAS is used for LC detection, fractions of the LC effluent are collected for subsequent analysis by GF-AAS. The chromatograms obtained with LC-GF-AAS appear slightly different from chromatograms obtained when using detectors that allow continuous sample aspiration. Figure 4.1.8 shows an example of a chromatogram obtained when a GF-AAS was used for LC detection.

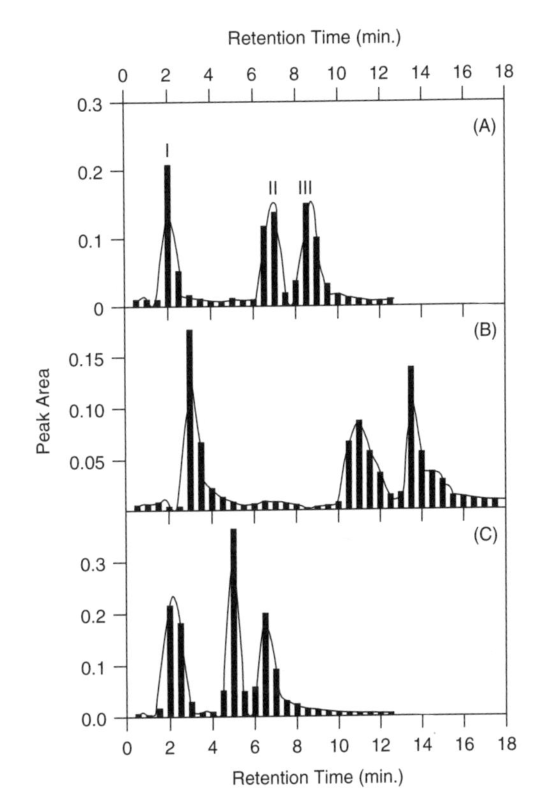

Figure 4.1.8. Chromatograms of separations obtained on reversed phase columns with GF-AAS detection. (a) Chromatogram obtained using a conventional LC column with collection of 500 μL fractions. (b) Chromatogram obtained using microbore LC column with collection of 40 μL fractions. (c) Chromatogram obtained using a micro LC column with collection of 5 μL fractions. I = *p*-arsanilic acid; II = 3-nitro-4-hydroxyphenylarsonic acid; III = 4-nitrophenylarsonic acid [42]. (Reprinted from *Journal of Chromatography*, Vol. 764, Pergantis *et al.*, Liquid chromatography and mass spectrometry, pp. 211–222, 1997, with permission from Elsevier Science.)

Cold vapor atomization is a special atomization technique used for the determination of mercury by AAS. Detection limits in the microgram per liter range can be realized with this technique. Mercury is oxidized to Hg^{2+} with sulfuric and nitric acid followed by reduction back to metallic mercury with $SnCl_2$. Gas is bubbled through the reaction mixture sweeping the metallic mercury into an absorption tube.

ICP-AES is frequently used for detection in elemental speciation analyses. The liquid sample stream is converted to an aerosol by the nebulizer. The aerosol then passes through a baffled spray chamber that allows only the smallest aerosol droplets to reach the plasma. Between 1 and 5 % of the original sample actually reaches the plasma. Detection limits for ICP-AES are typically in the range of $mg\,L^{-1}$ or high $\mu g\,L^{-1}$. These detection limits are often not low enough for many elemental speciation applications. To improve the sensitivity of ICP-AES, work has been done to improve sample transport efficiencies. Direct injection nebulization and hydride generation have been investigated to improve limits of detection.

ICP-MS is probably the most widely used LC detection method for elemental speciation analyses because of its superior sensitivity and because it is easily interfaced to LC systems. The use of an ICP-MS instrument as a detector for LC has been reviewed by Sutton and Caruso [46] ICP-MS systems are more expensive than FAAS and ICP-AES systems, but they offer detection limits in the nanogram per liter range. The sample introduction system for ICP-MS is nearly identical to the sample introduction system for ICP-AES, so like ICP-AES, ICP-MS suffers from poor sample transport efficiency. However, ICP-MS offers superior limits of detection because of the large signal-to-noise ratios that result from the use of a quadrupole mass analyzer. To improve sensitivity, work has been done to try to improve the sample transport efficiency. Techniques such as direct injection nebulization, ultrasonic nebulization, and thermospray nebulization have been used, with some success, to improve sensitivity. Micronebulizers such as the high efficiency nebulizer and the microconcentric nebulizer have been used to improve sensitivity when introducing sample at very low flow rates. Pneumatic nebulizers such as the glass concentric nebulizer and the cross flow nebulizer still remain the most commonly used nebulizers to introduce liquid sample into an ICP.

5 INTERFACING LIQUID CHROMATOGRAPHY WITH DETECTORS USED FOR ELEMENTAL SPECIATION

One of the advantages of utilizing LC for elemental speciation analyses is the ease with which LC systems can be interfaced to element-specific detectors. In detectors based on ICP-MS, ICP-AES, and FAAS sample is continuously aspirated into the plasma at a flow rate comparable to the flow rates typically utilized in LC analyses ($0.5-2.0\,\mathrm{mL\,min^{-1}}$). Interfacing the LC system to these detectors is easily achieved using a length of inert tubing such as polyetherether ketone (PEEK) tubing to connect the end of the LC column with the sample nebulizer. Care should be taken to minimize the length and internal diameter of the tubing to minimize band broadening.

Interfacing LC systems to detectors based on GF-AAS that introduce sample in discrete sample volumes is more difficult. Fractions of the column effluent may be collected for subsequent analysis by GF-AAS. Another approach is to completely volatilize the sample effluent prior to introduction into the graphite furnace.

Hydride generation is a technique that has been used to improve the sensitivity of FAAS, GF-AAS, ICP-AES, and ICP-MS in elemental speciation studies. If the species of interest are capable of forming volatile hydrides, the column effluent may be subjected to reaction with sodium borohydride converting the separated analytes into volatile hydrides which may then be swept to the flame, plasma, or graphite furnace. Only the elements arsenic, bismuth, germanium, lead, antimony, selenium, tin, and tellurium are capable of forming volatile hydrides. Hydride generation improves a method's sensitivity since the analytes are separated from much of the sample matrix. It also allows the sample to be introduced to the detector in gaseous form so energy is not required to volatilize the sample. Sample transport efficiencies are also dramatically improved. The use of hydride generation in chemical speciation with atomic spectrometry has been reviewed by Nakahara [47]. One limitation of hydride generation is that all of the species of interest for a particular element may not form volatile hydrides. For example, arsenate readily forms a volatile hydride, but arsenobetaine and arsenosugars, which are important environmental arsenic species, do not. To overcome this problem, after the species are separated chromatographically, they may be decomposed on-line prior to hydride generation. Le *et al.* [48] used on-line microwave decomposition of the LC column effluent to rapidly decompose organoarsenic compounds to arsenate. Once in the form of arsenate, volatile hydrides were formed that were detected by AAS.

6 CONCLUSIONS AND FUTURE DIRECTIONS

LC coupled with atomic spectrometric detection will continue to be one of the mainstays of elemental speciation analysis. LC offers the analyst tremendous flexibility and a technique that is generally compatible with and easily interfaced to atomic spectrometric detectors. Future research is likely to be seen in the areas of chiral LC, SEC, and micro LC. As more chiral stationary phases become commercially available, this technique will be utilized to a greater extent. SEC has already been used significantly in the area of elemental speciation, but as more focus is being placed on the role of metals in biological systems and an understanding of metalloproteins, SEC will be utilized to an even greater extent. Micro LC will become increasingly popular particularly as micronebulization techniques improve. Interfacing micro LC columns to atomic spectrometers has posed challenges in the past, but advances in the development of micronebulizers have made interfacing the two techniques easier. The reduced solvent consumption and sample requirements make micro LC an attractive technique.

Early work in elemental speciation involving LC utilized destructive detection methods that provided elemental information about analytes but not structural information. Coupling LC columns to mass spectrometers that utilize 'soft' ionization techniques allows the analyst to obtain structural information. Examples of 'soft' ionization techniques include electrospray ionization and tunable plasmas. Frequently unidentified metal-containing species are found in environmental and clinical samples during a speciation analysis. Often these species are unknown metabolites, and learning their identities may help researchers understand how an organism metabolizes a drug or reacts to a pollutant. Because of this, the use of detectors capable of providing structural information will continue to increase. These detectors are discussed in detail in subsequent chapters of this book.

It should be realized that frequently there are multiple LC techniques capable of solving an analytical problem, and rarely is there only one set of conditions capable of achieving a separation. To understand this point, one need only to look at a recent review on the use of LC for the speciation of organotin compounds [49]. Dozens of chromatographic conditions have been listed from the literature for the separation of organotin species. The versatility offered by LC will continue to make it one of the major separation techniques used for elemental speciation in the decade ahead.

7 REFERENCES

1. Toyo'oka, T. (Ed.), *Modern Derivatization Methods for Separation Sciences*, John Wiley & Sons, Ltd, Chichester, 1999.
2. Liu, W. and Lee, H. K., *J. Chromatogr. A*, **834**, 45 (1999).
3. Martin, A. J. P. and Synge, R. L. M., *Biochem. J.*, **35**, 1358 (1941).
4. Xu, H. and Lesage, S., *J. Chromatogr.*, **607**, 139 (1992).
5. Stäb, J. A., Rozing, M. J. M., van Hattum, B., Cofino, W. P. and Brinkman, U. A. T., *J. Chromatogr.*, **609**, 195 (1992).
6. Rivaro, P. and Frache, R., *Analyst*, **122**, 1069 (1997).
7. Cairns, W. R. L., Ebdon, L. and Hill, S. J., *Fresenius' J. Anal. Chem.*, **355**, 202 (1996).
8. Klinkenberg, H., van der Wal, S., Frusch, J., Terwint, L. and Beeren, T., *At. Spectrom.*, **11**, 198 (1990).
9. Rivas, C., Ebdon, L., Evans, E. H. and Hill, S. J., *Appl. Organomet. Chem.*, **10**, 61 (1996).
10. Dauchy, X., Cottier, R., Batel, A., Jeannot, R., Borsier, M., Astruc, A. and Astruc, M., *J. Chromatogr. Sci.*, **31**, 416 (1993).
11. Zoorob, G. K. and Caruso, J. A., *J. Chromatogr. A*, **773**, 157 (1997).
12. AL-Rashdan, A., Vela, N. P., Caruso, J. A. and Heitkemper, D. T., *J. Anal. At. Spectrom.*, **7**, 551 (1992).
13. Jiang, S. and Houk, R. S., *Spectrochim. Acta*, **43B**, 405 (1988).
14. Jiménez, O. and Marina, M. L., *J. Chromatogr. A.*, **780**, 149 (1997).
15. Ding, H., Wang, J., Dorsey, J. G. and Caruso, J. A., *J. Chromatogr. A.*, **694**, 425 (1995).
16. Suyani, H., Heitkemper, D., Creed, J. and Caruso, J., *Appl. Spectrosc.*, **43**, 962 (1989).
17. Wang, J., Tomlinson, M. J. and Caruso, J. A., *J. Anal. At. Spectrom.*, **10**, 601 (1995).
18. Thomas, P., Finnie, J. K. and Williams, J. G., *J. Anal. At. Spectrom.*, **12**, 1367 (1997).
19. Ackley, K. L., B'Hymer, C., Sutton, K. L. and Caruso, J. A., *J. Anal. At. Spectrom.*, **14**, 845 (1999).
20. Creed, J. T., Magnuson, M. L., Pfaff, J. D. and Brockhoff, C., *J. Chromatogr. A.*, **753**, 261 (1996).
21. Ulrich, N., *Fresenius' J. Anal. Chem.*, **360**, 797 (1998).
22. Inoue, Y., Sakai, T. and Kumagai, H., *J. Chromatogr. A*, **706**, 127 (1995).
23. Suyani, H., Creed, J., Davidson, T. and Caruso, J., *J. Chromatogr. Sci.*, **27**, 139 (1989).

24. Teräsahde, P., Pantsar-Kallio, M. and Manninen, P. K. G., *J. Chromatograph. A*, **750**, 83 (1996).
25. Mason, A. Z., Storms, S. D. and Jenkins, K. D., *Anal. Biochem.*, **186**, 187 (1990).
26. Dean, J. R., Munro, S., Ebdon, L., Crews, H. M. and Massey, R. C., *J. Anal. At. Spectrom.*, **2**, 607 (1987).
27. Matz, S. G., Elder, R. C. and Tepperman, K., *J. Anal. At. Spectrom.*, **4**, 767 (1989).
28. Lyon, T. D. B. and Fell, G. S., *J. Anal. At. Spectrom.*, **5**, 135 (1990).
29. Crews, H. M., Dean, J. R., Ebdon, L. and Massey, R. C., *Analyst*, **114**, 895 (1989).
30. Klueppel, D., Jakubowski, N., Messerschmidt, J., Stuewer, D. and Klockow, D., *J. Anal. At. Spectrom.*, **13**, 255 (1998).
31. Stalcup, A. M., Chiral separations, in *Kirk–Othmer Encyclopedia of Chemical Technology*, 4th edn, Supplement, John Wiley & Sons, Inc., New York, 1998, p. 133.
32. Méndez, S. P., Gónzalez, E. B., Fernández Sánchez, M. L. and Sanz Medel, A., *J. Anal. At. Spectrom.*, **13**, 893 (1998).
33. Sutton, K. L., Ponce de Leon, C. A., Ackley, K. L., Sutton, R. M. C., Stalcup, A. M. and Caruso, J. A., *Analyst*, **125**, 281 (2000).
34. Ponce de Leon, C. A., Sutton, K. L., Caruso, J. A. and Uden, P. C. *J. Anal. At. Spectrom.*, **15**, 1103 (2000).
35. Méndez, S. P., Gónzalez, E. B. and Sanz Medel, A., *J. Anal. At. Spectrom.*, **15** 1109 (2000).
36. Vissers, J. P. C., Claessens, H. A. and Cramers, C. A., *J. Chromatogr. A.*, **779**, 1 (1997).
37. Yang, F. J. (Ed.), *Microbore Column Chromatography, a Unified Approach*, Marcel Dekker, New York, 1989.
38. Tangen, A., Trones, R., Greibrokk, T. and Lund, W., *J. Anal. At. Spectrom.*, **12**, 667 (1997).
39. Shum, S. C. K., Pang, H. and Houk, R. S., *Anal. Chem.*, **64**, 2444 (1992).
40. Powell, M. J., Boomer, D. W. and Wiederin, D. R., *Anal. Chem.*, **67**, 2474 (1995).
41. Garraud, H., Woller, A., Fodor, P. and Donard, O. F. X., *Analusis*, **25**, 25 (1997).
42. Pergantis, S. A., Cullen, W. R., Chow, D. T. and Eigendor, G. K., *J. Chromatogr. A.*, **764**, 211 (1997).
43. Zoorob, G. K., McKiernan, J. W. and Caruso, J. A., *Mikrochim. Acta*, **128**, 145 (1998).
44. Olesik, J. W. and Moore, A. W., *Anal. Chem.*, **62**, 840 (1990).
45. Skoog, D. A., Holler, F. J. and Nieman, T. A., *Principles of Instrumental Analysis*, 5th edn, Harcourt Brace College Publishers, Philadelphia, PA, 1998.
46. Sutton, K. L. and Caruso, J. A., *J. Chromatogr. A*, **856**, 243 (1999).
47. Nakahara, T., *Bunseki Kagaku*, **46**, 513 (1997).
48. Le, X., Cullen, W. R. and Reimer, K. J., *Talanta*, **41**, 495 (1994).
49. Harrington, C. F., Eigendorf, G. K. and Cullen, W. R., *Appl. Organomet. Chem.*, **10**, 339 (1996).

4.2 Gas Chromatography and Other Gas Based Methods

J. Ignacio García Alonso and Jorge Ruiz Encinar

University of Oviedo, Spain

1 INTRODUCTION

There is a clear need today of measuring not only the total concentration of a certain element in the sample but also the concentration of the different chemical species in which this particular element may be distributed. The ability of a certain element to form different chemical species will be responsible for its geochemical distribution in the environment, its bio-availability and its toxicity towards different living organisms. Recently, the IUPAC [1] has defined the terms 'elemental speciation' (the distribution of defined chemical species of an element in a system) and 'chemical species' (specific form of an element defined as to isotopic composition, electronic or oxidation state, and/or complex or molecular structure)

Handbook of Elemental Speciation: Techniques and Methodology R. Cornelis, H. Crews, J. Caruso and K. Heumann
 ISBN: 0-471-49214-0

in trying to clarify those important concepts in modern analytical chemistry.

The big impact of speciation in modern analytical chemistry can be reflected not only in the vast amount of analytical methodologies developed in the last decade but also by the fact that new regulations are being implemented in different countries based on these studies [2]. This means that interest in elemental speciation will increase even more as these methodologies need to be implemented in routine chemical laboratories and included in quality assurance programs.

From an analytical point of view, speciation analysis is not an easy task. From the sampling to the final measurement step we have to take into account a series of problems which could make all our efforts worthless:

1. We need to keep the identity and concentration of the different species in the sample unchanged throughout the whole analytical procedure or, at least, provide means to correct for any changes in speciation.
2. We need to separate and positively identify the different chemical species present in the sample.
3. We need to determine very low levels of those species with adequate precision and accuracy.
4. We need to take into account matrix effects from complex biological and environmental samples which could affect most pretreatment processes used.

All the above mentioned factors make speciation analysis a complex task requiring several analytical steps which may include extraction, clean-up, derivatisation, preconcentration, separation and final measurement procedures. Most publications on speciation analysis have focused mainly on the last two analytical steps: separation and measurement. The preferred techniques make use of a potent chromatographic procedure (gas chromatography or high performance liquid chromatography) coupled to sensitive and selective (or specific) atomic detection techniques such as atomic absorption (AAS), atomic emission (AES) and mass spectrometry (MS) in combination with flames or, more recently, plasmas (ICP or MIP).

In this chapter we will focus on the use of gas chromatography (GC) for speciation analysis both for volatile compounds per se and for those nonvolatile compounds which can be derivatised and then separated by GC. We will describe different modes of GC and those derivatisation methods which are currently employed for speciation analysis. Finally, and due to its current interest, we will review the main characteristics and applications of the coupling of GC with inductively coupled plasma mass spectrometry (ICP-MS), discussing current and future applications.

2 SPECIES WHICH CAN BE ANALYSED BY GAS CHROMATOGRAPHY

The basic conditions governing the suitability of GC for the separation and detection of a given compound are its volatility and thermal stability. Within the field of speciation very few compounds fulfil those requirements directly and the analyst has to resort to chemical reactions to transform nonvolatile compounds (usually ionic) into volatile thermally stable compounds. Those reactions are known as derivatisation reactions and may include hydride generation, alkylation, etc. Many derivatisation reactions have been developed for GC and that has broadened the field of application of GC to the speciation of a large variety of compounds.

2.1 Volatile species

Within the field of elemental speciation several authors have identified volatile metal species in different samples. Feldmann and coworkers [3, 4] have identified volatile species such as dimethylmercury (Me_2Hg), dimethylselenium (Me_2Se), tetramethyltin (Me_4Sn), trimethylantimony (Me_3Sb), trimethylbismuth (Me_3Bi), methylated arsines (Me_xAsH_y, $x + y = 3$), dimethyltellurum (Me_2Te), tetraalkylated lead compounds (Et_xMe_yPb, $x + y = 4$) in sewage sludge gases and, more recently, molybdenum and tungsten hexacarboniles [5] ($Mo(CO)_6$ and $W(CO)_6$) in gases from municipal waste disposal sites. Most of these compounds are extremely toxic and their presence in these samples needs to be studied. Unfortunately, the quantification of these species is complex because of the lack of commercial standards or certified reference materials for most of these compounds. For that reason, Feldmann [6] developed a semiquantitative methodology consisting in the addition of nebulised water at the exit of the GC column in order to obtain the same response from the volatile compound and from the aqueous standards of those elements.

Volatile tetraalkyllead compounds have been used for years as antiknocking agents in gasoline. These compounds were detected in air samples at levels up to 14 ng Pb m^{-3} by sampling 70 L of air in a cryogenic trap. The authors concluded that their presence in air was due to gasoline evaporation rather that to its incomplete combustion [7]. Tetraethyllead is the standard antiknocking agent but other authors have detected the presence of mixed methyl-ethyl tetraalkyllead compounds in gasoline by GC-ICP-MS [8, 9]. Recently, the measurement of lead isotope ratios [10] in different organolead compounds present in gasoline and atmospheric particulate matter has been used to study lead pollution sources.

Volatile compounds of Se, Sn, Hg and Pb have been also detected in natural [11] and marine [12] waters. These studies demonstrated that the interaction between anthropogenic and natural sources led to the formation of volatile metallic and organometallic species in rivers, estuaries and the sea. Similar studies carried out by the same research group, using cryogenic trapping, showed the presence of volatile species of Hg, Sn, In, Ga, Se, P and As in air samples collected in urban and rural areas. The same authors detected volatile tin hydrides and methylated butyltin compounds in the bay of Arcachon and other polluted areas of France, Belgium and Holland [13]. The extremely low concentration levels of these compounds in the samples required special preconcentration techniques. In this case, cryogenic trapping was used after purging the compounds from 1 L of water with He or passing 100 L of air through adsorption columns immersed in liquid nitrogen. Thermal desorption GC-ICP-MS was used to achieve both species separation and adequate limits of detection. Previous environmental studies on volatile species in the atmosphere used room temperature trapping on solid sorbents [14, 15] and thermal desorption to the GC column [16], avoiding the use of organic solvents which could cause contamination.

Recently [17] the coupling of GC to a double focusing ICP-MS instrument allowed the speciation of sulfur compounds in human saliva incubated under anaerobic conditions. The sulfur compounds detected, H_2S, CH_3SH, $(CH_3)_2S$, $(C_2H_5)_2S$, $(CH_3)_2S_2$ and $(C_2H_5)_2S_2$, are generated by anaerobic microorganisms in the mouth and can be the cause of more than 80 % of the cases of bad breath. The use of high resolution in the mass spectrometer allowed the separation of the ^{32}S peak from its polyatomic interference $^{16}O_2^+$ and the detection of up to eight volatile sulfur compounds was facilitated by the use of a special GC column (SPB-1 Sulphur).

Other applications for the determination of volatile compounds, such as the detection of tetraalkyllead compounds in gasoline [18] or halogenated compounds in macro-algae [19] are

Table 4.2.1. Volatile elemental species detected according to the type of sample.

Sample type	Detected species	Technique used	Ref.
Gasoline	Tetraalkyllead compounds	GC-FAAS	18
Landfill gas	As, Se, Sn, Sb, Te, Hg, Pb, Bi, Mo, W compounds	GC-ICP-MS	5
Air	Tetraalkyllead compounds	GC-AAS	14
Air	Hg^0, MeHgCl, Et_2Hg, Me_2Hg	GC-AFS	15
Air	MeHgCl, Me_2Hg	GC-AED	16
Fermented saliva	H_2S, CH_3SH, $(CH_3)_2S$, $(C_2H_5)_2S$, $(CH_3)_2S_2$, $(C_2H_5)_2S_2$	GC-SF-ICP-MS	17
Macro-algae	CH_3I, CH_2I_2, $CHBr_3$, $CHBrCl_2$, CH_2BrCl, C_2H_5I	PT-GC-AED	19
Estuarine waters	Me_2Se, Me_2Se_2, Me_2Hg, Et_2Hg, Me_4Sn, Et_4Sn, Me_4Pb, Et_4Pb	CT-GC-ICP-MS	11

worth mentioning here. Table 4.2.1 summarizes the different volatile species detected according to the type of sample.

2.2 Nonvolatile elemental species

Most elemental species are ionic or polar compounds which show high boiling points and low thermal stability. Over the years, suitable derivatisation reactions have been developed for their separation and determination by GC. Those chemical reactions were selected in order to produce compounds of low polarity, low boiling points and high thermal stability with adequate chemical yields. In many cases, these derivatisation reactions can be coupled to a preconcentration and/or separation technique (e.g. liquid–liquid extraction), improving the sensitivity and selectivity of the analytical process. In the following pages we will review the most important derivatisation reactions for elemental speciation using GC.

2.3 Derivatisation reactions

In all speciation steps we need to preserve the relevant metal–carbon bonds so we can identify the original species present in the sample, and this is especially true for derivatisation reactions. In that sense, all derivatisation reactions used, prior to GC separations, follow one of the following processes: (i) conversion of inorganic or organometallic ions into covalent volatile compounds in aqueous media (e.g. hydride generation, ethylation), (ii) conversion of inorganic or organometallic ions into covalent volatile compounds in organic media using Grignard reagents (e.g. butylation) and (iii) conversion of ionic species into stable volatile chelates (dithiocarbamates, acetonates and trifluoroacetonates).

It is important to realize that all derivatisation procedures can be affected by matrix effects in the sample and, hence, recovery studies have to be carried out for different sample types under the selected experimental conditions. Unfortunately, there is a lack of commercially available derivatised analytes which makes the computing of derivatisation recoveries difficult [20, 21]. Most authors use internal standards to correct for derivatisation recoveries and analyte losses but this is clearly not the best solution. Other authors recommend the use of isotopically enriched species and isotope dilution procedures which do not require known or quantitative recoveries in any of the extraction, clean-up and derivatisation steps: if we can ensure that isotopic equilibrium between the original and spiked species in the sample is achieved, the recovery factors in any subsequent separation and derivatisation procedure will not affect the final concentration results because the final measurement is an isotope ratio that will be constant and independent of the number of molecules or atoms isolated to measure that ratio. Another advantage of the use of enriched isotopes is the fact that species transformations or degradation can be detected using several enriched isotopes [22].

2.3.1 Hydride generation

This well-known derivatisation reaction is usually applied to small inorganic and organometallic ions which form highly volatile covalent compounds. Different species containing the elements As, Sb, Hg, Sn, Pb, Bi, Cd, Se, Te and Ge have been derivatised using this technique [23]. The reaction takes place with the sample in acidic aqueous medium by adding sodium borohydride both as reductor and hydride source. Depending on the species and the element considered we have to optimise the concentration of reducing agent and the type and concentration of the acid used [24, 25]. Unfortunately, this procedure cannot be applied to all species of the same element due to thermodynamic and/or kinetic constraints or because of low stability of the hydrides formed. For example, this is the case for some alkylated lead species [26] and the high oxidation states of Se, As [27] and Sb. In the last case, these high oxidation state species have to be reduced previously, with the consequent loss of speciation information.

On the other hand, the methylated species of mercury [26] and germanium [28] form hydrides easily and the metal–carbon bond is preserved during the reaction which facilitates the identification and quantification of the species. Similar results were obtained for tin speciation: tin hydrides are readily formed and the tin–carbon bond is not broken during derivatisation. There have been numerous papers on tin speciation using hydride generation before GC separation. In most cases, hydride generation was coupled to a cryogenic trapping step in a chromatographic column immersed in liquid nitrogen. After the trapping step the liquid nitrogen was removed and a nichrome wire, wound around the column, was heated electrically to desorb the analytes and separate them in the GC column. AAS [29] and ICP-MS [30] have been used as suitable selective and sensitive detection methods. The system used by Segovia García *et al.* [30], which is basically an evolution of that published previously by Donard *et al.* [29], is presented in Figure 4.2.1. Basically, a given volume of sodium borohydride was added to the sample with the help of a programmed peristaltic pump. A flow of helium carrier gas transported the hydrides formed from the sample container to the GC column which was immersed in liquid nitrogen. For ICP-MS work it was important to remove the excess of hydrogen formed before the ICP torch and that was accomplished with the help of a three-way valve connected after the GC column during the preconcentration step. After the given purge and trap time had elapsed, then (i) the liquid nitrogen was removed, (ii) the solenoid valve just before the GC column and the three-way valve after the column were switched, (iii) the heater of the column was connected and (iv) the ICP-MS acquisition sequence was initiated. This procedure allowed the on-line derivatisation–preconcentration–separation–detection of butylated tin species with absolute detection limits between 50 and 200 pg for AAS and between 4 and 7 pg for ICP-MS.

Clark and Craig [31] and Sullivan *et al.* [32] suggested an interesting approach in which the hydride generation occurred in a special reactor packed with sodium borohydride and located inside the gas chromatograph. This 'reactive GC' approach was applied successfully for tin speciation with a considerable reduction in analysis time.

However, hydride generation suffer from severe matrix interferences when applied to real samples

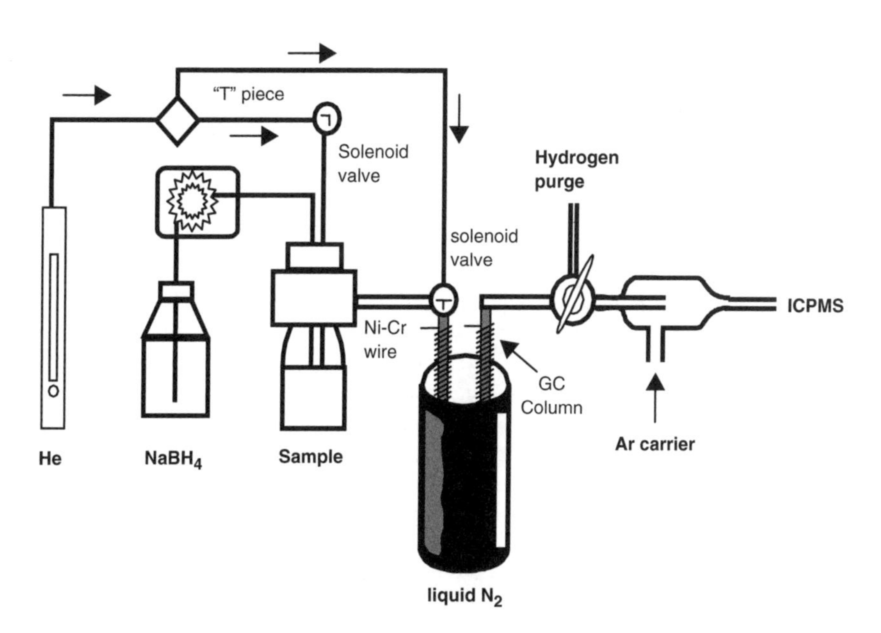

Figure 4.2.1. Hydride generation, cryogenic trapping, GC separation and ICP-MS detection for the speciation of butyltin compounds. Reprinted with permission from John Wiley & Sons, Ltd.

such as wastewaters, sediments and biota. These interferences cause both signal suppression and decreased precision in the measurements. Different authors have published studies on the effect of organic matter and transition metals on hydride generation yields [21, 33] but most studies have focused on those matrix effects for butyltin speciation studies [34] due to the usually complex matrices in which those compounds are determined. Organic solvents, pesticides and other organic compounds did not influence hydride reaction yields but humic substances caused low reproducibility due to the formation of foams during the reaction. More serious interferences were caused by transition metals in the reaction medium which decomposed the organometallic hydrides formed and sodium borohydride itself [26]. A possible solution to this problem was the addition of L-cysteine to the reaction medium [33]. This reagent formed complexes with the transition metals improving the sensitivity and reproducibility of the hydride generation reaction.

Other applications of hydride generation for elemental speciation in combination with GC have been published [35–38]. For example, the speciation of antimony in freshwater plant extracts was described by Dodd *et al.* [36]. These authors indicated that the problem of organostibine molecular rearrangement could be solved by adequate selection of the experimental procedure. Table 4.2.2 summarizes those and other applications of hydride generation in elemental speciation studies.

Table 4.2.2. Elemental species detected after hydride generation and separation by GC.

Sample type	Detected species	Instrumentation used	Ref.
Fish	TBT, DBT	GC-FPD	32
Estuarine waters	Sn^{4+}, Me_xSn^{4-x}, Bu_xSn^{4-x}	GC-QF-AAS	29
Mussels	MBT, DBT, TBT	GC-QF-AAS	35
Aquatic plants	Different Sb compounds	GC-MS	36
Natural and wastewaters	Ge^{4+}, $MeGe^{3+}$, Me_2Ge^{2+}	CT-ICP-MS	28
Harbour sediment	$MeSnCl_3$, MBT, DBT, TBT	CT-GC-QF-AAS	34
Soils	As(III), As(V), MMA, DMA	GC-SF-ICP-MS	37
Seawater	Hg^+, $MeHg^+$	GC-AFS	38

2.3.2 *Aqueous ethylation*

The ethylation reaction consist in the addition of one or more ethyl groups to inorganic or alkylated metal species to form the di- (Hg, Se), tri- (Bi) or tetraalkylated species (Sn, Pb) which are hydrophobic, volatile and thermally stable and, hence, suitable for GC separations. The ethylation reaction is simple and quantitative (except for inorganic ions), can be performed in the aqueous phase and does not destroy existing metal–carbon bonds. There are two ways in which ethylation can be performed: (i) using Grignard reagents [39, 40] (ethyl magnesium bromide) or (ii) using sodium tetraethylborate [41]. In the first case ethylation is performed in an anhydrous organic phase, while in the second case it is performed in the aqueous phase. The use of Grignard reagents is less prone to matrix interferences when compared to aqueous ethylation or hydride generation. This is because Grignard ethylation takes place in an organic phase after the separation of the analytes from the matrix by liquid–liquid extraction. Also, De la Calle-Guntiñas *et al.* [42] have reported that the derivatisation using Grignard reagents (ethylation and pentylation) provided higher reaction yields than aqueous ethylation using $NaBEt_4$. The main disadvantage of Grignard ethylation is the requirement of an anhydrous organic medium for the reaction to take place and that means that a previous liquid–liquid extraction procedure has to be applied. For the speciation of organotin compounds this liquid–liquid extraction was performed in a strong acidic medium [43], whereas organolead compounds were extracted in weak acid media [44]. Complexing agents used for extraction included tropolone [40, 45] (for organotin compounds), ditizone and diethyldithiocarbamate (DDTC). The fact that the derivatisation takes place in an inert atmosphere and that the excess of Grignard reagent has to be destroyed after derivatisation makes ethylation using Grignard reagents cumbersome and time consuming.

Aqueous ethylation using sodium tetraethylborate combines the advantages of working in the aqueous phase of hydride generation (no previous extraction into organic phase, speed, convenience of use) with the low matrix interferences of

Grignard reagents. De Diego *et al.* [46] compared the analytical performance characteristics of both aqueous ethylation and hydride generation for mercury speciation in samples of high salt and transition metal content. The results demonstrated that the presence of transition metals seriously affected hydride generation while the content of chloride influenced the ethylation yields. Other authors suggested that the two derivatisation techniques are complementary [47], as their performance depends on the type of sample measured. In general terms, ethylation using $NaBEt_4$ fits well in the actual trend of speciation analysis: to minimize the number of analytical steps, and so reduce sample handling and analysis time. This reduction in sample handling and reagent usage, which is provided by aqueous ethylation, leads to improved reproducibility and lower analytical errors (losses of analytes, contamination, etc).

Aqueous ethylation for trace metal speciation was first proposed by Rapsomanikis *et al.* [48] in 1986 for the speciation of methylated organolead compounds in waters using a home-made GC-AAS interface after purge and trap preconcentration. Later, this derivatisation procedure was applied for the speciation of, mainly, Sn, [49, 50, 51] Hg [48, 52, 53] and Pb [48, 54] compounds. An excellent review on the use of $NaBEt_4$ for trace metal speciation was published by Rapsomanikis [55]. In this review, the author explains the chemical reactions that take place and the applications of this reagent for trace metal speciation. We have also to take into account that, when analysing complex samples, a large excess of $NaBEt_4$ is usually required to compensate for the consumption of the reagent by other sample components [56].

In situ derivatisation using $NaBEt_4$ has also been applied to reduce analysis time [57]. In this case, a packed reactor inside the gas chromatograph containing $NaBEt_4$ was used in a similar way to that previously described for hydride generation [31, 32]. Another alternative for shorter analysis times consist in combining matrix extraction, derivatisation with $NaBEt_4$ and liquid–liquid extraction in a single analytical step [58]. All this process can be performed in less than 3 mins with the use of focused microwave systems [59].

Table 4.2.3. Organometallic species which have been detected after derivatisation using sodium tetraethylborate.

Sample type	Detected species	Instrumentation used	Ref.
Waters	Me_2Pb^{2+}, Me_3Pb^+	CT-GC-AAS	48
Waters	Hg, MeHg, EtHg	GC-AFS/MS/MIPAES	60
River sediment	MBT, DBT, TBT	GC-QF-AAS	50
Fish	Hg, MeHg	GC-FAPES	53
Sediments, biota and wastewaters	MBT, DBT, TBT	GC-FPD	61
Open-ocean seawater	MBT, DBT, TBT, MPhT, DPhT, TPhT	GC-ICP-MS (shield torch)	62
Sediment	MBT, DBT, TBT	GC-AAS/MS	57
Fish and marine sediments	MBT, DBT, TBT	GC-MIP-AES	63

Other applications of aqueous ethylation have been described [60, 61, 62, 63] and a summary of relevant applications of ethylation for elemental speciation is presented in Table 4.2.3.

The combination of ethylation with $NaBEt_4$ and solid-phase microextraction (SPME) for the detection of very low concentrations of organometallic compounds in waters is a rapidly growing field [64]. We will expand on the use of SPME for extraction/injection in combination with GC later in this chapter.

2.3.3 *Aqueous propylation*

The generalized use of $NaBEt_4$ as a derivatisation reagent for organotin speciation has not been extended to other organometallic species of Pb or Hg. This was limited by the fact that most Pb species of interest contained ethyl groups which, after derivatisation, would form the same species. For example, Et_4Pb, Et_3Pb^+, Et_2Pb^{2+} and Pb^{2+} will all form Et_4Pb after derivatisation with the corresponding loss of speciation information. The same can be said of Et_2Hg, $EtHg^+$ and Hg^{2+}, which would prevent the use of $EtHg^+$ as internal standard for the determination of both inorganic and methylmercury compounds. This problem can be solved by resorting to a different alkylation reagent. So, Grignard reagents based on propyl magnesium or butyl magnesium have been extensively used for organolead speciation [65, 66].

Another recent approach, which maintains the advantages of easy handling and aqueous derivatisation of $NaBEt_4$, and can be applied for both Pb and Hg speciation is the use of sodium tetrapropylborate ($NaBPr_4$). The synthesis of this derivatisation reagent was described in detail by De Smaele *et al.* [67], who applied it to the simultaneous speciation of Pb, Hg and Sn compounds in environmental samples. Later, other authors compared both derivatisation reagents for the speciation of mercury compounds [68]. In this paper, the advantages of $NaBPr_4$ over $NaBEt_4$ for the determination of both inorganic mercury and methylmercury consisted mainly in the ability to use ethylmercury as internal standard for quantification after selective GC-ICP-MS detection. The application of this reagent to the detection of organolead compounds in snow and road dust has been described [69].

The main drawback of the use of $NaBPr_4$ is the low chemical stability of this reagent in comparison with its ethylated analogue. In spite of this, the use of aqueous propylation has increased in recent years as the reagent can now be obtained commercially. Some of the applications described to date are summarized in Table 4.2.4.

2.3.4 *Grignard derivatisation*

Grignard derivatisation reactions, which use alkylmagnesium halides, can only be performed in a water-free organic phase and under inert atmosphere. That means that these reactions require a more complex experimental set-up but have the advantage of quantitative derivatisation in most cases. The organometallic species, usually present in an aqueous phase, have to be extracted into an organic solvent prior to derivatisation and this is done generally using chelating agents such as tropolone, DDTC, etc. Different alkyl groups have been used including methyl, propyl, butyl, pentyl and hexyl. However, the most generally used alkyl group has been the butyl group.

Table 4.2.4. Applications of tetrapropylborate to elemental speciation.

Sample type	Detected species	Instrumentation used	Ref.
Snow and road dust	TML, DML, TEL, DEL	GC-MIP-AES	69
Fish	Hg, MeHg	GC-ICP-MS	68
Sediments	MBT, DBT, TBT	GC-ICP-MS	67

Butylation reactions, using for example butylmagnesium chloride, are interesting and useful alternatives for the speciation of both Hg and Pb compounds in biological [68, 70, 71] and environmental samples [72]. Radojevic *et al.* [73] compared butylation and propylation efficiencies for organolead compounds using Grignard reagents. They concluded that propylation yields for dialkylated species (R_2Pb^{2+}) were higher than butylation yields. Other authors have also described lower butylation yields for these species [44].

The extraction of organolead compounds into the organic phase has been performed using DDTC at pH 9 in the presence of EDTA to prevent the co-extraction of inorganic lead. [70, 72, 74] In this way the extraction of inorganic lead, usually present in much higher concentration than the organolead compounds, is prevented. This is necessary as a large quantity of inorganic lead would consume the available butylation reagent, reducing the derivatisation yield for the other organolead compounds, and form huge amounts of the late eluting Bu_4Pb which could cause memory effects due to tailing and column contamination.

Butylation has also been evaluated for the speciation of mercury compounds. García Fernández *et al.* [68] compared aqueous ethylation and propylation with Grignard butylation for the determination of methylmercury in marine reference materials (DOLT-2, NRCC, Canada). Both aqueous propylation and Grignard butylation provided similar results as ethylmercury could be used as internal standard for both derivatisation reactions.

The use of pentylmagnesium halides has been mainly focused to the determination of organotin compounds in biological [45, 35, 75] and environmental samples [76]. However, this reagent has been also applied with good results to the speciation of both lead and mercury [77]. Stäb *et al.* [78] reported that derivatised organotin species were more prone to degradation during the destruction of the excess of Grignard reagent when the derivatisation alkyl group was smaller. So, larger alkyl groups, such as pentyl or hexyl provided more stable species. Also, the lower volatility of the pentylated species in comparison with those ethylated or propylated would facilitate the preconcentration of those species by evaporation of the solvent under an inert gas flow. There are two drawbacks which have been described for the pentylation of organotin compounds: first, the derivatised compounds would require higher elution temperatures and could condensate in the interface between the gas chromatograph and the detector [24]. Second, the derivatisation yields for the pentylation of di- and trialkyltin compounds were lower than when using shorter chain derivatisation reagents [76].

Other Grignard reagents, such as methylmagnesium [78, 79], propylmagnesium [80, 81] or hexylmagnesium halides [82] have also been proposed for elemental speciation. Cai *et al.* [82] applied supercritical fluid extraction (SFE) in combination with Grignard derivatisation. The main advantage of this approach is that extraction, derivatisation and clean-up can be performed in a single step without the need for organic solvents. The technique of SFE has been also applied for the extraction of mercury [83], lead [84] and arsenic [85] compounds in environmental samples.

Some relevant applications of Grignard derivatisation in elemental speciation are summarized in Table 4.2.5.

2.3.5 *Other derivatisation reactions*

Many other derivatisation reactions have been described for elemental speciation. Of special interest are those derivatisation reactions which can be performed in the aqueous phase. For example, the use of sodium tetraphenylborate has been proposed [60, 86] as an alternative to $NaBEt_4$ when information about $EtHg^+$ is required together with inorganic Hg and $MeHg^+$.

The determination of sulfur- and selenium-containing amino acids is another field where derivatisation reactions for GC separation are often employed. Studies reported by Clausen and Nielsen [87] demonstrated that L-selenomethionine is the Se species which can be more efficiently absorbed by the organisms and that explains the growing interest in the determination of Se-containing amino acids in foods and nutritional supplements. The derivatisation of amino acids for GC separation has been the subject of numerous publications. In general, it is necessary to derivatise both the carboxylic acid and the amino group to obtain thermally stable volatile compounds. This derivatisation can be performed in two sequential steps [88] consisting in the esterification of the carboxylic acid in an acidic medium using an alcohol (isopropanol or isobutanol) and the subsequent acylation of the amino group using an anhydride (trifluoroacetic or heptafluorobutiric). Later, the direct derivatisation of both the carboxylic acid and the amino group using alkyl chloroformates was proposed [89]. In our experience [90] the two-step procedure provided higher recoveries and a cleaner GC-MS chromatogram while the one-step

Table 4.2.5. Applications of Grignard reagents in elemental speciation.

Sample type	Derivatisation	Extraction	Detected species	Instrument	Ref.
Atmospheric particulates	BuMgCl	DDTC/hexane	DML, DEL, MEL	GC-ICP-MS	72
Rainwater	PrMgCl	DDTC/hexane	TML, DEL	GC-ICP-MS	80
Human urine	BuMgCl	DDTC/pentane–hexane	Pb^{2+}, TML, TEL	GC-MS	70
Marine products	EtMgBr	NaCl–HCl/Et_2O–hexane	TBT, DBT, TPhT	GC-FPD	39
Mussels	MeMgI	NaCl–HCl/Et_2O–hexane	MBT, DBT, TBT, MPhT, DPhT, TPhT	GC-MIP-AES/MS	78
Fish tissue	BuMgCl	DDTC/toluene	Hg, MeHg	GC-ICP-MS	68
Polyurethane foam	PrMgBr	HCl/toluene	DBT	GC-FPD	81
Sediments	HexMgBr	SFE	TBT, PDT, TPhT	GC-FPD	82

Table 4.2.6. Analytical characteristics of some derivatisation reactions used for the speciation of organometallic compounds.

Characteristic	Hydride generation	Aqueous alkylation	Grignard reagents
Reaction yield	High for small ions but lower for trialkylated species	High, except for monoalkylated and inorganic species	Very high
Reproducibility	Low due to foam formation	High	High
Interferences	Severe in samples with high organic or metallic content	Low matrix effects except in the presence of chloride	No interferences due to matrix separation
Stability of the reagents	High	Decompose slowly with light and humidity	Require an inert and dry atmosphere
Ease of handling	High due to the low reaction pH (pH ≅ 2)	High (pH ≅ 5)	Very low due to extraction into an organic phase, inert atmosphere and destruction of the excess of reagent
Speed of reaction	Instantaneous, allows on-line derivatisation	High, can be accelerated using MW irradiation	Low, the global process requires several reaction steps
Limitations	Only suitable for small organometallic species in simple matrices	$NaBEt_4$ not suitable for some applications, use of $NaBPr_4$ instead	None
Cost of reagents	Low	High	High

procedure formed one main product and a second subproduct showing the selenium isotopic pattern in the GC-MS spectrum.

2.3.6 Comparison of derivatisation reactions

The analytical characteristics of three derivatisation reactions currently used for the speciation of organometallic compounds are compared in Table 4.2.6. As can be observed in the table, the suitability of one or other reaction for a given application will depend mainly on the type of organometallic compounds analysed and on the matrix. In general terms, aqueous alkylation reactions, when both tetraethyl- and tetrapropylborate are available, seem to offer the best analytical characteristics. This is also justified by the increasing number of publications on elemental speciation using this type of derivatisation reaction.

2.4 Preconcentration and clean-up before GC separation

Elemental speciation in complex biological (human serum, fish tissue, ...) or environmental samples (sediments, waste waters, ...) requires, in many cases, the use of different preconcentration and/or clean-up procedures which could isolate the species of interest from the sample matrix. Samples with high fat content, sulfur-containing compounds, etc. could seriously affect both derivatisation and extraction yields and even prevent the detection of the sought compounds due to spectral interferences. Most preconcentration and clean-up procedures make use of classical approaches which are tedious, time consuming, require large quantities of organic solvents and are error prone due to contamination or analyte losses. The fact that most standard analytical procedures make use of classical clean-up and preconcentration methods has prevented further developments in this field [91]. However, in the last few years, many solid-phase extraction (SPE) and solid-phase microextraction (SPME) procedures have been published which provided improved analytical performance characteristics comparable with classical methods.

Procedures involving SPE either with C18 [40] or fluorisil [39] cartridges or filtration membranes [70] have been published. All these procedures are normally applied before the final derivatisation reaction. However, the clean-up of the organic extract prior to the final injection into the gas chromatograph has also been applied using, for example, silica gel columns [32, 40].

The last revolution in the field of organic sample preparation, which can also be applied for preconcentration and clean-up in elemental speciation, has been the introduction of SPME [92]. Since the pioneering work of Arthur and Pawliszyn [93] in 1990 the use of SPME procedures in elemental speciation, and in organic analysis in general, can be considered almost routine nowadays. The advantages of SPME for elemental speciation arise from its simplicity: in a first step the analytes are distributed between the fibre and the sample and, once equilibration has taken place, the analytes are thermally desorbed in the GC injector block. The whole SPME procedure provides several advantages in comparison with classical preconcentration and clean-up methods: (i) it does not require the use of organic solvents, (ii) it is fast and simple to perform, (iii) it allows *in situ* preconcentration, (iv) it can be automatised, (v) it can be selective by adequate selection of the fibre material, (vi) it has less breakthrough problems than SPE, (vii) it requires low sample volumes because of its high preconcentration factor and (viii) it does not require further clean-up. All these characteristics make SPME an excellent sample preparation technique.

In spite of the fact that SPME was first developed for the analysis of organic compounds, applications in elemental speciation were quickly developed [94–102] for the analysis of lead, mercury and tin species with excellent results. Most of these publications make use of nonpolar poly(dimethylsiloxane) fibres which require prior derivatisation of ionic organometallic species. In most cases derivatisation is done using $NaBEt_4$ in the same solution where SPME is performed. However, derivatisation can be performed in the fibre itself or in the GC injector after preconcentration [103].

The coupling of SPME with a gas chromatograph is relatively simple using traditional split/splitless injectors for capillary columns. The only modification required is the use of narrower bore glass liners, which provide a linear flow of gas along the fibre surface [103] and which facilitate the desorption of the analytes in the injector. The parameters which influence the retention of the analytes in the fibre include: (i) the nature of the fibre itself, (ii) the mode of operation (head space or immersed), (iii) the volume of sample, (iv) its pH and ionic strength, (v) the mode of stirring, (vi) adsorption temperature and (vii) adsorption time. After adequate optimisation of all these parameters, detection limits in the low $ng\,L^{-1}$ range have been accomplished for Pb, Hg and Sn organometallic species [98]. Selected applications of SPME to elemental speciation using GC are included in Table 4.2.7.

Stir bar sorptive extraction (SBSE) [104] is a new extraction procedure in which magnetic stir bars are coated with poly(dimethylsiloxane) and act as both stirrer and extraction phase. This procedure has been applied for elemental speciation showing that the extraction efficiency of stir bars is superior to that of SPME but they require special desorption equipment and cryofocusing of the analytes prior to GC separation [105]. All derivatisation procedures used with liquid–liquid extraction, SPE and SPME can be also used with SBSE, so we can expect an increase in the use of SBSE for elemental speciation in the near future.

Table 4.2.7. Applications of SPME for elemental speciation using GC as separation technique.

Type of sample	Derivatisation reagent	Detected species	Instrument	Ref.
Sea water	$NaBH_4$	MBT, DBT, TBT	GC-FPD	99
River water, fish	$NaBEt_4$	MeHg	GC-MS	94
Mussels and potatoes	$NaBEt_4$	TPhT	GC-ICP-MS	102
Tap water	$NaBEt_4$	Pb^{2+}, Et_4Pb	GC-ITMS/FID	96
Natural and wastewaters, sediments	$NaBEt_4$	MBT, DBT, TBT, MPhT, DPhT, TPhT	GC-FPD	100
Waters and sediments	$NaBEt_4$	TeET, TeBT, MBT, DBT, TBT	GC-FPD	101

3 SEPARATION TECHNIQUES USED

3.1 Cryogenic trapping and thermal desorption in packed columns

This analytical methodology has been extensively employed for the preconcentration and separation of both volatile (directly) and ionic (after derivatisation) species. The basic idea consists in the purge of the analytes, either derivatised or not, from the sample using a gentle flow of He and their trapping on a U-shaped glass column filled with chromatographic material and immersed in liquid nitrogen (−192 °C). After the trapping step, the liquid nitrogen is removed and the glass column is heated using a nichrome wire [25]. The analytes are then desorbed sequentially from the column and carried to the detector by another flow of helium. This experimental set-up is depicted in Figure 4.2.1 for hydride generation coupled to ICP-MS. The connections between the column and the detector have to be as short as possible to minimize dead volumes and peak broadening. Some authors recommend also the heating of the connection tubing between column and detector to prevent condensation of low volatility analytes and, hence, the existence of peak broadening effects. For the determination of volatile species in air or other gases it is also recommended to use a dry ice–acetone trap (−20 °C) to retain water vapour prior to the liquid nitrogen trap. The separation of the different species is based more on the differences in boiling points than on the properties of the chromatographic packing used. Normally apolar sorbents, such as Chromosorb, are used.

As we have said before, this system has been extensively used for the analysis of volatile species in gases (air, landfill gas, etc.) and waters allowing the separation of the different elemental species and the treatment of large volumes of sample. In comparison with the use of solid sorbents at room temperature cryogenic trapping offers two distinct advantages: (i) it is not selective, so many different species of various elements can be trapped simultaneously and (ii) unstable chemical species can be preserved for long periods of time before the desorption and analysis is performed. The latter characteristic is specially important for the analysis of gaseous samples [106]. In this way, different volatile species have been determined in natural waters [11, 13], landfill gases [5] and different atmospheres [106]. The low limits of detection reported using ICP-MS detection (down to femtogram levels) indicate the degree of development reached by this methodology. The coupling of the cryogenic trap to ICP-MS instruments offer different possibilities. While Pécheyran *et al.* [106] use a supplementary flow of argon to support the plasma, Feldmann and Cullen [5] employ a T piece to mix the helium carrier gas with a wet aerosol obtained using the ICP-MS nebuliser.

For the detection of nonvolatile species using this approach the only difference is the use of a reaction cell before the cryogenic trap. In this reaction cell the nonvolatile species are derivatised, converted into volatile species and transported by a flow of helium to the cryogenic trap (see Figure 4.2.1). For derivatisation, hydride generation has been mainly employed [107, 108] but also aqueous ethylation has been used [48] with satisfactory results for lead, mercury and tin species. However, it was observed that the detection of high molecular weight hydrides and ethylated compounds suffered from condensation and memory effects [109].

An interesting development in this approach is the possibilities of automation of the whole system [110]. For this purpose, a modified split/splitless injector using a programmed temperature vaporizer (PTV) was used to trap at −40 °C ethylated methyl- and butyltin species. To prevent band broadening, the whole chromatograph was cooled to preconcentrate the analytes at the top of the chromatographic column. The detection limits, using standard GC-MS instrumentation, were in the low $\mu g\,L^{-1}$ levels.

3.2 Gas chromatography with packed columns

Packed column GC was the first chromatographic technique to be applied for elemental speciation. It has important advantages over thermal desorption from either cryogenic traps or solid sorbents: (i) the interaction with the chromatographic packing influences the separation of the analytes together with the differences in volatility, (ii) the columns are stable and can be used many times, (iii) the analytes can be introduced dissolved in organic apolar solvents which allows the analysis of elemental species derivatised using Grignard reactions and (iv) the separation can be improved by temperature programming improving the chromatographic resolution.

However, packed column GC lacks the necessary resolution for the analysis of complex environmental and biological samples and cannot be routinely applied unless a selective or specific elemental detector is used. For elemental speciation using atomic detectors most of the first separations were described using this methodology. The packing material used was generally 5–10 % Carbowax or OV-101 supported on Chromosorb. These first studies will be discussed later in this chapter.

3.3 Gas chromatography with capillary columns

The use of capillary columns provides improved resolution as compared with packed or megabore columns. This resolution improvement can be clearly observed in Figure 4.2.2 [109] when moving from a packed column to a megabore column and finally to a capillary column for the separation of the same analytes. Additionally, the reduction in the carrier gas flow rate from 15–20 to 1–4 mL min^{-1} provides increased sensitivity due to the lower dilution factor in the mobile phase. However, we have to take into account that this increased chromatographic resolution will also depend on the connections between the column and the detector. For atomic detectors, adequate interfaces have to be built which do not produce extra band broadening and, hence, losses of theoretical plates in the coupling [109].

The main limiting factor of the use of capillary columns is their limited loading capacity. Typically, 1–2 μL of sample are injected in the column which, in most cases, represents a small percentage

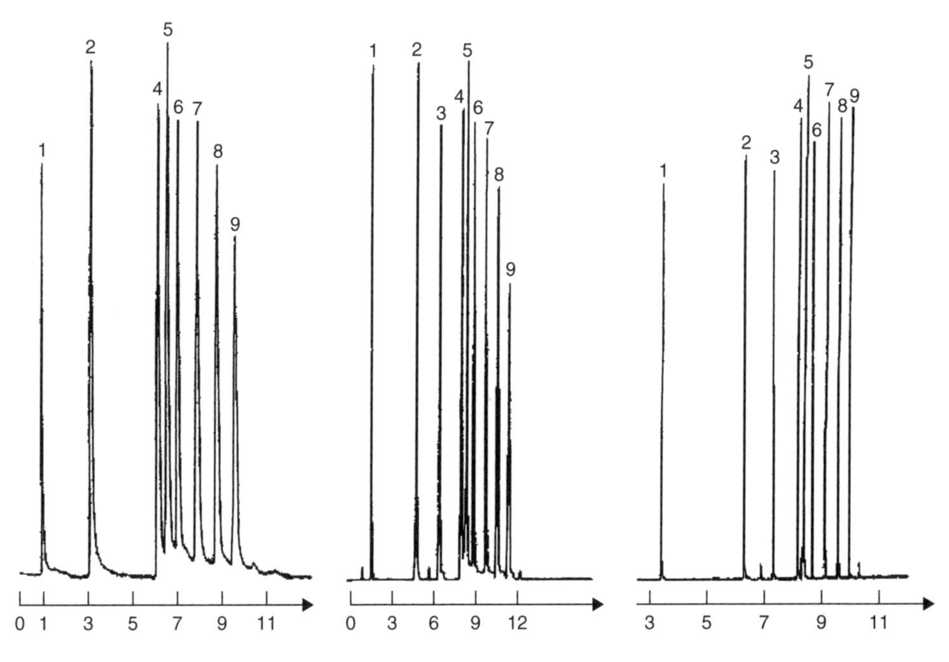

Figure 4.2.2. Improvement in chromatographic resolution from a packed column (left), to a megabore column (center) and to a capillary column (right) [109]. (Reprinted with permission from Elsevier Science.)

of the total sample volume. In speciation analysis this can be a problem as very low concentrations have to be measured usually. Different alternatives have been proposed to increase the sensitivity of capillary GC: (i) sample preconcentration by evaporation to dryness and reconstitution in a lower solvent volume, (ii) the use of electronic pressure control systems in large volume injectors [111] and (iii) the preconcentration of the analytes in temperature programmed high volume injectors which allow solvent purge at low temperature [62, 112]. However, the first of those procedures causes low reproducibility and the second is limited by the large amount of solvent reaching the detector.

The efficiency of capillary columns is increased by decreasing the internal diameter of the column, but the loading capacity decreases. This problem can be solved by resorting to multicapillary columns. Recently, multicapillary columns consisting in a bundle of ca. 1000 capillaries of 40 μm internal diameter have been commercialised [113]. The high efficiency of these columns allowed the use of carrier gas flows as high as $100\,mL\,min^{-1}$, reducing the analysis time to one-tenth of that required for a standard capillary column without reduction in loading capacity. Another advantage of multicapillary columns is their ability to work under isothermal conditions which simplifies the instrumentation used [114]. Columns of this type were first applied for elemental speciation by Rodriguez Pereiro and coworkers [115–119] using atomic detectors. The small length of these columns and the use of isothermal separations allowed the construction of a miniaturized speciation instrument coupled to a MIP-AES detector [120].

4 ICP-MS AS DETECTOR FOR GAS CHROMATOGRAPHY

4.1 Development of atomic gas chromatography detectors for elemental speciation

Detector traditionally used in the analysis of organic compounds by GC, such as the flame ionisation detector (FID) or the thermal conductivity detector (TCD), are not suitable for elemental speciation due to their lack of selectivity and sensitivity. The electron capture detector (ECD) has been applied for the detection of Pb [121], Sn [122], Hg [123], Se [124] and As [125] compounds but it cannot ensure specific detection. Atomic spectroscopy detectors (AAS, AES, AFS and elemental MS) are, on the other hand, perfectly suited for elemental speciation analysis by GC due to their high elemental sensitivity and selectivity. The fact that most elemental species are measured in complex samples at very low concentration levels requires the use of selective or, better, specific detectors which only respond to the element of interest without interferences from co-eluting compounds. The big advantage of the coupling of gas chromatographs to atomic detectors is that the influence of the matrix can be reduced or totally eliminated allowing near-specific detection.

Atomic absorption spectroscopy (AAS) is considered the most selective atomic spectroscopic technique for trace element speciation [126] because of the so-called key and lock mechanism [127]: only the element of interest will be able to absorb the radiation emitted by the hollow cathode lamp. This advantage was considered of utmost importance in earlier speciation studies. However, there is a clear disadvantage to this approach: we will be able to detect only those species which contain the element of interest. Nowadays it is considered necessary to be able to detect selectively and simultaneously different elements if required. Complex environmental and biological samples may contain species of different elements which could be interrelated. In this sense, atomic emission detectors, such as the flame photometric detector (FPD) or the microwave induced plasma atomic emission detector (MIP-AED) have been consolidated as suitable detectors for elemental speciation with GC. The combination of MIP-AED with simultaneous diode array detection is one of the most powerful combinations nowadays for elemental speciation.

In the first years of speciation studies, AAS established itself as the most popular GC detection technique for elemental speciation. The first way devised to couple those two techniques was to introduce the gaseous eluent from the GC column into the spray chamber of conventional AAS instruments using a short, heated transfer line. Later, to avoid dilution of the GC eluent, the analytes were introduced directly into the burner. The first coupling of GC and AAS was described by Kolb *et al.* [128] in 1966 for the determination of organolead compounds in gasoline. Since then, many applications have been published for the speciation of Pb [129], Cr [130], As, Ge, Sn and Se [131] and many others. In order to increase the residence time of the species in the flame and, hence, improve the sensitivity, Ebdon *et al.* [132] used a ceramic tube inside the flame where the analytes were 'trapped'. This idea was also followed in other laboratories [133, 134] but the big breakthrough came from the use of electrically heated tubes where electrothermal atomisation was achieved. These electrically heated atomizers were built either from graphite [135–137] or quartz [50, 138–140] and achieved temperatures between 1000 and 2000 °C. Because of the lack of flame gases which diluted the sample, sensitivities were ca. 100-fold better than these for conventional flame AAS.

Emission in flames has been applied for elemental speciation using mainly standard FPDs. The speciation analysis of tin using a FPD is well represented in the literature [141, 142] owing to its good sensitivity and selectivity for this element. Also, its widespread availability and ease of use made it the detector of choice in many modern speciation applications [61, 81, 143]. However, the general use of emission detectors for elemental speciation arrived with the development of analytical plasmas. Its high sensitivity and the possibility of detecting several elements simultaneously made atomic emission detection using plasmas an ideal tool for elemental speciation. Research on the coupling of microwave induced plasmas (MIP) and inductively coupled plasmas (ICP) to gas chromatographs for elemental speciation started in parallel with the development of AAS detectors.

The first description of an MIP used as elemental detector for GC was given by McCormack *et al.* [144] in 1965. In this application organic compounds were detected selectively using atomic emission. The coupling of an MIP to the GC is relatively simple using a heated transfer line which takes the column only millimetres from the plasma, so reducing the dead volume [145]. Another advantage is that the use of He as carrier gas in the chromatograph is compatible with the MIP as this gas is the most used plasma gas. The only disadvantage of this coupling is the need to bypass the plasma when the solvent front elutes or to switch on the plasma after the solvent front has eluted [146]. This is due to the low tolerance of the MIP towards organic solvents and any molecular gas in general. The first commercial GC-MIP-AED instrument was commercialised by Hewlett Packard (now Agilent) in 1989. This commercial system was applied by many laboratories for the speciation of Pb [147], Sn [148], Hg [149] and Se [150] in both environmental and biological samples. Also, many laboratories have performed their own 'home made' GC-MIP couplings with excellent results [151, 152]. Another GC-MIP system, the 'Automatic Speciation Analyser' has been described recently [120]. This system combines on-line sample preparation by microwave assisted extraction/derivatisation, cryogenic trapping and multicapillary GC separation coupled to a MIP-AES detector to automate the whole speciation procedure.

More details on the use of AAS and AES for elemental speciation studies are given in Chapter 5.1 of this book.

The use of the ICP as an emission source for elemental analysis first and as an ion source for elemental mass spectrometry later boosted research on the coupling of this source to GC for elemental speciation. In comparison with the MIP, the ICP source is more tolerant to molecular gases and provides better atomisation efficiencies. However, when the ICP was operated as an emission source, it provided worse sensitivity than the MIP due to the dilution of the GC eluent in the high carrier gas flow rate of the ICP. Also, the ICP provided lower excitation capabilities than the MIP for the detection of nonmetals, and some elements could not be detected at all due to the high plasma emission background because of the entrainment of air in the plasma (e.g. for H, C, O, N).

The development of the ICP as an ion source for elemental mass spectrometry in the 1980s and the growing research on suitable GC-ICP-MS interfaces resulted in a real alternative to the GC-MIP-AES detector for multielemental speciation. ICP-MS offers extremely high sensitivity and selectivity, true multielemental capabilities with fast scanning or simultaneous detectors and the possibility of measuring isotope ratios. This last capability, in combination with isotope dilution analysis, is the base of the most powerful analytical technique for the validation of speciation methodologies: speciated isotope dilution analysis. In the rest of this chapter we will focus on the coupling of GC to ICP-MS for elemental speciation studies.

4.2 Interfaces for GC-ICP-MS

In this section we will review the different interfaces described for the GC-ICP-AES and GC-ICP-MS coupling. Because of the similarities between both types of couplings, we will not distinguish between AES or MS detection.

The first interfaces were developed for the coupling of packed column GC to the ICP and we will describe those first. In those first studies, the end of the chromatographic column was connected to a T piece in which an additional flow of argon was introduced (about 0.9 L min^{-1}) to transport the analytes to the plasma. This additional argon flow was necessary to punch a hole in the central channel of the plasma as the carrier gas flow from the gas chromatograph, 10–20 mL min^{-1}, was insufficient for this purpose. The addition of an extra gas flow for the coupling with an ICP is a constant feature of most interfaces developed later. However, this caused dilution of the analytes and losses in sensitivity in comparison with MIP detectors. In these early studies different organic compounds were evaluated and the detection capabilities of the GC-ICP-AES coupling compared with other GC detectors [153–157]. Later, the end of the chromatographic column or the transfer line was connected directly to the torch, either at the base [158] or within the central channel [159] reducing the problems of dilution of the analytes. Chong and Houk [159], using a stainless steel transfer line, found that dead volumes could be minimized by adequate positioning of the transfer line within the central channel of the torch. The direct coupling of the transfer line to the torch has the inconvenience that the spray chamber and nebuliser have to be removed for the GC coupling. Peters and Beauchemin [160, 161] designed an interface which solved this problem. An automatic switching valve allowed the alternative use of the nebuliser or the GC interface without modifications to the instrumental set-up. This interface also used a setup described previously [162] which consisted in the addition of an external sheathing argon flow which prevented the dilution of the analytes and centred the eluent in the central channel of the torch, improving their transport through the plasma.

At the beginning of the 1990s the first publications on the coupling of capillary GC with ICP-MS were described. Kim *et al.* [8, 163] used a novel interface for the detection of Pb, Sn, Fe and Ni compounds. The interface consisted of a heated aluminium bar with a longitudinal slit in which the capillary column was introduced. A heating tape was used to maintain the high temperature throughout the interface. The last part of the chromatographic column was inside a stainless steel tube and introduced in the central channel of the torch almost to its end. The necessary argon make-up flow was introduced using a T piece. From the same laboratory, an improved version of the interface was described in which the argon make-up gas was previously heated and cold spots in the system eliminated. This interface was applied for the analysis of high boiling point compounds such as Fe, Ni and V porphyrins [164, 165] with high temperature capillary GC separation.

From the same period are the first papers published by Caruso's group [166, 167] for the coupling of low pressure ICP sources to GC for the detection of brominated organic compounds. The transfer line used consisted of a heated stainless steel tube, isolated with glass fibre tape. A short length of a 0.25 mm i.d. fused silica column was used to transfer the analytes to the torch. As with previous interfaces, a T piece was used to introduce a small make-up flow of, in this case, N_2 (at 100 mL min^{-1}) needed to keep the plasma. The sampling cone was made of aluminium and modified to be connected to the torch using an O-ring to keep the low pressure in the plasma. The exclusion of air and the use of lower gas flows resulted in lower polyatomic interferences in the detection of S and P by ICP-MS. Later, Evans *et al.* [168] used a similar interface to study plasma conditions in which both molecular and atomic information could be obtained. By modifying the forward RF power, the pressure in the plasma and the mixture of plasma gases used, atomic information (with picogram detection limits) and molecular spectra, similar to those provided by electron impact, could be obtained. These are the so-called 'tunable sources', which would offer both elemental (for quantification) and molecular information (for identification purposes).

Back in 1995 Prange and Jantzen [169] described a novel interface design based on a heated quartz transfer line in which the capillary column was inserted all the way to the base of the plasma. For the first time, detection limits in the femtogram range using ICP-MS detection were obtained. Also, the simultaneous detection of different Sn, Hg and Pb species was demonstrated. In this case, the argon make-up gas was previously heated to prevent peak broadening in the interface.

The suitability of a certain interface design for GC-ICP-MS coupling can be indicated by the number of publications in which this interface is used after its first description. This adjective of 'suitable' can be applied then to the interface described by De Smaele *et al.* [170] also in 1995. This interface consisted of a heated transfer line 2.5 m in length with three concentric tubes of stainless steel, Teflon and fused silica respectively. Initially, the part of the transfer line inside the plasma torch was not heated and that caused condensation of high molecular weight analytes present in the samples in this part of the transfer line. Later, after modifications to this part of the interface, the analytical characteristics of the interface improved [171, 172].

Gallus and Heumann [173] coupled GC to ICP-MS using a relatively simple and inexpensive interface. In this case, the capillary column was connected to a heated stainless steel transfer line through a six-way switching valve. The transfer line was introduced in the torch up to about 3 cm below the induction coil. By switching the valve, gaseous standards could be introduced in the plasma using a flow cell. Both the flow cell and the switching valve were inside the chromatographic oven. Mass bias, for isotope ratio work, could be corrected using this approach.

The interface described by Pritzl *et al.* [174] consisted of a special injector tube of 4 mm external diameter, constructed in stainless steel, which was attached to a heated transfer line (20–350 °C). A narrow tube, also in stainless steel, transversed the transfer line and injector tube almost to the base of the plasma. The argon make-up gas, preheated in the oven, was introduced coaxially with the column.

In the search for quantification of elemental species for which no standards were available different authors have designed interfaces in which the eluent from the column was mixed with an aqueous aerosol in the spray chamber [6] obtaining relatively stable plasma conditions. By nebulising aqueous standards of the element of interest a semiquantitative estimation of the concentration of the species in the samples was possible. Prohaska *et al.* [37] used a similar approach for the coupling of a double focusing ICP-MS instrument to the gas chromatograph for the analysis of different arsenic species. The last part of the chromatographic column was connected to the sample inlet of a standard concentric nebuliser and the nebuliser itself connected to the base of the torch by using a piece of PTFE tubing. Condensation of the analytes from the GC eluent in the gas phase would not affect the separation and detection as the analytes would be transported as an aerosol to the plasma.

A similar idea lies behind the interface described by Montes Bayón *et al.* [9] depicted graphically in Figure 4.2.3. The last 10 cm of the column are inserted in a heated copper tube and connected to a Swagelok T piece. The make-up argon gas is introduced through the side arm of the T piece at room temperature and flows externally to the

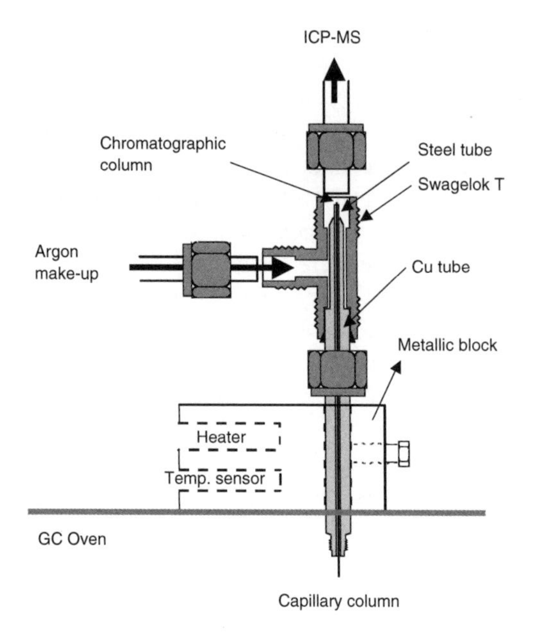

Figure 4.2.3. GC-ICP-MS interface developed by Montes Bayón *et al.* [9]. (Reproduced by permission of The Royal Society of Chemistry.)

heated copper tube inside the T piece. The condensation of the analytes in the gas phase after exiting the column would not cause band broadening as the high velocity external argon flow would prevent deposition of the analytes in the transfer tubing connecting to the torch. The authors use flexible room temperature PFA tubing for the connection between the interface and the torch. The main advantage of this interface design in comparison with those reported previously is its flexibility and the use of room temperature transfer tubing from the interface to the torch. Further publications from the same research group [72, 175] demonstrated that peak profiles using this interface were similar to those obtained using standard GC detectors such as FID. Only slight band broadening was observed for high molecular weight species, such as tetrabutyllead or tributylethyltin.

Krupp *et al.* [176, 177] were first to describe the coupling of a gas chromatograph to a multicollector ICP-MS instrument for precise and accurate isotope ratio measurements. Their transfer line consisted of a steel tube, electrically heated, inserted directly in the central channel of the torch. Using a T piece, a wet aerosol is introduced at the base of the torch. This aerosol is used for the nebulisation of Tl required for mass bias correction in their lead isotope ratio measurements. A summary of the general characteristics of different GC-ICP-MS interfaces is given in Table 4.2.8.

Table 4.2.8. General characteristics of transfer lines described for capillary GC-ICP-MS coupling.

Year	General characteristics	Ref.
1992	Rigid transfer line in Al ($\phi = 2.5$ cm). Four thermocouples for T control. Capillary column introduced almost to the plasma. Species with elevated molecular weight can be analysed.	164, 165
1993	Transfer line made of stainless steel (1 m, $\phi = 0.16$ cm) heated at 270 °C. Deactivated fused silica tube inside a water-cooled torch. Low pressure ICP.	167, 168
1995	Quartz transfer line heated at 240 °C permits the direct insertion of the analytes in the plasma.	169
1995	Transfer line in deactivated silica inside an steel tube (2.5 m, $\phi = 0.31$ cm). Ar make-up is heated inside the GC. Flexible.	171, 172
1996	Transfer line in stainless steel connected to the torch through a six-way valve inside the GC oven. Flow cell to introduce gaseous standards.	173
1996	Special stainless steel injector ($\phi = 4$ mm) to connect torch and heated steel transfer line (20–350 °C). Ar make-up gas previously heated inside GC oven and added coaxially to the column. Relatively versatile.	174
1997	Heated PTFE tube (1 m, $\phi = 0.3$ mm) to connect the capillary column to the torch. Nebulised standards are mixed using a T piece. Standardless quantification. Rigid.	6
1999	Capillary column introduced in the central channel of a concentric nebuliser which is connected to the torch by a PTFE tube. Nonheated, rigid.	37
1999	Similar to standard GC detectors. The eluent from the column is mixed with the Ar make up gas in a T piece and connected to the plasma with a piece of PFA tube. Flexible, PFA tube nonheated.	9, 72, 175
2001	Transfer line of heated steel tube. Ar make-up, preheated in the GC oven, goes inside the transfer line. A T piece at the base of the torch is used to introduce wet aerosols.	176, 177

4.3 Analytical characteristics of the GC-ICP-MS coupling

The coupling of a gas chromatograph to an ICP-MS instrument is not as simple and straightforward as the use of the ICP-MS instrument as detector for liquid chromatography where the exit of the chromatographic column can be connected directly to the nebuliser. As we have seen in the previous chapter special interfaces have to be constructed which increase the complexity of the coupling. However, there are several advantages which make GC-ICP-MS favourable in comparison with alternative HPLC-ICP-MS couplings [178]:

(i) High sample introduction efficiency. We can assume safely that 100 % of our sample will be transported to the plasma. In comparison with the sample introduction efficiency of typical nebuliser/spray chamber assemblies [2–3 %] used in HPLC coupled with ICP-MS, we can expect better sensitivity.
(ii) Efficient use of plasma energy. The analytes are carried to the plasma only by monoatomic gases, when He is used as carrier gas, so plasma energy can be used more efficiently for atomisation and ionisation. There is no need to volatilise and atomise solvent molecules as it is for liquid sample introduction.
(iii) Separation of the solvent from the analytes. The use of the GC allows separation of the solvent molecules from the analyte molecules during the chromatographic run. So, the analytes arrive in the plasma only in the company of the GC carrier gas and the make-up argon gas. That means that, when the analytes elute from the column, the plasma is stable and no distortion of the plasma by solvent molecules is observed [160].
(iv) Stable plasma conditions during the chromatographic run. The chromatographic separation takes place using temperature programming and not by gradient elution as is typical of HPLC separations. This means that the sensitivity remains constant during the separation, a stable baseline is obtained and, hence, low detection limits can be achieved.
(v) High chromatographic efficiency. In comparison with HPLC peak widths are much narrower and so signal to noise ratio is better using GC.
(vi) Low spectral interferences. The use of a dry plasma and the absence of any solvent lead to much less spectral interferences in comparison with solution nebulisation.
(vii) Low signal drift. The amount of material reaching the sampler and skimmer cones is very small so there is less chance of cone blocking.

All these characteristics make the GC-ICP-MS coupling ideal for ultratrace speciation analysis in comparison with alternative HPLC procedures. As an example, Table 4.2.9 compares absolute detection limits published for the speciation of organotin compounds using both HPLC [179] and GC [9, 62, 63, 180–185] as separation techniques and using different detectors. When comparing different GC detectors, the best detection limits are provided by the GC-ICP-MS coupling which goes down to the single femtogram range using shield torch instruments [62]. The comparison between HPLC-ICP-MS and GC-ICP-MS is also clear: more than three orders of magnitude of difference in sensitivity in two papers published by the same authors [9, 179].

Most publications using the GC-ICP-MS coupling are devoted to the speciation analysis of Sn, Hg, Pb, As and Se. Table 4.2.10 shows examples of the detection limits which can be achieved without special preconcentration procedures for the

Table 4.2.9. Absolute detection limits (pg as Sn) published for the detection of tin species using different detection techniques.

Technique	Absolute detection limits (pg as Sn)	Ref.
GC-MS	0.5–1	180
GC-QF-AAS	700–1200	181
GC-ECD	1–50	182
GC-MIP-AES	0.4	63
GC-ICP-AES	25	183
GC-FPD	3–160	184
GC-FPD (quartz induced lumines)	0.8	185
GC-ICP-MS	0.05–0.1	9
GC-ICP-MS (shield torch)	0.0007–0.0016	62
HPLC-ICP-MS	200	179

Table 4.2.10. Absolute detection limits (pg) achieved by capillary GC coupled to ICP-MS for the detection of organometallic species of different elements.

Element	Range (pg)	Ref.
Sn	0.05–0.1	9
Hg	0.09–0.83	68
Se	2.5	106
Pb	0.002–0.009	72
As	0.54	6
Bi	0.09	6
I, Br, Cl	16, 40, 158	168

detection of these and other species using capillary GC-ICP-MS. As can be observed, detection limits are at or below the picogram range for most elements except for the halogens.

In spite of the extremely low limits of detection offered by standard GC-ICP-MS, different authors have devised procedures for reducing even more those detection limits. In this sense, preconcentration systems such as purge and trap [186] and large volume injectors [62] have provided relative detection limits in the low $ng\,L^{-1}$ to $pg\,L^{-1}$ concentration range. In combination with shield torch, Tao *et al.* [62] obtained sub-femtogram detection limits for butyl- and phenyltin compounds. They observed that the use of a shield torch with a dry plasma offered a 100-fold increase in sensitivity in comparison with standard plasma conditions. In this way, concentrations of organotin compounds in the low $pg\,L^{-1}$ range were detected in open-ocean seawaters. However, this tremendous increase in sensitivity required ultraclean sample preparation methodologies to be employed [62]. For example, recent work carried out in our laboratory for the determination of butyltin compounds in coastal seawater samples by GC-ICP-MS [187] has shown that blank values for both TBT, DBT and MBT reduced well below $1\,ng\,kg^{-1}$ for 100 ml samples when sample preparation was carried out under clean room conditions (class 1000).

The optimisation of sensitivity using the GC-ICP-MS coupling has been approached from different angles. It is clear that the optimum plasma and ion lens conditions used for wet plasmas differ greatly from those under dry plasmas and, hence, adequate optimisation strategies have to be implemented. To avoid time-consuming optimisations by sequential injections of standards in the system, many authors employ xenon traces (1 %) added to the chromatographic carrier gas for optimisation. Gas flows, torch position, plasma and ion lens conditions are optimised by continuous monitoring of the ^{126}Xe signal. De Smaele *et al.* demonstrated the clear correlation between optimum conditions for Xe and those for organotin species [171]. This optimisation strategy was also employed by different authors [106, 119]. In other cases, volatile species of the element of interest are continuously fed to the plasma for optimisation, such as arsine [174] or mercury vapour [117] using a switching valve. Ruiz Encinar and coworkers [10, 175, 188] have applied a simple approach: they used polyatomic argon species, $m/z = 80$ ($^{40}Ar_2{}^+$), only for the optimisation of the ion lens after the optimisation of plasma conditions had previously been performed by nebulisation. These authors indicated drastic changes in the optimum values of different ion lenses of a quadrupole ICP-MS instrument by changing from wet to dry plasma conditions.

The GC-ICP-MS interfaces described in Table 4.2.8 do not provide means of diverting the eluting solvent front from the plasma. This is due to the robustness of the ICP in comparison with other plasmas and its high tolerance towards organic solvents. However, this mode of operation can cause carbon deposits in the sampling cone which, in turn, give rise to signal drift due to partial blocking of the sampling cone. When this effect was observed, the addition of a small flow of oxygen to the argon make-up gas (about $20\,mL\,min^{-1}$) and/or the increase of the RF power eliminated these carbon deposits [63, 119]. However, it has been observed recently [117] that the addition of oxygen caused a severe decrease in the sensitivity for mercury which was correlated with the flow of O_2 added. Many application published using GC-ICP-MS do not use an extra oxygen flow.

The selectivity of GC-ICP-MS for elemental speciation is excellent. In comparison with the complex emission spectra of GC-MIP-AES or GC-ICP-AES, the simplicity of the mass spectra using the ICP as an ion source is remarkable. In this sense, ICP-MS is very selective and the natural isotope abundances of the elements give a clear indication of the presence of a given elemental species in the sample [189]. New spectral interferences have not been described for the speciation of trace elements using GC-ICP-MS both because of lack of formation of new polyatomic ions and also because of the chromatographic separation from the solvent peak. Organic compounds in the plasma decompose completely and no new polyatomic ions containing carbon have been described using GC-ICP-MS. The only polyatomic species observed by GC-ICP-MS is $^{81}BrH^+$ which would interfere with ^{82}Se in the speciation of selenoaminoacids. Bromine impurities in the chloroform solvent gave a large solvent peak at mass 82 which was well separated from the selenium-containing aminoacids and hence did not interfere [90]. Other well-known spectral interferences in ICP-MS, such as $^{16}O_2{}^+$ on $^{32}S^+$, have also been observed by GC-ICP-MS but their magnitude was much lower than when using aqueous nebulisation [17].

The multi-isotopic and multi-elemental capabilities of the ICP-MS have also been exploited for elemental speciation using GC-ICP-MS. The only true limitation of this coupling is the fast transient chromatographic peaks, of ca. 2 s, which limit the number of isotopes which can be monitored simultaneously in sequential instruments. If we assume that we need at least ten data points to follow accurately a chromatographic peak and that this peak lasts for ca. 2 s, the optimum total integration time for all isotopes selected will be about 200 ms. In sequential instruments, such as quadrupole systems, this time will have to be divided between the number of selected isotopes reaching a practical limit of about ten isotopes monitored simultaneously (assuming that fast peak jumping and only one point per peak are used). This limit is set also by the fact that counting statistics worsen dramatically as the integration time per isotope is reduced to a few ms and this will also affect the achievable detection limit of the system. However, for simultaneous instruments, such as time of flight systems, the problem is the opposite. These instruments acquire ca. 20 000 full spectra per second which can be accumulated up to the selected integration time. If we assume that we can accumulate spectra for a maximum of 200 ms, to obtain adequate GC peak profiles, and that we have to divide this integration time between the total number of acquired data points in the mass spectrum (ca. 25 000 data points in the LECO ICP-TOF-MS instrument), the total integration time per points will be about 0.008 ms. That will make counting statistics very poor for GC-ICP-TOF-MS, but truly multi-isotopic and multi-elemental.

In spite of these capabilities, most multi-elemental or multi-isotopic speciation studies published using GC-ICP-MS concentrate only on the measurement of a few elements or a few isotopes. For example, the simultaneous speciation of Hg, Sn and Pb has been performed successfully [97]. The combination of multicapillary GC with ICP-MS allowed the determination of 11 species of these three elements in less than 2 min [119]. Multi-isotopic studies have also been published using both quadrupole and multicollector ICP-MS. In the first study [10] three Pb isotopes were measured for isotope ratios in the assessment of organolead sources in airborne particulate matter. In the second study, [176, 177] all four Pb isotopes plus ^{202}Hg and $^{203,205}Tl$ were measured simultaneously in ethylated NIST 981 lead isotopic standard.

Long-term stability of GC-ICP-MS systems has been a subject of worry for most people working in this field. Sensitivity drift is common in ICP-MS instruments and this can be worsened by carbon deposits in the cones. For this reason the use of internal standards is mandatory for GC-ICP-MS work. Different modes of internal standardization have been proposed for GC-ICP-MS:

(a) Methodological internal standards. As in any other analytical technique, the complexity and large number of analytical steps involved in speciation analysis require the use of adequate internal standards to compensate for errors in the whole speciation process. These internal standards are added to the sample at the beginning of the analytical procedure to compensate for nonquantitative recoveries, low derivatisation yields, dilution errors and so on. Obviously, these kind of internal standards have to behave like the analytes with similar chemical and physical properties. Some authors [63, 190] have indicated that, when the correction factors are elevated, it is recommended to apply simultaneously standard additions to the sample. Examples of this type of internal standards, which of course would also correct for drift in the GC-ICP-MS system, are tripropyltin (for butyltin compounds) or ethylmercury (for inorganic and methylmercury). The use of isotopically enriched elemental species [175, 188] would fall also in this category of methodological internal standards.

(b) Instrumental internal standards. There are two general procedures used to correct for instrumental variations in GC-ICP-MS. The first method consists in the addition to the final sample of a volatile compound, not present in the sample, but containing the metal under study [191]. This way of internal standardization would compensate for both sensitivity drift and injection precision. Examples of this type of internal standards are tetraalkylated Sn [169] or Pb [72] compounds and dialkylated Hg compounds [68]. The second method consists in the use of an impurity in the carrier gas (usually Xe) for signal normalization. This mode of internal standardization will compensate only for instrumental drift [51, 80, 102, 171, 172]. In many cases a combination of both modes of instrumental correction is used [102, 171].

4.4 Isotope ratio measurements with GC-ICP-MS

The analytical characteristic which makes the coupling of a gas chromatograph to the ICP-MS instrument unique in comparison with the other GC detectors indicated in Table 4.2.9 is the possibility of direct measurement of elemental isotope ratios. We have indicated in the previous chapter that the ideal internal standard would have to possess chemical and physical properties similar to the analytes. It is clear that another isotope of the same element would be the ideal internal standard and, hence, the measurement of elemental isotope ratios by GC-ICP-MS will open the way for the application of isotope dilution methodologies in elemental speciation. Another advantage of the use of isotope ratios is the improvement in measurement precision. For example, injection uncertainties of ca. 6–7 % were reduced to 2–3 % using Xe impurities in the carrier gas [80]. A similar injection uncertainty, in peak area mode [10, 175], was reduced to 0.5–1.5 % by measuring isotope ratios.

The first publication on the measurement of isotope ratios by GC-ICP-MS appeared in 1987 [159]. In that publication Chong and Houk measured isotope ratios for B, C, N, Si, S, Cl and Br in different organic compounds with instrumental precisions ranging between 0.37 and 18 %. They concluded that this technique could be applied to study reaction mechanisms and in metabolic studies using stable isotopic tracers but the precision would be inadequate for the study of natural isotope fractionation effects. Later, in 1993, Peters and Beauchemin [161] measured isotope ratios in organotin compounds and found isotope ratios in reasonably good agreement with the natural tin ratios. Precision was in the range 0.3–3.6 % in peak height mode. Nowadays, it is considered that peak area offers more precise and accurate isotope ratio measurements than peak height [10, 173].

The optimisation of the measurement parameters is critical in obtaining adequate isotope ratio measurements by GC-ICP-MS. When using sequential instruments (quadrupole, single collector double focusing) the selection of the right integration time will depend both on the ability to follow adequately the peak profiles and on optimising counting statistics [10]. In general, the fast transient signals generated by the GC are better followed using short integration times per isotope. However, under those circumstances counting statistics might be poor and isotope ratio precisions inadequate. On the other hand, for longer integration times counting statistics improve but ICP-MS is no longer able to follow the peak profiles accurately. Under those circumstances spectral skew in sequential instruments starts to be noticeable and isotope ratio precision is reduced [10]. These

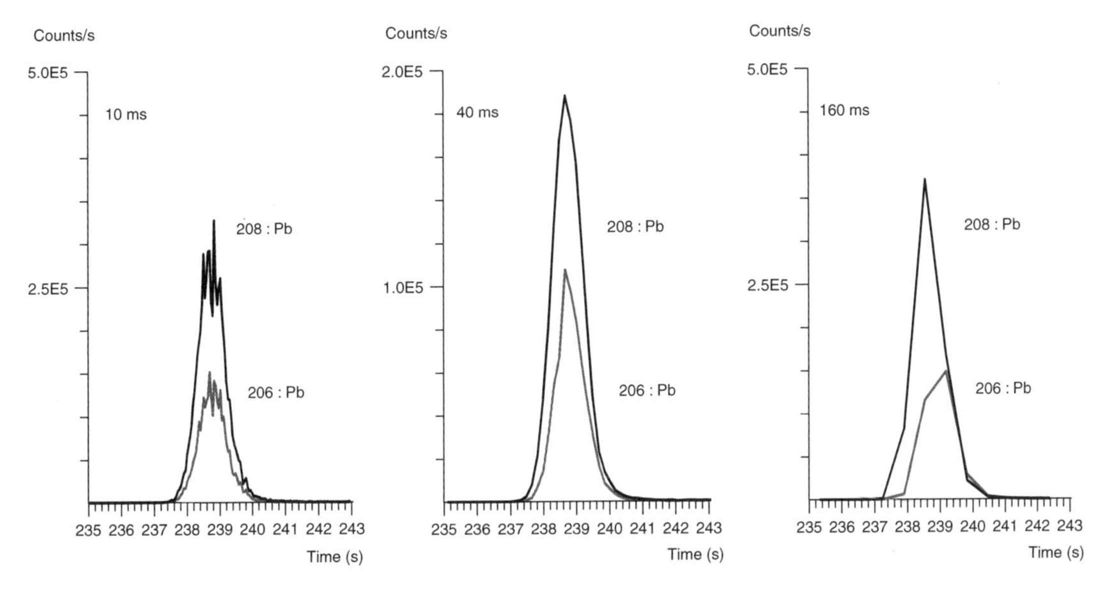

Figure 4.2.4. Influence of integration time on the GC-ICP-MS peak profiles measured for $MeEt_3Pb$, at masses 206 and 208, in leaded gasoline using a quadrupole instrument [10]. (Reproduced by permission of The Royal Society of Chemistry.)

two opposed effect are illustrated in Figure 4.2.4 for the measurement of lead isotope ratios in tetraalkyllead compounds in gasoline [10]. As can be observed, for a very short integration time per isotope (10 ms) peak profiles are very well defined but show high noise. For a long integration time (160 ms) peak profiles are poor and spectral skew is noticeable. A compromise between noise and spectral skew was obtained by keeping the total integration time between 150 and 200 ms [80, 188]. Under those conditions from two to eight isotopes could be monitored simultaneously using integration times from 25 to 100 ms [192].

In order to obtain accurate isotope ratio measurements using GC-ICP-MS we have to take into account factors such as detector dead time, mass bias, spectral interferences and chemical blanks. We will discuss briefly those factors and the way in which they can be corrected.

(a) Detector dead time is the time needed for the detection and electronic handling of single detection events. This effect is the cause of nonlinearity of pulse-counting detectors at high counting rates but does not affect analogue detectors [193]. When an isotope ratio is measured, counting losses will affect one isotope differently from the other and so the measured isotope ratio will be inaccurate and will depend on the concentration level being measured. There are different ways in which the detector dead time for a given instrument can be computed [194], but in most cases the measurement of isotope ratios by nebulisation of standards at different concentration levels is necessary. Once the detector dead time is known, the measured intensities on each point of the chromatogram have to be corrected using the equation:

$$N_{\mathrm{T}} = \frac{N_0}{1 - N_0\tau}$$

where N_0 and N_{T} are the observed and real count rates (counts s^{-1}) and τ is the detector dead time (s). The use of analogue measurement modes is less sensitive than pulse counting but does not require dead time correction [195].

(b) Mass bias, or mass discrimination, is a well-known effect in all ICP-MS instruments. It consist in the preferential transmission of heavier ions in the mass spectrometer which results in isotope ratio measurements which are biased by a constant factor [196]. The value of this mass bias factor depends mainly on the mass difference between the isotopes and can change from element to element,

Table 4.2.11. Isotope ratios measured for natural standards of MBT, DBT and TBT using GC-ICP-MS. Uncertainty expressed as standard deviation for three injections of 800 pg as tin.

Isotope ratio	Natural ratio	Experimental ratios measured			ΔM	Average relative error
		MBT (s)	DBT (s)	TBT (s)		
116/120	0.4463	0.4350 (0.0034)	0.4357 (0.0025)	0.4334 (0.0065)	4	−0.02596
117/120	0.2357	0.2296 (0.0020)	0.2287 (0.0027)	0.2313 (0.0026)	3	−0.02489
118/120	0.7434	0.7333 (0.0054)	0.7348 (0.0022)	0.7373 (0.0067)	2	−0.01110
119/120	0.2637	0.2599 (0.0001)	0.2610 (0.0030)	0.2586 (0.0040)	1	−0.01458
122/120	0.1421	0.1424 (0.0008)	0.1434 (0.0006)	0.1431 (0.0017)	−2	0.00624
124/120	0.1777	0.1824 (0.0027)	0.1809 (0.0032)	0.1806 (0.0039)	−4	0.02004

from instrument to instrument and from day to day. In order to illustrate this effect, Table 4.2.11 shows the experimental and theoretical tin isotope ratios measured by GC-ICP-MS for mono-, di- and tributyltin (MBT, DBT and TBT respectively) [188]. As can be observed, the relative error in the measured ratio, $(R_{\mathrm{exp}} - R_{\mathrm{theo}})/R_{\mathrm{theo}}$, is a function of the mass difference between the two isotopes, ΔM. By plotting the relative error versus the mass difference a linear function is usually obtained. The slope of this line is the mass bias factor, K. Once K is determined for a given GC-ICP-MS system by injecting isotopic standards, all subsequent isotope ratio measurements can be corrected [197]. For elements which do not show natural variations, such as tin, the natural element showing natural isotope composition can be used [198]. Following this approach, Ruiz Encinar *et al.* observed that the mass bias factor measured for organotin compounds drifted slightly after repeated injections of a natural abundance mixed butyltin standard during a run of 4 h [10, 175, 188]. This problem was solved by injecting a natural abundance standard every three samples and correcting each group of three samples using the average of the K values measured before and after the samples. A different approach was followed by Gallus and Heumann [173]. In this case, a system for the continuous introduction of a volatile natural selenium compound was devised and used for mass bias correction. Finally, Krupp *et al.* [176, 177] added natural thallium using on-line nebulisation for external mass bias correction of lead isotopes with satisfactory results. It is worth noting that mass bias is an instrumental effect and hence independent of the elemental species being monitored (see Table 4.2.11). This means that mass bias correction could be performed in the same chromatogram using a different elemental species added to the sample with well-known isotope composition [173, 175].

(c) Chemical blanks affect the measurement of isotope ratios in elemental speciation in a subtle way: only those species which show significant blank values will be affected. For example, high inorganic tin blanks will not affect the measurement of isotope ratios for butyltin compounds after ethylation [188]. To understand the effect of chemical blanks we need to consider two possibilities: isotope ratio measurements for source characterization, such as Pb source studies, or for isotope dilution analysis. In the latter case, chemical blanks are treated as other samples and the concentration in the blank computed by isotope dilution analysis and subtracted from that in the sample. When the measurements are performed for source characterisation chemical blanks are more difficult to take into account as the isotope composition of the blank will be slightly different from that in the sample. In those cases blank subtraction of absolute intensities at different masses might be the only alternative, but only when the blank values are much smaller that those in the sample. The reduction of chemical blanks to negligible levels by working with ultrapure reagents is advisable. Blank values affecting isotope ratio measurements have been reported for the speciation of Hg [197] and Pb [10] by GC-ICP-MS. For the speciation of butyltin compounds in sediments blank values have been reported to be negligible [175, 188] but high blank levels have been found for the analysis of seawater samples when high preconcentration factors are utilised [62, 187]. This may be due to their use as stabilizers in different plastics or to contamination of the laboratory atmosphere. From our experience, using only cleaned glass containers and working under clean room conditions are essential to measure low levels of butyltin compounds in seawater samples [187].

(d) Spectral interferences by GC-ICP-MS are less severe than when using conventional nebulisation. Isobaric interferences (e.g. ^{204}Hg and ^{204}Pb) are normally separated in the chromatographic column as the elution time of different lead and mercury compounds will be different and polyatomic interferences (e.g. ^{38}Ar^{40}Ar on ^{78}Se) will only increase the chromatographic baseline by a constant value. The integration of the GC peak will normally compensate for a small baseline value. Only in the cases when the baseline value is high and noisy in comparison with the analyte peak (e.g. $^{16}O_2$ on ^{32}S) will we need to resort to high resolution mass spectrometers [17] or collision cell instruments. No other serious spectral interferences have been reported for isotope ratio work using GC-ICP-MS.

4.4.1 Applications of isotope ratio measurements

There are many different fields in which isotope ratio measurements with GC-ICP-MS have proved useful. For example, in the study of isotope fractionation, the determination of isotope ratios in different elemental species can offer information which will be lost after total digestion of the sample. Many environmental, biological and geological processes produce compounds which might show natural or man-made isotope fractionation. Examples of those elemental species which might show changes in isotope ratios and can be detected by GC-ICP-MS are those of sulfur and lead [176]. Of course, the better the level of precision reached in the GC-ICP-MS coupling the larger this field of application will be.

The differentiation between sources of lead organometallic compounds was recently demonstrated by Ruiz Encinar *et al.* [10] using quadrupole ICP-MS. The measurement of lead isotope ratios by GC-ICP-MS of organolead standards allowed them to differentiate between lead sources. For example, Figure 4.2.5 shows some of the

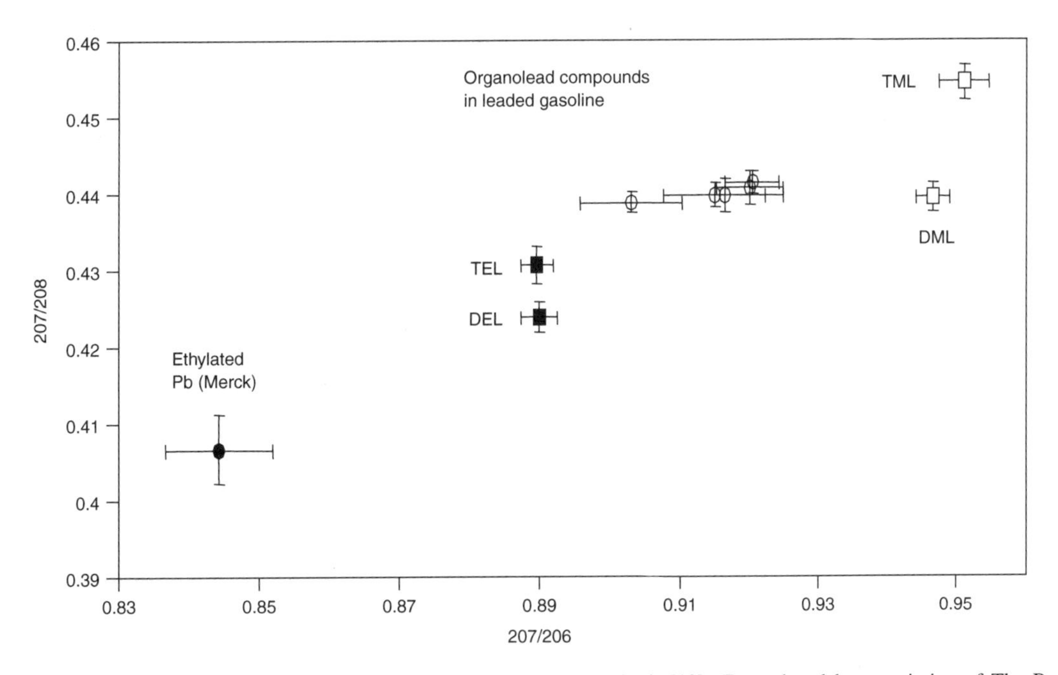

Figure 4.2.5. Lead isotope ratios measured for different organolead standards [10]. (Reproduced by permission of The Royal Society of Chemistry.)

results obtained: tetraalkyllead compounds present in gasoline (Me_4Pb, Me_3EtPb, Me_2Et_2Pb, $MeEt_3Pb$ and Et_4Pb) could be differentiated from butylated Et_3Pb^+ (TEL) or Me_3Pb^+ (TML) standards obtained from ABCR (Karlsruhe, Germany) which themselves showed a different lead signature. The reaction of the trialkylated species with iodine monochloride to form the corresponding dialkylated species (DEL and DML), which were then butylated to form the volatile species, did not produce drastic isotope changes. Finally, an inorganic lead ICP standard from Merck was ethylated using tetraethylborate and also showed a different lead signature by GC-ICP-MS (see Figure 4.2.5). The excellent precision results obtained by Krupp *et al.* using a multicollector ICP-MS [176, 177] for lead isotope ratio measurements by GC-ICP-MS are very promising for these type of studies.

Another field of application of isotope ratio measurements by GC-ICP-MS is in the study of artefacts during sample preparation. The use of certain sample preparation procedures may alter the speciation of a given element. In these cases, the use of isotopically labelled elemental species has allowed researchers to study these processes in depth. The isotopically labelled species can be followed throughout the whole analytical procedure and any change in the isotopic composition of another species will be shown by isotope ratio measurements. In this way, Hintelmann *et al.* [197] measured methylation velocities of inorganic mercury in sediments by using ^{199}Hg-enriched (92.57 at%) mercury nitrate. Later, the same authors studied methylation of inorganic mercury under different sample preparation conditions [199] such as distillation and acid and basic extractions. They concluded that methylation depended both on the type of sample (specially the organic matter content) and on the sample preparation procedure employed. In a similar study using inorganic ^{202}Hg (97 at%), García Fernández *et al.* [68] did not observe methylation of inorganic mercury during derivatisation processes involving $NaBEt_4$ or $NaBPr_4$. Similar results were obtained by Ruiz Encinar *et al.* [175] in the study of trans-butylation reactions for butyltin compounds during extraction and derivatisation with $NaBEt_4$. For this purpose, a highly pure DBT standard enriched in ^{118}Sn (98.44 at%) was synthesized and added to a certified sediment containing all three butyltin compounds. The measurement of isotope ratios by GC-ICP-MS demonstrated that only the DBT peak showed an altered 120/118 isotope ratio. The same authors observed that lead isotope ratios were unchanged during derivatisation with $NaBEt_4$ using enriched ^{204}Pb [10].

The applications of isotope dilution analysis for trace element speciation will be discussed later in this chapter.

4.5 Comparison of different ICP-MS instruments

Most of the GC-ICP-MS applications described to date have been performed on quadrupole ICP-MS instruments. This is mainly due to their relative higher abundance in analytical chemistry laboratories but we also have to take into account that their good sensitivity, speed and robustness make them suitable for most GC-ICP-MS applications. However, there are certain applications where other type of ICP-MS instruments would be a better choice.

Double focusing instruments offer higher sensitivities at low resolution power but their main advantage over quadrupole systems is in the speciation of spectrally interfered elements such as sulfur, phosphorus and silicon where the high continuous background makes speciation with quadrupole systems impossible. For example, Rodriguez Fernández *et al.* [17] measured volatile sulfur compounds in fermented saliva using a double focusing ICP-MS instrument as detector at mass 32 (95 at% abundance). Working at 3000 resolution power they could separate the sulfur peak from the polyatomic $^{16}O_2^+$. By using the most abundant sulfur isotope they could reach detection limits between 7 and 33 $\mu L\,L^{-1}$ (v/v) for 1 mL of sample. Figure 4.2.6 shows one of the chromatograms obtained for headspace sampling of saliva incubated at 37 °C for 48 h. Most of these sulfur compounds could be identified and they are thought to be related to bad breath problems [17].

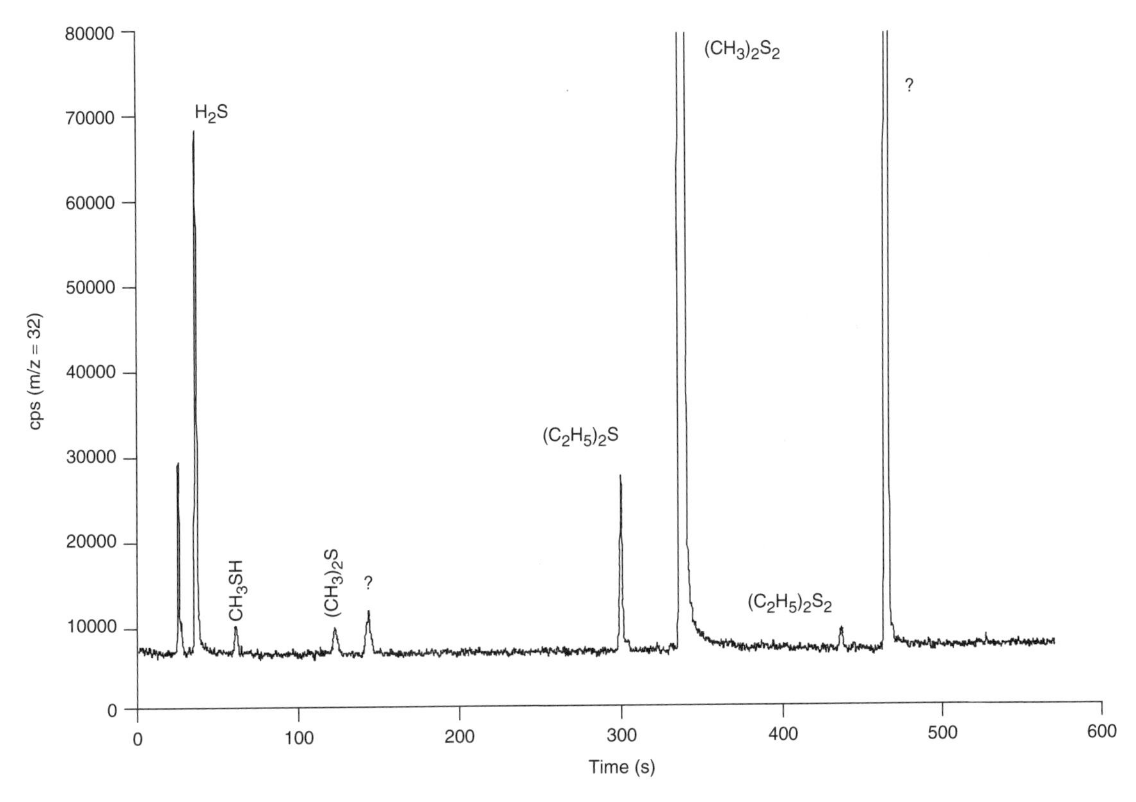

Figure 4.2.6. GC-ICP-MS chromatogram of volatile sulfur compounds in fermented saliva obtained with a double focusing instrument at mass 32 and resolution power of 3000 [17]. (Reproduced by permission of The Royal Society of Chemistry.)

The speciation of organosilicon compounds was also performed using an analogous system at resolution 3000 and mass 28 [200].

The high sensitivity capabilities of double focusing instruments was evaluated by Prohaska *et al.* [37] for arsenic speciation at mass 75 in low resolution. This instrument was evaluated for the detection of volatile arsenic compounds (arsine, mono-, di- and trimethylarsine) in a microcosm experiment. Similarly, high sensitivity applications for selenium speciation have been described [201].

A current limitation of quadrupole and single collector double focusing instruments is the sequential nature of their measurements. For certain applications, such as isotope ratio measurements or multielemental speciation, a simultaneous measurement of several isotopes or elements would be advantageous. This simultaneous measurement is offered nowadays by time of flight instruments and multicollector sector field instruments.

The expectations that arose with the introduction of time of flight ICP-MS instruments for elemental speciation have not been completely fulfilled. In principle, the measurement of many isotopes in fast transient signals, such as those obtained by GC separations, would be better performed on simultaneous instruments avoiding the well-known 'spectral skew' effect of sequential instruments. Also, for the precise measurement of isotope ratios, simultaneous instruments would be better suited. The main problem of TOF-ICP-MS instruments for elemental speciation is the need to acquire whole mass spectra continuously. When we need to focus on a few masses only, we would like to spend the maximum time integrating those masses and not to waste precious time on unwanted portions of the spectrum. In this way, the sensitivity of TOF instruments is ca. 5–10 times lower than quadrupole instruments and precision of isotope ratios for GC transient signals was between 1.2 and 2.9 % RSD [202, 203] and limited by counting statistics. However, when high concentrations were used and the measurements were performed in analogue mode isotope ratio precisions using GC-ICP-TOF-MS ranged between 0.3–0.5 % RSD [204]. For quadrupole instruments, where the use of time is more efficient, isotope ratio precisions were clearly below 1 % in counting mode [10, 175, 188]. The accuracy of isotope ratio measurements using TOF instruments was adequate in analogue mode but reduced seriously in counting mode when peaks larger than 20 000 cps reached saturation [204]. Of course, when truly multielemental speciation is required and no *a priori* information of the species present is given, such as the studies performed on landfill and other gases [3–6], TOF instruments could be the better choice.

The situation is completely different using multicollector instruments. In these cases we are continuously monitoring several isotopes, so time usage is optimal. Minor fluctuations in the plasma are also compensated so isotope ratios, even for fast transient signals, should be very precise indeed. To date, very few applications of GC-ICP-MS using multicollector instruments have been published [176, 177, 205] but we should expect much more in the near future. Isotope ratio precisions for major lead isotopes in tetraethyllead were in the range 0.02–0.07 % RSD. The accuracy of the measurements was also excellent for NIST 981 lead isotope standard (between 0.02 and 0.15 % deviation from the certified values). This isotope ratio accuracy was accomplished by on-line mass bias correction using thallium which was nebulised continuously and mixed with the GC eluent using a T piece [176, 177]. Detection limits using a collision cell single focusing multicollector [176] ranged from 2.9 fg (^{208}Pb) to 126 fg (^{204}Pb) and adequate isotope ratio precisions were obtained at concentrations as low as 0.5 ng mL^{-1} of Et_4Pb, as lead.

5 APPLICATION OF THE GC-ICP-MS COUPLING

5.1 Environmental applications

We consider it adequate to classify the environmental applications of GC-ICP-MS with respect to the nature of the analysed samples. We will consider first gaseous samples and the liquid and solid samples.

(a) Gaseous samples. The atmosphere can be considered the main route of dispersion of organometallic compounds in the environment. These compounds are present in the atmosphere as both volatile compounds and aerosols and can be formed by natural processes or generated by human activity. It is clearly established nowadays that we need to know the concentration of organometallic compounds in the atmosphere to understand the biogeochemical cycle of certain elements [106]. In this way, studies on the speciation of Sn, Se, As, P, In, Ga, Hg and Pb [106] and Pb [72] in atmospheric samples have been carried out by GC-ICP-MS. Also, characterisation of the source of different lead organometallic compounds in atmospheric particulate samples has recently been carried out by measuring lead isotope ratios [10]. In this study the authors compared lead isotope ratios in different organolead species measured for leaded gasoline, airborne particulates from both an urban and a rural area and an urban dust reference material (CRM 605). In addition, the samples were digested and the isotopic composition of total lead was measured by nebulisation using thallium as internal isotope standard. The results obtained are illustrated in Figure 4.2.7. As can be observed, the isotopic composition of organolead species in the atmosphere of Oviedo was similar to that found in leaded fuel but different from those in the reference material. On the other hand, total lead in Oviedo had an isotopic signature in between that of the rural area (industrial lead from a nearby aluminium smelter) and leaded fuel from gasoline showing the relative contribution of both sources to lead pollution in Oviedo.

The atmospheric transport or organolead compounds from central Europe to the artic was studied

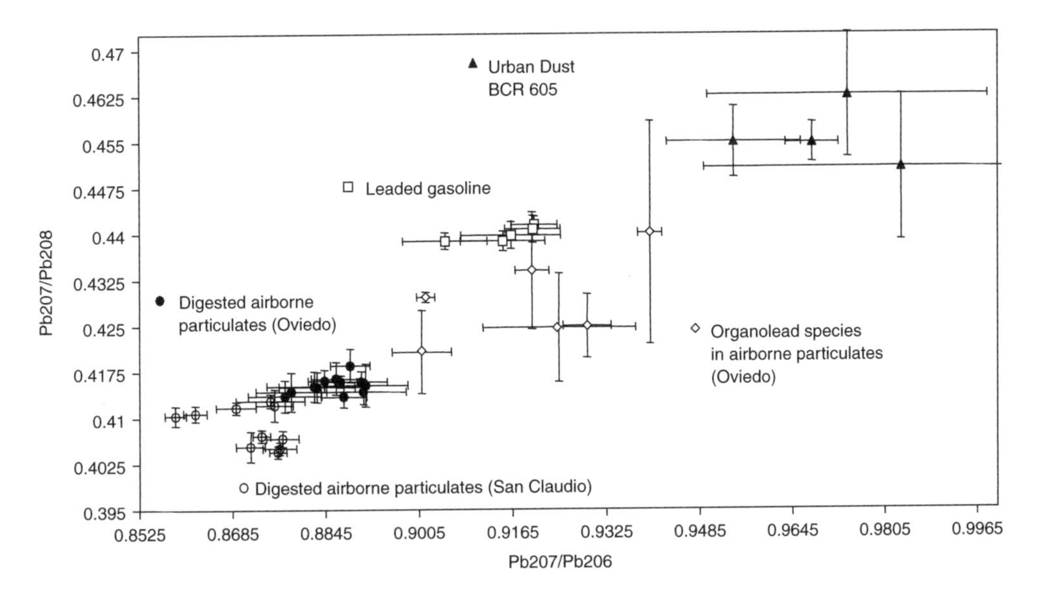

Figure 4.2.7. Lead isotope ratios measured for different atmospheric particulate samples [10]. (Reproduced by permission of The Royal Society of Chemistry.)

by Adams *et al.* [19] by measuring organolead compounds in arctic snow and ice. They observed that meteorological variations and solar light were the main factors influencing the transport and fate of organolead compounds in the atmosphere.

Volatile species from many different elements have been detected also in landfill gases [4, 5, 204]. In these studies Feldmann *et al.* have reported the presence of hydrides and methylated species of As, Se, Sn, Sb, Te, Hg, Pb and Bi. Even so, volatile Mo and W hexacarboniles were detected.

(b) Liquid samples. The low concentrations in which elemental species are found in waters require, in many cases, the use of preconcentration procedures. Solid–liquid extraction, solid-phase microextraction [62] and purge and cryogenic trapping [13] are the more used procedures. Using cryogenic trapping, Amoroux *et al.* [13] observed the presence of tin hydrides in seawater and concluded that chemical and biochemical methylation processes taking place in the sediments lead to the mobilization of tin species to the atmosphere. The same authors [11] studied the presence of volatile Se, Sn, Hg and Pb species in estuarine waters. Heisterkamp *et al.* [80] studied the presence of organolead compounds in rainwater, tap water and snow samples. Rainwater showed the highest concentrations of these compounds, dimethyllead being the more abundant (concentrations in the range of $\mathrm{ng\,l^{-1}}$). Open-ocean seawater was analysed by Tao *et al.* [62] in the search for organotin compounds. Both butylated and phenylated tin species were detected at the $\mathrm{pg\,l^{-1}}$ range in these samples. Recently, Rodríguez González *et al.* [187] determined the levels of butyltin compounds in coastal seawaters using isotope dilution GC-ICP-MS. Levels up to $80\,\mathrm{ng\,kg^{-1}}$ of TBT (as Sn) were found in marinas while levels below $1\,\mathrm{ng\,kg^{-1}}$ were found on beaches and in open areas.

(c) Solid samples. Most environmental solid samples studied in elemental speciation consist of sediments. Sediments act as the final sink for organometallic compounds and many biological transformations can occur there. The presence of elemental species in sediments is normally due to anthropogenic origin and so these species are usually adsorbed on the surface of sediment particles. That means that extraction methods used for elemental speciation in sediments rarely require total digestion of the sample [109]. Extraction methods used involve acidic leaching with the help of complexing agents or organic solvents such as methanol. Mechanical shaking, ultrasonic extraction, soxhlet extraction, distillation and microwave assisted extraction have all been used for elemental speciation in sediments. However, we have to take into account that extraction methods are the weakest link in the traceability chain for elemental speciation. The use of isotopically enriched elemental species has shown to be a viable alternative method to study and optimise sample preparation procedures [197, 206–208]. Tin, mercury and lead are the elements for which most speciation studies in sediments have been carried out by GC-ICP-MS [30, 51, 97, 169, 209].

GC-ICP-MS studies on soils have been also performed. Prohaska *et al.* [37] created a closed microcosm in which the biomobility of arsenic could be studied under different conditions.

5.2 Biological applications

Biological samples require total decomposition of the tissue as the elemental species are incorporated in the matrix. This decomposition step has to be performed with extreme care in order to keep the nature of the species unchanged. Controlled decomposition of biological tissues for elemental speciation has been performed in different ways: (i) enzymatic hydrolysis using different proteases and lipases, (ii) alkaline hydrolysis using KOH [102] or tetramethylammonium hydroxide (TMAH) [117, 119] or (iii) acid hydrolysis using HCl in a saline medium [210]. The use of microwave assisted extraction [117, 119] or ultrasonic extraction [210] has been proposed recently to accelerate the hydrolysis processes.

The elements for which most applications have been published in biological samples are Hg, Sn and Se. The high toxicity of methylmercury and its bioaccumulation in fish explains the interest for the speciation of this element in biological samples. In the case of tin, the toxicity of butyltin compounds for oysters and mussels is well known and, in the case of selenium, its biological window between essentiality and toxicity is very narrow. The increasing use of selenium supplements, in which selenium can be present as different selenium species, has been addressed by different authors. In this sense, Perez Mendez *et al.* have demonstrated that the coupling of a gas chromatograph to a quadrupole or a double focusing ICP-MS instrument resulted in one of the most sensitive and selective methods for the measurement of selenoaminoacids in food supplements. Also, the concept of chiral speciation was demonstrated by the separation of D- and L-selenoaminoacids using a chiral stationary phase [201, 211].

The detection of sulfur compounds in biological samples is an area of activity where many GC-ICP-MS applications can be foreseen. For example, the detection of volatile sulfur compounds in fermented saliva by double focusing ICP-MS after GC separation has been published [17] and GC-ICP-MS methods for the detection of methionine and other sulfur-containing aminoacids have been presented [212].

5.3 Isotope dilution analysis with GC-ICP-MS

If we consider that obtaining quality data in trace analysis for total elemental concentrations is a difficult task, we have to agree that the difficulties will be much greater when we try to determine different elemental species in which one element is distributed [213]. The complexity of the matrices, the different analytical steps which need to be performed (digestion, extraction, clean-up, preconcentration, derivatisation, separation and determination) and the possibility of species interconversion make elemental speciation a difficult and error-prone task [21]. The application of 'primary' analytical methods, such as isotope dilution analysis in combination with mass spectrometry [214] could be a powerful tool in the search for quality assurance in trace element speciation [215]. All the well-known advantages of isotope dilution analysis [216], namely the correction for incomplete recoveries or low derivatisation yields, correction for sensitivity drifts, ideal internal standard and high precision measurements, can be used advantageously in trace metal speciation. This concept was first described by Heumann for elemental speciation [213, 217]. In these papers, Heumann indicated two possible ways in which isotope dilution analysis could be applied in elemental speciation: (i) using a species-unspecific spike or (ii) using different species-specific spikes. The first approach is useful when the structure or identity of the species is unknown and/or no standard is available for quantification. Practically, the spike is added post-column after the separation of the species and can be used for quantification by converting the original chromatogram (counts s^{-1} versus time) into a mass flow chromatogram (ng s^{-1} versus time) using the isotope dilution equation. The integration of the chromatographic peaks will provide directly the amount of trace element in this elemental species [218, 219]. This first approach has been used only after HPLC or CE separations and does not offer all the advantages of isotope dilution analysis listed previously (only corrects for sensitivity drifts).

The full advantages of isotope dilution analysis for elemental speciation are realized only with the second approach: the use of one or more isotopically enriched species of the element of interest. This implies that we need to know the chemical structure of the species of interest and that we can obtain or synthesize those species isotopically enriched. For this approach the enriched spike(s) is(are) added at the beginning of the analytical procedure and that makes full use of all isotope dilution advantages. There are several examples of publications on isotope dilution analysis in combination with GC-ICP-MS (GC-ICP-IDMS) but these have focused only on three elements: Hg [197, 199, 220, 221], Se [173] and Sn [175, 187, 188, 207, 208].

Hintelmann *et al.* [199] studied the formation of methylmercury during the extraction of inorganic mercury from different environmental samples. They used both Hg(II) nitrate, enriched in ^{199}Hg, and Hg(II) chloride, enriched in ^{200}Hg, to compute methylation reactions, and methylmercury chloride, enriched in ^{201}Hg, to calculate the recovery of the extraction technique employed. The results obtained indicated important overestimations of $MeHg^+$ using distillation as a sample preparation technique. The conclusions of this study were that methylation of mercury depended on the type of sample, on the relative concentrations of inorganic and methylmercury and on the extraction technique employed.

In a different study, Snell *et al.* [220] used ^{198}Hg-enriched (96 at%) Hg(II) and $MeHg^+$ to develop analytical methodologies with quantitative recoveries when computation was performed using the isotope dilution equation. The use of standard addition calibration provided results of less quality. Recently, Demuth and Heumann [221] used the double spike technique to study methylation and demethylation reactions for mercury during derivatisation. Their results indicated that certain extraction conditions favoured the formation of Hg(0) and decomposition of the original species.

Gallus and Heumann [173] developed back in 1996 a methodology for the determination of Se(IV) and Se(VI) species by GC-ICP-IDMS. Selenite was derivatised to piazselenol (volatile) and this compound determined by GC-ICP-MS. Then, selenate was reduced to selenite and derivatised in the same way. The concentration of selenate was calculated by difference from that of selenite. They used ^{82}Se-enriched (62 at%) species, added to the sample prior to derivatisation or reduction, and measured the ratios 77/82 and 78/82 in the sample. The determinations using both isotope ratios led to comparable results. The detection limit of the method was estimated to be 0.02 ng ml^{-1} for both selenium species.

Ruiz Encinar and coworkers [175, 188, 207, 208] evaluated the possibilities of isotope dilution analysis for tin speciation in sediments using different isotopically enriched spikes. In the first study, [175] the synthesis of enriched dibutyltin (DBT), using ^{118}Sn (98.44 at%), was described and applied to the determination of DBT in certified reference materials with satisfactory results. Isotope equilibration was achieved by mechanical shaking of the sample and spike during 12 h using 4 ml of a 75/25 mixture of acetic acid and methanol. No interconversion of the different butyltin species in the sediment was detected using this extraction procedure.

In a second study [188] the synthesis of a ^{119}Sn-enriched (82.4 at%) mixture of MBT, DBT and TBT was described. This mixed spike was characterized by reverse isotope dilution analysis and applied to the simultaneous determination of the three butyltin species in certified sediments. In this case, isotope equilibration and extraction were performed as for the ^{118}Sn-enriched DBT and no interconversion reactions were detected during the reverse isotope dilution experiments. Isotope ratio precision on triplicate injections was, on average, 0.8 %, depending on peak size, and the precision of the whole procedure for independent triplicate analyses was between 1 and 1.5 %. This indicated that satisfactory results could be obtained even for single injection, single extraction experiments which could give a considerable reduction in time in comparison with other calibration strategies.

The third study, published recently by the same research group [207, 208], compared three extraction techniques: mechanical shaking, ultrasonic extraction and microwave assisted extraction for the speciation of butyltin compounds in sediments. The basic idea was to be able to compare the extraction efficiency of the three techniques for MBT, DBT and TBT and the possibility of interconversion reactions during extraction. For this purpose, a mixed spike was prepared using the enriched species synthesized previously [175, 188]. The final spike mixture used contained MBT and TBT, both enriched in ^{119}Sn, and DBT enriched in ^{118}Sn. The isotopic composition of this spike can be clearly seen in Figure 4.2.8. As can be observed, the abundance of the ^{120}Sn isotope, the main isotope in natural tin, is very low in all species.

By measuring the ratios 120/118 and 120/119 in spiked certified reference materials and assuming that the decomposition of butyltin compounds would follow a simple debutylation mechanisms, the authors could calculate the concentration of all three butyltin species in the sample and the corresponding decomposition factors.

The results showed that MBT was strongly bound to the sediment matrix, as previously suspected [222], and that mechanical shaking was not able to extract it quantitatively. On the other hand, all three extraction techniques tested could

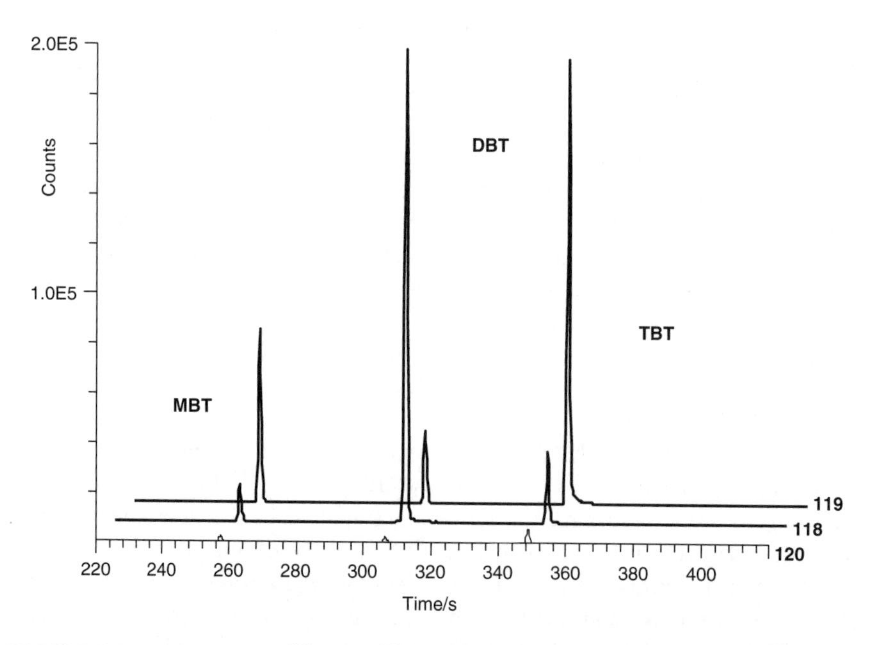

Figure 4.2.8. GC-ICP-MS chromatogram of a ^{118}Sn and ^{119}Sn double isotope spike for butyltin speciation [207]. (Reproduced by permission of The Royal Society of Chemistry.)

extract DBT and TBT easily. One interesting conclusion of this study was that microwave assisted extraction caused the decomposition of butyltin compounds with transformation factors up to 16 % depending on the butyltin species considered, the microwave power applied and the extraction time used.

The use of isotope dilution analysis in elemental speciation will continue to increase in the near future. The ability to compensate for nonquantitative recoveries will be of tremendous interest in studying extraction techniques where the recoveries are strongly influenced by the sample matrix, such as solid-phase microextraction. The analysis of butyltin compounds in natural and seawater, another area where high preconcentration factors are required, would benefit from the use of isotope dilution methodologies [187]. Also, the application of double spike techniques, or even triple spike techniques will help in establishing sound extraction procedures for elemental speciation and in the development of more reliable certified reference materials. Finally, the application of isotope dilution methodologies for the GC-ICP-MS determination of other elements and species, such as Se- and S-containing aminoacids, will be a reality in the near future.

5.4 Reference materials and quality control

Elemental speciation, until now restricted more or less to research laboratories, is coming slowly into environmental regulations in different countries. For example, tributyltin is regulated in seawater ($10\,ng\,l^{-1}$ as TBT) and drinking water ($63\,ng\,l^{-1}$ as TBT) in the USA. This trend is also followed by the EU where TBT has to be measured in continental waters (directive 76/464/EEC) [223]. Other elemental species, such as Cr(VI), $MeHg^+$, BrO_3^- and 'inorganic As', are already included in various 'black lists' and have to be measured in different environmental and biological samples (e.g. inorganic As in fish products). This means that sound speciation procedures will have to be implemented in routine or contract laboratories and that new reference materials for elemental speciation will be needed in the near future.

We all know that sample preparation procedures are still the 'Achilles heel' of modern speciation methodologies [224, 225]. If we need to develop new reference materials or reference methodologies to be used in contract laboratories we will have to pay much more attention to these sample preparation steps. It has been observed in previous certification campaigns [226] that certification of certain elemental species was impaired by the wide range of interlaboratory results or by the lack of stability of the material itself. Isotope dilution analysis with species-specific spikes will surely play a predominant role in this context.

There are several areas where isotope dilution analysis will help in the preparation of 'speciated' reference materials. First, homogeneity studies would be better performed using isotope dilution. The homogeneity of a given material cannot be certified to a value lower than the uncertainty in the measurement procedure used. Hence, if we could improve our measurement uncertainty by 5–10 times by using isotope dilution methodologies the better the homogeneity studies would be. The same can be said for stability studies where enriched species could be used to understand why a certain elemental species is degrading and what the degradation product is. This will help in devising alternative stabilization procedures which would not alter the speciation in the sample. Finally, the certification campaign would benefit from the use of speciated isotope dilution analysis as has been demonstrated previously for the preparation of trace element reference materials.

6 CONCLUSIONS

Trace metal speciation methodologies are very well developed in research laboratories. The coupling of a high efficient separation technique (HPLC, GC or CE) to a sensitive and selective elemental detection method (MIP-AES or ICP-MS) seems to be ideal for elemental speciation. However, we have to take into account that new elemental species are discovered continuously and many 'unknown' species are still found in the literature (see for example Figure 4.2.6). To identify these species we need to resort to detection techniques which are capable of producing 'molecular' information, such as electron impact MS, electrospray MS or new tunable GD-MS, MIP-MS or even ICP-MS, which are complementary to the elemental detection method.

Having said that, the speciation of elements which form volatile compounds or which can be derivatised to form volatile compounds is perfectly adapted for GC separation coupled to elemental detectors. Techniques such as MIP-AES or ICP-MS are the most sensitive and selective which can be used in detectors for GC. It is clear that the speciation of Hg, Sn and Pb in environmental and biological samples is better performed using GC-ICP-MS. This is clear when we compare the analytical characteristics of this coupling with those of any other combination published in the literature. Recently, GC-ICP-MS has started to be applied for the speciation of Se and S with excellent results. We have to expect more developments in the near future for the speciation of these elements.

The main characteristic of GC-ICP-MS, in which it clearly differs from other atomic detection techniques such as MIP-AES, is its ability to provide isotopic information. This extra information can be used in two ways: to investigate the sources of elemental species, as in the studies performed on lead and sulfur, or to use isotope dilution methodologies to improve the quality of the analytical information obtained. We should expect more developments in this field as the combination of GC with multicollector ICP-MS instruments has only started to be evaluated.

7 REFERENCES

1. Templeton, D. M., Ariese, F., Cornelis, R., Danielson, L. G., Muntau, H. van Leeuwen Lobinski, R. H. P., *Pure Appl. Chem.*, **72**, 1453 (2000).
2. Quevauviller, P., Astruc, M., Morabito, R., Ariese, F. and Ebdon, L., *Trends Anal. Chem.*, **19**, 180 (2000).
3. Feldmann, J. and Hirner, A. V., *Int. J. Environ. Anal. Chem.*, **60**, 339 (1995).
4. Feldmann, J., Koch, I. and Cullen, W. R., *Analyst.*, **123**, 815 (1998).
5. Feldmann, J. and Cullen, W. R., *Environ. Sci. Technol.*, **31**, 2125 (1997).
6. Feldmann, J., *J. Anal. At. Spectrom.*, **12**, 1069 (1997).
7. Radziuk, B., Thomassen, Y., van Loon, J. C. and Chau, Y. K., *Anal. Chim. Acta*, **105**, 255 (1979).
8. Kim, A. W., Foulkes, M. E., Ebdon, L., Hill, S., Patience, R. L., Barwise, A. G. and Rowland, S. J., *J. Anal. At. Spectrom.*, **7**, 1147 (1992).
9. Montes Bayón, M., Gutierrez Camblor, M., García Alonso, J. I. and Sanz-Medel, A., *J. Anal. At. Spectrom.*, **14**, 1317 (1999).
10. Ruiz Encinar, J., Leal Granadillo, I. A., García Alonso, J. I. and Sanz-Medel, A., *J. Anal. At. Spectrom.*, **16**, 475 (2001).
11. Amoroux, D., Tessier, E., Pécheyran, C. and Donard, O. F. X., *Anal. Chim. Acta*, **377**, 241 (1998).
12. Amoroux, D. and Donard, O. F. X., *Geophys. Res. Lett.*, **23**, 1777 (1996).
13. Amoroux, D., Tessier, E. and Donard, O. F. X., *Environ. Sci. Technol.*, **34**, 988 (2000).
14. Hewitt, C. N. and Harrison, R. M., *Anal. Chim. Acta*, **167**, 277 (1985).
15. Bloom, N. and Fitzgerald, W. F., *Anal. Chim. Acta*, **208**, 151 (1988).
16. Ballantine, D. S. and Zoller, W. H., *Anal. Chem.*, **56**, 1288 (1984).
17. Rodriguez Fernández, J., Montes Bayón, M., Pereiro, R. and Sanz-Medel, A., *J. Anal. At. Spectrom.*, **16**, 1051 (2001).
18. Robinson, J. W., Kiesel, E. L., Goodbread, J. P., Bliss, R. and Marshall, R., *Anal. Chim. Acta*, **92**, 321 (1977).
19. Adams, F. C., Heisterkamp, M., Candelone, J. P., Laturnus, F., Van de Velde, K. and Boutron, C. F., *Analyst*, **123**, 7671 (1998).
20. Quevauviller, P., Donard, O. F. X., Maier, E. A. and Griepink, B., *Mikrochim. Acta*, **109**, 169 (1992).
21. Morabito, R., Massanisso, P. and Quevauviller, P., *Trends Anal. Chem.*, **19**, 113 (2000).
22. Kingston, H. M., Huo, D., Lu, Y. and Chalk, S., *Spectrochim. Acta*, **53B**, 299 (1998).
23. Campbell, A. D., *Pure Appl. Chem.*, **64**, 227 (1992).
24. Dirkx, W. M. R., Lobinski, R. and Adams, F. C., in *Quality Assurance for Environmental Analysis*, Quevauviller, Ph., Maier, E. A. and Griepink, B. (Eds), Elsevier, Amsterdam, 1995, Chapter 15.
25. Donard, O. F. X. and Pinel, R., in *Environmental Analysis using Chromatography Interfaced with Atomic Spectroscopy*, Harrison, R. M. and Rapsomanikis, S. (Eds), Ellis Horwood, Chichester, 1989, Chapter 7.
26. Weber, J. H., *Trends Anal. Chem.*, **16**, 73 (1997).
27. Le, X. C., Cullen, W. R. and Reimer, J. K., *Anal. Chim. Acta*, **285**, 277 (1994).
28. Jin, K., Shibata, Y. and Morita, M., *Anal. Chem.*, **63**, 986 (1991).

29. Donard, O. F. X., Rapsomanikis, S. and Weber, J. H., *Anal. Chem.*, **58**, 772 (1986).
30. Segovia García, E., García Alonso, J. I. and Sanz-Medel, A., *J. Mass Spectrom.*, **32**, 542 (1997).
31. Clark, S. and Craig, P. J., *Appl. Organomet. Chem.*, **2**, 33 (1988).
32. Sullivan, J. J., Torkelson, J. D., Wekell, M. M., Hollingworth, T. A., Saxton, W. L., Miller, G. A., Panaro, K. W. and Uhler, A. D., *Anal. Chem.*, **60**, 626 (1986).
33. Martin, F. M. and Donard, O. F. X., *J. Anal. At. Spectrom.*, **9**, 1143 (1994).
34. Martin, F. M., Tseng, C.-M., Belin, C., Quevauviller, P. and Donard, O. F. X., *Anal. Chim. Acta*, **286**, 343 (1994).
35. Pannier, F., Astruc, A., Astruc, M. and Morabito, R., *Appl. Organomet. Chem.*, **10**, 471 (1996).
36. Dodd, M., Pergantis, S. A., Cullen, W. R., Li, H., Eigendorf, G. K. and Reimer, K. J., *Analyst*, **121**, 223 (1996).
37. Prohaska, T., Pfeffer, M., Tulipan, M., Stingeder, G., Mentler, A. and Wenzel, W. W., *Fresenius' J. Anal. Chem.*, **364**, 467 (1999).
38. Ritsema, R. and Donard, O. F. X., *Appl. Organomet. Chem.*, **8**, 571 (1994).
39. Ishizaka, T., Nemoto, S., Sasaki, K., Suzuki, T. and Saito, Y., *J. Agric. Food. Chem.*, **37**, 1523 (1989).
40. Fent, K. and Muller, M. D., *Environ. Sci. Technol.*, **25**, 489 (1991).
41. Honeycutt, J. B. J. and Riddle, J. M., *J. Am. Chem. Soc.*, **83**, 369 (1961).
42. de la Calle-Guntiñas, M. B., Scerbo, R., Chiavarini, S., Quevauviller, P. and Morabito, R., *Appl. Organomet. Chem.*, **11**, 693 (1997).
43. Maguire, R. J. and Huneault, H., *J. Chromatogr.*, **209**, 458 (1981).
44. Harrison, R. M. and Radojevic, M., *Environ. Technol. Lett.*, **6**, 129 (1985).
45. Gómez-Ariza, J. L., Morales, E., Giraldez, I., Beltran, R. and Escobar, J. A. P., *Fresenius' J. Anal. Chem.*, **357**, 1007 (1997).
46. De Diego, A., Tseng, C. M., Stoichev, T., Amoroux, D. and Donard, O. F. X., *J. Anal. At. Spectrom.*, **13**, 623 (1988).
47. Bloom, N. S., *Can. J. Fish. Aquat. Sci.*, **46**, 1131 (1989).
48. Rapsomanikis, S., Donard, O. F. X. and Weber, J. H., *Anal. Chem.*, **58**, 35 (1986).
49. Ashby, J. R. and Craig, P. J., *Appl. Organomet. Chem.*, **5**, 393 (1991).
50. Cai, Y., Rapsomanikis, S. and Andreae, M. O., *J. Anal. At. Spectrom.*, **8**, 119 (1993).
51. Ritsema, R., de Smaele, T., Moens, L., de Jong, A. S. and Donard, O. F. X., *Environ. Pollut.*, **99**, 271 (1998).
52. Rapsomanikis, S. and Craig, P. J., *Anal. Chim. Acta*, **248**, 563 (1991).
53. Jimenez, M. S. and Sturgeon, R. E., *J. Anal. At. Spectrom.*, **12**, 597 (1997).
54. Sturgeon, R. E., Wiilie, S. N. and Berman, S. S., *Anal. Chem.*, **61**, 1867 (1989).
55. Rapsomanikis, S., *Analyst*, **119**, 1429 (1994).
56. Chau, Y. K., Yang, F. and Brown, M., *Anal. Chim. Acta*, **338**, 51 (1997).
57. Ashby, J., Clark, S. and Craig, P. J., *J. Anal. At. Spectrom.*, **3**, 735 (1988).
58. Kuballa, J., Wilken, R.-D., Jantzen, E., Kwan, K. K. and Chau, Y. K., *Analyst*, **120**, 667 (1995).
59. Rodriguez Pereiro, I., Schmitt, V. O., Szpunar, J., Donard, O. F. X. and Lobinski, R., *Anal. Chem.*, **68**, 4135 (1996).
60. Cai, Y., Monsalud, S., Jaffé, R. and Jones, R. D., *J. Chromatogr. A*, **876**, 147 (2000).
61. Montigny, C., Lespes, G. and Potin-Gautier, M., *J. Chromatogr. A*, **819**, 221 (1998).
62. Tao, H., Rajendran, R. B., Quetel, C. R., Nakazato, T., Tominaga, M. and Miyazaki, A., *Anal. Chem.*, **71**, 4208 (1999).
63. Szpunar, J., Schmitt, V. O., Lobinski, R. and Monod, J. L., *J. Anal. At. Spectrom.*, **11**, 193 (1996).
64. Lespes, G., Desauziers, V., Montigny, C. and Potin-Gautier, M., *J. Chromatogr. A*, **826**, 67 (1998).
65. Lobinski, R., Boutron, C. F., Candelone, J. P., Hong, S., Szpunar, J. and Adams, F. C., *Anal. Chem.*, **65**, 2510 (1993).
66. Lobinski, R., Szpunar, J., Adams, F. C., Tesseidre, P. L. and Cabanis, J. C., *J. Assoc. Off. Anal. Chem.*, **76**, 1262 (1993).
67. De Smaele, T., Moens, L., Dams, R., Sandra, P., Van der Eycken, J. and Vandyck, J., *J. Chromatogr. A*, **793**, 99 (1998).
68. García Fernández, R., Montes Bayón, M., García Alonso, J. I. and Sanz-Medel, A., *J. Mass Spectrom.*, **35**, 639 (2000).
69. Heisterkamp, M. and Adams, F. C., *J. Anal. At. Spectrom.*, **14**, 1307 (1999).
70. Pons, B., Carrera, A. and Nerín, C., *J. Chromatogr. B*, **716**, 139 (1998).
71. Bulska, E., Emteborg, H., Baxter, D. C., Frech, W., Ellingsen, D. and Thomassen, Y., *Analyst*, **117**, 657 (1992).
72. Leal Granadillo, I. A., García Alonso, J. I. and Sanz-Medel, A., *Anal. Chim. Acta*, **423**, 21 (2000).
73. Radojevic, M., Allen, A., Rapsomanikis, S. and Harrison, R. M., *Anal. Chem.*, **58**, 658 (1986).
74. Chakraborti, D., De Jonghe, W. R. A., Van Mol, W. E., Van Cleuvenbergen, R. J. A. and Adams, F. C., *Anal. Chem.*, **56**, 2692 (1984).
75. Gui-Bin, J. and Qun-Fang, Z., *J. Chromatogr. A*, **886**, 197 (2000).
76. Stäb, J. A., Cofino, W. P., van Hattum, B. and Brinkman, U. A. T., *Fresenius' J. Anal. Chem.*, **347**, 247 (1993).
77. Liu, Y., Lopez-Avila, V., Alcaraz, M. and Beckert, W. F., *J. High Resolut. Chromatogr.*, **17**, 527 (1994).

78. Stäb, J. A., Brinkman, U. A. T. and Cofino, W. P., *Appl. Organomet. Chem.*, **8**, 577 (1994).
79. Meinema, H. A., Burger-Wiersma, T., Versluis-de Haan, G. and Ch. Gevers, E., *Environ. Sci. Technol.*, **12**, 288 (1978).
80. Heisterkamp, M., De Smaele, T., Candelone, J. P., Moens, L., Dams, R. and Adams, F. C., *J. Anal. At. Spectrom.*, **12**, 1077 (1997).
81. Nagase, M., Toba, M., Kondo, H. and Hasebe, K., *Analyst*, **123**, 1091 (1998).
82. Cai, Y., Alzaga, R. and Bayona, J. M., *Anal. Chem.*, **66**, 1161 (1994).
83. Quevauviller, P., Fortunati, G. U., Filippelli, M., Bortoli, A. and Muntau, H., *Appl. Organomet. Chem.*, **12**, 531 (1998).
84. Quevauviller, P., Ebdon, L., Harrison, R. M. and Wang, Y., *Appl. Organomet. Chem.*, **13**, 1 (1999).
85. Wenclawiak, B. W. and Krah, M., *Fresenius' J. Anal. Chem.*, **351**, 134 (1995).
86. Minganti, V., Capelli, R. and De Pellegrini, R., *Fresenius' J. Anal. Chem.*, **351**, 471 (1995).
87. Clausen, J. and Nielsen, S. A., *Biol. Trace Elem. Res.*, **15**, 125 (1988).
88. Lamkin, W. M. and Gehrke, C. W., *Anal. Chem.*, **37**, 383 (1965).
89. Husek, P. and Sweeley, C., *J. High Resolut. Chromatogr.*, **14**, 751 (1991).
90. Vázquez Peláez, M., Montes Bayón, M., García Alonso, J. I. and Sanz-Medel, A., *J. Anal. At. Spectrom.*, **15**, 1217 (2000).
91. Pawliszyn, J., *Trends Anal. Chem.*, **14**, 113 (1995).
92. Alpendurada, M. F., *J. Chromatogr. A*, **889**, 3 (2000).
93. Arthur, C. L. and Pawliszyn, J., *Anal. Chem.*, **62**, 2145 (1990).
94. Cai, Y. and Bayona, J. M., *J. Chromatogr. A*, **696**, 113 (1995).
95. Tutschku, S., Mothes, S. and Wennrich, R., *Fresenius' J. Anal. Chem.*, **354**, 587 (1996).
96. Górecki, T. and Pawliszyn, J., *Anal. Chem.*, **68**, 3008 (1996).
97. Moens, L., De Smaele, T., Dams, R., Van Den Broeck, P. and Sandra, P., *Anal. Chem.*, **69**, 1604 (1997).
98. Lespes, G., Desauziers, V., Montigny, C. and Potin-Gautier, M., *J. Chromatogr. A*, **826**, 67 (1998).
99. Jiang, G. B., Liu, J. Y. and Yang, K. W., *Anal. Chim. Acta*, **421**, 67 (2000).
100. Aguerre, S., Bancon-Montigny, C., Lespes, G. and Potin-Gautier, M., *Analyst*, **125**, 263 (2000).
101. Millán, E. and Pawliszyn, J., *J. Chromatogr. A.*, **873**, 63 (2000).
102. Vercauteren, J., De Meester, A., De Smaele, T., Vanhaecke, F., Moens, L., Dams, R. and Sandra, P., *J. Anal. At. Spectrom.*, **15**, 651 (2000).
103. Lord, H. and Pawliszyn, J., *J. Chromatogr. A*, **885**, 153 (2000).
104. Baltussen, E., Sandra, P., David, F. and Cramers, C., *J. Microcolumn Sep.*, **11**, 737 (1999).
105. Vercauteren, J., Peres, C., Devos, C., Sandra, P., Vanhaecke, F. and Moens, L., *Anal. Chem.*, **73**, 1509 (2001).
106. Pécheyran, C., Quetel, C. R., Lecuyer, F. M. M. and Donard, O. F. X., *Anal. Chem.*, **70**, 2639 (1998).
107. Seligman, P. F., Valkirs, A. O. and Lee, R. F., *Environ. Sci. Technol.*, **20**, 1229 (1986).
108. Schebek, L. and Andreae, M. O., *Environ. Sci. Technol.*, **25**, 871 (1991).
109. Lobinski, R. and Adams, F. C., *Spectrochim Acta Part B*, **52**, 1865 (1997).
110. Eiden, R., Schöler, H. F. and Gastner, M., *J. Chromatogr. A*, **809**, 151 (1998).
111. Munari, F., Colombo, P. A., Magni, P., Zilioli, G., Trestianu, S. and Grob, K., *J. Microcolumn Sep.*, **7**, 403 (1995).
112. Ceulemans, M., Lobinski, R., Dirkx, W. M. R. and Adams, F. C., *Fresenius' J. Anal. Chem.*, **347**, 256 (1993).
113. Cooke, W. S., *Today's Chem. Work*, **5**, 16 (1996).
114. Lobinski, R., Rodriguez Pereiro, I., Chassaigne, H., Wasik, A. and Szpunar, J., *J. Anal. At. Spectrom.*, **13**, 859 (1998).
115. Rodriguez Pereiro, I., Schmitt, V. O. and Lobinski, R., *Anal. Chem.*, **69**, 4799 (1997).
116. Wasik, A., Rodriguez Pereiro, I. and Lobinski, R., *Spectrochim. Acta Part B*, **53**, 867 (1998).
117. Slaets, S., Adams, F., Rodriguez Pereiro, I. and Lobinski, R., *J. Anal. At. Spectrom.*, **14**, 851 (1999).
118. Lobinski, R., Sidelnikov, V., Patrushev, Y., Rodriguez Pereiro I. and Wasik, A., *Trends Anal. Chem.*, **18**, 449 (1999).
119. Rodriguez Pereiro, I., Monicou, S., Lobinski, R., Sidelnikov, V., Patrushev, Y. and Yamanaka, M., *Anal. Chem.*, **71**, 4534 (1999).
120. Wasik, A., Rodriguez Pereiro, I., Dietz, C., Spuznar, J. and Lobinski, R., *Anal. Commun.*, **35**, 331 (1998).
121. Forsyth, D. S. and Marshall, W. D., *Anal. Chem.*, **55**, 2132 (1983).
122. Soderquist, C. J. and Crosby, D. G., *Anal. Chem.*, **50**, 1435 (1978).
123. Zarnegar, P. and Mushak, P., *Anal. Chim. Acta*, **69**, 389 (1974).
124. Uchida, H., Shimoishi, Y. and Toci, K., *Environ. Sci. Technol.*, **14**, 541 (1980).
125. Andreae, M. O., *Anal. Chem.*, **49**, 820 (1977).
126. Ebdon, L., Hill, S. and Ward, R. W., *Analyst*, **111**, 1113 (1986).
127. Ebdon, L., *An Introduction to Atomic Absorption Spectroscopy*, Heyden, London, 1982.
128. Kolb, B., Kemmer, G., Schlesser, F. H. and Wiedeking, E., *Fresenius' J. Anal. Chem.*, **221**, 166 (1966).
129. Chau, Y. K., Wong, P. T. S. and Goulden, P. D., *Anal. Chim. Acta*, **85**, 421 (1976).
130. Wolf, W. R., *Anal. Chem.*, **48**, 1717 (1976).

131. Hahn, M. H., Mulligan, K. J., Jackson, M. E. and Caruso, J. A., *Anal. Chim. Acta*, **118**, 115 (1980).
132. Ebdon, L., Ward, R. W. and Laethard, D. A., *Analyst*, **107**, 129 (1982).
133. Foster, R. C. and Howard, A. G., *Anal. Proc.*, **26**, 34 (1989).
134. Forsyth, D. S. and Cleroux, C., *Talanta*, **38**, 951 (1991).
135. Segar, D. A., *Anal. Lett.*, **7**, 89 (1974).
136. Parris, G. E., Blair, W. R. and Brinckman, F. E., *Anal. Chem.*, **49**, 378 (1977).
137. Apte, S. C. and Gardner, M. J., *Talanta*, **35**, 539 (1988).
138. Chau, Y. K., Wong, P. T. S. and Goulden, P. D., *Anal. Chem.*, **47**, 2279 (1975).
139. Chau, Y. K., Wong, P. T. S., Bengert, G. A. and Dunn, J. L., *Anal. Chem.*, **56**, 271 (1984).
140. Forsyth, D. S. and Marshall, W. D., *Anal. Chem.*, **57**, 1299 (1985).
141. Müller, M. D., *Anal. Chem.*, **59**, 617 (1987).
142. Dachs, J., Alzaga, R., Bayona, J. M. and Quevauviller, P., *Anal. Chim. Acta*, **286**, 319 (1994).
143. Gui-Bin, J., Qun-Fang, Z. and Bin, H., *Bull. Environ. Contam. Toxicol.*, **65**, 277 (2000).
144. McCormack, A. J., Tong, S. C. and Cooke, X. D., *Anal. Chem.*, **37**, 1470 (1965).
145. Rosenkranz, B. and Bettmer, J., *Trends Anal. Chem.*, **362**, 489 (1998).
146. Kato, T., Uehiro, T., Yasuhara, A. and Morita, M., *J. Anal. At. Spectrom.*, **7**, 15 (1992).
147. Witte, C., Szpunar, J., Lobinski, R. and Adams, F. C., *Appl. Organomet. Chem.*, **8**, 621 (1994).
148. Ceulemans, M., Witte, C., Lobinski, R. and Adams, F. C., *Appl. Organomet. Chem.*, **8**, 451 (1994).
149. Mena, M. L., McLeod, C. W., Jones, P., Withers, A., Mimganti, V., Capelli, R. and Quevauviller, P., *Fresenius' J. Anal. Chem.*, **355**, 456 (1995).
150. de la Calle Guntiñas, M. B., Ceulemans, M., Witte, C., Lobinski, R. and Adams, F. C., *Mikrochim. Acta*, **120**, 73 (1995).
151. Estes, S. A., Uden, P. C. and Barnes, R. M., *Anal. Chem.*, **53**, 1829 (1981).
152. Creed, J. T., Mohamad, A. H., Davidson, T M., Ataman, G. and Caruso, J. A., *J. Anal. At. Spectrom.*, **3**, 923 (1988).
153. Windsor, D. L. and Denton, M. B., *Appl. Spectrosc.*, **32**, 366 (1978).
154. Windsor, D. L. and Denton, M. B., *J. Chromatogr. Sci.*, **17**, 492 (1979).
155. Windsor, D. L. and Denton, M. B., *Anal. Chem.*, **51**, 1116 (1979).
156. Brown, R. M., Northway, S. J. and Fry, R. C., *Anal. Chem.*, **53**, 934 (1981).
157. Brown, R. M. and Fry, R. C., *Anal. Chem.*, **53**, 532 (1981).
158. Van Loon, J. C., Alcock, L. R., Pinchin, W. H. and French, J. B., *Spectrosc. Lett.*, **19**, 1125 (1986).
159. Chong, N. S. and Houk, R. S., *Appl. Spectrosc.*, **41**, 66 (1987).
160. Peters, G. R. and Beauchemin, D., *J. Anal. At. Spectrom.*, **7**, 965 (1992).
161. Peters, G. R. and Beauchemin, D., *Anal. Chem.*, **65**, 97 (1993).
162. Murillo, M. and Mermet, J. M., *Spectrochim. Acta*, **42B**, 1151 (1987).
163. Kim, A. W., Hill, S., Ebdon, L. and Rowland, S. J., *J. High Resolut. Chromatogr.*, **15**, 665 (1992).
164. Pretorius, W. G., Ebdon, L. and Rowland, S. J., *J. Chromatogr.*, **646**, 369 (1993).
165. Ebdon, L., Evans, H., Pretorius, W. G. and Rowland, S. J., *J. Anal. At. Spectrom.*, **9**, 939 (1994).
166. Story, W. C., Olson, L. K., Shen, W.-L., Creed, J. T. and Caruso, J. A., *J. Anal. At. Spectrom.*, **5**, 467 (1990).
167. Evans, E. H. and Caruso, J. A., *J. Anal. At. Spectrom.*, **8**, 427 (1993).
168. Evans, E. H., Pretorious, W., Ebdon, L. and Rowland, S., *Anal. Chem.*, **66**, 3400 (1994).
169. Prange, A. and Jantzen, E., *J. Anal. At. Spectrom.*, **10**, 105 (1995).
170. De Smaele, T., Verrept, P., Moens, L. and Dams, R., *Spectrochim. Acta*, **50B**, 1409 (1995).
171. De Smaele, T., Moens, L., Dams, R. and Sandra, P., *Fresenius' J. Anal. Chem.*, **355**, 778 (1996).
172. De Smaele, T., Moens, L., Dams, R. and Sandra, P., *LC-GC Int.*, **14**, 876 (1996).
173. Gallus, S. M. and Heumann, K. G., *J. Anal. At. Spectrom.*, **11**, 887 (1996).
174. Pritzl, G., Stuer-Lauridsen, F., Carlsen, L., Jensen, A. K. and Thorsen, T. K., *Int. J. Environ. Anal. Chem.*, **62**, 147 (1996).
175. Ruiz Encinar, J., García Alonso, J. I. and Sanz-Medel, A., *J. Anal. At. Spectrom.*, **15**, 1233 (2000).
176. Krupp, E. M., Pécheyran, C., Meffan-Main, S. and Donard, O. F. X., *Fresenius' J. Anal. Chem.*, **370**, 573 (2001).
177. Krupp, E. M., Pécheyran, C., Pinaly, H., Motelica-Heino, M., Koller, D., Young, S. M. M., Brenner, I. B. and Donard, O. F. X., *Spectrochim. Acta*, **56B**, 1233 (2001).
178. Vela, N. P., Olson, L. K. and Caruso, J. A., *Anal. Chem.*, **65**, 585A (1988).
179. García Alonso, J. I., Sanz-Medel, A. and Ebdon, L., *Anal. Chim. Acta*, **283**, 261 (1993).
180. Stäb, J. A., Cofino, W. P., Van Hattum, B. and Th. Brinkman, U. A., *Fresenius' J. Anal. Chem.*, **347**, 247 (1993).
181. Ashby, J. R. and Craig, P. J., *Sci. Total Environ.*, **78**, 219 (1989).
182. Arakawa, Y., Wada, O., Yu, T. H. and Iwai, H., *J. Chromatogr.*, **216**, 209 (1981).
183. Duebelbeis, D. O., Kapila, S., Yates, D. E. and Manahan, S. E., *J. Chromatogr.*, **351**, 465 (1986).
184. Tolosa, I., Bayona, J. M., Albaigés, J., Alencrasto, L. F. and Tarradellas, J., *Fresenius' J. Anal. Chem.*, **339**, 646 (1991).

185. Jiang, G. B., Ceulemans, M. and Adams, F. C., *J. Chromatogr. A*, **727**, 119 (1996).
186. Sato, K., Khori, M. and Okochi, H., *Bunseki Kagaku*, **45**, 575 (1996).
187. Rodríguez González, P., Ruiz Encinar, J., García Alonso, J. I. and Sanz-Medel, A., *J. Anal. At. Spectrom.*, **17**, 824 (2002).
188. Ruiz Encinar, J., Monterde Villar, M. I., Gotor Santamaría, V., García Alonso, J. I. and Sanz-Medel, A., *Anal. Chem.*, **73**, 3174 (2001).
189. Horlick, G. and Montaser, A., Analytical characteristics of ICP-MS, in *Inductively Coupled Plasma Mass Spectrometry*, Montaser, A. (Ed.), Willey-VCH, New York, 1998, pp. 503–614.
190. Quevauviller, P., Astruc, M., Ebdon, L. and Griepink, B., *Appl. Organomet. Chem.*, **8**, 639 (1994).
191. Abalos, M., Bayona, J. M., Campañó, R., Granados, M., Leal, C. and Prat, M. D., *J. Chromatogr. A*, **788**, 1 (1997).
192. García Alonso, J. I., Ruiz Encinar, J., Leal Granadillo, I. A., García Fernandez, R., Montes Bayón, M. and Sanz-Medel, A., Trace metal speciation in environmental analysis using inductively coupled plasma mass spectrometry, in *Advances in Mass Spectrometry, Vol. 15*, Gelpí, E. (Ed.), John Wiley and Sons, Ltd, Chichester, 2001, pp. 265–282.
193. Vanhaecke, F., Moens, L. and Taylor, P., Use of ICP-MS for isotope ratio measurements, in *Inductively Coupled Plasma Spectrometry and its Applications*, Hill, S. J. (Ed.), Sheffield Academic Press, Sheffield, 1999, pp. 145–207.
194. Nelms, S. M., Quétel, C. R., Prohaska, T., Vogl, J. and Taylor, P. D. P., *J. Anal. At. Spectrom.*, **16**, 333 (2001).
195. Ruiz Encinar, J., García Alonso, J. I., Sanz-Medel, A., Main, S. and Turner, P. J., *J. Anal. At. Spectrom.*, **16**, 315 (2001).
196. Heumann, K. G., Gallus, S. M., Radlinger, G. and Vogl, J., *J. Anal. At. Spectrom.*, **13**, 1001 (1998).
197. Hintelmann, H., Evans, R. D. and Villenueve, J. Y., *J. Anal. At. Spectrom.*, **10**, 619 (1995).
198. Rosman, K. J. R. and Taylor, P. D. P., *J. Anal. At. Spectrom.*, **14**, 5N (1999).
199. Hintelmann, H., Falter, R., Ilgen, G. and Evans, R. D., *Fresenius' J. Anal. Chem.*, **358**, 363 (1997).
200. Klemens, P. and Heumann, K. G., Determination of total silicon and silicon species by isotope dilution ICP-HRMS, Paper presented at the *Second International Conference on Double Focusing Sector Field ICP-MS*, Vienna, September 12–15, 2001.
201. Pérez Méndez, S., Montes Bayón, M., Blanco González, E. and Sanz-Medel, A., *J. Anal. At. Spectrom.*, **14**, 1333 (1999).
202. Costa Fernández, J. M., Bings, N. H., Leach, A. M. and Hieftje, G. M., *J. Anal. At. Spectrom.*, **15**, 1063 (2000).
203. Bings, N. H., Costa Fernández, J. M., Guzowski, J. P. Jr., Leach, A. M. and Hieftje, G. M., *Spectrochim. Acta*, **55B**, 767 (2000).
204. Haas, K., Feldmann, J., Wennrich, R. and Stärk, H-J., *Fresenius J. Anal. Chem.*, **370**, 587 (2001).
205. Donard, O. F. X., Metal speciation and trace metal cycling in the environment. Paper presented at the *Fourth Euroconference on Environmental Analytical Chemistry*, Visegrad, Hungary, September 14–19, 2000.
206. Huo, D., Lu, Y. and Kingston, H. M., *Environ. Sci. Technol.*, **32**, 3418 (1998).
207. Ruiz Encinar, J., Rodríguez González, P., García Alonso, J. I. and Sanz-Medel, A., *Anal. Chem.*, **74**, 270 (2002).
208. García Alonso, J. I., Ruiz Encinar, J., Rodríguez González, P. and Sanz-Medel, A., *Anal. Bioanal. Chem.*, **373**, 432 (2002).
209. Rajendran, R. B., Tao, H., Nakazato, T. and Miyazaki, A., *J. Anal. At. Spectrom.*, **15**, 1757 (2000).
210. Yu, Q., Quian, J. and Frech, W., *J. Anal. At. Spectrom.*, **15**, 1586 (2000).
211. Pérez Méndez, S., Blanco González, E. and Sanz-Medel, A., *J. Anal. At. Spectrom.*, **15**, 1109 (2000).
212. Montes Bayón, M., García Alonso, J. I., Vazquez Pelaez, M., Rodriguez de la Flor, R. and Sanz Medel, A., Coupling GC to double focusing ICP-MS to perform speciation of sulphur and selenium containing aminoacids. Paper presented at the *First International Conference on High resolution ICP-MS*, Norfolk, Virginia, May 25–28, 2000.
213. Heumann, K. G., in *Metal Speciation in the Environment*, Broekaert, J. A. C., Guçer, S. and Adams, F. (Eds), *NATO ASI Series*, Vol. G 23, Springer, Berlín, 1990, pp. 153–168.
214. De Bievre, P., *Anal. Proc.*, **30**, 328 (1993).
215. Sanz-Medel, A., *Spectrochim. Acta*, **53B**, 197 (1998).
216. Fasset, J. D. and Paulsen, P. J., *Anal. Chem.*, **61**, 643A (1989).
217. Heumann, K. G., *Int. J. Mass Spectrom. Ion Processes*, **118/119**, 575 (1992).
218. Sariego Muñiz, C., Marchante Gayón, J. M., García Alonso, J. I. and Sanz-Medel, A., *J. Anal. At. Spectrom.*, **16**, 587 (2001).
219. Rottmann, L. and Heumann, K. G., *Anal. Chem.*, **66**, 3709 (1994).
220. Snell, J. P., Stewart, I. I., Sturgeon, R. E. and Frech, W., *J. Anal. At. Spectrom.*, **15**, 1540 (2000).
221. Demuth, N. and Heumann, K. G., *Anal. Chem.*, **73**, 4020 (2001).
222. Ceulemans, M. and Adams, F. C., *Anal. Chim. Acta*, **317**, 161 (1995).
223. Quevauviller, P., *Spectrochim. Acta*, **53B**, 1261 (1998).
224. Quevauviller, P. and Morabito, R., *Trends Anal. Chem.*, **19**, 86 (2000).
225. Adams, F. C. and Slaets, S., *Trends Anal. Chem.*, **19**, 80 (2000).
226. Quevauviller, P., Astruc, M., Morabito, R., Ariese, F. and Ebdon, L., *Trends Anal. Chem.*, **19**, 180 (2000).

4.3 Capillary Electrophoresis in Speciation Analysis

Bernhard Michalke

Institute for Ecological Chemistry, Neuherberg, Germany

ABBREVIATIONS

CE	capillary electrophoresis
CZE	capillary zone electrophoresis
DIN	direct injection nebulizer
EOF	endoosmotic flow
ESI-MS	electrospray ionization mass spectrometer

Handbook of Elemental Speciation: Techniques and Methodology R. Cornelis, H. Crews, J. Caruso and K. Heumann
 ISBN: 0-471-49214-0

ICP-MS	inductively coupled plasma mass spectrometer
IEF	isoelectric focusing
ITP	isotachophoresis
LIF	laser induced fluorescence
LOD	limit of determination
MCN	microconcentric nebulizer
MECC	micellar electrokinetic chromatography

1 SPECIATION – INTRODUCTION

1.1 Importance of element speciation analysis

The quality and quantity of the various species of an element in a matrix are highly responsible for the mobility, bioavailability and finally the ecotoxicological or toxicological impact of that element rather than its total concentration [1–5]. Therefore, only a knowledge of the species enables an assessment of whether it is toxic, without (known) impact at a specific concentration, or whether it is essential. This stresses the necessity of speciation analysis to determine the speciation of an element in a specific matrix.

Definitions related to speciation have been given earlier in this book. However, it must be noted that the use of capillary electrophoresis (CE) in speciation analysis is closely linked to the analytical activity of identifying and measuring species, including identification of the binding partners of elements as well as providing methods for quality control [6].

2 POSITION OF CAPILLARY ELECTROPHORESIS IN SPECIATION

As yet there is no such instrument as an easy 'speciation analyzer'. Combination and hyphenation of separation technologies and element- or molecule-selective detection systems are generally the basis for speciation analyses. Separation and detection methods already established in other fields have to be combined partly in new ways and modified according to the particular speciation problems. Therefore, element species must be separated before being analyzed, either by nonselective detectors (e.g. UV) or by element- or molecule-selective detector such as ICP-MS or ESI-MS. One of the most powerful separation techniques is CE. It provides a most efficient separation of species and is often superior to liquid chromatography separation techniques. In addition, a single CE instrument even allows several different modes in separation: CZE, MECC, IEF, ITP and CEC. These separation mechanisms are commonly based on the application of a high voltage. However, their separation principles are quite different, providing completely different mechanisms of characterization and identification for element species. The latter is of particular significance, as species identification is rarely done by one single method but needs multidimensional strategies. In addition, this variability is most advantageous in finding separation solutions for nearly every specific separation problem. While CE supplements conventional HPLC methods, it shows unique promise for speciation purposes because it causes only a minor disturbance to the existing equilibrium between different species. There is no stationary phase, which has a huge surface area and gives various possibilities for undesired interactions. Therefore, species integrity is thought to be less easily affected than with HPLC.

The use of CE in speciation is complex: it may be used as a primary separation mechanism or as a secondary separation technique after e.g. HPLC, in a second dimension for identification. Often it is combined with nonselective (direct or indirect) UV detection. In whatever dimension it is employed, parallel runs in different separation modes on one sample are performed for a wide characterization. As an improvement fractionation may be performed at the outlet of the capillary to analyze further the separated element species by ICP-MS for the respective metals. This approach must be recognized as a preliminary stage for finally hyphenated techniques linked to selective detectors for either elements or whole molecules. The various combinations of CE techniques in different

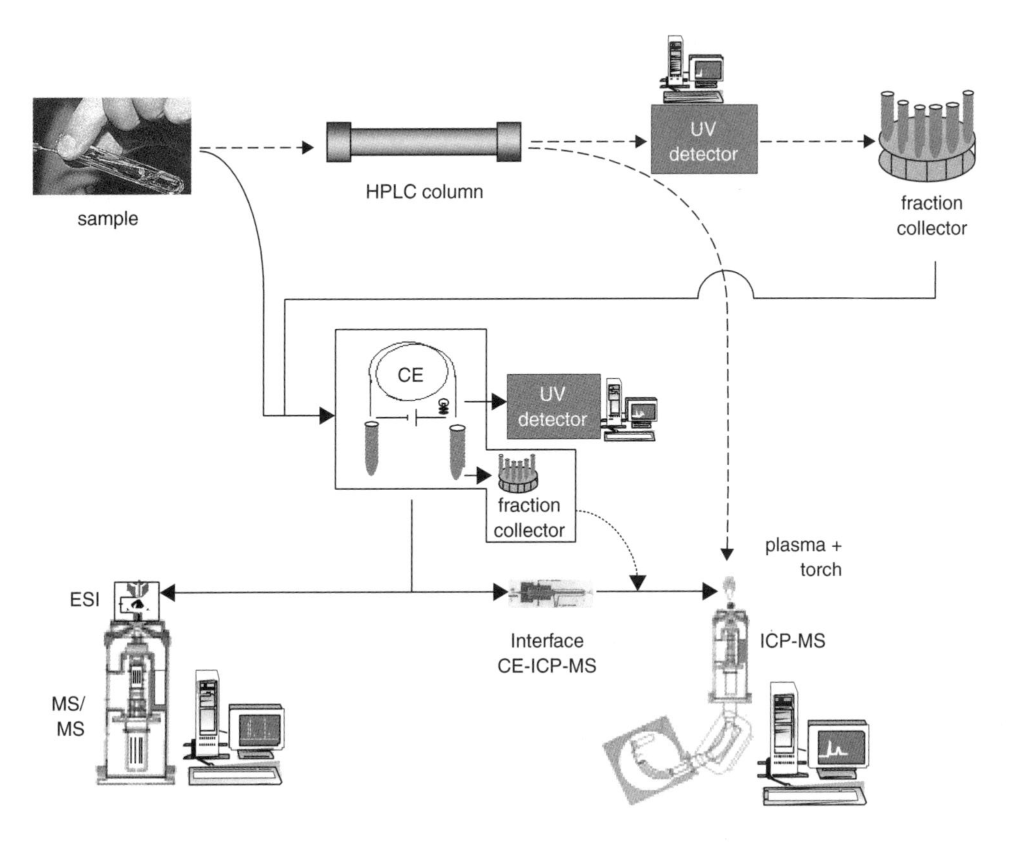

Figure 4.3.1. The position of capillary electrophoresis in element speciation analysis: CE can be used as a primary separation system with UV detection and fractionation for subsequent element determination. A more elegant way is its use in hyphenated systems, with either ICP-MS or ESI-MS. Then isotopic or molecular information can be obtained. CE can also be used as a secondary separation device, e.g. after HPLC separation, for quality control using the detection capabilities shown.

stages of multidimensional strategies and with different detectors, which supply information about isotopes and elements (ICP-MS) or molecules and structural compounds (ESI-MS/MS), make CE immensely valuable in speciation analysis. Figure 4.3.1 schematically shows the position of CE in speciation analysis.

1. In Figure 4.3.1 it is shown that CE may be used as the primary separation technique. The sample is introduced to a CE instrument and species are separated by one of the different separation modes. Detection may be done by direct UV or indirect UV if analytes are UV transparent. However, conductivity detectors or LIF are also employed.

2. CE may be also used for quality control and species identification as a second dimension separation device. Here, the first separation and characterization of species are performed by another method, such as HPLC, combined with fraction collection and subsequent element detection. In order to improve the reliability of speciation results, it is necessary to subject the HPLC fractions to other separation techniques based on different properties of the molecules for clear identification. Element concentrations are determined for specific HPLC peak fractions, which are further investigated by (different) CE methods. A nearly overall characterization is possible in this way, provided that no species alteration has occurred during the primary (HPLC) separation. It is assumed that the identified species and the elements determined in the fractions are indeed associated with one another. However, this assumption does not

necessarily hold in all cases, even if they are detected in the same fraction. To overcome this uncertainty CE may be run additionally with fraction collection at the end of the CE capillary. The fractions are again introduced into an element detector. Elements are shown to be attached to a specific species, both after HPLC and after CE separation. The species identification, as well as the knowledge about species–element association here, is based on orthogonal strategies, providing high testifying power.

3. A more elegant choice is to use a hyphenated system of CE with an element-selective detector such as ICP-MS. The high resolving power of CE and its capability of different separation technologies are here combined with element and/or isotope information. Combinations with ICP-OES have also been occasionally described, but cannot play an important role in speciation at physiologically or environmentally normal concentrations. For these applications the detection sensitivity of ICP-OES is insufficient. Specific problems in interfacing the two systems arise from the interface itself.

4. Finally, even with CE-ICP-MS not all problems in speciation are solved. Generally, combinations of separation and element detection systems identify species by means of comparing retention/migration times of analytes and standard compounds. When standards are missing identification is rarely possible. Even a characterization is difficult when employing hyphenated systems, as no fractions for subsequent investigations are available. Therefore, CE-ICP-MS should be complemented by a parallel setup of CE-ESI-MS. The electrospray ionization source is capable of handling low flow rates typical of CE and provides a smooth ionization, where the species is not altered (in many cases). Therefore, with this hyphenation the high resolving power of CE and its capacity for different separation technologies is combined with molecular (MS) and/or structural information (MS/MS).

3 PRINCIPLES OF METAL SPECIATION BY CAPILLARY ELECTROPHORESIS

3.1 Injection

The first method of sample introduction is to use a positive pressure at the inlet or, vice versa, a negative pressure (suction) at the outlet. Given a homogenous sample composition all components of the sample are injected at their respective concentration without any preference. Typical sampling volumes range between 5 and 50 nL. Hydrodynamical injection modes are based on a similar principle: either the sample at inlet is set higher than the outlet or the latter is positioned below the inlet. In both modes gravity is forcing the sample into the capillary during the time the height difference is maintained.

Another injection method is to apply a high voltage to the sample for a short injection time, typically around 5–15 s. This injection mode discriminates between charged and noncharged analytes or anions and cations. Conversely, there is the possibility of introducing selectively only analytes of a desired charge (positive or negative) by the application of an injection voltage with the appropriate polarity. This is called electrokinetic injection. A high charge density of the analyte helps for a preferred injection by the system and finally leads to a preconcentration during injection. However, it has been found that the injection reproducibility of slow moving species is much worse. Thus, electrokinetic injection is not considered to be a reliable method in elemental speciation [7]. It also lacks on a wide dynamic range.

3.2 Separation

CE uses the separation principle of differences in the electrically driven mobility of charged analytes, similar to conventional electrophoresis. Here, an electric field is applied along an open-tube column with a low inner diameter at high voltage, typically between 20 and 30 kV. This is the reason for its comparatively short analysis time and very high separation efficiency. Usually, 200 000–700 000 theoretical plates are achieved. The lack of a stationary phase having a large surface is considered as a further advantage of CE in speciation [8]. The molecules move with different velocities in the electric field. The borders of the analyte bands do not show laminar profiles, but for the CE show typically sharp vertically ones. This again improves resolution and also the signal to noise ratio and thus affects detection sensitivity. The latter is high for total analyte masses but low when looking at analyte concentrations. The reason is the very small total sample volumes, being in the range of 5–50 nL.

In contrast to conventional flat bed electrophoresis this technique is easily adapted for automation and quantitative analysis. The resulting electropherograms can be similarly processed like LC chromatograms. Usually a complete CE analysis is very fast and completed earlier than an LC separation of the same analytical problem.

3.3 Different separation modes

Development of the method and changing of analytical parameters are readily done by just changing the electrolyte system when replacing a few milliliters of one buffer system with other buffers. Thus, several different separation principles are available, distinguishing and separating the species according to their different physicochemical properties. Table 4.3.1 gives an overview of the separation modes, separation principle and target analytes (according to ref. 9).

Detection may be performed on the column by UV and LIF, or at the end of the capillary by conductivity or (preferably) element-selective and molecule-selective detectors. Combinations used for element speciation are discussed below. A scheme of a conventional CE system is given in Figure 4.3.2.

At pH values higher than 3 a laminar flow is built up, the more the pH takes on basic values. This flow is called endoosmotic flow (*EOF*). It is induced by the negatively charged inner capillary surface where the silanolic groups of the bare fused silica capillary attract positively charged ions. An 'electrical double layer' is set up where hydrated cations are moved to the cathode by the electric field and thus produce the EOF. Often the EOF is faster than the current-driven movement of anions in the opposite direction. A very fast EOF can result in an alteration in the separation of species. On the other hand, the EOF helps for a prolonged separation time and thus increased resolution when the electrically driven movement and the EOF are opposed. Usually, the EOF is directed towards the cathode as long as the capillary walls remain negatively charged. This means that only cationic metal species are moving by electrophoresis in the same direction as the EOF. However, without special measures to adjust the differences in electrophoretic mobilities, the simple co-electroosmotic mode offers efficient separations for only a limited number of cations. In contrast, many anionic metal complexes promise a good resolution due to inherent differences in electrophoretic mobilities. However, anionic electrophoretic mobilities and the EOF have different directions. Such a counter-electroosmotic migration substantially reduces the range of anionic analytes that can be separated for detection at the cathode end.

Table 4.3.1. Overview of the separation modes, separation principle and target analytes [9].

Separation mode	Abbreviation	Principle	Target analytes
Capillary zone electrophoresis	**CZE**	Charge density charge/mass ratio	Charged molecules, amino acids, proteins
Capillary isoelectric focusing	**cIEF**	Isoelectric point	Proteins, peptides
Capillary isotachophoresis	**cITP**	Analyte specific conductivity	Differently dissociated molecules
Micellar electrokinetic capillary chromatography	**MECC**	Hydrophobicity	Neutral molecules with different ability to enter charged hydrophobic micelles

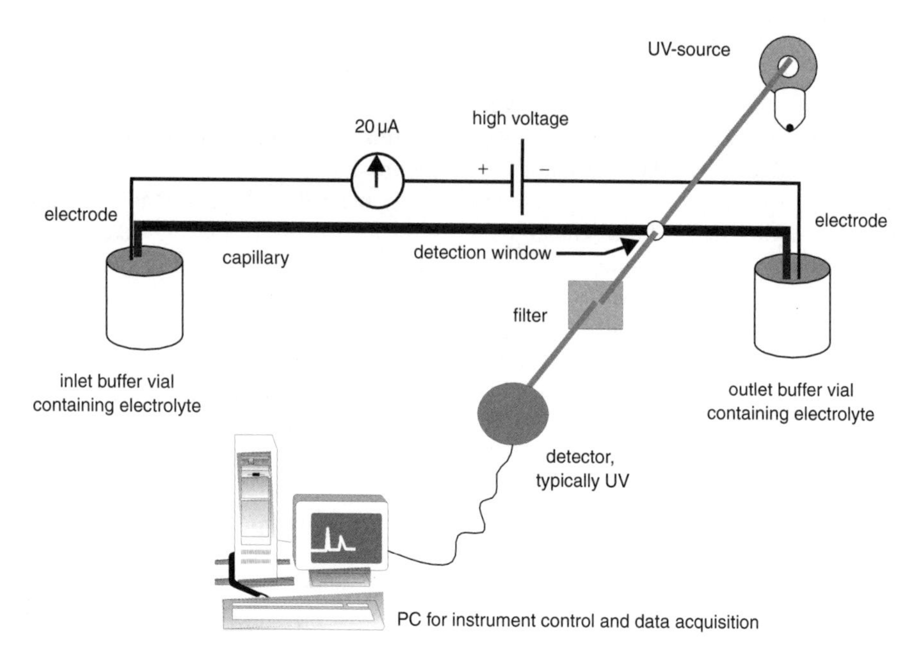

Figure 4.3.2. Principle scheme of capillary electrophoresis.

The EOF may be varied by applying coatings to the capillary wall: it can be slowed down, totally avoided or may be even reversed. These coatings may be of a dynamic nature by purging the capillary with suitable additives before each analysis, or bound permanently to the wall by covalent bondings. Uncoated silanolic groups are likely to adsorb basic molecules, preferentially basic proteins. Therefore, in order to prevent the precipitation of proteins and to suppress the EOF, capillaries with a polymer-coated inner surface are recommended for such separations.

The principle of CZE separation assumes that the difference in ionic mobilities has a decisive effect on the resolution. The ionic mobilities of analytes are related, in turn, to their charge densities, i.e. charge-to-size ratios. In MECC, which operates under the same conditions as CZE but with micellar electrolytes, the principle mechanism of separation depends on whether the analytes are charged or not. For charged analytes, both electrophoretic migration in the aqueous electrolyte and solubilization into micelles play a role. The separation of electrically neutral compounds, on the other hand, is dominated by the distribution between the aqueous electrophoretic and micellar phases only. Accordingly, the analyte's hydrophobicity governs the distribution ratio and determines its migration behavior. The separation of small metal ions is an area where CE is being used to an increasing extent [10]. The distinctive feature of metal cations, metal-complexed ions and metal oxoanions is their high charge-to-size ratio and hence high electrophoretic mobility. As a striking electrophoretic property, it should lead to rapid separations with high efficiency. However for metal cations, many of which are of nearly identical charge and hydrated ionic radius, the differences in mobility are not sufficient to provide good separations. Further technologies are given by isotachophoresis. This method is mostly achieved by the use of discontinuous buffer systems, distinguished by differences in conductivity. When the sample is positioned between a 'terminating buffer' at the inlet, having a very low conductivity and a leading electrolyte with a higher conductivity than the sample compounds, the analytes will be positioned according to increasing conductivity as soon as the high voltage is turned on. The major advantage of isotachophoresis in speciation analysis is its capability to be combined on line with CZE. This results in a prefocusing of the sample before it is separated by CZE. Therefore, the sample volume can be increased partly up to 300 nL (in 1.5 m capillaries) without loss in resolution and thus concentration detection limits may be improved.

3.4 Problems in speciation, problems and risk for species preservation, suitability of (complexing) buffers, metal release in IEF

One limitation is the very small sample volume to be analyzed, typically only a few nanoliters. This causes problems in ensuring a representative sample and sets high demands on homogeneity of the sample. Further, the concentration detection limits are generally about two orders of magnitude worse than those for LC separations. The detection capability of the UV detectors generally used principally follows the Lambert–Beer law, where the length of the light path through the sample governs the sensitivity. In CE this length is the inner diameter of the capillary, which is typically 25–75 μm. However, since capillary electrophoresis is hyphenated to powerful detectors such as ICP-MS, LODs of 0.05–1 $\mu g\,L^{-1}$ have been reported [11–13]. The high voltage itself can alter the integrity of elemental species. Similarly, the buffer/electrolyte composition and nature of additives can have detrimental effects on species stability [14].

The mobility of analytes is dependent on the actual strength of the electric field and the (pH-dependent) EOF. Unfortunately, the EOF may be influenced by the sample itself, which may have a high buffering capacity, thus locally changing the pH and EOF. Furthermore, the electric field is changing along the capillary due to differences in analyte conductivity, again being partly a function of sample composition. Therefore, the migration time of a specific compound may be shifted from standard to sample and from sample to sample. This is widely reported in the literature and known as 'migration time shifts'. The reproducibility within one sample is generally high, provided the buffer reservoirs are not depleted, but the intersample reproducibility is sometimes worse. Species identification by comparison with standard electropherograms is thus most questionable. The addition of an internal standard helps to correct species migration time according to standard electropherograms. Another possibility is the standard addition procedure of the compounds investigated. In the case that comigration of other analytes is excluded this clearly identifies the species.

The application of the high voltage itself may alter the structure of metal species. Loosely bound metals may be removed. Also, recomplexation is possible by several buffer components investigated, e.g. for borate and phosphate buffers. This has to be specifically considered when the net charge of the species is changed and subsequently it is moving to the inlet. It will no longer be detected anymore.

When using hyphenated techniques an additional suction flow may be forced on the 'open-tube' capillary. Suction is likely to be induced by a necessary sheath flow in the nebulizer interface or by the nebulization gas itself at the end of the capillary.

When cIEF is used an EOF must be strictly avoided [15]. Therefore, only coated capillaries are suitable for cIEF. This technique is considered to show excellent advantages in separating peptides and proteins (but is restricted to such molecules) concerning resolution and concentration detection limits. The reason is that the whole capillary is filled with the sample. The total sample volume therefore is around 500–1500 nL (depending on capillary length) compared to CZE methods with ca. 5–20 nL. The sample must be mixed with 'ampholytes' (approximately 2 % in the sample), which determine the pH gradient inside the capillary and thus the final position of metal–protein species in the capillary. Unfortunately, one must consider that the ampholytes may alter species and lead to recomplexation. This is especially stressed as peptides and proteins show no net charges at their respective isoelectric points and some loosely bound metals may then be removed by the high voltage.

When applying MECC, additives for micellation are needed in the electrolyte. Metal contaminations may be occurring at the polar surface of micelles, which mimics element species. Hydrophobic organic–metal complexes with loosely bound metals may be altered.

3.5 Separation of varying species types

3.5.1 *Element species in different oxidation states*

Free metal ions usually show similar electrophoretic mobilities and insufficient stability in electrolyte solutions. In addition, detectability is difficult using conventional detectors, due to the lack of detectable properties for metal ions. Complexation presents a valuable approach for performing speciation of metals with different oxidation states [16]. Complete conversion of metals into charged complexes, which takes place upon addition of a complexing reagent to a sample before introduction to the capillary, renders them different in charge density while alteration of the initial concentration ratio of different oxidation states is prevented. For speciation involving metal oxidation forms of opposite charges, precapillary complexation is a straightforward strategy to get the same charge and electrophoretic direction for the species. The exchange kinetics must be slow enough (compared with the time required for the separation) to obtain the individual peaks for each species [14]. To preserve the stability or chemical integrity of complexed metal species the number of electrolyte constituents should be kept to a minimum. Just for improved detection simple carrier electrolytes containing only protonated imidazole for indirect UV detection should be used.

3.5.2 *Analysis of organometallic compounds*

Since different species of one metal with alkyl or aryl substituents often have similar mobilities, the separation power of CE usually needs enhancement. Enhanced separation efficiency can be gained by selecting a suitable electrolyte additive, such as a weak complexing reagent or β-cyclodextrin, which is capable of forming complexes with organometallic compounds. Otherwise, organometallic–ligand complexes formed before introduction into the capillary can be used effectively.

3.5.3 *Analysis of elements bound to organic compounds such as proteins*

There are many metal-containing species of biological significance that have been subjected to CE speciation. In accordance with their charges, these species can be separated while moving towards the cathode. For preventing precipitation of proteins and to suppress the EOF, capillaries with a polymer-coated inner surface are recommended for such separations. CE can also be used for the analysis of the interaction between metals and proteins as assessed by mobility-shift assays. Important structural information about metalloproteins such as transferrin, concerning heterogeneity of the attached carbohydrate chains as well as the degree of metal saturation, can be obtained by both CZE and cIEF techniques.

4 DETECTION MODES AND THEIR ADVANTAGES AND PROBLEMS

4.1 UV detection and indirect UV detection (iUV)

One major area of research in applying CE for metal speciation is the development of sensitive detectors. The small diameter of separation capillaries leads to high efficiencies in CE separations but is also responsible for the major limitation in detection sensitivity. Although the mass sensitivity is very high, the concentration sensitivity is generally one or two orders of magnitude lower than that for HPLC. Since real-world samples contain metal species at the $\mu g\,L^{-1}$ level or lower, sensitive detectors must be coupled to the capillary in a CE system.

Conventional detectors are predominantly UV detectors. They are mostly considered as nonspecific detectors, because in only a few cases do elemental species show typical UV spectra. Species-selective electropherograms must be obtained by using a UV scanning detection mode. However, UV detection is nonspecific for most types of speciation analysis. This generally can lead to a 'false' detection of element species. Overlapping of peaks, resulting possibly in incorrect quantification, may not be apparent. The detection sensitivity is mostly too poor as many species do not show sufficient UV response. Therefore, their detection generally requires derivatization using a suitable chromophore prior to separation. According to experiences of the Standard Measurement and Testing program of the EU *'derivatization steps are often far from being controlled'*. The respective steps are *'rarely fully understood'* [17].

In contrast, indirect UV absorbance detection is presently the most versatile method to solve the problem of universal detection for metal speciation by CE. The key to this approach is the displacement of a highly absorbing electrolyte co-ion by the sample ions. When choosing a UV-active co-ion, a close mobility match to the analyte ions is required; otherwise, asymmetrical peak shapes are generated. Various cationic and anionic co-ions such as imidazole, pyridine and chromate have been successfully utilized for indirect detection of metal species in CE.

Unfortunately, even with indirect UV detection, the sensitivity is rarely adequate for monitoring naturally concentrated species in the real world. The addition of chromophores to the buffer may again affect species integrity in some cases. Another possibility is to link 'invisible' element species to complexing UV-active complexes. This has been shown to provide detection sensitivity, but with the risk of species alteration. Original species information is easily lost.

4.2.1 *A few examples of speciation using CE-UV/iUV*

Several attempts have described to speciate selenium compounds usually by direct UV detection. Unfortunately, the speciation methods mostly had to be limited to separate standards or standard-added real samples, as detection sensitivity was inappropriate. Albert *et al.* [18] separated four standard selenium compounds within 20 min at concentrations considerably higher than those found in e.g. biological samples. However, the high separation potential of the technique is demonstrated clearly (Figure 4.3.3). A similar

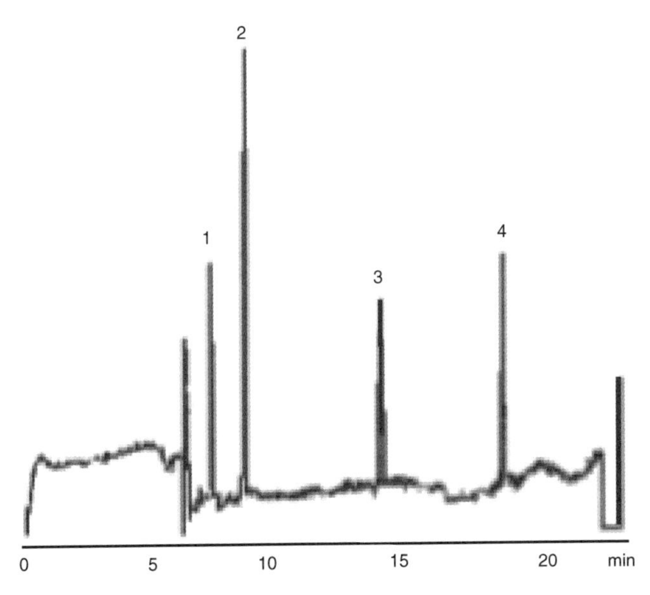

Figure 4.3.3. Electropherogram of Se compounds with the following CE conditions: fused silica capillary 52 cm × 50 μm ID; electrolyte, 80 mM phosphate buffer (pH 8.5); 2 mM TTAB; hydrodynamic injection 5 s; separation at −12 kV; direct UV detection at 200 nm. Peak 1 = Se(VI), peak 2 = Se(IV), peak 3 = SeC, peak 4 = SeM. Reprinted from *Analusis*, Albert, M. M. *et al.*, Vol. 21, pp. 403–407, 1993, copyright notice of Springer-Verlag [18].

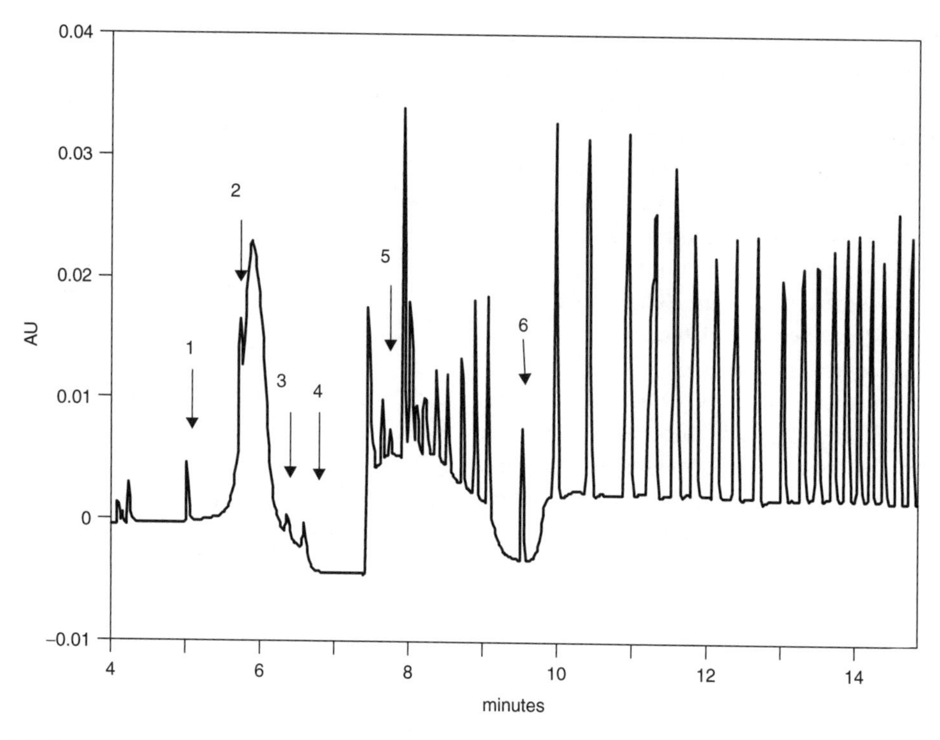

Figure 4.3.4. Electropherogram of a human milk sample for the following CE conditions: fused silica capillary 50 cm × 50 μm ID; electrolyte, 10 mM Na_2CO_3; pH 11.6; pressure injection 25 psi s; separation at −18 kV; direct UV detection at 200 nm. Peak 1 = Se-cystamine, peak 2 = Se(VI), peak 3 = SeC, peak 4 = Se(IV), peak 5 = SeM, peak 6 = Se-carrying glutathione (reprinted from [19]). Peak positions were determined by standard addition. Se(VI) and Se(IV) seem to comigrate with other compounds, as they were proven to be absent in the native sample.

approach was taken by Michalke [19], which is shown in Figure 4.3.4. Here human milk was analyzed using an alkaline electrolyte and direct UV detection. The arrows mark the positions where Se standards are observed when added to the sample. Here too, the impressive separation capability of CZE is shown. However, the nonspecific UV detector shows various signals that are not due to Se compounds. It is also demonstrated that some of the Se species are not resolved from major compounds (e.g. peak 2) or are very close to detection limit and thus difficult to quantify. It should be noted that subsequent analysis of this sample by hyphenation with ICP-MS excluded the presence of Se in compounds 2 and 4. Obviously, unknown but UV-active non-Se compounds were migrating at positions where added Se standards also appeared. The pitfalls of nonselective detection are clearly demonstrated here. Further speciation experiments have been performed, e.g. for As speciation, organolead species and organotins. The latter were determined by indirect UV detection at concentrations around $5\ mg\ L^{-1}$.

4.2.2 A few examples of speciation using CE-UV and with ETV-ICP-MS for quality control

The use of CE as a second separation system (two-dimensional) with conventional detection technique in multidimensional concepts with other separation techniques can be an easy means of quality control. It helps to increase the certainty of identification of element species. However, even in this case comigration is not excluded completely. Furthermore, the attribution of element concentrations in consecutive factions (e.g. from HPLC) to species, determined in the same fractions by CE, is still not always guaranteed. Complex schemes are necessary to overcome the remaining problems. A suitable method is to make a first identification by retention times in the first separation dimension (HPLC). Collected fractions are subsequently subjected to CE for further analysis to provide a two-dimensional identification. It has been proven [20] that in rare cases of very complex matrices comigration still takes place. Thus, in parallel another separation method by CE – either based on completely different electrolyte systems or using another separation mode, such as cIEF – was introduced. Coincidence of identification in each of the three separations finally provides a high certainty of identification. Figure 4.3.5 shows a flow chart of the two experiments performed in parallel.

To obtain additional information about the correct attribution of elements to identified species a fraction collection may be set at the outlet of the capillary. The CE fractions then are analyzed e.g. by ETV-ICP-MS, for the elements of interest. The element concentrations are compared to the corresponding concentrations in HPLC fractions of the 'first dimension separation'. Electrothermal vaporization is suggested for sample introduction into ICP-MS, because ETV is capable of handling very small sample volumes (few μL). This is necessary to avoid an unnecessary increase in dilution of analytes. In any case, the dilution of off-migration analytes from CE into the vial is immense!

For quantitative working (which is recommended for quality control) it is necessary to know the injection volume of CE exactly. Michalke and Schramel [21] introduced a suitable but time-consuming method: a standard compound of known concentration was injected into the capillary applying different injection times in

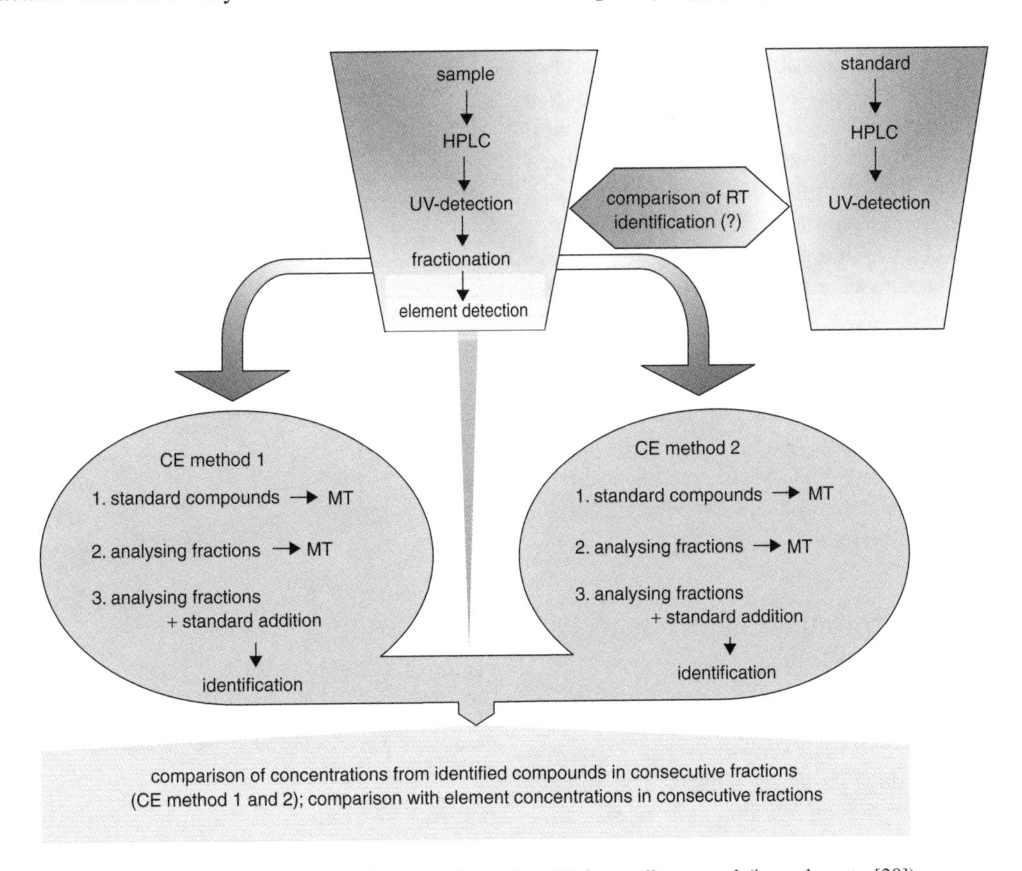

Figure 4.3.5. Flow chart of consecutive investigations when using CE for quality control (in analogy to [20]).

consecutive experiments. The capillary was then purged and the effluent collected into vials containing 100 μL of buffer. The capillary volume was 0.7 μL and thus made a negligible contribution to the final volume. The element was measured in the collection vial and the exact injection volume calculated according to the standard concentration, the measured concentration in vial and the vial volume (100 μL). Finally, a calibration curve was calculated from consecutive experiments with varying injection times. When using this setup platinum complexation by methionine as well as aging and degradation of the Pt–methionine complex was demonstrated. The whole procedure for quality control, species identification and element attribution was performed for Se species in the same experiment [21]: after separating Se compounds via SEC, two different CZE methods were applied for identification. Finally, fractions were collected at the end of the capillary and subjected to ETV-ICP-MS. Thus, Se was attributed to specific Se species after HPLC separation and in addition after a subsequent CZE separation. For quality control reasons fractions of each collected peak as well as of the inlet and outlet vials were monitored for Se. Only the identified Se species showed a measurable Se concentration. Figure 4.3.6 demonstrates the electropherogram of one CZE separation, which was made after an SEC separation. The identified Se-carrying glutathione gives the only fraction with Se.

A more elegant way is to use ICP-MS directly for on-line detection in capillary electrophoresis. The following section gives some information about the potential and about some problems arising from this detection system when used for CE detection in elemental speciation.

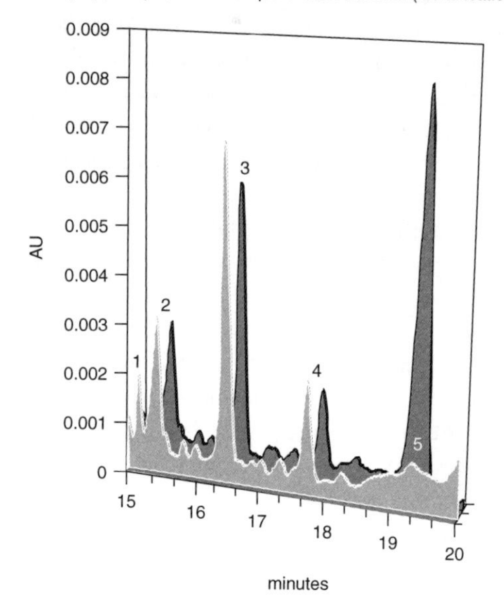

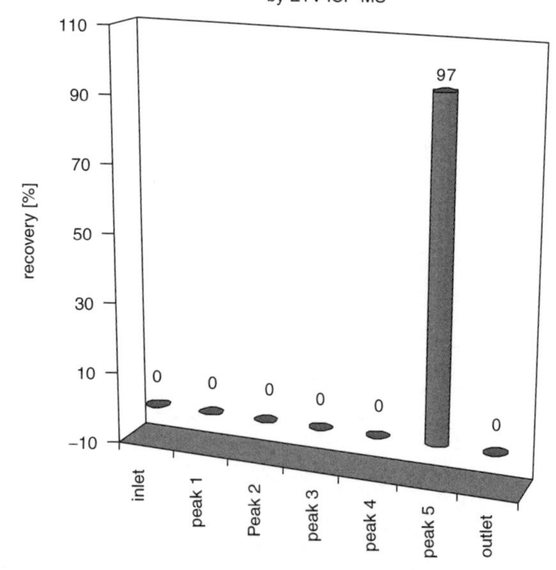

Figure 4.3.6. Analysis of an SEC fraction from human milk, containing Se-carrying glutathione: identification by standard addition and quantification of Se in CE fractions. Se-carrying glutathione was the only Se species in this SEC fraction.

4.3 Inductively coupled plasma mass spectrometry detection

The big advantages of this method are its multielement capability and high sensitivity [22]. Isotopic and elemental information is gained. The ionization source is an inductively coupled Ar plasma.

The sample introduction is performed by an interface, connecting the CE system with the ICP-MS instrument. Even in hyphenated systems detection limits for element concentrations in element species have been reported to be in the 10–100 ng L^{-1} range [11, 13, 23]. There are quadrupole systems available (qICP-MS) and highly resolving sector field ICP-MS (sf-ICP-MS).

qICP-MS is considerably cheaper, but detection limits generally are worse than those for sf-ICP-MS. The quadrupole mass filter provides a resolution of 300. Therefore, 'polyatomic interferences' are likely, especially in the mass range 40–80. A prominent interference is that one on ^{75}As. The ^{75}As isotope is a mono-isotope, which is easily interfered with the $^{40}Ar^{35}Cl^{+}$ cluster [24]. The latter is produced in the Ar plasma when chlorine is introduced by buffers or as a sample component. The highly resolving sf-ICP-MS can distinguish between the interference and the element isotope, as the mass resolution is 7500–10 000 (instead of 300 for qICP-MS). However, when using a high mass resolution the detection sensitivity is reduced. When employing (only) element-selective detectors such as an ICP mass spectrometer one has to realize that only the element in the species is detected, not the whole molecule. This gives the advantage that the separation of molecules with different elements does not need to be complete. The detector can distinguish them. However, as the molecule itself is not seen, identification is only possible by comparing retention times. In natural samples this cannot always be done with certainty. The hyphenation of CE to ESI-MS can help in this matter. For recognition of polyatomic interferences the monitoring of several isotopes of one element can be helpful. Only when the natural isotope ratio is determined in a peak, are interferences unlikely. Unsatisfactory sensitivity is still a problem in samples of the lowest concentration. It is recommended to monitor the most abundant isotopes of an element, except when these isotopes are major targets of interferences; as an example ^{80}Se is mentioned, being totally overlapped by the $^{40}Ar^{40}Ar$ cluster. Even the second most abundant isotope ^{78}Se suffers strong interference and thus the signal to noise is worse than for ^{77}Se, although the abundance of the latter is a factor of three lower.

On the other hand, an ICP mass spectrometer is a sequential detector, monitoring the programmed isotopes for several ms. When programming too much isotopes for determination in parallel the detector gets too slow for highly resolved and fast appearing peaks on one specific isotope. This causes a loss in chromatographic resolution of the hyphenated system.

4.4 The interfacing to ICP-MS

Much effort has been devoted to interfacing CE with inductively coupled plasma (ICP) mass spectrometry (MS). It has been demonstrated that such hyphenated CE techniques could provide not only sub-μg L^{-1} detection limits for the analysis of many types of environmental samples, but also the capability for multielement monitoring of various metal functionalities [11]. At present an efficient interface is still a challenge. In addition, the tiny amounts of sample result in concentration detection limits that are usually higher than of chromatographic methods

4.4.1 Requirements of the interface

The most critical point in hyphenating CE to ICP-MS is the interface itself. It has to fulfill several requirements. One is the closing of the electrical circuit from CE at the end of the capillary although the outlet of the capillary is connected to a nebulizer. Another problem is the low flow rate of CE, generally not matching the flow rates for an efficient nebulization. Thus, it must be supplemented either by an additional sheath flow or by increasing the flow through the capillary itself. The first solution results in an undesired, considerable dilution of analyte species. The second alters dramatically the separation, unless the capillary flow is increased after a suitable pause (e.g. a 20 min pause without a flow) only for transportation of the species bands to the detector (two-step mode). There is another undesired flow occurring when the nebulization gas is turned on ('suction flow') or a sheath flow builds up a backpressure to the capillary (reversed flow). As CE capillaries are principally 'open-tube systems' nebulizer suction or backpressure can cause flow rates of up to 1 m capillary length per minute (approx. 2 μL min^{-1}). This is twice the volume of frequently used capillaries (50 cm × 50 μm ID). Effective separation is then no longer achieved. The nebulization efficiency is often a function of exact and reproducible positioning of the end of the capillary at the nebulization gas stream. Owing to the low sample volumes analyzed this is a critical point in setting up a system. Finally, the interface must preserve the high resolution and separation power from CE when the species is transferred to the detector. Peak broadening or memory effects must be avoided to ensure that no signals are artifacts. For several commercial CE systems it may be a problem to maintain temperature control of the whole capillary up to the interface. Avoiding capillary heating, however, is a critical point for reproducibility, efficient separation and preventing sample degradation during separation. This problem may be overcome by a temperature-controlled liquid flow around the capillary, ensuring that the capillary is not overheated. Finally, the composition of the sheath electrolyte must be checked and has been investigated by several groups [25, 26]. It turned out that nitric acid was best suited to be used for this task, although the sheath electrolyte also has the function of an outlet electrolyte. However, as mostly +/− polarity was used H^+ ions did not enter the capillary and decrease the background electrolyte pH. Further, the use of an inorganic acid instead of a salt solution prevents the nebulizer from crusting and blocking.

4.4.2 Technical solutions

The closing of the electrical circuit of CE during nebulization is a primary problem needing a solution. The first published attempt was based on a Meinhard design and used a silver-painted end to the capillary, which was grounded [11]. A very low current of only few μA was measured. Most researchers have employed a coaxial electrolyte flow around the CE capillary. The grounded outlet electrode is in contact with this electrolyte flow in all cases. Usually a current of 10–30 μA was determined. Furthermore, the sheath flow was used to adapt the flow rate to a suitable nebulization efficiency. These interface models were based on a (modified) Meinhard design, on modified MCN nebulizers or a modified DIN. The optimization of nebulization efficiency was performed by an optimal adoption of the flow rate when using systems based on MCN or DIN or by an exact positioning of the CE the position of the capillary at the point of nebulization, employing a micrometer screw.

The reduction of dead volume and preservation of resolution were achieved using laboratory made special spray chambers. Michalke and Schramel [27] set up a spray chamber with an additional gas flow, which coats the inner surface and inhibits condensation. The mass transport into the ICP mass spectrometer was accelerated. Schaumlöffel and Prange [26] constructed a low-volume spray chamber similar to that used by Polec *et al.* [28] and the modified DIN [23] operated without spray chamber.

Suction through the capillary is often not considered in papers. However, if dealt with either it is quantified and alterations on separation estimated or attempts are made to avoid this undesired flow. There are two solutions based on capillary dimension: (a) using a long CE capillary (1.5 m) with standard inner diameter of 50 μm [12] or (b) a short (2 cm) but narrow interface capillary (25 μm ID) set at the end of the CE capillary [26].

Both are based on the law of Hagen and Poiseuille [29].

$$\frac{V}{t} = \frac{r^4}{8}\frac{\pi}{\eta}\frac{p_1 - p_2}{L} \qquad (4.3.1)$$

V = volume, which flows in the capillary at a time interval t,
r = radius of the capillary
p_1 = pressure at capillary's inlet,
p_2 = pressure at capillary's outlet, *if there is a suction* θ, *then* $p_2 < p_1$, *and* $\theta = p_1 - p_2$
η = viscosity of the buffer,
L = capillary length

The value of $\theta = p_1 - p_2$ is the force (suction) which pulls the electrolyte volume 'V' during the time interval t through the capillary. θ decreases with increased length or decreased inner diameter of the capillary. Another solution in the literature is to apply a negative pressure at the inlet during separation. However, exactly meeting the point of equilibrium between nebulizer suction and counter suction at the inlet is complicated to achieve. Other experiments aimed for a (fragile) equilibrium between nebulization gas flow (influencing the suction), and the sheath flow (feeding the suction flow). However, in this case the nebulization gas flow must fulfill two different tasks, which are efficient nebulization and controlling the dynamic flow. Logically, an optimized compromise between the two tasks is difficult to achieve. Finally, self-aspiration of the sheath flow has been suggested to overcome the suction flow but has not been widely used in the literature. If the suction flow is neglected separation usually is altered markedly. Before starting to analyze samples the suction flow should be checked and quantified. Attempts have been described in the literature to check the suction flow. Michalke *et al.* [12] checked it in three steps. (1) The capillary was filled with buffer, the high voltage turned on and the nebulizer gas off; this determined the CE current. (2) The capillary inlet was exposed to air and the nebulizer gas turned on for ca. 60 min. (3) The capillary inlet was again put into the buffer vial, the high voltage was turned on and nebulizer gas off; the CE current was determined again. When suction occurred, air was drawn into the capillary and interrupted the electrical circuit. In this case the current must be zero in step (3). Typically, no difference in current was seen between steps (1) and (3) for 1.5 m capillaries. Schaumlöffel and Prange [26] analyzed an Rb-containing standard solution by CZE and varying nebulization gas streams in consecutive runs. Here, too, no suction flow was seen, as the standard was monitored at 80 s during all runs, independently on the nebulizer gas flow rates.

Hyphenated systems are mostly operated in a conventional mode: ICP-MS is detecting the effluent from the capillary in real time. There is only one group of researchers who have switched from this mode to a 'two-step mode': Separation was performed in a very long capillary over 15–20 min, during which species were not leaving the capillary [12]. Then the separated analyte bands were moved to ICP mass spectrometer within 2 min by pressure at the inlet. Advantages were seen in reducing a suction flow using the long capillary and by keeping the sheath flow low ($10\,\mu L\,h^{-1}$), as it was not used for improvement of nebulization efficiency. It contributed only a little (nearly no species dilution) to the total flow during the pressure driven detection step ($1.5\,\mu L\ min^{-1} = 90\,\mu L\,h^{-1}$). When performing IEF hyphenated to ICP-MS, such a two-step procedure is obligatory. In this case the long capillary, being completely filled with sample and ampholytes, helps to improve concentration detection limits. Unfortunately, in a few CZE separations this pressure driven postseparation flow compromised the resolution.

Up to now interfaces for hyphenating CE to ICP-MS were laboratory-made or laboratory-modified systems based on commercially available nebulizer parts. Recently, at least two interfaces (from refs [12 and 26]) were made available by companies. In Figure 4.3.7 schemes are presented of the first published interface by Olesik, and of the two widely used interfaces due to Michalke or Schaumlöffel.

4.4.3 Potential of CE-ICP-MS

When the interface is working reliably no specific, 'coupling problems' occur and investigations can concentrate on the broad potential of this technique. The inaccessible advantages and potential of CE-ICP-MS are its high separation capability, the short analysis time and the high selectivity and sensitivity of detection. The ICP mass spectrometer accepts all buffers and modifiers without any problems, as the respective volumes reaching the plasma are in the nL to μL range only. This does not affect plasma stability. Therefore, online preconcentration methods, such as ITP combined with CZE, are easily possible, providing species separation that is still acceptable by using markedly increased sample volumes (= improved concentration detection limits). Buffer sandwiches or discontinuous buffer systems easily achieve improvements in separation. Here, the sample plug may be positioned in the middle of the capillary

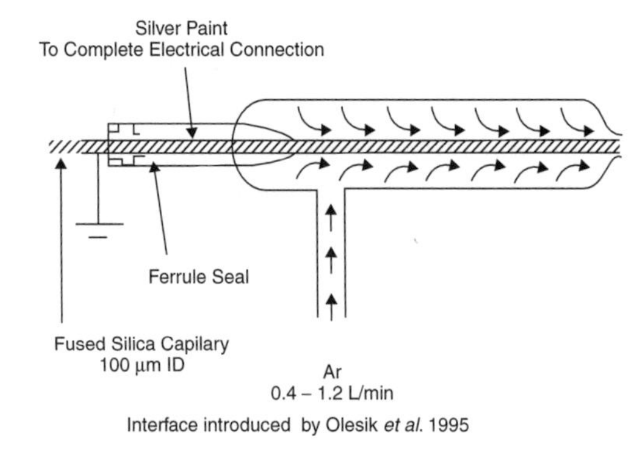

Pt - electrode (grounded)
screw (black) with silicone seal (white)
"O"-ring
"O"-ring
adapter to ICP-MS
CE capillary fixed to the nut by a screw (black) and a silicone seal (white)
nut for exact positioning of CE capillary
Ar gas stream ca. 0.85 L/min
Ar/H_2 gas stream 0.2 L/min
outlet buffer (electrolyte) 10 μL/h
spray chamber
nebulizer
Interface used widely by Michalke *et al.* 1996 - now

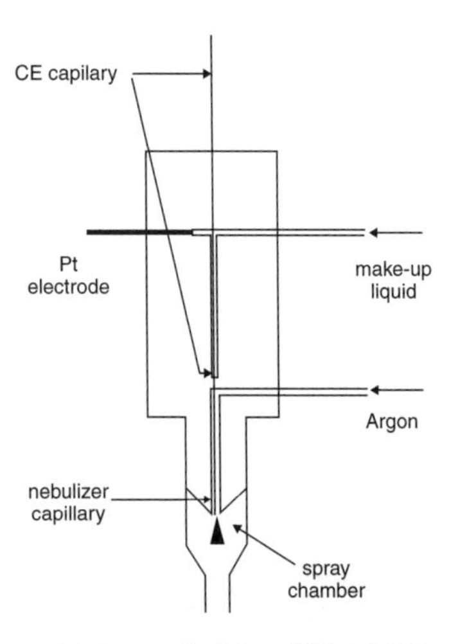

Figure 4.3.7. Schemes of some interfaces used for CE-ICP-MS. The first was published by Olesik *et al.* [11]. Reprinted with permission from Olesik *et al.*, *Anal. Chem.*, **67/1**, 1–12. Copyright 1995 American Chemical Society. The second used by Michalke *et al.* [12]. *Fresenius' Journal of Analytical Chemistry*, B. Michalke and P. Schramel, Vol. 357, pp. 594–599, 1997, copyright notice of Springer-Verlag. The third was published by Schaumlöffel *et al.* [26] Reproduced by permission of Wiley-VCH, STM-Copyright and Licenses.

and the pH of electrolytes is chosen in a way that some species appear anionic, others cationic. Thus, the former move towards the cathode, and the latter towards the anode. Before leaving the capillary either the buffer pH is changed to induce each species now to move to the ICP-MS instrument or a pressure driven detection step is started. The different separation modes allow a separation solution for nearly all elemental species and a wide characterization of the sample. The powerful ICP-MS detector provides elemental and isotope information, as well as multi element capability and low detection limits. Results are reported in the range 0.05–30 $\mu g\,L^{-1}$ depending on the species [11, 12, 13, 23, 26]. Species identification is possible via migration time and comparison with standard solutions. Further, there are no stationary phases that can impair species stability [30]. Several authors have already demonstrated applications to real-world samples from the real world with very different matrices and very low species concentrations.

4.4.4 Limitations of CE-ICP-MS

Many problems can be related to the attempt to decrease concentration detection limits to concentrations in the real world when using (partly inadequate) stacking and separation conditions. Difficulties were often related to chemical interactions of samples, electrolytes and the capillary or to detector interferences [31]. This is not surprising as species stability can easily be impaired by 'incorrect' CE conditions, predominantly complexing electrolytes, inadequate pH etc. [7, 14, 32]. A very serious problem is a total or partial sticking of a compound to the capillary. In this case quantifications are typically wrong and, most critical 'pseudo-species' are detected. Such peaks may suggest species within a sample, but are only artifacts. Since comparison of migration times of standards and samples performs species identification, such artifacts may appear at a specific migration time and thus may be 'identified' as a certain species. The well-known migration time variations according to differences in ionic strength of buffers or samples are a further problem for species identification [33]. Standard additions help to overcome this uncertainty [20]. But generation of new species during analysis is also recognized as a serious source of error [31, 32]. In these papers alkaline conditions helped to focus Sb species, but conversely altered some of them, as shown for Sb(III) tartrate standards. Three peaks appeared for one compound, proving decomposition. There are also problems known for the detection part. These problems can be summarized as variations in detection times according to sample/buffer viscosity when using a two-step procedure and interfered mass signals due to polyatomic or isobaric interferences [34]. The latter obliges one to analyze at least two isotopes per element to detect a possible violation of natural isotope ratio, thus pointing to interferences. However, as ICP-MS is a sequential detection system, the monitoring of too many isotopes in parallel may result in missing fast migrating peaks of one isotope. Detection limits of the CE-ICP-MS system up to now are either just suitable or still too high for several real-world samples. Thus, coupling to more sensitive detectors, e.g. HR-ICP-MS, is recommended and has already been partly achieved in the literature [25, 26, 35].

4.4.5 A few examples of speciation using CE-ICP-MS

Applications to real-world samples, e.g. eluates of tunnel dust and soil (platinum speciation [14]), plant extracts or sewage and fouling sludges (antimony, arsenic [32, 36]), or serum and human milk (selenium and iodine [13, 27]) or other body fluids (human cytosol) and tissues (metallothioneins, Cu, Cd; Co in cobalamins) have already been demonstrated. Lobinski [35] used CE-ICP-MS in parallel to CE-ESI-MS. Surprisingly, this author found worse LOD when using CE-ICP-MS than for CE-ESI-MS. An explanation was seen in the fact that detecting only Co in the molecule (ICP-MS detection) needs a higher mass detection sensitivity than does detecting the whole molecule (ESI-MS detection). The same group obtained a similar result for MT speciation by CE coupled to ICP-MS or ESI-MS [37]. However, these findings are contradictory to findings of other groups, where the superior detection sensitivity of ICP-MS overcompensates this effect.

As another example Figure 4.3.8 shows the analysis of Se species in a human milk sample,

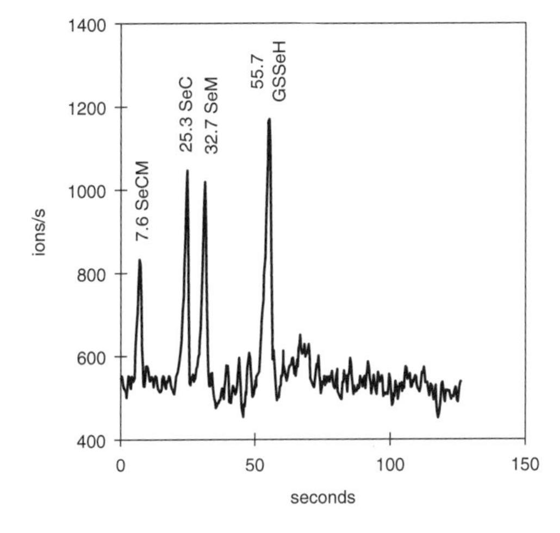

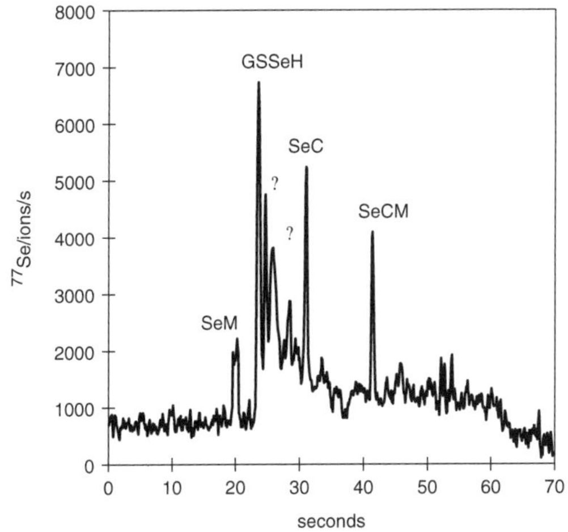

Figure 4.3.8. A comparison of two different CE separation modes hyphenated to ICP-MS for Se speciation in human milk. An advanced characterization is possible by employing two different CE modes. For CZE analysis the sample was preconcentrated by freeze-drying. The compounds are (CZE): SeCM, SeC, SeM and GSSeH.

performed by CZE and in parallel by cIEF, each coupled to ICP-MS. ^{78}Se and ^{77}Se were monitored (only ^{78}Se is shown). The electropherograms show multidimensional characterization by the use of different separation methods. The cIEF separation is superior as regards detection limits and resolution. Both methods detect the same major Se species, but cIEF is able to separate and detect at least three unknown species in addition. As species were analyzed in urine by van Holderbeke *et al.* [25] or in several other materials by Schlegel [36]. An $ArCl^+$ cluster did not interfere with detection of species in both papers. For monitoring all relevant As species in a single run a fast EOF was applied which pumped anionic, uncharged and a cationic species to the detector. One of the most frequent applications is the analysis of metallothioneins. Several groups have investigated the isoforms of this protein and its Cd and Cu bondings. As examples, Polec *et al.* [28], Mounicou *et al.* [37] and Prange *et al.* [38] are mentioned. The investigations of Polec *et al.* and Mounicou *et al.* used both ICP-MS and ESI-MS hyphenations for MT characterization and will be discussed in the corresponding section. Prange *et al.* analyzed MT samples in a volume of 22 μL for Cu, Zn, Cd and Pb, employing their laboratory developed interface and an sf-ICP-MS system. They compared human brain Cu-MT from healthy persons with that from patients with Alzheimer's disease. The two electropherograms are shown in Figure 4.3.9.

4.5 ESI-MS detection

Electrospray ionization is an ionization process that may preserve the whole species intact under optimal circumstances. ESI is suitable for extremely low flow rates. It is based on 'ion evaporation', where charged droplets of the analytes are transferred into gas phase. A volatile buffer consisting of a considerable amount of e.g. methanol supports this ion evaporation. In fact the high volatilization capability of CE electrolytes is necessary. The droplets are formed at the end of the CE capillary by the application of a high voltage, typically around 5 kV. The success of this detection method is based on the ability to produce multicharged ions from element species of high molecular weight such as metalloproteins and thus making the analysis of these compounds available, up to MW = 150 000–200 000 Da. The possibility of coupling this detector to LC or CE systems makes it additionally immensely valuable [39]. The soft ionization of element species finally gives the chance to preserve the whole molecule (element species) when it is transferred to the gas phase and is subsequently analyzed in the mass spectrometer [40]. Structural changes (mostly) do not take place as long as covalent bondings are present. In special cases (e.g. with selenium) stable element–organic molecules can be analyzed. As an example selenocystamine is discussed: the most prominent signal is the protonated species ($[SeCM \times H]^+$). The specific molecular mass at 249 is detected, but up to 19 peaks in the mass

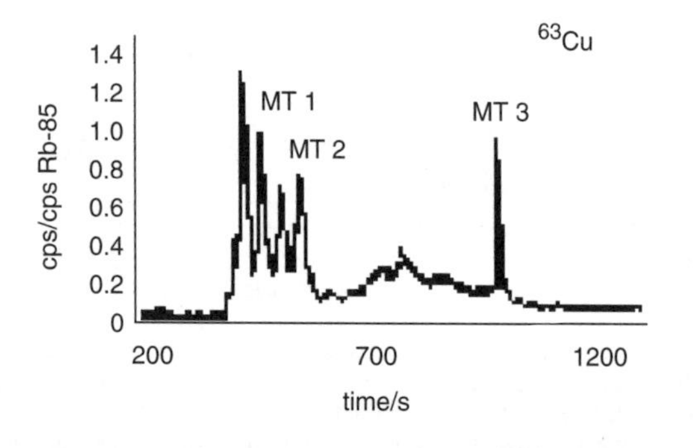

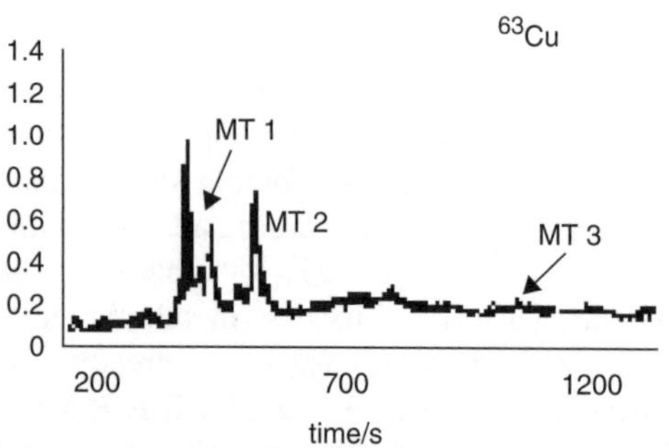

Figure 4.3.9. Analysis of metallothioneins in brain cytosol from healthy persons compared with those in brain cytosol from patients with Alzheimer's disease (reprinted from [38] by permission of John Libbey Eurotext).

range $m/z = 237-255$ should be monitored, which originate from the two Se atoms in the SeCM molecule and the specific isotope pattern of Se. Some of the Se isotopes, however, show very low abundances and thus may not be detected [41]. Here, the (structural) information of two Se atoms which are part of the SeCM molecule is available without any further analytical efforts.

When applying collision induced dissociation (CID) together with an MS/MS system further structural information can be gained. CID is possible when the parent ions (filtered by the first quadrupole) are accelerated by an electric field and then pass through a region of minimal gas pressure ($\leq 10^{-3}$ bar). This results in collision reactions that are governed by the kinetic energy from the acceleration of parent ions. Undesired ion–solvent clusters may be destroyed or the parent ions are fragmented into (molecule-specific) daughter ions, selected by a second quadrupole. The latter provides structural information about the element species (parent ion). No other detection technique is able to provide such detailed information about the molecular weight and even the structure of compounds analyzed.

4.6 Problems of ESI-MS in speciation

One problem comes from the ion–solvent clusters. During the transfer of gas-phase ions into the high vacuum (10^{-9} bar) a condensation of solvent molecules (e.g. methanol, water) to the gas-phase ions probably happens. The product is called an ion–solvent cluster. Production of ion–solvent clusters results in a splitting of one species into multiple signals, worsening detection limits and increasing spectral complexity. Electrolytic processes at the metallic ESI tip needle have been observed, resulting in the generation of new species or a transformation of species (e.g. by metal exchange). When analyzing free metal ions such as Cu(II), multiple signals from ion–solvent clusters are monitored (Figure 4.3.10). Most important, however, is the

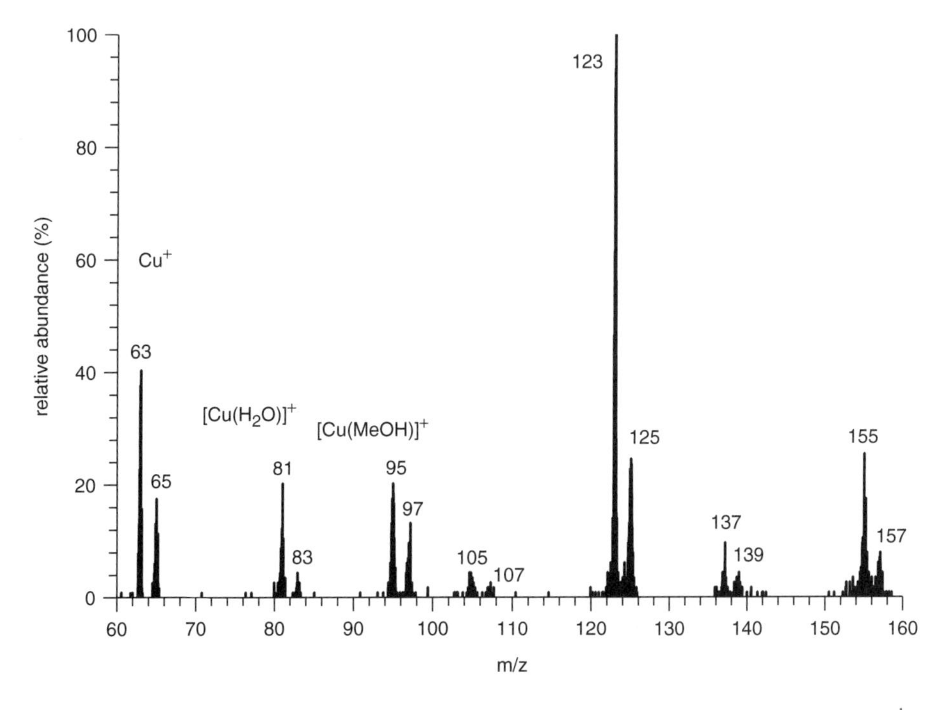

Figure 4.3.10. Pitfalls of electrospray ionization: an ESI-MS spectrum of $CuCl_2 \cdot xH_2O$ in H_2O (500 µg Cu L^{-1}) is presented. Multiple ion–solvent clusters are seen. Collision offset voltage = −25 V. Reprinted from *J. Chromatogr. A*, Vol. 819, O. Schramel *et al.*, pp. 231–242, Copyright (1998), with permission from Elsevier Science [48].

fact that native counter ions of the metal ion are replaced by H_2O and/or methanol, independent on the counter ion initially present (e.g. $[Cu(MeOH)]^+$ instead of $Cu^{2+}2Cl^-$). This causes a total loss of initial species information. Destruction of the ion–solvent cluster by CID is rarely possible due to a reduction of Cu(II) to Cu(I) and the generation of only a few Cu(II)–solvent clusters follows this. These Cu-based clusters are easily identified by the isotope pattern coming from $^{63}Cu/^{65}Cu$. Finally a charge reduction at the cluster may occur as found by Agnes and Horlick[42]:

$$Cu(MeOH)_n{}^{2+} \longrightarrow Cu(MeO)(MeOH)_{n-2}{}^+ + H(MeOH)^+ \quad (4.3.2)$$

4.7 CE-ESI-MS

4.7.1 Requirements of the ESI interface and solutions

ESI interfaces for CE are commercially available. The requirements of the interface are similar to those discussed earlier. The closing of the electrical circuit from CE during ion evaporation is provided by an electrolyte sheath flow. Effective ion production is possible using a suitable spray voltage easily controlled by instrument software. For older instruments a laboratory made device for reproducibly optimized positioning of the CE capillary to the ESI-tip is still necessary and has been described by Schramel *et al.* [41].

4.7.2 Potential of CE-ESI-MS

The potential of this hyphenated technique is defined by the advantages from CE discussed for species separation combined with the possibilities and species information coming from ESI-MS. In contrast to the CE-ICP-MS coupling, direct species detection is available here. When the element of interest imposes its specific isotope pattern on the total molecule mass, elemental information within the species is also possible: this means maximized species information is gained in one analytical effort. If the elemental pattern is not seen on the total molecular mass, structural information is provided when applying MS-MS mode. Species identification is gained via migration time *and* via m/z of species.

4.7.3 Limitations of CE-ESI-MS

The problems of CE-ESI-MS result from the limitations of CE discussed for species separation combined with the problems from ESI-MS. A big problem is that only volatile buffers are possible for ESI detection. Therefore, a free choice for separation electrolytes is not available. Mostly a compromise between separation and detection capability is necessary. Furthermore, Mounicou *et al.* [37] found detrimental effects on separation when the sheath buffer was different from the inlet buffer. This is a significant disadvantage compared with CE-ICP-MS. Elemental information is possible only in specific cases where the isotope pattern is imposed on the molecule mass (e.g. selenoamino acids). Unfortunately, the ionization process itself, as mentioned above, causes several severe problems in speciation. Summarizing, these problems are a total loss of species information by gas-phase ligand replacement, gas-phase intramolecular charge transfer, thermal decomposition on heated detector parts, ion-solvent cluster generation combined with worsening of detection limits and increasing problems of species identification by spectral complexity, generally worse detection limits, resulting in species of low concentration not being monitored. Application to real-world samples is thus rare.

4.7.4 A few examples of speciation using CE-ESI-MS

There are several examples employing CE-ESI-MS. Most of them have been in combination with ICP-MS technology and part of multidimensional concepts. In ref. [41] the possibilities and limitations of this technology were demonstrated using species of Sb, Cu and Se as examples. Nearly all other studies have analyzed metallothioneins

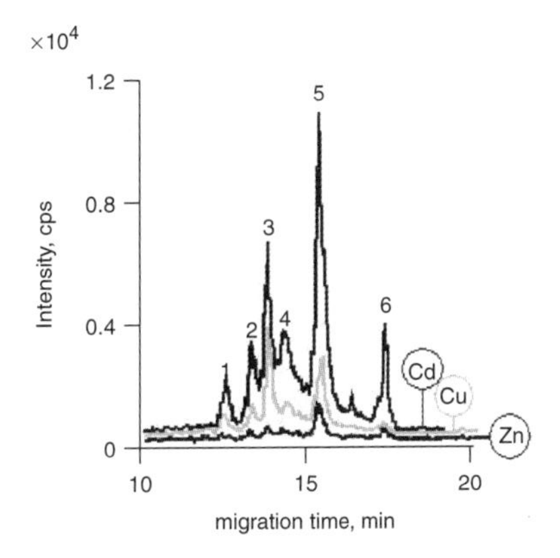

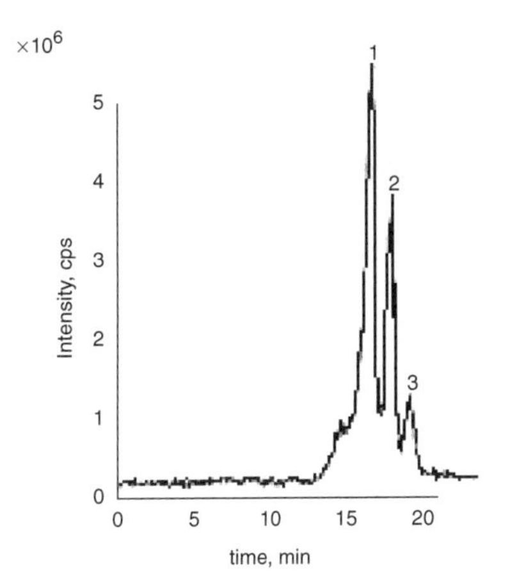

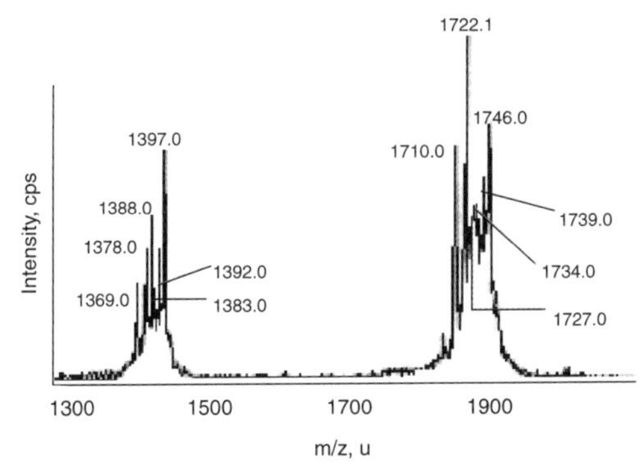

Figure 4.3.11. (a) An electropherogram of MT 1, analyzed by CZE-ICP-MS. Spectra for Cu, Cd and Zn are seen. (b) The same sample analyzed by CZE-ESI-MS, where the total ion current is monitored. The peaks were further investigated and the specific masses could be attributed to Cd_7-MT2 ($m/z = 1382$), accompanied by a mixed complex of Cd_7-MT2 Cd_6Zn-MT2, Cd_5Zn-MT2 and Cd_4Zn-MT2. Other subisoforms were also identified. Reprinted from K. Polec *et al.*, *Cell Mol. Biol.*, **46/2**, 221–235 (2000), reprinted with permission from CMB [28].

or selenocompounds, each of them stable during the ESI process. As mentioned above, impressive examples are published in refs [37] and [28], both studies from the same group. They both investigated metallothioneins first by CE-ICP-MS and subsequently by CE-ESI-MS for advanced identification of the metal–MT compounds. Comparing the two methods they found a poorer resolution in CE-ESI-MS and problems arising from their instrumental setup, which allowed no temperature control of the capillary outside the instrument. Mt isoforms were determined and subisoforms were additionally found, attributed to Cd, Cu and Zn. The Mt isoforms were then analyzed by CE-ESI-MS and subsequently their m/z values were determined. Figure 4.3.11 shows the respective electropherograms and mass identifications.

5 COMBINATION OF CE-ESI-MS AND CE-ICP-MS: MAXIMIZED SPECIES INFORMATION

Summarizing the advantages and limitations of both hyphenation techniques we can use the high resolution power of CE for the separation of metal species [43, 44] combined with molecular and structural information from ESI-MS and elemental and isotope information from ICP-MS detection.

Electrospray using soft ionization provides the following informations about element species:

- direct detection and quantification of species [45, 46];
- possibly elemental information by a characteristic isotopic pattern [47, 48];
- structural information by using the MS/MS technique [46].

Thus, information is gained about the element within the species, resulting in total species information and identification. Additionally, species identification is possible according to migration times and standard addition procedures.

Unfortunately, some disadvantages are known:

- Only volatile buffers are appropriate for the ESI process. Therefore, a compromise between optimal separation and sensitive detection is often unavoidable [46].

- Elemental information is achieved only indirectly by a characteristic isotope pattern. If the element does not show a very characteristic isotope pattern and/or the whole species is rather large (isotope pattern superimposed) no elemental information is gained.
- Possible losses of species information are known due to gas-phase ligand replacements or gas-phase intramolecular charge transfers [42, 48]. Electrolytic processes at the stainless steel capillary tip are not excluded.
- Detection limits are strongly compound dependent. For metal ions low values between 3 and 300 $\mu g\,L^{-1}$ are reported without CE coupling. However, for metal-containing amino acids and peptides detection limits of around 500 $\mu g\,L^{-1}$ are still typical with CE coupling. This is too poor for most real samples.

The use of an ICP mass spectrometer as element-specific detector for CE is a useful complementary technique to ESI-MS. The advantages of ICP mass spectrometer as a detector are:

- the very low detection limits of this hyphenated system ($<1\,\mu g\,L^{-1}$ [12, 23, 26]);
- direct element information and element quantification;
- All buffer systems and stacking procedures fit well to the detector, so no compromises between separation and detection are necessary [23, 27];
- Species identification via migration times.

On the other hand there are some limitations:

- Detection of only the element is possible. There is no detection of the species itself.
- Unknown species cannot be identified. They can only be characterized, e.g. via cIEF/ICP-MS [49].
- m/z signals can suffer interference from polyatomic interferences, resulting in pseudo-element signals [50, 51]. Detection limits worsen.

Considering all these factors, CE-ESI-MS provides maximum information, i.e. direct determination of the element in its specific form. CE-ICP-MS directly yields elemental information. Structural information can be obtained indirectly by means of CE data.

The combination of CE-ICP-MS and CE-ESI-MS can provide maximum species information.

6 CONCLUSION

Capillary electrophoresis is proving a very important and suitable tool for speciation investigations. Generally it acts as a separation method for elemental species before they are detected either nonselectively or in elemental- and/or molecule-selective ways. Its use in orthogonal multidimensional strategies helps to provide species information with increased certainty. Speciation analysis can often be performed where other methodical approaches cannot promise success. Especially in combination with detection techniques such as ICP-MS and ESI-MS its very high potential is realized and used for improved speciation of the elements. Nevertheless, the various limitations and problems – such as loss of species information or mimicking of unknown species by generating artifacts – must be carefully considered when using this technology. Finally, as CE can only play a (essential) part in speciation analysis together with other technologies, it should be used (only) in those fields where separation is difficult and cannot be achieved more easily and cheaply with other methods, and where species concentrations are high enough for detection without any (severe) problems. Thus the typical sample to be speciated by CE contains element species that are difficult to separate (uses the high resolution of CE) but in high species concentrations (low sample intake causes no detection problem).

7 REFERENCES

1. Mota, A. M. and Simaes Goncalves, M. L., Direct methods of speciation of heavy metals in natural waters, in *Element Speciation in Bioorganic Chemistry*, Caroli, S. (Ed.), John Wiley & Sons, Inc., New York, 1996, Chapter 2, pp. 21–87.
2. Morrison, G. M. P., Trace element speciation and its relationship to bioavailability and toxicity in natural waters, in *Trace Element Speciation: Analytical Methods*

and Problems, Batley, G. E. (Ed), CRC Press, Boca Raton, FL, 1989, Chapter 2, pp. 25–42.
3. Florence, T. M., *Trends Anal. Chem.*, **2**, 162 (1983).
4. Turner, D. R., *Met. Ions Biol. Syst.*, **18**, 137 (1984).
5. Templeton, D. M., Ariese, F., Cornelis, R., Danielsson, L.-G., Muntau, H., van Leeuwen, H. P. and Lobinski, R., *Pure Appl. Chem.*, **72**, 1453 (2000).
6. Michalke, B., *Fresenius' J. Anal. Chem*, **350**, 2 (1994).
7. Majidi, V. and Miller-Ihli, N. J., *Analyst*, **123**, 803 (1998).
8. Dunemann, L. and Begerow, J., *Kopplungstechniken zur Elementspeziesanalytik*, VCH, Weinheim, 1995.
9. Kuhn, R. and Hofstetter-Kuhn, S., *Capillary Electrophoresis: Principles and Practice*, Springer, Berlin, 1993.
10. Timerbaev, A. R., *Electrophoresis*, **18**, 185 (1997).
11. Olesik, J. W., Kinzer, J. A. and Olesik, S. V., *Anal. Chem.*, **67**, 1 (1995).
12. Michalke, B. and Schramel, P., *Fresenius' J. Anal. Chem.*, **357**, 594 (1997).
13. Michalke, B. and Schramel, P., *Electrophoresis*, **20**, 2547 (1999).
14. Michalke, B., Lustig, S. and Schramel, P., *Electrophoresis*, **18**, 196 (1997).
15. Bulletin 1641, *Coatings: The Key to Success in Capillary Electrophoresis of Proteins*, Bio-Rad, Hercules, CA, 1992.
16. Jen, J. F., Wu, M. H. and Yang, T. C., *Anal. Chim. Acta*, **339**, 251 (1997).
17. Quevauviller, Ph., Maier, E. A. and Griepink, B., Quality control of results of speciation analysis, in *Element Speciation in Bioorganic Chemistry*, Caroli, S. (Ed.), John Wiley & Sons Inc., New York, 1996, Chapter 6, pp. 195–222.
18. Albert, M. M., Demesmay, C. and Rocca, J. L., *Analusis*, **21**, 403 (1993).
19. Michalke, B., *Speziesanalytik in Umwelt und biomedizinischen Proben*, Habilitationsschrift, Technische Universität Graz, Austria, 1999.
20. Michalke, B., *Fresenius' J. Anal. Chem.*, **351**, 670 (1995).
21. Michalke, B. and Schramel, P., *J. Chromatogr. A*, **750**, 51 (1996).
22. Hill, S. J., Bloxham, M. J. and Worsfold, P. J. J., *Anal. At. Spectrom.*, **8**, 499 (1993).
23. Liu, Y., Lopez-Avila, V., Zhu, J. J., Wiederin, D. R. and Beckert, W. F., *Anal. Chem.*, **67**, 2020 (1995).
24. Hill, S. J., Ford, M. J. and Ebdon, L., *J. Anal. At. Spectrom.*, **7**, 719 (1992).
25. van Holderbeke, M., Zhao, Y., Vanhaecke, F., Moens, L., Dams, R. and Sandra, P., *J. Anal. At. Spectrom.*, **14**, 229 (1999).
26. Schaumlöffel, D. and Prange, A., *Fresenius' J. Anal. Chem.*, **364**, 452 (1999).
27. Michalke, B. and Schramel, P., *Electrophoresis*, **19**, 270 (1998).
28. Polec, K., Mounicou, S., Chassaigne, H., Lobinski, R., *Cell Mol. Biol.*, **46**, 221 (2000).
29. *Formeln Physik Chemie Mathematik*, Buch und Zeit Verlag GmbH, Köln, 1980.
30. Harms, J. and Schwedt, G., *Fresenius' J. Anal. Chem.*, **350**, 93 (1994).
31. Michalke, B. and Schramel, P., *J. Anal. At. Spectrom.*, **14**, 1297 (1999).
32. Michalke, B. and Schramel, P., *J. Chromatogr. A*, 1999, **834**, 341 (1999).
33. Bondoux, G., Jandik, P. and Jones, R. W., *J. Chromatogr. A*, **602**, 79 (1992).
34. *ICP-MS Interferenz Tabelle*, Finnigan MAT, Bremen, 1995.
35. Lobinski, R., *Appl. Spectrosc.*, **51**, 260A (1997).
36. Schlegel, D., Arsen-Speziationsanalytik mit Ionenchromatographie und Kapillarelektrophorese in Kopplung mit elementspezifischer Detektion, UfZ-Bericht 5/1999, ISSN 0948–9452, Leipzig, 1998.
37. Mounicou, S., Polec, K., Chassaigne, H., Potin-Gautier, M., Lobinski, R., *J. Anal. At. Spectrom.*, **15**, 635 (2000).
38. Prange, A, Schaumlöffel, D., Richarz, A. and Brätter, P., Speciation of metallothioneins in animal and human samples from nanoliter volumes, in *Metal Ions in Biology and Medicine*, Vol. 6, Centeno, J. A., Colley, Ph., Vernet, G., Finkelman, R. B., Gibb, H. and Etienne, J. C. (Eds), John Libbey Eurotext, Paris, 2000, pp. 430–432.
39. Smith, R. D., Loo, J. A., Barinaga, C. J., Edmonds, C. G. and Udseth, H. R., *J. Chromatogr.*, **480**, 211 (1989).
40. Cole, R. B., *Electrospray Ionization Mass Spectrometry – Fundamentals, Instrumentation and Applications*. John Wiley & Sons, Inc., New York, 1997.
41. Schramel, O., Michalke, B. and Kettrup, A., *Fresenius' J. Anal. Chem.*, **363**, 452 (1999).
42. Agnes, G. R. and Horlick, G., *Appl. Spectrosc.*, **48**, 655 (1994).
43. Jackson, P. E. and Haddad, P. R., *Trends Anal. Chem.*, **12**, 231 (1993).
44. Michalke, B., *Fresenius' J. Anal. Chem.*, **354**, 557 (1996).
45. Chassaigne, H. and Lobinski, R., *Anal. Chem. Acta*, **359**, 227 (1998).
46. Chassaigne, H. and Lobinski, R., *Fresenius' J. Anal. Chem.*, **361**, 267 (1998).
47. Corr, J. J. and Anacleto, J. F., *Anal. Chem.*, **68**, 2155 (1996).
48. Schramel, O. Michalke, B. and Kettrup, A., *J Chromatogr. A*, **819**, 231 (1998).
49. Michalke, B. and Schramel, P., *J. Chromatogr. A*, **807**, 71 (1998).
50. Seubert, A. and Meinke, R., *Fresenius' J. Anal. Chem.*, **348**, 510 (1994).
51. Michalke, B. and Schramel, P., *Analusis*, **26**, M51 (1998).

4.4 Gel Electrophoresis for Speciation Purposes

Cyrille C. Chéry

Laboratory for Analytical Chemistry, Ghent University, Belgium

ABBREVIATIONS

1D	one-dimensional
2DE	two-dimensional gel electrophoresis
AAS	atomic absorption spectroscopy
AES	atomic emission spectroscopy
C	degree of cross-linking
CE	capillary electrophoresis

Handbook of Elemental Speciation: Techniques and Methodology R. Cornelis, H. Crews, J. Caruso and K. Heumann
 ISBN: 0-471-49214-0

CHAPS	3-[(3-cholamidopropyl) dimethylammonio]-1-propanesulfonate
DTT	Dithiothreitol
ETV	electrothermal vaporisation
GE	gel electrophoresis
GF	graphite furnace
ICP-MS	inductively coupled plasma – mass spectrometry
IEF	isoelectric focusing
IPG	immobilized pH gradient
LA-ICP-MS	laser ablation – inductively coupled plasma – mass spectrometry
MALDI	matrix-assisted laser desorption/ionisation
MS	mass spectrometry
NAA	neutron activation analysis
PAGE	polyacrylamide gel electrophoresis
pI	isoelectric point
PIXE	particle induced X-ray emission
SDS	sodium dodecyl sulphate
SGE	slab gel electrophoresis
T	total acrylamide concentration
Tris	tris(hydroxymethyl)aminomethane
V h	volt hour

1 INTRODUCTION AND DEFINITION

1.1 Introduction; gel electrophoresis and speciation

Although far less used for speciation purposes than its parent method, capillary electrophoresis (CE), gel electrophoresis (GE), or more precisely slab gel electrophoresis (SGE), is a very promising method. It not only allows a rapid separation of a complex mixture, even if detection can be tedious, but it also allows various separation mechanisms according to the need. Some of the terms and principles of the two methods are comparable and it is thus advisable to take a look at the previous chapter on capillary electrophoresis in parallel to this one. Furthermore, although this chapter is more dedicated to gel slabs, the one-dimensional methods can be transposed to gel rods or columns, as is the case for isoelectric focusing.

One may wonder why gel electrophoresis is still interesting when capillary electrophoresis is so powerful. Let us give a striking image: gel electrophoresis would indeed belong to the past if this method was comparable in its figure of merit with thin layer chromatography and capillary electrophoresis with capillary chromatography; in other words, gel electrophoresis would be a basic quality control method and CE a method with unmatchable capacities, because of its high resolution and speed of analysis. However, gel electrophoresis cannot be considered as a cheap alternative to CE, if we pursue the comparison with chromatography, but as a method with other goals and a different output. It is an efficient and rapid separation method for complex mixtures; it allows detection with radiotracers; it can be two-dimensional; the amount of material is larger and allows an off-line identification by the means of trypsin cleavage and mass spectrometry; and, last but not least, the material is always available for further studies: gels stored for years can be used for identification since the compounds have not been degraded.

Although the application of gel electrophoresis for speciation purposes is just beginning, it is definitely worth applying on a large scale.

Some excellent books are recommended [1–3] to get a deeper and practical insight into gel electrophoresis and its applications. In this chapter, only the use of gel electrophoresis for elemental speciation purposes will be examined in detail.

1.2 Speciation

1.2.1 Applicability

The area covered by gel electrophoresis for speciation purposes is charged macromolecules to which any metal or metalloid is bound, covalently or not. Even if this chapter deals mainly with proteins, other macromolecules can be separated by this method, such as DNA or humic acids.

This method has been used in combination with numerous metals. The applicability is not limited to certain elements, but by practical considerations

Table 4.4.1. Applications of SGE to elemental speciation: a survey.

Element	Matrix	SGE separation	Detection	Remark	Reference
Co	Serum	Crossed immunoelectrophoresis	LA-ICP-MS	Stability of species?	[32]
Fe	Apotransferrin	Native	Autoradiography		[30]
	Bacterium	SDS-PAGE	PIXE		[34]
P	Milk	SDS-PAGE	NAA		[20, 22]
Pb	Humic acids	SDS-PAGE	LA-ICP-MS		[33]
Pt	Grass	SDS-PAGE	Voltammetry	Contamination from electrodes?	[15]
	Serum	Native 2DE	Autoradiography	Very promising method	[11]
Se	Soft tissues	2DE	Autoradiography		[25, 26]
	Yeast	2DE	Autoradiography	Stability checked	[14]
			ETV-ICP-MS	Stability checked	[36]
			LA-ICP-MS		[5]
	Glutathione peroxidase	SDS-PAGE	Mineralisation: HPLC-fluorescence		[16]
	Glutathione peroxidase	SDS-PAGE	GF-AAS		[18]
			ETV-ICP-MS	Stability checked	[36]
	Soft tissues	SDS-PAGE	Liquid scintillation		[19]
			LA-ICP-MS		[37]
			HG-AFS		[38]
V	Serum	Native	Autoradiography	Stability checked	[39]
Multi-element	Kidney	IEF	NAA	Stability of species?	[23]
	Liver	IEF	X-ray fluorescence		[40]

such as the limits of the detection method and the amount of material that can be brought on the gel.

The most representative applications of gel electrophoresis to speciation are summarized in Table 4.4.1.

1.2.2 Limitations

Although the detection of the metal is possible (see Section 3.2.), quantitation is still difficult. Up to now, gel electrophoresis has remained a semiquantitative method and this is also the case when it is used for speciation.

It should always be kept in mind that, although gel electrophoresis is a versatile separation method, speciation implies that the compound must be kept intact. The integrity of the analyte is fundamental and even dictates the separation process; therefore the choice of buffers, pH and electrodes, to mention the most important parameters, is crucial.

Some standard methods, which use detergents and denaturing agents, cannot be considered for speciation of protein–metal complexes because the basic structure of the complex is lost and the protein is stripped of its metal. Ideally, even if the metal is combined with the protein, the stability of the protein during the separation process has to be checked; indeed, artefacts can occur, due among other things to oxidation of residues of the protein, as in the case of selenoproteins.

Additional considerations have to be taken into account, such as possible contamination of the gel through the electrodes [4]. This is exemplified by the use of platinum electrodes. The large contact area between the electrode and the gel promotes the release of oxidized platinum into the gel, which, when platinum–protein complexes are studied, leads to the detection of an artefact. Contamination from the gels has never been reported and mineralisation of commercial gels followed by detection by ICP-MS has demonstrated that no high blank value has to be feared [5]. Lack of purity of chemicals can still be a problem and therefore home-made gels should be prepared with extreme caution and with the purest chemicals available.

1.3 Apparatus

Basically, an electrophoresis experiment requires a high voltage generator, typically up to 2000 V, a set of electrodes and a temperature-controlled separation surface or chamber. The last point is critical since heat is produced through the Joule effect, which can disturb the separation. For small gels, less than 10 cm^2, a Pelletier cooling is very efficient; for larger ones water cooling is the best solution. Several variations are available from this starting set, including submarine electrophoresis, vertical or horizontal electrophoresis. Other accessories may be necessary if the method is used intensively, such as staining trays. Detection for elemental speciation will be discussed separately.

1.4 Definitions

In an electric field, charged molecules or complexes will migrate to the electrode bearing the charge opposite to their global charge. This global charge is not necessarily the charge of the molecule, but rather the charge of the particle that is formed during electrophoresis, for example the charge of the protein in an SDS micelle or the charge of a metallo-complex. If a voltage V is applied between two electrodes separated by a distance L, a field E appears according to equation (4.4.1). The migration velocity (ν) of a particle in this field is proportional to the mobility (μ) of the particle and the field strength (E) (equation 4.4.2), where μ is an intrinsic parameter of the particle.

$$E = V/L \tag{4.4.1}$$

$$\nu = \mu E \tag{4.4.2}$$

The unit mainly used in electrophoresis is the volt hour (V h), since this value is proportional to the displacement d of the particle. Indeed, it can be proved that the velocity, ν, becomes rapidly constant and by combining equations (4.4.1) and (4.4.2):

$$d = \nu t = (\mu/L)Vt \tag{4.4.3}$$

Since μ is a characteristic of the particle and L of the system, the distance d to the electrode is related to a unit that has the dimension of a voltage multiplied with a time, which is traditionally expressed in V h.

This unit is particularly interesting if a separation has to be translated from one gel size to another. If a certain gel size is used to create and optimise a method and half the number of lanes are used afterwards, the number of V h is invariant, the parameters that have to be adjusted are simply current and power, both divided by two. In contrast, if the migration distance is divided by two, the number of V h is divided by two, although this time it may be a rough approximation.

Migration occurs in a liquid medium, namely the buffer, which is one of the key elements of the separation and especially of the stability of the metal bound to the macromolecule. This buffer, which is not necessarily a pH buffer but rather a good solvent for the particles, may be a single solution or a combination of two or three solutions if the separation process requires it (see Section 2.4).

The gel is the second key parameter for a good separation of the particles. It determines to a first approximation which separation mechanism occurs. Various gels are available, agarose and polyacrylamide being the most common ones. Generally, the choice is made between the two according to the size of the particles to be analysed. For larger particles, typically over 10 nm in diameter, agarose gel is preferred, especially for the analysis of DNA or RNA. Polyacrylamide gel is the polymer of choice for most proteins.

This polymer is obtained by copolymerisation of acrylamide and a cross-linking agent, usually N,N'-methylenebisacrylamide, which confers its three-dimensional structure on the gel. The pore size is defined by two parameters, C and T, both expressed in per cent. They are related to the polymerisation process and the quantities of monomer and cross-linking agent used, T being the mass of acrylamide per gel volume and C the percentage of cross-linking agent in the gel. For our purpose, it is only necessary to know that, if T increases, the pore size decreases (the more polymer per volume, the less free volume).

1.5 Typical applications

Whereas applications of gel electrophoresis are numerous, they are just emerging in the field of elemental speciation. They range from separation of DNA/RNA, humic acids, proteins, to dyes. A particular branch of gel electrophoresis is worth mentioning: two-dimensional gel electrophoresis (2DE). It is the most advanced and successful method, this success being due to the increasing importance of proteomics: the combination of PROTEin analysis and genOMICS aims at unravelling the mysteries of the expression of the genome into proteins. Thus proteomics aims to characterise all proteins present in a particular sample, e.g., *Saccharomyces cerevisiae*, human serum, *Eschericia coli* to quote just a few. Two-dimensional electrophoresis has become a widespread and very reliable analytical method for this purpose. To give a single example, 2DE enables the separation of more than 10 000 proteins on a single gel [6].

One further point which has to be mentioned about the application of gel electrophoresis is the widespread availability of ready-to-use gels and buffers, making it less and less necessary to make them oneself, which used to be most time consuming.

In the field of speciation by means of gel electrophoresis, most work has been done on proteins. That is why this chapter will concentrate on this point. But the examples that are quoted here can always be transposed to other types of samples, as long as the mechanism of separation is relevant (e.g., isoelectric focusing is of no use for a molecule without pI).

2 TECHNIQUES AND PROCEDURES

2.1 Basics

Even before choosing the procedures, the following question has to be addressed: is the metal covalently bound to the protein? If so, as is the case for selenium in some proteins, the species are relatively stable during the separation, and denaturing conditions can be used; this means conditions where only the primary structure is preserved. Otherwise, nondenaturing electrophoresis must be applied, even if this implies a loss of separation efficiency; nondenaturing is equivalent to native. The first decision is therefore whether a native or a denaturing procedure is to be used. The other questions to be answered are related to the kind of sample or the mixture of proteins to be separated: gradient or linear gel (Section 2.3), with or without stacking (Section 2.4). Any paired combination is feasible, giving eight theoretical associations.

2.2 Native/denaturing electrophoresis

The first and most straightforward method is native electrophoresis. This means that the proteins, without any modification to their secondary and tertiary structures, are submitted to electrophoresis. The buffer is chosen so that the protein is not denatured. Biochemists use this separation method when they are interested in the activity of the enzymes that are isolated, activity that would be lost if the proteins were denatured. However, this type of separation is subject to a major drawback, as no buffer system is suited to all separations. Firstly, no universal buffer exists for the separation of all proteins. In a buffer with a pH below 10, proteins with a pI of 11, thus positively charged, migrate to the cathode, and proteins with a pI below 10, negatively charged, migrate to the anode; in other words, they migrate in opposite directions and cannot be seen on a single gel. A solution would be to use extreme pHs but those are prohibited since they denature the proteins. Secondly, no buffer exists that allows the correct preservation of all metal–protein complexes, e.g., some buffers may affect a vanadium–protein complex without affecting a platinum complex.

For speciation purposes, however, the native method is compulsory when the metal is not covalently bound to the protein. Should the protein be denatured, the complexing site would be destroyed and the metal would be set free.

Various nondenaturing buffers have been proposed [7] and only one will be given as example in Section 2.5.1.

2.3 Restricting medium: gradient or linear gel

Basically, separation takes place in either a restricting medium or a free medium. Restricting means that the particles interact with the gel, either physically, e.g., because of their size, or, with an even broader definition, chemically, because of interactions of the proteins with a pH gradient or with antibodies.

For the separation according to size, except during stacking, a restricting medium is preferred. There is still a question to be answered, namely whether the gel must be used with a constant density or a gradient. A gradient is a continuous change in density of the gel in the direction of migration, or in other words, a continuous gradient of the pore size. The gel begins with large pore sizes and ends with restricted sizes, so that the friction force constantly increases in the gel, up to a point where the velocity of the protein is very low. A gradient also ensures the separation of a broad mass range, typically from 10 to 200 kDa, in comparison with smaller ranges for homogeneous gels. Thus, a gradient gel allows sharper and sometimes easier separations, especially when little is known about the range of molecular masses of interest. Homogeneous gels still have advantages, especially when one is interested in a particular mass or family of proteins, since they offer a kind of zoom process. Further, a more precise mass determination is possible, especially when Ferguson plots are used.

2.4 Stacking or sample concentration: discontinuous buffers

By stacking, one understands a process capable of concentrating the sample in the gel before the real separation occurs. This is particularly interesting if the analytes are present at low concentration. Indeed, stacking allows the proteins to be concentrated in sharp bands, a prerequisite for R_f measurements, i.e. migration distance, or speciation, when the metal concentration is low. The principle is isotachophoresis, or more precisely moving-boundary electrophoresis. The method is also widely known as discontinuous or disc electrophoresis because of the discontinuity of properties between stacking and separating gels in buffer and pH.

For clarity, the basics of the method will be summarized here, for a system migrating from the cathode to the anode. The gel is physically made of two zones, the first one, where the sample is applied, being the stacking gel, where the gel is not restricting; the second is made of the resolving gel, where separation occurs. A set of three buffers must be chosen, but even though this task is very tedious it is one of the best documented [7]. The goal is to obtain three ions with increasing mobilities from the cathode to the anode:

- a terminating ion, with a low mobility, at the cathode in our case;
- a leading ion, with the highest mobility of the ions, present in the gel and the anode;
- a common counter ion.

The proteins are applied at the cathode and, in a manner of speaking, are sandwiched between the two ions. When a voltage is applied, molecules range according to their mobilities, from the leading ion, the proteins, to the terminating ion. This occurs with a most interesting characteristic, a constant concentration in one band, dependent on the concentration of the leading ion. Through this effect the proteins are pre-separated in sharp bands (see Figure 4.4.1).

At the border between stacking and separating gel, a new force, the friction with the gel material, affects the macromolecules. The terminating ion does not interact with the gel, since its size is negligible in comparison with the pore size, and migrates further. At this stage, the proteins are surrounded by the terminating ion and migrate farther, but this time following the principles of zone electrophoresis.

2.5 Application

In order to give a practical idea of a separation by gel electrophoresis, two methods have been chosen: one-dimensional native electrophoresis and two-dimensional high resolution electrophoresis. Those separations illustrate the most extreme cases in trace element speciation with gel electrophoresis: 2DE is used when high separation capacity is needed but when the species can resist

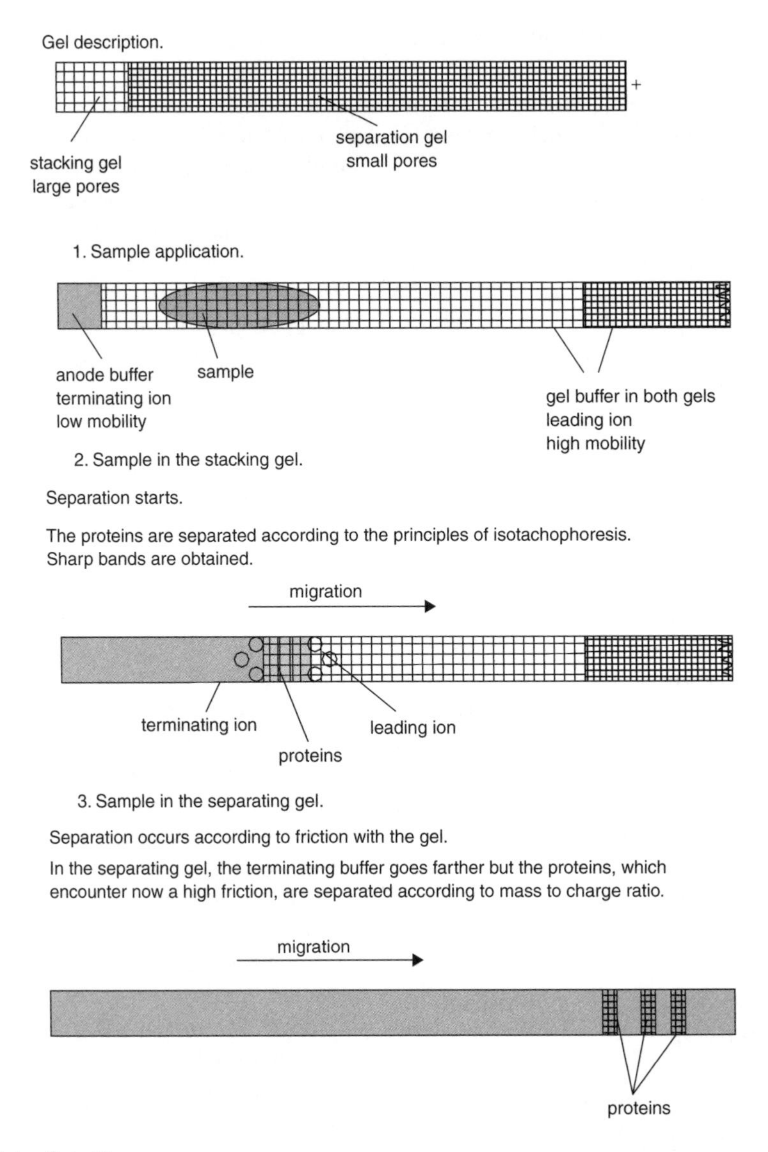

Figure 4.4.1. Principle of stacking.

denaturation, whereas nondenaturing electrophoresis is used when the complex is fragile and would not resist to a denaturing separation. These examples are largely inspired by dedicated and extended chapters in the books aforementioned; they also take into account the availability of commercial sets of gels and buffers.

Common features can be recognized in both examples. First of all, the stability of the species is a point that cannot be stressed enough and must be checked by an independent method. Since the gels may trap oxygen during the polymerisation, if the species are sensitive to oxidation, reducing agents can be used such as thioglycolate [8].

Secondly, the proteins must be brought into solution, in a sample buffer compatible with the separation method. The quantity of protein necessary for an optimum detection of the former is given by rules of thumb: in fact, this quantity primarily depends upon the detection method, silver-staining, rubies-staining, coomassie blue or ^{14}C. For example, a sensitive method such as silver-staining requires around 10 mg mL^{-1} as total concentration of proteins; for the visualisation of a specific protein, circa 10 ng protein are necessary, a quantity that is dependent not only on the protein itself, but also on the separation method. Indeed, in a one-dimensional experiment, the protein is spread over a whole band in comparison with a spot from 2DE, concentrated on 1 mm^2, therefore requiring less protein. One must also take into account the quantity of trace element in the sample and the detection of the metal. However, a limiting factor is that gel electrophoresis has also a maximum loading capacity above which no separation occurs. A compromise has to be found between the two parameters. If a high amount of protein has to be used to detect the trace element, the use of coomassie blue for the proteins can be considered, which is about 50 times less sensitive than silver-staining; should the latter be used, the risk is high that the whole gel would be darkened by the proteins.

2.5.1 Nondenaturing electrophoresis

Although one-dimensional electrophoresis does not imply that the method be nondenaturing, such a combination has been chosen for simplicity. The converse is true, nondenaturing electrophoresis being mostly one-dimensional up to this date.

The choice of the buffer system is the first step. A set of buffers optimum for the separation of proteins with acidic pI (below circa pI 8), and widely used, is a slight modification of the set described by Laemmli [9]:

- electrode buffer, glycine/Tris base, pH 8.3;
- gel buffer, Tris base/HCl, pH 6.4;
- polarity, separation towards anode, the sample is applied at the cathode;
- 100 mL of each solution is sufficient. Buffers strips and gel are rehydrated with adequate solution.

This set belongs to the class of discontinuous buffers, with Tris as the common ion, chloride the leading ion and glycine the trailing ion.

For other separations, such as separations of basic proteins (pI above 8), other sets have been optimised and can be found in authoritative reviews [7] or books [3].

The second step is the preparation of the sample. The sample, proteins in our example, must be dissolved in a buffer compatible with the method, they must be kept in solution and, of course, the species must be stable. In order to be compatible with the electrophoresis procedure, the sample should not contain too much salt. In fact, a sample with too high an ionic strength is one of the main causes of failure of a separation, since a high charge concentration causes a drop in the resistance of the solution. Indeed small ions migrate more easily and in such a case the proteins stay at their point of application. A further requirement is a good solvent that allows a smooth penetration of the sample in the gel. This step may be critical, especially for large or hydrophobic molecules, for which it is difficult to go from a free solution to a solution in a gel. Therefore, an optimal sample is the combination of the following:

- the desalted original sample;
- diluted in the gel buffer, which is the first solution with which the proteins will be in contact in the gel, with 10 % glycerol, to mimic the gel concentration and therefore facilitate the penetration in the gel;
- mild detergents, useful to keep the proteins in solution and prevent aggregation, but there is a real danger that the proteins could be denatured or the species degraded.

Parallel to the choice of the buffer, the problem of the stability of the compounds has to be tackled. As already mentioned, the choice of a buffer system is the key not only to a good separation of proteins but also to speciation. In order to apply the separation method to speciation, it is advisable to first test the stability of the compounds to be separated in the buffers. For example, if the species vanadium–protein has to be separated in a given buffer, experiments ought to be performed with the species in the buffer to check whether free metal is produced, i.e., whether the equilibrium between free vanadium and complexed vanadium

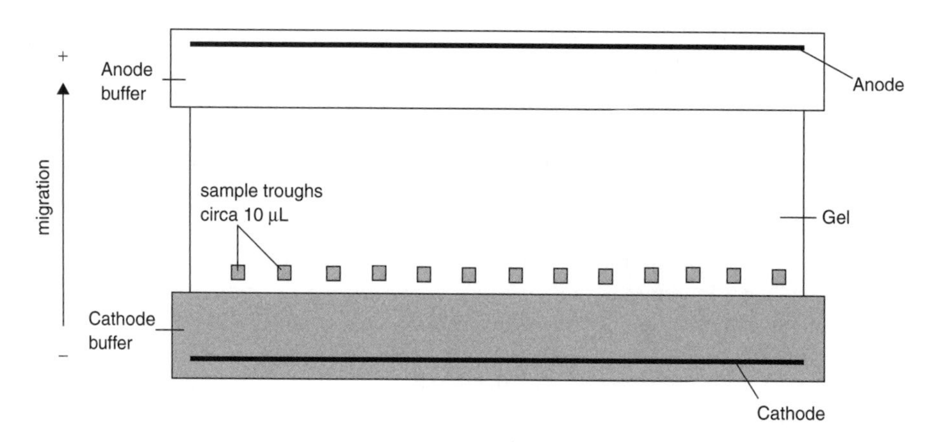

Figure 4.4.2. Typical set for flatbed separation.

is disturbed. Such experiments can be ultrafiltration or size-exclusion chromatography, any method that is able to separate the metal–complexes from the free element [39].

An example of incompatibility of a buffer set and a trace element is vanadium. Indeed, glycine can complex vanadium and strips proteins from the trace element to which it was bound [10]. Therefore, the use of buffers other than glycine or tricine has to be explored.

- Equipment: no equipment is standard for 1D electrophoresis. There is a large choice between flatbed, submarine or vertical systems, each of which has advantages over the other, but none influencing speciation.
- Gel: various gel sizes can be used, $5\,\text{cm} \times 5\,\text{cm}$ or $25\,\text{cm} \times 10\,\text{cm}$ gels, most of which are commercially available. If a specific gel is needed, preparation is possible in the laboratory, at low cost; once again, one should refer to the appropriate literature. Another possibility is to use a rehydrated gel, wash it thoroughly, dry and rehydrate it in the necessary buffer.

The buffers are laid directly under the respective electrode. The sample is brought on the gel either:

- directly, if the sample volume is small (1 μL);
- onto a sample strip (1–10 μL);
- in a sample trough (up to 15 μL) if these were foreseen during polymerisation.

See Figure 4.4.2 for a typical set with a flatbed system.

The separation programme is also dependent on the apparatus but common features are always identifiable. A low voltage (ca. $25\,\text{V}\,\text{cm}^{-1}$) is applied in the first step to let the proteins enter the gel material, i.e. the stacking gel. Afterwards the real separation begins.

For example, for a homogeneous gel, 25 cm (approximately 25 lanes) on 11 cm (separation length), stacking gel $T = 5\,\%$ (thus wide pores for the stacking effect), $C = 3\,\%$ (33 mm), resolving gel $T = 10\,\%$, $C = 2\,\%$ (77 mm), the programme can be written as [2]:

	Voltage (V)	Current (mA)	Power (W)	Duration (min)
1st step	500	10	10	10
2nd step	1200	28	28	50

The separation is stopped when the dye, indicating the front line, is at the anode. The subsequent steps are the detection of the trace elements and the visualisation of the proteins.

2.5.2 Two-dimensional gel electrophoresis (2DE)

Two-dimensional electrophoresis is the latest development of gel electrophoresis, as evident from the exponentially increasing number of articles about this method. As already mentioned, the reason for this success is due to proteomics, since a gel can map nearly all proteins present in a sample. For speciation analysis, 2DE is only applicable if the metal under study is covalently bound, since the method is denaturing in its most refined form, i.e. high resolution. Nondenaturing two-dimensional electrophoresis has been described [11], but is based on different separation principles from those mentioned here and little has been published until now. That is why this part concentrates on denaturing electrophoresis.

2DE unites two separation mechanisms which exist independently from one another: isoelectric focusing (IEF) and sodium dodecyl sulfate – polyacrylamide gel electrophoresis (SDS-PAGE). It is thus a separation according to pI in the first dimension and size in the second. In most of the contemporary publications, IEF is used as the first dimension and SDS-PAGE as the second dimension. Both methods can be used independently and what is mentioned here is true for both SDS-PAGE one-dimensional electrophoresis and IEF one-dimensional electrophoresis, with slight modifications for the latter. One should refer to books [2, 3, 12] or articles [6, 13] for a practical and state-of-the-art description of the method.

2.5.2.1 Isoelectric focusing (IEF)

This separation procedure relies on one of the major characteristics of proteins, their isoelectric point. Indeed, the charge of a protein is pH dependent, and at a characteristic pH this net charge is zero. It is possible to polymerise a gel with a pH gradient, termed an immobilized pH gradient (IPG), or to use a chemically created pH gradient, formed by free carrier ampholytes. They are both commercially available. The choice of the form of gradient is sometimes important for the quality of the separation. To begin with IEF separations, especially as the first dimension of 2DE, it is generally more secure to use the IPG technology, where the IPG strips, usually about 5 mm wide and 5–20 cm long, are stored in a dehydrated form.

If the acidic part of the gel is pointed towards the anode (+), a protein initially at the anode, which is positively charged below its pI, migrates towards (−) and thus towards its pI. At the pI, the charge is zero and the field does not influence the particle any more. This reasoning is conversely valid for a protein initially present at the cathode. Should the protein diffuse, below its pI, the charge is positive and it is repelled by the anode (+), while above it the charge is negative and it is repelled by the cathode (−). That is why the separation is usually named focusing, stressing the fact that the protein comes to a definite spatial point in the gel by a ping-pong mechanism.

Although not a requirement for IEF, it is better to denature the proteins at this stage, firstly to be compatible with the second dimension and secondly to bring hydrophobic proteins in solution. The quantity of protein is also a determining factor for detection and must be adjusted to the detection method, as already mentioned, but should not exceed 200 μg in a narrow strip, about 20 cm long. This solubilisation takes place in a mixture, hereafter referred to as sample solution, containing [13]:

- urea, a chaotropic agent, used to break the hydrogen bonds in and between proteins and to unfold them;
- a nonionic detergent, such as CHAPS, to bring the proteins in solutions without contributing to the ionic strength of the solution;
- a reducing agent, DTT, to break the disulfide bonds in proteins;
- possibly a protease inhibitor, depending on the sample, in order to prevent any proteolysis.

Once again, it must be checked whether the sample is stable in this solution. The same strategy as that mentioned in Section 2.5.1 is recommended. In particular, since the separation occurs in a pH gradient, control of the stability of the species is necessary at the extreme pHs, usually 3 and 11.

For our example, IPG, the strip must be rehydrated for at least 10 h, either in the sample solution or in a solution containing the same chemicals, except the analytes, the sample being added at the end of rehydration. After rehydration, a low voltage (in our example 300 V) is applied for 1 h to force the proteins into the gel and afterwards a high voltage is used to achieve the focusing (up to 8000 V, if it can be reached) for at least 6 h.

Between the two dimensions an equilibration step is necessary to change the solution in which the proteins will be separated in the second dimension, namely SDS.

2.5.2.2 *Sodium dodecyl sulfate – polyacrylamide gel electrophoresis (SDS-PAGE)*

The separation principle is based on the constant affinity of SDS for proteins, since 1.4 g SDS binds 1 g protein. From then on, the intrinsic charge of the protein is masked by the charge of SDS; thus there is a constant charge per gram of protein. Instead of a separation according to mass and charge, charge is here related to mass. This method has other advantages, when it can be applied to speciation, such as an enhanced solubility of the proteins or a real random coil of the chain, which allow an easier mass estimation of the particle.

For SDS-PAGE, the choice is once again left open between gradient or homogeneous; a gradient gel may be easier to begin with, since the mass range is broader, especially when little is known about the sample. Most of the time a stacking effect is used to enhance the resolution in the second dimension. The last point can be recognised in the setup of the buffers [9]:

- cathode buffer, glycine/Tris, $10\,g\,L^{-1}$ SDS;
- gel buffer, Tris/HCl, $10\,g\,L^{-1}$ SDS;
- anode buffer, Tris/HCl;
- glycine is the trailing and chlorine the leading ion.

The IPG strip is laid parallel and next to the cathode (−). Since the micelles are negatively charged, they migrate towards the anode.

The voltage programme is comparable with the others already mentioned, for a typical gel (25 cm × 11 cm):

(i) 200 V (50 mA and 30 W maximum). The voltage is low for an optimum sample entry, until the sample front is 5 mm away from the IPG strip.

(ii) The strip must be then removed and the cathode buffer put in on its place to avoid dehydration of the PA gel.

(iii) 600 V (50 mA and 30 W maximum). The separation is stopped when bromophenol blue has reached the anode.

Afterwards, the gel is ready for the detection of the trace element and the proteins.

2.5.2.3 *Example: 2DE of selenised yeast*

The aforementioned method was applied to the separation of an extract of yeast, enriched in selenium [14]. The radiotracer ^{75}Se was used to allow detection by means of the phosphor screen technology. A key point of the separation is the protection of the selenoamino acids against oxidation, by a chemical derivatisation. Without this precaution, the species are not stable during electrophoresis.

After separation, the proteins were fixed as mentioned later and the selenium-containing proteins detected with a phosphor screen for 1 week. After detection of the trace element, the gel was silver stained and the two pictures, autoradiogram and silver staining, can be compared (see Figure 4.4.3 and Figure 4.4.4 respectively).

Two practical points are worth mentioning. First of all, all gels and material are commercially available, which reduces the amount of work. However, if necessary, all these gels can be readily

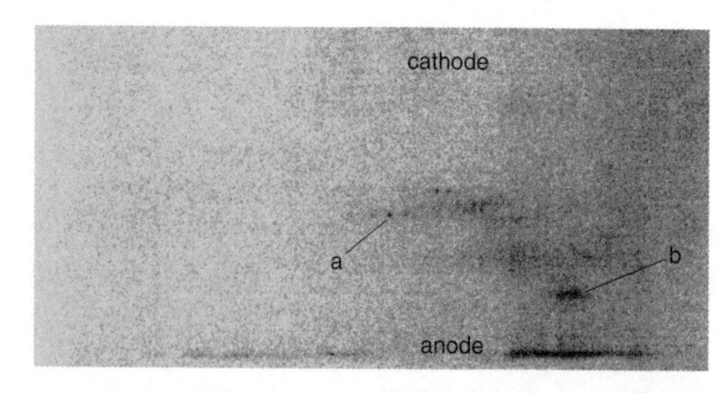

Figure 4.4.3. Autoradiogram of a 2DE gel of yeast enriched with ^{75}Se. a and b are two different spots, materialised to ease the comparison between silver staining and autoradiography. Reprinted from *Fresenius' Journal of Analytical Chemistry*, Two-dimensional gel electrophoresis, C. C. Chéry *et al.*, Vol. 371, pp. 775–781, 2001, copyright notice of Springer-Verlag [14].

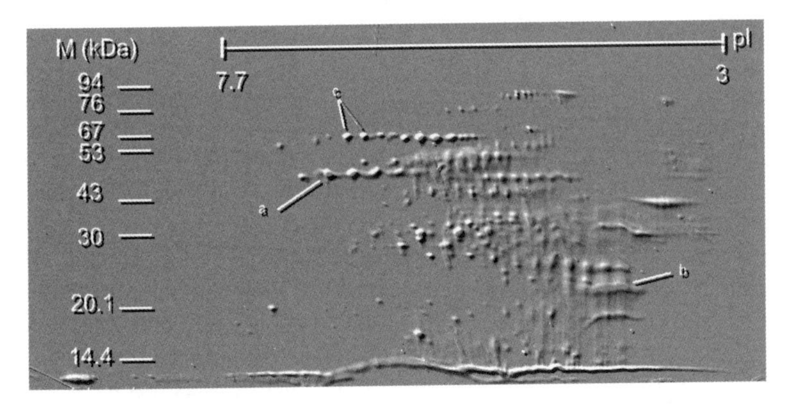

Figure 4.4.4. Silver staining of the gel used in Figure 4.4.3. Spots a and b are materialised for comparison with the autoradiogram. Note that the spots c do not appear on the autoradiogram, which proves that some proteins do not contain selenium. Reprinted from *Fresenius' Journal of Analytical Chemistry*, Two-dimensional gel electrophoresis, C. C. Chéry *et al.*, Vol. 371, pp. 775–781, 2001, copyright notice of Springer-Verlag [14].

made in the laboratory at low cost. Secondly, the whole separation requires less than 2 h of work, although it lasts 20 h in total. The first dimension is not labour intensive. Once the IPG is prepared, the separation occurs without external intervention (15 h); this can be done overnight. Preparation for the second dimension requires about 1 h and the run, although best supervised from time to time, does not require further intervention.

3 DETECTION OF TRACE ELEMENTS

3.1 General

Once again, the crucial question arises whether a species is stable during separation or detection. Indeed, a method widely used prior to the detection of proteins is precipitation. By soaking the gel with an acidic solution, the proteins precipitate and diffusion of the bands or spots is no longer a problem. The gel is soaked in a solution of ethanol, acetic acid and water and left to dry. It is in fact the first step of detection by staining. If the gel must be stored for a long period before analysis, glycerol can be added to this solution, otherwise the gel may shrink. However, such a treatment is prohibited if the species are labile and the only methods left are either drying or freezing the gel.

A second choice to be made is to use either a method that allows the detection of the trace elements in the entire gel or a method that requires that the gel be cut in subsamples. The former is more convenient but the latter allows quantitative analysis. Thus, the choice between the two sets of methods is dictated by strategy and need. One has to know which question is to be addressed: where is the trace element, or in which quantity is it present?

3.2 Detection of trace elements in subsamples of the gel

Although this procedure is more time consuming, since an additional manipulation is required before detection, analysis of subsamples is more reliable for quantitative analysis and more sensitive. Furthermore, they are sufficient when only a rough image of the distribution of trace elements is necessary. From the analytical point of view, all methods used for the detection of trace elements can in fact be considered; however, for simplicity, only the methods whose applications have already been published are presented here.

3.2.1 Liquid introduction system: inductively coupled plasma – mass spectrometry (ICP-MS), AAS and AES

In order to get a first image of the distribution of trace elements in gels, ICP-MS with a liquid introduction system is a method of choice. The gel must be excised and the pieces mineralised by microwave assisted digestion. Each fraction is then quantified by ICP-MS. This method is particularly tedious, but gives detection limits that are excellent for most elements. Any other method that relies on dissolved subsamples is, however, applicable, such as AAS, AES, voltammetry [15] or fluorescence [16].

3.2.2 Solid sample analysis: electrothermal vaporisation (ETV) – ICP; graphite furnace – atomic absorption spectrometry (GF-AAS)

These methods can be seen as improvements of the introduction system of the sample. Indeed, as solid sampling analysis, they do not require destruction of the pieces of gel and still offer excellent detection limits. The sample is brought in the oven, the organic material ashed and the trace elements brought to the plasma in the case of ETV [17, 36] or detected in the furnace for GF-AAS [18]. The disadvantages of the method are that a precise optimisation of the temperature programme and of the use of chemical modifiers are necessary and that the optimisation is valid for only one element. Still, the detection is simpler than with a liquid introduction system and can be quantitative if the system is adequately calibrated.

3.2.3 Nuclear analytical chemistry: scintillation counting, neutron activation analysis (NAA)

Scintillation counting can be used with samples radioactively labelled. According to the nuclear characteristics of the radiolabel, the method of choice will be either liquid scintillation counting or well-type NaI(Tl) detection. The former requires that the subsample be solubilised, whereas the latter is directly suitable for the solid piece of gel. Liquid scintillation has already been successfully applied to the detection of ^{75}Se after PAGE [19]. The advantage of the method is a higher sensitivity than the phosphor screen technology, but at the expense of a good resolution of the gel.

NAA [20–23] has been one of the first detection methods used for trace elements in gels. Even if this method implies a large investment and a heavy infrastructure, the method is interesting because it relies on a totally different principle from the spectrometric methods mentioned previously. After the gel is sliced, the elements it contains are activated in a nuclear reactor through neutron bombardment. After a cooling time, the elements are detected by recording the whole gamma spectrum with a Ge(Li) detector and by assigning the peaks specific to each element. Quantitation is once again possible. Theoretically, this method can be applied to a whole gel and an autoradiogram recorded (see later). However, the high neutron fluxes produce a high temperature and a high radiation which damage the gel; furthermore, a high background is obtained since any impurity and traces of the reagents in the gel are also activated.

3.3 Detection of trace elements in a whole gel

The methods presented here rely on concepts already presented in Section 3.2, with the exception of autoradiography, and are refinements of the sample introduction system.

3.3.1 Autoradiography

Autoradiography is the method of choice when the material submitted to electrophoresis is labelled with a radiotracer. A whole and precise image of the distribution of radioactivity is obtained when the autoradiography screen is laid on the gel, recording a kind of photographic picture of the radioactive material. Furthermore, it is not limited to gels, but can be applied to all thin materials, such as tissues.

Previously, X-ray films were used, but the contemporary technology relies on phosphor screens [24], which allow a more rapid detection with higher resolution. Further, a phosphor screen is theoretically infinitely reusable, as long as it has not been contaminated. After the signal has been read, the screen is erased by exposing it to visible light for a couple of minutes and is ready for use again.

The phosphorescent material is a suspension of $BaFBr:Eu^{2+}$ crystals in a polymer. This crystal, when excited by radiation, sends an electron to the conduction bands, resulting in a different chemical structure, $BaFBr^-$ and Eu^{3+}. When the excited crystal is exposed to light, typically from an He–Ne laser (633 nm), this energy is enough to destabilise the excited electron, which falls back to its ground state, emitting a photon at 390 nm. A laser is used because it allows an excellent spatial resolution. The luminescence is recorded with the position of the laser and finally, when the whole screen has been scanned, a precise image of the position and intensity of the original radioactivity is obtained, in the form of a digitised picture with the optical density as a function of the position.

Although both X-ray and phosphor screens primarily allow the detection of ^{14}C, ^{32}P and ^{35}S, pure beta-emitters, they also may be applied to other isotopes (^{75}Se [14, 25–27], ^{63}Ni [28], ^{109}Cd [29], ^{59}Fe [30], ^{65}Zn and ^{45}Ca [31]). The only drawback is that the detection efficiency is lower for some of these isotopes, depending on the nature of the radioactivity they emit. This means that a higher specific activity has to be used to get the same optical density as with the same activity of, e.g., ^{35}S, or that a longer exposure time is required.

From the last remark it is obvious that both parameters for detection by autoradiography are of importance, i.e. specific activity and exposure time. To get an idea of whether a sample can be detected by this method, a simple test is required: a dry gel can be rehydrated with a solution of known activity. By a simple weighing before and after rehydration, the amount of activity is known in the gel. The gel is left to dry, packed in plastic and exposed to the phosphor screen. Scanning of the gel will reveal if the original activity was high enough.

Let us describe the practical use of a phosphor screen. Just before use, the screen has to be erased since a background builds up after a while, even if the screen remains unused. After separation, on the condition that the species remains stable with treatment, the proteins are fixed in the gel by an acidic solution and the gel is left to dry. The staining should stop at this point, since a metal-based staining (e.g. silver staining) can quench the radiation emitted from the gel. In order to avoid contamination of the phosphor screen by the tracers in the gel, the gel is wrapped in a plastic foil, for example Mylar, the thinnest possible and only carbon-based to avoid loss of radiation through absorption of the material. A phosphor screen may also be stored at low temperature (less than $-20\,^{\circ}C$) while exposed to a gel, without apparent detriment to the quality of the picture. Thus, radiography also allows the detection of species that cannot be fixed in acid.

The phosphor screen is left for a period determined as mentioned beforehand, typically 1 day for ^{35}S and 2 days for ^{75}Se at similar specific activities. For longer periods, laying the screen in a lead coffer can improve the signal to background ratio by lowering the natural surrounding activity. The screen is then read by laser densitometry, erased and stored. Standard softwares are available for the image treatment of the gel.

3.3.2 Laser ablation – inductively coupled plasma – mass spectrometry (LA-ICP-MS)

This method is a further step toward a reliable direct introduction of the sample into the spectrometer. Other combinations are plausible, such as LA-AES, but the most powerful is presented here. LA-ICP-MS is a very promising technique for the detection of metals in gels after separation.

The detection method is itself an on-line hyphenation, between classical ICP-MS and a laser. The sample, the gel in other words, lies in an ablation cell; a spot is ablated by the laser and the fumes from the ablation are brought from the cell to the plasma in a tube by a continuous gas flow, generally argon. ICP-MS then gives the elemental composition of the protein present at the ablation site.

Various laser types are commercially available, from ArF to Nd:YAG lasers, with wavelengths respectively from 193 to 1064 nm. The ablation crater ranges from the μm range up to 200 μm, and is thus ideal to achieve a high resolution in the case of gel electrophoresis. Furthermore, the sample can be moved in two dimensions, as the latest equipment is computer controlled, allowing a precise screening of the gel. This last point is very advantageous for 2DE.

For a material such as polyacrylamide and the minimum protein spot size expected (about 200 nm in diameter), a low wavelength and very energetic beam are not necessary; fractionation, which means a different composition of the ablation composition from the original sample, is probably no problem, allowing the use of widespread and affordable lasers.

Up to now, this method has been used for only few elements in gels (Co in serum [32], Pb in humic acids [33]) but can be extended to any element detected by ICP-MS [5].

3.3.3 Particle induced X-ray emission (PIXE)

This technique relies on the excitation of the electron of the inner shell by a collimated beam of energetic charged particles, protons. After excitation, the atoms emit characteristic X-ray spectra that allow their detection, mostly by an energy dispersive analysis. However, this method requires a heavy investment since, e.g., a cyclotron is necessary to produce protons of a few MeV.

With an appropriate apparatus to translate the gel, the former can be scanned; the method has already been applied to one-dimensional gels, where a scanning in the direction of migration is the easiest [34].

3.3.4 Mass spectrometry (MS)

The applicability of MS to gel electrophoresis is well known and widespread, especially in the field of proteomics with matrix-assisted laser desorption/ionisation (MALDI). The combination of the two for speciation purposes has not yet been reported but no hindrance really exists, except for the low specificity of the method for metals. MS may be thought of as a detection method dedicated to organic molecules, in our case protein identification, but it has already been successfully applied to elemental speciation, for example to the speciation of arsenic or selenium [35]. The same can be applied with other elements, provided they give a typical loss, for example as for Se, or a recognisable isotopic envelope. Two examples to explain respectively the notions of typical loss and isotopic envelope: the loss of m/z 386 can only be attributed to Glu–Cys–(^{80}Se)–Gly in ref. [35], and not to its sulfur analogue; if subpeaks can be recognised in a major peak and if these peaks are separated by the same m/z as between isotopes of an element and with the same height ratio as in the natural abundance, the major peak can be attributed to a species containing this element.

4 CONCLUSION

Gel electrophoresis may seem tedious for elemental speciation purposes but its figures of merit make it worth giving it a try. To be applicable, one must be sure that the analytical data are representative of the elemental species in the original sample. Thus, attention must be paid to the stability of the compounds under investigation, even if this means additional tests.

The figures of merit for gel electrophoresis qualify this method as an integral part of speciation: high resolution, various separation mechanisms, the quantity of material after separation it yields, hyphenation with powerful tools such as MALDI and laser ablation – ICP-MS.

5 ACKNOWLEDGEMENTS

CCC is Research Assistant of the Fund for Scientific Research – Flanders (Belgium) (F.W.O.)

6 REFERENCES

1. Andrews, A. T., *Electrophoresis. Theory, Techniques, and Biochemical and Clinical Applications*, Oxford University Press, New York, 1988.
2. Westermeier, R., *Electrophoresis in Practice*, VCH, Weinheim, 1993.

3. Hames, B. D. (Ed.), *Gel Electrophoresis of Proteins*, Oxford University Press, Oxford, 1998.
4. Lustig, S., De Kimpe, J., Cornelis, R., Schramel, P. and Michalke, B., *Electrophoresis*, **20**, 1627 (1999).
5. Chéry, C. C., Günther, D., Cornelis, R., Vanhaecke, F. and Moens, L., *Electrophoresis*, submitted (2003).
6. O'Farrell, P. H., *J. Biol. Chem.*, **250**, 4007 (1975).
7. Chrambach, A. and Jovin, T. M., *Electrophoresis*, **4**, 190 (1983).
8. Bansal, M. P., Ip, C. and Medina, D., *J. Soc. Exp. Biol. Med.*, **196**, 147 (1991).
9. Laemmli, U. K., *Nature*, **227**, 680 (1970).
10. Lustig, S., Lampaert, D., De Cremer, K., De Kimpe, J., Cornelis, R. and Schramel, P., *J. Anal. At. Spectrom.*, **14**, 1357 (1999).
11. Lustig, S., De Kimpe, J., Cornelis, R. and Schramel, P., *Fresenius' J. Anal. Chem.*, **363**, 484 (1999).
12. Rabilloud, T. (Ed.), *Proteome Research: Two-Dimensional Gel Electrophoresis and Identification Methods*, Springer, Berlin, 2000.
13. Görg, A., Obermaier, C., Boguth, G., Harder, A., Scheibe, B., Wildgruber, R. and Weiss, W., *Electrophoresis*, **21**, 1037 (2000).
14. Chéry, C. C., Dumont, E., Cornelis, R. and Moens, L., *Fresenius' J. Anal. Chem.*, **371**, 775 (2001).
15. Messerschmidt, J., Alt, F. and Tölg, G., *Electrophoresis*, **16**, 800 (1995).
16. Vézina, D., Bélanger, R. and Bleau, G., *Biol. Trace Elem. Res.*, **24**, 153 (1990).
17. Wróbel, K., González, E. B., Wróbel, K. and Sanz-Medel, A., *Analyst*, **120**, 809 (1995).
18. Sidenius, U. and Gammelgaard, B., *Fresenius' J. Anal. Chem.*, **367**, 96 (2000).
19. Qu, X. H., Huang, K. X., Wu, Z. X., Zhong, S., Deng, L. Q. and Xu, H. B., *Biol. Trace Elem. Res.*, **77**, 287 (2000).
20. Stone, S. F., Zeisler, R., Gordon, G. E., Viscidi, R. P. and Cerny, E. H., *ACS Symp. Ser.*, **445**, 265 (1991).
21. Stone, S. F., Hancock, D. and Zeisler, R., *J. Radioanal. Nucl. Chem.*, **112**, 95 (1987).
22. Stone, S. F., Zeisler, R. and Gordon, G. E., *Biol. Trace Elem. Res.*, **26**, 85 (1990).
23. Jayawickreme, C. K. and Chatt, A., *J. Radioanal. Nucl. Ch.*, **124**, 257 (1988).
24. Johnson, R. F., Pickett, S. C. and Barker, D. L., *Electrophoresis*, **11**, 355 (1990).
25. Behne, D., Kyriakopoeulos, A., Weiss-Novak, C., Kalckloesch, M., Westphal, C. and Gessner, H., *Biol. Trace Elem. Res.*, **55**, 99 (1996).
26. Jamba, L., Nehru, B., Medina, D., Bansal, M. P. and Sinha, R., *Anticancer Res.*, **16**, 1651 (1996).
27. Lecocq, R. E., Hepburn, A., Lamy, F., *Anal. Biochem.*, **127**, 293 (1982).
28. Nielsen, J. L., Poulsen, O. M. and Abieildtrup, A., *Electrophoresis*, **15**, 666 (1994).
29. Scott, B. J. and Bradwell, A. R., *Clin. Chim. Acta*, **127**, 115 (1983).
30. Vyoral, D. and Petrák, J., *Biochim. Biophys. Acta*, **1403**, 179 (1998).
31. Scott, B. J. and Bradwell, A. R., *Clin. Chem.*, **29**, 629 (1983).
32. Neilsen, J. L., Abildtrup, A., Christensen, J., Watson, P., Cox, A. and McLeod, C. W., *Spectrochim. Acta B*, **53**, 339 (1998).
33. Evans, R. D. and Villeneuve, J., *J. Anal. At. Spectrom.*, **15**, 157 (2000).
34. Szökefalvi-Nagy, Z., Bagyinka, C., Demeter, I., Kovács, K. L. and Quynh, L. H., *Biol. Trace Elem. Res.*, **26**, 93 (1990).
35. McSheehy, S., Pohl, P., Szpunar, J., Potin-Gautier, M. and Łobiński, R., *J. Anal. At. Spectrom.*, **16**, 68 (2001).
36. Chéry, C. C., Chassaigne, H., Verbeeck, L., Cornelis, R., Vanhaecke, F. and Moens, L., *J. Anal. At. Spectrom.*, **17**, 576 (2002).
37. Fan, T. W.-M., Pruszkowski, E. and Shuttleworth, S., *J. Anal. At. Spectrom.*, **17**, 1621 (2002).
38. Chen, C. Y., Zhao, J. J., Zhang, P. Q. and Chai, Z. F., *Anal. Bioanal. Chem.*, **372**, 426 (2002).
39. Chéry, C. C., De Cremer, K., Dumont, E., Cornelis, R. and Moens, L., *Electrophoresis*, **23**, 3284 (2002).
40. Gao, Y. X., Chen, C. Y., Chai, Z. F., Zhao, J. J., Liu, J., Zhang, P. Q., He, W. and Huang, Y. Y., *Analyst*, **127**, 1700 (2002).

CHAPTER 5
Detection

5.1 Atomic Absorption and Atomic Emission Spectrometry

Xinrong Zhang and Chao Zhang

Tsinghua University, Beijing, China

1 INTRODUCTION

Speciation measurements of trace elements have become important because many studies have shown that the determination of total amounts in a sample, without distinguishing between its chemical species, is no longer adequate [1–4]. The main analytical challenges for trace element speciation are the very low concentration at which they often occur and the requirement for identification of the chemical forms in which the element is present. Both requirements are far from satisfactorily met by most of the existing commercial instrumentation available for inorganic and organic analysis [5, 6].

As indicated in ref. 6, elemental analysis has traditionally aimed at complete analyte recovery and high sensitivity, in order to measure the total amount of a specific element contained in a sample. Atomic absorption and emission spectrometric techniques (AAS, AES and ICP-MS) have been developed to achieve these goals. The methodology developed for element determination aims at complete matrix dissolution in order to optimize atomization and quantitation. With these techniques, however, little information can be gained

Handbook of Elemental Speciation: Techniques and Methodology R. Cornelis, H. Crews, J. Caruso and K. Heumann
 ISBN: 0-471-49214-0

with respect to the chemical forms and structures of the compounds in which the element is present.

Organic structure analysis, in contrast, has always been directed at molecular identification of the analyte. The analytical efforts were therefore focused on structural identification rather than on recovery and sensitivity issues. The techniques that have been developed for the identification of chemical structures include mass spectrometry (MS), nuclear magnetic resonance spectroscopy (NMR), infrared/ultraviolet spectrometry (IR/UV), etc. Separation techniques, such as gas or liquid chromatography, could be coupled for preseparation of the compounds before identification. As relatively high concentrations are usually available for structure identification, sensitivity would not be an important issue in organic analysis.

Speciation studies, however, require the detection of trace level quantities of the elements in the samples. At the same time, the measurements should provide information about the chemical forms or structures in which the element is present. From this point of view, we can immediately see the technical difficulties related to element speciation. Because of their low sensitivity, most popular techniques, (such as IR/UV, NMR and MS), which have been extensively and successfully used for the identification of the forms and structures in organic analysis, are usually no longer adequate for element speciation. Atomic spectrometric detectors, the powerful tools with high sensitivities for most metals, fail to provide information about the chemical forms in which an element is present. At this moment, there is hardly any commercial instrument available yet for element speciation.

In laboratories this problem has been solved by combining several analytical techniques. Most successful combinations result from the coupling of a separation technique and an element-selective detector [7–9]. In general, the detection limits of these hyphenated systems are strongly dependent on the selected detectors, although the detection limits of hyphenated techniques are inferior to those of atomic spectrometric detectors alone. This may be due to the relatively small sample volumes that can be introduced into the chromatographic system and the peak broadening that occurred during separation, and also because of low sample flow rates needed for compatibility with LC flows.

The element-selective detection techniques generally used for speciation purposes include atomic absorption spectrometry (AAS), atomic emission spectrometry (AES), atomic fluorescence spectrometry (AFS) and inductively coupled plasma – mass spectrometry (ICP-MS). In comparison with AES and ICP-MS, AAS is the more popular detection method. It has been successfully coupled to gas and liquid chromatography (GC/LC) in many laboratories [5, 7]. As the eluent from GC/LC can be easily introduced into flame AAS (FAAS), the interface between chromatography and FAAS is very simple. However, FAAS could not provide the necessary sensitivity for element speciation in most cases. Hydride generation atomic absorption spectrometry (HGAAS) coupled to chromatography is becoming a common technique for the speciation of those elements that can form hydrides. The major merit of this hyphenated technique is its high sensitivity. In addition, the matrix effect can be removed effectively before entering the AAS detector. Unfortunately, the technique is suitable only for a limited number of elements, such as arsenic, mercury, selenium, lead, etc. Elements that do not produce hydrides cannot be determined by HGAAS. Electrothermal atomic absorption spectrometry (ETAAS) combines high sensitivity with an extensive range of elements suitable for detection. Unfortunately, the sequential nature of the drying and ashing steps prior to atomization makes it difficult to couple onto the continuous flow of the HPLC effluent. Up to now, it remains a problem to design an interface for on-line coupling of ETAAS to a chromatographic system.

In comparison with AAS, plasma source AES offers advantages of multielement operation, easy coupling to chromatography and acceptance of the continuous flow of LC eluent. The important disadvantages are associated with the sensitivity of the plasma to organic solvents and the overall inefficiency of the nebulizer system. The poor tolerance of the plasma source for the common mobile phase, such as ion-pair reagents, limits the application of this technique to the speciation measurements of the trace level of analytes.

The present chapter aims to review the development of AAS and AES as element-selective detection methods in hyphenated techniques for element speciation. The important techniques that will be reviewed include flame AAS, HGAAS, ETAAS and ICP-AES. The important coupling techniques covered in the chapter will include chromatography and capillary electrophoresis. Considering that flow injection analysis (FIA) is easily coupled with AAS and AES, and that it could be used to differentiate between redox species such as Cr(III)/Cr(VI), As(III)/As(V) and to perform online preconcentrations and separations, this technique is concisely described in the present chapter.

2 FLAME AND HYDRIDE GENERATION AAS

Flame AAS is one of the most simple, cheap and reliable detectors that has been extensively coupled to chromatographic and other separating techniques for element speciation. The poor detection limit for trace elements remains the main problem with this method. This can be partly ascribed to the analyte transport inefficiency of the pneumatic nebulizer for sample introduction into the flame atomization system. The flame AAS nebulizer has less than 5 % efficiency for sample introduction. Many efforts have therefore been undertaken to improve its efficiency. Another way of improving the detection limit in elemental speciation is to develop preconcentration techniques, including column absorption and atomic-trapping, that have interfaced well with flame AAS for elemental speciation in recent years [10–12].

2.1 Technical developments in sample introduction

In order to increase the sample introduction efficiency, the aerosol chamber from a glass concentric nebulizer system originally developed for ICP, has been adapted for a nebulizer interface for FAAS [13]. This resulted in an almost 100 % analyte transport efficiency due to the improved flow characteristics and efficient desolvation [14, 15]. With this design the detector signal was ten times better than that obtained using a conventional nebulizer.

The use of thermospray (TS) in FAAS is another method of increasing sensitivity, because it greatly improves the efficiency of the sample introduction. A common design of TS apparatus includes a capillary through which the solution is pumped by an HPLC pump. The capillary is heated to a temperature of 100–200 °C either by thermal contact with a heated block or by passing an electric current through the capillary itself [16]. Maintaining a constant temperature is necessary to minimize differences in nebulization conditions with changes in sample flow rate or composition. As the solution to be nebulized progresses through the capillary, it first boils and then forms droplets. The advantage of TS is that the droplets are smaller than those produced by a pneumatic nebulizer, giving increased signals and lower detection limits. The use of TS has greatly improved the sensitivity of hyphenated techniques for element speciation. Chang and Robinson have developed a TS apparatus for interfacing HPLC and FAAS [17]. The instrument was used for cadmium speciation studies in urine. Their experimental data showed that a 75 μm orifice and 0.05 cm i.d. capillary produced sensitivity much higher than that using a commercial flame atomizer. The desolvation mechanism of TS was also studied by the same authors [18]. They believe that, with further modification of the TS design, even higher analytical sensitivity for ultratrace metal analysis and better compatibility for interfacing HPLC with FAAS and ICP-AES can be expected. The main limitation of coupling TS with FAAS is that the system tends to clog because of the capillary commonly used in a TS apparatus. As a result, TS-FAAS is poorly tolerant of the introduction of a high salt solution.

Berndt and Yanez developed a hydraulic high pressure nebulization (HHPN) technique with high temperature (300 °C) superheated liquids for sample introduction [19]. The liquid sample was nebulized providing aerosol yields of up to 90 % in flame AAS. This new nebulization method combines the advantages of HHPN and thermospray techniques (very small aerosol droplets, high aerosol yield, nebulization of saturated salt solutions) and could be easily interfaced between chromatographic and spectrometric setups.

In addition to the design improvements, various quartz adapters have also been tested to increase the sensitivity of FAAS detection [8]. Improvement factors of 2.0 to 3.5 have been observed for the different elements using the conventional adapter with an exit slot at 180° to the entrance slot. Using the adapter with slots at 120° to each other, yielded improvement factors of 4 to 5. It has been found that the sensitivity can be additionally enhanced by shifting the conventional quartz adapter with slots at 180° by 10–15°. Thus a 4–6 times lower characteristic concentration for arsenic has been obtained. Obviously, the S-type convection flow through the slots ensures more stable temperature conditions in the quartz tube. However, it should be noted that slotted quartz tubes are vulnerable to vitrification in the presence of alkali metal ions in buffers; therefore they are pretreated with La(III) for improving long-term performance. Following a suitable separation scheme the concentration of different species has been determined, e.g. As(III) and As(V) in wastewater and Tl(I) and Tl(III) in soil extracts [20, 21].

2.2 Separation and preconcentration

The chromatographic separation of species is very often associated with their preconcentration. For instance, after extraction and chromatographic separation of As(III) and As(V) in a water sample, the arsenic concentration in the eluent is 20 to 25 times higher than initially in the samples. After such a separation/preconcentration step researchers often use simple, cheap and reliable flame AAS methods for the detection of the concentrated species [20, 22–25]. The on-line scheme has been extensively studied by combination of flow-injection separation/preconcentration and flame AAS detection. Fe(II) and Fe(III) could be well separated by using a C18-modified silica column combined with FIA flame AAS. Fe(III) was passed straight to the AAS detector whereas Fe(II) was trapped as Fe(II)-ferrozine and then eluted with methanol [23]. A similar procedure was used for the speciation of Cr(III)/Cr(VI) by inserting a microcolumn packed with acidic alumina [24] or Se(IV)/Se(VI) by an anionic exchange column [25]. The detection limits are 0.6 ng for Se(VI) and 1 ng for selenite, respectively, using flame AAS detection.

Speciation of inorganic and methylmercury could also be carried out by coupling FIA preconcentration with flame AAS, although atomic fluorescence spectrometry offers higher sensitivity [5]. Figure 5.1.1 shows a typical manifold for the speciation of mercury species using FIA preconcentration. A mixed solution of inorganic mercury and methylmercury was flushed onto a microcolumn packed with sulphydryl cotton. Inorganic mercury was not retained and was reduced to elemental Hg° with $SnCl_2$ solution, whereas methylmercury was absorbed on the column. Although the methylmercury is present at very low trace levels in the original solution, the absorption procedure allows a high degree of preconcentration, proportional to the volume of sample processed through the column. After recording the inorganic mercury peak, the elution of methylmercury occurs upon acidification of the column using 0.1 mol L^{-1}

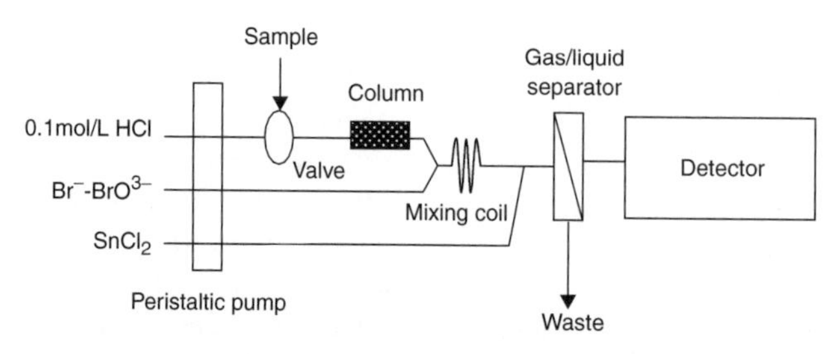

Figure 5.1.1. Schematic diagram of inorganic and methylmercury speciation by coupling FIA and AFS with an *on-line* preseparation technique using a sulphydryl cotton microcolumn [5].

HCl solution. On-line bromination and reaction of methylmercury with $SnCl_2$ generated Hg°, which was measured by the detector.

Preconcentration based on atom trapping has been developed in recent years to increase the sensitivity of certain elements in flame AAS for element speciation. In this technique, atoms are trapped on a cooled tube mounted above a conventional spray chamber–burner assembly. After a fixed sample collection time (conventional nebulization), the tube is rapidly heated to a temperature that causes atomization of the trapped analyte. Sensitivity enhancement depends upon the collection time and the analyte element. The technique works reasonably well for volatile elements such as arsenic, lead, cadmium and zinc. Matrix occlusion interferences limit this method to samples with a low dissolved solid content.

The combination of on-line hydride generation, atomic trapping and flame AAS has received considerable interest during the last decades. This hyphenated technique was first used in 1975 for the determination of methylated forms of selenium in freshwater environments [26]. Since then, it has been used for the speciation of arsenic, antimony, selenium, germanium, mercury and organic tin species. Table 5.1.1 lists the derivatizing conditions for the speciation of these elements using atomic trapping. The on-line integration of the different steps allows this approach to be used for a wide range of applications for samples such as air, water, sediments, and biological tissues, etc.

The introduction of cold trapping and chromatographic separation steps to hydride generation provides both high sensitivity and selectivity for real samples. This on-line hyphenated system provides derivatization by hydride generation or ethylation, preconcentration by cryotrapping, separation by packed column gas chromatography and detection by quartz furnace atomic absorption spectrometry. In addition to the direct gain in sensitivity achieved by hydride formation and cold trapping for elements such as As, Bi, Sb, Se, Sn, Ge and Te, the system provides the possibility of

Table 5.1.1. Derivatizing conditions for speciation using hyphenated techniques.

Species	Reagent	Derivatizing conditions	Sample pretreatment	Reference
As(III)	$NaBH_4$	2 mL 2 % aqueous $NaBH_4$	1–3 mL of 5 % potassium hydrogen phthalate pH 3.5–4	27
As(V) MMA DMA Trimethylarsine	$NaBH_4$	4 × 2 mL 2 % aqueous $NaBH_4$	pH 1–1.5 with 5 mL of saturated solution (10 % w/v) of oxalic acid in water	27
Sb(III) Sb(V)	$NaBH_4$	1 % $NaBH_4$ and 5.0 mol L^{-1} HCl solution	pH 4.0 with 50 mmol L^{-1} citrate solution	28
Se(IV) Se(VI)	$NaBH_4$	1.8870 g $NaBH_4$ in 0.5 % (w/v) NaOH and diluting with 250 mL of the same solution	with 25, 10 and 5 % (v/v) HCl : HBr solutions	29
Selenomethionine	$NaBH_4$	0.3 % $NaBH_4$ in 0.2 % NaOH		30
Ge $Me_xGe^{(4-x)+}$	$NaBH_4$	6 mL of 20 % $NaBH_4$ in 0.06 M NaOH per 100 mL of sample	5 mL of 1.9 M Tris-HCl + 10 mL of 300 gL^{-1} NaCl + 1 mL of 0.2 M EDTA per 100 mL of sample	31
$Me_xPb^{(4-x)+}$	$NaBEt_4$	3 mL of 0.43 % $NaBEt_4$ in water	pH 4.1	32
Hg^{2+}	$NaBH_4$	1 mL of 0.4 % aqueous $NaBH_4$	pH 4	33
$MeHg^+$	$LiB(C_2H_5)_3H$	0.1 % solution of $LiB(C_2H_5)_3H$ in THF	pH 4	33
TRISn[a] $Me_xSn^{(4-x)+}$	$NaBH_4$	2 × 1 mL of 1 % aqueous $NaBH_4$	pH 6.5 with 4 mL of 2 M Tris-HCl	34
TRISn $Me_xSn^{(4-x)+}$	$NaBH_4$	1 mL of 4 % $NaBH_4$ in 0.02 M NaOH	pH 2 with 0.2 mL of 5 M HNO_3	35
n-$Bu_xSn^{(4-x)+}$ Et_3Sn^+	$NaBH_4$	2 × 2.5 mL of 6 % aqueous $NaBH_4$	pH 1.6 with 2 mL of 5 M HNO_3	36

[a]TRISn, total recoverable inorganic tin; THF, tetrahydrofuran.

redox speciation of inorganic species of As (III and V), Sb(III and V) and Se(IV and VI). It is also very efficient for most low boiling alkylated species such as the mono-, di-, and trimethylated forms, monomethylarsonate (MMA) and dimethylarsinate (DMA); dimethylselenide, dimethyldiselenide and diethylselenide; methyl-, ethyl-, and especially butyltin (including tetrabutyltin); inorganic, methyl-, dimethyl-, and diethylmercury; as well as inorganic, methyl- and ethyllead species, etc. However, it should be noted that species such as arsenobetaine, arsenocholine, arsenosugars and selenomethionine or selenocysteine require a wet digestion procedure because these compounds have higher organometallic forms that are not directly amenable to gaseous derivatizing methods.

2.3 Chromatography coupled to flame AAS

Combining gas chromatography (GC) with atomic detectors produces powerful performance instruments for speciation analysis. This hyphenated technique has been extensively studied in recent years. Its main advantage over liquid chromatography is that the gas stream emerging from GC can be readily introduced to flame AAS so as to overcome the problems associated with the overall insensitivity of the flame system. The nebulizer of a flame AAS apparatus is only around 5–10 % efficient for liquid samples. If the gas emerging from the GC can be introduced bypassing the nebulizer, an increase in sensitivity can be achieved. The main disadvantages of GC coupled to flame AAS for element speciation are that many of the organometallic compounds are nonvolatile and therefore have to be derivatized prior to gas chromatographic determination. The samples must also be thermally stable and not break down at the oven temperature used in GC. The transfer lines linking the two instruments must also be heated in order to prevent condensation of the analyte. There must be no dead volume or cold areas in the lines. It was therefore applicable only for a limited number of determinations. Typical examples can be found in Table 5.1.2.

In comparison with GC coupled to flame AAS, coupling liquid chromatography to flame AAS is

Table 5.1.2. Element speciation using GC coupled to flame AAS.

Analyte	Chromatography	Detection limits (DL)	Reference
TRISn $Me_xSn^{(4-x)+}$	Chromosorb GAW-DMCS 45–60, 3 % SP2100	DL 20–25 pg for Sn. Reagent: $NaBH_4$	32
Organotin	Cryogenic trapping GC	Interferences were found to be a problem and the reasons why are discussed	37
Organotin	Ethylation GC	DL 2–4 ng/g for methyltin and butyltin	38
Organotin	Ethylation CT/GC	DL $MeSn^{3+}$ 135 pg	39
Tributyltin	Hydride generation GC	DL 0.1–1 ng L^{-1}	40
$Me_xPb^{(4-x)+}$	Chromosorb WAW-DMCS 80–100, 10 % SP2100	DL 9–10 pg as Pb Reagent: $NaBEt_4$	32
Me_4Pb	GC	DL 13 pg as Me_4Pb	41
Me_4Pb Me_3EtPb Me_2Et_2Pb $MeEt_3Pb$ Et_4Pb	Packed column, 3 % OV-101 on gas chromQ	DL 12–25 pg as Pb	42
As(III) As(V) MMA DMA	Glass beads (40 mesh)	DL 19–61 pg as As. Reagent: $NaBH_4$	43
Arsenite Arsenate MMAA DMAA TMAO	Hydride generation, trapping, GC	Absolute detection limits in the range of 0.1–0.5 ng for all compounds	44
As(III) As(V) DMA MMA TMAO	Hydride generation, GC	DL 200–400 and 2–10 ng L^{-1} for inorganic (As(III), As(V)) and methylated (DMA, MMA, TMAO) arsenic species, respectively	45
Hg^{2+} $MeHg^+$	Ethylation, cryogenic trapping, GC Chromosorb W-HP (60/80 mesh size, 10 % SP2100)	Potential artifacts and interfering compounds were studied during the speciation analysis	46
Methyl-, ethyl- and phenylmercury	Hydride generation, GC with a Supelco SPB-1 capillary column	DL 16 ng, 12 ng and 7 ng for methyl-, ethyl- and phenylmercury, respectively	47

MMA, monomethylarsonate; DMA, dimethylarsinate; TMAO, trimethylarsine oxide; TRISn, total recoverable inorganic tin; Me, methly; Et, ethyl.

more popular for element speciation because it offers several advantages over GC. One of the major advantages is that the analyte can be separated at ambient temperature with no need for derivatizing. This not only shortens the sample throughput time but it also reduces possible losses during the process. Also, there are more variable operational parameters; both the stationary and the mobile phase can be varied simultaneously to achieve better separation. A large variety of stationary phases are available and ion-exchange, normal and reverse phases, as well as gel permeation chromatography allow the separation of ions, organometallics of low and high molecular mass as well as metal–protein compounds. The eluent from HPLC can easily be introduced into the nebulizer for aerosol formation in order to evaporate in the flame for atomization. A typical example of HPLC coupled to FAAS for the speciation of methyl- and ethyltin compounds has been described by the ref. 48: An ODS Spherisorb S 5W column (250 mm × 3.0 mm i.d.) is used for the separation of the species. The mobile phases were acetone/pentane (3 + 2) at 1.0 ml min^{-1} for methyltin compounds and acetone/pentane (7 + 3) at 1.2 ml min^{-1} for ethyltin compounds. The AAS detection was carried out with an N_2O–C_2H_2 flame. The detection limits were 11–19 ng Sn for 50 μL sample injection. Applications using liquid chromatography coupled to flame AAS are summarized in Table 5.1.3.

The difficulty in coupling HPLC to FAAS is the balance of optimal flow rates between HPLC separation and AAS detection. The common flow rate for HPLC is around 1–1.5 mL min^{-1} but the flow rate for FAAS is much higher, causing starvation. Although an additional solvent can be introduced into the nebulizer at the end of the HPLC column [67], this leads to undesirable sample dilution. Another possibility is to attach a Teflon funnel [68] to the nebulizer or to introduce a small T-piece [69] into the transfer line.

An important disadvantage in interfacing HPLC with flame AAS is the efficiency of sample introduction; only around 5 % of the sample could be introduced into the flame for detection. Another disadvantage of this interface is the dispersion, not only in the HPLC column but also in the interface tube and FAAS detector, which decreases the sensitivity and resolution of the HPLC. The sensitivity could be greatly improved by using hydride generation (HG) for sample introduction, although this is limited only to the elements that can form hydrides, such as As, Sb, Se, Sn and Pb.

2.4 Chromatography coupled to hydride generation AAS

Figure 5.1.2 [59] shows a diagram of the chemical hydride generator. After HPLC separation, the eluent is introduced into the hydride generator and mixed first with hydrochloric acid, then with 1–5 % $NaBH_4$ solution. The gaseous hydrides formed in the reaction coil are separated in a gas–liquid separator, introduced by inert gas flow into the heated quartz absorption cell and detected by AAS.

The most popular use of HPLC-HGAAS has been for the speciation of reducible arsenic species [59, 70]. Four reducible arsenic species, e.g. As(III), As(V), monomethylarsonate (MMA) and dimethylarsinate (DMA) could be separated by using an anionic exchange column or a C18 column with ion-pair reagent of tetrabutylammonium ion as the mobile phase. The gas–liquid mixture upon derivatization can be introduced into the gas–liquid separator and the hydride can be detected by AAS. With this system, improved detection limits can be achieved through the removal of matrix interferences. The system is suitable for a wide range of samples, including urine and serum [71–73]. A similar set-up can be used for the speciation of antimony, tin, lead, selenium, etc. It could also be used for mercury, although Hg does not form a hydride. The eluting mercury compounds are converted into Hg° vapour by reduction with $SnCl_2$ and sweeping the vapour through a quartz-windowed cell aligned in the light path of AAS. A summary of several applications based on HPLC-HGAAS is given in Table 5.1.3.

A thermochemical hydride generation interface has been developed for HPLC-AAS by Blais and coworkers. This on-line interface is

Table 5.1.3. Application of hyphenated systems using liquid chromatography coupled to atomic absorption spectrometry.

Analyte	Chromatography	Detector	Comments	Reference
As	Dionex anion exchange 500 mm × 3 mm	HG-FAAS	Detection limit (DL) 3–16 ng mL^{-1}	49
Seven As species	IC-HPLC	FAAS	Hydrogen–argon entrained flame, with a slotted tube atom trap for signal enhancement. Analysis of aqueous extracts of soil samples from a polluted land site	50
Cd-metallothionein	HPLC, elution with linear gradient of Tris buffer	AAS	DL 5 μg g^{-1}	51
Zn^{2+} and Cd-metallothionein	HPLC	FAAS	New design of thermospray nebulizer was assessed	52
Cr^{3+}, Cr^{6+}	FI, microcolumn packed with activated alumina (acidic form)	FAAS	Recovery of 90–106 % for natural water	24
Fe^{2+}, Fe^{3+}	Flow injection (FI), C_{18}-modified silica column	FAAS	Fe^{3+} passes straight to detector, whereas Fe^{2+} is trapped as Fe^{2+}-ferrozine and then eluted with methanol	23
Hg	Develosil-ODS (30 μm) precolumn, and STR-ODS-H (5 μm) column	FAAS	DL 0.1 ng	53
Pb	μBondapak C_{18} column	FAAS using air–C_2H_2 flame	DL 10 ng	54
Pb species	Chelex 100. Spheron oxin. Amberlite XAC-2, C_{18} and cellulose sorbents modified with phosphoric acid and carboxymethyl groups	FAAS	DL 0.17 μg L^{-1} Cellulose sorbents were found to have the best retention characteristics	55
Sn	ODS Spherisorb S5W, 250 mm × 3.0 mm i.d.	FAAS using N_2O–C_2H_2 flame	DL 11–19 ng	48
Arsenic species	An anion exchange and a CAS1 ion exchange column connected in series	HGAAS	DL 1.6 ng mL^{-1} for As(III), As(V), MMA and *p*-APA[a] and 1.9 ng mL^{-1} for DMA, AsB and AsC	56
Arsenic species in serum of uraemic patients	Cation exchange liquid chromatography	UV photo/oxidation and HGAAS	DL 1.0, 1.3, 1.5 and 1.4 μg L^{-1} of arsenic for MMA, DMA, AsB and AsC	57
AsB, AsC, DMA	Cation exchange chromatography using a new solid-phase type based on the CBC technology	UV/MW reactor to the HGAAS	DL about 1 μg L^{-1} for each arsenic species	58
Arsenic species in human serum	Reversed phase ion-pair chromatography, polymer-based anion exchange chromatography and silica-based anion exchange chromatography	HGAAS	DL 0.49, 0.44, 0.92 and 0.40 μg L^{-1} for As(III), As(V), MMA and DMA in serum, respectively.	59
Inorganic arsenic and selenium	Separation of ions by HPLC using a phosphate buffer	HGAAS	Contaminated ground water samples	60
(a) Trace concentrations of As, Cd, Pb and Se (b) Various arsenic and selenium compounds	A reversed phase (C-18), ion-pair (tetrabutylammonium) HPLC procedure for the separation of four arsenic species	HGAAS or ETAAS	DL 0.004 μg L^{-1} for both As and Se	61
Inorganic and total mercury	HPLC	CVAAS	Mussel tissue	62

Table 5.1.3. *(continued)*

Analyte	Chromatography	Detector	Comments	Reference
Methylmercury and inorganic mercury	A reversed phase C-18 column	CVAAS	Quantitative recovery for both inorganic mercury and methylmercury from a spiked natural water sample	63
Sb(III) and Sb(V)	HPLC, Hamilton PRP-X100	HGAAS-ICP-MS	The detection limits were 5 and 0.6 ng per 100 μL sample for Sb(III) and Sb(V)	64
Sb(III) and Sb(V) in wastewaters	Two HPLC columns of different lengths (PRP-X 100, 250 mm × 4.1 mm i.d. and PRP-X 100, 100 mm × 4.1 mm i.d.)	HGAAS	DL 1.0 and 0.8 $\mu g L^{-1}$ for Sb(V) and Sb(III), respectively	28
Inorganic and organic antimony compounds	Dionex AS14 for the separation of Sb(V) and Sb(III); ION-120 column $TMSbCl_2$ and Sb(V)	FI-HGAAS	Detection limits of 0.4, 0.7, and 1.0 $\mu g L^{-1}$ for $TMSbCl_2$, Sb(III), and Sb(V)	65
Organic and inorganic selenium species in urine	A vesicle mediated HPLC	Microwave-HGAAS or ICP-MS	DL ranged between 1.0 and 5.3 $\mu g L^{-1}$	66

[a] *p*-APA, *p*-aminophenylarsenate; $TMSbCl_2$, trimethylantimony dichloride.

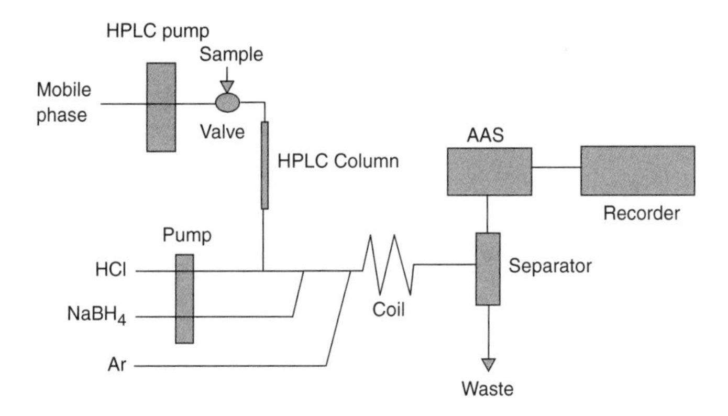

Figure 5.1.2. Diagram of HPLC-HGAAS system for As speciation.

based on thermospray nebulization of the HPLC methanolic eluent, pyrolysis of the analyte in a methanol–oxygen flame, gas-phase thermochemical hydride generation using excess hydrogen, and cool diffusion flame atomization of the product in a quartz cell mounted in the AAS optical beam. It has been used for the determination of arsenic species and selenonium compounds [74, 75]. The low cost, high reproducibility and relatively low detection limits of this system make it suitable for speciation study.

Mineralization may sometimes be necessary for organometallic compounds because some of these species cannot produce hydrides under reducing conditions. On-line digestion, based on microwave digestion or UV photolysis has been extensively studied in recent years. Examples include the speciation of organic arsenic, tin and selenium species [76]. These approaches allow highly sensitive determinations of analyte species that do not form volatile hydrides by hydride generation. Detection limits down to 1.5–2.0 $ng mL^{-1}$

can be achieved for organoarsenic compounds, such as arsenobetaine (AsB), arsenocholine (AsC), trimethylarsine oxide (TMAO), and tetramethylarsonium ion ($TMAs^+$) [77].

Although microwave-assisted digestion has gained wide acceptance as a rapid method for sample decomposition in speciation analysis [61, 78], it is not without problems. In-stream treatment with a mixture of oxidants at temperatures reaching 200–300 °C to achieve complete dissolution, may produce a high pressure in the tubing with the overheated solution. This is even more serious when $NaBH_4$ solution (to produce hydride) has been mixed in. A cooling system is therefore necessary after the microwave digestion, which causes post-column broadening in the coupled HPLC-AAS.

A simpler on-line digestion method for organometallic species has been developed by using a low power UV lamp as a source of reaction energy [57, 79, 80]. Organometallic compounds, such as arsenobetaine (AsB) and arsenocholine (AsC), which cannot produce hydrides directly by reaction with $NaBH_4$ in acidic medium, can be well digested in 1.5 % $K_2S_2O_8$ solution with the use of a 6 W UV lamp. This procedure allows the HPLC-HGAAS determination of AsB and AsC at levels of 1.5 and 1.4 ng mL^{-1} in serum of uraemic patients [57].

The main problem with UV on-line digestion is that a longer coil and a lower flow-rate have to be used to improve the efficiency. Both cause a broadening of the chromatographic peaks and consequently poorer separation of the species. To improve the separation, an argon segment-flow technique was developed for arsenic speciation using HPLC-HGAAS with UV-assisted digestion [57]. After separation of the arsenic species, argon was injected into the moving carrier stream of the mobile phase immediately after the column. This prevents physical dispersion of the analytes and controls the peak broadening, independently of mobile phase flow-rate, manifold geometry, coil length and diameter. Figure 5.1.3 compares the chromatograms of the unsegmented and segment-flow technique. The segment-flow method displays a marked improvement in resolution.

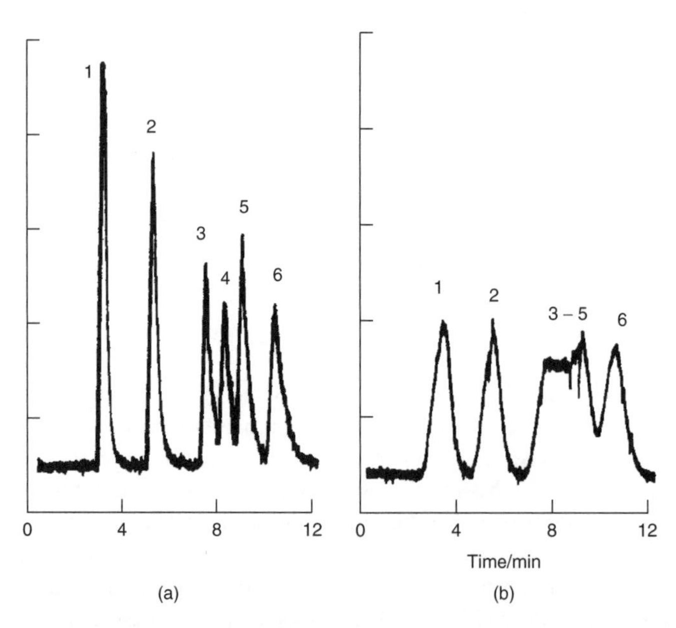

Figure 5.1.3. Segmented and unsegmented flow techniques (column: Dionex Ionpac CS 10, 4 mm × 250 mm; mobile phase: 100 mmol L^{-1} HCl–50 mmol L^{-1} NaH_2PO_4). (a) segmented flow; (b) unsegmented flow. peak 1, MMA; peak 2, DMA; peak 3, AsB; peak 4, TMAO; peak 5, AsC; peak 6, TMAs [57]. MMA, monomethylarsonate; DMA, dimethylarsinate; TMAO, trimethylarsine oxide; AsB, arsenobetaine; TMAs, tetramethylarsonium ion.

3 ELECTROTHERMAL AAS

Electrothermal AAS, which first appeared on the market about 1970, provides enhanced sensitivity because the entire sample is atomized in a short period and the average residence time of the atoms in the optical path is 1 s or more. A few microlitres of sample are first evaporated at low temperature and then ashed at a somewhat higher temperature in an electrically heated graphite tube. After ashing, the current is rapidly increased to several hundred amperes, which causes the temperature to soar to 2000–3000 °C. The atomization of the sample occurs in a few milliseconds to seconds. The absorption is measured in the region immediately above the heated surface. These procedures offer the advantage of high sensitivity for small amounts of sample. Unfortunately, the sequential nature of drying and ashing prior to atomization make it difficult to couple to the continuous flow of a chromatographic eluent. Another obstacle is the matrix effect of samples on ETAAS detection, which makes chemical modification of the matrix and background correction necessary. Until now the on-line coupling of chromatography to ETAAS detection has been rather limited compared to the various types of AAS. The need for sampling the effluent before injection of discrete aliquots has led to the development of two basic procedures of coupling: (1) collection of the effluent in fractions and analysis of each fraction by ETAAS; (2) periodic sampling of the effluent and injection into the furnace.

Procedures based on fraction collection imply their off-line analysis, as there is no physical connection between the chromatographic system and the detector. Thus the chromatograms are not obtained in real time. In spite of this disadvantage, the procedures based on ETAAS detection are simple and sensitive and have been extensively applied to the speciation of Al [81–83], As [20, 84, 85], Cd [86], Cr [87–89], Fe [90, 91], Se [92–94], Sn [95, 96], Sb [97–99], Zn [100], etc. in natural water, urine, blood, soil, sediment, and airborne particles [101].

Some researchers have designed an automated device for effluent sampling in real time. These designs allow a portion of the flowing eluent to be introduced periodically into the furnace for measurement. Two types of autosampler have been designed for this purpose: one is using a PTFE flow-through cell that is fixed into a cup of the autosampler of the AAS. Eluent from the column flows from the bottom of the cell and is drained by one channel of the peristaltic pump. Portions of the flowing effluent are taken periodically by the autosampler arm. The interface between the graphite furnace and the chromatographic system is shown in Figure 5.1.4(a) [102].

In a second approach, the effluent is stored as discrete fractions in the fraction collector. After complete elution of the fractions, they are injected off-line into the graphite atomizer. Figure 5.1.4(b) shows the interface between the graphite furnace and the chromatographic system [103]. The latter

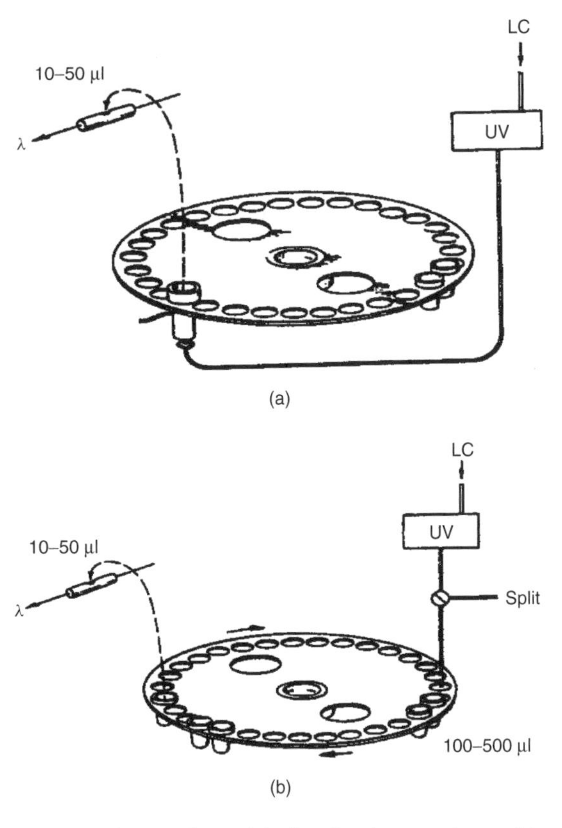

Figure 5.1.4. Two views of the interface between the graphite furnace and the chromatographic system: (a) flow-through cell [102]; (b) off-line injection [103].

method yields more data points per chromatographic peak, resulting in a better signal to noise ratio. Both methods were demonstrated to be suitable for the speciation of different elements. Chromatographic peaks are depicted by several bars, each of them representing the integrated signal for a certain collection period. These systems do not provide a continuous, on-line, real-time analysis. The resolution depends on the number of pulses detected.

An interface similar to thermospray (TS) nebulization in AAS has been proposed for HPLC-ETAAS by Nygren *et al.* [104]. They used a fused silica capillary heated through stainless steel tubing (200 °C) where the effluent is volatilized as aerosols. The aerosols are directly introduced into a glass carbon graphite furnace for detection. This glassy carbon material proved sufficiently temperature resistant up to at least 2000 °C. When the graphite furnace is operated at 2000 °C, a temperature gradient from 2000 °C at the furnace to the temperature below 1400 °C at the narrow part of the glassy carbon tube is formed to protect the fused silica capillary. The method offers the advantage of on-line and real-time analysis, but serious interference is caused by the matrix of eluent and samples since no drying and ashing occur prior to atomization. This technique has been used to determine di- and tributyltin species with a detection limit of 0.5 ng.

A technique for element speciation with an ETAAS detector based on 'permanent modification' has been developed in recent years [105]. This is, in fact, a very promising development in chemical matrix modification in view of increasing sample throughput with 'fast' programmes, reducing reagent blanks, preliminary elimination of unwanted modifier components, compatibility with on-line and *in situ* enrichment, etc. The technique was first studied by applying a single, manual injection of 50 μg Pd and 50 μg Ir on the integrated L'vov platform of the transversely heated graphite atomizer THGA tube, allowing up to 300 complete cycles of hydride trapping and atomization in hydride generation ETAAS determination of As, Bi and Se [106]. The vapour or hydride was introduced by the tip of the quartz capillary tube that was inserted automatically from the outlet of the gas–liquid separator at the centre of the graphite tube. The temperature of the graphite furnace during atomization was increased to 2000 °C over seconds using maximum power. This technique allows one to apply the 'fast' programmes, which could therefore be suitable for coupling the separation technique to ETAAS with increasing sample throughput. Table 5.1.4 shows examples of permanent modification in ETAAS for element speciation.

Despite the better overall sensitivity of ETAAS, coupling with chromatographic separation is still

Table 5.1.4. Examples of permanent modification in ETAAS for element speciation.

Analyte	Matrix	Modifier	Comments	Reference
Si	Serum	W-treated pyrocoated GT	Anion exchange HPLC-ETAAS for speciation; 600 μL fractions collected; 10 μL injections	107
Sn	Fruits, vegetables	Mg; Pd; $NH_4H_2PO_4$; Ti- or Zr-treated GT	Speciation of tricyclohexyltin hydroxide after $CHCl_3$ extraction; Zeeman STPF or Ti-treated tubes	108
As	Aquatic plant, biological tissues, urine, water	Ir-treated GT	On-line UV photo-oxidation or MWD with FI-HG-ETAAS for 'first-order' speciation; LOD 0.14 ng	109
Se	Soil	Pd-treated GT	GC–trapping–ETAAS for speciation of $(CH_3)_2$ Se, $(C_2H_5)_2$ Se and $(CH_3)_2Se_2$	110
Se	Aq. Solution	BN-coated GT vs. Pd-Mg $(NO)_2$ modifier addition	Se(IV), Se(VI) and selenomethionine studied	111
Se	Seawater	Ir-treated platform	Speciation protocol: Se(IV) determined without pretreatment; total Se after pre-reduction in 5 mol L^{-1} HCl	112
Sn	Water, sediments	110 μg of Zr or 240 μg of W and 2 μg of Ir permanent modifier	FI-HG-ETAAS for Sn(IV), monomethyltin, dimethyltin, trimethyltin, diethyltin and monobutyltin	113

troublesome. The interface is difficult, mainly owing to the discontinuous nature of the ETAAS. So far, ETAAS has been applied to the speciation studies for about 20 elements in various environmental matrices. Among them, Cr has been extensively studied in different environment samples. As, Se, Sb, Pb, Sn, Hg and Al are the elements that have been more or less widely analysed. Cu, Cd, Zn, Fe and Ni have been studied somewhat and little work has been done on Co, Si, V, Mg, Mo, and Tl.

4 PLASMA AES

Element speciation based on the plasma source has the advantage of accepting the continuous flow of the HPLC eluent and is therefore easier to hyphenate. Another advantage is its multielement capacity: Plasma AES can detect both metals and nonmetals. In the literature, different plasma sources have been proposed for element speciation coupled with chromatographic separation, e.g. microwave-induced plasma (MIP), direct current plasma (DCP) and inductively coupled plasma (ICP). The best one is still ICP because the continuous HPLC flow quenches the discharge of the MIP source, causing difficulties on the interface. DCP appears to be better than MIP since it provides a more stable plasma, especially with the introduction of mixed organic–aqueous eluents. However, with DCP direct interfacing, the detection was limited to levels above 100 ng mL^{-1} [114], several times higher than with an ICP source.

4.1 ICP source

The ICP source was first developed in the 1960s and is becoming the most important source for atomic spectrometry [115]. The argon-based plasma is compatible with aqueous aerosols and offers high energy for drying, dissociation, atomization and ionization of the analytes. The temperatures reached by an argon ICP vary from 4500 to 10 000 K, depending on the definition of 'temperature' (kinetic temperature, electron temperature, atomization temperature, ionization temperature) and the location inside the plasma. An ICP is therefore called a nonthermal equilibrium plasma. The temperature in the inner channel, used for analytical purposes, is about 5500–6500 K, high enough to destroy all molecular bonds and even to ionize many elements.

The sample introduction system in the ICP source is usually designed for liquids, but solids or gases are also possible. The nebulizer and the spray chamber constitute the most critical units of an ICP-chromatography set-up for element speciation. The standard configuration of an ICP includes a pneumatic nebulizer for the formation of aerosols and a spray chamber for the separation of the droplets by size. Only small droplets should be able to enter the plasma, otherwise it becomes unstable or extinguishes. Therefore, only 1–5 % of the sample reaches the plasma torch with the pneumatic nebulizer. The first requirement for a nebulizer is its compatibility with flow rate and eluent composition. Water-based eluents are often deleterious to cones because of their salt content. Eluents containing organic modifiers or organic eluents tend to affect plasma stability because of the increased solvent vapour pressure. Therefore, a frequent observation has been that there is poor tolerance of the ICP for the mobile phases commonly used in HPLC, particularly with ion-pairing or size exclusion LC separation. Ways to overcome these problems are directly related to improve rates of liquid consumption and efficiency of nebulization of the analyte into the plasma. Typical examples of nebulizers used in interfacing HPLC-ICP include cross-flow and concentric pneumatic nebulizers, ultrasonic nebulizers and thermospray nebulizers.

4.1.1 Interface based on the concentric and cross-flow pneumatic nebulizers

Both concentric and cross-flow nebulizers are widely used in ICP-AES. For the concentric nebulizer, the analyte solution is fed through a capillary surrounded by a second capillary, while the nebulizer gas flows through the space between them and produces the aerosol. The concentric nebulizers are robust and allow high operation stability because of their monolithic construction, but they tend to clog, especially with low aspiration gas flow rates. Another kind of nebulizer, cross-flow nebulizer is designed by feeding the solution through a vertically mounted capillary, nebulized by the gas flow from a horizontal capillary which ends close to the tip of the former one [116]. Although the cross-flow design is more tolerant of solutions with high salt contents, it is subject to periodic blockage by salt deposits and to tip blockage as a result of salting out, caused by a reduction in temperature.

Concentric and cross-flow nebulizers have been extensively utilized for element speciation by coupling HPLC to ICP-AES. For instance, a method was developed for arsenic speciation in marine organisms by coupling an anion exchange column with ICP-AES. The eluate was passed directly to a concentric nebulizer and detected by ICP-AES [117]. The difference between axial and radial viewing for the speciation of polar silicon compounds using HPLC-ICP-AES with a cross-flow nebulizer has also been documented [118]. It appeared that no improvement in detection limit was achieved using the axial system. However, since the flow rate is around $1\,mL\,min^{-1}$, the common design is not ideal for the purpose because of the dead volume or the liquid and gas injection areas. Therefore, coupling microcolumn HPLC and CE to ICP-AES by using a concentric nebulizer was developed by several authors. A CE-ICP interface using a modified concentric nebulizer has been described and applied to the separation and correlation of metal species in metallothioneins of rabbit liver [119]. By replacing the central tube of the concentric nebulizer, a modified concentric nebulizer was developed with the CE capillary as the interface to couple to the ICP-AES. The detection limit of Cr, based on peak area, is approximately $10\,\mu g\,L^{-1}$ [120].

Another type of concentric nebulizer for the simultaneous determination of both hydride forming and nonhydride forming elements by ICP-AES has been described [121]. The large droplets from a concentric nebulizer are trapped and react with a borohydride solution pumped into a small hydride generator fitted in the spray chamber. The hydrides that are formed and the sample aerosol enter the plasma simultaneously, while the nonvolatile reaction products and surplus reductant flow to the drain. This system provides a more than 20-fold improvement in detection limits for the hydride forming elements without degrading the performance for other elements.

Microconcentric nebulizers have attracted much attention as alternative sample introduction devices for ICP-AES. The main difference with respect to conventional concentric nebulizers is a conspicuous reduction in their critical dimensions. This reduction allows a more efficient gas–liquid interaction at low liquid flow rate, thus improving the nebulizer performance in terms of better efficiency of nebulization and transportation of sample. The main advantage of microconcentric nebulizers is that they can give rise to detection limits similar to or better than those obtained with conventional nebulizers, with liquid flow rates 10–40 times lower that result in increased transport efficiency and/or enhanced signal stability [122]. A study of five nebulizers associated to three spray chambers in conjunction with HPLC-ICP-AES found that best signal-to-noise ratios were obtained by using a microconcentric nebulizer and a cyclone spray chamber without affecting the chromatographic resolution. Response of the ICP to each species of As(III), As(V), DMA and MMA was strongly affected by the selection of the nebulizer and spray chamber [123].

The direct injection nebulizer is a microconcentric pneumatic nebulizer that allows direct nebulization of typically $100\,\mu L\,min^{-1}$ of analyte solution into the plasma. The direct injection nebulizer offers a 100 % analyte transport efficiency, reduced memory effect, rapid response time and good precision. The speciation of organic and inorganic selenium in a biological certified reference material, using microbore ion-exchange chromatography coupled to ICP-AES via a direct injection nebulizer has been reported. Separation for selenomethionine, selenite, selenate and selenocystine can be obtained within 5 min [124]. The speciation of chromium using HPLC-ICP-MS via a direct injection nebulizer with an absolute detection limit of 3 pg for both Cr(III) and Cr(VI) has been reported. However, the direct injection nebulizer requires a much more expensive setup than a common concentric pneumatic nebulizer.

4.1.2 *Interface based on ultrasonic nebulizers*

Sample introduction by utilizing ultrasonic nebulizers has become popular in recent years. In an ultrasonic nebulizer, a longitudinal acoustic wave is produced by an oscillator coupled to a transducer. The latter is oriented in such a way that the direction of wave propagation is perpendicular to the interface between the gas and the sample liquid to be nebulized. Aerosol formation occurs by the action of 'geysers' when the amplitude of the waves becomes sufficiently large [116]. Ultrasonic nebulizers are not subject to clogging, and yield detection limits that are currently 10–50 times better than those of pneumatic nebulizers, due to their high aerosol transport efficiencies (up to 30 %). Moreover, a system conjugating an ultrasonic nebulizer with flow-injection ICP-AES yielded a sensitivity enhancement factor of 225 for the speciation of vanadium(IV) and vanadium(V) in river water samples [125].

Applicability of an ultrasonic nebulizer coupled to different chromatographic techniques has been explored by several authors. Ion exchange chromatography coupled to ICP-AES by using an ultrasonic nebulizer was used for the speciation of arsenic(V) and monomethylarsonate in the $ng\,L^{-1}$ range [126]. Ultrasonic nebulization ICP-AES in conjunction with size exclusion chromatography was also studied. It found that speciation of Ca, Cu, Fe, Mg, Mn and Zn in human milk could be applied to assess the concentration range and binding pattern of the elements in the milk of 60 lactating mothers [127].

4.1.3 *Interface based on thermospray nebulizer*

The primary reason to employ alternative methods of nebulization such as thermospray as an interface between liquid chromatography and *ICP-AES* is to achieve better sensitivity and better limits of detection, since ICP-AES is not sensitive enough for trace element speciation in many samples, when compared to ICP-MS. Thermospray nebulization is accomplished by pumping a liquid sample at moderately high pressure through an electrothermally heated capillary. The aerosol droplet size produced by the thermospray is much smaller than those of pneumatic nebulization, which is favourable for desolvation, volatilization and atomization, as mentioned in Section 2.1.

In a review of new developments in thermospray sample introduction for atomic spectrometry from 1992 to 1997 [128], HPLC thermospray ICP-AES was summarized for speciation of tin [129], chromium [130], Se [131] etc. The general principles and operational characteristics of thermospray have also been discussed, together with a review of the applications of thermospray sample introduction with atomic spectrometry detection [132]. Speciation of Cr(III) and Cr(VI) was carried out by using thermospray sample introduction with ICP-AES and parameters such as control temperature, pH, and pump rate were also studied [133]. The direct speciation of selenite and selenate with thermospray sample introduction coupled to ICP-AES has also been developed [134].

4.2 Other plasma sources

Although the majority of sample introduction systems described in the literature are concerned with ICP-AES detection, other plasma sources have also been exploited for element speciation. Particularly, low power MIP has been of interest as an excitation source, because it provides good sensitivity for a number of elements and is inexpensive and easy to operate. Some limitations of the MIP discharge have been established, *i.e.* low tolerance of the introduction of even a limited amount of sample and instability when operated at low power. Hence, applications on this topic are limited to a few papers, which describe several designs of interfaces for efficient liquid introduction or special modifications of cavities in order to increase the plasma stability [125–137].

To overcome the problems of low tolerance for sample introduction and instability for plasma caused by the direct introduction of liquid samples, cold vapour and hydride generation techniques were applied to the speciation in LC-MIP-AES. These techniques offer the possibility for volatile analytes to enter the plasma without the interfering mobile phase. This was achieved by on-line coupling of liquid chromatography to low power argon microwave-induced plasma for the speciation of mercury and arsenic compounds with continuous cold vapour or hydride generation techniques [138]. Signal enhancements of around 100 % were found in micelles of cetyltrimethylammmonium bromide. The MIP generated at reduced pressure might have more resistance against molecular gases such as hydrogen produced in the hydride generator than atmospheric pressure plasmas [139]. For mercury, detection limits were found to be between $0.15\,\mu g\,L^{-1}$ for inorganic Hg and $0.35\,\mu g\,L^{-1}$ for methylmercury. For arsenic, the authors cite values between 1 and $6\,\mu g\,L^{-1}$.

In comparison with the hyphenation difficulties of liquid samples to MIP-AES, GC is advantageous for the coupling to MIP-AES. First, the analytes can be introduced quantitatively into the plasma in the gaseous form and no nebulization and drying is necessary. Secondly, GC separations can use helium as the carrier gas, which is ideally suited to helium MIPs. The interface can be built easily with a simple heated transfer line with low dead volume. On the other hand, GC requires the derivatizing of analytes prior to analysis, as the native species normally occur in the ionic state and lack the necessary volatility and thermal stability. Table 5.1.5 summarizes the typical examples of GC-MIP-AES for element speciation.

Direct current plasma (DCP) appears to offer certain advantages with regard to chromatography coupling. This source involves a low voltage (10–50 V), high current (1–35 A) discharge, stabilized by the flowing inert gas, usually argon. The interface includes a quartz jet tube to convey the GC effluent directly into the DCP plume. The DCP remains stable in a high solvent background and its design does not require a venting valve, especially with the introduction of the mixed organic–aqueous eluents. GC coupled to DCP emission spectroscopy has been used for the determination of organotins in fish and shellfish [147]. A similar device was employed for the determination of methylmercury [148]. However, with DCP direct interfacing, there is the problem of high detection limits, which is the reason for the scarcity of publications on element speciation using DCP. Speciation analysis with different plasma sources has been reviewed [149].

Table 5.1.5. GC-MIP-AES for element speciation.

Species	Samples	Comments	Detection limit	Reference
Organolead compounds	Snow samples	*In situ* propylation with sodium tetrapropylborate derivatization	$0.15–0.21\,ng\,kg^{-1}$ (As, Pb)	139
Organolead compounds	Tapwater and peat	*In situ* butylation with tetrabutylammonium tetra-butylborate for derivatization	sub-$ng\,L^{-1}$ range	140
Mercury	Certified reference materials	Purge-and-trap multicapillary GC	$0.01\,pg\,mL^{-1}$ for methylmercury	141
Butyltin organomercury tetraalkyllead	Sediment and gasoline	Multicapillary GC	sub-$pg\,L^{-1}$ range	142
Organolead compounds	Gasoline	Multicapillary GC	$<1\,ng\,mL^{-1}$	143
Mercury species	Natural water	Packed column GC for large volume injections	$8\,pg\,L^{-1}$	144
Mercury species	Canal waters	Sulphydryl cotton microcolumn for preconcentration	$10\,ng\,L^{-1}$ for methyl- and ethylmercury; $16\,ng\,L^{-1}$ for inorganic mercury	145
Organotin compounds	Sediment samples	Microwave-assisted leaching	$2\,ng\,g^{-1}$	146

5 CONCLUSION

AAS and AES have been coupled to different separation techniques for elemental speciation. The potential of these hyphenated systems is strongly dependent on the selection of the detectors, the interfaces and the design for sample introduction. It remains a problem to develop an interface for on-line coupling ETAAS to chromatographic systems, although this type of detector offers high sensitivity with an extensive range of elements for detection when compared with other detectors such as flame AAS, HGAAS and ICP-AES. Commercial instruments for element speciation should also be developed, since home-made hyphenated instruments have severely restricted the application of these methods in routine environmental and clinical laboratories for element speciation.

6 REFERENCES

1. Cornelis, R., *Ann. Clin. Lab. Sci.*, **26**, 252 (1996).
2. Cornelis, R., Crews, H., Donard, O., Ebdon, L., Pitts, L. and Quevauviller, P., *J. Environ. Monit.*, **3**, 97 (2001).
3. Bernhard, M., Brinckman, F. E. and Sadler, P. J., *The Importance of Chemical Speciation in Environmental Processes*, Springer, Berlin, 1986.
4. Caroli, S., *Element Speciation in Bioinorganic Chemistry*, John Wiley & Sons Inc., New York, 1996.
5. Donard, O. F. X. Hyphenated techniques applied to the speciation of organometallic compounds in the environment, in *Environmental Analysis, Applications and Quality Assurance*, Barcelo, D. (Ed.), Elsevier, New York, Chapter 16, 1993.
6. Donard, Trace metal speciation: finally, correctly addressing trace metal issues, in the Report on the consultants' meeting to: *Outline of a Co-ordinated Research Project on Trace Element Speciation to Enhance Technology Transfer in This Field by the Development of Suitable Methodologies and/or the Production of Suitably Characterized Reference Materials*, International Atomic Energy Agency, Vienna, November 23–26, 1998.
7. Ellis, L. A. and Roberts, D. J., *J. Chromatogr. A*, **774**, 3 (1997).
8. Havezov, I., *Fresenius' J. Anal. Chem.*, **355**, 452 (1996).
9. Cornelis, R., De Kimpe, J. and Zhang, X. R., *Spectrochim. Acta, Part B*, **53**, 187 (1998).
10. Matusiewicz, H., *Spectrochim. Acta B*, **52**, 1711 (1997).
11. Pasullean, B., Davidson, C. M. and Littlejohn, D., *J. Anal. At. Spectrom.*, **10**, 241 (1995).
12. Garbos, S., Bulska, E. and Hulanicki, A., *At. Spectrosc.*, **21**, 128 (2000).
13. Gustavsson, A., *Spectrochim. Acta, Part B*, **42**, 118 (1997).
14. Nygren, O., Nilsson, C. A. and Gustavsson, A., *Spectrochim. Acta, Part B*, **44**, 589 (1989).
15. Gustavsson, A. and Nygren, O., *Spectrochim. Acta, Part B*, **42**, 883 (1987).
16. Saverwijns, S., Zhang, X. R., Vanhaecke, F., Cornelis, R., Moens, L. and Dams, R., *J. Anal. At. Spectrom*, **12**, 1047 (1997).
17. Chang, P. P. and Robinson, J. W., *Spectrosc. lett.*, **30**, 193 (1997).
18. Chang, P. P. and Robinson, J. W., *Spectrosc. lett.*, **29**, 1469 (1996).
19. Berndt, H. and Yanez, J., *Fresenius' J. Anal. Chem.*, **355**, 555 (1996).
20. Russeva, E., Havezov, I. and Detcheva, A., *Fresenius' J. Anal. Chem.*, **347**, 320 (1993).
21. Tsakovsky, S., Ivanova, E. and Havezov, I., *Talanta*, **41**, 721 (1994).
22. Naghmush, A. M., Pyrzynska, K. and Trojanowicz, M., *Talanta*, **42**, 851 (1995).
23. Krekler, S., Frenzel, W. and Schulze, G., *Anal. Chim. Acta*, **296**, 115 (1994).
24. Sperling, M., Xu, S. K. and Welz, B., *Anal. Chem.*, **64**, 3101 (1992).
25. Kolbl, G., Kalcher, K. and Irgolic, K. J., *Anal. Chim. Acta*, **284**, 301 (1993).
26. Chau, Y. K., Wong, P. T. S. and Goulden, P. D., *Anal. Chem.*, **47**, 2279 (1975).
27. Braman, R. S., Johnson, D. L., Craig, C., Foreback, C. C., Ammons, J. M. and Bricker, J. L., *Anal. Chem.*, **49**, 621 (1977).
28. Satiroglu, N., Bektas, S., Genc, O. and Hazer, H., *Turk. J. Chem.*, **24**, 371 (2000).
29. Gallignani, M., Valero, M., Brunetto, M. R., Burguera, J. L., Burguera, M. and Petit de Peña, Y., *Talanta*, **52**, 1015 (2000).
30. Chatterjee, A., Shibata, Y. and Morita, M., *Microchem. J.*, **69**, 179 (2001).
31. Hambrick, G. A., Freolich, P. N., Andreae, M. O. and Lewis, B. L., *Anal. Chem.*, **56**, 421 (1984).
32. Rapsomanikis, S., Donard, O. F. X. and Weber, J. H., *Anal. Chem.*, **58**, 35 (1986).
33. Craig, P. J., Mennie, D., Ostah, N., Donard, O. F. X. and Martin, F., *Analyst*, **117**, 823 (1992).
34. Braman, R. S. and Tompkins, M. A., *Anal. Chem.*, **51**, 12 (1979).
35. Andreae, M. O. and Byrd, J. T., *Anal. Chim. Acta*, **156**, 147 (1984).
36. Randall, L., Donard, O. F. X. and Weber, J. H., *Anal. Chim. Acta*, **184**, 197 (1986).
37. Martin, F. M., Tseng, C. M., Belin, C., Quevauviller, P. and Donard, O. F. X., *Anal. Chim. Acta*, **286**, 343 (1994).

38. Shawky, S., Emons, H. and Durbeck, H. W., *Anal. Commun.*, **33**, 107 (1996).
39. Martin, F. M. and Donard, O. F. X., *Fresenius' J. Anal. Chem.*, **351**, 230 (1995).
40. Ritsema, R., *Mikrochim. Acta*, **109**, 61 (1992).
41. Bergmann, K. and Neidhart, B., *Fresenius' J. Anal. Chem.*, **356**, 57 (1996).
42. Radziuk, B., Thomassen, Y., Van Loon, J. C. and Chau, Y. K., *Anal. Chim. Acta*, **105**, 255 (1979).
43. Howard, A. G. and Comber, S. D. W., *Mikrochim. Acta*, **109**, 27 (1992).
44. Molenat, N., Astruc, A., Holeman, M., Maury, G. and Pinel, R., *Analusis*, **27**, 795 (1999).
45. Guerin, T., Molenat, N., Astruc, A. and Pinel, R., *Appl, Organomet. Chem.*, **14**, 401 (2000).
46. Tseng, C. M., De Diego, A., Wasserman, J. C., Amouroux, D. and Donard, O. F. X., *Chemosphere*, **39**, 1119 (1999).
47. He, B. and Jiang, G. B., *Fresenius' J. Anal. Chem.*, **365**, 615 (1999).
48. Burns, D. T., Glockling, F. and Harriot, M., *Analyst*, **106**, 22 (1981).
49. Ricci, G. R., Shepard, L. S., Colovos, G. and Hester, N. E., *Anal. Chem.*, **53**, 610 (1981).
50. Hansen, S. H., Larsen, E. H., Pritzl, G. and Cornett, C, *J. Anal. At. Spectrom.*, **7**, 629 (1992).
51. Pan, A. H., Wang, Z. and Ru, B., *Biomed. Chromatogr.*, **5**, 193 (1991).
52. Chang, P. P. and Robinson, T. W., *J. Environ. Sci. Health, Part A*, **28**, 1147 (1993).
53. Munaf, E., Haraguchi, H., Ishii, D., Takeuchi, T. and Gota, M., *Anal. Chim. Acta*, **235**, 309 (1990).
54. Messman, J. D. and Rains, T. C., *Anal. Chem.*, **53**, 1632 (1983).
55. Naghmush, A. M., Pyrzynska, L. and Trojanowicz, M., *Talanta*, **42**, 851 (1995).
56. Han, H. B., Liu, Y. B., Mou, S. F. and Ni, Z. M., *J. Anal. At. Spectrom.*, **8**, 1085 (1993).
57. Zhang, X. R., Cornelis, R., De Kimpe, J. and Mees, L., *Anal. Chim. Acta*, **319**, 177 (1996).
58. Sur, R., Begerow, J. and Dunemann, L., *Fresenius' J. Anal. Chem.*, **363**, 526 (1999).
59. Zhang, X. R., Cornelis, R., De Kimpe, J. and Mees, L., *J. Anal. At. Spectrom.*, **11**, 1075 (1996).
60. Rassler, M., Michalke, B., Schramel, P., Schulte-Hostede, S. and Kettrup, A., *Int. J. Environ. Anal. Chem.*, **72**, 195 (1998).
61. Tyson, J. F., *J. Anal. At. Spectrom.*, **14**, 169 (1999).
62. Rio-Segade, S. and Bendicho, C., *Ecotox. Environ. Safe.*, **42**, 245 (1999).
63. Rio-Segade, S. and Bendicho, C., *Talanta*, **48**, 477 (1999).
64. Smichowski, P., Madrid, Y., Guntinas, M. B. D. and Camara, C., *J. Anal. At. Spectrom.*, **10**, 815 (1995).
65. Krachler, M. and Emons, H., *J. Anal. At. Spectrom.*, **15**, 281 (2000).
66. LaFuente, J. M. G., Dlaska, M., Sanchez, M. L. F. and Sanz-Medel, A., *J. Anal. At. Spectrom.*, **13**, 423 (1998).
67. Yoza, N. and Ohashi, S., *Anal. Lett.*, **6**, 595 (1973).
68. Slavin, W. and Schmidt, G. J., *J. Chromatogr. Sci.*, **17**, 610 (1979).
69. Ebdon, L., Hill, S. J. and Jones, P., *Analyst*, **110**, 515 (1985).
70. Cornelis, R., Zhang, X. R., Mees, L., Christense, J. M., Byrialsen, K. and Dyrschel, C., *Analyst*, **123**, 2883 (1998).
71. Zhang, X. R., Cornelis, R., De Kimpe, J., Mees, L. and Lameire, N., *Clin. Chem.*, **44**, 141 (1998).
72. Zhang, X. R., Cornelis, R., De Kimpe, J., Mees, L. and Lameire, N., *Clin. Chem.*, **43**, 406 (1997).
73. Zhang, X. R., Cornelis, R. and Mees, L., *J. Anal. At. Spectrom.*, **13**, 205 (1998).
74. Blais, J. S., Momplaisir, G. M. and Marshall, W. D., *Anal. Chem.*, **62**, 1161 (1990).
75. Blais, J. S., Huyghues-Despointes, A., Momplaisir, G. M. and Marshall, W. D., *J. Anal. At. Spectrom.*, **6**, 225 (1991).
76. Howard, A. G., *J. Anal. At. Spectrom.*, **12**, 267 (1997).
77. Ouyang, J., Shi, Y. and Zhang, X. R., *Chin. J. Anal. Chem.*, **27**, 1151 (1999).
78. Le, X. C., Cullen, W. R. and Reimer, K., *Talanta*, **41**, 495 (1994).
79. Howard, A. G. and Hunt, L. E., *Anal. Chem.*, **65**, 2995 (1993).
80. Tsalev, D. L., Sperling, M. and Welz, B., *Spectrochim. Acta, Part B*, **55**, 339 (2000).
81. Bantan, T., Milacic, R., Mitrovic, B. and Pihlar, B., *Fresenius' J. Anal. Chem.*, **365**, 545 (1999).
82. Bantan, T., Milacic, R., Mitrovic, B. and Pihlar, B., *J. Anal. At. Spectrom.*, **14**, 1743 (1999).
83. Wrobel, K., Gonzalez, E. B. and Sanz-Medel, A., *Trace Elem. Med.*, **10**, 97 (1993).
84. Van Cleuvenbergen, R. J. A., Van Mol, W. E. and Adams, F. C., *J. Anal. At. Spectrom.*, **3**, 169 (1988).
85. Burguera, M. and Burguera, J. L., *J. Anal. At. Spectrom.*, **8**, 229 (1993).
86. Fernandez, F. M., Tudino, M. B. and Troccoli, O. E., *J. Anal. At. Spectrom.*, **15**, 687 (2000).
87. Kelko-Levai, A., Varga, I., Zih-Perenyi, K. and Lasztity, A., *Spectrochim. Acta, Part B*, **54**, 827 (1999).
88. Vassileva, E., Hadjiivanov, K., Stoychev, T. and Daiev, C., *Analyst*, **125**, 693 (2000).
89. Horvath, Z., Lasztity, A., Varga, I., Meszaros, E. and Molnar, A., *Talanta*, **41**, 1165 (1994).
90. Bermejo, P., Pena, E., Dominguez, R., Bermejo, A., Fraga, J. M. and Cocho, J. A., *Talanta*, **50**, 1211 (2000).
91. Vanlandeghem, G. F., Dhaese, P. C., Lamberts, L. V. and Debroe, M. E., *Anal. Chem.*, **66**, 216 (1994).

92. Yan, X. P., Sperling, M. and Welz, B., *Anal. Chem.*, **71**, 4353 (1999).
93. Robles, L. C., Feo, J. C., de Celis, B., Lumbreras, J. M., Garcia-Olalla, C. and Aller, A. J., *Talanta*, **50**, 307 (1999).
94. Gammelgaard, B. and Larsen, E. H., *Talanta*, **47**, 503 (1998).
95. Astruc, A., Dauchy, X., Pannier, F., Potin-Gautier, M. and Astruc, M., *Analusis*, **22**, 257 (1994).
96. Bermejo-Barrera, P., Soto-Ferreiro, R. M., Dominguez-Gonzalez, R. and Bermejo-Barrera, A., *Ann. Chim.*, **86**, 495 (1996).
97. de la Calle Guntinas, M. B., Madrid, Y. and Camara, C., *J. Anal. At. Spectrom.*, **8**, 745 (1993).
98. Garbos, S., Rzepecka, M., Bulska, E. and Hulanicki, A., *Spectrochim. Acta Part B*, **54**, 873 (1999).
99. Farkasovska, I., Zavadska, M. and Zemberyova, M., *Chem. Listy*, **93**, 173 (1999).
100. Guenther, K. and Waldner, H., *Anal. Chim. Acta*, **259**, 165 (1992).
101. Das, A. K. and Chakraborty, R., *Fresenius' J. Anal. Chem.*, **357**, 1 (1997).
102. Laborda, F., Vicente, M., Mir, J. M. and Castillo, J. R., *Fresenius' J. Anal. Chem.*, **357**, 837 (1997).
103. Vickrey, T. M., Howell, H. E. and Paradise, M. T., *Anal. Chem.*, **51**, 1880 (1979).
104. Nygren, O., Nilsson, C. A. and Frech, W., *Anal. Chem.*, **60**, 2204 (1988).
105. Tsalev, D. L., Slaveykova, V. I., Lampugnani, L., D'Ulivo, A. and Georgieva, R., *Spectrochim. Acta, Part B*, **55**, 473 (2000).
106. Shuttler, I., Feuerstein, M. and Schlemmer, G., *J. Anal. At. Spectrom.*, **7**, 1299 (1992).
107. Wrobel, K., Blanco Gonzalez, E., Wrobel, K. and Sanz-Medel, A., *Analyst*, **120**, 809 (1995).
108. Giordano, R., Ciaralli, L., Gattorta, G., Ciprotti, S. and Costantini, S., *Microchem. J.*, **49**, 69 (1994).
109. Willie, S. N., *Spectrochim. Acta Part B*, **51**, 1781 (1996).
110. Jiang, G.-b., Ni, Z.-m., Zhang, L., Li, A., Han, H.-b. and Shan, X.-q., *J. Anal. At. Spectrom.*, **7**, 447 (1992).
111. Krivan, V. and Kueckenwaitz, M., *Fresenius' J. Anal. Chem.*, **342**, 692 (1992).
112. Cabon, J. Y. and Erler, W., *Analyst*, **123**, 1565 (1998).
113. Tsalev, D. L., D'Ulivo, A., Lampugnani, L., Di Marco, M. and Zamboni, R., *J. Anal. At. Spectrom.*, **11**, 979 (1996).
114. Morita, M., Uehiro, T. and Fuwa, K., *Anal. Chem.*, **53**, 1806 (1981).
115. Montaser, A., *Inductively Coupled Plasma Mass Spectrometry*, Wiley-VCH, New York, 1998.
116. Boumans, P. W. J. M., *Inductively Coupled Plasma Emission Spectrometry*, John Wiley & Sons, Inc., New York, 1987.
117. Amran, M. B., Lagarde, F. and Leroy, M. J. F., *Mikrochim. Acta*, **127**, 195 (1997).
118. Ebdon, L., Foulkes, M., Fredeen, K., Hanna, C. and Sutton, K., *Spectrochim. Acta, Part B*, **53**, 859 (1998).
119. Deng, B. Y. and Chan, W. T., *Electrophoresis*, **22**, 2186 (2001).
120. Chan, Y. Y. and Chan, W. T., *J. Chromatogr. A* **853**, 141 (1999).
121. Huang, B. L., *Can. J. Appl. Spectrosc.*, **39**, 117 (1994).
122. Hettipathirana, T. D. and Davey, D. E., *J. Anal. At. Spectrum.*, **13**, 483 (1998).
123. Chausseau, M., Roussel, C., Gilon, N. and Mermet, J. M., *Fresenius' J. Anal. Chem.*, **366**, 476 (2000).
124. Emteborg, H., Bordin, G. and Rodriguez, A. R., *Analyst*, **123**, 245 (1998).
125. Wuilloud, R. G., Wuilloud, J. C., Olsina, R. A. and Martinez, L. D., *Analyst*, **126**, 715 (2001).
126. Ochsenkuhn-Petropulu, M. and Schramel, P., *Anal. Chim. Acta*, **313**, 243 (1995).
127. Bocca, B., Alimonti, A., Coni, E., Di Pasquale, M., Giglio, L., Bocca, A. P. and Caroli, S., *Talanta*, **53**, 295 (2000).
128. Conver, T. S., Yang, J. F. and Koropchak, J. A., *Spectrochim. Acta, Part B*, **52**, 1087 (1997).
129. Van Berkel, W. W., Balke, J. and Maessen, F. J. M. J., *Spectrochim. Acta, Part B*, **45**, 1265 (1990).
130. Roychowdhury, S. B. and Koropchak, J. A., *Anal. Chem.*, **62**, 484 (1990).
131. Laborda, F., de Loos-Vollebregt, M. T. C. and de Galan, L., *Spectrochim. Acta, Part B*, **46**, 1089 (1991).
132. Zhang, X. H., Chen, D., Marquardt, R., John, A. and Koropchak, J. A., *Microchem. J.*, **66**, 17 (2000).
133. Zhang, X. H. and Koropchak, J. A., *Anal. Chem.*, **71**, 3046 (1999).
134. Yang, J. F., Conver, T. S. and Koropchak, J. A., *Anal. Chem.*, **68**, 4064 (1996).
135. Hansen, G. W., Huf, F. A. and DeJong, H. J., *Spectrochim. Acta, Part B*, **40**, 307 (1985).
136. Kollotzek, D., Oechsle, D., Kaiser, G., Tschoepel, P. and Toelg, G., *Fresenius' J. Anal. Chem.*, **318**, 485 (1984).
137. Rosenkranz, B. and Bettmer, J., *Trends Anal. Chem.*, **19**, 138 (2000).
138. Costa-Fernandez, J. M., Lunzer, F., Pereiro-Garcia, R. and Sanz-Medel, A., *J. Anal. At. Spectrom.*, **10**, 1019 (1995).
139. Heisterkamp, M. and Adams, F. C., *J. Anal. At. Spectrom.*, **14**, 1037 (1999).
140. Heisterkamp, M. and Adams, F. C., *Fresenius' J. Anal. Chem.*, **362**, 489 (1998).
141. Pereiro, I. R., Wasik, A. and Lobinski, R., *Anal. Chem.*, **70**, 4063 (1998).
142. Pereiro, I. R., Wasik, A. and Lobinski, R., *J. Chromatogr. A*, **795**, 359 (1998).

143. Pereiro, I. R. and Lobinski, R., *J. Anal. At. Spectrom.*, **12**, 1381 (1997).
144. Hanstrom, S., Briche, C., Emteborg, H. and Baxter, D. C., *Analyst* **121**, 1657 (1996).
145. Mena, M. L., McLeod, C. W., Jones, P., Withers, A., Minganti, V., Capelli, R. and P. Quevauviller, P., *Fresenius' J. Anal. Chem.*, **351**, 456 (1995).
146. Pereiro, I. R., Wasik, A. and Lobinski, R., *Fresenius' J. Anal. Chem.*, **363**, 460 (1999).
147. Krull, I. S., Panaro, K. W., Noonan, J. and Erickson, D., *Appl. Organomet. Chem.*, **3**, 295 (1989).
148. Panaro, K. W., Erickson, D. and Krull, I. S., *Analyst*, **112**, 1097 (1987).
149. Lobinski, R. and Adams, F. C., *Spectrochim. Acta, Part B*, **53**, 1865 (1997).

5.2 Flow Injection Atomic Spectrometry for Speciation

Julian F. Tyson

University of Massachusetts Amherst, MA, USA

1 FLOW INJECTION ANALYSIS

The need for greater automation of chemical analysis procedures is a major driving force in analytical chemistry research and development. Automation brings a number of advantages in terms of the analytical figures of merit, such as improved long-term precision, as well as in the fiscal figures of merit such as reduced cost per analysis. One way of viewing flow injection (FI) is as a procedure for the automation of serial determinations. Indeed, the early publications in the FI literature concentrate on this aspect of the technique. Explanations of the basic principles were often made in terms of FI being 'segmented flow without the bubbles in slightly narrower, nonwettable tubing'.

Since the term 'flow injection analysis' was first used, in 1975, to describe a procedure for chemical analysis, controlled fluid flow as an integral part of an analytical measurement procedure is now so widespread in current analytical methodology that it is difficult to provide a concise definition of FI. Definitions which are based on concepts such as 'the gathering of information from a concentration gradient formed when a well-defined zone of fluid, dispersed in a continuous unsegmented stream of a carrier, flows through a suitable detector', to paraphrase a definition provided by Ruzicka and Hansen [1], who are widely acknowledged to be two of the technique's inventors, would nowadays be considered to include high performance liquid chromatography (HPLC), capillary zone electrophoresis (CZE), and – as gases are fluids – gas chromatography (GC). The scope of what is considered to be flow injection can be ascertained from the material collected at the flow analysis database [2], in which there are now over 10 000 citations.

Handbook of Elemental Speciation: Techniques and Methodology R. Cornelis, H. Crews, J. Caruso and K. Heumann
 ISBN: 0-471-49214-0

Flow injection was also at the forefront of what has turned out to be another major theme of current analytical chemistry research and development, the drive for miniaturization of chemical measurement systems. It was not long before the theory of the characteristics of fluid flow and mixing in closed conduits was being examined, and the benefits of even further decreases in working scales became apparent. If mass transfer processes in liquids are to be driven by diffusion, then rapid events are only achieved by keeping the relevant distances small.

FI is a microscale operation. That is, the volumes of sample solution involved are in the microliter range, the flow rates involved are in the $\mu L\,s^{-1}$ range, masses of solid reagents are in the mg range, the volumes of reactors are in the μL to mL range. As FI is also a low pressure operation, the tubing used for conduits is typically 0.5–0.8 mm in diameter and not more than a few meters in length, hence fluids can be propelled by peristaltic (or syringe) pumps working at pressures of a few hundred kPa.

Dispersion is controlled by the hydrodynamics of such flow systems which are dominated by laminar flow. In a closed circular pipe, the flow is characterized by a parabolic velocity gradient ranging from zero at the walls to twice the average linear velocity at the center. For most systems, some dispersion is desirable as this is the mechanism by which sample solution and reagent solution are mixed. Typically, rapid radial mixing is required with minimum longitudinal mixing. These features can be promoted by inducing the appropriate secondary flow patterns which are functions of coiling, confluence point geometry and the insertion of packing materials into the flow lines. For a given manifold design, sample volume, and flow rate(s), the dispersion processes are highly reproducible as are residence times. Thus all samples and standards are subjected to exactly the same chemistry for exactly the same time, and thus it is not necessary for any process to be independent of residence time; as long as processes which affect the magnitude of the signal ultimately measured by the spectrometer progress to an extent which gives a signal above that equivalent to the detection limit, then the manifold may be used for quantitative analysis.

The key features of FI are often summarized as sample injection, controlled dispersion and reproducible timing. Many of the flow-based procedures for the preconcentration of trace elements for determination by atomic spectrometry (AS) would nowadays be considered under the general heading of FIAS, even though the procedure did not involve sample 'injection' as such, rather the procedure would be based on the introduction of a relatively large volume of sample by the continuous introduction of solution for a controlled time. Once separation from the matrix components had occurred, the procedure might well be based on the handling of a limited zone of analyte within the flow manifold.

Twenty five years or so after the appearance of the first papers, it is easily seen that FI as a means of automating chemical analysis procedures has underpinned the commercial success of the technique. However, the concept of handling fluids by pressure-driven flow in narrow, nonwettable conduits proved to be extremely versatile – capable of 'automating' many procedures in addition to mixing of homogeneous aqueous solutions – and thus the original invention gave rise to a considerable number of research themes in a considerable number of research laboratories around the world.

2 FLOW INJECTION AND ATOMIC SPECTROMETRY

The scale of flow injection is compatible with normal operation of the various atomic spectrometers as detectors for the flow systems: flame atomic absorption spectrometry (FAAS) requires introduction at $5\,mL\,min^{-1}$, inductively coupled plasma (ICP) optical emission spectrometry (OES) or mass spectrometry (MS) requires introduction at $1\,mL\,min^{-1}$, and graphite furnace (GF) AAS requires about $20\,\mu L$ of sample solution.

Although it got off to a relatively slow start, the concept of using FI as a sample handling system for atomic spectrometry has proved to be a particularly fruitful research topic, and the flow injection atomic spectrometry (FIAS) combination could easily be considered a major category of analytical chemistry methodology in the same way as the acronym HPLC represents a major collection of current analytical chemistry methodology. In 1998, Sturgeon [3] wrote, 'without doubt, the greatest impact on sample processing and introduction for atomic spectrometry has derived from the fields of FI technology and microwave radiation.' The combination of flow analysis techniques with atomic spectrometry detection (FIAS) now forms a significant subdiscipline of the flow analysis field. Several textbooks have appeared, the most recent of which [4] included a chapter on cryofocusing (for 'metal and metalloid speciation in the environment'), and a chapter on chromatographic separations (for 'trace element speciation in biological systems'); thus the boundaries between 'flow injection' and other flow-based analytical methods are becoming blurred. The transient signals produced by FI sample introduction can model the response of an atomic spectrometer to the transient concentration profiles in HPLC eluents [5, 6].

The literature is regularly reviewed as Atomic Spectrometry Updates in the *Journal of Analytical Spectrometry*; the Updates are entitled *Advances in Atomic Emission, Absorption and Fluorescence Spectrometry, and Related Techniques* and *Atomic Mass Spectrometry*, published in the June and August issues, respectively, cover the developments relating to FIAS. The relevant literature is large. Each year several hundred papers appear in the original primary literature concerned with some aspect of the combination of FI with atomic spectrometry. Many of these are what might be termed 'application' papers, i.e. the contents are the description of a new method for the determination of X (some analyte or analytes) in Y (some matrix). There are relatively few publications in which a genuinely new FI technique is described. Many of the application papers are concerned with overcoming some limitation of the atomic spectrometry technique. Such limitations are often related to detection capability, or inaccuracies due to the presence of some potentially interfering matrix component. Thus a considerable amount of research effort is being devoted to the development of FI procedures for the preconcentration and/or separation of the analyte(s) from the matrix. Categorizing the current research activity in this fashion does not relate immediately to the issues of speciation.

3 SPECIATION OF THE FIAS LITERATURE

However, one possible categorization of the literature relating to FIAS topics is to divide published work into two categories: those papers in which some aspect of speciation is featured, and those papers in which some aspect of speciation is not featured. In the first category would be work in which the principal use of FI was (a) to transport the sample solution to the spectrometer, (b) to preconcentrate all of the analyte, or (c) to separate all of the analyte from the matrix. Solutions containing high quantities of dissolved solids or suspended particles can be handled by FI. There are a surprising variety of ways, not all of which would appear to have been discovered yet, for the dilution of sample solutions en route to the spectrometer, so that off-range samples can be handled without removal from the autosampler tray, dilution and reanalysis. It is also possible to add reagents, such as ionization buffers or releasing agents, by the mixing of sample and reagent solutions in a suitably designed manifold.

In the second category, would be work in which some selectivity had been created via selective introduction to the instrument: most likely by selective chemical reaction in the FI manifold. The selectivity induced would be based on a separation, and thus the use of FIAS for speciation purposes is founded on the implementation of reactions for separation in a flow injection manifold where the key processes are controlled by fluid flow in closed, fixed-geometry conduits, reactors and other vessels.

The separations that are typically implemented in a FI system might be termed 'nonchromatographic'. They are usually binary in character, i.e. the analytes are separated into two groups on the basis of the process. These processes include, for example, precipitation, liquid–liquid extraction, solid-phase extraction (retention on a solid reagent), chemical vapor generation (conversion to a gaseous derivative), and dialysis. Continuous-flow sample delivery with such techniques was the subject of a book [7], which covers many of the techniques that are used for speciation analysis, though the speciation was not a major topic of the book, nor indeed was flow injection. A slightly more recent book [8] approached the topic of separation from the FI perspective. The book also covered the closely related topic of preconcentration. Again, speciation did not figure very prominently in the table of contents or in the index. Although atomic spectrometry featured prominently among the instrumental techniques used for quantification, other spectroscopic and electrochemical techniques were also covered. In Welz's 'attempt at a forecast' of where speciation analysis is going [9], FI-based procedures are featured prominently. The extent to which various instrumental techniques have been used with FI is apparent from a relevant review article [10] in which the literature up to 1995 was reviewed under the title Chemical speciation by flow-injection analysis. As the authors of this review point out, FI has the ability to help overcome a general problem encountered in speciation analyzes involving separations, namely the redistribution of species following the disturbing of the equilibrium by the separation, as the time between reaction and measurement can be kept short (maybe only a few seconds).

There are several reports of procedures in which FIAS is used as part of a method which results in the production of information about the speciation of the element, or elements, of interest, but in which the diagnostic chemistry is not performed in the FI manifold. For example, As(III) was selectively determined in fish tissue by distillation of $AsCl_3$ followed by FI-HG-AAS [11]. Such procedures have been considered outside the scope of this chapter, which deals with speciation methodology in which the chemistry which forms the basis of the speciation has been performed on line in the FI manifold.

4 OVERVIEW OF CHEMICAL REACTIONS FOR SPECIATION

To a first approximation, the response of any atomic spectrometer is independent of the chemical form of the element(s) introduced, and thus there is little scope for using the response of the spectrometer as the basis of a measurement of chemical speciation. Selectivity towards the different chemical forms of the element of interest are therefore produced by reaction chemistry designed to physically separate fractions of the analyte prior to measurement. In principle, any chemical reaction or process that produces such separation has the potential for application as a speciation procedure; however, to be useful in the context of helping to solve problems by the provision of reliable information about chemical composition, it is preferable that the separation be easily related to well-defined chemical species. There are, though, schemes of speciation which are quite widely used, but which are based on separations for which the species separated are not well defined. For example there are schemes for the estimation of various forms of elements in soils based on leaching with solutions of increasing complexing or solubilizing ability [12, 13]. In 1985, Van Loon wrote [14], 'I treat such procedures with some degree of skepticism because past experience suggests that it is not possible in most cases to restrict the extraction to a particular soil fraction. However, in spite of these problems agricultural scientists find these procedures helpful'. It is possible [15] to implement such procedures in a FI format.

This separation basis for speciation means that the analysis, typically, proceeds in several stages, each of which requires a fresh portion of the sample material: following suitable pretreatement, the total element will be determined, then a fresh sample portion will be obtained and a measurement made of the element that is separated under the conditions selected. This might be repeated with a different set of conditions selective for another analyte species or a different separation scheme might be used. Finally, a fraction will be quantified by calculation as the difference between the total and the sum of the various speciated fractions measured.

As with other analytical methods in which reaction chemistry precedes instrumental measurement, the species quantified are only defined in terms of the chemistry which forms the basis of the procedure. For example, it is common to describe the reaction in which acidified potassium antimonyl tartrate, ammonium molybdate and ascorbic acid are added to the sample in the appropriate order followed, possibly, by heating and preceded, possibly, by the addition of masking agents as 'the determination of phosphate'. What, in fact, is measured are all species that will react to give a blue color under the conditions used; such species might include other phosphorus-containing oxo-anions, and oxo-anions of other elements. Most speciation procedures are based on the goal of being able to related what is measured to the concentration of a defined chemical entity as this, in general, leads to information which aids in understanding the problem being studied. However, this is not the only goal of speciation analysis. An important question about materials that might be ingested (either deliberately as in the case of food and drink) or accidentally (as in the case of soil near pressure-treated decks) concerns the bioavailability of potentially harmful elements. In assessing the 'arsenic status' of a soil, it may be more important to know how much arsenic is bioavailable than it is to know how much arsenic is present as arsenite, arsenate, monomethyl arsonic acid and so on. Devising simple chemical tests that are reliable indicators of bioavailability is not a trivial problem, and is an active area of research at present.

All of the separation procedures commonly used in analysis can be adapted to the FI format. Even distillation can be performed in an FI system [11, 16] though there appears to be little interest in such systems at present. The only well-established chemical separation procedure that has not yet been adapted to the FI format is fire assay. There is some variation in the design of manifold for the various separation procedures and the way such manifolds are interfaced with the spectrometer (not always a trivial exercise in the case of graphite furnace atomizer instruments); however, it is possible to identify a prototypical FI manifold for separating analyte species either from each other or from other sample components. This is shown in Figure 5.2.1. The sample solution is either injected into a carrier (C), or is pumped continuously, and merged with a reagent R. It might be necessary to add more than one reagent (for example the addition of a complexing agent followed, after a suitable reaction time, by the addition of an extracting solvent). The sample components

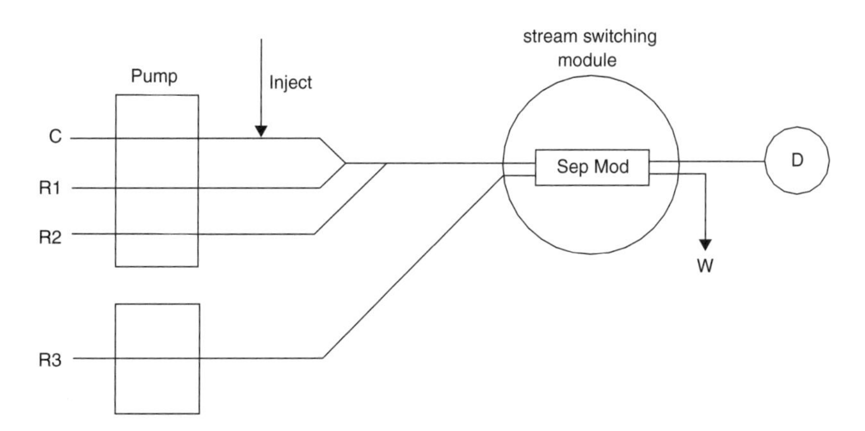

Figure 5.2.1. Generalized FI manifold for speciation studies. C, carrier stream; R1, R2, R3, reagents; Sep Mod, separation module; D, detector; W, waste. Typically R1 and R2 are selective for some component of the sample which is injected via a suitable valve at the location shown by the 'Inject' arrow. The reagent R3 would be involved in the release of sample components retained in the separation module.

pass into the separation module (whose exact nature depends on the nature of the separation). There are several possible next stages: (1) selected sample species can pass to the detector with other matrix components of the sample, (2) selected sample species can pass to the detector while other matrix components of the sample pass to waste, (3) sample species are retained, matrix components pass to the detector or to waste, then sample species are released from the separation module, (4) sample species are retained, matrix components pass to the detector or to waste, then sample species are sequentially released from the separation module, or (5) a combination of (1) or (2) with (3) or (4). Although the figure is drawn as though the separation module was mounted in one particular orientation, it may well be that it is located in a stream switching module that allows it to be moved from one flow line to another and for the direction of flow to be reversed.

The most widely used separation modules, for speciation and other applications, are (a) the stripping coil and gas–liquid separator and (b) the solid-phase extraction cartridge (often referred to as a 'minicolumn') or coil. Less widely used devices are (a) the liquid–liquid extraction coil and phase separator, (b) the particulate filter, and (c) the dialyzer. In one or two cases there is some ambiguity as to the nature of the separation process. For example, when ammonium pyrrolidine dithiocarbamate is added to an aqueous solution of trace metals, complexes are formed with can be retained on a C-18 cartridge or on the walls of an open tubular reactor. As these complexes are generally thought to be insoluble, it is not entirely clear whether the retention is by solid phase extraction or filtration. More detailed illustrations of the implementation these separation schemes are given in Figure 5.2.2.

Most of the speciation schemes implemented with the aid of FI technology are based on differences in thermodynamic properties of the participants in the reactions which form the basis of the separation. For example, hydride generation (HG) is quite widely used to discriminate between different oxidation states of inorganic selenium. For most samples, the soluble inorganic selenium exists as either selenite (SeO_3^{2-}) or selenate (SeO_4^{2-}) in which selenium is in the 4+ state or the 6+ state, respectively. Extensive experimental evidence supports the contention that, when a solution of selenate in acid is mixed with an alkaline solution of borohydride (BH_4^-) so that the resulting solution is acid, no volatile selenium species is formed; whereas if the experiment is repeated with selenite a substantial percentage (approaching 100) of this species is converted to hydrogen selenide (H_2Se). Thus HG provides the basis for the sequential determination of these two species. In the case of solid-phase extraction, the basis of separation is the formation of a product with the immobilized reagent that is more stable than (a) the solution-phase reactant form of the target species and (b) all other possible products with other species in the solution phase. Thus when the solution and solid are separated, only one species is retained by the solid reagent. For example, an anion exchange column will retain negatively charged forms of chromium from aqueous solution (provided the conditions are appropriate) and positively charged forms will remain in solution. This procedure therefore provides one possible basis for the separation of chromium in the +6 oxidation state (as these species are negatively charged) from chromium in the +3 state (species which are positively charged).

Features of chemical reactions such as ionic strength effects, pH effects and buffer capacity must also be borne in mind. The position of the equilibrium of reactions involving ionic species can be significantly affected by ionic strength. Samples which have substantial inorganic matrices (such as seawaters) may have ionic strengths orders of magnitude greater than those of the calibration standards. Some samples may have high acid concentrations as a result of the pretreatement procedure which overwhelm the buffer capacity of the solution used for pH adjustment. Some samples may have a buffer capacity high enough to resist that of the buffer solution.

There are some schemes which are based on kinetic considerations. For example, in the

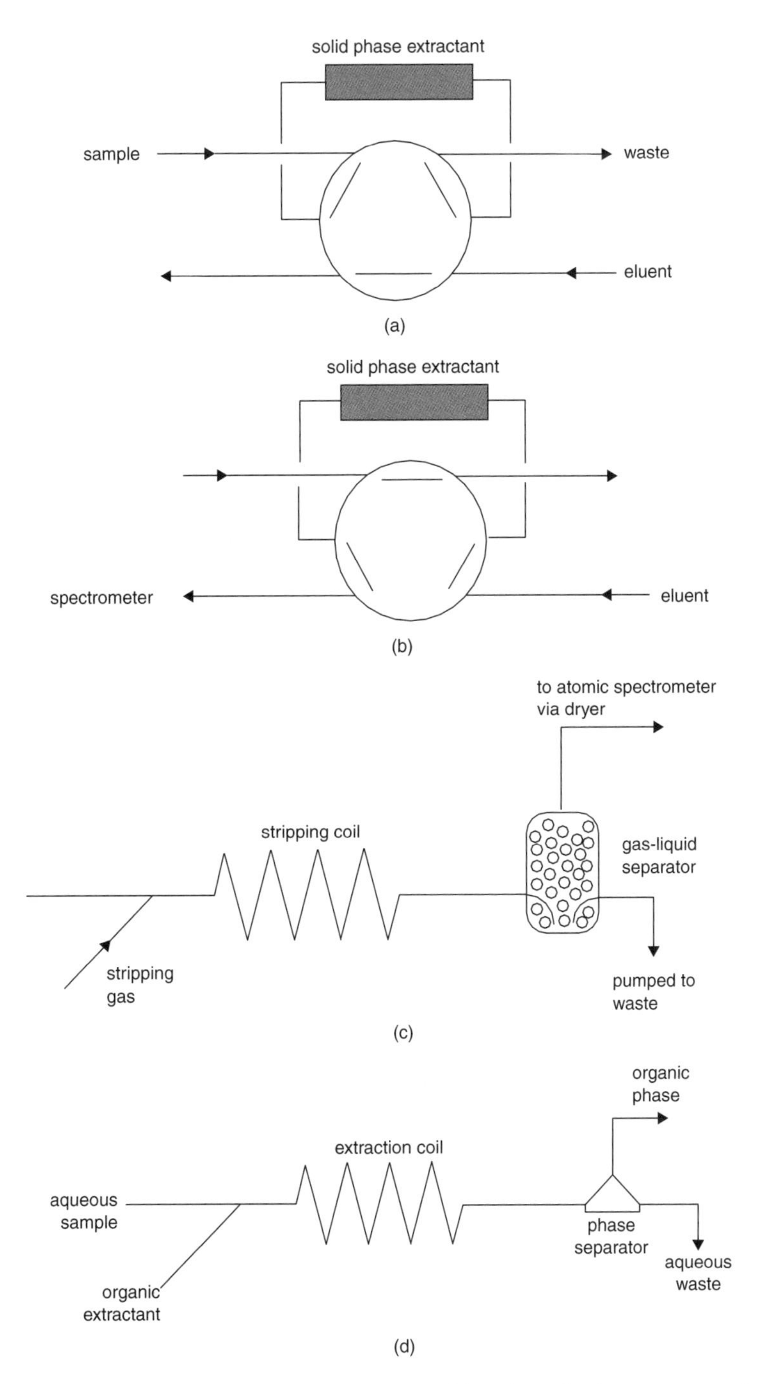

Figure 5.2.2. FI separation devices. (a) Solid-phase extractant mounted in column in six-port rotary valve in 'load' position; (b) valve switched to elute position; (c) chemical vapor generation manifold showing gas–liquid separator filled with glass beads; (d) liquid–liquid extraction manifold showing conical phase separator in which less dense organic solvent is removed at the upper outlet; (e) dialysis (or supported liquid membrane) separating donor and acceptor solutions; (f) in-line filter for collection of precipitate [can be mounted in loop of injection valve as shown for (a) and eluted by back-flushing as shown in (b)].

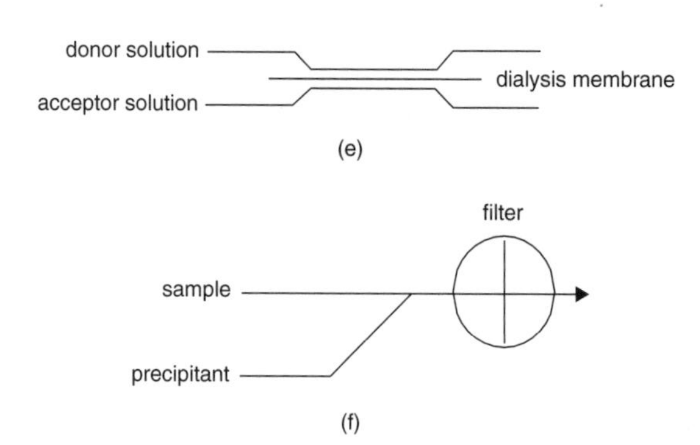

Figure 5.2.2. (*continued*)

speciation of aluminum in natural waters, the fraction identified as being potentially the most toxic to aquatic organisms contains monomeric, inorganic hydrated ions. These species are often quantified on the basis of the speed of the reaction of aluminum species with chelating reagents, such as 8-hydroxyquinoline (8-HQ). This idea was proposed prior to the availability of FI techniques, but may be readily implemented in an FI manifold as the combination of controlled injection volume and flow rate allows the contact time between analyte species and reagent to be precisely controlled. Thus if the 8-HQ is immobilized in a short column the so-called 'fast reacting' or 'labile' aluminum can be determined by measurement of the aluminum that is retained on the passage of the injected slug though such a column.

As residence times in FI systems are on the order of seconds, reactions which are exploited as the basis of speciation measurements are usually fast. Some reactions which are quite widely used are not fast enough at room temperature, and to implement these in a flow injection system requires some additional features to be incorporated into the manifold or its operation. To some extent, residence times can be increased by slowing the flow or increasing the length of the manifold tubing or both. There are some practical limitations to these approaches arising from the limitations of peristaltic pumps. Firstly it is difficult to pump reliably (i.e. with good precision and no pulsations) at very low flow rates ($\mu L\ min^{-1}$), and secondly increasing the tube length increases the back pressure and it can become impossible to pump the solutions though the manifold. Residence times can be increased by stopping the flow for a desired period. This requires computer control over the pumping device, but is probably the best approach if this stopped-flow mode is used. Another approach to this problem is to increase the rate of the desired reaction by raising the temperature. This can be done by immersing the reaction coil in a conventional heating bath or by irradiation with microwaves. This latter approach provides an additional possibility for speciation as the energy is controllable. Thus it is possible to make a measurement of the species introduced to the spectrometer arising from the chemical reactions occurring with the microwaves on and a measurement of the species produced with the microwaves off. For example, (selenate + selenite) can be determined by HG if the sample contains hydrochloric or hydrobromic acids and the microwaves are 'on' (as selenate is reduced to selenite). When the microwaves are off, only selenite is detected as selenate is not 'borohydride active'.

The kinetics of mass transfer processes should also be borne in mind. If a solid-phase extraction procedure is to be used as the basis of separation in an FI system, then the amount of material retained on a small column in an FI system will depend on the extent to which the species in solution get close to the immobilized reagent species on the surface of the solid support. If the support material creates large interstitial spaces and the analyte species are dilute and a high volumetric flow rate is used, the retention efficiency may be very poor (only a few percent of the target material is retained). Thus a considerable inaccuracy is incurred if the concentration of the nonretained element is equated with species A and the element subsequently eluted is equated with species B. Diffusion in liquids is slow; to use it as a mechanism of mass transfer means that the distances involved must be small (or residence times must be long). One of the reasons that HPLC is HP is that mass transfer by diffusion in the interstitial spaces is effective because the interstitial spaces are small. This is achieved at the expense of flow rate as the back pressure for a column of useful dimensions is high. High pressure pumping (with HPLC pumps) is not normally considered a feature of FI. Residence times can be manipulated as discussed in the previous paragraph.

Some speciation schemes depend on the kinetics of heat transfer to effect temporal separation of volatile species after cryotrapping. This procedure is normally applied following generation of volatile derivatives of several species in the sample. For example, inorganic arsenic, monomethylarsonic acid and dimethylarsinic acid can be derivatized with borohydride to arsine, monomethyl arsine and dimethyl arsine, all three of which can be trapped at liquid nitrogen temperatures. When the trap is warmed, the three compounds elute in order of lowest boiling (arsine) first, highest boiling (dimethylarsine) last. In an interesting variation on this experiment, Burguera *et al.* trapped the arsines in a coil inside a microwave oven [17] and remobilized them by microwave heating when the arsines were volatilized in the reverse order, i.e. the compound with highest boiling point (but highest dipole moment) was volatilized first.

5 HYDRIDE GENERATION

The basic principles of hydride generation (HG) were comprehensively covered in a recent book by Dedina and Tsalev [18], which despite the title does cover HG for atomic fluorescence and emission as well. There has been a substantial body of work published since this book was published (maybe as many as a thousand papers) whose contents have been reviewed in the regular Atomic Spectrometry Update articles published bimonthly in the *Journal of Analytical Atomic Spectrometry*. The salient features of FI-HG were covered by Dedina in a chapter [19] in a recent FI book [4]. In addition to the general advantageous features of FI compared with batch procedures, there is an additional feature of FI-HG that is beneficial: kinetic discrimination over interferences caused by the reaction of borohydride with matrix components. In general, the primary HG reaction is fast compared with the competing interfering reactions, and improved tolerance to interferences is obtained in an FI system as the residence time of the hydride in the liquid phase is very short. Much of the work published recently concerning the determination of the 'hydride-forming' elements is concerned with some aspect of speciation, though a substantial number of papers describe HPLC separation with element specific detection. By far the greatest interest at present is in the determination of arsenic and selenium species. Other elements of interest include tin, lead, and antimony. FI-HG has been used for the determination of bismuth, cadmium, and germanium. Mercury species may also be determined via HG, though the determination of inorganic mercury is normally considered not to involve the formation of a hydride. A commonly used reagent is tin(II) in hydrochloric acid solution which reduces Hg(II) species to elemental Hg, which can be blown out of solution with a stream of argon or nitrogen. Inorganic mercury also reacts with borohydride, BH_4^-, to release elemental mercury vapor. If mercury hydride is formed it is considered to be unstable with respect to elemental mercury and hydrogen and to have a very short lifetime.

Hydride generation may be used as the basis of a speciation scheme by virtue of selective derivatization. The most common reagent is borohydride, BH_4^-. Not all forms of the so-called 'hydride-forming' elements are 'borohydride active' (i.e. give a volatile derivative, which may be easily atomized), when mixed with a borohydride solution followed by acidification.

Howard [20] has reviewed speciation based on reaction with borohydride, covering discrimination based on (a) pH control, (b) cryotrapping, and (c) HPLC. Tsalev, in a recent comprehensive review [21] surveyed HG procedures for speciation. HG may also be used in conjunction with other separation schemes, such as solid-phase extraction (in the case of FI-based speciation) or HPLC. In both cases, consideration has to be given to the borohydride activity of the separated species and further chemical transformation may be necessary in the case of species which are inactive. Most procedures of this sort are based on the conversion of the borohydride-inactive species to the simple inorganic anion of lowest oxidation state. This may involve reduction, by agents such as chloride, bromide, iodide, or ascorbic acid in the case of inorganic ion oxidation state adjustment, or combinations of oxidation and reduction when organocompounds containing bonds between carbon and the element of interest are to be determined. Oxidants such as alkaline persulfate, acid dichromate or permanganate have been used. All of these types of reactions have been promoted by microwave heating [22, 23], and there is evidence that the desired conversions can be obtained under less aggressive chemical conditions than are needed when conventional heating is applied. It is also possible to promote some of these reactions with UV light [20], an attractive proposition for a FI procedure as flow-through UV photoreactors are commercially available.

5.1 Antimony

An HG-inductively coupled plasma atomic emission spectrometry (ICP-AES) procedure for the determination of antimony(III) and antimony(V), based on the kinetics of the reduction reaction of antimony(V) with L-cysteine, was devised by Feng *et al.* [24]. They found that the degree of conversion was a linear function of time for a period of up to 10 min. Their method was based on measurements made 2 and 8 min after the addition of the reducing agent. Potential interferences from transition metals were removed with a column of Chelex-100. Ulrich [25] devised an ICP-AES method in which the reactivities of various antimony species [inorganic Sb(III), Sb(V) and trimethylstilboxide (TMeSbO)] towards fluoride formed the basis of the separation. In the presence of fluoride and iodide, only trimethylstilboxide was detected; in the presence of fluoride both Sb(III) and TMeSbO were detected. In the presence of iodide, Sb(V) was reduced to Sb(III) which was detected along with TMeSbO. The limits of detection were around $1\,\mu g\,L^{-1}$ for each species. Deng *et al.* showed [26] that the partial contribution of the signal from Sb(V) obtained in their FI-HG-AFS procedure could be eliminated by the addition of 8-hydroxyquinoline added originally to suppress the interferences by transition metals in the analysis of lake waters and sediments.

5.2 Arsenic

Nielsen and Hansen [27] devised an AAS procedure for the determination of As(III) and As(V) in water samples. Total As was measured following on-line reduction with ascorbic acid and potassium iodide in a knotted reactor immersed in an oil bath at 140 °C. Without the heating bath and reducing agents, and with mild acid conditions, only As(III) was determined. The procedure was applied to the analysis of a certified drinking water material. A similar procedure was developed by Krenzelok and Rychlovsky [28] who reduced As(V) to As(III) with potassium iodide (in 6 M HCl) alone at 60 °C. A membrane gas–liquid separator was used. Willie [29] devised what he called a 'first-order' speciation procedure with detection by ETAAS following trapping of the volatile arsenic-containing species on the interior of the atomizer. Conditions were used under which volatile derivatives were formed from As(III), As(V), MMA, and DMA, but not from arsenobetaine, arsenocholine or tetramethylarsonium. On-line UV oxidation or microwave heating in the presence of alkaline persulfate converted all species to borohydride-active forms. The procedure was applied to the analysis of several water and urine reference materials. A somewhat similar method was proposed by Cabon and Cabon [30] for the analysis of seawater. They showed that As(III) was not stable with respect to oxidation to As(V) in acidified seawater and slowly converted to As(V) in coastal waters at the natural pH. The method was extended [31] by cryogenically trapping the volatile hydrides from As(III) and As(V), generated under differing conditions, and MMA and DMA on a chromatographic phase followed by sequential release. It was claimed that 'evidence for the biotransformation of arsenic in seawater was clearly shown'. Burguera *et al.* [17] remobilized the arsines trapped in a liquid-nitrogen-cooled coil by irradiation with microwaves. These four species can, apparently, be differentiated on the basis of reaction conditions [32]; whereas the addition of L-cysteine produced equal sensitivities for each species [33]. This was thought to be due to the reduction of the As(V) species (arsenate, MMA and DMA) to organosulfur–As(III) species, which would be identical with or respond identically to the species formed when arsenite and L-cysteine react. This reagent has also been studied by Tsalev *et al.* [34] who showed that the species produced on reaction with L-cysteine could be separated by HPLC. They also showed [35] that UV-assisted oxidation with persulfate would transform arsenite, MMA, DMA, arsenobetaine, arsenocholine, trimethylarsine oxide and tetramethylarsonium to arsenate. The reaction was used for the determination of total As by FI-HG-AAS, and for post-column derivatization following HPLC separation.

Electrochemical HG has the same sort of selectivity as borohydride and so can be applied for speciation. Pyell *et al.* [36] devised a system that was selective for As(III) and, with the addition of a cryotrap, was selective for arsine over the various methylated arsine species.

There are several reports of procedures in which the quantification was by FI-HG-atomic spectrometry following various pretreatments in the batch mode. Perhaps the most important of these is the procedure developed by Chaterjee *et al.* [37] in support of the studies of the contamination of well-water in West Bengal, described by the researchers as 'the biggest arsenic calamity in the world'.

5.3 Selenium

An AAS procedure for the determination of Se(IV) and Se(VI) in seawater was developed by Fernandez *et al.* [38] in which the Se(VI) was reduced in the FI manifold to Se(IV) on addition of concentrated hydrochloric acid and passing though a coil immersed in a heating bath at 140 °C. Hydride generation was performed at 0 °C by placing the relevant part of the manifold in an ice bath. Stripeikis *et al.* [39] measured relevant reaction kinetic parameters for such a system and claimed to have the best throughput, $60\,h^{-1}$, of any FI system described so far. Microwave heating is also suitable for the conversion of Se(VI) to Se(IV) by hydrochloric acid, and such a procedure was developed by Burguera and coworkers [40] for the determination of these species in citrus fruit juices and geothermal waters. They later modified the procedure [41] so that reduction was effected by a mixture of hydrochloric and hydrobromic acids; in comparison to the procedure which uses only hydrochloric acid, lower concentrations of the acids (10 %) could be tolerated. He *et al.* [42] determined these selenium species in seawater by a similar procedure, but in which quantification was by AFS. The AFS detection limits of $5\,ng\,L^{-1}$, were about 50 times better than those of the AAS methods. Moreno *et al.* [43] compared on- and off-line procedures for the extraction (and reduction) of selenium species from wastewaters and sewage sludges, and concluded that accurate analyzes by the on-line method required calibration by the method of standard additions. Varade and Luque de Castro [44] showed that subcritical water (250 °C and 200 bar) extracted both inorganic and organic compounds from sludges, but that the organic compounds did not survive the extraction. The on-line reduction to Se(IV) was helped by microwave irradiation.

It has been reported [45, 46] that UV irradiation converts both Se(VI) and various organoselenium compounds to Se(IV). This has been used for the speciation of selenium species in seawater by various off-line pretreatments [45] and for HPLC post-column derivatization [46]. This post-column derivatization has also been achieved by microwave-assisted reaction with a mixture of potassium bromate and hydrobromic acid [47].

Clearly most researchers thought that organoselenium compounds are not borohydride active; however, recently published evidence [48] suggests that selenomethionine, selenoethionine and trimethylselenonium do react with borohydride to produce methyl selenol and dimethyldiselenide, ethyl selenol and diethlyldiselenide, and dimethylselenide, respectively. It has also been shown that trimethylselenonium can be determined by HG-AAS with the same sensitivity as for selenite [49]. The reasons for these differences in findings over the reactivity of these organoselenium compounds are not yet clear, but may have something to do with (a) the acidity of the solution or (b) the presence of matrix components capable of reacting with borohydride.

5.4 Mercury

Gallignani *et al.* [50] showed that inorganic mercury could be determined in the presence of methylmercury by the use of stannous chloride as the reductant to generate mercury vapor. Total mercury was determined after on-line microwave-assisted oxidation with persulfate. For determination by AAS, the limit of detection was $0.1\,\mu g\,L^{-1}$. Rio-Segade and Bendicho devised somewhat similar procedures for distinguishing between inorganic mercury and methylmercury. In extracts containing both species only inorganic mercury was determined when stannous chloride was used as the reductant [51]. When on-line oxidation with potassium peroxodisulfate (persulfate) in sulfuric acid was triggered by heating [52], total mercury was measured; with the heating off, only inorganic mercury was determined. In the former procedure, mercury species were extracted from fish tissue into hydrochloric acid with the help of ultrasound. However, it has been reported [53] that such a procedure can convert methylmercury to inorganic mercury, but as the results obtained for the analysis of three fish reference materials were in agreement with the certified values there may be conditions under which no conversion occurs. Tao *et al.* [54] obtained accurate results for marine tissue reference materials following dissolution of mercury species in tetramethylammonium hydroxide solution. Inorganic mercury was selectively determined with stannous chloride after the addition of L-cysteine. Burguera *et al.* [55] determined mercury species in fish-egg oil. A surfactant was added to generate an emulsion which was much easier to pump in the FI system than the original viscous oil. Inorganic mercury was determined on the addition of borohydride (when presumably methylmercury was not detected) and total mercury after oxidation with persulfate. However, for the 22 samples examined no inorganic mercury was detected whereas the organic mercury content ranged from 2 to $3\,\mu g\,L^{-1}$. There is conflicting evidence in the literature concerning the product of the reaction of methylmercury with borohydride. Some researchers [e.g. 49] find that mercury vapor is produced, whereas others [e.g. 56, 57] find that methylmercuryhydride is produced. Recent work [58] suggests that the concentration of both the borohydride and the sodium hydroxide, used to stabilize the borohydride solution, are important factors.

5.5 Other elements

5.5.1 *Bismuth*

Bismuth species in urine were determined by AAS following ethylation with tetraethylborate [59]. The procedure was free from interferences from other components in urine and was applied after a simple $1 + 1$ dilution. The limit of detection was $2\,\mu g\,L^{-1}$ and the method was used to follow the urinary clearance of therapeutic doses of bismuth subcitrate

5.5.2 *Cadmium*

Post-column HG with ICP-MS detection allowed the detection of cadmium-containing metallothioneins in eel liver and kidney separated by vesicle-mediated, reversed-phase HPLC [60]. The procedure was validated by the examination of rabbit liver cadmium metallothioneins.

5.5.3 *Lead*

Organolead compounds in the concentration range $0.25-8\,mg\,L^{-1}$ were selectively determined in the presence of inorganic lead by FI-HG-ETAAS [61]. A carrier stream containing hydrochloric acid and ethylenediaminetetraacetic acid suppressed the inorganic lead signal for concentrations up to $10\,mg\,L^{-1}$.

5.5.4 *Tin*

Post-column UV photooxidation in the presence of persulfate at 95–100 °C converted a variety of alkyltin compounds to forms that could be detected by HG-AAS [62] with better than 80 % recovery. However, tetrabutyltin was only 15 % recovered.

6 SOLID-PHASE EXTRACTION

More papers are published about the implementation of a solid-phase extraction procedure in an FI mode with direct coupling to an atomic spectrometer than about any other FIAS topic. Most of these publications are concerned not with speciation analysis, but with either separation of analyte(s) from a potentially interfering matrix or with preconcentration, or both. Many of the methods developed are designed to extend the capabilities of flame AAS (FAAS).

The possible applications for speciation analysis are based on the selective retention of species by a column packed with a suitable solid-phase extractant. Unretained species may either be discarded to waste or be directed to the spectrometer for determination. Species retained on the column may then be eluted with an appropriate reagent and measured by the spectrometer. In some cases, it is possible to sequentially elute more than one retained species.

Terminology can be a little confusing in this area as some authors refer to the procedure described in the previous paragraph as 'extraction chromatography'. In chromatographic terms, species which are not retained have a capacity factor of 0. Capacity factor (often given the symbol k') is the number of column volumes, over and above the void volume, needed to elute a component from the column. Species which are retained have very large capacity factors, maybe approaching infinity. The term 'chromatography' will not be used in relation to these solid-phase extraction procedures unless it is clear that in the procedure described the species were being eluted under conditions which produced $k' > 0$ and $\ll$ infinity.

6.1 Arsenic

A procedure in which the complex between As(III) and ammonium diethyldithiophosphate was selectively retained on a C-18 column was devised by Pozebon *et al.* [63]. The retained species was eluted with $120\,\mu L$ of ethanol into an autosampler cup for transferal ($30\,\mu L$ plus $10\,\mu L$ of 0.1 % palladium nitrate solution) into a graphite furnace atomizer. Total arsenic was determined after reduction with potassium iodide and ascorbic acid and hence As(V) by difference. The working range was $0.3-3\,\mu g\,L^{-1}$. Tyson [64] developed a procedure in which the hydride was generated from the surface of an anion-exchange resin by the sequential retention of analyte and borohydride followed by the passage of a discrete volume of acid through the column. A speciation procedure for As(III), As(V) and methylated arsenicals was devised based on control of pH of the carrier stream and some off-line oxidations. At pH 2.3 only arsenate is not fully protonated and hence this was the only species retained by the column. Reaction with hydrogen peroxide in nitric acid oxidized As(III) to As(V) and thus As(III) could be found by difference. Further oxidation with alkaline persulfate in a sealed vessel in a microwave oven converted all species to As(V) and thus the methylated arsenic species were found, again, by difference. The limit of detection was $4\,ng\,L^{-1}$ for a 10 mL sample volume. Burguera *et al.* [65] trapped As(V), MMA and DMA on a combined cation–anion exchange column and then sequentially removed them by eluting first with 0.006 M trichloroacetic acid [to remove As(V)], then with 0.2 M trichloroacetic acid (to remove MMA) and finally with 5 M ammonium hydroxide solution (to remove DMA). As(III) was not retained and was determined by difference after a separate determination of total As after an off-line, microwave-assisted, acid decomposition. The species in the various eluents were determined by HG-AAS after 'reduction' with a solution containing 15 % (m/V) potassium iodide and 5 % (m/V) ascorbic acid. The limits of detection were around 0.1–0.3 $\mu g L^{-1}$ in a 2 mL urine sample. A similar approach has been used by Yalcin and Le [66] who retained (a) DMA on a resin-based, cation exchange resin, followed by elution with 1 M HCl, (b) MMA and As(V) on a silica-based, anion exchanger, followed by elution of the MMA with 0.06 M acetic acid and of the As(V) with 1 M HCl, and (c) all four species on alumina. As(III) was not retained on the cation or anion exchange materials. The procedure was applied to the determination of the species in drinking water down to concentrations around 0.05 $\mu g L^{-1}$.

6.2 Chromium

All of the published FIAS work is concerned with measurement of Cr(III) and Cr(VI). Sperling and coworkers [67] developed an ETAAS procedure in which Cr(VI) was selectively retained as the complex with diethyldithiocarbamate (DDC) on C-18 silica and then eluted with ethanol. Total Cr was determined after oxidation of Cr(III) to Cr(VI) with persulfate, and hence Cr(III) was determined by difference. The limits of detection were 20 $ng L^{-1}$. They also developed a procedure for FAAS [68] in which an alumina column was rendered selective to each of the species in turn by varying the pH of the carrier solution. At pH 2 Cr(VI) was retained, and at pH 7, Cr(III) was retained. The eluents were ammonia solution (0.5 M) and nitric acid (1 M), respectively. The limits of detection were 1 $\mu g L^{-1}$. More recently, titanium dioxide has been used in a similar fashion. Vassileva [69] developed a procedure for the retention of each species on two separate columns followed by determination by ICP-AES, and Yu *et al.* [70] selectively retained Cr(VI) at pH 2 with detection of the unretained Cr(III) by ICP-MS. However, the Cr(VI) was not eluted for subsequent determination, rather a total Cr value was obtained by direct introduction so that Cr(VI) was determined by difference. The limits of detection were 70–80 $ng L^{-1}$.

Several other solid-phase extraction materials have been used for one or other of the species. Naghmush *et al.* [71] investigated the performance of various functionalized cellulose absorbents, a chelating resin and some ion exchange resins. For FAAS the detection limits were around 1 $\mu g L^{-1}$. Jiminez *et al.* [72] collected Cr(III) on Amberlite IR-20 cation exchange resin and Cr(VI) on Amberlite IRA-400 anion exchange resin. The FAAS detection limits were 10 and 1 $\mu g L^{-1}$ for Cr(III) and Cr(VI), respectively. Kelko-Levai *et al.* retained Cr(III) on 'IDAEC' and Cr(VI) on the anion exchanger diethylaminoethyl (DE)-cellulose [73]. After elution the elements were quantified either by ETAAS or by total reflection X-ray spectrometry.

The method of Sperling *et al.* [67], based on retention of the DDC complexes on C-18, was recently modified by Rao *et al.* [74] so that both species were retained: Cr(VI) at pH 1–2 and Cr(III) at pH 4–9 [in the presence of Mn(II) which was reported to enhance the FAAS signal for Cr(III) by a factor of 10]. The eluent was methanol and the overall enrichment factor was 500 for a 5 min preconcentration, allowing determination of both species in the concentration range 0.2–200 $\mu g L^{-1}$. Cespon-Romero *et al.* [75] devised a procedure in which Cr(III) was selectively retained on a poly(aminophosphonic acid) chelating resin followed by elution with hydrochloric acid (0.5 M). Total Cr was determined after off-line reduction of Cr(VI) with ascorbic acid. The limit of detection was 0.2 $\mu g L^{-1}$ for a 6.6 mL sample volume.

6.3 Selenium

Selenium(IV) and (VI) were preconcentrated on an alumina column (activated by the carrier stream of 0.01 M nitric acid) by workers in Camara's group [76]. Following elution with 2 M ammonia solution the Se(IV) was determined by on-line HG-AAS. Total selenium was determined by reduction prior to introduction to the column, and thus Se(VI) was determined by difference. For 25 mL of sample, a preconcentration factor of 50 was obtained with detection down to around 6 ng L^{-1}. Pyrzynska *et al.* [77] devised an off-line procedure based on the same preconcentration, but with successive elution of the Se(IV) with 1 M ammonia and the Se(VI) with 4 M ammonia, and determination by ETAAS. The limit of detection of detection for Se(IV) was 50 ng L^{-1}. A sequential-elution, on-line procedure was devised by Bryce and coworkers [78]. The two selenium species were retained on a strong anion-exchange material followed by elution of Se(IV) with 2 M formic acid and of Se(VI) with 6 M HCl. On-line reduction to Se(IV) was aided by passage through a microwave oven. Determination was by HG-AFS and for a sample volume of 600 μL and an elution volume of 350 μL, the detection limit was 40 ng L^{-1}. Yan *et al.* [79] selectively retained Se(IV) as the pyrrolidine dithiocarbamate complex on C-18, from 4.2 mL of sample, followed by elution with 26 μL of ethanol directly into a graphite furnace atomizer pretreated with iridium. The detection limit was 4 ng L^{-1}.

Carrero and Tyson [80] retained Se(IV) together with borohydride on an anion exchange resin followed by HG directly from the solid phase by the passage of a slug of acid. For a sample volume of 9 mL the detection limit was 100 ng L^{-1}. The work has been further adapted [81] for the determination by HG-ETAAS with in-atomizer trapping, achieving a detection limit of 4 ng L^{-1} for a sample volume of 20 mL. In this revised procedure, the selenium was loaded first onto the column followed by an appropriate amount of borohydride. No signal was obtained from any Se(VI) retained in the column.

6.4 Other elements

The so-called 'fast reacting' aluminum species were retained on an Amberlite XAD-2 (nonionic) column following a reaction time of 3 s with 8-hydroxyquinoline in a procedure devised by Fairman and Sanz-Medel [82]. The retained Al was eluted with 1 M HCl and determined by ICP-AES. The procedure also incorporated a preconcentration of up to 18-fold giving a limit of detection of 2 μg L^{-1}. It was shown that the nontoxic AlF^{2+} species was not included in the 'fast reacting' retained aluminum. The method was also adapted for field sampling and for determination by ICP-MS [83].

Hulanicki and coworkers have developed procedures for the selective retention of antimony(III) prior to determination by GFAAS [84] or FAAS [85]. In the first procedure, the Sb(III) was retained on C-18 as the chelate with ammonium pyrrolidine dithiocarbamate followed by elution with ethanol directly into the graphite furnace atomizer. Total antimony was determined after reduction with L-cysteine. The limit of detection was 7 ng L^{-1}. The stability of Sb(III) and Sb(V) spiked into several natural matrices was investigated, and it was found that, while both species were stable in urine, Sb(III) was not stable in tapwater. In the second method [85], Sb(III) was retained on a 'DETA sorbent with grafted diethylenetriamine groups' then eluted with nitric acid directly into a flame atomizer to give a detection limit of 0.9 μg L^{-1}. Total antimony was determined by GF-AAS and thus Sb(V) was determined by difference. Antimony species added to a well water which reached rocks formed during the oligocene period was determined.

An on-line UV decomposition procedure was devised by Comber *et al.* [86] to investigate the speciation of copper in natural waters. Complexation with three model ligands, glycine, NTA and ETDA, was investigated. With the lamp off, only 84 %, 45 % and 2 % respectively of the copper was collected by a column of Chelex-100 chelating resin, but when the lamp was turned on, over 90 % of the copper was collected. A procedure for the speciation of copper and manganese in cow's milk has been devised by Abollino *et al.* [87] in which four operationally defined 'species' were measured. The separations were based on (a) precipitation with casein (species associated with proteins), (b) species retained on an anion exchange column, (c) species retained on a chelating cation exchange column and (d) species not associated with proteins not retained by either column. The initial casein precipitation was performed off line, but the two ion exchange resins were incorporated into an FI manifold, in which the sample could be directed to one or other of the columns followed by back-flushing with 2 M HCl and determination by ICP-AES. Total copper and casein-precipitated copper were determined by GF-AAS.

Several procedures for the differentiation of iron(II) and iron(III) have been devised, and the earlier ones are summarized in ref. 10. More recently Bagheri *et al.* [88] retained Fe(III) selectively on 2-mercaptobenzimidazole loaded onto silica gel; Fe(II) was not retained. The Fe(III) was removed with thiocyanate solution. Total iron was determined after oxidation of Fe(II) with hydrogen peroxide. The limit of detection was around $1\,\mu g\,L^{-1}$.

A procedure for separation of tetralkyllead and the sum of inorganic lead and organolead species having a smaller number of alkyl groups was developed by Naghmush *et al.* [89]. All of the species were retained on Cellex P, a functionalized cellulose sorbent. The two groups of retained species were eluted with ethanol and nitric acid, and determined by FAAS. The limit of detection for inorganic lead was $0.2\,\mu g\,L^{-1}$ for a sample of 50 mL loaded at $7\,mL\,min^{-1}$. A somewhat more complicated procedure which provided greater resolution among a similar set of analytes was devised by Valcarcel and coworkers [90]. Inorganic lead was precipitated as the chromate which was continuously collected then dissolved in nitric acid and the lead determined by FAAS. The trialkyl cations trimethyllead and triethyllead, in the filtrate, were retained on a column of the fullerene C-60 following derivatization with diethyldithiocarbamate. This retention was selective depending on the conditioning of the column: either *n*-hexane or isobutyl methyl ketone was used.

Despite the considerable interest in the determination of mercury species, there has been very little work published in which an FI solid-phase extraction procedure has formed the basis of the distinction between species. As long ago as 1992, Wei and McLeod [91] devised a method in which methylmercury was selectively retained on sulfhydryl cotton. However, descriptions of further usage of this method are somewhat ambiguous. The original research group applied the procedure to field sampling of the Manchester Ship Canal [92], but also suggested that the material could be loaded with both inorganic Hg as well as organomercury species to form a possible reference material [93]. Cai *et al.* [94] described a procedure in which methyl- and ethylmercury were retained on sulfhydryl cotton followed by elution and conversion to the bromides for subsequent determination by GC with AFS detection. Apparently inorganic mercury was not retained. However, Yu and Yu recently described a procedure [95] in which both inorganic mercury and lead species were retained on the material, whereas Kwokal and Branica devised a procedure [96] for the preconcentration of methylmercury. Both of these methods were applied to the analysis of waters. Frech and coworkers preconcentrated both inorganic and organomercury as the dithiocarbamate complexes prior to elution and determination (after butylation) by GC [97], or by HPLC with post-column hydride generation [98].

Organotin compounds were retained on a silica-based C-18 column as the first step in a GC procedure for the determination of these compounds in water developed by Szpunarlobinska *et al.* [99]. The species were ethylated on the column by the passage of a solution of tetraethylborate, eluted with methanol and injected onto a GC column for separation followed by detection by microwave-induced plasma atomic emission spectrometry (MIP-AES). Grotti *et al.* [100] interfaced the HPLC separation of butyltin compounds with graphite furnace AAS via a HG manifold.

Vanadium(IV) and (V) were determined [101] in river water by a procedure in which V(IV) was selectively retained on a column of Amberlite XAD-7 as the complex with 2-(5-bromo-2-pyridylazo)-5-diethylaminophenol while V(V) was masked with 1,2-cyclohexanediaminetetraacetic acid. The retained vanadium was eluted with nitric acid and determined by ICP-OES. In the absence of the masking agent both oxidation states were retained and so V(V) was determined by difference. The limit of detection for a 10 mL sample was 20 ng L^{-1} for an instrument equipped with an ultrasonic nebulizer.

7 OTHER PROCEDURES

Adsorption of insoluble derivatives on the inner wall of open tubular reactors forms the basis of preconcentration procedures for a number of elements. When coupled with chemistry that is selective for a particular oxidation state of an element, such a procedure has the potential to provide speciation information. Nielsen and Hansen [102] devised such a procedure for the determination of Cr(VI) by ETA-AAS. The Cr(VI) complex with ammonium pyrrolidine dithiocarbamate was retained on a knotted PTFE tubular reactor and then eluted with 55 μL of ethanol directly into the graphite furnace atomizer. A preconcentration for 60 s at 5 mL min^{-1} gave a signal enhancement of 20-fold and a detection limit of 4 ng L^{-1}. The method has also been adapted for implementation in the sequential injection mode [103]. Gaspar and Posta showed [104] that PEEK tubing also collected this complex. They obtained a detection limit by FAAS of 2 μg L^{-1} for a sample volume of 5 mL. In addition to the usual water samples, they applied the method to the analysis of cigarette ash. Yan *et al.* [105] retained the Fe(III) pyrrolidinedithiocarbamate (PDC) complex on the interior of a PTFE reactor from a solution containing 0.07–0.4 M HCl, prior to elution with 1 M nitric acid and detection by ICP-MS. When the acidity of the sample solution was decreased to 0.001–0.004 M HCl, both Fe(II) and Fe(III) were retained and thus Fe(II) could be determined by difference. For loading at 5 mL min^{-1} for 30 s, a detection limit of 80 ng L^{-1} was obtained. The enhancement factor was 12, the retention efficiency was 80 %, and the throughput was 21 h^{-1}. This procedure was an adaptation of an earlier method [106] for the determination of inorganic arsenic species by ICP-MS. The As(III)–PDC complex was retained on PTFE tubing, from solutions whose acidity ranged from 0.01 to 0.7 M with respect to nitric acid, prior to elution with 1 M nitric acid. Neither As(V) nor mono-nor dimethylarsenic species were retained. After reduction of As(V) to As(III) with L-cysteine, total inorganic arsenic was determined (again, the methylated forms were not retained) and hence As(V) was determined by difference. The detection limit was 20–30 ng L^{-1}.

In addition to the retention of organic complexes on the interior of such reactors, it has been shown possible to collect metal hydroxide precipitates and to exploit co-precipitation both as a means of preconcentration and as the basis of a speciation scheme. Zou *et al.* [107] selectively retained Cr(III) by coprecipitation with lanthanum hydroxide, prior to dissolution in 0.5 M HCl and determination by FAAS. For a loading period of 110 s the limit of detection was 0.8 μg L^{-1} and the method was applied to the analysis of water and human hair. Nielsen *et al.* [108] collected Se(IV) with the same chemistry; but, after elution with HCl, the selenium was determined by HG-AAS. The detection limit was an impressive 5 ng L^{-1}.

FI liquid–liquid extractions for speciation have not been developed to any extent, probably because of the difficulty of achieving reliable phase separation. Nielsen *et al.* [109] extracted the PDC complex of Cr(VI) into MIBK, 55 μL of which was then delivered to a graphite furnace for determination by AAS to give a detection limit of 3 ng L^{-1}. Phase separation was achieved in a small conical PTFE vessel with a stainless steel base. The procedure was applied to various water samples including the wastes from incineration and desulfurization plants.

In principle, it is possible to connect flow injection detectors in series (as long as the only destructive detector is the last in the sequence), though there do not appear to be many procedures developed based on this concept. Girard and Hubert [110] described a procedure for the determination of chromium species in which a molecular absorption detector was connected in series with a flame atomic absorption spectrometer. Cr(VI) was detected by visible absorption spectrometry of the 1,5-diphenylcarbazide complex and total chromium was detected by AAS.

The method was applied to the analysis of stainless steel welding dust. It was found that for a material with 30 000 mg L^{-1} of Cr in the solid, of which 25 000 mg L^{-1} was 'extractable', the Cr(VI) content was 22 500 mg L^{-1}. The method was based closely on a procedure first described by Lynch *et al.* [111] in 1984.

FI chemical vapor generation with cryotrapping can be used as the first stage in a procedure in which the derivatives are subsequently separated by GC and detected with an atomic spectrometric detector. A recent example of this methodology [112] demonstrates that considerable temporal separation of the two stages is possible as the derivatives of arsenic, germanium, mercury, and tin with borohydride were formed in an FI system and cryotrapped on board ship with subsequent analysis in the laboratory by GC with ICP-MS detection. Alkylarsenic compounds other than the methylated species were tentatively identified in samples taken from the Rhine estuary.

8 COMPARISON WITH OTHER SPECIATION PROCEDURES

The capability of FI-based speciation procedures is somewhat limited when compared with the performance of procedures based on high performance liquid or gas chromatography coupled with element-specific detection, in the sense that FI procedures are only capable of providing information about a limited number of analyte species. FI procedures become rather cumbersome when the goal of the analysis is to quantify four (or more) components. However, if the goal is to provide information about a limited number of components based on, for example, the classifications 'inorganic', 'organic', 'toxic', 'nontoxic', 'fast reacting', and so on, then FI procedures offer the advantages of speed, simplicity and relatively low cost. While a fully automated, computer-controlled FI system will be considerably more expensive than a manually operated system, the capital and running costs will be considerably less than those of a high performance liquid chromatograph. As other analytical performance parameters, such as detection limit are a function of the detection technique; to a first approximation these are the same for the two approaches. However, it should be borne in mind that HPLC procedures cause a substantial on-line dilution (maybe by as much as a factor of 100) and may thus be inferior to procedures in which there is preconcentration step (such as by trapping a generated hydride on the interior of a graphite furnace). It should not be assumed that useful speciation information can only be obtained by the combination of some high resolution separation technique coupled with plasma-source mass spectrometric detection.

9 REFERENCES

1. Ruzicka, J. and Hansen, E. H., *Flow Injection Analysis*, 2nd edn, John Wiley & Sons, Inc., New York, 1988, p. 380.
2. Chalk, S. J., *The Flow Analysis Database*, http://www.fia.unf.edu/fad.lasso (accessed June 2001).
3. Sturgeon, R. E., *J. Anal. At. Spectrom.*, **13**, 351 (1988).
4. Sanz-Medel, A. (Ed.), *Flow Analysis with Atomic Spectrometric Detectors*, Elsevier, Amsterdam, 1999.
5. Beauchemin, D., *J. Anal. At. Spectrom.*, **13**, 1 (1988).
6. Dempster, M. A. and Marcus, R. K., *J. Anal. At. Spectrom.*, **14**, 43 (1999).
7. Valcarcel, M. and Luque de Castro, M. D., *Non-chromatographic Continuous Separation Techniques*, Royal Society of Chemistry, Cambridge, 1991.
8. Fang, Z., *Flow Injection Separation and Preconcentration*, VCH, New York, 1993.
9. Welz, B., *Spectrochim. Acta, Part B*, **53**, 169 (1998).
10. Campanella, L., Pyrzynska, K. and Trojanowicz, M., *Talanta*, **43**, 825 (1996).
11. Oygard, J. K., Lundebye, A. K. and Julshamn, K., *J. AOAC Int.*, **82**, 1217 (1999).
12. Woolson, E. A., Axley, J. H. and Kearney, P. C., *Soil Sci. Soc. Am. Proc.*, **37**, 254 (1973).
13. Iu, K. L., Pulford, I. D. and Duncan, H. J., *Anal. Chim. Acta*, **106**, 319 (1979).
14. Van Loon, J. C., *Selected Methods of Trace Metal Analysis*, Wiley-Interscience, New York, 1985, p. 285.
15. Lin, S. L., Man, H. and Shu, Z., *Lab. Robotics Automat.*, **12**, 86 (2000).
16. Pilhar, B. and Costa, L., *Anal. Chim. Acta*, **114**, 275 (1980).
17. Burguera, J. L., Burguera, M., Rivas, C. and Carrero, P., *Talanta*, **45**, 531 (1998).
18. Dedina, J. and Tsalev, D., *Hydride Generation Atomic Absorption Spectrometry*, John Wiley and Sons, Ltd, Chichester, 1995.

19. Dedina, J., Flow methods in gas–liquid separations, in *Flow Analysis with Atomic Spectrometric Detectors*, Sanz-Medel, A. (Ed.), Elsevier, Amsterdam, 1999.
20. Howard, A. G., *J. Anal. At. Spectrom.*, **12**, 267 (1997).
21. Tsalev, D. L., *J. Anal. At. Spectrom.*, **14**, 147 (1999).
22. Jin, Q., Liang, F., Zhang, H., Zhao, L., Huan, Y. and Song, D., *Trends Anal. Chem.*, **18**, 479 (1999).
23. Burguera, M. and Burguera, J. L., *Quim. Anal.*, **15**, 112 (1996).
24. Feng, Y. L., Narasaki, H., Chen, H. Y. and Tian, L. C., *Anal. Chim. Acta*, **386**, 297 (1999).
25. Ulrich, N., *Anal. Chim. Acta*, **417**, 201 (2000)
26. Deng, T. L., Chen, Y. W. and Belzile, N., *Anal. Chim. Acta*, **432**, 293 (2001).
27. Nielsen, S. and Hansen, E. H., *Anal. Chim. Acta*, **343**, 5 (1997).
28. Krenzelok, M. and Rychlovsky, P., *Coll. Czech. Chem. Commun.*, **63**, 2027 (1998).
29. Willie, S. N., *Spectrochim. Acta, Part B*, **51**, 1781 (1996).
30. Cabon, J. Y. and Cabon, N., *Anal. Chim. Acta*, **418**, 19 (2000).
31. Cabon, J. Y. and Cabon, N., *Fresenius' J. Anal. Chem.*, **368**, 484 (2000)
32. Rude, T. R. and Puchelt, H., *Fresenius' J. Anal. Chem.*, **350**, 44 (1994).
33. Le, X. C., Cullen, W. R. and Riemer, K. J., *Anal. Chim. Acta*, **285**, 277 (1994).
34. Tsalev, D. L., Sperling, M. and Welz, B., *Talanta*, **51**, 1059 (2000).
35. Tsalev, D. L., Sperling, M. and Welz., B., *Spectrochim. Acta, Part B*, **55**, 339 (2000).
36. Pyell, U., Dworkschak, A., Nitcshke, E. and Neidhart, B., *Fresenius' J. Anal. Chem.*, **363**, 495 (1999).
37. Chatterjee, A., Das, D., Mandal, B. K., Chowdhury, T. R., Samanta, G. and Chakraborti, D., *Analyst*, **120**, 643 (1995).
38. Fernandez, M. G. C., Palacios, M. A. and Camara, C., *Anal. Chim. Acta*, **283**, 386 (1993).
39. Stripeikis, J., Costa, P., Tudino, M. and Troccoli, O., *Anal. Chim. Acta*, **408**, 191 (2000).
40. Burguera, J. L., Carrero, P., Burguera, M., Rondon, C., Brunetto, M. R. and Gallignani, M., *Spectrochim. Acta, Part B*, **51**, 1837 (1996).
41. Gallignani, M., Valero, M., Brunetto, M. R., Burguera, J. L., Burguera, M. and de Pena, Y. P., *Talanta*, **52**, 1051 (2000).
42. He, Y. Z., Azouzi, H., Cervera, M. L. and de la Guardia, M., *J. Anal. At. Spectrom.*, **13**, 1291 (1988).
43. Moreno, A. E., Perez-Conde, C. and Camara, C., *J. Anal. At. Spectrom.*, **15**, 681 (2000).
44. Varade, C. M. R. and Luque de Castro, M. D. L., *J. Anal. At. Spectrom.*, **13**, 787 (1998).
45. Cabon, J. Y. and Erler, W., *Analyst*, **123**, 1565 (1998).
46. Vilano, M., Padro, A., Rubio, R. and Rauret, G., *J. Chromatogr. A*, **819**, 211 (1998).
47. La Fuente, J. M. G., Dlaska, M., Sanchez, M. L. F. and Sanz-Medel, A., *J. Anal. At. Spectrom.*, **13**, 423 (1998).
48. Chatterjee, A., Shibata, Y., Yoneda, M., Banerjee, R., Uchida, M., Kon, H. and Morita, M., *Anal. Chem.*, **73**, 3181 (2001).
49. Chatterjee, A. and Shibata, Y., *Anal. Chim. Acta*, **398**, 273 (1999).
50. Gallignani, M., Bahsas, H., Brunetto, M. R., Burguera, M., Burguera, J. L. and de Pena, Y. P., *Anal. Chim. Acta*, **369**, 57 (1988).
51. Rio-Segade, S. and Bendicho, C., *J. Anal. At. Spectrom.*, **14**, 263 (1999).
52. Rio-Segade, S. and Bendicho, C., *Spectrochim. Acta, Part B*, **54**, 1129 (1999).
53. Capelo, J. L., Lavilla, I. and Bendicho, C., *Anal. Chem.*, **72**, 4979 (2000).
54. Tao, G. H., Willie, S. N. and Sturgeon, R. E., *J. Anal. At. Spectrom.*, **14**, 1929 (1999).
55. Burguera, J. L., Quintana, I. A., Salager, J. L., Burguera, M., Rondon, C., Carrero, P., de Salager, R. A. and de Pena, Y. P., *Analyst*, **124**, 593 (1999).
56. Puk, R. and Weber, J. H., *Anal. Chim. Acta*, **292**, 175 (1994).
57. Palmer, C. D., Ph.D. Dissertation, University of Massachusetts, Amherst, 2001.
58. Palmer, C. D., Rio-Segade, S. and Tyson, J. F., work in progress.
59. Mota, J. P. V., de la Campa, M. R. F. and Sanz-Medel, A., *J. Anal. At. Spectrom.*, **13**, 431 (1998).
60. Infante, H. G., Sanchez, M. L. F. and Sanz-Medel, A., *J. Anal. At. Spectrom.*, **15**, 519 (2000).
61. Bettmer, J. and Cammann, K., *Appl. Organomet. Chem.*, **8**, 615 (1994).
62. Tsalev, D. L., Sperling, M. and Welz, B., *Spectrochim. Acta, Part B*, **55**, 339 (2000).
63. Pozebon, D., Dressler, V. L., Neto, J. A. G. and Curtius, A. J., *Talanta*, **45**, 1167 (1998).
64. Tyson, J. F., *J. Anal. At. Spectrom.*, **14**, 169 (1999).
65. Burguera, J. L., Burguera, M. and Rivas, C., *Quim. Anal.*, **16**, 165 (1997).
66. Yalcin, S. and Le, X. C., *J. Environ. Monitor.*, **3**, 81 (2001).
67. Sperling, M., Xin, X. F. and Welz, B., *Analyst*, **117**, 629 (1992).
68. Sperling, M., Xu, S. K. and Welz, B., *Anal. Chem.*, **64**, 3101 (1992).
69. Vassileva, E., *Analysis*, **28**, 878 (2000).
70. Yu, J. C., Wu, X. J. and Chen, Z. L., *Anal. Chim. Acta*, **436**, 59 (2001).
71. Naghmush, A. M., Pyrzynska, K. and Trojanowicz, M., *Anal. Chim. Acta*, **288**, 247 (1994).
72. Jiminez, M. S., Martin, L., Mir, J. M. and Castillo, J. R., *At. Spectrosc.*, **17**, 201 (1996).
73. Kelko-Levai, A., Varga, I., Zih-Perenyi, K. and Lasztity, A., *Specrochim. Acta, Part B*, **54**, 827 (1999).
74. Rao, T. P., Karthikeyan, S., Vijayalekshmy, B. and Iyer, C. S. P., *Anal. Chim. Acta*, **369**, 69 (1998).

75. Cespon-Romero, R. M., Yebra-Biurrun, M. C. and Bermejo-Barrero, M. P., *Anal. Chim. Acta*, **327**, 37 (1996).
76. Larraya, A., Cobofernandez, M. G., Palacios, M. A. and Camara, C., *Fresenius' J. Anal. Chem.*, **350**, 667 (1994).
77. Pyrzynska, K., Drzewicz, P. and Trojanowicz, M., *Anal. Chim. Acta*, **363**, 141 (1998).
78. Bryce, D. W., Izquierdo, A. and Luque De Castro, M. D., *J. Anal. At. Spectrom.*, **10**, 1059 (1995).
79. Yan, X. P., Sperling, M. and Welz, B., *Anal. Chem.*, **71**, 4353 (1999).
80. Carrero, P. E. and Tyson, J. F., *Analyst*, **122**, 915 (1997).
81. Carrero, P. E. and Tyson, J. F., *Spectrochim. Acta, Part B*, **53**, 1931 (1998).
82. Fairman, B. and Sanz-Medel, A., *Fresenius' J. Anal. Chem.*, **355**, 757 (1996).
83. Fairman, B. and Sanz-Medel, A., *J. Anal. At. Spectrom.*, **10**, 281 (1995).
84. Garbos, S., Rzepecka, M., Bulska, E. and Hulanicki, A., *Spectrochim. Acta, Part B*, **54**, 873 (1999).
85. Garbos, S., Bulska, E. and Hulanicki, A., *At. Spectrosc.*, **21**, 128 (2000).
86. Comber, M. H. I., Eales, G. J. and Nicholson, P. J. D., *J. Automat. Chem.*, **14**, 5 (1992).
87. Abollino, O., Aceto, M., Bruzzoniti, M. C., Mentasti, E. and Sarzanini, A., *Anal. Chim. Acta*, **375**, 299 (1998).
88. Bagheri, H., Gholami, A. and Najafi, A., *Anal. Chim. Acta*, **424**, 233 (2000).
89. Naghmush, A. M., Pyrzynska, K. and Trojanowicz, M., *Talanta*, **42**, 851 (1995).
90. Baena, J. R., Gallego, M. and Valcarcel, M., *Spectrochim. Acta*, **54**, 1869 (1999).
91. Wei, J. and McLeod, C. W., *Talanta*, **39**, 1537 (1992).
92. Jian, W., Mena, M. L., McLeod, C. W. and Rollins, J., *Int. J. Environ. Anal. Chem.*, **57**, 99 (1994).
93. Mena, M. L. and McLeod., C. W. *Mikrochim. Acta*, **123**, 103 (1996).
94. Cai, Y., Jaffe, R., Alli, A. and Jones, R. D., *Anal. Chim. Acta*, **334**, 251 (1996).
95. Yu, F. and Yu, T. L., *Spectrosc. Spectr. Anal.*, **20**, 898 (2000).
96. Kwokal, Z. and Branica, M., *Croat. Chem. Acta*, **73**, 97 (2000).
97. Emteborg, H., Baxter, D. C. and Frech, W., *Analyst*, **118**, 1007 (1993).
98. Xin, X. F., Frech, W., Hoffmann, E., Ludke, C. and Skole, J., *Fresenius' J. Anal. Chem.*, **361**, 761 (1998).
99. Szpunarlobinska, J., Ceulemans, M., Lobinski, R. and Adams, F. C., *Anal. Chim. Acta*, **278**, 99 (1993).
100. Grotti, M., Rivaro, P. and Franche, R., *J. Anal. At. Spectrom.*, **16**, 270 (2001).
101. Wuilloud, R. G., Wuilloud, J. C., Olsina, R. A. and Martinez, L. D., *Analyst*, **126**, 715 (2001).
102. Nielsen, S. and Hansen, E. H., *Anal. Chim. Acta*, **366**, 163 (1998).
103. Nielsen, S. and Hansen, E. H., *Anal. Chim. Acta*, **422**, 47 (2000).
104. Gaspar, A. and Posta, J., *Anal. Chim. Acta*, **354**, 151 (1997).
105. Yan, X. P., Hendry, N. J. and Kerrich, R., *Anal. Chem.*, **72**, 1879 (2000).
106. Yan, X. P., Kerrich, R. and Hendry, N. J., *Anal. Chem.*, **70**, 4736 (1998).
107. Zou, H. F., Xu, S. K. and Fang, Z. L., *At. Spectrosc.*, **17**, 112 (1996).
108. Nielsen, S., Sloth, J. J. and Hansen, E. H., *Analyst*, **121**, 31 (1996).
109. Nielsen, S. C., Sturup, S., Spliid, H. and Hansen, E. H., *Talanta*, **49**, 1027 (1999).
110. Girard, L. and Hubert, J., *Talanta*, **43**, 1965 (1996).
111. Lynch, T. P., Kernoghan, N. J. and Wilson, J. N., *Analyst*, **109**, 839 (1984).
112. Tseng, C. M., Amouroux, D., Brindle, I. D. and Donard, O. X. F., *J. Environ. Monitor.*, **2**, 603 (2000).

5.3 Detection by ICP-Mass Spectrometry

Frank Vanhaecke

Laboratory for Analytical Chemistry, Ghent University, Belgium

Gunda Köllensperger

Analytical Chemistry Research Group, BOKU Universität für Bodenkultur Wien, Vienna, Austria

1 INDUCTIVELY COUPLED PLASMA-MASS SPECTROMETRY

1.1 Introduction

Inductively coupled plasma-mass spectrometry (ICPMS) is a remarkably powerful and versatile technique for (ultra)trace element determination, characterized by extremely low limits of detection (LODs), a wide linear dynamic range, multielement capabilities, surveyable spectra and a high sample throughput. As such, ICPMS cannot be used for elemental speciation, as all molecules introduced into the high temperature ion source (an Ar ICP) are broken down into atoms, which are subsequently ionized. However, if the different species of interest can be separated from one another *before* their introduction into the plasma, ICPMS can be used as a highly sensitive and element-specific on-line multielement detector.

1.2 Operating principle

Although the use of laser ablation or – to a lesser extent – electrothermal vaporization permits the

Handbook of Elemental Speciation: Techniques and Methodology R. Cornelis, H. Crews, J. Caruso and K. Heumann
 ISBN: 0-471-49214-0

direct analysis of solid samples, ICPMS is mainly intended for the analysis of (aqueous) sample solutions. Traditionally, such a sample solution is converted into an aerosol by means of a pneumatic nebulizer. In order to ensure a stable plasma and an efficient atomization and ionization in the ICP, the larger droplets ($d > 10\,\mu m$) are removed from the sample aerosol by means of a spray chamber. Subsequently, the aerosol is swept by the Ar nebulizer or carrier gas into the ICP, which can be considered as an extremely hot (ionization temperature approximately 7500 K) electrical flame, generated at the end of a quartz torch. During their residence in this ICP, the droplets are desolvated and the sample molecules are broken down into atoms, which are subsequently excited and ionized. For the majority of elements, the efficiency of ionization in an Ar ICP exceeds 90 % (Figure 5.3.1), and in spite of their high first ionization potential, important metalloids and nonmetals, such as As, Se, S or Cl, are still sufficiently ionized to allow sensitive determination.

Since the ICP is operated under atmospheric pressure, whereas in the mass spectrometer a high vacuum is required, an interface between both components is necessary (Figure 5.3.2). This interface consists of two successive, coaxial and water-cooled cones, with a small central aperture – the sampling cone and the skimmer. As a result of the difference in pressure between the expansion chamber – the region between the sampling cone and the skimmer – and the ion source, a fraction of the ICP is extracted into the interface region and undergoes supersonic expansion [3]. Because of the sudden drop in particle density, ion–electron recombination or other reactions are avoided, such that the composition of the extracted gas is 'frozen' and hence, is representative of that of the ICP. The majority of the extracted gas is subsequently evacuated by means of a vacuum pump, but a central

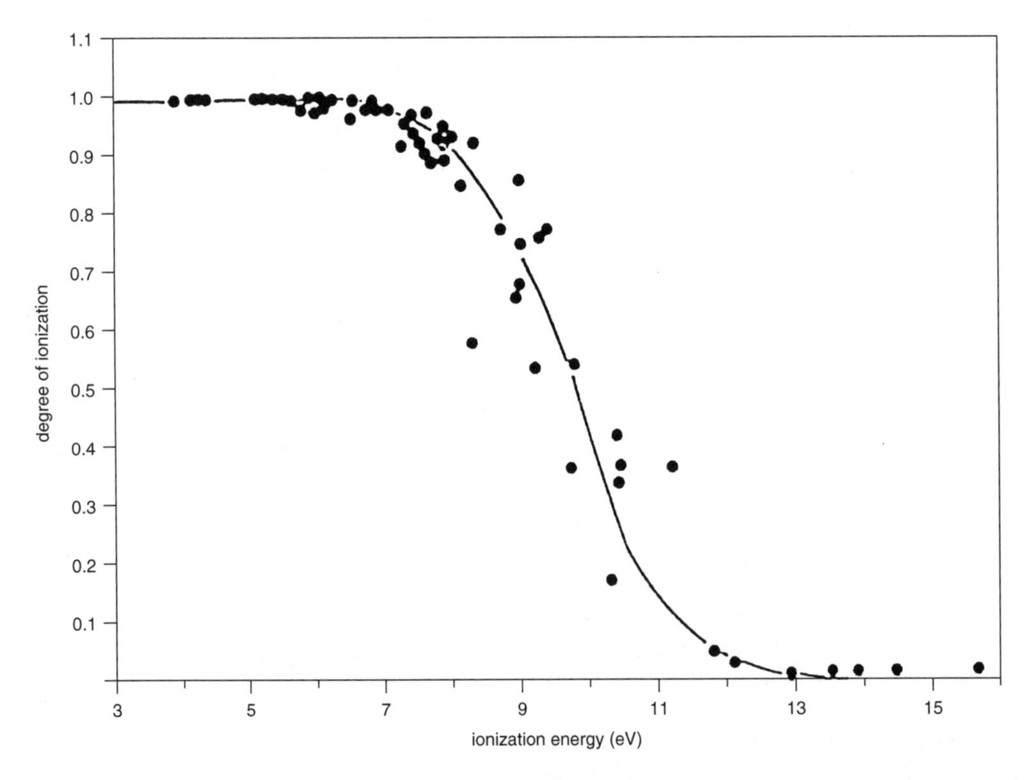

Figure 5.3.1. Ionization efficiency in the ICP as a function of the analyte's first ionization energy [1]. Reprinted from *An Introduction to Analytical Atomic Chemistry*, L. Ebdon, E. H. Evans (Ed.), A. Fisher and S. J. Hill, 1998, © John Wiley & Sons Limited. Reproduced with permission.

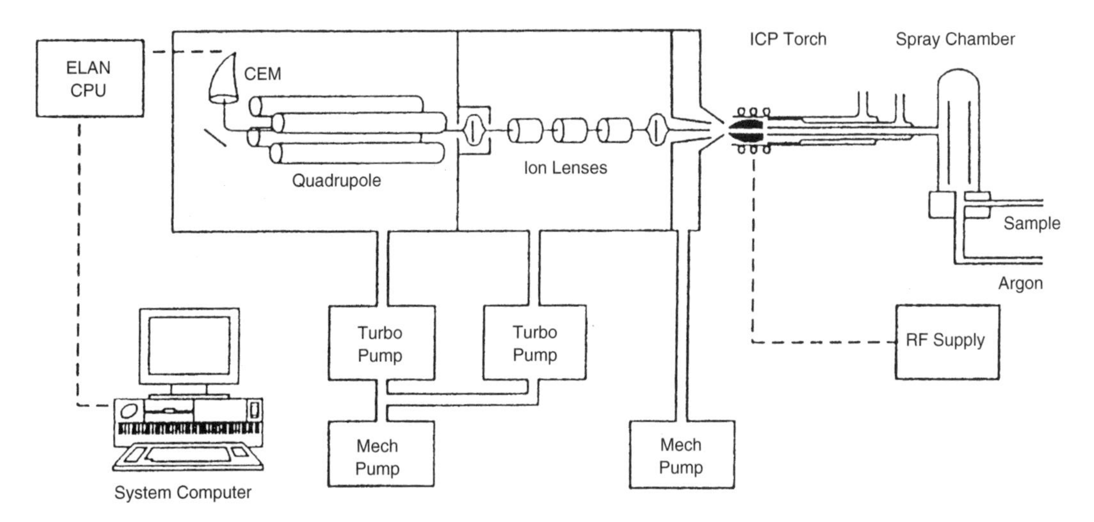

Figure 5.3.2. Schematic diagram of a quadrupole-based ICPMS instrument [2]. Reproduced by permission of PerkinElmer Sciex.

beam enters a higher vacuum region by passing through the skimmer aperture. This beam consists of ions, electrons and neutral particles. A negatively charged extraction lens selectively attracts the positive ions, which are subsequently transported and introduced as efficiently as possible into the mass analyzer. The latter is accomplished by the ion lens system, the construction and complexity of which differ markedly from one type of instrument to another.

Originally, all ICP-mass spectrometers available were equipped with a quadrupole filter for mass analysis. Although nowadays instruments equipped with a double-focusing sector field mass spectrometer, a time-of-flight analyzer or even an ion trap are also commercially available, the majority (approximately 90 %) of instruments used worldwide are still equipped with a quadrupole filter. Such a quadrupole filter acts as a bandpass mass filter and transmits only those ions with a mass-to-charge ratio within a narrow mass window (approximately 1 u). Of course, by changing the voltages applied to the quadrupole rods, the position of the mass window can be selected and the operator has the choice between *scanning* of the entire mass spectrum or a smaller mass region on the one hand or monitoring the intensity of a limited number of pre-selected analyte signals (*peak jumping or hopping*) on the other.

Finally, the ions transmitted by the mass analyzer have to be detected. The majority of instruments use an electron multiplier for this purpose, although some manufacturers equip their instrumentation with a Daly-type detector. An electron multiplier is an extremely sensitive detection device, permitting each individual ion to be counted. The intensity of each output signal is compared with that of a threshold to suppress the background noise. Photon noise on the other hand is reduced by using a photon or shadow stop and/or by mounting the detector or the entire mass analyzer off axis. In order to cope with higher count rates, electron multipliers can also be operated in analog mode, where ions are no longer counted, but an output signal, the intensity of which is proportional to the intensity of the ion beam striking the detector surface, is measured instead.

1.3 Figures of merit

LODs of course vary among instruments (governed by background noise and sensitivity) and from one element to the other (governed by mass number, ionization potential and the isotopic abundance of the isotope monitored). When not limited by a very high ionization potential (e.g., Cl, Br) or an increased background due to blank contamination and/or interfering ions (see later), low- to sub-ng L^{-1} LODs can be obtained. Owing to the extremely low background observed with sector field ICPMS (<0.2 counts s^{-1} vs. <1–10 counts s^{-1} for quadrupole-based instruments) and its high ion transmission efficiency, the instrumental LOD with this type of instrumentation is $\leq$1 pg L^{-1}. In many instances, it is no longer the instrumental detection power, but the purity of the reagents and recipients used that limits the LODs attainable. Additionally, modern ICPMS instruments offer a linear dynamic range of 8 to 9 orders of magnitude. Short-term RSDs on signal intensities of $\leq$2 % are typical, while the isotope ratio precision is $\geq$0.05 %. Although the robustness of ICPMS instruments has been improved during the past decade, it is not advisable to introduce solutions with a total content of dissolved solids exceeding 1–2 g L^{-1}.

While quadrupole-based ICPMS instruments show only unit mass resolution (peak width approximately 0.5 u over the entire mass spectrum), sector field ICPMS also permits measurements at a higher mass resolution ($R_{maximum} \geq$ 10 000). Hence, in many instances, spectral overlap of signals from ions showing the same *nominal* mass can be avoided, permitting lower LODs to be obtained for traditionally problematic elements. This topic will be discussed into more detail in a later section.

With quadrupole-based ICPMS instrumentation, the complete mass range – from Be to U – can be scanned in approximately 0.1 s. Owing to hysteresis of the magnetic sector, the scanning speed with a sector field mass spectrometer is lower and scanning the full mass range takes about 0.5 s. With a recently introduced newly developed magnetic field regulator however, the scanning speed can be enhanced to a value almost similar to that of quadrupole-based instrumentation. With TOF-ICPMS on the other hand, up to 30 000 full mass spectra can be recorded every second. The sensitivity of TOF-ICPMS instrumentation, however, is inferior to that of quadrupole-based ICPMS by 1–2 orders of magnitude. Additionally, a recent investigation showed that, even when working with transient signals only lasting a few seconds, the 'simultaneous' monitoring of up to 20 mass-to-charge ratios – a number hardly ever exceeded in speciation work – is feasible with quadrupole-based instrumentation [4].

Finally, the purchase price (2002) of a quadrupole-based or TOF-based ICPMS instrument is in the order of 200 000 €; that of a sector field instrument is about twice as high.

1.4 Spectral interferences

1.4.1 Introduction

Very soon after the introduction of ICPMS, it became clear that nonspectral (matrix-induced signal suppression or enhancement) and spectral (overlap of the signals of ions showing a difference in mass <0.5 u) interferences were its most prominent disadvantages. While, in most cases, nonspectral interferences could be fairly easily coped with – e.g., by means of sample dilution, the use of (a) carefully selected internal reference(s) or application of standard additions or isotope dilution instead of (an) external standard(s) for calibration – the occurrence of spectral interferences proved to be more troublesome. Despite the high temperature in the ICP, molecular ions, originating from the plasma gas (Ar), entrained air, the solvent and/or the matrix, occur in ICPMS and their signal may complicate the mass spectrum and analyte quantification to a large extent. Especially for complex matrices and in the lower mass range ($\leq$80 u), obtaining accurate results or sufficiently low LODs is therefore not self-evident. As an illustration, Table 5.3.1 presents a number of elements – the majority of which are of interest from a biomedical and/or environmental point of view – and the molecular ions potentially giving origin to spectral overlap with the signal from the most abundant analyte isotope.

Self-evidently, throughout the years, efforts have been made to avoid, reduce or correct for the effects of spectral interferences. In the first place, whenever possible, problems should be avoided by appropriate selection of the nuclide(s) monitored, although selecting an isotope with a lower

Table 5.3.1. Examples of spectral interferences *potentially* encountered using ICPMS. For each element, only the major isotope (highest isotopic abundance) is given. The *actual* occurrence of spectral overlap depends on the matrix composition.

Element	Major isotope and isotopic abundance (%)	Molecular ions (potentially) causing spectral overlap at low mass resolution
Al	^{27}Al, 100 %	$^{12}C^{14}NH^+$, $^{13}C^{14}N^+$
Si	^{28}Si, 92.2 %	$^{14}N_2{}^+$, $^{12}C^{16}O^+$
P	^{31}P, 100 %	$^{14}N^{16}OH^+$
S	^{32}S, 95.0 %	$^{16}O_2{}^+$
K	^{39}K, 100 %	$^{38}ArH^+$
Ca	^{40}Ca, 96.9 %	$^{40}Ar^+$
Sc	^{45}Sc, 100 %	$^{12}C^{16}O_2H^+$, $^{28}Si^{16}OH^+$, $^{29}SiO^+$
Ti	^{48}Ti, 73.8 %	$^{32}S^{16}O^+$, $^{31}P^{16}OH^+$
V	^{51}V, 99.8 %	$^{35}Cl^{16}O^+$, $^{37}Cl^{14}N^+$
Cr	^{52}Cr, 83.8 %	$^{40}Ar^{12}C^+$, $^{35}Cl^{16}OH^+$
Fe	^{56}Fe, 91.7 %	$^{40}Ar^{16}O^+$, $^{40}Ca^{16}O^+$
Cu	^{63}Cu, 69.2 %	$^{40}Ar^{23}Na^+$, $^{31}P^{16}O_2{}^+$
Zn	^{64}Zn, 48.6 %	$^{32}S^{16}O_2{}^+$, $^{31}P^{16}O_2H^+$
As	^{75}As, 100 %	$^{40}Ar^{35}Cl^+$
Se	^{80}Se, 49.6 %	$^{40}Ar_2{}^+$

isotopic abundance obviously leads to a deterioration in sensitivity and detection power. Especially in elemental speciation work, where total element concentrations that are already low are further distributed over different species, such a reduction in sensitivity is often not acceptable. Also the conditions of sample preparation and species separation should be selected such that the occurrence of spectral interferences is avoided to the largest possible extent. In relatively simple cases, mathematical correction has also been used successfully, although very rapidly one may run into rather complex calculation schemes, requiring the use of dedicated computer programs. Sometimes, even blank correction, preferably using a matrix-matched blank, may be sufficient. However, very often, more effective, technically more complex and unfortunately more expensive measures are required for coping with spectral overlap.

1.4.2 *Cool plasma conditions*

With all modern ICPMS instruments, the plasma can be operated under cool plasma conditions [3, 5] – obtained by using a low RF power and an increased nebulizer gas flow rate. In order to also enable operation of the ICP under cool plasma conditions with instruments equipped with a load coil that is not electrically balanced, insertion of a grounded metal plate in between the coil and the ICP torch is necessary to capacitively decouple both components [5]. Otherwise, the occurrence of intense secondary discharges counteracts the desired effect. The use of cool plasma conditions leads to a substantial reduction in the intensity of Ar^+ and Ar-based ions and hence makes the determination of ultratrace amounts of traditionally difficult elements, such as K, Ca and Fe, possible. Unfortunately, the use of these cool plasma conditions also brings about important disadvantages: elements characterized by a high ionization potential are no longer efficiently ionized, the signal intensity of other types of molecular ions (e.g., oxide ions) increases and matrix effects become more pronounced. Hence, cool plasma conditions are only useful for the determination of relatively light elements in fairly clean and simple matrices and are predominantly used in the semiconductor industry for the analysis of high purity chemical reagents. Their use in speciation work has only seldom been reported.

Vanhaecke *et al.* [6] evaluated the merits of cool plasma conditions to cope with the spectral overlap of the signals of ArC^+ and $ClO(H)^+$ with $^{52}Cr^+$ and $^{53}Cr^+$ analyte signals. This overlap hampered accurate quantification of Cr(III) and Cr(VI) species, separated from one another using a microbore anion exchange HPLC column, in industrial process solutions. They came to the conclusion that, while the intensity of ArC^+ could be sufficiently suppressed, the ratio $ClO(H)^+/Cr^+$ deteriorated on switching to cool plasma conditions. Therefore, these authors preferred to use a higher mass resolution (see later) instead. Despite the important loss in signal intensity inherent to an increase in mass resolution, sub-$\mu g\,L^{-1}$ LODs were obtained. The increased tendency of oxide formation observed under cool plasma conditions was used to advantage by Divjak and Goessler [7] in sulfur speciation. Sulfide, sulfite, sulfate and thiosulfate were separated from one another using an anion exchange HPLC column, while detection was accomplished using ICPMS. Since the signal of $^{32}S^+$ suffers from a major spectral overlap by the O_2^+ signal, the instrument was tuned for a maximum SO^+/S^+ ratio and the $^{32}SO^+$ ion signal was monitored for analytical purposes instead. In the case of highly saline samples, the determination of sulfide was reported to be hindered by severe signal suppression.

1.4.3 Aerosol desolvation

Aerosol desolvation reduces the occurrence of oxide- and Cl-based ions. Since the introduction of ICPMS, cooled spray chambers – originally by means of a thermostated water jacket, later on also by means of Peltier elements – have been used to limit the introduction of water (vapor) into the plasma. More efficient ways of sample introduction, e.g., thermospray or ultrasonic nebulization, require the use of more effective desolvation systems [8]. Basically, these units consist of a heated part, in which the solvent is vaporized and a cooled part, in which the solvent vapor is condensed and removed. Membrane desolvation is a more novel approach. In a membrane desolvator, the Ar carrier gas loaded with sample aerosol is directed through a cylinder, the walls of which are made out of a microporous material. Heated Ar (the sweep gas) flows around this central cylinder in the opposite direction to the carrier gas flow. Because of the elevated temperature, the solvent is vaporized and the gaseous solvent molecules can penetrate the semipermeable membrane and are carried off by the aforementioned sweep gas. In addition to reducing the level of oxide- and Cl-based molecular ions, application of a membrane desolvator also permits the direct analysis of (volatile) organic solvents [9]. In the latter case, addition of O_2 (at a low gas flow rate) to the plasma is advisable to avoid carbon deposition on the torch and the interface.

1.4.4 Multipole collision cell

Several manufacturers produce ICPMS instrumentation equipped with a multipole collision cell. Such a collision cell consists of six (hexapole) or eight (octopole) rods to which an RF-only voltage is applied. In order to cope with the occurrence of Ar-based molecular ions, which give rise to spectral overlap, H_2 can be added to the cell. As a result of the occurrence of selective ion–molecule reactions, e.g., charge, atom and proton transfer, the signal intensities of Ar^+ and of Ar-containing molecular ions such as ArC^+, ArO^+, $ArCl^+$ and Ar_2^+ are suppressed by three or more orders of magnitude, such that trace levels of Ca, Cr, Fe, As and Se can be accurately determined [10, 11]. Application of He as a thermalization gas results in both an improvement in the ion transmission efficiency (collisional focusing) and an enhanced reaction efficiency.

Marchante-Gayón *et al.* [12] used reversed phase and ion-pairing HPLC to separate the Se species present in human urine from one another. By introducing H_2 as a reaction gas and He as a thermalization gas in the hexapole collision cell of the ICPMS detector, the signal intensity of the argon dimer (Ar_2^+) ion, which normally precludes the use of the most abundant Se isotope ^{80}Se (isotopic abundance: 49.6 %), could be suppressed by orders of magnitude. This experimental set-up was used for urinary Se speciation before and after intake of commercially available nutritional supplements.

1.4.5 Dynamic reaction cell

The dynamic reaction cell is a similar approach, but owing to the use of a quadrupole unit instead of a hexapole or octopole configuration the cell can be simultaneously used as a bandpass mass filter. As a result, the lifetime of newly created and unwanted species can be limited, such that they neither give rise to a signal in the mass spectrum nor take part in further reactions [13]. As a result, a larger selection of gases, e.g., NH_3, CH_4, or NO_2, can also be used, such that the application range is not limited to Ar^+ and Ar-containing molecular ions. Spectral overlap due to the occurrence of an oxide ion can, e.g., be avoided by converting the latter into a dioxide or higher oxide ion [14] and similar atom transfer reactions have also been shown to be successful for overcoming isobaric overlap [15]. Since hexapole or octopole arrangements do not show mass filtering capabilities, an extension of the application range of collision cells is aimed at via discrimination between analyte ions and newly (in-cell) created ions on the basis of their energy [16].

By using CH_4 as a reaction gas in a dynamic cell, Sloth and Larsen [17] were able to suppress the signal intensity of the $^{40}Ar_2^+$ ion by approximately five orders of magnitude. Hence, monitoring of the major isotope of Se (^{80}Se) was enabled and sensitive detection of selenoamino acids, separated by cation exchange HPLC, could be accomplished (absolute limit of detection: approximately 5 pg as Se).

1.4.6 High mass resolution

Finally, the most universal strategy to cope with spectral overlap is created by using a double-focusing sector field mass spectrometer instead of the more traditional quadrupole filter in ICPMS instrumentation. Sector field mass spectrometers offer a far higher mass resolution than do quadrupole filters, such that ions differing in mass by only a fraction of a mass unit can still be separated from one another (Figure 5.3.3) and straightforward quantification becomes self-evident, despite the presence of molecular or doubly charged ions of the same *nominal* mass-to-charge ratio. The maximum mass resolution ($\geq$ 10 000) offered by present-day sector field ICPMS instruments is sufficient to overcome the large majority of spectral interferences known, the only limitation of this approach being the substantial loss in ion transmission efficiency observed on increasing the mass resolution.

Rottmann and Heumann [19] were the first to report on the use of a sector field ICPMS instrument operated at a higher mass resolution as a detector in a speciation study. This study aimed at obtaining insight into the association of metals with different fractions of the dissolved organic matter in natural waters and applied a combination of an HPLC system, equipped with a size exclusion column, and an ICPMS instrument for this purpose. Interference-free determination of Fe, whose accurate quantification is hampered at low mass resolution due to spectral overlap of the signals of $^{56}Fe^+$ and $^{40}Ar^{16}O^+$, was accomplished at a resolution setting of 3000. Cabezuelo *et al.* [20] used anion exchange fast protein liquid chromatography coupled to sector field ICPMS for studying Al species in blood serum of both ureamic patients and healthy subjects. Accurate determination of low Al concentrations was hindered by the occurrence of $^{13}C^{14}N^+$, $^{12}C^{15}N^+$ and $^{12}C^{14}NH^+$. These molecular ions mainly originated from the ammonium acetate mobile phase used. At a higher mass resolution setting ($R = 3000$) however, Al could be measured interference free. Hence, despite the fact that its basal level in normal serum is below $5\,\mu g\,L^{-1}$ (total element concentration), speciation of Al was accomplished. Vanhaecke *et al.* [6] used a higher mass resolution to ensure accurate results in Cr speciation work. Cr(III) and Cr(VI) were established to co-elute from the anion exchange column with Cl^- and HCO_3^-, respectively, such that the signals of both $^{52}Cr^+$ and $^{53}Cr^+$ suffered from spectral overlap, due to the occurrence of $^{40}Ar^{12,13}C^+$, $^{35}Cl^{16}OH^+$ and $^{37}Cl^{16}O^+$ ions. At a resolution setting of 3000, both Cr signals could be measured interference free.

1.5 Nonspectral interferences

While nonspectral interferences – matrix-induced signal suppression or enhancement – can often be fairly easily coped with when total element concentrations have to be determined, the situation is more complicated in the context of elemental speciation.

Often, the elements to be speciated are present at a low level only and as the total element content is further 'distributed' over a number of species, dilution is in many cases unacceptable. For complex samples, also matrix matching – imitation of the sample matrix by adding high purity chemicals to the standard solutions – is not self-evident. When aiming at the determination of total element concentrations, both nonspectral interferences and signal instability and/or drift can often be corrected for by using a carefully selected internal reference. To all blank, sample and standard solutions, an

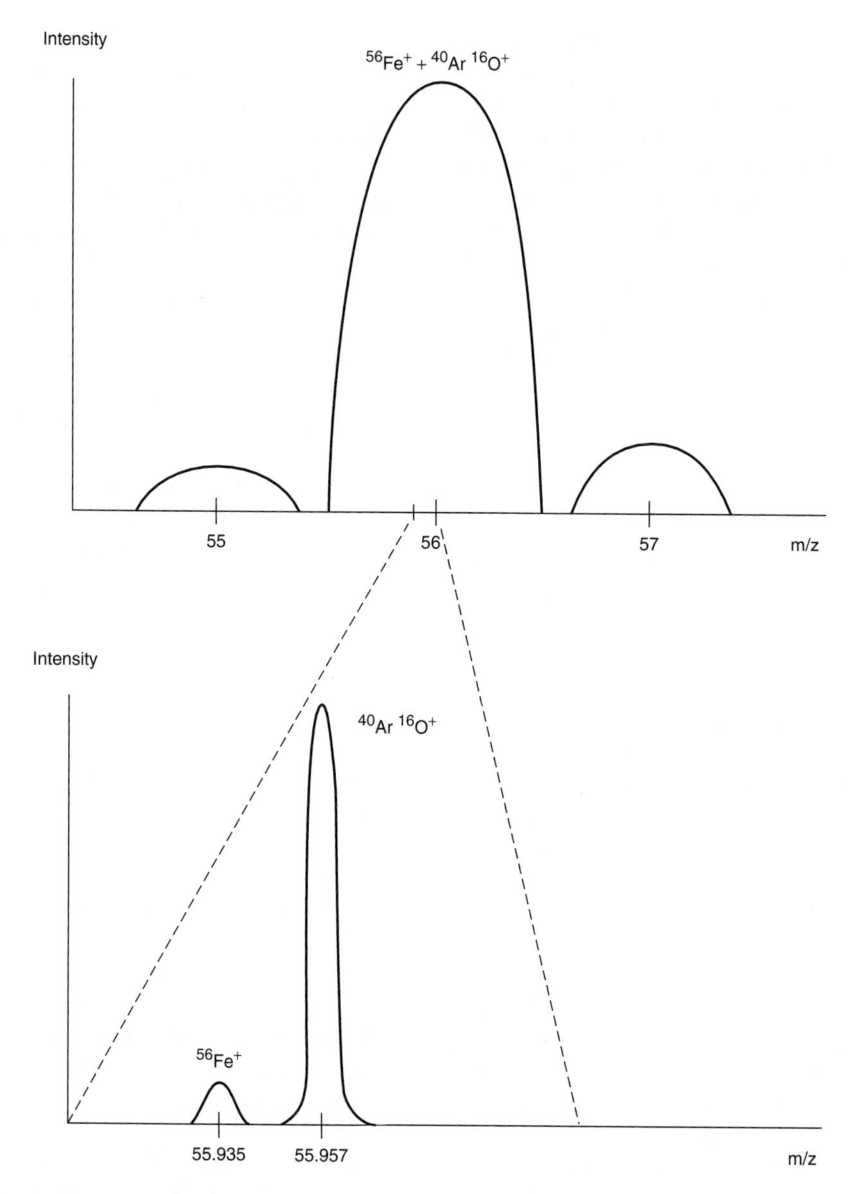

Figure 5.3.3. Schematical representation of the mass spectrum at $m/z = 56$, as obtained using (a) a quadrupole-based ICPMS instrument and (b) a sector field ICPMS instrument operated at a higher mass resolution [18]. Reproduced by permission of International Scientific Communications.

equal amount of an internal reference element is added and it is assumed that this element undergoes the same suppression or enhancement as the analyte element(s). All calculations are subsequently carried out using the ratio of the signal intensity of the analyte element to that of the internal reference. Usually, accurate correction is obtained by using an internal reference element with a mass number close to that of the analyte elements, although some matrices (the example of C-containing matrices being notorious) selectively enhance the signal intensity of specific elements (e.g., As, Sb and Se) by affecting the ionization process [21]. As a result of the use of a separation technique, adding an internal reference element to the sample solution does not offer a solution for nonspectral interferences, as the prerequisite of a 'common fate' of this internal reference element and the sample components is not fulfilled. Saverwyns *et al.* [22] solved this problem to a large extent by mixing the effluent of their HPLC column – used to separate Cr(III) and Cr(VI) from one another – with a continuous flow of a Co standard solution. For GC-ICPMS, using an internal reference is even less obvious. De Smaele *et al.* [23] used H_2 with 1 % Xe as a carrier gas and continuously monitored the $^{126}Xe^+$ signal as an internal reference. Chen and Houk [24] pointed out that the signals of some polyatomic ions can also be used as an internal reference for analytes at a nearby mass-to-charge ratio. Especially in speciation work, this approach may be useful.

Finally, when using the method of standard additions or isotope dilution as a means of calibration (see later), nonspectral interferences are automatically corrected for as sample and standard are affected to exactly the same extent.

2 CALIBRATION

2.1 Traditional approaches

Self-evidently, when using ICPMS for detection in speciation work, the 'traditional' calibration approaches of internal standardization, external standardization and the method of standard additions can be used.

Internal standardization is the simplest approach, in which the concentration of the species involved is estimated from the signal intensity of a single compound with known concentration. With this approach, the results obtained should not be considered as being more than semiquantitative, because in a chromatographic separation process, not only the content of the trace element(s) which is (are) the subject of the speciation study, but also that of the matrix (which may also contain other substances present at much higher concentration levels), will vary as a function of time. This may lead to a time-dependent variation in the degree of nonspectral interference. In addition, as an entire separation process may take a considerable time (e.g., several minutes), signal drift may also further compromise the results obtained. An advantage of internal standardization, however, is that not for every species detected a corresponding standard is required. In many instances, acquiring these standards is not obvious and in the case of complex molecules from biochemical (e.g., proteins) or environmental (e.g., humic acids) origin often impossible.

When the identity and structure of all of the species of interest are known to the analyst and the corresponding standards are commercially available or can be synthesized, external calibration becomes possible. As a result of the wide linear range exhibited by ICPMS, single point calibration is often used instead of a calibration line. It should be realized that since sample and standard are measured at a different moment, while the standard will normally show a different matrix composition, the accuracy of quantitative results is often still not guaranteed. Application of an internal reference signal permits one to correct for the varying sensitivity of the detector – as a result of variations in the composition of the column effluent and/or as a function of time – and can hence improve the reliability of the data produced. The use of such an internal reference was discussed in Section 1.5. Feldmann [25] even described the use of such a Rh^+ internal reference signal (obtained by continuous nebulization of a Rh standard solution into the ICP) for semi-quantitative determination of elemental species, introduced into the ICP after separation by GC and for which no standards were (commercially) available.

Adding the species-specific standards to the sample (standard additions) brings about an automatic correction for nonspectral interferences as sample and standard undergo the same matrix-induced suppression or enhancement. Signal drift or instrument instability between measurement of the sample and of the sample to which standard has been added however, is still not corrected for. Simultaneous application of an internal reference signal can hence further ameliorate the quality of the results obtained.

2.2 Isotope dilution

As a mass spectrometric technique, ICPMS also provides the analyst with isotopic information, such that also isotope dilution (ID) can be used for calibration, provided that the element studied shows at least two isotopes, the signal of which is free from spectral overlap. In this approach, a known amount of tracer – i.e. the chemical compound to be determined, in which the element studied is characterized by an isotopic composition sufficiently different from the natural one – is added to a known amount of sample. From the change induced in the isotopic composition of the analyte element by this spiking process (Figure 5.3.4), the concentration of the target species can be accurately calculated by using the following equations:

$$R_{\text{sample}} = \frac{{}^{1}n_{\text{sample}}}{{}^{2}n_{\text{sample}}},$$

$$R_{\text{tracer}} = \frac{{}^{1}n_{\text{tracer}}}{{}^{2}n_{\text{tracer}}},\ R_{\text{blend}} = \frac{{}^{1}n_{\text{blend}}}{{}^{2}n_{\text{blend}}}$$

$$n_{\text{sample}} = \frac{\theta 1_{\text{tracer}}}{\theta 1_{\text{sample}}} \frac{R_{\text{sample}}}{R_{\text{tracer}}} \left[\frac{R_{\text{tracer}} - R_{\text{blend}}}{R_{\text{blend}} - R_{\text{samp}}} \right] n_{\text{tracer}} \tag{5.3.1}$$

$$n_{\text{sample}} = \frac{\theta 2_{\text{tracer}}}{\theta 2_{\text{sample}}} \left[\frac{R_{\text{tracer}} - R_{\text{blend}}}{R_{\text{blend}} - R_{\text{samp}}} \right] n_{\text{tracer}} \tag{5.3.2}$$

where ${}^{1}n$ is the signal intensity for the lighter isotope, ${}^{2}n$ is the signal intensity for the heavier isotope, $\theta 1$ is the isotopic abundance of the lighter isotope and $\theta 2$ is the isotopic abundance of the heavier isotope.

Equation (5.3.1) is used if the tracer is enriched in the lighter isotope, whereas equation (5.3.2) is preferred when the tracer is enriched in the heavier isotope. If all ratios (R_{sample}, R_{tracer} and R_{blend}) are measured experimentally, the ratios obtained can be introduced into the appropriate equation as such. When tabulated values are used for R_{samp} and/or R_{tracer} instead, the experimental results have to be corrected for mass discrimination [26]. It is useful to mention in this context that, although most elements show an isotopic composition that

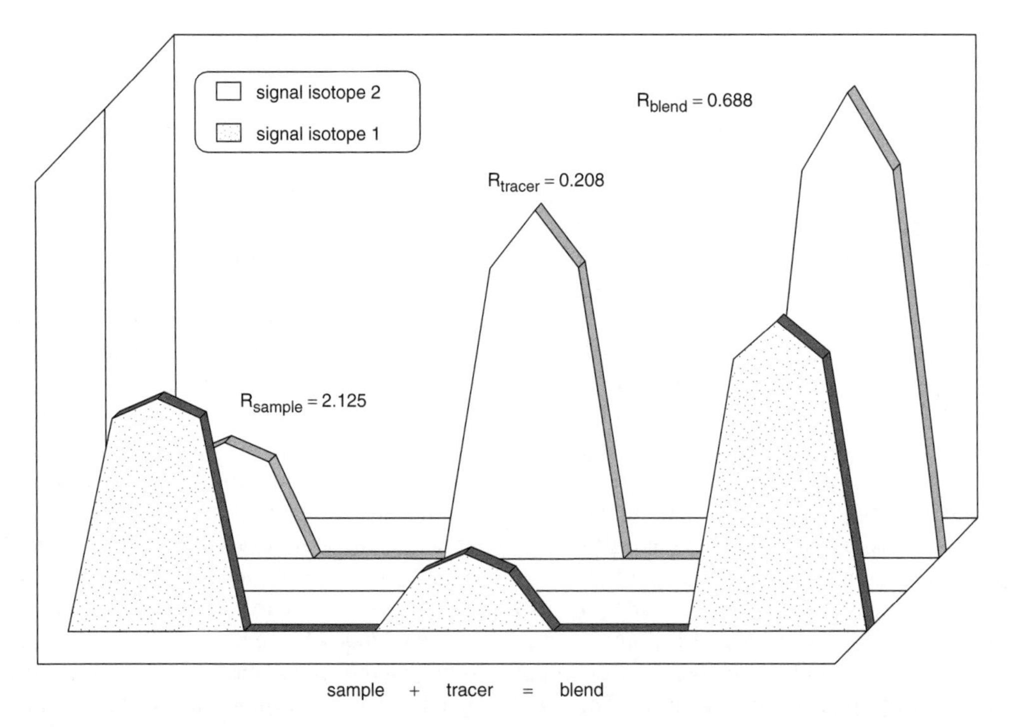

Figure 5.3.4. Schematic representation of the principle of IDMS.

is constant, some show natural (e.g., Sr, Pb) or man-made (e.g., Li and U) variations. For these elements, experimental determination of the isotopic composition in the sample and in commercial standards, respectively, is required. Finally, the information required for the tracer – n_{tracer} and θ(enriched isotope) or $^{\mathrm{enriched}}n_{\mathrm{tracer}} = \theta$(enriched isotope) n_{tracer} – can be obtained from the corresponding certificate or, if necessary, via reverse ID. The latter is an ID procedure, whereby the enriched tracer is considered as the sample and a standard of natural isotopic composition as the tracer.

ID shows distinct advantages over other calibration techniques. Once isotopic equilibration has been established, analyte losses cause no further deterioration in the final analysis result because they do not affect the isotope ratio. Additionally, isotope ratios are practically unaltered by nonspectral interferences, instrument instability or signal drift either, such that these phenomena are also automatically corrected for. Actually, variations in the matrix composition and/or measurement parameters may have a small effect on the mass discrimination, but in the context of ID this effect is of no significance. Hence, ID has the capability of providing the analyst with more accurate and more precise analysis results.

A limitation of ID as a calibration approach typical of elemental speciation work is the lack of commercially available isotopically enriched standards. Therefore, several groups have been obliged to synthesize species-specific tracers in house.

Heumann *et al.* [27] have used ID for the determination of iodide and iodate in mineral waters. These species were separated from one another using ion exchange HPLC. As I is monoisotopic (^{127}I), a long-lived radionuclide (^{129}I) had to be used in the production of species-specific tracers. If no organoiodine species were present, the sum of the iodide and iodate results was seen to be in excellent agreement with the total concentration of I. Ebdon *et al.* [28] used ID for the determination of trimethyllead TML – a degradation product of the corresponding tetraalkyllead, added as an anti-knocking agent to petrol – in artificial rainwater. For this purpose, ^{206}Pb-enriched TML was synthesized from 'a radiogenic lead' reference material (NIST 983), by converting the latter into $PbCl_2$, which was subsequently derivatized using a MeMgI Grignard reagent. TML was separated from inorganic Pb and triethyllead (TEL) by means of reversed phase ion-pairing HPLC. The results obtained agreed well with the corresponding reference values.

The use of species-specific tracers, enabling the use of ID in elemental speciation is not limited to HPLC-ICPMS and several authors have reported on the use of this approach in GC-ICPMS applications as well.

Encinar *et al.* [29] synthesized a mixture of isotopically enriched mono- (MBT), di- (DBT) and tributyltin (TBT) tracers by butylation of elemental Sn, enriched in ^{119}Sn, using BuCl and Et_3N and I_2 as catalysts. The concentrations of the three target species in the mixed tracer were determined by reverse ID. Commercial MBT, DBT and TBT standards (with Sn of natural isotopic composition) were used for this purpose. The entire process included derivatization of the target species using $NaBEt_4$, separation using capillary gas chromatography and detection by ICPMS. Subsequently, this mixed tracer solution could be used for the simultaneous determination of TBT (mainly originating from the use of antifouling paints) and its degradation products DBT and MBT in sediment samples. Provided that complete extraction of the target species from the sediment samples is accomplished, such that complete isotopic equilibration is guaranteed, ID permits more reliable results to be obtained. Once the above-mentioned prerequisite is fulfilled, other sources of error – incomplete derivatization, analyte losses during extraction of the derivatized target species in an organic solvent and/or the measurement itself – no longer affect the final result. The accurate results obtained for two certified reference materials illustrated the utility of this approach.

Snell *et al.* [30] produced isotopically enriched $(CH_3)_2Hg$, CH_3HgCl and $HgCl_2$ and used these as species-specific tracers for ID purposes in the analysis of natural gas condensates by means of

GC-ICPMS. Even in the case of species interconversion, which is normally detrimental in speciation work, species-specific ID offers a solution. Hintelmann *et al.* [31], e.g., established that the formation of additional CH_3Hg^+ from inorganic Hg during sample preparation may lead to an overestimation of the 'original' CH_3Hg^+ content in environmental samples. This alteration of the distribution of Hg over its various chemical forms during the pre-treatment could be unequivocally demonstrated by adding an isotopically enriched stable tracer of Hg^{2+} to reference materials that were subjected to sample preparation and subsequent analysis by means of HPLC-ICPMS or GC-ICPMS. Species-specific ID was suggested to cope with this problem, because isotopically enriched standards are expected to undergo the same changes as the species to be determined [32].

Species-specific ID, however, is only possible if the identity of the species is known and their composition and structure is sufficiently simple to permit synthesis of a corresponding tracer. In order to preserve the advantages offered by ID to the largest possible extent when the aforementioned conditions are not fulfilled, Heumann and coworkers introduced the use of species-unspecific ID in HPLC applications [19, 33, 34]. In this approach, the species of interest are *first* separated from one another using an appropriate form of HPLC and the HPLC effluent is subsequently mixed with a continuous flow of enriched tracer (in a Y junction) prior to its introduction into the ICP (Figure 5.3.5). Hence, in this approach, isotopic equilibration is only obtained in the ICP, where due to the high temperature all compounds are broken down into atoms, irrespective of their original chemical form. Finally, the experimental set-up also contains a flow injection valve, permitting injection of a standard solution of natural isotopic composition for calibration of the tracer flow (reverse ID).

As is shown in Figure 5.3.6 for Cu, with this approach, the isotope ratio of interest is monitored as a function of time. When no copper-containing species are eluted from the column, the measured ratio is that of the enriched spike (approximately

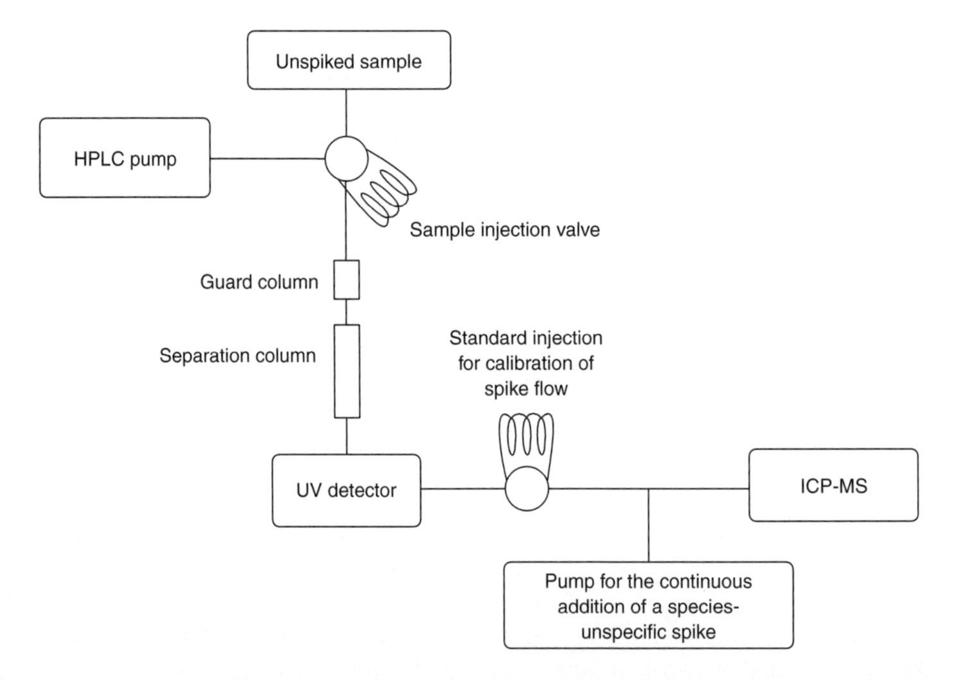

Figure 5.3.5. Schematic diagram of the instrumental set-up for HPLC-ICPMS using species-unspecific ID for calibration [19]. Reprinted with permission from L. Rottmann and K. G. Heumann, *Analytical Chemistry*, **66**, 3709. Copyright (1994) American Chemical Society.

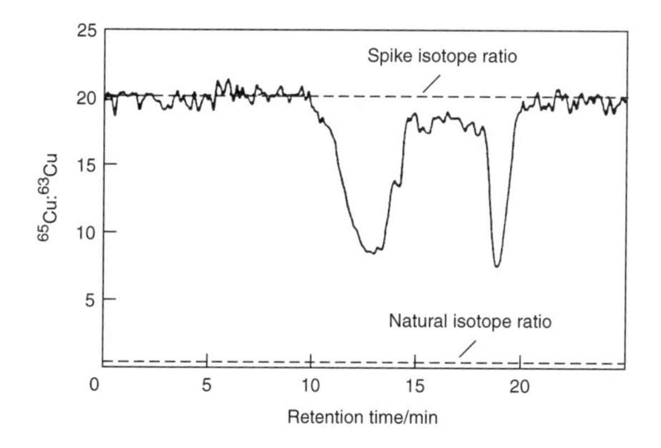

Figure 5.3.6. $^{65}Cu/^{63}Cu$ isotope ratio as a function of the retention time during analysis of a river water sample using a combination of an HPLC system, equipped with a SEC column and ICPMS, with species-unspecific ID for calibration [27]. Reproduced by permission of the Royal Society of Chemistry.

20 for the ^{65}Cu-enriched $Cu(NO_3)_2$ spike used in this example). Cu-containing fractions of the eluent, however, shift the isotope ratio of interest in the direction of the natural isotope ratio, as can be seen from the two 'negative' peaks in the chromatogram. Finally, the original chromatogram, showing the variation of the Cu isotope ratio as a function of time, can be converted into a chromatogram, which directly displays the amount of Cu present in the effluent as a function of time. This approach is, of major interest, for e.g., the quantification of metal complexes with proteins in body fluids or with humic substances in waters of different origin, the complex composition and structure of which do not permit synthesis of a species-specific spike. In the case of species-unspecific ID, nonspectral interferences, signal drift and instrument instability are automatically corrected for; losses during the sample pre-treatment and/or species interconversion on the other hand are not corrected for since isotopic equilibration is only accomplished in the ICP.

This approach was, e.g., used to study metal interactions with dissolved organic materials in natural aquatic systems. For this study, an HPLC system, equipped with a size exclusion column (SEC) was coupled on line with a UV detector and an ICP-mass spectrometer. Simultaneous registration of the presence of UV-absorbing organic matter and the metals in the column effluent permitted conclusions to be drawn about the interactions between the metals under investigation and humic substances, which form the major part of dissolved organic matter in waters of different origin. In addition, the ID approach allows accurate quantitative information to be obtained. Later on, simultaneous ICPMS detection of both the metals of interest and the content of dissolved organic carbon (DOC) was made possible by also adding ^{13}C-enriched benzoic acid as a species-unspecific tracer [35].

3 USE OF THE ICP AS A SOFT IONIZATION SOURCE

With commercially available ICPMS instrumentation, the ICP is operated at atmospheric pressure and at RF powers exceeding 1000 W (except in case of operation under cool plasma conditions, which requires a reduction of the RF power by some hundreds of watts). Molecules introduced into such a plasma are broken down into atoms, which are subsequently ionized, such that all structural information is lost. During the past couple of years however, efforts have also been made to couple low power low pressure ICP ion sources to an MS detection system [36, 37]. With such instrumentation, both molecular and atomic spectra can be obtained, depending on the RF power used (5–90 W). The molecular spectra obtained at low RF power resemble electron impact spectra, such that application of existing library spectra can facilitate species identification on the basis of the structural information provided. In addition to the RF power, the plasma pressure and composition (pure He and mixed Ar/He plasmas are used) and the introduction of reaction gases have been demonstrated to influence the molecular fragmentation. These ICPMS devices with low pressure low power plasmas have been combined with GC, to separate organo-Br, -Hg -Pb and -Sn compounds. Both qualitative and quantitative information could be obtained with this set-up. In spite of the promising results, no low power low pressure ICPMS instrumentation is commercially available.

4 HIGH PERFORMANCE LIQUID CHROMATOGRAPHY (HPLC)-ICPMS

4.1 Coupling

HPLC-ICPMS offers an unmatched performance for the detection and determination of nonvolatile metallospecies [38]. As a result of the different separation strategies that can be used – (1) normal phase and reversed phase partition, (2) ion exchange (3) size exclusion, (4) affinity and (5) adsorption chromatography – HPLC is extremely versatile. The interaction of the mobile phase with the sample constituents has a direct effect on the distribution coefficients (except in the case of size exclusion chromatography, SEC) and contributes to this versatility. The choice of an HPLC technique primarily depends on the research objective. Currently, reversed phase (RP) partition HPLC is one of the most frequently used approaches. It implies distribution of the target compounds between a nonpolar stationary phase and a relatively polar mobile phase.

An important advantage in the coupling of HPLC with ICPMS is the compatibility of the chromatographic effluent flow rate and the liquid flow rate required for stable pneumatic nebulization. Moreover, HPLC is operated at room temperature. Hence, interfacing of the two techniques is straightforward. However, problems may arise upon introduction of HPLC effluents into the ICP [38, 39]. LC mobile phases generally consist of some combination of organic solvents, salts in buffer solutions and/or ion-pairing reagents. Generally, ICPMS requires more dilute buffers to be used and only tolerates lower concentrations of organic solvents than ICPOES. In ion exchange chromatography, buffer concentrations often exceed 0.1 M. Such concentrations are likely to cause short-term signal suppression or enhancement and can cause blockage of the nebulizer and/or the sampling cone as well as erosion of the sampling cone and the skimmer [39]. These irreversible changes in the interface aperture configuration lead to unwanted sensitivity losses. In RPC, the organic solvent must be coped with, whereas the buffer concentration is seldom a problem. High loads of organic solvents negatively influence the performance because of the decreased plasma stability – even plasma extinction can be observed in extreme cases – and deposition of carbon on the torch and sampling cone [39]. Water-cooled spray chambers and introduction systems equipped with a desolvation unit are therefore used to reduce the amount of solvent introduced into the plasma. An increase in forward RF power can improve the plasma stability, but is accompanied by an increase of the reflected power, which is harmful to the RF generator in the long run [40]. The deposition of carbon on the sampling cone, causing an elevated noise level and excessive signal drifts, can in most cases be minimized by addition of oxygen (ca. 3 % V/V) to the nebulizer gas flow, although this may lead to a reduced sampling cone lifetime [39, 40]. Finally, it has to be considered that mobile phases may give rise to a more complex mass spectrum, leading to more spectral interferences, particularly at a mass-to-charge ratio <80.

Figure 5.3.7 shows the experimental set-up of a versatile HPLC-ICPMS system, based on a conventional pneumatic nebulizer interface [41]. The interface itself is simple: a piece of narrow bore tubing connects the outlet of the LC column with the liquid flow inlet of a pneumatic nebulizer (Meinhard nebulizer or PFA low flow nebulizer). Post-column addition of an internal reference element allows correction for changes in the plasma conditions and/or other fluctuations in sensitivity. Post-column effluent split, accomplished by a variable micro splitter valve, is optional as a remedy to reduce the amount of mobile phase introduced into the ICP, although at the expense of a loss in sensitivity. The dead volume is reduced to the largest possible extent by using capillaries with small diameters, micro flow-splitters and miniature T-pieces. To enhance the overall performance of the system, a software-controlled four-port valve, which is installed after the separation column, allows switching from chromatographic effluent to 1 % HNO_3 in-between measurements to rinse the sample introduction system.

A major limitation of the hyphenation of HPLC and ICPMS, is the low analyte transport efficiency to the plasma (usually 5 % or less), inherent to

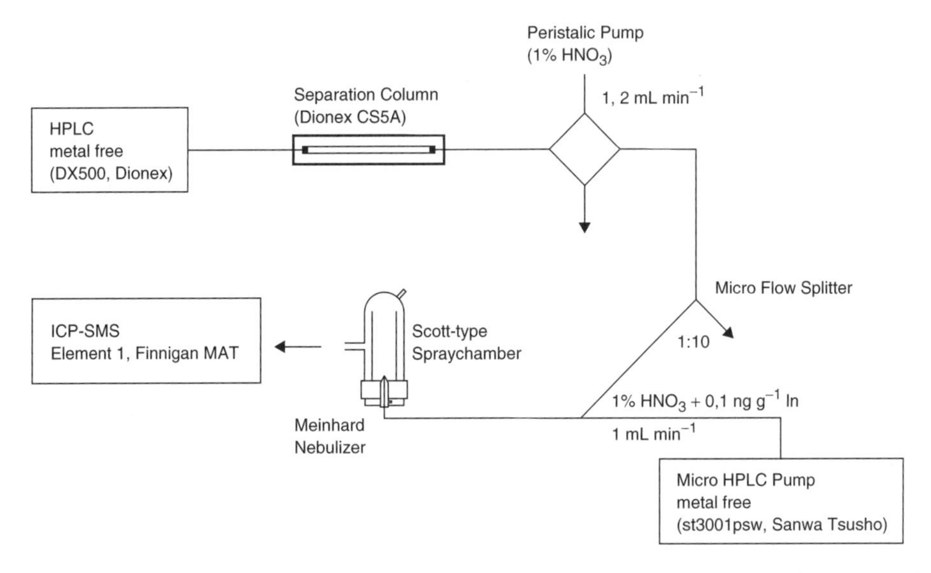

Figure 5.3.7. Experimental set-up of an HPLC-ICPMS system: flow splitting and make-up flow addition are used for reduction of nonspectral interferences [41]. Reproduced by permission of the Royal Society of Chemistry.

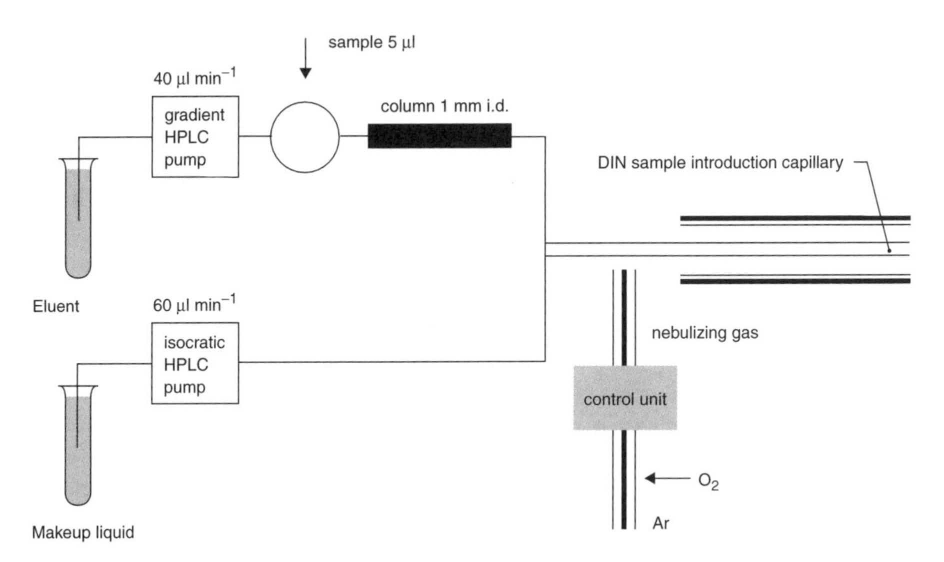

Figure 5.3.8. Schematic representation of a DIN-based interface for the coupling of microbore HPLC to ICPMS [42]. Reproduced by permission of the Royal Society of Chemistry.

pneumatic nebulization [40]. Hence, an increase in transfer efficiency would result in a corresponding improvement in LODs. The use of other types of nebulizers (e.g., an ultrasonic nebulizer) can increase the analyte transport efficiency, but can also give rise to an additional extra-column dead volume and is therefore not recommendable. One option that shows promise for certain applications is the use of a direct injection nebulizer (DIN) [42] (Figure 5.3.8). This DIN is a microconcentric pneumatic nebulizer, which is positioned inside the central tube of the ICP torch. Also the more recently developed direct insertion high efficiency nebulizer (DIHEN) allows the direct and 100 %

efficient introduction of $\sim 100\,\mu L\,min^{-1}$ flow rates into the ICP. The DIN interface offers a low dead volume, short wash-out times and virtually no memory effects. By using this introduction technique, peak broadening can be minimized and transport efficiency can approach 100 % with mobile phase flow rates up to $0.1\,mL\,min^{-1}$ for microbore columns. Lobinski *et al.* [42] reported that carbon deposits were less important for this type of interface than for conventional pneumatic nebulizer interfaces. Moreover, the system showed better salt compatibility. As a drawback, the effect of organic solvents on the plasma stability was substantial and necessitated addition of oxygen, while the degree of plasma robustness may be reduced. Oxide ion formation levels in the plasma were not reported in this study, but are also known to be considerably higher.

4.2 Illustrative applications

HPLC-ICPMS has been used for the speciation of a variety of biochemically and/or environmentally interesting elements. Most studies have focused on the assessment of the environmental and/or health risks by the determination of potentially dangerous species of a single target element [38, 39] in different environmental compartments (air, natural water, soils, sediments and biota) or in human body fluids.

As a result of the large difference in toxicity between its different chemical forms, many publications have reported on the speciation of As using HPLC-ICPMS. Already in 1993, Larsen *et al.* [43] used HPLC-ICPMS for the determination of eight arsenic compounds in human urine. In a first phase (Figure 5.3.9.(a)), dimethylarsinate (DMA), As(III), monomethylarsonate (MMA) and As(V) were separated from one another and from the positively charged As species (eluting in the void volume) using an anion exchange resin. In a second phase (Figure 5.3.9.(b)), the other As compounds – aresenobetaine (AsB), trimethylarsine oxide (TMAO), arsenocholine (AsC) and the tetramethylarsonium ion (TMAs) – were separated from one another and from the negatively charged species using a cation exchange resin. More recently, it was demonstrated that, when using an anion exchange resin that also shows some nonpolar activity (nonpolar sites in addition to ion exchange sites), AsB – which is uncharged at the pH value of the mobile phase used – could be determined within the same chromatographic run as DMA, As(III), MMA and As(V), while only the cationic arsenic compounds elute in the void volume [44]. Under these conditions, one measurement of a urine sample is sufficient to draw meaningful conclusions concerning (professional) exposure to inorganic As, without risk of misinterpretation due to the intake of food of marine origin (a major source of the nontoxic AsB). Also the chromatographic resolving power and the LODs ($\leq 0.05\,\mu g\,L^{-1}$) of anion exchange chromatography – ICPMS for As speciation have been improved considerably [45, 46].

Recently, other more explanatory HPLC-ICPMS studies have aimed at the elucidation of the mechanisms of biotransformation of inorganic metal ions and simple inorganic species [38]. A trend in the field of speciation by HPLC-ICPMS is the detection and identification of ligand – metal complexes in biological samples. Metalloproteins, trace metal complexation in blood and blood plasma, selenoproteins in human and animal body fluids and tissues, metallodrugs and their interaction with proteins are subjects of investigation in these studies [47].

An interesting HPLC-ICPMS study on the interaction of *cis*-$[Pt(NH_3)_2Cl_2]$ (cisplatin) with 5′-guanosine monophosphate (5′-GMP) was reported by Hann *et al.* [48]. The anti-tumoral activity of platinum drugs is based on their coordinative binding with lone pairs of electrons of DNA bases. As a consequence, the structure as well as the functionality of the DNA is modified and cell replication is inhibited. Hence, investigation of the species formed by the interaction of cisplatin with DNA bases is a key to understanding the activity of this drug. The combination of high performance ion chromatography with sector field ICPMS provided unambiguous stoichiometrical information on the major GMP adduct. Cisplatin was incubated with 5′-GMP under physiological conditions

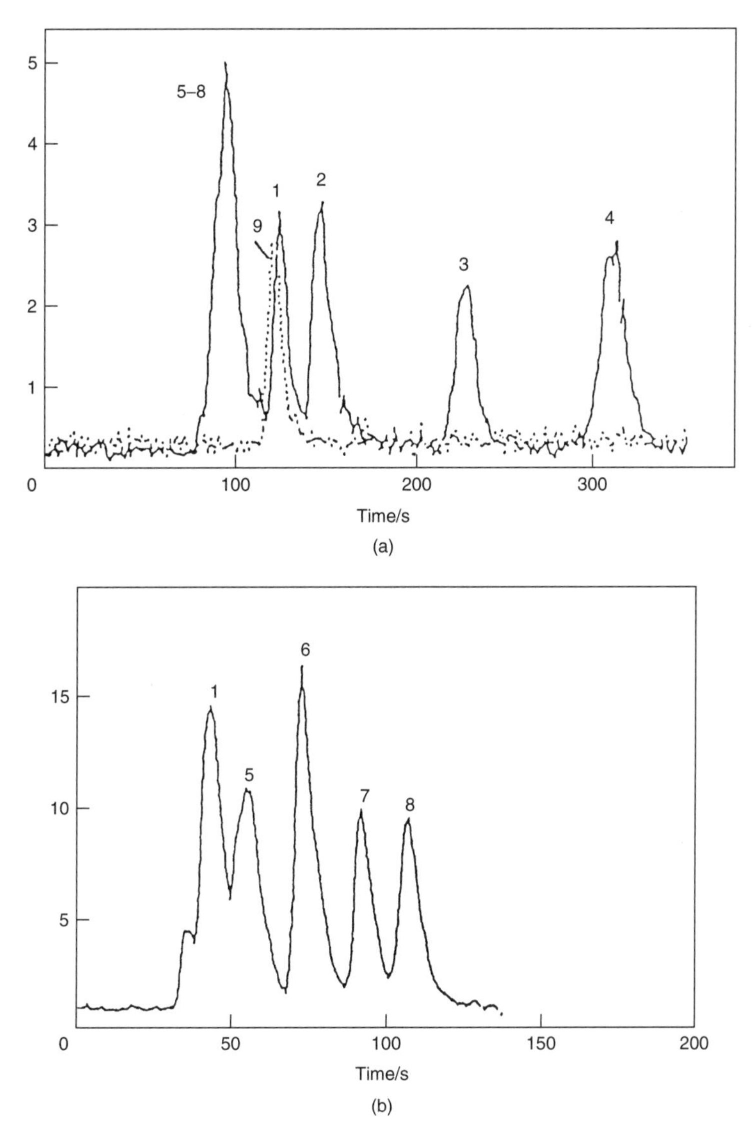

Figure 5.3.9. (a) Chromatogram obtained for a fourfold diluted urine sample, spiked with 1, DMA; 2, As(III); 3, MMA; 4, As(V); 5, AsB; 6, TMAO; 7, AsC; 8, TMAs and 9, $Sb(OH)_6^-$ (used as an internal reference) using an HPLC system (equipped with an anion exchange resin) coupled on-line to an ICP-mass spectrometer. (b) Chromatogram obtained for a fourfold diluted urine sample, spiked with 1, DMA; 5, AsB; 6, TMAO; 7, AsC and 8, TMAs using an HPLC system (equipped with a cation exchange column) coupled on-line to an ICP-mass spectrometer [43]. Reproduced by permission of the Royal Society of Chemistry.

for 180 h. The simultaneous detection of P and Pt by sector field ICPMS and the determination of the instrumental response (sensitivity) for both elements resulted in an P/Pt elemental ratio of 2/1, corresponding to the molar ratio in the *bis* adduct *cis*-$[Pt(NH_3)_2(GMP)_2]^{2-}$. Higher mass resolution proved to be mandatory for the accurate determination of ^{31}P. Additionally, the time-dependent reaction course of the cisplatin-5′–GMP system could be followed as can be observed in Figure 5.3.10. The concentration decrease of 5′-GMP and the formation of adducts was monitored on the basis

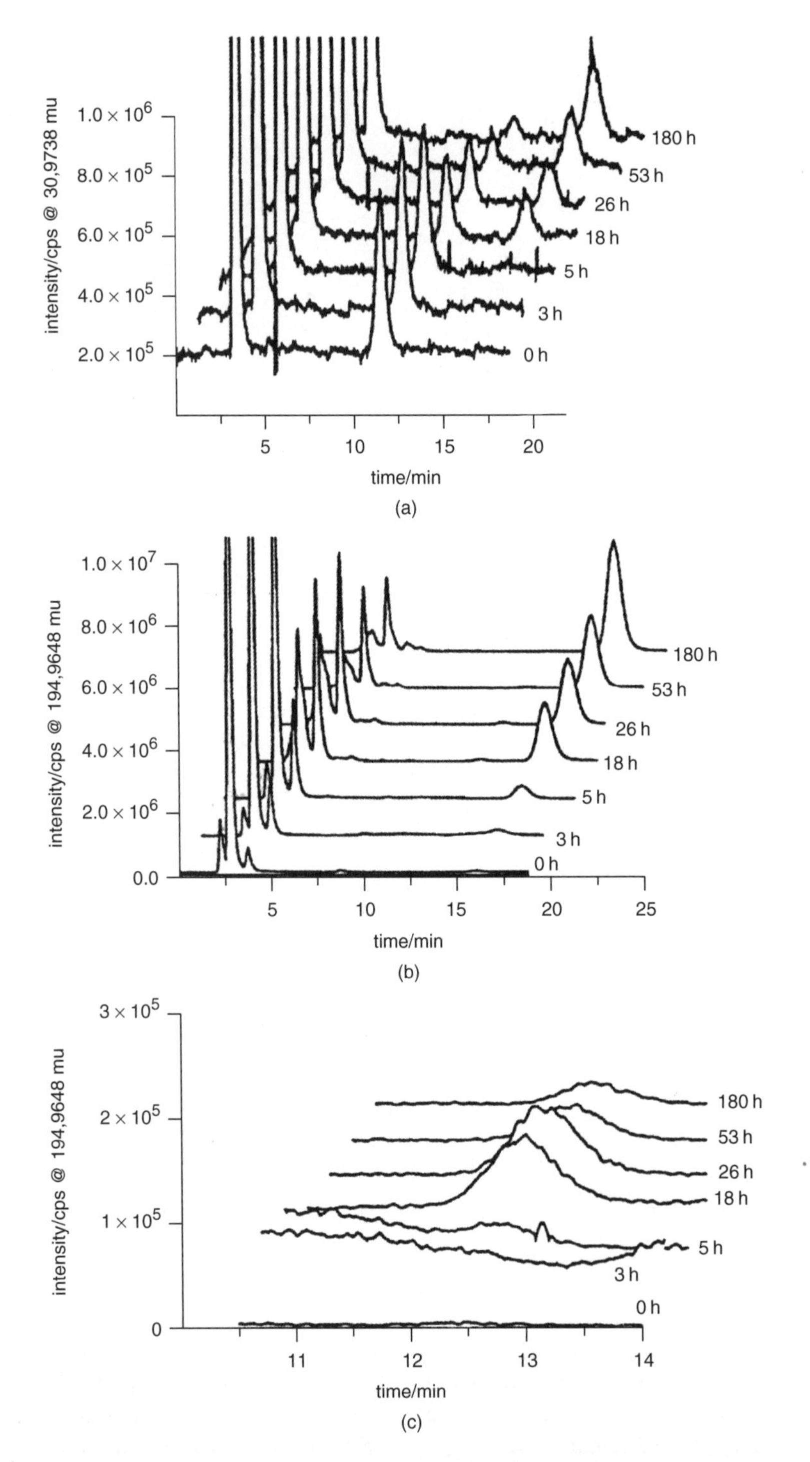

Figure 5.3.10. Time-dependent monitoring of reaction of *cis*-[$Pt(NH_3)_2Cl_2$] (cisplatin) with 5′-guanosine monophosphate (5′-GMP): P (a) and Pt (b) chromatograms obtained after different incubation times (bisadduct observed at a retention time of 15.5 min). (c) shows the intermediate monoadduct (observed at a retention time of 12.0 min) [48]. Reproduced from *Fresenius' Journal of Analytical Chemistry*, **370**, 581, 2001, copyright notice of Springer-Verlag.

of UV and ICPMS detection. An intermediate monoadduct was observed together with the major product, the bisadduct *cis*-$[Pt(NH_3)_2(GMP)_2]^{2-}$.

Other interesting HPLC-ICPMS applications exploit the multielement capability of ICPMS. Wang *et al.* [49] used size exclusion chromatography coupled to sector field ICPMS to study metal–protein binding. They investigated the association of various elements with proteins of human and bovine serum. V, Co, Cd and Mo were found in two distinct protein fractions. Mn was observed in three protein fractions and in the low molecular weight fraction. The latter fraction also contained some lanthanides. Most lanthanides, however, were bound to proteins in the mass range of 70–90 kDa. Alkali metals and Tl were found as free metals not bound to proteins. Jakubowski *et al.* [50] studied the metabolism of platinum in biological systems. Pt, S and C (^{13}C) were monitored simultaneously. By means of this approach, 90 % of the Pt found in a grass sample could be assigned to inorganic platinum. The remaining 10 % occurred in four different organic fractions.

A review by Szpunar and Lobinski [47] reported on the potential of HPLC-ICPMS for the study of biomacromolecular complexes. Bidimensional HPLC with ICPMS detection, i.e. separation of metallopolypeptides by size-exclusion HPLC and subsequent signal identification by anion exchange HPLC-ICPMS, was discussed. Species identification was accomplished on the basis of the matching of their retention times with those of standards. Species of interest are biomacromolecular metal complexes found in plants (e.g., polysaccharides, phytochelatins), biological fluids (e.g., proteins, porphyrins) or animal tissues (e.g., metallothioneins). The authors emphasized that, at this point of research, limitations in terms of separation selectivity and signal identification are becoming increasingly conspicuous because of the unavailability of standards. The parallel use of electrospray MS (ESMS) was proposed as a promising alternative. The gap between the LODs offered by ICPMS and ESMS, respectively, and the difficulty of unambiguous attribution of a peak in the mass spectrum to a species are considered as current limitations.

Up to now, HPLC-ICPMS analysis is almost exclusively research, and most of the published methods have not, or have not sufficiently, considered some of the fundamental requirements of routine analysis [51]. However, for selected applications, such as the separation of Cr(III) and Cr(VI) and the simultaneous separation of different oxidation states of As and Se, so-called speciation kits are commercially available. These kits, especially developed for use with ICPMS, consist of a microbore HPLC column and a guard disk for column protection [52].

5 GAS CHROMATOGRAPHY (GC)-ICPMS

5.1 Coupling

Compared with HPLC-ICPMS, GC-ICPMS offers a higher resolving power and 100 % introduction efficiency, allows a more stable plasma and gives rise to fewer spectral interferences as a result of the plasma being dry and finally leads to less sampling cone and skimmer wear. Of course, GC-ICPMS can only be used for the separation and detection of sufficiently volatile and thermally stable compounds or compounds that can be derivatized into a volatile form. Also the coupling of a gas chromatograph with an ICP-mass spectrometer is somewhat more complicated as a heated transfer line is required, such that condensation of the species that have been separated from one another in the gas chromatograph and hence peak broadening can be avoided. Additionally, typical effluent flow rates with GC are in the order of $1\,mL\,min^{-1}$, while for ICPMS, a carrier gas flow rate in the order of $1\,L\,min^{-1}$ is required to obtain an annular plasma. As a result, addition of a make-up gas is required.

Recently, a transfer line enabling coupling of GC to ICPMS was commercially introduced [53]. Many research groups, however, have developed their own metallic or quartz heated transfer line. De Smaele *et al.* [54] reported the development of a transfer line permitting rapid coupling and decoupling of a capillary gas chromatograph

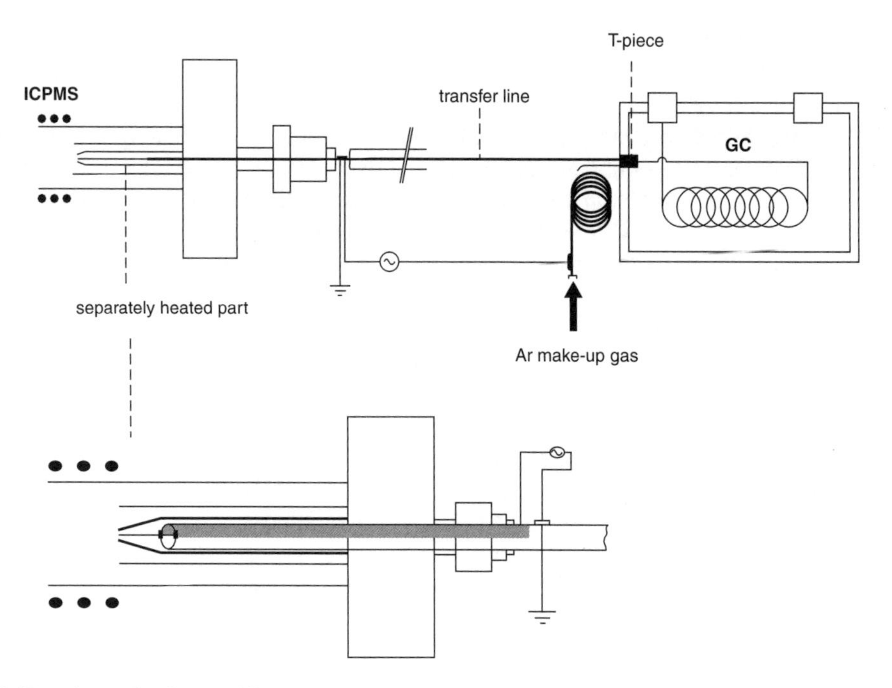

Figure 5.3.11. Heated transfer line enabling hyphenation of CGC with ICPMS without significant band broadening [54]. Reprinted from *Spectrochimica Acta*, Vol. 50, De Smaele *et al.*, A flexible interface for ..., p. 1409, 1995 with permission from Elsevier Science.

to an ICPMS instrument. Their transfer line (Figure 5.3.11) basically consists of two resistively heated stainless steel tubes. The first stainless steel tube contains the deactivated fused silica capillary that transports the GC effluent into the center of the ICP. The Ar make-up gas is resistively heated in the second stainless steel tube and is subsequently introduced into the first stainless steel tube, where it flows around the fused silica capillary. This set-up ensures an equable temperature all over the transfer line. Finally, the whole transfer line is thermally isolated. To avoid peak broadening, even under the most demanding circumstances, the last part of the transfer line was resistively heated separately. Therefore, this part was lengthwise cut in two halves, which were electrically isolated from one another over their entire length with polyimide tape, except for an electrical contact at the very end.

The transfer line developed by Montes Bayòn *et al.* [55] is interesting, because its design is markedly different from that of the majority of transfer lines described in the literature. In this design, the last part of the GC capillary is inserted into a short piece of heated metallic tube. This concentric assembly is inserted into a metallic T-piece (Figure 5.3.12). Unheated Ar make-up gas is introduced into the T-piece via the perpendicular side arm. Since this make-up gas is forced to flow through the narrow opening between the metallic tube and the T-piece, it acquires a higher velocity and is hence able to prevent condensation of the separated species on the walls of the T-piece or within the flexible nonheated PTFE tubing used for connection to the ICPMS instrument. This interface was reported to perform well in GC-ICPMS analysis, aiming at the determination of organometallic compounds of Hg, Pb, S, Se and Sn.

5.2 Illustrative applications

The major application field of GC-ICPMS involves the determination of organometallic compounds of Sn, Hg and Pb in environmental matrices.

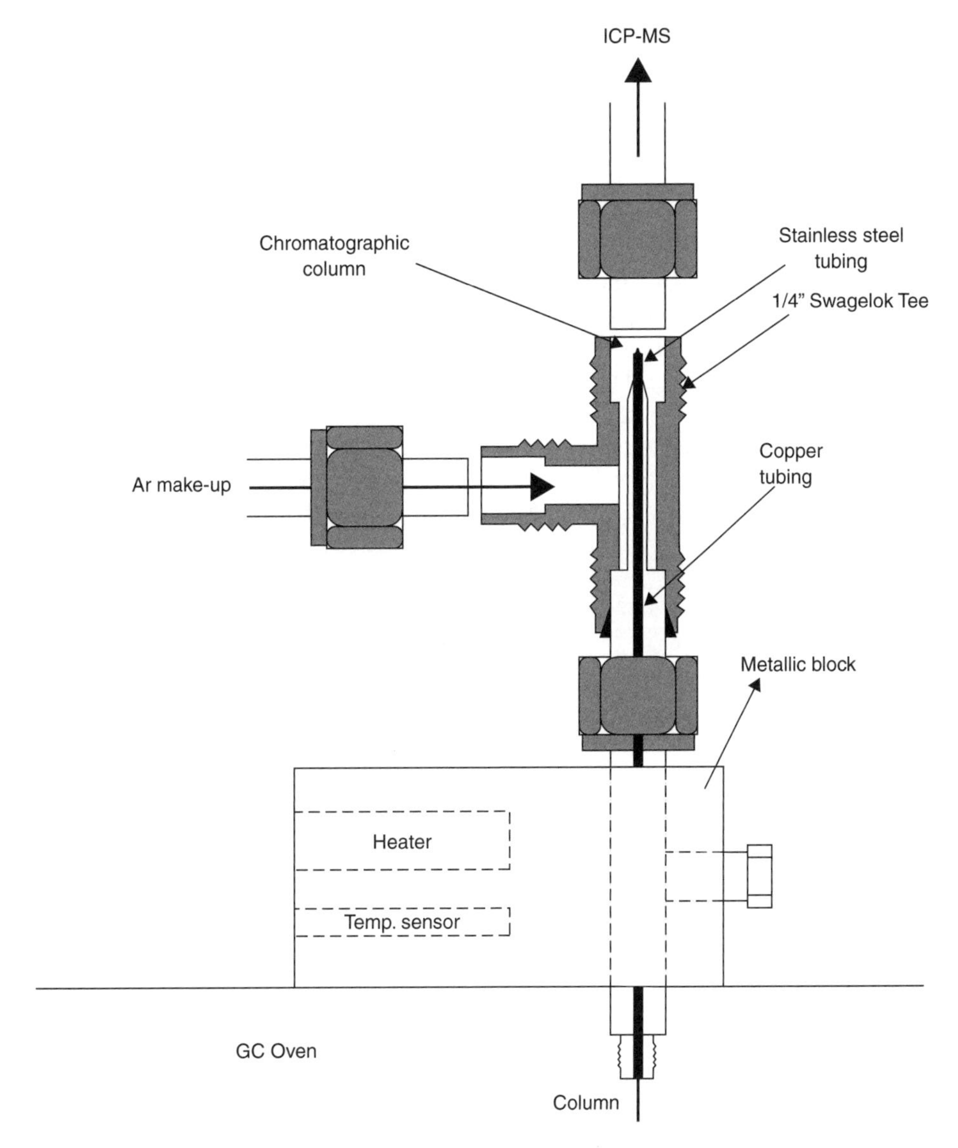

Figure 5.3.12. Unheated transfer line for hyphenation of CGC and ICPMS [55]. Reproduced by permission of the Royal Society of Chemistry.

Ritsema *et al.* [56], e.g., reported on the use of GC-ICPMS for the determination of butyltin compounds in harbor sediments, sampled in two marinas along the Dutch coast. For many years, TBT was added as the active component to antifouling paints, intended to prevent the growth of algae and mussels on the hulls of ships and in docks. As a result of the growing awareness of the outspoken toxicity of TBT, and to a lesser extent of its degradation products DBT and MBT, its use was regulated more strictly and prohibited for pleasure boats in many countries, including the Netherlands. To investigate the effects of this ban, the content of butyltin compounds in sediments sampled between 1992 and 1995 was determined. After being leached from the sediments using a

MeOH–acetic acid mixture, the species of interest were ethylated using $NaBEt_4$ and the derivatized species were extracted into hexane. These extracts were subsequently subjected to GC-ICPMS analysis. As no decreasing trend was observed as a function of the sampling time between 1992 and 1995, while high TBT/DBT ratios were found in all sediment extracts, it was concluded that sediments can act as a reservoir of TBT. TBT is hence slowly released into the harbor water, such that intolerably high levels are still expected for a long time. Vercauteren *et al.* [57] used GC-ICPMS for the determination of traces of the pesticide fentin (triphenyltin) in environmental samples. The analyte of interest was extracted from the matrices investigated (potatoes and mussels) by means of headspace solid-phase microextraction (SPME) after digestion using TMAH or a mixture of KOH and EtOH and *in situ* derivatization using $NaBEt_4$. Leal-Granadillo *et al.* [58] investigated the possibility of determination of organolead compounds in airborne particulate matter using GC-ICPMS after their extraction into an organic solvent and derivatization by means of a Grignard reaction. Femtogram LODs (as Pb) were accomplished for trimethyl- (TML), dimethyl- (DML), triethyl- (TEL) and diethyl-Pb (DEL). Analysis of airborne particulate matter sampled in the Spanish city Oviedo revealed the presence of several organo-Pb compounds. Armstrong *et al.* [59] demonstrated a LOD <1 pg (as Hg) for methylmercury (MM) using GC-ICPMS. The accuracy of the results obtained was demonstrated by the agreement between experimental results and the corresponding certified contents of MM in marine tissue certified reference materials. Finally, the approach developed was used for the determination of MM in 'real-life' samples of ringed seal and beluga whale.

Multielement approaches have also been described in the literature. Jantzen and Prange [60] carried out an extensive study on the occurrence of organometallic species of Sn, Hg and Pb in sediments along the River Elbe. The species of interest were extracted from the matrix in an organic solvent after *in situ* derivatization with $NaBEt_4$ and were subsequently subjected to GC-ICPMS analysis. Peycheran *et al.* [61] collected air samples with a cryogenic trap and subsequently analyzed these using GC-ICPMS for their content of volatile metal species. Inorganic Hg and tetraalkyl-Pb compounds were found to be the major species in samples originating from the Bordeaux urban environment (France). The quantitative results obtained for the organometallic compounds of Pb illustrated the beneficial effect of the decreased use of leaded petrol. The efficiency and high sample throughput of the method permits the influence of meteorological factors and automotive traffic parameters to be studied into more detail. By using a combination of *in situ* purging and cryogenic trapping, this method could also be used for analyzing natural waters [62]. Low to sub-pg L^{-1} LODs were reported for organometallic compounds of Hg, Pb, Se and Sn (Figure 5.3.13). The method was used to investigate the occurrence of these species in estuaries in France, Belgium and the Netherlands.

Finally, in a number of instances, organometallic compounds of other elements have also been studied. Gallus and Heumann [63] determined the selenite (directly) and selenate (= total Se − selenite) content in water samples by means of GC-ICPMS. As selenite is a nonvolatile species, it was converted into a volatile piazoselenol by reaction with 1,2-diamino-4-trifluoromethylbenzene prior to its determination. Feldmann *et al.* [64] provided evidence for the biomethylation of Bi – an element widely used in alloys, cosmetics and pharmaceutical products – to trimethyl-Bi (TMB) using GC-ICPMS for the analysis of gases of sewage sludge digesters and landfill gases. Peycheran *et al.* [65] demonstrated the utility of GC-ICPMS for monitoring the workspace air in the semiconductor industry in view of professional exposure to volatile species of As, In and P. The same approach can also be used to check the purity of volatile reagents used in the production of semiconductors. Finally, Grüter *et al.* [66] recently reported on a GC-ICPMS-based method, capable

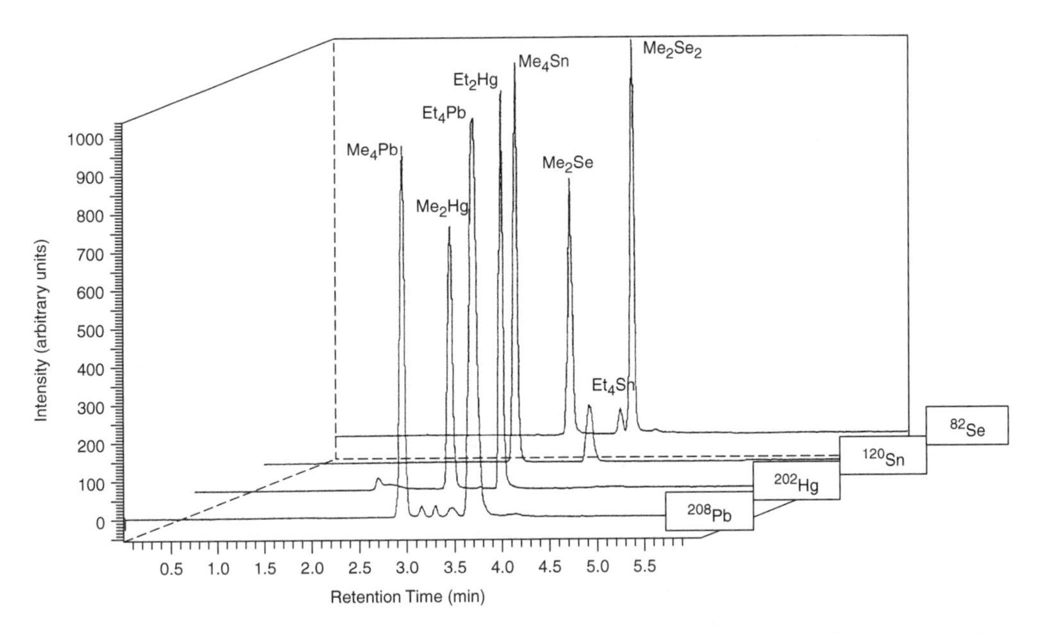

Figure 5.3.13. Multielement chromatogram for an aqueous standard solution, obtained after purging and cryogenic trapping followed by GC-ICPMS [62]. Reprinted from *Analytica Chimica Acta*, Vol. 337, Amouroux *et al.*, Sampling and probing volatile . . ., p. 241, 1998 with permission from Elsevier Science.

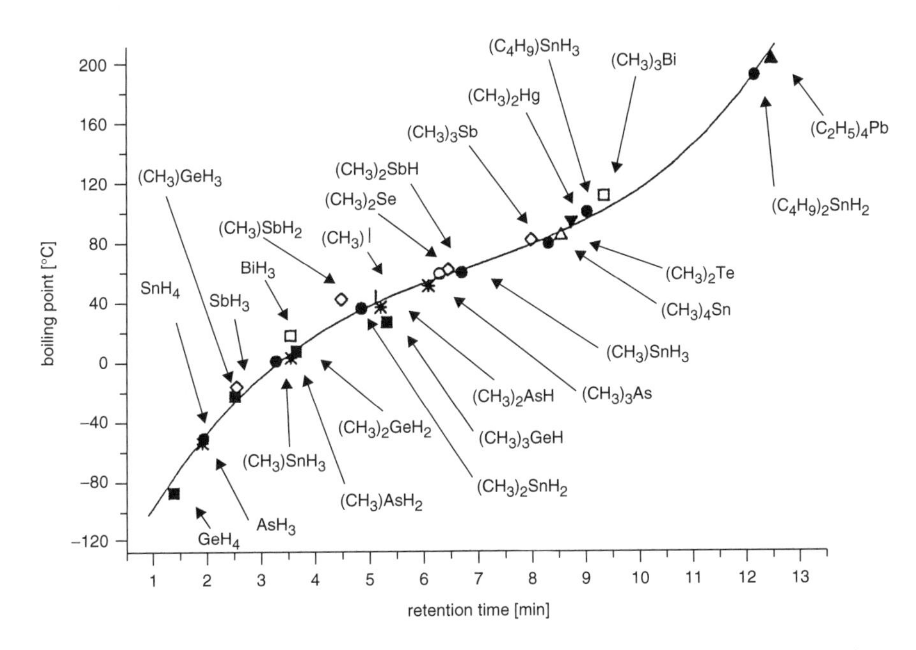

Figure 5.3.14. Relationship between boiling point and elution time of volatile organometal(loid) compounds for a purge-and-trap GC-ICPMS set-up [66]. Reprinted from *Journal of Analytical Chemistry*, A new HG/LT-GC/ICP-MS multielement speciation technique for real samples in different matrices, U. M. Grüter, J. Kresimon and A. V. Hirner, Vol. 368, p. 67, 2000, copyright notice of Springer-Verlag.

of determination of the metal(loid)–organic compounds of twelve elements – As, Bi, Ge, Hg, I, Mo, Pb, Sb, Se, Sn, Te and W – with LODs below 1 pg. Derivatization was accomplished by means of hydride generation, and the volatile species were subsequently collected in a cryogenic trap. By heating the trap slowly from – 196 to 150 °C, and introducing the sequentially revolatilized compounds into a GC, complete separation could be ensured. A clear correlation between boiling point and retention time could be established (Figure 5.3.14). Soil samples from municipal waste deposits were subjected to analysis with this instrumental set-up, in view of the possibility of methylation caused by the microbiological activity and the reducing conditions often typical for these matrices. For nine elements, organic species were found with concentrations in the upper ng (as element) per kg level. In a waste deposit that had been closed for a longer time (15 years), a lower extent of biomethylation was established. This can probably be explained by an almost complete decomposition of organic material, leading to reduced bioactivity.

6 CAPILLARY ELECTROPHORESIS (CE)-ICPMS

6.1 Coupling

CE-ICPMS combines a high separation efficiency with the sensitivity of ICPMS and can be considered as the most important recent trend in the field of hyphenated techniques [39]. CE techniques have important advantages over more conventional separation approaches. CE is able to separate cationic, anionic and neutral species, it shows potential to handle labile complexes (e.g., noncovalently bound) and colloidal systems, it is rapid and the columns are relatively simple and cheap. Hence, a variety of compounds with metal functionality are amenable to speciation by CE-ICPMS. The method represents an interesting alternative to HPLC-ICPMS, when 'gentle' separation schemes are required to preserve the true chemical information in a real-life sample. Moreover, since the technique is characterized by an extremely low sample consumption, it is ideally suited when the amount of sample available is limited, a situation frequently encountered in biological, biomedical or nuclear research.

As only very small sample volumes (typically 1–100 nL) are introduced in CE, it is generally difficult to obtain satisfactory LODs in terms of concentration for most species. Accordingly, development of highly sensitive and selective detectors has been very important and challenging since CE came into existence. Olesik *et al.* [67] published the first paper on CE-ICPMS in 1995. The first hyphenation was accomplished by using a conventional concentric nebulizer in combination with a conical spray chamber. It was evident from this first publication that the key to the analytical success of CE-ICPMS is the design of the interface. Establishing efficient sample transport without degradation of separation resolution proved to be a major challenge. The basic requirements of the hyphenation can be summarized as follows. (1) The interface must include an electrical connection at the CE capillary exit end, enabling application of an electrical field gradient along the CE capillary as a driving force for species separation. A stable electric current is crucial for reproducible separations. (2) The interface must adapt the flow rate of the electro-osmotic flow (EOF) inside the CE capillary (nL min^{-1} range) to the liquid flow rate required by the nebulizer. (3) The interface must prevent nebulizer suction from causing laminar flow inside the capillary. Nebulizer suction was identified [67] as the principal factor jeopardizing separation resolution in CE-ICPMS. (4) Because ICPMS is a post-column detection method, some dead volume is inevitable. The challenge is to make the impact of this dead volume as small as possible.

During the last few years, several interfaces for coupling of CE with ICPMS have been developed with various degrees of success. These constructions are mostly based on Meinhard [68, 69], microconcentric [70–72] or ultrasonic nebulizers [73], but also interfaces based on cross-flow [70], high efficiency [74] or direct injection nebulizers [75] have been developed. Almost all

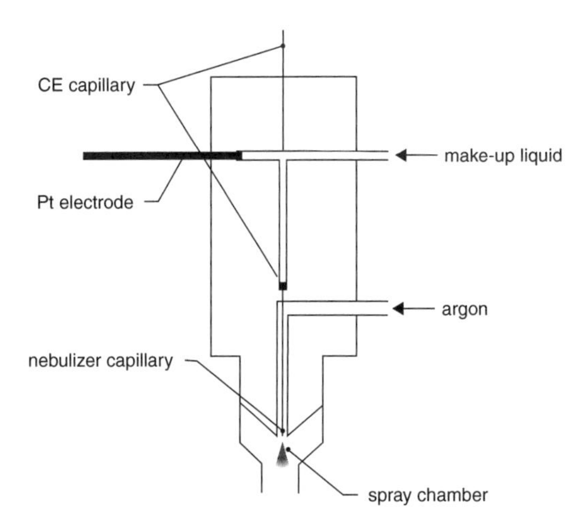

Figure 5.3.15. Schematic representation of a commercially available CE-ICPMS interface based on a modified microconcentric nebulizer [76]. Reprinted from *Journal of Analytical Chemistry*, D. Schaumlöffel and A. Prange, Vol. 364, p. 452, 1999, copyright notice of Springer-Verlag.

interfaces developed to date use (either home-built or commercially available) spray chambers with a low dead volume, such as the cyclonic or the conical type to reduce memory effects and peak broadening. Schaumlöffel and Prange [76] developed an interface based on a modified microconcentric nebulizer, which is now commercially available (Figure 5.3.15). Flow rates between 2 and $12\,\mu L\,min^{-1}$ were obtained in the self-aspiration mode for this modified nebulizer. A conducting make-up buffer flows concentrically around the CE capillary exit to provide the necessary electrical connection as well as to supplement the low EOF in order to achieve stable nebulizer operation. A cross-shaped piece connects the nebulizer with the CE capillary (vertical fittings). The two horizontal fittings are intended for the platinum electrode and for the make-up liquid, respectively. The make-up liquid is transported by self-aspiration of the nebulizer and is in electrical contact with the platinum electrode. Mixing occurs at the end of the CE capillary. The nebulizer is plugged into a low volume spray chamber of about 5 mL for minimizing band broadening of the CE signals. A Teflon tube of 60 cm length connects this interface to the ICPMS torch. The possibility to adjust each parameter (e.g., the capillary position) in an exact and reproducible way is one of the key advantages of this interface. The authors report a dilution factor of only 5–20 for the CE analytes, depending on CE capillary dimensions and voltage.

A cheap, stable and dependable interface can also be obtained by combining a commercially available microconcentric nebulizer with a low volume spray chamber [70–72] such as the cyclonic spray chamber (Figure 5.3.16). Laminar flow in the CE capillary can be counterbalanced by using a liquid sheath flow. Either the sheath flow is pumped at matched flow rates or the nebulizer is operated in the self-aspiration mode, with a leveled sheath liquid reservoir. As a drawback, this use of a sheath flow compromises sensitivity to a higher degree compared to the commercially available design, since the nebulizer accommodates flow rates in the $100\,\mu L\,min^{-1}$ range, such that

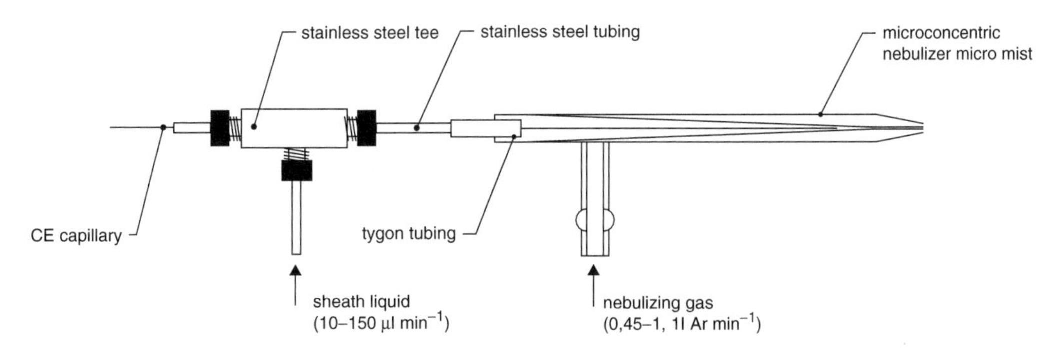

Figure 5.3.16. Schematic representation of a CE-ICPMS interface based on a commercially available microconcentric nebulizer (Micromist, Glass Expansion, Switzerland), used in combination with a cyclonic spray chamber [71]. Reproduced by permission of the Royal Society of Chemistry.

the sample is significantly diluted (up to factor of 1000). Lower flow rates compromise nebulization efficiency and result in memory effects. Majidi and Miller-Ihli [70] reported the occurrence of spikes and peak tailing in the electropherogram due to poor transport and mixing efficiency of the CE effluent with sheath buffer prior to nebulization. With higher flow rates again spikes were observed. This was explained by release of analytes from the wall of the low volume spray chamber. Application of a negative pressure to the inlet vial is an alternative approach for counterbalancing the nebulizer suction [72].

Other interesting interfaces based on direct sample injection have been developed. The aim was to improve sample transfer efficiencies up to 100 %, without adverse effects derived from the spray chamber, such as analyte–chamber interactions, band broadening and/or sample losses. A DIN interface was first implemented by Liu *et al.* [75]. The CE capillary was placed concentrically inside the DIN sample introduction capillary so that the liquid sample was directly nebulized into the central channel of the plasma torch. The DIN's low dead volume offered faster sample introduction and shorter wash-out times. Majidi *et al.* [77] reported the use of a DIHEN for CE-ICPMS. With a sample uptake rate of $85\,\mu L\,min^{-1}$, the LODs could be improved by a factor of 2 compared with a cross-flow nebulizer interface. Recently, Bendahl *et al.* [78] developed a demountable DIHEN operating at low sample uptake rates ($10\,\mu L\,min^{-1}$) for CE-ICPMS hyphenation (Figure 5.3.17). As a drawback, both DINs and DIHENs are known to be difficult to install and optimize on ICPMS instruments. Moreover, the DIN requires a gas displacement pump for sample delivery and a coaxial carrier gas in addition to the nebulizer gas and its cost is also a consideration. Tangen *et al.* [79] found severe problems due to the interference of the high voltage used for the CE separation with the RF power supply of the ICPMS plasma with a DIN interface. Finally, it has to mentioned that the absence of a spray chamber may also have negative effects, such as an increased oxide ion formation.

6.2 Illustrative applications

To date, the application of CE-ICPMS to real samples has been reported in only a few papers. Most research has focused on hyphenation-related problems and aimed at the construction of a robust interface and improvement of the LODs. The analytical figures of merit have been evaluated on the basis of different applications. Van Holderbeke *et al.* [71] showed the potential of CE-ICPMS for arsenic speciation. Polec *et al.* [80] evaluated an interface based on a self-aspirating micronebulizer for a study of metal binding by recombinant and native metallothioneins. Ackley *et al.* [81] tested the method for analysis of metalloporphyrins standards. Michalke and Schramel [82] investigated the capability of CE-ICPMS for Sb speciation. On the basis of iodine speciation, Michalke [83] addressed problems related to the conflicting situation of improving LODs by increased sample volumes and preconcentration by stacking procedures on the one hand and preserving separation efficiency and species stability on the other. Detection of pseudospecies artefacts as a result of chemical interaction of samples, electrolytes and the capillary and spectral interferences was stated as major problem in CE-ICPMS.

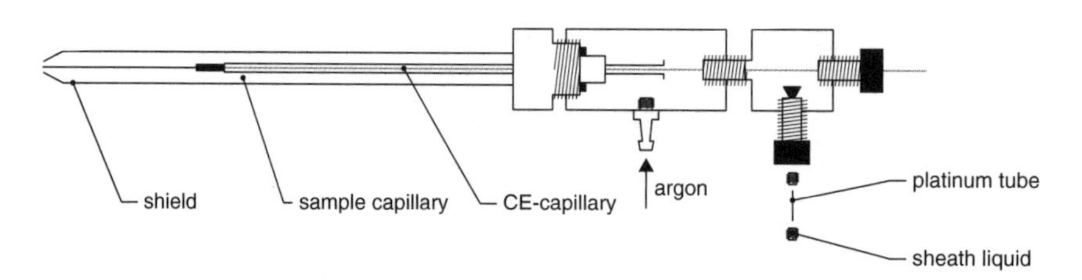

Figure 5.3.17. Demountable DIHEN for interfacing CE to ICPMS [78]. Reproduced by permission of the Royal Society of Chemistry.

Michalke and Schramel [84] applied CE-ICPMS to Se speciation in human serum and breast milk of early lactation. Se speciation is of growing interest, due to the toxic and essential properties exhibited by this element. CE separation of six Se species – including both inorganic and organic species – was accomplished. In parallel, CE separations based on isoelectrical focusing have been developed to discover the isoelectric points of the organic Se species. LODs of the CE-ICPMS system were found to be just acceptable or still too high for the samples investigated. Therefore, preconcentration was necessary to detect Se species in human milk. In accordance with other studies, no inorganic Se was observed in this type of samples. Ten different Se species were detected in human serum. Unfortunately, the signals were close to the detection limit, while identification of the species was not included in this study.

In a recent paper, Prange *et al.* [85] investigated metallothionein isoforms in human brain cytosol by CE-ICPMS. For the first time, CE-ICPMS was used in comparative studies on the distribution of isoforms of metallothioneins in brain samples taken from subjects with Alzheimer's disease and from a control group. The isoforms were separated by CE and the elements Cu, Zn, Cd and S were detected by sector field ICPMS. For accurate determination of some of these elements, the use of a higher mass resolution was a prerequisite. Defatting cytosol with subsequent acetonitrile precipitation for protein elimination was found to be the optimum sample preparation procedure. Figure 5.3.18 shows the electropherogram of a cerebellum cytosol for the elements Cu, Zn and Cd after acetonitrile precipitation. The authors suggest that the peaks observed correspond to the species known as MT-1, MT-2 and MT-3 on the basis of migration time comparison with reference MT from rabbit liver. Similar patterns of MT isoforms were found for all brain regions of the subjects investigated. For subjects with Alzheimer's disease, the levels of MT-1 and MT-3 were reported to be reduced in all temporal and occipital samples, while the Cu pattern was observed to be affected. Furthermore, the detection of sulfur by

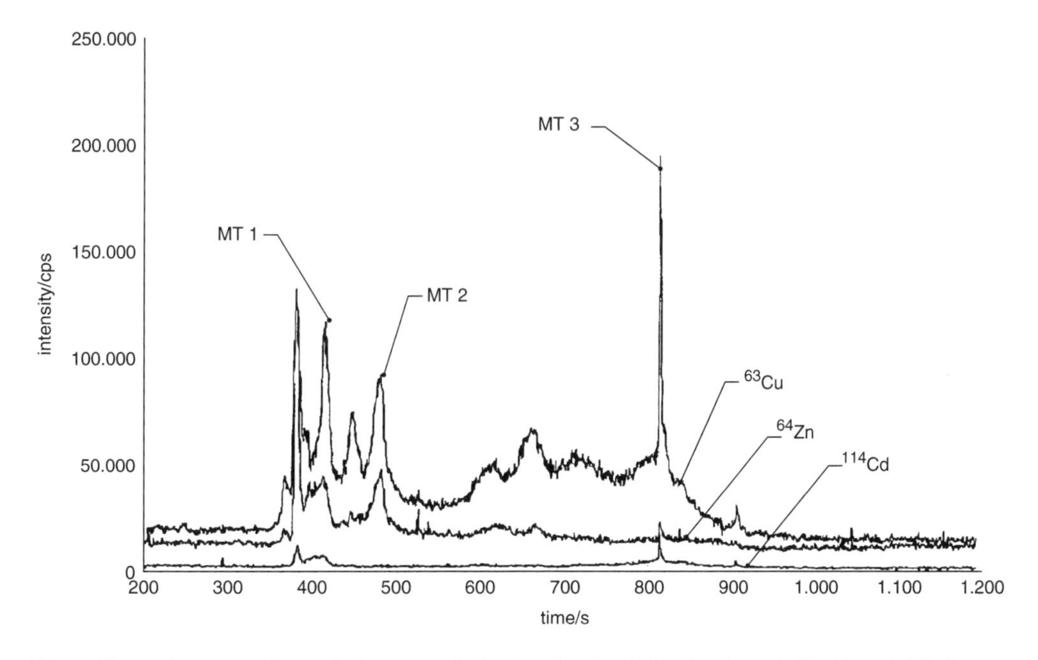

Figure 5.3.18. Electropherogram of a cerebellum cytosol showing the signals for the elements Cu, Zn and Cd after acetonitrile precipitation. The authors suggest that the peaks observed correspond to MT-1, MT-2 and MT-3 on the basis of migration time comparison with that for reference MT from rabbit liver [85]. Reprinted from *Journal of Analytical Chemistry*, A. Prange *et al.*, Vol. 371. p. 764, 2001, copyright notice of Springer-Verlag.

ICPMS and quantification by ID have been investigated as a new method for the quantification of MT isoforms. After metal quantification, the molar ratios of sulfur to metals can be used to characterize the composition of the metalloprotein complex of the MT isoforms. Thus, a suggestion for their stoichiometric formulae can be given.

7 SUPERCRITICAL FLUID CHROMATOGRAPHY (SFC)-ICPMS

7.1 Coupling

SFC can be regarded as a hybrid technique between gas and liquid chromatography, combining some of the best features of each, i.e. the high diffusion coefficient of GC with the solubility properties of LC [38]. Compounds which are traditionally difficult to separate by GC, such as thermally labile, nonvolatile and high molecular mass compounds can be separated by SFC with relative ease. Compared to LC, SFC is faster due to the lower viscosity of the mobile phase and high diffusion coefficients of the analyte. When using SFC, the use of organic solvents and the relatively long on-column residence times observed with many LC techniques can be avoided [40]. Critical temperatures and critical pressures of the mobile phases used in SFC lie within the usual HPLC chromatographic conditions, such that little instrument modification is required. Pressure control is the primary variable for the chromatographic separation. Carbon dioxide is so far the most commonly used mobile phase.

In SFC-ICPMS, the mobile phase changes from the supercritical fluid to the gaseous state before entering the plasma. Technically, the decompression is accomplished in the interface by implementing a restrictor connected to the end of the analytical column [39]. This process of expansion is subject to the Joule–Thomson effect and results in a net cooling. Hence, the restrictor zone must be sufficiently heated. Generally, coupling SFC to ICP torches requires similar conditions to those for GC, where a heated transfer line connects the oven with the torch. Nebulizer and spray chamber typically used for liquid sample introduction into the plasma can be eliminated. In summary, the main factors to be considered for coupling SFC and ICPMS are as follows. (1) Sufficient heat should be provided to the restrictor (temperature > 150 °C) in order to avoid cluster formation and wall condensation. (2) Plasma response to the supercritical fluid is critical. (3) As with any other coupled technique, a sufficiently high analyte transport efficiency is crucial for successful hyphenation.

There is a marked preference in selecting either packed microcolumns or capillary columns over packed SFC columns for SFC-ICPMS, since the high mobile phase flow rates with the latter (compared to that with capillary systems) may cause severe perturbations in the plasma [86]. Interfacing capillary SFC and ICPMS instruments was first accomplished by Shen *et al.* [87]. Most of the SFC-ICPMS studies reported to date use this functional design (Figure 5.3.19). It provides a heated transfer line connecting the SFC oven with the ICPMS torch. The restrictor is mounted in a heated copper tube (200 °C). This assembly

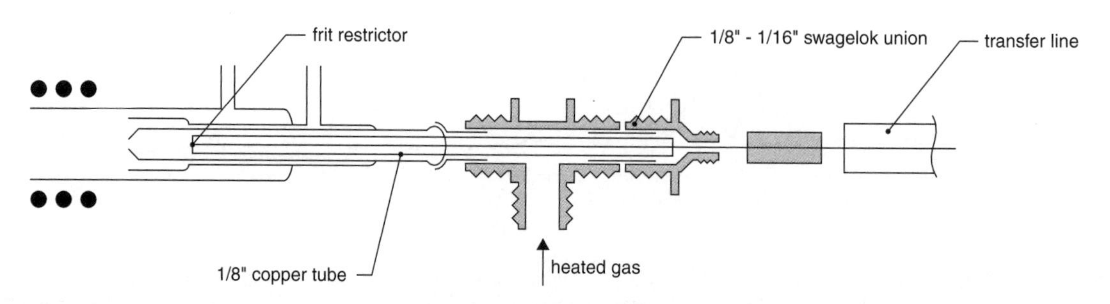

Figure 5.3.19. Schematic representation of an SFC-ICPMS interface [87]. Reprinted with permission from W. Shen *et al.*, *Analytical Chemistry*, **63**, 1491. Copyright (1991) American Chemical Society.

is inserted into the central tube of a regular ICP torch. The tip of the restrictor is positioned flush with the end of the central tube of the ICP torch. A three-way Swagelock union was used for introduction of a heated Ar make-up gas to transport the analyte into the plasma. The make-up gas flow rate approximates the normal nebulizer gas flow rate used in ICPMS. Plasma and auxiliary gas flows are not altered. The combination of heating the transfer line and restrictor zone and providing a heated make-up gas flow ensures a proper restrictor temperature. Shen *et al.* evaluated the performance of this instrumental set-up on the basis of the separation of tetraalkyltin compounds. The authors critically discussed the background spectral features of Ar ICPMS with CO_2 as an SFC eluent. Major background interferences were identified as $^{12}C^+$, $^{12}C^{16}O_2{}^+$ and $^{40}Ar^{12}C^+$. Also, the effect of the mobile phase CO_2 on the sensitivity – due to plasma quenching – was investigated. At CO_2 flow rates typical for capillary SFC ($<1\,mL\,min^{-1}$), the effects of CO_2 on the plasma were found to be minimal and thus normal ICP operating conditions could be used. At higher flow rates, prolonged CO_2 introduction resulted in carbon deposition on the sampler and skimmer.

7.2 Illustrative applications

Speciation studies using SFC-ICPMS have recently been reviewed by Vela and Caruso [86]. The number of publications and applications has been limited, in part due to the popularity of and the wide availability of detailed information on LC and GC separations [40]. So far, most SFC-ICPMS studies have focused on a better understanding of the processes occurring in the transfer line or the interface. Another factor to consider is the fact that the most common mobile phase for SFC, CO_2, is not ideal for most organometallic compounds. Elution problems and strong interaction with the stationary phase are problems related to the mismatching in terms of polarities of mobile phase and analyte compounds. The use of organic modifiers or formation of nonpolar metal complexes from polar organometallic compounds have been studied as solutions for these problems.

The potential of ICPMS as an element-selective detector for SFC was initially demonstrated with organotin compounds [87]. Tetrabutyltin, tributyltin chloride, triphenyltin chloride, and tetraphenyltin have been separated in a single chromatographic run with LODs in the subpicogram range. The analytical figures of merit of SFC-ICPMS and SFC-FID were compared systematically [88]. The resolution obtainable with SFC-FID was not always observed in SFC-ICPMS. Temperature fluctuations in the transfer line and the restrictor zone – affecting the mobile phase density and velocity – were identified as the main source of peak broadening. As an additional consequence, variations in retention times were observed. However with SFC-ICPMS, LODs were improved by one order of magnitude compared to SFC-FID.

Carey *et al.* [89] evaluated the performance of SFC-ICPMS for the determination of organochromium compounds. Owing to the spectral interferences (ArC^+ at m/z 52), nitrous oxide replaced carbon dioxide as the mobile phase. For the determination of ionic chromium and different Cr oxidation states, it was necessary to complex the chromium prior to injection into the SFC system. β-Ketonate compounds were chosen as complexing agents. Problems with the detection of thermally labile chromium complexes were encountered. This was related to the manner in which the restrictor was heated.

8 ALTERNATIVE APPROACHES

The large majority of elemental speciation studies using ICPMS are accomplished by means of on-line coupling of a chromatographic or electrophoretic column. However, some alternative and often creative methods have been described that deserve proper attention.

Jin *et al.* [90] accomplished speciation of Ge by trapping the hydrides of inorganic, monomethyl-(MMGe) and dimethyl-Ge (DMGe) – formed by reaction with $NaBH_4$ and removed from the reaction vessel by a flow of He – in a cooled trap. Subsequently, re-volatilization of the compounds as a function of their boiling point could be accomplished, such that fractionated introduction into the

ICP enabled quantitative species-specific results to be obtained. Sub-pg LODs were achieved. All cited Ge compounds were detected in natural and waste waters, while in the latter sample, other, unknown Ge species were also observed.

Willie *et al.* [91] used graphite furnace – ICPMS to determine the contents of inorganic mercury and methylmercury (MM) in fish tissue. This approach is also based on the different volatilities of the target compounds. In a first measurement, Hg is vaporized from the samples of interest (solubilized using TMAH) and introduced into the ICP, such that the total content of this element could be determined. For the second measurement, iodoacetic acid, sodium thiosulfate and acetic acid are added to the sample. During the drying step of the temperature program, MMI is already removed from the furnace, such that during the vaporization step, only the inorganic fraction of the Hg is introduced into the ICP and measured. An even more elegant approach was developed by Gelaude *et al.* [92]. In this approach, a carefully optimized temperature program enabled the vaporization of MM and inorganic Hg to be separated in time, such that both compounds could be determined within the same measurement. Quantification was accomplished by means of species-unspecific ID using a permeation tube, containing elemental Hg enriched in ^{200}Hg. Analysis of certified reference materials demonstrated the suitability of this approach.

Feng *et al.* [93] reported on the simultaneous determination of inorganic As(III) and As(V) using hydride generation – ICPMS. Since the reaction conditions can be selected such that As(III) is readily converted into the corresponding hydride, while As(V) is not converted at all, it is possible to either determine the As(III) content, or the total As content, the latter after complete pre-reduction of As(V) to As(III). Hence, determination of both As(III) and As(V) (by subtraction) is possible, but requires two measurements of the same sample. Feng *et al.* modified this approach by adding L-cysteine as a reducing agent, which *slowly* converts As(V) into As(III). By measuring the As^+ signal intensity at two moments in time after the start of the reaction, the contents of both As(III) and As (V) can be calculated, such that one measurement suffices for each sample. A similar approach was developed for the simultaneous determination of Sb(III) and Sb(V) [94].

Finally, a very promising approach was described by Nielsen *et al.* [95], who used a combination of gel electrophoresis and laser ablation ICPMS for studying the binding of Co to blood serum proteins. For this purpose, human serum (enriched with Co) was subjected to immunoelectrophoresis. By rastering the agarose gel obtained in this separation process with an IR laser and analyzing the material ablated from the gel by ICPMS, a Co distribution map was obtained. By comparing this to the protein distribution map, obtained on staining the proteins with Coomassie Blue, the main Co-binding serum proteins could be identified. Evans and Villeneuve [96] used a similar approach to study the association of Pb to humic and fulvic acids. For this purpose, isotopically enriched Pb was added to the sample and the aforementioned organic compounds were subsequently separated from one another as a function of their molecular size on a polyacrylamide gel. Finally, the dried gel plates were subjected to LA-ICPMS analysis to find out which organic compounds Pb was bound to.

9 REFERENCES

1. Fisher, A. and Hill, S. J., Inductively coupled plasma mass spectrometry, in *An Introduction to Analytical Atomic Spectrometry*, Ebdon, L. and Evans, E. H. (Eds), John Wiley & Sons, Ltd, Chichester, 1998, Chapter 5, pp. 115–136.
2. Elan 5000 Users Manual, 0993–8429, May 1992, Rev. B.
3. Niu, H. and Houk, R. S., *Spectrochim. Acta, Part B*, **51**, 779 (1996).
4. Resano, M., Verstraete, M., Vanhaecke, F. and Moens, L., *J. Anal. At. Spectrom.*, **16**, 1018 (2001).
5. Sakata, K. and Kawabata, K., *Spectrochim. Acta, Part B*, **49**, 1027 (1994).
6. Vanhaecke, F., Saverwyns, S., De Wannemacker, G., Moens, L. and Dams, R., *Anal. Chim. Acta*, **419**, 55 (2000).
7. Divjak, B. and Goessler, W., *J. Chromatogr. A.*, **844**, 161 (1999).
8. Minnich, M. G. and Houk, R. S., *J. Anal. At. Spectrom.*, **13**, 167 (1998).
9. Brenner, I. B., Zander, A., Plantz, M. and Zhu, J., *J. Anal. At. Spectrom.*, **12**, 273 (1997).
10. Feldmann, I., Jakubowski, N. and Stuewer, D., *Fresenius' J. Anal. Chem.*, **365**, 415 (1999).

11. Feldmann, I., Jakubowski, N., Thomas, C. and Stuewer, D., *Fresenius' J. Anal. Chem.*, **365**, 422 (1999).
12. Marchante-Gayón, J. M., Feldmann, I., Thomas, C. and Jakubowski, N., *J. Anal. At. Spectrom.*, **16**, 45 (2001).
13. Tanner, S. D. and Baranov, V. I., *At. Spectrosc.*, **20**, 45 (1999).
14. Simpson, L. A., Thomsen, M., Alloway, B. J. and Parker, A., *J. Anal. At. Spectrom.*, **16**, 1375 (2001).
15. Bandura, D. R., Baranov, V. I. and Tanner, S. D., *Fresenius' J. Anal. Chem.*, **370**, 454 (2001).
16. Shaw, P., Spence, B., Lee, K. and Nelms, S., Speed and versatility with CCT^{ED} ICP-MS. How fast switching and energy discrimination can get the highest quality data in the shortest possible time. Paper presented at the *2002 Winter Conference on Plasma Spectrochemistry*, Scottsdale, AZ, USA, January 6–12, 2002.
17. Sloth, J. J. and Larsen, E. H., *J. Anal. At. Spectrom.*, **15**, 669 (2000).
18. Lapitajs, G., Greb, U., Dunemann, L., Begerow, J., Moens, L. and Verrept, P., *Int. Lab.*, May, 21 (1995).
19. Rottmann, L. and Heumann, K. G., *Anal. Chem.*, **66**, 3709 (1994).
20. Cabezuelo, A. B. S., Bayòn, M. M., Gonzalez, E. B., Garcia Alonso, J. I. and Sanz-Medel, A., *Analyst*, **123**, 865 (1998).
21. Vanhaecke, F., Riondato, J., Moens, L. and Dams, R., *Fresenius' J. Anal. Chem.*, **355**, 397 (1996).
22. Saverwyns, S., Van Hecke, K., Vanhaecke, F., Moens, L. and Dams, R., *Fresenius' J. Anal. Chem.*, **363**, 490 (1999).
23. De Smaele, T., Moens, L., Dams, R. and Sandra, P., *Fresenius' J. Anal. Chem.*, **355**, 778 (1996).
24. Chen, X. S. and Houk, R. S., *J. Anal. At. Spectrom.*, **10**, 837 (1995).
25. Feldmann, J., *J. Anal. At. Spectrom.*, **12**, 1069 (1997).
26. Heumann, K. G., Gallus, S. M., Rädlinger, G. and Vogl, J., *J. Anal. At. Spectrom.*, **13**, 1001 (1998).
27. Heumann, K. G., Rottmann, L. and Vogl, J., *J. Anal. At. Spectrom.*, **9**, 1351 (1994).
28. Ebdon, L., Hill, S. J. and Rivas, C., *Spectrochim. Acta, Part B*, **53**, 289 (1998).
29. Encinar, J. R., Monterde Villar, M. I., Santamaria, V. G., Garcia Alonso, J. I. and Sanz-Medel, A., *Anal. Chem.*, **73**, 3174 (2001).
30. Snell, J. P., Stewart, I. I., Sturgeon, R. E. and Frech, W., *J. Anal. At. Spectrom.*, **15**, 1540 (2000).
31. Hintelmann, H., Falter, R., Ilgen, G. and Evans, R. D., *Fresenius' J. Anal. Chem.*, **358**, 363 (1997).
32. Hintelmann, H. and Evans, R. D., *Fresenius' J. Anal. Chem.*, **358**, 378 (1997).
33. Rottmann, L. and Heumann, K. G., *Fresenius' J. Anal. Chem.*, **350**, 221 (1994).
34. Heumann, K. G., Gallus, S. M., Rädlinger, G. and Vogl, J., *Spectrochim. Acta, Part B*, **53**, 273 (1998).
35. Vogl, J. and Heumann, K. G., *Anal. Chem.*, **70**, 2038 (1998).
36. O'Connor, G., Ebdon, L., Evans, E. H., Ding, H., Olson, L. K. and Caruso, J. A., *J. Anal. At. Spectrom.*, **11**, 1151 (1996).
37. Waggoner, J. W., Milstein, L. S., Belkin, M., Sutton, K. L., Caruso, J. A. and Fannin, H. B., *J. Anal. At. Spectrom.*, **15**, 13 (1999).
38. Szpunar, J., Witte, C., Lobinski, R. and Adams, F., *Fresenius' J. Anal. Chem*, **351**, 351 (1995).
39. Sutton, K., Sutton, R. C. and Caruso, J. A., *J. Chromatogr. A.*, **789**, 85 (1997).
40. Vela, N. P., Olson, L. K. and Caruso, J. A., *Anal. Chem.*, **65**, 585 A (1993).
41. Hann, S., Prohaska, T., Köllensperger, G. and Stingeder, G., *J. Anal. At. Spectrom.*, **15**, 721 (2000).
42. Lobinski, R., Pereiro, I. R., Chassaigne, H., Wasik, A. and Szpunar, J., *J. Anal. At. Spectrom.*, **13**, 859 (1998).
43. Larsen, E. H., Pritzl, G. and Hansen, S. H., *J. Anal. At. Spectrom.*, **8**, 557 (1993).
44. Saverwyns, S., Zhang, X., Vanhaecke, F., Cornelis, R., Moens, L. and Dams, R., *J. Anal. At. Spectrom.*, **12**, 1047 (1997).
45. Kavanagh, P., Farago, M. E., Thornton, I., Goessler, W., Kuehnelt, D., Schlagenhaufen, C. and Irgolic, K. J., *Analyst*, **123**, 27 (1998).
46. Lintschinger, J., Schramel, P., Halatak-Rauscher, A., Wendler, I. and Michalke, B., *Fresenius' J. Anal. Chem.*, **362**, 313 (1998).
47. Szpunar, J. and Lobinski, R., *Pure Appl. Chem.*, **71**, 899 (1999).
48. Hann, S., Zenker, A., Galanski, M., Bereuter, T. L., Stingeder, G. and Keppler, B. K., *Fresenius' J. Anal. Chem*, **370**, 581 (2001).
49. Wang, J., Houk, R. S., Dreessen, D. and Wiederin, D. R., *J. Biol. Inorg. Chem.*, **364**, 546 (1999).
50. Jakubowski, N., Thomas, C., Klueppel, D. and Stuewer, D., *Analysis*, **26**, M37 (1998).
51. Welz, B., *Spectrochim. Acta, Part B*, **53**, 169 (1998).
52. Vanhaecke, F. and Moens, L., *Fresenius' J. Anal. Chem*, **364**, 440 (1999).
53. Agilent GC-ICP-MS Interface Technology, Agilent Technologies application note, 5988-3071EN (2001).
54. De Smaele, T., Verrept, P., Moens, L. and Dams, R., *Spectrochim. Acta, Part B*, **50**, 1409 (1995).
55. Montes Bayòn, M., Gutiérrez Camblor, M., Garcia Alonso, J. I. and Sanz-Medel, A., *J. Anal. At. Spectrom.*, **14**, 1317 (1999).
56. Ritsema, R., De Smaele, T., Moens, L., de Jong, A. S. and Donard, O. F. X., *Environ. Pollut.*, **99**, 271 (1998).
57. Vercauteren, J., De Meester, A., De Smaele, T., Vanhaecke, F., Moens, L., Dams, R. and Sandra, P., *J. Anal. At. Spectrom.*, **15**, 651 (2000).
58. Leal-Granadillo, I. A., Garcia-Alonso, J. I. and Sanz-Medel, A., *Anal. Chim. Acta*, **423**, 21 (2000).
59. Armstrong, H. E. L., Corns, W. T., Stockwell, P. B., O'Connor, G., Ebdon, L. and Evans, E. H., *Anal. Chim. Acta*, **390**, 245 (1999).

60. Jantzen, E. and Prange, A., *Fresenius' J. Anal. Chem.*, **353**, 28 (1995).
61. Peycheran, C., Lalere, B. and Donard, O. F. X., *Environ. Sci. Technol.*, **34**, 27 (2000).
62. Amouroux, D., Tessier, E., Peycheran, C. and Donard, O. F. X., *Anal. Chim. Acta*, **377**, 241 (1998).
63. Gallus, S. M. and Heumann, K. G., *J. Anal. At. Spectrom.*, **11**, 887 (1996).
64. Feldmann, J., Krupp, E. M., Glindemann, D., Hirner, A. V. and Cullen, W. R., *Appl. Organomet. Chem.*, **13**, 739 (1999).
65. Peycheran, C., Quétel, C. R., Lecuyer, F. M. M. and Donard, O. F. X., *Anal. Chem.*, **70**, 2639 (1998).
66. Grüter, U. M., Kresimon, J. and Hirner, A. V., *Fresenius' J. Anal. Chem.*, **368**, 67 (2000).
67. Olesik, J. W., Kinzer, J. A. and Olesik, S. V., *Anal. Chem.*, **67**, 1 (1995).
68. Lu, Q., Bird, S. M. and Barnes, R. M., *Anal. Chem.*, **67**, 2949 (1995).
69. Michalke, B. and Schramel, P., *J. Chromatogr. A*, **750**, 51 (1996).
70. Majidi, V. and Miller-Ihli, N. J., *Analyst*, **123**, 803 (1998).
71. Van Holderbeke, M., Zhao, Y., Vanhaecke, F., Moens, L., Dams, R. and Sandra, P., *J. Anal. Atom. Spectrom.*, **14**, 229 (1999).
72. Taylor, K. A., Sharp, B. L., Lewis, D. J. and Crews, H. M., *J. Anal. Atom. Spectrom.*, **13**, 1095 (1998).
73. Kirlew, P. W., Caruso, J. A. and Castillano, M. T. M., *Spectrochim. Acta B*, **53**, 221 (1998).
74. Kinzer, J. A., Olesik, J. W. and Olesik, S. V., *Anal. Chem.*, **68**, 3250 (1996).
75. Liu, Y., Lopez-Avila, V., Zhu, J. J., Wiederin, D. R. and Beckert, W. F., *Anal. Chem.*, **67**, 2020 (1995).
76. Schaumlöffel, D. and Prange, A., *Fresenius' J. Anal. Chem.*, **364**, 452 (1999).
77. Majidi, V., Qvarnström, J., Tu, Q., Frech, W. and Thomassen, Y., *J. Anal. At. Spectrom.*, **14**, 1993 (1999).
78. Bendahl, L., Gammelgaard, B., Jons, O., Farver, O. and Hansen, S. H., *J. Anal. At. Spectrom.*, **16**, 38 (2001).
79. Tangen, A., Lund, W., Josefson, B. and Borg, H., *J. Chromatogr. A.*, **826**, 87 (1998).
80. Polec, K., Szpunar, J., Palacios, O., Gonzalez-Duarte, P., Atrian, S. and Lobinski, R., *J. Anal. At. Spectrom.*, **16**, 567 (2001).
81. Ackley, K. L., Day, J. A. and Caruso, J. A., *J. Chromatogr. A.*, **888**, 293 (2000).
82. Michalke, B. and Schramel, P., *J. Chromatogr. A.*, **834**, 341 (1999).
83. Michalke, B., *J. Anal. At. Spectrom.*, **14**, 567 (1999).
84. Michalke, B. and Schramel, P., *J. Chrom. A.*, **807**, 71 (1998).
85. Prange, A., Schaumlöffel, D., Brätter, P., Richarz, A. N. and Wolf, C., *Fresenius' J. Anal. Chem.*, **371**, 764 (2001).
86. Vela, N. P. and Caruso, J. A., *J. Biochem. Biophys. M.*, **43**, 45 (2000).
87. Shen, W., Vela, N. P., Sheppard, B. S. and Caruso, J. A., *Anal. Chem.*, **63**, 1491 (1991).
88. Vela, N. P. and Caruso, J. A., *J. Chromatogr.*, **641**, 337 (1993).
89. Carey, J. M., Vela, N. P. and Caruso, J. A., *J. Chromatogr. A.*, **622**, 329 (1994).
90. Jin, K., Shibata, Y. and Morita, M., *Anal. Chem.*, **63**, 986 (1991).
91. Willie, S. N., Grégoire, D. C. and Sturgeon, R. E., *Analyst*, **122**, 751 (1997).
92. Gelaude, I., Dams, R., Resano, M., Vanhaecke, F. and Dams, R., *Anal. Chem.*, **74**, 3833 (2002).
93. Feng, Y.-L., Chen, H.-Y., Tian, L.-C. and Narasaki, H., *Anal. Chim. Acta*, **375**, 167 (1998).
94. Feng, Y.-L., Narasaki, H., Tian, L.-C. and Chen, H.-Y., *At. Spectrosc.*, **21**, 30 (2000).
95. Nielsen, J. L., Abildtrup, A., Christensen, J., Watson, P., Cox, A. and McLeod, C. W., *Spectrochim. Acta, Part B*, **53**, 339 (1998).
96. Evans, R. D. and Villeneuve, J. Y., *J. Anal. At. Spectrom.*, **15**, 157 (2000).

5.4 Plasma Source Time-of-flight Mass Spectrometry: A Powerful Tool for Elemental Speciation

Andrew M. Leach, Denise M. McClenathan and Gary M. Hieftje

Indiana University, Bloomington, IN, USA

1 INTRODUCTION

To generate speciation information, two types of data must be obtained about a sample: which elements are present and what the state or immediate environment of those elements is. The most common methods used to obtain this information involve a combination of separation techniques and elemental detectors. These systems differentiate species on the basis of selective retention (chromatography), electrophoretic mobility (electrophoresis), or volatility (electrothermal vaporization) prior to optical or mass spectral elemental detection. A more novel method for the generation of speciation information that is unique to mass spectrometry is the use of a modulated or pulsed ionization source that provides both atomic and molecular or fragmentation information about the sample in a sequential and repetitive fashion.

Handbook of Elemental Speciation: Techniques and Methodology R. Cornelis, H. Crews, J. Caruso and K. Heumann
 ISBN: 0-471-49214-0

Restricted measurement time is a common feature of virtually all speciation techniques. Separation methods produce analyte peaks with limited widths that arrive sequentially at the detector. Similarly, modulated ionization sources produce rapidly alternating windows of analyte information. The measurement of such transient signals can be extremely difficult with conventional mass-spectral systems.

Historically, the most common mass spectrometers used in elemental analysis are sequentially scanned systems such as the quadrupole mass filter or the scanned sector-field mass spectrometer. With scan-based systems, elemental information is generated by selectively allowing ions of a single mass-to-charge ratio (m/z) to traverse the mass spectrometer and be detected. To produce a spectrum, the mass-selection device is scanned in a sequential manner. In most laboratory situations in which analysis time is not limited, scanned systems provide excellent sensitivity and good spectral resolution. However, when measurement time is restricted, the fact that only one m/z is measured at a time produces an inverse relationship between the best achievable sensitivity (and thus precision) and the maximum number of elements and isotopes that can be monitored. Equation (5.4.1) describes the relationship between sensitivity and mass coverage range for scanned systems when analysis time is limited:

$$t = \sum_{n} \frac{c}{r} \qquad (5.4.1)$$

In this equation, t is the available analysis time, which is related, for example, to the length of a chromatographic peak, c is the number of signal (ion) counts needed for a given m/z to achieve a desired level of precision and sensitivity, n is the number of m/z to be monitored, and r is the count rate (ions s^{-1}) for a given m/z. If the analysis time t is held constant, the only way to enhance the sensitivity or precision of a measurement is to increase the number of counts (c) collected for an individual m/z; consequently, fewer elements (n) can be monitored.

An additional problem associated with the measurement of transient signals by means of scanned detection systems is the quantitation error known as *spectral skew* [1]. Transient signals, such as those produced by most separation techniques, exhibit a temporally changing concentration profile. Spectral skew is the relative enhancement or suppression of signals from adjacent m/z that is produced by time-dependent changes in analyte concentration that are on the same time scale as the scan time of the detection system. The combination of the ever-increasing speed of modern separation systems and the instrumentally restricted maximum scan rate of quadrupoles and sector-field instruments will cause spectral skew to become an increasingly troublesome difficulty.

Trapping mass analyzers such as the quadrupole ion trap and Fourier transform (FT) ion cyclotron resonance mass spectrometer eliminate not only spectral skew but also the tradeoff among sensitivity, precision and mass coverage range, since they simultaneously extract all m/z from the ionization source [2]. Additionally, the ability to selectively modify or remove trapped m/z through ion chemistry makes trap-based analyzers highly attractive. However, limited ion capacity (10^4 to 10^6 ions) has restricted the sensitivity and dynamic range of these systems. Furthermore, when operated in either a scanned or simultaneous (FT or image-current) detection mode, trapping spectrometers offer relatively low duty cycles. Their restricted spectral generation rate limits the applicability of trapping systems to the measurement of transient signals.

Sector-field instruments equipped with multiple discrete detectors or detector arrays are another class of mass spectrometer that simultaneously extracts selected m/z and avoids the disadvantages of scanned systems for the analysis of transient signals. Both detector geometries benefit from a 100 % duty cycle and practically unlimited spectral generation rates. However, multidetector systems are typically limited to monitoring less than ten elements and isotopes per analysis [3]. Detector-array systems are able to monitor large sections or a complete elemental mass spectrum simultaneously, but to date have been limited by the performance of available detector technology [4].

2 TIME-OF-FLIGHT MASS SPECTROMETRY

Time-of-flight mass spectrometry (TOFMS) is one of the oldest and perhaps simplest forms of mass analysis. First proposed in 1948 by Cameron and Eggers [5], TOFMS separates ions based on the fundamental relationship between kinetic energy (*KE*), mass (m) and velocity (v) shown in (equation 5.4.2):

$$KE = \frac{1}{2}mv^2 \quad (5.4.2)$$

The acceleration of a population of ions through a potential field so that all ions receive the same kinetic energy will result in a mass-dependent velocity dispersion, with light ions achieving higher velocities than heavier ions. Following this acceleration process, ions are allowed to separate based on their velocity difference within a field-free drift region commonly known as a flight tube. Detection of the ion arrival times (related to velocity) at a detector positioned at the end of the flight tube produces a mass spectrum.

Time-of-flight mass spectrometry has recently seen a dramatic increase in application, fueled by bioanalytical research that exploits the theoretically infinite mass range afforded by TOFMS for the measurement of large peptides and DNA fragments. Although the ability to measure high-mass ions is relatively unimportant in elemental mass spectrometry (commonly limited to m/z 1–238), other characteristics have motivated atomic spectroscopists to utilize TOFMS [6–8]. Because ions of different mass are detected as a function of time, TOFMS is not, strictly speaking, a simultaneous detection technique. However, since all of the ions that produce a given time-of-flight mass spectrum are extracted simultaneously from the ionization source, the tradeoff between sensitivity and mass coverage range inherent to scanned systems is eliminated, and spectral skew is avoided. Additionally, the limited atomic mass range, moderate acceleration potentials (−2000 V), and short flight lengths (1 m) can produce complete elemental mass spectra in less than 50 μs. With spectral generation rates greater than 20 kHz, TOFMS is a nearly ideal detection system for transient signals.

3 MASS RESOLUTION OF TOFMS

Resolving power ($m/\Delta m$) in TOFMS is proportional to ion flight time and calculated as $R = T/2\,\Delta t$, where T is the total flight time from the extraction region to the detector and Δt is the width of the analyte peak. The factor of 2 in the denominator is a result of the squared relationship between mass and velocity. The ultimate resolving power of TOFMS systems is limited by several factors related to the spatial and velocity distributions of the ion packet extracted for mass analysis. Figure 5.4.1 shows a diagram of a simple two-stage acceleration TOFMS instrument. Ions

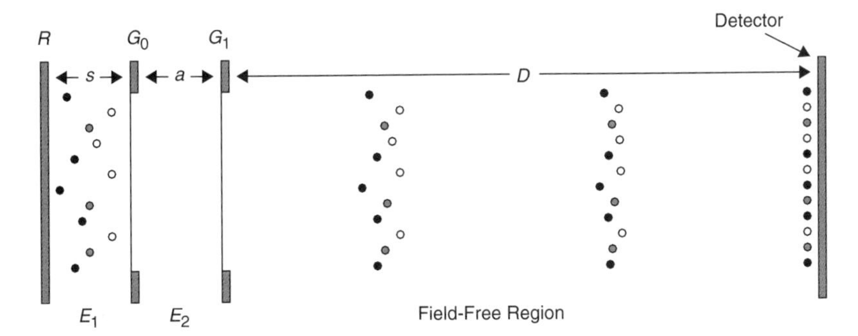

Figure 5.4.1. Diagram of a two-stage acceleration time-of-flight mass spectrometer. The extraction region is defined by the repeller plate R and gridded electrode G_0, having a width of s and a potential difference of E_1. The acceleration region is defined by gridded electrodes G_0 and G_1, having a width of a and a potential difference of E_2. The field-free drift region has a length of D that ends at the ion detector.

enter the mass spectrometer and fill the extraction region defined by the repeller plate (R) and a gridded electrode (G_0). While ions are filling the extraction region, R and G_0 are held at the same voltage, ordinarily ground potential. At a set delay time, the repeller is pulsed to a high positive voltage (for the analysis of positive ions). This event produces a linear potential field between R and G_0 and causes ions within the extraction region to move toward G_0. Following the extraction region is a second, constant-acceleration stage defined by two gridded electrodes, G_0 and G_1. Accelerated ions then enter a field-free drift region where mass separation occurs, ending at the detector.

3.1 Compensation for an initial spatial distribution

If a packet of isomass ions is accelerated through the same potential difference, the ions will achieve an equal velocity and arrive at the detector with a time spread corresponding to their initial spatial distribution in the direction of the flight tube. This situation will obviously compromise resolution. To compensate for the length of the extraction region, a technique known as *space focusing* was conceived and developed by Wiley and McLaren [9]. Application of a voltage gradient across the extraction region produces a relationship between the initial position of an ion and the potential through which that ion is accelerated. With a positive voltage applied to the repeller plate, positive ions of a given m/z that are positioned close to the repeller are accelerated to a greater final velocity than ions closer to the G_0 electrode. The difference in velocity of isomass ions causes the ions with elevated velocities to overtake the slower ions within the flight tube. The position where the spatial distribution of the extracted ion packet is minimized is known as the *space-focus plane*. Conveniently, the position of the space-focus plane is the same for ions of all m/z.

Optimal mass resolving power requires compensation for the initial spatial distribution of ions within the extraction region. Equation (5.4.3) describes the basic premise of space focusing, where two ions, one positioned in the middle of the extraction region ($s_{1/2}$) and one positioned a distance Δs from $s_{1/2}$, produce a distribution of flight times.

$$t(s_{1/2} \pm \Delta s) - t(s_{1/2}) = \text{minimum} \qquad (5.4.3)$$

To achieve the highest possible mass resolution the temporal spread of ion arrival times at the detector must be minimized. Equation (5.4.3) can be analyzed with a Taylor series expansion to produce equation (5.4.4):

$$\sum_{n=1}^{\infty} \frac{t^n(s_{1/2})}{n!}(\pm\Delta s)^n - t(s_{1/2}) = t'(s_{1/2})(\pm\Delta s) + \frac{1}{2}t''(s_{1/2})(\pm\Delta s)^2 + \cdots \qquad (5.4.4)$$

In this power series, t' and t'' denote first- and second-order coefficients. To achieve enhanced resolving power, successively higher-order coefficients must be included in calculating the location of the optimal space-focus plane. First-order space focusing occurs when $\Delta t = 0$ for the t' term, while second-order space focusing is achieved by simultaneous solution of the t' and t'' terms.

Analysis of the equations of motion for ions within the TOF system shows that the ratio of the field strengths within the extraction (E_1) and acceleration (E_2) regions can be used to control the quality and position of the space focus plane [10]. With the two-stage acceleration system shown in Figure 5.4.1, a solution for first-order space focusing finds the position of lowest spatial dispersion (D) to be:

$$D = s\left[\left(1 - \frac{g}{f}\right)\left(\sqrt{1+2f}\right)^3 + g\left(2 + \frac{1}{f}\right)\right] \qquad (5.4.5)$$

where s and a are the widths of the extraction and acceleration regions, respectively, $g = a/s$ and $f = (E_2/a)/(E_1/s)$. Higher mass resolution can be accomplished with second-order space focusing if specific operational conditions are observed. Second-order space focusing is achieved when $E_1 = E_2$, where a space focus plane is produced at $D = s + 2a$.

At this point, it should be noted that the quadratic relationship between ion velocity and initial position within the extraction region prohibits the complete solution of the Taylor series expansion with the previously described two-stage linear field system. However, use of a quadratic extraction field would produce a linear relationship between ion velocity and initial position and thus allow ideal space focusing [11].

Additionally, second-order space focusing typically produces a focal plane close to the extraction region, by which location little mass dispersion has been achieved. Since resolution is directly related to total flight time, only low resolving powers are obtained at this primary space-focus plane. However, as will be discussed in the following section, the space focus plane is ideally suited to act as a virtual source for an ion mirror that re-images the plane at a greater flight length.

A full mathematical description of space focusing is beyond the scope of the present text. For a more in-depth review of space-focusing theory, the reader is referred to the literature [9, 10, 12–14].

3.2 Compensation for initial energy distribution

The equations of motion for space focusing are commonly solved by assuming that the extracted ions have no initial energy and thus an energy distribution of zero. However, in reality, most ionization sources used for elemental analysis are operated at atmospheric pressure, which requires the use of a vacuum interface to couple the source with the mass spectrometer. Transmission from the ionization source to the mass spectrometer generates an isokinetic (equal-velocity) ion beam and thus a mass-dependent energy [15]. The velocity of the ion beam is dictated by the most abundant species, argon for an inductively coupled plasma (ICP), but will exist in a Maxwellian distribution corresponding to the temperature of the ionization source. Additionally, potentials imparted by the ionization source, such as the offset potential of an ICP, will augment the distribution of ion velocities within the TOFMS extraction region. The existence of this energy distribution will produce a spread of final post-acceleration isomass ion velocities and thus will result in degraded resolution.

The most common method used to compensate for an initial ion energy distribution is an electrostatic ion mirror, also known as a reflectron. Developed by Mamyrin and coworkers [16, 17], a properly designed reflectron can significantly reduce timing errors caused by initial energy distributions of up to 10 % of the total acceleration potential. Ion mirrors are potential ramps consisting of a series of resistively separated ring electrodes (Figure 5.4.2). High-energy ions penetrate farther into the reflectron than ions of lower energy and thus follow a longer path to the detector. This energy-dependent flight length causes the faster ions to travel a longer distance and to arrive at the detector coincident with slower ions that traversed a shorter path. A second function of the ion mirror is to re-image the original space focus plane. The bent ion path of a reflectron-equipped TOF instrument can effectively double the flight length of the system without an increase in the instrument size. This enhanced flight length will result in an additional gain in mass resolving power.

A special resolution consideration arises from the possibility that ions might be moving toward the repeller (away from the detector) prior to their extraction for mass analysis. These ions would

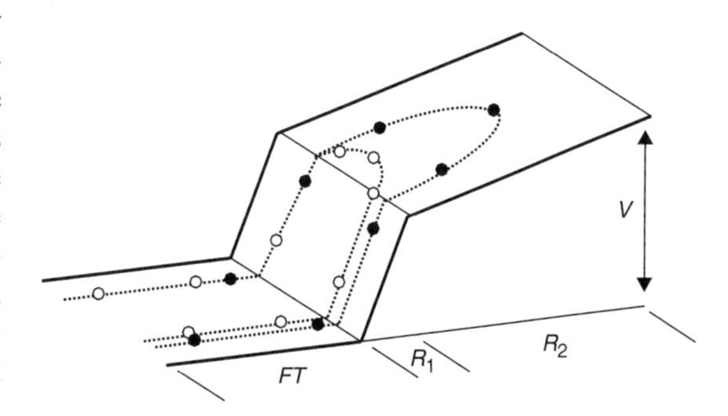

Figure 5.4.2. Energy compensation with a two-stage reflection ion mirror. The reflectron is a voltage (V) ramp composed of a retardation region (R_1) and a reflection region (R_2). Ions with different energies disperse within the flight tube (FT). High-energy ions (●) penetrate the reflectron to a greater extent than lower energy ions (○) and thus travel a greater distance within the ion mirror. High-energy ions catch up with lower-energy ions in the second passage though the flight tube, to arrive simultaneously at the detector.

possess a negative initial velocity with respect to their final velocity imparted by the TOF extraction and acceleration processes. Upon initiation of the repeller pulse these ions would first decelerate prior to being re-accelerated toward the detector. This *turnaround time* error is difficult to overcome and can result in degraded mass resolution.

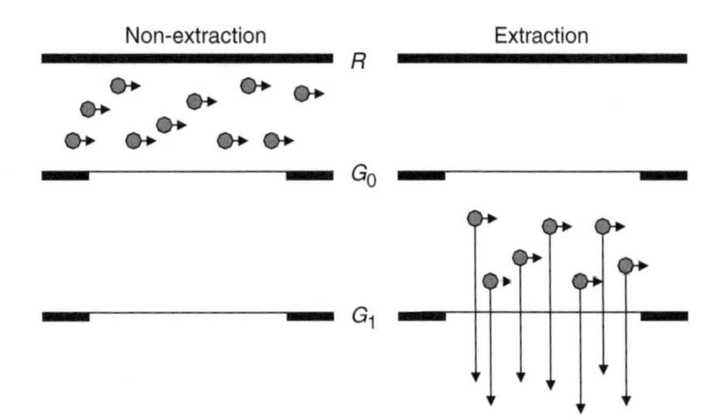

Figure 5.4.3. Diagram of an orthogonal acceleration TOFMS system. The extraction region is defined by the repeller plate (R) and a gridded electrode (G_0), commonly held at ground potential. The acceleration region extends from G_0 to a second gridded electrode (G_1), held at flight-tube potential. While ions are entering the extraction region the repeller is held at ground potential. Extraction of ions for mass analysis is initiated when the repeller is pulsed to a positive potential, injecting ions into the flight tube at a 90° angle with respect to their original axis of propagation. Arrows denote the relative direction and magnitude of ion velocity vectors, but are not drawn to scale.

4 TOFMS INSTRUMENTATION

Although several time-of-flight geometries have been explored, two systems, orthogonal acceleration (*oa*) and axial acceleration (*aa*) have enjoyed the widest acceptance. These systems differ in the relative angle between the axis of the primary ion beam from the ionization source and the axis of the TOF flight tube. Each system possesses advantages relative to the other and will be discussed from a historical standpoint here.

4.1 Orthogonal acceleration time-of-flight mass spectrometry

Hieftje and coworkers [18–21] performed the initial experiments that coupled an ICP source to a time-of-flight mass spectrometer with an instrument based on orthogonal acceleration. Figure 5.4.3 shows a schematic diagram of the orthogonal acceleration process. In this system, the primary ion beam enters the extraction region between the repeller plate and the G_0 electrode. At a set time, the repeller is pulsed to a positive voltage and the ions are accelerated into the flight tube, positioned at a 90° angle to the original axis of ion propagation. Because the most significant initial ion energy distribution is oriented perpendicular to the TOF axis, orthogonal extraction systems can achieve relatively high mass resolving power (>2000 FWHM). Additionally, the width of the primary ion beam in the direction of the flight tube can be restricted by optics positioned prior to the extraction region. As a result, less stringent demands are placed upon space focusing, so resolution is improved [20].

Although the initial velocity component of the primary ion beam, oriented perpendicular to the field-free region, does not significantly limit mass resolving power, this energy produces many of the instrumental difficulties experienced with *oa*-TOFMS systems. The isokinetic ion beam extracted from the ionization source results in a mass-dependent energy perpendicular to the TOFMS axis. Light ions have low initial energy and are easily redirected by the repeller pulse. However, heavier ions possess more energy and thus follow a more curved path to the detector. Differences in the flight path of ions result in beam broadening that can require the use of an extended extraction zone and larger, more expensive detectors that have elevated susceptibility to noise. Initial experiments with ICP-*oa*-TOFMS attempted to correct for this mass-dependent ion trajectory by means of steering plates positioned within the flight tube [19]. Adequate mass dispersion has occurred at the steering plates to allow the application of a voltage ramp to the plates that redirects all m/z along roughly the same flight path. Although this compensation technique reduced beam divergence, some loss in mass resolving power was experienced.

Guilhaus [22] has demonstrated a simpler technique, known as spontaneous drift, that allows ions to follow their natural, mass-dependent path. Although this method results in a broadened ion beam and can require the use of enlarged reflectrons and detectors, spontaneous drift avoids the slight degradation of mass resolving power caused by steering plates. A smaller detector can be used with the spontaneous drift technique if a longer extraction region is employed [23]. By taking light ions from the leading edge of the ion packet while detecting heavy ions from the trailing edge, the spatial distribution within the extraction region can be used to compensate for the mass-dependent ion trajectory.

4.2 Axial acceleration time-of-flight mass spectrometry

To address several of the limitations of *oa*-TOFMS, a second geometry has been developed that accelerates ions coaxially to their initial axis of propagation from the ionization source. A diagram of the axial acceleration process is shown in Figure 5.4.4. Similar to *oa*-TOFMS, a two-stage acceleration system is employed. However, in an axial system, the repeller plate is replaced with a gridded electrode. Ions pass through the repeller electrode to enter the TOF extraction region. This simple alteration results in several important changes to the operational characteristics of an *aa*-TOFMS system.

With an *aa*-TOFMS system, the axis with the largest initial ion energy spread is positioned colinear with the TOFMS axis. This coaxial geometry provides both benefits and disadvantages when compared to an orthogonal acceleration system. Since the isokinetic energy of the primary ion beam is directed along the TOFMS axis, the mass-dependent trajectory to the detector experienced with orthogonal systems is significantly reduced [7]. The lower energy distribution perpendicular to the TOFMS axis also lessens beam divergence within the field-free region and results in greater ion transmission efficiency to the detector. Conversely, the increased energy distribution directed along the TOF axis places greater demands on the energy-compensation ability of the reflectron ion mirror. Additionally, to improve the efficiency of ion utilization in a TOFMS, it is desirable to make the length of the extraction zone as long as possible. Because this length is in the direction of the flight tube in an *aa*-TOFMS, greater care is needed to achieve the best space focusing.

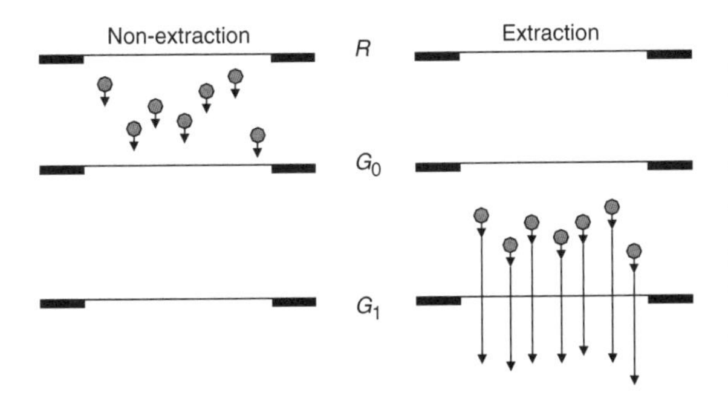

Figure 5.4.4. Schematic diagram of an axial acceleration TOFMS system. The extraction region is defined by the gridded repeller electrode (R) and a second gridded electrode (G_0), commonly held at ground potential. The acceleration region extends from G_0 to a third gridded electrode (G_1), held at flight-tube potential. Ions enter the extraction region through the repeller (held at ground potential). Extraction of ions for mass analysis is initiated when the repeller is pulsed to a positive potential, injecting ions into the flight tube at a 0° angle with respect to their original axis of propagation. Arrows denote the relative direction and magnitude of ion velocity vectors, but are not drawn to scale.

4.3 Other considerations for TOFMS with continuous ionization sources

The open geometry of TOF systems requires that unwanted ions be strictly controlled to reduce noise and interferences. Ions that reside in the extraction region during a repeller pulse are accelerated into the flight tube for analysis. The arrival of these ions at the detector is temporally referenced to the extraction pulse and proportional to mass. Ions that pass through the extraction region when the repeller is at ground potential will also be accelerated into the flight tube and can result in detector events. However, the flight time of a nonextracted ion cannot be calculated since there is no referenced start time. Because nonextracted ions will regularly enter the acceleration region, these ions will be seen as a continuous background signal.

Adventitiously, extracted and background ions can be differentiated on the basis of their energy [24]. Extracted ions will attain an energy proportional to the sum of the repeller and acceleration potentials. In contrast, background ions will receive energy only from the acceleration region since they enter that region at a time when a repeller pulse is not being applied. Placement of an energy barrier with a potential slightly higher than the acceleration voltage immediately prior to the detector will stop background ions while passing extracted ions. As shown in Figure 5.4.5, a simple energy barrier can be constructed from a series of three gridded electrodes, of which the exterior electrodes are held at flight-tube potential while the center electrode provides the energy discrimination.

Use of an energy discrimination system can significantly reduce continuum ion background. However, although ions can be stopped by a potential field, neutral species will be unaffected.

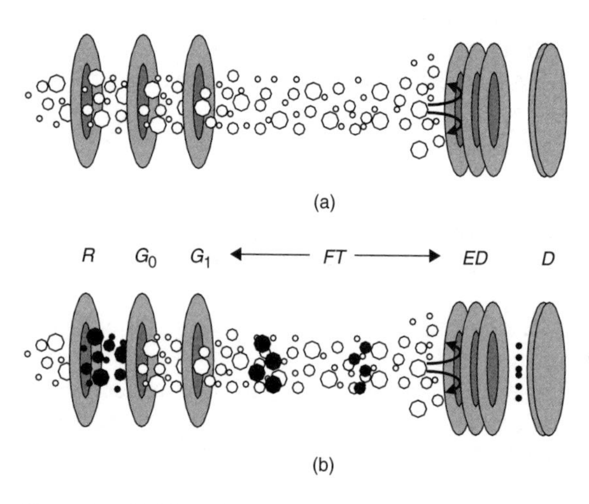

Figure 5.4.5. Effect of energy discrimination system on continuum background ions. Ion optics consists of: repeller (R), grounded electrode (G_0), acceleration electrode (G_1), flight tube (FT), energy discrimination electrodes (ED), and detector (D). The middle ED electrode is held at a positive potential to reject continuum background ions (○). (a) Repeller potential at ground. None of the ions entering the flight tube has sufficient energy to overcome ED barrier. (b) Repeller electrode is pulsed to a positive potential to start time-of-flight analysis sequence. Ions (●) present within the extraction region (defined by R and G_0) during this pulse gain additional energy and are subsequently able to pass the ED barrier to result in detector events.

Ions that are neutralized after acceleration, most likely through charge exchange with residual gas in the flight tube, will retain a significant portion of their initial energy. If charge exchange occurs between the reflectron and the detector, the high-energy neutrals that are produced will traverse the energy discrimination barrier and result in detector events and elevated background.

To limit the production of high-energy neutrals, the pressure inside the flight tube must be as low as possible and the number of ions (especially Ar^+) admitted to the flight tube must be restricted. Time-of-flight mass spectrometry is an inherently pulsed technique. However, most of the ionization methods used for elemental analysis produce ions in a continuous fashion. To reduce the number of unwanted ions presented to the mass spectrometer, the primary ion beam can be modulated.

The position of the flight tube at a 90° angle relative to the axis of the primary ion beam makes the orthogonal extraction process a nearly ideal modulation approach. In theory, all ions enter the extraction region from one side and exit through the other. Only when the repeller plate is pulsed to a positive potential are ions redirected into the flight tube for mass analysis. However, ion scattering occurs; also, the radial velocity of some ions in the primary ion beam is sufficient to cause unwanted ions to gain entrance to the flight tube [24]. These undesired ions can then undergo charge exchange and result in elevated background noise.

To further restrict the number of unwanted ions within the TOFMS system, an ion gate can be positioned ahead of the extraction region (Figure 5.4.6). In a fashion similar to that used in the energy discrimination system, a modulation gate can be constructed of three gridded electrodes. The two exterior electrodes (M_1 and R) are maintained at ground potential while the center electrode (M_2), held at a positive potential, stops the passage of ions. During a defined temporal window, the gate electrode is brought to a low potential to transmit ions. Returning the gate electrode to its original elevated potential then produces an ion packet. Other modulation gates incorporate steering plates or lenses to deflect the

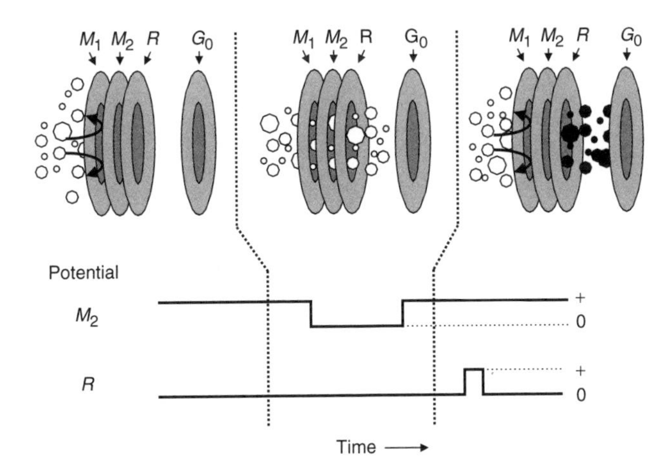

Figure 5.4.6. Temporal potential sequence of a three-electrode pre-acceleration modulation system. Modulation electrode M_1 remains at ground potential throughout analysis. With modulation grid M_2 at a positive potential all ions (○) are prevented from entering the TOF extraction region (defined by repeller R and ground electrode G_0). When M_2 is dropped to ground potential, ions pass through the modulation system to fill the extraction region. After a set delay, M_2 is returned to its original potential, impeding the transmission of additional ions and generating an ion packet (●). A subsequent delay allows ions within the modulation system to enter the TOF extraction region. Finally, R is pulsed to a positive potential to start the TOF analysis process.

primary ion beam [20]. Although beam modulation can be beneficial in an orthogonal extraction system, modulation is critical to the performance of axial systems, in which the primary ion beam is oriented directly into the flight tube.

While the simultaneous extraction of all m/z has proven to be one of the keys to the success of TOFMS for elemental analysis, it imparts some difficulties. The primary ion beam of most conventional plasma sources is composed predominantly of ionized support gas and solvent vapor. Argon in an ICP can be greater than 200-fold more abundant than analytes present at the $\mathrm{mg\,kg^{-1}}$ level. Additionally, ambient gas and solvent ions including N^+, O^+, OH^+, N_2^+, NO^+, O_2^+ and several argon-containing polyatomics can be present at high concentrations. For ultratrace measurements, the disparity between matrix and analyte concentrations is great. Therefore, matrix ions must be removed from the extracted ion packet to avoid detector saturation while maintaining sufficient detector gain for analyte measurements.

The most common technique used to remove unwanted ions from the extracted ion packet is selective ion deflection. Although ion deflection can be accomplished via a number of instrumental configurations, the basic operation is the same. Ions are extracted into the TOF drift region and allowed to undergo partial mass separation. A deflection system as simple as two parallel steering plates is positioned at the first space focus plane. Application of a potential pulse to one plate will selectively alter the trajectory of ions within the deflection region, causing them to miss the detector. A sequence of pulses can be used to remove more than one m/z window.

The performance of an ion deflection system can be evaluated on the basis of several criteria, most notably deflection efficiency (the ratio of ions in a selected mass window with and without the application of a deflection pulse) and deflection-window width (important for minimizing the effects of ion deflection on adjacent m/z). The temporal width of a deflection window is defined by the combination of the physical dimensions of the deflection system in the time-of-flight axis and the width of the voltage pulse. Deflection efficiency is dictated by the applied potential used to steer the unwanted ions, and the length of time the selected ions experience the deflection field (proportional to deflection window width). As should be evident from the previous statements, the efficiency and temporal width of deflection systems are usually inversely related. Traditional parallel-plate systems provide good deflection efficiency, ranging from 10^3 to 10^4, but are limited in temporal narrowness by the physical dimensions of the plates. As a result, they often result in the simultaneous deflection of several m/z even at relatively low masses ($<40\,m/z$). An alternative deflection system described by Vlasak *et al.* [25] consists of a series of closely spaced parallel wires; pulses of opposite potential are simultaneously applied to adjacent wires. This system is commonly referred to as a comb deflector due to its physical appearance. The temporal fidelity achieved with a comb deflector is significantly better than that of parallel-plate systems due to the narrow region (in the time-of-flight axis) in which ions experience the applied deflection pulse. However, as expected, the deflection efficiency of comb deflectors is typically worse than parallel-plate systems, ranging from 10^2 to 10^3.

5 PERFORMANCE OF ICP-TOFMS SYSTEMS

Although time-of-flight mass spectrometers have been used with a number of plasma ionization sources including glow discharges (GD) and microwave plasma torches (MPT), this section will focus on ICP-based systems due to their recent success. Two commercial ICP-TOFMS instruments are already available, the orthogonal acceleration Optimass 8000 (GBC Scientific Equipment Pty. Ltd., Dandenong, VIC, Australia) and the axial acceleration Renaissance (LECO, St. Joseph, MI, USA). This section compares the characteristics of several instruments, both academic and commercial.

5.1 Sensitivity, noise and limits of detection

Theoretically, it is expected that the in-line geometry of the axial acceleration system should exhibit lower beam divergence and better ion-transmission efficiency than the orthogonal system. In turn, increased transmission efficiency would result in improved sensitivity. However, the sensitivity of both commercial instruments is found to be in the range of 500 to 10 000 cps $\mu g\,L^{-1}$ per m/z (range caused by heavy-mass bias in both instruments) [26, 27]. The similar sensitivities of the two acceleration geometries can be explained by the relative size of each instrument's extraction region. To gain sensitivity in an orthogonal acceleration system the length of the extraction region can be increased to improve the instrument duty cycle (percentage of ions utilized). However, a longer extraction region in an axial acceleration instrument will place greater demands on the space-focusing abilities of the typical two-stage acceleration system and might result in degraded resolution.

The folded flight path produced by the presence of a reflectron minimizes photon-related noise in both acceleration geometries. The orthogonal acceleration system demonstrates very low detector background with less than one count per second per m/z for most elements [26]. As would be expected, the axial acceleration instrument exhibits slightly higher background (1–10 cps per m/z) due to the greater likelihood of high-energy neutrals within the flight tube [27]. The combination of comparable sensitivity and reduced noise provided the orthogonal acceleration instrument with somewhat better limits of detection than the axial system. Depending upon operating conditions, the axial instrument exhibited detection limits in the single $ng\,L^{-1}$ range, while the orthogonal system was typically in the sub-$ng\,L^{-1}$ regime. The limits of detection for both instruments should improve as the instruments mature.

5.2 Mass resolving power and abundance sensitivity

In general, the mass resolving power of ICP-TOFMS instruments improves with increased m/z. Although the temporal separation between adjacent m/z peaks is greater for low masses, higher resolution is achieved for heavy ions due to longer flight times. The orthogonal geometry produced resolving powers (FWHM) of 500 for ^{6}Li and 2200 for ^{238}U [26]. The axial system specifies a lower resolving power (but at the more stringent criterion of 10 % peak height) of 615 for ^{209}Bi due to the increased demands on space focusing and elevated initial energy spread [27]. Work performed on a similar axial system in this laboratory has resulted in resolving powers (FWHM) ranging from 410 for ^{7}Li to 1140 for ^{209}Bi [28]. Although both instruments achieved resolution significantly greater than convention quadrupole-based systems (operated in the first Mathieu stability region), resolution was inferior to sector-field instruments and insufficient for separation of important ICP-MS isobars such as ^{52}Cr with ^{40}Ar^{12}C ($R = 2400$) and ^{56}Fe with ^{40}Ar^{16}O ($R = 2500$).

Abundance sensitivity is the effect of adjacent m/z on one another, calculated in terms of equivalent concentrations. High abundance sensitivity means that trace measurements can be made in the presence of high concentrations of an element at an adjacent m/z. The orthogonal acceleration TOFMS produced abundance sensitivities of 2.8×10^6 on the low-mass side and 1.4×10^4 on the high-mass side [26]. A noncommercial axial system provided abundance sensitivities of 6.7×10^5 on the low-mass side and 1.7×10^4 on the high-mass side [29]. Both geometries achieved low-mass side abundance sensitivities similar to quadrupole instruments, but were worse on the high-mass side due to peak tailing. It is worth noting that the abundance sensitivity measured with quadrupole systems is typically better for the high-mass side than for the low-mass side (opposite to the trend seen with TOF-MS instruments).

5.3 Isotope ratio measurement

Multiplicative noise in the ionization source is a major cause of degraded isotope-ratio precision, especially when the source fluctuations are on the same time scale as the scan rate of the mass spectrometer [30]. Although faster scan rates improve ratio precision, all m/z must be sampled from the plasma simultaneously to compensate fully for ionization-source multiplicative noise. Mahoney *et al.* [6] demonstrated that the simultaneous extraction capability of ICP-TOFMS could provide isotope ratio precision on the order of 0.056 % RSD. Good agreement with values predicted by counting statistics suggests that further improvements in precision are possible with longer integration times. The commercial orthogonal acceleration instrument provided isotope-ratio precision that was as good as 0.2 % RSD (depending upon concentration and integration time, as expected) [26]. At higher concentrations, Vanhaecke and coworkers [31] reported precision better than 0.05 % RSD with a commercial axial acceleration system. The measurement precision of both commercial instruments was shown to improve based on counting statistics up to integration times between 30 and 60s [26, 27]. Beyond these integration lengths other noise sources appear to be dominant in the determination of isotope-ratio precision.

Instrumental mass bias and, even more important, its stability, are important factors that often limit isotope ratio accuracy. In general, mass bias for the two commercial ICP-TOFMS instruments is found to drop with increased m/z [26, 32] and to be similar in magnitude to that reported for quadrupole- and magnetic sector-based instruments [33]. Mass-bias fluctuation and drift have been reported with both commercial ICP-TOFMS systems. Sturgeon *et al.* [26] attributed minor short-term variations in mass bias experienced with the orthogonal acceleration instrument to changes in ion-optic voltages. With the axial acceleration system, Emteborg and coworkers [32] reported sudden, unexplained shifts in mass bias that degraded the long-term stability of the instrument. However, the authors noted that the rapid data acquisition speed of TOF-MS instruments limited the need for long-duration measurements. As with other types of ICP-MS instruments, proper calibration will allow accurate isotope ratio measurements to be performed with ICP-TOFMS systems.

5.4 Mass analyzer comparison

Table 5.4.1 displays a comparison of the figures of merit typically achieved by the three types of mass analyzers currently competing in the commercial ICP-MS market. Instruments based on

Table 5.4.1. Comparison of mass analyzer figures of merit.

	ICP-QMS[a]	ICP-SFMS[a,b]	ICP-TOFMS
Sensitivity (cps μg L^{-1})	10^5	10^6	10^3-10^4 per m/z
Background (cps)	1–10	<0.1	1–10 per m/z
Resolving power[c]	Unit Mass	Up to 10 000	Up to 1500
Abundance sensitivity[d]	$10^6/10^7$	$10^6/10^6$	$10^6/10^4$
Isotope ratio precision[e] (% RSD)	0.1	0.05	0.02
Ion sampling mode	Sequential	Sequential	Simultaneous
Spectral generation rate	2000–3000 $m/z\ s^{-1}$ 10 full spectra s^{-1}	350 $m/z\ s^{-1}$ 1.7 full spectra s^{-1}	6 000 000 $m/z\ s^{-1}$ [f] 20 000 full spectra s^{-1}
Cost	$	$$	$

[a] Ref. 34.
[b] Ref. 35.
[c] 10 % valley definition.
[d] Low-mass side/high-mass side.
[e] 10 min integration time.
[f] Based on 300 m/z measured at 20 000 Hz.

quadrupole mass filters have historically dominated research in the field of ICP-MS. Quadrupole systems are in general robust and of relatively low cost while offering good sensitivity. Low resolving power, typically unit-mass resolution, and relatively high noise have been the main disadvantages of ICP quadrupole mass spectrometers (QMS) [34]. Sector-field mass spectrometers (SFMS) combine excellent sensitivity and low noise to produce the best limits of detection currently achievable with ICP-MS, often less than 0.1 pg g^{-1} [34, 35]. Additionally, most ICP-SFMS systems can be operated in a high-resolution mode ($R > 7500$) that eliminates the majority of common elemental isobaric interferences. Criticism of sector-field instruments typically stems from their high cost and slow scan speeds. Fast spectral generation rates and the potential of high-resolution measurements have resulted in the recent interest in the use of time-of-flight mass spectrometers for elemental analysis. However, the widespread acceptance of ICP-TOFMS is currently hindered by somewhat higher detection limits. It is important to note that ICP-TOFMS is the least mature of the three classes of mass analyzer compared here, so it can be expected that the performance of time-of flight instruments will rapidly improve.

6 HYPHENATED TOFMS SPECIATION ANALYSIS

The most obvious method for the generation of speciation information is to couple a separation system with an element-selective detector. Several outstanding reviews on the topic of hyphenated speciation analysis have recently been published [36–39]. The following section highlights hyphenated speciation systems that employ plasma source time-of-flight mass spectrometry (PS-TOFMS).

6.1 Gas chromatography

Capillary gas chromatography (CGC) is a convenient method for the separation of volatile species based on their selective partitioning between stationary and mobile phases. Separations typically require several minutes with analyte peak widths as narrow as hundreds of milliseconds. To provide adequate temporal resolution, single m/z sampling rates of at least 10 Hz are typically desired. On this time scale, multielemental analysis of complex mixtures with conventional scan-based mass analyzers can be experimentally difficult.

Leach and coworkers [40] demonstrated the use of an ICP-*aa*-TOFMS instrument for the speciation analysis of a simple mixture of organometallic compounds. Although this study did not analyze complex mixtures, the figures of merit for the use of ICP-TOFMS with gas chromatography were explored. The high spectral generation rate of TOFMS provided excellent temporal resolution of fast transient signals. Simultaneous extraction of all m/z provides the capability to perform isotopic analysis on GC peaks. Figure 5.4.7 illustrates the detection of all ten tin isotopes during the separation of tetramethyl- and tetraethyltin. Calculation of the $^{118}Sn/^{120}Sn$ isotopic distribution for ten injections containing 5 pg of tin per compound produced an isotope ratio accuracy of 0.28 % with a precision of 2.88 % RSD. This precision was found to be in good agreement with counting statistics. Detection limits in the tens of femtograms (calculated as the metal) per injection with a dynamic range of six orders of magnitude were demonstrated for three organotin and organolead compounds.

A helium microwave plasma torch (MPT) was coupled to an *oa*-TOFMS system for the detection of halogenated hydrocarbons separated by gas chromatography [41]. Figure 5.4.8 shows the detection of several chlorine-containing hydrocarbons at m/z 12 for carbon and m/z 35 for chlorine. The simultaneous extraction of all m/z by

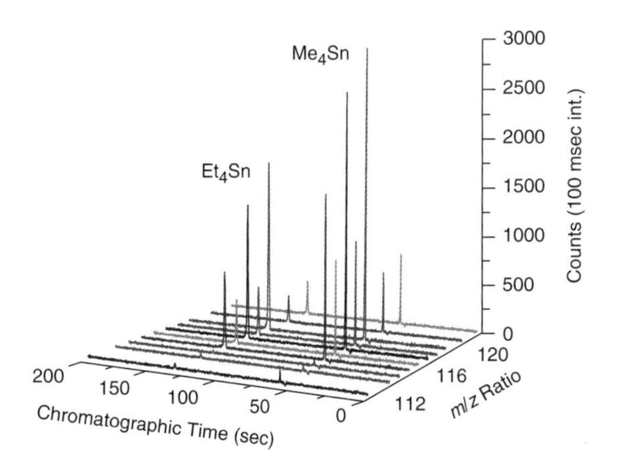

Figure 5.4.7. Multi-isotopic detection during the separation of tetramethyltin (Me_4Sn) and tetraethyltin (Et_4Sn). Injection contained 50 pg of tin per compound. Reproduced by permission of the Royal Society of Chemistry.

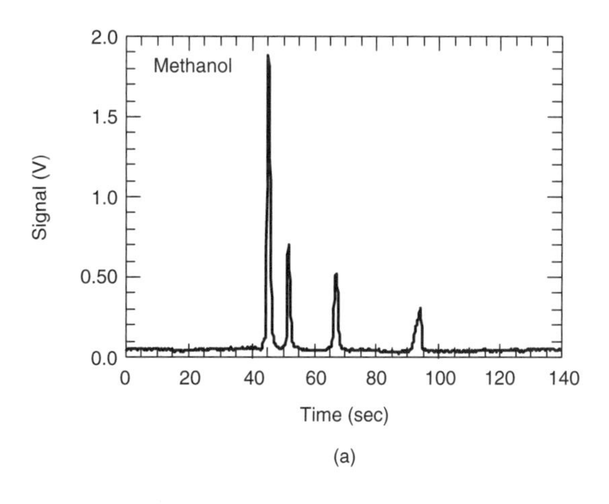

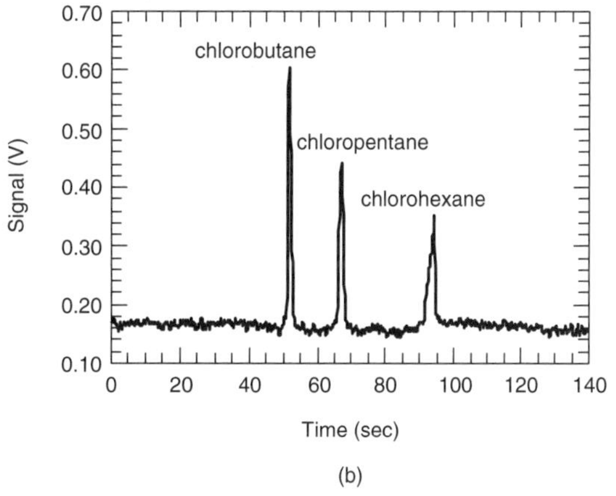

Figure 5.4.8. Isotope-specific chromatograms of halogenated hydrocarbons (chlorobutane to chlorohexane) in methanol. Twin boxcar averagers used for data collection. (a) Signal from ^{12}C. (b) Signal from ^{35}Cl. Reprinted with permission from B. W. Pack, J. A. C. Broekaert, J. P. Guzowski, J. Poehlman, and G. M. Hieftje, *Anal. Chem.*, **70**, 3957. Copyright (1998) American Chemical Society.

TOFMS allowed the direct identification of analyte peaks by their distinctive empirical formulas (Figure 5.4.9). Detection limits for the GC-MPT-TOFMS were found to be in the low femtomole range for all analytes.

In a similar study, Guzowski and Hieftje [42] demonstrated the use of a simple gas-sampling glow discharge (GSGD) coupled with an *oa*-TOFMS for the detection of halogenated GC eluents. This low-pressure ionization source provided a cost-effective alternative to bulkier atmospheric-pressure sources such as the ICP. Additionally,

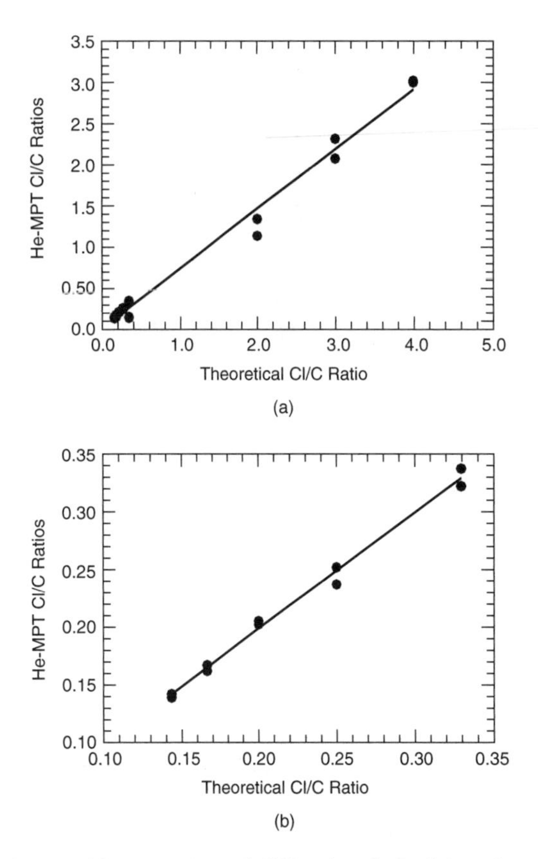

Figure 5.4.9. Comparison of Cl/C ratios obtained from flow-cell MPT-TOFMS measurements to values derived from empirical formulas. (a) Ratios obtained from both aromatic and aliphatic species. Compounds used and corresponding Cl/C ratios: chlorotoluene (Cl : C 0.143), chlorobenzene (Cl : C 0.167), dichlorobenzene (Cl : C 0.333), chloroheptane (Cl : C 0.143), chlorohexane (Cl : C 0.167), chloropentane (Cl : C 0.200), chlorobutane (Cl : C 0.250), chloropropane (Cl : C 0.333), methylene chloride (Cl : C 2.0), chloroform (Cl : C 3.0), and carbon tetrachloride (Cl : C 4.0). Correlation coefficient, $r = 0.994$. (b) An expanded view of the low Cl : C ratio region in (a) which consists of a homologous series of aliphatic halogenated hydrocarbons (chloropropane to chloroheptane). Reprinted with permission from B. W. Pack, J. A. C. Broekaert, J. P. Guzowski, J. Poehlman, and G. M. Hieftje, *Anal. Chem.*, **70**, 3957. Copyright (1998), American Chemical Society.

reduced-pressure operation lessened the likelihood of air entrainment, resulting in simplified mass spectra. Speciation analysis was performed on a range of halogenated hydrocarbons with atomic detection limits in the low- to mid-femtogram per second regime. A dynamic range greater than three orders of magnitude was reported. As will be discussed in the next section, this GSGD could be operated in both atomic and molecular ionization modes to provide further speciation information.

The recent development of multicapillary columns [43, 44] and ultrafast GC [45] will likely foster the application of PS-TOFMS to the speciation analysis of volatile compounds. With these methods producing peak widths as short as 10 ms, the ability of scan-based analyzers to monitor multielemental separations will quickly be overwhelmed. For example, a minimum of five data points should be collected per analyte peak to provide acceptable temporal fidelity [46]. Although commercial quadrupole systems offer scan rates as high as 3000 amu s^{-1} [34], measurements will be limited to fewer than six elements or isotopes for a 10 ms transient. Additionally, spectral skew will then severely compromise the accuracy of the quantitative information generated by scanned systems.

6.2 Liquid chromatography

Conventional liquid chromatography is widely used in the field of speciation analysis for the separation of nonvolatile species. However, PS-TOFMS has experienced limited application to the element-selective detection of LC eluents due to the relatively slow speed of such separations. Ferrarello *et al.* [47] demonstrated the separation of several metallothionein-like proteins with an ICP-*aa*-TOFMS instrument. This study showed the potential of PS-TOFMS for fast protein speciation. The authors noted that the simultaneous extraction capability of ICP-TOFMS would be attractive for isotope ratio or isotope dilution experiments with LC separations. The recent development of ultrahigh pressure LC (pressures between 20 and 40 kpsi), which generates peak widths on the order of 100 to 1000 ms in duration, should result in the further application of PS-TOFMS to LC speciation [48].

6.3 Capillary electrophoresis

High separation efficiency and minimal required sample volume has led to the increased popularity of capillary electrophoresis for elemental speciation analysis. Costa-Fernandez *et al.* [49] demonstrated the use of CE for the rapid separation of a mixture of cations in a cyanide buffer. In addition to V(V), Cr(VI), Ni(II) and Cu(II), speciation of Co(II) and Co(III), and As(III), As(V), and dimethylarsinic acid was performed (see Figure 5.4.10). The entire analysis was completed in less than 70 s, with peak widths on the order of 1–3 s FWHM. Although the separation did not fully resolve several of the analytes, element-selective detection provided the ability for identification and quantitation. This study examined several experimental variables including buffer concentration, flow rates and separation voltages. Additionally, suction flow associated with the use of pneumatic nebulization was addressed.

Like other separation techniques, capillary electrophoresis is constantly being driven to faster time scales. Microfluidic technology that allows the complete manipulation and separation of analytes on a single microchip-sized substrate provides fast, efficient, low-volume measurements [50]. Although microfluidic CE has yet to be used for elemental speciation with PS-MS, it seems likely in the near future.

6.4 Electrothermal vaporization

In electrothermal vaporization (ETV), a small volume of sample solution, typically ranging from 1 to 20 μL, is heated within a graphite furnace. Application of a temperature ramp to the furnace causes analytes to be released as a function of volatility. Mahoney and coworkers [51] used this element-selective volatility to provide an orthogonal means of enhancing the effective spectral resolution of ICP-TOFMS. With an *oa*-TOFMS system that exhibited a nominal resolution of 2000, several isobars ($^{112}Cd/^{112}Sn$, $^{113}Cd/^{113}In$, $^{114}Cd/^{114}Sn$) that would require resolution as high as 300 000 could easily be separated.

Although ETV is not commonly associated with speciation analysis, differences in compound volatility can be used to provide analyte separation. Guzowski *et al.* [52] demonstrated the speciation of two iron compounds with ETV-GSGD-TOFMS (see Figure 5.4.11). A furnace temperature of

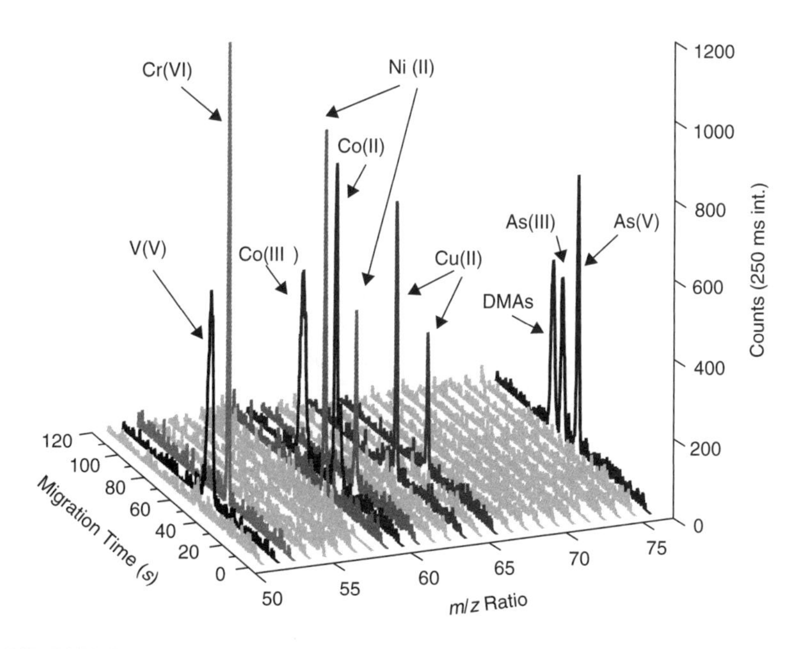

Figure 5.4.10. CE-ICP-TOFMS separation and multielemental detection of V(V), Cr(VI), Co(II), Co(III), Ni(II), Cu(II), As(V), As(III) and DMAs cyanide complexes. Injection contained 100 pg of each metal analyte per compound (400 pg of metal per arsenic species). Reproduced by permission of the Royal Society of Chemistry.

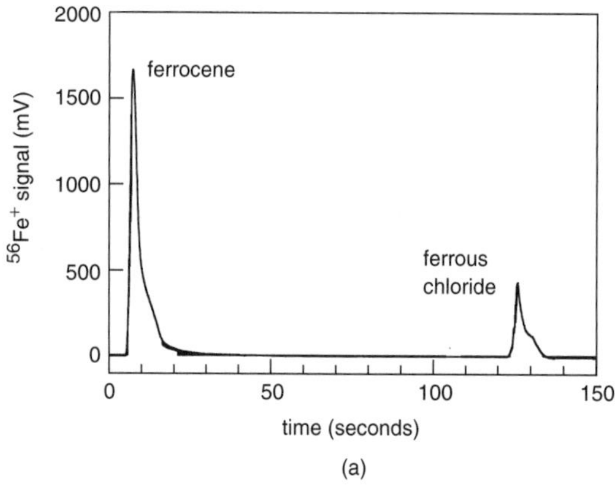

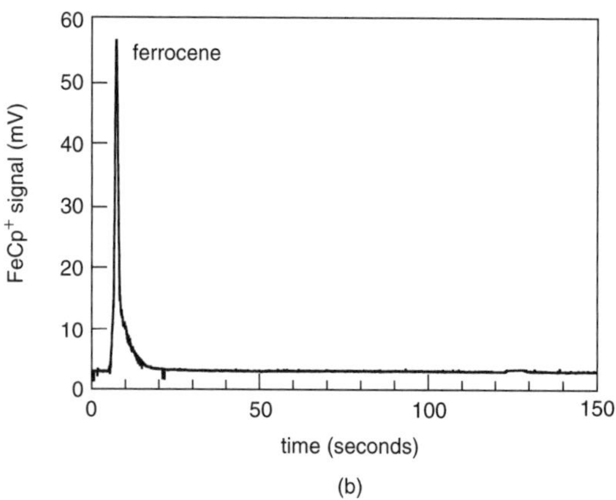

Figure 5.4.11. Chemical speciation of two iron compounds, ferrocene (5 µg) and ferrous chloride (250 ng), achieved with ETV and with a GSGD operated in the modulated mode (10 Hz). Helium gas sampling glow discharge: atomic current 40 mA, molecular current 15 mA, and discharge pressure 5.10 Torr helium. ETV parameters: ash 400 °C (120 s); ramp 300 °C s^{-1}; atomize, 2200 °C (4 s); and transfer line flow rate 90 ml min^{-1} helium, heated at 130 °C. The inorganic salt was initially added to the cell, and the solvent evaporated at 300 °C, then the organic matrix was added after the cell had cooled. (a) Atomic mode; signal collection at m/z 56, $^{56}Fe^+$. Response observed for both organometallic and inorganic components. (b) Molecular ionization mode; fragmentation signal collected at m/z 120, $FeCp^+$. The ferrocene sublimes at a temperature significantly lower than the appearance temperature of the inorganic salt. Reprinted from *Spectrochim. Acta, Part B*, Vol. 55B, J. P. Guzowski, J. A. C. Broekaert, and G. M. Hieftje, pp. 1295–1314, Copyright (2000), with permission from Elsevier Science.

400 °C readily volatilized ferrocene. Elevation of the furnace temperature to 2200 °C caused the release of iron chloride from the furnace into the plasma. Measurement of atomic and molecular information provided the unique identification of both analytes without the need for standards.

7 MODULATED IONIZATION SOURCE TOFMS SPECIATION ANALYSIS

As was mentioned previously, ionization sources can be operated under different conditions to provide distinct analyte mass spectra. 'Hard' ionization conditions generate atomic information, while 'softer' conditions result in molecule fragmentation or formation of the molecular ion. Operated independently or in combination with a hyphenated separation technique, the combination of atomic and molecular information allows the unique identification of analytes.

7.1 Two-state modulated systems

Guzowski *et al.* [53] have demonstrated that a gas sampling glow discharge can be operated in at least two distinct ionization modes. Use of helium as a discharge gas at pressures between 1 and 10 Torr, currents between 10 and 130 mA, and potentials between 200 and 4000 V produced relatively clean atomic mass spectra. With a lower pressure (0.1–3.0 Torr) argon discharge operated at currents of 3–20 mA and potentials of 30–300 V, a more diffuse plasma was formed. This reduced-pressure discharge acts as a softer ionization source to produce molecular fragmentation similar to that in conventional 70 eV electron impact (EI) systems. In the atomic mode, the GSGD provided excellent sensitivity with limits of detection for several halogenated hydrocarbons in the tens of picograms per second range. Operation in the molecular mode made possible the use of EI libraries for the unique identification of analytes based on their distinctive fragmentation patterns.

In later studies by the same group [54] it was discovered that the different ionization modes of the GSGD could be exploited with a single discharge gas at a constant pressure. This simple experimental alteration greatly enhanced the attractiveness of the GSGD. With a single discharge gas (helium) at a pressure of 5.5 Torr, a change in the current–voltage characteristics of the plasma could greatly affect the ionization properties. Operation at a current of 25 mA and a potential of −350 V generated atomic ions. With a current of 15 mA and a potential of +210 V, the surface area of the discharge was increased by greater than a factor of 10, to produce a diffuse plasma with soft ionization capabilities. Because current and voltage were the only two experimental factors that dictated the GSGD ionization properties, simple electronics could be used to rapidly switch the source from atomic to molecular mode at rates up to 100 Hz. Figure 5.4.12 shows the atomic and molecular mass spectra of bromoform generated by the switched GSGD. In the atomic spectrum the two isotopes of bromine are clearly visible. The molecular spectrum exhibits good agreement with the spectrum from a 70 eV EI source. Notably absent from the GSGD fragmentation spectrum is the molecular ion, suggesting that the GD is slightly more energetic than a conventional EI source.

Combined with gas chromatography, the switched GSGD provides high sensitivity and structural identification of transient analyte peaks. Figure 5.4.13 demonstrates how the molecular

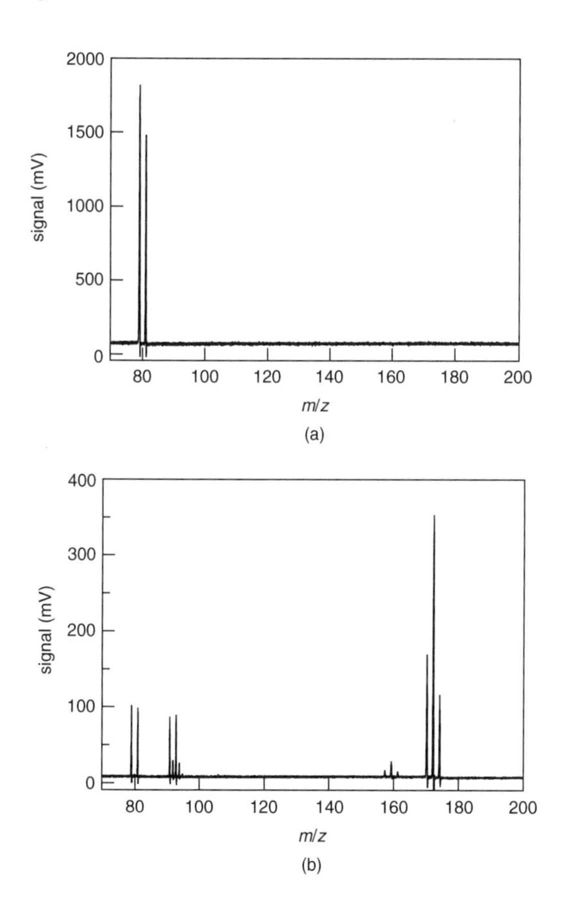

Figure 5.4.12. Atomic and molecular mass spectra for bromoform vapor swept into the GSGD while the plasma was modulated at 10 Hz. Data were collected on a digital oscilloscope (1000 transients averaged). (a) Atomic ionization mode; discharge conditions, 5.50 Torr helium, 30 mA, 350 V. (b) Molecular fragmentation mass spectrum; discharge conditions, 5.50 Torr helium, 20 mA, 250 V. Reproduced by permission of the Royal Society of Chemistry.

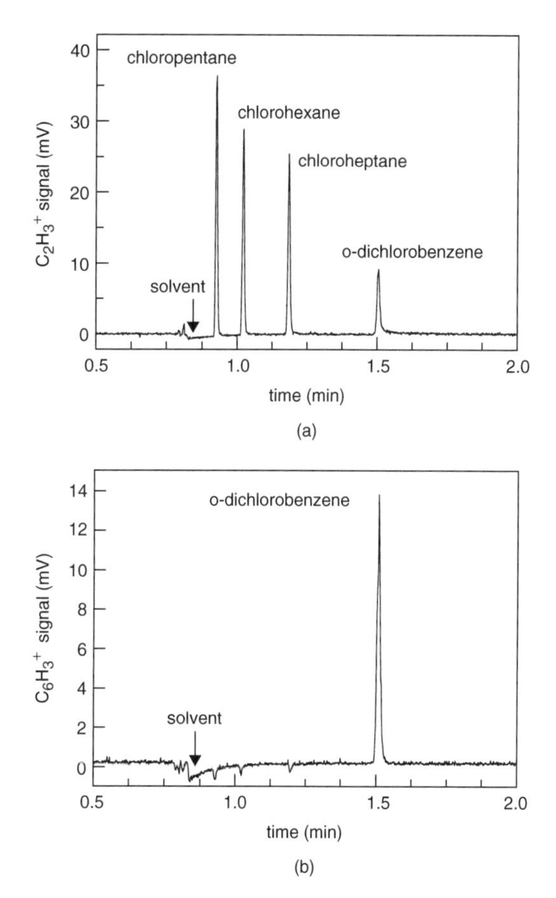

Figure 5.4.13. Chromatogram of chlorinated hydrocarbons with the ionization source operated in the static molecular-fragmentation mode with boxcar averagers used for data collection. Plasma conditions: pressure, 5.50 Torr; current, 20 mA 250 V. (a) m/z 27 ($C_2H_3^+$) signal; a common fragment for these analytes. (b) m/z 75 ($C_6H_3^+$) signal illustrating the selectivity of this method. Reproduced with permission from J. P. Guzowski, Jr., and G. M. Hieftje, *Anal. Chem.*, **72**, 3812. Copyright (2000) American Chemical Society.

mode of the GSGD complements the elemental detection of the atomic ionization conditions by monitoring fragments unique to specific analytes such as $C_6H_3^+$ from *o*-dichlorobenzene [42]. To detect the modulated signals generated by the GSGD, particularly when coupled with GC, the fast spectral generation rate (15 000 complete mass spectra per second) of TOFMS was essential.

7.2 Single-state modulated systems

The pulsed glow discharge has been suggested as a source nearly ideally suited for use with TOFMS [55–58]. Higher instantaneous power provides greater sensitivity than in DC discharges without overheating the source cathode. Additionally, the ionized discharge gas that is extracted from the source can be temporally discriminated against to improve signal to noise.

Steiner *et al.* [59] exploited the time-dependent energy properties of a millisecond pulsed glow discharge to produce a source capable of providing both elemental and molecular information (Figure 5.4.14). Immediately upon initiation of the GD pulse, the source potential accelerates electrons toward the cathode. These fast electrons strike gas atoms (argon) to produce a high-energy discharge capable of electron-impact like ionization and molecular fragmentation. After approximately 0.5 ms the discharge stabilizes and assumes pseudo-steady-state conditions during which both discharge gas and cathode vapor are ionized. After termination of the discharge pulse, electrons recombine with ionized argon to produce long-lived metastable atoms. These metastables generate significantly less fragmentation and produce molecular ions through Penning ionization pathways.

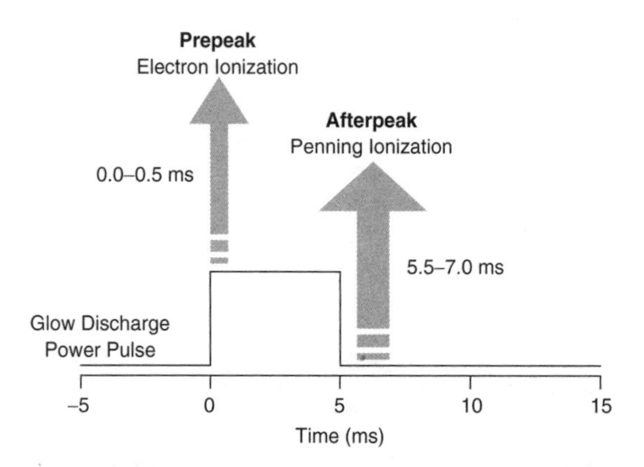

Figure 5.4.14. Depiction of glow discharge pulse sequence illustrating the location and duration of each analytical temporal region (prepeak and afterpeak). Reproduced by permission of the Royal Society of Chemistry.

Majidi and coworkers [60] later extended the use of modulated sources for the generation of concurrent elemental and molecular information to microsecond pulsed discharges. Comparison of Figures 5.4.15 (a) and (b) shows that 80 μs after plasma ignition, the fragmentation pattern produced for *p*-xylene is in fairly good agreement with that from a 75 eV EI spectrum. The mass spectrum produced 470 μs after plasma ignition is dominated by the molecular ion. The authors noted that although identical fragmentation patterns are expected for other isomers of xylene, use of a separation technique prior to the GD would provide additional information needed for the unique identification of each analyte.

8 CONCLUSIONS

Plasma source time-of-flight mass spectrometry is a relatively new addition to the arsenal of techniques available to analytical chemists. At present, several figures of merit of PS-TOFMS, most notably sensitivity, lag behind those of more conventional mass-analyzer configurations. Continuing improvement in vacuum interface geometry and space-focusing capability should cause this gap to decrease. Although the mass resolving power of most PS-TOFMS systems is currently not sufficient to resolve many of the common isobars associated with elemental analysis, recent development in the field of ion guide collision cells promises to push resolving power above 3000 in the near future by reducing the energy distribution of sampled ions through collisions with buffer gas molecules [28, 61]. Additionally, ion chemistry in collision cells with reactive gases provides a means to selectively eliminate specific isobar interferences.

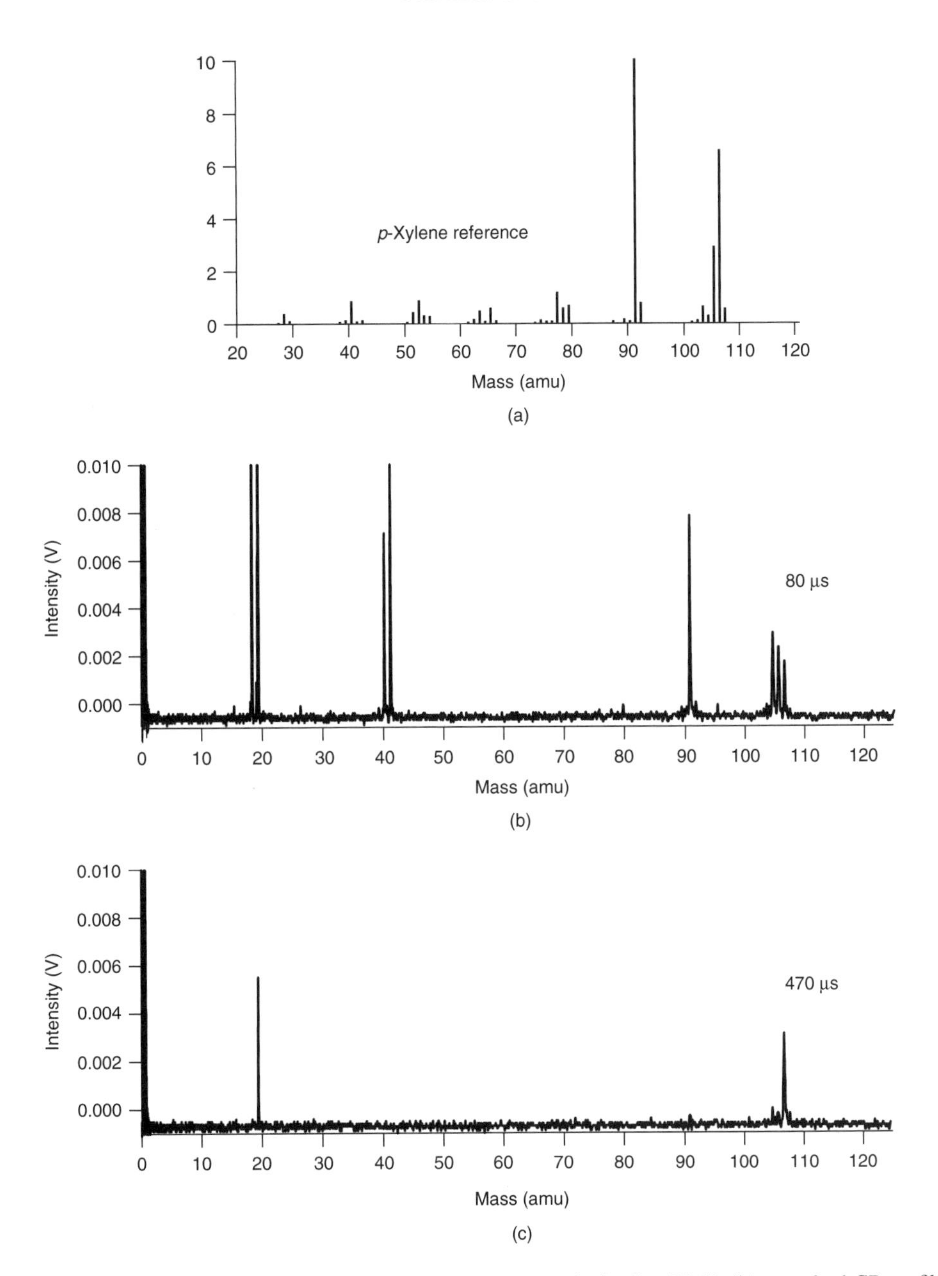

Figure 5.4.15. Mass spectra of *p*-xylene: (a) NIST EI reference spectrum obtained at 75 eV; (b) μs-pulsed GD profile obtained 80 μs after the plasma ignition; (c) μs-pulsed GD profile obtained 470 μs after the plasma ignition. Reproduced by permission of the Royal Society of Chemistry.

The greatest strengths of TOFMS that make it extremely attractive for the elemental analysis of complex systems are the simultaneous extraction of all m/z and high spectral generation rate. The combination of these two traits allows PS-TOFMS to monitor fast transient signals without loss of temporal fidelity or corruption of quantitative data by spectral skew. Coupled with a variety of ionization sources including the ICP, MPT and GD, TOFMS has proven to be a capable system for the detection of hyphenated separations. As the speed of chromatographic and electrophoretic methods increases, the use of TOFMS in speciation analysis will likely experience a dramatic rise. Additionally, the development of rapidly modulated ionization sources will dictate the use of detectors such as TOFMS that can monitor the large quantity of spectroscopic information generated.

9 ACKNOWLEDGEMENTS

This research was supported in part by the US Department of Energy through grant DOE DE-FG02-98ER14 890, by the LECO Corporation, and by ICI Technologies.

10 REFERENCES

1. Holland, J. F., Enke, C. G., Allison, J., Stults, J. T., Pickston, J. D., Newcome, B. and Watson, J. T., *Anal. Chem.*, **55**, 997 (1983).
2. Duckworth, D. C. and Barshick, C. M., *Anal. Chem.*, **70**, 709 (1998).
3. Halliday, A. N., Christensen, J. N., Lee, D.-C., Hall, C. M., Luo, X. and Rehkamper, M., Multiple-collector inductively coupled plasma mass spectrometry, in *Inorganic Mass Spectrometry*, Barshick, C. M., Duckworth, D. C. and Smith, D. H. (Eds), Marcel Dekker, New York, 2000, pp. 291.
4. Solyom, D. A., Burgoyne, T. W. and Hieftje, G. M., *J. Anal. At. Spectrom.*, **14**, 1101 (1999).
5. Cameron, A. E. and Eggers, D. F., *Rev. Sci. Instrum.*, **19**, 605 (1948).
6. Mahoney, P. P., Ray, S. J. and Hieftje, G. M., *Appl. Spectrosc.*, **51**, 16 (1997).
7. Myers, D. P., Ray, S. J. and Hieftje, G. M., Inorganic time-of-flight mass spectrometry, in *Inorganic Mass Spectrometry*, Barshick, C. M., Duckworth, D. C. and Smith, D. H. (Eds), Marcel Dekker, New York, 2000, pp. 447.
8. Guilhaus, M., *Spectrochim. Acta, Part B*, **55**, 1511 (2000).
9. Wiley, W. C. and McLaren, I. H., *Rev. Sci. Instrum.*, **26**, 1150 (1955).
10. Piseri, P., Iannotta, S. and Milani, P., *Int. J. Mass Spectrom. Ion Processes*, **153**, 23 (1996).
11. Flory, C. A., Taber, R. C. and Yefchak, G. E., *Int. J. Mass Spectrom. and Ion Proc.*, **152**, 169 (1996).
12. Cotter, R. J., *Time-of-Flight Mass Spectrometry. Instrumentation and Applications in Biological Research*, ACS Professional Reference Books, Washington DC, 1997.
13. Boesl, U., Weinkauf, R. and Schlag, E. W., *Int. J. Mass Spectrom. Ion Processes*, **112**, 121 (1992).
14. Sanzone, G., *Rev. Sci. Instrum.*, **41**, 741 (1970).
15. Niu, H. and Houk, R. S., *Spectrochim. Acta, Part B*, **51**, 779 (1996).
16. Karataev, V. I., Mamyrin, B. A. and Shmikk, D. V., *Sov. Phys.-Tech. Phys.*, **16**, 1177 (1972).
17. Mamyrin, B. A., Karataev, V. I., Shmikk, V. and Zagulin, V. A., *Sov. Phys. JETP*, **37**, 45 (1973).
18. Myers, D. P. and Hieftje, G. M., *Microchem. J.*, **48**, 259 (1993).
19. Myers, D. P., Li, G., Yang, P. and Hieftje, G. M., *J. Am. Soc. Mass Spectrom.*, **5**, 1008 (1994).
20. Myers, D. P., Li, G., Mahoney, P. P. and Hieftje, G. M., *J. Am. Soc. Mass Spectrom.*, **6**, 400 (1995).
21. Myers, D. P., Li, G., Mahoney, P. P. and Hieftje, G. M., *J. Am. Soc. Mass Spectrom.*, **6**, 411 (1995).
22. Guilhaus, M., *J. Am. Soc. Mass Spectrom.*, **5**, 588 (1994).
23. Guilhaus, M., Selby, D. and Mlynski, V., *Mass Spectrom. Rev.*, **19**, 65 (2000).
24. Mahoney, P. P., Ray, S. J., Hieftje, G. M. and Li, G., *J. Am. Soc. Mass Spectrom.*, **8**, 125 (1997).
25. Vlasak, P. R., Beussman, D. J., Davenport, M. R. and Enke, C. G., *Rev. Sci. Instrum.*, **67**, 68 (1996).
26. Sturgeon, R. E., Lam, J. W. H. and Saint, A., *J. Anal. At. Spectrom.*, **15**, 607 (2000).
27. Tian, X., Emteborg, H. and Adams, F. C., *J. Anal. At. Spectrom.*, **14**, 1807 (1999).
28. Leach, A. M. and Hieftje, G. M., *Int. J. Mass Spectrom*, **212**, 49 (2001).
29. Ray, S. J. and Hieftje, G. M., *J. Am. Soc. Mass Spectrom*, submitted (2003).
30. Furuta, N., *J. Anal. At. Spectrom.*, **6**, 199 (1991).
31. Vanhaecke, F., Moens, L., Dams, R., Allen, L. and Georgitis, S., *Anal. Chem.*, **71**, 3297 (1999).
32. Emteborg, H., Tian, X., Ostermann, M., Berglund, M. and Adams, F. C., *J. Anal. At. Spectrom.*, **15**, 239 (2000).
33. Heumann, K. G., Gallus, S. M. and Vogl, J., *J. Anal. At. Spectrom.*, **13**, 1001 (1998).
34. Montaser, A., *Inductively Coupled Plasma Mass Spectrometry*, Wiley-VCH, New York, 1998.
35. Jakubowski, N., Moens, L. and Vanhaecke, F., *Spectrochim. Acta, Part B*, **53B**, 1739 (1998).
36. Sutton, K., Sutton, R. M. C. and Caruso, J. A., *J. Chromatogr. A*, **789**, 85 (1997).
37. Lobinski, R., *Appl. Spectrosc.*, **51**, 260 (1997).
38. Lobinski, R., Pereiro, I. R., Chassaigne, H., Wasik, A. and Szpunar, J., *J. Anal. At. Spectrom.*, **13** (1998).
39. Lobinski, R. and Szpunar, J., *Anal. Chim. Acta*, **400**, 321 (1999).
40. Leach, A. M., Heisterkamp, M., Adams, F. C. and Hieftje, G. M., *J. Anal. At. Spectrom.*, **15**, 151 (2000).
41. Pack, B. W., Broekaert, J. A. C., Guzowski, J. P., Poehlman, J. and Hieftje, G. M., *Anal. Chem.*, **70**, 3957 (1998).
42. Guzowski, J. P., Jr. and Hieftje, G. M., *Anal. Chem.*, **72**, 3812 (2000).

43. Pereiro, I. R., Schmitt, V. O. and Lobinski, R., *Anal. Chem.*, **69**, 4799 (1997).
44. Pereiro, I. R., Wasik, A. and Lobinski, R., *J. Chromatogr.*, **795**, 359 (1998).
45. van Deursen, M. M., Beens, J., Leclercq, P. A. and Cramers, C. A., *J. Chromatogr. A*, **878**, 205 (2000).
46. Leclercq, P. A. and Cramers, C. A., *Mass Spectrom. Rev.*, **17**, 37 (1998).
47. Ferrarello, C. N., Bayon, M. M., de la Campa, R. F. and Sanz-Medel, A., *J. Anal. At. Spectrom.*, **15**, 1558 (2000).
48. Wu, N., Collins, D., Lippert, J. A., Xiang, Y. and Lee, M. L., *J. Microcolumn Separations*, **12**, 462 (2000).
49. Costa-Fernandez, J. M., Bings, N. H., Leach, A. M. and Hieftje, G. M., *J. Anal. At. Spectrom.*, **15**, 1063 (2000).
50. Kutter, J. P., *Trends Anal. Chem.*, **19**, 352 (2000).
51. Mahoney, P. P., Ray, S. J., Li, G. and Hieftje, G. M., *Anal. Chem.*, **71**, 1378 (1999).
52. Guzowski, J. P., Broekaert, J. A. C. and Hieftje, G. M., *Spectrochim. Acta, Part B*, **55B**, 1295 (2000).
53. Guzowski, J. P., Broekaert, J. A. C., Ray, S. J. and Hieftje, G. M., *J. Anal. At. Spectrom.*, **14**, 1121 (1999).
54. Guzowski, J. P., Jr. and Hieftje, G. M., *J. Anal. At. Spectrom.*, **15**, 27 (2000).
55. Hang, W., Yang, P., Wang, X., Yang, C., Su, Y. and Huang, B., *Rapid Commun. Mass Spectrom.*, **8**, 590 (1994).
56. Harrison, W. W. and Hang, W., *J. Anal. At. Spectrom.*, **11**, 835 (1996).
57. Hang, W., Baker, C., Smith, B. W., Winefordner, J. D. and Harrison, W. W., *J. Anal. At. Spectrom.*, **12**, 143 (1997).
58. Harrison, W. W., Hang, W., Yan, X., Ingeneri, K. and Schilling, C., *J. Anal. At. Spectrom.*, **12**, 891 (1997).
59. Steiner, R. E., Lewis, C. L. and Majidi, V., *J. Anal. At. Spectrom.*, **14**, 1537 (1999).
60. Majidi, V., Moser, M., Lewis, C., Hang, W. and King, F. L., *J. Anal. At. Spectrom.*, **15**, 19 (2000).
61. Guzowski, J. P. and Hieftje, G. M., *J. Anal. At. Spectrom.*, **16**, 781 (2001).

5.5 Glow Discharge Plasmas as Tunable Sources for Elemental Speciation

R. Kenneth Marcus

Clemson University, Clemson, SC, USA

1 INTRODUCTION

The very essence of the elemental speciation experiment is the identification and quantification of the various chemical entities present in a given analytical specimen. In the *vast* majority of these determinations, the first step in the process is some form of chemical separation used to isolate the components in chemical space. This separation can occur as a selected complexation, immobilization, precipitation, or some form of liquid, gas, or supercritical fluid chromatography. More recently, capillary electrophoresis and field flow fractionation have been employed for separation. Once the chemical species have been separated, the second step of the speciation experiment involves the detection and quantification of the target central metal ion. Most often, either inductively coupled plasma optical (atomic) emission and mass spectrometries (ICP-OES/MS) or flame atomic absorption (FAA) is used as the detector element. Note that in the latter case, the identity of the analyte must be known *a priori*.

The typical speciation experiment involves liquid or gas chromatography followed by atomic spectroscopy. Chromatography alone cannot identify specific chemical species. Identities can only be inferred by correlating analytical retention times to standard mixtures, if direct overlaps in composition exist. While ICP-OES and MS are excellent means of making trace metal determinations, the essence of the methods requires that any analyte-containing entities be in the free atom state. It must be kept in mind that there are cases where speciation does not involve metal ions explicitly, so capabilities in elemental detection must extend to nonmetals including 'gaseous' elements and the halides. In any case, if only elemental analysis is possible, then there is no direct evidence of the initial analyte's identity generated in either the separation or the detection process. Thus, the typical speciation experiment involves qualitative analysis by inference, based on relative chemical reactivity or chromatographic retention times for chemical entities containing the monitored element. Clearly, the most effective way to perform elemental speciation is to employ detection means that provide direct evidence of the chemical form of each constituent. It is possible that chemical separation will never be circumvented as an integral part of the

Handbook of Elemental Speciation: Techniques and Methodology R. Cornelis, H. Crews, J. Caruso and K. Heumann
 ISBN: 0-471-49214-0

experiment, but there will definitely always be a need for chemically specific means of detection.

Beyond the ability for the aforementioned atomic spectroscopic methods to provide element-specific information, chemical speciation in the areas of biological or environmental analysis requires far greater amounts of information. There are two more relevant levels of specificity that can be related to the chemical identity of chromatographically separated species. The first level of complexity is the empirical formulae of compounds, basically the relative number of atoms of each element in the eluting molecule. For example, the empirical formula for tetraethyl lead ($Pb(CH_2CH_3)_4$ is PbC_8H_{20}. A determination of the empirical formula for this compound is not specific to this particular molecule though.

The greatest level of specificity comes from methods that allow the unambiguous identification of a complete analyte entity. For complete determinations, there are few spectroscopic methods that allow unambiguous identification of any molecule. Mass spectrometry is one such method, but only when aspects of the complete molecular entity (i.e., molecular weight) and selective information related to molecular structure (i.e., fragmentation patterns) exist. When both of these aspects are used in gaining speciation information via mass spectrometry such methods outweigh conventional atomic spectrometry sources that, by definition, decompose samples down to their atomic form to report elemental information. Scientifically desirable methods are those that provide element specific information, which is complemented by relative atomic ratios as well as complete molecular weight signatures. Thus the detector element must inherently be capable of delivering different amounts of energy to the analyte to produce atomic, fragment, and molecular signals. Conventional spectrochemical sources operating at atmospheric pressure are not capable of providing this range of information for a number of reasons. Alternatively, low pressure glow discharge (GD) sources do indeed have such capabilities. GD sources, operating at pressures of $\sim$1 Torr in rare gas environments and at powers of less than 60 W have proven the basic capabilities of producing optical emission spectra which are reflective of an analyte's empirical formula and mass spectra that are composed of the desired information.

Described here are the design aspects and analytical characteristics of GD sources employed in elemental speciation studies. Sources designed for receiving either gaseous or liquid samples will be presented. In addition, sampling in either the optical emission or mass spectrometry modes will be demonstrated for both sample forms. While there are no commercially available instruments in which glow discharge sources are employed to explicitly monitor liquid or gaseous streams, it is believed that work of the sort described here will go far toward bringing these capabilities to wider acceptance [1].

2 GLOW DISCHARGES AS SPECIATION DETECTORS

GD sources offer a number of interesting possibilities as speciation detectors for gaseous and liquid sample analysis. Many of the key points are described below with reference to how they are different from spectrochemical sources that are more widely employed in this area [2–4].

- *Inert gas environment* – Because they operate at reduced pressures and in inert environments, the plasma is not inherently contaminated by atmospheric species such as water, oxygen, hydrocarbons, and fragments thereof. As a result, GD sources can be employed for elemental analysis across the entire periodic table (save the discharge gas), which is paramount in the ability to perform empirical formula determinations of organometallics by optical emission or mass spectrometry. In terms of mass spectrometric analysis, the absence of background gases eliminates possibilities of deleterious side reactions.
- *Low temperature environment* – Specifically for the case where mass spectrometry is the desired sampling mode, the gas-phase temperatures and densities must be sufficiently low so as not to induce dissociation in molecular species. Thermal excitation of vibrational modes in analyte molecules can serve either to dissociate

species directly or to make them more prone to fragment in the course of subsequent ionization events. The kinetic temperatures of most GD sources lie in the range of 100–500 K, as opposed to ~2500 K in combustion flames and ~10 000 K in ICP sources. Of course, high temperatures are desirable in cases where only elemental analysis is required and are the *de facto* goal in all optical emission experiments.

- *Low power operation* – While it is a natural assumption that low operation powers will yield low gas phase temperatures, the desire for low power operation has another role in terms of designing speciation detectors. In all forms of chromatography there is a desire to match the size, cost, and experimental complexity of the detector element with that of the separation stage. GD devices operating at average powers of less than 60 W (i.e., less than a light bulb) can be constructed to fit in the palm of your hand. Therefore, these are *sources that can be coupled to chromatographs* rather than a *chromatograph coupled to the source*. This may be a matter of semantics, but is very important when one wishes to have technology adapted by chromatographers, environmental chemists, and nutritionists.
- *Analytical versatility* – Given that there are many forms of chromatographic separation and sample introduction, one must consider the range of sample forms that a potential speciation detector can receive. In regards to GD sources, this is an issue of sample introduction as opposed to the spectrochemical source *per se*. Glow discharges readily accept vapor-phase sample forms; thus gas chromatography and vapor generation are viable means of species separation. On the other hand, direct liquid injection is not a viable option. As will be discussed in subsequent sections of this chapter, liquid introduction can be accomplished in a straightforward manner. In the end, a single GD source can be designed to accept eluents from either gaseous or liquid streams.

In the following sections, a number of analytical applications of glow discharge devices in the field of elemental speciation will be described. The discussion will be restricted to GD experiments that yield higher levels of chemical information than simple determinations of metal ion content in flowing streams. In the case of optical emission sampling, the nonmetal elements provide information toward elucidating empirical formulae. In mass spectrometric analysis, the ability to produce information on both molecular and elemental constituents must be realized. Based on examples appearing in the literature to date, GD plasmas should play a very prominent role in the rapidly expanding area of chemical speciation.

3 GAS SAMPLING GLOW DISCHARGES

All GD sources have in their very nature an aspect of operation with a flowing gas stream. Commercial GD-OES and GDMS systems rely on a stream of argon (typically) passing through the source as a means of providing the discharge gas. The argon flow also serves to flush debris from the system and, in the case of MS analysis, to carry analyte ions from the source volume into the mass analyzer stage. In the case of solid specimen analysis, argon is the discharge gas of choice because of its high sputtering efficiency combined with high-lying metastable levels. Metastable levels of approximately 11.5 eV play an important role in the excitation and ionization characteristics in the discharge negative glow region. In the case of gaseous sample introduction, sputtering is not required; on the contrary it is detrimental. Therefore, the discharge gas can be chosen solely on the basis of achieving the desired excitation/ionization characteristics, with helium providing the most energetic of GD plasmas. In general, gaseous sample introduction is achieved by simply transporting the analyte vapor to the source in a flow of He gas, as is the case in most gas chromatography separations.

3.1 Optical emission detection

The underlying principles for the use of GD-OES in chemical speciation were presented before the term had actually become common. The potential capabilities of the use of GD sources as detectors for elemental speciation were first demonstrated in 1989 by Puig and Sacks who used a heated injector to introduce hydrocarbon vapors into a hollow cathode discharge (HCD) and monitored the optical emission of the halides F, Cl, and Br [5]. The discharge source operated at pressures of 20–40 Torr He and at currents of up to 100 mA. Transient signals resulting from the introduction of 250 μL samples suggested that chromatographic integrity could be retained while yielding detection limits in the range of 0.6–17 ng level for the halides introduced as small halocarbon molecules. Similarly, Winefordner and coworkers extended the analyte set to include I and S [6]. Those researchers suggested future applications in gas chromatography as well.

The initial studies explicitly pointing to the use of GD-OES as a speciation tool (i.e., the ability to generate empirical formulae) were presented in a series of papers by Hieftje and coworkers [7–10]. That early work followed the concept of McLuckey and coworkers who developed an atmospheric pressure sampling glow discharge for organic mass spectrometry [11]. The gas sampling glow discharge (GSGD) incorporated a metallic capillary to deliver volatilized organic compounds through the cathode region of a planar GD source [7]. Those studies showed that small molecules could be effectively dissociated such that empirical formulas of halogen-versus-carbon content could be derived in a straightforward fashion. Detection limits on the 5 ng s^{-1} were achieved for chlorine, with subsequent studies also illustrating the use of hydride generation for determinations of As in solution [8].

A detailed study of the effects that organic vapor introduction had on the discharge operating characteristics for a Grimm-type source geometry was presented by Hieftje *et al.* [10]. It was found that the discharge voltage increased upon injection of the organic compounds, which were manifest in changes in the spectral background. This points to the importance of obtaining background-corrected signals on the chromatographic time scale. Three different introduction geometries were explored regarding the exit of the silica capillary: (1) flush with the face of the plane cathode, (2) a conical depression in the cathode surface, and (3) a 2 mm deep, 1 mm diameter hollow cathode. Spatially-resolved optical measurements clearly indicate that the plasma energetics are localized within the cavities of geometries (2) and (3), with a very diffuse distribution observed in the plane cathode case. These studies also considered the temporal stability and noise structure of the plasma. The Cl(II) emission from dichloromethane introduced continuously over a 30 min period showed a variation of only 0.9 % RSD. The associated analyte noise power spectrum was 'white' up to a frequency of 800 Hz. The detection limit for chlorine introduced as dichloromethane was determined to be 20 ng s^{-1}, higher than the previous studies which used more open geometries [7]. The degradation in detection limit was attributed to the higher spectral background present in the restricted Grimm-source geometry. Linear response curves were demonstrated for the C/Cl content for a range of halocarbons, with acceptable responses noted as well for the cases of C/H and H/Cl ratios.

In the area of source development, Sanz-Medel and coworkers have published an excellent series of papers on the use of glow discharge devices as optical emission detectors for gaseous sample introduction [12–17]. Their work differed from that of the previously cited authors in that radio frequency (RF) powered plasma sources were employed. In the case of optical emission detection, RF powering affords a more energetic plasma in terms of electron energies and densities [18]. They have described two basic designs for gaseous sample introduction, which are depicted in Figure 5.5.1. Early studies involved a modification of the so-called 'Marcus' geometry normally used for solids analysis [12]. In this approach (Figure 5.5.1(a)), the gaseous sample was introduced through the limiting (anode) orifice (shown in greater detail on the right-hand side of the figure), tangential to the cathode surface. In this way, the eluent was delivered directly into the plasma negative glow region. Different introduction points for the plasma make-up gas have also been investigated. The make-up gas is

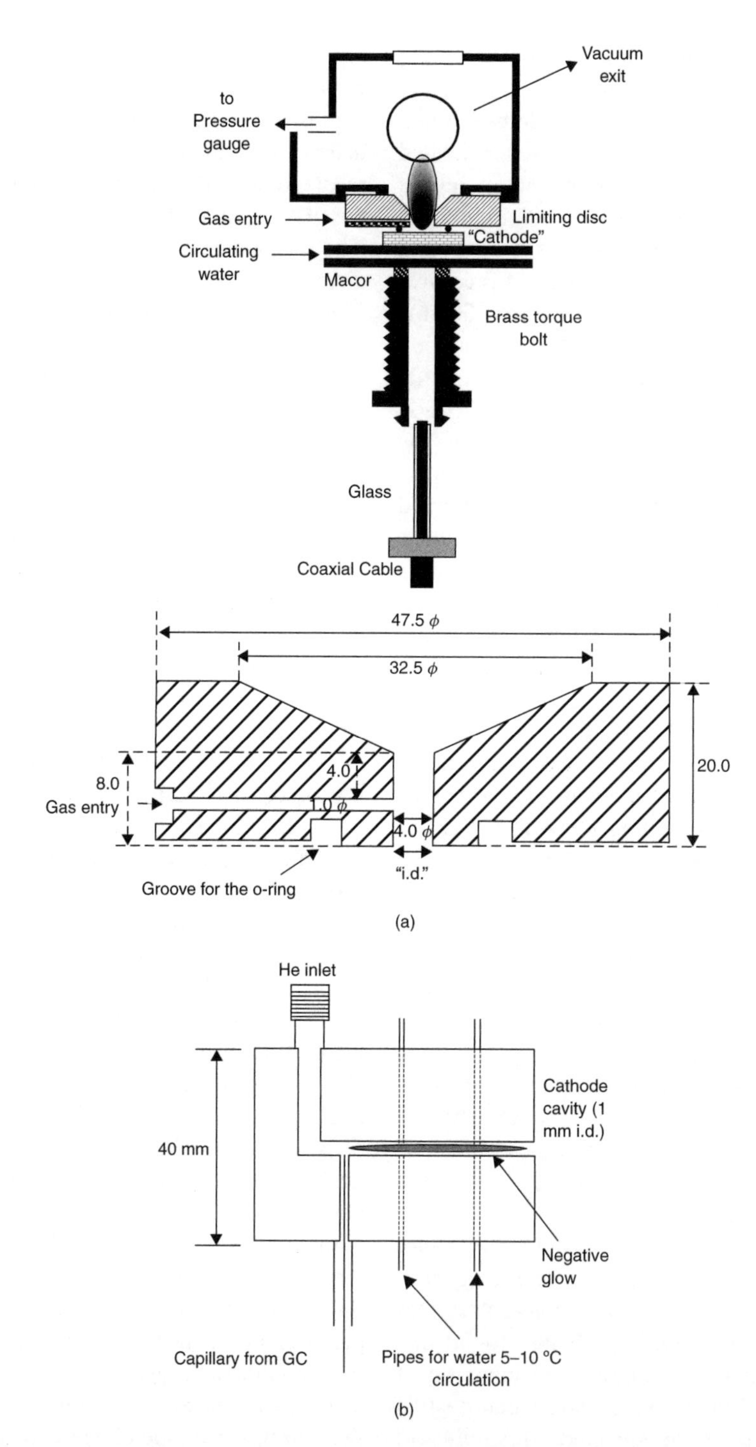

Figure 5.5.1. Basic discharge geometries employed by Sanz-Medel and coworkers for RF-GD-OES analysis of gaseous samples. (a) Introduction through limiting orifice of Marcus-type source geometry. Reproduced from reference 12 with permission from the American Chemical Society. (b) Hollow cathode geometry. Reproduced from reference 15 with permission from the Royal Society of Chemistry.

required because the He flow rate from the sample introduction is not sufficiently high to maintain a stable plasma. Even the earliest studies produced detection limits for nonmetals (C, Cl, Br, and S), introduced in the form of organic vapors through an exponential dilution flask, that were superior to those of more widely used methods. Further demonstration of the methodology involved more thorough evaluations of the plasma operation parameters as well as alternative methods of sample introduction.

Figure 5.5.2 illustrates the versatility of the RF-GD-OES sampling method through the three basic forms of sample introduction employed by the Oviedo group; exponential dilution [12], volatile vapor generation (either cold vapor, hydride, or oxidative) [13, 14], and gas chromatography [15, 16]. In one application, volatile chlorine (Cl_2) was oxidatively generated from chloride in solution and detected via Cl(I) emission in the NIR region of the spectrum (837.6 nm) [13]. Both RF and direct current (DC) modes of plasma operation were investigated. Detection limits for the RF mode (0.1 ng mL^{-1}) were found to be superior to those for the DC, and better than any previously reported values, without sacrifice in other figures of merit. In addition, empirical formulas for a range of chloro-hydrocarbons and a pair of sulfur compounds were accurately determined. Gas chromatography separations were employed for mercury determinations in fish tissue extracts [15]. Detection limits for methyl-, ethyl-, and inorganic mercury ranged from 1–3 ng mL^{-1} when the Hg(I) 253.6 nm transition was monitored. Unfortunately, only chromatographic retention time was used to identify species, rather than the use of nonmetals to

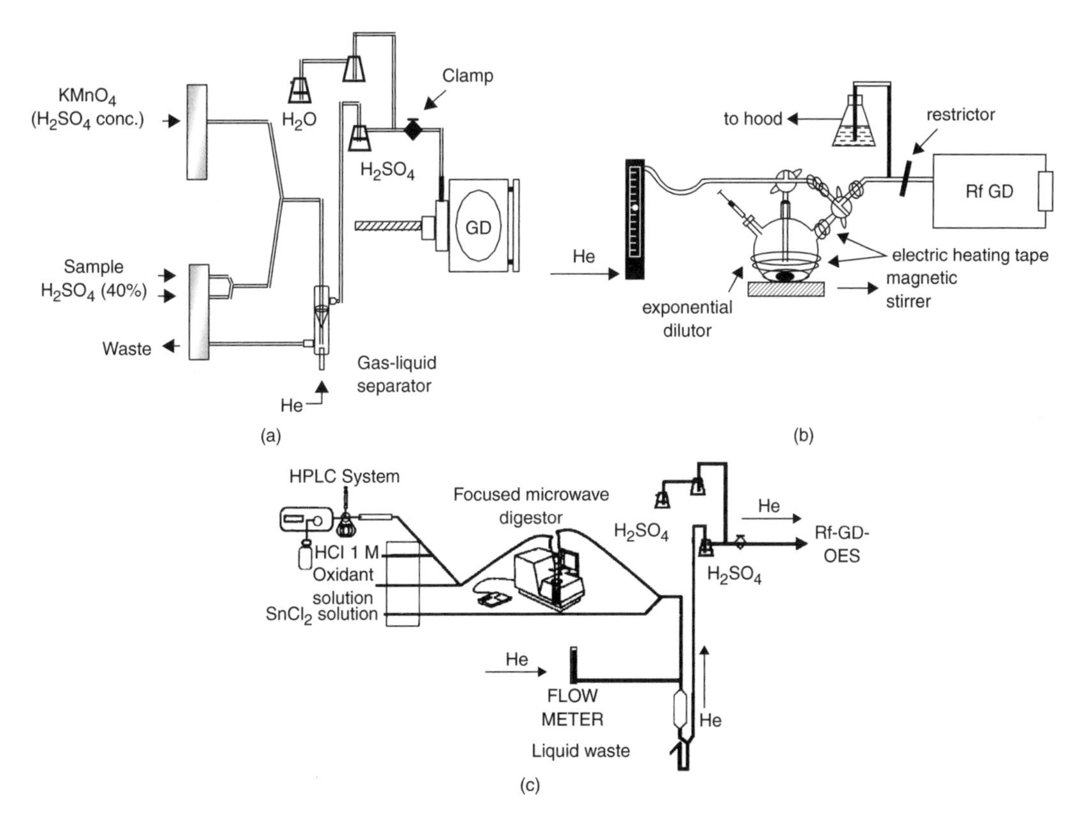

Figure 5.5.2. Examples of different modes of gaseous sample generation for introduction to an RF-GD-OES source. (a) Oxidation of chloride to chlorine [13]. Reproduced with permission from the Royal Society of Chemistry. (b) Exponential dilution of organic vapors [12]. (c) HPLC separation followed by microwave oxidation to generate Hg vapor [16]. Reproduced with permission from Maik Nauka, Russia.

determine empirical formula. A different approach to introducing the same materials involved the use of liquid chromatography to separate the organic and inorganic mercury species, followed by in-line microwave digestion, and finally cold vapor generation of Hg^0 that was introduced into the plasma [17]. Liquid chromatography separated the various Hg species in time, prior to more or less generic vapor generation. Both $NaBH_4$ and $SnCl_2$ were evaluated as the reducing agents to produce the mercury vapor, with realized detection limits of 1.8 and 0.2 ng mL^{-1}, respectively. The sample preparation and separation steps were viewed by the authors to be more useful and rapid than the aforementioned GC methodology. In comparison with more common ICP methods, the GD-OES technique '... has favorable detection limits, lower instrumentation costs, and very low plasma gas consumption' [16].

The group of Sanz-Medel has also looked at different source geometries for the use of GD-OES for gas specimen analysis. It has long been known that hollow cathode geometries yield plasmas that are of higher density and greater energy than those employing planar (flat) cathode geometries [19]. Figure 5.5.1(b) depicts the RF-HC-GD-OES source developed in that laboratory [15]. Different from the planar source arrangement, the gas eluent and the entirety of the He discharge are introduced in the base of the plasma. Arsenic and antimony hydride generation was used as the sample introduction platform for the comparison studies [14]. While the DC powered flat cathode geometry produced lower limits of detection than the other possible modes and geometries, it was determined that the RF-HC combination was more analytically useful based on low detection limits, greater sensitivity for low sample volumes, and greater temporal fidelity of signal transients. The RF-HC-GD-OES source was used as a GC detector for organic and inorganic mercury species [15]. In this instance, the species in solution were isolated by first performing a Grignard reaction to form the mercury chlorides, followed by extraction in a mixture of diethyldithiocarbamate in toluene. Similar to many plasma sources used for GC detection, the authors point out that the eluting solvent vapor can extinguish the plasma, and so the discharge was not initiated until the solvent front had passed. Very high quality chromatograms were realized in the separation of fish tissue reference materials. Here again, the HC geometry yielded far lower detection limits (5–10×) than the flat cathode geometry, which in turn were lower than those for GC with MIP-OES detection. A recent report from this group has illustrated the use of solid-phase microextraction (SPME) as a very powerful means of sample preconcentration, while also eliminating solvent effects in performing tin and lead speciation experiment by RF-(HC)GD-OES [20].

3.2 Mass spectrometric detection

Similar to the case of GD-OES analysis of gaseous samples, the use of GDMS for elemental speciation applications has its roots very much in early works by McLuckey and coworkers in the development of the ASGD source for organic mass spectrometry [11]. Those studies illustrated well the fact that a reduced pressure GD source can produce fragmentation patterns that are very similar in nature to those of electron impact (EI) sources. This is an important feature as only EI sources are reproducible to the point of the generation of 'universal' spectral libraries. Of course, in the generic case of elemental speciation, one desires the ability to observe molecular ions as well as chemically significant fragmentation patterns. This is the strength of EI and is in fact realized for the most part with the low pressure GD sources applied to speciation.

The first dedicated effort in the use of GDMS for elemental speciation was described by Caruso and coworkers at the University of Cincinnati [21]. That work involved designing a cubic discharge cell that was mounted in the torch position of a commercial ICP-MS instrument. Figure 5.5.3(a) depicts the design of the RF-GDMS source used for elemental speciation based on a direct insertion probe (DIP) as described by Duckworth and Marcus [25]. A high-purity aluminum disk (~6 mm diameter) served as the cathode. The fused silica GC capillary was passed through a heated steel capillary mounted on a flange of

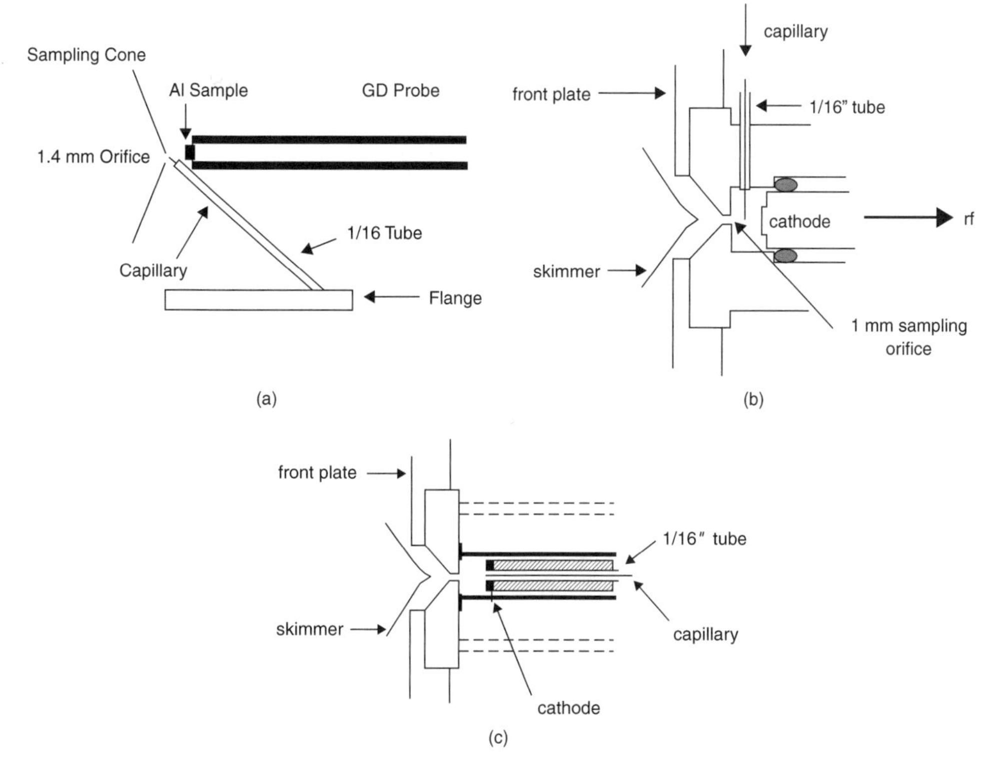

Figure 5.5.3. Three configurations of RF-GDMS sources used for gas chromatography detection. (a) Open cell geometry using a six-way cube [21]. (b) Reduced-volume source with movable probe [22]. (c) Coaxial introduction of gas eluent through cathode [23]. Reproduced with permission from the Royal Society of Chemistry.

the cube, such that it terminated within 1 mm of the ion exit orifice. Placement of the cathode approximately 1 cm from the orifice ensured that the eluting species entered directly into the plasma negative glow for ionization and extraction to the mass analyzer. Those early studies clearly showed that the extent of fragmentation for alkyltin and lead species was dependent on the discharge conditions of power and pressure, with generally very good overlap observed with EI spectral libraries. Detection limits for Sn of 1 pg were shown to be very competitive with more established methods.

Two subsequent design modifications were made to the RF-GD source used for elemental speciation. The second design by Caruso and coworkers involved a dramatic reduction in the plasma volume as depicted in Figure 5.5.3(b) [23]. Based on the structure of the RF plasma, higher discharge pressures ($\sim$25 Torr He) were used with this source than the initial design ($<$1 Torr). Cool on-column injections could be employed readily with this geometry, with no need for solvent venting in many instances. Comparison between chromatograms obtained with a flame ionization detector and that from the GDMS detection for a mixture of alkyltin compounds showed very good correlation in terms of peak shape and resolution. The roles of carrier gas flow rate, discharge pressure, and cathode–anode separation distance were evaluated for three organotin compounds, which showed that each compound responded in the same way. This would indicate that there should be minimal matrix effects. The RF-GD mass spectra were shown to be very similar again to EI libraries, as illustrated in Figure 5.5.4(a) for tetramethyltin. By the same token, the mass spectra of related alkyltin compounds produce

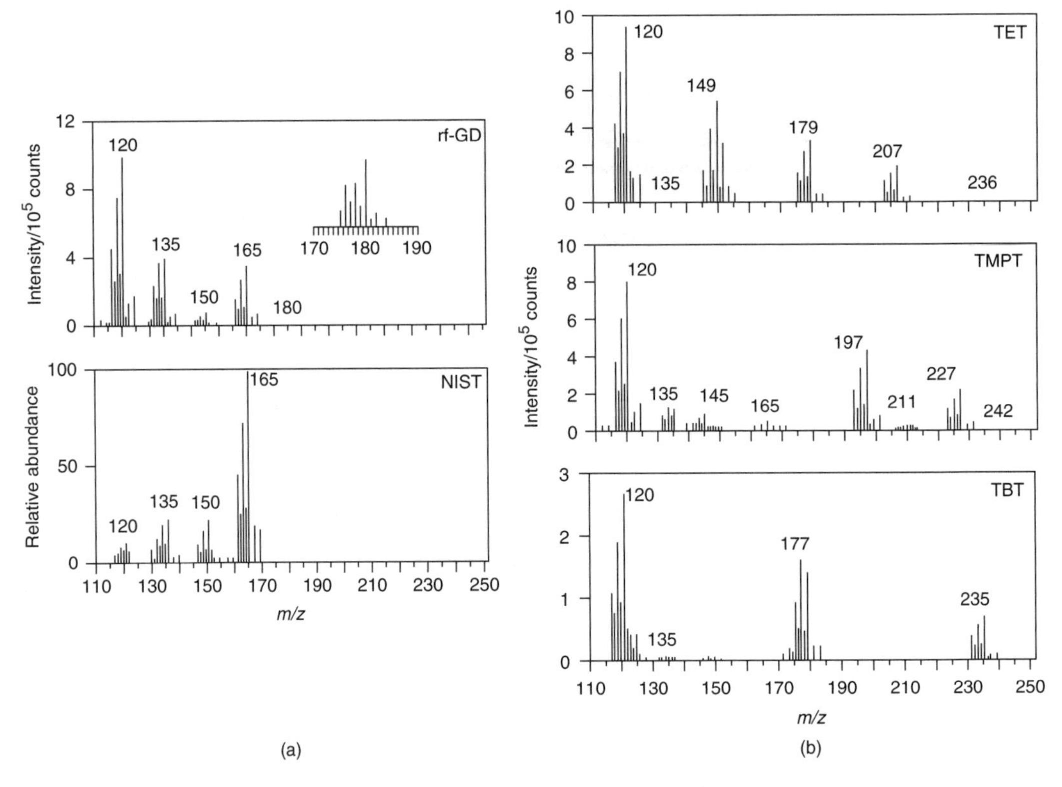

Figure 5.5.4. Mass spectra of TMT. (a) RF-GD ionization source (RF power = 30 W, pressure = 26 Torr He, sampling distance = 2.5 mm) and NIST reference library [21]. (b) RF-GD mass spectra of eluting TET, TMT, and TBT [23]. Reproduced with permission from the Royal Society of Chemistry.

very distinct, yet easily interpreted, fragmentation spectra (Figure 5.5.4(b)). A very important set of experiments was performed to elucidate how differences in discharge conditions (i.e., power and pressure) and sampling distance are manifest in the degree of fragmentation for the tetramethyltin compound. Interestingly, while discharge conditions effect the absolute signal intensities, there are no appreciable differences in degree of dissociation. On the other hand, changes in sampling distance do yield different spectral structures. Simply put, while the bare metal ion signal ($^{120}Sn^+$) definitely decreases as a function of distance, there is lesser degrees of fragmentation at short distances (i.e., a greater fraction of molecular species). Table 5.5.1 lists the determined figures of merit for the RF-GDMS analysis of alkyltin compounds. As can be seen, the figures are quite excellent, with detection limits on the subpicogram level. The responses across the family of compounds are really quite uniform, with the only deviation is for tetrabutyltin, the least volatile of the molecules.

One final variation in the source designs investigated by the Caruso group involved something akin to a hollow cathode geometry, with the GC capillary actually passing through the center of the DIP and terminating at the cathode surface, as shown in Figure 5.5.3(c) [24]. With this geometry, the eluting compounds pass coaxially through the RF-GD plasma. In this way, variations of the sampling distance also effect the time each analyte is present in the discharge. The analytical figures of merit for this geometry are included as well for tetramethyltin in Table 5.5.1. While there are few substantive differences, one very interesting improvement is seen in the >3X larger slope (i.e., sensitivity) realized with the latter source geometry. Different as well, increases in discharge

Table 5.5.1. Figures of merit of RF-GDMS analysis of alkyltin compounds introduced by gas chromatography [22, 23].

Parameter	TMT[21]	TET[21]	TMPT[21]	TBT[21]	TMT[22]
Linear range studied (decades)	2.5	2.5	2.5	2.5	2
Slope/counts s^{-1} pg^{-1}	681	615	379	262	2237
Correlation coefficient	0.9950	0.9874	0.9713	0.9991	0.997
Log–log slope	1.011	1.126	1.187	0.9196	0.973
MDA[a] pg^{-1}	1.2	2.0	2.6	87	1.0
Detection limit pg^{-1}	0.6	0.6	1.2	13	0.6
RSD (%)	<5 %	<5 %	<5 %	<5 %	3.5

[a] Minimum detectable amount (MDA) based on peak height, all others based on peak area. Reproduced with permission from the Royal Society of Chemistry.

pressure did indeed produce greater amounts of fragmentation, enhancing the relative amounts of atomic ions. Thus, there is indeed an amount of tunability that can be achieved in the use of RF-GDMS for elemental speciation. In a subsequent study principally using low pressure inductively coupled plasmas, the mixing of Ar and He discharge gases was also shown in the case of the RF-GD to allow a greater amount of control of discharge fragmentation of organometallic compounds [25]. This tunability arises from the fact that the two gases have appreciably different metastable energies.

A very enlightening series of papers has illustrated the potential of using different regimes of glow discharge powering to effect the production of mass spectra that are either 'atomic' or 'molecular' in nature. Continuing along the lines described previously for their work with the gas sampling glow discharge (GSGD) employed for optical emission detection [7–10], Hieftje and coworkers have coupled that source to a time-of-flight (TOF) mass spectrometer to allow greater ion throughput and simultaneous analysis across the desired mass regions [26–30]. A 1.6 mm o.d. stainless steel capillary was used to introduce analyte vapors into the discharge, where the planar cathode was machined to effect a small hollow in which the plasma was struck [26]. The authors showed that the extent of analyte fragmentation could be controlled via the use of either He or Ar discharge gases, with the former yielding 'atomic' spectra permitting assignments of empirical relationships and the latter producing 'molecular' spectra. The majority of the studies involved the evaluation of how discharge parameters affected 'atomic' ion signals. Unfortunately, the responses of the signals for the different fragment ion species were not the same, such that measured empirical formulae were very dependent on discharge current and pressure. Detection limits for a range of chloro- and bromohydrocarbons were in the general range of 40–90 pg s^{-1} for the molecular ions, and perhaps ~2 × lower in the case of halogen detection.

Given the experimental complications of changing discharge gases during the course of a single gas chromatogram, Hiefje and coworkers have developed a GSGD that produces either atomic or molecular species on-the-fly through the use of a DC, 'switched' discharge source [27, 28]. The switching between the 'atomic' and 'molecular' modes was affected by operation at high (30 mA) and low (20 mA) discharge currents at a 50 % duty cycle at frequencies of up to 150 Hz. As illustrated in Figure 5.5.5, the pulsing of the

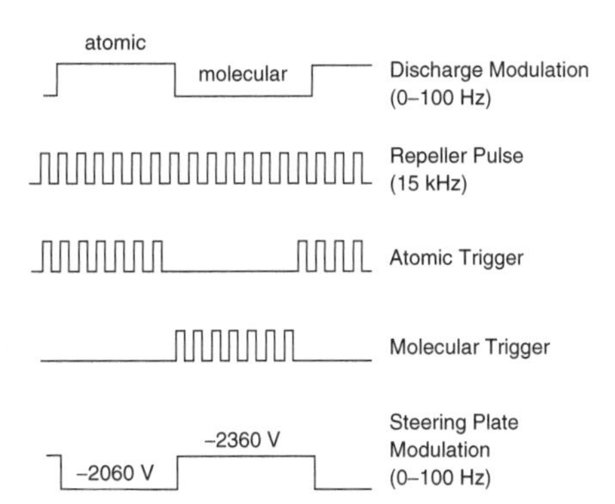

Figure 5.5.5. Pulse sequences employed in a 'switched' GDMS source for the acquisition of atomic and molecular mass spectra on a TOF mass analyzer [28]. Reproduced with permission from the Royal Society of Chemistry.

discharge occurs at such a rate that many TOF mass spectra can be accumulated during each half-cycle [28]. Different discharge configurations and powering schemes were investigated using bromoform introduced via an exponential dilution flask at the model compound. Ultimately, a slight depression formed in the cathode used to effect a hollow cathode geometry. Figure 5.5.6 illustrates the versatility of the switched source operation in the case of the GC separation of a series of *n*-chlorohydrocarbons [27]. As seen in the top two chromatograms, the respective signals for the ^{12}C and ^{35}Cl ions correlate very well, with the Cl^+ signal showing much better signal to noise characteristics, particularly when the much lower mole fraction of chlorine is considered. The chromatogram extracted from the propyl group ($C_3H_5^+$) signals also showed excellent termporal agreement with the atomic chromatograms, with very good signal-to-noise as well. Calibration curves were shown to be linear on a log–log scale ranging from the picogram to nanogram levels with detection limits ranging from 1 to 18 pg s^{-1} for halocarbons in both the atomic and molecular modes, with the former generally being on the single pg s^{-1} level. Mass spectra acquired under molecular conditions were shown to correlate with conventional EI source.

In addition to the use of direct vapor and gas chromatographic sample delivery, the Indiana University group illustrated the use of electrothermal vaporization (ETV) as a means of introduction of inorganic and organic compounds [29]. In this approach, the appearance temperature of a given analyte provides insight into the chemical identity. For example, metals of different volatility appear in temporal order of their melting points (e.g., Zn → Sn → Cd). Here too, the GSGD was operated

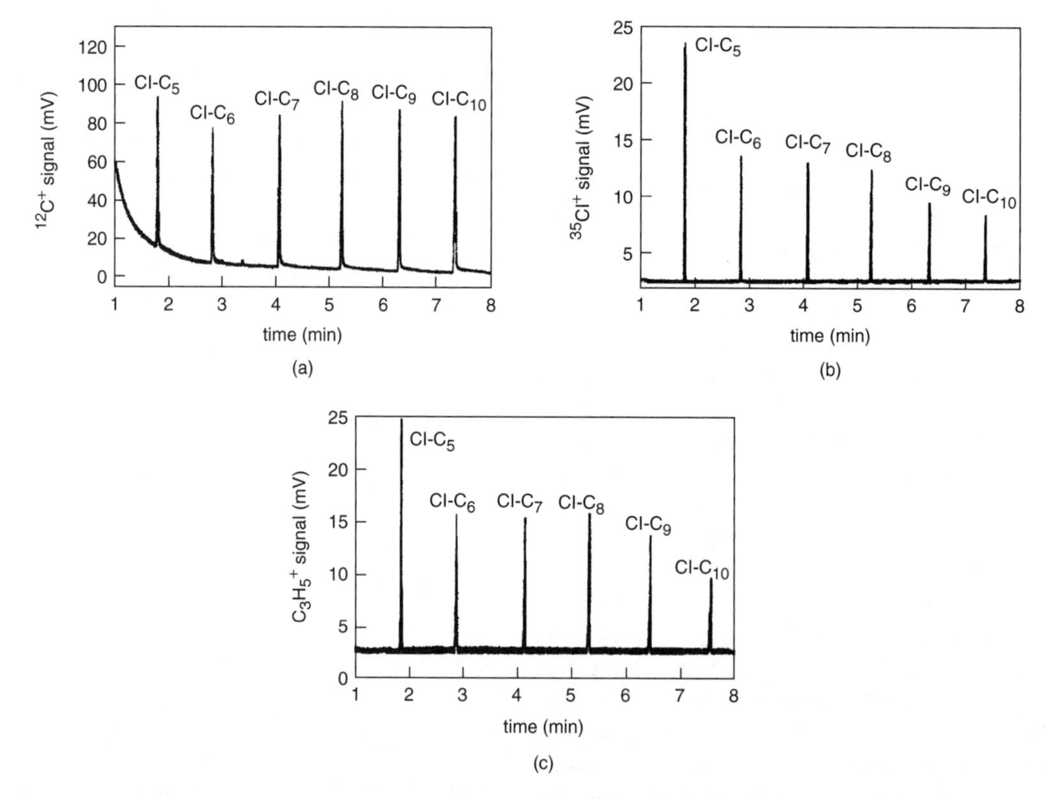

Figure 5.5.6. Gas chromatography separation of a number of *n*-chloroalkanes employing the atomic ($^{12}C^+$ and $^{35}Cl^+$ in A and B respectively) and molecular ($C_3H_5^+$ in C) modes of GSGD operation [27]. Reproduced with permission from the American Chemical Society.

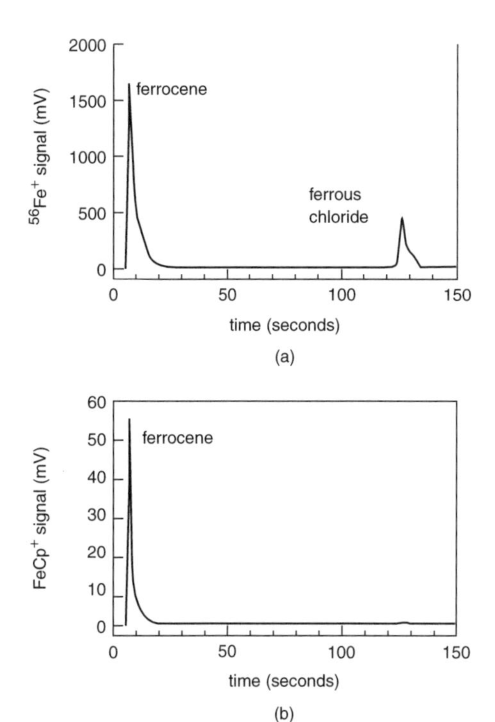

Figure 5.5.7. Speciation of organic (ferrocene) and inorganic (ferrous chloride) iron species by selective volatilization and use of a switched-source GSGD and TOF-MS to monitor (a) atomic iron at $^{56}Fe^+$ and (b) the molecular analyte at $FeCp^+$ ($m/z = 120$ Da) [28]. Reproduced with permission from Elsevier Science.

in the switching mode to optimize 'atomic' signals and also identify 'molecular' species. Figure 5.5.7 illustrates the differentiation of organic iron (ferrocene) from inorganic iron (ferrous chloride). In the top figure, the mass spectra produced in the atomic portion of the waveform are monitored for $^{56}Fe^+$. The transient signal plot shows two Fe species of different volatility. On the other hand, monitoring of $m/z = 120$ ($FeCp^+$) clearly indicates that the first peak corresponds to an iron atom with a cyclopentadiene ligand (i.e., ferrocene after loss of one cyclopentadiene unit). The GD experiment here clearly provides greater species information than would have been obtained in the more conventional coupling of ETV with ICP-MS.

Majidi and coworkers have taken the concept of a switched glow discharge into a very different realm by the use of microsecond time scale pulsing of the plasma [31]. These experiments build upon a number of papers in the use of pulsed-source GDMS in elemental analysis of solids. Harrison and coworkers have shown that on–off pulsing of the GD produces mass spectra that represent two different modes of ionization [32, 33]. In the discharge-on period, the mass spectra are indicative of the situation where electron ionization is the dominant mechanism. In the post-pulse regime (within a few microseconds of plasma cessation), the mass spectra clearly indicate that ionization occurs via a Penning-type collision with remnant metastable discharge gas atoms. It must be kept in mind that the EI process is energetic enough to ionize sputtered atoms and other gaseous molecules, while the Penning process involves a fixed energy dictated by the discharge gas species (i.e., ~11.5 eV for Ar). The use of microsecond pulse lengths provides great temporal selectivity in ion sampling as the post-pulse region can be sampled selectively in terms of pulsing of the repeller gate to the orthogonal TOF mass analyzer. In essence, the temporal evolution of plasma energetics can be selectively used to yield mass spectra of different amounts of fragmentation.

Figure 5.5.8 illustrates the concept of temporal sampling to achieve different levels of species information [31]. The mass spectrum of ethylbenzene acquired by pulsing the repeller 45 μs after the ignition of the discharge pulse (which has a 20 μs width) displays very much more fragmentation than the reference EI spectrum. While the flight time from the plasma source to the repeller gate is not provided, it is assumed that these ions reflect more the ionization conditions of the plasma-on regime. On the other hand, the spectrum acquired from the ion population passing the repeller 305 μs after plasma ignition indicates 'softer' ionization conditions than standard 70 eV EI. A number of alkylated aromatic compounds show similar temporal qualities. Use of the temporal characteristics of these compounds and the common inorganic thermometric species tungsten hexacarbonyl ($W(CO)_6$) provides a picture of the ionization conditions as a function of delay time that is presented in Figure 5.5.9. As can be seen, the energy begins very high, ultimately decreasing to the metastable level of the Ar discharge.

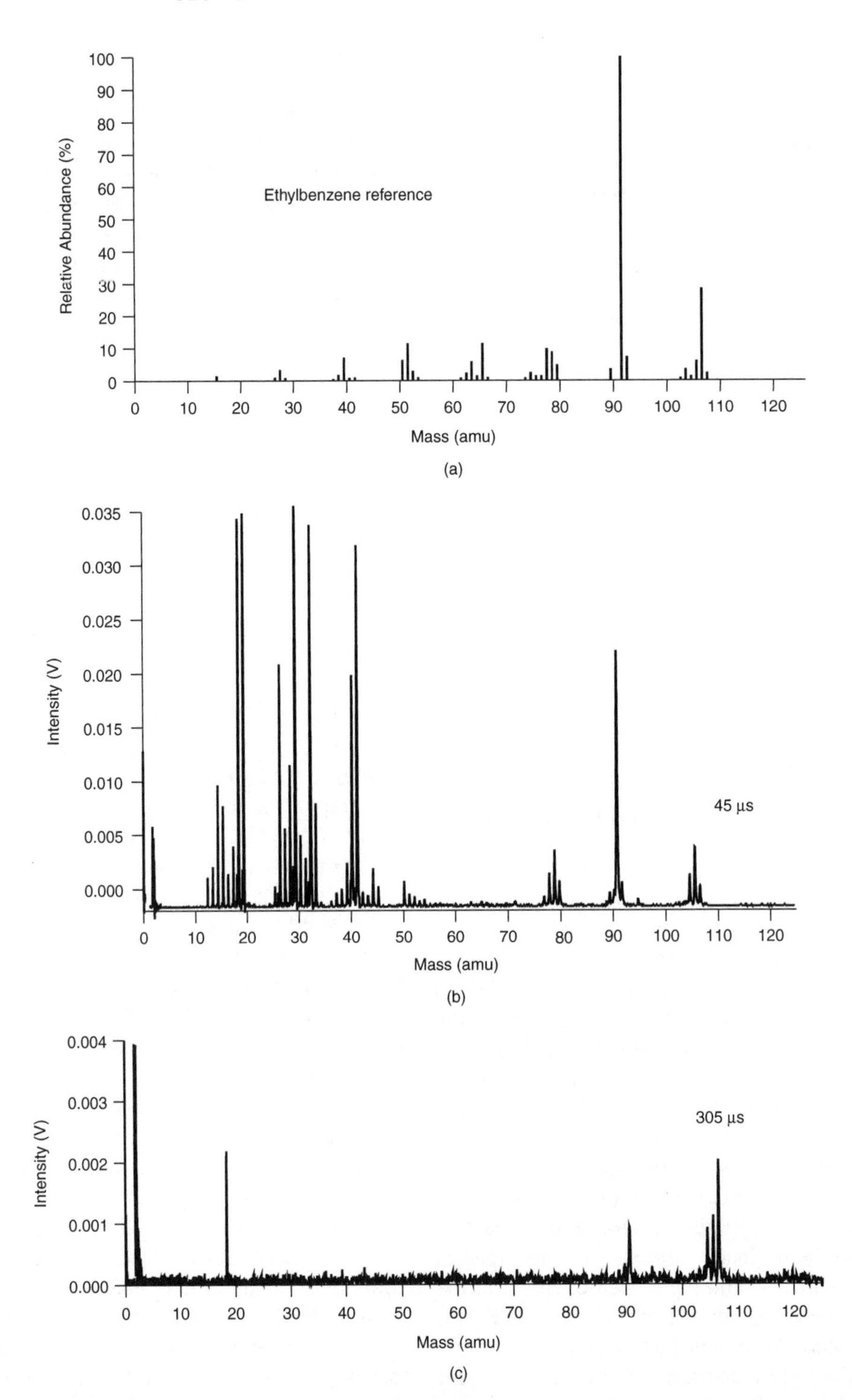

Figure 5.5.8. Mass spectra of ethylbenzene. (a) NIST EI reference spectrum. (b) μs-pulsed sampled 45 μs after plasma ignition (25 μs after termination). (c) *μ*s-pulsed sampled 305 μs after pulse plasma ignition (285 μs after termination) [31]. Reproduced with permission from the Royal Society of Chemistry.

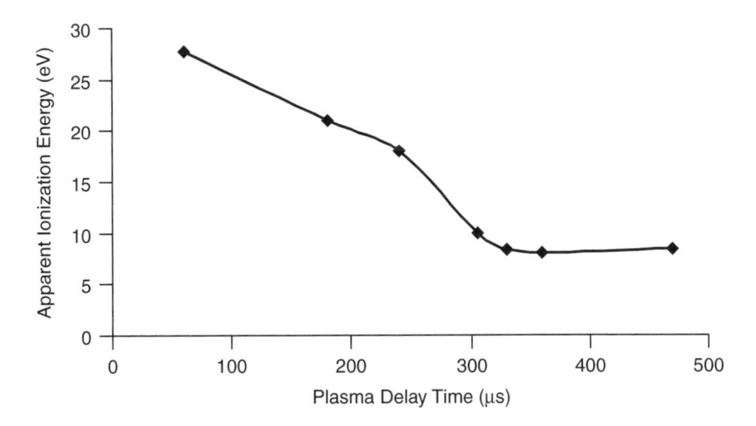

Figure 5.5.9. Measured plasma ionization energy as a function of time for μs-pulsed GD based on introduction of hydrocarbon and $W(CO)_6$ vapors [30]. Reproduced with permission from the Royal Society of Chemistry.

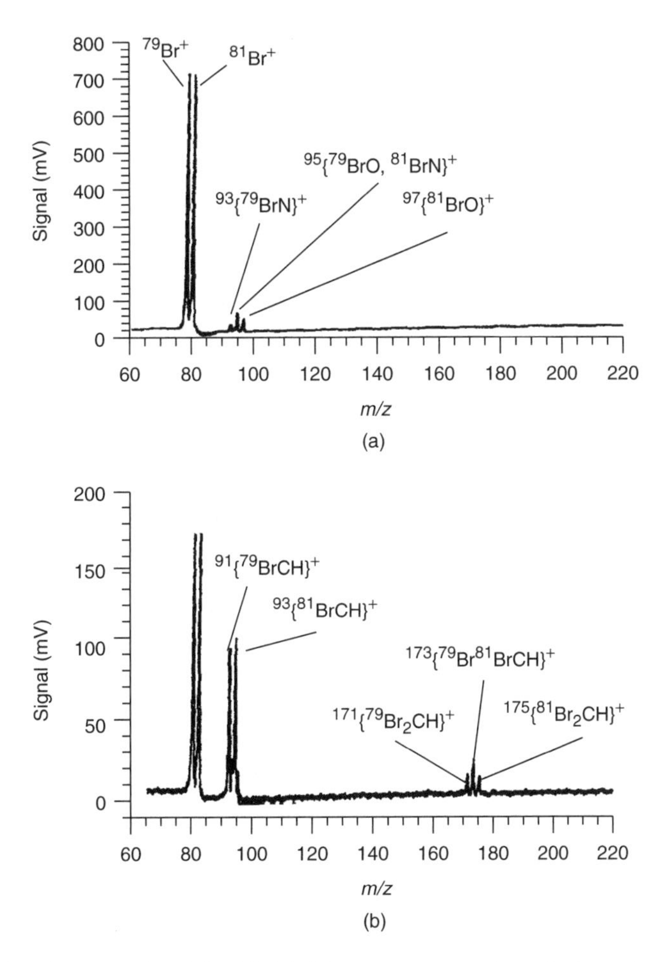

Figure 5.5.10. Atomic and molecular modes of the MPT-ASGD tandem source shown for bromoform introduction. (a) Atomic mode with both MPT and ASGD powered. (b) Molecular mode with ASGD powered (no MPT) [30]. Reproduced with permission from Elsevier Science.

This approach to speciation offers a great deal of experimental flexibility, though perhaps too much with regards to practical implementation. Future studies were suggested involving the use of other discharge gases and the addition of chemical ionization agents to the discharge.

One other approach has been used in the concept of switched sources for elemental speciation of gaseous samples. Ray and Hieftje have recently described a tandem microwave plasma torch–atmospheric sampling glow discharge (MPT-ASGD) apparatus mounted on a quadrupole mass analyzer [30]. In this approach, the gaseous sample is subject to the collisional/ionization environment of the MPT and then that of the ASGD. As such, three modes of operation are possible: MPT alone, ASGD alone, or the MPT-ASGD in series. In this implementation, the GD was maintained by a positive potential on the sampling aperture, with the ions transmitted to the MS effectively through the cathode that was actually the second-stage skimmer cone of the instrument. This was termed reverse-biased operation. In practice, use of the MPT alone provided very little direct ionization. Figure 5.5.10 depicts the mass spectra obtained in the two other operational modes. As expected, the combined sources provided exclusively atomic spectra (relative to the parent halocarbon analyte) as analyte molecules were subjected to the two collisionally active plasmas. Finally, use of the ASGD by itself yielded mass spectra that were similar in structure to EI libraries, as would be expected given its relatively low current operation. The authors noted that reverse-biased ASGD operation produced ions of very high kinetic energies, which can be detrimental to throughput and mass resolution. This is the same situation noted by Harrison and coworkers in early hollow cathode GDMS, where sampling ions through the cathode fall lead to many mass spectrometric difficulties [34]. Other difficulties presented included an electrical 'communication' between the plasmas and the need to separately ignite the MPT with a Tesla coil. The latter point makes switching on chromatographic time scales impossible in the current method. The authors do suggest that there may be a variety of other 'tandem source' combinations which may be effective in the area of elemental speciation by mass spectrometry.

4 LIQUID SAMPLING GLOW DISCHARGES

As alluded to early in this chapter, a typical approach to chemical speciation is the online detection of species separated by liquid chromatography followed by ICP-MS. This of course alleviates the possibility of direct species identification. Based on the work described in the previous sections, the coupling of gaseous sample introduction method to low pressure GD sources is a relatively straightforward line of reasoning. On the other hand, the introduction of solution phase samples into glow discharge sources is similar to the introduction of liquid samples into the low pressure ion sources used in organic mass spectrometry in terms of the detrimental effects of solvent vapors to the plasma operation. Electron impact and chemical ionization sources are incapable of affecting desolvation and the presence of residual solvent vapors causes severe depression of analyte ion signals and leads to high levels of spectral interference [35]. By analogy, it is not surprising that liquid chromatography/mass spectrometry (LC/MS) interfaces, the moving belt and the particle beam [36–39], have also been implemented for sample introduction into GD devices. The moving belt and particle beam devices fall into the category of 'transport-type' LC/MS interfaces. While achieved by different means, these interfaces include aspects of on-line sampling, desolvation, solvent vapor removal, and analyte delivery into low pressure environments. In LC/MS applications, analytes are delivered solvent free to ion sources operating at pressures $<10^{-4}$ Torr, at solution flow rates in the case of conventional LC separations in the range $0.2–2.0\,\mathrm{mL\,min^{-1}}$. Thus, use of these interfaces with GD sources would provide an environment wherein the analyte should have no memory of its solution phase heritage. While the moving belt has been used in a simple elemental analysis, it has not been used to perform speciation-types of experiments.

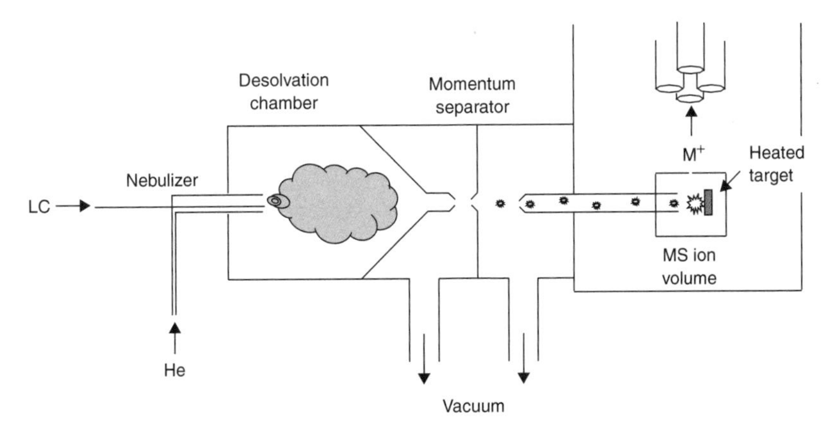

Figure 5.5.11. Basic components of a particle beam LC/MS interface.

A more widely used transport-type LC/MS interface is the particle beam (PB), shown schematically in Figure 5.5.11. PB interfaces, as first described by Willoughby and Browner [38], include some form of a nebulizer housed in a heated chamber where desolvation of the aerosol takes place, followed by a multistage momentum separator. The separator affects particle enrichment by removal of nebulizer gases and solvent vapors through the vacuum. In simple terms, mixtures of gas and particulate matter passing through small ($\leq$1 mm diameter) orifices across pressure differentials produce an expansion based on the relative momenta of the stream components. Lightweight species (gases) tend to expand radially and are skimmed from the beam, while more massive analyte particles continue to pass through successive orifices. Use of two or more stages of differential pumping yields a true beam of particles, free from residual gases. Complete removal of the solvent vapors allows particle delivery to ionization volumes operating at pressures of $\sim 10^{-4}$ Torr where they are flash vaporized into the gas phase. The PB interface provides a means of sampling liquids at flow rates of up to $\sim$1.5 mL min^{-1}, while allowing ionization by electron impact, chemical ionization and many other methods. Marcus and coworkers have used this approach to introduce particles into GD sources for analysis by optical emission and mass spectrometries as depicted in Figure 5.5.12.

Similar to the observations with transport-type interfaces for organic LC/MS, the noted difficulties with the MB interface eventually led the exploration of new interface approaches for solution introduction, specifically the PB interface [37, 38]. The ability to work across a wide range of solution phase compositions and flow rates, while still delivering dry analyte particles to the mass spectrometer ionization volume are the advantages of the PB approach. Strange and Marcus first described the use of a PB interface for the introduction of liquids into GD plasmas [40]. In that work, Cu and Al disks were employed as the cathode in the discharge and the target at which the particle beam impinged. The observed optical emission spectra indicated that there was little or no water solvent carried over from the momentum separator at solution flow rates of 0.1–1.0 mL min. Scanning electron micrographs of particles collected at the cathode surface indicated that particles ranging from 0.5–10 μm in diameter were delivered to the source.

4.1 Optical emission detection

A series of reports by You and Marcus described the design aspects, the sample introduction characteristics, and the analytical performance of the particle beam-hollow cathode-optical emission spectrometry (PB-HC-OES) system [41–44]. The HC geometry was adopted to obtain a more energetic plasma as well as to provide a heated surface (the wall) to vaporize the introduced particle.

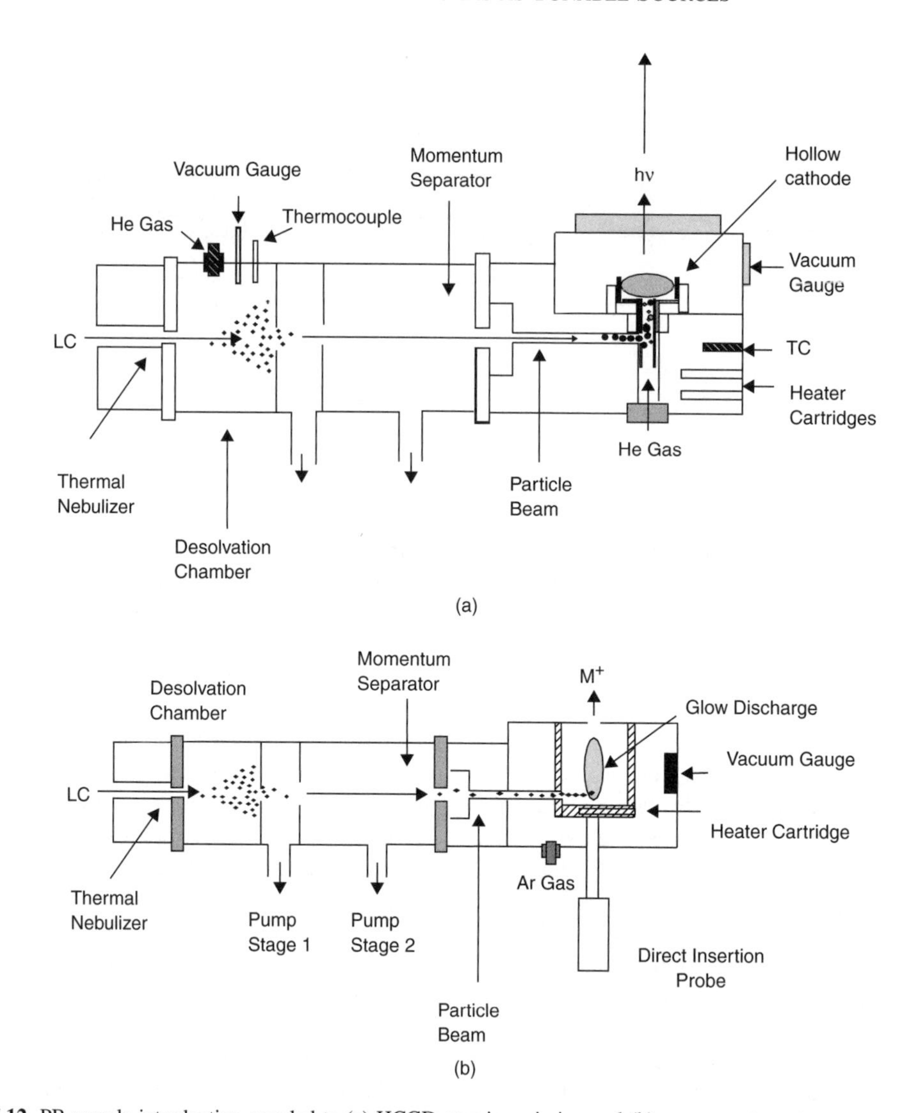

Figure 5.5.12. PB sample introduction coupled to (a) HCGD atomic emission and (b) mass spectrometry sources.

As expected, the response in optical emission of the introduced analyte species depended on the discharge operating conditions of discharge gas pressure and identity and current [41]. The results were in general agreement with those obtained with other HC sources, e.g., He was found to be a better excitation environment than Ar, and the optimum discharge gas pressure was in the 2–4 Torr range. Furthermore, the analyte emission intensity was found to be roughly proportional to the discharge current (up to 100 mA). A cursory evaluation of the roles of the nebulizer tip temperature, the solution flow rate, and the cathode block (vaporization) temperature was also presented. Analyte response profiles for flow injection introduction showed little or no dispersion or tailing, suggesting that the memory effects were low and that the device could be effectively used as a chromatographic detector. For the case of analytical calibration curves obtained for Na and Cs (nitrates) with neat aqueous solutions, detection limits of 0.05 and 0.1 $\mu g\, mL^{-1}$ for 200 μL injections were achieved, which are about two orders of magnitude lower than those in the case of the planar cathode PB-GD geometry.

Detailed studies have been performed to understand the nebulization and particle transport characteristics in PB-HC-OES [42, 43]. The respective roles of the nebulizer tip temperature, the He nebulizer gas flow rate, and the use of a supplemental He gas flow in the desolvation chamber were studied as a function of the analyte solution flow rate. Each of these parameters was found to directly contribute to the nebulization and vaporization processes in a more or less straightforward manner. It was important for the use of the device for LC applications to find that the Cu(I) analyte emission was insensitive to the water:methanol solvent composition, ranging from 100 % water to 100 % methanol. As with other pneumatic nebulizer systems, they obtained the highest analyte responses with the use of small (~50 μm) inner diameter silica capillaries and high He nebulizer gas flow rates. Calibration curves for aqueous Cu, Pb, Fe and Mg (including the aforementioned HCl addition) exhibited very good linearity, and the detection limits were found to range from 12 to 25 ng mL^{-1} for 200 μL injections.

One of the key advantages of the use of low pressure, inert gas plasmas lies in the fact that there is no continuous emission background from atmospheric species such as C, N, O, H, etc. Of course, these are also components of most liquid sample matrices and LC mobile phases. Therefore, the determination of such elements depends on the use of highly efficient solvent removal methods, such as the PB interface. The use of standard flame and furnace atomic absorption and ICP sources for such determinations is less attractive due to low desolvation efficiencies and constant atmospheric background. The access to the above 'gaseous' elements enables some level of molecular information to be obtained in elemental speciation experiments. Namely, it is possible to determine atomic ratios of these elements and to elucidate of the empirical formula of a compound eluting from an LC column. The application of PB-GD-OES to empirical formula determinations, and by extension to elemental speciation was demonstrated. Basically, a plot of the ratio of the H(I) optical emission signal to that of C(I) as a function of the molar ratios of those elements for a range of aliphatic amino acids was quite linear [44]. A similar response curve for the H(I)/N(I) ratio is shown in Figure 5.5.13, illustrating the ability to distinguish between a number of aromatic amino acids as well [45]. It can be seen that the agreement is excellent. Table 5.5.2 depicts some of the analytical figures of merit for the application of LC/PB-GD-OES for the analysis of a pair of organomercury compounds as well as a number of amino acids. Elemental detection

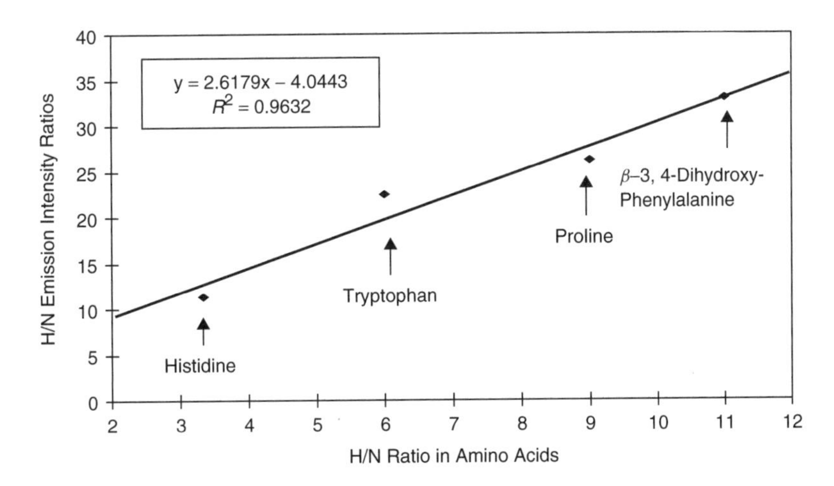

Figure 5.5.13. Comparison of experimentally obtained H(I) and N(I) optical emission intensity ratios with the empirical formula values (H/N) for a range of aromatic amino acids determined by PB-HC-OES [45]. Reproduced with permission from the Royal Society of Chemistry.

Table 5.5.2. Analytical response characteristics for organomercury compounds by PB-HC-AES [45]. Reproduced with permission from the Royal Society of Chemistry.

Analyte	Wavelength (nm)	r^2	Detection limit (in 200 μL injections)
Hg in thimerosal	435.8	0.9979	2.9×10^{-11} mol thimerosal (0.03 ppm Hg)
Hg in merbromin	435.8	0.9995	4.1×10^{-11} mol merbromin (0.04 ppm Hg)
C in thimerosal	538.0	0.9834	3.5×10^{-8} mol thimerosal (19 ppm C)
C in merbromin	538.0	0.9934	5.2×10^{-8} mol merbromin (63 ppm C)
Na in thimerosal	589.0	0.9943	1.4×10^{-10} mol thimerosal (0.02 ppm Na)
Br in merbromin	614.9	0.9997	1.3×10^{-10} mol merbromin (0.1 ppm Br)
H in thimerosal	656.3	0.9877	2.2×10^{-9} mol thimerosal (0.1 ppm H)

limits for H, C, and N in amino acid specimens can be performed in the range from ~0.1 to 3 ppm for 200 μL injection volumes. This corresponds to molecular detection limits of 10^{-9} M! It is envisioned that this approach has great potential for applications in elemental speciation.

4.2 Mass spectrometric detection

In order to obtain the most comprehensive information in elemental speciation, one would want unambiguous molecular weight and structural information of compounds eluting from the separation column. Thus, ideally, one would have a mass spectrometry ion source capable of accepting LC flows and still produce meaningful molecular and elemental mass spectra. Traditional 'organic' ionization sources such as electron impact (EI) and chemical ionization (CI), while excellent at providing information on the molecular level, are virtually useless for the ionization of free metal atoms and small inorganic complexes. By the same token, ICP-MS is a very powerful means of performing elemental analysis of liquid specimens, but by its nature all molecules are decomposed down to the atomic level. One accordingly needs an ionization source of modest energy such that organic compounds can still be kept intact, while some atomic species yield simple 'elemental' mass spectra.

Gibeau and Marcus have used the same approach as described above, namely a PB interface and a GD ionization source, to realize a versatile LC detector [46]. Different from the OES application, the beam of particles impinges directly on a disk-type cathode where they are vaporized/sputtered into the gas phase for subsequent ionization. While not nearly so efficient as an HC source (to be pursued in the future), work with this geometry has demonstrated that the basic approach provides the sorts of information desired in elemental speciation. As in the case of OES, experiments with amino acids have been done to characterize this technique. In Figure 5.5.14, the PB-GDMS spectra obtained for 200 μL injections of 150 ppm solutions for a set of selenoamino acids are shown [47]. These spectra are very straightforward to interpret and provide the exact type of information required in comprehensive speciation experiments. All other 'small' molecules examined to date have yielded mass spectra that are *very* similar to those generated with traditional EI sources, and in fact identification with the aid of standard databases is possible. One observes signals representative of the molecular weight of the molecule, and the successive loss of organic functional groups. The mass spectra obtained with 'elemental' solutions are very simple in structure, with little or no evidence of oxide species or the like. The sensitivity of the method at an early stage of development is comparable with that obtained when coupling PB interfaces to EI sources, and the detection limits for both organic molecules and elemental species are at the single nanogram level. Surprisingly, this value is in line with commercial ICP-MS instrumentation. A separation and speciation of mixtures of inorganic lead and organolead compounds show the flexibility of the LC/PB-GDMS technique, as with the GD ionization source spectra are produced which accurately represent both types of species [47]. Also, separation and

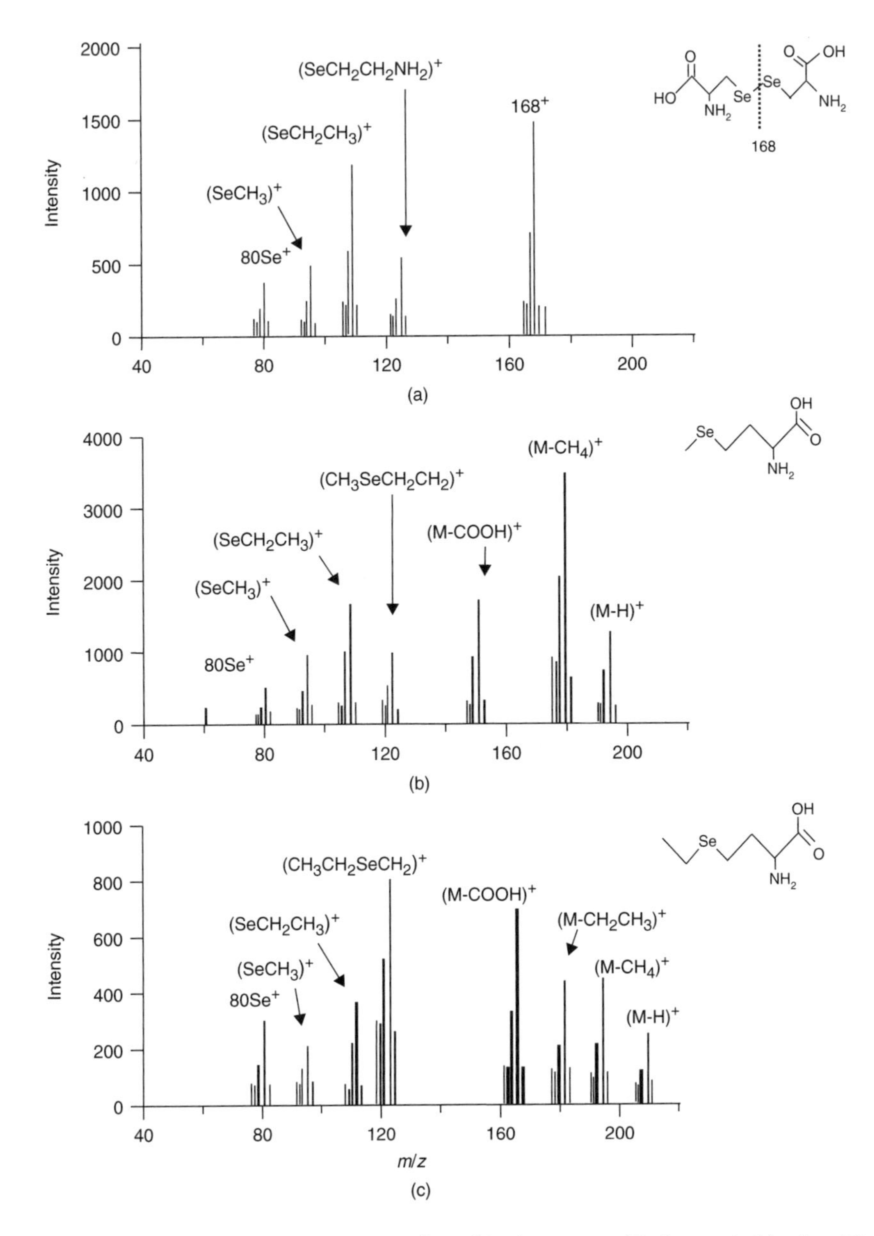

Figure 5.5.14. PB-GD mass spectra of (a) seleno-DL-cystine, (b) seleno-DL-methionine, and (c) seleno-DL-ethionine [47]. Reproduced with permission from Elsevier Science.

identification of mixtures of polyaromatic hydrocarbons (PAHs), steroids, and selenoaminoacids has been carried out. It is believed that this methodology holds a great promise in providing more comprehensive speciation information than any other MS source.

5 CONCLUSIONS

The preceding and following chapters of this volume set out very clearly the challenges and opportunities that exist in the very important area of elemental speciation. Issues of sampling, preservation, sample preparation, separation and detection are integrally linked. Unfortunately, the part of the experiment that allows the actual identification of species is probably the least evolved of them all. Existing, commercially available methods have been *adapted* from their initial forms to be very selective and sensitive detectors for elemental (predominately metal) species. To deliver the target information though, detectors *designed* explicitly for these jobs provide the best opportunity to assemble the required systems. Low pressure, low power GD devices offer the possibility to obtain the desired information in the biochemical and environmental analysis communities. Small, low cost platforms that enable the ready coupling of a variety of sample introduction forms have a far greater chance of acceptance by the nonplasma community. Ultimately, though, it is the information content provided by GD-OES and GDMS that must answer the call of the speciation community. Based on studies presented to date from a handful of laboratories, the promise is there and awaits the support of the instrumentation community to realize its suggested potential.

6 REFERENCES

1. Marcus, R. K., Evans, E. H. and Caruso, J. A., *J. Anal. At. Spectrom.*, **15**, 1 (2000).
2. Chapman, B. N., *Glow Discharge Processes*, Wiley-Interscience, New York, 1980.
3. Marcus, R. K., (Ed.) *Glow Discharge Spectroscopies*, Plenum, New York, 1993.
4. Baude, S., Broekaert, J. A. C., Delfosse, D., Jakubowski, N., Fuechtjohann, L., Orellana-Velado, N. G., Pereiro, R. and Sanz-Medel, A., *J. Anal. At. Spectrom.*, **15**, 1516 (2000).
5. Puig, L. and Sacks, R., *Appl. Spectrosc.*, **43**, 801 (1989).
6. Ng, K. C., Ali, A. H. and Winefordner, J. D., *Spectrochim. Acta*, **46B**, 309 (1991).
7. Starn, T. K., Periero, R. and Hieftje, G. M., *Appl. Spectrosc.*, **47**, 1555 (1993).
8. Periero, R., Starn, T. K. and Hieftje, G. M., *Appl. Spectrosc.*, **49**, 615 (1995).
9. Broekaert, J. A. C., Starn, T. K., Wright, L. J. and Hieftje, G. M., *Spectrochim. Acta*, **48B**, 1207 (1993).
10. Broekaert, J. A. C., Starn, T. K., Wright, L. J. and Hieftje, G. M., *Spectrochim. Acta*, **53B**, 1723 (1998).
11. McLuckey, S. A., Glish, G. L., Asano, K. G. and Grant, B. C., *Anal. Chem.*, **60**, 2220 (1988).
12. Centineo, C., Fernandez, M., Pereiro, R. and Sanz-Medel, A., *Anal. Chem.*, **69**, 3702 (1997).
13. Rodriguez, J., Pereiro, R. and Sanz-Medel, A., *J. Anal. At. Spectrom.*, **13**, 911 (1998).
14. Orellana-Velado, N. G., Pereiro, R. and Sanz-Medel, A., *J. Anal. At. Spectrom.*, **13**, 905 (1998).
15. Orellana-Velado, N. G., Pereiro, R. and Sanz-Medel, A., *J. Anal. At. Spectrom.*, **15**, 49 (2000).
16. Martinez, R., Pereiro, R., Sanz-Medel, A. and Bordel, N., *Fresenius' J. Anal. Chem.*, **371**, 746 (2001).
17. Orellana-Velado, N. G., Fernandez, M., Pereiro, R. and Sanz-Medel, A., *Spectrochim. Acta*, **56B**, 113 (2001).
18. Pan, X., Hu, B., Ye, Y. and Marcus, R. K., *J. Anal. At. Spectrom.*, **13**, 1159 (1998).
19. Slevin, P. J. and Harrison, W. W., *Appl. Spectrosc. Rev.*, **10**, 201 (1975).
20. Orellano-Velado, N. G., Pereiro, R. and Sanz-Medel, A., *J. Anal. At. Spectrom.*, **16**, 376 (2001).
21. Olson, L. K., Belkin, M. and Caruso, J. A., *J. Anal. At. Spectrom.*, **11**, 491 (1996).
22. Olson, L. K., Belkin, M. and Caruso, J. A., *J. Anal. At. Spectrom.*, **12**, 1255 (1997).
23. Belkin, M., Waggoner, J. W. and Caruso, J. A., *Anal. Commun.*, **35**, 281 (1998).
24. Milstein, L. S., Waggoner, J. W., Sutton, K. L. and Caruso, J. A., *Appl. Spectrosc.*, **54**, 1286 (2000).
25. Duckworth, D. C. and Marcus, R. K., *J. Anal. At. Spectrom.*, **7**, 711 (1992).
26. Guzowski, J. P., Jr., Broekaert, J. A. C., Ray, S. J. and Hieftje, G. M., *J. Anal. At. Spectrom.*, **14**, 1121 (1999).
27. Guzowski, J. P., Jr. and Hieftje, G. M., *Anal. Chem.*, **72**, 3812 (2000).
28. Guzowski, J. P., Jr. and Hieftje, G. M., *J. Anal. At. Spectrom.*, **15**, 27 (2000).
29. Guzowski, J. P., Jr., Broekaert, J. A. C. and Hieftje, G. M., *Spectrochim. Acta*, **55B**, 1295 (2000).
30. Ray, S. J. and Hieftje, G. M., *Anal. Chim. Acta*, **445**, 35 (2001).
31. Majidi, V., Moser, M., Lewis, C., Hang, W. and King, F. L., *J. Anal. At. Spectrom.*, **15**, 19 (2000).
32. Klingler, J. A., Savickas, P. J. and Harrison, W. W., *J. Am. Soc. Mass Spectrom.*, **1**, 138 (1990).
33. Harrison, W. W. and Hang, W., *Fresenius J. Anal. Chem.*, **355**, 803 (1996).
34. Bruhn, C. G., Bentz, B L. and Harrison, W. W., *Anal. Chem.*, **51**, 673 (1979).
35. Chapman, J. R., *Practical Organic Mass Spectrometry: A Guide for Chemical and Biochemical Analysis*, John Wiley & Sons, Ltd, Chichester, 1993.
36. Scott, R. P. W., Scott, C. G., Munroe, C. G. M. and Hess, J., Jr., *J. Chromatogr.*, **99**, 395 (1974).
37. Games, M. B., *Adv. Mass Spectrom.*, **10B**, 323 (1986).
38. Willoughby, R. C. and Browner, R. F., *Anal. Chem.*, **56**, 2626 (1984).
39. Creaser, C. S. and Stygall, J. W., *Analyst*, **118**, 1467 (1993).

40. Strange, C. M. and Marcus, R. K., *Spectrochim. Acta*, **46B**, 517 (1991).
41. You, J., Fanning, J. C. and Marcus, R. K., *Anal. Chem.*, **66**, 3916 (1994).
42. You, J., Depalma, P. A., Jr. and Marcus, R. K., *J. Anal. At. Spectrom.*, **11**, 483 (1996).
43. You, J., Dempster, M. A. and Marcus, R. K., *J. Anal. At. Spectrom.*, **12**, 807 (1997).
44. You, J., Dempster, M. A. and Marcus, R. K., *Anal. Chem.*, **69**, 3419 (1997).
45. Dempster, M. A. and Marcus, R. K., *J. Anal. At. Spectrom.*, **15**, 43 (2000).
46. Gibeau, T. E. and Marcus, R. K., *Anal. Chem.*, **72**, 3833 (2000).
47. Gibeau, T. E. and Marcus, R. K., *J. Chromatogr.*, **A 915**, 117 (2001).

5.6 Electrospray Methods for Elemental Speciation

Hubert Chassaigne

European Commission – Joint Research Centre, Geel, Belgium

1 INTRODUCTION

In terms of analytical developments, the demonstration of analytical craft and skills of an analyst to determine a particular elemental distribution in a sample are aimed at the detection of unknown elemental species, their identification and/or structural elucidation [1, 2]. Exciting potential opportunities are offered by electrospray (ES and related techniques) for a soft ionization of metal-containing species and by tandem mass spectrometry (MS/MS) for a precise determination of molecular weight and structural characterization of molecules at trace levels in complex matrices.

The evolution in speciation analysis is partially due to the advent and spread of electrospray tandem quadrupole and quadrupole time-of-flight (Q-TOF) mass spectrometers in analytical laboratories and the coupling of ES MS/MS with high resolution separation techniques, such as high performance liquid chromatography (HPLC – MS/MS) [1, 3, 4]. Microbore and capillary LC are compatible with the majority of electrospray MS interfaces [5] and provide high sensitivity for elemental trace analysis.

Speciation affects the bioavailability and toxicity of elements and so is important in toxicology and nutrition [6, 7]. Both essential and nonessential elements are taken into the body with foodstuffs and some elements may be biologically incorporated in food itself. Classic speciation analyses put strong emphasis on the evaluation of risk induced by the contamination of foodstuffs [2]. The greatest area of interest in the biological field concerning trace element speciation probably relates to its influence on the bioavailability of

Handbook of Elemental Speciation: Techniques and Methodology R. Cornelis, H. Crews, J. Caruso and K. Heumann
 ISBN: 0-471-49214-0

essential elements and the availability and toxicity of toxic metals [8]. Organometallic species and metal complexes arising in plants and animals from biotransformation of metals and metalloids are important in understanding the mechanisms of their biological effects.

Electrospray ionization is the method of choice for organometallic species, proteins, oligopeptides and their metal complexes. This chapter is illustrated with a number of examples related to trace element speciation analysis in biochemistry. The compounds of interest are endogenous and exogenous metal and metalloid species in foodstuffs (Se species in yeast and As in seaweeds), metal complexes with peptides and oligopeptides in plants (Cd-induced phytochelatins) and metal–protein complexes (Cd and Zn complexes with metallothioneins) in animal tissues.

2 ELECTROSPRAY AND RELATED IONIZATION TECHNIQUES

2.1 Electrospray

Droplets are generated when a high voltage is applied to a liquid stream; this technique is known as electrospray (ES) [9, 10]. The part of the ionization source in which the ES process takes place is operated at atmospheric pressure (Figure 5.6.1). The liquid sample is delivered via a fused silica capillary inserted into a metal needle. The needle is held at a potential difference (typically 3–5 kV) relative to the mass spectrometer's entrance orifice. The typical flow range operated with this technique is 1–100 μL min^{-1}. The voltage on the needle causes the spray to be charged as it is nebulized. The droplets evaporate in a region maintained at a vacuum of about 1 mbar causing the charge to increase on the droplets. Larger droplets explode into smaller droplets and so on until protonated analyte molecules are released into the gas phase [11]. Ions in the partial vacuum of the ion block are extracted and focused electrostatically (by ion lenses) into a quadrupole or a hexapole system which efficiently transports ions into the mass analyser. Pure ES in the context of MS is accomplished without a nebulizing gas.

A new development in ES source geometry is the orthogonal sampling technique. The spray is directed perpendicularly to the spectrometer's entrance hole (not shown). By modifying the geometry, the source design is claimed to result in increased ruggedness and sensitivity compared with the early on-axis sampling systems,

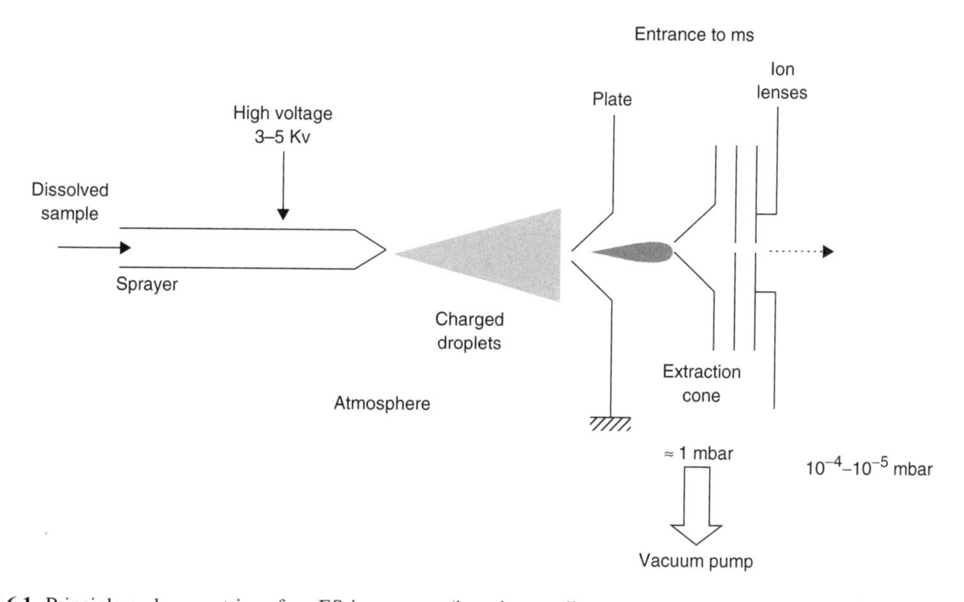

Figure 5.6.1. Principle and geometries of an ES ion source (in-axis sampling system).

because fewer large droplets, neutral components and particles may enter the vacuum chamber of the mass spectrometer and contaminate the mass analyser. Ions are extracted orthogonally from the spray into the sampling cone aperture leaving large droplets, involatile materials, particulates and unwanted components to collect in the vent port.

2.2 Pneumatically assisted electrospray

An alternative ES technique uses a gas to assist nebulization and desolvation of liquids in ES. This method is called pneumatically assisted ES (Figure 5.6.2). The source is an ambient temperature ion source (without the application of any heat) that typically accepts flow rates from 1 to 200 $\mu L\,min^{-1}$ without flow splitting. In this way quasimolecular ions can be generated from very labile and high molecular weight compounds, without any degradation.

New pneumatically assisted ES exploits a dual orthogonal sampling technique to deliver two stages of sampling, one for ruggedness (contamination avoidance) and a second for sensitivity in mass spectrometry (Figure 5.6.2):

- In the first stage the spray is directed perpendicularly past the sampling cone as in the simple orthogonal source geometry. Ions in solution are emitted into the gas phase.
- The second orthogonal step enables the volume of gas (and ions) sampled from atmosphere to be increased by a factor of 2–4 compared with conventional sources. The jet passes orthogonally to the second aperture so the flow into it is significantly decreased. Ions in the partial vacuum of the ion block are extracted electrostatically into the quadrupole or hexapole, which efficiently transports ions to the mass analyser.

Background noise is significantly reduced by dual orthogonal sampling, contributing to the

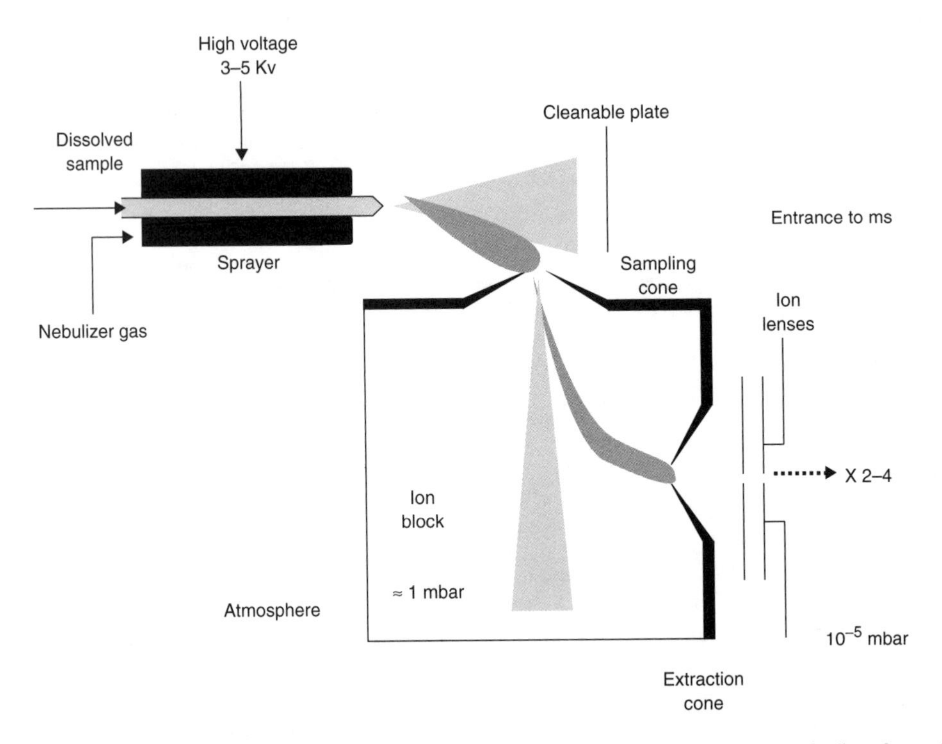

Figure 5.6.2. Principle of pneumatically assisted ES for labile and high molecular mass compounds (interface with a dual orthogonal sampling technique).

very good limits of detection and the high sensitivity obtained.

Additional modifications of the ion source are proposed to increase the sample utilization efficiency and thus the sensitivity obtained in MS. Heated pneumatically assisted ES sources achieve high sensitivity for quantification at high flow rates (up to 1000 $\mu L\,min^{-1}$) (not shown). The advantages are:

- superior ionization efficiencies without thermal degradation of labile compounds for improved sensitivity;
- enhanced stability at high flow rates and high aqueous mobile phases for improved results and reliability with gradient HPLC methods;
- compatible with a wide range of flow rates, separation techniques and solvent compositions for increased flexibility.

2.3 Microelectrospray and nanoelectrospray

Other developments resulting from the miniaturization of the ES source due to demands for low analyte consumption in the direct introduction mode and hence low flow rates in chromatography. They became known as microelectrospray (flow rate ca 100–200 $nL\,min^{-1}$) and nanoelectrospray (since the flow rate is ca 10–20 $nL\,min^{-1}$) (not shown). Other terms have been coined such as microspray and so on. The nanoelectrospray technique was introduced in 1994 [12, 13]. This in turn can provide more concentrated solutions from small amounts of sample and/or long analysis times because of the low flow rates. The samples are loaded into pulled capillaries that have been coated with metal. The flow typically starts when a voltage is applied (although a little pressure is sometimes needed) and then continues until the sample is depleted or the metal sputters away.

The advantages of the miniaturization of the sprayer are:

- near 100 % sample utilization;
- minimized contamination of the instrument;
- elimination of cross contamination due to disposable spray capillaries;
- ability to spray from purely aqueous as well as purely organic solvent;
- flexibility for a variety of on-line electrospray couplings (e.g. capillary LC).

An example of application in speciation analysis was the identification of arsenic-containing compounds (arsenosugars) at the picogram level using nanoelectrospray MS [14]. The application of the method to real samples reveals the potential of the technique for trace analysis.

3 SAMPLE INTRODUCTION AND ANALYSIS IN TANDEM MASS SPECTROMETRY

3.1 Sample introduction in electrospray

With recent design improvements to provide flexibility, ease-of-use, and fast installation, the sources are now ideal for any other ES approach such as coupling to HPLC (nanoscale to analytical) techniques. However, typical LC eluents do not readily lend themselves to direct coupling with MS. The coupling of reversed-phase HPLC with ES is the most compatible and has been widely described in the literature [15, 16]. It is also generally agreed that the greatest sensitivity in ES and pneumatically assisted ES is achieved with the lowest liquid flow and the smallest diameter HPLC column [17].

Before beginning an LC/MS analysis, the mass spectrometry should be optimized for the specific application in speciation analysis. The best method of tuning is either by infusion or by direct loop injection of the sample solution. In this way, the effect of different mobile phase compositions on sample ionization can be tested in positive and/or negative ionization mode, and suitable solvents can thus be selected.

For direct analysis of samples in ES, 50/50 H_2O–AcN is typically used as the mobile flow phase. Although 100 % water can be used in ES, better sensitivity is obtained with some organic solvent present in the water. Even 5–10 % of MeOH or AcN increases the stability of the nebulization process and the desolvation step. When using high percentages of water it is usually necessary to increase the nebulization gas and/or the source temperature to aid desolvation.

If the pH of the mobile phase needs to be reduced to enhance LC separations, then acetic and formic acids are suitable. Formic acid is more acidic and less is required to reach a desired pH. Normally, less than 1 % of acid would be added, but often with LC more needs to be added to reach the required pH. Care should be taken during negative ion analysis, as the addition of acid to the mobile phase can suppress ionization. Weakly acidic compounds will not form deprotonated ions in acidic conditions.

Inorganic acids (e.g. H_2SO_4, H_3PO_4) should not be used with LC-MS even though there are examples in the literature of very low levels of HCl ($<0.005\,\%$) being utilized. Care should be taken if employed, as HCl is corrosive, and even at low concentrations, can cause corrosion. This assumes that the ionization process is not degraded by the addition of such acids. The ion source would require more regular cleaning, however, if these additives were used.

Buffers such as phosphate, Tris, and Hepes cannot be used in ES. Even trace levels of these interfere with the ES process. Only volatile buffers such as ammonium acetate (CH_3COONH_4) or ammonium formate ($HCOOHN_4$) can be used. To improve chromatography without degrading the MS performance it is best to use as little CH_3COONH_4 as possible up to a maximum concentration of 0.1 M. In many cases CH_3COONH_4 can be used to replace phosphate buffers which are highly incompatible with LC-MS systems.

Excess Na^+, K^+, and detergents (such as sodium dodecyl sulfate) are very bad for ES and frequently result in no data. Detergents, by their nature, are concentrated at the surface of a liquid. This causes a problem in ES as the ionization relies on the evaporation of ions from the surface of a droplet. The detergent congregates at the surface of the droplet thus suppressing other ions.

3.2 Tandem quadrupole MS/MS system

The ES source operates at atmospheric pressure, so a quadrupole analyzer, which does not employ high voltages, is easier to interface to the ES source. Tandem quadrupole mass spectrometry makes use of a mass analyser (quadrupole MS 1 in Figure 5.6.3) to select a particular m/z value (usually the molecular ion) for a CID (collision-induced dissociation) in a collision cell (quadrupole or hexapole). The fragment ions, called product ions, that form in the collision cell are mass analysed (by the quadrupole MS 2) to record a product ion spectrum. This spectrum contains structural information about the ion selected in the tandem MS experiment.

A tandem quadrupole MS/MS system has a mass-to-charge ratio (upper) limit typically around m/z 3000, in both MS and MS/MS

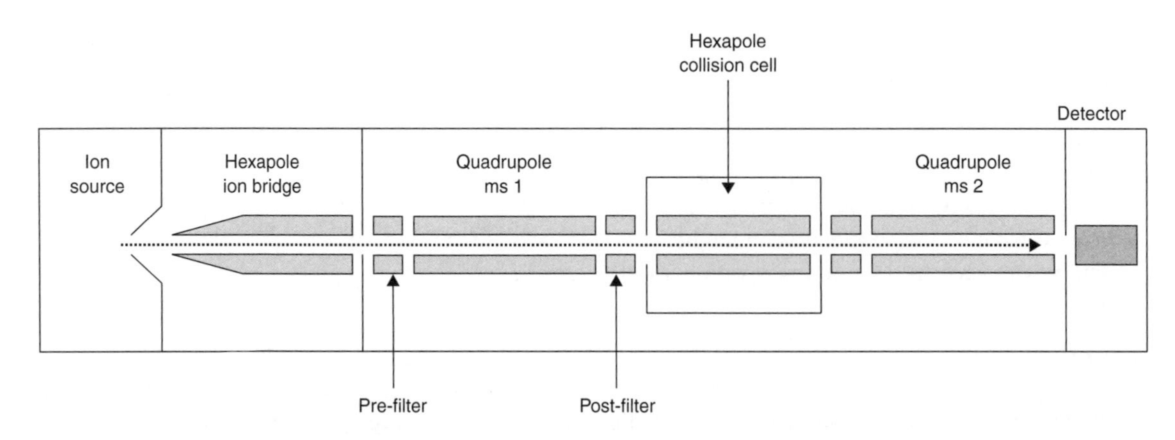

Figure 5.6.3. Tandem quadrupole mass spectrometer (MS/MS system).

modes. The mass accuracy is ca 0.01 % over the entire mass range. The typical resolution power of such an instrument is 3000–4000 (FWHM). This system brings unparalleled performance for quantification and identification in the speciation field. The tenfold increase in sensitivity, enhanced robustness, and ease of use of the innovative new ion sources (that accept higher flows and the presence of salt contents in the samples) combined with new collision cells (e.g. linear accelerator for a better fragmentation efficiency) deliver higher productivity with greater confidence in the results.

3.2.1 MS mode

In the MS operating mode, the first quadrupole MS 1 (see Figure 5.6.3) is swept over a given mass range and MS 2 is operated in RF-only mode. The sensitivity (signal-to-noise ratio) can be significantly increased by scanning a narrow mass range or by using the single ion monitoring mode to focus on particular masses of interest.

The ES mass spectra are characterized by quasi molecular ions, e.g. $[M + H]^+$, $[M + Na]^+$ etc. In the simplest case, one or more protons can be attached to the analyte molecule leading to the formation of a singly charged ion (generally for metal- and metalloid-containing species with a mass up to 1000 Da) or multicharged ions (e.g. in the case of polypeptides and proteins and their complexes with metals). In the MS mode, no identification can be performed. The only information obtained is the molecular mass and the isotopic pattern, allowing the confirmation of known compounds and their complexes.

The MS spectrum obtained in the case of low molecular mass compounds (up to 1000 Da) is very simple and contains only singly charged molecular ions (singly protonated molecular ions). However, for species containing an element having several isotopes, such as organoselenium compounds, an envelope corresponding to protonated molecules with the most abundant isotopes can be observed. Figure 5.6.4(a) shows the spectrum obtained for a common selenoamino acid encountered in biological samples, selenomethionine [18]. Its shows an m/z envelope within which the isotopic pattern of selenium can be identified (protonated molecular ion $[M + H]^+$ at m/z 198 containing the most abundant ^{80}Se isotope). A partial fragmentation can be obtained despite the relatively low extraction energy used (potential of the orifice or the extraction cone). Thus a fragment M-17 corresponding to the loss of an OH group from the carboxylic group is observed in Figure 5.6.4(a).

ES is particularly useful for obtaining the molecular weight of analytes that would normally exceed the normal mass range of sector and quadrupole instruments. The most obvious feature of an ES spectrum is that the molecular ions of large molecules carry multiple charges, which reduce their mass-to-charge ratio compared with a singly charged species. This allows mass spectra to be obtained for large molecules.

For polypeptides and proteins and their complexes with metals, the formation of multiply protonated molecular ions occurs [10, 19, 20]. Figure 5.6.5(a) shows an MS spectrum obtained for a class of metalloproteins, Cd-induced metallothioneins isolated from animal tissues (isoform MT-2 from rabbit liver) [21, 22]. Two different envelopes of multicharged peaks can be seen in the spectrum. Assuming that adjacent peaks in the ion envelope differ by only one charge and that the charging is due to protonation, the relation between a multiply charged peak at m/z and the relative molecular mass is shown in Figure 5.6.5(a) [23]. The higher the number of m/z signals that can be seen for a given compound in the mass spectrum, the more precise is the result of the molecular mass determination. Three sub-isoforms of MT-2 (which differ by a few amino acids in their composition) with very close molecular masses can be identified in the preparation. They were identified as MT-2a, MT-2b and MT-2c (Figure 5.6.5(a)) with molecular masses of 6125.5, 6146.0, and 6155.5 Da, respectively. Apart from the metallothioneins, the sample also contains another protein (multicharged peaks +15, +14 and +13) with the molecular mass 15 570.0 Da, identified as superoxide dismutase.

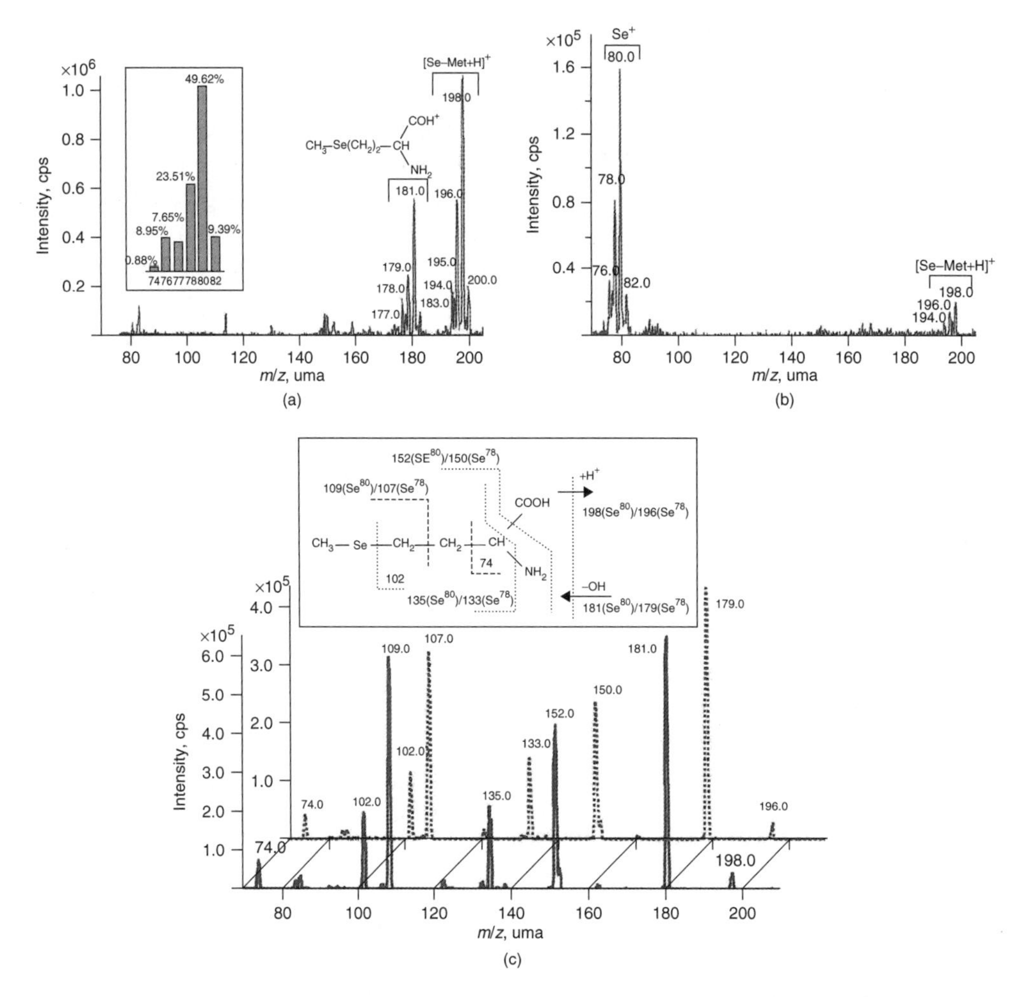

Figure 5.6.4. Illustration of the potential of ES MS/MS in speciation analysis using different data acquisition modes. Characterization of a selenium-containing compound (selenomethionine): (a) MS spectrum, (b) SCID mode for elemental analysis, (c) MS/MS spectrum. Reprinted from *Trends in Analytical Chemistry*, Vol. 19, H. Chassaigne, V. Vacchina and R. Lobinski, p. 300, Copyright (2000), with permission from Elsevier Science.

3.2.2 MS/MS mode

In the MS/MS mode, mass spectrometer 1 is used as a filter to transmit the quasi molecular ion $[M + H]^+$ (see Figure 5.6.3) to the collision chamber where it is fragmented by collision with neutral gas molecules in a process referred to as collision-induced dissociation (CID). The collision gas is typically nitrogen and the collision energies used can be in the range 10–50 eV depending on the mass of the compounds and to obtain optimum fragmentation. The mass of mass spectrometer 1 is fixed and mass spectrometer 2 is swept over a given mass range to determine the ions that result from the fragmentation of the precursor molecular ion. The resulting fragment ions are analysed by the second mass analyser (see quadrupole mass spectrometer 2 in Figure 5.6.3) allowing their mass determination and information on the molecule to be obtained.

In particular, for species containing an element having more than one stable isotope, such as selenomethionine, valuable information can be obtained by fragmenting the two protonated molecules containing the adjacent most abundant isotopes (m/z 196 and 198 corresponding to ^{78}Se and ^{80}Se, respectively) (Figure 5.6.4(c)) [18]. Selenium-containing fragments are separated by the distance of two mass units whereas fragments that do not contain selenium will remain at the same m/z value, thus facilitating the interpretation of the mass spectra. However, peaks obtained by fragmentation of the molecular signals due to different isotopes may vary by 0.1–0.5 Da with a quadrupole instrument, which makes it difficult to identify the elemental species in the CID MS spectra [24].

Some studies clearly show the limitations of a triple quadrupole instrument for speciation studies and especially in the case of monoisotopic elements (e.g. arsenic). Since arsenic has only one isotope, the molecular ions observed in the MS mode may suffer from a severe risk of overlaps in the case of real matrices which would mean that the positive identification of arsenic-containing compounds exclusively on that basis would be unreliable [25]. However, a deeper insight into the structure of the compounds can be gained by analysing product ions resulting from the CID of the protonated molecular ions.

ES MS/MS is also widely used to determine the sequence of polypeptides and proteins. Peptides of limited molecular mass (up to 2000 Da) can be subject to sequence analysis by MS/MS [26]. Peptides fragment primarily at the amine bonds to produce a 'ladder' of sequence ions. The charge

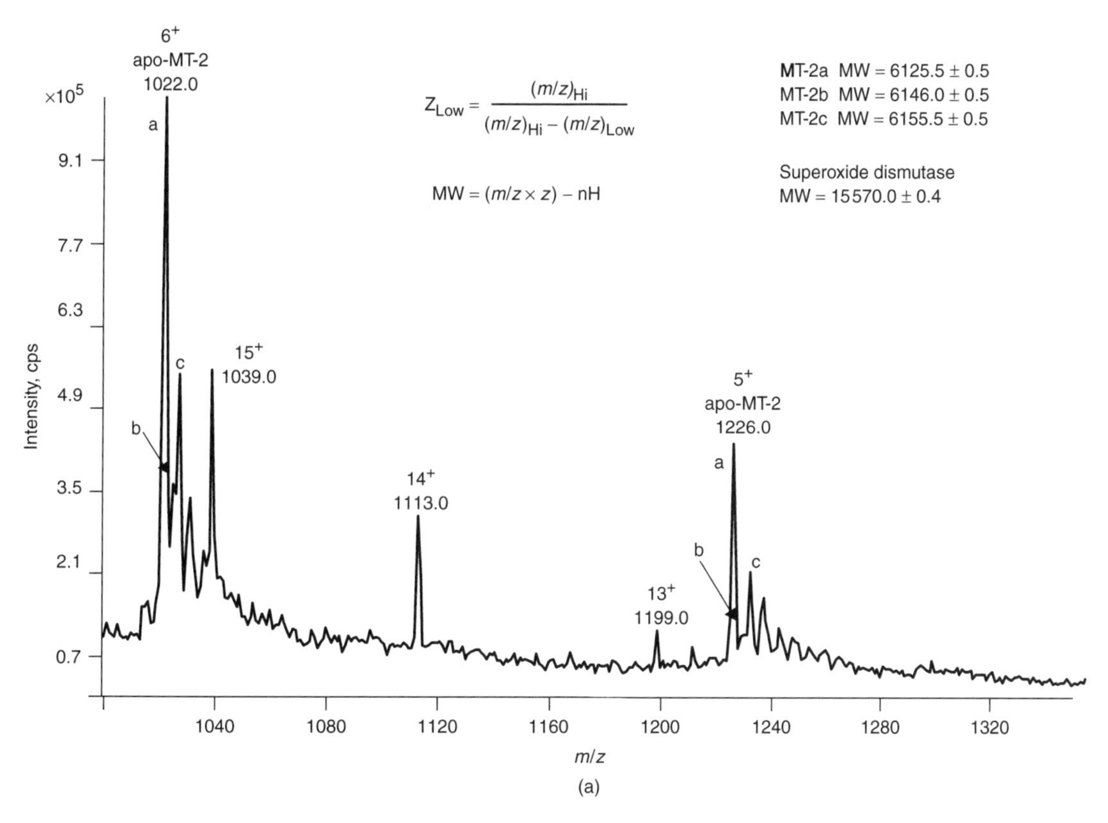

Figure 5.6.5. Potential of ES MS/MS for the analysis of proteins and peptides. (a) Determination of the molecular mass of Cd-induced metallothionein MT-2 (rabbit liver) using the multiply charged ion envelope. Three sub-isoforms are identified as MT-2a, MT-2b, and MT-2c. Reprinted from the *Journal of Analytical Chemistry*, Vol. 361, H. Chassaigne, R. Lobinski, pp. 267–273, Figure 4, 1998, copyright notice of Springer-Verlag. (b) CID mass spectra of a Cd-induced peptide isolated from plants (phytochelatin PC_4) Reprinted from *Trends in Analytical Chemistry*, Vol. 19, H. Chassaigne, V. Vacchina and R. Lobinski, p. 300, Copyright (2000), with permission from Elsevier Science.

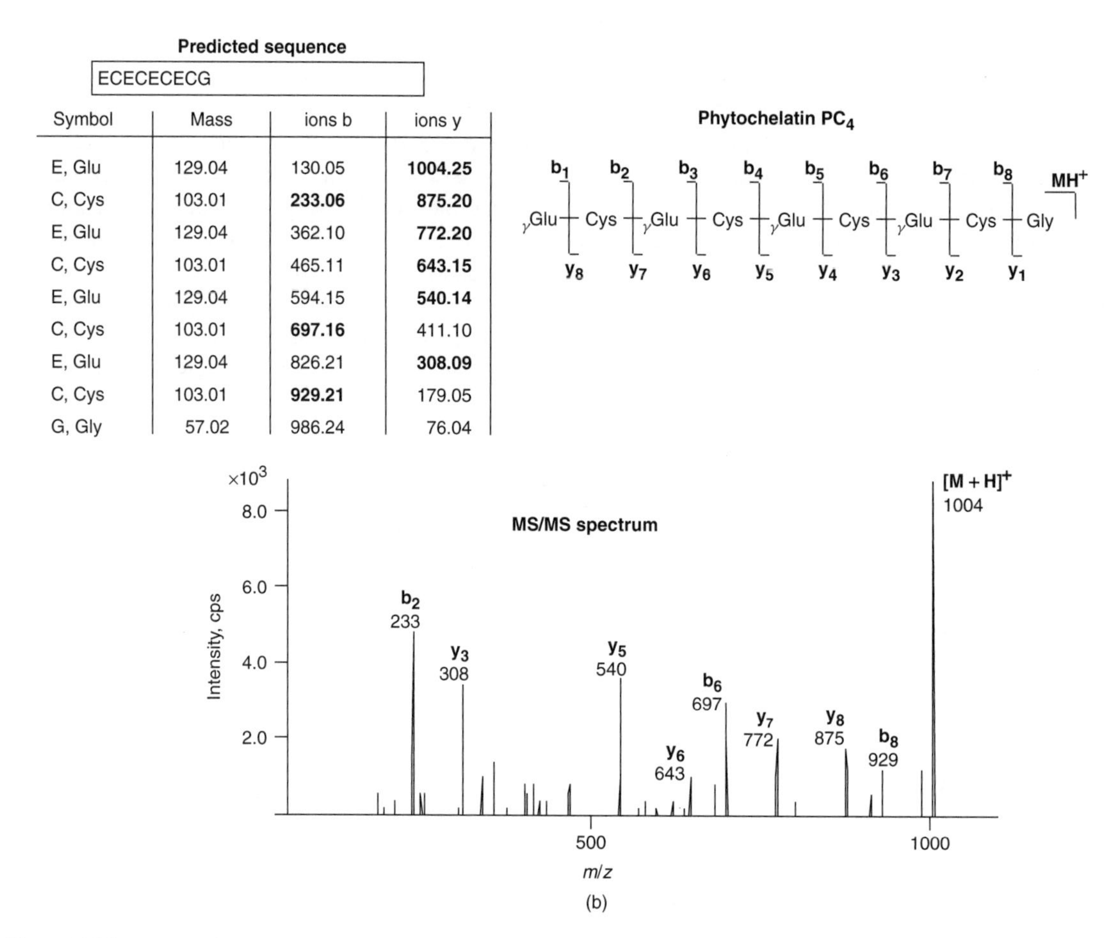

Predicted sequence

ECECECECG

Symbol	Mass	ions b	ions y
E, Glu	129.04	130.05	**1004.25**
C, Cys	103.01	**233.06**	**875.20**
E, Glu	129.04	362.10	**772.20**
C, Cys	103.01	465.11	**643.15**
E, Glu	129.04	594.15	**540.14**
C, Cys	103.01	**697.16**	411.10
E, Glu	129.04	826.21	**308.09**
C, Cys	103.01	**929.21**	179.05
G, Gly	57.02	986.24	76.04

Figure 5.6.5. (*continued*)

can be retained on the amino terminus (type b ion) or on the carboxy terminus (type y ion). Thus a complete series of ions from both types allows the determination of the amino acid sequence. The potential of this mode is illustrated in the case of peptides such as cadmium-induced phytochelatins in plants (phytochelatin PC_4 with MW 1004 Da) (Figure 5.6.5(b)) [18].

3.2.3 In-source collision-induced dissociation mode

The ES interface can be tuned to provide fragment ions that unlock the structure of eluting molecules and this leads to the formation of elemental ions of interest in speciation analysis. In-source fragmentation can be controllably induced by the voltage applied to the extraction cone or orifice in the interface (see Figure 5.6.2). Source collision-induced dissociation (SCID) is promoted by increasing this voltage, resulting in a CID mass spectrum which is a unique fingerprint of the analyte. In the example of a selenoamino acid (selenomethionine) (see Figure 5.6.4(b)), the protonated ion $[M+H]^+$ at m/z 198 can be broken down by increasing the ionization energy. At a sufficiently high ionization energy, elemental Se cations can be obtained from the sample [18, 27]. The detection limits for the elements, however, are 2–3 orders of magnitudes higher than in the case of inductively coupled plasma mass spectrometry (ICP MS) [18].

The type of information obtained in this operational mode of ES mass spectrometry is of paramount importance in speciation analysis. In the case of complex samples when a number of concomitant ions can interfere with the identification of the element isotopic pattern (e.g. for selenium) and especially for a monoisotopic elemental species (e.g. for arsenic), the SCID may be an alternative.

3.2.4 Quantitative analysis

For quantification, the tandem quadrupole mass spectrometer can be operated in the different acquisition modes as previously described:

- Full scan in MS (scanning over a defined mass range).
- SIM (selected ion monitoring) mode. The mass spectrometer sensitivity (signal-to-noise ratio) can be significantly increased, by narrow mass scanning or by selected ion monitoring.
- Full scan in MS/MS.
- MRM (multiple reaction monitoring) mode. MRM allows MS/MS to be used for multiple analytes in direct analysis and coeluting analytes in HPLC. MRM allows a few parent ions to be isolated and dissociated under optimum conditions with monitoring of all the corresponding product ions.

An example of quantification using a quadrupole instrument was given by Madsen *et al.* [28]. The identity of arsenic-containing species (arsenosugars) was confirmed by LC-ES MS with a variable fragmenter voltage which provided simultaneous elemental and molecular detection. LC-ES MS was further used to quantify four arsenosugars, producing values within 5–14 % (depending on the compound) of the ICP MS data.

3.3 Quadrupole time-of-flight MS/MS system

A quadrupole time-of-flight (Q-TOF) instrument is an MS/MS system combining the simplicity of a quadrupole (mass spectrometer 1) with the ultrahigh efficiency of a TOF mass analyser (mass spectrometer 2) (Figure 5.6.6). The system exploits a TOF mass analyser to achieve simultaneous detection of ions across the full mass range. This is in contrast to conventional instruments (tandem quadrupole system) that must scan over one mass at a time. A Q-TOF instrument offers up to 100 times more sensitivity than tandem quadrupole instruments when acquiring full product ion (MS/MS) mass spectra.

A Q-TOF system has a mass-to-charge ratio (upper) limit typically in excess of m/z 20 000, in both MS and MS/MS modes, enabling the analysis of very large molecules as multiply charged ions. The high resolving power (5000–10 000 FWHM) enables improved mass measurement accuracy for small molecules, charge state identification of multiply charged ions and greater differentiation of isobaric species. The inherent stability of the reflectron TOF analyser routinely delivers excellent mass measurement accuracy (0.0002–0.0005 %) for molecules of low molecular mass (up to 1000 Da). Exact mass measurements enable the masses of molecular and/or product ions (MS/MS) to be confirmed for known compounds. For unknowns the number of plausible structures may be restricted to a small number with the aid of additional chemical information (metallic or nonmetallic element of interest in the structure).

For identification, the Q-TOF hybrid MS/MS system is a powerful qualitative tool providing high full-scan sensitivity and excellent resolution of intact compounds as well as product ion spectra. The system can easily determine the elemental composition of various compounds.

For a Q-TOF instrument the acquisition modes for quantification are:

- Full scan in MS with the reflectron TOF analyser (mass spectrometer 2 in Figure 5.6.6).
- Full scan in MS/MS. CID mode of molecular ions selected by mass spectrometer 1 and simultaneous detection of fragment ions across the full mass range by mass spectrometer 2 (Figure 5.6.6).

The performances of a Q-TOF mass spectrometer has been evaluated in terms of accuracy and precision for the identification of low

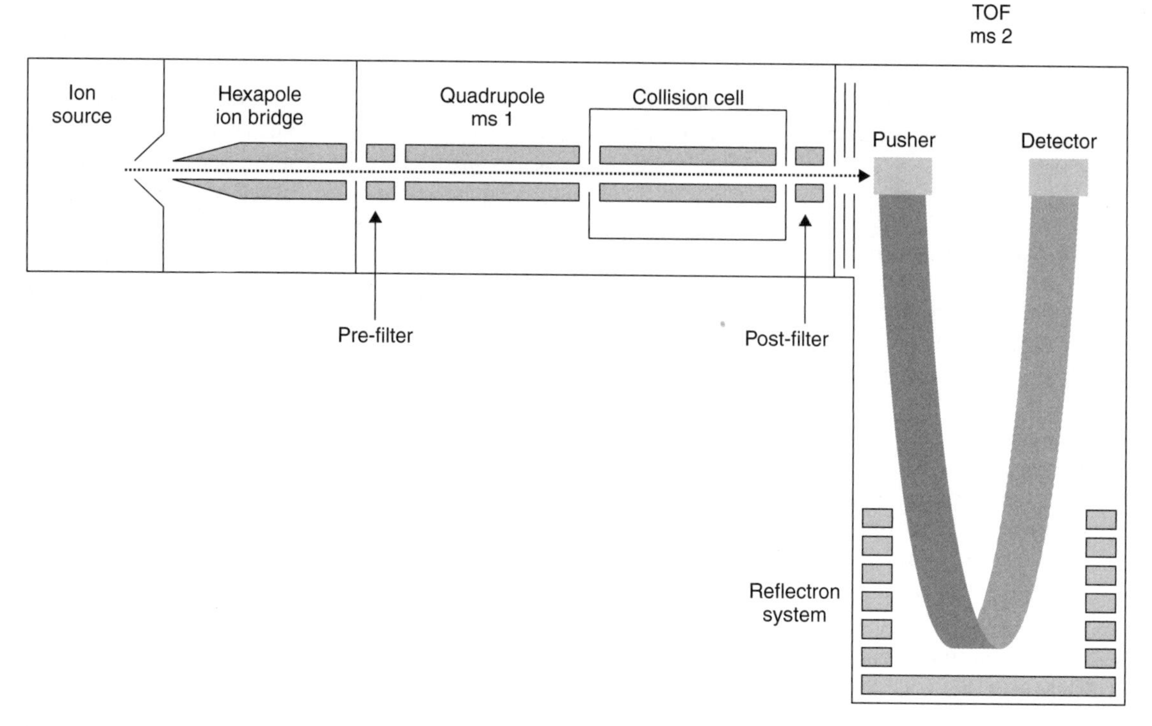

Figure 5.6.6. Geometry of a quadrupole TOF mass spectrometer system (MS/MS system).

molecular weight selenium compounds [24, 29]. The accuracy of mass measurements, evaluated with standards (selenomethionine, selenoethionine and selenocystine) was found to be between 0.005 and 0.01 % (0.01–0.02 Da) [24]. Previous work in this field was done with a triple quadrupole mass analyser [30, 31]. A triple quadrupole instruments offers 0.1–0.5 mass resolution which makes the recognition of the selenium pattern in a mass spectrum often ambiguous, especially when a foreign compound gives a peak at one of the molecular masses within the isotopic cluster. The precision of the molecular mass measurement in the MS/MS mode was evaluated as 0.004 % for selenomethionine and selenocystine [24].

Another study shows the performance of a Q-TOF instrument for the characterization of arsenic compounds in biological samples [32]. The mass accuracy obtained with this instrument enables the characterization of arsenosugars in a complex sample. These results offer new perspectives for the rapid recognition, and subsequent identification, of unknown arsenic species in crude extracts, without the need for extensive purification or previously characterized standards.

4 APPLICATIONS IN ELEMENTAL SPECIATION ANALYSIS

The technique was shown to be capable of producing gas-phase ions of highly labile and nonvolatile compounds, such as peptides, proteins and oligosaccharides [18]. This ionization process is so gentle that noncovalently bound metal complexes, such as complexes of Cd and As [27], were also shown to be desolvated and were studied by MS. However, the sample must be soluble in solvents (AcN, MeOH, water, ...) and the species stable at very low concentrations. Examples are discussed to illustrate the potential of ES in speciation analysis.

4.1 Selenium speciation in yeast

The fact that selenium (Se) is essential for human beings was demonstrated only recently in some regions where the intake of selenium is very low. Certain forms of cancers [33, 34] and cardiovascular diseases [35] have in some studies been associated with a low intake of selenium by people. Hence, supplements have been used for improving the Se status in the human body and to prevent disease to progress. However, the chemical form and concentration in which Se is introduced in to the body are of primary importance. Further investigations have suggested that selenized yeast produce selenomethionine along with other organoselenium compounds. Selenized yeast makes the selenium both bioavailable and provides it in a form which the body may use beneficially. Since the bioavailability and the toxicity of Se are closely correlated with its chemical form and concentration, the information on Se speciation is vital [36, 37]. For these reasons the demand for accurate and sensitive methods for Se speciation in nutritional supplements, e.g. Se-enriched yeast, has rapidly increased.

Selenocompounds cannot be analysed by ES MS in yeast extracts directly because of the presence of a matrix composed of high molecular weight compounds and salts suppressing the signal [24]. A sequential extraction and a fractionation of the extract by preparative size exclusion chromatography is often required to isolate the low molecular weight fractions [38]. Further purification of low molecular weight compounds by HPLC (e.g. anion exchange) may be necessary [39, 40]. An approach was proposed by Casiot *et al.* [41] who isolated the major selenocompounds in yeast extracts and identified them on the basis of the CID pattern of the protonated molecular ions corresponding to the adjacent Se isotopes by ES tandem quadrupole MS/MS. Thus, selenomethionine and Se-adenosylhomocysteine were successfully identified [36, 41]. In contrast, previous work was carried out with a triple quadrupole system which offers poor resolution and often makes the recognition of the Se pattern in a mass spectrum ambiguous [24].

Figure 5.6.7(a) shows the mass spectrum obtained for a chromatographic fraction [18, 30]. The spectrum reveals the presence of an unresolved ion cluster, matching the characteristic abundance of the Se isotope pattern, centred at m/z 433. The MS mode gives valuable information on the molecular mass of the species and the presence of one Se in the molecule. Information on the identity of the compound can be obtained by fragmenting the quasi molecular ion by CID. The MS/MS spectra obtained for the fragmentation of the ions at m/z 431 and 433 (corresponding to the two most abundant Se isotopes ^{78}Se and ^{80}Se, respectively) are shown in Figure 5.6.7(b). The comparison of the two spectra allows the differentiation between fragments that do not contain Se (m/z signal in both spectra at the same value) and fragments that contain Se (m/z signal in both spectra appear with a difference of 2 Da). The structure proposed for this compound was Se-adenosyl-homocysteine [30, 31].

In recent work, the purification of selenium compounds from the yeast extract has been simplified for further characterization of low molecular weight Se compounds by Q-TOF mass spectrometry [24]. In quadrupole MS of this fraction only the most abundant cluster centred at m/z 433 could be seen [30] whereas a number of other clusters, e.g. m/z 182, 196, 298, are observed in the MS spectrum of low molecular weight compounds (Figure 5.6.8(a)). When performing elemental analysis calculations with the instrument, the compound at m/z 433.0350 can be assigned the formula $C_{14}H_{20}N_6O_5Se$ [24]. In Figure 5.6.8(b) the observed and the calculated isotopic pattern of the molecule are compared. An insight into the structure of the compound can be obtained by CID (data not shown). The data are similar to those obtained with a triple quadrupole MS/MS system (see Figure 5.6.7(b)). The advantage of the Q-TOF analyser compared with the triple quadrupole system is the rapidity of spectra combined with high resolution, which enables the rapid fragmentation of all the molecular ions within the cluster using a small amount of sample. ES Q-TOF, especially when in tandem mode, provides novel information regarding the identity of selenocompounds in selenized yeast [24].

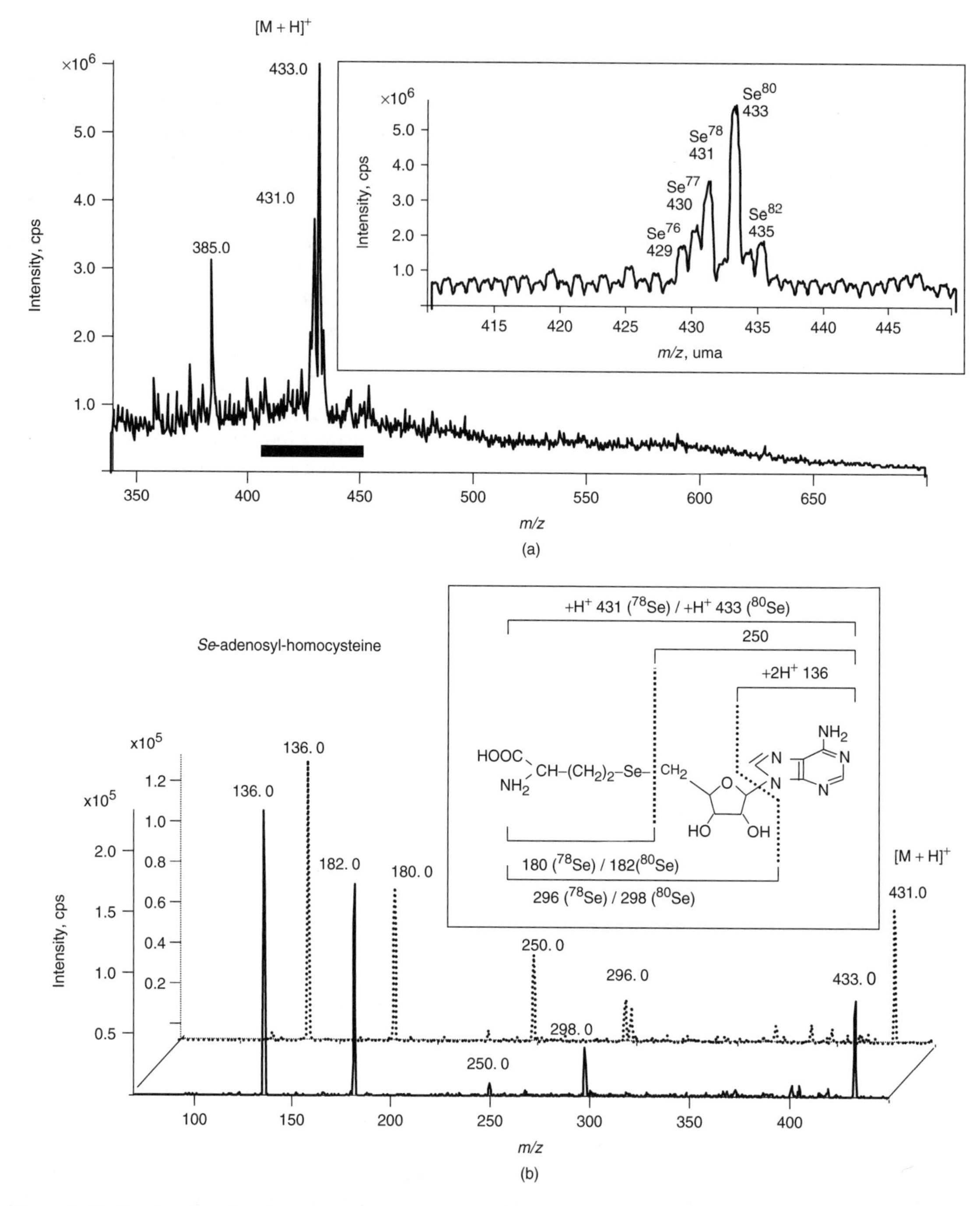

Figure 5.6.7. Identification of a selenocompound in a selenized yeast extract by ES quadrupole MS/MS. (a) ES MS spectrum of a fraction collected in the RP HPLC – ICP MS chromatogram of a water yeast extract. Reproduced by permission of the Royal Society of Chemistry. (b) ES MS/MS spectra of the selenium-containing ions at m/z 431 and 433. The fragmentation pattern of the identified compound, Se-adenosylhomocysteine, is shown in the inset. Reprinted from *Trends in Analytical Chemistry*, Vol. 19, H. Chassaigne, V. Vacchina and R. Lobinski, p. 300, Copyright (2000), with permission from Elsevier Science.

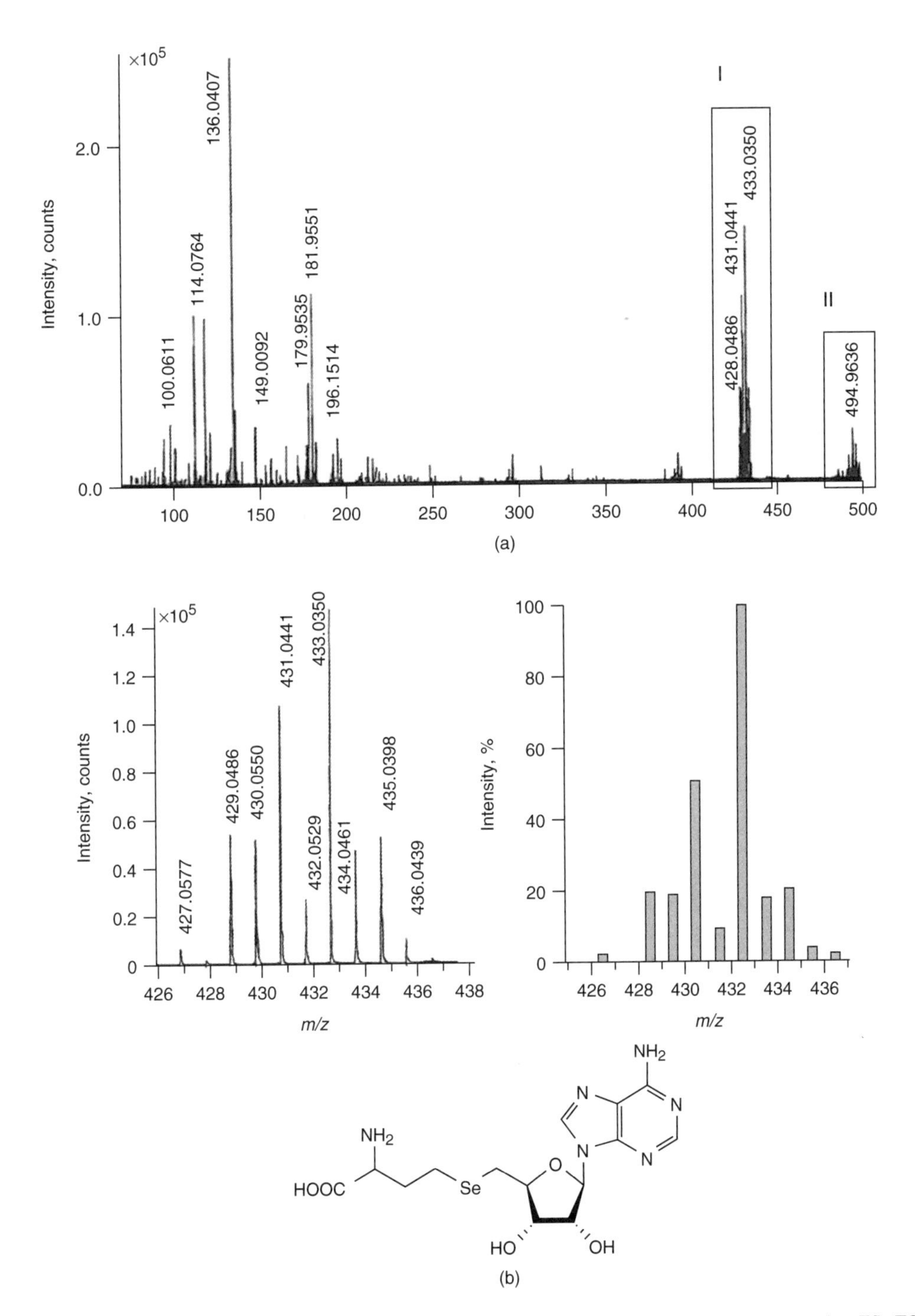

Figure 5.6.8. Confirmation of the elemental composition of a selenium compound in yeast extract by ES TOF MS (Se-adenosylhomocysteine). (a) ES MS spectrum of a yeast fraction collected after preparative anion exchange chromatography – ICP MS. (b) Zoom of the *m/z* 433.0350. Se-containing species compared with a theoretical pattern for the species shown. Reproduced by permission of the Royal Society of Chemistry.

4.2 Arsenosugars in seaweeds

Arsenic biotransformations by marine life are known to give rise to a wide range of organoarsenic compounds. Arsenobetaine (AsB) is the most abundant species in marine animals whereas shellfish and algae may contain a class of arsenic-containing ribofuranosides (so-called arsenosugars) [14, 42, 43]. This class of compounds has been raising growing concern since recent reports indicating the possibility of metabolizing arsenosugars to the carcinogenic dimethylarsinic acid (DMAA) by the human body [44]. As arsenosugar standards are not available, the characterization of such compounds in seaweeds requires their isolation [45] and purification before their analysis by mass spectrometry.

The poor potential of tandem quadrupole mass spectrometry in the case of arsenic-containing species is hampered by the fact that this element is monoisotopic so it is difficult to attribute a peak in the mass spectrum to an arsenic species. However, a confirmation of the identity of the arsenosugars present can be obtained by the fragmentation of the pseudomolecular ions and mass spectrometry of the resulting fragments.

The matrix suppression of ES prevents its application for crude extracts but the technique may be used for signal identification in fractions highly purified by multidimensional chromatographic techniques, e.g. size exclusion HPLC and ion exchange HPLC [46–48]. Figure 5.6.9 shows the CID mass spectra of the protonated molecules of four arsenosugars (A–D) extracted from algae and purified by complementary chromatographic techniques [25]. As indicated in the earlier studies [46], the common feature of the MS/MS spectra of arsenosugars is the presence of an ion signal at *m/z* 237 corresponding to the oxonium ion of the dimethylarsinoylpentose moiety. Another characteristic ion is that at *m/z* 195, which results from the break-up of the furane ring and indicates the attachment of the dimethylarsinoyl moiety to the 5′ position of the furane ring [25]. Arsenosugars A and C lose the SO_3 moiety readily; the result is the presence of an M^+-80 fragment in the case of sugar C or of an M^+-98 fragment (*m/z* 295) in the case of arsenosugar A. The fragments at *m/z* 97 and 80 have been assigned to OSO_3H and SO_3, respectively. The fragmentation never leads to the bare As^+ and or to the AsO^+ ion common in the mass spectra of simpler organoarsenic species (such as arsenobetaine), even when high extraction energy is used [25].

In recent studies, the use of accurate, high resolution MS has been shown to be a powerful analytical technique of great promise for the identification of unknown arsenic compounds in crude algal extracts [32]. Q-TOF mass spectra of a fractionated extract (by cation exchange chromatography) were obtained over the *m/z* range 0–400 on an instrument calibrated to a mass accuracy of 0.0002 %. Figure 5.6.10(a) shows the presence of a peak at *m/z* 329.1. However, since the mass spectrum contains a number of peaks and arsenic has only one stable isotope (in contrast to selenium), the selection of this particular *m/z* value has required an assumption that arsenosugars were present in the mixture. When performing elemental analysis calculations with such an instrument, increasing the mass accuracy of the measurement reduces the number of possible elemental combinations for the ion being investigated. This peak may correspond to the protonated molecule of arsenosugar B (see Figure 5.6.9(b)). CID tandem MS was performed on the *m/z* 329.1 ion (Figure 5.6.10(b)). The resultant mass spectrum of the fragment ions was also subject to a search using the elemental composition tool, which confirmed the composition of arsenosugar B. ES Q-TOF MS with high mass accuracy allows the number of peaks that could correspond to arsenic to be considerably reduced, facilitating recognition of unknown arsenic-containing compounds and representing a useful application on a more widely available instrument.

4.3 Cadmium-induced phytochelatins in plants

Like all organisms, plants present a dilemma in that metals such as Cu and Zn are essential trace metal nutrients taking part in redox reactions, electron transfers, a multitude of enzyme-catalyzed

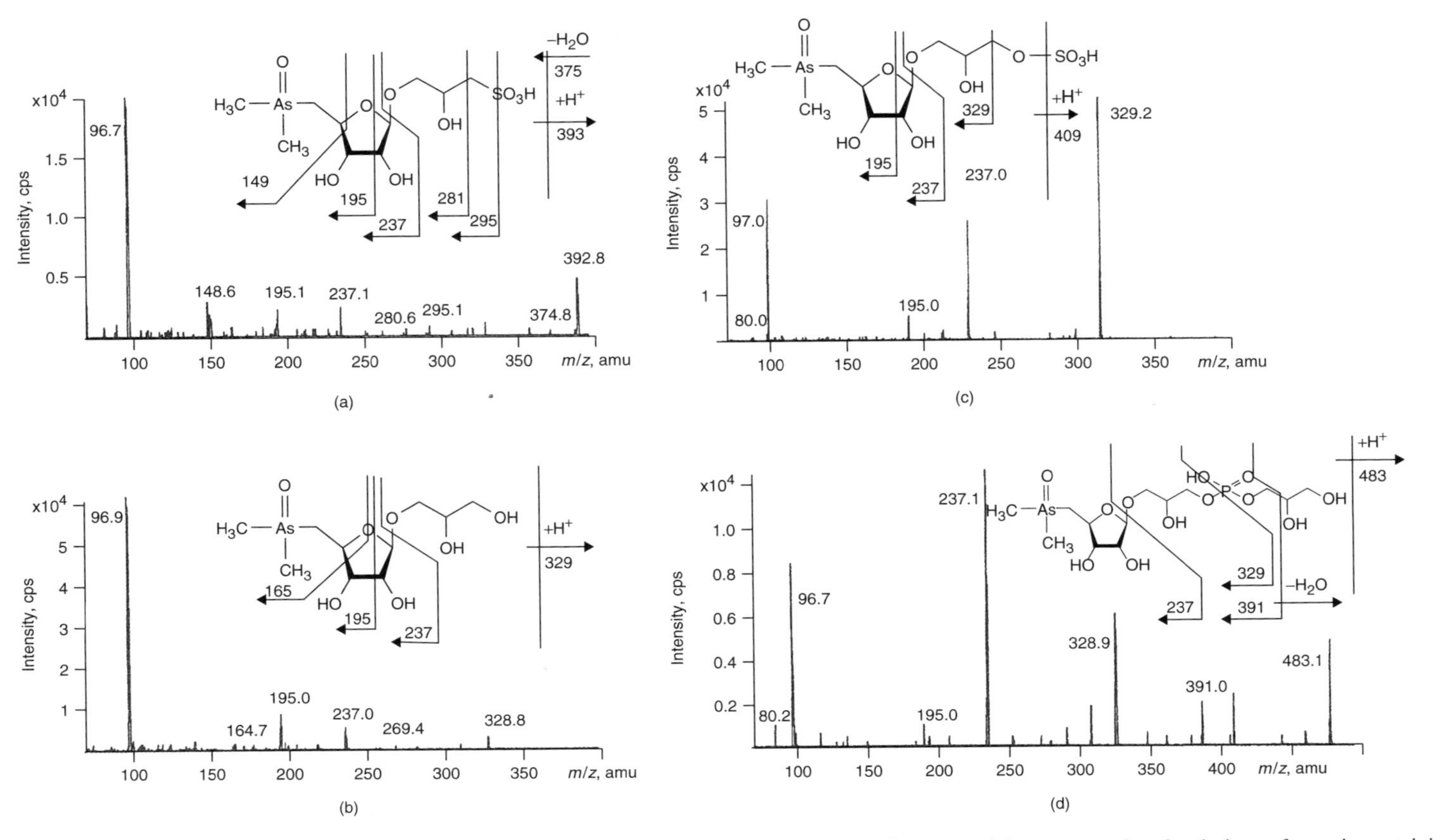

Figure 5.6.9. Identification of arsenic compounds in algal extracts by ES quadrupole MS/MS. MS/MS spectra of the protonated molecule ions of arsenic-containing ribofuranosides (arsenosugars) extracted from algae and purified by complementary chromatographic techniques. (a) Arsenosugar A; (b) arsenosugar B; (c) arsenosugar C; (d) arsenosugar D. Reprinted from *Analytica Chimica Acta*, Vol. 410, S. McSheehy, M. Marcineck, H. Chassaigne, and J. Szpunar, p. 71, Copyright (2000), with permission from Elsevier Science.

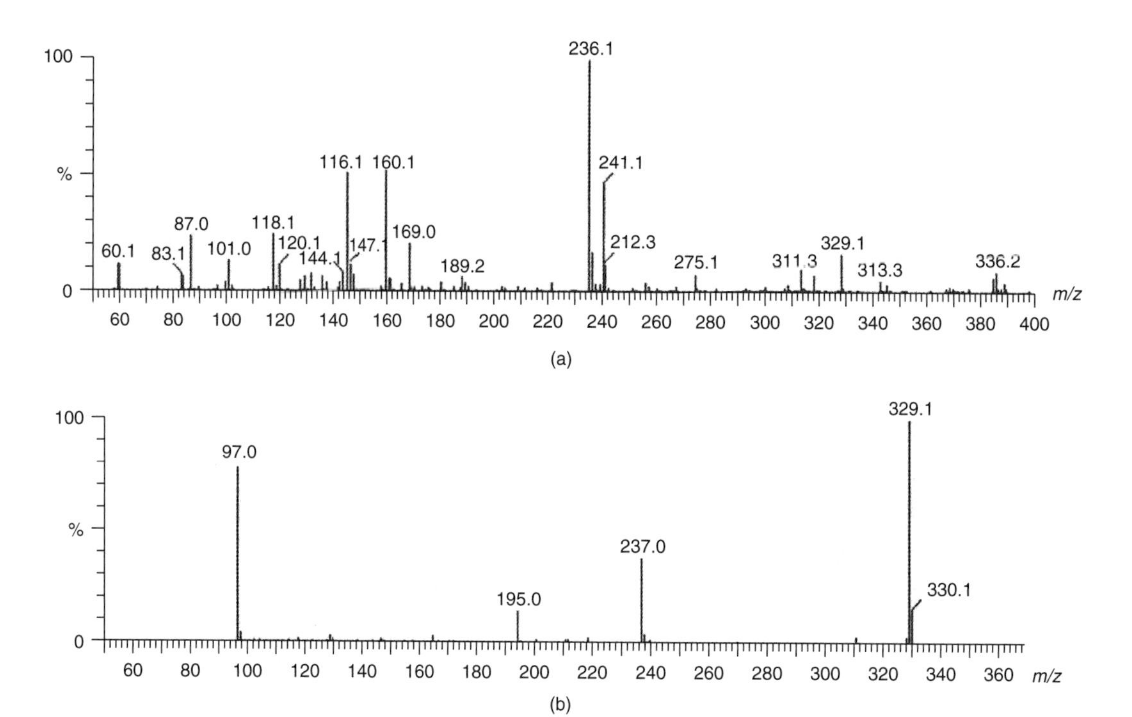

Figure 5.6.10. Identification of arsenic compounds in algae extracts by ES quadrupole TOF MS/MS. (a) MS spectrum of an extract roughly purified by cation exchange preparative chromatography. (b) ES MS/MS spectrum of the protonated molecule ion of the arsenic-containing ion at *m/z* 329.1. The fragmentation pattern allows the confirmation of the identity of the compound as arsenosugar B. Reproduced by permission of The Royal Society of Chemistry.

reactions, and structural function in nucleic acid metabolism. However the same metals present at high concentrations, and even low concentrations of the more potent ions of Cd, Hg, etc., are strongly poisonous, resulting in growth inhibition and death of the organism [49]. In order to survive plants must have developed efficient and specific mechanisms by which heavy metals are taken up and transformed into a physiologically tolerable form, providing the essential elements for the plant's metabolic function [49]. Some studies have revealed that the majority of higher plant species detoxify the metals by chelating them to peptides of the family of phytochelatins [50].

Phytochelatins (PC) are metal-binding peptides which are enzymatically synthesized from glutathione (GSH). These peptides have been shown to be induced in plants by various metals such as Cd, Cu and several other metals [49, 51]. In addition to PCs which possess the typical $(\gamma\text{Glu-Cys})_n$-Gly ($n = 2$–11) sequence, several variant structures have also been detected which differ in the C-terminal amino acid and are called isophytochelatins (iso-PC) [52]. Homologues of glutathione with C-terminal linked Glu instead of Gly and isophytochelatins (Glu) (iso-PC(Glu), $(\gamma\text{Glu-Cys})_n$-Glu, $n = 2$–3) were recently isolated from maize plants exposed to cadmium [53, 54]. In addition, desglycine phytochelatins (desGly-PC, $(\gamma\text{Glu-Cys})_n$), which lack the C-terminal amino acid residue, were first discovered in maize [52].

The classical approach used for the determination of PC extract from plant is based on reversed-phase HPLC with on-line derivatization of the sulfhydryl groups with Ellman's reagent (DTNB) and spectrophotometric detection. Reversed-phase chromatography of Cd-exposed and control maize root extract was used for isolation and purification of phytochelatins and phytochelatin-related peptides (chromatograms not shown) [55]. In the

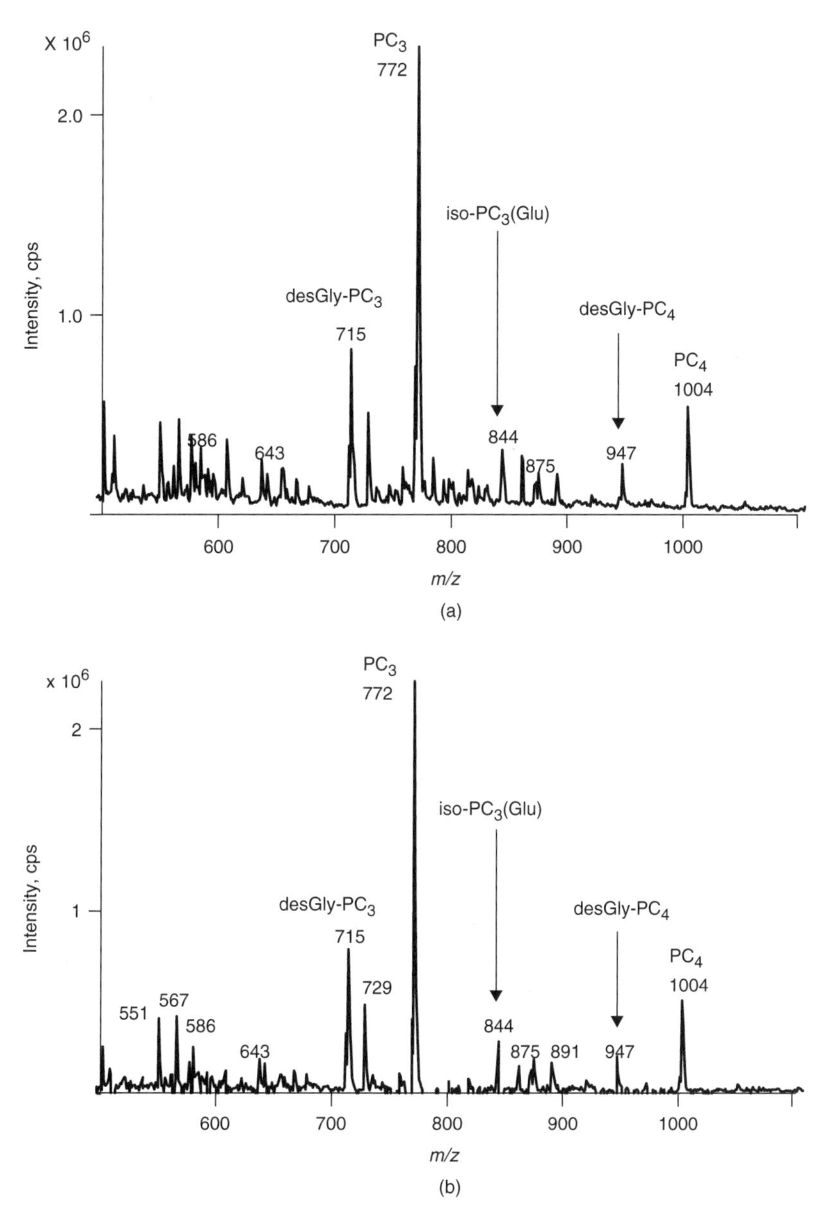

Figure 5.6.11. Identification of Cd-induced phytochelatins in plant root extracts (maize) by ES quadrupole MS. (a) MS spectrum of a fraction collected in reversed-phase HPLC – UV of Cd-exposed plant root extracts. (b) MS spectrum after substraction of the control (blank). Reprinted from *Phytochemistry*, Vol. 56, H. Chassaigne, V. Vacchina, T. M. Kutchan, and M. H. Zenk, p. 657, Copyright (2001), with permission from Elsevier Science.

ES MS spectrum of the chromatographic fraction shown in Figure 5.6.11, several minor peaks are observed that can be attributed to peptides which may be naturally present in plants or to Cd-induced peptides. The peaks at m/z 715, 772, 947 and 1004 may correspond to the molecular ions of desGly-PC_3, PC_3, desGly-PC_4 and PC_4, respectively. The spectrum also shows the presence of the peak at m/z 844 that is attributed to the molecular ion of iso-PC_3(Glu). The sequence of this last form

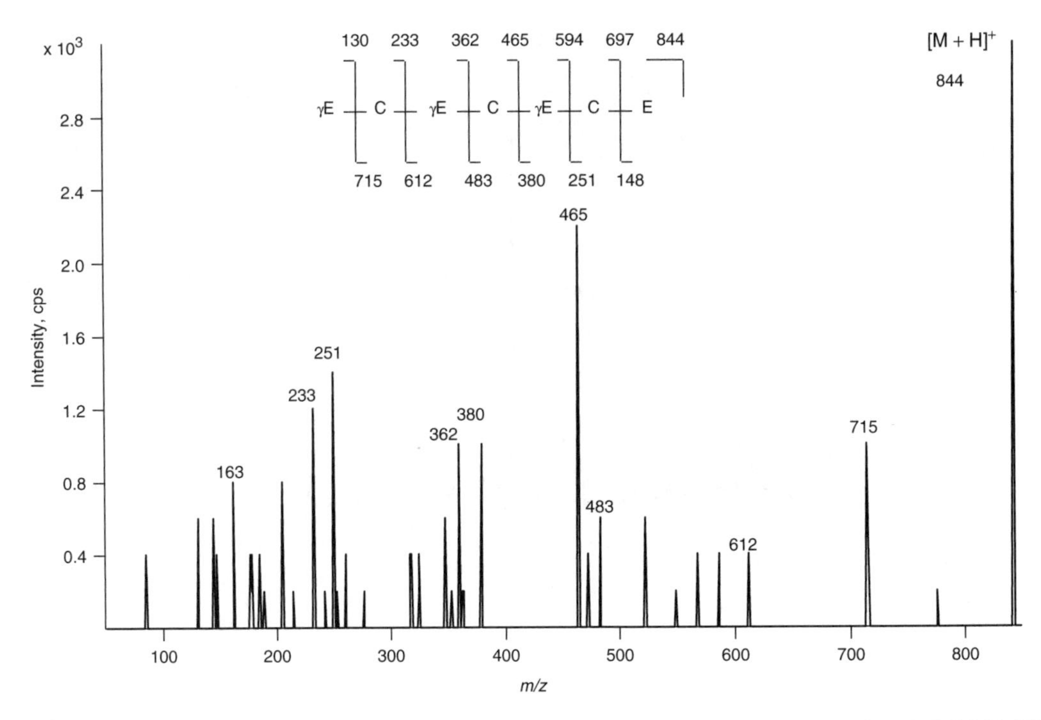

Figure 5.6.12. Identification of Cd-induced phytochelatins in plant root extracts (maize) by ES quadrupole MS/MS. MS/MS spectrum of a peak corresponding to a PC-related peptide detected in the MS spectrum at *m/z* 844 (cf. Figure 5.6.11). The fragmentation pattern allows the identification of the compound, as iso-phytochelatin, iso-PC_3(Glu). Reprinted from *Phytochemistry*, Vol. 56, H. Chassaigne, V. Vacchina, T. M. Kutchan, and M. H. Zenk, p. 657, Copyright (2001), with permission of Elsevier Science [55].

was confirmed by fragmenting the molecular ion in the CID mode (Figure 5.6.12). The complete series of y-type ions was observed and a good sensitivity was obtained with the triple quadrupole instrument [55]. However, ES Q-TOF with higher mass accuracy and resolution should represent the future in peptide and oligopeptides research and especially in the speciation field, facilitating recognition of unknown compounds in crude or partially purified extracts.

4.4 Cadmium complexes with metallothioneins in animal tissues

Characterization of macromolecules involved in the sequestration of heavy metals as well as in the metalloregulation in animals has been attracting considerable interest. Metallothioneins (MTs) are cysteine-rich, metal-binding proteins involved in the detoxification of metals and metabolism of essential metals, and are found in various organs of animals [56]. Certain MTs exist as different isoforms coded from multiple genes whose individual expression can only be determined by fastidious techniques (RNA assay using specific DNA). The methodology developed for analysis of translated proteins is also a challenge since sequence variation between MT isoforms encountered in mammalian species can vary from one to a few amino acids (in 60–62 amino acids).

Figures 5.6.13(a) and (b) show a typical reversed-phase chromatogram of the rabbit liver MT-2 isoform sample (commercially available) obtained with on-line inductively coupled plasma MS and ES MS detection, respectively. The identification on the basis of the retention time in HPLC is impossible because of the unavailability of standards of particular protein isoforms with sufficient purity. The major peak is split in two,

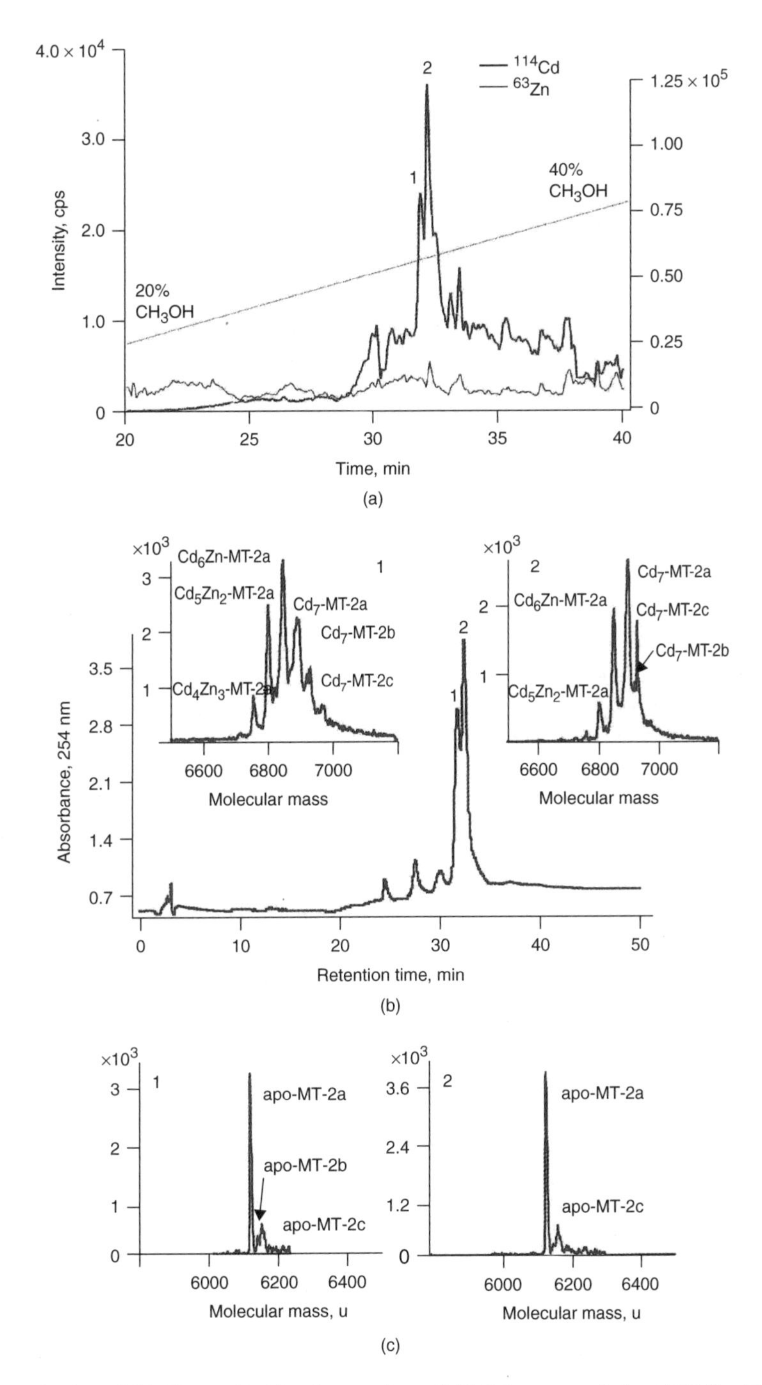

Figure 5.6.13. Characterization of Cd, Zn-metallothionein complexes (MT-2) by reversed-phase HPLC with ES quadrupole MS detection. (a) HPLC – ICP MS chromatogram with Cd and Zn specific detection of the MT-2 sample. (b) HPLC – ES MS chromatogram under the same chromatographic conditions. Reconstructed mass spectra taken at the maxima of the peaks 1 and 2 are shown in the insets. (c) Reconstructed mass spectra at the maxima of the peaks 1 and 2 after acidification of the column effluent. Reprinted from *Trends in Analytical Chemistry*, Vol. 19, H. Chassaigne, V. Vacchina, and R. Lobinski, p. 300, copyright (2000), with permission of Elsevier Science [18].

which might suggest the existence of two sub-isoforms (which differ by one or a few amino acids in their composition) of the MT-2 isoform. The mass spectra taken at the maximum of peaks 1 and 2 (spectra in the insets to Figure 5.6.13(b)) give more information on the identity of the eluted compounds. Since the separation is carried out at pH 7.0 to ensure the stability of metal–MT complexes, several artefacts due to the formation of mixed metal complexes (Cd, Cd–Zn) coexist and make the interpretation of the spectra difficult.

This problem can be solved by a post-column acidification whose purpose is to remove the metals from the metallothionein complexes and to detect the apo-MT [57]. Figure 5.6.13(c) (after post-column acidification) confirms the complexity of the mass spectra of peaks 1 and 2 caused by the presence of different metallo species that eclipse the occurrence of species with different amino acid composition. It is clearly seen that the two major peaks in the spectra belong to the MT-2 and MT-2c sub-isoforms, with respective masses of 6126.0 and 6150.0 Da. The difference between the mass spectra of peaks of metallo and epo-metallo proteins allows the determination of the stoichiometry of the separated compounds (see spectra in the insets to Figure 5.6.13b).

5 CONCLUSION

The main advantages of ES and related ionization techniques in elemental speciation analysis are the following:

- soft ionization method;
- suitable for analyzing large biomolecules;
- suitable for analyzing polar and even ionic compounds (e.g. metal complexes);
- enables coupling of MS/MS and HPLC.

The latest developments of the technology concern the better sensitivity obtained and the lower vulnerability of the ionization source to the matrix composition or the eluent composition in HPLC. New pneumatically assisted ES combined with a dual orthogonal sampling technique allow contamination avoidance and improve the sensitivity obtained in MS. Additionally, the ionization and desolvatation processes can be supported by a heated probe for analysis at high flow rates and to achieve a high sensitivity. The microelectrospray and nanoelectrospray techniques extend the capacities of the ionization technique for the analysis of very small quantities of compounds and for nanoscale LC separations (capillary HPLC).

ES MS/MS allows a precise determination of the molecular mass and the elucidation of the molecular structure of chemical species, and for this reason is turning out to be a necessary tool for the identification and characterization of unknown metal- and metalloid-containing compounds. In this chapter we have highlighted the performances and the limitations in terms of sensitivity, mass measurement precision and resolution of a triple quadrupole mass spectrometer preventing the identification of new species. This study also evaluates the potential of an ES Q-TOF mass spectrometer for the investigation of new compounds in speciation analysis. ES TOF MS with higher resolution and mass accuracy allows the detection of new species in a complex mixture without the need for fastidious purification. When it is used in tandem Q-TOF mode, novel information regarding the identity of compounds is obtained.

6 ACKNOWLEDGEMENTS

H. Chassaigne acknowledges a postdoctoral fellowship from the European Commission–Joint Research Centre.

7 REFERENCES

1. Cornelis, R., De Kimpe, J. and Zhang, X., *Spectrochim. Acta*, **53**, 187 (1998).
2. Lobinski, R., *Fresenius' J. Anal. Chem.*, **369**, 113 (2001).
3. Bettmer, J., *Anal Bioanal Chem*, **372**, 33 (2002).
4. Sanz Medel, A., *Spectrochim. Acta*, **53**, 197 (1998).
5. Haskins, W. E., Wang, Z., Watson, C. J., Rostand, R. R., Witowski, S. R., Powell, D. H. and Kennedy, R. T., *Anal. Chem.*, **73**, 5005 (2001).
6. Taylor, A., Branch, S., Halls, D. J., Owen, L. M. W. and White, M., *J. Anal. At. Spectrom.*, **15**, 451 (2000).
7. Taylor, A., Branch, S., Fisher, A., Halls, D. J. and White, M., *J. Anal. At. Spectrom.*, **16**, 421 (2001).
8. Templeton, D. M., *Fresenius' J. Anal. Chem.*, **363**, 505 (2001).

9. Yamashita, M. and Fenn, J. B., *J. Phys. Chem.*, **88**, 4451 (1984).
10. Fenn, J. B., Mann, M., Meng, C. K., Wong, S. F. and Whitehouse, C. M., *Mass Spectrom. Rev.*, **9**, 37 (1990).
11. Kebarle, P. and Tang, L., *Anal. Chem.*, **65**, 972A (1993).
12. Wilm, M. S. and Mann, M., *Int. J. of Mass Spectrom. Ion Process.*, **136**, 167 (1994).
13. Emmett, M. R. and Caprioli, R. M., *J. Am. Soc. Mass Spectrom.*, **5**, 605 (1994).
14. Pergantis, S. A., Wangkarn, S., Francesconi, K. A. and Thomas-Oates, J. E., *Anal. Chem.*, **72**(2), 357 (2000).
15. Wachs, T., Conboy, J. C., Garcia, F. and Henion, J. D., *J. Chromatogr. A*, **29**, 357 (1991).
16. Niessen, W. M. A. and Tinke, A. P., *J. Chromatogr. A*, **703**, 357 (1995).
17. Whitehouse, C. M., Dreyer, R. N., Yamashita, M. and Fenn, J. B., *Anal. Chem.*, **57**, 675 (1985).
18. Chassaigne, H., Vacchina, V. and Lobinski, R., *Trends Anal. Chem.*, **19**, 300 (2000).
19. Fenn, J. B., Mann, M., Meng, C. K., Wong, S. F. and Whitehouse, C. M., *Science*, **246**, 64 (1989).
20. Smith, R. D., Loo, J. A., Ogorzalek-Loo, R. R., Busman, M. and Udseth, H. R., *Mass Spectrom. Rev.*, **10**, 359 (1991).
21. Chassaigne, H. and Lobinski, R., *Fresenius' J. Anal. Chem.*, **361**, 267 (1998).
22. Chassaigne, H. and Lobinski, R., *Anal. Chem.*, **70**, 2536 (1998).
23. Kellner, R., Mermet, J. M., Otto, M. and Widmer, H. M., *Analytical Chemistry*, Wiley-VCH, Weinheim, 1998.
24. McSheehy, S., Szpunar, J., Haldys, V. and Tortajada, J., *J. Anal. At. Spectrom.*, **17**, 507 (2002).
25. McSheehy, S., Marcinek, M., Chassaigne, H. and Szpunar, J., *Anal. Chim. Acta*, **410**, 71 (2000).
26. Yates, J. R. III, McCormack, A. L., Link, A. J., Schieltz, D., Eng, J. and Hays, L., *Analyst*, **121**, 65R (1996).
27. LeBlanc, J. C. Y., *J. Anal. At. Spectrom.*, **12**, 525 (1997).
28. Madsen, A. D., Goessler, W., Pedersen, S. N. and Francesconi, K. A., *J. Anal. At. Spectrom.*, **15**, 657 (2000).
29. Lindemann, T. and Hintelmann, H., *Anal. Bioanal. Chem.*, **372**, 486 (2002).
30. Casiot, C., Vacchina, V., Chassaigne, H., Szpunar, J., Potin-Gautier, M. and Lobinski, R., *Anal. Commun.*, **36**, 77 (1999).
31. McSheehy, S., Pohl, P., Szpunar, J., Potin-Gautier, M. and Lobinski, R., *J. Anal. At. Spectrom.*, **16**, 68 (2001).
32. Pickford, R., Miguens-Rodriguez, M., Afzaal, S., Speir, P., Pergantis, S. A. and Thomas-Oates, J. E., *J. Anal. At. Spectrom.*, **17**, 173 (2002).
33. Clark, L. C., Combs, G. F., Turnbull, S. W., Slate, E. H., Chalker, D. K., Chow, J., Davis, L. S., Glover, R. A., Graham, G. F. and Gross, E. G., *J. Am. Med. Ass.*, **276**, 1957 (1996).
34. Ip, C., Birringer, M., Block, E., Kotrebai, M., Tyson, J. F., Uden, P. C. and Lisk, D. J., *J. Agric. Food Chem.*, **48**, 2062 (2000).
35. Neve, J., *J. Cardiovasc. Risk*, **3**, 42 (1996).
36. Kotrebai, M., Birringer, M., Tyson, J. F., Block, E. and Uden, P. C., *Anal. Commun.*, **36**, 249 (1999).
37. Templeton, D. M., Ariese, F., Cornelis, R., Danielsson, L. G., Muntau, H., Van Leeuwen, H. P. and Lobinski, R., *Pure Appl. Chem.*, **72**, 1453 (2000).
38. Chassaigne, H., Chery, C. C., Bordin, G. and Rodriguez, A. R., *J. Chromatogr. A*, **976**, 409 (2002).
39. McSheehy, S., Pannier, F., Szpunar, J., Potin-Gautier, M. and Lobinski, R., *Analyst*, **127**, 223 (2002).
40. Larsen, E. H., Hansen, M., Fan, T. and Vahl, M., *J. Anal. At. Spectrom.*, **16**, 1403 (2001).
41. Casiot, C., Szpunar, J., Lobinski, R. and Potin-Gautier, M., *J. Anal. At. Spectrom.*, **14**, 645 (1999).
42. Shibata, Y., Morita, M. and Fuwa, K., *Adv. Biophys.*, **28**, 31 (1992).
43. Francesconi, K. A. and Edmonds, J. S., *Oceanogr. Mar. Biol. Annu. Rev.*, **31**, 111 (1993).
44. Le, X. C., Cullen, W. R. and Reimer, K. J., *Clin. Chem.*, **40**, 617 (1994).
45. Francesconi, K. A., Edmonds, J. S. and Stick, R. V., *J. Chem. Soc. Perkin Trans.*, **1**, 1349 (1992).
46. Corr, J. J. and Larsen, E., *J. Anal. At. Spectrom.*, **11**, 1215 (1996).
47. McSheehy, S. and Szpunar, J., *J. Anal. At. Spectrom.*, **15**, 79 (2000).
48. McSheehy, S., Pohl, P., Lobinski, R. and Szpunar, J., *Analyst*, **126**, 1055 (2001).
49. Zenk, M. H., *Gene*, **179**, 21 (1996).
50. Gekeler, W., Grill, E., Winnacker, E. L. and Zenk, M. H., *Z. Naturforsch.*, **44**, 361 (1989).
51. Rauser, W. E., *Plant Physiol.*, **109**, 1141 (1995).
52. Grill, E., Winnacker, E. L. and Zenk, M. H., *FEBS Lett.*, **197**, 115 (1986).
53. Meuwly, P., Thibault, P. and Rauser, W. E., *FEBS Lett.*, **336**, 472 (1993).
54. Meuwly, P., Thibault, P., Schwan, A. L. and Rauser, W. E., *Plant J.*, **7**, 391 (1995).
55. Chassaigne, H., Vacchina, V., Kutchan, T. M. and Zenk, M. H., *Phytochemistry*, **56**, 657 (2001).
56. Stillman, M. J., Shaw, C. F. and Suzuki, K. T., *Metallothionein Synthesis, Structure and Properties of Metallothioneins, Phytochelatins and Metalthiolate Complexes*, VCH, New York, 1992.
57. Chassaigne, H. and Lobinski, R., *J. Chromatogr. A*, **829**, 127 (1998).

5.7 Elemental Speciation by Inductively Coupled Plasma-Mass Spectrometry with High Resolution Instruments

R. S. Houk

Iowa State University, Ames, IA, USA

Handbook of Elemental Speciation: Techniques and Methodology R. Cornelis, H. Crews, J. Caruso and K. Heumann
 ISBN: 0-471-49214-0

1 INTRODUCTION

This chapter provides information pertinent to use of high resolution mass spectrometers with an ICP source. The use of these instruments for analysis of biological materials in general has been reviewed recently [1]. The magnetic sector is by far the most widely used instrument for high resolution measurements with the ICP, and most of the chapter deals with these devices. The terms 'high resolution,' 'sector field,' and 'magnetic sector' have become synonymous in the ICP lexicon, although there are other ways to achieve high resolution, and sectors are used at low resolution where possible. Compared to quadrupole instruments, scanning magnetic sector instruments offer the following advantages: (a) higher sensitivity[1] at unit mass resolution, (b) sufficient resolution to separate many chemically different ions at the same nominal m/z value, e.g., $^{56}Fe^+$ ($m/z = 55.9349$) from $^{40}Ar^{16}O^+$ ($m/z = 55.9567$)[2], (c) the ability to identify an ion conclusively by accurate m/z measurements[3], and (d) lower instrument background (1 count s^{-1} or less). Whether the higher sensitivity and lower background translate into better detection limits depends on the blank signals for the element(s) of interest. In the author's opinion, the ability to conclusively identify the atomic analyte ion based on high resolution and accurate m/z measurements makes sector instruments particularly attractive for speciation measurements. The potential for better detection limits is also invaluable, because the analyte compounds are typically diluted during the chemical separation.

The other general type of magnetic sector instrument, the multicollector, is also discussed in this chapter. These instruments are capable of highly precise isotope ratio measurements. Usually, the magnetic field is fixed, and ions of different m/z ratios are measured in separate detectors. Noise from the ICP is cancelled when the ratio of signals is measured, so these devices should provide high ratio precision for analytes in the transient chromatographic peaks observed in speciation studies [2, 3].

Thus, both general types of magnetic sector instrument have potential uses in speciation. A concise description of the operating principles of magnetic sector instruments is given, followed by a summary of their value and limitations when used with on-line chemical separations, the usual mode of providing speciation information with ICP-MS. Particular ways to enhance the performance of sector instruments, such as solvent removal and reduction of the plasma potential, and representative applications are also summarized.

2 MASS ANALYSIS WITH MAGNETIC AND ELECTROSTATIC SECTORS

Much of the discussion in this section follows that in Roboz [4]. More advanced treatments can be found in other references [5, 6].

2.1 Mass analysis in radial magnetic field

Figure 5.7.1 shows a positive ion of mass m and charge z moving with velocity $\mathbf{v}$ in a vacuum through a uniform magnetic field of strength $\boldsymbol{B}$. The magnetic field is oriented perpendicular to the plane of the paper; one magnet pole piece is above the plane of the figure, the other pole piece is below. Movement of the ion charge through the magnetic field generates a magnetic force $\boldsymbol{F}_m$ that acts in a direction perpendicular to the direction of $\mathbf{v}$. The magnetic force thus deflects the ion in a curved path. As the ion moves further through the magnetic field, the orientation of $\mathbf{v}$ changes, and $\boldsymbol{F}_m$ always acts perpendicular to the direction of $\mathbf{v}$ at each position in the field. Thus, the ion continues to move in a curved path as long as it is in the magnetic field. The magnitude of the ion velocity remains the same, but the direction of $\mathbf{v}$

[1] Sensitivity refers to analyte signal per unit concentration, i.e., the slope of a calibration curve. The minimum detectable amount of analyte is more properly referred to as the detection limit.

[2] Throughout this chapter, the m/z values cited for a singly charged ion have had the mass of one electron removed from the sum of the masses of the various atoms.

[3] The differences in the binding energies of the nuclei ^{56}Fe, ^{40}Ar and ^{16}O and the numbers of protons, neutrons and electrons in each atom lead to slightly different m/z values for $^{56}Fe^+$ and $^{40}Ar^{16}O^+$. Tables of accurate mass values are available from various handbooks and manufacturers.

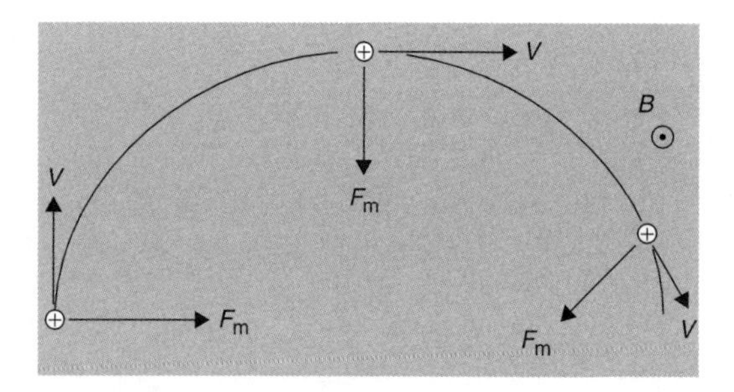

Figure 5.7.1. Deflection of positive ion moving at velocity **v** through a region of uniform magnetic field, strength ***B***. The magnetic field lines are oriented into the plane of the page. The magnetic force $\boldsymbol{F}_{\mathrm{m}}$ acts perpendicular to the direction of **v**.

changes, so the magnetic interaction does induce acceleration and represents a force in that sense.

The ion undergoes uniform circular motion such that the radius of curvature r_{m} balances the magnetic force at the applied field strength

$$\boldsymbol{F}_{\mathrm{m}} = \boldsymbol{B} ze\mathbf{v} = mv^2/r_{\mathrm{m}} \qquad (5.7.1)$$

To induce the ion to travel from the source into the magnetic field, it is usually accelerated through a potential difference V, which gives it a kinetic energy

$$KE = \tfrac{1}{2}mv^2 = ze\mathbf{v} \qquad (5.7.2)$$

where e = electron charge. Note the important difference between capital V for accelerating voltage and lower case **v** for velocity.

These two equations can be combined to give the following general form for ion motion in a magnetic field

$$m/z = B^2r_{\mathrm{m}}^2/2V \qquad (5.7.3)$$

where the constant e has been combined with z and will be accounted for shortly.

Some comment about units is also in order. The standard unit for B is Tesla (T); $1\,\mathrm{T} = 10^4\,\mathrm{G} = 10^8$ lines of force $\mathrm{m}^{-2} = 1$ weber m^{-2}, where G = gauss. The symbol B is sometimes also referred to as magnetic flux density. Some older literature uses the symbol H instead of B.

A convenient reduced form of equation (5.7.3) is as follows

$$m/z = 4.83 \times 10^{-5} B^2r_{\mathrm{m}}^2/V \qquad (5.7.4)$$

where m/z is measured relative to $^{12}C = 12.00000\ldots$, $z = +1, +2, -1, -2\ldots$, r_{m} is in cm, B is in gauss, and V is in volts. A singly charged ion ($z = 1$) at $m = 100$ in a magnetic field $B = 10^3$ gauss with $V = 2000\,\mathrm{V}$ moves with a radius of curvature of 64 cm. To separate and detect such an ion, the flight tube through the magnetic field is constructed with this value of r_{m}.

Equations (5.7.2) and (5.7.3) also show that there are several ways to obtain a spectrum:

(1) Scan B while holding r_{m} and V constant.
(2) Scan V while holding r_{m} and B constant.
(3) Keep B and V constant and detect ions of various m/z values at different values of r_{m} with a position-sensitive detector.

Generally, a particular instrument designed to change m/z value by either methods (1) or (2) is used for scanning applications, while one meant to measure different ions at different radial positions is referred to as a multicollector. Equation (5.7.3) also shows that the m/z scale is not linear in either B, V, or r_{m}. Peaks for heavier ions are spaced more closely. The instrument software usually displays the spectrum as if the m/z scale were linear, however.

2.2 Kinetic energy analysis in radial electric field

Equations (5.7.2) and (5.7.3) show that a spread of ion kinetic energy (reflected in a spread of V values) causes a variety of m/z values to be observed at particular values of B and r_{m}. To improve mass resolution, a method for selecting ions with a narrow spread of kinetic energy is often used. A simplified schematic diagram of such an electrostatic analyzer is given in Figure 5.7.2. Again, an ion is accelerated through potential difference V into the gap between two curved metal plates. The electric field between the plates is V'/d, where d is the plate spacing. The ion experiences a radial electrostatic force F_{e} that is oriented toward the center of curvature of the plates. This force deflects the ion in a curved path

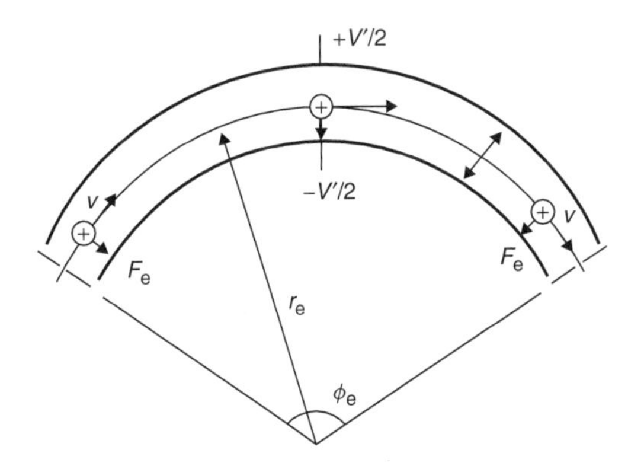

Figure 5.7.2. Deflection of ion moving inside a radial electrostatic condenser, gap thickness d, radius of curvature r_e, angle of deflection ϕ_e, radial electric field $E = V'/d$. The electric force $\boldsymbol{F}_e$ acts perpendicular to the direction of ion velocity v.

between the sector plates. Again, the condition for uniform circular motion is

$$\mathbf{F}_e = mv^2/r_e = ze\mathbf{E} \tag{5.7.5}$$

where $E =$ radial electric field $= V'/d$. This expression can be combined with equation (5.7.2) for the ion kinetic energy to derive the following general expression

$$r_e = 2V/E \tag{5.7.6}$$

Thus, ions of a given kinetic energy V (actually zeV) can be selected by adjusting E (i.e., the applied voltages V'). Mass does not appear in this expression, so an electrostatic analyzer such as that in Figure 5.7.2 does not provide a mass spectrum. It is typically used in combination with a magnetic analyzer to select ions of the desired range of kinetic energies to improve the resolution. Alternatively, a multipole collision cell can be used to reduce the spread of kinetic energy by collisional energy transfer, as discussed in more detail below.

2.3 Focusing properties of electrostatic and magnetic fields

Electrostatic and magnetic analyzers can tolerate ion beams of a finite width or angular divergence and still provide good performance. These devices

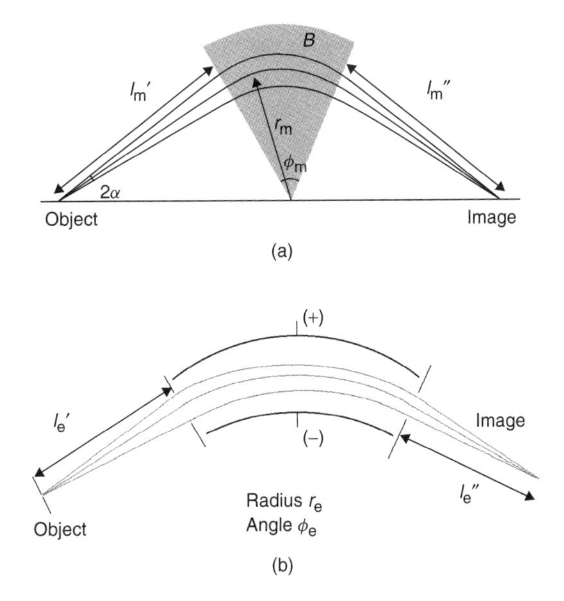

Figure 5.7.3. Focusing effect of magnetic sector (a, subscript m) and electrostatic sector (b, subscript e). The image distance is l' and the focused image is formed at l''. See also equations (5.7.7)–(5.7.12).

have focusing properties illustrated in Figure 5.7.3. For either the electrostatic analyzer (subscript e, Figure 5.7.3(b)) or the magnetic analyzer (subscript m, Figure 5.7.3(a)), ions injected directly on center travel along the path dictated by the equations described above. An ion injected above this central path travels a longer distance in the field and is deflected slightly more, whereas an ion injected below the central path is deflected less extensively. Such off-center ions can be focused to a point outside the field at positions l_e'' or l_m''. These focal positions are the usual positions for entrance or exit slits. In the simple cases (perpendicular ion entry, source, center of curvature and image all on same plane for magnetic analyzer) shown in Figure 5.7.3, their locations can be predicted from the following expressions:

$$f_e^2 = (l_e' - g_e)(l_e'' - g_e) \tag{5.7.7}$$

$$f_e = \frac{r_e}{\sqrt{2}\,\sin(\sqrt{2}\phi_e)} \tag{5.7.8}$$

$$g_e = \frac{r_e}{\sqrt{2}\,\tan(\sqrt{2}\phi_e)} \tag{5.7.9}$$

$$f_m^2 = (l'_m - g_m)(l''_m - g_m) \quad (5.7.10)$$

$$f_m = r_m / \sin \phi_m \quad (5.7.11)$$

$$g_m = r_m / \tan \phi_m \quad (5.7.12)$$

The focal properties of more complex geometries and corrections for the fringe fields (especially that of the magnet) can be determined [7].

These considerations govern the selection of parameters used in construction of the instrument, such as angle of deflection ϕ, radius of curvature r and slit positions l' and l''. For example, the intermediate slit between the electrostatic and magnetic analyzer is located at a position that provides a focused image for both, i.e., at the object position l'' for the first analyzer and the image position l' for the second.

2.4 Double focusing principle

The focusing properties of both the electrostatic and magnetic field can be used in combination to provide very narrow ion beams (i.e., high spectral resolution) without a disastrous loss of transmission. Velocity focusing occurs when ions of the same m/z value but different kinetic energy recombine at the same point. Curve g of Figure 5.7.4 represents the set of such velocity focusing points plotted together into a velocity focusing curve for ions of various m/z values. At the same time the broadening caused by the initial angular divergence in the ion beam is minimum at certain points on the direction focusing curve r.

The condition for double focusing is that the velocity dispersion suffered by the ions in the

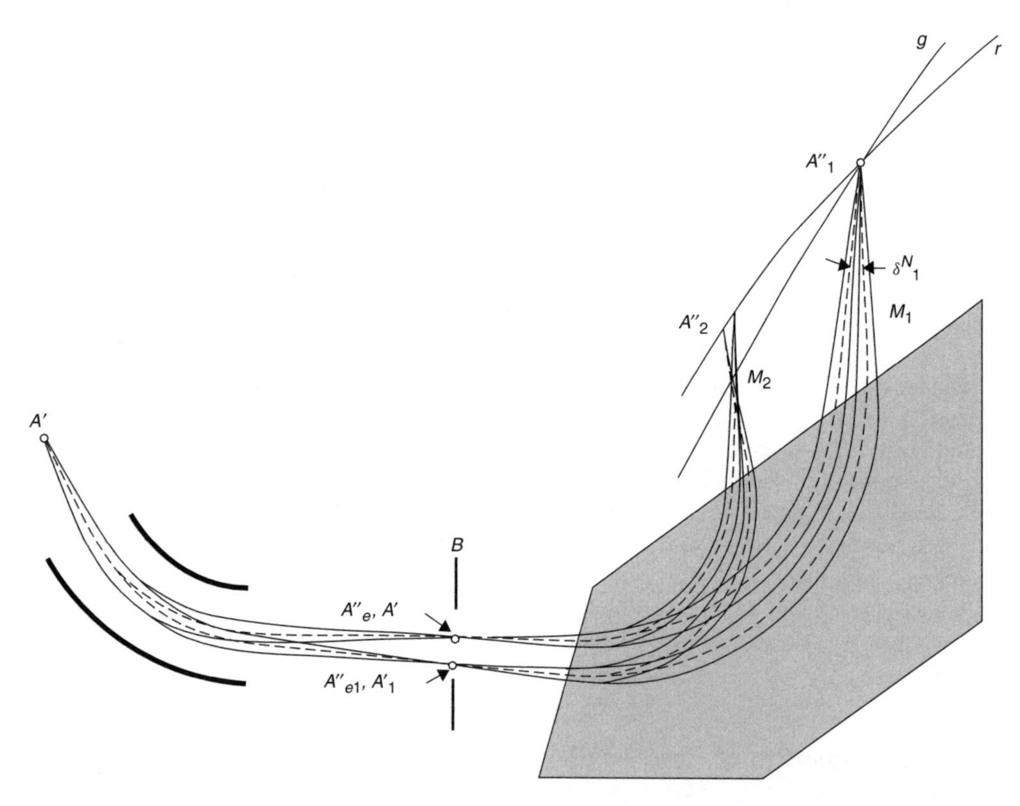

Figure 5.7.4. Double focusing mass analyzer. For ions of various m/z values, the direction focusing curve is r, where the angular aberration due to the finite angular width of the beam is minimum. The velocity focusing curve is g, where ions of the same m/z value but different kinetic energy dispersed by the electrostatic analyzer recombine. For the particular settings of accelerating voltage and magnetic field strength, ions of m/z value M_1 are separated from others with both velocity and direction focusing at point A''_1. Reproduced from ref. 4 with permission from Springer-Verlag.

electrostatic analyzer is exactly compensated by the magnetic field, while the direction-focusing of both fields and the mass dispersion of the magnetic field are maintained. For the instrument shown in Figure 5.7.4, this double focusing condition is met for ions of $m/z = M_1$ at point A_1'', where the two focusing curves intersect.

At the value of B that brings the ion beam for M_1 to the double focusing point, the lighter ion M_2 has a smaller radius of curvature r_m and leaves the magnetic sector at a position where the direction focusing curve diverges from the velocity focusing curve. The ion M_2 thus produces a broader image because it is not observed at the single position where the direction and velocity focusing compensate precisely. The values of V and/or B could be adjusted to bring the paths for ion M_2 to the double focusing point A_1'', which is one way to scan the m/z value transmitted by such instruments.

2.5 Slit widths and resolution

Resolution R is defined in the usual way as $R = m/\Delta m$, the m/z value measured divided by the separable mass difference. It can be defined with various values for the allowable valley between peaks, relative to the peak heights, i.e., 5 % valley, 10 % valley, 50 % valley, etc. The ratio $m/\Delta m$ is sometimes also called resolving power.

The resolution necessary to separate a given pair of ions can be estimated as follows. Suppose we wish to resolve $^{75}As^+$ ($m/z = 74.9210$) from $^{40}Ar^{35}Cl^+$ ($m/z = 74.9307$), a common problem ion. The resolution required is

$$R = m/\Delta m \sim 75/(74.9307 - 74.9210) = 7800$$

If the peaks were perfect triangles of equal height, they would be separated to baseline at this resolution setting. If $ArCl^+$ is much more abundant than As^+, substantially better resolution would be needed to account for tailing of the peaks. Note also that the atomic ion is the lightest one at $m/z = 75$, which is usually the case at m/z values below $m/z \sim 100$ [8].

The resolution is affected by factors such as the widths of the slits and ion beam, kinetic energy spread of the ion beam, and aberrations, i.e., imperfections due to fringe fields and other causes. Roboz [4] gives several general equations that illustrate the relation between these parameters:

$$R = \frac{1}{(S_1 + S_2)/r_m + \Delta V/V} = \frac{r_m}{S_1 + S_2 + \beta r_m} \quad (5.7.13)$$

where $\Delta V/V$ is the kinetic energy spread ΔV relative to the accelerating voltage V and β is a number that describes the magnitude of aberration. This equation shows that the resolution increases at smaller slit widths, smaller kinetic energy spread, and larger accelerating voltage.

For a double focusing instrument, it can be shown that the resolution is roughly related to the radius of curvature of the electrostatic analyzer [4]:

$$R \sim ar_e/S \quad (5.7.14)$$

where the entrance and exit slits are of equal width S and the proportionality constant a is approximately 1–2. Thus, an instrument with an electrostatic analyzer of radius 10 cm with 10 μm slits would be expected to have a resolution of the order of 10 000, roughly what is achieved with ICP-MS instruments.

The relation between slit width and peak shape is also important. If the slits are wide open (Figure 5.7.5 top), the beam is narrower than the slit. The resulting peak has an extended flat section, corresponding to transmission of the entire beam through the last analyzer and exit slit to the detector. Such a peak shape provides the best precision, as the signal at the top of the peak is not strongly affected by small variations in accelerating voltage or magnetic field strength. It is also straightforward to hop back and forth between the centers of such flat-topped peaks, in the same fashion that the voltages applied to a quadrupole can be varied to hop between peaks.

To achieve higher resolution, the slits are made narrower than the inherent beam width (Figure 5.7.5 bottom). Ideally, the peaks would be triangular, but there is typically some curvature on

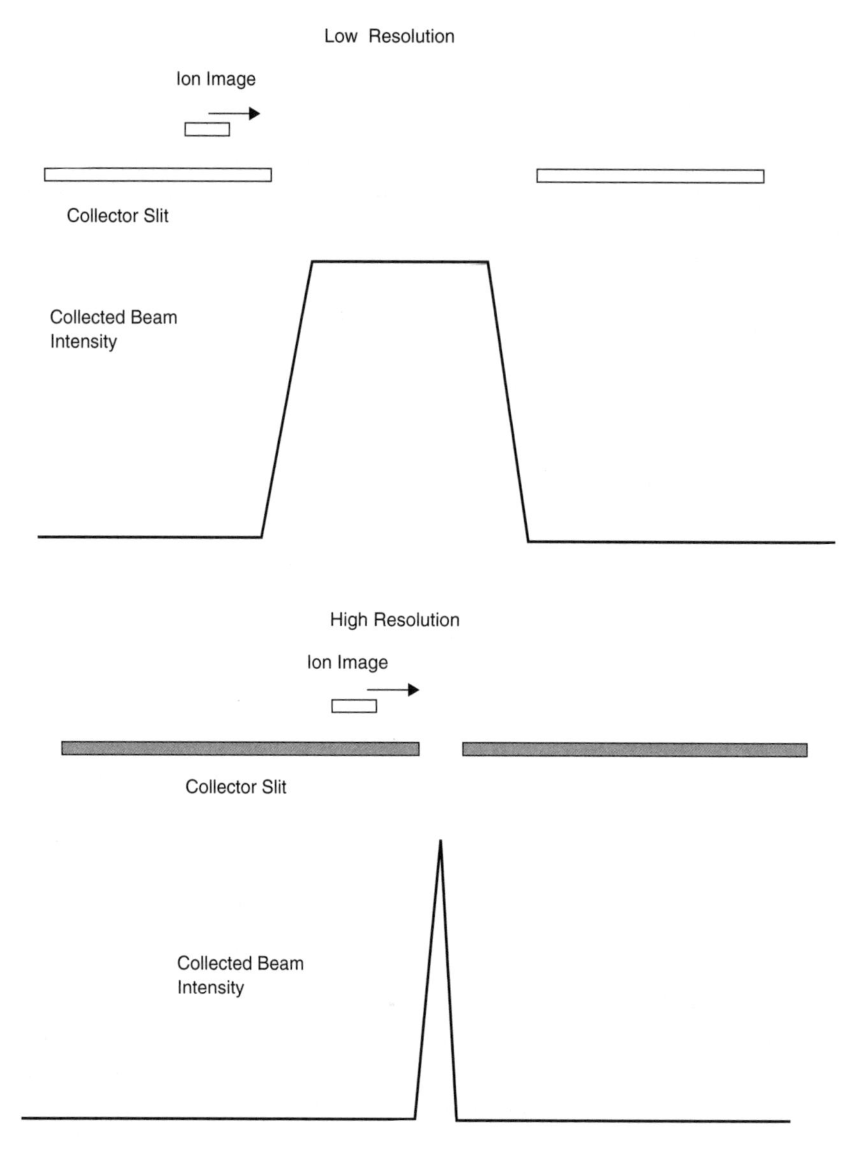

Figure 5.7.5. Relation between beam size, slit width and peak shape at low and high resolution settings. Courtesy of Thermo Elemental.

the sides. Naturally, both peak height and area are sacrificed to achieve higher resolution. It is also more difficult to control the accelerating voltage and magnetic field strength with sufficient accuracy to hop directly onto the sharp tip of a peak in high resolution. Traditionally, these instruments are scanned when operated in high resolution. Recent improvements in magnet stability allow the width of the scanned region to be only a narrow strip in the center of the peak, so that little time is spent elsewhere except near the center of the peak.

Examples of peaks actually measured on a Finnigan Element are shown in Figure 5.7.6. Note the different sensitivity scales. For a given element and isotope, the peak height decreases roughly tenfold when the resolution is changed from 300 to 4000 (slit widths ~20 μm), followed by another tenfold loss from 4000 to 8000. The slits are only

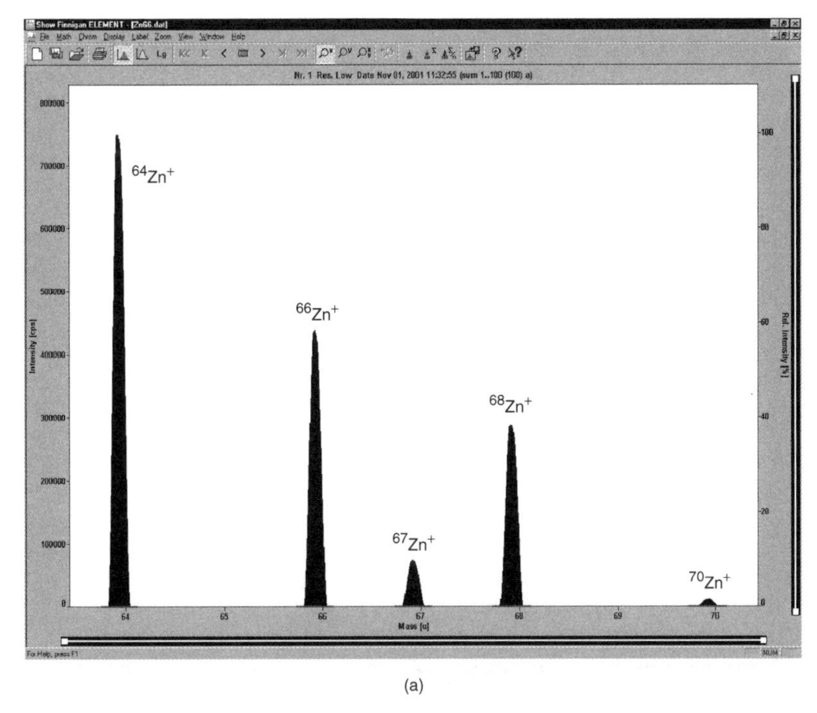

(a)

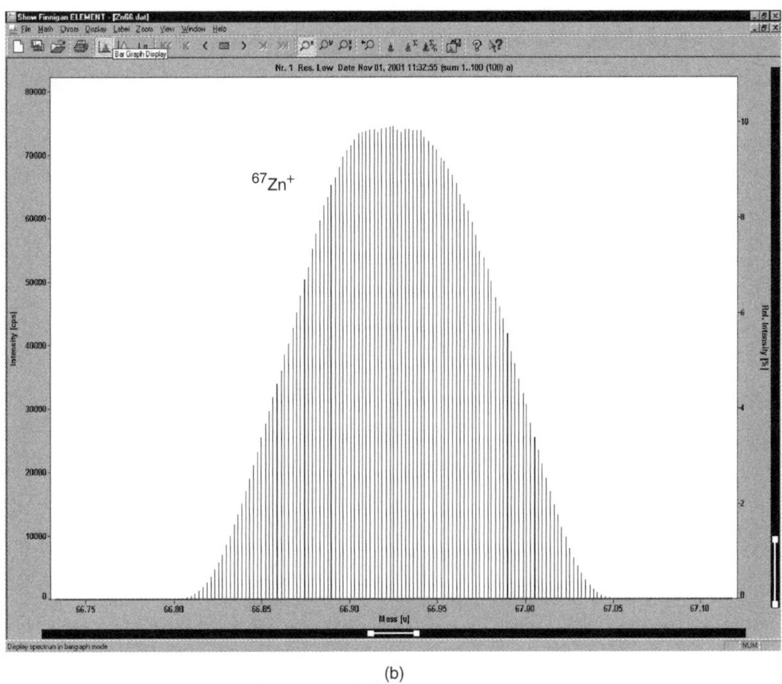

(b)

Figure 5.7.6. Actual peaks from Finnigan Element. (a) Zn^+, low resolution, $R \sim 300$, $10\,\mu g\,L^{-1}$ Zn. (b) Expanded view of $^{67}Zn^+$ showing flat top of peak. (c) Separation of V^+ from ClO^+, medium resolution, $R \sim 4000$, $10\,\mu g\,L^{-1}$V. (d) Separation of As^+ from $ArCl^+$, high resolution, $R \sim 10\,000$, 10 ppb As in 1 % aqueous HCl. The accurate m/z value for $^{75}As^+$ is 74.9216, very close to the peak centroid. (e) Separation of $^{78}Se^+$ from $^{38}Ar^{40}Ar^+$, high resolution, $R \sim 10\,000$, $10\,\mu g\,L^{-1}$ Se.

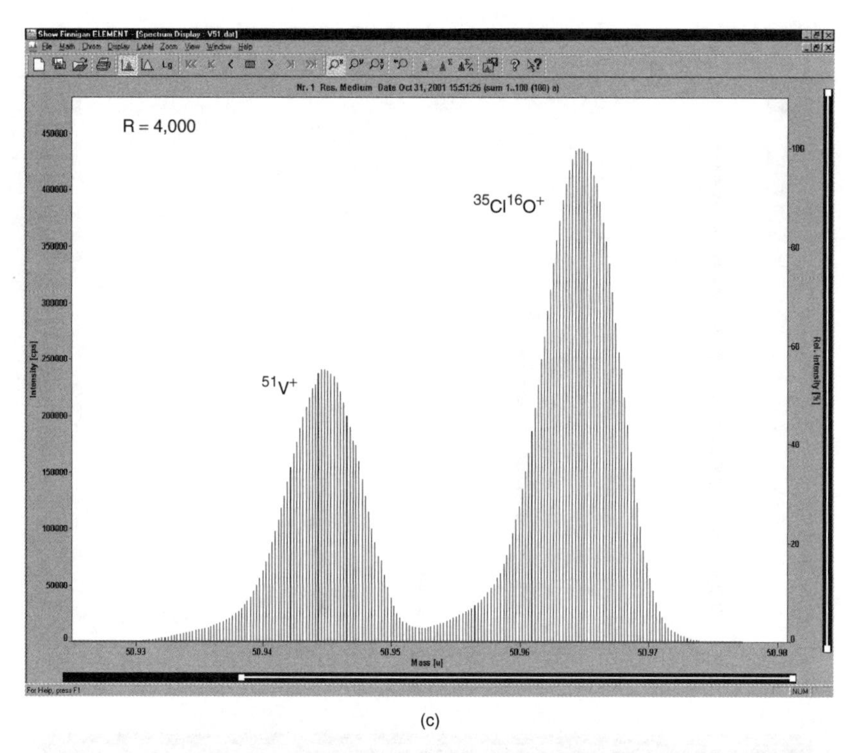

(c)

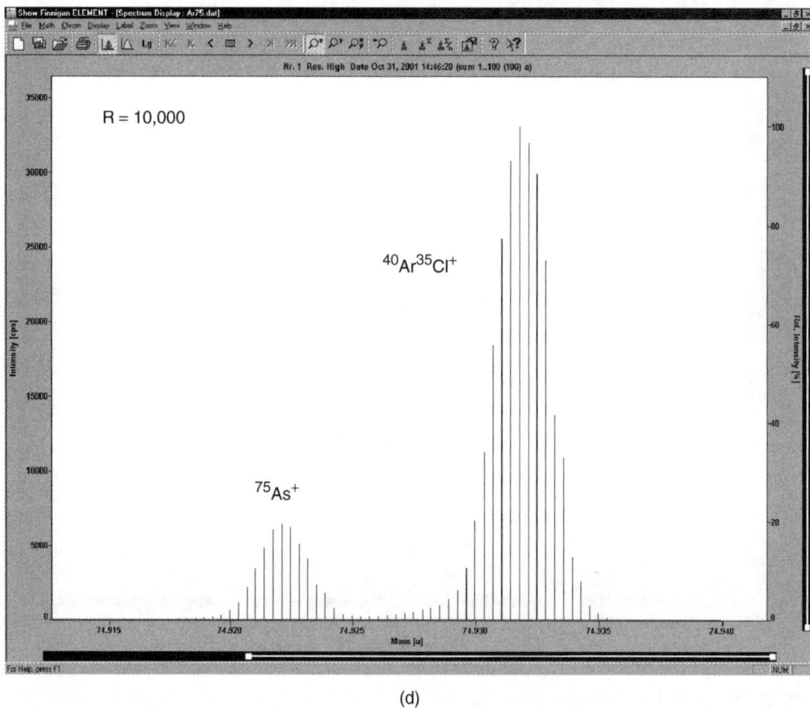

(d)

Figure 5.7.6. *(continued)*

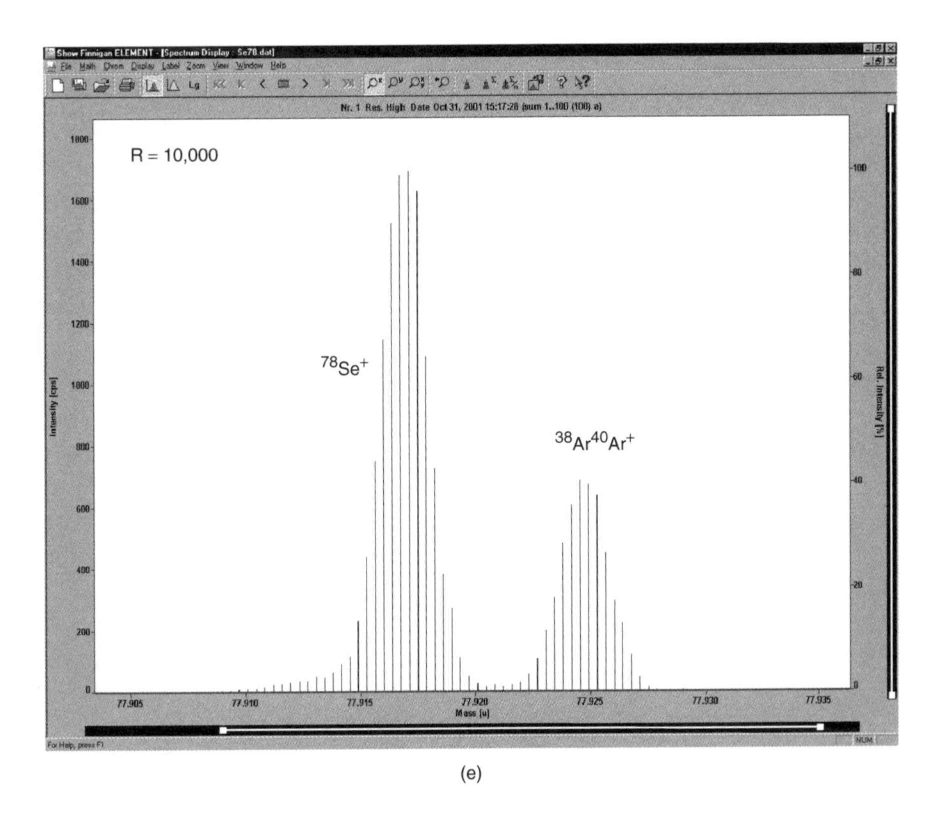

(e)

Figure 5.7.6. *(continued)*

10 μm wide at the high resolution setting, which places stringent requirements on their manufacture, mounting and physical condition and on the focusing conditions needed to put the ion beam onto the slit.

3 USE OF MAGNETIC SECTORS WITH IONS FROM AN ICP

3.1 Acceleration of ions

With a quadrupole instrument, the sampler and skimmer are typically grounded. The ions leave the skimmer with the kinetic energy gained from entrainment in the flow of argon, plus a contribution from the plasma potential. The kinetic energy possessed by an ion is given roughly by the plasma potential plus the difference in voltage between the region where it was formed and the region of interest [9]. For quadrupoles, the ion kinetic energy need be only a few electronvolts, whereas magnetic sectors perform best when the ions are accelerated to several thousand electronvolts. The method by which the ions are accelerated is therefore important.

In the initial ICP magnetic sector experiment by Bradshaw *et al.* [10], the ions were accelerated by application of a high positive voltage to both the sampler and skimmer (Figure 5.7.7(a)). The flight tube through the analyzers is then kept grounded. Naturally, the sampler and skimmer must be isolated electrically from the rest of the vacuum system. Some care is also necessary to prevent arcing from the high voltage interface through the vacuum tubing to the interface pump, which is typically operated at or near the interface voltage using isolation transformers. Several present manufacturers use this approach.

The reader may wonder why positive ions are not simply repelled by the high positive voltage (+4 to +8 kV) applied to the sampler and skimmer.

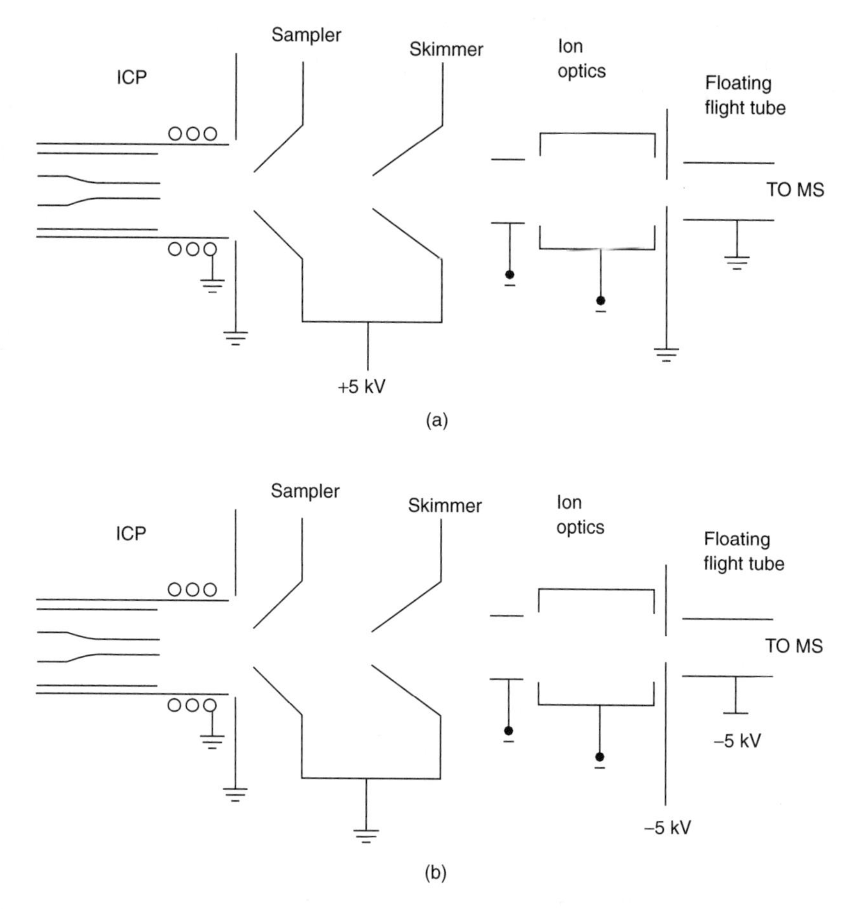

Figure 5.7.7. Voltage arrangements for extracting ions into magnetic sector mass analyzer. (a) Sampler and skimmer float at positive accelerating voltage, while flight tube is grounded. (b) Sampler and skimmer are grounded while flight tube floats below ground. Reprinted from *Spectrochimica Acta, Part B*, Vol. 51, Niu and Houk, pp. 779–815, Copyright (1996), with permission from Elsevier Science.

There are several reasons. A space charge sheath forms around the inner edge of the sampler and skimmer, which shields the plasma from the potential applied to the cones. Some say the plasma potential also floats up near the accelerating voltage. Ions pass through the sampler and skimmer mostly by virtue of the gas flow rather than the applied potentials.

An alternate way to accelerate the ions is to leave the sampler and skimmer grounded and apply a negative voltage to the flight tube (Figure 5.7.7(b)). This is the approach taken in the Finnigan Element and Neptune instruments [11, 12]. Finnigan claims that the high voltage can be switched more rapidly by applying it to the flight tube than to the interface, which permits faster peak switching with the grounded interface.

3.2 Effect of load coil configuration

Present magnetic sector ICP-MS instruments use a load coil that is grounded at the downstream end (Figure 5.7.8). On some instruments, a C-shaped metal shield (also called a guard electrode or plasma screen) can be inserted between the coil and the outer tube of the torch (Figure 5.7.8). The shield removes the capacitive coupling between the load coil and the plasma and reduces the plasma potential [9, 13]. It is usually grounded.

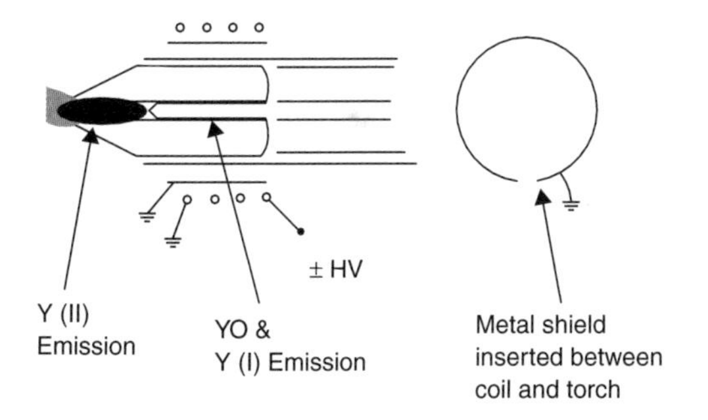

Figure 5.7.8. Shielded ICP torch. The usual initial radiation zone from red YO and Y(I) emission and blue analytical zone from Y(II) emission are also shown.

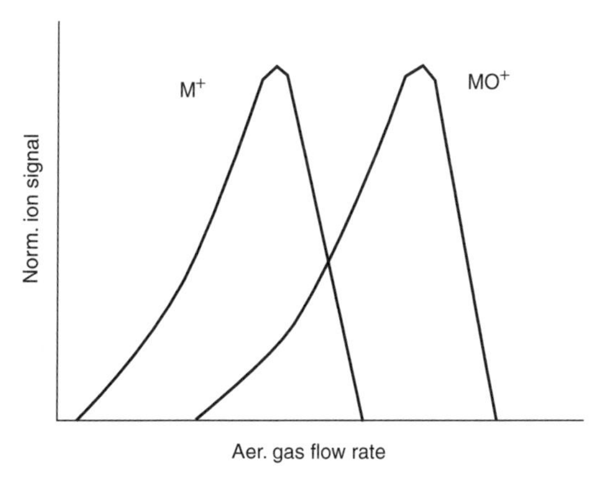

Figure 5.7.9. Plots of ion signal versus aerosol gas flow rate showing different zones of abundance for M^+ and MO^+. Both plots have been normalized to the same height; the maximum signal for MO^+ is usually much lower than that for M^+, depending on the element and the solvent loading. Adapted from ref. 16 with permission.

There are various reports on the effects of the shield with sector instruments. On Finnigan Element devices, which use the grounded interface (Figure 5.7.7(b)), grounding the shield improves the sensitivity roughly tenfold. Appelblad *et al.* [14] report that operating with the shield in place but floating allows selection of a value for the aerosol gas flow rate that minimizes matrix effects by mimicking the 'cross-over point' seen in ICP emission spectrometry. Some magnetic sector instruments that use the high voltage interface (Figure 5.7.7(a)) do not report sensitivity improvements using the shield [15].

3.3 'Cool' plasma

On any of the sector instruments, grounding the shield allows use of so-called 'cool' plasma conditions. The usual adjustable plasma conditions are power, aerosol gas flow rate, and sampling position. At a particular sampling position and power, a plot of M^+ ion signal versus aerosol gas flow rate generally gives a pyramidal shape (Figure 5.7.9) [16]. A similar shaped plot (normalized to the same maximum) displaced to higher aerosol gas flow rate is seen for the generally undesirable MO^+ ions.

The variation of signal with these operating parameters is interrelated. At higher power, the aerosol gas flow rate that generates maximum M^+ sensitivity is displaced to lower aerosol gas flow rate. If the plasma is retracted further away from the sampler, lower power and/or higher aerosol gas flow rate are required to maximize sensitivity.

The 'zone model' [16, 17] has been described to explain these effects. Figure 5.7.8 also shows the zones characteristic of the various species formed as the sample travels through the ICP. If a concentrated solution containing yttrium (or some other element whose MO species emit in the visible) is introduced into the plasma, the plume comprising emission from MO and neutral M atoms is often called the initial radiation zone (IRZ). The IRZ protrudes further downstream as power decreases and aerosol gas flow rate increases. This can be thought of as 'pushing' the MO and M species further downstream in the plasma until they are heated sufficiently to become atomized and ionized.

At the power, aerosol gas flow rate, and sampling position where M^+ sensitivity is maximum, the sampler is typically just 1 to 2 mm downstream from the tip of the IRZ. Physically, the plume of M^+ ions expands rapidly, and only ions from a small cross-section of the plasma just in front of the sampler traverse both sampler and skimmer [9]. These conditions yield best M^+ sensitivity for all instruments. The background spectrum under such conditions is dominated by atomic ions O^+ and Ar^+ and ions with a few protons like H_2O^+ and

ArH^+. However, if the power is reduced and/or aerosol gas flow rate is increased, with the shield grounded, the background spectrum changes to one dominated by molecular ions such as H_3O^+, NO^+ and $O_2{}^+$. This is the 'cool' plasma condition [18]. Atomic background ions are much less prevalent, and so are their reaction products ArH^+, ArN^+, ArO^+ and $Ar_2{}^+$. The changes to plasma conditions that induce the 'cool' plasma would also cause an increase in plasma potential and secondary discharge, so the 'cool' plasma can be generated only if the shield is in place and grounded or the voltage applied to the load coil is inherently balanced.

The primary use of the 'cool' plasma is to remove the worst of the polyatomic ions that plague measurement of K^+, Ca^+, and Fe^+. There are also reports that operation under 'cool' conditions decreases atomic ion background, especially for volatile elements such as Li, Na, and Pb. Although the traditional use of the 'cool' plasma has been to allow measurement of such elements with quadrupole instruments, the high resolution capability of scanning magnetic sectors is of further value. For example, measurements on a Finnigan Element show low levels of 'new' background ions such as $(H_3^{17}O^+)(H_2^{16}O)_2$ and $(H_2D^{16}O^+)(H_2^{16}O)_2$ at $m/z = 56$ (Figure 5.7.10). The protons ($m = 1.0078$) make these ions easy to resolve from $^{56}Fe^+$ ($m/z = 55.9349$, slightly under 56). These background ions are probably also present in quadrupole instruments and are often ascribed to 'Fe blank'. The ability to separate the interferences conclusively, identify ions definitively by accurate m/z measurements, and measure the true blank due to atomic ions are big advantages of magnetic sector instruments.

There are several drawbacks to the 'cool' plasma. Elements such as Ce and W that form strongly bound oxides are seen only as MO^+ or $MO_2{}^+$ ions, and elements with first ionization energies much above that of NO (9.25 eV) are not ionized efficiently in 'cool' mode, unless they are seen as oxides. The latter category includes As and Se, two key elements in speciation. Matrix effects are generally more severe in 'cool' mode. There have been relatively few publications on the 'cool' plasma for speciation with magnetic sector instruments [19].

3.4 Space charge effects

In the plasma, the charge on the positive ions is balanced by an equal number of electrons. The extracted ion beam remains quasineutral until it leaves the skimmer. Preferential loss of the light, highly mobile electrons results in a beam of positive ions leaving the skimmer with total current of the order of 1 mA. The maximum current I_{max} of ions at mass-to-charge ratio m/z that can be focused through a lens of length L and diameter D is given by

$$I_{max} \sim 0.9(z/m)^{1/2}(D/L)^2 V^{3/2} \qquad (5.7.15)$$

where V is the ion kinetic energy inside the lens (volts), m/z is given relative to $^{12}C = 12$, and I_{max} is in μA [20]. It is important to note that m/z is the value for the major ion comprising the bulk of the ion beam, usually considered to be 40 for Ar^+ in the ICP. For example, it is not correct to say that equation (5.7.15) shows that space charge effects are less severe for heavy analytes. Space charge effects (i.e., defocusing and loss of ions) can still be significant at current values less than this calculated maximum.

For a quadrupole instrument, $V \sim 200$ V or less, for which I_{max} is 100 μA or less. Since this value is well under the expected ion current of ~1 mA, low voltage extraction optics are susceptible to the space charge effect. For a sector instrument with accelerating voltage of 4000 and 8000 V, I_{max} is 9 mA and 25 mA, respectively, higher than the expected ion current through the skimmer. Thus, space charge effects are likely to be less severe in sector instruments. Whether they are totally absent or unimportant is another matter. Space charge defocusing can still be significant at currents well below the limit calculated from equation (5.7.14). The effects are exacerbated if the beam is focused to a fine image, and sharp focusing is much more important in sector instruments than in quadrupoles. Certainly mass bias and matrix effects are still present with sector

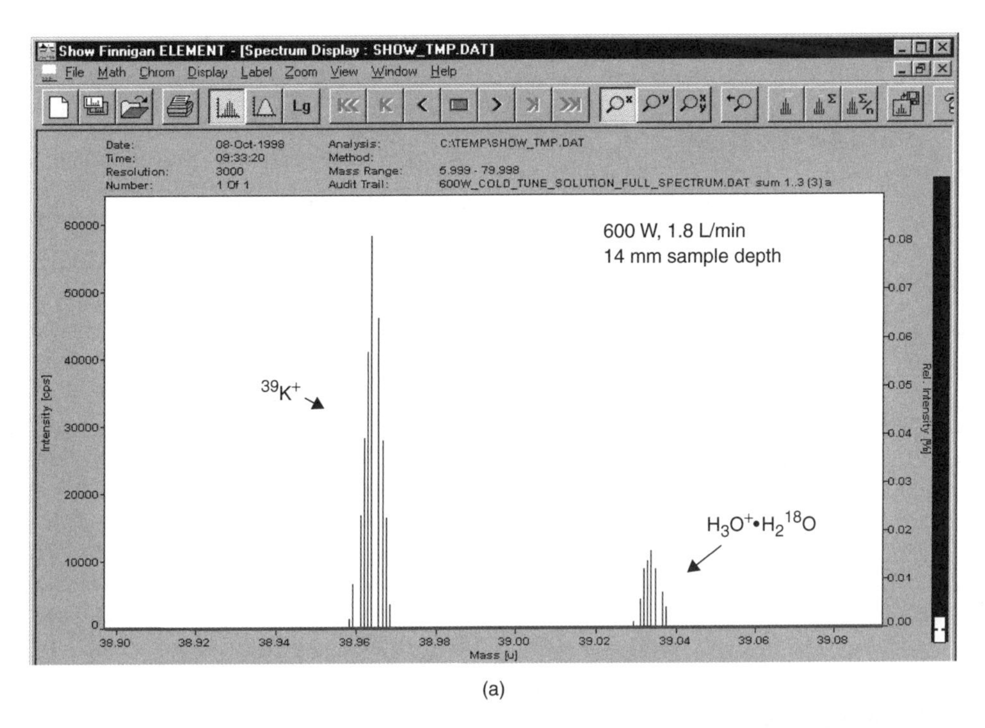

(a)

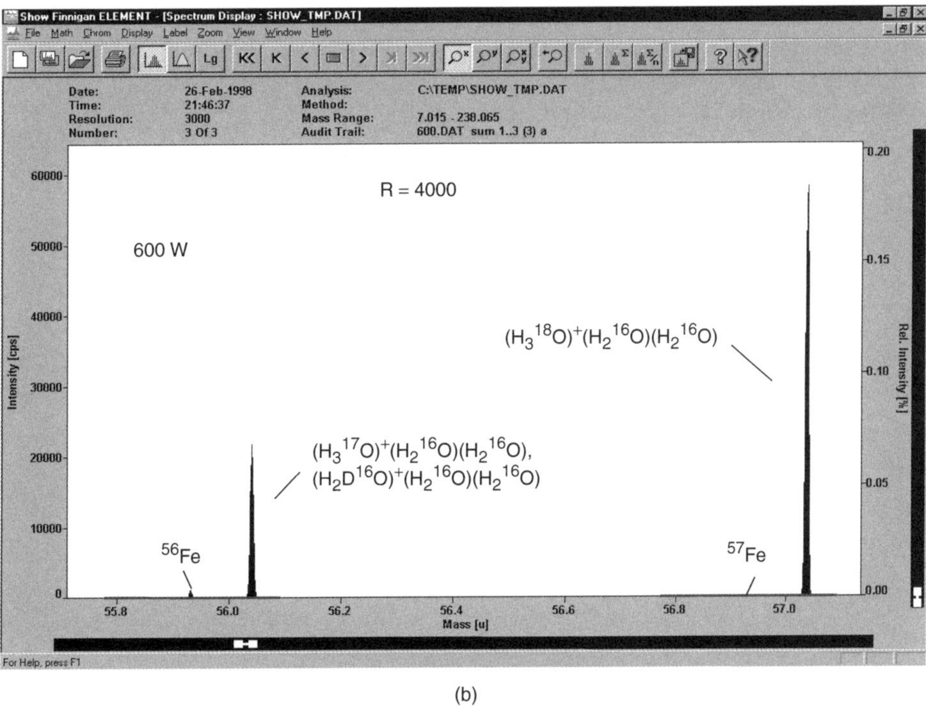

(b)

Figure 5.7.10. Background spectra at $m/z = 39$, 56 and 57 in cool plasma mode, Finnigan Element, $R = 4000$. Note water cluster ions and their separation from $^{39}K^+$, $^{56}Fe^+$ and $^{57}Fe^+$. Data provided by D. R. Wiederin, Elemental Scientific Inc.

instruments, although evidence has been presented that they have rather different fundamental causes than the effects seen with low voltage, quadrupole instruments [21]. The entire matter of precisely how a beam is formed from the plasma emanating out of the skimmer is still an issue for basic study [22].

3.5 Beam shaping with quadrupole or hexapole lenses

The beam leaving the skimmer has a circular cross-section, while the entrance and exit slits are rectangular. Ions in the outer sections of the circular beam will be lost unless the shape of the beam is changed. This can be done with the set of quadrupole lenses shown in Figure 5.7.11. There are two sets of four metal rods with DC voltages applied as shown. The positive voltages on the horizontal poles squeeze the ions in that direction, while the negative voltages allow the beam to expand in the vertical direction. A combination of two or more of such lenses can produce the desired focusing properties in both dimensions [23].

Hexapole lenses operate on the same principle, except that there are six poles, of course. Note that these quadrupole or hexapole lenses are DC devices; do not confuse them with the RF multipole collision cells described below.

3.6 Collision cells

In recent years, collision cells[4] have been supplied with quadrupole ICP-MS instruments for either removal of interfering ions or conversion of M^+ analyte ions to more easily measured species. Oxidation of As^+ to AsO^+ to avoid $ArCl^+$ interference is an important example of the latter process [24]. The operating principles for these devices have been described [25, 26]. Ions pass through an RF quadrupole, hexapole, or octopole that contains a collision gas. The pressure is sufficient that the ions undergo several collisions while inside the cell. A chemically unreactive gas (e.g., He) can be used to thermalize the ions, and/or a reactive one (such as H_2 or NH_3) can be used to induce a desired chemical reaction. The term 'chemical resolution'

[4] Some workers distinguish between 'collision cells' that are not mass analyzers and use relatively lower pressure, fewer collisions and have more likelihood of endothermic reactions from 'reaction cells' that are quadrupole mass analyzers and use higher pressure and exothermic ion–molecule reactions. The present discussion does not distinguish between these various devices.

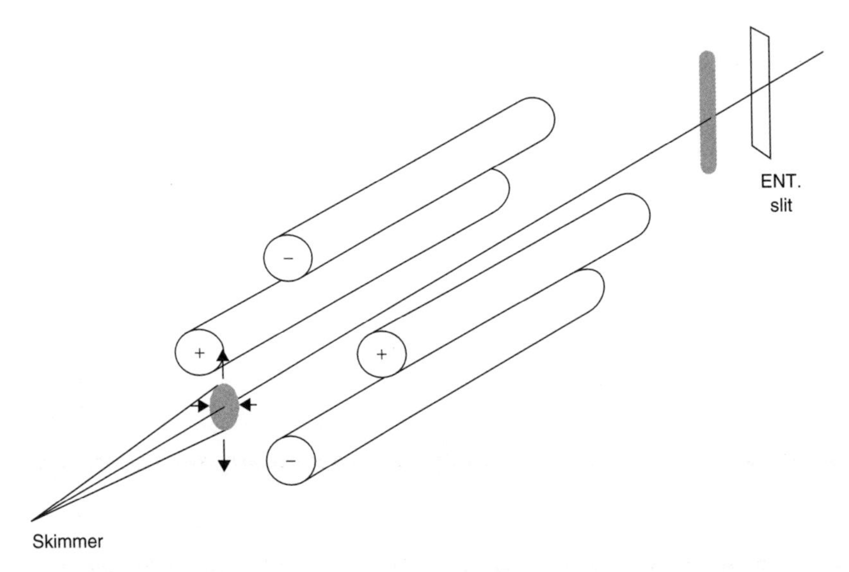

Figure 5.7.11. DC quadrupole lens used between skimmer and electrostatic or magnetic sector. The beam leaving the skimmer is round, while the applied voltages 'squeeze' the beam into a rectangular shape that matches the slit.

is sometimes used to describe the higher selectivity imparted by the collision cell. Some means of distinguishing desired analyte ions from unwanted product ions generated in the ion–molecule reaction is usually necessary.

These same principles can be used with magnetic sector analyzers. Micromass also uses an RF hexapole collision cell to reduce the kinetic energy spread of ions, which eliminates the need for an electrostatic analyzer, at least for their low resolution multicollector device (Figure 5.7.12). For each elastic, nonreactive collision with a stationary neutral of mass m_2, an ion of mass m_1 with initial kinetic energy E_1 loses energy to E_1' according to equation (5.7.16) [27–29]:

$$E_1' = E_1 \left[\frac{m_1^2 + m_2^2}{(m_1 + m_2)^2} \right] \quad (5.7.16)$$

These collisions also scatter ions, but the RF field applied to the hexapole rods helps contain them, so transmission and sensitivity do not suffer too much. Polyatomic ions have a larger cross-section for kinetic energy loss than atomic ions at the same nominal m/z value. Yamada *et al.* [30] exploit this property to show that subsequent kinetic energy analysis can be used to discriminate against polyatomic ions. Bandura *et al.* [31] and Vanhaecke *et al.* [32] show that such collisions can also be used to reduce high-frequency noise in the ion signal from fluctuations in the plasma. This improves the precision in isotope ratios measured with quadrupole instruments, and should also help improve precision in scanning sector devices. Of course, the same collisional processes used with quadrupole instruments to remove interferences, convert the analyte M^+ ion to one more readily measured, or reduce noise from the ICP could also be implemented with sector devices.

4 EXAMPLES OF SCANNING, HIGH RESOLUTION SECTOR INSTRUMENTS

It is not feasible to describe all the presently available devices in detail. Several of the more widely used models are discussed to illustrate the workings of these instruments.

4.1 Finnigan Element

This instrument is based on the work of Giessman, Jakubowski and associates [11, 12]. Of the various magnetic sector instruments, this is the only one the author has used personally. As mentioned previously, this system uses the grounded interface (Figure 5.7.7). After the skimmer, the ions are accelerated through a graphite electrode at high negative potential. The flight tube is kept below ground potential by whatever voltage is needed to transmit the ions of interest, up to 8 kV. There are quadrupole lenses for beam shaping and focusing. The sectors are arranged in a reverse Nier–Johnson geometry (Figure 5.7.13). The 60° magnetic analyzer is followed by the 90° electrostatic analyzer, exit slit, and detector. Thus, the ions are mass analyzed in the magnetic sector first, then ions of a selected band of kinetic energies are transmitted through the exit slit to the detector. There are three fixed slit settings for low resolution ($R = 300$), medium resolution ($R = 4000$), and high resolution ($R = 8000$ or better). The slit settings can be changed quickly during an acquisition, so medium or high resolution can be used only at m/z values where they are needed.

It is faster to scan m/z by changing accelerating voltage (V in equation 5.7.3), so the magnetic field is usually kept at the desired value while ions in a certain range are measured. This can be done by electrostatic peak switching at low resolution or electrostatic scanning at medium or high resolution. The magnetic field is then changed to a value in the next desired m/z range, and these ions are measured by changing accelerating voltage.

4.2 VG Axiom

This instrument is descended from the original ICP sector MS work of Bradshaw *et al.* [10], which

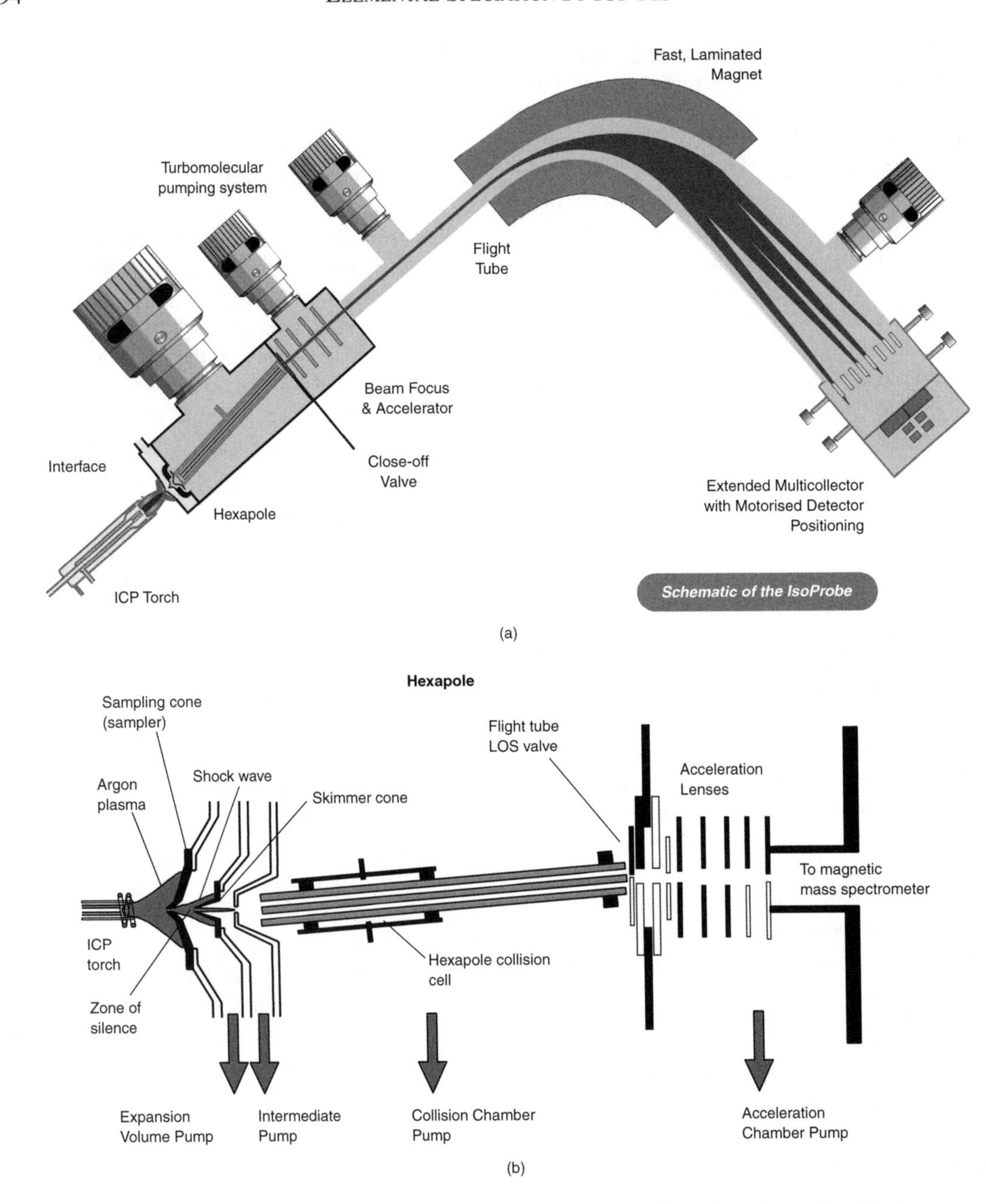

Figure 5.7.12. Micromass Isoprobe multicollector instrument with hexapole collision cell used to reduce kinetic energy spread on ion beam. (a) overall instrument; (b) detail of interface and collision cell. Courtesy of Micromass.

began interest in the concept. Marriott *et al.* [33] have described the Axiom. The high voltage interface (Figure 5.7.7(a)) is used. The ion optical system is shown in Figure 5.7.14. The electrostatic analyzer is first, which is called forward geometry. Scanning measurements are done with a single electron multiplier at the double focusing position after the magnet. The slit widths can be selected between a wide variety of values, so the resolution is continuously selectable between 300 and 12 000.

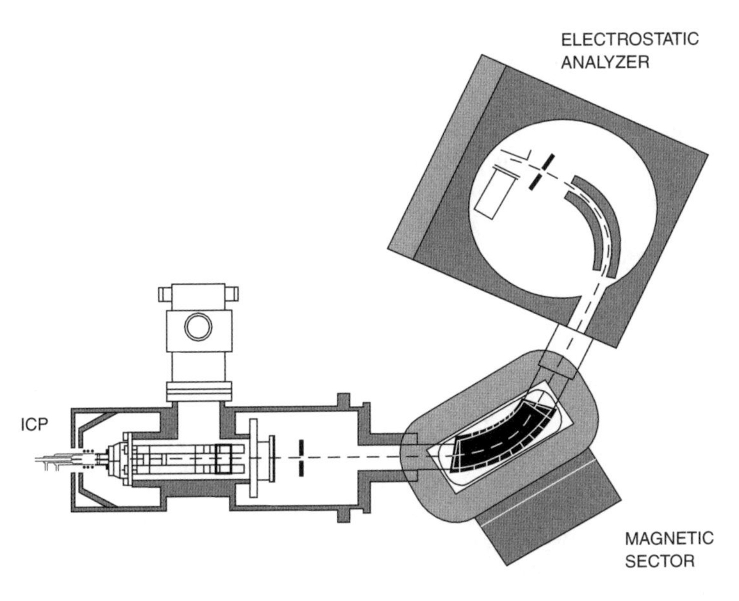

Figure 5.7.13. Thermo Finnigan Element instrument, reverse Nier-Johnson geometry. Courtesy of Thermo Elemental.

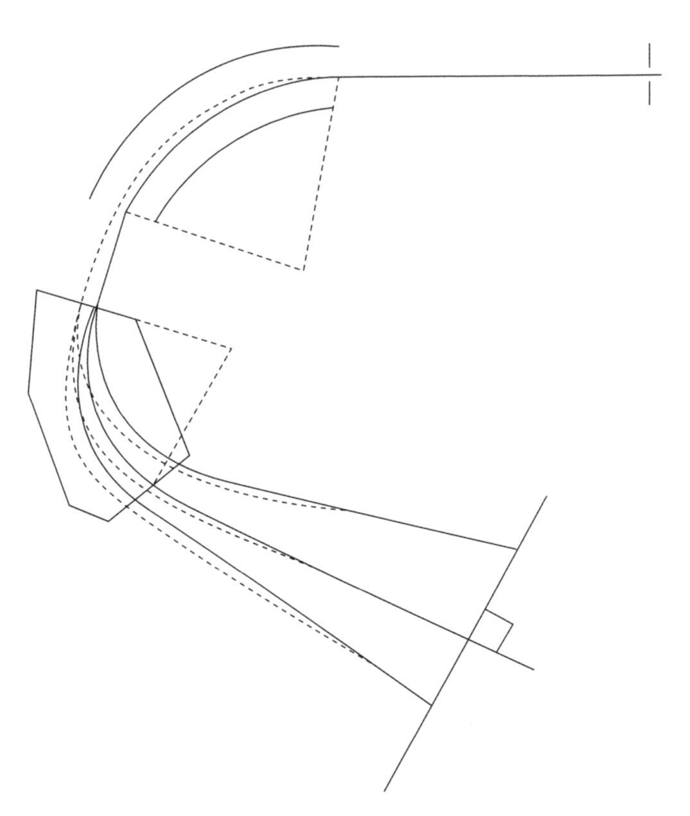

Figure 5.7.14. Ion optics of Axiom instrument from Thermo Elemental. The ions pass from the source at upper right through an electrostatic analyzer first, followed by the magnetic sector. The exit optics are such that either a single electron multiplier or an array of Faraday cups can be used at the detector plane with good focusing. Reproduced from ref. 33 with permission of the Royal Society of Chemistry.

5 EXAMPLES OF MULTICOLLECTOR INSTRUMENTS

For high precision isotope ratio measurements of metallic elements for geochemical studies, the best possible precision is desired, so thermal ionization MS is the norm. Sample throughput would be higher, multielement studies would be possible, and certain problem elements could be measured more readily if highly precise isotope ratios could be obtained by ICP-MS.

Scanning instruments measure the various isotopes of interest sequentially. The ion beam from the ICP fluctuates due to effects such as passage of large droplets and oscillation of the plasma at audio frequencies [34, 35]. Fast scanning or peak hopping helps compensate for the effects of this noise when ratios are measured, and isotope ratios can be measured with precision of $\sim$0.02 % relative standard deviation (RSD) on scanning sector instruments. However, nearly all of this noise is eliminated if ion signals at different m/z values can be measured truly simultaneously. Walder *et al.* [36] did the first multicollector measurements on ions from an ICP, using an ICP extraction interface and beam shaping optics added onto a spectrometer designed primarily for thermal ionization. These studies demonstrated that ICP-MS could attain similar precision to thermal ionization and led to the development of present ICP multicollector devices.

5.1 VG Axiom with multicollector array

The electrostatic analyzer must come first with a multicollector. The ion optical system of the Axiom is designed such that ions of a substantial range remain in reasonable focus outside the m/z value corresponding to perfect double focusing (mass M_1 in Figure 5.7.4). The two focal curves g and r lie nearly together on either side of point A_1'', so ions of a range of masses of about 10 % of M_1 are reasonably well focused at the same values of B and V. Thus, this instrument can also be used in multicollector mode. The electron multiplier is replaced by a set of separate Faraday cup detectors (like those shown in Figure 5.7.12(a)). Ions at up to eight adjacent m/z values are measured truly simultaneously with this arrangement. There is a central slit for an electron multiplier for measurement of low abundance isotopes.

Such simultaneous measurements yield isotope ratios whose precisions are, in the short term at least, limited only by counting statistics and the stability of the mass bias corrections. The response of a Faraday cup is very stable, but it provides no amplification like that of an electron multiplier, so measurements with multicollectors require more analyte than do scanning instruments.

5.2 Micromass Isoprobe

As shown in Figure 5.7.12, this instrument does not have an electrostatic analyzer. A collision cell is used to 'thermalize' ions, i.e., to reduce their kinetic energies and kinetic energy spread, before they enter the magnetic sector. Instead of isolating ions of a range of kinetic energies and rejecting the others, the operating concept of the collision cell is to utilize a larger fraction of the ions than is the case with an electrostatic sector.

A spectrum for the various isotopes of Tl and Pb is shown in Figure 5.7.15. Note the flat topped, highly stable peaks, one of the desirable characteristics of these multicollector instruments. This device is also capable of resolution of $\sim$4000.

5.3 NU Plasma

This company provides a relatively compact device capable of resolution of $\sim$500 (Figure 5.7.16). The zoom lens adjusts the separation between ion beams so that the spacing between collectors need not be changed mechanically when the magnet mass is changed. In addition to the usual Faraday cup detectors, ions at up to three m/z values can be monitored using electron multipliers with an ion deflector assembly. A larger version capable of high mass resolution is also available.

5.4 Finnigan Neptune

This instrument (Figure 5.7.17) has the ICP, interface and ion lenses from an Element added onto an

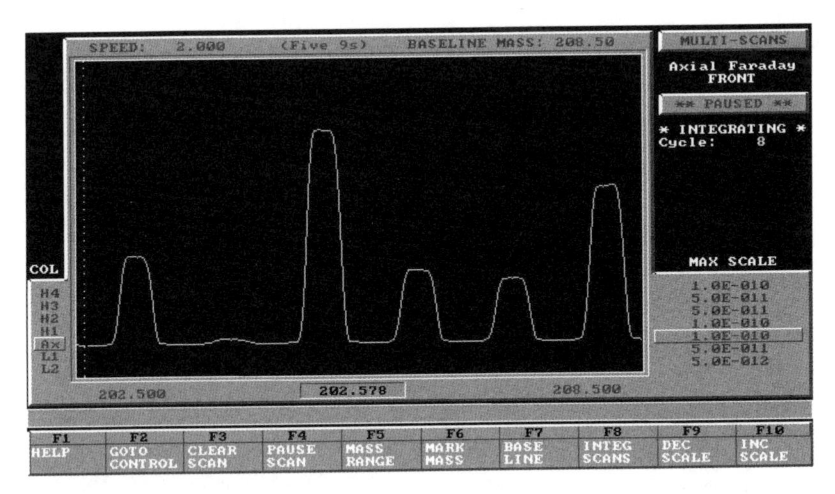

Figure 5.7.15. Scan of mass spectral peaks for Tl^+ ($m/z = 203$ and 205) and Pb^+ ($m/z = 204, 206, 207, 208$) from Micromass Isoprobe. Reproduced courtesy of Micromass.

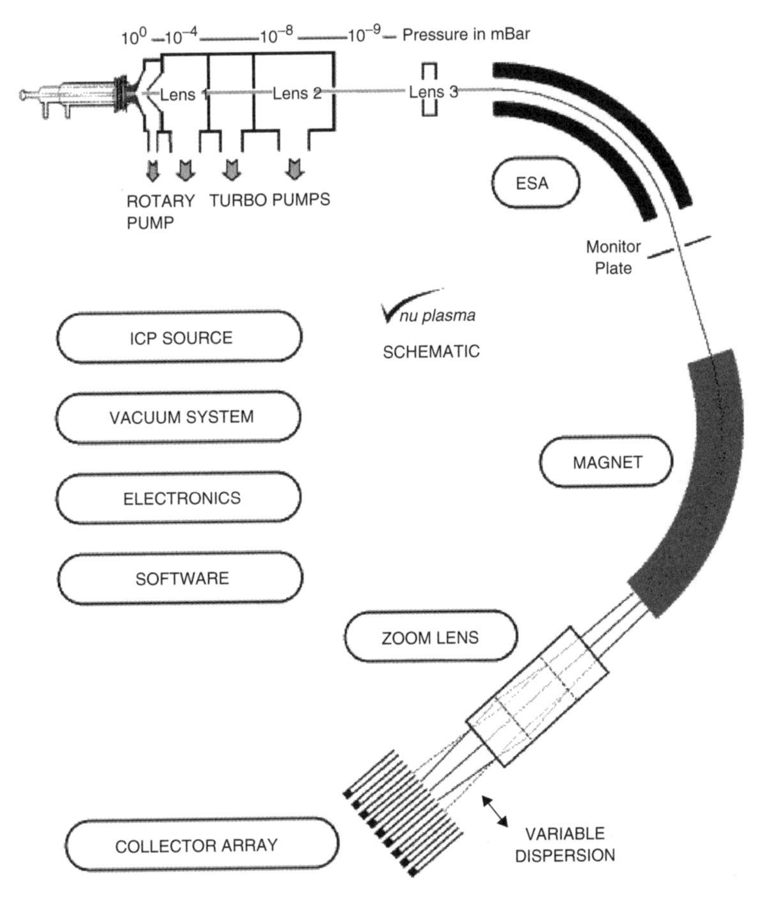

Figure 5.7.16. NU Plasma multicollector instrument with zoom lens and multiple electron multiplier detectors. Figure provided courtesy of NU Instruments.

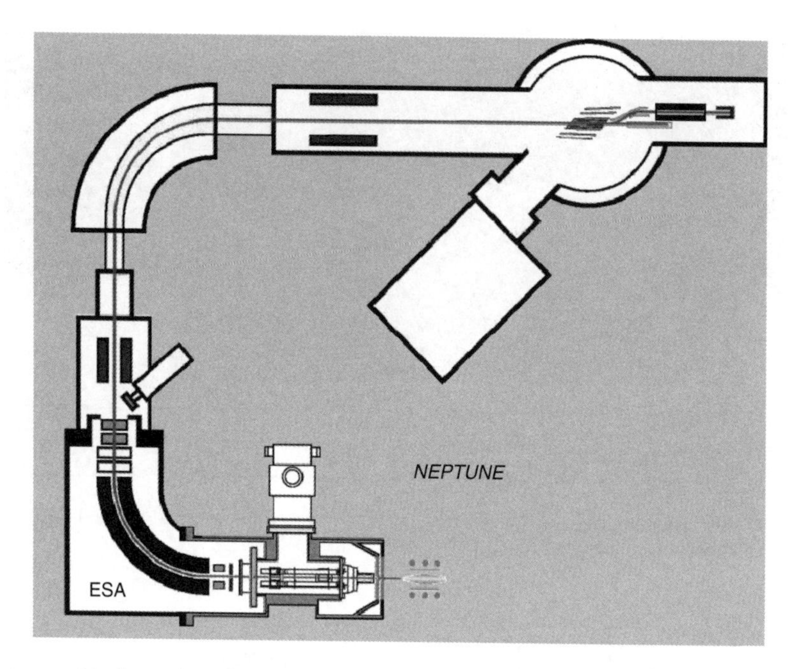

Figure 5.7.17. Neptune multicollector instrument. Note the magnification provided by having the image distance roughly twice the object distance, which facilitates operation of the multicollector array. Figure provided courtesy of Thermo Elemental.

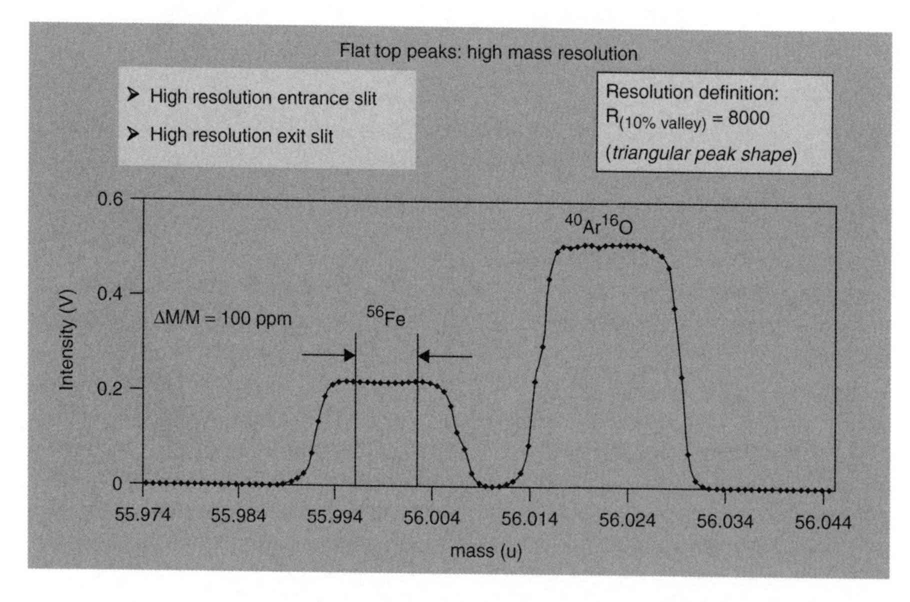

Figure 5.7.18. High resolution separation of $^{56}Fe^{+}$ from $^{40}Ar^{16}O^{+}$ in multicollector mode. Figure provided by Thermo Elemental.

electrostatic and magnetic analyzer from the Triton instrument for thermal ionization. The large mass analyzer has a dispersion of 81 cm, and a zoom lens is used. This device can provide flat topped peaks at a resolution 3000 for simultaneous multi-collection at up to three m/z values, as shown for $^{56}Fe^{+}$ in Figure 5.7.18. Resolution up to 8000 can be achieved with sharply pointed peaks.

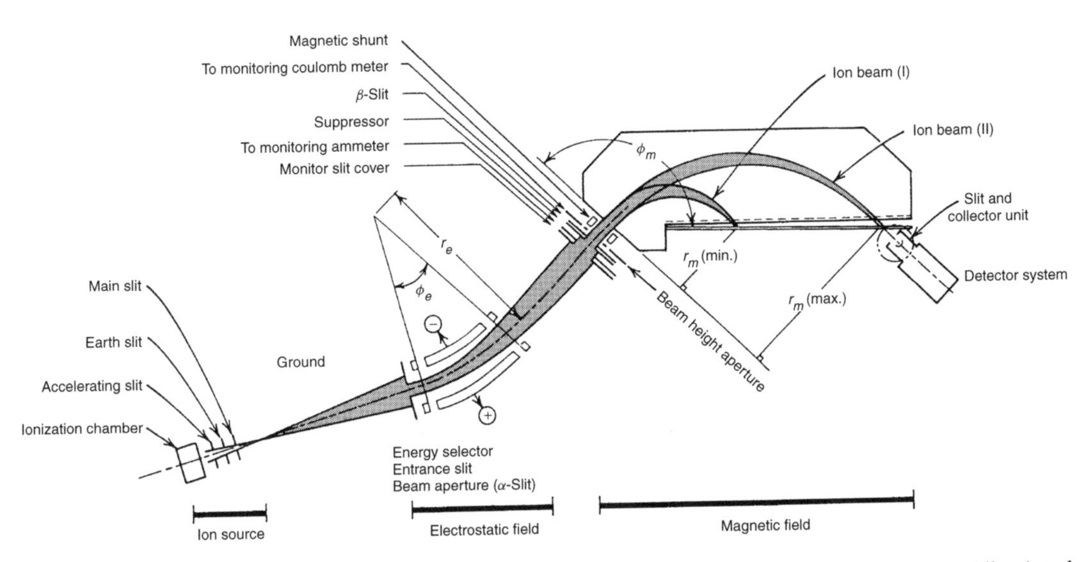

Figure 5.7.19. Mattauch–Herzog geometry. The ion beam leaves the electrostatic analyzer as a parallel beam. All m/z values within the upper m/z limit of the device come back into focus along a single plane near the boundary of the magnetic field. Reproduced from ref. 4 with permission from Jeol.

5.5 Mattauch–Herzog instrument

The multicollector instruments described previously are meant to monitor ions at a set of eight to ten adjacent m/z values. The Mattauch–Herzog geometry can measure all m/z values at once. In this geometry (Figure 5.7.19), ϕ_e is 31.83°. The focusing equations (5.7.7)–(5.7.9) can be solved for this angle to show that the ions leave the electrostatic analyzer as a parallel beam ($l''_e \rightarrow \sim$). The magnetic sector satisfies the equation

$$\sin \phi_m = \sqrt{2} \sin \sqrt{2}\phi_e \qquad (5.7.15)$$

The ion path is an S shape; the direction of deflection in the electrostatic analyzer is opposite to that in the magnetic sector [4].

Under these conditions, the direction and velocity focusing curves are straight lines that coincide. Thus, ions of the entire m/z range achieve double focusing along the same plane, as shown in Figure 5.7.19. This arrangement is well suited to a planar, array detector and was used with photographic plates in the old spark source MS. An ICP-MS instrument of this sort with an electrical array detector would be capable of truly simultaneous measurement of the entire m/z range, provided the detector is large enough to observe the whole range with sufficient dispersion. Work along these lines has been described by Cromwell *et al.* [37, 38] and is proceeding under the direction of Hieftje and Denton [39–41].

6 SPECIATION MEASUREMENTS WITH SECTOR INSTRUMENTS

6.1 Dry sample introduction

The composition of the solvent that feeds the nebulizers is a matter of concern in ICP-MS. A variety of schemes for solvent removal can be implemented with any ICP-MS instrument:

(1) cooled spray chamber (Figure 5.7.20).
(2) traditional desolvation [42] (Figure 5.7.21(a)), i.e., heating the aerosol followed by cooling at or near the freezing point of the solvent.
(3) additional solvent removal by membrane or cryogenic desolvation (Figures 21(b) and (c)).

There are obvious advantages to removing solvent for ICP-MS, many of which are of particular advantage in high resolution measurements. Other factors such as temperature remaining the same,

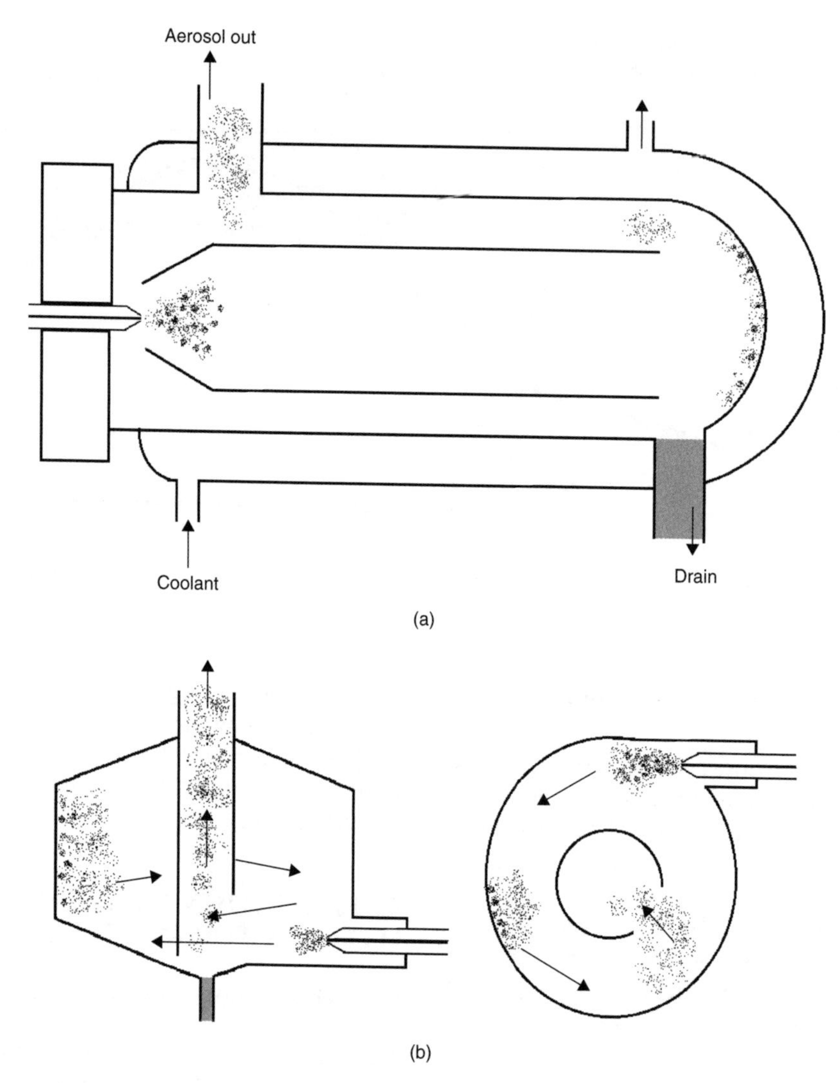

Figure 5.7.20. Cooled spray chambers for solvent removal: (a) cooled double pass Scott chamber; (b) Cyclone chamber, side and top views. In both chambers, most of the large droplets are deposited at the bends, while fine droplets pass out to the plasma.

lower levels of polyatomic ions containing oxygen atoms would be expected from simple mass balance considerations,

$$MO^+ \rightleftharpoons M^+ + O$$

assuming most of the O atoms come from the solvent. The magnitude of the atomic mass defect varies with mass [8], and some important polyatomic ion interferences are just within the resolution capabilities of the instruments. For example, separation of $^{140}Ce^{16}O^+$ from $^{156}Gd^+$ requires nominal resolution of at least 7200, with the accompanying large sacrifice in sensitivity. Thus, using plasma conditions that reduce the abundance of MO^+ is useful. Removal of organic solvents minimizes carbon deposition and helps maintain a properly hot plasma, and solvent removal obviates noise from passage of wet droplets through the plasma. In certain cases, removing solvent leads to enhancements in sensitivity with sector instruments, which is welcome even if the basic reasons are not clear. In general, the plasma potential is

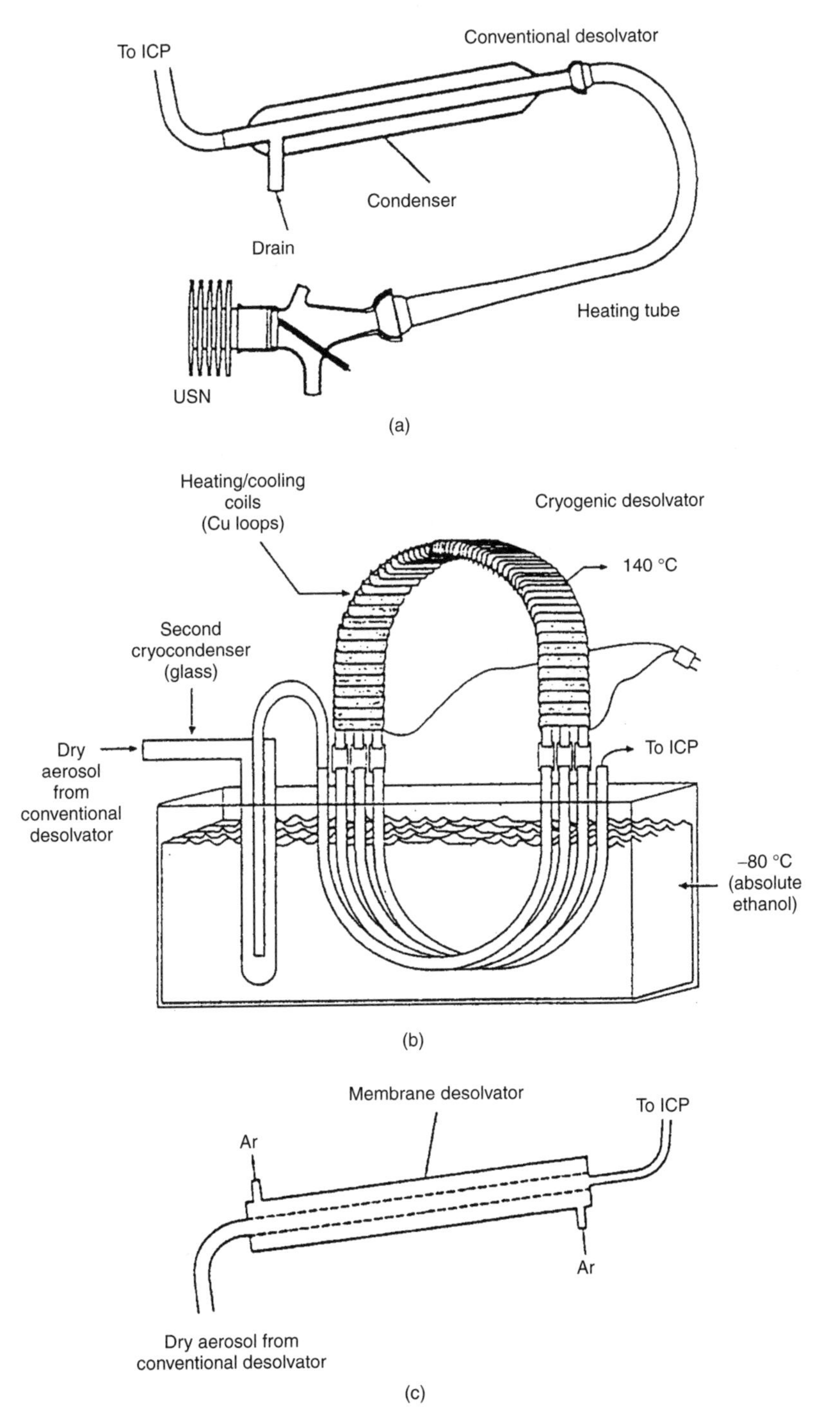

Figure 5.7.21. Additional solvent removal options: (a) conventional desolvation with heated chamber and condenser at or near the freezing point of the solvent; (b) cryogenic desolvation. The aerosol undergoes repeated heating and cooling steps. (c) Membrane desolvation. Reproduced from ref. 42 with permission of the Royal Society of Chemistry.

lower when the plasma is dry, so ions from a dry plasma would be expected to have lower spreads of kinetic energy, which could improve transmission and sensitivity.

Generic disadvantages of solvent removal are band broadening, longer rinse out time, and worse memory effects, especially for volatile elements such as Hg, B, Os and I. Volatile species can also be lost during desolvation. Such memory or loss problems are generally less severe if the droplets are not heated to dryness.

First we discuss solvent removal options with conventional nebulizers that operate at liquid flow rates of 0.5 mL min^{-1} or more and have droplet transport rates to the plasma of ~1–2 %. These devices normally require spray chambers to remove the very large droplets and to keep the total solvent load below the maximum the plasma can tolerate. The Scott-type double pass chamber (Figure 5.7.20(a)) [43] remains common, although the cyclone chamber (Figure 5.7.20(b)) [44] shows moderately faster rinse out behavior and is also seeing extensive use recently.

Either of these spray chambers can be cooled to temperatures near the freezing point of the solvent. The aerosol leaving the spray chamber can also be heated and then either cooled again or passed through a membrane desolvator. The sample constituents are injected into the plasma as dry particles with some solvent vapor remaining. If the aerosol is to be heated anyway, a conical spray chamber that allows higher initial droplet transport can be tolerated. The higher droplet production rate and transport from the spray chamber are the primary reasons behind the sensitivity advantage of ultrasonic nebulizers [45].

It must be admitted that solvent removal beyond simply cooling the spray chamber is not commonplace in speciation measurements, with either sector or quadrupole instruments, largely because of the perceived problem of dead volume after the aerosol is produced. In the author's experience, it is much more important to minimize dead volume in the liquid connection between column and nebulizer. For well-behaved species, i.e., those not overly volatile or prone to memory effects, whether the additional dead volume and rinse out time is a problem depends on the time duration of the peaks provided by the chemical separation. Species that are volatile can be readily lost or suffer long memory effects when their aerosols are heated.

On-line hydride generation is another option for solvent removal for elements that form volatile hydrides (e.g., As, Se, Pb, Sn) or for Hg. It is important to verify that the various chemical forms are indeed converted to hydrides with reproducible, hopefully complete, recovery. On-line microwave digestion just before the hydride generation cell has been used for this purpose [46].

6.2 Micronebulizers

These devices operate at low liquid flow rates, 100 μL min^{-1} or less for this discussion. Aerosol is transported more efficiently out of the spray chamber with micronebulizers, which compensates for the lower liquid uptake. Thus, sensitivity can be as good as or better than that obtained from conventional nebulizers. It is also easier to remove solvent; indeed, for water at liquid flow rates of 40 μL min^{-1} or less, the solvent evaporates naturally anyway [47]. At lower uptake rates, the aerosol can be fed directly into a membrane desolvator, without the intermediate condenser. Since the total solvent load can be lower, the plasma can tolerate higher concentrations of organic solvent, and there should be less variation of sensitivity as solvent composition changes during gradient elution. The trend in LC separations is toward lower flow columns anyway, the outlets of which can be fed to micronebulizers with less extensive splitting. Those micronebulizers with very low liquid uptake rates (e.g., 20 μL min^{-1} for the PFA-20, Elemental Scientific, Inc., or even 5 μL min^{-1} for the cross-flow device recently described by Tsunoda *et al.* [48]) are readily adapted to capillary electrophoresis (CE) separations since it is easy to compensate for their low natural suction. For these reasons, use of micronebulizers for speciation is expected to increase in the near future.

6.3 Gradient elution

As with other ICP-MS devices, the sensitivity generally varies with the solvent composition, mainly because the characteristics of the aerosol change with solvent composition. Changing the fraction of organic solvent present also results in formation of polyatomic ions from the solvent; the sector instrument has the desirable feature that many of these polyatomic ions can be separated from M^+ ions using high resolution. Nevertheless, these effects complicate the use of solvent gradients in LC.

Browner and coworkers [49] use an oscillating capillary nebulizer (OCN) for LC measurements with solvent gradients. The droplets are produced by vibration of a capillary tip, not by the usual pneumatic process. The droplets are produced mechanically, and the droplet characteristics are not greatly affected by the solvent composition.

Other measures that facilitate use of gradient elution are as follows:

(1) Restrict the range of solvent compositions used.
(2) Measure the change in sensitivity during the gradient by adding analyte to the solvents, then apply correction factors at the appropriate retention times when analyzing samples.
(3) Add another solvent stream post-column that is the inverse of the stream through the column [50]. For example, if a gradient from 10 % to 30 % methanol–water goes through the column, mix in a stream whose composition goes from 90 % to 70 % methanol–water, then nebulize the mixed stream. Of course, a second gradient pump is required. Dilution and post-column band broadening are also expected.
(4) Continuous introduction of an internal standard post-column, followed by measurement of the ratio of analyte signal to that for the internal standard. As demonstrated by Heumann *et al.* [51] and Garcia Alonso *et al.* [52], the internal standard can be an enriched minor isotope of the analyte, the signal from which will be affected by changes in solvent composition to the same degree as that for the analyte. The m/z windows measured for both analyte and internal standard must either be free of polyatomic ions, or the polyatomic ion background must remain constant. Again, the high resolution capability helps separate these interferences and facilitates accurate measurements using organic solvents or other additives, as demonstrated below.
(5) Remove solvent (Figures 5.7.20 and 5.7.21) and/or add oxygen to the aerosol gas flow to minimize carbon deposition on the sampler and skimmer. Of course, extra oxygen will enhance the abundance of oxide ions.

6.4 Scan speed and data acquisition issues

As with any scanning instrument, the rate of scanning or hopping from peak to peak is important when monitoring a transient signal. First we consider single collector instruments that operate with a fixed value of r_m. As shown in equation (5.7.3), the m/z value can be changed by changing either the magnetic field B or the accelerating voltage V. The magnet takes longer to stabilize upon change than does the accelerating voltage, and the accelerating voltage must be at least a certain value to ensure proper extraction and transmission of ions. Thus, measurements are commonly made in one particular m/z range of interest by changing V while keeping B constant. The magnet setting is then changed to a different value for the next set of analytes, with more measurements made by changing V, and so on.

For the flat-topped peaks seen in low resolution (Figures 5.7.6(a) and (b)), the beam is much narrower than the slit, so it is relatively easy to change B and/or V and still have the beam pass through the slit. Thus, the B and V values can be adjusted to hop directly onto the m/z value of interest with little dead time, even when the magnetic field is changed. The 'dead time' involved in such hops is of the order of a few milliseconds, comparable to that of a quadrupole MS.

With the sharply pointed peaks seen in medium or high resolution (Figure 5.7.6(b) and (c)), it is much harder to 'find' the peak top reliably, so the m/z value is usually scanned, even when the intent is to monitor only one m/z value. It then becomes necessary to select experimental parameters to optimize data collection. The use of such scans

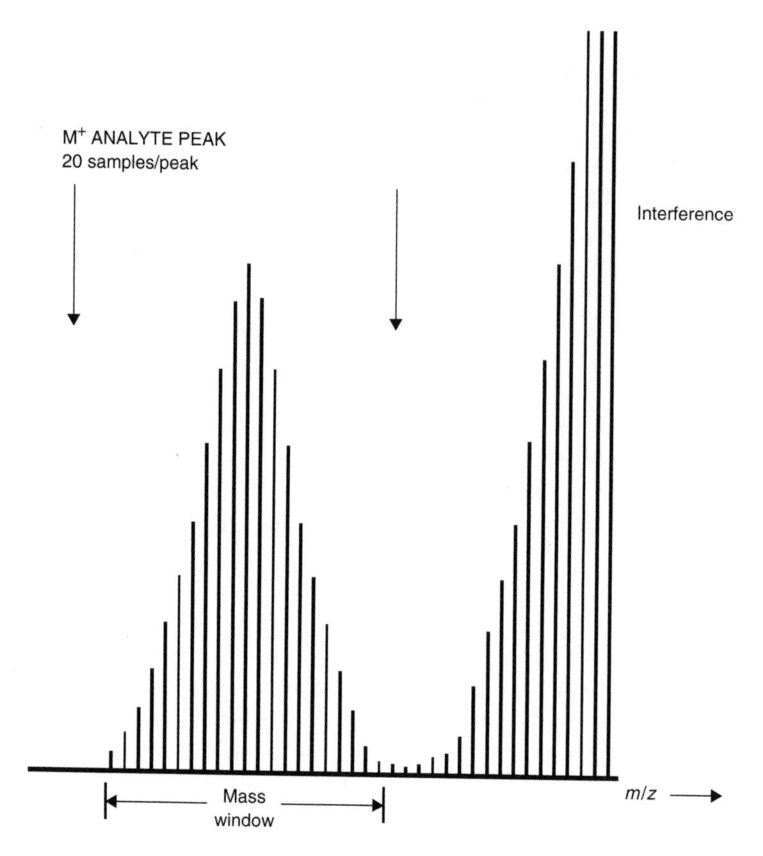

Figure 5.7.22. Schematic diagram of medium or high resolution scan showing data acquisition parameters. Each vertical line represents one m/z position, i.e., a slightly different setting of accelerating voltage at a given magnetic field setting.

in high resolution does limit the number of m/z values that can be monitored with a desired signal-to-noise ratio during a transient signal.

A description of a typical set of scan parameters is in order here. The terminology is taken from that used for the Finnigan Element; other devices have different jargon but follow the same concepts. Consider the schematic high resolution mass scan shown in Figure 5.7.22. The mass window is the percentage of the nominal peak width chosen in which to acquire data; a value of 100 % could be used to encompass only $^{56}Fe^+$ and not $^{40}Ar^{16}O^+$. The samples per peak value gives the number of settings of accelerating voltage V per peak, represented by the 20 lines for M^+ in Figure 5.7.22. Sample time is the amount of time spent at each value of V, i.e., the time the mass analyzer spends on each line shown. This value is 5 ms or longer, with small values necessary for fast transients or to monitor more elements per chromatographic peak. There is also a settling time needed for the m/z value to stabilize, typically 1 ms if only V is changed, 300 ms if the magnet setting is changed. Thus, each line shown in Figure 5.7.22 would be measured in 6 ms, for a total time of 120 ms spent measuring just the Fe^+ peak each time through the sequence of peaks selected. On new Element instruments, the mass window can be selected to be as little as 10 % of the peak width reproducibly in medium or high resolution. All the signals for each line, or that for a chosen fraction of the lines, within the specified mass window can be added together when the chromatogram is plotted.

In the author's experience, there are two measures that can be taken to speed up data acquisition: avoid changing the slit width during each cycle (this imposes a 1 s delay), and minimize the number of magnetic field settings where possible. On the Element, at least six m/z values at one or two B settings can be monitored on chromatographic peaks lasting 10 s with little loss of peak definition.

Multicollector instruments are generally just kept at a fixed magnetic field setting, as the usual objective is the best possible precision. It has been demonstrated recently that these devices are capable of high precision during GC peaks [2, 3]. With the recent advent of zoom optics for the collectors, as described above, ions from various m/z ranges can be imaged onto the detectors without large-scale mechanical adjustments of the spacing. Mattauch–Herzog instruments with array detectors should be able to monitor as many m/z values as desired, if the dispersion is enough to separate the peaks sufficiently onto a detector of practical size.

6.5 Blanks

The high sensitivity of magnetic sector ICP-MS devices does not automatically translate into better detection limits. The blank level, i.e., the signal from the element of interest in the solvent, often restricts the detection limit. In most experiments, small peaks are observable for all the elements in the blank. The organic solvents and organic buffer additives are particularly prone to such contamination.

The usual clean sample handling procedures need to be followed in speciation measurements as well. One key point is whether the contaminant element binds to the molecules in the sample. If not, the baseline in the chromatogram merely becomes elevated, as shown for Cr in human serum in Figure 5.7.23 [53]. The Cr^+ in this baseline can be subtracted from that which is in the chromatographic peaks, so the Cr bound to

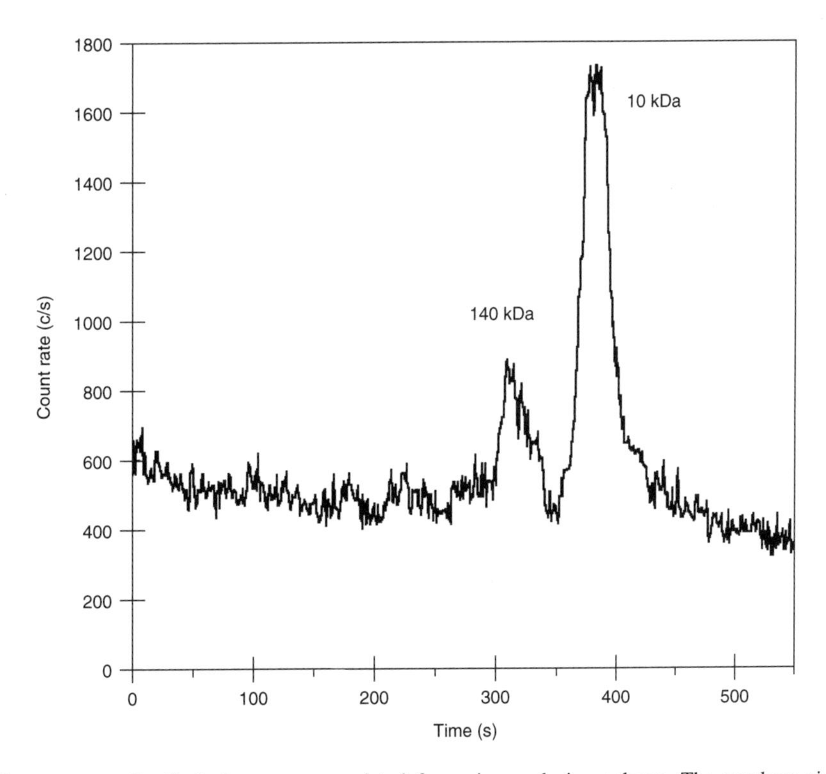

Figure 5.7.23. Chromatogram for Cr in human serum eluted from size exclusion column. The numbers give the estimated molecular weights of the proteins based on calibration with protein standards. Reproduced from ref. 53 with permission from the American Chemical Society.

proteins can still be estimated. Of course, a high background level also carries more noise with it, and much of the Cr apparently bound to proteins may actually come from contamination, which is hard to rule out completely.

In LC experiments, some measures that help reduce trace element contamination are as follows:

(1) Conventional columns and pumps contain stainless steel exposed to the eluent, so use metal-free pumps and columns lined with glass or some other non-metallic material.
(2) Sanz-Medel and coworkers [1, 54, 55] use a small scavenger column between the pump and injector. The scavenger column is packed with Chelex 100 or ion exchange resin and removes metals from the eluent stream just before it is used for the separation.
(3) In the author's experience, Teflon tubing is cleaner than PEEK.
(4) Clean containers are particularly important in many biological applications. One way to minimize contamination is to use large sample containers so that the ratio (sample volume/wall area of container) is relatively large, which minimizes the contribution of the element coming from or going into the container wall. This is often not an option for the small volumes available for enzyme or protein solutions.
(5) Some instruments are installed completely into clean room facilities. Others find it useful to keep the sample handling area enclosed with a positive pressure of filtered air to exclude dust.

7 CAPABILITIES AND REPRESENTATIVE APPLICATIONS WITH LC SEPARATIONS

This discussion is meant to illustrate the capabilities of sector instruments for speciation measurements. It is not a full review of all such studies.

7.1 High resolution and accurate mass measurements

As stated previously, the ability to separate the atomic ion of interest from polyatomic ions at the same nominal m/z value provides very high confidence that the ions measured are the ones desired. An example is shown in Figure 5.7.24. The objective is to measure chromium speciation in serum by size exclusion chromatography with ICP-MS. Although there is a substantial Cr^+ background, additional Cr from the sample elutes in two chromatographic peaks. The spectral scan in Figure 5.7.25 shows that only a small percentage of the total signal at $m/z = 52$ is from $^{52}Cr^+$. The measured m/z value for $^{52}Cr^+$ agrees well with the expected value ($m/z = 51.9405$). This identification is simplified by the fact that, in this m/z range, the atomic ion is the lightest one likely in each particular m/z window [8].

Most of the remaining ions in Figure 5.7.24 are $^{40}Ar^{12}C^+$ (hereafter just referred to as ArC^+), and there is a hint of yet a third ion on the low mass tail of ArC^+. A low resolution measurement would ascribe much too high a value for chromium concentration in this chromatographic peak. Application of a correction for ArC^+ based on that coming from the solvent would improve the situation but still ignores the possibility that the carbon from the eluting protein could increase the amount of C^+ in the plasma and the corresponding level of ArC^+ seen in the spectrum. Use of a solvent gradient with an organic modifier is another reason why the ArC^+ signal can change during the chromatogram. In the author's opinion, spectral separation of the M^+ analyte ion from the polyatomic ions is the most general solution to such overlap problems, especially when there are two or more polyatomic ions at the nominal m/z value of interest.

7.2 High sensitivity

In low resolution mode the sensitivity from a sector instrument is approximately 100 times higher than that obtained using quadrupole instruments. Of course, the sensitivity for the element of interest in the blank is also higher, so the detection limits are not automatically 100 times better. This sensitivity enhancement is of particular value in cases where the blank is low and/or where the sample is diluted extensively during the chromatographic separation. An example is

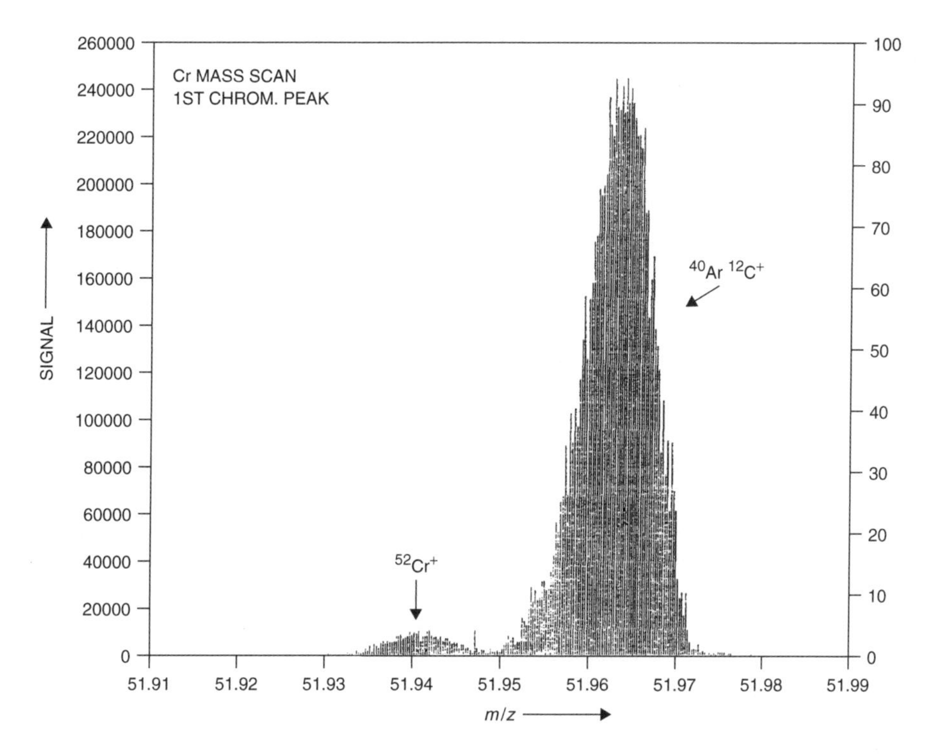

Figure 5.7.24. Mass scan during first chromatographic peak from Figure 5.7.23 showing spectral resolution of $^{52}Cr^+$ from $^{40}Ar^{12}C^+$. Reproduced from ref. 53 with permission from the American Chemical Society.

shown in Figure 5.7.25. Uranium and thorium can be readily observed at ambient levels in serum without prior chemical preconcentration, even though the separation dilutes the analyte by a factor of ~30. The retention times show that these elements are bound to proteins. Even if the sample had been contaminated, the additional uranium and thorium are still attached to proteins [53].

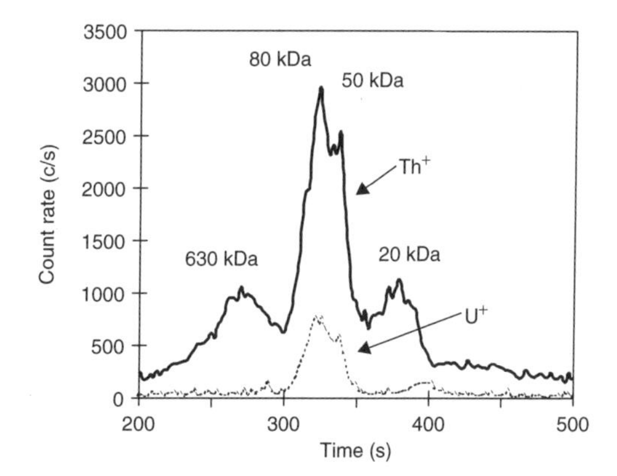

Figure 5.7.25. Selected ion chromatogram for U and Th in human serum, low spectral resolution. The concentrations are estimated to be approximately $1\,ng\,L^{-1}$ U and $3\,ng\,L^{-1}$ Th. Reproduced from ref. 53 with permission from the American Chemical Society.

7.3 Selected applications

The data shown in Figures 5.7.23–5.7.25 illustrate results obtained in the author's group with size exclusion separations and a magnetic sector ICP-MS device. Other such studies include multielement speciation in serum [56] and liver [57]. Harrington *et al.* [58] report the use of this separation method for quantification of Fe species in meat. Size exclusion chromatography has several desirable features: separations can be done at physiological pH, organic solvents that would be expected to unfold or denature proteins are usually not used, and the relation between retention time and molecular weight can be determined by calibration with known proteins. Thus, at least the molecular weights of unknown proteins can be estimated without isolated standards of the same compounds. Compared to other chromatographic methods, size exclusion generally provides poorer chromatographic resolution, however.

Sanz-Medel and coworkers have used fast protein LC separations with a magnetic sector instrument to measure speciation of several elements, especially Al in serum [54, 55]. A resolution of 4000 is sufficient to separate $^{27}Al^+$, the only Al isotope, from polyatomic ions at $m/z = 27$. Aluminum in transferrin measured by LC-ICP-MS yields the same twin chromatographic peaks as a transferrin standard, which shows that the bulk of the Al in serum is bound to transferrin. This group also studied other elements using the same chromatographic method [59] and measured metal distribution patterns in mussels using size exclusion separations [60].

Jakubowski *et al.* [61, 62] use a reversed phase, isocratic separation with a hydraulic high pressure nebulizer [63] and solvent removal to measure selenium-containing amino acids, selenomethionine, selenoethionine, selenocystamine, and selenocystine. Performance at either low or high mass resolution was compared. As expected, low resolution gave better Se^+ sensitivity, but background ions from the organic solvent were found at high resolution. Examples were $^{12}C_6H_5{}^+$ at $m/z = 77$ and $^{12}C_4H_2^{16}O_2{}^+$ and $^{12}C_4H_6^{16}O^+$ at $m/z = 82$. The latter m/z position is often thought to be 'clean' for Se. Fortunately, the H atoms ($m = 1.0078$) displace the masses of these organic ions well above those of the atomic Se^+ isotopes, and these polyatomic ions were readily separated from the corresponding Se^+ isotopes at $R = 1400$. Detection limits were 2–4 pg at low resolution and 0.4–2 pg at $R = 1400$.

Jakubowski's group [64] then applied this LC-ICP-MS method to the measurement of Se compounds in herring gull eggs. A Se-selective chromatogram obtained after injection of such an extract is shown in Figure 5.7.26. The top frame shows the original chromatogram from a 1:5 diluted extract. At least six separate Se compounds are apparent. There is little problem monitoring Se in these samples; the extracts contain a total of about 57 $\mu g\,L^{-1}$ Se. The chromatogram in the bottom frame was obtained from a diluted extract spiked with four standard compounds at 2 $\mu g\,L^{-1}$ Se each. These spikes elute with four of the peaks seen in the original chromatogram, but compounds 3, 4, 5, and 6 remain unidentified. The problem of identification of the species responsible for chromatographic peaks that do not elute with standard compounds remains endemic to ICP-MS. In its usual mode of operation, the ICP destroys structural information about the compounds. Jakubowski *et al.* [65] have also investigated Pt speciation in plants using size exclusion chromatography. Sulfur was also measured to indicate when general proteins elute.

Measurement of phosphorylation of proteins and peptides is another important potential application of LC-ICP-MS. The sensitivity of molecular mass spectrometric methods differs for different compounds, and the sensitivity generally decreases as the extent of phosphorylation increases. These molecular methods can be used to determine molecular weight and sequence, and this information can then be combined with measurement of $^{31}P^+$ and $^{32}S^+$ to both quantify the protein and determine the extent of phosphorylation. Interferences such as NOH^+ at $m/z = 31$ can be separated at $R = 4000$.

Wind *et al.* [66, 67] recently reported such measurements. Capillary LC (flow rate $\sim 4\,\mu L\,min^{-1}$) was interfaced to a microflow nebulizer (PFA 100) and a Finnigan Element. The low liquid flow rate reduces changes in sensitivity with eluent composition, but these effects still occur. To use gradient elution, the $^{31}P^+$ and $^{32}S^+$ sensitivity was measured while continuously introducing standard phosphate and cysteine in the eluent during a gradient (Figures 5.7.27(a) and (b)). The ratio of P^+/S^+ sensitivity measured for the standard mixture at each retention time was used to correct the signals for each element during subsequent separation of proteins or peptides using the same solvent gradient (Figure 5.7.27(c)). The measured P/S ratio could then be compared with the known or assumed ratio for caseins and for peptides with known phosphorylation, with good agreement. Detection limits were ~0.1 pmol.

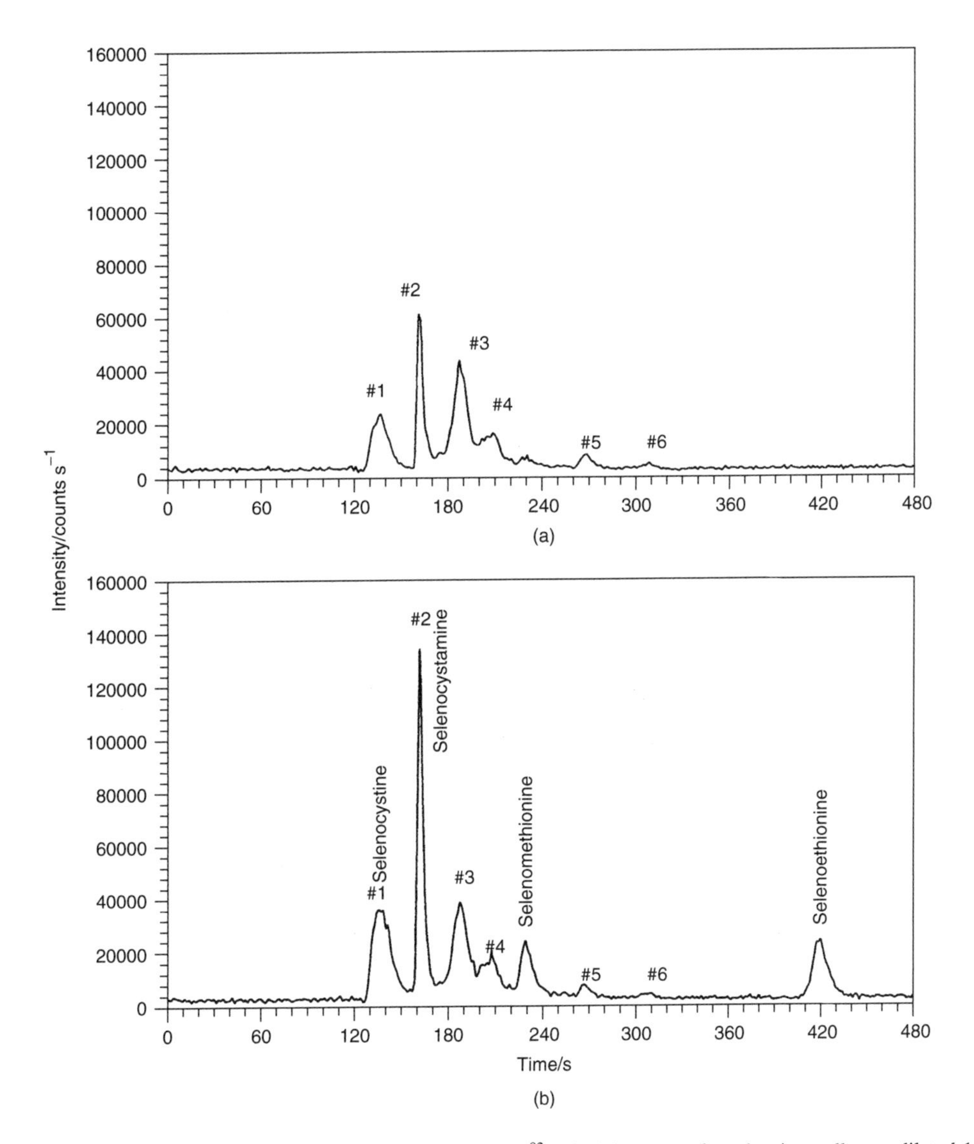

Figure 5.7.26. Selected ion chromatogram measured at $R = 400$ for ^{82}Se in (a) extracts from herring gull eggs diluted 1 : 5, and (b) diluted extract spiked with standard organoselenium compounds at $2\,\mu g\,L^{-1}$. Reproduced from ref. 64 with permission of the Royal Society of Chemistry.

This methodology will prove very useful in studies of post-translational modification of proteins and peptides.

LC-ICP-MS measurements are also valuable for analysis of DNA. Wang *et al.* [53] found that various metal cations bound to DNA fragments, even toxic metals such as Pb and Cd (Figure 5.7.28). Chromium(III) and chromate also bound to DNA, which suggested that the $CrO_4{}^{2-}$ anion was first reduced to a Cr cation, probably Cr^{3+}. The carcinogenic role of chromate is related to its ability to oxidize DNA, either directly or through intermediates. Jakubowski's group [68] quantified DNA adducts with styrene oxide using LC-ICP-MS in combination with electrospray ionization. The ICP-MS monitored $^{31}P^{+}$ at $R = 3000$. The P^{+} sensitivity was the same for P from DNA as for inorganic phosphate, so standards of the actual adducts were

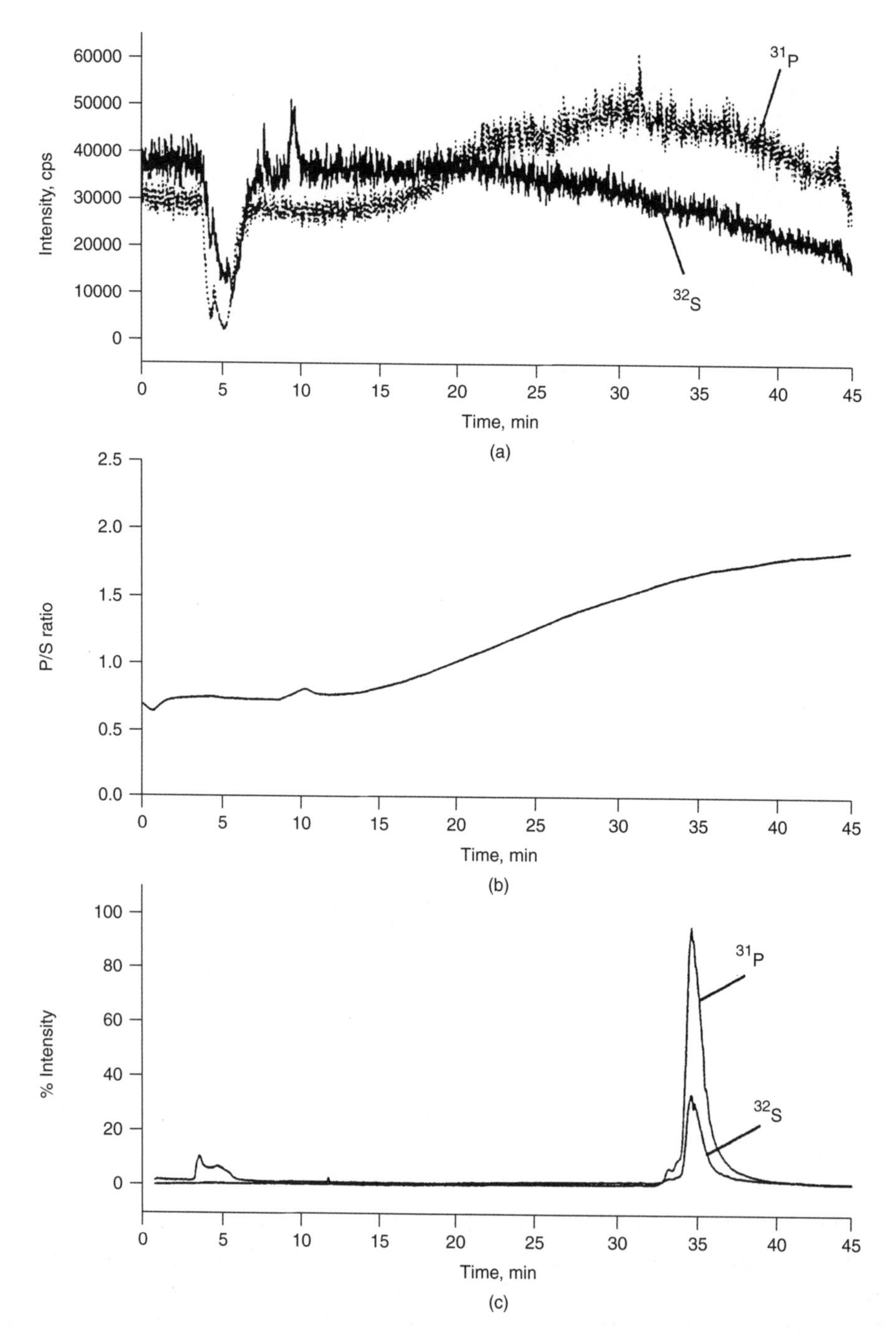

Figure 5.7.27. Determination of P/S ratio during LC separation. (a) Variation of sensitivity with eluent composition during gradient elution. (b) P/S signal ratio measured from (a). (c) Elution of α-casein with correction derived from (b). Reproduced from ref. 67 with permission from the American Chemical Society.

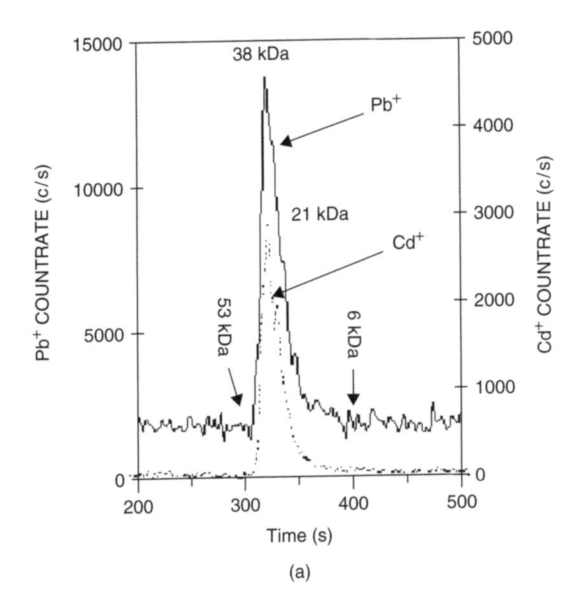

Figure 5.7.28. Selective ion chromatograms for ^{208}Pb and ^{114}Cd in mixture of DNA restriction fragments separated by size exclusion chromatography. The baseline at $m/z = 208$ represents Pb eluting from the column, tubing, etc. Additional Pb and Cd elutes with the DNA fragments of the approximate molecular weights shown. Reproduced from ref. 53 with permission from the American Chemical Society.

not necessary, which is a general advantage of atomic spectroscopic methods. The detection limit was 20 pg, or 14 modifications in 10^8 bases, which is comparable with that of other methods for detection of DNA damage.

Finally, Hassellov *et al.* [69] used field flow fractionation with ICP-MS to study trace metal adsorption onto colloidal particulates. Suspended particles of various sizes were separated and nebulized into a Finnigan Element. The major elements Si, Al, minor elements Fe and Mn, and adsorbed trace elements Cs, La and Pb could be followed into various particle size fractions.

8 OTHER SEPARATION TECHNIQUES

8.1 CE separations

Another potential niche for the high sensitivity provided by sector instruments is for CE, where the absolute amount of sample injected is very low. An ICP-MS device is a mass flow sensitive detector, so it struggles in very low flow applications, and it needs all the sensitivity it can get for CE. The process of nebulizing the CE effluent must not disturb the flow in the capillary, which is a challenge because of the suction generated by most nebulizers. Various workers have described methods to compensate for suction [70]. Using the high sensitivity of sector instruments, Prange and Schaumloffel [71] and Tsunoda *et al.* [72, 73] have recently described special nebulizer arrangements for CE separations that generate little or no suction and do not disturb the quality of the CE separation. These developments exploit the fact that, at very low liquid flow rate, the nebulized droplets evaporate quickly into solid particles, which can be transported to the ICP with very high efficiency.

An example is shown in Figure 5.7.29 from Prange and Schaumloffel [71]. An Element is used to monitor four elements (As, Se, Sb and Te) in a single injection. Two magnet settings are necessary. Despite the sequential nature of detection and the need to wait for the magnet to stabilize, these four elements can be monitored readily in electrophoretic peaks that are very sharp, 10 s peak widths in some cases. A proper interface between the CE capillary and the nebulizer is essential in this measurement. Recent studies have demonstrated separation of various isoforms of the metal storage protein metallothionein, for which the high efficiency and soft nature of CE are critical [74]. In another use of sector instruments with CE for speciation, Sanz-Medel and coworkers [75] report measurement of Hg^{2+}, $MeHg^+$ and $EtHg^+$ using a simple T-interface.

8.2 Gel electrophoresis

Electrophoresis in a flat configuration is, of course, also possible and offers the advantage of separation in two dimensions, typically isoelectric point (pI) and molecular weight. Several workers have used this format for metal speciation. The bands can be cut out of the gel, then dissolved and analyzed as solutions [76]. Laser ablation has been used to

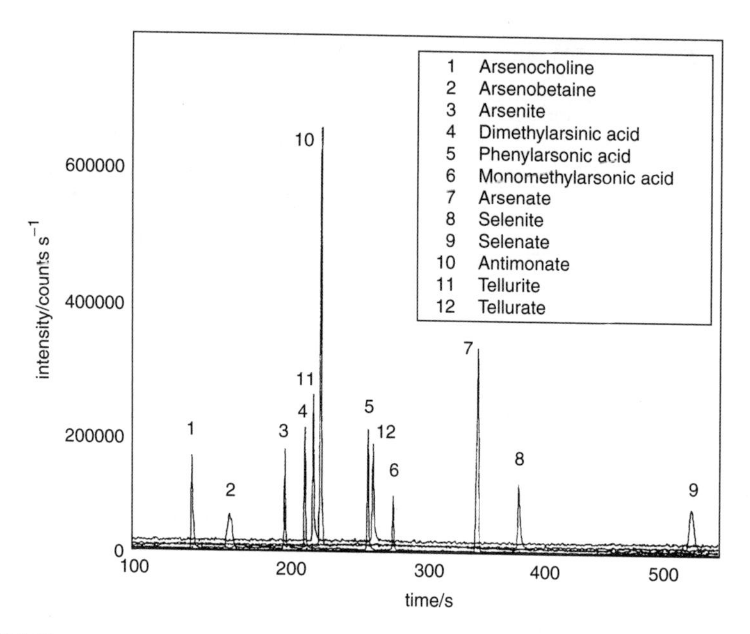

Figure 5.7.29. CE-ICP-MS separation of 12 species from four elements in one sample injection. Two magnet settings were used. The concentrations (as element) are As, Sb, Te 100 $\mu g L^{-1}$ each, and Se 1 $mg L^{-1}$. Reproduced from ref. 71 with permission of the Royal Society of Chemistry.

locate bands that contain the elements of interest without removing them from the gel [77].

8.3 GC separations

Rather quietly, GC-ICP-MS has become widely used in some important real applications, particularly for monitoring of Pb and Sn species in the marine environment. Such species are generally extracted from the samples, after derivatization if necessary. This sample preparation step also removes most interferences.

Compared with liquid phase separations, capillary GC has several general advantages:

(1) There is no nebulizer, so all the analyte can be introduced into the plasma.
(2) There is no solvent, with the accompanying advantages of a dry plasma.
(3) The chromatographic resolution and ability to separate many components are often superior for GC.
(4) The separation is tuned by changing the column temperature, which does not induce a large change in sensitivity of the ICP-MS device.

Naturally, GC is only useful for compounds that are volatile and thermally stable or can be converted into derivatives that have these properties.

There have been several publications on GC-ICP-MS with sector instruments. Krupp *et al.* [2, 3] converted Pb to Et_4Pb and introduced this compound through a capillary GC interface into a multicollector instrument. Good precision was obtained despite the short duration of the GC peaks (~3 s) because no scanning was necessary. The GC was connected to the ICP by a heated interface. They also simultaneously introduced a nebulized standard solution containing Tl for instrument optimization and mass bias corrections on-the-fly, i.e., as the Pb signals were also being measured.

Although not strictly speciation, Evans *et al.* [78] also report measurements of Hg isotope ratios using transient Hg signals from Hg vapor trapped from coal samples. Aerosols containing Tl were mixed with the Hg vapor for mass bias corrections. Four different multicollectors and three other ICP-MS instruments were tested for this application.

GC separations have also been used with scanning sector instruments. Sanz-Medel and coworkers [79] used an Element to measure ^{32}S in compounds that cause bad breath. This peak was readily resolved from $^{16}O_2^+$ at the medium resolution setting of the instrument. Of the many possible compounds in breath samples, this is a sensitive, fast and unambiguous way to find those that contain sulfur. They use a simple interface with flexible tubing so that the GC can be moved independently of the ICP-MS. Prohaska *et al.* [80] measured arsenic species from soils by both LC and GC separations. The GC capillary was simply fed into the usual concentric nebulizer, which was connected directly to the injector tube of the ICP.

9 HIGH RESOLUTION WITH MASS ANALYZERS OTHER THAN MAGNETIC SECTORS

Although magnetic sectors are by far the most common devices for high resolution measurements with an ICP, other types of mass analyzer have been or could be used for such measurements. These are mentioned briefly below, in the event that they develop further into widespread usage.

9.1 High resolution quadrupoles

The standard quadrupole operates in the first stability region and is normally limited to unit mass resolution. Douglas and coworkers [81] have described the use of quadrupoles in alternate stability regions for either high resolution or unit mass resolution of high kinetic energy ions. Compared with work in the first stability region, the quadrupole control must provide higher power to reach these new regions, which limits the mass range, but this is only a minor limitation for atomic ions such as those from an ICP. Resolution up to 8800, sufficient to separate $^{56}Fe^+$ from $^{40}Ar^{16}O^+$ to baseline with 10 % of the original ion signal remaining, has been demonstrated.

Amad and Houk [82] achieved resolution of 22 000 with a multiple pass quadrupole operated in the first stability region, although this concept has not yet been applied to ions from an ICP.

9.2 Fourier transform ion cyclotron resonance

Eyler and coworkers [83] have published one paper on use of FT-ICR-MS with ions from an ICP. This type of instrument has the highest resolution of any mass analyzer and is capable of separating different atomic ions at the same nominal mass. In more recent work, mass resolution up to 600 000, sufficient to separate $^{40}Ca^+$ from $^{40}Ar^+$ to baseline, was achieved [84]. As yet, the sensitivity is much lower than that obtained from beam-type instruments.

9.3 Time-of-flight MS

One of the most important recent developments in organic MS has been the capability to do high resolution and accurate mass measurements with TOF mass analyzers. These instruments are the basis of the very successful GC-TOF and quadrupole-TOF instruments. With orthogonal ion extraction [85], these devices now have a high duty cycle for ions produced from continuous sources such as electrospray, and they are also very effective at measuring product ions produced in tandem MS experiments.

There are presently two commercially available TOF-MS instruments with an ICP source, from LECO and GBC. Although the present ICP-TOF-MS devices were probably not designed with this in mind, resolution sufficient to separate many of the worst cases of polyatomic ion overlap ought to be possible, with the potential duty cycle advantages of TOF for transient samples. In a recent paper, Adams and coworkers [86] describe the use of TOF-MS for multielement speciation in cytosols.

10 HIGH RESOLUTION MEASUREMENTS WITH PLASMAS OTHER THAN THE ICP

At first glance, it should be possible to use any of the alternate plasmas with a magnetic sector instrument, although few such measurements have been reported as yet. Nam *et al.* [33] used a He ICP with a home-made Mattauch–Herzog instrument, partly in the hope that the array detector would perform better with He^+ at one end of it than with Ar^+ in the middle. As usual with a He plasma, Ar^+, NO^+, or other species persisted as major ions anyway.

11 CONCLUSION

Magnetic sector instruments provide both high sensitivity and selectivity for speciation measurements. If ions in the same m/z range are monitored so that the magnetic field setting is not changed, the scan speed of a sector is comparable to that of quadrupoles. A longer settling time ($\sim$300 ms) is needed if the magnetic field setting is changed. Their spectra acquisition rate is much slower than that of TOF instruments, but they provide much higher basic signal-to-noise ratios than present TOF devices. It is expected that usage of sector instruments in speciation measurements will increase with the newer devices that are dependable and simple to use. Multicollector instruments also have a role in speciation measurements, as isotope ratios can be measured on transient peaks with high precision without scanning.

12 ACKNOWLEDGEMENTS

The Ames Laboratory is operated for the US Department of Energy by Iowa State University under Contract W-7405-Eng-82. This work was supported in the Chemical and Biological Sciences Program by the Director of Science, Office of Basic Energy Sciences, Division of Chemical Sciences. The author thanks Daniel R. Wiederin for providing Figure 5.7.10, David B. Aeschliman for measuring the spectra in Figure 5.7.6, the latter worker and Jill Ferguson for critical comments on the manuscript, and Jin Wang, Dawn Dreessen and Regine Schoenherr for making many of the speciation measurements reported. Early speciation work by our group was performed on a magnetic sector MS provided by CETAC Technologies, Omaha, NE. Chuck Douthitt provided valuable literature surveys.

13 REFERENCES

1. Marchante-Gayon, J. M., Sariego Muniz, C., Garcia Alonso, J. I. and Sanz-Medel, A., Analysis of biological materials by double focusing ICP-MS, in *Advances in Atomic Spectrometry*, Vol. 7, Sneddon, J. (Ed.), Elsevier, Amsterdam, 2001; *Anal. Chim. Acta*, **400**, 307 (1999).
2. Krupp, E., Pecheyran, C., Pinaly, H., Montelica-Heino, M., Koller, D., Young, S. M. M., Brenner, I. B. and Donard, O. F. X., *Spectrochim. Acta, Part B*, **56**, 1219 (2001).
3. Krupp, E. M., Pecheyran, C., Meffan-Man, S. and Donard, O. F. X., *Fresenius' J. Anal. Chem.*, **370**, 573 (2001).
4. Roboz, J., *Introduction to Mass Spectrometry; Instrumentation and Techniques*, Wiley-Interscience, 1968, New York, Chapters 2 and 3.
5. Septier, A., *Focusing of Charged Particles*, Academic Press, New York, 1967.
6. Wollnik, H., *Optics of Charged Particles*, Academic Press, San Diego, CA, 1987.
7. Kerwin, L., Ion Optics, in *Mass Spectrometry*, McDowell, C. A. (Ed.), McGraw-Hill, New York, 1963.
8. Becker, J. S. and Dietze, H.-J., *J. Anal. At. Spectrom.*, **12**, 881 (1997).
9. Houk, R. S. and Niu, H. S., *Spectrochim. Acta, Part B*, **51**, 779 (1996).
10. Bradshaw, N., Hall, E. F. H. and Sanderson, N. E., *J. Anal. At. Spectrom.*, **4**, 801 (1989).
11. Giessman, U. and Greb, U., *Fresenius' J. Anal. Chem.*, **350**, 186 (1994).
12. Feldmann, I., Tittes, W., Jakubowski, N., Stuewer, D. and Giessmann, U., *J. Anal. At. Spectrom.*, **9**, 1007 (1994).
13. Gray, A. L., *J. Anal. At. Spectrom.*, **1**, 247 (1986).
14. Appelblad, P. K., Rodushkin, I. and Baxter, D. C., *J. Anal. At. Spectrom.*, **15**, 359 (2000).
15. Freedman, P., NU Plasma, personal communication, 2000.
16. Vaughan, M.-A., Horlick, G. and Tan, S., *J. Anal. At. Spectrom.*, **2**, 765 (1987); Horlick, G. and Montaser, A., Analytical characteristics of ICPMS, in *ICPMS*, Montaser, A. (Ed.), Wiley-VCH, New York, 1998, p. 527.
17. Vanhaecke, F., Dams, R. and Vandecasteele, C., *J. Anal. At. Spectrom.*, **8**, 433 (1993).
18. Jiang, S.-J., Stevens, M. A. and Houk, R. S., *Anal. Chem.*, **60**, 1217 (1988); Sakata, K. and Kawabata, K., *Spectrochim. Acta, Part B*, **49**, 1027 (1994); Tanner, S. D., *J. Anal. At. Spectrom.*, **10**, 905 (1995).
19. Vanhaecke, F., Saverwyns, S., De Wannemacker, G., Moens, L. and Dams, R., *Anal. Chim. Acta*, **419**, 55 (2000).
20. Gillson, G., Douglas, D. J., Fulford, J. E., Halligan, K. W. and Tanner, S. D., *Anal. Chem.*, **60**, 1472 (1988); Tanner, S. D., *Spectrochim. Acta, Part B*, **47**, 809 (1992).
21. Nonose, N. and Kubota, M., *J. Anal. At. Spectrom.*, **16**, 551 (2001).
22. Sakata, K., Yamada, N. and Sugiyama, N., *Spectrochim. Acta, Part B*, **56**, 1249 (2001).

23. Wollnik, H., *J. Mass Spectrom.*, **34**, 991 (1999).
24. Bandura, D. R., Tanner, S. D., Baranov, V. I., Koyanagi, G. K., Lavrov, V. V. and Bohme, D. K., Ion–molecule chemistry solutions to the ICP-MS analytical challenges, in *Plasma Source Mass Spectrometry: the New Millennium*, Holland, G. and Tanner, S. D. (Eds), Royal Society of Chemistry, Cambridge, UK, 2001, p. 141.
25. Eiden, G. C., Barinaga, C. J. and Koppenaal, D. W., *Rapid Commun. Mass Spectrom.*, **11**, 37 (1997).
26. Tanner, S. D., Baranov, V. I. and Bandura, D. R., *Spectrochim. Acta, Part B*, **57**, 136 (2002).
27. Douglas, D. J., *J. Am. Soc. Mass Spectrom.*, **9**, 101 (1998).
28. Tanner, S. D., Baranov, V. I. and Bandura, D. R., Reaction chemistry and collisional processes in multipole devices, in *Plasma Source Mass Spectrometry: the New Millennium*, Holland, G. and Tanner, S. D., (Eds), Royal Society of Chemistry, Cambridge, UK, 2001.
29. Bandura, D. R., Baranov, V. I. and Tanner, S. D., *Fresenius' J. Anal. Chem*, **370**, 454 (2001).
30. Yamada, N., Takahashi, J. and Sakata, K., *J. Anal. At. Spectrom.*, **17**, 1213 (2002); *Spectrochim. Acta, Part B*, submitted (2002).
31. Bandura, D. R. and Tanner, S. D., *At. Spectrosc.*, **20**, 69 (1999); Bandura, D. R., Baranov, V. I. and Tanner, S. D., *J. Anal. At. Spectrom.*, **15**, 921 (2000).
32. Moens, L. J., Vanhaecke, F., Bandura, D. R., Baranov, V. I. and Tanner, S. D., *J. Anal. At. Spectrom.*, **16**, 991 (2001).
33. Marriott, P., Fletcher, R., Cole, A., Beaumont, I., Lofthouse, J., Bloomfield, S. and Miller, P., *J. Anal. At. Spectrom.*, **13**, 1021 (1998).
34. Olesik, J. W., *Appl. Spectrosc.*, **51**, 158A (1997).
35. Houk, R. S., Winge, R. K. and Chen, X., *J. Anal. At. Spectrom.*, **12**, 1139 (1997).
36. Walder, A. J., Koller, D., Reed, N. M., Hutton, R. C. and Freedman, P. A., *J. Anal. At. Spectrom.*, **8**, 1037 (1993); Walder, A. J. and Furuta, N., *Anal. Sci. (Jpn)*, **9**, 675 (1993); Halliday, A. N., Lee, D.-C., Christenson, J. N., Walder, A. J., Freedman, P. A., Jones, C. E., Hall, C. M., Yi, W. and Teagle, D., *Int. J. Mass Spectrom. Ion Process.*, **146/147**, 21 (1995).
37. Cromwell, E. and Arrowsmith, P., *J. Am. Soc. Mass Spectrom.*, **7**, 458 (1996).
38. Nam, S., Montaser, A. and Cromwell, E. F., *Appl. Spectrosc.*, **52**, 161 (1998).
39. Burgoyne, T. W., Hieftje, G. M. and Hites, R. A., *J. Anal. At. Spectrom.*, **12**, 1149 (1997); *Anal. Chem.*, **69**, 485 (1997); Solyom, D. A., Burgoyne, T. W. and Hieftje, G. M., *J. Anal. At. Spectrom.*, **14**, 1101 (1999).
40. Solyom, D. A., Gron, O. A., Barnes IV, J. H. and Hieftje, G. M., *Spectrochim. Acta, Part B*, **56**, 1717 (2001).
41. Barnes IV, J. H., Sperline, R., Denton, M. B., Barinaga, C. J., Koppenaal, D. and Hieftje, G. M., *Anal. Chem.*, **74**, 5317 (2002).
42. Minnich, M. G. and Houk, R. S., *J. Anal. At. Spectrom.*, **13**, 167 (1998).
43. Scott, R. H., Fassel, V. A., Kniseley, R. N. and Nixon, D. R., *Anal. Chem.*, **46**, 75 (1974).
44. Wu, M., Madrid, Y., Auxier, J. A. and Hieftje, G. M., *Anal. Chim. Acta*, **286**, 155 (1994).
45. Tarr, M., Zhu, G. and Browner, R. F., *Appl. Spectrosc.*, **45**, 1424 (1992).
46. Gonzalez LaFuente, J. M., Marchante-Gayon, J. M., Fernandez Sanchez, M. L. and Sanz-Medel, A., *Talanta*, **50**, 207 (1999).
47. Olesik, J., Hensman, C., Rabb, S. and Rago, D., Sample introduction, plasma–sample interactions, ion transport and ion–molecule reactions: fundamental understanding and practical improvements in ICP-MS, in *Plasma Source Mass Spectrometry: the New Millennium*, Holland, G. and Tanner, S. D. (Eds), Royal Society of Chemistry, Cambridge, UK, 2001, p. 5.
48. Li, J., Umemura, T., Odake, T. and Tsunoda, K., *Anal. Chem.*, **73**, 1416 (2001).
49. Wang, L., May, S. W., Browner, R. F. and Pollock, S. H., *J. Anal. At. Spectrom.*, **11**, 1137 (1997).
50. Elder, R. C., University of Cincinnati, personal communication, 1988.
51. Heumann, K. G., Rottmann, L. and Vogl, J., *J. Anal. At. Spectrom.*, **9**, 1351 (1994).
52. Garcia Alonso, J. I., Encinar, J. R., Muniz, C. S., Marchante Gayon, J. M. M. and Sanz Medel, A., Isotope dilution analysis for trace metal speciation, in *Plasma Source Mass Spectrometry: the New Millennium*, Holland, G. and Tanner, S. D. (Eds), Royal Society of Chemistry, Cambridge, UK, 2001, p. 327.
53. Wang, J., Houk, R. S., Dreessen, D. and Wiederin, D. R., *J. Am. Chem. Soc.*, **120**, 5793 (1998).
54. Lopez Garcia, A., Blanco Gonzalez, E. and Sanz-Medel, A., *Mikrochim. Acta*, **112**, 19 (1993).
55. Belen Soldado Cabazuelo, A., Montes-Bayon, M., Blanco Gonzalez, E., Garcia Alonso, J. I. and Sanz-Medel, A., *Analyst*, **123**, 865 (1998).
56. Wang, J., Houk, R. S., Dreessen, D. and Wiederin, D. R., *J. Biol. Inorg. Chem.*, **4**, 546 (1999).
57. Wang, J., Dreessen, D., Wiederin, D. R. and Houk, R. S., *Anal. Biochem.*, **288**, 89 (2001).
58. Harrington, C. F., Elahi, S., Merson, S. A. and Ponnampalavanar, P., *Anal. Chem.*, **73**, 4422 (2001).
59. Montes-Bayon, M., Soldado Cabazuelo, A. B., Gonzalez, E. B., Garcia Alonso, J. I. and Sanz-Medel, A., *J. Anal. At. Spectrom.*, **14**, 947 (1999).
60. Ferrarello, C. N., del Rosario Fernandez de la Campa, M., Sariego Muniz, C. and Sanz-Medel, A., *Analyst*, **125**, 2223 (2000).
61. Thomas, C., Jakubowski, N., Stuewer, D., Klockow, D. and Emons, H., *J. Anal. At. Spectrom.*, **12**, 1221 (1998).
62. Feldmann, I., Jakubowski, N., Stuewer, D. and Thomas, C., *J. Anal. At. Spectrom.*, **15**, 371 (2000).
63. Jakubowski, N., Feldmann, I., Stuewer, D. and Berndt, H., *Spectrochim. Acta, Part B*, **47**, 119 (1992); Berndt, H., *Fresenius' J. Anal. Chem.*, **331**, 321 (1988); Jakubowski, N.,

Jepkens, B., Stuewer, D. and Berndt, H., *J. Anal. At. Spectrom.*, **9**, 193 (1994).
64. Jakubowski, N., Stuewer, D., Klockow, D., Thomas, C. and Emons, H., *J. Anal. At. Spectrom.*, **16**, 135 (2001).
65. Klueppel, D., Jakubowski, N., Messerschmidt, J., Stuewer, D. and Klockow, D., *J. Anal. At. Spectrom.*, **13**, 255 (1998).
66. Wind, M., Edler, M., Jakubowski, N., Linsheid, M., Wesch, H. and Lehman, W. D., *Anal. Chem.*, **73**, 29 (2001).
67. Wind, M., Wcclıst, H. and Lehmann, W. D., *Anal. Chem.*, **73**, 3006 (2001).
68. Siethoff, C., Feldmann, I., Jakubowski, N. and Linscheid, M., *J. Mass Spectrom.*, **34**, 421 (1999).
69. Hassellov, M., Lyven, B. and Beckett, R., *Environ. Sci. Technol.*, **33**, 4528 (1999).
70. Majidi, V. and Miller-Ihli, N., *Analyst*, **123**, 863 (1998); Day, J. A., Sutton, K. L., Somon, R. S. and Caruso, J. A., *Analyst*, **125**, 819 (2000).
71. Prange, A. and Schaumloffel, D., *J. Anal. At. Spectrom.*, **14**, 1329 (1999); *Fresenius' J. Anal. Chem.*, **364**, 452 (1999).
72. Li, J., Umemura, T., Odake, T. and Tsunoda, K., *Anal. Chem.*, **73**, 1416 (2001).
73. Li, J., Umemura, T., Odake, T. and Tsunoda, K., *Anal. Chem.*, **73**, 5992 (2001).
74. Wang, Z. and Prange, A., *Anal. Chem.*, **74**, 626 (2002).
75. Silva da Rocha, M., Soldado, A. B., Blanco-Gonzalez, E. and Sanz-Medel, A., *J. Anal. At. Spectrom.*, **15**, 513 (2000).
76. Lustig, S., De Kimpe, J., Cornelis, R., Schramel, P. and Michalke, B., *Electrophoresis*, **20**, 1627 (1999); Lustig S., De Kimpe, J., Cornelis, R. and Schramel, P., *Fresenius' J. Anal. Chem.*, **363**, 484 (1999).
77. Alt, F., Messerschmidt, J. and Weber, G., *Anal. Chim. Acta*, **359**, 65 (1998); Becker, J. S., Boulyga, S. F., Becker, J. S., Damoc, N. -E. and Przybylski, M. *Int. J. Mass Spectrom.*, submitted (2003).
78. Evans, R. D., Hintelmann, H. and Dillon, P. J., *J. Anal. At. Spectrom.*, **16**, 1064 (2001).
79. Rodriguez-Fernandez, J., Montes-Bayon, M., Pereiro, R. and Sanz-Medel, A., *J. Anal. At. Spectrom.*, **16**, 1051 (2001).
80. Prohaska, T., Pfeffer, M., Tyulipan, M., Stingeder, G., Mentler, A. and Wenzel, W. W., *Fresenius' J. Anal. Chem.*, **364**, 467 (1999).
81. Ying, J.-F. and Douglas, D. J., *Rapid Commun. Mass Spectrom.*, **10**, 649 (1996); Du, Z., Douglas, D. J. and Konenkov, N. V., *J. Anal. At. Spectrom.*, **14**, 1111 (1999).
82. Amad, M. H. and Houk, R. S., *Anal. Chem.*, **70**, 4885 (1998); *J. Am. Soc. Mass Spectrom.*, **11**, 407 (2000).
83. Milgram, K. E., White, F. M., Goodner, K. L., Watson, C. H., Koppenaal, D. W., Barinaga, C. J., Smith, B. J., Winefordner, J. D., Marshall, A. G., Houk, R. S. and Eyler, J. R., *Anal. Chem.*, **69**, 3714 (1997).
84. Watson, C. H., Houk, R. S. and Eyler, J. R., unpublished results.
85. Guilhaus, M., Selby, D. and Mlynski, V., *Mass Spectrom. Rev.*, **19**, 65 (2000).
86. Infante, H. G., Van Campenhout, K., Blust, R. and Adams, F. C., *J. Anal. At. Spectrom.*, **17**, 79 (2002).

5.8 On-line Elemental Speciation with Functionalised Fused Silica Capillaries in Combination with DIN-ICP-MS

J. Bettmer

Johannes Gutenberg-Universität Mainz, Germany

1 INTRODUCTION

Hyphenated techniques are usually necessary tools for metal speciation analysis. The combination of a chromatographic or electrophoretic separation unit with an element-selective detection system can provide qualitative as well as quantitative information about the different forms in which an element occurs. Many papers have reviewed the frequently used techniques and their applications [1–11].

As the detection method in metal speciation ICP-MS (inductively coupled plasma-mass spectrometry) has become more and more popular owing its excellent properties concerning detection limit, sensitivity, selectivity and the simplicity to couple both gas and liquid chromatographic systems to it. Another important advantage is the possibility of monitoring different isotopes of one element which has been proved a success for identifying artefact formations and degradation processes during analysis [12–14]. In particular, isotope-labelled species nowadays contribute to a better understanding for such processes combined with the possibility of applying isotope dilution for accurate and precise quantification [13, 15].

This paper aims to present an alternative method for the direct elemental speciation using ICP-MS. The application of an on-line separation during sample transport to the plasma unit will be demonstrated for the examples of chromium and mercury speciation. Instead of chromatographic columns, chemically modified capillaries as used in capillary electrophoresis (CE) allow a selective retention of one of the relevant species.

2 EXPERIMENTAL SET-UP

For introducing the liquid sample into the plasma a direct injection nebulizer (DIN) was used. In

Handbook of Elemental Speciation: Techniques and Methodology R. Cornelis, H. Crews, J. Caruso and K. Heumann
 ISBN: 0-471-49214-0

particular, for elements showing memory effects such as mercury this kind of sample introduction is favourable because of the minimisation of this negative effect. The sample transport by the direct injection nebulizer was obtained through fused silica capillaries (400–700 mm × 0.15 mm o.d., 50 μm i.d., encased in a Teflon tube). The nebulizer gas was connected to the carrier gas supply (range 0–2 L min^{-1}) while the nebulizer auxiliary gas was connected to the blend gas supply (range 0–2 L min^{-1}) of the ICP-MS system. The liquid transport was performed using a gas displacement pump with Milli-Q water as carrier solution. A metal-free injection valve was a component of the pump module of the DIN which allowed time-controlled injections. An additional argon supply (150 psi) was necessary for the DIN pump for the loading of the loop with the sample. For the DIN two different sample introduction capillaries are available, with 50 and 75 μm i.d. For these investigations the 50 μm i.d. capillaries were chosen for a more efficient interaction between capillary surface and analytes. The general operation conditions for the ICP-MS in combination with the DIN are shown in Table 5.8.1, whereas the schematic adaptation of the DIN to the plasma torch is presented in Figure 5.8.1. For these studies ICP-QMS was used in the time-resolved mode and peak height measurements were chosen for data evaluation.

Table 5.8.1. ICP-MS (HP 4500) operating conditions with direct injection nebulizer (DIN).

Parameter	Value (DIN)
RF power	1320 W
Reflected power	0–8 W
GDP pressure (DIN)	300–420 psi
Nebulizer pressure (DIN)	75 psi
Gas flow rate	
Plasma	13–14 L min^{-1}
Auxiliary	0.8–1.5 L min^{-1}
Carrier	0.4–0.7 L min^{-1}
Capillary tip position	0.3 mm in front of nebulizer body
Sample depth	9.5 mm
Sample cone	Ni, 1.0 mm orifice
Skimmer cone	Ni, 0.4 mm orifice
Dwell time	300 ms

3 PREPARATION AND PROCEDURE OF CAPILLARY MODIFICATIONS

In the field of capillary electrophoresis the surface modification of fused silica capillaries is a common technique for the suppression or reversal of the electroosmotic flow [16]. Different techniques, mostly based on the chemical binding of silane

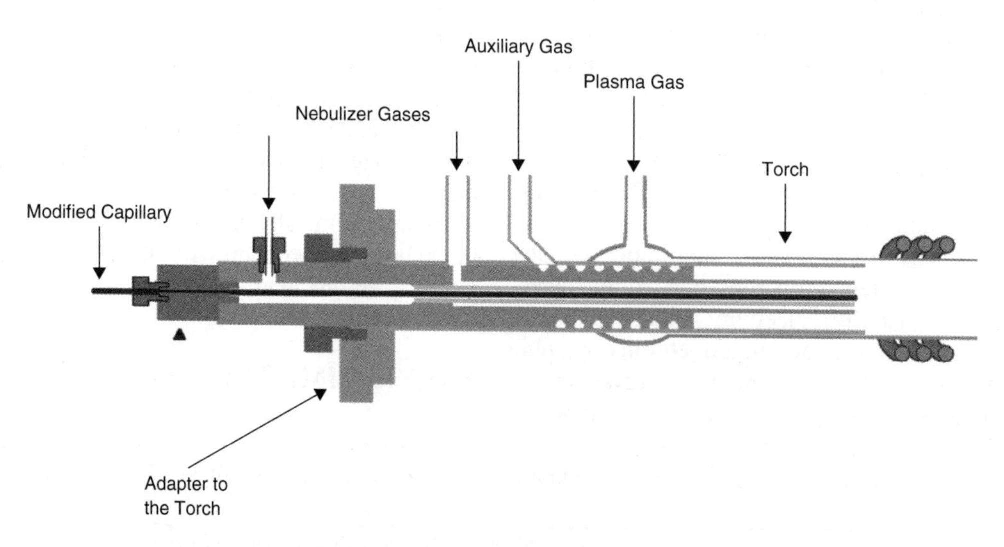

Figure 5.8.1. Adaptation of the DIN to the ICP torch.

reagents with the desired functional groups to the surface have been carried out. This technique was adapted to our purpose of modifying the inner surface of a fused silica capillary with a cationic and anionic exchanger material.

The activation of the inert surface was obtained by rinsing the capillaries (50 µm i.d. × 360 µm o.d., length 40–70 cm) with 1 M NaOH. A minimum of 30 min was necessary for a stable activation of the capillary inert surface. Afterwards the capillary was flushed with Milli-Q water (15 min) in order to remove the majority of sodium ions and dried by purging with nitrogen (5 bar) at 80 °C for 1 h.

After the activation of the capillary surface the cationic exchanger coating was obtained by rinsing the capillaries with the coating reagent. The coating solution consisted of 5 % (v/v) 2-(4-chlorosulfonylphenyl)-ethyltrichlorosilane in methylene chloride. The capillary was rinsed with this solution for 2 h at a temperature of 60 °C with a

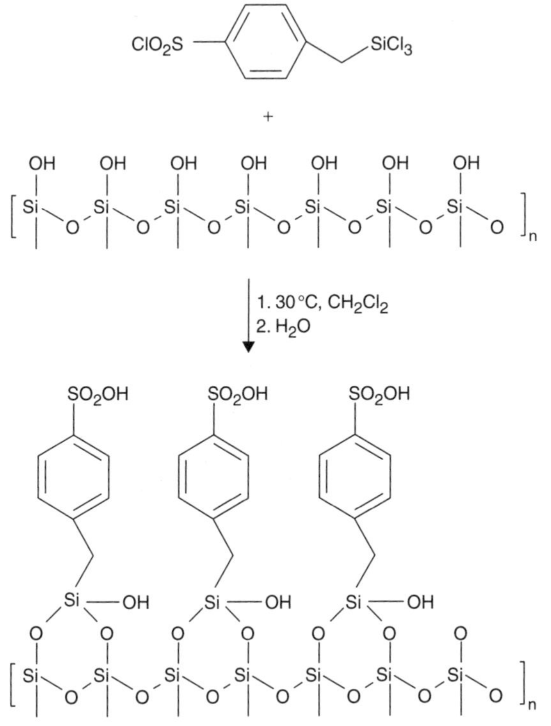

ClO_2S — $SiCl_3$

+

1. 30 °C, CH_2Cl_2
2. H_2O

Figure 5.8.2. Scheme of the coating procedure (cation exchanger).

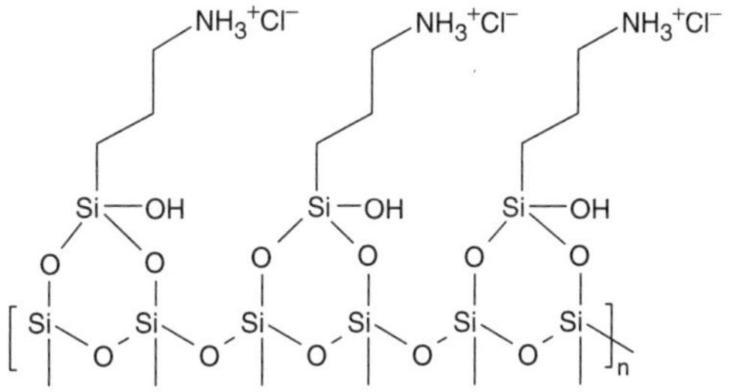

$NH_3^+Cl^-$ $NH_3^+Cl^-$ $NH_3^+Cl^-$

Figure 5.8.3. Schematic surface of a modification with an anion exchanger.

pressure of 2 bar. This dynamic coating procedure was stopped every 15 min so that the capillary was coated for 5 min in a static mode. After the coating the reagent not bound to the surface was removed with methylene chloride (20 min) and with water (15 min) at room temperature. Before use the capillaries were purged with nitrogen (5 bar) at 80 °C for 2 h. Figure 5.8.2 presents the scheme of this coating procedure for the example of a cation exchanger material [17].

In the case of an anion exchanger material, 3-aminopropyltrimethoxysilane (Figure 5.8.3) was used as the bifunctional reagent. The coating procedure was analogously performed as described for the cation exchanger.

4 CHARACTERISATION OF THE MODIFIED CAPILLARIES

4.1 Stability and capacity

The stability of the coatings was investigated to optimise the efficiency of the capillaries without destroying the surface while removing retained species. Special attention was paid to the pH values of the introduced samples. These solutions were prepared on a daily base to guarantee the best possible accuracy of these data. Testing the possible pH range was necessary because for some of the exchange resins no pH working range has been determined so far. For this investigation short pieces of 1.5–2.5 cm were connected between the nebulizer and the delivering pump flushed with solutions of different pH ranging from 1.2 to 10.5. For the sulfur-containing columns the m/z

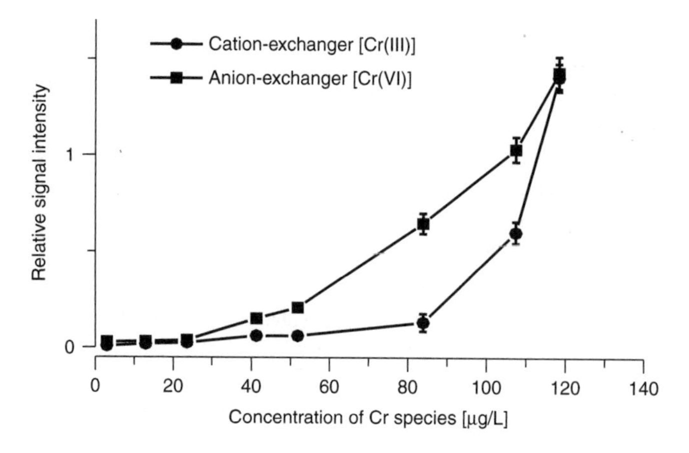

Figure 5.8.4. Capacity studies of the modified capillaries.

ratios 32 (S^+) and 64 ($SO_2{}^+$) were constantly measured. For the nitrogen-containing columns m/z 54 (ArN^+) was monitored. Although these masses have a naturally a high background level a destruction of the coating could be recognised by a significant increase of these levels. It turned out that the coatings investigated suffered destruction at a pH lower than 1.8 and higher than 9.0. Hence, for the later experiments usually pH 2 or higher was used as the most acidic solution to remove retained species from the capillary. The stable pH ranges were later investigated for the most efficient preconcentration or retention of the species.

The next step involved the evaluation of the breakthrough capacity of the columns. The used coatings vary in their ion exchange strength. Therefore it was necessary to guarantee that the coatings were not overloaded. Figure 5.8.4 shows a capacity study for chromium species. Each data point represents a series of five injections (10 μL) with given concentration. The breakthrough limit for these capillaries can be set between 40 and 50 $\mu g\,L^{-1}$.

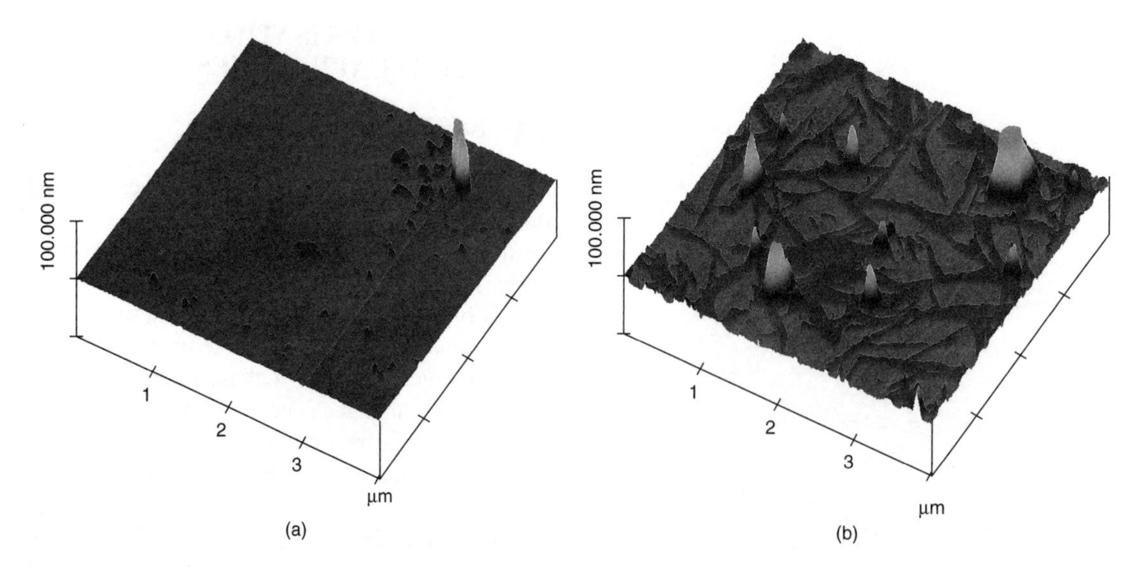

Figure 5.8.5. (a) AFM picture of an untreated surface; (b) AFM picture of an activated surface; (c) AFM picture after modification with an anion exchanger; (d) AFM picture after modification with a cation exchanger.

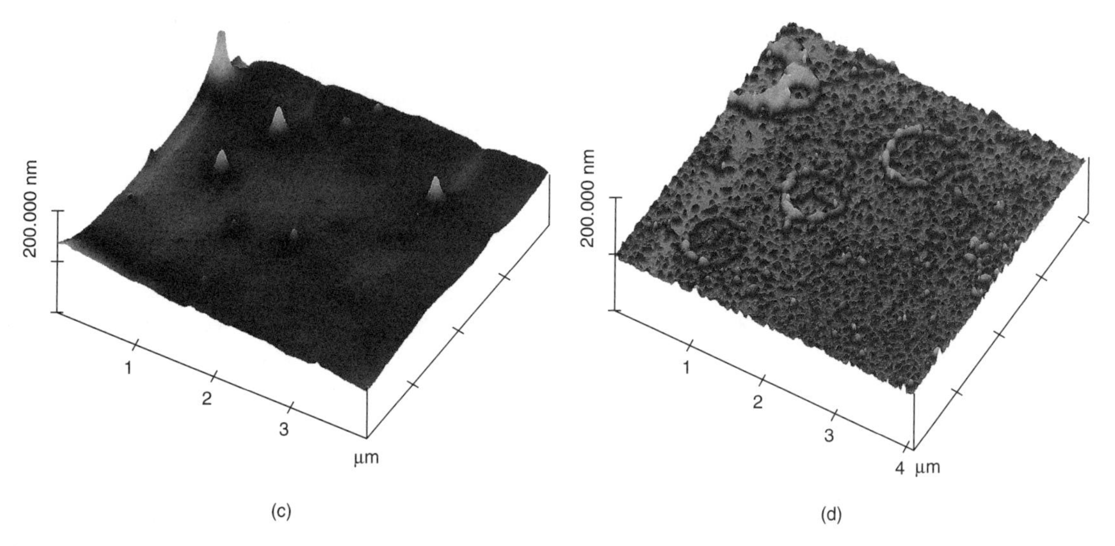

Figure 5.8.5. *(continued)*

4.2 Atomic force microscopy

To be sure that every group has been derivatized, and then to be able to perform quantitative retention of the species, a check of the internal surface of the capillaries has been made via AFM (atomic force microscopy).

Figure 5.8.5(a) shows the internal surface of silica fused capillaries as they are before any treatment. The surface is smooth, while some impurities are present and represented as light peaks. Figure 5.8.5(b) shows the surface after flushing it with NaOH: the darker parts of the surface did not react, while the light structured parts represent the removal of SiO_2 layers. Finally, the activated surface has been rinsed with the derivatizing reagents: Figures 5.8.5(c) and (d) show the result of this final step. While using 3-aminopropyltrimethoxysilane the surface still looks smooth and even (Figure 5.8.5(c)), and derivatization occurred on the major part; otherwise a large part seemed to remain unchanged when 2-(4-chlorosulfonylphenyl)-ethyltrichlorosilane was used (Figure 5.8.5(d)).

Impurities are anyway still present in every step of the modification, and might be responsible for a nonquantitative coating of the capillaries. The second reason for this observation is that the inability to activate the surface completely may lead to a nonquantitative retention of interfering species during the analysis, so that the exchange capacity is limited.

5 APPLICATIONS OF THE MODIFIED CAPILLARIES IN ELEMENTAL SPECIATION

5.1 Modified capillaries for the separation of spectral interferences in ICP-QMS

In ICP-QMS analysis spectral interferences have to be considered in order to guarantee the quality of analytical measurements. A new way of chemical separation of elements responsible for the existence of spectral interferences such as cluster ions or doubly charged ions was developed by the use of the modified capillaries. Anionic and cationic exchanger coatings have proven to retain selectively anions and cations, respectively, which could strongly interfere with the determination of selected elements. Examples were shown for platinum group elements such as Pt, Pd, and Rh (matrix separation concerning the interfering elements Cu, Pb, Rb, Sr, and Hf) and for As (monoisotopic ^{75}As is interfered with by the formation of $^{40}Ar^{35}Cl$) [17–19].

5.2 Chromium speciation

The differentiation between the two dominant redox species of chromium (Cr(III) and Cr(VI)) has gained importance during the last decades. Since the carcinogenic potential of hexavalent chromium has been realised, many analytical methods have been developed. They mainly use hyphenated techniques, the coupling of a chromatographic separation and mostly element-selective detection unit. In the preceding chapters techniques for applying ICP-MS as detector have been mentioned.

In contrast to conventional methods our development is without the adaptation of a chromatographic system to the ICP-MS detector. The system achieves the separation of both species by the selective retention of one species depending on the kind of modified capillary. In the case of retaining Cr(VI), which is favourable in the case of higher Cr(III) concentrations, the use of an anion exchanger material (illustrated as an example in Figure 5.8.3) is preferred. For Cr(III) preconcentration a cation exchanger is consequently applied.

For developing a separation method the most influential factor, the pH, was optimised in order to retain one of the species, while the other can quantitatively be eluted. The optimum pH for the cationic exchanger is 5.5, while the use of the anionic exchanger for the retention of Cr(VI) needs a pH of about 6.0. An example of the preconcentration of Cr(VI) is given in Figure 5.8.6, demonstrating that the relevant species will be quantitatively retained in the capillary after six injections. However, flushing the capillary with nitric acid releases the retained species and the signal intensity for this accumulated chromium was compared with an injection of 10 $\mu g\,L^{-1}$ Cr(III). As shown, these two results are comparable and emphasise the possibilities given by this method.

In order to prove its reliability the method was applied for the determination of chromium in a reference material CRM 545 (welding dust). The sample preparation was carried following basically the instructions given in the EC certification report [20]. In brief: the dust was weighed, diluted in a 2 % NaOH–3 % Na_2CO_3 solution and agitated in a heated ultrasonic bath for 20 min. Prior to separation and analysis the solution was diluted by factor of 20 000 and the pH was adjusted to 5.5 using dilute hydrochloric acid.

In Figure 5.8.7 a typical time-resolved picture of the species-selective flow injection procedure is demonstrated. Six replicate injections of the dissolved sample show the signal for Cr(VI) while the injection of nitric acid releases the amount of Cr(III) initially preconcentrated in the capillary.

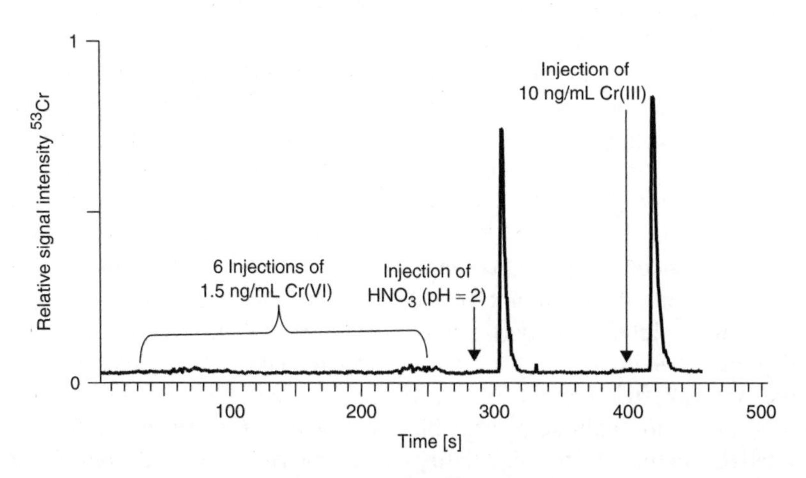

Figure 5.8.6. Retention and elution of Cr(VI) using an anion exchanger capillary.

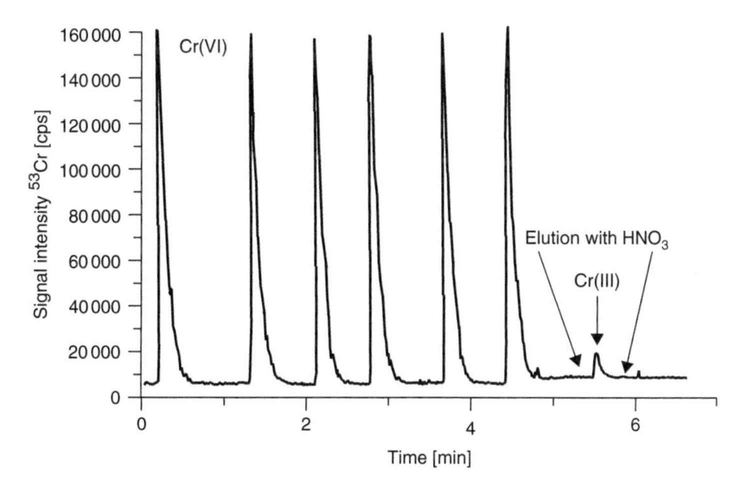

Figure 5.8.7. Time-resolved measurement of CRM 545 (six replicate injections of the dissolved sample [21]).

Table 5.8.2. Determination of chromium species in welding dust CRM 545.

	This work ($g\,kg^{-1}$)	Certified ($g\,kg^{-1}$)
Total leachable chromium	38.3 ± 0.9	39.5 ± 1.3
Cr(VI)	38.1 ± 1.6	40.2 ± 0.6
Cr(III)	0.27 ± 0.08	Not certified

The results obtained with this newly developed method are summarised in Table 5.8.2 and show excellent agreement with certified values. Recent publications have reported small amounts of Cr(III) in the sample which could also be observed during our investigations.

These results show the applicability of the developed method for the differentiation of Cr(III) and Cr(VI) without the necessity of a chromatographic system [21]. The detection limits obtained are $8\,ng\,L^{-1}$ for Cr(III) and $31\,ng\,L^{-1}$ for Cr(VI), respectively.

5.3 Mercury speciation

Mercury is one of the elements for which analytical developments have been pursued most intensely in order to speciate the organic forms of mercury, especially methylmercury. Hyphenated techniques have been applied to solve this analytical problem. In most cases a gas or liquid chromatographic separation unit was coupled to an element-selective detection system in order to obtain qualitative and quantitative information about the different binding forms.

This work aims to apply modified capillaries to a screening method of mercury compounds. The differentiation of organic and inorganic species as a preliminary diagnostic tool allows the estimation of the existence and concentration of the organic species. In the case that organic compounds could be detected a chromatographic separation could provide information about each single species.

For method development the behaviour of the most important organic species, methylmercury ($MeHg^+$) was compared with the behaviour of inorganic mercury (Hg^{2+}). With the use of the cationic exchanger capillary modified with 2-(4-chlorosulfonylphenyl)-ethyltrichlorosilane, the retention behaviour of both compounds was observed as a function of the pH of the injected solution. Figure 5.8.8 demonstrates that the signal intensities of both compounds showed a local minimum at a pH of 5.5. Our investigations proved that the signal suppression of Hg^{2+} was caused by its complete retention inside the capillary. In the case of methylmercury the minimum in signal intensity was derived from a signal suppression in the plasma. No methylmercury could be found after rinsing the capillary with acid in order to elute retained compounds. This result made the differentiation between the two species after

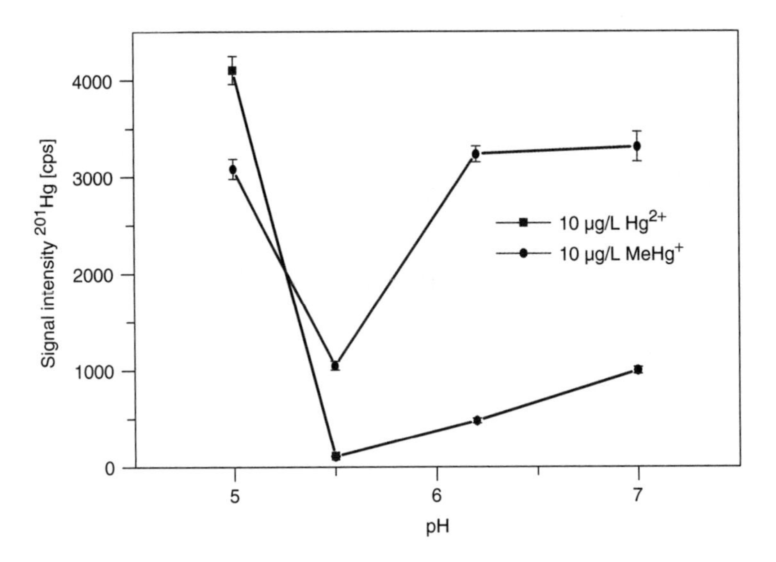

Figure 5.8.8. Influence of pH on the signal intensities of Hg^{2+} and $MeHg^{+}$.

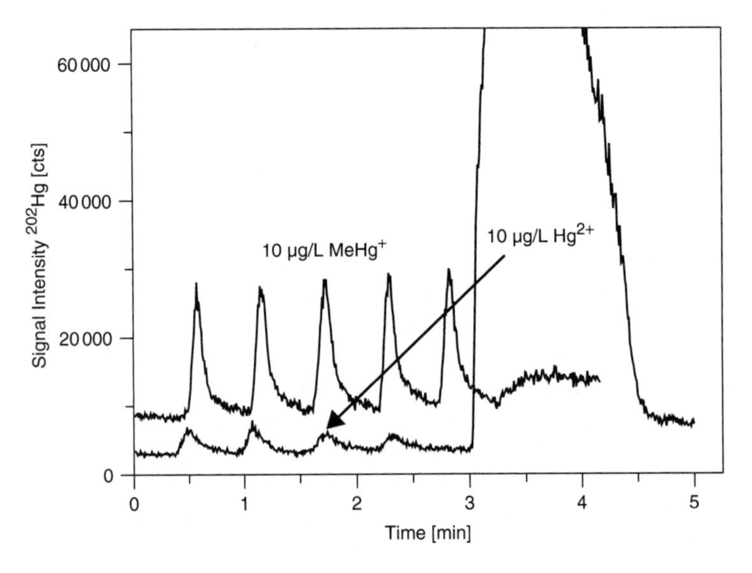

Figure 5.8.9. Injection and elution profiles of Hg^{2+} and $MeHg^{+}$ (five replicate injections of the analytes and elution with 10^{-3} M HCl).

pH adjustment possible – whereas other alkylated species like dimethylmercury behave like $MeHg^{+}$. Typical injection signals for both species are shown in Figure 5.8.9. Elution of retained Hg^{2+} could be obtained by injecting 10^{-3} M HCl. The signal forms demonstrate that the use of the direct injection nebulizer (DIN) minimises the memory effects of mercury substantially.

Under optimised conditions the selectivity was investigated. A comparison of signal intensities from $MeHg^{+}$ and Hg^{2+} injections was carried as demonstrated in Figure 5.8.10. A concentration excess of 100 ($[Hg^{2+}]/[MeHg^{+}]$) shows no signal for inorganic mercury indicating that $MeHg^{+}$ can be selectively detected at levels higher than 1 % of total mercury content in a sample. This

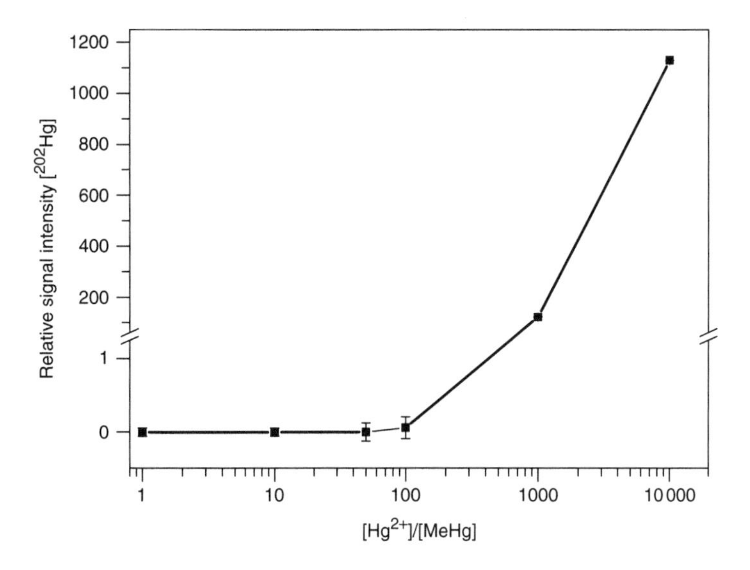

Figure 5.8.10. Selectivity of $MeHg^+$ versus Hg^{2+}.

circumstance is particularly likely in biological and marine samples, for which this method can be easily applied to prove the existence of organomercury compounds.

The method was applied for the analysis of a tuna sample which was foreseen as a reference material for total mercury content (9/97). The sample was dissolved in 25 % (w/v) TMAH (tetramethylammonium hydroxide) by microwave digestion (15 min, 45 W), and after filtration and dilution the solution was adjusted to pH of 5.5. The results obtained were compared with the results obtained using a hyphenated technique (capillary cold trap-GC-AAS, CCT-GC-AAS [22]). They were in a good agreement regarding the methylmercury content (Table 5.8.3) and emphasised the potential of this method as a screening method for organomercury compounds. The detection limit for methylmercury was 100 ng L^{-1}, sufficient for its determination in biological samples.

Table 5.8.3. Determination of mercury species in reference material 9/97 (tuna fish).

	Concentration (mg kg^{-1})
Modified DIN-ICP-MS	
Concentration (organic mercury)	2.6 ± 0.2
Comparative investigations using CCT-AAS	
Concentration ($MeHg^+$)	2.4 ± 0.4

6 CONCLUSIONS AND FUTURE DEVELOPMENTS

These investigations showed that a lot of information about elemental species can be obtained by the use of modified capillaries for sample introduction in ICP-MS, realised for the first time for metal speciation analysis. On the example of redox species of chromium a differentiation between Cr(III) and Cr(VI) is possible using the exchange capabilities of the modified surfaces. In the case of mercury a differentiation between inorganic and organic species is possible using a modification with a cation exchanger. This can be useful as a screening method. For further information about the exact species present an analysis using chromatographic separation is necessary.

Further investigations are under way, especially to evaluate a similar method for the screening of organolead compounds. Furthermore, the application of more selective modifications, e.g. complex-forming reagents, is being studied to form species-selective interactions to the capillary surface. Another aim to be pursued is the application of isotope dilution for precise and accurate quantification of the species of interest.

7 ACKNOWLEDGEMENTS

The author thanks the European Commission for the financial support (SMT-4-CT98-2233, MOSIS: Modified Sample Introduction Systems) and the project partners C. Cámara, R. Garcia-Sanchez (Universidad Complutense de Madrid, Spain), L. Ebdon (University of Plymouth, UK), B. Rosenkranz, H.-G. Riepe (Westf. Wilhelms-Universität Münster, Germany) and J. Schmitt (Johannes Gutenberg-Universität Mainz, Germany) for excellent cooperation.

8 REFERENCES

1. Harrison, R. M. and Rapsomanikis, S. (Eds), *Environmental Analysis Using Chromatography Interfaced with Atomic Spectroscopy*, Horwood, Chichester, 1989.
2. Batley, G. E. (Ed.), *Trace Element Speciation: Analytical Methods and Problems*, CRC Press, Boca Raton, FL, 1989.
3. Uden, P. (Ed.), *Element-specific Chromatographic Detection by Atomic Emission Spectroscopy*, American Chemical Society, Washington, DC, 1992.
4. Kramer, J. R., Allen, H. E. (Eds), *Metal Speciation: Theory, Analysis and Application*, Lewis, Chelsea, 1988.
5. Krull, I. S. (Ed.), *Trace Metal Analysis and Speciation*, Elsevier, Amsterdam, 1991.
6. Szpunar-Lobinska, J., Witte, C., Lobinski, R., Adams, F. C., *Fresenius' J. Anal. Chem.*, **351**, 351 (1995).
7. Donard, O. F. X., Martin, F. M., *Trends Anal. Chem.*, **11**, 17 (1992).
8. Chau, Y. K., Zhang, S. Z., Maguire, R. J., *Analyst*, **117**, 1161 (1992).
9. Rosenkranz, B., Bettmer, J., *Trends Anal. Chem.*, **19**, 138 (2000).
10. Ebdon, L., Hill, S., Ward, R. W., *Analyst*, **112**, 1 (1987).
11. Vela, N. P., Olson, L. K., Caruso, J. A., *Anal. Chem.*, **65**, 585 A (1993).
12. Hintelmann, H., Falter, R., Ilgen, G. and Evans, R. D., *Fresenius' J. Anal. Chem.*, **358**, 363 (1997).
13. Demuth, N. and Heumann, K. G., *Anal. Chem.*, **73**, 4020 (2001).
14. Encinar, J. R., Alonso, J. I. G. and Sanz-Medel, A., *J. Anal. At. Spectrom.*, **15**, 1233 (2000).
15. Nusko, R. and Heumann, K. G., *Fresenius' J. Anal. Chem.*, **357**, 1050 (1997).
16. Hjerten, S., *J. Chromatogr.*, **347**, 191 (1985).
17. Riepe, H.-G., Gómez, M., Cámara, C. and Bettmer, J., *J. Mass Spectrom.*, **35**, 891 (2000).
18. Riepe, H.-G., Gómez, M., Cámara, C. and Bettmer, J., *J. Anal. At. Spectrom.*, **15**, 507 (2000).
19. Garcia, R., Gómez, M., Palacios, M. A., Bettmer, J. and Cámara, C., *J. Anal. At. Spectrom.*, **16**, 481 (2001).
20. BCR Information, *The Certification of the Contents of Cr(VI) and Total Leachable Cr in Welding Dust Loaded on a Filter (CRM 545)*, EUR 18026 EN (1997).
21. Rosenkranz, B., Riepe, H.-G., Bettmer, J. and Ebdon, L., *J. Anal. At. Spectrom.*, in press.
22. Dietz, C., Madrid, Y., Cámara, C. and Quevauviller, P., *Anal. Chem.*, **72**, 4178 (2000).

5.9 Speciation Analysis by Electrochemical Methods

Raewyn M. Town

The Queen's University of Belfast, Northern Ireland

Hendrik Emons

EC Joint Research Center IRMM, Geel, Belgium

Jacques Buffle

CABE Geneva, Switzerland

Handbook of Elemental Speciation: Techniques and Methodology R. Cornelis, H. Crews, J. Caruso and K. Heumann
 ISBN: 0-471-49214-0

SYMBOLS

A	electrode area
a_i	activity of species i
AC	alternating current
AdSV	adsorptive stripping voltammetry
ACV	AC voltammetry
ASV	anodic stripping voltammetry
α_c	cathodic transfer coefficient
c	bulk concentration
$c_{L,T}$	total ligand concentration
c_L	free ligand concentration
$c_{M,T}$	total metal concentration
c_M	free metal concentration
C_d	differential capacity
CE-EC	capillary electrophoresis with electrochemical detection
CV	cyclic voltammetry
D	diffusion coefficient
δ	steady-state diffusion layer thickness
DC	direct current
DPV	differential pulse voltammetry
E	potential
E_{pa}	anodic peak potential
E_{pc}	cathodic peak potential
E_{eq}	equilibrium potential
E_{ref}	potential of the reference electrode
E°	standard potential
$E^{o\prime}$	formal redox potential
$E_{1/2}$	half-wave potential
ΔE_p	peak potential difference
EPPS	N-2-hydroxyethylpiperazine-N'-3-propanesulfonic acid
F	Faraday constant
f	frequency
HEPES	N-2-hydroxyethylpiperazine-N'-2-ethanesulfonic acid
HEPPS	N-2-hydroxyethylpiperazine-N'-3-propanesulfonic acid
HMDE	hanging mercury drop electrode
i	current (faradaic i_f)
i_c	capacity current
i_l	diffusion-limited current
i_p	peak potential
i_{pa}	anodic peak current
i_{pc}	cathodic peak current
ISE	ion selective electrode
k_0	standard rate constant
k_c	cathodic rate constant
L	lability criterion parameter
LCEC	liquid chromatography with electrochemical detection
LSV	linear sweep voltammetry
μ	reaction layer thickness, $(D_M/k_a')^{1/2}$
n	number of electrons transferred
Ox	oxidized form of the reactant
PIPES	piperazine-N-N'-bis(2-ethanesulfonic acid)
r	electrode radius
R	gas constant
Red	reduced form of the reactant
R_s	solution resistance
Q	charge
N	number of moles of electrolyzed material
SCE	saturated calomel electrode
SWV	square wave voltammetry
SV	stripping voltammetry
t	time
t_m	measuring time
t_p	pulse time

T	temperature
TES	N-tris(hydroxymethyl)methyl-2-aminoethanesulfonic acid
TMFE	thin mercury film electrode
v	scan rate
WE	working electrode
Superscript	
b	bulk
Subscript	
ox	oxidized species
red	reduced species

1 INTRODUCTION

It is well established that the total concentration of an element provides no information about its bioavailability, toxicity, transport properties, or residence time within a given system. The quantitative analysis of chemical species in real-world samples represents a complex process. It is a far more challenging task than that of determining total concentrations. The ideal method would be one that can be deployed *in situ* with the analytical signal being directly interpretable in terms of the species present in the medium. However, in practice, measurements of element speciation often involve a number of consecutive steps, including general sampling, sample preparation, possibly some additional species separation such as chromatography, the measurement step of species quantification (detection) and data evaluation. IUPAC has published recommended guidelines for nomenclature of chemical speciation and fractionation of elements [1]. This chapter presents a description of detection methods which exploit electrochemical processes at electrodes.

In comparison with many analytical techniques, electrochemistry is unique in that it is based on interfacial phenomena. It finds its main application in the investigation of dissolved species but can also be used in certain cases for the direct measurement of solids. Electroanalytical techniques have certain features that are advantageous for speciation analysis. In contrast to atomic spectrometry or ICP-MS they belong to the low energy excitation techniques which is the reason why they are species selective rather than element selective. Many of them exhibit excellent detection limits coupled with a wide dynamic range. Measurements can generally be made on very small samples, typically in the microliter volume range.

If the main targets of speciation analysis are grouped into redox states, metal(loid) complexes and organometal(loid) compounds, analytes in all three areas can be determined by electroanalysis.

2 OVERVIEW OF ELECTROANALYSIS

2.1 Fundamentals

Electrochemical methods are based on the measurement of electrical signals associated with molecular properties or interfacial processes of chemical species. Owing to the direct transformation of the desired chemical information (concentration, activity) into an electrical signal (potential, current, resistance, capacity) by the methods themselves, they provide an easier and cheaper access to automation, computer control and data handling in comparison with the other analytical methods requiring an additional transducer to obtain electrical signals. Two major difficulties in the application of electroanalytical techniques to complex real-world samples have been the lack of selectivity of electrochemistry and the susceptibility of the electrode surface to fouling by surface active material in the sample. The fundamentals of electrochemical methods which are suitable for the quantitative determination of chemical species are described on various levels in a number of textbooks and monographs [2–7]. However, the relatively high level of theoretical knowledge and interdisciplinary understanding necessary to apply most of the techniques to new samples of complex composition or to develop electroanalytical procedures seems to be a barrier for the broader use of these methods. Moreover, the great number of different techniques can make the proper selection for solving a certain problem difficult not only for the newcomer but even for more experienced scientists. In addition, there are still various names for the same technique in the literature. The reader is referred to IUPAC recommendations for classification and nomenclature of electroanalytical techniques [8, 9]. In the following, we will concentrate on those methods with a broader applicability for speciation analysis.

Electroanalytical techniques can be classified in various ways. An approach mainly based on the character of the measured signals and their excitation is shown in Figure 5.9.1. Only major techniques were selected for this 'family tree'. Electrochemistry can be divided into 'ionics' and 'electrodics'. Conductance measurements are amongst the important analytical methods that are based on electrochemical properties of chemical species in the bulk phase. Because of the rather nonselective nature of the measurement it is rarely used to determine certain species directly in the sample solution, but it provides an easily obtainable parameter to estimate the total number of ions in liquid samples. Conductometry does find useful application in studying counterion association with polyelectrolytes where the data obtained are complementary to those provided by voltammetric methods [10, 11]. More important for speciation analysis is the application of conductivity detectors in chromatographic systems.

The majority of electroanalytical methods belong to the field of electrodics. That means the analytical signal is produced at an interface, mostly formed at a metal/solution contact, by heterogeneous processes with the participation of charged species. Measurements in the state of electrochemical equilibrium of electrochemical cells, i.e. without any overall current flowing, are the basis of potentiometry. The importance of these methods for species analysis in liquids cannot be overestimated particularly due to the exciting developments in the field of ion-selective electrodes in the 1960s and 1970s.

Dynamic electrochemical techniques involve externally initiated electrolysis (oxidation or reduction) at the surface of an electrode. Quantitation of

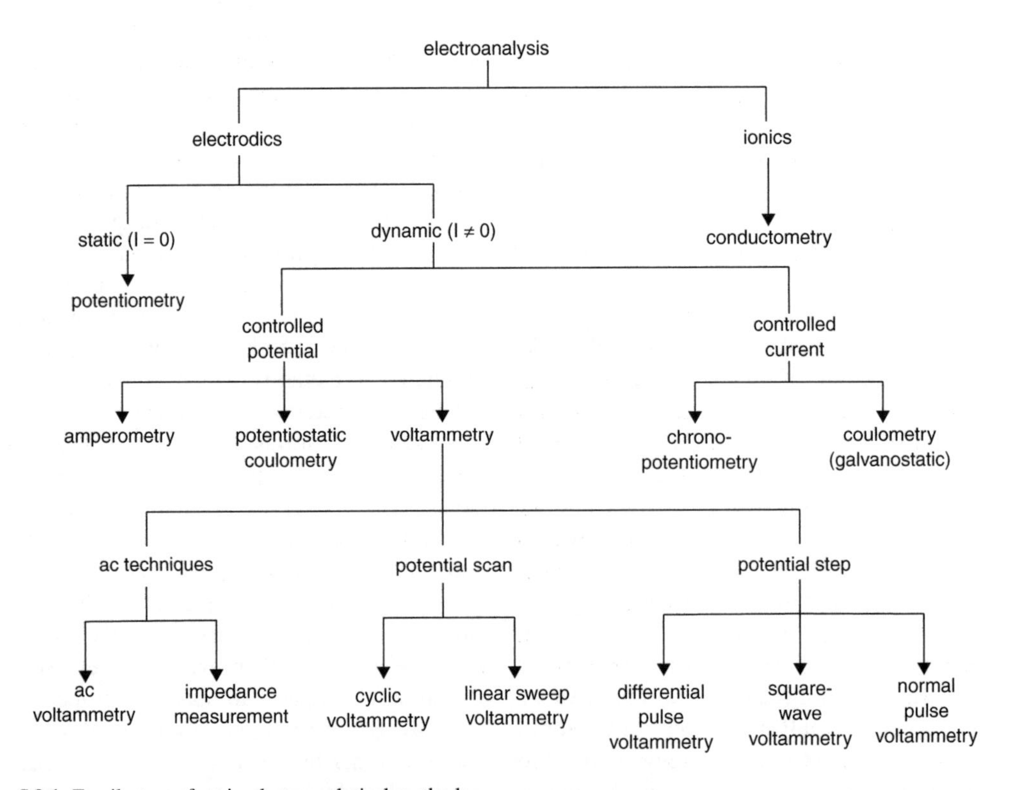

Figure 5.9.1. Family tree of main electroanalytical methods.

a species (the analyte) in the sample is achieved by measuring the current (or charge) generated by the surface redox process.

The basic system of electrochemical experiments is the electrochemical cell. It consists of at least two independent electrodes which are immersed in an electrolyte solution. The electrodes are electrically connected through both the solution and external wires via a measuring device. The current flowing through this electrical circuit can result either from the occurrence of spontaneous electrode reactions (potentiometry) or from heterogeneous redox reactions which are externally driven by the application of a potential difference at the electrodes (electrolytic cell).

The fundamental relation between reactive species at electrodes and the electrode potential under equilibrium conditions is expressed by the Nernst equation (5.9.1):

$$E_{eq} = E^0 + RT/nF \ln\left(\prod a_{ox}{}^{\nu} / \prod a_{red}{}^{\nu}\right) \quad (5.9.1)$$

or by inclusion of the activity coefficients into the standard electrode potential E^0 of the electrochemical cell reaction (leading to a formal potential $E^{0\prime}$):

$$E_{eq} = E^{0\prime} + RT/nF \ln\left(\prod c_{ox}{}^{\nu} / \prod c_{red}{}^{\nu}\right) \quad (5.9.2)$$

where R, T, n, and F have their usual meaning, a_{ox} and a_{red} are the activities of Ox and Red, respectively, c_{ox} and c_{red} are the corresponding concentrations, and ν is the stoichiometric coefficient of the reaction.

The potential of the indicator (or working) electrode is measured or applied (for electrolytic cells) with respect to a reference electrode which serves as a 'calibration point' for the potential difference.

For electroanalytical purposes the mass transfer of the analyte is driven either by diffusion alone or by a combination of diffusion and convection. Convection must be avoided or carefully controlled during the experiments because it influences the thickness of the diffusion layer δ (Figure 5.9.2) which must be constant during the analytical measurement.

Another layer has to be considered at the electrode/solution interface: the so-called electrical double layer. On the solution side it contains the first few species layers at the electrode where the electron transfer actually occurs and where the greatest potential difference appears and is composed mainly of electrically oriented solvent dipoles and adsorbed electrolyte ions. Moreover, a more extended region of 1–3 nm thickness (depending on the electrolyte concentration) is included which is characterized by a potential

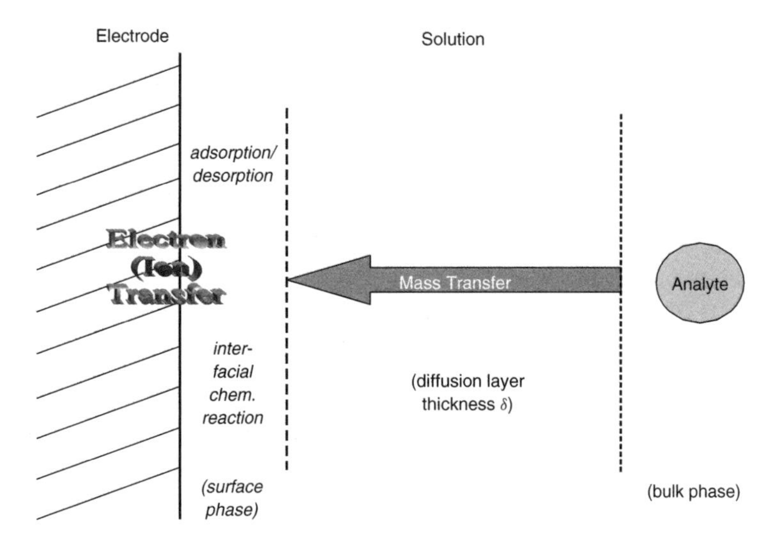

Figure 5.9.2. Steps of an electrochemical reaction.

gradient and consequently a different spatial distribution of charged species in comparison to the bulk phase of solution. From the point of view of analytical measurements one has to take into account that there is a charge separation in this interfacial region which gives rise to a capacity current i_c according to

$$i_c = C_d (dE/dt) \quad (5.9.3)$$

where C_d is the differential double layer capacity. It is potential dependent and varies usually between 10 and 100 $\mu F cm^{-2}$. The capacity current i_c is part of the overall measured current (although it is not related to the electron transfer process) and contributes in most cases only to the background signal of the analytical measurement. Therefore, i_c should be separated from the faradaic current i_f, which carries the concentration-dependent analytical information, or it should be kept at least very small and constant for trace analysis. The latter approach can be fulfilled with the use of microelectrodes; i_f and i_c can be separated in potential step techniques (e.g. pulse techniques) by exploiting their different time-decay characteristics, by taking the inverse of the time derivative of potential in chronopotentiometry [12] and by using the difference in phase of i_c and i_f in AC techniques.

The electron transfer process is usually the process of interest for most analytical determinations and gives rise to the so-called faradaic current. It is based on Faraday's law:

$$Q = nFN \quad (5.9.4)$$

where Q is the charge of electricity, F is the Faraday constant, N is the number of moles of material electrolyzed, and n is the number of electrons per molecule involved in the electrolysis process. Many electroanalytical techniques measure current (i), which is obtained from the above equation by differentiation with respect to time (t) to give

$$i = dQ/dt = nF(dN/dt) \quad (5.9.5)$$

which shows that the current is a measure of the rate of electrolysis at the electrode surface. The analytical use of electrochemistry relies on the proportionality of the rate of electrolysis, as measured by current, and the concentration of the species undergoing electrolysis at the working electrode. An important aspect of dynamic electrochemistry is the potential that is applied to the electrochemical cell. The applied potential provides the driving force for the electrolysis reaction upon which the analysis is based. The rate constant of the electron transfer depends exponentially on the electrode potential E and is described for a reduction by

$$k_c = k_0 \exp[-\alpha_c nF(E - E^{0\prime})/RT] \quad (5.9.6)$$

where k_0 and α_c are reaction-specific kinetic constants called standard rate constant and cathodic transfer coefficient, respectively. Obviously, electroanalysis has to obey and can exploit the sensitive control of the reaction rate by the electrode potential. However, both the electron transfer kinetics and the analyte transport, usually by diffusion, have to be considered together for the design and evaluation of electroanalytical measurements.

2.2 Instrumentation

In general electroanalysis requires only relatively few and inexpensive instruments. Potentiometry is performed with a potentiometer (pH/mV meter) with high input impedance and the dynamic techniques are using a potentiostat.

2.2.1 Potentiostats

The potentiostat applies a defined potential across the working electrode and the reference electrode for the typical three-electrode configuration of modern electrochemical cells. Moreover, it measures the current at the working electrode. Commercially available instrumentation has been reviewed [13]. There is increasing interest in the development of portable battery-powered potentiostats for field use [14, 15] and multiplex instrumentation for measurements with multichannel array electrodes [16, 17].

2.2.2 *Cell designs*

The importance of mechanical, geometric, and hydrodynamic factors in electrode design have been summarized [18, 19]. Flow-through cells allow automation of electroanalytical determinations and are essential for *in situ* measurements. Systems range from on-line (e.g. shipboard) flow-injection type cells [20] to fully immersible units in which pressure compensation is an integral component to allow measurements at depth and to overcome problems associated with the liquid junction of the reference electrode [21]. Key design aspects of such cells have been discussed [19]. Special considerations are required for fabrication of microanalytical systems [22]. In all cases, the construction materials should be carefully chosen to minimize adsorption of sample components [23].

2.2.3 *Electrodes and electrode materials*

2.2.3.1 *Potentiometry*

The measuring electrode (also called indicator electrode) for potentiometric speciation analysis consists of an appropriately chosen ion-selective electrode as discussed in Section 2.3.1. The various types of ISEs have been reviewed [24, 25]. The potential difference is measured versus a reference electrode (usually saturated calomel electrode = SCE, or Ag/AgCl electrode) which provides an invariant potential.

2.2.3.2 *Dynamic techniques*

Common three-electrode cells contain a reference electrode, a counter electrode (CE, represented typically by a Pt wire or plate, sometimes also a glassy carbon electrode, etc.) and the working electrode (WE). A survey of available types of reference electrodes, including those suitable for miniaturization and elimination of internal solution, has recently been published [19].

The analyte of interest will react (mostly by electrolysis) at the WE whereas the counter reaction will take place at the CE. By that a current flow through the reference electrode is avoided. The working electrode should have a reproducible surface morphology and area, and a low residual current.

Mercury remains the electrode material of choice for detection of metals due to its large hydrogen overvoltage and its remarkable reproducibility. However, solid electrode substrates including Au, Pt, Ag, and various forms of carbon also find applications. Various reviews have surveyed the range of electrodes used in voltammetric analysis [19, 26, 27].

Two types of mercury electrodes are commonly used for stripping voltammetry (see Section 2.3.3): the hanging mercury drop electrode (HMDE) and the thin mercury film electrode (TMFE). Commercial HMDE systems are able to produce mercury drops with very reproducible dimensions (the internal diameter of the capillaries is typically 60–200 μm). Narrow bore glass capillaries have been used for production of renewable micromercury drops [28, 29]. Thread electrodes, in which a mercury column is enclosed within a hydrophilic dialysis membrane tube (ca 150 μm diameter), have been proposed for use in flow cells [30]. TMFEs are produced by electrochemical deposition of a thin mercury coating onto an electrically conductive substrate, e.g. glassy carbon, platinum, gold or iridium. The film can be preformed or deposited *in situ* after adding Hg^{2+} to the sample solution; *in situ* formation should, however, be avoided when element speciation in the sample is of interest. Many workers have preferred the TMFE over the HMDE, because it exhibits greater sensitivity, lower detection limits and sharper stripping peaks as a result of the greater electrode surface/volume ratio and shorter diffusion distances for the deposited metal. In addition, a TMFE can be rotated to produce more reproducible solution hydrodynamic conditions than can be achieved by solution stirring. However, the HMDE is more reproducible and easier to operate with respect to the repetitive formation of the electrode surface and is better defined from a theoretical point of view. The deposition of mercury onto carbon produces droplets of the metal, rather than a true film [31]. There are numerous publications on pretreatment protocols for the carbon surface [32] and various deposition procedures [33] which are purported to result in useful mercury 'films'. Some metal supports, e.g. gold and platinum, form amalgams which have different electrochemical properties and thus lower the reproducibility of the procedure. It has been shown that iridium offers an attractive alternative as a conductive support for the TMFE due to its good wettability by mercury that allows formation of a perfect mercury surface (rather than the separate droplets formed on glassy carbon), and its very low solubility in mercury that prevents the drawbacks of amalgam formation associated with Pt, Au, and Ag substrates [34, 35].

Microelectrodes. Microelectrodes are characterized by possessing at least one geometrical dimension in the low μm range [36]. Single microelectrodes have been fabricated as disk, band, fiber or microdrop and offer several advantages for speciation measurements in real-world samples, including the very small influence of ohmic drop which facilitates application in low ionic strength media [37] (e.g. freshwaters), their greater stability and reproducibility, their insensitivity to bulk solution convection [38] (a consequence of spherical diffusion), and the potential for analyses at high spatial resolution arising from their small physical dimensions. There is an increasing number of reports of application of microelectrodes to speciation analysis [15, 37, 39–41]; measured concentrations correspond to the mobile and labile metal fraction [19].

Care must be taken in fabrication of microelectrodes to ensure reproducible results. For example, perfect electrical contact, Ir–glass sealing, and Ir-disk morphology were found to be the key aspects for obtaining reproducible mercury-coated Ir microelectrodes [42]. Mercury deposits on such electrodes can be used for several days without renewal of the mercury layer and the electrodes themselves have a lifetime of at least several years [42].

Various microelectrode arrays, typically based on photolithographic procedures, for example the deposition of Ir followed by electroplating with mercury, have also been reported for electroanalysis of trace metals [17, 43, 44].

Modified electrodes. Chemically modified electrodes designed for diverse applications have gained increasing interest in recent years [45]. For element speciation purposes, modification procedures are typically aimed towards either (i) increased selectivity towards target species [46, 47] or (ii) protection of the electrode surface from fouling due to adsorption of organic materials in the sample matrix. It is presumed that a thin semipermeable layer will exclude potential fouling materials by size and/or charge exclusion. For example, cellulose acetate may protect against adsorption of proteins [48] and humic acids [49]; Nafion is purported to prevent adsorption of humic substances [50]. None of these approaches are ideal because the coatings themselves are not inert towards the target elements, e.g. Nafion is an ion-exchange polymer [51], and they are often difficult to prepare in a reproducible and controlled manner.

Recently, gel integrated microelectrodes (GIME) have been reported in which a microelectrode is integrated into a relatively thick agarose gel layer. The GIME has been shown to possess many advantages in terms of long-term operation, antifouling properties, and reproducibility. In particular, it has been shown to selectively exclude colloids and macromolecules from the electrode surface while being inert towards the target elements and ensuring diffusion-controlled transport due to its anticonvective properties [52]. The GIME is commercially available (Idronaut, Italy).

2.3 Measuring techniques

The following sections discuss a variety of electroanalytical techniques that differ in the mode of excitation signal-response characteristics.

2.3.1 *Potentiometry*

In potentiometry, the equilibrium potential of an indicator electrode is measured against a selected reference electrode using a high impedance voltmeter, i.e. effectively at zero current. The measured signal is a function of the activity of such species in solution that influence the potential at the indicator electrode (see equation 5.9.1). A broad range of ion-selective electrodes (ISE) has been developed by placing a separating membrane on the tip of a more or less conventional second reference electrode. The potential measured is the difference of potential across this membrane which is influenced by the activity of the species on either side of the membrane. If the activity of the ionic species x with the charge n_x remains constant in the membrane the measured potential difference is

$$E - E_0 = \Delta E = (RT/n_x F) \ln a_x{}^b \qquad (5.9.7)$$

Therefore, a potential variation of $59/n_x$ mV should be observed per decade of variation in bulk activity $a_x{}^b$ at 298 K. In this way, activities typically in the range of 10^{-1}–10^{-6} mol dm^{-3} can be measured. Recently, much lower detection limits of the order of 10^{-11} mol dm^{-3} have been claimed by metal ion buffering in the internal reference solution of polymer membrane ISEs [53]. A perfect ISE would respond to only one kind of ionic species in solution containing any ion at any concentration. In practice, this goal cannot be achieved and one has to consider interference effects of other ions, as quantified by their selectivity coefficients.

Various recommended methods have been proposed for the calibration of ISEs [54] and for the determination of selectivity coefficients [55, 56].

Ion-selective electrodes can be classified according to their membrane characteristics as shown in Figure 5.9.3. Examples for types of ISEs are summarized in various review articles [57] and IUPAC has reported recommended nomenclature [58].

2.3.2 *Fixed potential methods – amperometry*

The simplest dynamic electroanalytical technique is the application of a fixed potential to an appropriately chosen working electrode and measurement of the current due to electrolysis of the analyte. If this potential is conveniently chosen or steady-state convection in the measuring cell is employed, the magnitude of the current is directly proportional to the concentration of the species of interest. This technique is also termed amperometry and plays a role as flow-stream detection technique for speciation analysis (see below).

2.3.3 *Voltammetry*

More information from dynamic potential-controlled measurements is available if the potential is changed in a well-defined manner during the experiment. The measurement of corresponding current–potential curves is called voltammetry (derived from *volt–ampere–metrology*) and can be exploited to characterize electrochemical systems also qualitatively as well as for increasing the selectivity. It should be noted that polarography

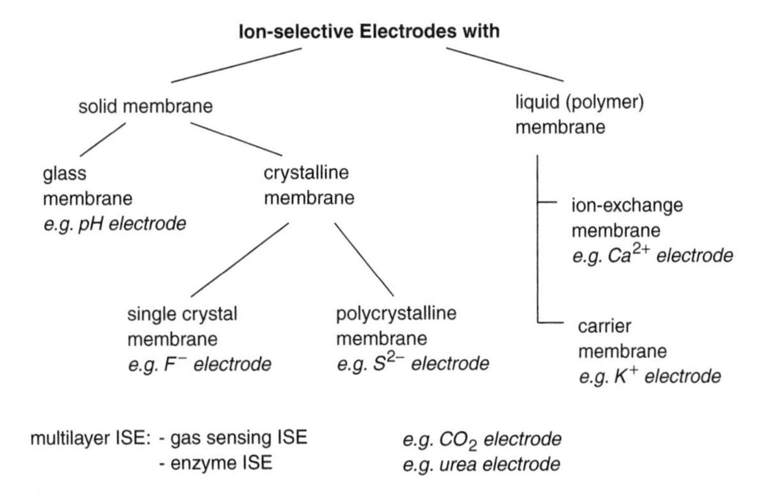

Figure 5.9.3. Ion-selective electrodes classified according to their membrane composition.

is frequently used in the literature instead of voltammetry for current–potential measurements. But the term 'polarography' is recommended now for the specific case of voltammetric measurements at working electrodes with constantly or periodically changing surfaces, mainly the dropping mercury electrode (DME). In the following the general term voltammetry includes polarography, and selected voltammetric techniques with analytical importance for speciation are briefly introduced. The characteristic parameters for these methods for reversible and irreversible systems are tabulated in the literature [7, 18].

2.3.3.1 *Cyclic voltammetry (CV)*

Cyclic voltammetry is perhaps the most versatile electroanalytical technique for the study of electroactive species in quiescent solution. It can be used to observe rapidly their chemical reactions and redox behavior over a wide potential range, i.e. in an extended energy region.

The excitation signal for CV is a linear potential scan (or sweep) with a triangular waveform, as shown in Figure 5.9.4 together with a corresponding response curve. The important parameters of a cyclic voltammogram are the magnitudes of the anodic peak current (i_{pa}) and cathodic peak current (i_{pc}) as well as the anodic peak potential (E_{pa}) and cathodic peak potential (E_{pc}). It should be mentioned that the measurement of peak currents is often complicated by problems in establishing correct baselines, particularly for more complicated systems.

The data for the oxidation and reduction peaks are suited to characterize redox reactions and partially also of preceding or following chemical processes. If the species undergo a reversible electrochemical reaction at the electrode without any chemical complications, i.e. electron transfer occurs very fast with respect to the other steps especially mass transport, the measurement of E_{pa} and E_{pc} allows the estimation of the formal redox potential

$$E^{0'} = \frac{1}{2}(E_{pa} + E_{pc}) \quad (5.9.8)$$

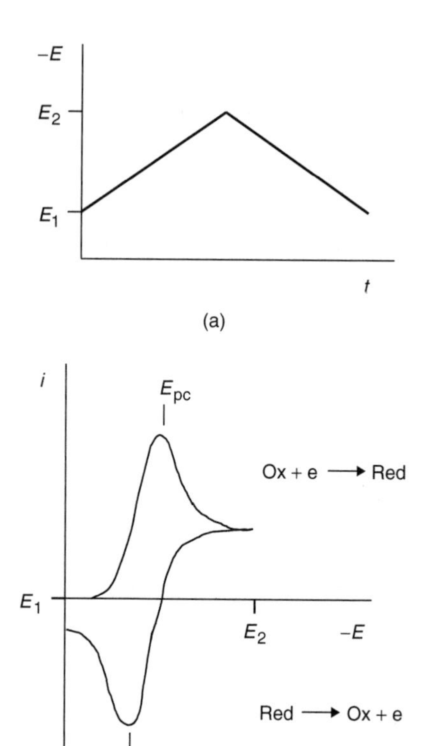

Figure 5.9.4. Cyclic voltammetry: principal excitation (a) and response (b) curves.

Diagnostic criteria for such reversible processes are fulfilled if

$$\Delta E_p = E_{pa} - E_{pc} = 59\,\mathrm{mV}/n \quad \text{at } T = 298\,\mathrm{K} \quad (5.9.9)$$

$$i_{pa}/i_{pc} = 1 \quad (5.9.10)$$

$$i_p \propto v^{1/2} \quad (5.9.11)$$

where v represents the potential scan rate (dE/dt).

In addition, the peak potentials should be independent of the scan rate. The peak current for a reversible couple at a normal-sized planar electrode is described (at 298 K) by the Randles–Sevcik equation:

$$i_p = (2.69 \times 10^5)n^{3/2}AD^{1/2}v^{1/2}c \quad (5.9.12)$$

where i_p is in amperes, the diffusion coefficient D is in $cm^2 s^{-1}$, v is in $V s^{-1}$ and the bulk concentration of the reactant c is in $mol cm^{-3}$. Despite the linear correlation between the peak current and the concentration, cyclic voltammetry is rarely used for quantitative analysis. Its detection limit is often insufficient (in the range of 10^{-5} $mol dm^{-3}$) and the accurate determination of i_p is complicated by nonlinear baselines.

In practice, most systems show some degree of irreversibility in their CV. As a result of a slower electron transfer or coupled chemical reactions, the oxidation and reduction signals are diminished, broader peaks appear, the peak potential difference increases and the most marked feature of a CV of a totally irreversible system is the absence of a reverse peak. The equations (5.9.8)–(5.9.12) are not applicable for such systems.

2.3.3.2 Linear sweep voltammetry (LSV)

This method can be described as one half of a CV experiment. The excitation signal is a linear potential ramp usually in the range of 10–1000 $mV s^{-1}$ which causes a peak-shaped current–potential curve as response. The peak current depends on the bulk concentration of the electroactive species and is described for a reversible process at a planar electrode by equation (5.9.12). In general, detection limits for many systems are about 10^{-5} $mol dm^{-3}$ due to the capacitive current. Application of LSV is limited due to the relatively high detection limit and the often insufficient peak resolution in systems containing multiple redox species. However, LSV still finds important application in stripping voltammetry (see below). In addition this excitation technique is used for the development of amperometric/voltammetric detection schemes in flow analysis. In this case, so-called hydrodynamic voltammograms are recorded by applying the LSV excitation (Figure 5.9.5(a)) to an electrochemical cell with stirred or flowing analyte solution. The resulting current–potential curve (Figure 5.9.5(b)) has a sigmoid shape which is known from classical DC polarography at the dropping mercury electrode. The mass transport by controlled convection causes the establishment

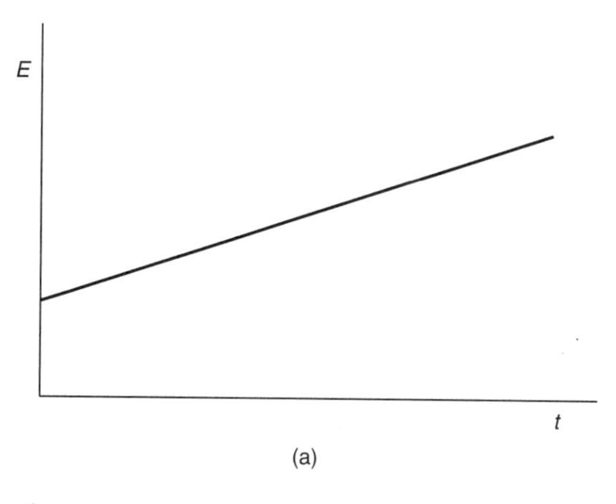

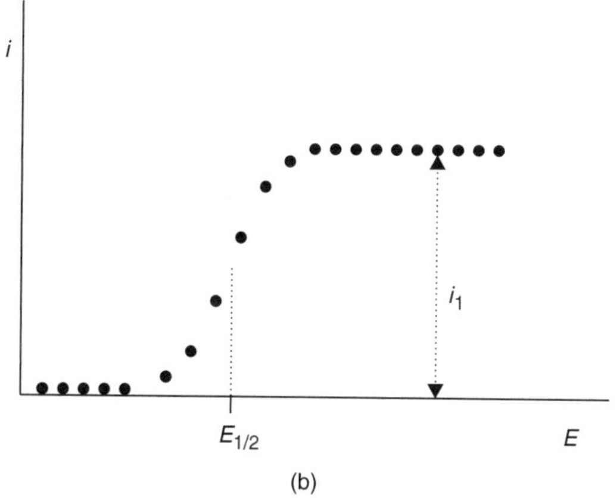

Figure 5.9.5. Hydrodynamic voltammogram: principal excitation (a) and response (b) curves.

of a diffusion layer with constant thickness in the vicinity of the working electrode. Therefore, the current reaches a plateau that corresponds to the condition of the analyte being reduced as rapidly as it is transported to the surface by diffusion in the solution near the electrode. The potential at which the current is half of its plateau value is termed the half-wave potential ($E_{1/2}$) and the plateau current is termed the limiting current i_l. $E_{1/2}$ and i_l are the two paramount parameters in the analytical application of electrochemistry. i_l is used for quantitation by means of its relationship to the bulk concentration of the analyte

$$i_l \propto c \tag{5.9.13}$$

and $E_{1/2}$ is used for qualitative identification through its relationship to the formal electrode potential $E^{0\prime}$ of the redox couple

$$E_{1/2} = E^{0\prime} - (RT/nF)\ \ln\ (D_{\mathrm{Ox}}/D_{\mathrm{Red}}) - E_{\mathrm{ref}} \quad (5.9.14)$$

where D_{Ox} and D_{Red} are the diffusion coefficients for the oxidized and reduced forms of the couple, respectively, and E_{ref} is the potential of the reference electrode.

2.3.3.3 Normal (NPV) and reverse (RPV) pulse voltammetry

NPV can be described as a combination of chronoamperometric measurements which register i–t curves of potential step experiments. In chronoamperometry the electrode potential is often immediately changed from a region without any electrode reaction to the diffusion-limited range and the resulting current–time response at a planar electrode in quiescent solution is given by the Cottrell equation

$$i = nFAD^{1/2}c/(\pi^{1/2}t^{1/2}) \quad (5.9.15)$$

For NPV increasing potential step heights are applied as indicated by the excitation signal in Figure 5.9.6(a). A typical normal pulse voltammogram exhibits a sigmoidal current–potential wave (Figure 5.9.6). The current is measured only for a few milliseconds (t_{m}) at the end of each pulse to reduce the capacitive current. The initial potential E_{i} is normally fixed in a range without faradaic reactions and the pulse time t_{p} should be short (commonly about 50 ms). The sampled current on the plateau i_{l} can be calculated for a reversible reaction according to

$$i_{\mathrm{l}} = nFAD^{1/2}c^{\mathrm{b}}/\left[\pi^{1/2}(t_{\mathrm{p}} - t_{\mathrm{m}})^{1/2}\right] \quad (5.9.16)$$

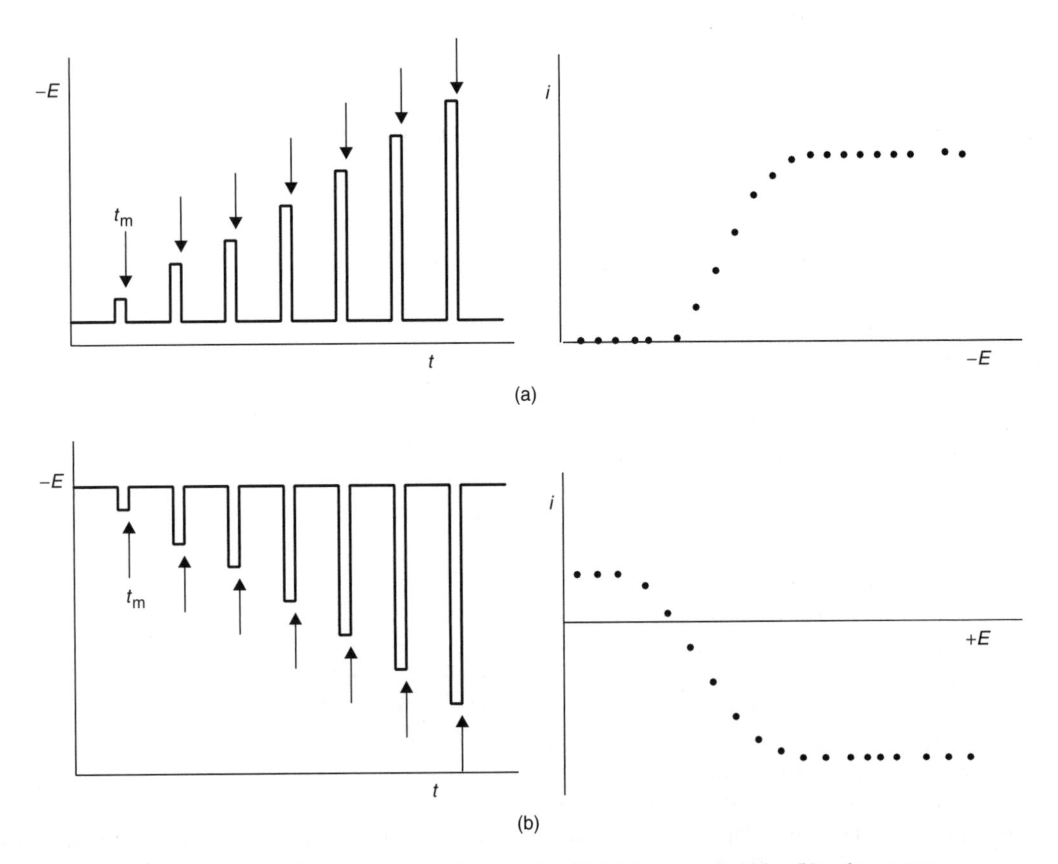

Figure 5.9.6. Principal excitation and response curves for normal pulse (a) and reverse pulse (b) voltammetry.

For polarographic experiments the electrode is held at a base potential at which negligible electrolysis occurs for most of the lifetime of each mercury drop. After a fixed waiting period following drop formation the potential is rapidly switched to a value E for ca 50 ms and the current is sampled near the end of this pulse time. This potential cycle is repeated with successive mercury drops with E being shifted progressively to more reducing values. If the waiting time between the pulses is chosen properly (typically 1–4 s) to allow restoration of the initially uniform character of the concentration distribution in solution at E_i, the limiting current for NPV is larger than that for sampled DC polarography by a factor of about 6. Therefore, detection limits for NPV are in the range of 10^{-6} mol dm^{-3}.

In reverse pulse voltammetry (RPV) the base potential is selected to be in the limiting current region for the reduction (or oxidation) and increasingly oxidizing (or reducing) values are applied during the pulse time (Figure 5.9.6(b)).

NPV at the DME is very sensitive to adsorption processes, which give rise to maxima at the onset of the polarographic wave, the magnitude of which increases with decreasing pulse time. This phenomenon and its use as a tool for detecting the presence of adsorbing substances have been described in detail [59]. In contrast, the current measured at the DME by RPV is less susceptible to adsorption of electroinactive ligands because at extremely reducing potential values the diffusion-limited current depends only on the properties of the bulk solution and at extremely oxidizing potentials it depends only on the properties of the amalgam [60]. Nevertheless, the current–potential curves of RPV can also be affected by adsorption [61] and thus caution should be exercised in application of the DeFord–Hume approach for calculation of stability constants from such experiments (Section 3.1).

2.3.3.4 *Differential pulse voltammetry (DPV)*

An improvement of the i_f/i_c ratio for simple electrode reactions can be achieved with differential pulse voltammetry. The excitation signal (Figure 5.9.7(a)) consists of a staircase (or ramp) potential with a pulse train. In contrast to the former methods the output is now the difference between individual currents measured before the pulse is applied and at its end, respectively. Therefore, the response is a peak-shaped signal (Figure 5.9.7(c)). Typical parameters for DPV measurements of simple faradaic reactions are pulse heights ($\Delta_p E$) between 10–100 mV, a pulse width (t_p) of approximately 50 ms, times between pulses of 0.5–5 s and 'scan rates' between 1 and 10 mV s^{-1}.

The peak current i_p can be estimated for a reversible faradaic reaction according to

$$i_p = nFAc(D/\pi t_p)^{1/2} \frac{1-\sigma}{1+\sigma} \qquad (5.9.17)$$

with

$$\sigma = \exp(nF\Delta_p E/2RT) \qquad (5.9.18)$$

The DPV peak potential E_p lies close to the voltammetric half-wave potential $E_{1/2}$ for small $\Delta_p E$:

$$E_p = E^{0\prime} + RT/nF \ln(D_{red}/D_{ox})^{1/2} - \Delta_p E/2 = E_{1/2} - \Delta_p E/2 \qquad (5.9.19)$$

The $\Delta_p E$ values of ca 50 mV used in most practical analyses result from a compromise between maximum peak current (i_p increases with $\Delta_p E$ according to equation 5.9.17) and sufficient peak resolution (peak width increases as the pulse height grows larger). The detection limit of DPV is often of the order of 10^{-7} mol dm^{-3}. The capacitive background in DPV is flat in the range −0.2 to −1.0 V versus SCE due to the fact that the double layer capacity does not vary much in this range. Therefore, DPV curves are often easier to evaluate than wave-shaped curves.

It is important to note that the concentration-dependent signal, i.e. the peak current, can be significantly lower for irreversible reactions than that predicted by equation (5.9.17). Such processes are causing also broader peaks and equations (5.9.17)–(5.9.19) are not applicable. One has to consider that the time scale of DPV experiments is usually shorter than that for

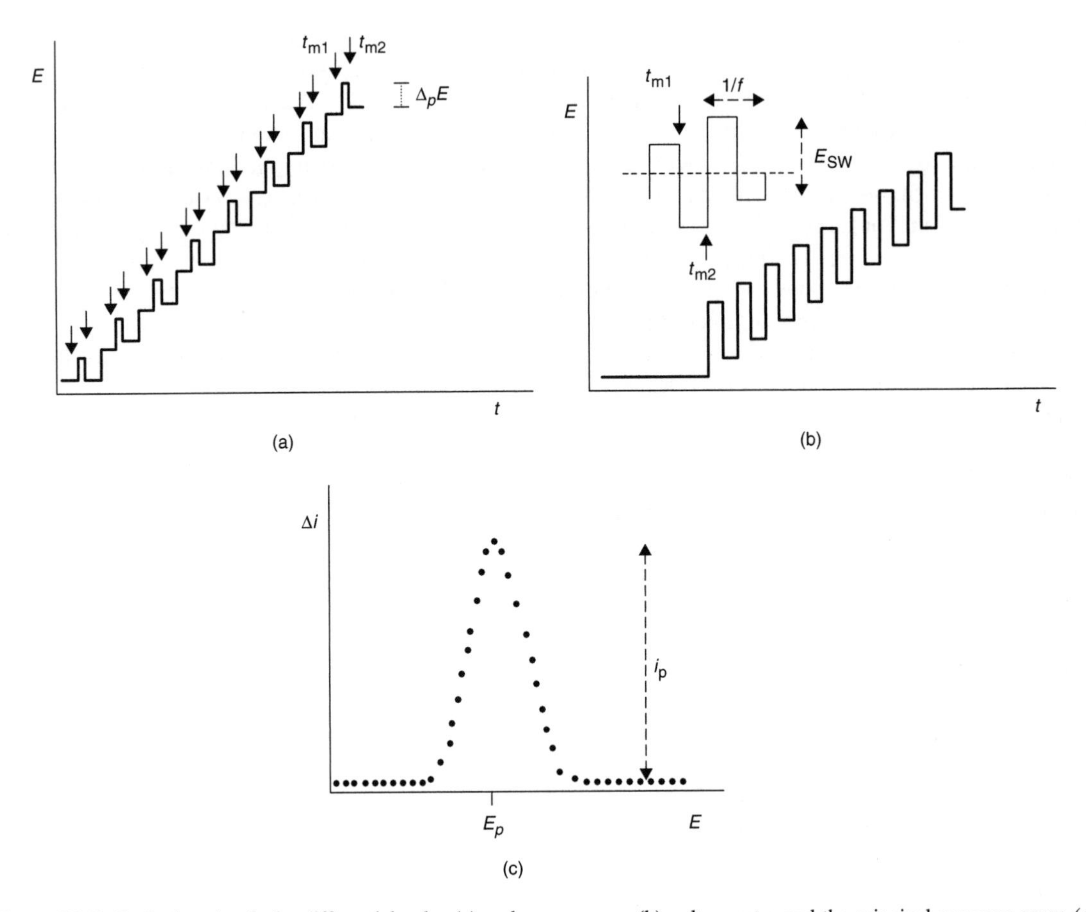

Figure 5.9.7. Excitation signals for differential pulse (a) and square wave (b) voltammetry and the principal response curve (c).

linear sweep voltammetry. Therefore, the degree of reversibility toward the pulse methods may differ from that shown toward LSV and kinetic effects can play a more important role for DPV measurements.

2.3.3.5 Square wave voltammetry (SWV)

The excitation signal of this large-amplitude differential technique consists of a symmetrical square wave superimposed on a staircase, as shown in Figure 5.9.7(b). The current is sampled at the end of the forward pulse as well as at the end of the reverse pulse during each square wave cycle and the difference between the two values is plotted versus the staircase potential. The resulting response curve is comparable to DPV (Figure 5.9.7(c)) and as for DPV the peak height of the signal can be increased by application of larger amplitudes of the excitation function, which causes on the other hand also a peak broadening.

Most experimental parameters are comparable with those of DPV. However, the potential scan rate determined by the square wave frequency (5–500 Hz) is much faster. Detection limits for favorable systems are usually slightly better, i.e. in the range of 10^{-8} mol dm^{-3}. SWV is often more sensitive than DPV because both forward and reversed currents are measured.

From an analytical point of view a major advantage of SWV is its speed, which allows the recording of a complete voltammogram within a few seconds. Certainly, the shorter measuring time has only a negligible effect on analysis time and sample throughput for practical applications, because the total analytical procedure requires much more time than the actual measurement. However, there are several analytical aspects, where faster voltammetry is desired. The working electrode surface is exposed for less time to the detector reactions as well as to interference processes, which could result in less pronounced surface alterations especially at solid electrodes. Such a fast technique can be used to obtain three-dimensional i–E–t plots in flow injection or chromatographic systems during the residence time of the injected sample in the flow cell detector. Moreover, a kinetic discrimination can be achieved against less reversible interfering reactions such as the oxygen reduction. The theoretical background for SWV is much more well established than that for DPV [62].

2.3.3.6 AC voltammetry (ACV)

There are several electrochemical methods that are based on the concept of impedance. For analytical purposes the most important is AC voltammetry, where a constant sinusoidal AC potential is superimposed upon a DC potential ramp (Figure 5.9.8(a)). Typically, the AC potential has a frequency f of 10–1000 Hz and a peak-to-peak amplitude of 4–20 mV. The role of the DC potential is to set mean surface concentrations for both redox states of the reactant, which then face an excitation signal of low amplitude. The current flowing through the cell contains both AC and DC components. The registered responses are either the total AC current as a function of the DC potential, as shown in Figure 5.9.8(b), or preferably the current components in phase (or out of phase) with the AC potential as a function of the DC potential. The latter method is called phase-selective AC voltammetry and is based on the different electrical behavior of ohmic resistors and capacitors in AC circuits. Therefore, it allows the effective discrimination between faradaic (in phase) and capacity (out of phase) currents and leads to detection limits of the order of 5×10^{-7} mol dm^{-3} for reversible systems.

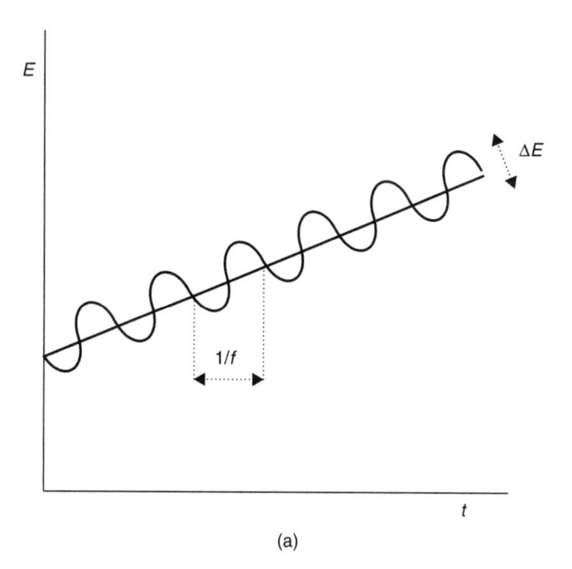

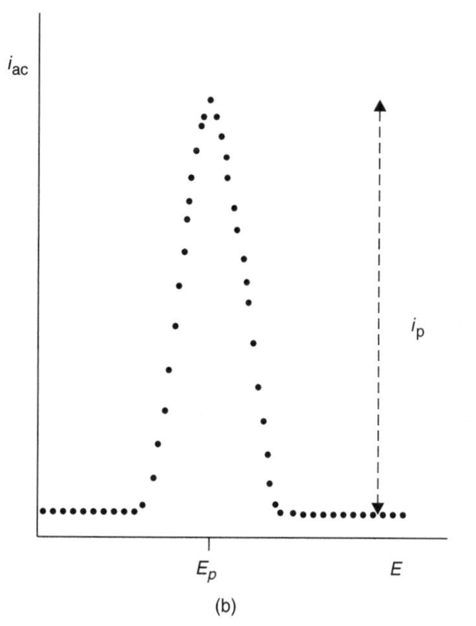

Figure 5.9.8. AC voltammetry: principal excitation (a) and response (b) curves.

The bulk concentration can be determined from the peak current of the response curve. The corresponding correlation for reversible systems is

$$i_p = n^2F^2A(2\pi fD)^{1/2}\ \Delta Ec/4RT \qquad (5.9.20)$$

Equation (5.9.20) is only valid if the AC time window is much shorter than the DC one and small

AC amplitudes are applied, i.e. ΔE should be less than $10/n$ mV.

It is important to note that AC techniques are very sensitive with respect to slow electron-transfer kinetics resulting in smaller signals for such processes than predicted by equation (5.9.20). This feature can, however, be exploited for kinetic discrimination against certain electroactive interferences such as dissolved oxygen. It may be advantageous for the analysis of fast-reacting analytes to measure the AC current at a frequency of $2f$ instead of the fundamental. This so-called second-harmonic AC voltammetry can provide better separation from capacitive current leading to lower detection limits for fast electron-transfer reactions.

The AC voltammetry can also be used to analyze electroinactive compounds which are surface active by measuring their adsorption signals [63–65]. Such tensammetric measurements should be performed in a potential range without interferences from faradaic reactions.

2.3.3.7 *Anodic stripping voltammetry (ASV)*

Voltammetric stripping (SV) techniques are now widely recognized as powerful tools for trace analysis of metal ions and certain organic compounds in solutions. They offer excellent detection limits (down to 10^{-12} M for certain metals!) coupled with inherent species selectivity. These features arise from the two-step nature of this technique: preconcentration of the analyte at the electrode, followed by generation of the analytical signal by stripping from the electrode.

The trace determination of many metal ions can be performed by anodic stripping voltammetry (ASV). Figure 5.9.9 illustrates determination of Pb^{2+} at a HMDE by ASV. The accumulation process comprises the reduction of metal ions at constant potential for several minutes, mostly facilitated by convection. Thus, Pb^{2+} is electrochemically extracted as elemental lead into the mercury electrode forming an amalgam. The resulting lead concentration in mercury is substantially higher than that in the solution of metal ions being analyzed because of the much smaller electrode volume in comparison with the solution volume. After

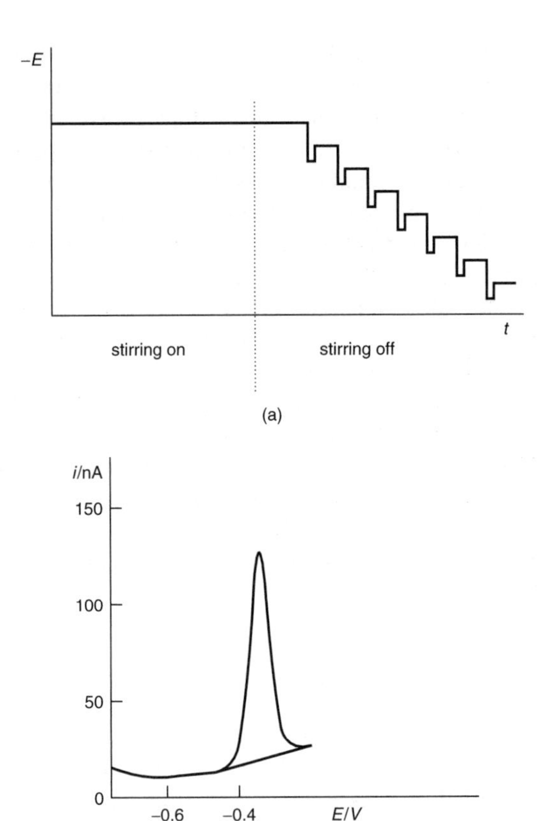

Figure 5.9.9. Anodic stripping voltammetry: excitation signal (a) and measuring curve (b) for the determination of Pb^{2+} at an HMDE with DPV detection.

discontinuing the stirring, the potential is changed to more positive potentials by linear sweep, differential pulse or square wave voltammetry. This causes the oxidation of the amalgamated lead back into the solution registered by the current peak in Figure 5.9.9(b). The peak height is determined by the concentration of Pb in the mercury electrode which is in turn proportional to the amount of Pb^{2+} in the sample if appropriate experimental parameters such as electrode area, deposition time, stirring etc. are held constant and there are no secondary effects such as peak broadening due the presence of heterogeneous ligands (see below). The necessary accumulation time depends on the analyte concentration and can reach up to 20 min at the 10^{-9} mol dm^{-3} level. With this ASV approach about 15 amalgam-forming metals can be determined, including Tl, Cd, Zn, Cu, Bi, In and Ga.

The trace analysis of other metal ions (Hg, Au, As, Se) can be performed after their electrolytic deposition as a metal film on bare solid electrodes made from carbon or gold.

2.3.3.8 Stripping chronopotentiometry (SCP)

Stripping chronopotentiometry (also referred to as 'potentiometric stripping analysis') [9], analogous to SV, is a two-step technique. The first, deposition, step is identical to that for ASV, but reoxidation of the accumulated metal is then achieved by application of a constant oxidizing current or by constant flux of a chemical oxidant (usually Hg^{2+} or O_2). The electrode potential is recorded as a function of time during the stripping step, and the analytical signal is the time taken for reoxidation (the transition time, τ). The analytical signal can be enhanced simply by application of a smaller stripping current, or by a lower flux of chemical oxidant (a major advantage over voltammetric stripping techniques), with the practical detection limit being determined by the presence of redox active impurities, e.g. dissolved oxygen, in the sample (which represents the major limitation compared to voltammetric techniques, particularly SWV and ACV). The instrumentation required for SCP is simpler than that for voltammetric methods.

There are many claims in the literature (based on empirical observations) that SCP is less susceptible to interferences from adsorption of organic compounds on the electrode surface than is SV [66]. A more rigorous understanding of this behavior has recently been developed: determination of transition times from the area under peaks in the inverse of the time derivative of potential (dt/dE) versus E plots is the correct strategy for effective elimination of charging currents [12]. The area under the baseline corresponds to the time necessary for charging, which is thus effectively eliminated from the analytical signal by this approach. Some workers report poor baselines for dt/dE versus E plots, ascribe this to 'adsorption', and apply some arbitrary 'background correction' protocol [67]. However, correct interpretation requires knowledge of the stripping time regime under which measurements are made. When high stripping currents are used with a HMDE, the accumulated metal is not completely stripped from the electrode during the transition plateau and poor baselines result from the ongoing faradaic processes that follow this incomplete depletion. When experimental conditions are chosen such that measurements are made in the complete depletion regime, a limit of detection directly comparable with that for DP-SV is achieved with the advantage that discrimination against capacitive charging is achieved by an approach that avoids the adsorption complications associated with pulse SV waveforms [12].

2.3.3.9 Adsorptive stripping voltammetry (AdSV)

In recent years a fast growing number of nonelectrolytic accumulation procedures using the adsorption of the species of interest at the electrode surface has been developed for metal ion complexes as well as for an increasing number of organic compounds. Tabulations of experimental conditions for a range of elements have been reported [68, 69]. This so-called adsorptive stripping voltammetry (AdSV) allows the trace determination of metals such as Al, Fe, Co, Ni, Mo, V, Cr, Ti, U, La, which cannot be measured by ASV due to nonfavorable reversible reactions or the absence of amalgam formation at mercury electrodes. The technique involves addition of surface-active ligands (L_a) with an affinity for the species of interest M, followed by accumulation of the resulting ML_a complexes as a monolayer on the electrode surface (either in open-circuit, or by application of an appropriate accumulation potential), and finally quantification (stripping) by a reducing potential scan (for which a range of waveforms can be employed, e.g. DPV, SWV). The analytical signal obtained during the stripping step may arise from reduction of either the element or the ligand in the adsorbed complex, or from catalytic effects [70–72]. This approach can also improve the analytical signal for other metals which are conventionally measured by ASV. AdSV measurements are particularly susceptible to interference from other surface-active material in the sample which will compete with the chelate complexes for coverage of the electrode surface.

Note that the term 'cathodic stripping voltammetry (CSV)', is often used to denote AdSV. We recommend that this terminology be avoided because it can cause confusion with other stripping methods which use a change of the oxidation state during the analyte accumulation followed by its reductive determination [9].

Application of SV and SCP to measurements of very low concentrations requires use of special guidelines for trace analysis. All precautions should be obeyed to prepare carefully the standards and to avoid contamination of the sample or loss of analyte, e.g. by adsorption at cell walls. Moreover, such two-step procedures at the trace level are subject to several interferences. The adsorption of foreign surface-active compounds can influence not only the accumulation during the adsorptive stripping mode, but also the measuring step of the ASV for metals, particularly when pulse modes are employed. Complications can also arise from the formation of intermetallic compounds in the electrode or electroactive interferences with comparable redox potentials to the analyte. It has been frequently shown that such problems can be avoided by the careful development of an appropriate stripping method, for which both electrochemical and chemical parameters were optimized taking into account the sample matrix [14, 73, 74].

2.3.4 Electrochemical detection in liquid chromatography and flow-injection analysis

Electrochemistry offers a number of advantages for the trace determination of certain species in liquid chromatography (LCEC) and flow-injection analysis (FIAEC). One commonly used electrochemical detector is a thin-layer cell in which the working electrode is positioned in a thin channel through which the mobile phase flows. A detector of this type is shown schematically in Figure 5.9.10, which illustrates also the

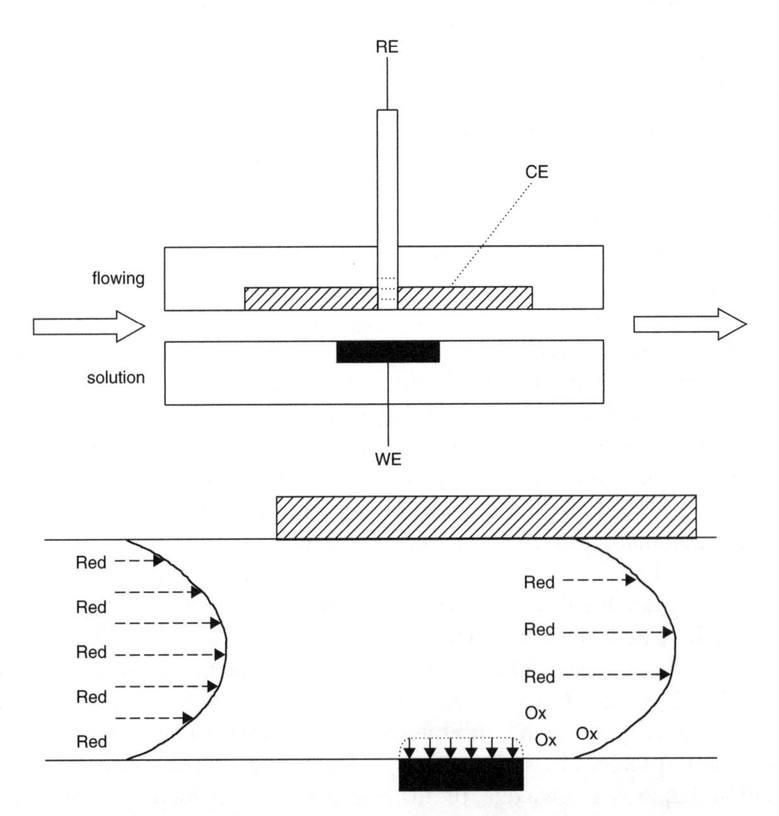

Figure 5.9.10. Scheme of an electrochemical thin-layer cell and the corresponding flow-through detection.

detection of a 'plug' of analyte Red that has eluted from the column being detected by oxidation to Ox as it sweeps over the electrode in the thin channel. For maximum sensitivity the potential of the electrode is held on the limiting current region of a hydrodynamic voltammogram (see Figure 5.9.5) for oxidation of Red. The resulting chromatogram shows a peak current response for the detection of Red. Its concentration can be quantified by measuring either the peak height or integrating the peak and measuring the charge, both of which are proportional to the bulk concentration.

Thin-layer detectors of this type are mostly operated as fixed potential (amperometric) devices and the current is measured as a function of time. Typically about 5 % of the analyte is electrolyzed as it sweeps past the electrode. Amperometric detectors can achieve detection limits as low as 10^{-8}–10^{-9} mol dm^{-3} of injected analyte. Coupled with an injection sample of 10 μL, they can detect as little as 10^{-16} mol (or 0.1 fmol). Amperometry is advantageous in comparison to the voltammetric techniques because the charging current is minimized by operating at a fixed potential.

Interest in voltammetric detectors in which the potential is scanned to provide a voltammogram, however, is increasing because they provide qualitative information about the species being detected in the form of the half-wave potential. This can be especially important in speciation studies as an additional parameter to be used in conjunction with chromatographic retention time for qualitative identification of a species. As for voltammetric techniques, the development of acceptable voltammetric detectors cannot be based on linear scan voltammetry due to the poor sensitivity of this mode. Pulse techniques such as differential pulse and square wave voltammetry are the most commonly used methods to diminish charging-current contributions. Recently, microelectrodes have been applied in electrochemical detectors to alleviate charging-current problems [75]. Coulostatic detectors have also been designed to obtain voltammetric information [76].

3 PRINCIPLES OF SPECIATION BY ELECTROANALYSIS

3.1 Thermodynamic aspects

The most straightforward information for determination of thermodynamic equilibria parameters (K values) is provided by analytical signals from ISEs (i.e. potentiometric E values). In this technique equilibrium is assumed to exist throughout the measurement system ($i \rightarrow 0$) and the analytical signal depends only on the free metal ion activity (which is fixed by its equilibrium across all binding sites in the system). Voltammetric signals are also sensitive to K, but since the signal depends on the flux of species to the electrode surface, correct interpretation requires concomitant consideration of kinetic factors (D, k_d) (see below).

ISEs provide a direct measure of the free hydration ion activity, $\{M\} = c_{M,T}\gamma/\alpha$, where γ is the activity coefficient for M, and α its degree of complexation ($= c_{M,T}/c_M$). α is related to the stability constants for metal complex species; for the simplest case of a single well-characterized simple ligand forming a single complex, ML, $\alpha = 1 + Kc_L$. Determination of α, and thus K values, requires knowledge of $c_{M,T}$, γ, and the E_o and slope (calibration) parameters for the ISE (equation 5.9.7). At constant ionic strength, equation (5.9.7) can be written as

$$\Delta E = s \log c_{M,T} - s \log \alpha \qquad (5.9.21)$$

where $s = 2.303RT/nF$.

Complexation parameters are usually determined from titrations of ligand with metal (or vice versa) [18]. For increasing $c_{M,T}$ at constant $c_{L,T}$, α for each point in the titration can be calculated from equation (5.9.21). The most useful part of the titration curve is the region in which $c_{M,T} \ll c_{L,T}$, under these conditions $c_L \approx c_{L,T}$ and thus $\alpha = 1 + Kc_{L,T}$ for the simple case. In the case of heterogeneous ligands (typical of real-world samples), various data interpretation models are applied to extract K values (or their distribution) from the titration curves [14]. Despite their limitations (in particular interference from other ions and low sensitivity), ISEs are a useful tool for studying metal ion complexation, e.g. the influence of parameters such as the metal ion loading ($c_{M,T}/c_{L,T}$) on α, which is an important factor for natural heterogeneous ligands, is more readily tested with ISE potentiometry than via other methods [77].

For voltammetric methods and labile complexes (see Section 3.3 for definitions), α can be calculated from current and potential values obtained in the absence and presence of ligand via the DeFord and Hume expression

$$\ln \alpha = \frac{nF}{RT}(E_{1/2} - E_{1/2}{}^{L}) + \ln\left(\frac{i_l}{i_l{}^{L}}\right) \quad (5.9.22)$$

with $i_l{}^{L}/i_l = \sqrt{\overline{D}/D_M}$ on macroelectrodes (see Section 3.2.3 for definitions), where i_l denotes the polarographic limiting current, $E_{1/2}$ the half-wave potential, superscript L denotes values in the presence of ligand(s) and $\overline{D}$ is the mean diffusion coefficient for M and the complex. Under conditions where each ligand concentration is such that $c_{L,T} \gg c_{M,T}$, α is related to the stability constants, K_i, for metal complexes ML_i, by

$$\alpha = 1 + \sum_i K_i c_{L_i}$$

and $\overline{D}$ is given by: $\overline{D} = D_M/\alpha + \Sigma D_{ML_i} f_i$, where f_i is the fraction i of complex ML_i with respect to $c_{M,T}$. These expressions are directly applicable to other voltammetric techniques by substitution of E_p and i_p for $E_{1/2}$ and i_l.

For a system containing M and a single metal complex species, ML

$$\frac{i_l{}^{L}}{i_l} = \sqrt{\frac{1}{\alpha} + \frac{D_{ML}}{D_M}\left(\frac{\alpha - 1}{\alpha}\right)} \quad (5.9.23)$$

(For microelectrodes (Section 3.2.3) the same equation holds but without the square root.)

There are three useful limiting cases of equation (5.9.23):

(i) small simple ligands for which $D_{ML} = D_M$, thus $i_l{}^{L}/i_l = 1$, and $\ln \alpha = nF(E_{1/2} - E_{1/2}{}^{L})/RT$

(ii) $\alpha \gg D_M/D_{ML}$ typical of strong complexation by small ligands. In this case $i_l{}^{L}/i_l = \sqrt{D_{ML}/D_M}$ and α can be computed from equation (5.9.22) if D_{ML} is known.

(iii) $D_M/D_{ML} \gg \alpha - 1$ typical of weak complexation by large ligands. In this situation ML is essentially immobile and the current is determined primarily by the free metal ion, and

$$\frac{i_l{}^{L}}{i_l} = \sqrt{\frac{1}{\alpha}} = \sqrt{\frac{c_M}{c_{M,T}}}$$

Application of this methodology has been discussed in more detail for different types of metal complex species [18, 19]. The DeFord–Hume expression (equation 5.9.22) has been extended to the case for heterogeneous ligands [78]; various data interpretation models are applied to extract K values (distributions) from the voltammetric curves for such systems (e.g. Freundlich isotherm).

Many publications on element speciation assume that the system under consideration is at equilibrium even at the electrode surface, i.e. that the complexes are labile (see Section 3.2) However, this depends strongly on the measurement time scale of the technique and must be tested and verified in each case [14, 75]. A more complete understanding of element speciation must incorporate knowledge of the interconversion rate of species with respect to diffusion rate, and thus a study of chemical and physical kinetic properties is required.

3.2 Kinetic aspects

With the exception of potentiometry, electroanalytical techniques are nonequilibrium (dynamic) techniques. Thus sound theoretical concepts are required to relate the analytical signals to the underlying properties of the system under study. The important concepts involved in voltammetric measurements are redox reversibility, chemical lability, and physical mobility.

Reversible behavior results when the charge transfer rate, k_0, is much higher than the diffusion rate which corresponds to $k_0 \gg \overline{D}/\delta$ for macroelectrodes (linear diffusion) and $k_0 \gg \overline{D}/r$ for microelectrodes (spherical diffusion, r = electrode radius). Interpretation of data is simplified for reversible systems and this condition is assumed throughout the following discussion. Tests for reversibility using various electrochemical methods are tabulated (see also Section 2.3.3.1) [18, 19].

The dynamic behavior of metal complexes at the voltammetric interface has been discussed in detail by several authors [18, 19, 79–81], and only a brief outline is included here. Complexes are considered as dynamic when (i) their mobility is high enough, i.e. they can move towards the electrode surface by diffusion, at a rate that is nonnegligible compared with that of free M, and (ii) their lability is also high enough, i.e. they can associate/dissociate a large number of times during their diffusive transport. Since the latter depends on the measurement time scale of the technique used, lability also depends on this time scale. Based on these concepts, the following definitions are used:

- Complexes are said to be *immobile*, when their diffusion rate (mobility) is negligible compared with that of the free metal ion, M.
- Complexes are said to be *inert* when the number of association/dissociation steps is negligible during the time scale of interest (usually the time of diffusion through the diffusion layer, which also corresponds to the measurement time of the technique).
- Complexes are said to be *labile*, when they dissociate/associate a very large number of time during their diffusive transport to (or from) the electrode.
- Complexes are *semi-labile* when they exhibit behavior borderline between inert and labile complexes.
- Complexes which are both mobile and either labile or semi-labile are called *dynamic*.

It is most important to note that only dynamic complexes can contribute to the measured voltammetric current.

The above definitions can be defined on a mathematical basis as follows. Consider the simple reaction

$$\mathrm{M} + \mathrm{L} \underset{k_\mathrm{d}}{\overset{k_\mathrm{a}}{\longleftrightarrow}} \mathrm{ML} \qquad (5.9.24)$$

At one extreme, the inert situation arises when $k'_\mathrm{a}t$, $k_\mathrm{d}t \ll 1$, where t is the measurement time. This means that complex dissociation during this time is negligibly small and the flux (i.e. the current) is thus controlled by the diffusion of free metal species in the bulk solution.

The flux $J(t)$ in solution containing free metal ions, M, and a single dynamic complex, ML, has been computed in the presence of an excess of ligand L, for semi-infinite diffusion as the sole transport process towards the electrode and the condition that this latter is a perfect sink for M (i.e. the potential is negative enough to ensure a complete reduction of M). For a dynamic system at a macroelectrode, $J(t)$ is given by [82–84]:

$$J(t) = \frac{{k_\mathrm{d}}^{1/2}{D_\mathrm{M}}^{1/2}c^*_\mathrm{T}(1+\varepsilon K')^{3/2}}{\varepsilon^{3/2}K'(1+K')} \times \exp(\Lambda^2 t)\mathrm{erfc}(\Lambda t^{1/2}) \qquad (5.9.25)$$

where $c^*_\mathrm{T} = c^*_\mathrm{M} + c^*_\mathrm{ML}$, $\varepsilon = D_\mathrm{ML}/D_\mathrm{M}$, and

$$\Lambda = \frac{{k_\mathrm{d}}^{1/2}(1+\varepsilon K')}{\varepsilon^{3/2}K'(1+K')^{1/2}} \qquad (5.9.26)$$

$$k'_\mathrm{a} = k_\mathrm{a}c^*_\mathrm{L}, \text{ and } K' = k'_\mathrm{a}/k_\mathrm{d} = Kc^*_\mathrm{L}.$$

A lability criterion (based on the magnitude of the term $\Lambda t^{1/2}$) is used to describe the contribution of metal complex species to the overall flux of metal at the electrode surface. The lability of metal complex species will decrease with increasing values of K' and decreasing values of ε for a given measurement time scale. Two kinetic limiting cases can be identified:

(i) $\Lambda t^{1/2} \gg 1$, labile case. The flux is diffusion controlled and reduces to:

$$J(t) = \frac{\overline{D}^{1/2}c^*_\mathrm{T}}{(\pi t)^{1/2}} \qquad (5.9.27)$$

It corresponds to purely diffusion controlled coupled transport of M and ML (i.e. the association/dissociation kinetics are fast compared to diffusion).

(ii) $\Lambda t^{1/2} \ll 1$, nonlabile case. The flux is entirely controlled by the chemical kinetics of dissociation of ML (i.e. electron-transfer kinetics are slow relative to diffusional transport), and for large $K'(\varepsilon K' \gg 1)$, the flux is given by:

$$J(t) = k_d c_T{}^* \left(\frac{D_M}{k'_a}\right)^{1/2} \quad (5.9.28)$$

The case of inert complexes can be seen as the limit of equation (5.9.28), when $J(t)$ is negligibly small compared to the equivalent flux which would be obtained in absence of L.

This approach has been extended to analysis of the steady-state case (δ as the variable) [79, 85], chemically heterogeneous systems that involve a dissociation rate constant distribution [86], and to micro uptake surfaces at which spherical diffusion must be considered [79].

3.2.1 Dependence of measured parameters on lability

Characteristic changes in the current–potential curves are observed according to the lability of the metal complex species being measured relative to the curves for free metal ions. This is shown schematically for DC polarography (or NPV) and for DPV (or ACV) in Figure 5.9.11.

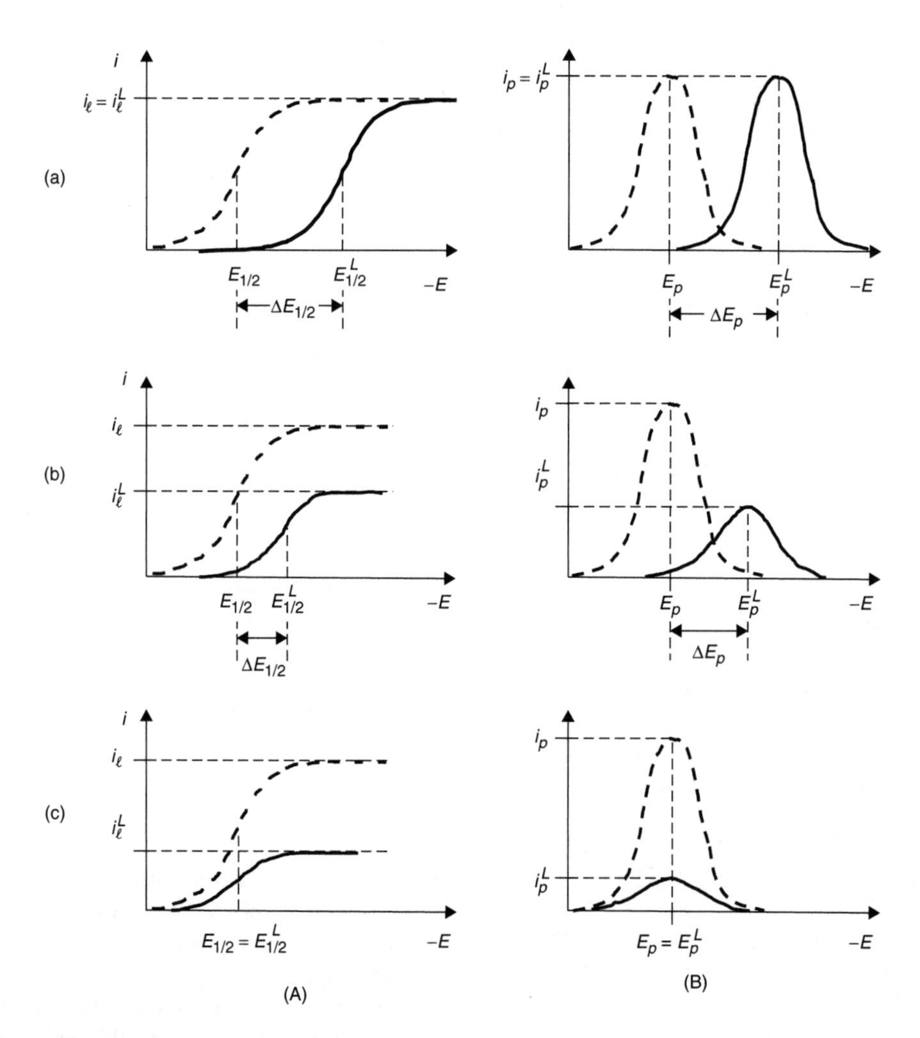

Figure 5.9.11. Comparison of $i = f(E)$ curves obtained by (A) DC polarography or NPV and (B) AC polarography or DPV, for (a) labile, (b) nonlabile, and (c) inert metal complex species (solid lines) with those for the free metal ion (dashed lines).

The corresponding curves for other electroanalytical methods follow a similar pattern [18]. As explained above, inert complexes do not dissociate during the measurement time and thus do not contribute to the flux (and thus current). The current measured for such systems is thus proportional to the free metal ion concentration and there is no change in the potential characteristics. In contrast, dynamic fully labile complexes do contribute to the measured current, the potential is shifted relative to the ligand-free case and the magnitude of this shift is a measure of the thermodynamic stability of the complexes. Furthermore, for labile complexes, with $D_{ML} \approx D_M$, the limiting current is the same in the presence and absence of ligand; a reduction is observed when $D_{ML} < D_M$. Dynamic nonlabile complexes display intermediate behavior, i.e. some reduction in limiting current (but less than that for inert complexes) and some shift in E (but less than that for labile complexes).

3.2.2 Dependence of lability on measurement time scale

As discussed above and shown in particular in the lability criteria expression, the lability of a given complex is not an intrinsic property, but rather an operationally defined concept which depends on the measurement time scale of the analytical technique employed. Typical time scales, t_m, for electroanalytical techniques are shown in Figure 5.9.12.

3.2.3 Lability at microelectrodes

In recent years there has been an upsurge in the application of microelectrodes [36] to element speciation measurements. Rational interpretation of experimental data obtained at these electrodes requires a sound theoretical understanding of processes occurring at the microsurface. Transition from a macro- to a microelectrode influences both the nature of the flux to the surface and also the extent of lability (and thus bioavailability, Section 5.2) of metal complex species. Below is a brief discussion of the application of the lability concepts to microsurfaces. More details are given in recent publications [87, 88].

Diffusion regimes range from a linear profile at macroelectrodes, to spherical one at a microelectrode. The limit between these regimes is determined by the size of the electrode radius relative to the diffusion layer thickness, i.e. the ratio r/δ. The thickness of the diffusion layer in quiescent solution at a planar electrode is given by $\delta = \sqrt{\pi \overline{D} t}$, while under convective conditions (electrode rotation or solution stirring at rotation rate ω) $\delta \gg (\overline{D}/\omega)^p$ ($p \approx 0.3$–0.5).

Irrespective of the electrode size, the current (or metal flux) to the electrode surface at constant potential is given by

$$i = nFA\overline{D}c^b\left(\frac{1}{\delta} + \frac{1}{r}\right) \qquad (5.9.29)$$

For a macroelectrode, i.e. a planar electrode at which linear diffusion occurs, $1/r \ll 1/\delta$; i is thus primarily controlled by δ and therefore sensitive to solution convection or time, since δ depends on time in purely diffusive transport.

In contrast, for microelectrodes $1/r \gg 1/\delta$, therefore i is determined principally by r and consequently is independent of both time and solution convection (a major advantage for *in situ* applications, Section 4).

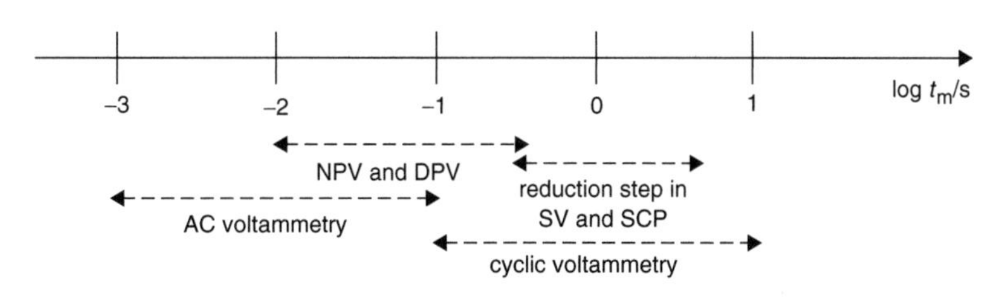

Figure 5.9.12. Effective measurement times for electroanalytical techniques.

Because lability is relative to transport rate, and diffusive transport depends on microelectrode size, it follows that lability of complexes will themselves depend on electrode size. This aspect has been discussed by de Jong and van Leeuwen [84].

3.3 Considerations for sample preparation and experimental conditions

3.3.1 *Measuring solutions*

The electroanalytical methods described above are used for measurements in solution. Therefore, all aspects of the preparation of nonliquid samples for speciation analysis, which are discussed in Chapter 2.1 of this Handbook, have to be carefully considered. Moreover, the specific demands of electroanalysis may give rise to additional problems. Many of the techniques require a sufficient electrical conductivity of the measuring solution provided by ionic components that are not participating in any electrochemical process influencing the analytical signal. The addition of such a supporting electrolyte, however, can change the original species distribution in solution. The use of microelectrodes is a way to overcome this problem, since they can perform measurements in solution with high resistivity. Indeed, due to their small size, the measured current is small, as is the corresponding ohmic drop. Usually, the pH of the measuring solution cannot be modified, since otherwise, the whole of dynamic complexation equilibria will be shifted. When the pH can be changed, great care must be taken to choose appropriate buffers, as they usually also contain potential ligands that can change the speciation of metal ions by complexation or adsorption at interfaces (electrodes, cell wall). The complexing ability and surface activity of several commonly used buffers have been characterized and compared [89, 90]. During the reduction of metal ions, the potential is sufficiently negative to also reduce the oxygen present in the solution. When elimination of oxygen is not sufficient, its reduction leads to an increase in pH at the electrode surface that may drastically change the speciation if the solution is not sufficiently buffered. Therefore, the sample preparation and manipulation must be designed very carefully in accordance with the chemical properties of the analyte species and the requirements of the measuring method. As a general rule, sample handling should be minimized.

3.3.2 *Adsorption effects*

Adsorption of foreign material, in particular colloidal and macromolecular organic compounds [91], from the sample at the working electrode usually drastically interferes with speciation analysis [18, 19]. When the test metal species itself is adsorbed, special signals can arise. Pulse and AC modes of polarography and voltammetry are useful tools to check for the presence of such adsorption processes [59]. The impact of adsorption will depend on the time scale of the technique, the concentration of adsorbing material, and the electrical potential at the interface [18, 92].

Many of the applications of electrochemical techniques to speciation analysis (Section 4) involve measurements in matrices containing significant amounts of organic compounds that have a propensity for adsorption on electrode surfaces. Various approaches have been proposed to overcome interference from adsorption of organic compounds ranging from protective coating of the electrode surface (Section 2.2.3), to removal of the organic matter, e.g. by addition of fumed silica [93]. Most of these methods have some drawbacks, and any methodology which may perturb the initial solution composition should be avoided.

3.3.3 *Avoiding saturation of ligands at the electrode surface*

During the stripping step of ASV, the surface concentration of oxidized metal ions is far greater than that in the bulk solution and thus secondary reactions may be induced at the electrode surface [94]. An appropriate choice of experimental conditions, including the use of a sufficiently large excess of ligand, is required to minimize the impact of these effects [95]. The use of a medium exchange has been proposed to overcome this problem [96], but reproducible results have proved difficult to achieve in practice. The construction of stepwise stripping voltammograms (so-called 'pseudopolarography'), in which the magnitude of the stripping signal is recorded as a function of deposition potential, may be a more appropriate approach to obtaining speciation information than attempts to interpret single ASV measurements [97] provided conditions are carefully chosen to avoid the problem of ligand saturation during the stripping step.

Note that a large excess of ligand over metal also simplifies data interpretation (Section 3.2). Under these conditions a constant ratio between complexed and free metal can be assumed over the diffusion layer and thus the mean diffusion coefficient operating from the bulk solution to the electrode surface can be taken as constant.

4 APPLICATIONS

During recent years speciation analysis has been mainly directed at the quantification of either various redox states of an element, its organometallic species or metal complexes in equilibrium with each other. These various analytes of interest possess different chemical stability and volatility that has to be considered for the selection and optimization of the analytical procedures. In principle, the ideal method for any application is one that can be deployed *in situ* with minimal sample perturbation. At present, such measurements are in the minority, but rapid advances in this field are anticipated.

The vast majority of literature to date on the application of electroanalytical methods to determination of speciation has been directed at speciation of metal complexes in solution and reports very empirical data, e.g. percent 'ASV-labile' metal, with experimental details often very poorly defined. Such data have no real meaning in themselves, and certainly cannot be extrapolated to conditions other than those under which they were determined (e.g. to predict the impact of a change in pH, metal ion loading, etc.).

As is evident from the discussion in Section 3, (and see refs [18, 19]) reliable determination of useful speciation parameters (stabilities of metal complex species and kinetic association/dissociation rate constants) from electroanalytical measurements requires knowledge of the time scale and diffusion regime of the technique employed. Furthermore, in real-world samples, the stability and rate constant parameters are typically represented by a (continuous) distribution of values. This factor, together with other possible interferences (Section 3.3) means that correct interpretation of electroanalytical measurements can be difficult, and all experimental conditions must be fully reported to facilitate utility of results to others. Despite these difficulties, the sensitivity of electroanalytical techniques to many important speciation parameters makes them very powerful tools for many applications.

Examples of applications of electroanalytical methods to determination of element speciation in various media are given below.

4.1 Aquatic systems

There are several reviews on application of electroanalytical methods to element speciation in aquatic systems [14, 19, 98, 99]. Redox speciation is an important topic where voltammetry/polarography and/or ASV could in principle be used to distinguish between Fe(III)/(II), Cr(VI)/(III), Tl(III)/(I), Sn(IV)/(II), Mn(IV)/(II), Sb(V)/(III), As(V)/(III), and Se(VI)/(IV). On the other hand, there are only a few papers dealing with the determination of organometallic species by direct electrochemical detection. For instance, Schwarz *et al.* [100]. have described the analysis of butyltin species in surface water from a harbor by AdSV with tropolone. The three butyltin compounds are electroactive and can be detected down to 0.5–5 $\mu g\,dm^{-3}$. A further LOD improvement by a factor of 10–20 has been obtained by pre-concentration of the analytes at a solid-phase extraction column. Bond *et al.* [101] have found an interaction of the reduction processes of butyltin species (without tropolone complexation) at the mercury electrode which pointed to the necessity of additional separation steps for speciation analysis (see below).

ISE. An ISE has been directly applied to the determination of the free metal concentration in seawater [102]. Most studies have used ISEs to follow metal titrations of water samples; various models are then applied to the data for extraction of apparent binding site affinities and corresponding complexation capacities [14, 103].

Laboratory and on-site polarography of O_2, Fe(II), Mn(II) and S(−II). It has been shown, more than 20 years ago, that DC polarography, sampled DC polarography, DP polarography and CV can be used on board a ship to determine the speciation of Fe(II), Mn(II), and S(−II) in eutrophic waters. This subject has been reviewed in several publications [19, 104, 105]. The method allows determination of the total dissolved S(−II) (which is usually dominated by S(−II) and its protonated forms) and the total dissolved Fe(II) and Mn(II) (which in most waters are dominated (95 %) by the aquated Fe^{2+} and Mn^{2+} ions) [106]. In addition, a special peak occurs in the presence of adsorbable small polynuclear FeS complexes. The polarography of the Fe(II)/S(−II) system has been studied in detail both in the laboratory and on site [107]. By taking the difference between the colorimetric measurement of Fe(II) and S(−II) after 0.45 μm filtration, and the direct DPP signals for these ions, the concentration of colloidal FeS can also be obtained. The formation of colloidal FeS is time dependent and can be followed by this method. It also allows them to be discriminated readily from the colloidal forms of Mn(IV) and Fe(III) oxides. Detailed concentration profiles of these various species and their seasonal evolution with time have been reported by using this method and their redox interactions were discussed [107]. Recently this approach has been used to measure specifically the free Mn^{2+} by an *in situ* voltammetric probe, at depths in the range −80 to −95 m below the surface of Lake Lugano (Tessin, CH), and to discriminate it from colloidal Mn upsurged by storm events from the sediment present 2–3 m below [19].

AdSV. AdSV is probably one of the most widely used methods for measurement of metal complexation parameters (stability constants and complexation capacities) in natural waters [108]. This approach is based on competition between the added ligand and complexants present in the sample for the elements of interest, and therefore cannot be employed in a truly *in situ* form, although 'on-line' procedures have been proposed [109, 110]. It should be noted that the pH of the sample is often adjusted in order to achieve optimal complexation conditions for the added ligand, e.g. the following buffers (at ca. 0.01 mol dm^{-3} concentration) have been used in analysis of seawater samples by AdSV: Tris [111], EPPS [111, 112], HEPES [113, 114], HEPPS [115, 116, 117], borate [118, 119, 120], PIPES [121]; and in freshwaters: TES [122], EPPS [123], PIPES [124], HEPPS [125], HEPES [126–128], acetate [129]. Any pH adjustment (and reagent addition) will undoubtedly perturb the sample equilibria. A different ligand and/or conditions are usually required for each element of interest. The ligands and conditions employed for determination of a range of elements in environmental and biological matrices by AdSV have been tabulated [69, 70, 130].

Interpretation of AdSV data remains very empirical and data are of questionable value due to a number of artifacts whose importance is still poorly identified. For example, it is assumed that the added ligand reaches equilibrium with the sample components even though very stable complexes are purported to be detectable by AdSV ($\log K \approx 15$) which may have very low rates of association/dissociation, necessitating very long equilibration times with the added ligand. In addition, many of the ligands used for AdSV are unstable towards oxidation (e.g. catechol) and thus their concentration cannot be assumed to be constant over the time course of the experiment. An important interference is competitive adsorption of the added complex and natural surface-active sample components, e.g. humic substances, on the electrode surface. This aspect is usually poorly characterized and made worse by the fact that analyses typically involve the use of a range of concentrations of added ligand (to determine the so-called 'complexation capacity') such that the effective competition is not equal across all measurements. These 'titration' protocols are routinely used in determinations of metal complexation parameters (log K and complexation capacity) by AdSV in both seawaters [113, 117, 120] and freshwaters [124, 125, 128]. Considering this factor, attempts to develop multielement AdSV protocols, involving several added ligands, must be interpreted with caution [69].

The analytical detection window of AdSV depends on the concentration of added ligand and its stability constant with the element of interest [131]. When this factor is taken into account, the so-called 'strong' binding ligands reported by many workers are seen to be merely a consequence of the heterogeneity of ligands in these systems [132].

Microelectrodes. Microelectrodes are necessary for speciation measurements on low ionic strength freshwaters when no perturbation of the water is allowed, even the addition of a non complexing electrolyte. In addition, their signal is not influenced significantly by colloidal complexes with size larger than ~4 nm, so that they provide a more well-defined distinction between truly dissolved and colloidal (size limit ≈1 nm) species [19]. For example, measurements by square wave modulated stripping voltammetry with an agarose gel-coated mercury-coated iridium microelectrode showed that most of the Pb(II) and Cd(II) (80–90 %) in river waters heavily loaded with suspended particles was associated with colloidal material and enabled determination of binding site concentrations and corresponding stability constants for metals and proton [41].

Separation/detection. Several approaches have been reported for the exploitation of the capability of electrochemical techniques to detect very low analyte concentrations after species separation [133]. An off-line procedure for the separation of dibutyltin and triphenyltin by ion-exchange chromatography followed by ASV detection of the tin species has been described and tested with the extract of a sediment reference material [134]. Conductometric detection in capillary electrophoresis (CE) was used for the on-line determination of arsenic and selenium species (As(III), As(V), DMA, Se(IV), Se(VI)) with limits of detection of about 50 $\mu g\,dm^{-3}$ [135]. The method has been applied to water samples from a tailing of tin ore processing. The advantages of microelectrodes for amperometric CE detection [136] have been used for the speciation of mercury (Hg^{2+}, monomethylmercury, monoethylmercury) [137]. After electrophoretic separation the Hg species are reduced at a gold microelectrode at −0.2 V. The procedure provides detection limits of 0.2 $\mu g\,dm^{-3}$ (Hg^{2+}) and 5 $\mu g\,dm^{-3}$ (Me–Hg) with a dynamic range of three orders of magnitude. Also extracts of sediments have been analyzed by this approach. Field-portable CE instrumentation with electrochemical detection has been reported [138].

In situ. A more realistic understanding of element behavior in aquatic systems requires continuous real-time monitoring. Various voltammetric probes have been reported for *in situ* deployment in waters [19, 139]. It has been shown that microelectrodes are very important for such applications, for various reasons. Most of them are prototypes with a limit of detection of about 10^{-8} mol dm^{-3}, which is useful mostly for polluted waters. A commercial one, however, is available [19] with a detection limit down to 10^{-10} – 10^{-11} mol dm^{-3}. Although use of SW-ASV can minimize oxygen interference by kinetic discrimination of metal and oxygen currents (the direct SWV mode does not eliminate the O_2 component efficiently), in low pH buffer capacity freshwaters oxygen removal is anyway necessary to prevent pH changes induced by reduction of O_2 during the deposition step in ASV. An on-line system based on permeation of oxygen through silicone tubing surrounded by an enzymatic cross-linked O_2-scavenging gel has been recently developed and successfully deployed *in situ* coupled to a submersible voltammetric probe for determination of trace metals in oxygen-saturated freshwater [140].

Recently, a field-deployable instrument for the speciation analysis of As(III) and As(V) in potable water has been described [141]. It is based on ASV of As(III) on a gold electrode in 4.5 mol dm^{-3} HCl. Unfortunately, the As(V) content can only be calculated by difference from the total As concentration determined after oxidation of the sample.

4.2 Water/sediment systems

Processes occurring at the sediment–water interface involve fluxes of material over very small dimensions (10–100 μm) which are often redox sensitive [142]. Nonperturbing *in situ* techniques are thus the method of choice and miniaturized potentiometric and amperometric electrodes with tip sizes down to 10 μm have been developed. Detailed descriptions of such electrodes is available [143]. Potentiometric glass and liquid membrane ISEs have been developed in particular for recording pH and pCO_2 concentration gradients [144, 145], Ca^{2+} and NH_4^+ [146], while amperometric chemical microsensors have been developed for O_2 [147] and H_2S [148], and amperometric biosensors have been developed for BOD (biological oxygen demand), NO_3^- and CH_4. Luther and coworkers have used a mercury-coated gold microelectrode to measure submillimeter scale profiles of O_2, Fe(II) and S(−II) at the sediment–water interface [149] by using the same polarographic waves as those described above for water column analysis. In this environment, however, fouling, especially by sulfides, can be problematic for unprotected electrodes and in particular calibration in synthetic solutions can give rise to large errors when applied to measurements inside the sediment. This has been highlighted by means of a voltammetric system employing an individually addressable microsensor array (150 μm spacing between each microdisk electrode, each with a radius of 5 μm) [41]. This device was used to measure in one run Pb and Cd (both 5 $\mu mol\,dm^{-3}$) concentration gradients across an artificial sediment (silica beads)/water interface, with 200 μm resolution, and to follow their evolution with time during the diffusion of the metals from the water to the solid phase. Results could be interpreted in terms of molecular diffusion, complexation of the metals by the silica particles, and porosity and tortuosity of the 'sediment' phase [150]. The results obtained with this 'synthetic' well-controlled sediment strongly suggest that interpretation of direct measurements in real sediments must be made with caution.

4.3 Biological matrices

The effects and toxicity of an element and its metabolic behavior depend on its physicochemical form within an organism. But speciation analysis in the complex organic matrix of biological samples, even in liquids, is a very challenging task, and to date the vast majority of publications in this area have used chromatographic methods, usually coupled with atomic or mass spectrometric detection [151–153]. For example, a review on metal speciation in biological fluids cited 151 references, only one of which referred to use of voltammetry [154].

In addition to the *ex situ* analysis of biological body fluids direct *in vivo* measurements of chemical species are becoming more feasible due to the development of microanalytical tools and methods. Electroanalytical methods are in principle well suited for miniaturization [155] and have been among the first analytical *in vivo* applications. But one has to consider that a device developed for *in vivo* use must be biocompatible (nonperturbing of the local environment) and resistant to interference from biofouling such as protein adsorption.

ISE. Miniaturized ISEs have been developed for *in vivo* use and applied to e.g. monitoring of K^+ levels during cardiac events [156]. To date applications have been limited to major cations (e.g. K^+, Na^+, Ca^{2+}) with the intention to measure directly the 'free' ions [157].

ASV. There are some empirical reports of using ASV (typically at polymer-coated electrodes) for trace metal determination in body fluids such as urine, blood, and sweat [158], and in foods such as wine [159] and milk [160]. However, rigorous interpretation in terms of speciation has not been attempted.

SCP. As for ASV, application of SCP has been largely empirical with no real attempts to determine speciation parameters. The purported advantage of SCP over SV methods is that less sample pretreatment is required. For instance, the determination of labile and total copper and lead concentrations in wines has been reported [161].

5 NEW CONCEPTS AND PROSPECTS

5.1 Selectivity

One of the key requirements of speciation analysis consists in the selective detection of the various analytes in complex matrices. Electroanalytical methods are often very powerful for single-species analysis, but it is difficult to determine a range of species of the same element simultaneously. Moreover, many of the real-world samples of environmental, food and health control contain surface-active and/or electroactive matrix interferences. Therefore, two strategies for improving the selectivity in electroanalysis are followed: development of electrochemical sensors for selective single-species analysis, and the *on-line* coupling of chromatographic/electrophoretic separation with electrochemical detection. The first route is described in detail in Chapter 5.11 of this Handbook. The developments of LCEC and CE-EC are currently undergoing rapid changes because of the progress in micro- and nanotechnologies. It allows not only the miniaturization of the working electrode or the complete detection unit (microcells etc.), but even that of the whole analytical system after sampling (μTAS, micro total analysis system) [162]. The combination of microstructuring and microfluidics opens a new horizon also for the application of interfacial detection techniques based on electrochemical principles. For instance, capillary electrophoresis is well suited for the efficient separation of differently charged redox species in liquid samples and the analytes can subsequently be quantified by amperometry/voltammetry at a microelectrode or an array if they are electroactive. It is envisaged that new lab-on-a-chip developments will not only revolutionize the analysis of biomolecules such as DNA, but also the speciation analysis of dissolved metal(loid) species in the future by very efficient combinations of separation and detection methods.

5.2 Metal speciation dynamics and bioavailability

Much of the work on the relationship between metal speciation and bioavailability persistently adopts a thermodynamic approach, notably the widely used free ion activity model (FIAM) [163]. There is a plethora of disparate empirical reports on the relationship between electrochemically determined metal and that which is bioavailable. This situation has arisen from the lack of an appropriate theoretical framework for rational data interpretation. A few publications [164–166], however, have shown that the FIAM is limited to cases in which mass transfer is not flux determining. In general, dynamic aspects must be taken into account by quantifying the role of association/dissociation rate parameters for complexes in the medium in the supply of free metal towards the consuming biological interface [167]. This theoretical analysis has identified the conditions under which the metal species detected by voltammetric methods can be appropriately compared with that which may be available for biouptake; in particular the significance of measurements performed at microelectrodes vis-à-vis biouptake by microorganisms with sizes of the order of the operational diffusion layer thickness (see Section 3) [79, 89].

5.3 Spatial resolution

Most real-world samples are spatially heterogeneous, and indeed knowledge of this aspect may be key to understanding their behavior/functionality. Realistic stratification or heterogeneity data can only be obtained when the dimensions of the sensing element are much smaller than the heterogeneity dimensions of the medium. In addition, the total physical dimension of the sensing system should be such that it is nonperturbing of the medium; this is of particular concern for solid samples. For valid data interpretation, the relationship between the dimensions related to the system being probed and those of the sensor-related processes must be established.

Electrochemical measurements at microelectrodes have pushed forward into spatial resolution at the micrometer and even nanometer scale during recent years. Two directions of development are being followed: the placement (and often also movement) of single microelectrodes with micropositioning devices (leading to techniques such as scanning electrochemical microscopy [168, 169]), and measurements at microelectrode arrays with individually addressable electrodes. An example for the latter approach has been reported with the simultaneous recording of 64 complete voltammograms at an iridium-based microelectrode array [170]. In addition, the capability of electrochemical detection to be applicable in ultrasmall sample volumes (down to 1 pL) [171] offers the opportunity to obtain data about the spatial distribution of chemical species in a matrix after locally resolved microsampling.

5.4 Linking theoretical and experimental developments

Progress in understanding of element speciation requires concomitant theoretical and experimental developments. A sound theoretical basis can point the way for rational design of improved analytical systems. It is of fundamental importance that the analytical signals should be directly linked to the underlying properties of the system under study in a rigorous scientifically sound manner. A major disadvantage of many of the sample pretreatment and electrode modification protocols which are claimed to give an 'improved electrochemical response' is that the possibility of any such rigorous link is lost, and results are thus rendered meaningless or at best only empirical.

Therefore, the present opportunities to study interfaces, in particular solid–liquid interfaces, at a molecular/atomic level should be used for a better understanding of the structure–property relations of electrodes as the key component of electrochemical detection systems. On this basis tailor-made sensing surfaces can be designed for electrochemical speciation analysis. In addition, a point of view that is more oriented towards speciation dynamics will be necessary and measurements at very different time scales can conveniently be performed using electrochemical techniques (see Section 2.3).

5.5 Instrumentation

There is a need for further development of *in situ* sensors to allow measurements to be made under the most relevant conditions and to obviate the need for tedious, and probably perturbing, sampling procedures [172]. Electrochemical microtechnologies can contribute to the creation of such in-field monitors for speciation analysis in the environment, in particular for water analysis. But the development of bedside monitors for species determinations in health care units can also be envisaged. Another opportunity will be the integration of amperometric/potentiometric flow-stream cells into detection units, consisting also of optical measuring devices, for modular multidetector speciation analyzers.

Overall it has to be realized that the experimental and theoretical potential of electrochemistry for speciation analysis has not been extensively exploited in many respects. In principle, electrochemical detection could contribute more to the analysis of redox states and their dynamics, as well as to probing electrochemical potentials at interfaces relevant for species immobilization or transformation. These approaches, however, require the combination of different scientific and technical disciplines, including the corresponding multidisciplinary teaching and further developments of reliable electrodes and measuring devices.

6 REFERENCES

1. Templeton, D. M., Ariese, F., Cornelis, R., Danielsson, L. G., Muntau, H., van Leeuwen, H. P. and Lobinski, R., *Pure Appl. Chem.*, **72**, 1453 (2000).
2. Bond, A. M., *Modern Polarographic Methods in Analytical Chemistry*, Marcel Dekker, New York, 1980.
3. Bard, A. J. and Faulkner, L. R., *Electrochemical Methods. Fundamentals and Applications*, 2nd edn, John Wiley & Sons, Inc., New York, 2000.
4. Southampton Electrochemistry Group, *Instrumental Methods in Electrochemistry*, Ellis Horwood, Chichester, 1985.
5. Kissinger, P. T. and Heineman, W. R. (Eds), *Laboratory Techniques in Electroanalytical Chemistry*, 2nd edn, Marcel Dekker, New York, 1996.
6. Heyrovský, J. and Kuta, J., *Principles of Polarography*, Academic Press, New York, 1966.
7. Galus, Z., *Fundamentals of Electrochemical Analysis*, Ellis Horwood, Chichester, 1976.
8. IUPAC Commission on Electroanalytical Chemistry, *Pure Appl. Chem.*, **45**, 81 (1976).

9. Fogg, A. G. and Wang, J., *Pure Appl. Chem.*, **71**, 891 (1999).
10. van Leeuwen, H. P., Cleven, R. F. M. J. and Valenta, P., *Pure Appl. Chem.*, **63**, 1251 (1991).
11. van Leeuwen, H. P., *Colloids Surf.*, **51**, 359 (1990).
12. Town, R. M. and van Leeuwen, H. P., *J. Electroanal. Chem.*, **509**, 58 (2001).
13. Bond, A. M. and Švestka, M., *Collect. Czech. Chem. Comm.*, **58**, 2769 (1993).
14. Williams, G. and D'Silva, C., *Analyst*, **119**, 2337 (1994).
15. Bond, A. M., Luscombe, D. L., Tan, S. N. and Walter, F. L., *Electroanalysis*, **2**, 195 (1990).
16. Hintsche, R., Albers, J., Berndt, H. and Eder, A., *Electroanalysis*, **12**, 660 (2000).
17. Tercier-Waeber, M.-L., Pei, J., Buffle, J., Fiaccabrino, G. C., Koudelka-Hep, M., Riccardi, G., Confalonieri, F., Sina, A. and Graziottin, F., *Electroanalysis*, **12**, 27 (2000).
18. Buffle, J., *Complexation Reactions in Aquatic Systems. An Analytical Approach*, Ellis Horwood, Chichester, 1988.
19. Buffle, J. and Tercier-Waeber, M.-L., *In situ* voltammetry: concepts and practice for trace analysis and speciation, in *In Situ Monitoring of Aquatic Systems. Chemical Analysis and Speciation*, Buffle, J. and Horvai, G. (Eds), John Wiley & Sons, Ltd, Chichester, 2000, Chapter 9, pp. 279–405.
20. Achterberg, E. P. and Braungardt, C., *Anal. Chim. Acta*, **400**, 381 (1999).
21. Tercier-Waeber, M.-L., Buffle, J., Confalonieri, F., Riccardi, G., Sina, A., Graziottin, F., Fiaccabrino, G. C. and Koudelka-Hep, M., *Meas. Sci. Technol.*, **10**, 1202 (1999).
22. Fiaccabrino, G. C., de Rooij, N. F., Koudelka-Hep, M., Hendrikse, J. and van den Berg, A., Microtechnology for the development of *in situ* microanalytical systems, in *In Situ Monitoring of Aquatic Systems. Chemical Analysis and Speciation*, Buffle, J. and Horvai, G. (Eds), John Wiley & Sons, Ltd, Chichester, 2000, Chapter 12, pp. 571–610.
23. Díaz-Cruz, H. M., Esteban, M., van den Hoop, M. A. G. T. and van Leeuwen, H. P., *Anal. Chem.*, **64**, 1769 (1992).
24. Pranitis, D. M., Teltingdiaz, M. and Meyerhoff, M. E., *Crit. Rev. Anal. Chem.*, **23**, 163 (1992).
25. Thomas, J. D. R., *Pure Appl. Chem.*, **73**, 31 (2001).
26. Kalcher, K., Kauffmann, J.-M., Wang, J., Švancara, I., Vytøas, K., Neuhold, C. and Yang, Z., *Electroanalysis*, **7**, 5 (1995).
27. Bond, A. M. and Scholz, F., *Z. Chem.*, **30**, 117 (1990).
28. Town, R. M., Tercier, M.-L., Parthasarathy, N., Bujard, F., Rodak, S., Bernard, C. and Buffle, J., *Anal. Chim. Acta*, **302**, 1 (1995).
29. Novotný, L., *Fresenius' J. Anal. Chem.*, **362**, 184 (1998).
30. Jayaratna, H. G., Bruntlett, C. S. and Kissinger, P. T., *Anal. Chim. Acta*, **332**, 165 (1996).
31. Wu, H. P., *Anal. Chem.*, **66**, 3151 (1994).
32. Gross, M. and Jordan, J., *Pure Appl. Chem.*, **56**, 1096 (1984).
33. Cara, R. G.-M., Sánchez-Misiego, A. and Zirino, A., *Anal. Chem.*, **67**, 4484 (1995).
34. Kounaves, S. P. and Buffle, J., *J. Electroanal. Chem.*, **216**, 53 (1987).
35. Kounaves, S. P. and Buffle, J., *J. Electroanal. Chem.*, **239**, 113 (1988).
36. Štulík, K., Amatore, C., Holub, K., Mareček, V. and Kutner, W., *Pure Appl. Chem.*, **72**, 1483 (2000).
37. Ciszkowska, M. and Stojek, Z., *J. Electroanal. Chem.*, **466**, 129 (1999).
38. Matysik, F.-M. and Emons, H., *Electroanalysis*, **4**, 501 (1992).
39. Daniele, S., Baldo, M.-A., Ugo, P. and Mazzocchin, G.-A., *Anal. Chim. Acta*, **219**, 19 (1989).
40. Feinberg, J. S. and Bowyer, W. J., *Microchem. J.*, **47**, 72 (1993).
41. Pei, J., Tercier-Waeber, M.-L. and Buffle, J., *Anal. Chem.*, **72**, 161 (2000).
42. Tercier, M.-L., Parthasarathy, N. and Buffle, J., *Electroanalysis*, **7**, 55 (1995).
43. Feeney, R. and Kounaves, S. P., *Electroanalysis*, **12**, 677 (2000).
44. Le Drogoff, B., El Khakani, M. A., Silva, P. R. M., Chaker, M. and Ross, G. G., *Appl. Surf. Sci.*, **152**, 77 (1999).
45. Kutner, W., Wang, J., L'Her, M. and Buck, R. P., *Pure Appl. Chem.*, **70**, 1301 (1998).
46. Kalcher, K., *Electroanalysis*, **2**, 419 (1990).
47. Arrigan, D. W. M., *Analyst*, **119**, 1953 (1994).
48. Dam, M. E. R., Thomsen, K. N., Pickup, P. G. and Schrøder, K. H., *Electroanalysis*, **7**, 70 (1995).
49. Christensen, M. K. and Hoyer, B., *Electroanalysis*, **12**, 35 (2000).
50. Capelo, S., Mota, A. M. and Gonçalves, M. L. S., *Electroanalysis*, **7**, 563 (1995).
51. Bagel, O., Degrand, C. and Limoges, B., *Anal. Chem.*, **71**, 3192 (1999).
52. Tercier, M.-L. and Buffle, J., *Anal. Chem.*, **68**, 3670 (1996).
53. Sokalski, T., Ceresa, A., Zwickl, T. and Pretsch, E., *J. Am. Chem. Soc.*, **119**, 11347 (1997).
54. Buck, R. P. and Coşofret, V. V., *Pure Appl. Chem.*, **65**, 1849 (1993).
55. Umezawa, Y., Umezawa, K. and Sato, H., *Pure Appl. Chem.*, **67**, 507 (1995).
56. Ren, K., *Fresenius' J. Anal. Chem.*, **365**, 389 (1999).
57. Whitfield, M., *Ion Selective Electrodes for the Analysis of Natural Waters*, AMSA Handbook No. 2, Australian Marine Association, Sydney, 1971.
58. Buck, R. P. and Lindner, E., *Pure Appl. Chem.*, **66**, 2527 (1994).

59. van Leeuwen, H. P., Buffle, J. and Lovriæ, M., *Pure Appl. Chem.*, **64**, 1015 (1992).
60. Galceran, J., Salvador, J., Puy, J., Mas, F., Giménez, D., Esteban, M., *J. Electroanal. Chem.*, **442**, 151 (1998).
61. Galvez, J. and Park, S.-M., *J. Electroanal. Chem.*, **263**, 269 (1989).
62. Osteryoung, J. and O'Dea, J. J., Square-wave voltammetry, in *Electroanalytical Chemistry*, Bard, A. J. (Ed.), Vol. 14, Marcel Dekker, New York, 1986, pp. 209–308.
63. Jehring, H., *Elektrosorptionsanalyse mit der Wechselstrompolarographie*, Akademie-Verlag, Berlin, 1974.
64. Emons, H., Werner, G., Haferburg, D. and Kleber, H.-P., *Electroanalysis*, **1**, 555 (1989).
65. Emons, H., Schmidt, Th. and Stulik, K., *Analyst*, **114**, 1593 (1989).
66. Ostapczuk, P., Stoeppler, M. and Dürbeck, H. W., *Fresenius' Z. Anal. Chem.*, **332**, 662 (1988).
67. Jagner, D., Sahlin, E. and Renman, L., *Talanta*, **41**, 515 (1994).
68. Kalvoda, R., *Fresenius' J. Anal. Chem.*, **349**, 565 (1994).
69. Paneli, M. G. and Voulgaropoulos, A., *Electroanalysis*, **5**, 355 (1993).
70. van den Berg, C. M. G., *Anal. Chim. Acta*, **250**, 265 (1991).
71. Emons, H., Schmidt, Th. and Werner, G., *Anal. Chim. Acta*, **228**, 55 (1990).
72. Leon, C., Emons, H., Ostapczuk, P. and Hoppstock, K., *Anal. Chim. Acta*, **356**, 99 (1997).
73. Wang, J., *Stripping Analysis*, VCH, Deerfield Beach, 1985.
74. Brainina, Kh. and Neyman, E., *Electroanalytical Stripping Methods*, John Wiley & Sons, Inc., New York, 1993.
75. You, T., Yang, X. and Wang, E., *Electroanalysis*, **11**, 459 (1999).
76. Acworth, I. N., Naoi, M., Parvez, H. and Parvez, S. (Eds), *Coulometric Electrode Array Detectors for HPLC*, VSP, Utrecht, 1997.
77. Buffle, J., Tessier, A. and Haerdi, W., Interpretation of trace metal complexation by aquatic organic matter, in *Complexation of Trace Metals in Natural Waters*, Kramer, C. J. M. and Duinker, J. C. (Eds), Martinus Nijhoff/Dr W. Junk, The Hague, 1984, pp. 301–316.
78. Filella, M., Buffle, J. and van Leeuwen, H. P., *Anal. Chim. Acta*, **232**, 209 (1990).
79. van Leeuwen, H. P., *J. Electroanal. Chem.*, **99**, 93 (1979).
80. van Leeuwen, H. P., Cleven, R. and Buffle, J., *Pure Appl. Chem.*, **61**, 255 (1989).
81. van Leeuwen, H. P. and Buffle, J., *J. Electroanal. Chem.*, **296**, 359 (1990).
82. de Jong, H. G., van Leeuwen, H. P. and Holub, K., *J. Electroanal. Chem.*, **234**, 1 (1987).
83. de Jong, H. G. and van Leeuwen, H. P., *J. Electroanal. Chem.*, **234**, 17 (1987).
84. de Jong, H. G. and van Leeuwen, H. P., *J. Electroanal. Chem.*, **235**, 1 (1987).
85. van Leeuwen, H. P., *Anal. Proc.*, **28**, 66 (1991).
86. Pinheiro, J. P., Mota, A. M. and van Leeuwen, H. P., *Colloids Surf. A*, **151**, 181 (1999).
87. Galceran, J., Puy, J., Salvador, J., Cecilia, J. and van Leeuwen, H. P., *J. Electroanal. Chem.*, **505**, 85 (2001).
88. Pinheiro, J. P. and van Leeuwen, H. P., *Environ. Sci. Technol.*, **35**, 894 (2001).
89. Soares, H. M. V. M., Conde, P. C. F. L., Almeida, A. A. N. and Vasconcelos, M. T. S. D., *Anal. Chim. Acta*, **394**, 325 (1999).
90. Soares, H. M. V. M. and Conde, P. C. F. L., *Anal. Chim. Acta*, **421**, 103 (2000).
91. Damaskin, B. B., Petrii, O. A. and Batrakov, V. V., *Adsorption of Organic Compounds on Electrodes*, Plenum, New York, 1971.
92. Pinheiro, J. P., Mota, A. M., Gonçalves, M. S. and van Leeuwen, H. P., *Environ. Sci. Technol.*, **28**, 2112 (1994).
93. Kubiak, W. and Wang, J., *J. Electroanal. Chem.*, **258**, 41 (1989).
94. Buffle, J., *J. Electroanal. Chem.*, **125**, 273 (1981).
95. Buffle, J., Mota, A. M. and Gonçalves, M. L. S., *Port. Electrochim. Acta*, **3**, 293 (1985).
96. Florence, T. M. and Mann, K. J., *Anal. Chim. Acta*, **200**, 305 (1987).
97. Town, R. M. and Filella, M., *J. Electroanal. Chem.*, **488**, 1 (2000).
98. Florence, T. M., Electrochemical techniques for trace element speciation in waters, in *Trace Element Speciation: Analytical Methods and Problems*, Batley, G. E. (Ed.), CRC Press, Boca Raton, FL, 1989, Chapter 4, pp. 77–116.
99. Taillefert, M., Luther, G. W. and Nuzzio, D. B., *Electroanalysis*, **12**, 401 (2000).
100. Schwarz, J., Henze, G. and Thomas, F. G., *Fresenius' J. Anal. Chem.*, **352**, 479 (1995).
101. Bond, A. M., Turoczy, N. J. and Carter, R. J., *Anal. Chim. Acta*, **310**, 109 (1995).
102. Zirino, A., van der Weele, D. A., Belli, S. L., De Marco, R. and Mackey, D. J., *Mar. Chem.*, **61**, 173 (1998).
103. Verweij, W. and Ruzic, I., *Croat. Chem. Acta*, **70**, 419 (1997).
104. Davison, W., Buffle, J. and De Vitre, R., *Pure Appl. Chem.*, **60**, 1535 (1988).
105. Davison, W., *Anal. Proc.*, **28**, 59 (1991).
106. DeVitre, R. R., Buffle, J., Perret, D. and Baudat, R., *Geochim. Cosmochim. Acta*, **52**, 1601 (1988).
107. Davison, W., Buffle, J. and DeVitre, R. R., *Anal. Chim. Acta*, **377**, 193 (1988).
108. Town, R. M. and Filella, M., *Aquat. Sci.*, **62**, 252 (2000).
109. Achterberg, E. P., Colombo, C. and van den Berg, C. M. G., *Cont. Shelf Res.*, **19**, 537 (1999).
110. Achterberg, E. P. and van den Berg, C. M. G., *Mar. Pollut. Bull.*, **32**, 471 (1996).

111. Apte, S. C., Gardner, M. J. and Ravenscroft, J. E., *Mar. Chem.*, **29**, 63 (1990).
112. Gardner, M. J. and Ravenscroft, J. E., *Chem. Spec. Bioavail.*, **3**, 22 (1991).
113. van den Berg, C. M. G., *Anal. Chim. Acta*, **164**, 195 (1984).
114. Braungardt, C., Achterberg, E. P. and Nimmo, M., *Anal. Chim. Acta*, **377**, 205 (1998).
115. van den Berg, C. M. G. and De Luca Rebello, A., *Sci. Tot. Environ.*, **58**, 37 (1986).
116. Bruland, K. W., Rue, E. L., Donat, J. R., Skrabal, S. A. and Moffett, J. W., *Anal. Chim. Acta*, **405**, 99 (2000).
117. Donat, J. R. and Bruland, K. W., *Mar. Chem.*, **28**, 301 (1990).
118. van den Berg, C. M. G. and Donat, J. R., *Anal. Chim. Acta*, **257**, 281 (1992).
119. Campos and C. M. G. van den Berg, M. L. A. M., *Anal. Chim. Acta*, **284**, 481 (1994).
120. Donat, J. R. and van den Berg, C. M. G., *Mar. Chem.*, **38**, 69 (1992).
121. Gledhill, M., van den Berg, C. M. G., Nolting, R. F. and Timmermans, K. R., *Mar. Chem.*, **59**, 283 (1998).
122. Fischer, E. and van den Berg, C. M. G., *Anal. Chim. Acta*, **432**, 11 (2001).
123. Gardner, M. and Ravenscroft, J., *Chemosphere*, **23**, 695 (1991).
124. Gardner, M., Dixon, E. and Comber, S., *Chem. Spec. Bioavail.*, **12**, 1 (2000).
125. Witter, A. E., Mabury, S. A. and Jones, A. D., *Sci. Tot. Environ.*, **212**, 21 (1998).
126. Jin, L. and Gogan, N. J., *Anal. Chim. Acta*, **412**, 77 (2000).
127. Xue, H., Oestreich, A., Kistler, D. and Sigg, L., *Aquat. Sci.*, **58**, 69 (1996).
128. Xue, H. and Sunda, W. G., *Environ. Sci. Technol.*, **31**, 1902 (1997).
129. Jones, M. J. and Hart, B. T., *Chem. Spec. Bioavail.*, **1**, 59 (1989).
130. Zuhri, A. Z. A. and Voelter, W., *Fresenius' J. Anal. Chem.*, **360**, 1 (1998).
131. van den Berg, C. M. G., *Anal. Proc.*, **28**, 58 (1991).
132. Town, R. M. and Filella, M., *Limnol. Oceanogr.*, **45**, 1341 (2000).
133. Buchberger, W., *Fresenius' J. Anal. Chem.*, **354**, 797 (1996).
134. Ochsenkühn, K. M., Ochsenkühn-Petropoulou, M., Tsopelas, F. and Mendrinos, L., *Fresenius J. Anal. Chem.*, **369**, 633 (2001).
135. Schlegel, D., Mattusch, J. and Wennrich, R., *Fresenius J. Anal. Chem.*, **354**, 535 (1996).
136. Matysik, F.-M., *Electroanalysis*, **12**, 1349 (2000).
137. Lai, E. P. C., Zhang, W., Trier, X., Georgi, A., Kowalski, S., Kennedy, S., MdMuslim, T. and Dabek-Zlotorzynska, E., *Anal. Chim. Acta*, **364**, 63 (1998).
138. Kappes, T., Schnierle, P. and Hauser, P. C., *Anal. Chim. Acta*, **393**, 77 (1999).
139. Wang, J., Larson, D., Foster, N., Armalis, S., Lu, J., Rongrong, X., Olsen, K. and Zirino, A., *Anal. Chem.*, **67**, 1481 (1995).
140. Tercier-Waeber, M.-L. and Buffle, J., *Environ. Sci. Technol.*, **34**, 4018 (2000).
141. Huang, H. and Dasgupta, P. K., *Anal. Chim. Acta*, **380**, 27 (1999).
142. Tessier, A., Carignan, R. and Belzile, N., Processes occurring at the sediment–water interface: emphasis on trace elements, in *Chemical and Biological Regulation of Aquatic Systems*, Buffle, J. and De Vitre, R. R. (Eds), Lewis, Boca Raton, FL, 1994, Chapter 4, pp. 139–197.
143. Buffle, J. and Horvai, G. (Eds), In Situ *Monitoring of Aquatic Systems. Chemical Analysis and Speciation*, *IUPAC Series on Analytical and Physical Chemistry of Environmental systems*, John Wiley & Sons, Ltd, Chichester, 2000.
144. Cai, W.-J. and Reimers, C. E., Sensors for *in situ* pH and pCO_2 measurements in seawater and at the sediment–water interface, in In Situ *Monitoring of Aquatic Systems. Chemical Analysis and Speciation*, Buffle, J. and Horvai, G. (Eds), John Wiley & Sons, Ltd, Chichester, 2000, Chapter 3, pp. 75–119.
145. Müller, B., Buis, K., Stierli, R. and Wehrli, B., *Limnol. Oceanogr.*, **43**, 1728 (1998).
146. De Beer, D., Potentiometric microsensors for *in situ* measurements in aquatic environments, in In Situ *Monitoring of Aquatic Systems. Chemical Analysis and Speciation*, Buffle, J. and Horvai, G. (Eds), John Wiley & Sons, Ltd, Chichester, 2000, Chapter 5, pp. 161–194.
147. Glud, R. N., Gundersen, J. K. and Ramsing, N. B., Electrochemical and optical oxygen microsensors for *in situ* measurements, In Situ *Monitoring of Aquatic Systems. Chemical Analysis and Speciation*, Buffle, J. and Horvai, G. (Eds), John Wiley & Sons, Ltd, Chichester, 2000, Chapter 5, pp. 19–73.
148. Kühl, M. and Steuckart, C., Sensors for *in situ* analysis of sulfide in aquatic systems, In Situ *Monitoring of Aquatic Systems. Chemical Analysis and Speciation*, Buffle, J. and Horvai, G. (Eds), John Wiley & Sons, Ltd, Chichester, 2000, Chapter 5, pp. 121–159.
149. Luther, G. W., Reimers, C. E., Nuzzio, D. B. and Lovalvo, D., *Environ. Sci. Technol.*, **33**, 4352 (1999).
150. Pei, J., Tercier-Waeber, M.-L., Buffle, J., Fiaccabrino, G. C. and Koudelka-Hep, M., *Anal. Chem.*, **73**, 2273 (2001).
151. Florence, T. M., Trace element speciation in biological systems, in *Trace Element Speciation: Analytical Methods and Problems*, Batley, G. E. (Ed.), CRC Press, Boca Raton, FL, 1989, Chapter 9, pp. 319–341.
152. Caroli, S. (Ed.), *Element Speciation in Bioinorganic Chemistry*, John Wiley & Sons, Inc., New York, 1996.
153. Caruso, J. A., Sutton, K. L. and Ackley, K. L. (Eds), *Elemental Speciation*, Elsevier, Amsterdam, 2000.
154. Das, A. K., Chakraborty, R., Cervera, M. L. and de la Guardia, M., *Mikrochim. Acta*, **122**, 209 (1996).

155. Emons, H., Electrochemical microanalysis, in *Electrochemical Microsystems Technologies*, Schultze, J. W., Osaka, T., Datta, M. (Eds), Taylor and Francis, London, 2002, pp. 371–383.
156. Buck, R. P., Coşofret, V. V., Lindner, E., Ufer, S., Madaras, M. B., Johnson, T. A., Ash, R. B. and Neuman, M. R., *Electroanalysis*, **7**, 846 (1995).
157. Lindner, E. and Buck, R. P., *Anal. Chem.*, **72**, 336A (2000).
158. Hoyer, B. and Florence, T. M., *Anal. Chem.*, **59**, 2839 (1987).
159. Azenha, M. A. G. O. and Vasconcelos, M. T. S. D., *J. Agr. Food Chem.*, **48**, 5740 (2000).
160. Vidal, J. C., Viñao, R. B. and Castillo, J. R., *Electroanalysis*, **4**, 653 (1992).
161. Green, A. M., Clark, A. C. and Scollary, G. R., *Fresenius' J. Anal. Chem.*, **358**, 711 (1997).
162. Manz, A., Graber, N. and Widmer, H. M., *Sens. Actuators B*, **1**, 244 (1990).
163. Morel, F. M. M. and Hering, J. G., *Principles and Applications of Aquatic Chemistry*, John Wiley & Sons, Inc., New York, 1993.
164. Whitfield, M. and Turner, D., Critical assessment of the relationship between biological thermodynamic and electrochemical availability, in *Chemical Modeling in Aqueous Systems*, Jenne, E. A (Ed.), *ACS Symp. Ser.* **93**, American Chemical Society, Washington, DC, 1979, Chapter 29, pp. 657–680.
165. Tessier, A., Buffle, J. and Campbell, P. G. C, Uptake of trace metals by aquatic organisms, in *Chemical and Biological Regulation of Aquatic Systems*, Buffle, J. and DeVitre, R. R (Eds), Lewis, Ann Arbor, MI, 1994, Chapter 6, pp. 199–232.
166. van Leeeuwen, H. P., *Environ. Sci. Technol.*, **33**, 3743 (1999).
167. van Leeuwen, H. P., *J. Radioanal. Nucl. Chem.*, **246**, 487 (2001).
168. Bard, A. J., Fan, F. R. F., Kwak, J. and Lev, O., *Anal. Chem.*, **61**, 132 (1989).
169. Wittstock, G., Emons, H., Ridgway, T. H., Blubaugh, E. O. and Heineman, W. R., *Anal. Chim. Acta*, **298**, 285 (1994).
170. Pei, J., Tercier-Waeber, M.-L., Buffle, J., Fiaccabrino, G. C. and Koudelka-Hep, M., in *Electroanalysis*, Emons, H. and Ostapczuk, P. (Eds), Forschungszentrum Jülich GmbH, Jülich, 2000, p. B 04.
171. Clark, R. A., Hietpas, P. B. and Ewing, A. G., *Anal. Chem.*, **69**, 259 (1997).
172. Batley, G. E., Collection, preparation, and storage of samples for speciation analysis, in *Trace Element Speciation: Analytical Methods and Problems*, Batley, G. E. (Ed.), CRC Press, Boca Raton, FL, 1989, Chapter 1, pp. 1–24.

5.10 Future Instrumental Development for Speciation

Andrew N. Eaton and Fadi R. Abou-Shakra

Micromass UK Ltd, Manchester, UK

1 INTRODUCTION

It is evident from the IUPAC definition given at the start of this book that speciation is really a combination of two distinct aims. Firstly to identify all of the species in a sample which contain a given element or elements, and secondly to quantify them. The plethora of papers published on the subject of speciation suggests two things to the authors: that there is a tremendous amount of interest in speciation, and that none of the analytical techniques that are currently available provides a complete solution to the challenges involved.

Elemental analysis techniques such as atomic absorption spectrophotometry (AAS), atomic fluorescence spectrometry (AFS), inductively coupled plasma atomic emission spectrometry (ICP-AES) and inductively coupled plasma mass spectrometry (ICP-MS), and others, provide excellent element specific information. A discussion of the relative merits and demerits of each of these is outside the scope of this chapter, and for the purposes of this work, such techniques will be collectively referred to as 'elemental detection techniques.'

The ability of these techniques to provide element specific detection is their strength, *and also their major weakness for speciation work*, since, by definition, they measure the amount of an element present irrespective of the ligands to which it is complexed or the compound – the exact opposite of the information which we are trying to glean.

The usual approach to circumvent this problem has been to use a separation system, usually chromatography, to separate the various compounds prior to analysis. Until recently, elemental detection techniques were only capable of acquiring steady state signals. However, systems are now becoming commercially available with data acquisition systems which allow them to be coupled to the gamut of separation systems available. With speciation analysis being driven by former elemental analysts who already have elemental

Handbook of Elemental Speciation: Techniques and Methodology R. Cornelis, H. Crews, J. Caruso and K. Heumann
 ISBN: 0-471-49214-0

detection techniques, such instruments are likely to feature heavily in the future of speciation, and the requirements for integrating such systems with the appropriate separation system will be discussed later in the chapter.

However, coupling a separation system, such as a chromatograph, to an elemental detector gives us only half of the solution, as there remains the issue of identifying the species present in the chromatogram. Most workers involved in such research take the approach of spiking the sample with an authentic standard of one of the expected element-containing species, and by noting which peak in the chromatogram gets bigger, an identity can be assigned. There are, however, a number of drawbacks to this approach.

There is the possibility of two or more species co-eluting at a given retention time. If both species contain the element of interest, then the elemental detector cannot distinguish between the two species. The only safe options are to re-run the sample using a different separation, the chances of the same two peaks co-eluting under two different sets of conditions being very small, or to use techniques such as isotope dilution which involve significant manual sample preparation and repeated analyses. This increases the method development time and, at least, doubles the analysis time. In addition, it is necessary to spike the sample with each individual standard, or isotopic tag, and run each spiked solution through the separation. For samples containing even a modest number of species, this represents a huge amount of repetition and a greatly increased method development time. While this may be acceptable for a research project, it would definitely not be practical in a high throughput, routine analysis laboratory, even if this spiking and re-running process were made completely automatic, something which no commercially available system can do at present.

This technique relies upon the availability of authentic standards for each species in the sample. If the sample is relatively simple, this may not be a problem, but as the technique is applied to more and more complex samples, it becomes impossible to predict all of the possible species, let alone obtain authentic standards for each. A good example of this is in the work by Marchante-Gayón *et al.* [1] on the determination of selenium species in nutritional commercial supplements. Although many of the peaks in their chromatograms could be identified by spiking with authentic standards, such as selenite, selenate, selenocystine, selenomethionine and selenoethionine, there remain a number of peaks in each chromatogram which do not correspond to any of these compounds and therefore cannot be identified. Worse still, the technique offers no possible clues whatsoever as to how to proceed with the identification of such unknowns.

Elemental detection therefore is not the whole solution and it becomes essential to look to other techniques or combinations of techniques to provide the necessary structural elucidation capability.

2 REQUIREMENTS FOR ELEMENTAL DETECTION SYSTEMS COUPLED TO CHROMATOGRAPHY TECHNIQUES

In general, and provided that the chromatography used is adequate to the task, the detection and quantification of a specific metal ion as it elutes from the column should be a very simple task. The usual parameters, such as detection limit, and stability, apply, just as they do in all such analyses, but the task is not a particularly demanding one. Having identified those peaks in the chromatogram which contain the metal, the process of identifying the species comes into play, and for this the elemental detection techniques are useless and alternative methodologies and techniques are required, especially as samples of increasing complexity are investigated.

2.1 Control of chromatograph

In the majority of work published on the coupling of chromatography systems to elemental detectors, there has been only minimal control of the chromatograph by the elemental detector data system, or vice versa, usually just the provision of sending a trigger signal to either start the chromatographic run, or to start the acquisition. This means that both parts of the system have to be set up independently of each other, with the result that the entire set of experimental conditions are not stored in any single location. In research projects, this is not a serious problem, but when speciation becomes routine and such analysis results are being used as evidence in legal proceedings, it is essential to have a clear audit trail for all of the method and results. For this reason alone, it will become necessary to control the entire system from a single set of software.

Control will need to take the form of defining the gradient to be used, if any, starting and stopping the pump, injecting samples and controlling the autosampler, and recording all of these settings in addition to those of the elemental analyser.

2.2 Data acquisition

The use of chromatography imposes on the elemental detection system a requirement to be able to monitor continuously variable signals, depending upon the nature of the technique used.

For essentially single element techniques such as AAS or 'reaction cell' type ICP-MS instruments, there are no special requirements other than the ability to constantly monitor and store the output from the detector, the time interval between successive measurements being relatively unimportant. Similarly, techniques which provide simultaneous detection of a number of elements such as simultaneous ICP-AES, or ICP-MS with a time-of-flight (TOF) analyser, also have no significant restriction on the data acquisition rate, other than the ability to continuously monitor and store the output from each detector.

The difficulty comes for multielement scanning techniques, such as sequential ICP-AES and ICP-MS. Here the time taken to perform each measurement and to move the analyser to the next measuring position, termed its 'duty cycle', means that there is often a trade-off between the number of elements which can be determined, and the resulting number of data points per chromatogram peak which can be acquired.

Table 5.10.1. Typical peak widths for various common separation techniques.

Separation technique	Peak width	Time per scan for ten points/peak
Capillary electrophoresis	2–5 s	200 ms
HPLC	5–20 s	500 ms
GC	2–3 s	200 ms
Fast GC	<100 ms	10 ms

While a large number of points per peak increases the quality of the peak shape, it is rarely the case that this is the most important consideration. Often it is more desirable to have a number of elements (analytes plus internal standard) and, if possible, multiple lines or isotopes of these elements, and this requirement usually means restricting the number of points per peak. For quantitative purposes, a minimum of ten points per peak is required in order to define the peak adequately, because, unlike spectroscopic peaks, the chromatography peak width may vary quite significantly.

In the main, the time available to take measurements is governed by the choice of chromatography used, which dictates the width of the resulting peaks. Table 5.10.1 shows a range of chromatographic techniques and typical peak widths, and the maximum time in which a scan can be performed, assuming ten points per peak.

In order to combine a chromatography technique with a particular elemental detector, it is essential that the acquisition system of the detector can acquire and store a scan fast enough to maintain the minimum number of points per peak for the type of chromatography used.

2.3 Data processing

The processing of chromatographic data is very different to processing spectra. Firstly, the data system needs to be able to display chromatograms, based on successive element scans, and to display such chromatograms using the total signal obtained for all peaks in the spectra, using single-element or line signals, and combinations of lines. An example of this, based on ICP-MS data for metabolites of a platinum-containing drug in urine, is shown in Figure 5.10.1.

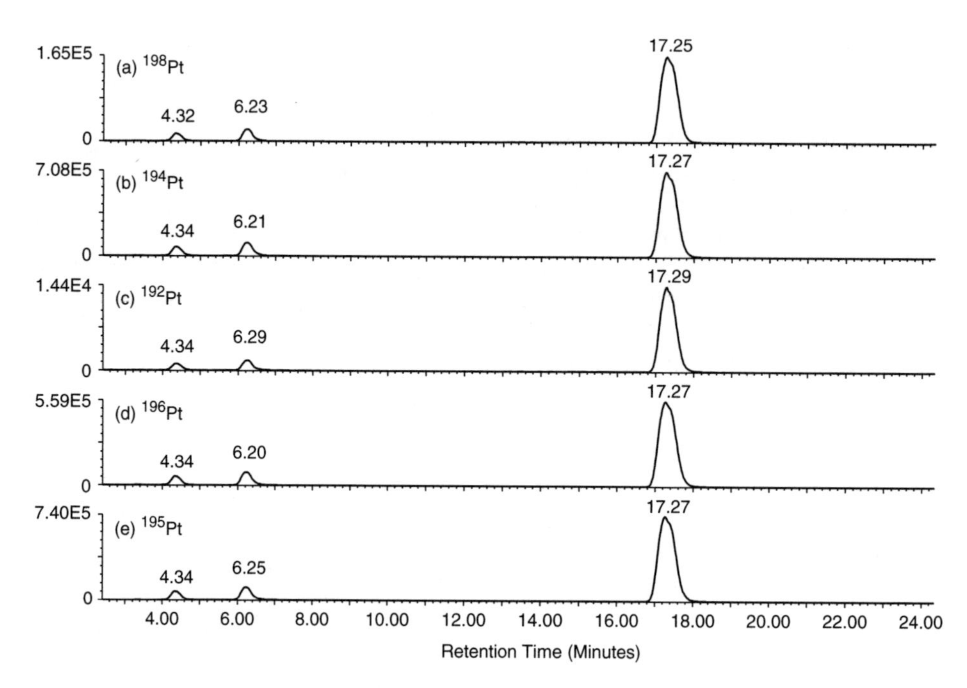

Figure 5.10.1. ICP-MS mass chromatograms showing the separation of three unknown metabolites present in urine after taking a drug which contained platinum: (a) mass chromatogram for m/z 198; (b) mass chromatogram for m/z 194; (c) mass chromatogram for m/z 192; (d) mass chromatogram for m/z 196; (e) mass chromatogram for m/z 195.

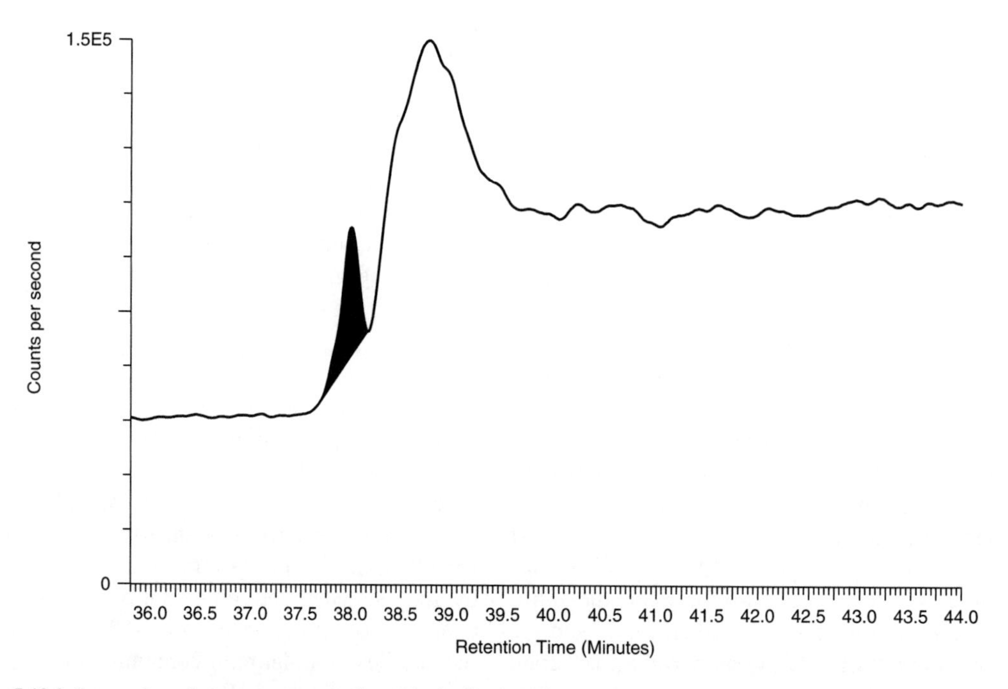

Figure 5.10.2. Integration of chromatogram peaks with sloping baseline.

Another difference is in the way peaks are integrated. Unlike elemental spectra, the baseline in a chromatogram can vary significantly during a run, as can be seen in Figure 5.10.2. Therefore, for routine work, the data system must be capable of automatically handling sloping baselines if accurate quantitation is to be carried out.

There are many specialist chromatography data processing packages available which carry out this type of data processing, usually designed for organic mass spectrometry applications, and these could be used to carry out this task, provided that the elemental detector data files are readable, or can be made readable, by the software. However, this increases the time and effort required to carry out the analysis, thus reducing throughput.

Most elemental data systems offer a quality control option, whereby data are processed during a run and checked against user defined quality control (QC) standards, and immediately re-run, or the instrument re-calibrated, if there is a problem. This facility is essential to maintaining throughput in routine laboratories. To carry out such QC checking, the data processing must, of necessity, be carried out on the instrument data system. Since, as far as the authors are aware, only one of the commercially available elemental detector data systems is capable of carrying out both chromatography data processing and QC checking, this is an area which will have to develop in the future, if elemental detectors are to be used for speciation in a routine manner.

3 STRUCTURE ELUCIDATION

It is clear that very different analytical techniques are required to identify the species, than are used only to measure the amount of the element present: the two requirements being somewhat mutually exclusive.

Although not exclusively so, the majority of the elements of interest are contained within organic compounds or bound by organic ligands. It is logical therefore to look at organic analysis techniques, such as nuclear magnetic resonance spectrometry (NMR), infrared spectrometry (IR), ultraviolet spectrophotometry (UV) and mass spectrometry (MS), which have been developed specifically to obtain structural information or elemental composition for organic compounds.

With the exception of mass spectrometry, all of these techniques provide information on the functional groups present and the types of bond which a compound contains, but they cannot positively identify the actual component under analysis. Only mass spectrometry provides both structural information, through fragmentation studies, and the possibility to identify the molecular weight of the compound, thereby allowing it to be positively identified in a single run. Organic mass spectrometry has developed dramatically in the last 10 years, primarily as a result of the development of the electrospray ionisation (ESI) technique, which is both a soft ionisation technique, allowing molecular weight information to be gleaned from even very labile compounds, and is applicable across a wide range of polarities of compounds, thereby increasing the range of amenable analyte compounds.

The technique of electrospray, as applied to speciation, is discussed elsewhere in this volume, and so will not be covered here. However, any discussion about the future of instrumentation for speciation needs to consider the implications of developments in organic mass spectrometry.

Organic mass spectrometry development had been driven largely by the requirements of the pharmaceutical and biotechnology industries to identify very quickly large numbers of compounds present in complex biological samples such as urine and plasma. In this respect, the complexity of the samples can be compared to those of environmental samples, where degradation of compounds by the biosystem leads to very complex mixtures.

The primary tool in such analysis is tandem mass spectrometry, or MS/MS which uses two mass analysers, the first to isolate a parent ion, chosen on the basis of its molecular weight, which is then fragmented, by means of a collision cell, and the fragments mass analysed by the second mass analyser. The instrumentation used, until recently, has been the 'triple quad', which uses two quadrupole analysers and an RF multipole (quadrupole, hexapole or octapole) encased in a gas tight canister to form the collision cell.

Recent variants on this approach have been the use of a quadrupole ion trap. Here, a single trap is used to retain all of the ions, to selectively eject all but the parent ion, fragment it, and eject the fragment ions to the detector according to mass, thereby producing a mass spectrum. Theoretically, this approach allows multiple collection, fragmentation and mass analysis, termed 'MS^n' (MS to the n), although the complexity involved in setting up such analyses has seen this technology confined to the research laboratories, rather than being used instead of the over 500 triple quad systems which are sold each year throughout the world.

Another more widely applicable variant on the triple quad has been to replace the second quadrupole with an orthogonal TOF analyser. The advantage of the TOF analyser is that it produces full mass scans, thereby counting all of the ions which pass through it, unlike a quadrupole mass filter which allows ions of a single mass through at any time. By admitting all of the ions to the detector, rather than just one mass, for a mass range of 500 Da, the sensitivity could theoretically be as much as 500 times greater, although a factor of 200 × is commonly found in practice.

4 COMBINATIONS

An increasing number of workers have taken the approach of combining more than one technique in order to obtain both structural and quantitative information. Typically, elemental detectors are used in parallel with organic mass spectrometer. The former offer in general the sensitivity, selectivity and specificity required to identify and possibly quantify the presence of a species containing the element of interest. The latter on the other hand, can be used to provide accurate mass information or structural information via parent–daughter fragmentation studies. An example of such an attempt is the work of Nicholson *et al.* where ICP-MS is used in parallel to ESI-TOF-MS [2].

The considerations that must be taken into account to ensure a successful use of a combination of techniques are: solvent compatibility, sensitivity of the techniques, and instrument control.

4.1 Solvent compatibility

A quick review of the current literature on the separation of metal/semimetal-containing species for detection by element-sensitive detectors shows that the great majority of the mobile phases used contain high levels of salts or ion-pairing reagents. This will lead a substantial reduction in the sensitivity for ESI based systems. However, advances in ICP generators and matching circuitry have made it possible to use highly organic mobile phases without affecting the stability and the performance of the ICP source. This in turn should open the road for the use of more ESI friendly chromatographic conditions, which would enable simultaneous ICP/ESI – MS detection on the same eluents. Figure 5.10.3, shows the separation of diphenyltin, dibutyltin, triphenyltin, and tributyltin using HPLC ICP-MS using a 65 % acetonitrile mobile phase.

This is paralleled with advances on the ESI front with the use of nebuliser assisted electrospray and 'Z-type' orthogonal interfaces which mean that those systems are currently more tolerant of salts in the mobile phase.

Another exciting advance is in the use of superheated water as a mobile phase [3]. This method is based upon the observation that water at temperatures in excess of 100 °C and which is kept in a liquid from as a result of a high back pressure can function as a suitable solvent for reverse-phase chromatography, without the need for salts or organic modifiers. This technique is currently in its infancy. However, should it prove to be successful for elemental speciation it would offer the ideal solution for the compatibility issues between the hard and the soft ionisation sources.

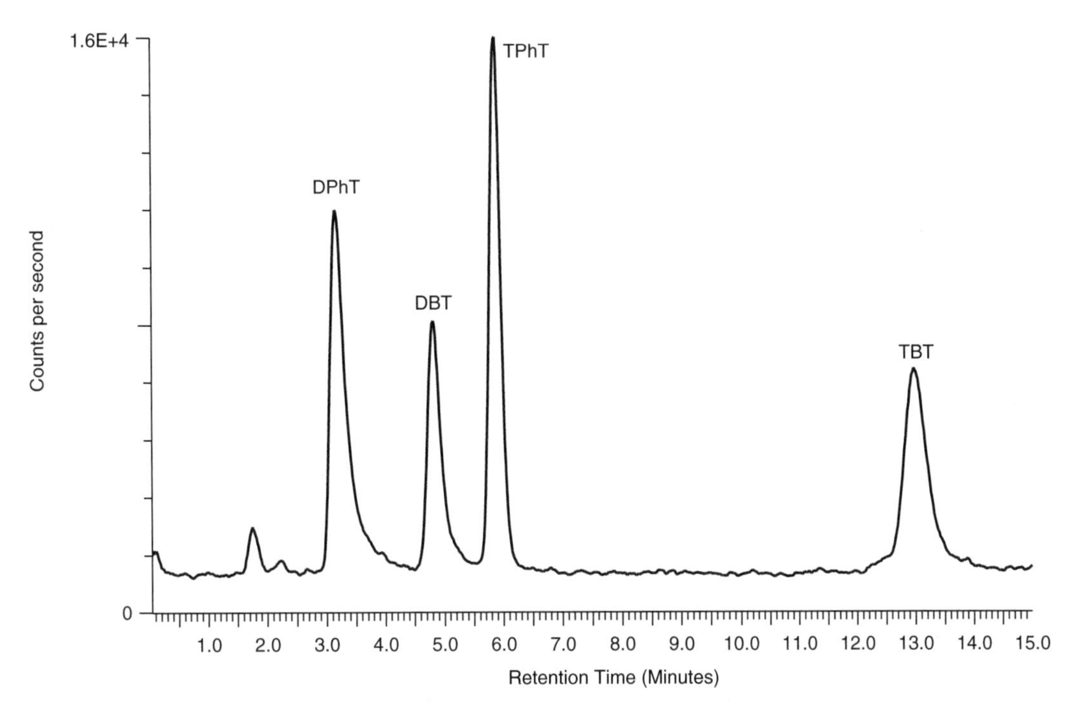

Figure 5.10.3. Separation of diphenyltin, dibutyltin, triphenyltin, and tributyltin using HPLC ICP-MS with a 65 % acetonitrile mobile phase.

4.2 Sensitivity of the techniques

Co-eluting material, which may not significantly affect the elemental detector, can be a major source of a high background signal intensity that limits the detection capabilities of an ESI-MS system. However, the use of MS-MS can provide a more selective way for identifying the species of concern and has been reported to offer a significant improvement in signal to background ratio. Advanced acquisition systems currently enable time scheduled acquisitions whereby the mass spectrometer is set to carry out specific MS-MS (precursor–product) monitoring at times coinciding with the retention time of the species of interest, the ability to trigger these experiments upon the detection of such peaks on the elemental system would be a very powerful tool indeed.

In addition, it is well known that with ESI, molecular clusters are formed during the ionisation and extraction stage. This does in turn reduce the sensitivity of the instrument by reducing the number of ions of a species that are available to the mass spectrometer. Declustering techniques, via excitational or collisional dissociation are offered as a standard in modern ESI-MS instrument. Therefore, method development work is required to ensure that a declustering mechanism is used to eliminate adducts of the species of interest. This declustering mechanism, however, must not be strong enough to cause any fragmentation of the species.

4.3 Instrument control

Parallel operation of elemental and structural detectors can be made more efficient using a single control system. The main benefit of such a system lies in its ability to detect the presence of a peak on the elemental detector and to subsequently trigger specific experiments such as MS-MS on the second detector, thereby improving its detection/identification power. Furthermore, additional information from the elemental detector regarding the presence of other elements such as S, P, Cl, etc. in the detected peak, can be used to speed up library searching in order to identify the subsequently detected peak on the structural type detector.

Finally for routine analytical work the use of a single control system can dramatically improve throughput through ease of use. It does also provide an easier platform for compliance with QC/QA and other regulatory protocols.

4.4 Multiple combinations

Although the combination of an elemental detector and organic mass spectrometry yields most of the necessary information required for speciation, it is likely that the complexity of systems under investigation will increase and more complex organic structures will be studied. This will mean that the fragmentation and accurate mass information may not be sufficient to fully determine the structure of the organic compounds. In such cases, multiple hyphenation is a possibility and Nicholson *et al.* [2] have shown that it is possible to couple ICP-MS, ESI-MS, NMR and diode array detectors to the eluent from a single HPLC instrument. The additional structural information provided, primarily by the NMR, greatly increases the chances of being able to correctly assign a structure and obtain quantitative measurements in a single run.

5 NEW TECHNIQUES

With the ultimate aim being a combination of an elemental-specific quantification technique and a structural identification technique, in a single system, several workers are investigating new techniques largely, although not exclusively, centred on new ion sources for mass spectrometry.

5.1 Alternatives to the ICP

Many groups are working on alternatives to the ICP [4]. These include development of the helium ICP and microwave induced plasma (MIP) sources, and also the use of high power nitrogen microwave induced plasmas [5, 6]. While the main focus of this work has been to overcome the problems associated with the argon plasma, primarily the formation of interfering argides in ICP-MS, in recent years the He-MIP has been applied to the analysis of halogens, usually coupled to gas chromatography which uses a helium mobile phase.

These alternative plasmas, however, are not intended to address the requirement for an instrument which combines elemental and molecular information in a single run, nor is there any indication that they have the potential to do so.

One area of research which does show potential is the development of the low pressure, low power plasma [7–9], and several workers have developed what is termed a 'tuneable plasma', (Marcus *et al.* provide a recent review [10]). Such systems, coupled to mass spectrometry, have been applied to the analysis of organometallics and organohalides introduced via a gas chromatograph, and have been used to produce either elemental or molecular information, yielding spectra which are very similar to those produced by an electron impact (EI) source mass spectrometer.

While showing significant potential, these tuneable plasmas suffered from linearity and sensitivity problems. Some of these have been alleviated by the use of reagent gases to modify the ionisation process [11]. This technique enabled the detection of molecular spectra for chlorobenzene, iodobenzene and dibromobenzene, with detection limits of 100, 140 and 229 pg, respectively.

It must be stated that work with such tunable plasmas has concentrated on obtaining both elemental and molecular information, but under different conditions; therefore each sample has to be run twice. Therefore the only advantage offered by this solution over the use of dedicated elemental and ESI analyser is a lower capital cost. In addition, so far the work has concentrated on the analysis of volatile species introduced by a gas chromatograph. It is questionable whether such tunable plasmas will be robust enough to function with aqueous samples in HPLC solvent systems [12]. This is a very serious limitation to the applicability of the technique.

5.2 Glow discharge

One of the most exciting areas of research has emerged from developments in the glow discharge source. Glow discharge has largely been applied to the analysis of conducting solid samples, although the development of the radio frequency glow discharge (RF-GD) has further extended the range to nonconducting materials [13].

You *et al.* [14] have demonstrated the use of a particle beam interface organic mass spectrometer fitted with an RF-GD source instead of the conventional EI source, and were able to produce both elemental and molecular species in the same spectra, although with poorer detection limits compared to conventional elemental analysis techniques such as ICP-MS.

Majidi *et al.* [15, 16] observed that 'glow discharge plasmas are steady state but highly heterogeneous plasmas with a number of dynamic processes occurring simultaneously.' They used a microsecond pulsed power supply, instead of a DC source, to produce a transient glow discharge plasma. As a result, the nature of the plasma is highly time dependent, and therefore the effect of different amounts of energy imparted to the analyte can be determined by sampling from the plasma at the appropriate time during or after the pulse. They used a TOF mass spectrometer to give them the ability to 'time gate' different parts of the plasma, thereby achieving anything from 'soft' chemical ionisation, giving molecular ion information, through EI type fragmentation to full elemental ionisation. Results have been presented for a number of volatile organic compounds, such as ethylbenzene, toluene and *p*-xylene, and the organometallic tungsten hexacarbonyl.

It could be postulated that this apparatus, coupled with a particle beam interface such as the one used by You *et al.* may result in an instrument that meets the requirement for a technique capable of measuring both elemental and molecular information in a single run, although much work would be needed before the system could be commercialised.

6 CONCLUSIONS

The fact that there is a scientific need to develop new instrumentation for speciation analysis is hopefully now well established. The current state of play involving many different approaches suggests that we are a long way from our single 'Speciator' instrument. It would be foolish to try to predict at this stage which area will dominate in the future, as this will depend not only upon scientific factors.

The extent to which speciation becomes a routine technique will depend on the need for it. Although we would like to think that the scientific requirement was enough, in the real world, instrument companies will only develop a product for which they can see a well defined market. This will not happen until the majority of Western countries have established environmental legislation which requires speciation, rather than total element measurements to be made. This is happening, but legislation proceeds at a snail's pace and it will be a number of years before it becomes the standard rather than the norm.

7 REFERENCES

1. Marchante-Gayón, J. M., Thomas, C., Feldmann, I. and Jakubowski, N., *J. Anal. At. Spectrom.*, **15**, 1093 (2000).
2. Nicholson, J. K., Lindon, J. C., Scarfe, G. B., Wilson, I. D., Abou-Shakra, F., Sage, A. B. and Castro-Perez, *J. Anal. Chem.*, **73**, 1491 (2001).
3. Wilson, I. D., *Chromatographia*, **52**, S28 (2000).
4. Evans, E. H., Giglio, J. J., Castillano, T. M. and Caruso, J. A., *Inductively Coupled and Microwave Induced Plasma Sources for Mass Spectrometry*, Barnett, N. W. (Ed.), Royal Society of Chemistry, Cambridge, 1995.
5. Deutsch, R. D. and Hieftje, G. M., *Appl. Spectrosc.*, **39**, 214 (1985).
6. Ohata, M. and Furuta, N., *J. Anal. At. Spectrom.*, **13**, 447 (1998).
7. Waggoner, J. W., Belkin, M., Sutton, K. L., Caruso, J. A. and Fannin, H. B., *J. Anal. At. Spectrom.*, **13**, 879 (1998).
8. Rodriguez, J., Pereiro, R. and Sanz-Medel, A., *J. Anal. At. Spectrom.*, **13**, 911 (1998).

9. Belkin, M., Waggoner, J. W. and Caruso, J. A., *Anal. Commun.*, **35**, 281 (1998).
10. Marcus, R. K., Evans, E. H. and Caruso, J. A., *J. Anal. At. Spectrom.*, **15**, 1 (2000).
11. O'Connor, G., Ebdon, L. and Evans, E. H., *J. Anal. At. Spectrom.*, **12**, 1263 (1997).
12. Pack, B. W., Broekaert, J. A. C., Guzowski, J. P., Poehlman, J. and Hieftje, G. M., *Anal. Chem.*, **70**, 3957 (1998).
13. Marcus, R. K., Harville, T. R., Mei, Y. and Shick, C. R. Jr., *Anal. Chem.*, **66**, 902A (1994).
14. You, J., Dempster, M. A. and Marcus, R. K., *Anal. Chem.*, **69**, 3419 (1997).
15. Steiner, R. E., Lewis, C. L. and Majidi, V., *J. Anal. At. Spectrom.*, **14**, 1537 (1999).
16. Majidi, V., Moser, M., Lewis, C. L., Hang, W. and King, F. L., *J. Anal. At. Spectrom.*, **15**, 19 (2000).

5.11 Biosensors for Monitoring of Metal Ions

Ibolya Bontidean and Elisabeth Csöregi

Lund University, Sweden

Wolfgang Schuhmann

Ruhr Universität Bochum, Germany

1 INTRODUCTION

Out of the 90 naturally occurring elements, 21 are nonmetals, 16 are light metals and the remaining 53 are heavy metals. Most of the metals are transition elements with densities above $5\,g\,cm^{-3}$ and have incompletely filled *d* orbitals. Some of the cations formed by these metals have the ability to form potentially redox-active complexes with suitable ligands. Therefore metal cations play an important role as trace elements in many sophisticated biochemical reactions. At higher concentrations, however, metal cations form unspecific complexes and have toxic effects in the cell. Of the 53, only 17 metals are available to the living cell and form soluble cations, thus showing biological influence: Fe, Mo, Mn, Zn, Ni, Cu, V, Co, W, Cr, As, Ag, Sb, Cd, Hg, Pb and U. Some of their main properties are presented in Table 5.11.1 [1].

The toxicity of metals is based not only on their oxidation state, but also on the form in which they occur, i.e. whether it is elemental, inorganic or organometallic. For example organomercurials are more toxic than Hg^{2+}, ethylated lead is extremely toxic and the same is valid for arsenic. Some metals (e.g. Ni, Co, Zn, Cu) are essential to microorganisms as trace nutrients, in contrast to others (e.g. Hg, Cd, Pb) which are extremely toxic even at trace levels; however, all metals are toxic in μM to mM concentrations.

Because of the extreme toxicity metals display for various forms of life, and their broad distribution in nature, detection of metals has evolved from being an analytical task into a necessity in various areas such as medicine, food industry, environment, etc. Recognizing the importance of metal monitoring, several methods have been developed over the years. Powerful methods such as atomic absorption and emission spectroscopy [2] or inductively coupled plasma mass spectroscopy [2, 3] have been described and are commercially available. These methods show good selectivity, sensitivity, reliability and accuracy; however, they often require very expensive instrumentation operated by highly skilled personnel.

Electrochemical detection methods including ion-selective electrodes, polarography and other

Handbook of Elemental Speciation: Techniques and Methodology R. Cornelis, H. Crews, J. Caruso and K. Heumann

ISBN: 0-471-49214-0

Table 5.11.1. Metals with biological influence and their oxidation forms.

Metal	Toxicity	Available inorganic ionic forms
V	Mostly toxic	V(V) in vanadate
Cr	Cr(III) has physiological role in man, Cr(VI) is toxic	Cr(VI) in chromate and Cr(III)
Mn	Very low toxicity, except in human where it acts on the central nervous system	Any oxidation state between Mn(II) and Mn(VII), Mn(II) predominant
Fe	Biologically most important metal, not toxic	Fe(II) and Fe(III)
Co	Co(II) has medium toxicity, but Co dust causes lung disease	Almost only Co(II), rarely Co(III)
Ni	Medium toxicity, may cause nickel allergy to humans	Ni(II) and very unstable Ni(III)
Cu	Toxicity is based on the ability of Cu to easily interact with radicals	Cu(II) and Cu(I)
Zn	Very important physiological function, 'no life without Zn', very low toxicity	Only as Zn(II)
As	Well-known toxin even at trace concentrations	As(V) in arsenate and As(III) in arsenite
Mo	Molybdate is the biologically most important oxyanion	Mo(VI) in molybdate
Ag	Very toxic due to its strong complex with sulfur	Ag(I) in Ag_2S
Cd	Toxic due to thiol binding and protein denaturation	Cd(II) in CdS
Hg	Metal with the strongest toxicity, no beneficial function	Hg(II) in HgS
Pb	Toxic for humans, acts on the nervous system, on blood pressure and on reproduction	Pb(II)
U	Radioactive toxin, no beneficial function is known	U(VI) in UO_2^{2-}

voltammetric methods [4] are less expensive but their main disadvantage is their inability to detect metals at extremely low concentration.

As environmental concentrations of metals are reduced, increasingly sensitive analytical methods are required to monitor their distribution. Moreover, none of the above mentioned methods is able to selectively detect the amount of metal which is bioavailable and therefore likely to present a risk to living organisms.

In this respect, biosensors are useful analytical tools since they are able to monitor that part of the total metal concentration that is available for the biological component used as recognition element. A large variety of biological recognition elements and transducers have been used in biosensor construction for metal detection. Different biosensor architectures are described briefly below and their main characteristics (dynamic range, DR; limit of detection, LOD) are presented in Table 5.11.2.

2 WHOLE CELL BIOSENSORS

Whole cell biosensors can be constructed using certain microorganisms, e.g. bacteria, yeasts, fungi, lichens, mosses, and water plants due to their ability to accumulate metals [5]. The advantages of using intact cells as sensing element in a biosensor are: (i) microorganisms are usually more tolerant to assay conditions than isolated biomolecules, due to mechanisms that enables them to regulate their internal composition, (ii) microorganisms provide information about the bioavailability of the analyte because the analyte must be taken up before the response is produced, (iii) microorganisms are relatively cheap since large quantities of these living organisms can be prepared comparably inexpensively.

Among the disadvantages of using microorganisms are: (i) increased response times of the biosensors (ii) difficulties of the regeneration of the sensor (iii) variation of the response when using cells from different batches (iv) influence of the biosensor response by culture age, temperature, cell density and aeration during induction. Some of the microorganisms have learned to survive and grow in environments containing metals, developing resistances to metals such as Zn, Cu, Ni, Co, Hg, Cr, and Pb.

Table 5.11.2. Metal biosensors and their properties.

Biological molecule			Transducer type	Operating conditions	M^{2+}	LOD	DR	Ref.
Whole cell	Mosses	*Sphagnum Sp.*	Stripping Differential pulse Voltammetry	Acetate pH 6.0, IS 0.7, 10% moss, carbon paste electrodes	Pb^{2+}	$2\,\mu g\,L^{-1}$	$5-125\,\mu g\,L^{-1}$	[45]
	Bacteria	*Nitrosomonas*	Amperometric	Oxygen uptake rate is measured	Cd^{2+} Cu^{2+}	$8.3\,mg\,L^{-1}$ $173\,mg\,L^{-1}$		[46]
		E. coli + *mer* promoter + *lux* genes from *Vibrio fischeri*	Optical detection	Bioluminescence is measured at 28 °C, 30 min response time under aeration	Hg^{2+}	0.1 μM	20 nM–4 μM	[8]
		E. coli + *lacZ* gene	Optical detection	Electrochemical assay of β-galactosidase activity	Cd^{2+}	nM		[47]
		Synechococcus smt + *lux* from *Vibrio fischeri*	Optical detection	Luminescence is measured	Zn^{2+}		0.5–4 μM	[48]
		E. coli + *lux* genes from *Vibrio fischeri*	Optical detection		Hg^{2+} Cu^{2+}	0.1 μM 0.1 μM		[49]
		R. silverii + *lux* operon from *Vibrio fischeri*	Optical detection	23 °C, 0.2% acetate, 20 mM MOPS, pH 7.0, $20\,\mu g\,mL^{-1}$ tetracycline	Cu^{2+} Zn^{2+} Cd^{2+} Cr^{6+} Pb^{2+} Tl^{+} Ni^{2+}	2 μM 5 μM 5 μM 1 μM 1 μM	2–40 μM 5–250 μM 5–200 μM 1 μM–40 μM 1 μM–40 μM	[10, 50]
		R. silverii + *lux* operon from *Vibrio fischeri*	Optical detection	Microorganisms immobilized in polymer matrices, 25 °C	Cu^{2+}	1 μM		[6]
		E. coli + *mer-lux* plasmid	Optical detection	Luciferase activity is detected	Hg^{2}	1–10 000 nM		[51]
		E. coli + *lux* operon	Optical detection	30 °C, M9 medium	Hg^{2+} Cu^{2+}	10 nM 1 μM		[52]
		E. coli + firefly luciferase gene	Optical detection	Luminescence is measured in microtiter plates after 60 min, at 30 °C	Hg^{2+}	0.1 fM	0.1 fM–0.1 μM	[7]
		Staphylococcus aureus + firefly luciferase gene	Optical detection	Luminescence is measured in scintillation counter after 60 min;	AsO_4^{3-} Cd^{2+}	1 μM 1 μM	1–5 μM 1–20 μM	[12]
		Staphylococcus aureus + firefly luciferase gene	Optical detection	Luminescence is measured in microtiter plates, 30 °C	Cd^{2+} Pb^{2+} Hg^{2+}		10 nM–1 μM 33 nM–330 μM 33–100 nM	[14]
		Bacillus subtilis + firefly luciferase	Optical detection	Luminescence is measured in microtiter plates, 30 °C	Cd^{2+} Zn^{2+}		3.3 nM–1 μm 1–33 μM	[14]
	Yeast		Fluorescence detection	Light emitted at 509 nm is measured when the system is excited at 395 nm	Cu^{2+}		0.1 μM	[53]

(continued overleaf)

Table 5.11.2. (*continued*)

Biological molecule		Transducer type	Operating conditions	M^{2+}	LOD	DR	Ref.
Tissue	Cucumber leaves	Amperometric detection	Effect of metal on the hydrolysis of cysteine is measured in the presence of L-cysteine desulfhydrolase	Pb^{2+}		30 nM	[54]
				Cd^{2+}		10 nM	
Protein	Apophytochelatin	UV spectrophotometric detection at 215 nm	270 mM apophytochelatin is used	Cd^{2+}		1–6 mg L^{-1}	[43]
	MerR-LacZα:M15 complex	Spectrophotometric detection	Microtiter plates coated with BSA–divinyl-sulphone–glutathion were treated with Hg^{2+} concentrations and after washing the protein was bound to it	Hg^{2+}	µg L^{-1} level		[44]
	glutathione	UV spectrophotometric detection at 215 nm	160 mM glutathion is used	Cd^{2+}		1–8 mg L^{-1}	[43]
		pH measurement	Protein crosslinked with glutaraldehyde and entrapped behind a dialysis membrane	Cd^{2+}		10–80 mg L^{-1}	[43]
	GST-SmtA	Capacitive	Protein immobilized with carbodiimide on a thiol modified gold electrode	Cu^{2+}	1 fM	1 fM–1 mM	[11, 74]
				Hg^{2+}	1 fM	1 fM–1 mM	
				Cd^{2+}	1 fM	1 fM–1 mM	
				Zn^{2+}	1 fM	1 fM–1 mM	
				Pb^{2+}	1 fM	1 fM–1 mM	
	MerR	Capacitive	Protein immobilized with carbodiimide on a thiol modified gold electrode	Cu^{2+}	1 fM	1 fM–1 mM	[11, 74]
				Hg^{2+}	1 fM	1 fM–1 mM	
				Cd^{2+}	1 fM	1 fM–1 mM	
				Zn^{2+}	1 fM	1 fM–1 mM	
	MerP	Capacitive	Protein immobilized with carbodiimide on a thiol modified gold electrode	Hg^{2+}	1 fM	1 fM–1 mM	[68]
	Phytochelatins	Capacitive	Protein immobilized with carbodiimide on a thiol modified gold electrode	Cu^{2+}	1 fM	1 fM–10 mM	[70]
				Hg^{2+}	1 fM	1 fM–10 mM	
				Cd^{2+}	10 fM	10 fM–10 mM	
				Zn^{2+}	<1 fM	<1 fM–10 mM	
				Pb^{2+}	1 fM	1 fM–10 mM	
Antibody	Antibody against Cd-EDTA complex	Spectrophotometric detection	Microtiter plates were coated with Cd-EDTA-BSA conjugate and then the antibody was added, HEPES buffer pH 7.0–7.2	Cd^{2+}	7 µg L^{-1}	10–2000 µg L^{-1}	[42]
Enzyme	Urease	ISFET	Inhibition of urease immobilized on different membranes, 0.02 M HEPES, 25 °C, batch mode	Cu^{2+}		1–10 mg L^{-1}	[31]
				Hg^{2+}		0.25–5 mg L^{-1}	
				Cd^{2+}		3–10 mg L^{-1}	
				Pb^{2+}		2–10 mg L^{-1}	

Table 5.11.2. (*continued*)

Biological molecule	Transducer type	Operating conditions	M^{2+}	LOD	DR	Ref.
	ISFET	Inhibition of urease immobilized in a Nafion film, 20 °C,	Hg^{2+} Cu^{2+}	1 μM 3 μM		[39]
	Thermometric detection	Acid urease is immobilised on controlled-pore glass	Cu^{2+}		5–100 μM	[37]
	Ammonia sensor	Inhibition of urease, cuvette test with ammonia sensitive coating on the wall, 0.1 N maleate buffer pH 6	Cu^{2+} Hg^{2+} Zn^{2+} Pb^{2+}	0.25 mg L−1 0.07 mg L^{-1} 50 mg L^{-1} 100 mg L^{-1}	0.4–0.7 mg L^{-1} 0.07–1 mg L^{-1} 50–70 mg L^{-1} 100–350 mg L^{-1}	[38]
	Ammonia sensor	Enzyme reactor with urease inhibited by mercury, enzyme immobilized on glass beads	Hg^{2+}		0–15 nM	[29]
	pH sensor	Inhibition of urease immobilized with thymol blue covalently bound to aminopropyl glass at the tip of an optical fiber	Cu^{2+} Hg^{2+}	2 μg L^{-1}		[34]
	Conductometric detection	Enzyme on interdigitated gold electrodes, residual activity of urease is measured, 5 mM Tris-HNO_3 pH 7.4, 50 mM urea	Hg^{2+} Cu^{2+} Cd^{2+} Pb^{2+} Co^{2+}		1–50 μM 2–100 μM 5–200 μM 0.02–5 mM 10–500 μM	[36]
	Conductometric detection	Inhibition of urease is monitored with a standing acoustic wave device	Hg^{2+}	20 μg L^{-1}		[35]
	Fluorimetric detection at 340/485 nm	Flow system, enzyme immobilized on controlled pore glass, 0.005 M phosphate buffer pH 6.5	Hg^{2+}		0.5–100 μg L^{-1}	[32]
	Potentiometric detection	Urease entrapped in PVC membrane at the surface of iridium oxide electrode	Hg^{2+}			[40]
	Fluorescence detection at 340/455 nm	Flow system, inhibition of urease detected using *o*-phthalaldehyde	Hg^{2+}	2 μg L^{-1}		[33]
	IrTMOS	Ammonia detection by IrTMOS, 0.05 M Tris-HCl, pH 8.3	Hg^{2+}	0.005 μM		[30]

(*continued overleaf*)

Table 5.11.2. *(continued)*

Biological molecule	Transducer type	Operating conditions	M^{2+}	LOD	DR	Ref.
L-Lactate dehydrogenase	Amperometric detection	Enzyme coimmobilized with L-lactate oxidase on the top of an oxygen electrode	Hg^{2+} Cu^{2+} Zn^{2+}	1 μM 10 μM 25 μM		[55]
Glycerophosphate oxidase	Oxygen electrode	Inactivation of enzyme by metal ions, enzyme immobilized by reticulation in gelatine film or covalent binding on a membrane	Hg^{2+}	μM	20–500 μM	[56]
Pyruvate oxidase	Oxygen electrode		Hg^{2+}	10 nM		[56]
L-Lactate dehydrogenase	Amperometric detection	Enzyme coimmobilized with L-lactate oxidase on the top of an oxygen electrode	Hg^{2+} Cu^{2+} Zn^{2+}	1 μM 10 μM 25 μM		[55]
Cholinesterase	Voltammetric detection	Flow system, enzyme immobilized on nitrocellulose film with glutaraldehyde	Pb^{2+} Cu^{2+} Cd^{2+}	5 μM 50 nM 5 μM		[41]
Alkaline phosphatase	Spectrophotometric detection	Chemiluminescence from enzyme catalyzed hydrolysis of a phosphate derivative of 1,2-dioxetane is measured	Zn^{2+}	0.17 mg L^{-1}		[19]
Horseradish peroxidase	Amperometric detection	Inactivation of the enzyme is measured	Hg^{2+}	0.1 μg L^{-1}		[55]
	Spectrophotometric detection	Inhibition of enzyme immobilized on solid supports is measured	Hg^{2+}	0.1 ng L^{-1}	Four orders of magnitude	[57]
Invertase	Amperometric detection	Enzyme inhibition by mercury is measured	Hg^{2+}		10–60 μg L^{-1}	[58]
	Amperometric detection	Inhibition of enzyme immobilized on a membrane is measured	Hg^{2+}	1 μg L^{-1}		[59]
Alcohol, sarcosine or glycerol-3-P oxidase	Amperometric detection	H_2O_2 production is measured with a Ru/graphite working electrode at +700 mV versus. Ag/AgCl	Hg^{2+} Ni^{2+} Cu^{2+} V^{5+}		0.05–0.5 mg L^{-1}	[60, 61]
Urease + acetylcholin-esterase	ISFET	Inhibition of the enzymes by HM is measured	Hg^{2+} Cu^{2+} Cd^{2+} Co^{2+}	μM range		[62]
Acetylcholin-esterase	Amperometric detection	Inhibition of enzyme by metal ions is measured	Cu^{2+} Cd^{2+} Fe^{2+} Mn^{2+}	0.01 pM 1 pM 10 pM 100 pM		[63]

Table 5.11.2. *(continued)*

Biological molecule		Transducer type	Operating conditions	M^{2+}	LOD	DR	Ref.
Apoenzyme	Alkaline phosphatase	Calorimetric detection	Enzyme immobilized on epoxide acrylic beads, 100 mM TRIS-HCl pH 8.0	Zn^{2+} Co^{2+}		0.01–1.0 mM 0.04–1.0 mM	[17]
	Alkaline phosphatase	Spectrophotometric detection	Flow injection system, change in absorbance at 405 nm is measured	Zn^{2+} Co^{2+}	sub-μM	0.1–10 μM 1–200 μM	[15, 18]
		Potentiometric detection	Flow-through ISFET, pH shift detected	Zn^{2+}		0.01–1.0 mM	[16]
		Optical detection	Chemiluminescence from enzyme catalyzed hydrolysis of a phosphate derivative of 1,2-dioxetane is measured	Zn^{2+}	0.5 μg L^{-1}	0.5–50 μg L^{-1}	[19]
	Ascorbate oxidase	Calorimetric detection	Flow system, enzyme immobilized on porous glass beads	Cu^{2+}		1–50 μM	[25]
	Ascorbate oxidase	Spectrophotometric detection	Absorbance at 265 nm is measured	Cu^{2+}		0.1–10 μM	[27]
		Amperometric detection	Polarographic oxygen electrode is used	Cu^{2+}		0.5–2 μM	[26]
	Carbonic anhydrase	Fluorescence lifetime detection	The affinity of the apoenzyme for different HM is used	Cu^{2+} Zn^{2+} Cd^{2+} Co^{2+} Ni^{2+}		pM nM nM	[24]
		Calorimetric detection	Flow system, enzyme immobilized on porous glass beads	Zn^{2+} Co^{2+}		25–250 μM 50–200 μM	[20, 21]
		Optical detection at 326/460 and 560 nm	Recognition of metal ion by apoenzyme transduced by the dansylamide fluorescent probe	Zn^{2+}		40–1000 nM	[22, 23]
	Galactose oxidase	Calorimetric detection		Cu^{2+}		5–20 mM	[28]
		Amperometric detection	Detection with oxygen electrode	Cu^{2+}		0.1–10 mM	[26]
	Alkaline phosphatase + ascorbate oxidase	Amperometric detection	Enzymes coimmobilized on a polymer membrane attached to a polarographic oxygen electrode	Cu^{2+} Zn^{2+}		2–100 μM 2–200 μM	[64]
	Tyrosinase	Amperometric detection	Flow system with oxygen electrode	Cu^{2+}		up to 0.05 mM	[65]

The general approach to constructing biosensors based on intact cells is to fuse an inducible promoter from one of the metal resistance operons to a reporter gene that includes genes for bioluminescent proteins such as luciferase. The light produced by luminescent proteins can be measured with photometers or luminometers. When the bacteria emit light, optical fibers can be used to transmit the light to the detection device [6]. Several biosensors were developed using the promoter–reporter gene concept. Different bacteria were genetically modified with *lux* genes from *Vibrio fischeri* or *luc* genes from firefly luciferase used as reporter genes fused to promoter regions of operons responsible for resistances to various metals: *mer* operon for detection of mercury [7–9] and copper [6, 10, 11], *ars* operon for detection of arsenic [12, 13], or *cad* operon for detection of cadmium [12, 14].

3 ENZYME AND APOENZYME BASED BIOSENSORS

The use of enzymes and apoenzymes as recognition elements in biosensors for metal detection is based on the fact that the metal may act as catalyst (cofactor) or inhibitor. There are enzymes that have binding selectivity for the metal ions which participate as cofactors in catalysis, a fact that has been exploited, e.g. in biosensors based on apo forms of alkaline phosphatase [15–19], carbonic anhydrase [20–24], ascorbate oxidase [25–27], and galactose oxidase [26, 28].

A more frequently used approach is based on the inhibition of the enzyme activity in the presence of metals allowing the correlation of the decrease in enzyme activity with the metal concentration. Most such biosensors are based on the inhibition of urease [29–40], but also carbonic anhydrase [23], cholinesterase [41], alkaline phosphatase [19] were used for this purpose. As can be derived from Table 5.11.2 enzymes or apoenzymes have been coupled to various transducers such as amperometric and potentiometric electrodes, optical fibers, conductometric or piezoelectric devices. As compared to cell-based biosensors, heavy metal biosensors using enzymes or apoenzymes as biological recognition element are easier to construct and can be regenerated to a certain extent.

4 ANTIBODY OR PROTEIN BASED BIOSENSORS

There have been a few attempts to use antibodies [42] or proteins [43, 44] as recognition elements for metal biosensors. Since immunoassays are quick, inexpensive, easy to perform, and portable, these assays are becoming increasingly accepted for environmental applications. Proteins, e.g. phytochelatins or metallothioneins, can be used as biological components for metal binding by their immobilization at the surface of an appropriate transducer, e.g. pH-sensitive field-effect transistor, resonating piezoelectric crystal or optical devices. These proteins selectively bind metal ions via thiolate complex formation. Changes within the layer of the immobilized protein (e.g. release of protons, changes in mass and optical properties) are transformed into measurable signals by the transducer.

A common drawback of the abovementioned biosensors, excepting whole cell and protein based ones, is the impossibility to detect bioavailable concentrations of metals. Moreover, cell based biosensors can only detect metals down to micromolar level, while the protein based capacitive biosensors are sensitive to concentrations in the femtomolar range. A more detailed description of such biosensor design is presented below.

5 PROTEIN BASED CAPACITIVE BIOSENSOR

Proteins from three different classes of metal binding proteins have been used as biological component of a biosensor, namely synechococcal metallothionein, SmtA [66], mercury resistance proteins, MerR [67] and MerP [68], and a phytochelatin [69, 70].

The SmtA protein is a metallothionein (MT) from the cyanobacterium *Synechococcus*. Metallothioneins are small proteins which sequester metal ions in a 'cage' structure. In animal metallothioneins there are two domains, each of which can sequester three or four metal ions. Metal binding is associated with a large conformational change in the protein, as the sulfhydryl groups of cystein residues coordinate the metal ions. The SmtA protein was overexpressed as a glutathione-S-transferase-metallothionein fusion protein, which was subsequently used as the biological recognition element of a biosensor.

MerR is the regulatory protein responsible for inducible expression of mercury resistance proteins [67]. Hg(II) binds to the dimeric MerR protein, and there is genetic evidence that this results in a conformational change in the protein [71, 72].

MerP is a 72 amino acid periplasmic mercury binding protein containing two cysteine residues, Cys-14 and Cys-17, both necessary for specific binding of Hg^{2+}. MerP contains a GMTCXXC metal binding domain that also can be found in Menkes copper transporting ATPase, a protein responsible for Menkes syndrome, which is a lethal X-chromosome hereditary disease of copper starvation. The amino acid sequence similarity for the two proteins is shown below and the exact matches were denoted with stars (*):

```
Menkes ATPase   LTQETVINID   GMTCNSCVQS   IEGVISK-KPG   VKSIRVSLAN
MerP            ATQTVTLAVP   GMTCAACPIT   VKKALSK-VEG   VSKVDVGFEK
                **           ****   *     *** *         *  *
```

Due to its similarity with Menkes copper transporting ATPase, a thorough study of MerP protein is important to elucidate the mechanism of metal transport in the cell, identifying the residues responsible for the specific metal binding. The developed MerP based biosensor seems to be a promising tool in this respect [68].

Plants, algae and some fungi are capable of synthesizing, on exposure to metals, thiol-rich peptides such as (^{1}-glutamylcysteinyl)$_n$-glycine with $n = 2$ to 11, also known as phytochelatins. Phytochelatins (PCs) seem to be involved in detoxification and homeostasis of trace metals in plants and thus serve functions analogous to metallothioneins in animals. Phytochelatins are enzymatically synthesized by a specific ^{1}-glutamylcysteine-dipeptidyl transpeptidase (phytochelatin synthase), which is activated by the presence of metal ions and uses glutathione as a substrate. Phytochelatins bind metal ions by thiolate coordination yielding intracellular metal complexes.

These different types of metal-binding proteins have been coupled to a highly sensitive capacitive transducer. The principle of capacitance measurements is based on the modulation of the electrical double layer and was described earlier [11, 73–75]. The total capacitance measured is given by the sum of the capacitances of the different layers covering the electrode surface: (i) the self-assembled nonconductive layer needed for protein binding, (ii) the protein layer, and (iii) the ionic space charge formed by the hydrated ions of the buffer.

When metal ions bind to the protein, a conformational change in the protein's structure invokes a change of the capacitance which is detected and correlated with the metal concentration. The assumed conformational change is schematically depicted in Figure 5.11.1.

Proteins were covalently immobilized on the surface of gold electrodes after carbodiimide activation of the carboxylic acid head groups of a thioctic acid monolayer. The protein-modified biosensors were then used as working electrode in a three(four)-electrode electrochemical flow cell. A Pt foil was used as auxiliary electrode, and a Pt wire and a commercial Ag/AgCl electrode served as quasi and real reference electrodes. The importance of using a second Ag/AgCl reference electrode was explained previously [73].

Measurements were made by applying a potential pulse of 50 mV at the working electrode after injection of 250 μL sample solutions into the 10 mM borate buffer carrier flow, kept at a pH value of 8.75. The current transients invoked by the application of the potential pulse are recorded, and the decrease in current is evaluated according

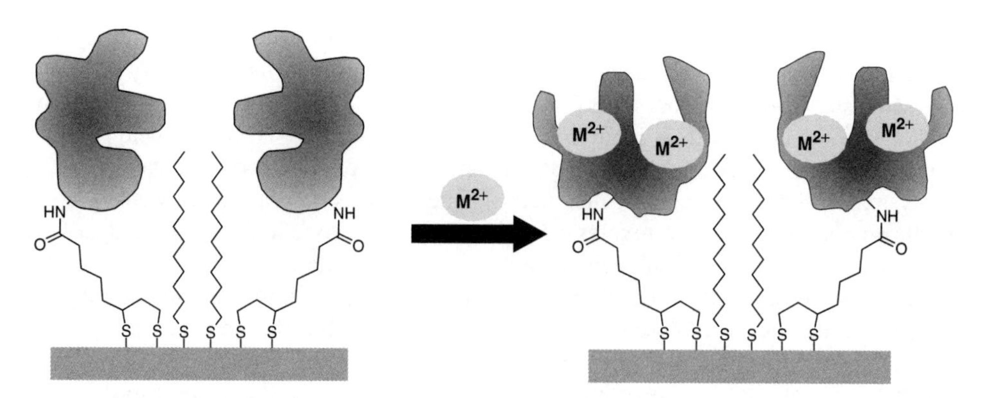

Figure 5.11.1. Schematic representation of conformational changes occurring upon binding of metal ions to proteins.

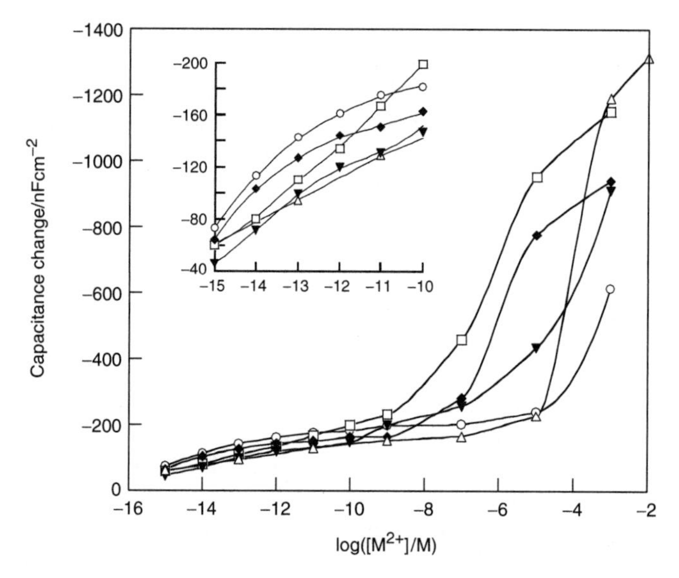

Figure 5.11.2. Typical calibration curves obtained for Pb (△), Hg (◆), Cd (○), Zn (▼), and Cu (□) obtained with a GST-SmtA based capacitive biosensor (10 mM borate buffer, pH 8.75, flow rate 0.5 ml min^{-1}, room temperature).

to the following equation:

$$i(t) = \frac{u}{R_s} \exp\left(-\frac{t}{R_s C_t}\right)$$

where $i_{(t)}$ is the current at time t, u is the amplitude of the potential pulse applied, R_s is the resistance between the gold and the reference electrodes, C_t is the total capacitance over the immobilized layer and t is the time elapsed after the potential pulse was applied.

Typically, all these protein based capacitive biosensors detected metals with high sensitivity and a very broad detection range starting at fM level up to mM concentrations. However, differences were noticed depending on the protein used and the metal ion to be detected. Typical calibration curves obtained for the determination of different metal ions using the GST-SmtA based electrode are shown in Figure 5.11.2, while Figure 5.11.3 depicts the signals obtained for the determination of Hg^{2+} with four different biosensors.

The SmtA-based sensors show highest sensitivity and extended linear range for Hg^{2+} at lower

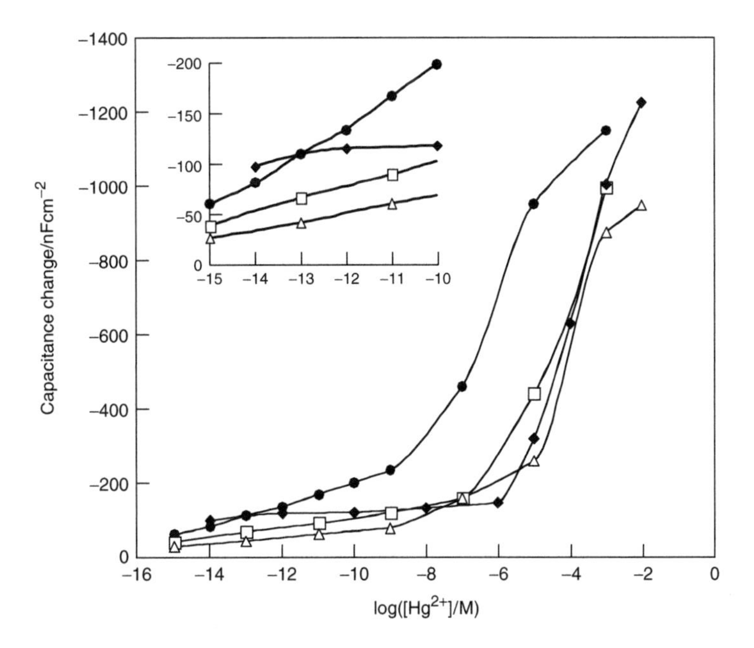

Figure 5.11.3. Calibration curves obtained for the determination of Hg^{2+} using GST-SmtA (•), MerR (□), MerP (◆) and phytochelatin (△) based capacitive biosensors.

Table 5.11.3. Concentrations at which the shape of the capacitive signal changes for capacitive biosensors based on GST-SmtA, MerR and phytochelatin.

Protein	Metal ion				
	Cd	Hg	Cu	Zn	Pb
GST-SmtA	10^{-9} M	10^{-10} M	10^{-6} M	10^{-7} M	10^{-7} M
MerR	10^{-6} M	10^{-7} M	10^{-5} M	10^{-6} M	No results
Phytochelatin	10^{-6} M	10^{-7} M	10^{-5} M	10^{-5} M	10^{-4} M

concentrations and a broad selectivity pattern, sensing Pb^{2+}, Hg^{2+}, Cd^{2+}, Zn^{2+}, and Cu^{2+} ions (Figure 5.11.2), while the MerR and phytochelatin based ones were more selective towards one specific metal ion, Hg^{2+} and Zn^{2+}, respectively (results not shown).

Biosensors based on all four proteins (GST-SmtA, MerR, MerP and PC) are able to measure Hg^{2+}, with the GST-SmtA based one being the most sensitive. The shape of the calibration curves may be explained by conformational changes related to the biological roles of the different proteins, namely binding and regulating the transport of Hg^{2+} in the animal cells (SmtA, MerP, MerR) or plant cells (PC). Obviously, depending on the metal concentration, there are two distinct responses of the biosensors. It can be assumed that in the low concentration range the signal is due to the titration of the cysteine and histidine residues in the amino acid sequence of the proteins with the metal ion. The accentuated capacitance changes occurring at higher concentrations (see Table 5.11.3) may be correlated with the conformational changes in the protein's structure. For all electrode types, the regeneration of the electrodes is possible by injecting EDTA.

6 CONCLUSIONS

Biosensors have been shown to be a versatile tool for monitoring metal ions with high sensitivity and selectivity. However, their bioanalytical characteristics (e.g. lifetime, linear range, sensitivity, selectivity) vary significantly for the different sensor types and fields of application. Therefore, one should always consider the particular application area before choosing the most adequate sensor configuration. On the other hand, biosensors based on metal-binding proteins are able to detect *bioavailable* concentrations of metals with a very high sensitivity and within an extremely broad concentration range. Considering present progress in biomolecule engineering, a further improvement in their selectivity is expected, and thus, they seem to be very promising tools for further fundamental and applied studies.

7 ACKNOWLEDGEMENTS

The European Commission (CEMBA contract EVK1-1999-0008) and the Swedish Research Council (NFR) supported this work financially. Professor Nigel Brown (University of Birmingham, UK) and Professor Ashok Mulchandani (University of California, Riverside, CA, USA) are acknowledged for their kind gift of SmtA, Mer type proteins and phytochelatins, respectively.

8 REFERENCES

1. Nies, D. H., *Appl. Microbiol. Biotechnol.*, **51**, 730 (1999).
2. Jackson, K. W. and Chen, G. R., *Anal. Chem.*, **68**, R231 (1996).
3. Burlingame, A. L., Boyd, R. K. and Gaskell, S. J., *Anal. Chem.*, **68**, R599 (1996).
4. Anderson, J. L., Bowden, E. F. and Pickup, P. G., *Anal. Chem.*, **68**, R379 (1996).
5. Wittmann, C., Riedel, K. and Schmidt, R. D., Microbial and enzyme sensors for environmental monitoring, in *Handbook of Biosensors and Electronic Noses: Medicine, Food and the Environment*, Kress-Rogers, E. (Ed.), CRC Press, Boca Raton, FL, 1997, pp. 299.
6. Leth, S., Maltoni, S., Simkus, R., Mattiasson, B., Corbisier, P., Klimant, I., Wolfbeis, O. S. and Csöregi, E., *Electroanalysis*, **14**, 35 (2002).
7. Virta, M., Lampinen, J. and Karp, M., *Anal. Chem.*, **67**, 667 (1995).
8. Tescione, L. and Belfort, G., *Biotechnol. Bioeng.*, **42**, 945 (1993).
9. Selifonova, O., Burlage, R. and Barkay, T., *Appl. Environ. Microbiol.*, **59**, 3083 (1993).
10. Corbisier, P., van der Lelie, D., Borremans, B., Provoost, A., de Lorenzo, V., Brown, N. L., Lloyd, J. R., Hobman, J. L., Csöregi, E., Johansson, G. and Mattiasson, B., *Anal. Chim. Acta*, **387**, 235 (1999).
11. Bontidean, I., Lloyd, J. R., Hobman, J. L., Wilson, J. R., Csöregi, E., Mattiasson, B. and Brown, N. L., *J. Inorg. Biochem.*, **79**, 225 (2000).
12. Corbisier, P., Ji, G., Nuyts, G., Mergeay, M. and Silver, S., *FEMS Microbiol. Lett.*, **110**, 231 (1993).
13. Tauriainen, S., Virta, M., Chang, W. and Karp, M., *Anal. Biochem.*, **272**, 191 (1999).
14. Tauriainen, S., Karp, M., Chang, W. and Virta, M., *Biosens. Bioelectron.*, **13**, 931 (1998).
15. Satoh, I. and Aoki, Y., *Denki Kagaku*, **58**, 1114 (1990).
16. Satoh, I. and Masumura, T., Flow injection biosensing of zinc(II) ions with use of an immobilized alkaline phosphatase reactor, in *Technical Digest of the 9th Sensor Symposium*, Sasaki, A. (Ed.), Tokyo, Japan, 1990, p. 197.
17. Satoh, I., *Biosens. Bioelectron.*, **6**, 375 (1991).
18. Satoh, I., *Ann. N. Y. Acad. Sci.*, **672**, 240 (1992).
19. Kamtekar, S. D., Pande, R., Ayyagari, M. S., Marx, K. A., Kaplan, D. L., Kumar, J. and Tripathy, S., *Anal. Chem.*, **68**, 216 (1996).
20. Satoh, I., Ikeda, K. and Watanabe, N., Microanalysis of zinc(II) ion by using an apoenzyme thermistor, in *Proceedings of the 6th Sensor Symposium*, Takahashi, K. (Ed.), Tokyo, Japan, 1986, p. 203.
21. Satoh, I., Continuous biosensing of heavy metal ions with use of immobilized enzyme-reactors as recognition elements, in *Proceedings of MRS International Meeting on Advanced Materials*, Karube, I. (Ed.), Pittsburgh, PA, Vol. 14, 1989, p. 45.
22. Thompson, R. B. and Jones, E. R., *Anal. Chem.*, **65**, 730 (1993).
23. Thompson, R. B., Maliwal, B. P., Feliccia, V. L., Fierke, C. A. and McKall, K., *Anal. Chem.*, **70**, 4717 (1998).
24. Thompson, R. B., Patchan, M. W. and Ge, Z., U.S. patent, 1999, p. 25.
25. Satoh, I., Kimura, S. and Nambu, T., Biosensing of copper(II) ions with an apoenzyme thermistor containing immobilized metalloenzymes in flow system, in *Digest of Technical Papers of the 4th International Conference on Solid-State Sensors and Actuators, Transducers '87*, Matsuo, T. (Ed.), Tokyo, Japan, 1987, p. 789.
26. Satoh, I., Kasahara, T. and Goi, N., *Sens. Actuat. :B*, **1**, 499 (1990).
27. Satoh, I. and Nambu, T. Flow-injection photometric biosensing of copper(II) ions with the use of an immobilized ascorbate oxidase column, in *Technical Digest of the 10th Sensor Symposium*, Nakamura, T. (Ed.), Tokyo, Japan, 1991, p. 77.
28. Satoh, I., *Ann. N. Y. Acad. Sci.*, **613**, 401 (1990).
29. Ögren, L. and Johansson, G., *Anal. Chim. Acta*, **96**, 1 (1978).

30. Winquist, F., Lundström, I. and Danielsson, B., *Anal. Lett.*, **21**, 1801 (1988).
31. Sakai, H., Kaneki, N., Tanaka, H. and Hara, H., *Sens. and Mater.*, **2**, 217 (1991).
32. Bryce, D. W., Fernández-Romero, J. M. and Luque de Castro, M. D., *Anal. Lett.*, **27**, 867 (1994).
33. Narinesingh, D., Mungal, R. and Ngo, T. T., *Anal. Chim. Acta*, **292**, 185 (1994).
34. Andres, R. T. and Narayanaswamy, R., *Analyst*, **120**, 1549 (1995).
35. Liu, D. H., Yin, A. F., Chen, K., Ge, K., Nie, L. H. and Yao, S. Z., *Anal. Lett.*, **28**, 1323 (1995).
36. Zhylyak, G. A., Dzyadevich, S. V., Korpan, Y. I., Soldatkin, A. P. and Elskaya, A. V., *Sens. Actuators B*, **24–25**, 145 (1995).
37. Preininger, C. and Danielsson, B., *Analyst*, **121**, 1717 (1996).
38. Preininger, C. and Wolfbeis, O. S., *Biosens. Bioelectron.*, **11**, 981 (1996).
39. Volotovsky, V., Nam, Y. J. and Kim, N., *Sens. Actuators B*, **42**, 233 (1997).
40. Krawczynski, T. K. V., Moszczynska, T. and Trojanowicz, M., *Biosens. Bioelectron.*, **15**, 681 (2000).
41. Budnikov, H. C., Medyantseva, E. P. and Babkina, S. S., *J. Electroanal. Chem.*, **310**, 49 (1991).
42. Khosraviani, M., Pavlov, A. R., Flowers, G. C. and Blake, D. A., *Environ. Sci. Technol.*, **32** 137 (1998).
43. Hilpert, R., Zenk, M. H. and Binder, F., Biosensors for the detection of heavy metal ions, in *GBF Monogr.*, 367 (1989).
44. Klein, J. and Mattes, R., *Anal. Biochem.*, **260**, 173 (1998).
45. Ramos, J. A., Bermejo, E., Zapardiel, A., Pérez, J. A. and Hernández, L., *Anal. Chim. Acta*, **273**, 219 (1993).
46. Verschure, L., Gernaey, K. and Verstraete, W., *Water (Wijnegem, Belg.)*, **14**, 163 (1995).
47. Biran, I., Babai, R., Levcov, K., Rishpon, J. and Ron, E. Z., *Environ. Microbiol.*, **2**, 285 (2000).
48. Erbe, J. L., Adams, A. C., Taylor, K. B. and Hall, L. M., *J. Ind. Microbiol.*, **17**, 80 (1996).
49. Holmes, D. S., Dubey, S. K. and Gangolli, S., *Environ. Geochem. Health*, **16**, 229 (1994).
50. Corbisier, P., Thiry, E., Masolijn, A. and Diels, L., Construction and development of metal ion biosensors in *Proceedings of the 8th Bioluminescence Chemiluminescence International Symposium*, (1994) p. 151.
51. Lyngberg, O. K., Stemke, D. J., Schottel, J. L. and Flickinger, M. C., *J. Ind. Microbiol. Biotechnol.*, **23**, 668 (1999).
52. Corbisier, P., Thiry, E. and Diels, L., *Environ. Toxicol. Wat. Qual.*, **11**, 171 (1996).
53. Liu, Y., Ensor, M. and Daunert, S., Copper ion detection using genetically engineered yeast, in *Book of Abstracts of The 215th ACS National Meeting*, Dallas, TX, 1998, 032-ANYL, Part 1.
54. Kremleva, N. V., Medyantseva, E. P., Budnikov, G. K. and Bormotova, Y. I., *J. Anal. Chem.*, **54**, 151 (1999).
55. Fennouh, S., Casimiri, V., Geloso-Meyer, A. and Burstein, C., *Biosens. Bioelectron.*, **13**, 903 (1998).
56. Gayet, J. C., Haouz, A., Geloso-Meyer, A. and Burstein, C., *Biosens. Bioelectron.*, **8**, 177 (1993).
57. Shekhovtsova, T. N. and Chernetskaya, S. V., *Anal. Lett.*, **27**, 2883 (1994).
58. Bertocchi, P., Ciranni, E., Compagnone, D., Magearu, V., Palleschi, G., Pirvutoiu, S. and Valvo, L., *J. Pharm. Biomed. Anal.*, **20**, 263 (1999).
59. Amine, A., Cremsini, C. and Palleschi, G., *Mikrochim. Acta*, **121**, 183 (1995).
60. Compagnone, D., Bugli, M., Imperiali, P., Varallo, G. and Palleschi, G., *NATO ASI Ser.*, **38**, 220 (1997).
61. Compagnone, D., Palleschi, G., Varallo, G. and Imperiali, P., Amperometric biosensors for the determination of heavy metals, in *Proc. SPIE–Int. Soc. Opt. Eng.*, **2504**, 141 (1995).
62. Starodub, N. F., Torbicz, W., Starodub, V. M., Kanjuk, M. I. and Ternovoj, K. S., *Transducers '97*, **2**, pp. 1383 (1997).
63. Stoytcheva, M., *B. Soc. Chim. Belg.*, **103**, 147 (1994).
64. Satoh, I., *Sens. Actuatuators B*, **13**, 162 (1993).
65. Mattiasson, B., Nilsson, H. and Olsson, B., *J. Appl. Biochem.*, **1**, 377 (1979).
66. Turner, J. S. and Robinson, N. J., *J. Ind. Microbiol.*, **14**, 119 (1995).
67. Hobman, J. L. and Brown, N. L., Metal ions in biological systems, in '*Mercury and its Effects on Environment and Biology*, Vol. 34, Sigel, A. and Sigel, H. (Eds), Marcel Dekker, New York, 1997 p. 527.
68. Bontidean, I., Hobman, J. L., Brown, N. L. and Csöregi, E., *Anal. Chem.*, submitted (2003).
69. Bae, W., Chen, W., Mulchandani, A. and Mehra, R. K., *Biotechnol. Bioeng.*, **70**, 518 (2000).
70. Bontidean, I., Ahlqvist, J., Mulchandani, A., Chen, W., Bae, W., Mehra, R. K., Mortari, A. and Csöregi, E., *Biosens. Bioelectron.*, in press (2003).
71. Parkhill, J., Ansari, A. Z., Wright, J. G., Brown, N. L. and O'Halloran, T. V., *EMBO J.*, **12**, 413 (1993).
72. Parkhill, J., Lawley, B., Hobman, J. L. and Brown, N. L., *Microbiology (Reading, UK)*, **144**, 2855 (1998).
73. Berggren, C. and Johansson, G., *Anal. Chem.*, **69**, 3651 (1997).
74. Bontidean, I., Berggren, C., Johansson, G., Csöregi, E., Mattiasson, B., Lloyd, J. R., Jakeman, K. J. and Brown, N. L., *Anal. Chem.*, **70**, 4162 (1998).
75. Berggren, C., Bjarnason, B. and Johansson, G., *Instrum. Sci. Technol.*, **27**, 131 (1999).

5.12 Possibilities Offered by Radiotracers for Method Development in Elemental Speciation Analysis and for Metabolic and Environmentally Related Speciation Studies

Rita Cornelis

Laboratory for Analytical Chemistry, University of Ghent, Belgium

1 INTRODUCTION

The development of methods for chemical speciation analysis can be facilitated to a great extent by incorporating a suitable radioisotope of the element into the system and measuring the radiation of the isolated species [1]. The most basic assumption is that the radiotracer behaves exactly in the same way as the stable isotopes of the element it characterises. Therefore, the radiotracer has to be incorporated under the identical form as the species endogenous to the system on the basis of complete isotopic exchange.

The separation steps are common to the nonradioactive procedures, and the only difference lies in the final detection of the analyte. The advantage in ease of detection of trace element amounts of radioactive isotopes is unsurpassed for gamma rays, and useful in many a case of positron and beta emitters. Detection of gamma rays consists of a simple measurement of the radioactivity without requiring the type of sample preparation and calibration that is necessary with other techniques. Even when the method seems effortless, such as electrothermal atomic absorption spectrometry (ET-AAS), inductively coupled plasma atomic emission spectrometry (ICP-AES), inductively coupled plasma mass spectrometry (ICP-MS) or electrochemistry, it will be more tedious than for gamma spectrometry. This is due to the fact that the emission of the radiation by the radionuclide is independent of physical and chemical influences.

Handbook of Elemental Speciation: Techniques and Methodology R. Cornelis, H. Crews, J. Caruso and K. Heumann
 ISBN: 0-471-49214-0

Additionally, the method of using radiotracers circumvents the hazard of detecting contamination from other sources of the trace element studied, as only the radioactive tracer is measured.

A major warning is needed. The sample to be analysed for its different species should never be subjected to neutron or particle activation in a nuclear reactor, cyclotron or accelerator prior to speciation analysis. This is totally inappropriate because when a chemical compound undergoes nuclear bombardment, the chemical valency and the chemical bonds might change through the Szilard–Chalmers effect, as this causes the radioactive atom in the process of its formation to break loose from its molecule with loss of the original chemical species. In 1934 Szilard and Chalmers showed that after the neutron irradiation of ethyl iodide most of the iodine activity formed could be extracted from the ethyl iodide with water. They used a small amount of iodine carrier (nonradioactive), reduced it to iodide, and finally precipitated it as AgI. Evidently the iodine–carbon bond was broken when an ^{127}I nucleus was transformed by neutron capture to ^{128}I [2].

A detailed study of the Szilard–Chalmers effect on the decomposition of six different organoarsenic compounds during neutron irradiation has been reported by Šlejkovec *et al.* [3]. For aqueous solutions of monomethylarsonic acid, dimethylarsinic acid, arsenobetaine, arsenocholine, tetramethylarsonium ion, trimethylarsine oxide the degree of decomposition was high (>80 % for 10 min of irradiation at $3.8 \times 10^{16}\, s^{-1}\, m^{-2}$, yielding mainly As(V)) whereas irradiation of solid arsenobetaine for 60 min gave low decomposition yields (<10 %). What makes these experiments even more interesting is that the irradiation resulted in very high specific activities for the decomposition products in the samples irradiated in aqueous solution. For As(V) specific activities about 1000 times higher than those expected from direct irradiation of As(V) were found (>3800 $kBq\, \mu g^{-1}$).

Whereas radiotracer techniques are simple, it is not practicable to envisage neutron activation analysis of the different fractions after separation of the species, owing to its time-consuming nature. First of all, the induced radioactivity is no longer specific for the radioisotope of the element, but is now due to the mixture of isotopes that became radioactive in the process of neutron activation. As a consequence elaborate radiochemical separation would be needed to separate the radioisotope of the species under investigation from the matrix radioactivity. Secondly the number of fractions after, e.g., a chromatographic separation, are about fiftyfold as numerous, making it an endless task.

Radiotracers are most useful in two ways:

(1) During the exploratory phase of method development for extraction, chromatographic and electrophoretic techniques.
(2) For *in vitro* and *in vivo*, and environmentally related studies about mobility, storage, retention, metabolism and toxicity of trace element species. It allows to follow the behaviour and transformation of a newly administered radiotracer, as the 'cold' or nonradioactive share of the element cannot be measured. Such radiotracer experiments can often be done using a carrier-free radioisotope, this means all isotopes added are radioactive, and they will not significantly add to the mass already present. This is an additional advantage as the original concentration of the trace element in the system remains unaffected.

2 MODES OF RADIOACTIVE DECAY RELEVANT FOR SPECIATION PURPOSES [4, 5]

Radioisotopes undergo radioactive decay by emitting alpha, beta or gamma rays.

Alpha rays are monoenergetic and consist of an He nucleus (mass 4, consisting of two neutrons and two protons); beta emitters emit either a positron (β^+) or an electron (β^-), and simultaneously a neutrino, beta rays are not monoenergetic, the total decay energy being divided over the particle and the neutrino; gamma emitters emit discrete gamma energies as electromagnetic waves.

Alpha emitters are mainly found among the heavier elements. As they need much more caution to handle and require more elaborate measurement techniques, they are not commonly used as tracers for elemental speciation purposes. It may be interesting to mention that all elements found in natural sources with atomic number greater than 83 (bismuth) are radioactive. They belong to the uranium, thorium and actinium series.

Many radioisotopes undergo decay by successive β- and γ-emission. A limited number of radioisotopes are solely β emitters, such as ^{3}H, ^{14}C, ^{32}P.

The decay of a radioisotope follows the exponential law, $N_t = N_0 e^{-\lambda t}$, where N_t is the (large) number of radioactive atoms at time t, N_0 is the number present at time $t = 0$ and λ is a constant characteristic of the particular radioisotope, related to the half-life ($t_{1/2}$) or the time required for an initial large number of atoms to be reduced to half that number by radioactive decay ($t_{1/2} = \ln 2/\lambda = 0.693/\lambda$). Radioactive decay is considered to be a purely statistical process. In practical work the number of atoms is not measured. The radioactivity emitted by a radioisotope follows the equation $A_t = A_0 e^{-\lambda t}$. Even the radioactivity is usually not measured in an absolute way. In order to do an absolute measurement it would be necessary to determine the detection efficiency, which depends on the nature of the detection instrument, the sensitivity of recording the particular radiation in that particular instrument, and the geometrical arrangement of sample and detector. This is circumvented by doing relative measurements by measuring a 'standard' solution at the beginning of the experiment and then comparing the radioactivity of all the subsequent samples in identical geometry and density of solution to that of the 'calibrant'. This procedure is suitable for the purpose for radiotracer based method development in elemental speciation studies.

3 ELEMENTS SUITABLE FOR RADIOTRACER STUDIES

Although a radioisotope exists for almost every element in the periodic table, only a limited number are suitable for studies in this way. Firstly the half-life of the isotope must be adequate for the duration of the experiment. As a rule, half-lives of less than a few hours are not acceptable for this purpose. Whenever more than one isotope of the same element is eligible, the isotope with the shortest half-life is to be preferred as this reduces the problem of waste disposal. This option can, of course, only be considered when this isotope is measurable under the experimental conditions with as high a sensitivity as possible.

Gamma-emitters are more suitable because self-absorption of the radiation in the medium is either negligible or can be easily corrected for. Any liquid or solid form is convenient for measurement with a NaI(Tl) scintillation detector or a Ge detector unit (see next paragraph).

Electron and positron emitters are second choice because they have a poor penetration range (the radiation is absorbed to a large extent by the medium and the walls of the counting vial), and therefore require labourious sample preparation. The detection method that may be envisaged for a flat sample is a Geiger–Müller counter or a proportional counter. In case of speciation work these historic detection systems would never be considered because they are too impractical for serial analysis. The other detection method that will be considered is liquid scintillation counting, where the sample is intimately mixed with a special scintillation cocktail. As this technique is prone to quenching and luminescence error, which adversely affect the counting efficiency, suitable correction protocols must be included for each set of experiments. The sample preparation is rather time consuming and the corrections may add significantly to the overall error of the measurement.

Alpha emitters are rarely used for trace experiments, because they are too radiotoxic and have very poor penetration (the radiation has a short range and can be stopped by the species of which they form an integral part, by the medium and by the walls of the vial in which the sample is collected). The exceptions are the actinides and ^{210}Po, a decay product of radiolead, which emit useful alpha particles. These radioisotopes need to be carefully monitored in the environment. Alpha particle detection can be carried out by 4π counting, 2π counting, scintillation counting, alpha particle spectrometry with an ionisation chamber or with the more common silicon surface barrier semiconductor, all requiring extensive sample preparation prior to counting. Elements with potentially useful radionuclides are given in Figure 5.12.1.

Today a major drawback in doing radiotracer research is that fewer and fewer radioisotopes are commercially available. Many nuclear facilities are closed. The demand for radioisotopes, is dwindling, except for those routinely used for medical diagnostic purposes. Whenever a special radioisotope is needed, the manufacturers may nevertheless be willing to produce it, albeit at exorbitant costs.

The most common production mode for radioisotopes will be the irradiation of the parent isotope in a nuclear reactor at a high neutron flux. The parent isotope undergoes (n,γ) reaction and gives rise to a radioisotope of the same element, but with the atomic weight increased with 1 unit mass. Examples are: ^{63}Cu(n,γ)^{64}Cu, ^{50}Cr(n,γ)^{51}Cr, ^{59}Co(n,γ)^{60}Co, ^{74}Se(n,γ)^{75}Se, ... [4]. Isotopes produced in this way will not be carrier free; this means they will consist of a mixture of mainly inactive carrier (nonradioactive, naturally occurring isotopes of the element) and the radioactive isotope. The specific activity of the radioisotope is defined by the counting rate per total mass of the element. In case of ultratrace element studies, it is necessary to know the amount of carrier added. Irradiation of the element enriched in the parent isotope of the radioactive daughter increases its specific activity, albeit at an additional cost. This was done, e.g., by Parent *et al.* [6], for the production of ^{191}Pt. The irradiated Pt was enriched in ^{190}Pt up to 4.19 %, which means about 400 times higher than the natural isotopic abundance ($\theta = 0.01\,\%$). It is also possible to make noncarrier added radioisotopes with fast neutrons via (n,p) reaction, e.g., ^{64}Zn(n,p)^{64}Cu [7]. On the assumption that the zinc target is free from copper impurities, and that no stable copper atoms are added during the post-irradiation radiochemical separation originating from copper impurities in the reagents and recipients, this is an efficient way to produce a carrier-free copper isotope.

An interesting alternative to produce an isotope with a high specific activity is to irradiate a compound where the element is bound to a carbon atom. Due to recoil, the radioactive element is knocked off and set free (similar to the arsenic example, given previously). To continue the example of copper, via recoil of ^{64}Cu from organocopper compounds high specific activity copper radiotracer may be produced.

The other interesting way to produce a radioisotope is through particle activation, followed by a radiochemical separation of the radioisotope

1	2											13	14	15	16	17	18
Ia	**IIa**											**IIIa**	**IVa**	**Va**	**VIa**	**VIIa**	**0**
H																	
		3	4	5	6	7	8	9	10	11	12		C				
Na		**IIIb**	**IVb**	**Vb**	**VIb**	**VIIb**		**VII**		**Ib**	**IIb**			P	S	C1	
K	Ca	Sc	Ti	V	Cr	Mn	Fe	Co	Ni	Cu	Zn	Ga	Ge	As	Se	Br	Kr
Rb	Sr	Y	Zr	Nb	Mo	Tc	Ru	Rh	Pd	Ag	Cd	In	Sn	Sb	Te	I	Xe
Cs	Ba	La	Hf	Ta	W	Re	Os	Ir	Pt	Au	Hg	T1	Pb	Bi	Po		

Lanthanides	Ce	Pr	Nd	Pm	Sm	Eu	Gd	Tb	Dy	Ho	Er	Tm	Yb	Lu
Actinides	Th	Pa	U	Np	Pu	Am								

Figure 5.12.1. Elements with potentially useful radionuclides.

from the matrix and from other interfering isotopes. An example is the production of ^{48}V through the nuclear reaction $^{48}Ti(p,n)$ ^{48}V, and the separation of the ^{48}V from the Ti matrix and interfering Sc isotopes. The advantage here is that the radioisotope is in principle carrier free, which means that all V atoms present are radioactive [8]. This assumes, however, that there was no V impurity either in the Ti target, or in the reagents used to do the radiochemical separation of the V. This will never be true. Nevertheless the radioisotope may be claimed to be carrier free, because the mass of V attributable to the nonradioactive contaminant may be orders of magnitude below or similar to the mass of the radiotracer.

Although it is feasible to do tracer experiments on a mixture of radionuclides with suitable nuclear characteristics, only one radioisotope is normally used, so that the isolation and counting of the species can be straightforward. Sometimes it may be very tempting to use two radioisotopes of the same element to follow the behaviour of two important species, e.g., two oxidation states. At the start of the experiment, e.g., an aquatic medium, each oxidation state of the element is labelled with its specific isotope. Depending on the chemical environment of the medium, the duration since the start of the experiment, the temperature, ..., the oxidation state will be stable, reduced or oxidized. This can be unravelled with a specific radiotracer per oxidation state. It requires radioisotopes with suitable nuclear characteristics: comparable half-life and nonoverlapping photopeaks. This situation, however, is seldom met. The element As is the exception, with a choice of three radioisotopes: ^{74}As ($t_{1/2} = 17.78$ days, γ: 634.8 keV), ^{76}As ($t_{1/2} = 26.4$ h, γ: 559.1 keV) and ^{77}As ($t_{1/2} = 38.8$ h, γ: 239.0 keV). The individual isotope can then be converted into a specific species, e.g., oxidation state (arsenate, arsenite, and in living organisms also into methylated species (monomethylarsenic acid or dimethylarsonic acid).

A dual radiotracer technique, with an emphasis of probing artefacts, was used in a case study of technetium and spinach [9]. The authors wanted to study the uptake of Tc by the plant, but were also on the alert for possible errors during their homogenisation procedure (possible losses and redox conversions). For the uptake by the plant they used $^{99}Tc^{VII}O_4^-$ (spike in the substrate (^{99}Tc: $t_{1/2} = 2.13 \times 10^5$ years, β^-: 294 keV)) that was transformed into ^{99}TcX by the plant (plant-formed species). For testing the homogenisation procedure they added another isotope, either $^{95m}TcO_4^-$ (^{95m}Tc: $t_{1/2} = 61$ days, γ: 204.1, 582.1, 835.1 keV) or $^{99m}TcO_4^-$ tracers (^{99m}Tc: $t_{1/2} = 6.02$ h, γ: 140.6 keV). The tracing of the different isotopes allowed them to correct for procedural losses and redox conversions. Similar radioisotope matches can be obtained for other elements, e.g., Mn, Co, Se, Sn, Sb and I.

In some cases researchers use a 'substitute' tracer, i.e., a radiolabel foreign to the compound, such as ^{125}I-labelled (^{125}I : $t_{1/2} = 59.4$ days, decay mode: electron capture) or a ^{11}C-labelled (^{11}C : $t_{1/2} = 20.3$ min, positron emitter) methyl group on a protein.

A small group of elements have natural radioisotopes. The most important ones are ^3H, ^{14}C, ^{40}K, ^{210}Pb, ^{226}Ra, ^{232}Th, ^{235}U and ^{238}U, all of which have very long half-lives, requiring special low radioactivity measurement facilities.

4 RADIONUCLIDE MEASUREMENTS [5]

As a rule radioactivity measurements will have to be normalised to equal counting times, corrected for decay during counting and decay against the counting of the reference.

The mathematical equation to correct for decay during counting ($t_{\text{counting time}}$), for decay versus time zero ($t_{\text{waiting time}}$) and to normalise per time unit is the following:

$$\text{Normalised radioactivity} = \text{measured radioactivity} \times \frac{\lambda \exp(+\lambda t_{\text{waiting time}})}{1 - \exp(-\lambda t_{\text{counting time}})}$$

with λ being the decay constant ($\lambda = 0.693/t_{1/2}$), $\exp(+\lambda t_{\text{waiting time}})$ the correction for radioactive decay between the arbitrary time zero and the start of the measurement, and $1 - \exp(-\lambda t_{\text{counting time}})$ the correction for decay during the measurement. The time unit of the normalised radioactivity will be that of the decay constant, e.g., when the half-life is expressed in hours, then this equation gives the normalised radioactivity per hour.

Besides the measurement of the radioactive sample, it is necessary to measure the radioactivity due to the background radioactivity of the surroundings and the electronic noise of the equipment. The net radioactivity is that measured for the sample minus that of the background. This background activity is always present and due, among others, to naturally occurring ^{40}K in concrete and other building materials, to remnants of fission products from fall-out of atomic bombs and/or nuclear disasters, and to electronic noise.

The counting time will be adjusted to the radioactivity of the sample (A^*) and to its ratio versus the background radioactivity (B). The minimum is a counting rate equal to that of the background measured during the same time, but if feasible a counting rate ten times that of the background is advisable. The precision of the counting rate depends on the counting statistics of the sample and on that of the background measurement. It is calculated in the following way:

The radioactivity A^* has an error due to counting statistics equal to $\sqrt{A^*}$ (square root of A^*), i.e., with a confidence level of 1 σ (65 %) and twice this value for 2 σ. The net radioactivity has a counting error equal to $\sqrt{(A^* + B)}$ (square root of the sum of the activity measurements A^* and B) with a confidence level of 1 σ (65 %) and twice this value for 2 σ. Relative counting errors exceeding 30 % are the upper limit.

This section will consider detection of γ emitters, and that of positron and electron emitters.

4.1 Gamma-detection

4.1.1 NaI(Tl) scintillation detector

The most important detector for γ counting, offering a high sensitivity, is the thallium activated NaI crystal scintillation detector. The detection is based on the fluorescence produced in the NaI (Tl activated) monocrystal, coupled to a photomultiplier, a power supply and an amplifier–analyser system. When the ionising radiation passes through the scintillator it produces photons in the UV–VIS range. The number of photons produced is proportional to the energy absorbed in the scintillator. Light falling on the photomultiplier is converted into a number of electrons that is proportional to the number of photons falling on the photocathode. The final electrical pulse is amplified and analysed. It is proportional to the energy of the detected γ-ray.

The resolution of the NaI(Tl) detector is about 7 % (full width at half-maximum (FWHM) of the 662 keV photopeak of ^{137}Cs. This means that this γ-peak is completely resolved from γ-rays with energies below 562 keV and above 762 keV, at equal intensity of the γ-peaks.

4.1.2 Semiconductor detectors

In a semiconductor the atoms are arranged closely together in a crystal lattice. At absolute temperature (0 K) the electrons fill up completely the lowest energy levels, called the valence band. At any other higher temperature there will be some thermal excitation of electrons from the valence band to the conduction band, leaving some empty places or 'holes', carrying a positive charge. The electron in the conduction band is free to move and if an electrical field is applied, it moves towards the positive electrode. When radiation is absorbed in the crystal electron–hole pairs are created and the collection of these charge carriers gives rise to an output signal proportional to the amount of energy absorbed. The detection efficiency of a semiconductor can be equal to or an order of magnitude lower than that of an NaI(Tl) scintillation detector, depending on the size of the crystal. The resolution, however, is superior. The FWHM of the 1.33 Mev photopeak of ^{60}Co is 2 keV or below. This means that γ-energies of 4 keV below or 4 keV above this energy are completely resolved. The most common semiconductor detectors are made of high purity germanium.

4.2 Beta detection

4.2.1 Liquid Scintillation Detector

In case of β^--emitters a special type of detector is needed, because the shielding of, e.g., a solid scintillator, prevents most of the radiation from penetrating. Therefore the radioactivity is intimately mixed with the scintillator, to which the energy of excitation is transferred. The scintillation vial is 'viewed' by one or two photomultipliers, power supplies and an amplifier–analyser system. The efficiency of detection is very high, and can reach 100 % for very low energy β-emitters, e.g., tritium. This type of measurement, however, is very prone to quenching of the fluorescence depending on the composition of the aqueous phase. There exist, however, adequate methods to correct for this drawback.

4.3 Autoradiography

This method is suitable for the detection of β-particles and γ-rays in flat samples, e.g., in a gel used for electrophoresis or a section of organic tissue cut with a microtome in cytology.

There exist two different techniques: the classical autoradiography, using X-ray films and since the late seventies autoradiography using phosphor technology.

4.3.1 X-ray film autoradiography

Provided that the X-ray film makes close contact with the flat sample, autoradiography makes excellent spatial resolution. The exposure of the film may take many weeks. Silver halide crystals in the emulsion respond directly to the β-particles and γ-rays emitted from the sample. Each emission converts several silver ions from a particular silver halide crystal to silver atoms to produce a stable latent image. When the film is subsequently developed these few silver atoms catalyse the reduction of the entire silver halide crystal (grain) to metallic silver to produce an autoradiography image of the radioisotope distribution. Under carefully controlled conditions, a densitometer can be used to quantify image formation. The dynamic range of the film is limited to about two orders of magnitude, and is complicated by the characteristic sigmoidal density *versus* log exposure response curve [10].

In order to enhance the effect an intensifying screen can be placed at the opposite side of the film. Use of, e.g., a hyperfilm, i.e., a plastic base with an emulsion on both sides, sandwiched between the sample and the intensifying screen, exposes the film not only to the direct radiation but also to the fluorescence emitted by the intensifying screen when hit by remaining radiation after crossing the film. Atoms of the intensifying screen are excited through this radiation, which is followed by fluorescence and conversion of silver ions to silver atoms.

4.3.2 Phosphor imaging [10]

Phosphor imaging is a newer (late seventies) and faster method, replacing the traditional X-ray films (and eliminating developing chemicals) by reusable storage phosphor screens that are 10 to 250 times more sensitive than film. The sample is placed in close contact with a phosphor screen for hours up to weeks, depending on the nature (β^--, γ-rays), the half-life and the number of radioactive atoms. The atoms of the screen that receive the radiation of the radioisotopes are excited. At the end, the screen is scanned by a laser beam. As the screen is scanned luminescence is collected from the excited areas by a fibre optic bundle, channelled to a photomultiplier and converted to an electrical signal. The amount of light is proportional to the amount of radioactivity in the sample in a wide linear dynamic range, covering five orders of magnitude and a sensitivity 10 to 250 times that of X-ray film. This phenomenon can be digitised and results in a two-dimensional picture of the location of the radioactivity. In the case of gelelectrophoresis this picture is then combined with that of the proteins, made visible through staining (e.g., silver staining, or Coomassie blue).

The results give a quantitative estimate of the presence of the trace element species.

5 GENERAL SOURCES OF ERROR IN RADIOACTIVITY MEASUREMENT

The proper function of the detector must be under adequate control to maintain stability and detect deviations. Although the dynamic range of the detection is very large, there is a practical limitation due to deviation from normality at high counting rates. A change in geometry (putting the sample at a greater distance from the detector) is often sufficient to eliminate errors of this type.

A major, unexpected source of error in using radiotracers can be the presence of a radioactive impurity. This has once been the situation for a ^{51}Cr ($t_{1/2} = 27.7$ days, γ: 320 keV) source provided by a commercial supplier. At the time of delivery only the 320 keV γ peak of ^{51}Cr could be measured on the intrinsic Ge detector, but after decay of about 280 days the radioactivity of this source was reduced 1000-fold and became equal to that of the ^{57}Co contaminant ($t_{1/2} = 271.8$ days, main γ 121.6 keV), which was originally masked by the Compton radiation caused by the ^{51}Cr 320 keV γ ray. A regular check with a high resolution Ge detector of the radiotracer in different fractions precludes such errors.

6 APPLICATION OF RADIOTRACERS TO SOLVE SPECIFIC CHEMICAL SPECIATION PROBLEMS

As said previously the procedures to separate the species are the same as those in use with most of the other analyte detection systems described in this book, but the compounds are measured through the radioactivity emitted by the labelled species, on the assumption that they underwent complete isotopic exchange with the nonradioactive share of the molecules. In contrast to most of the other systems, the detection of the radioactivity occurs mostly off line, which means the technique is nonhyphenated. The separation techniques that are liable to profit most from the ease of radioactivity detection are liquid chromatography, liquid–liquid extraction, ultrafiltration and gel electrophoresis. Applications in method development are documented in the life and environmental sciences. A minor share of applications concerns species that carry natural radioactivity, or radioactivity due to nuclear fission experiments or nuclear disasters.

6.1 Life sciences

Radiotracer examination of a particular step of an analytical procedure is a most useful tool in method validation. Hereafter follow a couple of examples.

A very delicate step in the measurement of arsenic by hydride generation-atomic absorption spectrometry (HG-AAS) is the reproducible and quantitative on-line transformation of the arsenic compounds from the aqueous aliquots into hydrides by $NaBH_4$ in the reaction coil followed by removal of the generated hydrides in a gas–liquid separator. In the coil the reduction of the traces of arsenate, arsenite, monomethylarsonic acid (MMA) and dimethylarsinic acid (DMA) can be adequately led to completion. In order to evaluate two different gas–liquid separators used in arsenic speciation by HG-AAS, van Elteren *et al.* [11] applied an ^{74}As(III) spike to establish the yield of hydride generation with respect to both gas–liquid separators and concluded that only one of the two types was satisfactory. The hydride procedure prior to gas–liquid separation was essentially quantitative for all species investigated, but only the 'classical separator' in which gas and liquid are separated by gravity was reliable to strip the arsine completely from the solution, independent of back-pressure. The second type of gas separator where gas diffuses through a permeable tube did not allow the complete stripping from the arsine and back-pressure worsened things.

An area that is prone to many artefacts is that of the metalloproteins in biological fluids and tissues. The easiest case is without any doubt that of a covalently bound metal. Nevertheless, the tedious optimisation study of the separation parameters, retention behaviour and recovery of metalloprotein complexes in serum by SEC was greatly facilitated by using metalloprotein complexes labelled with radiotracers (^{59}Fe: $t_{1/2} = 44.51$ days, γ: 1099 and 1291 keV and ^{65}Zn: $t_{1/2} = 244.3$ days, γ: 1115.5 keV) [12].

Another example concerns the separation of the very labile, noncovalently bound protein–vanadium complexes in human serum. De Cremer *et al.* [8] went through a long process of finding a suitable column and elution media for ^{48}V-labelled proteins (^{48}V $t_{1/2} = 15.98$ days, γ: 984 keV) that guaranteed the preservation of the original species. Ultrafiltration was a most useful tool for preliminary investigations about the stability of the complexes in the various media tested, because it allowed in a simple procedure to find out if the radioactivity remained with the proteins, which are bigger than the cut-off of the filter, or dissociated from the protein in that particular medium and ended up in the filtrate [13]. Similarly during the chromatographic experiments, it was easy to find out if the vanadium eluted with the expected mass of the protein in case of SEC, or as free vanadate. It could be shown that affinity chromatography was not applicable because the medium immediately destroyed the vanadium–protein binding.

One step beyond the development and optimisation of analytical methods with the aid of a radiotracer, is its use for *in vitro* and *in vivo* studies, to determine the distribution of added tracer to a living system or to a particular compound. Besides testing the procedures on 'real samples' it also provides basic knowledge on the behaviour of the species in a living system, which paves the way to subsequent 'cold' studies. The terminology 'cold' and 'warm' designates respectively 'stable' and 'radioactive' isotopes.

In vitro experiment consists in adding a radiotracer, e.g., ^{51}Cr radiotracer as $CrCl_3$ in physiological medium to serum, leave it to incubate for, e.g., 1 day at 37 °C. *In vivo* labelling is then, e.g., the intraperitoneal injection of ^{51}Cr radiotracer as $CrCl_3$ in physiological medium into a rat, wait an appropriate time to allow the compound to distribute over the organism, and measure the absorbed radiolabel in different body compartments (blood, soft and hard tissues). In both cases it is important to study under which form the ^{51}Cr is present. Of particular interest will be the fraction of ^{51}Cr bound to proteins. Ultrafiltration, gel filtration and ion exchange chromatography are the relevant tools to use [14]. The radiolabelling allowed to study the kinetics of newly added Cr to be investigated. Such an experiment with serum reveals how in case of Cr the ^{51}Cr binds mainly to transferrin and to a lesser extent to albumin [15].

A warning is needed. The application of a compound, whether *in vivo* or *in vitro*, whether radioactive or not, to study the distribution of added tracer to a living system or a particular substance can entail many artefacts. During the initial step of incubation, the tracer should be enabled to become part of the system or medium so as to reach equilibrium. Care should be taken to administer doses reflecting physiological levels of the element, so that fortuitous linkages to those components with many binding sites are avoided. Whenever feasible, carrier-free radioisotopes should be administered.

Radiotracer labelling has been widely used to study the behaviour of many essential and toxic elements in body fluids and tissues. Here follows a number of examples to illustrate the potential of radiotracer method.

Recently a study was conducted to determine the behaviour of $^{114m}InCl_3$ intraperitoneally injected in rats (^{114m}In : $t_{1/2} = 49.5$ days, γ: 192 keV), how it distributed over the body, and what kind of protein bound In species could be discerned in the different body fluids and tissues [16]. It may be interesting to follow the complete scheme, starting with the *in vivo* organ distribution. The results are given in Table 5.12.1, listed as percentage of the radioactivity per gram organ (wet weight) and percentage per organ (wet weight). Large amounts of indium are stored in kidney, liver and spleen. This is typical for ionic indium. Other studies applying colloidal indium oxide, reported that indium accumulates preferentially in liver, spleen and the reticuloendothelial system [17].

Table 5.12.1. *In vivo* organ distribution of ^{114m}In in rats expressed as per cent per gram organ of total activity in all measured organs and as per cent per organ of total activity in all measured organs [16]. Reproduced by permission of the Royal Society of Chemistry.

Organ	% per g organ	% per organ
Liver	16.5 ± 4.5	69.2 ± 7.3
Kidney	17.1 ± 2.8	10.8 ± 2.8
Heart	2.7 ± 0.5	0.7 ± 0.3
Testes	2.8 ± 0.6	2.3 ± 0.6
Thyroid glands	3.7 ± 0.5	0.5 ± 0.1
Lung	4.4 ± 1.1	2.0 ± 0.9
Spleen	21.8 ± 2.1	4.2 ± 0.6
Stomach	6.6 ± 2.1	3.0 ± 1.1
Bladder	13.4 ± 5.6	0.9 ± 0.3
Small intestine	6.1 ± 2.9	6.4 ± 3.1
Bone	4.9 ± 1.0	[a]

[a]Total bone mass not measured [16].

Table 5.12.2. *In vivo* intracellular distribution of ^{114m}In in kidney, liver and spleen. **A**, pellet of nucleus; **B**, pellet of mitochondria; **C**, pellet of lysosomes and peroxisomes; **D**, pellet of ribosomes and small vesicles; **E**, homogenate containing cytoplasma. Results denoted as percentage of total activity within each tissue homogenate [16]. Reproduced by permission of the Royal Society of Chemistry.

	Kidney	Liver	Spleen
% in A	9.7 ± 4.1	6.3 ± 0.3	3.4 ± 1.4
% in B	22.5 ± 1.0	14.1 ± 2.2	18.7 ± 2.9
% in C	5.7 ± 2.6	1.7 ± 0.5	7.7 ± 1.6
% in D	15.3 ± 1.0	4.6 ± 0.4	11.8 ± 1.5
% in E	46.8 ± 3.6	73.3 ± 3.3	58.4 ± 5.3

One step further, the intracellular distribution of the trace element in tissues of animals was investigated after *in vivo* treatment with a radiolabel. The example with indium continues. Van Hulle *et al.* [16] studied the partition of nontoxic trace doses of carrier free ^{114m}In as $InCl_3$ in the Wistar rat. Table 5.12.2 shows the results for the *in vivo* intracellular distribution of ^{114m}In in kidney, liver and spleen by differential centrifugation. In all cases the cytosolic fraction accounts for the highest activity of ^{114m}In, followed by the mitochondrial fraction. Some reflections have to be made about the outcome of such intracellular distribution studies. Although the fractions are described as nucleus, mitochondria, lysosomes,. . ., none of these fractions contains solely one specific organelle. The presence of other organelles can be determined by the analysis of their specific enzymes, the so-called marker enzymes. The distribution of the organelles has to be carried out separately on a small aliquot of each fraction, the remaining part being used for trace element analysis. At the end, a matrix of x equations with the x unknown concentrations of the trace element in each organelle has to be solved. When this work is undertaken using nonradiolabelled tissue, significant, even unsurmountable, problems due to the possible contamination of the cell fractions with the trace element under study by the many reagents and recipients can be anticipated. As the working conclusions may be approximately the same as those drawn from the simplified model of supposedly pure organelle fractions, no effort was made to work in the strict way. Working with *in vivo* labelled tissue, where a complete isotopic exchange with the natural isotopes has been achieved, circumvents the problem related to contamination hazards. The only instance where it may be feasible to work with a nonlabelled compound, is when it does not occur in nature and cannot be formed in the organism. For instance the study of the distribution of cisplatin is feasible in the cold way. To add to the degree of difficulty of this work, little is known about the preservation of the original trace metal – organelle binding throughout the tedious intracellular fractionation, a problem common to 'cold' and 'warm' procedures. It is not surprising that apparently very few attempts have been made to study speciation of 'cold' trace elements in this way.

When studying blood, the first aspect to consider is the partition of the species between serum and packed cells. In the case of indium the major share of the ^{114m}In activity in the blood is located in serum, 90.2 ± 4.1 % of total indium content of blood. Next, serum is subjected to different types of chromatographic separations. Figure 5.12.2 shows the elution profile of ^{114m}In in rat serum separated by SEC. There is only one peak of indium and it coincides with the elution of transferrin. Figure 5.12.3 shows the chromatogram of the *in vivo* speciation of ^{114m}In in blood lysate. The In elutes slightly ahead of haemoglobin, so further investigations are needed to determine whether or not haemoglobin is the main carrier of indium. The separation of rat urine

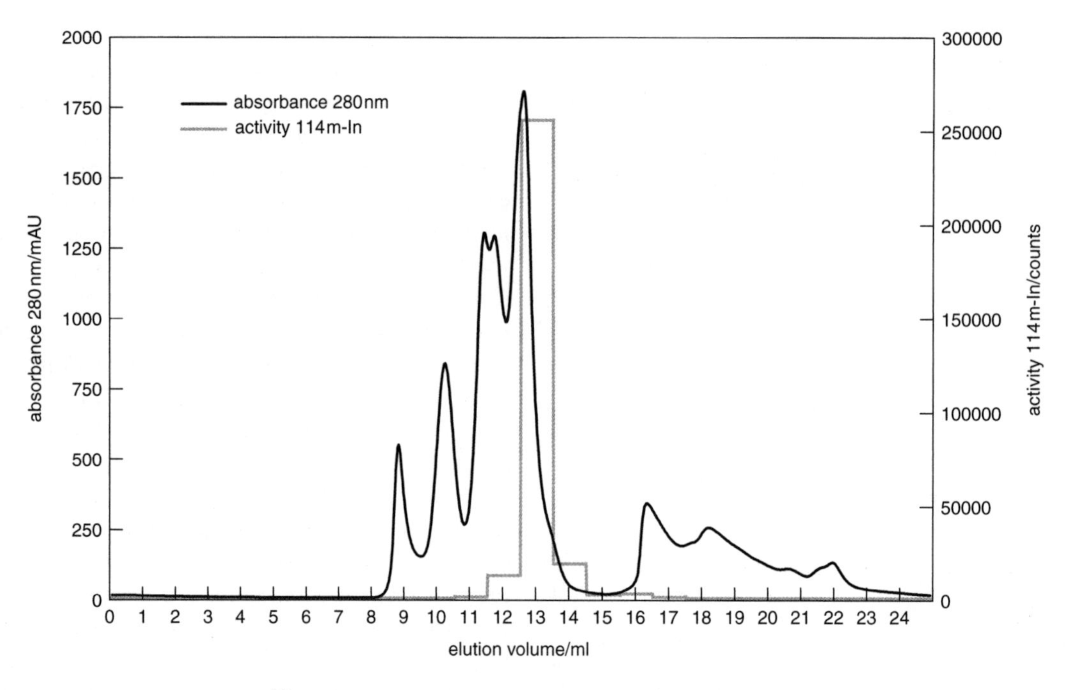

Figure 5.12.2. Elution profile of [^{114m}In]InCl$_3$/rat serum separated with size exclusion chromatography on Asahipak 520-GS 7G. Buffer: 10 mM hepes +5 mM NaHCO$_3$ + 0.15 M NaCl, pH 7.4; UV absorption at 280 nm [16] (reproduced by permission of The Royal Society of Chemistry).

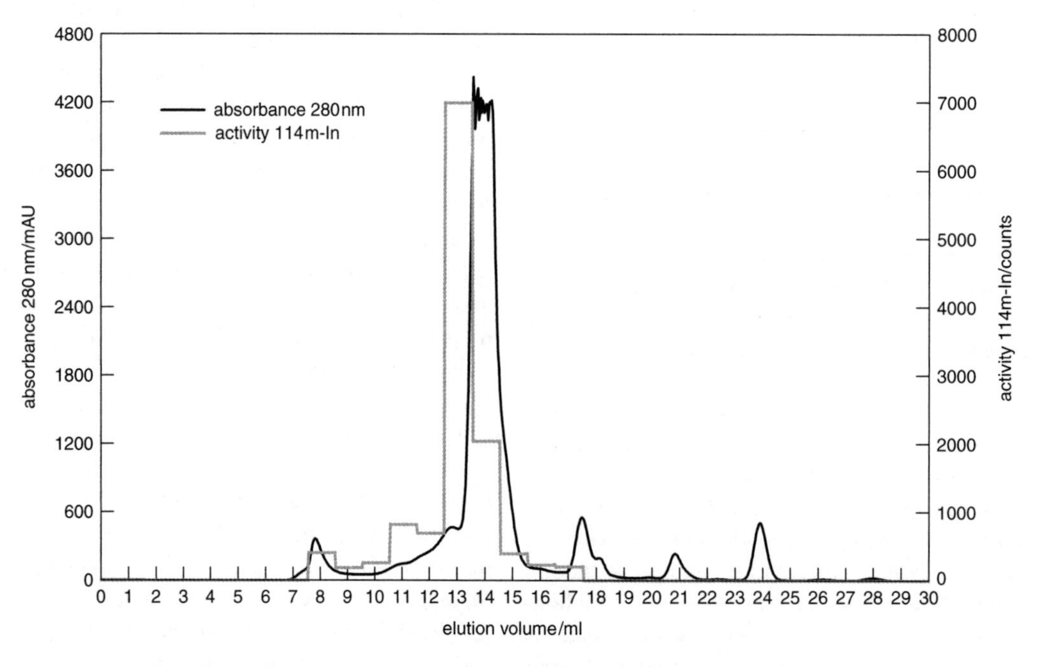

Figure 5.12.3. Elution profile of [^{114m}In]InCl$_3$/rat blood cell lysate separated with size exclusion chromatography on Superose 12 HR. Buffer: 15 mM hepes + 0.15 M NaCl, pH 7.2; UV absorption at 280 nm [16] (reproduced by permission of The Royal Society of Chemistry).

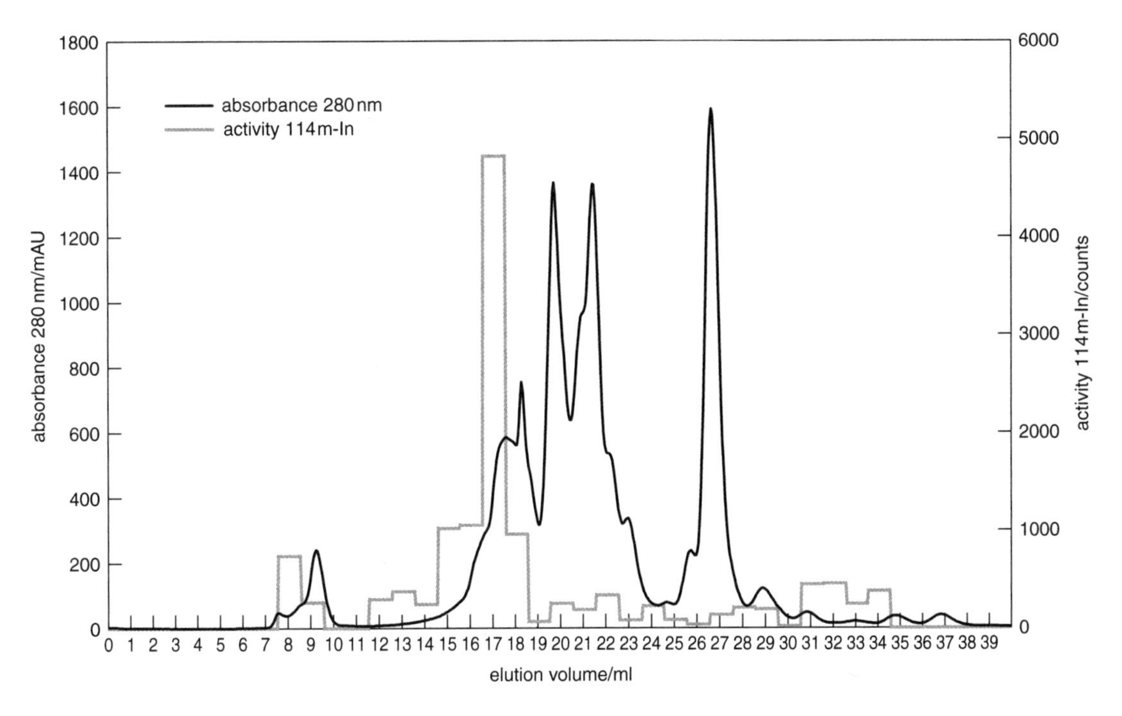

Figure 5.12.4. Elution profile of $[^{114m}In]InCl_3$/rat urine separated with size exclusion chromatography on Superdex Peptide. Buffer: 15 mM hepes +0.15 M NaCl, pH 6.5; UV absorption at 280 nm [16] (reproduced by permission of The Royal Society of Chemistry).

by SEC is shown in Figure 5.12.4. In urine, indium is mostly bound to the low molecular mass fraction, with a maximum corresponding to a molecular mass of 300–4000 Da. The results of liver cytosol by SEC are shown in Figure 5.12.5. It appears that indium preferentially or exclusively binds to the high molecular mass (HMM) fraction. The large amount of ^{114m}In might be partly due to the presence of serum transferrin in the organ after homogenisation. As the elution peak is much broader than expected, there is ground to think that there is one or more other HMM components present in the fractions. Only in the case of kidney cytosol (figure not shown), does a considerable amount of indium appear in the low molecular weight (LMM) fraction. Further research consists in identifying the HMM and LMM compounds carrying the element In.

Very extensive research on the *in vivo* behaviour of radiotracer Se in Wistar rats has been done by Behne and coworkers [18] over the last 15 years using ^{75}Se tracer with very high specific activity allowing the determination of the species in the pmol to fmol range. The wide array of biochemical separation techniques, combined with γ-scintillation spectrometry and also autoradiography in the case of gel electrophoresis allowed them to reveal most interesting findings about the diverse Se species in the organism, as well as about their transformation in the living system. They concluded that all of the selenium present in the mammalian organism is protein bound, and therefore speciation is mostly concerned with the determination of the different selenium-containing proteins. A method was developed for the determination of selenocysteine and selenomethionine in the selenium-containing proteins. The identification of specific selenoproteins was achieved by analysis of their selenoamino acid residues and by studies on their characteristics and possible biological functions. This is being followed by the development of methods for the quantitative analysis of the selenoproteins in question in the tissues of animals and man.

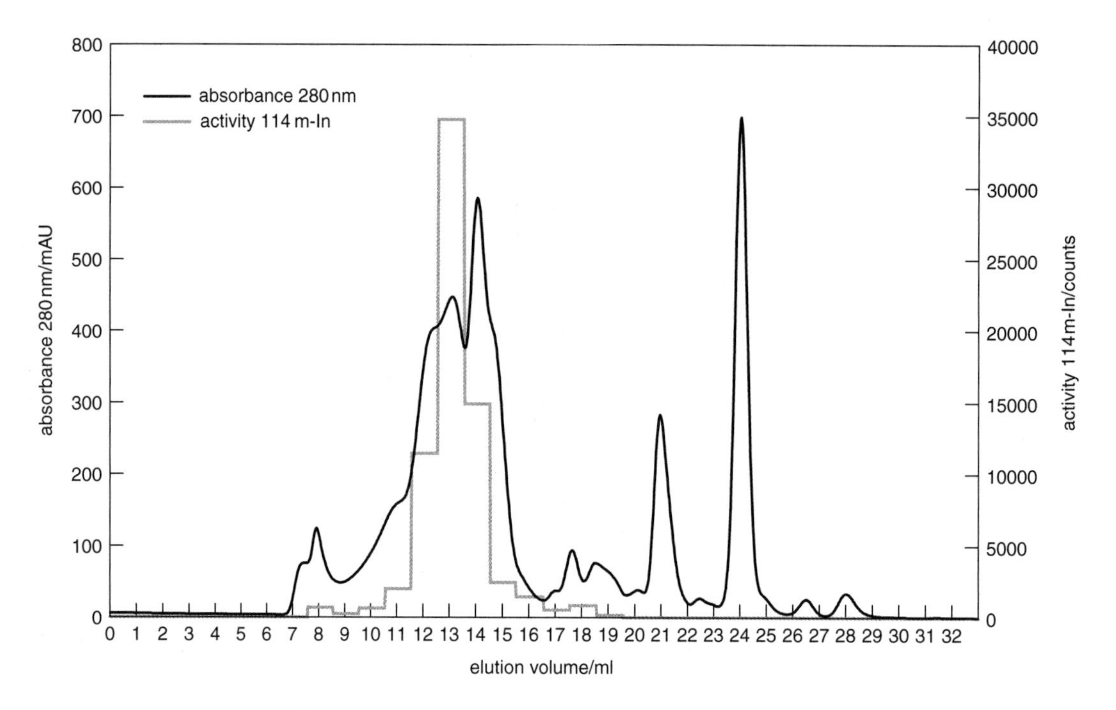

Figure 5.12.5. Elution profile of [^{114m}In]InCl$_3$/rat liver cytosol separated with size exclusion chromatography on Superose 12 HR. Buffer: 15 mM hepes +0.15 M NaCl, pH 7.2; UV absorption at 280 nm [16] (reproduced by permission of The Royal Society of Chemistry).

Method development for the identification of the different chemical forms of Se proteins in ^{75}Se-enriched yeast was successfully done by Chéry *et al.* using 2D electrophoresis, followed by phosphor imaging [19] (see Chapter 4.4).

Another study of which the successful outcome is solely due to the use of radiotracer, concerns the basal metabolism of intraperitoneally injected carrier-free ^{74}As-labelled arsenate in rabbits. [20]. It was very interesting to follow the process of the reduction of arsenate to arsenite, its binding to transferrin in serum, to haemoglobin in the red blood cells, and its transformation into monomethylarsonic acid, and next to dimethylarsinic acid, and the binding of these different components to tissue proteins. In healthy rabbits and also in man the addition of first one and then a second methyl group to the arsenic is a natural detoxification process. This process is, however, seriously inhibited when either toxic doses of inorganic arsenic (arsenate or arsenite) are administered, or in case of uraemia.

Therefore the next step consisted in studying the effect of toxic doses of arsenate and observe many variations in the metabolic pattern of inorganic arsenic [21]. In a further scenario the arsenate metabolism was studied in uraemic rabbits, showing the dramatic inhibiting effect of this pathology on arsenate metabolism, including the inhibitory effect on the detoxifying methylation mechanism transforming inorganic arsenic species into far less toxic methylated compounds [22].

6.2 Environmental sciences

In environmental sciences there is the search for the chemical species and where this appears to be impossible, it becomes the science of fractionation of elements, which means classification of an analyte or a group of analytes from a certain sample according to physical (e.g., size or solubility) or chemical (e.g., bonding, reactivity) properties [23].

The investigation of the different chemical forms of elements in natural waters, sediments and sludges is a matter of great difficulty. The species belong to chemically very different categories [24] and include different oxidation states (e.g., Cr(III) and Cr(VI)), inorganic compounds and their complexes (hydroxicomplexes, e.g., $Fe(OH)_3$, $Al_{13}O_4(OH)_{24}{}^{7+}$, complexes with $SO_4{}^{2-}$, Br^-, F^-, $CO_3{}^{2-}$, organic complexes, macromolecular compounds and complexes (e.g., the element being bound to humates, clays). To mimic the behaviour of elemental species as they occur in nature is not evident. It is difficult to create a laboratory scale system as complex as prevails in nature. Therefore the experimental set-up is usually simplified. The radiotracer must be of the highest specific activity and it has to be assumed that the tracer added in the ionic form will exchange with the natural system. In practice it seems that at least some exchange takes place, but that it is a slow process with a half-time of hours, up to days and weeks.

A radiolabel proved to be very useful for investigating the stability of Cr(III) and Cr(VI) species in water in a feasibility study to produce a reference material for intercomparison of speciation methodology. Unfortunately, only one useful radioisotope of Cr exists: ^{51}Cr ($t_{1/2} = 27.7$ days, γ: 320 keV). The absence of a specific isotope for each oxidation state is a problem common to any other conventional analytical method, where the measurement will detect any form of Cr present. The changes that might be anticipated in such a solution are the conversion of Cr(VI) into Cr(III) and the hydrolysis of Cr(III). Dyg *et al.* [25] studied the effect of different parameters on the stability of Cr(III) and Cr(VI) in water. These included a study of the stability as a function of time, possible losses caused by adsorption, temperature dependence and choice of the material of the container. The method used to differentiate between Cr(III) and Cr(VI) was based on the cationic behaviour of the Cr(III) and the anionic behaviour of Cr(VI) as chromate. In this work the Cr(III) and Cr(VI) were separated through extraction of Cr(VI) with Amberlite LA-1 or LA-2 diluted with isobutylmethylketone. The behaviour of the radiolabelled Cr(III) and Cr(VI) were checked in separate runs, with detection of the ^{51}Cr label.

Radiotracers in controlled laboratory experiments are a very tempting tool to study the behaviour of metal ions and other contaminants in natural waters, because they allow researchers in an uncomplicated manner to learn about the role of individual parameters on kinetics and sorption to particulate material. This was adopted by McCubbin and Leonard [26] for the element Th with the aid of ^{234}Th ($t_{1/2} = 24.1$ days, β^-: 198 keV, weak γ's) but the outcome of their experiments was only moderate as they concluded that caution should be exercised in extrapolating information obtained from investigations using radiotracers to predict trace element behaviour in natural waters.

Another interesting example deals with copper in the environment. The group of van Elteren investigated a chromatographic technique to investigate the lability of copper complexes under steady-state conditions using high specific activity ^{64}Cu ($t_{1/2} = 12.7$ h, γ: 511 keV) [27]. Along the same lines they examined the usefulness of solid-phase extraction (SPE) cartridges for copper speciation screening [28]. In the latter study five SPE cartridges were used to extract copper species from a sample. Each cartridge is expected to extract a specific (group of) copper species, depending on the chemical properties of the extracting phase. In theory this would lead to the extraction of free copper ions (Cu^{2+}) and labile inorganic copper complexes (like $CuCl^+$) by chelex cartridges, cationic species by SCX cartridges, anionic species by SAX cartridges and hydrophobic species by C18 and RP cartridges. However, in practice it may be possible that unwanted secondary interactions take place, making the cartridge less selective than desired. These artefacts were visualized by radiotracer studies. The composition of the test samples and calculated speciation with regard to Cu species at pH 6.0 are given in Table 5.12.3. The results are given in Figures 5.12.6–5.12.8, proving that secondary interactions must play a role in the extraction of copper species. Both Chelex and SCX cartridges are expected to retain free copper ions, while they should let negatively charged Cu–EDTA and Cu–TACTDD pass through. Indeed, it is found that free copper is almost completely retained on both

Table 5.12.3. Composition of test samples and calculated speciation with regards to Cu species at pH 6.0 [28].

Sample	Sample composition	calculated speciation
Test 1	1×10^{-6} mol L^{-1} Cu	76 % Cu^{2+} 21 % $CuCl^{+}$ 2 % $Cu(OH)_2(aq)$ 1 % $CuCl_2$
Test 2	1×10^{-6} mol L^{-1} Cu 7.5×10^{-7} mol L^{-1} EDTA	25 % Cu^{2+} 75 % $Cu{-}EDTA^{2-}$
Test 3	1×10^{-6} mol L^{-1} Cu 7.5×10^{-7} mol L^{-1} TACTDD	25 % Cu^{2+} or inorganic Cu 75 % $Cu{-}TACTDD^{-}$

EDTA: ethylenediaminotetraacetic acid, disodium salt.
TACTDD: 1,4,8,11-tetraazacyclotetradecan-5,7-dione.

cartridges as can be seen in Figure 5.12.6. However, Figure 5.12.7 and 5.12.8 indicate that part of Cu–EDTA and Cu–TACTDD complexes is also retained on these columns. In the absence of secondary reactions 25 % of the total copper should be retained, but the chelex cartridge retains more than 50 % and the SCX cartridge even more: about 70 % in case of Cu–EDTA and almost all copper in case of Cu–TACTDD. The SAX cartridge shows some secondary interactions as well. It retains 63 % of the total copper when only Cu^{2+} is present, while it should retain any cationic species. It seems to perform better with Cu–EDTA, as about 80 % of the total copper is retained when 75 % is expected. For Cu–TACTDD the performance is worse, only 42 % being retained. The C18 and RP cartridges show opposite behaviour with respect to free copper ions. While the C18 cartridge retains only 32 %, the RP cartridge retains more than 70 % of the total copper. The presence of Cu–EDTA reduces these values by a factor of two (17 % and 35 %, respectively), while with Cu–TACTDD similar responses are found, both cartridges retain about 50 % of the total copper.

In particular, free copper ions seem to be extracted effectively by cartridges that should not have any affinity for this species: SAX, C18 and RP. In general, these results make clear that secondary interactions must play a role in the extraction of copper species. With respect to this the C18 cartridge performs best,

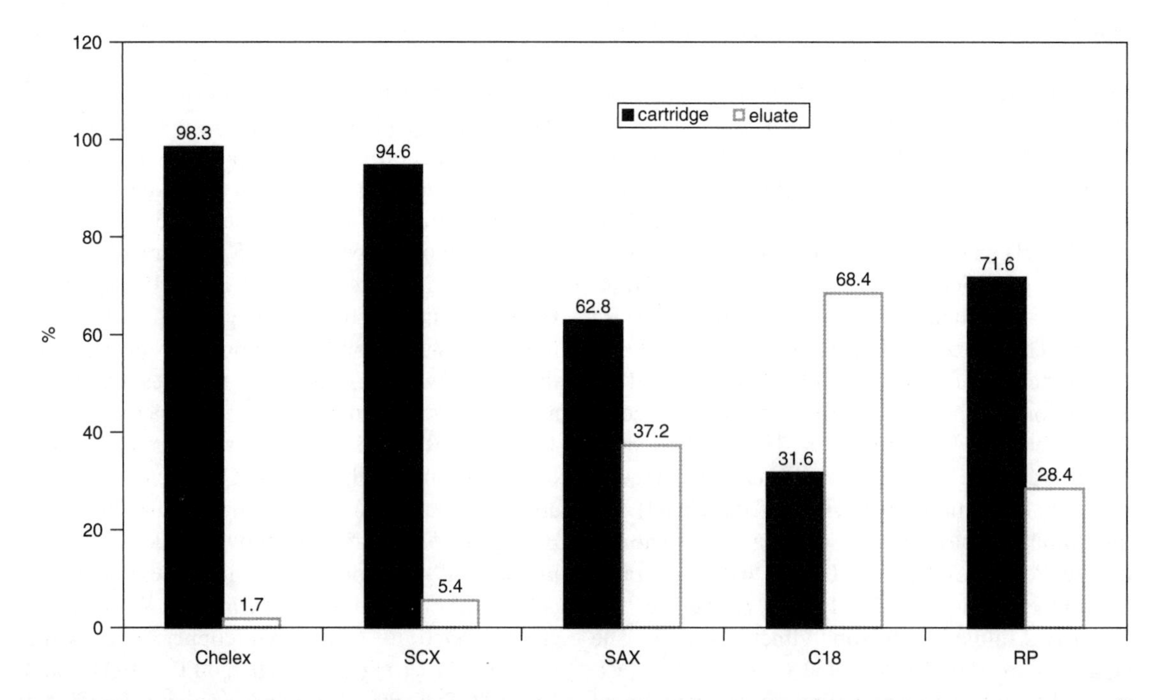

Figure 5.12.6. Extracted and eluted amounts of copper species (as percentage of total copper) of test sample 1, containing Cu^{2+} and inorganic Cu species [28] (reproduced by permission of Akadémiai Kiadó, Budapest).

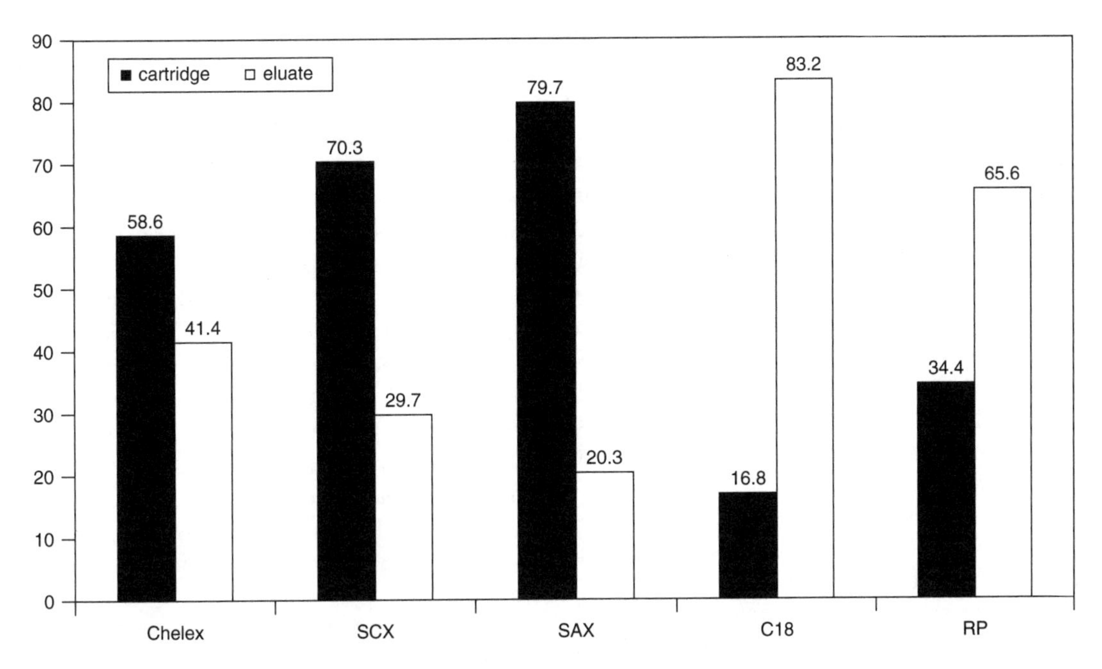

Figure 5.12.7. Extracted and eluted amounts of copper species (as percentage of total copper) of test sample 2, containing 75 % Cu–EDTA [28] (reproduced by permission of Akadémiai Kiadó, Budapest).

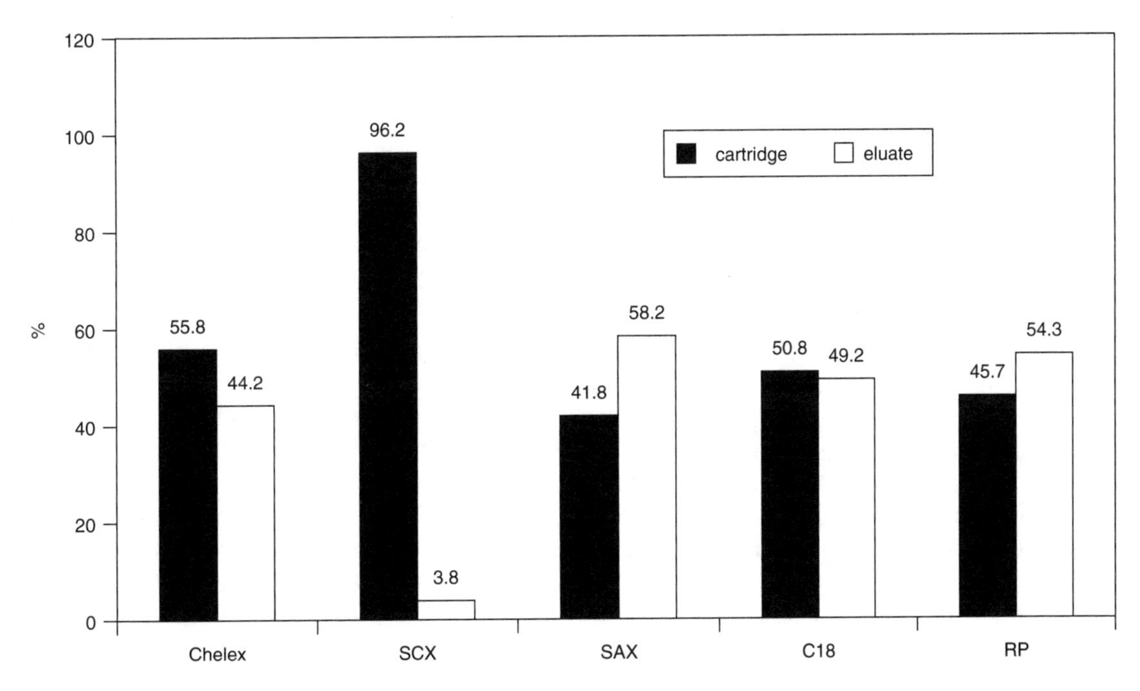

Figure 5.12.8. Extracted and eluted amounts of copper species (as percentage of total copper) of test sample 3, containing 75 % Cu–TACTDD [28] (reproduced by permission of Akadémiai Kiadó., Budapest).

followed by SAX and RP. The anionic species Cu–EDTA and Cu–TACTDD seem to undergo secondary interactions as well. They are not completely passed through Chelex, SCX, C18 and RP cartridges and Cu–TACTDD is not completely retained on the SAX cartridge. It seems that Cu–EDTA behaves more like an anionic species than Cu–TACTDD which has a more hydrophobic character. This study clearly shows the advantages of applying a highly specific activity radiotracer in speciation studies. The high specific activity of the radiotracer allowed the researchers to add only a negligible amount of mass to a sample, which ensures that initial chemical equilibria are maintained.

At the beginning of this section, fractionation was mentioned. Part of such work consists in applying sequential extraction steps. There exists a procedure for soils, sediments and sludges developed on a European scale in the frame of the EC Standards, Measurement and Testing programme (SM&T). These procedures do not provide information about chemical species, but fractionate the metals according to operationally defined selective dissolution of geochemical phases. Gilmore *et al.* [29] studied the readsorption and redistribution of lead in the SM&T sequential extraction procedure using a spike of ^{212}Pb tracer ($t_{1/2} = 10.64$ h, γ: 238.6, 300.0 keV).

The possibilities offered by radiotracers to study dynamic speciation are really unique. They allow researchers to study not only the equilibrium distribution of different species, but also the kinetics of their interconversion. This particular aspect of speciation analysis is nicely illustrated by the work of Achterberg *et al.* [30] on species kinetics and heterogeneous reactivity of dissolved copper in natural organic-rich freshwaters under steady-state conditions, *i.e.*, with minimum disturbance of existing equilibria, using high specific activity $^{64}Cu^{2+}$. Study sites with contrasting suspended particulate matter (SPM) characteristics were investigated. The analytical protocol allowed the differentiation between the following Cu species: SPM-associated Cu, dissolved reactive (free and labile) Cu, and organically complexed Cu. The data obtained were successfully evaluated by compartmental analysis, which showed the importance of organically complexed Cu in freshwaters, and the dominant role of the interactions between organically complexed Cu and SPM in SPM-rich water. The kinetic ^{64}Cu measurements indicated that the attainment of equilibrium between dissolved reactive and organically complexed Cu took about $<1-2$ h, and 4–15 h for the interaction between dissolved organically complexed and SPM-associated Cu. The kinetic study was augmented by voltammetric measurements of the dissolved (stable) Cu equilibrium speciation conditions in the natural waters. These measurements showed that the waters contained very low cupric ion concentrations ($10^{-12}-10^{-15}$ M), with more than 99.9 % of the dissolved Cu complexed by strong organic ligands (conditional stability constants: $10^{13.4}-10^{15.4}$).

Last but not least there are the extensive studies on the speciation of radionuclides in the environment [31]. The presence of trace amounts of fission radioisotopes such as ^{90}Sr, ^{99}Tc, ^{137}Cs and actinides in nature as a consequence of fall-out of the fission products of ^{235}U and ^{239}Pu used in nuclear weapon programmes, the storage of nuclear waste and also from the Chernobyl legacy (April 26, 1986) has been the subject of many studies. The Chernobyl accident can be considered as the ultimate global scale unauthorised radiotracer experiment of the past century, giving workers in the nuclear field ample opportunity to collect environmental samples to follow the pathways of certain elements in the geosphere. The relatively long half-lives (up to thousands of years) of many of the fall-out radionuclides are well suited for use in the study of these long-term processes.

The ultimate fate and effects of these long-lived radionuclides in the ecosystem are mainly dependent on their chemical forms, which determine their partitioning and chemical transport in various locations. These parameters together with the physical transport of water masses and associated suspended particles, play an important role in the availability of the radiation to living matter. Many studies cover this domain [32–37].

The assumption that the behaviour of the radionuclides can be inferred from their stable analogues does not necessarily hold true, as has been observed by Joshi [33] for the behaviour of Co, Cs and Pb in water from Lake Ontario. The concentrations of ^{60}Co, ^{137}Cs and ^{210}Pb are several orders of magnitude lower than those of the stable isotopes or members of the same subgroup of elements. Joshi also pointed out that the chemical forms of the stable analogues of several radionuclides in the Great Lakes have not yet been investigated. For the actinides, true stable analogues do not exist. This very important issue of the actinides in the environment will be dealt with in great detail in a special chapter of the next volume of the handbook.

Interesting radionuclide and mobility studies in Norwegian and Soviet soils were studied by Oughton *et al.* [34] 3 years after the Chernobyl accident. A sequential extraction procedure has been applied to study the speciation of Chernobyl-derived radionuclides (^{137}Cs and ^{90}Sr) in soils from Norway, and from Byelorussia and the Chernobyl region in the Ukraine. They used six different extraction media, starting with very mild (water), NH_4OAc (exchangeable), to NaOAc (carbonates), to $NH_2OH.HCl$ (easy reducible Fe/Mn oxides), H_2O_2 (oxidizable organic matter), to nitric acid (acid digestible), the remaining part being defined as residue. Figure 5.12.9 shows the relative distribution of ^{137}Cs ($t_{1/2} = 30.17$ years, γ: 662 keV) and Figure 5.12.10 for ^{90}Sr ($t_{1/2} = 28.64$ years, β^- : 546 keV) in the sequential extraction fractions for four types of soil samples collected in Norway. Most of the ^{137}Cs was strongly associated with soil components (retained in the residue), whereas ^{90}Sr was more mobile, up to 70 % being found in the easily extractable fractions. Amano *et al.* [36] have reported on the transfer capability of long-lived Chernobyl radionuclides from surface soil to river water in dissolved forms.

Sanada *et al.* [37] have studied the accumulation of ^{137}Cs, ^{90}Sr, $^{239+240}Pu$, and ^{241}Am in sediment of the Pripyat River in the exclusion zone of the Chernobyl nuclear plant and addressed the dissolution of these radionuclides from the sediment 12 years after the accident. The peak area of the concentrations of the radionuclides is

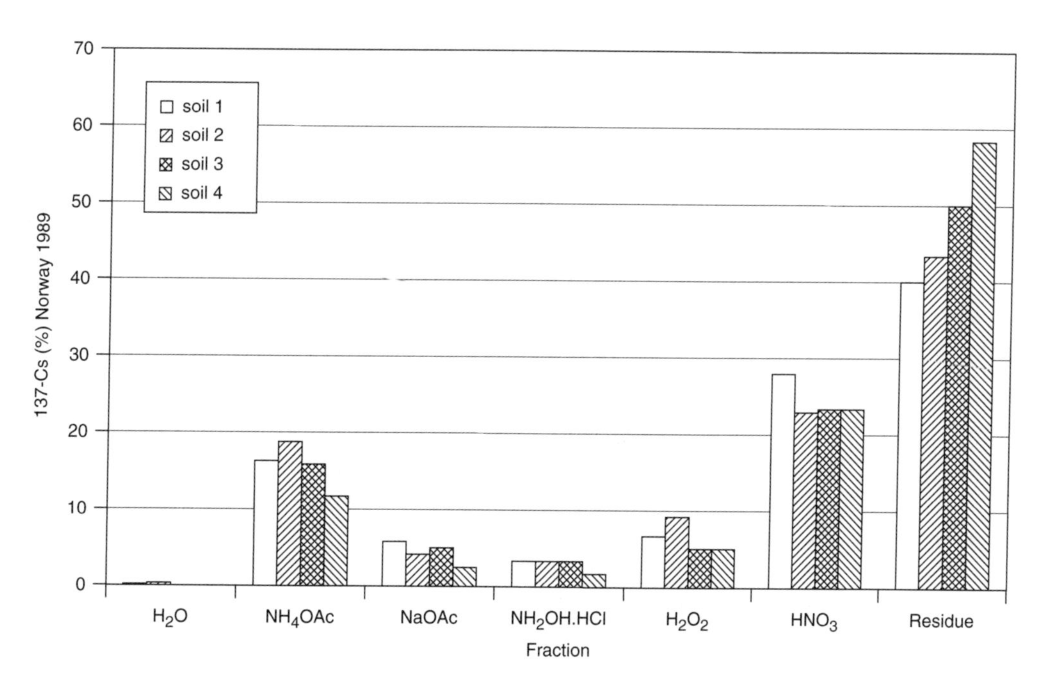

Figure 5.12.9. Relative distribution of ^{137}Cs in sequential extraction fractions. Samples ($n = 4$) were collected from Lierne, Norway, July 1989 [34] (reproduced by permission of The Royal Society of Chemistry).

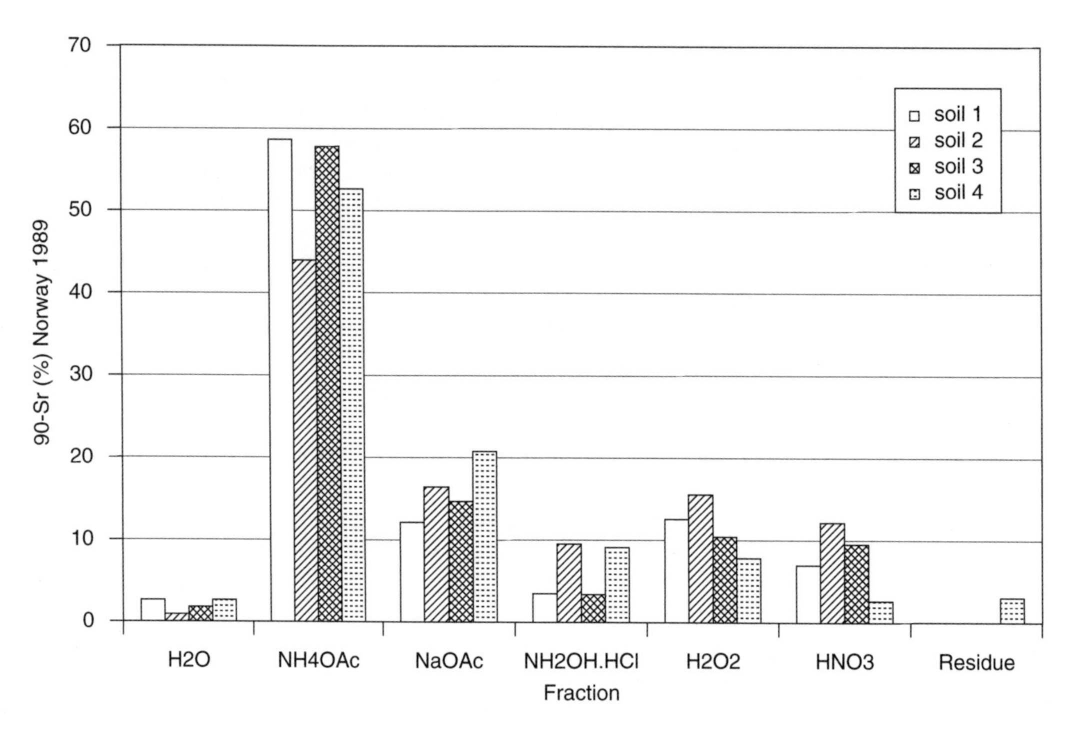

Figure 5.12.10. Relative distribution of ^{90}Sr in sequential extraction fractions. Samples ($n = 4$) were collected from Lierne, Norway, July 1989 [34] (reproduced by permission of The Royal Society of Chemistry).

at 20–25 cm depth. The composition was compared with that of the radioparticles released during the accident. An analysis using a selective sequential extraction technique was applied for the radionuclides in the sediments. It comprised acid soluble, reducible, oxidisable and residual fractions. The results are given in Figure 5.12.11. ^{137}Cs and $^{239+240}$Pu were concentrated in the residual phase, suggesting an effective fixation of these particles after the dissolution of the fuel particles to ambient environmental matrices. The higher distribution of ^{90}Sr over the acid soluble, reducible and oxidisable phases suggested a higher potential mobility for this element compared with ^{137}Cs and $^{239+240}$Pu. These results suggest that the possibility of release of ^{137}Cs and $^{239+240}$Pu from the bottom sediment was low compared with ^{90}Sr. The authors concluded that the potential dissolution and subsequent transport of ^{90}Sr from river bottom sediment should be taken into account with respect to the long-term radiological influence on the aquatic environment.

7 CONCLUSIONS

The previous examples are not exhaustive. They are meant to give an idea of how radiotracers are most handy in method development for elemental speciation analysis and in research aiming at metabolic and environmentally related studies.

Analytical chemists have been very resourceful in developing separation and measurement systems for many element species. Every time a proper analytical method has been developed, it requires validation to make sure that the data produced by different laboratories are comparable and traceable [38]. Radiolabelled species are most useful to assist in controlling and validating separate steps or the whole procedure. Unfortunately the use of radioactivity has fallen into disgrace with the general public and the politicians, so that many research centres have had to close their nuclear facilities. This slows down and may even halt further use of radiotracers for method development

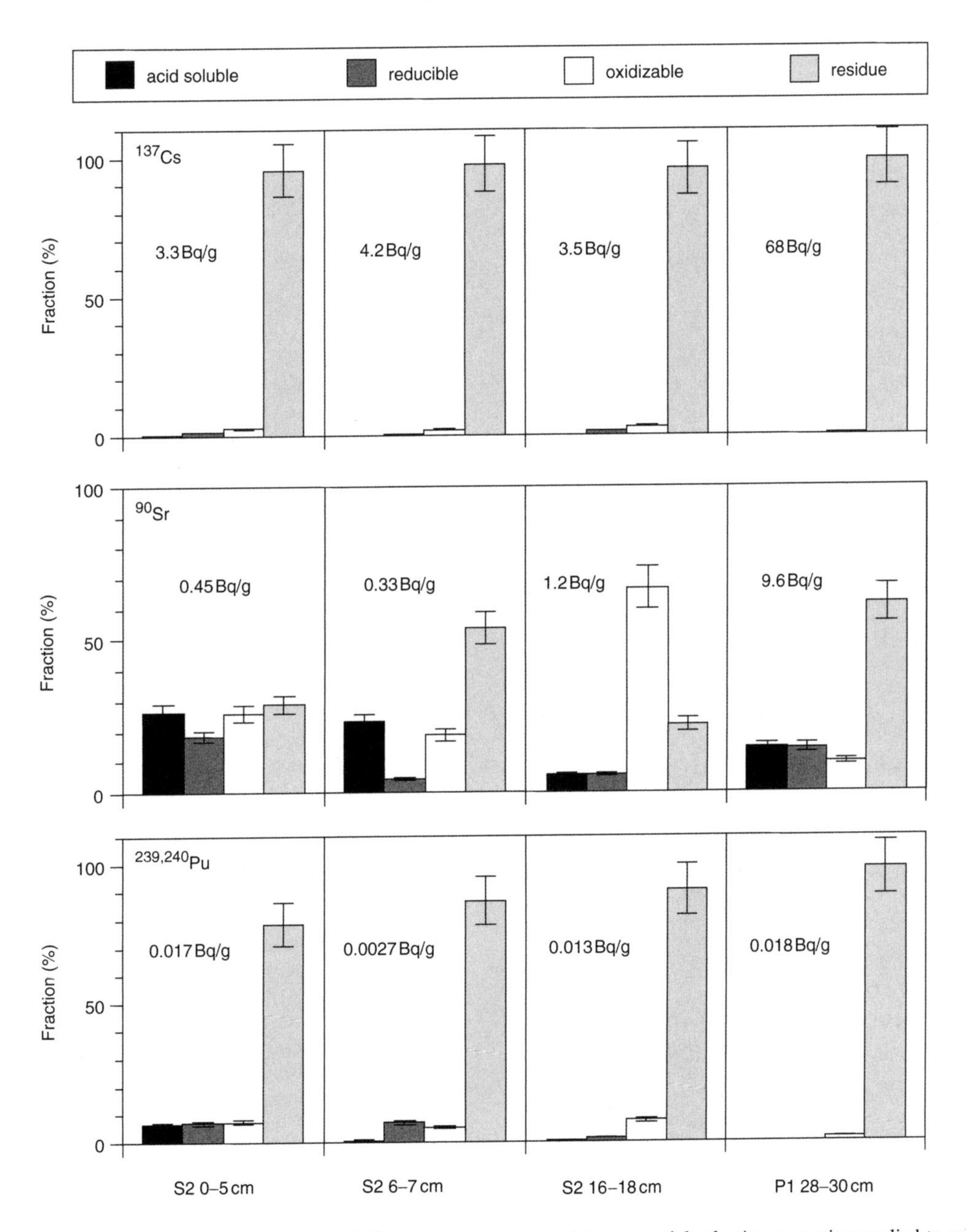

Figure 5.12.11. Fractions of radionuclides over different phases segregated by sequential selective extraction applied to selected samples of bottom sediment of the Pripyat river near the Chernobyl Nuclear Power Plant. The value is total radionuclide concentration [37] (Reprinted from *Applied Radiation and Isotopes*, Vol. 56, Sanada *et al.*, pp. 751, 2002, with permission from Elsevier Science).

in elemental speciation analysis and for metabolic and environmentally related speciation studies. Pessimism, however, is not a good attitude for further development.

Overall, the integration of the use of radiotracers offers an invaluable, complementary, analytical tool in elemental speciation analysis in the life and environmental sciences.

8 REFERENCES

1. Cornelis, R., *Analyst*, **117**, 583 (1992).
2. Friedlander, G., Kennedy, J. W., Macias, E. S. and Miller, J. M., *Nuclear and Radiochemistry*, 3rd edn, Wiley-Interscience, New York, 1981.
3. Šlejkovec, Z., van Elteren, J. T., Byrne, A. R. and de Goeij, J. J. M., *Anal. Chim. Acta*, **380**, 63 (1999).
4. Adams, F. and Dams, R., *Applied Gamma-ray Spectrometry*, Pergamon, Oxford, 1970.
5. Brune, D., Forkman, B. and Persson, B., *Nuclear Analytical Chemistry*, Chartwell-Bratt, 1984.
6. Parent, M., Strijckmans, K., Cornelis, R., Dewaele, J. and Dams., R., *Nucl. Instrum. Methods. B*, **86**, 355 (1994).
7. van Elteren, J. T., Kroon, K. J., Woroniecka, U. D. and de Goeij, J. J. M., *Appl. Radiat. Isotopes*, **51**, 15, 1999.
8. De Cremer, K., Cornelis, R., Strijckmans, K., Dams, R., Lameire, N. and Vanholder, R., *J. Chromatogr. B*, **757**, 21 (2001).
9. Harms, A. V., van Elteren, J. T., Wolterbeek, H. Th. and de Goeij, J. J. M., *Anal. Chim. Acta.*, **394**, 271 (1999).
10. Johnston, R. F., Pickett, S. C. and Barker, D. L., *Electrophoresis* **11**, 355 (1990).
11. van Elteren, J. T., Das, H. A. and Bax, D., *J. Radioanalyt. Nucl. Chem.*, **174**, 133 (1993).
12. Raab, A. and Brätter, P., *J. Chromatogr. B*, **707**, 17 (1998).
13. De Cremer, K., De Kimpe, J. and Cornelis, R., *Fresenius, J. Anal. Chem.*, **363**, 519 (1999).
14. Cornelis R, R., De Kimpe, J. and Zhang, X., *Spectrochim. Acta, Part B*, **53**, 187 (1998).
15. Borguet, F., Cornelis, R., Delanghe, J., Lambert, M. C. and Lameire, N., *Clin. Chim. Acta*, **238**, 71 (1995).
16. Van Hulle, M., De Cremer, K., Cornelis, R. and Lameire, N., *J. Environ. Monit.*, **3**, 86 (2001).
17. Castronovo, F. P. and Wagner, H. N., *J. Nucl. Med.*, **14**, 677 (1973).
18. Behne, D., Hammel, C., Pfeifer, H., Röthlein, D., Gessner and H., Kyriakopoulos, A., *Analyst*, **123**, 871 (1998).
19. Chéry, C. C., Dumont, E., Cornelis, R. and Moens L., *Fresenius' J. Anal. Chem.*, **371**, 775 (2001).
20. De Kimpe, J., Cornelis, R., Mees, L. and Vanholder, R., *Fund. Appl. Toxicol.*, **34**, 240 (1996).
21. De Kimpe, J., Cornelis, R., Wittevrongel, L. and Vanholder, R., *J. Trace Elem. Bio. Med*, **12**, 193 (1998).
22. De Kimpe, J., Cornelis, R., Mees, L., Vanholder, R. and Verhoeven, G., *J. Trace Elem. Bio. Med*, **13**, 7 (1999).
23. Templeton, D. M., Ariese, F., Cornelis, R., Danielsson, L.-G., Muntau, H., van Leeuwen, H. P. and Łobiski, R., *Pure Appl. Chem.*, **72**, 1453 (2000).
24. Bowen, H. J. M., Page, E., Valente, I. and Wade, R. J., *J. Radioanal. Chem.*, **48**, 9 (1979).
25. Dyg, S., Cornelis, R., Griepink, B. and Quevauviller, Ph., *Anal. Chim. Acta*, **286**, 297 (1994).
26. McCubbin, D. and Leonard, K. S., *Sci. Total Environ.*, **173**, 259 (1995).
27. Van Doornmalen, J., van Elteren, J. T. and de Goeij, J. J. M., *Analyt. Chem.*, **72**, 3043 (2000).
28. Van Doornmalen, J., Kroon, K. J., van Elteren, J. T. and de Goeij, J. J. M., *J. Radioanalyt. Nucl. Chem.*, **249**, 349 (2001).
29. Gilmore, E. A., Evans, G. J. and Ho, M. D., *Anal. Chim. Acta*, **439**, 139 (2001).
30. Achterberg, E. P., van Elteren, J. T. and Kolar, Z. I., *Environ. Sci. Technol.*, **36**, 914 (2002).
31. Salbu, B. and Steinnes, E., *Analyst*, **117**, 243 (1992).
32. von Gùnten, H. R. and Benes, P., *Radiochim. Acta*, **69**, 1 (1995).
33. Joshi, S. R., *Sci. Total Environ.*, **100**, 61 (1991).
34. Oughton, D. H., Salbu, B., Riise, G., Lien, H., Ostby, G. and Noren, A., *Analyst*, **117**, 481 (1992). Konoplev, A. V., Bulgakov, A. A., Popov, V. E. and Bobovnikova, T. I., *Analyst* **117**, 1041 (1992).
35. Salbu, B., Krekling, T. and Oughton, D. H., *Analyst* **123**, 843 (1998).
36. Amano, H., Matsunaga, T., Nagao, S., Hanzawa, Y., Watanabe, M., Ueno, T. and Onuma, Y., *Org. Geochem.*, **30**, 437 (1999).
37. Sanada, Y., Matsunaga, T., Yanase, N., Nagao, S., Amano, H., Takada, H. and Tkachenko, Y., *Appl. Radiat. Isotopes* **56**, 751 (2002).
38. Ebdon, L., Pitts, L., Cornelis, R., Crews, H., Donard, O. F. X. and Quevauviller, P., *Trace Element Speciation for Environment, Food and Health*, Royal Society of Chemistry, Cambridge, 2001.

CHAPTER 6

Direct Speciation of Solids

6.1 Characterization of Individual Aerosol Particles with Special Reference to Speciation Techniques

H. M. Ortner

Darmstadt University of Technology, Germany

Handbook of Elemental Speciation: Techniques and Methodology R. Cornelis, H. Crews, J. Caruso and K. Heumann
 ISBN: 0-471-49214-0

ACRONYMS

AAS	Atomic absorption spectrometry
AES	Auger electron spectrometry (this is the common acronym in topochemical analysis. Unfortunately it is identical with the acronym for atomic emission spectrometry in bulk analysis. In this paper, AES is used exclusively for Auger electron spectrometry)
AFM	Atomic force microscopy
BSE	Backscatter electron
CPC	Condensation particle counter
3D	Three dimensional
DMA	Differential mobility analyser
EBSD	Electron backscatter diffraction
EELS	Electron energy loss spectrometry
EDX	Energy-dispersive X-ray (fluorescence spectrometry)
EFTEM	Energy filtering transmission electron microscopy
EPMA	Electron probe microanalysis
ESEM	Environmental scanning electron microscopy
ESI	Electron spectroscopic imaging
EXAFS	Extended X-ray absorption fine structure
FEG	Field emission gun
FWHM	Full width at half-maximum
GC	Gas chromatography
GIF	Gatan imaging filter
HPLC	High performance liquid chromatography
HR	High resolution
ICP-OES,MS	Inductivity coupled plasma – optical emission spectrometry, mass spectrometry
ICSD	Inorganic Crystal Structure Data Base
JCPDS	Joint Committee of Powder Diffraction Standards
LA-ICP-MS	Laser ablation-inductively coupled plasma-mass spectrometry
LAMMS	Laser micro(probe) mass spectrometry
LC	Liquid chromatography
LDI-TOF-MS	Laser desorption/ionization-time of flight-mass spectrometry
MRP	Mass resolution power
MSC	Molecular sieve chromatography
PIXE	Proton induced X-ray emission (spectrometry); μ-PIXE is frequently used for PIXE with focused proton beams
PSE	Periodic system of the elements
PTFE	Poly(tetrafluoroethylene)
SAED	Selected area electron diffraction
SE	Secondary electron
SEM	Scanning electron microscopy
SIMS	Secondary ion mass spectrometry
SMPS	Scanning mobility particle sizing
STEM	Scanning transmission electron microscopy
TEM	Transmission electron microscopy

TLC	Thin layer chromatography
TXRF	Total reflection X-ray fluorescence (spectrometry)
UHV	Ultrahigh vacuum
WDX	Wavelength-dispersive X-ray (fluorescence spectrometry)
XANES	X-ray absorption near edge structure (spectrometry)
XPS	X-ray induced photoelectron spectrometry
XRS	X-ray spectrometry
Z	Atomic number of an element
ZAF	Atomic number – absorption – fluorescence correction procedure in EPMA

1 INTRODUCTION: THE SIGNIFICANCE OF SINGLE PARTICLE CHARACTERIZATION IN SCIENCE AND INDUSTRY

The characterization of particles, especially aerosol particles, is of great importance to a broad variety of scientific and industrial fields [1].

- In atmospheric sciences, the characterization of individual aerosol particles, their size distribution and chemical composition is of great relevance to modelling atmospheric processes and for environmental control purposes [2, 3]. Compared to conventional bulk techniques, individual particle analysis provides additional and complementary information concerning origin, formation, transport and chemical reactions [4–6].
- Occupational health monitoring relies on particle collection and subsequent particle characterization to evaluate health hazards for workers exposed to dusts from foundries, calcination ovens, powder handling, milling etc. [4, 7].

Particle characterization is an important source of information for compound identification for cleanroom control [6, 8]: in microelectronic compound fabrication or in ultratrace analytical laboratories for quality control of ultrapure chemicals, it is essential not only to monitor particle concentrations in air – which is done routinely – but also to identify the particles for possible elimination of sources of particulate contamination [6]. The identification of particles which caused malfunction of highly integrated microelectronic devices is also very important [8].

- Powders are the basis of powder technologies. Usually, single particle characterization is not necessary in powder technologies and bulk characteristics are determined in routine quality control. However, single particle characterization is essential for the determination of so called ‘heterogeneous impurities’ in raw and intermediate products in powder metallurgy [9, 10]. Heterogeneous impurities are particulate impurities in the raw and intermediate products of powder metallurgy with particle diameters above approximately 5 μm. They are introduced into the raw or intermediate powders at various stages of production as e.g. ore or gangue particles, abraded particles from grinding, milling and mixing operations or by careless powder manipulation or storage (e.g. cigarette ash, textile fibres, hair, rubber particles, aerosol dust, etc.). Owing to the sintering process in powder metallurgy, such impurities are not homogenized as in melting operations. Hence, the particles remain unaffected or react partially with the matrix material to form inclusions which usually act as centres for crack formation if the material is mechanically stressed. In fine wire drawing or foil production, wires break or the foils become perforated at the place of a heterogeneous inclusion. The analytical determination of heterogeneous impurities is therefore an important procedure in the quality control of raw and intermediate products of powder metallurgy [9, 10].
- The characterization of wear particles, e.g. in polymer extrudates or in motor lubricants [11], is another important area of technological relevance.
- The quality control of composite particles designed for coating operations by plasma spraying or of particles used in powder technology and metallurgy is a further very important field [12].

- The characterization of complex particles is of relevance in some applications, e.g. for pigments. Heterogeneous catalysts are usually also applied in particulate form and their characterization is accomplished using diverse methods of solid state analysis [13]. Another important group of particles of relevance in analytical chemistry are chromatographic materials for column fillings for gas chromatography (GC), high performance liquid chromatography (HPLC), molecular sieve chromatography (MSC) (e.g. macroreticular or pellicular resins or materials for the preparation of thin layer chromatography).
- A rather exotic but nevertheless important role is played by cosmic [14–16] and terrestrial particles in the degradation of material surfaces exposed to space in the low Earth orbit (LEO) range, i.e. in an altitude of 400–600 km above the Earth's surface. The most alarming observation was that more than 80 % of these particles are of anthropogenic origin, i.e. man made debris (e.g. paint particles from space shuttles, particulate residues from solid state rockets etc.). The differentiation between cosmic and anthropogenic particles is only possible by isotope ratio measurements of particles or what remains of such particles on degraded material surfaces [15, 17]. Because of speeds of typically several km s^{-1} on impact, practically no particles survive collisions with space-exposed material surfaces and what remains is a thin layer of recondensed matter after being vaporized during impact. This layer can only be qualitatively analysed by secondary ion mass spectrometry (SIMS). Material degradation by the combined effect of erosion by particles and oxidative processes on such eroded surfaces by the atomic oxygen which is still present in the LEO region is becoming a serious problem for satellite lifetimes in this region [15, 18].
- A similarly exotic field is the forensic characterization of the diverse particulate material from crime scenes, which is used as physical evidence and is a major task in forensic science [19].
- Last but not least, volcanic dust particles and interplanetary dust particles are studied intensely in geology: explosive volcanic eruptions can eject vast amounts of gaseous and particulate material into the atmosphere within a brief period. Much of this material remains in the troposphere for a considerable length of time [20]. These particles can greatly reduce the radiation of the sun reaching the Earth's surface and thus decisively influence weather patterns. In addition, interplanetary dust particles are studied intensely since they often stem from the very beginning of the formation of our solar system 4.5 billion years ago, and some of them might even stem from pre-solar system times [16, 17].

Taking into account this broad range of interest in particle characterization, a multimethod approach for the comprehensive characterization of particles has been developed for a particle diameter range from 10 nm to 100 μm, which is relevant in the above fields of study.

A thorough evaluation of the particle size distribution as well as of the lifetime of particles of varying size in different domains of the Earth's atmosphere showed that particles smaller than 100 nm in diameter exhibit very short lifetimes mainly due to agglomeration [20]. Furthermore, most sources of particles introduced into the atmosphere emit particles of 100 nm and more in diameter, many natural sources emit particles even in the 1 μm range [21]. This is the reason why our main scheme of particle characterization is designed for particles >100 nm. However, particles with diameters down to 10 nm are of relevance for health hazard studies to evaluate the toxicity of dust in many industries [22]. In this case, the application of transmission electron microscopy (TEM) is necessary for particle size distribution measurements in the nm range and for phase and/or compound identification of such particles. Alternatively, high resolution scanning electron microscopy (HR-SEM) can also be applied [1, 21].

Nanometer sized particles have also become technologically important in materials science for the production of materials with new and interesting properties. They have also revolutionized established technologies and led to the development of materials with greatly improved high-tech properties [23]. The characterization of such particles is also accomplished by the combined use of TEM and HR-SEM.

The large diameter end of aerosols is again limited by the increasingly short lifetimes of large particles in the range of 25–100 μm in diameter depending – as can be expected – on particle density and morphology. This upper limit is equally significant for health hazard investigations, since coarse particles are usually trapped in our respiratory system before the particles can reach the deeper bronchial system and the lungs. For various reasons, this is also the upper limit of particles of technological interest (heterogeneous particles in powder metallurgy, wear particles, harmful particles in microelectronics technology) [8–10].

2 SEMIQUANTITATIVE PARTICLE EVALUATION + MORPHOLOGY OFTEN = SPECIATION

It should be emphasized that besides special methods of single particle speciation, which are still less common and will be discussed later, the primarily described procedure of the routinely used HR-SEM and electron probe microanalysis (EPMA) should be regarded as a most important step towards single particle speciation. In many cases semiquantitative single particle analysis by EPMA, together with an often typical particle morphology as studied by HR-SEM, unambiguously answer the question of the compound making up the particle. In other cases speciation is possible by selected area electron diffraction (SAED) in the transmission electron microscope or by the modern technique of energy filtering transmission electron microscopy (EFTEM). In still other cases only the appropriate combination of such techniques will give a conclusive answer to particle speciation, especially of compound particles which occur very often in aerosols.

Figure 6.1.1 demonstrates this possibility of an HR-SEM identification of a particle found in an aerosol from the largest aluminium production facility of Norway [21]. Such cryolite coated Al_2O_3 particles were found to make up about 50 % (in particle numbers) of aerosols collected in the electrolysis hall at the Norsk Hydro Aluminium plant, Haugesund, Norway. They are the most likely reason for the so-called 'potroom asthma', which occurs in such sites of the electrolytic production, of aluminium metal [23].

Figure 6.1.2(a) shows a TEM micrograph of a platinum catalyst particle, which was collected together with nickel-containing particles in the 'reduction hall' of the secondary nickel refinery at Monchegorsk (Penninsula of Kola, Russia) [24]. Figure 6.1.2(b) gives the corresponding EDX spectrum. The copper lines stem from the copper grid of the TEM polycarbonate foil on which the particles were collected. The $C\,K_\alpha$ line is rather intense and stems from the soot which is obviously combined with the finely dispersed Pt. Figure 6.1.2(c) shows the electron diffraction grainy ring pattern. This is indicative of the very small grain size of the Pt particles, of the order of tens of nanometres. It also unambiguously identifies the particle as platinum.

Figure 6.1.1 demonstrates the principal possibility of compound identification by qualitative EDX analysis together with particle morphology which yields additional valuable information, e.g. related to the toxicity of certain aerosol particles. This information is even superior to mere speciation and this is the reason why the whole scheme of particle characterization that has been developed is elucidated. Figure 6.1.2 demonstrates the very valuable combination of direct speciation by electron diffraction in the transmission electron microscope. In addition to mere speciation, it yields a complete compound identification of crystalline phases which includes the relevant crystal structure. Again, this information is in many cases important since identical compounds with varying crystal structure might be of varying toxicity as is the case for nickel sulfides. Their toxicity falls in the following compound sequence [25]: α-Ni_3S_2, β-NiS $>$ NiO $\gg$ metallic Ni $\gg$ amorphous NiS.

2.1 The key to a profound particle characterization: a multimethod approach

Table 6.1.1 presents an overview of the multimethod approach to particle characterization of

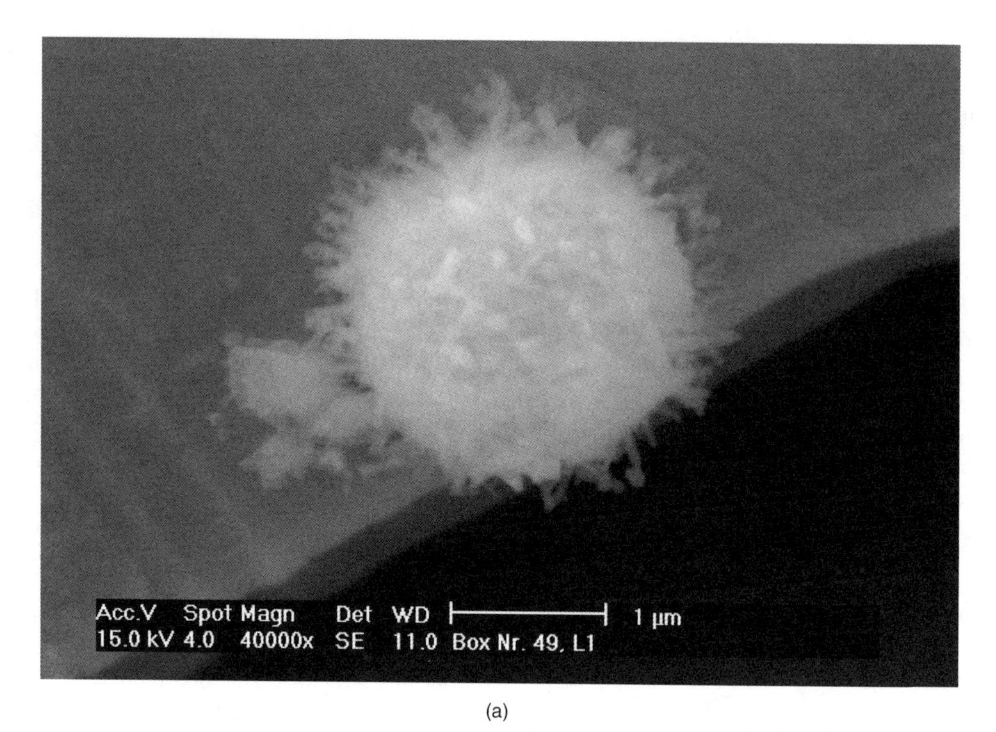

(a)

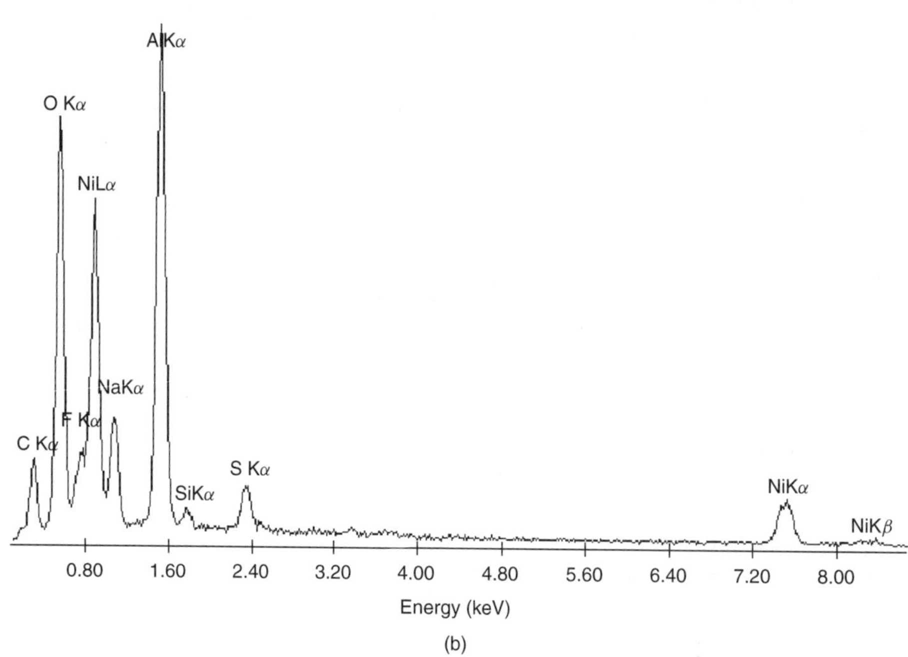

(b)

Figure 6.1.1. HR-SEM study of aerosol particles collected in the hall of one of the largest aluminium production sites of Norsk Hydro Aluminium at Haugesund, Norway. (a) Al_2O_3 particle coated with $Na_3[AlF_6]$. (b) qualitative EDX spectrum. Main components: Al, O. Coating: Na, Al, F. Further elements present: C, Si, S; the Ni lines stem from the sample holder.

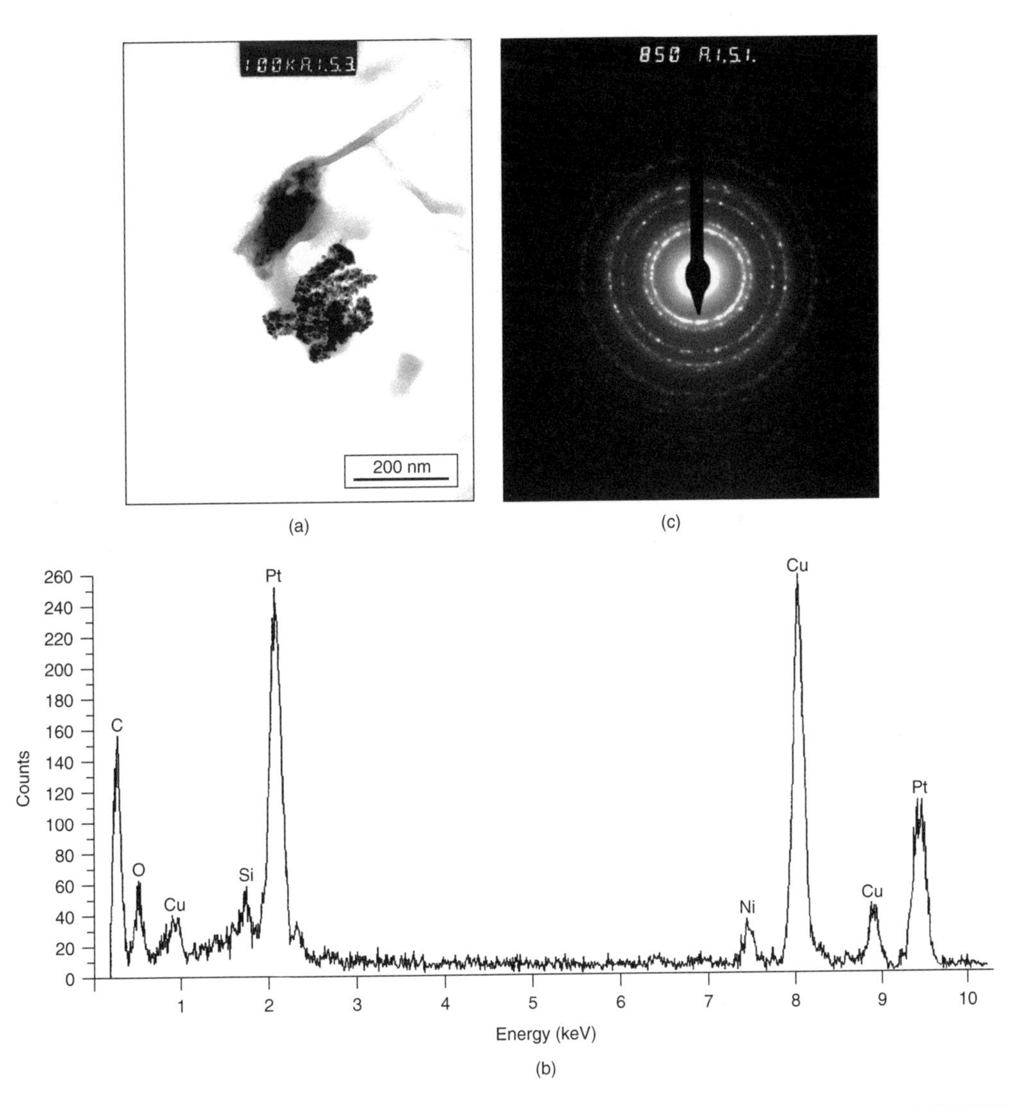

Figure 6.1.2. TEM characterization of a platinum-containing particle, an example of particle identification by TEM/EDX/electron diffraction. (a) Diffraction contrast micrograph. (b) Qualitative EDX analysis: the Cu lines stem from the filter material. (c) Electron diffraction ring pattern typical for Pt metal.

solid aerosols which was developed by us [1]. Of course, such a scheme does not work without certain limitations: since practically all topochemical methods applied are working under vacuum conditions (some of them even under ultrahigh vacuum (UHV) conditions), the inspection of droplets is not feasible and moisture and volatile compounds are thereby removed. Hence, many aerosol particles undergo a certain change before they can be inspected.

Figure 6.1.3 demonstrates the possibility of speciation and – more than that – the evaluation of the crystal structure and thus the phase identification by TEM. This is especially important for cases such as nickel compounds: the determination of carcinogeneity of such particles depends not only on compound identification but also on the determination of the crystal structure. Besides Millerite, Heazlewoodite (Ni_3S_2), NiO and $NiSO_4 \cdot H_2O$ have been found in aerosols of Monchegorsk [24].

Table 6.1.1. Scheme of single particle characterization with special reference to speciation.

Sampling	Particle collection with a three- or five-stage impactor onto high purity glassy carbon disks or by a one-stage impactor for very low aerosol concentrations. Carbon coating by evaporation (if necessary).
SEM survey (preferably by HR-SEM)	• Selection of samples with the appropriate particle density (if sampling with varying sampling times was performed). • First survey of particle size distribution. • First survey of particle morphologies. • Check on particle homogeneity or heterogeneity and on particle aggregates. • Qualitative analysis of individual particles by EDX for selection of elements to be studied by EPMA.
Semiquantitative EPMA investigation (down to 0.5 μm particle diameter)	• Elemental mapping (usually by WDX) for elements selected by the SEM survey. • Semiquantitative particle evaluation yielding the following information: – elemental composition of particles (including C, N, O but no trace contents); – number and size distribution of particles of specific composition; – systematic assignment of size distribution and morphology to identified classes of particles; – assessment of elemental homogeneity or heterogeneity of individual particles.
Quantitative bulk characterization by TXRF for main, minor and trace elements	• Quantitation is accomplished by use of Sc as internal standard for low Z elements and of Y for high Z elements.
ESEM work	• For particle inspection under moisture saturated conditions and investigations concerning morphological changes with decreasing humidity. • For the investigation of insulating particles without coating.
AFM work	• For quantitative volumetric particle investigations, under varying ambient conditions in the tapping mode.
EBSD	• Only applicable for well-developed crystals larger than 5 μm and with flat surfaces (e.g. platelets). Phase identification is then possible for single particles.
Explicit methods of particle speciation	
Valence bond X-ray spectrometry by EPMA-WDX	• Single particle speciation is often possible for low Z elements and for particles > 5 μm in diameter. Time consuming.
TEM/SAED	• For particles smaller than 0.5 μm in diameter, compound identification is possible for crystallized species usually performed in combination with elemental analysis by EDX and/or EELS.
EFTEM or ESI	• For particles smaller than 100 nm or for particle domains not thicker than 100 nm, mapping of elemental binding states is feasable with single nm-lateral resolution in favourable cases as long as the particle endures the high intensity of the electron bombardment.
XPS	• Most commonly used method of solid-state speciation. Synchrotron-XPS allows single particle characterization.
AES	• Speciation possible especially for low Z elements with excellent lateral and depth resolution (both in the low ten nm range).
Mößbauer spectrometry	• Unfortunately limited to the mg range (with special apparatus). No single particle speciation possible. Restricted to iron compounds and some other compounds.
Micro-Raman spectrometry	• Functional group analysis in organic particles, ≥1 μm possible with modern laser Raman microprobes.
μ-EXAFS and μ-XANES	• Requires synchrotron X-ray radiation. EXAFS yields a series of valuable information also for amorphous particles such as: – type(s) of nearest neighbour(s); – distances to nearest neighbours; – coordination number. XANES gives information on the valency state and/or chemical state of the probed atom (e.g. oxidic or metallic etc.) With appropriate X-ray optics, mapping of this information with resolution in the low ten μm range is feasible.

Table 6.1.1. *(continued)*

Trace and isotopic characterization of single particles by mass spectrometric methods and PIXE	
SIMS	The most sensitive topochemical method. The NANOSIMS has been constructed with dedication to particle analysis (cosmic particles) with diameters in the tens of nm range. Speciation is sometimes possible from the observed mass pattern. 3D SIMS is possible for particles with certain geometries (e.g. platelets). Lateral resolution with conventional sector-field mass spectrometers is in the single μm range, depth resolution in the low 10 nm range.
LAMMS and LA-ICP-MS	LAMMS is extensively used for single particle characterization. Speciation is feasible in favourable cases and especially for inorganic compounds. Isotopic composition is also determinable, also with LA-ICP-MS which is the more sensitive technique of the two.
LDI-TOF-MS	Can be used for the on-line characterization of single particles from a sampled air stream. Very fast. Quantitation possible in favourable cases.
μ-PIXE	Single particle trace analysis with focusing nucleoprobes possible. Lateral resolution around 5 μm.

2.1.1 Particle collection with a cascade impactor on glassy carbon disks

Taking into account the broad range of interest in particle characterization as outlined above, a multimethod approach for the comprehensive characterization of particles was developed for a particle diameter range from 10 nm to 100 μm, which is relevant in the above fields of study. A thorough evaluation of particle size distributions and of lifetimes of particles of varying size in different domains of the Earth's atmosphere showed that particles smaller than 100 nm in diameter exhibit very short lifetimes, mainly due to agglomeration [26]. Furthermore, most sources of particle emission emit particles of 100 nm and more in diameter. Many natural sources introduce particles into the atmosphere even in the 1 μm range [26]. In contrast, the presently and generally accepted ranges of particle sizes of relevance from a health related point of view are the following [27]:

- *Inhalable fraction:* is defined as the mass fraction of total airborne particles which is inhaled through the nose and/or mouth. It comprises particles smaller than approximately 100 μm in aerodynamic diameter.
- *Thoracic fraction:* is the mass fraction of inhaled particles penetrating the respiratory system beyond the larynx. It is given by a cumulative lognormal curve with a median aerodynamic diameter of 10 μm and a geometric standard deviation of 1.5.
- *Respirable fraction:* is the mass fraction of inhaled particles which penetrates to the unciliated airways of the lung (alveolar region). It is given by a cumulative lognormal curve with a median aerodynamic diameter of 4 μm and a geometric standard deviation of 1.5.
- *'High risk' respirable fraction:* is a definition of respirable fraction for the sick and infirm, or children. It is given by a cumulative lognormal curve with a median aerodynamic diameter of 2.5 μm and a geometric standard deviation of 1.5.

These are the reasons why our main scheme of particle characterization is designed for particles larger than 100 nm. In contrast, particles with diameters down to 10 nm are of great relevance for health hazard studies to evaluate the toxicity of dust in many industries [22] and of diesel soot [27]. It should be mentioned that the influence of fine and ultrafine particles on human health is attracting the greatest current attention. Often, adverse health effects seem to be linked with smaller particles [28]. In addition, it seems that nearly all of the mass emitted by modern diesel engines is in the nanometre diameter range [28]. The application of TEM is necessary for particle characterization in the nanometre range and for phase and/or compound characterization of such particles, as will be discussed later.

Alternatively, HR-SEM can also be applied which is experimentally much faster and easier. Particle size distribution measurements in

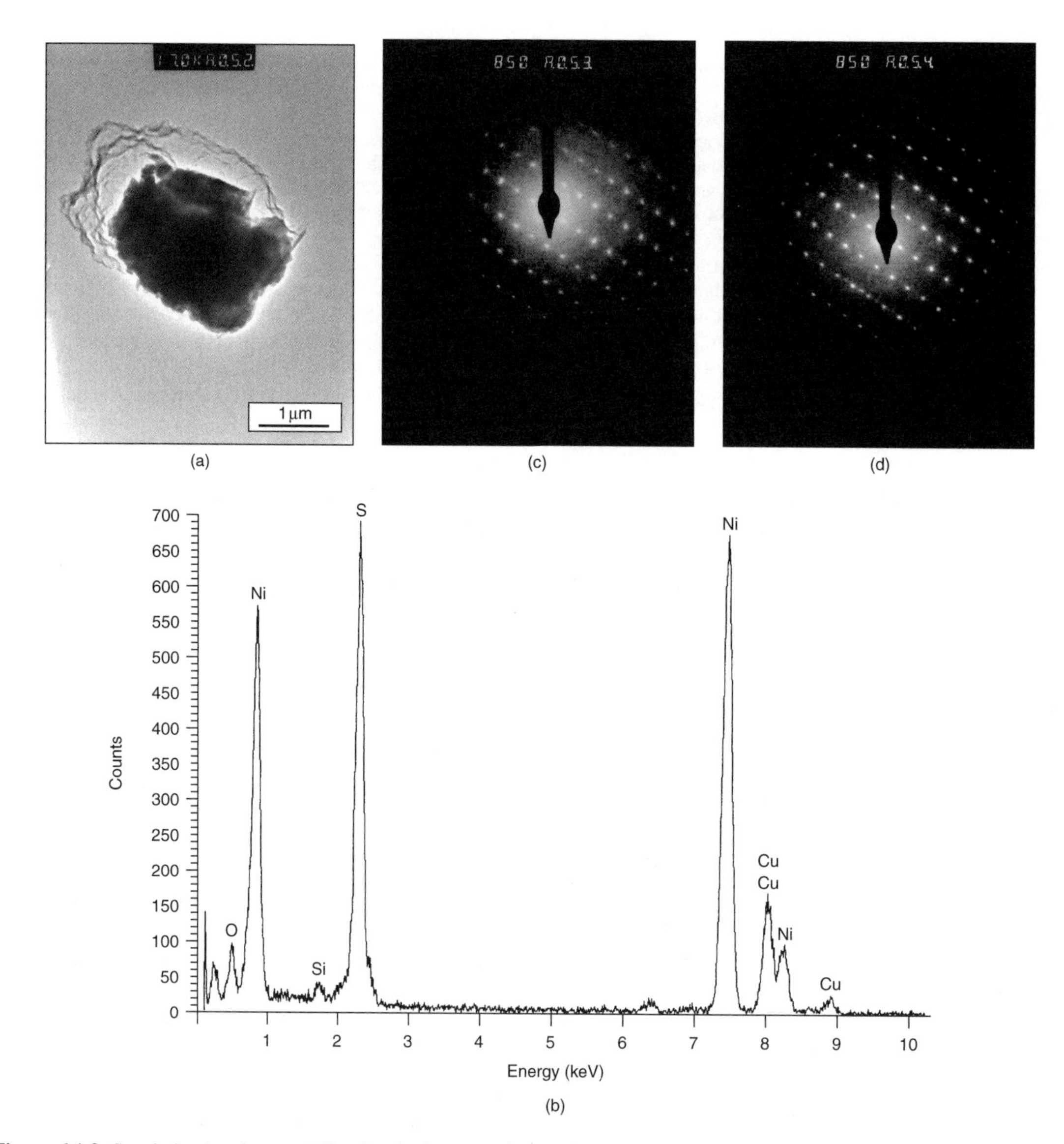

Figure 6.1.3. Speciation by electron diffraction in the transmission electron microscope: (a) diffraction contrast micrograph of an NiS particle; (b) qualitative EDX analysis; (c) electron diffraction, zone axis (21 − 2); (d) Electron diffraction, zone axis (211).

the nanometre range are today usually performed by scanning mobility particle sizing (SMPS). The SMPS instrument of TSI Inc., USA, is based on droplet growth by condensation nucleation, followed by particle size distribution analysis of the droplets grown into the micrometre range [29]. It consists of a differential mobility analyser (DMA) and a condensation particle counter (CPC).

Aerosol collection with a one-stage impactor has been used for particles with a minimum diameter of 100 nm for relatively long periods (hours) in order to obtain the necessary sensitivity for element detection in the pg m^{-3} range in clean-room atmospheres [6]. Collection times must, of course, be shorter (minutes or even seconds) for heavily contaminated atmospheres in industry in the range of mg m^{-3} for health hazard evaluation [30].

2.1.2 Electrostatic particle collection for nanometre particles

Particles with diameters below 100 nm cannot be deposited by impaction. Electrostatic deposition methods are in use for this purpose [29, 31, 32]. Electrostatic deposition is a well-known technique for collecting airborne particulates as well as artificially produced particles for subsequent investigation with various analytical systems [32]. For this technique high deposition efficiencies up to 100 % have been reported for particulates in a size range of 3.5 nm to 3 μm [32]. The theory of electrostatic deposition predicts high deposition efficiency for larger and smaller particles as well [33]. A suitable commercially available instrument is also offered by TSI, as was mentioned above [29, 31, 32].

2.1.3 Topochemical characterization of solid aerosols by SEM and EPMA

2.1.3.1 Scanning electron microscopy

The first step in our scheme of particle characterization is always an SEM survey, in which important information is collected on the following aspects:

(a) *Particle abundance.* This information is important for optimizing the particle collection parameters for an appropriate number of particles per unit area of the glassy carbon disks. In the case of remote sampling, this is not possible and samples are collected with various sampling times and the optimally loaded ones are selected for further investigations.

(b) *Size distribution.* This is an important parameter of its own. It also yields information on whether EPMA can be applied as next step or whether TEM or HR-SEM would have to be selected in the case of particles with diameters below 0.5 μm.

(c) *Particle morphologies.* In many cases, certain particle morphologies already give the possibility to identify certain particle types, e.g. soot particles exhibit such a typical morphology that they can usually be identified in this step. The same is true for most biological particles (pollen) salt particles from the sea and many mineral particles.

(d) *Particle homogeneity or heterogeneity.* For many aerosol particles, especially for aggregates, their heterogeneous structure often becomes evident.

(e) *Sample charging.* In most practical cases, it is not necessary to coat the glassy carbon disks with the collected particles because of the sufficient electrical conductivity of the glassy carbon. However, if larger amounts of highly insulating particles are present, coating with carbon is performed.

Several particles (the number depends on the extent of morphological and constitutional variation of the inspected sample) are then qualitatively analysed by EDX for the selection of elements which need to be mapped by EPMA.

It should be emphasized that particle morphologies and morphological details of particles in the nanometre range can only be properly studied by SEM instruments with a field emission gun or an LaB_6 electron source [34]. This is the reason why HR-SEM instruments are used throughout our aerosol investigations. It is also essential to use a modern thin window EDX detector since classical EDX detectors are not capable of analysing elements with an atomic number $Z < 11$. Oxygen and carbon are always included in our WDX element mapping. Since B and F are usually not expected in particles, their presence is only checked in special cases. This leaves only nitrogen mapping questionable for elements of the second period of the periodic system of elements (PSE). Like B and F, it is only mapped in special cases.

Environmental scanning electron microscopy uses instruments that are able to work under low pressures. These have been under development for quite some years. They have now reached maturity especially with the Philips ESEM series [35, 36]. Water vapour can be introduced as a gas into the specimen chamber so that saturated chamber conditions can be reached and maintained. Water can even be condensed onto as well as removed from the sample in a controlled manner. This allows the morphological and analytical investigation of samples under moist conditions which would create morphological artifacts in a normal SEM instrument under vacuum. In order to create a water saturated atmosphere at low pressure, a Peltier cooling stage is available which can, of course, also be used for the removal of water from samples under vacuum by freeze drying. Furthermore, a specimen chamber gas handling system allows the introduction of various gases at low pressure. Air can be introduced, e.g. for low or high temperature oxidation studies (in dry or moist air) since a hot stage is also available for temperatures up to 1500 °C. Such conditions are, of course, especially interesting for material science studies. The evolved gases of such experiments can be introduced into a quadrupole mass spectrometer via a small capillary and analysed. Gas concentrations of $<1\,\mu g\,mL^{-1}$ up to 100 % are thus determinable.

This successful operation of scanning electron microscopes at low pressure is only possible with some constructional developments [35, 36]:

(a) The beam gas path length, i.e. the distance that the primary electron beam has to overcome from the high vacuum system of the electron optics column to the sample, is kept to a minimum. This is also important for the lateral resolution of X-ray detection which is hampered by primary electrons being scattered by gas molecules.
(b) Secondary electron detectors have to be modified to so-called 'gas amplification detectors', whose operation is analogous to the flow proportional gas detector used in WDX. Backscattered electrons suffer negligible energy loss in the gas atmosphere and retain sufficient energy to activate large area scintillators without post-specimen acceleration.
(c) A great advantage of operation at elevated pressure is the automatic discharge of the negatively charged surface of insulators due to gas ions that are attracted by this charge and 'neutralize' it. The gas ions above the sample are generated by the primary electron beam and by secondary electron ionization processes. This enables the morphological study of insulating samples without a coating. In a subsequent analytical investigation under vacuum for light elements there is no interference of a conductive coating.

Variations of the morphology and volume of particles with e.g. varying humidity can also be advantageously investigated by *atomic force microscopy* (AFM) in the tapping mode [37]. Tapping mode means that the needle of the AFM instrument is raised from and lowered onto the sample with a relatively high frequency (kHz) while the sample is slowly moved along in the x, y plane. In this way, particles are generally not moved from their original position by the transgressing needle. Particle morphologies can thus be determined without interference of the needle which is usually led over the particle in the noncontact mode.

Since no comparative studies are yet available with the ESEM and the AFM it is difficult to say whether the AFM approach bears advantages over respective ESEM observations. Certainly, quantitative volumetric observations of expanding or shrinking particle volumes under varying ambient conditions are only possible by AFM. Single sub-μm particles have been investigated with respect to changing humidity in the surrounding atmosphere [37]. Volume calculations allowed monitoring of these changes on a quantitative basis.

2.1.3.2 Electron backscatter diffraction, EBSD

The evolution of this technology has essentially taken place in the 1980s. Today it is available as an add-on package to a scanning electron microscope [38]. The essential features of EBSD are its unique capabilities to yield crystallographic data by imaging in real time with a spatial resolution of 0.5 μm, combined with the regular capabilities of SEM, such as capacity for large specimens and option of chemical analysis. In other words, EBSD can automatically determine crystallographic data of each grain that is analysed by electron diffraction to yield so-called Kikuchi patterns from which these data can be obtained. Hence, EBSD could also be used for phase identification of each analysed particle [38]. This was the onset to our efforts to apply EBSD also to the characterization of aerosol particles. Unfortunately, these efforts were unsuccessful

[24]. The reason for this is that EBSD needs well-developed crystals at least 1 μm in diameter (due to the spatial resolution of EBSD of 0.5 μm). Another restriction is the fact that EBSD only works well for flat and polished surfaces. For the usually very irregularly shaped aerosol particles it is generally not possible to obtain the respective Kikuchi patterns. In addition, aerosol crystallites very frequently contain many imperfections, which consequently blur the evolution of a Kikuchi pattern. It is due to these circumstances that EBSD cannot be applied generally to phase identification of aerosol crystallites unless these are big enough (>1 μm in diameter) and well crystallized.

2.1.3.3 EPMA mapping

A *qualitative* method for the characterization of a great number of individual particles based on element distribution maps has recently been developed [39]. Using this procedure, the size, shape and qualitative chemical composition of each particle can be deduced from a combination of secondary electron (SE) and backscattered electron (BSE) images and element distribution maps for those elements which have been chosen by the above SEM survey. The knowledge of the qualitative chemical composition of individual particles is sufficient for many applications. For example, we have recently investigated the significance of iron in atmospheric processes [40]. The iron-bearing particles were characterized by our analysis procedure and were classified into several categories (i.e. metal, oxide, silicate particles). We are also using this procedure for source apportionment of aerosols, where characteristic elements for each source are known from bulk measurements by total reflection X-ray fluorescence (spectrometry) (TXRF) (see below).

In addition to the qualitative determination, a *semiquantitative* estimate of the chemical composition of each particle can also be obtained [39]. For this purpose, the count rates for the measured elements of each particle are derived from the respective element distribution maps. These are corrected for matrix and geometric effects using the particle *ZAF* procedures of Armstrong [41, 42]. In contrast to the qualitative procedure, spectrometer defocusing cannot be neglected in semiquantitative analysis, even at high magnifications. The correction is performed using algorithms which were also developed [43].

This semiquantitative approach turned out to be necessary e.g. for particle characterization in occupational health monitoring. In the course of the evaluation of samples collected at the largest nickel refinery in the world at Monchegorsk on the Kola peninsula (Russia), it became apparent that the differentiation of particles containing various amounts of nickel oxides and nickel sulfides in contrast to nickel sulfate particles was only feasable by semiquantitative characterization [30]. In contrast to a previously developed similar procedure [44] we use WDX rather than EDX in order to include the very important elements C, N and O. Recently, the EDX procedure [44] has been further developed to use windowless EDX detectors in EPMA [45]. In addition, Monte Carlo calculations were developed to improve the very difficult matrix correction procedures for the soft X-rays of the elements of the second period of the PSE [45]. Whether or not this leads to data comparable with the WDX EPMA procedure has not yet been investigated. It is doubtful because the second great drawback of the use of EDX detectors still remains: the essentially worse spectral resolution which inevitably leads to serious line interferences of spectral lines of metallic elements of higher Z with the analyte lines of the second period elements to be determined. Unfortunately, the EPMA mapping method is not capable of analysing particles smaller than 0.5 μm in diameter for instrumental and physical reasons. Therefore, other methods and instrumentation have to be applied for smaller particles.

2.1.4 Quantitative bulk characterization on glassy carbon disks by TXRF for main, minor and trace elements

It is beyond the scope of this chapter to give an overview of the wide range of methods used today for the bulk characterization of particulate samples (e.g. atomic absorption spectrometry (AAS), X-ray spectrometry (XRS), inductively coupled plasma-optical emission spectrometry (ICP-OES), inductively coupled plasma-mass spectrometry (ICP-MS), etc.) [46]. Only the use of TXRF is outlined here. TXRF is a trace and microanalytical technique and matches the specific requirements of impactor-collected particulate samples in an ideal way: it covers the small area where particles are deposited below the respective impactor jets and allows a safe multielement quantification by internal standardization. TXRF has been used extensively for quantitative aerosol analysis [47, 48], and is one of the most important methods for this purpose. Quantitative determination of the elemental composition of the collected aerosol is carried out with two TXRF instruments on the glassy carbon carriers of selected purity. Quantitation is accomplished by use of Sc and Y as internal standards for elements of low and high Z, respectively. Details of the procedure are described in refs 5 and 6. Since the application of the internal standard solution changes the aerosol particles it has to be applied as the last step of the usual procedure after the morphological and EPMA survey.

3 EXPLICIT METHODS OF PARTICLE SPECIATION

We are now arriving at methods which can give direct information on particle speciation. However, these methods are less frequently used in routine aerosol characterization than those discussed up to now.

3.1 Valence band X-ray spectrometry by EPMA-WDX

This method is already well established for the identification of binding states in solids, especially for elements of low and medium Z [49]. The basis is the 'chemical shift' of X-ray lines if the onset of the electron jump to an inner orbital vacancy lies in the valence band of the respective atom. Generally the line shift is small (about 1–3 eV) in comparison with the peak width of X-ray lines (full width at half-maximum (FWHM) ≈40 eV). In geology and mineralogy, a method has been developed to determine the Fe(II)/Fe(III) ratio in solid samples by precision measurement of the variation of the position and shape of the Fe L_α and Fe L_β lines [50]. The L_α/L_β intensity ratio might also vary. This method has the great advantage over other established methods for solid-state speciation (XPS, AES) that it exhibits much greater detection sensitivity and/or lateral resolution (compared with XPS only in this case). Bulk speciation can be carried out directly on samples collected on glassy carbon disks. For particles >5 μm, single particle speciation is also possible but time consuming as Höflich *et al.* were able to demonstrate recently [51]. Unfortunately, many particles exhibited instability by the long measuring times which are necessary for X-ray profile precision measurements.

3.2 TEM of particles smaller than 0.5 μm in diameter

3.2.1 Experimental procedure

Samples for TEM investigations are usually obtained on the last (fifth) stage of the cascade impactor with a Formvar foil placed on the glassy carbon disk. The Formvar foil is a polycarbonate filter reinforced by a copper grid. Very fine particles which cannot be collected by impaction are either deposited electrostatically or they are deposited by suction with a small pump on a Formvar foil which is placed on a ceramic filter support. In some cases, for 'concentrated' industrial aerosols, mere exposure of the Formvar foil to the contaminated atmosphere for a few minutes is sufficient. The Formvar foil is carbon coated after sampling and prior to TEM inspection. Particles can be characterized in only a limited number since automation comparable to that outlined above for the EPMA procedure is not yet available. Particles thicker than 50 nm cannot be penetrated by the primary electron beam. However, thinner portions of such particles can usually be inspected at their edges.

3.2.2 *Information content of TEM investigations*

The following information can be obtained by TEM investigations:

(a) *Particle size distribution* and partly *morphology* can be studied by imaging with the transmitted beam.

(b) *Phase identification* is possible by selected area electron diffraction (SAED). If point patterns are obtained for single crystals, the observed reflections can be converted into the respective lattice constants. The latter are used for compound identification by search in the Inorganic Crystal Structure Data Base (ICSD) using the procedure 'PIEP' which was developed by Miehe [52]. Other suitable programmes have also been developed [53]. In the case of very small crystallites and nanocrystalline particles, diffraction rings are obtained. Since atomic number dependences of the scattering power for electrons and X-rays are similar, the relative intensities in the corresponding electron and X-ray powder diffractograms are also similar. It is, therefore, possible in such cases to carry out a compound identification with the Joint Committee of Powder Diffraction Standards (JCPDS) data bank for X-ray powder diffraction data.

(c) *Compound identification* for amorphous particles is possible in simple cases (if no multiphase particles are present) by the determination of the elemental composition by EDX or by electron energy loss spectrometry (EELS) or, preferably, by a combination of the two methods. In favourable cases, speciation is possible because EELS peaks often contain binding information in the respective near edge fine structures [54]. EDX and EELS are complementary since EELS is most sensitive for low Z elements whereas EDX is more powerful for elements of medium and high Z by the nature of the underlying physical process. Xhoffer *et al.* have given a more detailed account of EELS for single particle analysis [48].

3.2.2.1 Energy filtering TEM (EFTEM) or electron spectroscopic imaging (ESI)

There are some very promising developments in TEM-EELS which will have a substantial influence on single particle characterization: up to now, EEL spectra were usually acquired from a small selected area (point analysis). Now, the two-dimensional acquisition of any spectral feature of an EEL spectrum has become possible with what is called energy filtering transmission electron microscopy (EFTEM) or electron spectroscopic imaging (ESI) [55, 56]. Energy filtering devices for TEM have become commercially available only recently and the Gatan imaging filter (GIF) can be attached to almost any 100–400 kV TEM instrument [57]. With ESI, energy filtered images are acquired which can then be combined to show the distribution of elements in the specimen with nanometre resolution [55]. This procedure has advantages over the scanning transmission electron microscopy (STEM)–parallel EELS combination since acquisition times for high resolution measurements of the latter are essentially longer than those for ESI [58]. In favourable cases, ESI allows the mapping of elemental binding states with single nanometre lateral resolution [55, 56].

3.3 X-ray induced photoelectron spectrometry (XPS)

XPS is the most commonly used method of solid-state speciation with excellent depth resolution in the single nanometre range [34, 59]. Lateral resolution, however, is limited to the low millimetre range in older instrumentation. Modern XPS instruments can achieve a lateral resolution in the lower micrometre range and can thus be used for single particle XPS characterization [1, 3, 4]. The best lateral resolution in the single micrometre range is achieved with XPS instrumentation adjoined to electron accelerators, e.g. in Grenoble or Hamburg (Hasylab). The application of XPS to environmental particulate samples has been discussed extensively by Xhoffer *et al.* [48]. It should be mentioned that XPS is a UHV method and only vacuum-stable particles can be studied. On the other hand, the danger of particle disintegration by X-rays is considerably lower than for electron probe methods discussed so far. On account of its excellent depth resolution, XPS in combination with SEM/EDX or EPMA/WDX allows researchers to distinguish between bulk composition and thin coatings of larger particles.

3.4 Auger electron spectrometry (AES)

AES can be used especially for the speciation of low Z elements with excellent lateral and depth resolution (in the 10 nm range). Auger signal shapes are frequently sensitive to relevant binding situations and allow e.g. the differentiation of graphitic, amorphous and carbidic carbon [34].

3.5 Mössbauer spectrometry

This method has been specifically applied for iron speciation in large, integral aerosol samples [60]. Iron is one of the most abundant elements in solid and aqueous atmospheric samples [61]. It is usually introduced into the atmosphere as soil dust, fly ash from power plants and waste incineration facilities, from exhaust of combustion engines and generally from industrial operations [62]. A thorough study of the significance of iron for atmospheric redox reactions and on the presence of iron in the above mentioned sources was recently carried out [63]. Magnetite, hematite, goethite and iron(III) silicates were found in the inspected samples [60, 63]. Speciation by Mössbauer spectrometry is also possible for a number of other elements [64].

3.6 Micro-Raman spectrometry

With the advent of well focused laser beams as sources for the excitation of Raman spectra and parallel developments in relevant instrumentation, the analysis of discrete particles $\geq 1\ \mu m$ in diameter has become possible by Raman spectrometry yielding valuable information on functional groups in organic particulates [34]. Although additional vibrations might be caused by a well-defined geometry of the inspected particles, this is usually of no concern for irregularly shaped particle assemblies which are generally present in aerosol samples. A more detailed discussion of micro-Raman spectrometry for particulate samples can be found elsewhere [46].

3.7 X-ray absorption techniques: μ-EXAFS and μ-XANES

These methods have been in use in materials characterization and biological studies for quite some time [34]. However, only recently has their lateral resolution been improved to the micrometre level by the use of polychromatic lenses in combination with an Si (111) channel-cut monochromator [65–67]. While EXAFS uses an energy region extending from 50 to as much as 1500 eV above the K-absorption edge of the probed element, XANES analyses the region within $\pm$ 50 eV of the absorption edge. EXAFS yields the following information [34]:

- nearest neighbour(s) distance(s);
- type(s) of nearest neighbour(s);
- coordination number;
- Debye–Waller factors, which are indicative of the degree of vibrational and static disorder.

XANES probes the shape of the absorption edge. This will change with the varying binding situation of the probed atom. Variations in the valence state can thus be determined and chemical state mapping has become feasable. The XANES spectrum is used as a 'fingerprint' spectrum for the element of interest in the material under examination. The EXAFS spectrum is mathematically manipulated in the following way [68]:

(a) Isolation of the fine structure from the general absorption.
(b) Conversion to a reciprocal space representation (k-space).
(c) Application of a Fourier transform that converts the k-space EXAFS spectrum into a kind of radial distribution function, which is called a radial structure function (RSF). The RSF

describes the local structure about the atom that absorbs the X-rays. Peaks in the RSF represent the coordination shells of atoms.

A disadvantage of these methods is the necessity to use very intense X-ray beams which are only available at large synchrotrons, e.g. the European Synchrotron Radiation Facility (ESFR) at Grenoble (France) or the HASYLAB facilities at DESY in Hamburg (Germany). A great advantage of these methods is the fact that they can be performed under ambient conditions provided that the X-ray energy in the probed region of the absorption edge of the element investigated is high enough ($Z > 11$). Only recently have these methods been used for particle characterization and work is in progress for Ni-containing particles [66].

4 TRACE AND ISOTOPIC CHARACTERIZATION OF SINGLE PARTICLES BY MASS SPECTROMETRIC METHODS AND PIXE

4.1 Secondary ion mass spectrometry (SIMS)

Relatively few methods are available for the determination of trace elements in single particles [1, 7, 48]. SIMS is known to be the most sensitive topochemical method [34, 69]. It has been successfully used for the trace characterization of single particles [15, 18]. Since it uses minimal sample amounts during analysis, it can almost be considered a nondestructive technique. A further unique characteristic of SIMS is its ability to determine isotopic abundances and isotopic ratios in single particles. This is important, e.g., for distinguishing between cosmic and Earth debris particles in erosion studies of space-exposed materials [14, 15, 18]. Owing to its good lateral resolution in the beam scanning mode (better than 0.5 μm), particles down to 0.5 μm in diameter can be characterized with conventional instrumentation. The newly designed NANOSIMS instrument can analyse particles with diameters in the 10 nm range [70]. From numerous trace element measurements of Stadermann [15] on cosmic particles, it can be deduced that at least the order of magnitude of trace constituents can be reliably determined by SIMS in single particles.

It is interesting that isotopic ratio measurements of single grains of cosmic dust by SIMS have yielded important information on pre-solar matter and on nucleosynthesis as the theory of element formation in stars [71]. In favourable cases, conclusions can be drawn on speciation from the pattern of various masses in a SIMS mass spectrum [34].

4.2 Laser micro(probe) mass spectrometry (LAMMS) and laser ablation-inductively coupled plasma-mass spectrometry (LA-ICP-MS)

LAMMS has been used extensively for single particle characterization and is a powerful tool especially for the characterization of poorly vacuum resistant organic particles [44, 48]. Equivalent in performance to SIMS, LAMMS is an off-line method because particles must be collected and mounted on a substrate. Detection of trace metals is feasible at the $\mu g\, g^{-1}$ level and speciation of inorganic compounds (especially those containing nitrogen and sulfur) is possible. Distinction between surface and volume compositions of a particle is also possible, in addition to the detection of trace organics (especially aromatics). However, large pulse-to-pulse and also particle-to-particle variations of the ion signal intensity inhibit quantification [72].

LA-ICP-MS is probably a more powerful technique owing to the much more efficient ionization of material in the inductively coupled plasma. Laser ablation of the sample, in contrast, should only produce a very fine aerosol which is quantitatively transported into the inductively coupled plasma by an argon carrier gas.

4.3 Laser desorption/ionization-time-of-flight-mass spectrometry (LDI-TOF-MS)

Laser desorption/ionization (LDI) coupled with time-of-flight MS (TOF-MS) was used by Johnston and Wexler [72] for the on-line characterization of single particles from a sampled air stream. Several particles can be sampled and analysed per second, allowing the compilation of large data sets in a very short time. This allows an improvement in the precision of analysis of particles exhibiting similar size and composition by averaging the recorded spectra. The average composition of each group is then quantitatively determined by comparison with spectra of relevant standard particles having known size and composition. Since quantitation is frequently synonymous with speciation, the latter is also possible with this method in favourable cases.

4.4 Proton induced X-ray emission (spectrometry) (PIXE)

PIXE is another method with a trace characterization capability for particles with detection limits of the order of $1-10\,mg\,g^{-1}$ [48, 73]. Other variants of nuclear microprobe techniques are still restricted to a few laboratories worldwide but can provide a wealth of information for individual particles [48]. μ-PIXE uses a well-focused proton beam and has been used for the single particle characterization of giant North Sea aerosols and other samples for major, minor and trace element analysis [74].

4.5 Three-dimensional SIMS characterization of large particles

In recent years it has been shown that three-dimensional (3D) imaging is a powerful new application of SIMS [75]. This method combines the surface imaging capabilities of SIMS with depth profiling, creating layer-by-layer images of elemental distributions as the primary ion beam sputters deeper and deeper into the sample. The amount of data produced during a typical measurement of this type can easily reach several hundred megabytes or even exceed one gigabyte. This is the reason why 3D SIMS has only become practicable with the advent of fast means of management of large amounts of data [34]. With suitable modern imaging software it is possible to convert this information into easy to understand 3D visualizations of elemental distributions within a given sample volume near the original surface [76]. When the images are created by rastering the primary ion beam, the lateral resolution achievable is only limited by the beam diameter, which can be significantly smaller than 1 μm. The depth resolution is only limited by the thickness of the ion beam mixing layer (at best several nanometres).

In general, particles represent one of the least suited categories of samples for analysis by 3D SIMS. Not only do particles often exhibit a heterogeneous composition, leading to a series of artifacts, but also their morphologies often make meaningful 3D SIMS impossible without special sample preparation and methods of corrections (e.g. eliminating effects of varying sputter rates in changing matrix compositions) [76]. Another important requirement for SIMS measurements, a flat sample surface at the beginning of analysis, is generally not complied with. However, in favourable cases, the 3D SIMS technique can be successfully applied to particle characterization [1]. Although SIMS primarily gives information on elemental (and isotopic) distributions these are very valuable for a further interpretation with regard to speciation.

5 CONCLUSION

The main incentives to this chapter were the following. An overview of modern methods of single particle characterization and speciation was, of course, the first aim. However, there is additionally a wealth of topochemical methods of particle characterization which can be used in combination to yield a profound means of particle characterization and simultaneously also give valuable answers with respect to speciation. Crystal structures of crystalline aerosol particles may also contribute significantly to the toxicity of certain aerosol compounds such as nickel sulfides. Therefore a corresponding TEM inspection seems essential, since only electron diffraction patterns will yield the necessary information on crystal structures with the necessary lateral and depth resolution useful for single particle analysis. It is the multimethod approach outlined here which produces valuable synergistic effects in an in-depth interpretation of relevant results. Of course, it is neither possible nor meaningful to use all the methods described together to provide solutions to questions related to particle characterization. However, certain methodological combinations have proved to be very successful in our experience, e.g. the combination of the bulk analytical method TXRF with the topochemical methods of HR-SEM, EPMA and TEM. The procedure developed by us for semiquantitative particle characterization by EPMA has revealed complex compositions of particles collected in the metallurgical industries [21, 24, 30, 77]. This has called for further characterization of the phase composition and crystal structures of such particles by TEM-EDX-SAED-EELS. Some relevant examples have been presented. SIMS, in contrast, can reveal the 3D compositional structure of larger particles with certain geometries [76] or the isotopic pattern of trace elements of particles, e.g. to distinguish between terrestrial and cosmic particles in near-Earth space – an important issue in modern space technology [14–16].

It is to be expected that certain combinations of the methods of solid-state characterization addressed here will lead to many new insights into the very diverse fields of science and technology for which particle characterization was shown here to be of relevance. The field of instrumental solid state particle characterization including speciation is a truly interdisciplinary one since such instrumentation is generally found in institutions of material science and/or solid state chemistry and physics.

6 ACKNOWLEDGEMENTS

The results presented here were obtained on the instrumentation of various working groups, mainly at the Institute of Material Science of the Darmstadt University of Technology but also at other institutions. I would, therefore, like to thank all our respective collaborative partners for excellent cooperation. Table 6.1.2 gives an overview on our numerous collaborative partners in particle characterization.

Table 6.1.2. Collaborative partners in single particle characterization.

Dr. Frank Stadermann	McDonnel Center for the Space Sciences, Washington University, St. Louis, MO, USA
Prof. Dr. Stephan Weinbruch Dr. Ing. Martin Ebert	Dept. of Environmental Mineralogy, Institute of Material- and Geosciences, Darmstadt University of Technology, Germany
Prof. Dr. G. Hohenberg Dr. Ing. Michael Wentzel Dipl.-Ing. Barbara Zelenka	Dept. of Combustion Engines, Institute of Mechanical Engineering, Darmstadt University of Technology, Germany
Dr. Peter Hoffmann Dipl. Ing. Burkard Höflich	Dept. of Chemical Analytics, Institute of Material- and Geosciences, Darmstadt University of Technology, Germany
Prof. Dr. Yngvar Thomassen, Asbjörn Skogstad	National Institute of Occupational Health, Oslo, Norway
Dr. V. P. Tchashchin Dr. M. Tchashchin	Kola Research Laboratory for Occupational Health, Kirovsk, Russia
Prof. Dr. Evert Nieboer	Dept. of Biochemistry and Occupational Health, McMaster University, Hamilton, Ontario, Canada and Institute of Community Medicine, University of Tromsö, Tromsö, Norway
Dr. G. Helas	Dept. of Biogeochemistry, Max Planck Institute for Chemistry, Mainz, Germany
Dr. Ing. W. Schulmeyer	Plansee AG, Dept. of Refractory Metals, Reutte, Tirol, Austria

7 REFERENCES

1. Ortner, H. M., Hoffmann, P., Stadermann, F. J., Weinbruch, S. and Wentzel, M., *Analyst*, **123**, 833 (1998).
2. Pruppacher, H. R. and Klett, J. D., *Microphysics of Clouds and Precipitation*, Reidel, Dordrecht, 1978.
3. Andreae, M. O., in *World Survey of Climatology*, Vol. 16: *Future Climates of the World: A Modelling Perspective*, Henderson-Sellers, A. (Ed.), Elsevier, Amsterdam, 1995, p. 347.
4. Willeke, K. and Baron, P. A. (Eds), *Aerosol Measurement*, Van Nostrand Reinhold, New York, 1993.

5. Ebert, M., Dahmen, J., Hoffmann, P. and Ortner, H. M., *Spectrochim. Acta, Part B*, **52**, 967 (1997).
6. Ebert, M., Hoffmann, P., Ortner, H. M. and Dahmen, J., *GIT Fachz. Lab.*, **10/96**, 982 (1996).
7. Ortner, H. M., *J. Environ. Monit.*, **1**, 273 (1999).
8. Grasserbauer, M. and Werner, H. W. (Eds), *Analysis of Microelectronic Materials and Devices*, John Wiley & Sons, Ltd, Chichester, 1991.
9. Ortner, H. M. and Wilhartitz, P., *Fresenius' J. Anal. Chem.*, **337**, 686, (1990).
10. Ortner, H. M., *The Influence of Trace Elements on the Properties of Hard Metals*, COST 503, Material Sciences Series: Powder Metallurgy – Powder Based Materials, Vol. IV, Valente, I. (Ed.), European Communities, Brussels, 1997, p. 70.
11. Hunt, T. M., *Handbook of Wear Debris Analysis and Particle Detection in Liquids*, Elsevier, London, 1993.
12. Ortner, H. M., Petter, H. and Birzer, W., in *Proceedings of the First Plasma-Technique Symposium*, Vol. **2**, Eschnauer, H., Huber, P., Nicoll, A. R. and Sandmeier, S. (Eds), Plasma Technik, Switzerland, 1988, p. 237.
13. Nietmantsverdriet, J. W., *Spectroscopy in Catalysis*, VCH, Weinheim, 1993.
14. Stadermann, F. J. and Jessberger, E. K., in *Proceedings of the European Conference on Space Debris*, ESA SD-01, 1993, p. 185.
15. Stadermann, F. J., Measurement of isotopic and elemental abundancies in single interplanetary dust particles by secondary ion mass spectrometry, Ph.D. Dissertation, University of Heidelberg, 1990 (in German).
16. McDonnel, A. M. (Ed.), *Cosmic Dust*, John Wiley & Sons, Ltd, Chichester, 1978.
17. Allamandola, L. J., Sandford, S. A. and Wopenka, B., *Science*, **237**, 56 (1987).
18. Stadermann, F. J. and Olinger, C. T., *Meteoritics*, **27**, 291 (1992).
19. Walls, H. J., *Forensic Science*, 2nd edn, Sweet and Maxwell, London, 1968.
20. Rampino, M. R., in *World Survey of Climatology*, Vol. 16: *Future Climates of the World: a Modelling Perspective*, Henderson-Sellers, A. (Ed.), Elsevier, Amsterdam, 1995, p. 95.
21. Höflich, B. L. W. and Ortner, H. M., Final Report: *Characterization of Single, Respirable Particles In Aluminium Potrooms at Norsk Hydro Aluminum*, 2001.
22. Wilson, R. and Spengler, J. D. (Eds) *Particles in Our Air: Concentrations and Health Effects*, Harvard School of Public Health, 1996.
23. Siegel, R. W., in *Material Science and Technology*, Vol. 15: *Processing of Metals and Alloys*, Cahn, R. W. (Ed.), VCH, Weinheim, 1991, p. 583.
24. Morlang, A., Phase analysis of single particles by electron diffraction in SEM and TEM, Diploma Thesis, Institute of Material Sciences, Darmstadt University of Technology, 2000, (in German).
25. Sundermann, F. W., Jr., in *Nickel in the Human Environment*, Sundermann, F. W., Jr (Ed.), International Agency for Research on Cancer, Lyon, 1984, p. 127.
26. Jaenicke, R., in *Atmosphärische Spurenstoffe*, Jaenicke, R. (Ed.), VCH, Weinheim, 1987, p. 321.
27. Harrison, R. M. and Van Grieken, R. (Eds.), *Atmospheric Particles*, IUPAC Series on Analytical and Physical Chemistry of Environmental Systems, Vol. 5, John Wiley & Sons, Ltd, Chichester, 1998.
28. Cahier, H., Carbonaceous combustion aerosols, in *Atmospheric Particles*, Harrison, R. M. and Van Grieken, R. (Eds), *IUPAC Series on Analytical and Physical Chemistry of Environmental Systems*, Vol. 5, John Wiley & Sons, Ltd, Chichester, 1998, p. 296.
29. Koropchak, J. A., Sadain, S., Yang, X., Magnusson, L.-E., Heybroek, M., Anisimov, M. and Kaufman, S. L., *Anal. Chem.*, **71**, 386A (1999).
30. Höflich, B. L. W., Wentzel, M., Ortner, H. M., Weinbruch, S., Skogstad, A., Hetland, S., Thomassen, Y., Chaschin, V. P. and Nieboer, E., *J. Environ. Monit.*, **2**, 213 (2000).
31. Model 3089 Nanometer Aerosol Sampler, Advance Product Information, TSI, St. Paul, MN, USA 2000.
32. Dixkens, J. and Fissan, H., *Aerosol Sci. Technol.*, **30**, 438 (1999).
33. Hincks, W., *Aerosol Technology*, John Wiley & Sons, Inc., New York, 1982.
34. Ortner, H. M., in *Analytiker Taschenbuch*, Vol. 19, Günzler, H. *et al.* (Eds), Springer, Berlin, 1998, p. 217.
35. The XL ESEM series, Company Brochure of Philips Electron Optics, Eindhoven, Netherlands.
36. The XL 30 SFEG Scanning Electron Microscope, Technical Data Sheet, Company Brochure of Philips Electron Optics, Eindhoven, Netherlands.
37. Köllensperger, G., Friedbacher, G., Kotzick, R., Niessner, R. and Grasserbauer, M., *Fresenius' J. Anal. Chem.*, **364**, 296 (1999).
38. Randke, V., *Electron Backscatter Diffraction*, Guide Book Series, Oxford Instruments, High Wycombe, England, 1996.
39. Weinbruch, S., Wentzel., M., Kluckner, M., Hoffmann, P. and Ortner, H. M., *Mikrochim. Acta*, **125**, 137 (1997).
40. Hoffmann, P., Dedik, A. N., Ensling, J., Weber, S., Sinner, T., Gütlich, P. and Ortner, H. M., *J. Aerosol Sci.*, **27**, 325 (1996).
41. Armstrong, J. T., in *Electron Probe Quantitation*, Heinrich, K. F. J. and Newbury, D. E. (Eds), Plenum, New York, 1991, p. 261.
42. Armstrong, J. T. and Buseck, P. R., *X-ray Spectrom.*, **14**, 172 (1985).
43. Kluckner, M., Brandl, O., Weinbruch, S., Stadermann, F. J. and Ortner, H. M., *Mikrochim. Acta*, **125**, 129 (1997).
44. Jambers, W., De Bock, L. and Van Grieken, R., *Analyst*, **120**, 681 (1995).
45. Chul-Un Ro, Osan, J. and Van Grieken, R., *Anal. Chem.*, **71**, 1521 (1999).

46. Smith, R.-U. (Ed.), *Handbook of Environmental Analysis*, 2nd edn, AOAC, Mc Lean, VA, 1995.
47. Klockenkämper, R., in *Total Reflection X-ray Fluorescence Analysis*, Winefordner, J. D. (Ed.), *Chemical Analysis Series*, Vol. 140, John Wiley & Sons, Inc., New York, 1997, p. 151.
48. Xhoffer, C., Wouters, L., Artaxo, P., Van Put, A. and Van Grieken, R., in *Environmental Particles I*, Vol. 1, IUPAC Series on Environmental Analytical Chemistry and Physical Chemistry, Buffle, J. and van Leeuwen, H. P. (Eds), Lewis, Boca Raton, FL, 1992, p. 107.
49. Meisel, A., Leonhardt, G. and Szargan, R., *X-ray Spectra and Chemical Binding*, Springer Series in Chemical Physics, Vol. 37, Springer, Berlin, 1989.
50. Höfer, H. E., Brey, G. P., Schulz-Dobrick, B. and Oberhäusli, R., *Eur. J. Mineral.*, **6**, 407 (1994).
51. Höflich, B. L. W., Rausch, A., Weinbruch, S., Helas, G., Ortner, H. M. and Ebert, M., Speciation of sulfur in individual aerosol particles by precisive measurement of chemical shift of the SK_α X-ray line in the electron microprobes, submitted to *Mikrochim. Acta.*
52. Miehe, G., in *Referate, 3. Jahrestagung der Deutschen Gesellschaft für Kristallographie*, Oldenbourg, Munich, 1995, p. 51.
53. Zaefferer, S., *J. Appl. Crystallogr.*, **33**, 10 (2000).
54. Bischoff, E., Campbell, G. H. and Rühle, M., *Fresenius' J. Anal. Chem.*, **337**, 469 (1990).
55. Hofer, T. and Warbichler, P., *Ultramicroscopy*, **63**, 21 (1996).
56. Hofer, T., Warbichler, P. and Grogger, W., *Ultramicroscopy* **59**, 31 (1995).
57. Kriwanek, O.-L., Gubbens, A. J., Bellby, N. and Meyer, C. E., *Microsc. Microanal. Microstruct.*, **3**, 187 (1992).
58. Grogger, W., Hofer, T. and Kothleitner, G., *Mikrochim. Acta*, **125**, 13 (1997).
59. Grasserbauer, M., Dudek, H. J. and Ebel, M. F., *Angewandte Oberflächenanalyse*, Springer, Berlin, 1985.
60. Weber, S., Hoffmann, P., Ensling, J., Dedik, A. N., Weinbruch, S., Miehe, G., Gütlich, P. and Ortner, H. M., *J. Aerosol Sci.*, **31**, 987 (2000).
61. Sedlak, D. L. and Hoigné, J., *J. Atmos. Environ.*, Part A, **27**, 2173, (1993).
62. Seinfeld, J. H., *Atmospheric Chemistry of Air Pollution*, John Wiley & Sons, Inc., New York, 1986.
63. Hoffmann, P., Dedik, A. N., Deutsch, F., Ebert, M., Hein, M., Hoffmann, H., Lieser, K. H., Ortner, H. M., Schwarz, M., Sinner, T., Weber, S., Weidenauer, M. and Weinbruch, S., in *Dynamics and Chemistry of Hydrometeors*, Jaenicke, R. (Ed.), Deutsche Forschungsgemeinschaft, Wiley-VCH, Weinheim, 2001, p. 440.
64. Gütlich, P., Linck, R. and Trautwein, A., *Mössbauer Spectroscopy and Transition Metal Chemistry*, Springer, Berlin, 1978.
65. Janssens, K., Vince, L., Wie, F., Proost, K., Vekemans, B., Vittiglio, G., Yan, Y. and Falkenberg, G., Feasibility of (trace-level) micro-XANES at Beamline L, in *HASYLAB Jahresbericht 1999*, Materlik, G. *et al.* (Eds), Hamburg, 1999.
66. Salbu, B., Janssens, K., Irekling, T., Simionovici, A., Drakopoulos, M., Raven, C., Snigireva, I., Snigirev, A., Lind, O. C., Oughton, D. H., Adams, F. and Kashparov, V. A., Micro XANES for characterization of fuel particles, in *ESRF Highlights 1999*, European Synchrotron Radiation Facility 2000, Grenoble, France, p. 24.
67. Janssens, K. H. A. (Ed.), *Microscopic X-ray Fluorescence Analysis*, John Wiley & Sons, Ltd, Chichester, 2000, p. 370.
68. Fariborz Goodarzi and Huggins, F. E., *J. Environ. Monit.*, **3**, 1 (2001).
69. Ortner, H. M. and Wilhartitz, P., *Mikrochim. Acta II*, 177 (1991).
70. Schuhmacher, M., Rasser, B., De Chambost, E., Hillion, F., Mootz, Th. and Migeon, H. N., *Fresenius' J. Anal. Chem.*, **365**, 12 (1999).
71. Bernatowicz, T. J. and Walker, R. M., *Phys. Today*, **50**, 25 (1997).
72. Johnston, M. V. and Wexler, A. S., *Anal. Chem.*, **67**, 721A (1995).
73. Injuk, J., Breitenbach, L., Van Grieken, R. and Wätjen, U., *Mikrochim. Acta*, **114/115**, 313 (1994).
74. Injuk, J. and Van Grieken, R., *Spectrochim. Acta, Part B*, **50**, 1787 (1995).
75. Gara, S., Stingeder, G., Tian, C., Hutter, H., Führer, H. and Grasserbauer, M., in *Secondary Ion Mass Spectrometry: SIMS VIII*, Benninghoven, A., Janssen, K. T. F., Thümpner, J. and Werner, H. W. (Eds), John Wiley & Sons, Ltd, Chichester, 1992, p. 537.
76. Stadermann, F. J. and Ortner, H. M., in *Secondary Ion Mass Spectrometry: SIMS X*, Benninghoven, A., Hagenhoff, B. and Werner, H. W. (Eds), John Wiley & Sons, Ltd, Chichester, 1997, p. 325.
77. Gunst, S., Weinbruch, S., Wentzel, M., Ortner, H. M., Skogstad, A., Hetland, S. and Thomassen, Y., *J. Environ. Monit.*, **2**, 62 (2000).

6.2 Direct Speciation of Solids: X-ray Absorption Fine Structure Spectroscopy for Species Analysis in Solid Samples

Edmund Welter

Hamburger Synchrotronstrahlungslabor (HASYLAB) at Deutsches Elektronen-Synchrotron (DESY), Hamburg, Germany

1 INTRODUCTION

Many of the known analytical methods for elemental speciation analysis are restricted to fluid samples, for instance the whole range of chromatographic methods. Consequently these methods are applicable to solid samples only if it is possible to dissolve the analyte without destruction of the species information. This is often a tedious and error-prone operation. An analytical method circumventing this problem by enabling the direct species determination in solids is thus highly desirable. Such a technique is provided by X-ray absorption fine structure (XAFS) spectroscopy.

Another positive property of XAFS spectroscopy is that it yields the species information without being significantly disturbed by the matrix due to its high degree of element selectivity. This is in contrast to other methods that enable the determination of the chemical speciation only in a *pure* or at least *highly concentrated* solid sample. Examples for the latter methods are X-ray diffraction (XRD) and IR spectroscopy. Their applicability is limited to more or less pure compounds because they cannot differentiate between analytical signals from the analyte and the matrix, a fact that makes it impossible to identify single species of the analyte in the sample.

The basic theoretical background, the experimental implementation and further enhancements of the applicability by introduction of new experimental techniques such as measuring XAFS spectra with μm spatial resolution and X-ray Raman spectroscopy will be discussed in the

Handbook of Elemental Speciation: Techniques and Methodology R. Cornelis, H. Crews, J. Caruso and K. Heumann
 ISBN: 0-471-49214-0

following. Some examples of contemporary use in species analysis will be given to demonstrate the potential of the method.

2 BASIC XAFS THEORY, THE REASON WHY XAFS SPECTROSCOPY IS ELEMENT *AND* SPECIES SELECTIVE

XAFS theory is discussed in detail in several monographs [e.g. 1–3]. In the following only the basic elements required to understand the analytically interesting element and species selectivity of XAFS spectroscopy can be discussed.

As in other absorption spectroscopical methods, for example in the UV/Vis or IR region, in X-ray absorption spectroscopy (XAS) the dependence of the absorption coefficient μ on the wavelength of the incoming X-ray beam is measured (equation 6.2.1). In X-ray spectroscopy the abscissa is usually scaled in energy units (eV), not wavelength or frequency.

$$\mu d = \ln \frac{I_1}{I_2} \tag{6.2.1}$$

with μ being the absorption coefficient in units of cm^{-1}, d the sample thickness in cm and I_1, I_2 the photon beam intensities in front of and behind the sample.

Figure 6.2.1 shows a simplified X-ray absorption spectrum of a sample that contains Fe, Cu and a small amount of Zn. Unlike the absorption spectra in UV/VIS or IR spectroscopy the X-ray absorption spectrum shows no peaks but edges at which the absorption coefficient μ increases abruptly. The reason for the difference between absorption spectra in the UV/Vis and in the X-ray region is that in XAS an electron is excited to the continuum whereas the transitions in optical UV spectroscopy take place between energetically well defined orbitals, see Figure 6.2.2.

The edge position corresponds to the energy that is necessary to lift a core electron from an inner shell to the continuum. This energy is specific for every element, thus making XAFS spectroscopy an element-selective method. The edges that are most

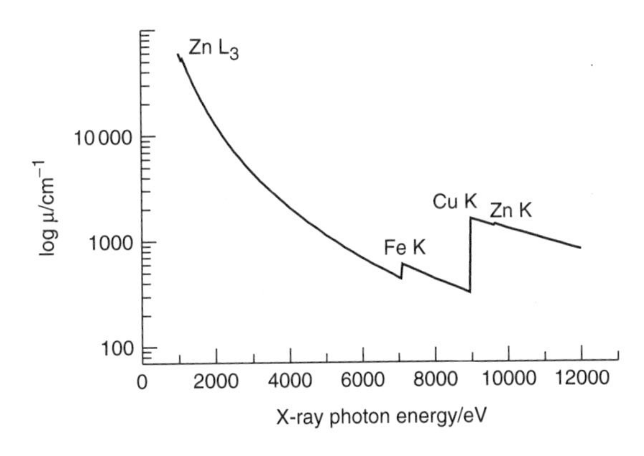

Figure 6.2.1. Absorbance coefficient (μ) over X-ray photon energy calculated for a sample with the hypothetical sum formula $Fe_2Cu_{20}Zn$ (spectrum calculated by use of XOP program package) [4].

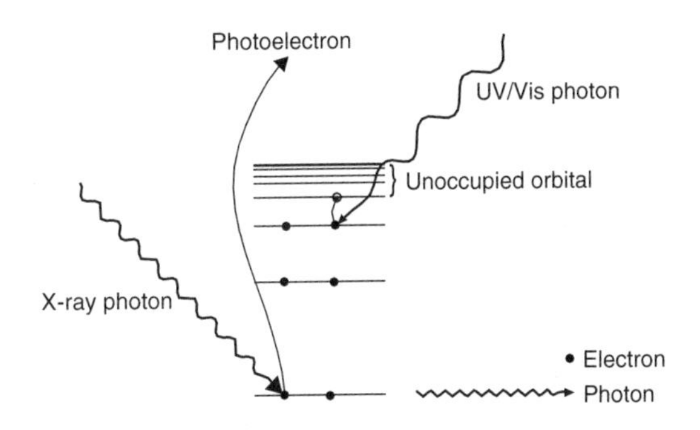

Figure 6.2.2. Schematic visualization of the reasons for the different shape of UV/Vis (peaks) and X-ray absorption spectra (edges). UV/Vis spectroscopy probes the small but discrete energy differences between the highest occupied and the lowest unoccupied atomic or molecular orbitals. Absorption of an X-ray photon lifts a core shell electron to the continuum.

often used are the K-edges. They result from the excitation of a K-shell electron to the continuum. For elements of higher Z the L-edges, normally the L_3-edge, are also used for XAFS spectroscopy. K-edge positions range from some 100 eV (K-edge spectra of low Z elements such as C, N and O) to ~115 keV (K-edge of uranium). The distances between edges of consecutive elements are of the order of some 100 eV for the 3d transition metals and ~3000 eV for the K-edges of the actinides.

Small perturbations of the core ground states due to the redox state lead to a 'chemical shift' of the edge position depending on the oxidation state of the absorbing atom. This chemical shift

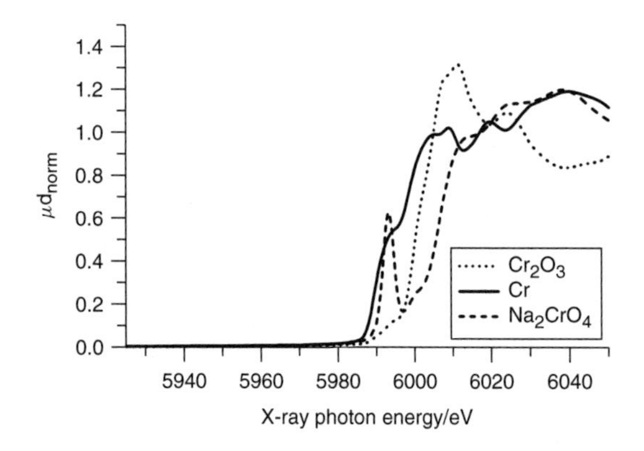

Figure 6.2.3. XANES spectra of three different Cr compounds with different Cr oxidation states, showing the chemical shift of the edge position. Note the large pre-edge peak of the Cr(VI) compound; this peak can be used to identify and quantify Cr(VI) in mixtures.

is normally of the order of some eV. This is exemplified in Figure 6.2.3, which shows the edge region of the XAS of three chromium compounds (Cr, Cr_2O_3 and Na_2CrO_4) that contain Cr in three different oxidation states. In general the edge position of the more highly oxidized compounds is shifted to higher energies, because the core electrons are bound more strongly in these compounds. However, Figure 6.2.3 not only shows the small shifts in the edge position caused by the oxidation state of the absorbing element. It can also be seen that the region above the edge is not as smooth as it is in the simplified spectrum in Figure 6.2.2 and instead shows well visible oscillations of the absorption coefficient above the edge. These oscillations can be found up to 1000–2000 eV and even higher above the edge depending on the element, the chemical compound of the element and experimental conditions. XAFS spectroscopy is based exactly on these oscillations of the absorption coefficient.

The fine structure of the absorption edge was first detected in 1920, but only with the invention of synchrotron radiation as a source of very intense X-rays to XAFS spectroscopy in 1974 did it become possible to measure XAFS spectra within a reasonable time and in diluted samples. The region above the edge is normally divided into two subregions, the first 50–100 eV above the edge are called X-ray absorption near edge structure (XANES), the region above the XANES region is called extended X-ray absorption fine structure (EXAFS).

A first idea for the understanding of the occurrence of the fine structure can be drawn from the comparison of XAS spectra from a monoatomic gas such as Kr and a diatomic gas such as Br_2 [5]. The spectrum from the monoatomic gas shows only some structures in the actual edge and absolutely no EXAFS oscillations, whereas the spectrum from a diatomic gas shows both some structures in the actual edge and oscillations in the EXAFS region. This finding indicates that the EXAFS oscillations might result from effects caused by neighboring atoms whereas the structure in the edge is caused by 'internal' effects in the absorber atom. The latter is in fact a result of transitions of a core shell electron to higher unoccupied orbitals, whereas the oscillations in the EXAFS region are caused by a different effect that will be qualitatively discussed in the following.

The fine structure, XANES as well as EXAFS, contains much more information about the sample than simply detecting the oxidation state of the absorber atom. It contains information about nature, number and distance of the next neighboring atoms. Figure 6.2.4 shows the XAS spectra of four different Pb compounds. It is clearly seen that the fine structure above the edge is unique for every single Pb compound. The reason for the occurrence of the fine structure oscillations and

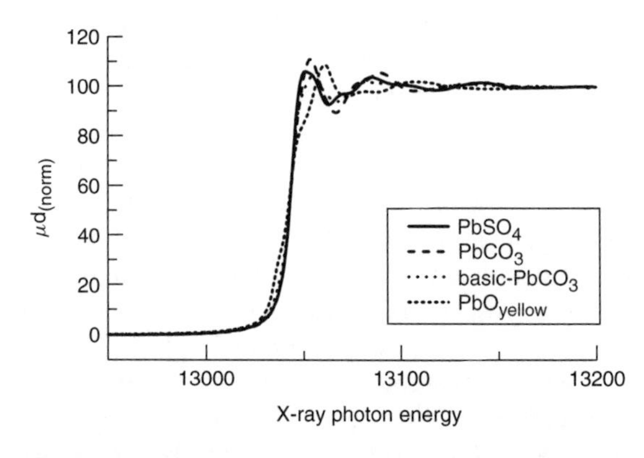

Figure 6.2.4. Normalized near-edge XAFS spectra of four inorganic Pb(II) compounds; spectra like these are used as reference spectra for fingerprint analysis of Pb-containing samples.

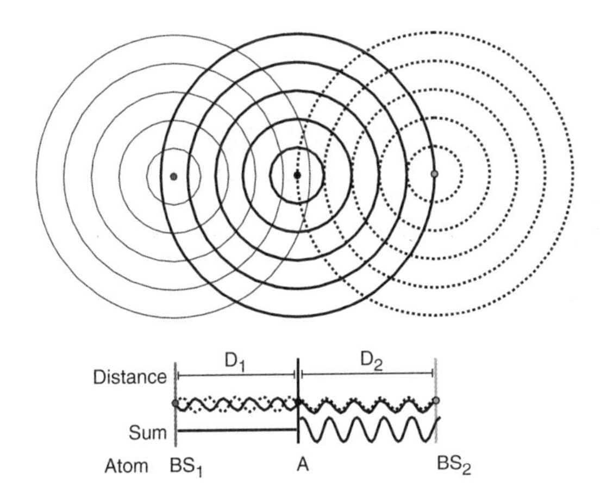

Figure 6.2.5. Qualitative explanation for the occurrence of the XAFS oscillations. Three atoms are shown, the absorbing atom A and two backscattering neighbor atoms (BS_1 and BS_2) at two different distances D_1 and D_2 with D_2 being larger than D_1 and $(D_2 - D_1)$ equal to half the wavelength of the outgoing photoelectron wave (note that the sinusoidal curves neglect the influence of the sample specific phase shifts during the scattering process).

their species selectivity can be understood qualitatively by looking at Figure 6.2.5.

If the energy of the absorbed X-ray photon $(E_{h\nu})$ is higher than the energy (E_0) that is necessary to lift a core shell electron to the continuum, an electron from an inner shell is lifted to the continuum. This photoelectron leaves the atom with a kinetic energy (E) given by:

$$E = E_{h\nu} - E_0 \tag{6.2.2}$$

To understand the occurrence of the fine structure oscillations it is necessary to look at the wave properties of the photoelectron. The wavelength (λ) of the outgoing spherical photoelectron wave is linked to the energy of the absorbed X-ray photon $(E_{h\nu})$ by:

$$\lambda = \frac{2\pi}{k} \tag{6.2.3a}$$

where

$$k = \sqrt{\frac{2m}{\hbar^2}(E_{h\nu} - E_0)} \tag{6.2.3b}$$

with m = mass of the electron, $E_{h\nu}$ and E_0 in joules.

Equation (6.2.2) shows that the energy of the photoelectron increases with increasing energy of the incident X-ray photons during the XAFS scan. This leads to a variation of the wavelength of the photoelectron wave (equations 6.2.3a and b). The outgoing photoelectron wave is partially reflected by the neighboring atoms. The interference between outgoing and reflected electron waves, which can either be constructive or destructive at the origin, causes the oscillations of the absorption cross-section, which is measured as a function of the energy of the incoming photon beam. A measurable reflection of the outgoing photoelectron wave is caused only by the nearest neighbor atoms. That means the fine structure probes the immediate chemical environment, the short range order (typically 2–3 nearest neighbors), of the absorbing atom species and gives information about the nature, distance and number of the nearest neighbors of the absorber atom.

The restriction to the determination of the short range order means that XAFS spectroscopy is also applicable in amorphous samples, whereas XRD probes the average position on lattices and is therefore normally performed on crystalline solid samples.

3 EXAFS DATA EVALUATION

The process most frequently used to evaluate the information contained in the EXAFS spectra is a multistep process, which is based on the 'short range, single electron, single scattering theory'. The model presented before is the basis of this theory. Major steps of the evaluation procedure are in the order they have to be done [6]:

- Background correction (subtraction of the background absorption that is caused by Rayleigh scattering, Compton scattering and by preceding absorption edges).
- Normalization of the edge jump to 1.
- Conversion of the initial energy axis to the photoelectron wave vector scale ('k-scale') using equation (6.2.4).

$$k = 0.152\sqrt{E - E_0} \tag{6.2.4}$$

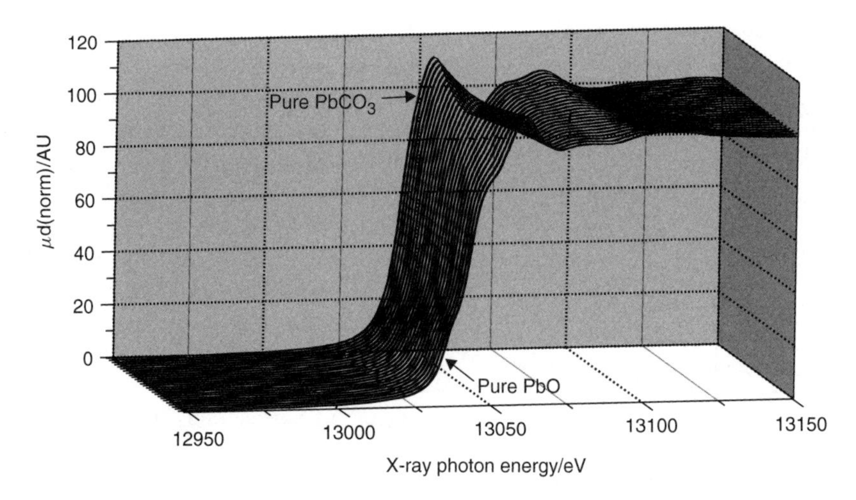

Figure 6.2.6. Linear combinations of normalized PbO and $PbCO_3$ XAFS spectra. The spectrum of pure PbO is shown in front that of pure $PbCO_3$ as the last spectrum. Linear combinations like these can be compared with experimental spectra from samples containing mixtures of compounds to determine the relative amounts of the constituents.

where E and E_0 are given in eV. Equation (6.2.4) can be calculated from equation (6.2.3b) by employing the values of m and $\hbar$ and converting the energy units from joules to electronvolts.
- Fourier transformation of the spectrum.
- Fitting (modeling) of theoretical spectra by use of several computer codes.

All steps are performed by use of special computer programs, see for instance refs 7 and 8.

This classical EXAFS data analysis is most useful, however, to analyze the spectra of pure compounds of the absorber atom or of single compounds of an element in a matrix consisting of other elements. It is not applicable to mixtures of several components of the same absorber atom. The calculated results like the measured spectra would be weighted linear combinations of the respective parameters of the different substances. This is exemplified in Figure 6.2.6, which shows the spectra of pure $PbCO_3$ and pure PbO framing spectra of mixtures of the two compounds. The spectra shown were calculated using linear combinations of the two pure substance spectra. Mixed spectra as they are often obtained from natural samples make the standard evaluation procedure inapplicable for species analysis in these type of samples.

Alternative evaluation strategies for XAFS spectra which can be used in samples that contain more than one species, will be presented with the examples for the use of XAFS spectroscopy in species analysis, especially in those from environmental species analysis.

4 THE XAFS EXPERIMENT

4.1 Transmission

The principal experimental setup that is needed to measure XAFS spectra is shown in Figure 6.2.7. It is basically the same as in any other absorption

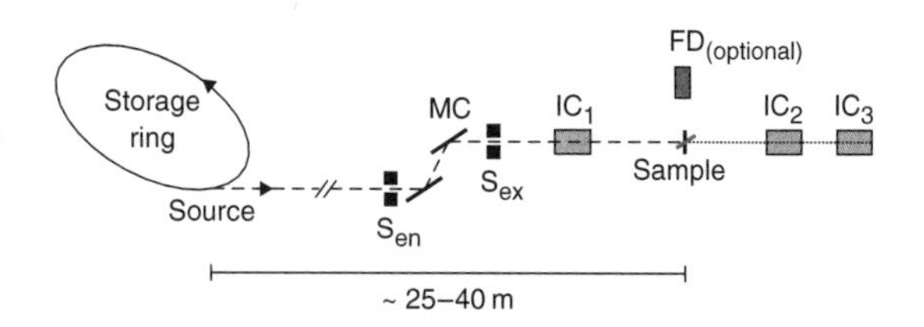

Figure 6.2.7. Schematic drawing of a synchrotron radiation XAFS experiment in transmission (black sample) and in fluorescence mode (grey sample and additional fluorescence detector) at a bending magnet beamline (S_{en}, monochromator entrance slit; S_{ex}, monochromator exit slit; MC, monochromator, IC1 – IC3, ionization chambers 1–3; FD, fluorescence detector). IC3 is used to measure an energy reference simultaneously.

spectroscopy method. One small difference is that the sample is usually positioned *between* a detector that measures the incoming intensity and a detector that measures the transmitted intensity. The basic elements of an XAFS experiment are the X-ray source, a monochromator, two detectors that measure the incident and the transmitted intensity and some kind of sample holder device to handle the sample itself.

In almost all XAFS experiments synchrotron radiation is used as the X-ray source, although there are concepts for laboratory XAFS spectroscopy equipment using X-ray tubes. The main reasons for the use of synchrotron radiation are the higher flux and the easy tunability of these sources. The small divergence of the synchrotron beam adds to the advantages for spectroscopy.

Synchrotron radiation is emitted during the acceleration of charged particles such as electrons or positrons. The first sources of synchrotron radiation were the bending magnets that are used to force the charged particles that are accelerated in a synchrotron ring into the orbital track. The synchrotron light from bending magnets is emitted in a narrow cone tangentially to the circumference of the synchrotron ring. The dipole radiation emitted by a synchrotron source covers a wide energy (or wavelength) range, starting in the infrared and going up to the hard X-ray and even γ-ray region.

The demand for higher flux led to the invention of special devices for the production of even more intense light. These insertion devices are called 'wigglers' and 'undulators'. Wigglers increase the flux by adding the dipole radiation of several alternating magnets which force the charged particles into a sinusoidal trajectory. At undulators the emission of photons at every dipole occurs – in contrast to wigglers where the emission of synchrotron radiation happens independently at every dipole – with a fixed phase relation. The emission spectrum of wigglers shows a broad energy spread like the emission spectrum of a bending magnet. In the case of an undulator the emission spectrum is overlaid by several small emission lines with very high intensity. This emission characteristic makes undulators less suited for EXAFS scans, which cover an energy range much broader than the emission lines. If one accepts the higher experimental effort associated with the use of undulators they are nevertheless very valuable sources for experiments which need an extremely high photon flux, for instance the registration of XAFS spectra from highly diluted samples.

However in many cases the flux of the source is not the limiting factor for an XAFS experiment. That is the reason why even today many XAFS experiments are still performed at bending magnet beamlines. They simply offer a sufficiently high intensity to measure spectra with good signal/noise ratio from pure or moderately diluted samples.

An XAFS scan requires monochromatic (mono-energetic) X-ray light, which must be scanable over a range of 1000–2000 eV. The monochromatic X-ray light is produced by the use of crystal monochromators. The two-crystal design shown in Figure 6.2.8 is most widespread. The major advantage of this design compared to a single-crystal design is that the monochromatic beam leaves the monochromator horizontally, with only a small vertical offset to the white beam. The material in most widespread use for the analyzer crystals is Si, because crystals of the required size and purity are available relatively cheaply. Furthermore, Si crystals have appropriate plane distances and a good reflectivity for measurements in the X-ray region. Crystals for different energy regions and spectral resolutions can be produced by cutting monocrystalline silicon along certain crystal planes, thus producing crystals with different distances (d) of the crystal planes. The transmitted wavelengths can be calculated using Bragg's law given by

$$n\lambda = 2d \sin\theta \qquad (6.2.5)$$

with $n = 1, 2, 3, \ldots$, λ = wavelength, d = crystal lattice spacing and θ = incidence angle.

Transmitted radiation with $n > 1$ is called 'higher harmonic'; this higher harmonic radiation must be excluded, because it produces a large background. There are two common techniques to achieve this. First, the use of mirrors. This strategy makes use of the fact that the angle under which total reflection occurs on a surface rapidly decreases with wavelength, so higher harmonics

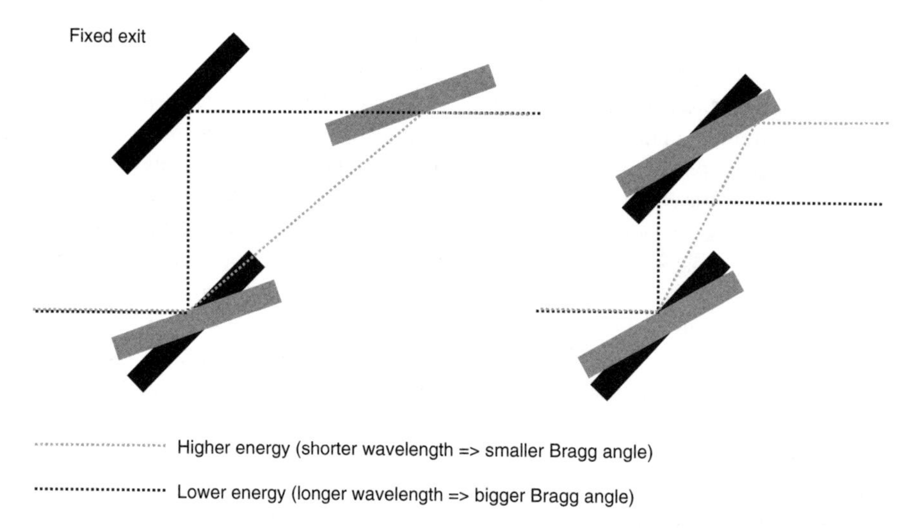

Figure 6.2.8. Principal setup of the two double crystal monochromator designs in most widespread use for XAFS spectroscopy. Energy scans are performed by rotating the crystals around axis 1, thus changing the angle θ (Bragg angle) under which the incident beam hits the first crystal. The fixed exit setup (left) needs two goniometers for the parallel rotation of both crystals and a linear movement for the translation of the second crystal. The alternative design (right) needs only one goniometer on which both crystals can be mounted but makes it necessary to move the sample and the exit slit accordingly.

with half the wavelength can be excluded almost completely using appropriate glancing angles. The second technique is to detune the double crystal monochromator slightly, which means tilting one of the crystals slightly, so that both crystals are no longer perfectly parallel. This works because the bandwidth of the transmitted radiation is much greater for the first harmonic than for that of the higher harmonics. Furthermore, the use of certain crystal planes such as Si(111), where the second order reflection is prohibited, further reduces higher harmonic contents [9].

The energy resolution of a crystal monochromator is determined by two main factors. Firstly the width of the reflected Bragg peak, which is a function of the chosen crystal material, of the chosen reflex and of the quality of the crystal. The second important factor is the divergence of the incident beam, because it determines the range of incident angles on the crystal. The divergence of laboratory sources such as X-ray tubes is normally decreased by use of collimators. Because of the small divergence of the synchrotron beam this is not necessary for XAFS measurements for which an energy resolution of 0.5 to some eV is sufficient. A higher spectral resolution would on the contrary be counterproductive, because the higher the resolution of the monochromator is, the lower is the transmitted flux and consequently the statistical quality of the data.

The spatial emission properties of synchrotron radiation also make it possible to work without focusing optical elements. Without focusing optical elements, a typical size of the beam on the sample would be 10 mm horizontal and 1 mm vertical. The beam shape is finally defined by slit systems, which also have the function to minimize stray light. The beam size of an unfocused beamline can be reduced using slits, but again only for the price of statistical quality. Focussing mirrors (toroidal or elliptical) are used in many XAFS beamlines to increase the flux density on the sample and to cut off higher harmonic (n in equation (6.2.5) >1) radiation.

4.2 Fluorescence yield

XAFS measurements in transmission work well in pure or highly concentrated samples. With increasing dilution the problem that the small edge of interest is located on a high background absorption becomes more and more important. This can be seen in Figure 6.2.1 where a small Zn K-edge is positioned on the much larger background which is mainly caused by Cu and Fe.

Therefore, diluted samples cannot be measured with the basic transmission mode XAFS experiment. Fortunately there are alternatives to the direct measurement of the absorption. One analytical signal, whose height is proportional to absorbance but much less affected by the background, is the fluorescence. Fluorescence photons are generated during the relaxation process of the ionized absorber atom. There are several possible relaxation processes but the emission of a fluorescence photon is the most important relaxation process for all elements with $Z > 32$ (Ge) [10] (for lighter elements the emission of an Auger electron is the preferred relaxation process). Fluorescence photons are emitted when an electron from a higher shell decays down to the hole in the K or L shell that is left after the emission of the photoelectron. The energy of the fluorescence photon is equal to the energy difference between the two orbitals, and is thus characteristic of the emitting element and always lower than the energy of the absorbed incident X-ray photon.

To measure XAFS spectra in fluorescence mode means that the intensity of the respective fluorescence line, which is proportional to the number of *absorbed* photons, is measured while the energy of the incident beam is scanned over the edge and the EXAFS region of the absorbing element as is done in transmission mode. Figure 6.2.9 shows the development of the Pb L X-ray fluorescence spectra during a scan of the incident energy between 12 800 and 13 200 eV. Clearly visible are the Pb $L_{\alpha 1,2}$ and the Pb $L_{\beta 2.15}$ emission lines that appear after the energy of the incident photons is high enough (13 035 eV) for the excitation of an L shell electron to the continuum. The intensity of these lines shows the first XAFS oscillations. The third line, which is visible before the Pb emission lines appear and which disappears under the Pb $L_{\beta 2.15}$ line, is caused by elastically scattered incident photons.

The fluorescence technique obviously makes a higher experimental effort necessary. The major

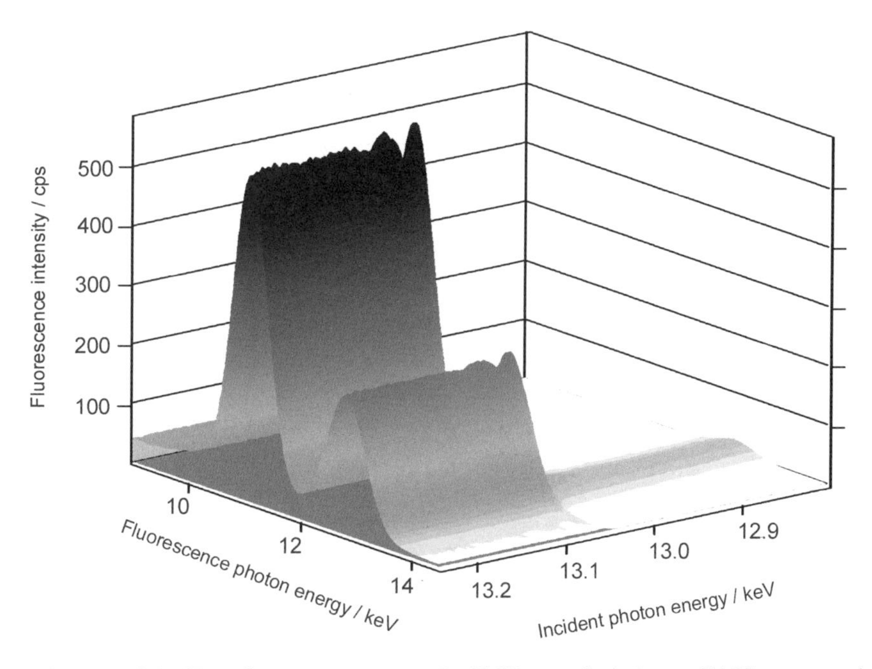

Figure 6.2.9. Development of the X-ray fluorescence spectra of a $PbCO_3$ sample during an XAFS scan over the Pb L_3-edge. The peaks are: Pb $L_{\alpha 1,2}$ (10 500 eV), Pb $L_{\beta 2,15}$ (12 622 eV) and elastically scattered photons very weak but visible before the edge (equal to incident beam energy). Spectra measured with a Ge detector optimized for high count rates.

addition to the standard transmission experiment is the detector system which consists of the actual detector and the signal processing electronics. It has to fulfill two major, partially contradictory, requirements. Firstly it must offer a good energy resolution, in most real samples at least 1000 eV, preferably much better. Secondly the detector system must handle high count rates (>100 kcps), because otherwise the time needed for measurements with satisfactory signal/noise ratio would be unacceptably long.

Today energy-dispersive semiconductor detectors which are optimized for high count rates are used routinely for this task. They make it possible to measure at count rates of 100 kcps with an energy resolution of 250–500 eV. With these detector systems the separation of the emission lines from the different elements in the sample (analyte/matrix) is normally large enough to measure the intensity of the fluorescence line of the analyte with often only negligible interference by the matrix-caused emission. Figure 6.2.10 shows raw Cr K XAFS spectra measured on a soil sample that contained 200 $\mu g\, g^{-1}$ Cr. While the transmission spectrum shows a small edge sitting on a high background the fluorescence spectrum shows a much better signal/background ratio and the first oscillations of the XANES region, although it is plagued by statistical noise.

It must be mentioned that fluorescence detection of XAFS spectra suffers from some inherent problems itself. The first problem is that of statistical noise. Because the count rates are much smaller than they are in transmission mode the data often suffer from statistical noise. The reason for the low count rates is simply a geometrical problem. A detector with a diameter of the entrance window of 2.5 cm, positioned in a distance of 5 cm, covers only 1.6 % (~0.2 sr) of the total solid angle. Approaches to solve this problem are obvious: the number of detectors can be increased and the distance between sample and detector can be decreased. Both approaches lead to an increase of the solid angle covered and a corresponding increase in the total count rate.

Another problem that must be considered is called 'self-absorption'; it leads to a decrease in the amplitude of the XAFS oscillations compared

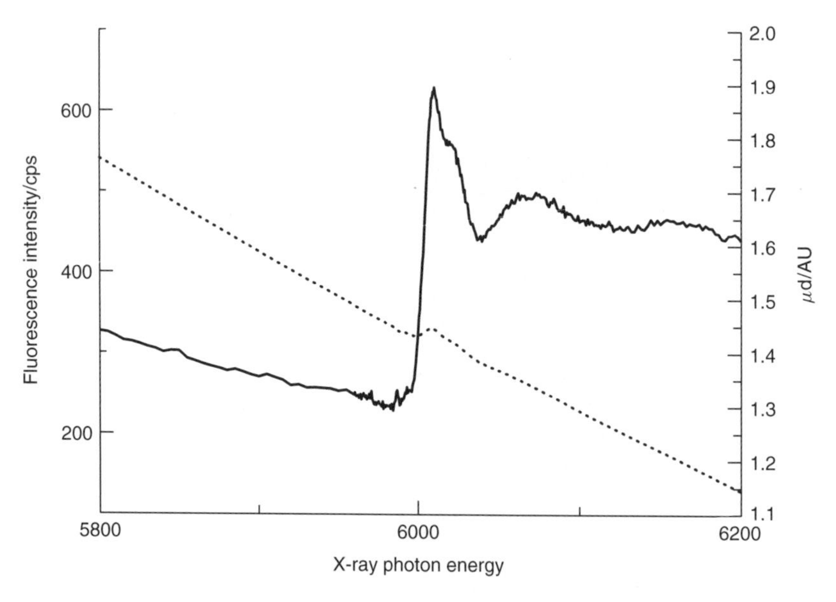

Figure 6.2.10. Comparison of a raw fluorescence and a raw transmission Cr XAFS spectrum measured simultaneously on a soil sample that contained 200 $\mu g\, g^{-1}$ Cr. Note the absence of a pre-edge peak, indicating the absence of Cr(VI). The spectra were measured at a bending magnet beamline with a five pixel Ge-detector for fluorescence detection. They demonstrate the much higher signal/background ratio achieved with the fluorescence technique.

with those measured in transmission mode. The magnitude of this effect depends among other factors on the concentration of the absorber in the sample [11, 12]. Especially if the standards are measured in transmission mode or have strongly differing sample composition, the spectra must be corrected before they can be used in any of the XAFS spectra evaluation procedures. Since the factors that lead to the self-absorption phenomenon are known the spectra can be corrected mathematically [12].

4.3 Sample preparation and handling

Sample preparation is probably the one step of an analytical method bearing the highest risk of loss of species information. Notably the ability to investigate wet samples is of utmost importance for species analysis. As briefly mentioned before it is one of the major advantages of XAFS spectroscopy that it can be performed in the original sample. The XAFS experiment is in most cases, dependent on the energy of the absorption edge, very flexible with respect to sample properties such as size, form, humidity or temperature. The XAFS spectra of most elements can be measured in wet samples under ambient conditions. The critical factor is the energy of the absorption edge; the lower the energy, the smaller is the penetration depth of the photons. At energies below 5–6 keV measurements are better performed under reduced pressure. This corresponds to the Cr K edge (5989 eV). However, even in these cases it is possible to find solutions, for instance by building special sample holders that make it possible to keep at least the actual sample under ambient conditions.

Either way it is necessary to choose a suitable sample thickness. If using transmission mode the sample thickness should be chosen so that a μd value between ~2.5 and 3.0 is achieved just above the edge. If fluorescence detection is used the sample should – but does not necessarily has to be – thicker to obtain the highest possible count rates.

Nevertheless, one restriction which might interfere with species analysis should be kept in mind. The samples must be homogenous (thickness, elemental and species composition) over the irradiated area. If this is not the case the sample has to be homogenized or the beam size has to be reduced to a length scale on which the sample is homogenous (see μ-XANES section). Homogenization of solid samples is normally done by grinding. In practice this can produce problems for species analysis, because the resulting enormous increase of surface can lead to artifacts, for example changes in oxidation state. In all cases where this risk exists care must be taken to avoid these changes in sample composition, for instance by simply working under a protective atmosphere.

Once ground the sample might be mixed with a light element bonding agent, for instance polyethylene or boron nitride powder, and pressed into pellets of appropriate thickness in a suitable press. Another possibility is to spread a thin layer of the powder on an adhesive tape, or mix the powder with glue.

The homogeneity of paste-like or liquid samples is normally less problematic. These samples can be handled in small containers with thin entrance and exit windows made from a light element material of appropriate mechanical strength.

5 EXAMPLES OF CONTEMPORARY USE IN SPECIATION ANALYSIS

A large number of applications of XAFS spectroscopy for analytical investigations that can be subsumed under the term 'species analysis' can be found in the literature. The chosen examples can by no means be a complete overview of this field of research. They are intended to give an idea of the many possible fields of application and to give some indications of applications. The examples are taken from two fields of research, environmental analysis and catalysis research.

5.1 Environmental analysis

Chemical speciation of metals plays an important role in the assessment of contaminated environmental compartments. Examples are the long-term behavior of heavy metals in soils and sediments or the safe disposal of waste containing heavy metals or radioactive elements. The typical sample material for these investigations is a solid diluted sample.

Two different approaches that use XAFS spectroscopy can be distinguished: firstly the physico/chemical characterization of single important components, for example the interaction of metals with humic acids or naturally occurring minerals; secondly the identification of chemical compounds in a (complex) natural sample. While it is often suitable to use the described classical procedure of EXAFS spectra interpretation in the first case this is not a possible strategy in the second case. The reason is that in the latter case the samples usually contain more or less complex mixtures of different compounds of the element of interest.

An example for the first case is the determination of the binding form of actinides and fission products in humic acids and certain clay and iron minerals. These investigations are aimed at the development of safe nuclear waste disposal and will serve as a way to predict the behavior of these elements in the environment if ever released [13]. In one of these studies the binding form of uranyl ions (UO_2^{2+}) to wood degradation products was investigated. For this purpose known complexing wood degradation products were added to solutions containing uranyl ion under strict control of pH, temperature and redox potential. The binding form of the uranium to these substances was determined by EXAFS spectroscopy and the results were compared with model calculations [14].

An example of an investigation of the chemical transformations that occur during the sorption of metals to clay minerals is a study of the sorption of Ni to clay minerals and aluminum oxides [15]. The kinetics of this process was investigated using time-resolved XAFS spectroscopy. On a time scale of several hundred hours major changes of the chemical binding form could be detected and were followed by the changes of the fine structure in the XAFS spectra. In either case the classic strategy was chosen to evaluate the spectra from these well-defined synthetic samples.

In contrast to the investigations on synthetic model compounds it is often not possible to interpret the EXAFS spectra in the usual way when investigating natural samples, because the sample contains more than one compound of an element. Then one can try to investigate the spectra by use of a fingerprint approach. This is done by measuring the spectra of pure reference compounds which might be constituents of the sample and calculating the best fitting linear combination of these reference compound spectra [16–19]. The working principle of this method can be seen in Figure 6.2.11, which shows some of the spectra already presented in Figure 6.2.6 together with a spectrum measured for a mixture of PbO and $PbCO_3$ that contained 25 % of the lead as PbO and the remaining 75 % as $PbCO_3$.

Often it is more useful to work with the near edge region than the actual EXAFS region for this type of analysis. This is because this region of the spectrum exhibits the largest features. The major steps of such a fingerprint analysis are shown in the flow chart in Figure 6.2.12. Several factors influence the reliability of the results drawn from the fingerprint method. In particular, when the actual edge region is used, a very careful energy calibration that eliminates the small shifts (typically 0.5–1.5 eV) of the measured edge position between two spectra is essential, because at this point μd changes drastically within small energy differences.

By far the greatest risk of erroneous results using the fingerprint method arises from the selection of inappropriate reference compounds. Too many reference compounds make it possible to find a good fit, although compounds contained are in fact missing. It is therefore mandatory to be well informed about the chemical behavior of the analyte and the possible reactions with matrix components, to minimize the number of reference compounds tested. Particularly problematic here are the numerous poorly defined compounds that are formed by interaction of the analyte with matrix components such as clay minerals and humic acids. These substances must be synthesized in the laboratory under strict control and adjustment to natural

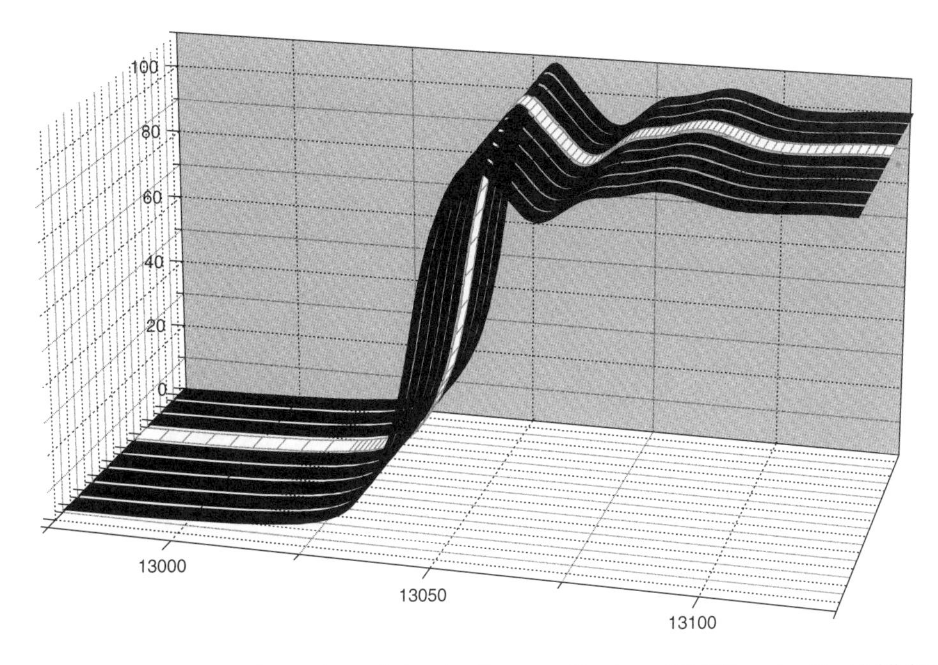

Figure 6.2.11. Working principle of a fingerprint analysis using linear combinations of XAFS spectra. The highlighted spectrum is measured on a sample that contained a mixture of $PbCO_3$ (72.9 % of total Pb) and PbO (27.1 % of total Pb). The surrounding spectra are linear combinations with weights of 24 % or 26 % for PbO and 76 % or 74 % for $PbCO_3$.

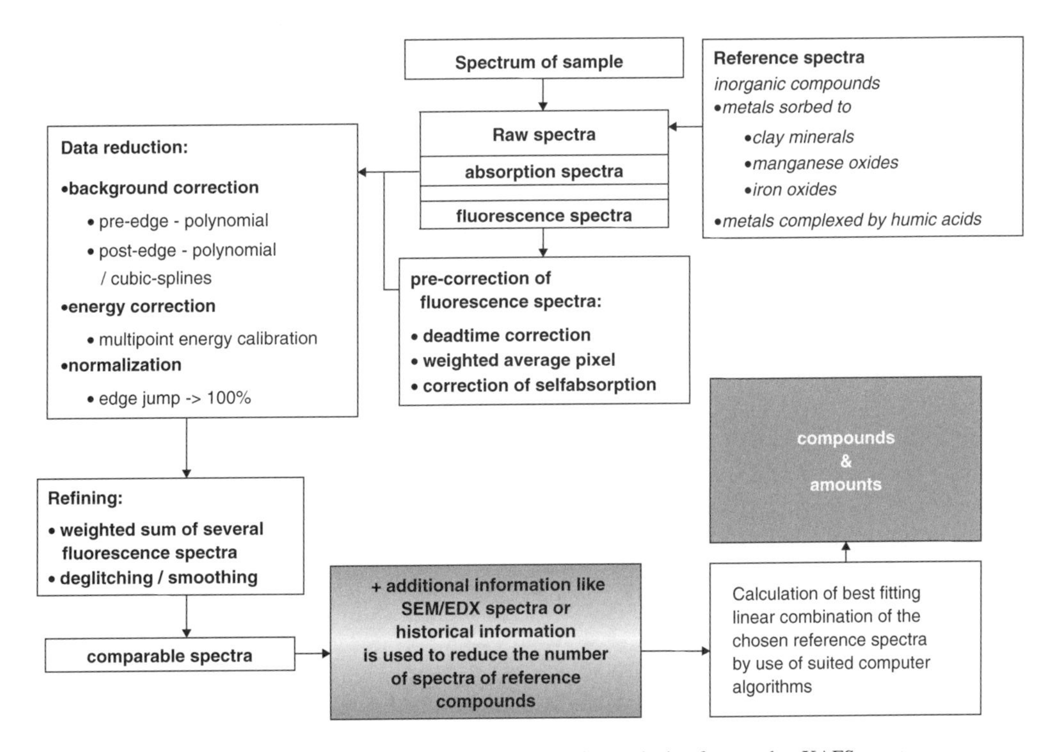

Figure 6.2.12. Flow chart depicting the major steps during a fingerprint analysis of near-edge XAFS spectra.

conditions of parameters such as pH, redox potential, temperature etc.

At this point the analysis of 'natural' samples strongly interferes with the characterization of single compounds. Preparing adsorbates of metals on clay minerals as reference materials for example makes it necessary to take the results of the investigations on the kinetics of the sorption process, which were mentioned before, into account [15].

Sometimes it is possible to work around the problem to choose a suitable not too large set of reference compounds by use of certain statistical methods. Principal component analysis (PCA) [20] is a method that was previously used for this purpose. Fundamental to this method is the availability of an adequate number of samples which contain differing amounts of the same chemical compounds. The number of samples must be at least one more than the number of compounds. Under this suppositions it is possible to calculate from the set of sample spectra virtual main compounds. The number of significant main compounds corresponds to the minimum number of real compounds that are contained in at least one of the samples. One major advantage of this technique is that up to this point PCA yields its information without the use of reference compound spectra.

In a further step called 'target transformation' PCA not only provides information about the number of main components, it also tests whether a certain pure compound is a constituent of at least one of the mixtures. Thus it is possible to limit the number of reference compound spectra used in the fingerprint method to those that are identified as being contained in at least one of the samples.

PCA has been used in several recent studies of chemical binding forms. Examples are the determination of iron binding forms in eight coal samples from different mines [21] or the identification of changes in the speciation of Cu during a chemical reaction [22]. In the first study at least four main components were found to contribute to the spectra of the mixtures. FeS_2 could be identified by target transformation as one of these species.

The Cu spectra for the latter example were measured in transmission mode on a copper metal foil that was placed in a high temperature furnace ($T > 700\,K$) under a continuous gas flow. The metal was oxidized by varying concentrations of oxygen to obtain either Cu_2O or Cu and the changes of the spectra followed over a longer period of time. The occurrence of Cu_2O during the oxidation could be verified by using PCA and target transformation.

5.2 Catalysis research

Another field of science where analysis of speciation in solids is important and in which XAFS spectroscopy is frequently employed is catalysis research. Following the alterations of chemical speciation of the catalytically active side of a catalyst yields valuable insight into the mechanism of the catalytic process. It is used as a basis for a more precisely aimed development of new catalysts. In the context of this chapter it is used as an example of the ability of XAFS spectroscopy to perform species analysis in solid samples *in situ* during an ongoing chemical reaction.

The high flexibility in the design of XAFS samples and sample holders enabled by the large penetration depth of X-rays is of special importance for *in situ* measurements during catalytic reactions. This will be demonstrated here using examples from investigations where XAFS spectroscopy was coupled with other analytical methods for special investigations. All these examples are not standard techniques at any beamline but will demonstrate the high flexibility of XAFS spectroscopy with respect to the sample environment. Catalysis research benefits especially from the fact that XAFS spectra can be measured under physico/chemical conditions (temperature, pressure, composition) similar to those in real use. Examples can be found in the literature where XAFS spectra were measured under high pressure [23], high temperature [24] or a reactive atmosphere [25].

Tracing the formation and transformation of chemical species during chemical reactions makes it necessary to decrease the time needed for the measurement of a single spectrum. The time needed for an average XAFS scan at a bending magnet beamline is 15–60 min, which is much too long to trace chemical reactions. Special experimental techniques make it possible to decrease the time needed to no more than some seconds or even ms, especially when the scan range is limited to the edge region (ΔE about 100 eV) which is often sufficient and high flux sources are used. Two different techniques are used to achieve such short measuring times, as follows.

In the first approach a polychromatic incident beam is used. With an optical element that disperses the transmitted polychromatic X-ray light spectroscopically after it has passed the sample it is possible to map the whole spectrum on a detector array, analogous to the well-known diode array detectors in UV/Vis spectroscopy. A drawback of this technique is that the spectra cannot easily be measured in fluorescence mode. This limits the use of the so-called dispersive EXAFS (DEXAFS) technique to higher concentrated samples.

The second approach is to 'simply' increase the scan speed of a conventional sequential XAFS scan. This makes it necessary to register the detector signals continuously thus avoiding the overhead caused by the stepwise registration of the spectra, as is usually employed in classic XAFS spectroscopy. This technique is named QEXAFS (quick scanning EXAFS). The major problem of this technique is obvious: the high scan speed leads to poorer counting statistics resulting in noisier spectra. QEXAFS experiments, particularly those with very high repetition rates, must be performed at high flux beamlines on insertion devices. However, using fluorescence detection and a scan range of 100 eV it was possible to measure the Cu Kα XANES spectra of samples containing 2.5 mmol L^{-1} Cu in water with repetition rates of 9 Hz with reasonable quality [26].

The combination of XAFS spectroscopy with other analytical methods offers the opportunity to get additional information, for example thermodynamical data, simultaneously. For this purpose XAFS was combined in one experiment with DSC [27]. Reactions of the general type

$$M{-}OOC{-}CH_2{-}X \longrightarrow MX + 1/n[{-}OOC{-}CH_2{-}]_n$$

were studied, with M being an alkali metal or silver and X a halogen. A commercial DSC apparatus that was mounted in the beam was used. The sample powders were mixed with inert boron nitride, so that the desired edge jump could be achieved. This mixture was placed in thin walled Al crucibles. A hole with 1 mm diameter allowed the beam to pass through the sample to measure the metal spectra. QEXAFS scans were then measured at increasing temperatures. They showed clearly the change in the binding form of the metal, coinciding with the DSC signal.

Other examples are the combination with FTIR spectroscopy [28], which was used to investigate the molecular structure of $RbReO_4$, or with X-ray diffraction (XRD) [29, 30], which adds long range order information to the short range order information about the immediate chemical environment of the absorber atom that is yielded from XAFS spectroscopy.

6 LIMITATIONS

The limitations of a method are probably more useful for a decision on whether a certain method can be helpful to solve an analytical problem or not than an enumeration of positive properties. As an analytical method that is (most often) performed on solid samples XAFS spectroscopy is affected by all the problems that are typical of this type of sample, especially representativeness of the samples for the bulk material, homogeneity of the actual sample etc. These problems will not be further discussed here, because they are not specific to this experimental technique.

Beside the positive aspects there are several factors limiting the usefulness of XAFS spectroscopy for elemental speciation analysis in solid samples. Some limitations, such as the general need for homogenous samples, the limited availability of synchrotron radiation source beamlines for

XAFS measurements or the problem of choosing suitable reference substance spectra for the fingerprint method, have already been mentioned.

For all samples which contain only one unknown species of an element the most reliable information is yielded by the 'classical' EXAFS evaluation procedure. A detailed discussion of error sources and how to minimize the errors introduced by the evaluation procedure can be found in the literature [31].

The fingerprint approach is not limited to a pure compound but works more reliably with a small number of compounds. Using this approach only main compounds with shares of more than $\sim$5 % of the total amount can be identified with reasonable certainty [17].

Although the actual detection limits depend strongly on the matrix properties it can be said that XAFS spectroscopy is not a technique that is suited for trace analysis. If the measurements are performed in transmission mode the detection limit is around $20\,g\,kg^{-1}$. The use of fluorescence detectors decreases the detection limit depending on the matrix properties by a factor between 10 and 1000 and in extreme cases even further. In general the detection limit is lower in samples that contain a heavy element as analyte in a light element matrix. This can be demonstrated with some examples. For U in water it was possible to measure EXAFS spectra in solutions containing only $1\,mmol\,L^{-1}$ of U [14]. The sorption process of Np at the α-FeOOH/water interface was investigated in samples that contained some $10\,\mu g\,g^{-1}$ Np [32].

Figure 6.2.10 shows a raw Cr K edge spectrum measured with fluorescence detection in a sample that contained $200\,\mu g\,g^{-1}$ of Cr in soil. This spectrum was measured at a bending magnet beamline (Beamline A1 at HASYLAB at DESY, Hamburg, Germany) using an energy-dispersive Ge detector.

An experimental setup such as the one that is used for X-ray Raman spectroscopy (see Figure 6.2.16) can also be used to increase the signal/noise ratio in fluorescence detection of 'conventional' XAFS spectroscopy. The analyser crystal used splits spectroscopically the radiation emitted from the sample. Thus it is possible to exclude from the actual detector the unwanted photons that are emitted by the matrix. The use of such a secondary monochromator makes it possible to make full use of modern synchrotron sources with very high flux, because the detector and electronics must no longer count all the photons but only those originating from the element of interest.

In the following, two newer experimental developments that can be helpful to overcome some of the limitations mentioned will be presented. The need for homogenous samples can often be overcome by reduction of the sample and beam size to a length scale on which the sample is homogenous. The necessity to measure the XAS spectra of lighter elements such as C, N and P under high vacuum conditions can be overcome by a technique that uses higher energy radiation to gather the desired information.

7 μ-XANES

The distribution of an analyte in a solid sample is often not homogenous or the distribution of different chemical binding forms is not homogenous. In these cases analytical methods with a high spatial resolution in the μm range can be a way to yield valuable additional information.

A technical prerequisite to perform XAFS spectroscopy with a spatial resolution in the μm range (μ-XANES or μ-EXAFS) is the availability of focusing X-ray optics. With the development of effective X-ray optics during the last decade it has become possible to perform X-ray absorption spectroscopy on a μm scale. The μ-focus technique minimizes the need for homogenous samples, while at the same time raising the question

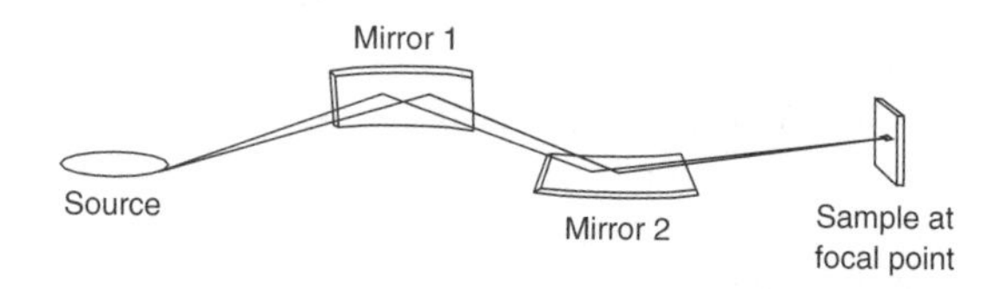

Figure 6.2.13. Scheme of a widely used setup of two elliptical focusing mirrors that enables horizontal and vertical focusing of an X-ray beam under grazing incidence (Kirkpatrick–Baez geometry). A setup like this is used at several μ-focus beamlines.

of whether the results are representative of the bulk material.

Three types of optical devices are used for focusing to μm spot sizes: first refractive optics elements such Fresnel zone plates and lenses, second two different types of reflective optics. The first of the latter are toroidal or elliptical mirrors which are operated under grazing incidence. Figure 6.2.13 shows as an example the 'Kirkpatrick–Baez' setup that is often used to produce a small focal point with X-ray mirrors. Complex experimental setups like this have to be employed, because X-ray photons are reflected only under grazing incidence. The second approach using reflective optics are special glass capillaries which guide and concentrate the X-ray photons. Each of these methods has its special merits and the decision which of the systems is used must be taken individually, according to the X-ray source, sample properties and scientific question.

If a μ-focus beam is available two different experimental strategies can be employed, the mapping technique in which a larger area of the sample is scanned and a two-dimensional map of the sample (surface) is produced and secondly the analysis of single particles. In either case the sample must be mounted on an x/y scanning device with sufficient spatial resolution. The position of the beam is normally fixed since it is a much higher experimental effort to change the position of the beam.

7.1 Mapping technique

Problems with the representativeness of the results yielded by μ-XANES can be partly overcome by mapping of larger areas on the sample. The same approach is chosen in the case of elemental mapping, using for example electron microprobes and X-ray fluorescence detection. μ-XANES adds the ability to discriminate between different chemical forms of the element under investigation. The major advantage of this approach, however, is that it markedly increases the relevance of the results obtained by adding the information about the spatial distribution of the analyte and its chemical forms to the information about the chemical form of the analyte at one spot, which is achieved with a nonscanning technique. The chemical parameter which is easiest to map using μ-XANES is the distribution of different oxidation states of the analyte on the sample surface.

To perform oxidation state mappings the elemental distribution, e.g. the corresponding fluorescence intensity, is measured at several different energies of the incident monochromatic beam. The corresponding energies of the incident beam are chosen so that certain oxidation states of the analyte are excited preferentially [33]. In case of Cr, images recorded at incident beam energies of about 5997, 6001 and 6007 eV would be used to distinguish between elemental Cr, Cr(III) and Cr(VI); see Figure 6.2.3. The distribution of the different oxidation states can then be calculated from the resulting elemental mappings.

This technique can be used for the determination of redox states and mineralogical associations of toxic or essential species in natural and contaminated soils, sediments, waste encapsulation materials or minerals. An example is the depiction of annual ring-like structures of alternating oxidation states of manganese in Mn/Fe nodules from the Baltic sea [34]. Another example is the investigation of the redox chemistry of metals at the root – soil interface of plants and its role in agriculturally relevant plant diseases, for instance the measurement of Se diffusion and reduction at the water–sediment boundary that was performed recently [35]. Further interesting applications would be measurements of the distribution of elements in different oxidation states in microelectronic devices or the mapping and chemical characterization of metals within single cells and at the binding domains of bio films. These films are really ubiquitous at water/mineral surfaces and are believed to play an important role in many different processes such as the binding of metals in soils and sediments or the corrosion of fresh and wastewater supply systems.

7.2 Single particle technique

The single particle technique yields essentially the same information as can be drawn from the standard XAFS technique, but with much higher spatial resolution. Three important reasons for using this experimental approach and not a bulk technique are as follows.

(1) *For small samples such as single dust particles or colloidal particles.* Examples are the characterization of single colloidal particles which are believed to play an important role in the transport of toxic metals in groundwater aquifers [36] or the determination of the redox state of the Earth's interior based on valence determinations on microcrystals within diamonds or volcanic glasses. One example of the latter is the investigation of Fe inclusions in volcanic glasses which is of interest in geochemistry [37]. Focusing was achieved using a Kirkpatrick–Baez mirror system. The samples investigated originated from two different volcanic sites, the glass inclusions had an Fe content of 10.86–9.31 %. The beam size was 15 μm × 15 μm. Only the actual edge region, −30 to +40 eV around the edge, was measured. Using standards with known Fe(II)/Fe(III) ratios for the calibration it was possible to determine the Fe(II)/Fe(III) ratio in the samples. In general it can be said that, if the sample contains regions (particles) that contain higher concentrations of the analyte than the bulk, the analyte/matrix signal ratio can be increased by the μ-XAFS technique compared with investigations on bulk samples.

(2) *Different binding forms of the analyte are contained in discrete particles.* If the analyte is contained in discrete particles in different binding forms the problem that arises from the overlay of different XAFS spectra can be avoided so that the 'classical' EXAFS spectrum interpretation becomes possible. An example is the identification of different uranium oxidation states in nuclear fuel particles, emitted during different phases of the Chernobyl incident [38].

(3) *High radioactivity or toxicity of the samples makes working with small amounts of sample material desirable or necessary.* An example from a recent investigation is a μ-XAFS study of sorbed Pu on tuff [39]. It was performed at an undulator beamline, using Kirkpatrick–Baez optics to focus the beam. The spatial resolution was limited by the beam size of 4 μm × 7 μm.

Further applications originate from all areas of research where small samples and/or inhomogeneous samples are involved.

8 X-RAY RAMAN SPECTROSCOPY

Elements such as C, N and O have an enormous importance in chemistry. Unfortunately XAFS spectroscopy is not easily usable for the determination of the local structure around light absorber atoms. This is because of the small energy of the absorption edges, less than 1000 eV. The penetration depth of X-ray photons with such a low energy is very small, so that the samples have to be very thin. For the same reason the measurements on light elements have to be performed under high vacuum conditions, thus making it very difficult if not impossible to measure wet samples under realistic conditions without species alterations.

In X-ray Raman spectroscopy (or nonresonant X-ray Raman scattering spectroscopy) a sample is irradiated with a fixed energy beam. The scattered radiation is investigated spectroscopically. In contrast to resonant edge absorption spectroscopy, which is normally used to measure XAFS spectra, the photon energy can be much higher than the edge energy. In the case of a nonresonant edge absorption the scattered photon will lose part of its energy. This energy loss corresponds to the energy of the respective absorption edge. The process is equivalent to electron energy loss spectrometry (EELS), which is a well-known analytical technique in transmission electron microscopy, or the classical Raman spectroscopy in the IR region. Figure 6.2.14 depicts the X-ray Raman process schematically.

A plot of the distribution of the scattered photons over the energy axis shows the same oscillations that can be found in the conventionally measured absorption spectrum. The major advantage of this technique, compared with the resonant measurement of the XAS spectra in the soft X-ray region, is as mentioned above that the X-ray photon energy can be chosen so high (~10 keV) that low Z elements in thick samples can be measured under environmental conditions.

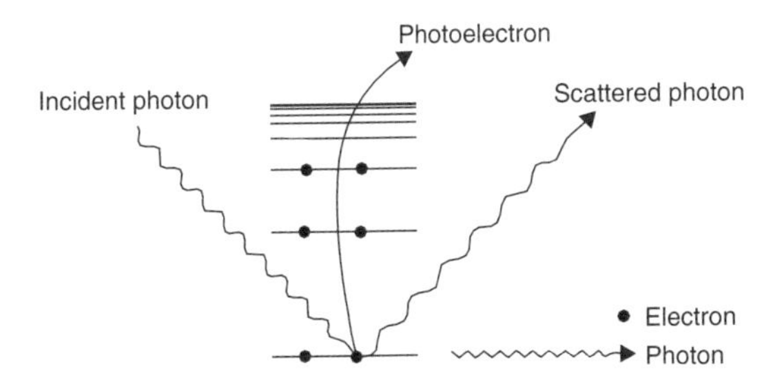

Figure 6.2.14. Schematic visualization of the X-ray Raman process. The incident photon is inelastically scattered whereby it loses a part of its energy. The photoelectron leaves the atom with an energy that is equal to this loss minus the threshold energy E_0, which is necessary to lift a core shell electron to the continuum. The outgoing photoelectron wave undergoes the same scattering processes as in XAFS spectroscopy.

For these elements X-ray Raman spectroscopy will become to a valuable alternative way to obtain the desired information. Up to now the major drawback has been the extreme weakness of the Raman scattering process. This makes it difficult to achieve good spectra with high resolution and a good signal/noise ratio with contemporary X-ray sources within reasonable time spans.

Figure 6.2.15 shows examples of the K-edge spectra from boron and nitrogen in different BN samples measured at a wiggler beamline [40]. The flux during these measurements was 10^{13} photons s^{-1}, the spectral width (FWHM) was 1.1 eV at 6 keV. The time needed for data collection of one spectrum was some hours. In an earlier study [41] measuring times between 24 h and 3 days were used to obtain X-ray Raman spectra of pure carbon compounds (diamond and graphite) at a bending magnet beamline. The flux was estimated to be 10^{11} photons s^{-1}, the linewidth of the exciting monochromatic X-rays was estimated to be 2 eV (FWHM). Today this technique is obviously still very time consuming, but with the predictable increase of the spectral flux from X-ray sources in the near future it will be possible to measure these spectra within reasonable time.

Examples for possible applications of X-ray Raman spectroscopy include the following originating from three totally different fields of research where the chemical form of a low Z element is of interest:

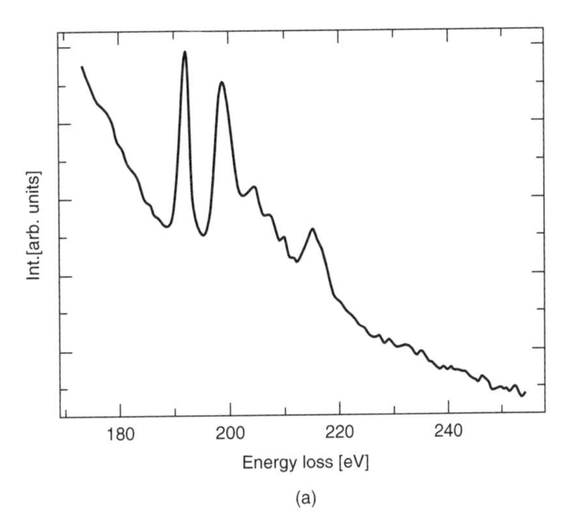

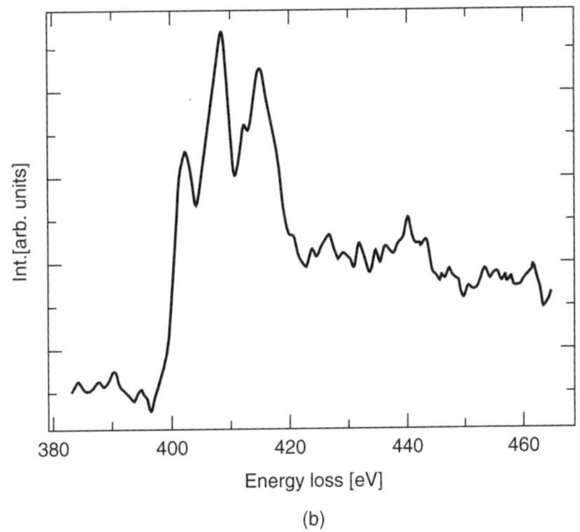

Figure 6.2.15. K-edge X-ray Raman spectra from (a) boron and (b) nitrogen measured in a BN sample, spectra were measured at a wiggler beamline (BL16X of Photon Factory, KEK, Tsukuba, Japan), from ref. 40. Reproduced by permission of the American Institute of Physics and the authors.

- Measurement of the XAS spectra of light elements in semiconductor materials. At present these studies have to be performed in the soft X-ray region. Materials of interest are III–V semiconductor materials such as Mg- or Si-doped GaN [42] or AlGaN and InGaN alloys [43].
- An interesting experiment would be the combination of X-ray Raman spectroscopy with the imaging techniques mentioned above. This

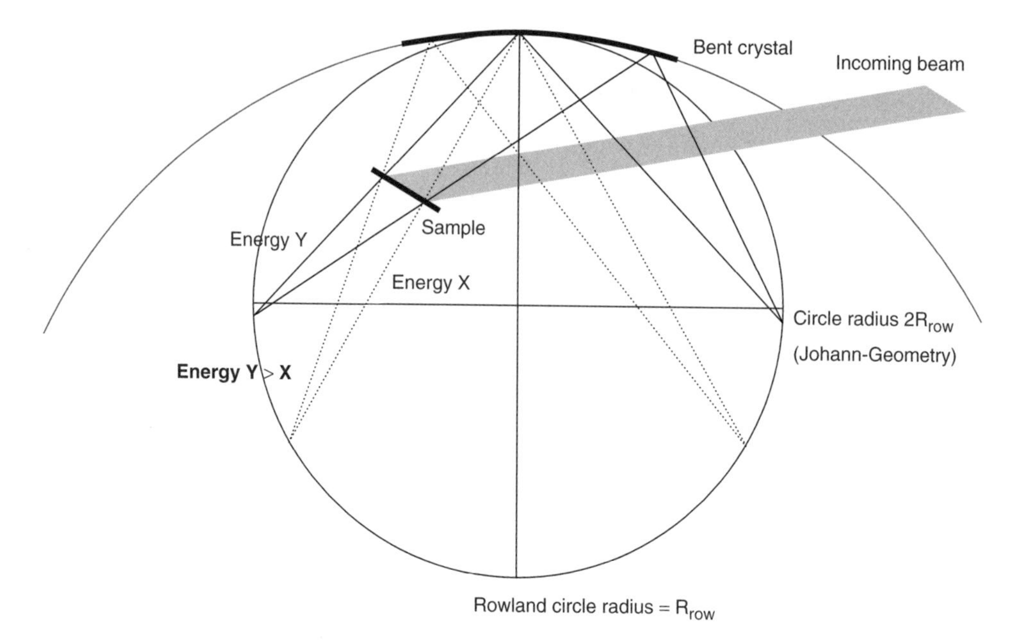

Figure 6.2.16. Schematic drawing of the ray paths in an experimental setup (Johann geometry) that employs a focusing bent crystal for the spectroscopic dispersion of the X-ray photons emitted from the sample. A setup like this can be used for the measurement of X-ray Raman spectra or to increase the signal/background ratio in a 'conventional' fluorescence mode XAFS scan.

would enable a three-dimensional mapping of light elements and their chemical binding forms in thick particles where the penetration depth of edge energy photons is much too small. Again, a scientifically and economically interesting goal for this method is semiconductors, especially B- and N-containing III–V semiconductors (see above).

- The use of X-ray Raman spectroscopy in catalysis research will make it possible to perform *in situ* experiments and look from the other side, not from the reactive (metal) center of a catalyst but from the substrate molecule on the chemical reaction mechanism.

Figure 6.2.16 shows the principle experimental setup that can be used for X-ray Raman experiments. It is based on a focusing, spherically or cylindrically bent analyser crystal and a two-dimensional detector such as a CCD chip. Both are mounted on the circumference of a Rowland circle. The crystal is bent to a radius twice that of the Rowland circle. This so-called Johann geometry enables measurements with high energy resolution (better 1 eV) for the X-ray Raman experiment and focuses the spectroscopically dispersed light on the detector, thus increasing the count rate by increasing the used solid angle.

9 SUMMARY AND OUTLOOK

Summarizing the foregoing it can be concluded that XAFS spectroscopy is an analytical method which yields all the necessary information to undertake species analysis in solid samples from a large number of different scientific fields. Particularly advantageous is the negligible amount of sample preparation that is required owing to the flexibility of the XAFS experiment with respect to sample properties such as humidity, size, shape, state of aggregation etc. However, XAFS spectroscopy requires homogenous samples and suitable synchrotron radiation sources or beamlines and the detection limits are often higher than the limits that are set in legal guidelines.

Recent and future experimental developments such as the μ-methods and X-ray Raman scattering techniques described will most probably extend the possible applications in species analysis markedly. The development of sources with even higher flux together with improved detectors will enable a further reduction of the detection limits.

At the moment a larger number of conventional XAFS beamlines and μ-XAFS beamlines are available for experiments. All types of synchrotron sources, from bending magnets to undulators, are used at one or the other storage ring for XAFS spectroscopy beamlines. Bending magnet beamlines have the advantage of easier operation and that more beamtime is available.

Undulator or wiggler beamlines, however, are the first choice for well-planned experiments on highly diluted samples. Because of the complex experimental techniques and data evaluation the newcomer in this field would be well advised to start with his experiments at a bending magnet beamline. All synchrotrons offer user support to help in planning and conducting experiments and choosing the best suited beamline for a particular application. The high experimental effort that is associated with the measurement of XAFS spectra appears to be appropriate in all cases where the analyte cannot be separated from the matrix without destruction of the species information or where measurements must be performed *in situ*.

10 REFERENCES

1. Teo, B. K., *EXAFS: Basic Principles and Data Analysis*, Springer, Berlin, 1986.
2. Koningsberger, D. C. and Prins, R. (Eds.), *X-Ray Absorption: Principles, Applications, Techniques of EXAFS, SEXAFS and XANES*, John Wiley & Sons, Inc., New York, 1988.
3. Fay, M. J., Proctor, A., Hoffmann, D. P. and Hercules, D. M., *Anal. Chem.*, **60**, 1225A (1988).
4. Sánchez del Río, M. and Dejus, R. J., XOP: Recent developments, *SPIE Proc.*, **3448** (1998); see also: http://www.esrf.fr/computing/scientific/xop/intro.html.
5. Kincaid, B. M. and Eisenberger, P., *Phys. Rev. Lett.*, **34**, 1361 (1975).
6. Sayers, D. E., Lytle, F. W. and Stern, A. E., *Phys. Rev. Lett.*, **27**, 1024 (1971).
7. Ankudinov, A. L., Ravel, B., Rehr, J. J. and Conradson, S. D., *Phys. Rev. B*, **58**, 7565 (1998).
8. Stern, E., Newville, M., Ravel, B., Yacoby, Y. and Haskel, D., *Physica B*, **208–209**, 117 (1995).
9. Koch, E. E. (Ed.), *Handbook on Synchrotron Radiation*, Vol. 1, North-Holland, Amsterdam, 1983.
10. Lide, D. R. (Ed.), *CRC Handbook of Chemistry and Physics: A Ready-reference Book of Chemical and Physical Data*, 80th edn., CRC Press, Boca Raton, FL, 1999.
11. Iida, A. and Noma, T., *Jpn. J. Appl. Phys.*, **32**, 2899 (1993).
12. Tröger, L., Arvanitis, D., Baberschke, K., Michaelis, H., Grimm, U. and Zschech, E., *Phys. Rev. B*, **46**, 3283 (1992).
13. Scheidegger, A. M., Strawn, D. G., Lamble, G. M. and Sparks, D. L., *Environ. Sci. Technol.*, **30**, 548 (1996).
14. Denecke, M. A., Reich, T., Pompe, S., Bubner, M., Heise, K. H., Nitsche, H., Allen, P. G., Bucher, J. J., Edelstein, N. M. and Shuh, D. K., *J. Phys. IV Fr.*, **7**, C2–637 (1997).
15. Scheidegger, A. M., Strawn, D. G., Lamble, G. M. and Sparks, D. L., *Geochim. Cosmochim Acta*, **62**, 2233 (1998).
16. Manceau, A., Boisset, M. C., Sarret, G., Hazemann, J. L., Mench, M., Cambier, P. and Prost, R., *Environ. Sci. Technol.*, **30**, 1540 (1996).
17. Welter, E., Calmano, W., Mangold, S. and Tröger, L., *Fresenius' J. Anal. Chem.*, **364**, 238 (1999).
18. Szulczewski, M. D., Helmke, P. A. and Bleam, W. F., *Environ. Sci. Technol.*, **31**, 2954 (1997).
19. Hesterberg, D., Sayers, D. E., Zhou, W., Plummer, G. M. and Robarge, W. P. *Environ. Sci. Technol.*, **31**, 2840 (1997).
20. Malinowsky, E. R. and Howery, D. G., *Factor Analysis in Chemistry*, John Wiley & Sons, Inc., New York, 1980.
21. Wassermann, S. R., *J. Phys. IV Fr.*, **7**, C2–203 (1997).
22. Hilbrandt, N., Wasserman, S. R. and Martin, M., *Solid State Ionics*, **101–103**, 431 (1997).
23. Frenkel, A. I., Wang, F. M., Kelly, S., Ingalls, R., Haskel, D., Stern, E. A. and Yacoby, Y., *Phys. Rev. B*, **56**, 10 869 (1997).
24. Koningsberger, D. C. and Vaarkamp, M., *Physica B*, **208 & 209**, 633 (1995).
25. Kappen, P., Grunwaldt, J. D., Hammershoi, B. S., Tröger, L. and Clausen, B. S., *J. Catal.*, **198**, 56 (2001).
26. Lützenkirchen-Hecht, D., Grundmann, S. and Frahm, R., *J. Synchrotron Radiat.*, **8**, 6 (2001).
27. Epple, M., Tröger, L. and Hilbrandt, N. *Synchrotron Radiat. News*, **10**, 11 (1997).
28. Wilkin, O. M. and Young, N. A., *J. Synchrotron Radiat.*, **6**, 204 (1999).
29. Topsoe, H., *Stud. Surf. Sci. Catal.*, **130**, 1 (2000).
30. Grunwaldt, J.-D., Molenbroek, A. M., Topsoe, N. Y., Topsoe, H. and Clausen, B. S., *J. Catal.*, **194**, 452 (2000).
31. Krappe, H. J. and Rossner, H. H., *Phys. Rev. B*, **61**, 6596 (2000).
32. Combes, J. M., Chisholm-Brause, C. J., Brown, G. E., Parks, G. A., Conradson, S. D., Eller, P. G., Triay, I. R.,

Hobart, D. E. and Miejer, A., *Environ. Sci. Technol.*, **26**, 376 (1992).
33. Sutton, S. R., Bajtz, S., Delaney, J., Schulze, D. and Tokunaga, T., *Rev. Sci Instrum.*, **66**, 1464 (1995).
34. Kersten, M. and Wroblewski, Th., 'Two dimensional XAFS Topography of Amorphous Oxyhydroxide Layers in Mn/Fe nodules', in *HASYLAB Annual Report 1998*, HASYLAB at DESY, Hamburg, Germany, 1998, pp. 875–876; also available on the World Wide Web: http://www-hasylab.desy.de/science/annual_reports/1998/part1/contrib/26/2052.pdf.
35. Tokunaga, T. K., Sutton, S. R., Bajt, S., Nuessle, P. and McCarthy, G. S., *Environ. Sci. Technol.*, **32**, 1092, (1998).
36. Kammer, F. v. d. and Förstner, U., *Water Sci. Technol.*, **37**, (6–7) 173 (1998).
37. Mosbah, M., Duraud, J. P., Metrich, N., Wu, Z., Delaney, J. S., San Miguel, A., *Nucl. Instrum. Methods B*, **158**, 214 (1999).
38. Salbu, B., *Radiat. Prot. Dosim.*, **92**, 49 (2000).
39. Duff, M. C., Newville, M., Hunter, D. B., Bertsch, P. M., Sutton, S. R., Triay, I. R., Vaniman, D. T., Eng, P. and Rivers, M. L., *J. Synchrotron Radiat.*, **6**, 350 (1999).
40. Watanabe, N., Hayashi, H., Udagawa, Y., Takeshita, K. and Kawata, H., *Appl. Phys. Lett.*, **69**, 1370 (1996).
41. Tohji, K. and Udagawa, Y, *Phys. Rev. B*, **39**, 7590 (1989).
42. Katsikini, M., Moustakas, T. D., Paloura, E. C., *J. Synchrotron Radiat.*, **6**, 555 (1999).
43. Katsikini, M., Fieber-Erdmann, M., Holub-Krappe, E., Korakakis, D., Moustakas, T. D. and Paloura, E. C. *J. Synchrotron Radiat.*, **6**, 558 (1999).

CHAPTER 7

Calibration

7.1 Calibration in Elemental Speciation Analysis

K. G. Heumann

Institut für Anorganische and Analytische Chemie, Mainz, Germany

1 INTRODUCTION

Similar to methods for the determination of total element concentrations, calibration methods for elemental speciation analysis can be classified into categories of absolute and relative methods [1]. An absolute method produces a result that is direct traceable to SI units. Gravimetry, titrimetry, and coulometry are such absolute determination methods. Absolute methods can be evaluated without any comparative measurement, but they depend on physicochemical constants such as Faraday's constant in the case of coulometry. In the absence of interferences, the amount of an elemental species determined by coulometry can be calculated using the measured amount of charges if the corresponding redox reaction is well defined. A well-known example of a coulometric determination of an elemental species is the anodic oxidation of As(III) to As(V) at a platinum electrode. The basis of titrimetric determinations of elemental species is a well-defined chemical reaction, where the amount of the analyte is related to the stoichiometric coefficients of the reaction. Titrimetric redox reactions are well established so that different oxidation states of an element can be determined, e.g. the determination of Fe(II) by permanganate in acidic solutions. The selective precipitation of sulfate, even in the presence of other sulfur species, is a well-known example of gravimetric elemental species determinations from student courses.

All these absolute methods must be sufficiently selective if other compounds are present in the sample besides the elemental species to be determined. For coulometric determinations

Handbook of Elemental Speciation: Techniques and Methodology R. Cornelis, H. Crews, J. Caruso and K. Heumann
 ISBN: 0-471-49214-0

selectivity can be achieved in many cases by applying a fixed potential difference to the electrodes, which does not allow a redox reaction for other compounds. Selectivity is normally an especially critical point in gravimetry because many compounds can usually be precipitated by the same precipitator. Possible interferences must therefore be eliminated before precipitation takes place. For example, gravimetric determination of nitrate is possible by precipitation as nitron nitrate. Interferences by nitrite, often also present in samples where nitrate occurs, can be eliminated by selective reduction of nitrite to elementary nitrogen, e.g. by amidosulfuric acid. Coprecipitation and adsorption of other species can also occur during gravimetric determinations. These interferences are usually difficult to prevent.

However, all absolute methods are extremely sensitive with respect to possible interferences so that they are normally only useful for the determination of pure samples. In addition, gravimetric and titrimetric methods can usually not be applied at trace levels, where it is very often necessary to analyze elemental species. Thus, absolute methods are not important at the moment for elemental speciation analysis. On the other hand, determinations of pure stock solutions of elemental species standards, which are used for calibration of relative methods, are often carried out by titrimetric methods.

Relative methods are those where detection of the elemental species in a sample to be analyzed is achieved by comparison with a set of calibration samples of known content. Reference materials (RMs) or certified reference materials (CRMs) are applied for calibration with identical, or at least similar, matrix composition. In contrast to CRMs that are certified for their total element concentration, there are only a few CRMs available for elemental species analysis. Aspects which need to be taken into consideration when calibrating by using CRMs, are summarized in Chapter 7.2. Most of the analytical methods applied today for elemental speciation are relative methods so that the corresponding type of calibration procedure with a set of calibrants is fully discussed in the following text.

Because of the practical advantages of linear calibration graphs they are always favored in analytical chemistry. Linear calibration graphs can be obtained by measuring only a few calibration standards and, in addition, are easily described by a simple mathematical function:

$$S = kc + b, \tag{7.1.1}$$

where S is the signal response of the detection method, c the concentration of the calibrant, k the calibration factor, and b the intersection with the y-axis. The k value reflects the sensitivity, and sensitivity is therefore better the higher the k value is for a given concentration range.

The quality of a calibration is, in principle, controlled by the repeatability of the measurement, the trueness of the standards used, and the validity of the comparison between the calibrant and the sample. Whereas the repeatability influences the precision of the analytical result, the two last factors mentioned influence the accuracy. The validity of the comparison between the calibrant and the sample is one of the most critical points in calibration techniques for trace analyses. Therefore different calibration modes are used depending on how critical this comparison is. External calibration modes, where the sample and the corresponding calibrant are separately measured, do not fit identical measuring conditions for the standard and the sample in the same way as internal calibration techniques. The following different modes of internal calibration have been used, up to now, for elemental speciation analysis:

(a) For the standard addition technique each sample is split into several subsamples and an increasing but known amount of the analyte is added to the different subsamples. This calibration technique usually eliminates possible influences of the matrix composition on the signal intensity of the detection system [2]. If this is a critical point during analysis, the standard addition method can successfully be applied for the determination of elemental species. Some of the possible errors during analytical steps prior to the final detection, for example, by extraction of an analyte from solid samples, can usually not be corrected for this calibration mode unless total equilibration between analyte and standard is guaranteed.

(b) The species-specific isotope dilution technique uses a known amount of a spike, which contains the elemental species to be determined in a different isotopic composition from that of the sample. For isotope dilution the spike is added to the sample and homogeneously mixed prior to all other analytical sample treatment steps. Because the isotope ratio of the isotope-diluted sample must then be measured, mass spectrometry has to be used as detection method. Isotope dilution mass spectrometry (IDMS) fits excellently the principles of an internal calibration because isotopes are, within an uncertainty of usually less than 0.1 %, identical in their chemical behavior and they can easily be detected by mass spectrometry. In addition, only a ratio of the amount of isotopes and not an absolute amount of the analyte must be measured, which allows loss of substance during sample pretreatment steps without any influence on the analytical result. This is due to the fact that, even in the case of loss of substance, the isotope ratio of the isotope-diluted elemental species does not change. These characteristics of IDMS normally lead to highly accurate results [3]. The isotope dilution technique is also a one-point calibration method, which saves time compared with other calibration methods. These advantages are the reason why an increasing number of elemental speciation analyses are carried out by IDMS. Under certain conditions other chemical forms of an element, not identical with the elemental species to be determined, can be used (species-unspecific IDMS; see Section 4.2.2).

2 SPECIAL FEATURES OF CALIBRATION FOR ELEMENTAL SPECIES ANALYSIS

A lack of availability of calibrants, problems with the stability of elemental species standards, possible species transformations during the sample treatment procedure, and a total separation of one species from the other in the case of a species-unspecific detection, are special features of calibration in elemental species analysis compared with the determination of total element concentrations in a sample. Whereas most of the relevant inorganic elemental species are commercially available today, e.g. selenite and selenate or iodide and iodate, there is often a lack of organoelemental species such as dimethylthallium. However, most of the anthropogenic organoelemental species distributed worldwide, e.g. tributyltin and tetraethyllead, are available on the market. There is a total lack of isotopically labeled elemental species so that in all these cases synthesis of the corresponding spike compounds must be carried out.

Literature procedures are available for the synthesis of most of the stable elemental species which are of actual interest in elemental speciation analysis. Nevertheless, an exact characterization has to be carried out after synthesis of an elemental species which is used as a calibrant. With regard to the use of such a calibrant the species-specific purity is an important topic to be taken into account. When synthesizing elemental species with natural isotopic composition substantial amounts of educts can usually be applied. In contrast to this only small amounts, usually in the milligram range, must be applied for the synthesis of isotopically labeled elemental species because of the costs of the isotope-enriched initial material. In these cases the chemical procedure must be optimized with respect to a high synthesis yield for the isotope-enriched compound but not for the other reactant(s). A couple of descriptions for the synthesis of isotopically labeled elemental species can be found in the literature. Corresponding procedures for ^{82}Se-enriched selenite and selenate as well as those for ^{129}I-enriched iodide and iodate are described by Heumann and coworkers [4, 5]. A ^{206}Pb-enriched trimethyllead spike was first used by Ebdon's group for the determination of the corresponding lead species in rainwater by species–specific IDMS [6]. Recently, procedures for the synthesis of isotopically labeled monomethylmercury have also been published [7, 8], where methylation of isotope-enriched Hg^{2+} ions was carried out by reaction with a Grignard reagent in organic solvent and with methylcobalamin in aqueous solution, respectively.

The stability of elemental species standard solutions used for calibration is a much more critical point than for standard solutions used for total element determinations. In addition to problems observed already with trace element standards, where adsorption at walls of storage vessels or evaporation of the solvent may influence the concentration, decomposition of the elemental species can also lead to time-dependent variations of the concentration in solutions. Different storage conditions, such as composition of the solvent, pH value, oxidants, temperature, but also the material of the storage vessel, are important parameters which can cause a possible decomposition of the elemental species. The storage conditions for elemental species standard solutions must therefore be carefully checked and they are more critical the lower the concentration level is.

For example, it was found that an aqueous monomethylmercury solution was stable for months when it was stored in a closed PFA vessel at 4 °C in the dark [8, 9], which could not be confirmed for PE vessels. A mixture of Cr(III) and Cr(VI) remains stable over months if it is stored in PFA bottles at 5 °C in an HCO_3^-/CO_2 buffer solution at pH 6.4 under a CO_2 blanket [10, 11]. Different stabilities were found for iodide and iodate solutions in the concentration range of a few $\mu g\,L^{-1}$ during their storage in distilled water. Whereas iodate was stable over months, the iodide concentration decreased by about 10 % after 5 days and by more than 25 % after 1 month [12], so that iodide solutions must be freshly prepared for calibration purposes. The instability of iodide is also the reason why iodine doping of food is often carried out with iodate and not with iodide. Contrary to the iodide instability, bromide, also as a mixture with bromate, was found to be stable for months in an HCO_3^-/CO_2 buffer solution [13].

Transformation of the elemental species during the sample treatment procedure can be another important source of error in elemental speciation. If calibration by a species-specific standard is carried out for the detection step, possible transformations of the corresponding elemental species at other analytical steps prior to detection are not reflected in the result obtained. The total analytical procedure should therefore be validated for possible species transformations to be sure that species transformation do not play a role. Such a validation of analytical methods can be carried out by applying the isotope dilution technique [8] (see Section 4.2.3). During the last few years doubts, especially on the accuracy of methylmercury determinations, arose when Hintelmann *et al.* observed additional formation of this mercury species during its water vapor distillation from sediments [14]. By determining the distilled methylmercury after ethylation and subsequent analysis with a GC-ICP-MS system, they found that substantial amounts of this species can be formed during the distillation process from inorganic Hg^{2+} ions. In contrast, Demuth and Heumann observed formation of elementary mercury from methylmercury during ethylation of this species in the presence of halide ions [8]. Species transformations have been best investigated for methylmercury, but they can also occur for other alkylated elemental species, e.g. of lead and tin.

Because most of the detection methods applied today in elemental speciation analysis, e.g. atomic absorption spectrometry, atomic emission spectrometry, and ICP-MS, respectively, are not species specific in their signal response, a complete separation of the different species one from the other must be carried out prior to detection. In addition, the recovery must be determined for each of the species analyzed if less than 100 % of the elemental species is isolated after the different sample treatment steps. However, under certain conditions electroanalytical methods enable direct detection of elemental species in solution even if other species of the same element are present (see Chapter 5.9). An interesting example is the determination of monomethylcadmium in aquatic samples in the presence of inorganic Cd^{2+} by differential pulse anodic stripping voltammetry, which allows the determination of this biogenically produced cadmium species at concentration levels down to about 0.5 $ng\,L^{-1}$ [15].

3 EXTERNAL CALIBRATION

External calibrations can be carried out with a species-specific or a species-unspecific standard solution. In the case of a species-specific calibration the chemical form of the calibrant is identical with the analyte to be determined, whereas for species-unspecific calibration the calibrant is different in its chemical form. Species-unspecific calibration can never be used for a complete analytical procedure including, for example, extractive or chromatographic separations. There will always be a fractionation between the different species during such a separation procedure which prevents calibration of these analytical steps by an elemental species not identical in its chemical form with the analyte.

Under certain conditions species-unspecific calibration is therefore only acceptable for the detection step. However, in this case it must be guaranteed that the response of the detector is identical for the different elemental species in the sample and in the standard solution. It can be assumed that detector systems operating at high temperatures, such as those with inductively coupled plasma excitation, should not show significant differences in the detector response for various elemental species. Table 7.1.1 shows a comparison of the ICP-MS response for a 0.5 % HNO_3 acidic solution containing inorganic lead ions (Pb^{2+}) and trimethyllead (Me_3Pb^+), respectively, for different nebulizer systems [16]. Both solutions are identical in their lead content ($12\,\mu g\,L^{-1}$). As can be seen from the results listed in Table 7.1.1 the same detector response (counts per second, cps) is obtained for both lead species within the given standard deviations using a cross-flow and a μ-flow nebulizer system, respectively. From this it follows that under these conditions the same amount of lead is introduced into the plasma torch, using either trimethyllead or inorganic lead, and that there is also no difference in the ionization efficiency for both species. A similar result, with identical response for both species, was also found for an iodide and iodate solution at the $4\,\mu g\,L^{-1}$ concentration level by applying a quadrupole ICP-MS with a cross-flow nebulizer [12].

Table 7.1.1. ICP-MS response for a Me_3Pb^+ and an inorganic Pb^{2+} solution of identical lead content measured with different nebulizer systems (0.5 % HNO_3 acidic solutions containing $12\,\mu g$ Pb L^{-1}; measured isotope ^{208}Pb).

Nebulizer system	Pb species	ICP-MS response (cps)
Cross-flow[a]	Pb^{2+}	6400 ± 400
	Me_3Pb^+	6360 ± 380
μ-flow[b]	Pb^{2+}	8960 ± 220
	Me_3Pb^+	9000 ± 210
Ultrasonic with membrane desolvator[a]	Pb^{2+}	13 850 ± 1200
	Me_3Pb^+	475 ± 55
	Blank (0.5 % HNO_3)	255 ± 50

Quadrupole ICP-MS instrument used: [a]ELAN 5000; [b]HP 4500.

Identical absorbances were measured by flame atomic absorption spectrometry for the above mentioned trimethyllead and inorganic lead solutions containing identical lead contents. This also did not change when the temperature of the flame was varied using different air–acetylene gas mixtures. In contrast, when using an ultrasonic nebulizer with a membrane desolvator (CETAC U-6000AT) as the introduction system for ICP-MS, a totally different result was observed. Whereas a high ICP-MS response was obtained for inorganic Pb^{2+} ions, the response for Me_3Pb^+ was only a few counts per second above the HNO_3 blank (Table 7.1.1). The membrane desolvator obviously eliminates methylated lead effectively from the sample before reaching the plasma so that introduction of alkylated heavy metal species by such a system is not a good choice because fractionation between different elemental species can easily occur.

From determinations of the total element concentration it is well known that the matrix can strongly influence the detector response. It has therefore to be proved if such matrix effects are also relevant for different elemental species. From the results summarized in Table 7.1.1 it follows that a 0.5 % HNO_3 acidic solution, for example, does not cause differences in the detector response for the two lead species analyzed (see results for cross-flow and μ-flow nebulizer). However, this is not necessarily also valid for all other matrices and/or concentration ranges.

Figure 7.1.1 represents the signal intensity of Pb^{2+} and Me_3Pb^+, respectively, measured with

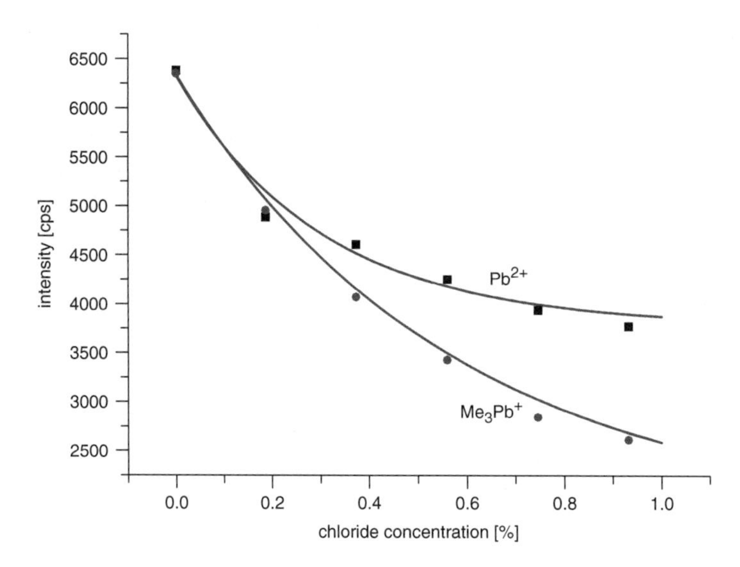

Figure 7.1.1. ICP-MS response of two different lead species containing the same lead content (12 μg Pb L^{-1}) dependent on the matrix influence of a diluted seawater sample (the concentration of the diluted seawater is represented by its chloride concentration).

an ICP-MS instrument (ELAN 5000), equipped with a cross-flow nebulizer, and its dependence on the salt concentration of a diluted synthetic seawater sample [16]. The salt concentration is expressed as its chloride content, with 1.8 % for undiluted seawater. Me_3Pb^+ can be biogenically produced in the ocean so that the corresponding matrix is relevant for trimethyllead determinations. For both lead species the signal intensity decreases substantially with increasing chloride concentration, affecting Me_3Pb^+ much more than Pb^{2+}. This leads to a difference of 15 % in the signal intensity at a chloride concentration of about 0.9 % between both species by a total response reduction of about 60 % for Me_3Pb^+ compared with the species solution without any matrix. From this it follows that calibration of species with a calibrant, different in its chemical form, can be strongly influenced by the matrix and also depends on its concentration. Summarizing the possibilities of species-unspecific calibration, it must be stated that the validity of such a calibration must always be proved for different matrix concentrations even if an identical response of different species is obtained for the matrix-free solution.

On-line coupling of chromatographic separation with atom spectrometric detection methods is one of the most powerful tools for speciation. In all these cases transient signals are produced which must be compared for calibration with corresponding signals of a standard solution. For quantification either the peak height or the peak area can be used for calculating the analytical result. Precision and accuracy are usually better when peak areas are evaluated whereas evaluation of the peak height is much simpler. In the case of symmetric peaks the peak height fits much better the result by peak areas compared with asymmetric peaks.

4 INTERNAL CALIBRATION

4.1 Standard addition method

Calibration by the standard addition method eliminates matrix interferences during detection and can also be used to correct for possible losses of the analyte during sample treatment procedures if sample and standard have been allowed to equilibrate. Because of possible matrix effects, which affect the detector response to different elemental species in a different way (see Section 3), species-specific standard addition should usually be applied for detector calibration. Calibration of the other analytical steps prior to detection can only be carried out by a species-specific standard. This means that the calibrant, added in increasing amounts to the different subsamples, should be identical in its chemical form with the elemental species to be determined. Under these conditions analytical results, obtained with the standard addition method, are usually more precise and accurate than those obtained by external calibration. One disadvantage of this calibration method is the fact that at least three to five aliquots (subsamples) need to be prepared because an increasing amount of the calibrant must be added to these different subsamples.

Figure 7.1.2(a) shows a standard addition calibration curve obtained for the determination of bromate in a mineral water sample by detection of the ^{79}Br isotope with a quadrupole ICP-MS after species separation by anion exchange chromatography, which is shown in Figure 7.1.2(b) for the original sample. The standard addition calibration curve resulted in a bromate concentration of $1.64\,\mu g\,L^{-1}$, whereas $14.6\,\mu g\,L^{-1}$ was obtained for bromide by applying an analogous calibration procedure also for this bromine species [13]. Linear calibration curves by the standard addition method are represented by equation (7.1.1). The corresponding mathematical expression for the bromate

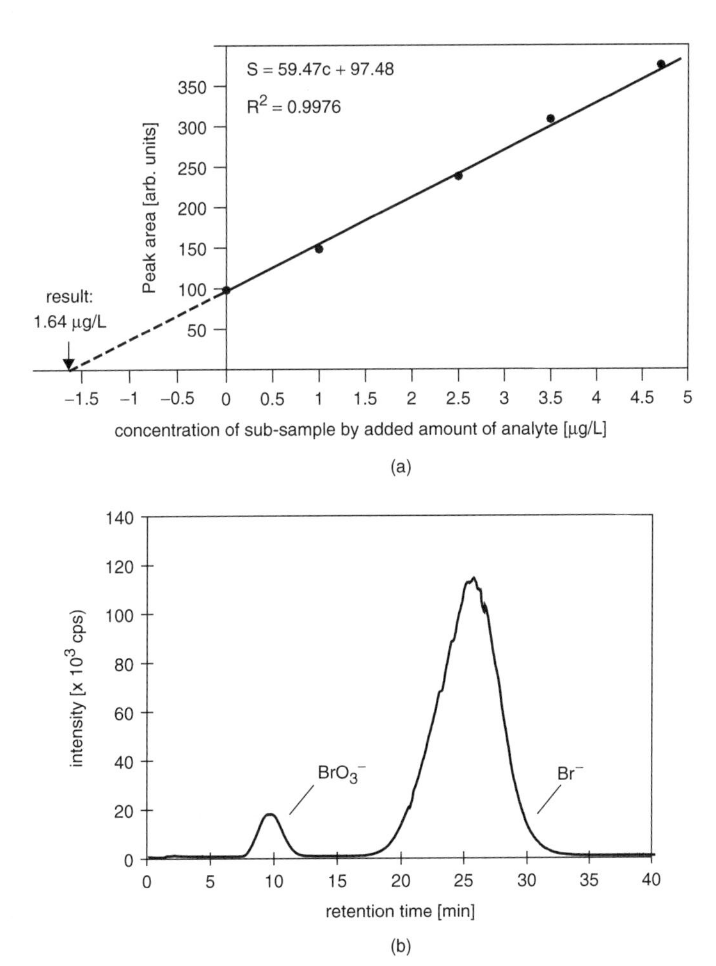

Figure 7.1.2. Determination of bromate by ICP-MS with the standard addition method in a mineral water sample after separation by anion exchange chromatography: (a) calibration curve; (b) chromatogram of separated bromate and bromide detected by ICP-MS.

determination discussed is given in Figure 7.1.2(a). The quality of the calibration curve of the standard addition method (linear regression) should always be judged by the correlation coefficient R, or better by its squared value R^2, which is given for the bromate determination in Figure 7.1.2(a). The correlation coefficient can vary between -1 and $+1$ and values close to -1 or $+1$ indicate a good fit of the regression curve. For a good calibration the R^2 value should be significantly larger than 0.95. The relative uncertainty of the result obtained by the standard addition method is best if the slope of the calibration curve is around 45° and becomes high for extremely gradual or steep curves.

4.2 Mass spectrometric isotope dilution technique

4.2.1 Fundamentals

Isotope dilution mass spectrometry is internationally accepted as a definitive method of proven high accuracy and precision where the possible sources of error are understood and usually also under control [3, 17]. One of the greatest advantages of IDMS is that loss of the analyte has no effect on the result after the isotope dilution process has taken place, which also means that no recovery must be determined for sample treatment procedures of the isotope-diluted sample. The isotope dilution step must guarantee total equilibration between the isotopically labeled spike and the sample species. This is usually no problem in aqueous systems, where many of the analyses for elemental species in the environment take place, or for samples totally dissolved, e.g. by an acid. IDMS of elemental speciation analysis can be particularly limited in cases where sample matrices cannot be totally dissolved because of possible species transformations during this process. The principles of IDMS are described in various textbooks [3, 18].

In principle, all different types of mass spectrometers can be used for the isotope dilution technique. However, in the case of using ionization methods where molecules or molecular fragments are produced and analyzed, as is true for applying electron impact or electrospray ionization (ESI), the natural isotopic pattern of the molecules can cause severe interferences. This is a special problem for elemental species containing large organic molecules, e.g. for metal complexes of biomolecules, because the $^{12}C/^{13}C$ pattern of such compounds causes many isotopic peaks. In this case mathematical corrections must be applied to obtain reliable analytical IDMS results. It is therefore much easier to apply ionization methods where preferably atomic ions are produced and detected, such as thermal ionization mass spectrometry (TI-MS) or ICP-MS. In addition, these ionization methods are mostly more sensitive than those producing molecular ions. The disadvantage is that the elemental species to be determined must be known and well defined because no structural information is obtained by these ionization techniques. The combined use of different ionization techniques will therefore certainly become a future trend for the quantification and characterization of elemental species of unknown composition. For example, ICP-IDMS can quantify the elemental species by a selected element in the molecule and ESI-MS may be able to identify its structure and composition.

TI-IDMS and ICP-IDMS have usually been applied to determine elemental species. Because these mass spectrometric methods are not able to differentiate between different chemical forms of an element, a total separation of all elemental species must be carried out prior to the mass spectrometric detection. In this case isotope-labeled spikes, identical in their chemical composition with all elemental species to be determined, must first be added to the sample. After equilibration of the elemental species of the sample with those of the spike, which is best done in a homogeneous solution, separation of the isotope-diluted species can be carried out. Then, the isotope ratio R of the spike isotope over a reference isotope of the element, used for labeling the elemental species, is measured for all separated fractions of the different elemental species. The schematic Figure 7.1.3 shows the principles of such an isotope dilution technique using a copper species as an example.

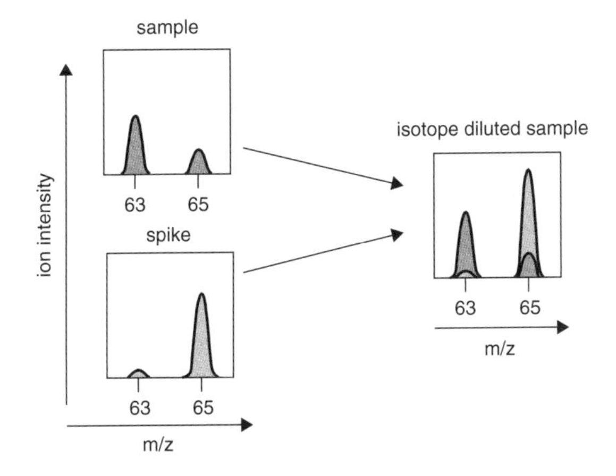

Figure 7.1.3. Schematic figure of the principles of isotope dilution mass spectrometry by the example of measuring the copper isotopes in a copper containing species.

The isotope ratio R of the isotope-diluted sample is the only quantity that needs to be measured for calibration by the species-specific IDMS technique. If the mass spectrometer produces isotope peaks with a flat top, as is the case for magnetic sector field instruments, the peak heights of the spike and reference isotope are used for calculation of the isotope ratio. In all other cases, e.g. by applying quadrupole instruments, the peak areas of the isotopes must be measured for the determination of the isotope ratio. As can be seen from Figure 7.1.3, the ion intensities of the lighter (^{63}Cu; in the following equations index 1) and the heavier (^{65}Cu; index 2) isotopes of the isotope-diluted sample are identical to the sum of the sample portion and spike portion used. Hence, equation (7.1.2) enables the isotope ratio R to be calculated:

$$R = (N^{S}h_2^{S} + N^{Sp}h_2^{Sp})/(N^{S}h_1^{S} + N^{Sp}h_1^{Sp}) \quad (7.1.2)$$

where $N^{S,Sp}$ is the number of molecules of elemental species in the sample and in the spike, and $h_{1,2}$ are the isotopic abundances of isotope 1 and isotope 2. After transformation of equation (7.1.2) for the content G_S of the elemental species in the sample one obtains

$$G_S = 1.6610^{-24}(M/W^{S})N^{Sp}[(h_2^{Sp} - Rh_1^{Sp})/(Rh_1^{S} - h_2^{S})] \quad [\mathrm{g\,g^{-1}}] \quad (7.1.3)$$

where M is the molecular weight of the elemental species and W^S is the sample weight.

The isotope abundances in the elemental species of the sample are usually identical with the well-known natural isotopic compositions, which are listed in corresponding IUPAC tables [19]. The isotope abundances of the spike solution and its concentration are determined separately where the concentration is usually obtained by a reverse IDMS technique using a standard solution of the elemental species with natural isotopic composition. Possible mass bias effects of the applied mass spectrometric techniques can best be eliminated by analyzing the isotopic compositions of the sample and the spike under the same conditions as the isotope ratio R, or they have to be corrected by independent measurements with standards [20]. Calibration by IDMS should also take into account an optimization of the ratio of the amount of sample species over spike species to minimize the error multiplication factor, which influences the precision of the IDMS result. Depending on the enrichment of the spike isotope this ratio should normally be between 0.1 and 10 [3].

4.2.2 Species-specific and species-unspecific calibration

Whereas TI-IDMS was the preferred technique before 1995 [3–5, 11], the first ICP-IDMS analyses of elemental species appeared in 1994 [6, 21]. Today, the number of investigations using ICP-IDMS for elemental species analysis exceeds those with TI-IDMS. This is due to several advantages of ICP-MS compared with TI-MS even if the selectivity is much better for TI-MS because of significantly reduced problems with spectrometric interferences. The multielement capability of ICP-MS is one of these advantages. However, the much more complicated sample preparation technique for TI-IDMS with its relatively high time consumption as well as the fact that only ICP-MS offers on-line coupling with separation techniques are the major reasons why ICP-IDMS is now more frequently used. The great difference in the sample treatment procedures between TI-IDMS and ICP-IDMS is demonstrated in Figure 7.1.4, showing

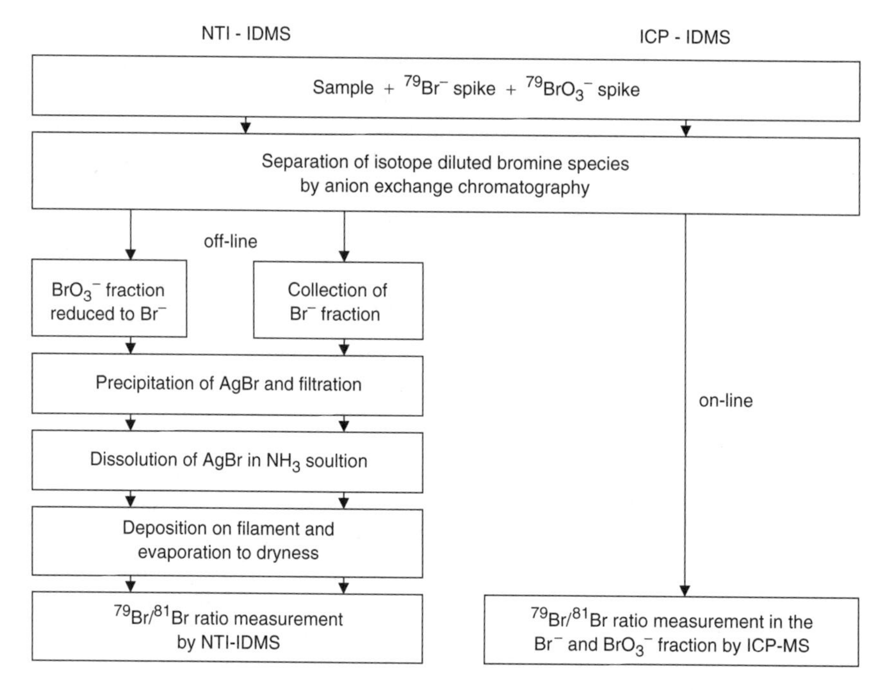

Figure 7.1.4. Comparison of the sample pretreatment techniques for the determination of bromide and bromate in an aquatic sample by negative thermal ionization isotope dilution mass spectrometry (NTI-IDMS) and inductively coupled plasma isotope dilution mass spectrometry (ICP-IDMS), respectively.

the determination of bromide and bromate by species-specific calibration with the isotope dilution technique [13]. Whereas the isotope-diluted fractions of bromate and bromide, separated by an anion exchanger column (Figure 7.1.2(b)), are introduced on-line into the ICP-MS instrument without any additional sample treatment, various off-line steps of elemental species/matrix separation and species isolation must be carried out in order to measure the bromine isotope ratio of the two species by negative thermal ionization mass spectrometry (NTI-MS).

Volatile and thermally stable elemental species are usually well defined with respect to their chemical composition and structure and can best be separated by gas chromatography. Species-specific IDMS calibration using on-line coupling of GC with ICP-MS is therefore preferably applied for this type of elemental species. The determination of dimethylselenide and dimethyldiselenide is a relevant example [22]. For GC-ICP-IDMS determinations the isotope dilution step is always carried out in the beginning of the analytical procedure so that the isotope-diluted sample passes the GC separation column. Isotope effects during this separation are usually very small so that they can be ignored. In addition, some nonvolatile elemental species can easily be converted into a volatile compound by derivatization which also offers the possibility of analyzing them by a GC-ICP-IDMS coupling system. For example, monomethylmercury ($MeHg^+$) is converted into the volatile ethylated compound MeEtHg by sodium tetraethyloborate [8, 14] and selenite is specifically converted into piazselenol by 1,2-diaminobenzenes [23].

Besides the species-specific spiking mode, a species-unspecific spiking mode is also possible under certain conditions. This is the only way to apply IDMS calibration to either elemental species where the chemical composition and structure is not exactly known (so that an isotope-labeled spike cannot be synthesized) or to all species where the synthesis of a labeled compound is too complicated. These are the reasons why metal complexes of humic substances or those of biomolecules have only been quantified, up to now, by a species-unspecific spiking mode when using IDMS calibration in connection with a HPLC-ICP-MS coupling system [24–26]. When applying the species-unspecific spiking mode, where the spike may exist in any chemical form, the isotope dilution step cannot be carried out before a complete separation of the different species has taken place. No loss of substance of the different species is therefore allowed up to this analytical step. It must also be guaranteed that the sample and the spike species do not produce different ICP-MS responses and that there is no discrimination between the sample and spike species by the sample introduction system (see Section 3; Table 7.1.1 and Figure 7.1.1).

Figure 7.1.5 shows the schematic diagram of an HPLC-ICP-IDMS system which is applied for species-specific but also for species-unspecific calibration [21]. An HPLC system, including a high pressure pump, sample injection valve, guard and separation columns, and a UV flow-through detector, is used for coupling liquid chromatography with ICP-MS. All different types of separation columns can be applied, such as size exclusion chromatography, normal and reversed-phase chromatography, and ion chromatography. If the species-specific isotope dilution mode is applied, the isotope-diluted sample is injected into the system, separated and then the separated elemental species are directly introduced into the ICP-MS instrument for measuring the isotope ratio R of all separated fractions. In this case the isotope ratio of a peak containing only a single elemental species has identical values at all peak positions. It is therefore possible to determine the isotope ratio, which can be directly converted into a concentration using equation (7.1.4) with the corresponding isotope intensities at a single peak position or, for better precision and accuracy, by measuring the total peak area of the reference and spike isotope. If IDMS calibration is carried out in the species-unspecific spiking mode, the spike solution is continuously added by a pump to the separated fractions after they have passed the UV detector. For precise species determinations it is very important that the pump produces a constant spike flow without pulsation. The spike flow is calibrated by a

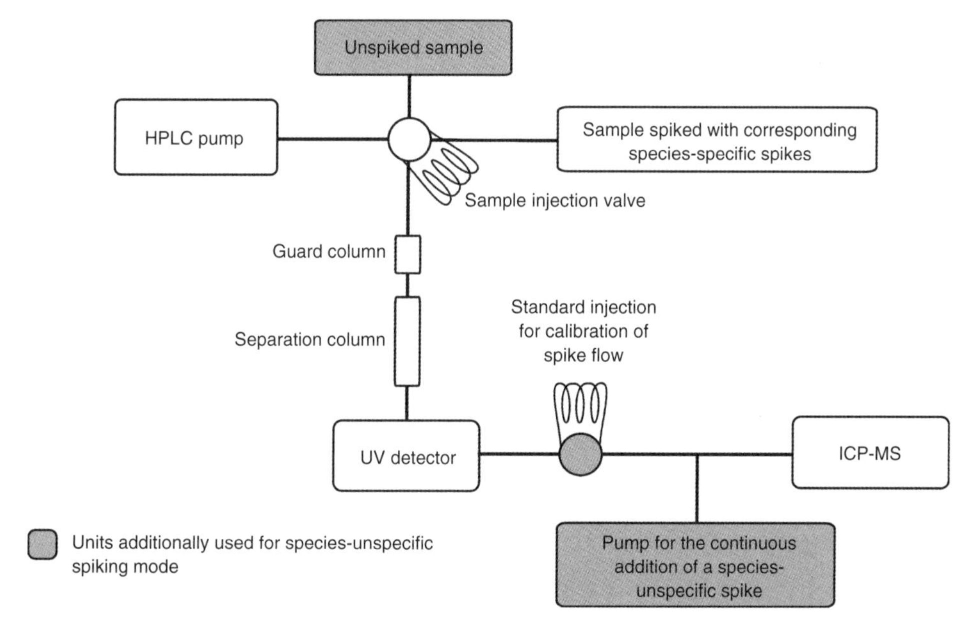

Figure 7.1.5. Schematic figure of an HPLC-ICP-IDMS system for elemental species analysis by species-specific and species-unspecific spiking modes.

corresponding standard solution injected by a separate valve into the system. The separated elemental species and the spike solution are completely mixed within a Y-shaped connection capillary and then the isotope diluted fractions are introduced into the ICP-MS instrument.

The isotope ratio *R* is measured at the appropriate retention time as represented in the top part of Figure 7.1.6 for a separated copper species when using a ^{65}Cu-enriched spike solution. So long as no copper species are eluted from the separation column the isotope ratio of the spike solution is measured. If a separated copper species appears the measured isotope ratio shifts towards the natural isotopic composition, which is proportional to the amount of the copper species. Using the calibrated spike flow and the known eluent flow the isotope ratio chromatogram can be directly converted into a mass flow chromatogram, which is shown at the bottom of Figure 7.1.6. More details about this HPLC-ICP-IDMS system and the mathematical conversion of the isotope ratio chromatogram into a mass flow chromatogram are given in ref. 21. This type of calibration offers the possibility of carrying out analysis by IDMS even for elemental species which are not sufficiently characterized. It also allows the determination of 'real-time concentrations' of chromatographically separated elemental species. These advantages make HPLC-ICP-IDMS one of the most powerful analytical tools for the quantification of trace amounts of elemental species.

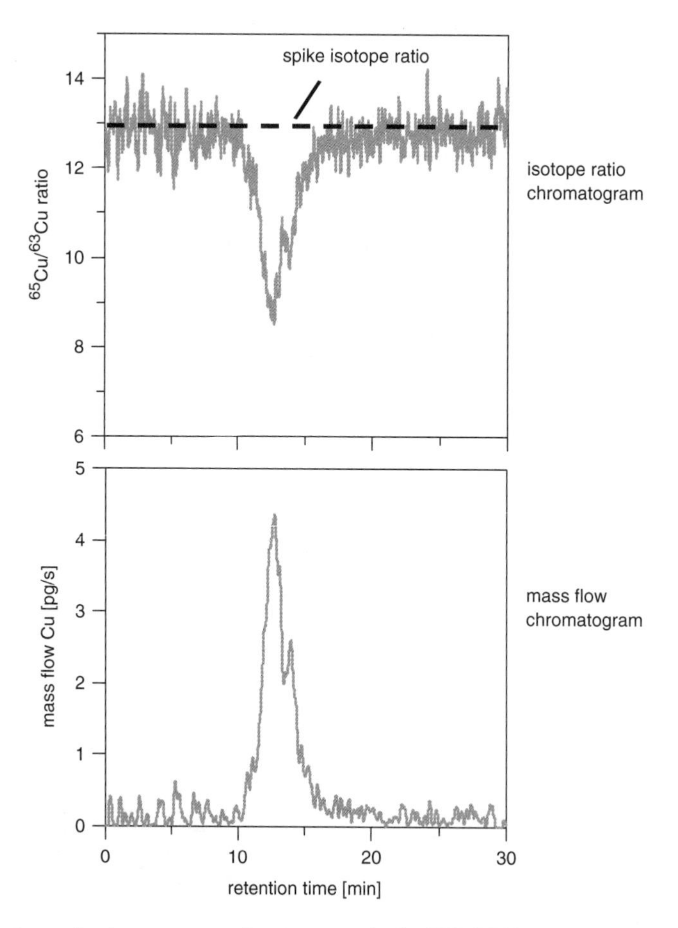

Figure 7.1.6. Measured isotope ratio chromatogram of a copper species by HPLC-ICP-IDMS with the species-unspecific spiking mode and its transformation into a mass flow chromatogram.

The HPLC-ICP-IDMS system represented in Figure 7.1.5 was used in connection with an ion chromatographic column and the species-specific calibration mode, for example, to determine iodide and iodate in aquatic samples [26]. Fractions of heavy metal complexes with humic substances, separated by size exclusion chromatography, have been quantified by species-unspecific IDMS calibration [21, 24, 26]. Recently, capillary electrophoresis was also successfully coupled with ICP-IDMS using species-unspecific spiking for the determination of sulfur in metallothionein fractions [25]. The corresponding mass flow chromatogram of a rabbit liver sample is represented by Figure 7.1.7. In this case the known amino acid sequence of the metallothionein fractions MT1 and MT2 also allowed calculation of the amount of metallothionein in these fractions by measuring the amount of sulfur.

Recently, electrothermal vaporization (ETV) was coupled with ICP-IDMS using species-unspecific calibration for the determination of mercury species in biological samples [27]. This ETV-ICP-IDMS system allows the direct determination of mercury species by solid sampling without dissolution of the sample, which is a great step forward with respect to a time-effective species analysis at a low contamination level. The temperature program which was used evaporated monomethylmercury at a much lower temperature than inorganic mercury so that these two mercury species could be separated by the ETV system. Species-unspecific calibration was carried out with ^{200}Hg-enriched elementary mercury (Hg^0) by mixing a well-defined gas flow of the elementary mercury spike, produced by an exactly tempered permeation tube, with the gaseous mercury species emitted by the ETV system. When methylmercury and inorganic mercury in a biological reference material (TORT-2) were measured with this ETV-ICP-IDMS system, the results obtained agreed very well with the certified values.

4.2.3 Validation of analytical procedures by the isotope dilution technique

An intensive discussion of possible systematic errors in analyses for elemental species arose after transformation of mercury species during sample treatment was first identified by using the isotope dilution technique with isotopically enriched Hg^{2+} ions [14]. By determining the distilled $MeHg^+$ species after ethylation with $NaBEt_4$ and subsequent analysis by GC-ICP-MS, it was found that substantial amounts of methylmercury can be formed during the distillation process from Hg^{2+}. The resulting overestimation of methylmercury in sediments was as high as 80 %. This clearly demonstrated that validation of analytical methods must always be carried out to guarantee accurate results in elemental species analysis. In contrast to

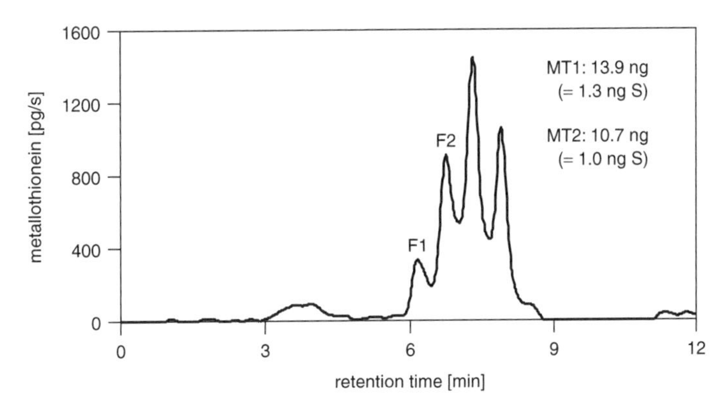

Figure 7.1.7. Mass flow chromatogram of sulfur from separated metallothioneins of a rabbit liver by CE-ICP-IDMS (MT1 and MT2 are known metallothionein fractions with 20 cysteine units; the mass flow of sulfur could therefore be used to calculated the corresponding amount of protein; F1 and F2 are unknown metallothionein fractions and the first peak at about 3.5 min migration time may be due to a sulfur impurity).

other atom spectrometric methods only mass spectrometry is able to verify such a transformation of elemental species during the sample treatment. The isotope dilution technique has therefore been increasingly applied during the last few years to validate analytical procedures for possible species transformations. As a result it was also found that methylmercury is transformed to elementary mercury during ethylation by $NaBEt_4$ in the presence of halide ions [8, 28].

Speciation of Cr(VI) and Cr(III) in solid environmental samples is a great challenge because bidirectional species transformation is possible during recommended sample treatment procedures, e.g. those by EPA method 3060A [29]. To determine the degree of transformation of Cr(VI) into Cr(III) and also the opposite reaction during sample pretreatment, a double-spiking IDMS method was developed using a ^{53}Cr(VI)-enriched and a ^{50}Cr(III)-enriched spike. Recently possible transformations of butyltin species were also investigated using the IDMS technique [30].

Even if the IDMS technique is able to identify possible transformations of elemental species during sample treatment procedures and is therefore able to offer validation of analytical methods, the question nevertheless remains whether under these conditions IDMS calibration results in accurate analytical data or not. The determination of $MeHg^+$ in a river water sample by applying a GC-ICP-IDMS system is used to show the corresponding result [8]. After the sample was spiked with ^{201}Hg-enriched $MeHg^+$ (isotope ratio $^{201}Hg/^{202}Hg = 6.45$), monomethylmercury was converted into the volatile MeEtHg by reaction with $NaBEt_4$. Under these conditions Hg^{2+} ions were also ethylated to become volatile Et_2Hg. After a purge and trap procedure the volatile mercury species were then separated using a capillary GC column. The corresponding chromatograms of the isotopes ^{201}Hg (spike isotope) and ^{202}Hg (reference isotope), detected with a quadrupole ICP-MS instrument, are represented at the top of Figure 7.1.8. The $MeHg^+$ peak shows an isotope ratio $^{201}Hg/^{202}Hg$ somewhat between that of the spike solution and the natural isotopic ratio of 0.44, which is due to the presence of monomethylmercury in the river water sample. Also the Hg^0 peak does not show the natural isotopic composition, which is a clear indication that transformation of methylmercury into elementary mercury took place. By applying $NaBPr_4$ for propylation instead of ethylation of $MeHg^+$, no transformation of methylmercury into elementary mercury could be observed (bottom of Figure 7.1.8). However, by evaluating the measured isotope ratio of the $MeHg^+$ peak for ethylation and propylation using equation (7.1.3) identical analytical results within the limits of error of $(3.8 \pm 0.1)\,ng\,mL^{-1}$ and $(3.6 \pm 0.1)\,ng\,mL^{-1}$, respectively, were obtained. This means that transformation of methylmercury during the ethylation process has no effect on the analytical result. This is due to the fact that loss of methylmercury during the ethylation process takes place only after a total equilibration between the spike and sample species, which demonstrates the great advantage of an IDMS calibration for elemental species analysis.

5 CONCLUSION

External as well as internal calibration can be applied for elemental species analysis. Even if some atom spectrometric detectors can also be calibrated under certain conditions by elemental species not identical with the species to be determined, the best way to obtain accurate analytical results is calibration with the elemental species of the analyte. This is absolutely necessary in all cases where separation techniques are involved in the analytical method such as for hyphenated techniques most frequently applied in elemental speciation. Internal calibration usually results in more precise and accurate results than does external calibration and can be carried out either by the standard addition or by the IDMS method. IDMS has the advantages of being a one-point calibration method and that loss of the analyte after the isotope dilution step usually has no effect on the result. However, IDMS needs the application of an isotopically enriched spike so that for more frequent

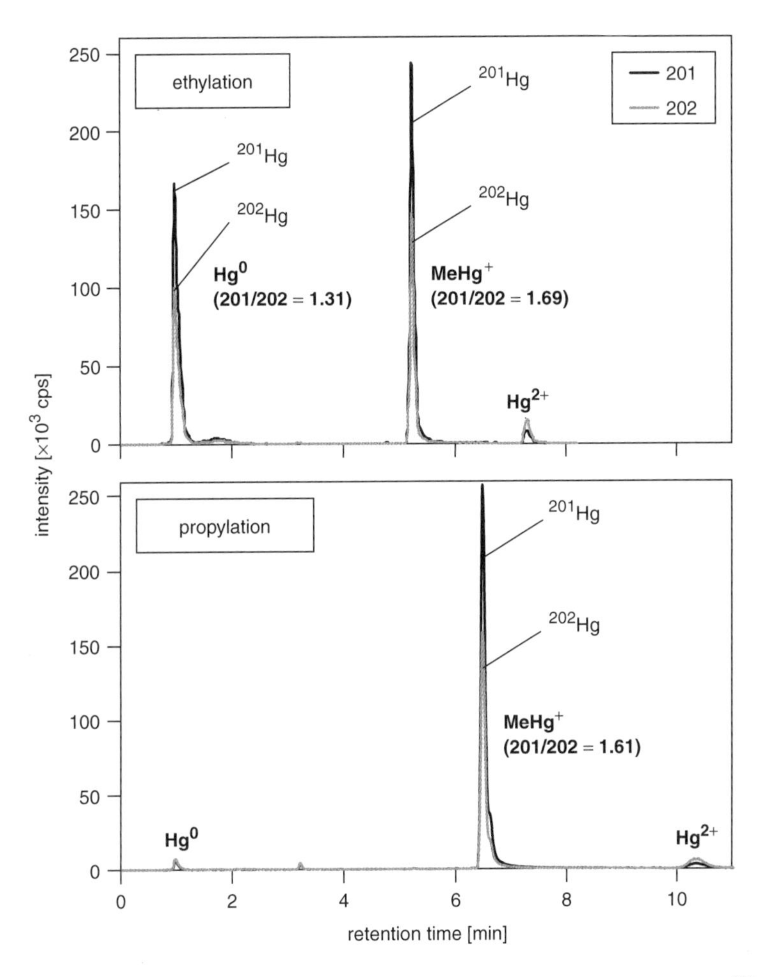

Figure 7.1.8. GC-ICP-IDMS chromatograms of a river water sample for methylmercury determination by ^{201}Hg-enriched $MeHg^+$ after derivatization with sodium tetraethylborate and sodium tetrapropylborate, respectively. (For all elemental species deviating from the natural isotopic ratio $^{201}Hg/^{202}Hg = 0.44$ the corresponding value is given in brackets.)

applications today's total lack of commercially available isotope-labeled elemental species must be overcome.

6 REFERENCES

1. Marshal, A., *Calibration of Chemical Analysis and Use of Certified Reference Materials*, Draft ISO Guide 32, ISO/REMCO N 262, International Organization of Standardization, Geneva, 1993.
2. Skoog, D. A., West, D. M. and Holler, F. J., *Analytical Chemistry – an Introduction*, Saunders College Publishing, Philadelphia, PA, 1994, p. 431.
3. Heumann, K. G., Isotope dilution mass spectrometry in *Inorganic Mass Spectrometry*, Adams, F., Gijbels, R. and van Grieken, R. (Eds), John Wiley, & Sons, Inc., New York, 1988, pp. 301–376.
4. Heumann, K. G. and Grosser, R., *Fresenius' J. Anal. Chem.*, **332**, 880 (1989).
5. Reifenhäuser, C. and Heumann, K. G., *Fresenius' J. Anal. Chem.*, **336**, 559 (1990).
6. Brown, A. A., Ebdon, L. and Hill, S. J., *Anal. Chim. Acta.*, **286**, 391 (1994).
7. Snell, J. P., Steward, I. I., Sturgeon, R. E. and Frech, W., *J. Anal. At. Spectrom.*, **15**, 1540 (2000).
8. Demuth, N. and Heumann, K. G., *Anal. Chem.*, **73**, 4020 (2001).
9. Leermakers, M., Lansens, P. and Baeyens, W., *Fresenius' J. Anal. Chem.*, **336**, 655 (1990).
10. Dyg, S., Cornelis, R., Griepink, B. and Verbeeck, P., Stability study of Cr(III) and Cr(VI) in water for

production of an aqueous chromium reference material, in *Metal Speciation in the Environment*, Broekaert, J. A. C., Gücer, S. and Adams, F. (Eds), NATO ASI Series G: Ecological Sciences, Vol. 23, Springer, Berlin, 1990, pp. 361–376.
11. Nusko, R. and Heumann, K. G., *Anal. Chim. Acta*, **286**, 283 (1994).
12. Vogl, J., Iodine species analysis in aquatic systems by HPLC/ICP-MS and application of on-line isotope dilution technique, Diploma Thesis, University of Regensburg, Germany, 1994.
13. Diemer, J. and Heumann, K. G., *Fresenius' J. Anal. Chem.*, **357**, 74 (1997).
14. Hintelmann, H., Falter, R., Ilgen, G. and Evans, R. D., *Fresenius' J. Anal. Chem.*, **358**, 363 (1997).
15. Pongratz, R. and Heumann, K. G., *Anal. Chem.*, **68**, 1262 (1996).
16. Helfrich, A. and Heumann, K. G., unpublished.
17. De Bièvre, P., *Fresenius' J. Anal. Chem.*, **350**, 277 (1994).
18. Smith, D. H., Isotope dilution mass spectrometry, in *Inorganic Mass Spectrometry – Fundamentals and Applications*, Barshick, C. M., Duckworth, D. C. and Smith, D. H. (Eds), Marcel Dekker, New York, 2000 pp. 223–240.
19. Rosman, K. J. R. and Taylor, P. D. P., *Pure Appl. Chem.*, **70**, 217 (1998).
20. Heumann, K. G., Gallus, S. M., Rädlinger, G. and Vogl, J., *J. Anal. At. Spectrom.*, **13**, 1001 (1998).
21. Rottmann, L. and Heumann, K. G., *Fresenius' J. Anal. Chem.*, **350**, 221 (1994).
22. Gallus, S. M., Development of a GC/ICP-MS system for selenium species determinations by isotope dilution mass spectrometry, PhD Thesis, University of Regensburg, Germany, 1998.
23. Gallus, S. M. and Heumann, K. G., *J. Anal. At. Spectrom.*, **11**, 887 (1996).
24. Heumann, K. G., Gallus, S. M., Rädlinger, G. and Vogl, J., *Spectrochim. Acta*, **B53**, 273 (1998).
25. Schaumlöffel, D., Prange, A., Marx, G., Heumann, K. G. and Brätter, P., *Anal. Bioanal. Chem.*, **372**, 155 (2002).
26. Heumann, K. G., Rottmann, L. and Vogl, J., *J. Anal. At. Spectrom.*, **9**, 1351 (1994).
27. Gelaude, I., Dams, R., Resano, M., Vanhaecke, F. and Moens, L., *Anal. Chem.*, **74**, 3833 (2002).
28. Lambertsson, L., Lundberg, E., Nilsson, M. and Frech, W., *J. Anal. At. Spectrom.*, **16**, 1296 (2001).
29. Huo, D. and Kingston, H. M., *Anal. Chem.*, **72**, 5047 (2000).
30. Alonso, J. I. G., Encinar, J. R., González, P. R. and Sanz-Medel, A., *Anal. Bioanal. Chem.*, **373**, 432 (2002).

7.2 Reference Materials

Philippe Quevauviller

European Commission, Brussels, Belgium

1 INTRODUCTION

The harmonisation of measurements and technical specifications is a continuous process, which is reflected in the proliferation of quality assurance guidelines, accreditation systems, proficiency testing schemes, etc. Besides the external means to control the quality of measurements, the availability of reference materials is one of the major tools for laboratories to monitor the performance of their analytical work. In several branches of the economic sector, in particular industry, these tools have been in use for more than a century (since 1901), their production being market driven. The needs to support industry by producing quality control tools has been recognised by major organisations such as the US National Institute for Standards and Technology, formerly National

Handbook of Elemental Speciation: Techniques and Methodology R. Cornelis, H. Crews, J. Caruso and K. Heumann
 ISBN: 0-471-49214-0

Bureau of Standards, at the beginning of the 20th century [1]. In other fields, such as environmental monitoring, and health and safety services, the reference material needs have been highlighted much later, and their production has often been supported by public funding. The first attempt to produce biological and environmental reference materials for chemical measurements was made in the 1950s and the first biological RM (elements in kale) was made available in 1960 [1]. In the 1970s and 1980s, a number of organisations started or developed programs designed to provide biological, environmental and food reference materials for a wide range of chemical parameters. An example of a successful interlaboratory program is a succession of research and development programs set up by the European Commission since 1973 for improving the quality of chemical measurements and producing Certified Reference Materials (CRMs) covering various sectors, as a result of laboratory and regulatory needs [2].

The need to determine chemical forms of elements rather than their total contents has been highlighted decades ago and has been discussed extensively in the literature [3] and in the present book. Similarly to other measurement fields, speciation analysis requires the availability of suitable reference materials for the purpose of verification of accuracy and quality assurance needs [4]. Measurements that represented a real start in speciation science were those that had a link with identified toxicity risks, namely the determination of methylmercury in biological tissues and organotins in environmental matrices. This awareness was naturally associated with expressed needs from the laboratories with respect to quality control tools, and organisations such as the National Institute for Environmental Sciences, NIES (Japan) and the National Research Council, NRCC (Canada) started work on the organisation of interlaboratory studies and CRM production for these compounds during the 1980s. Later on (1988 and after), the European Commission, through the BCR program (French acronym for Community Bureau of Reference), launched a series of projects aiming at improving the quality of speciation measurements for chemical forms of Al, As, Hg, Pb, Se and Sn in various biological and environmental matrices, along with extractable forms of trace metals in soils and sediments [4].

Resulting from the experience gained, this chapter summarises main aspects related to the preparation, certification and use of reference materials for speciation studies as reviewed from the literature [3–7].

2 DEFINITIONS AND CATEGORIES OF REFERENCE MATERIALS

2.1 Definitions

The ISO Guide 30 gives the following definitions for reference materials [8]:

- *Reference Material (RM):* a material or substance one or more of whose property values are sufficiently homogeneous and well established to be used for the calibration of an apparatus, the assessment of a measurement method, or for assigning values to materials.
- *Certified Reference Material (CRM):* a reference material, accompanied by a certificate, one or more of whose property values are certified by a procedure which establishes traceability to an accurate realisation of the unit in which the property values are expressed, and for which each certified value is accompanied by an uncertainty at a stated level of confidence.

The added value of the CRM in fact lies in the guarantee given for the certified property and its associated uncertainty. Recent discussions [9] have highlighted the need to clarify this definition, stressing that RMs are to be used solely for interlaboratory studies (including proficiency testing) and reproducibility evaluation (through control charting), while CRMs are to be used exclusively for the evaluation of accuracy (and, in specific cases, for calibration purposes).

The sections below examine the various types of reference materials that are currently encountered in chemical analytical work (including speciation), namely pure substances and calibrating materials, matrix-matched reference materials, and operationally defined reference materials. In this chapter, the terms RM (Reference Material) and CRM (Certified Reference Material) will be used.

2.2 Pure substances and calibrating materials

Reference materials can consist of pure substances or solutions for use in calibration and/or identification of given parameters, or aimed at testing part or totality of an analytical procedure (e.g. raw or purified extracts, spiked samples etc.). Examples given in the literature describe the preparation of various materials falling into this category that were used for the purpose of quality control of speciation analyses. To name but a few, calibrants of pure organotins, methylmercury chloride and trimethyllead were prepared in the framework of certification campaigns organised by BCR [4]. They were not aimed at producing pure substances for commercial purpose, but rather to ensure a firm comparability basis for measurements performed by laboratories participating in corresponding certifications (see below), i.e. for verifying that no systematic errors due to calibration were left undetected. Other examples are fish extracts (either raw or spiked with MeHg) and urban dust spiked with trimethyllead [10] that were prepared in the framework of interlaboratory studies to test the performance of separation and detection methods prior to certification work [4]. Pure substances have also been prepared to check derivatisation reactions in the framework of certification of organotin compounds, i.e. secondary organotin calibrants (e.g. ethylated, pentylated etc. forms of butyltin compounds) for verifying derivatisation yields [11].

Other materials, of known composition, are used for calibrating certain types of measurement instruments. Matrix-matched reference materials (see section below) are not intended for calibration purposes but some techniques, e.g. X-ray fluorescence spectrometry, require such materials for quantification purposes. This applies to total inorganic analyses, however, and no examples concerning speciation analysis are presently known. In the case of CRMs, calibrating solutions are prepared gravimetrically (by specialised laboratories) and examples are known for speciation, e.g. synthesis of pure arsenobetaine and preparation/certification of pure calibrating solution of this compound [12].

2.3 Matrix-matched reference materials

Matrix-matched reference materials represent, in principle, as much as possible the matrix analysed by the laboratory. Due to the requirements for homogeneity and stability (see further discussions below) of a reference material, it is never possible to exactly match the matrix of a fresh sample. It may be hoped that the changes that are introduced to the matrix during the production of the RM are not too significant for the purpose, but in principle one has to be aware that even a matrix-matched reference material in the best case only comes close to a sample of the respective matrix. In the case of RMs, the materials may be prepared by the laboratory for internal quality control purposes (e.g. establishment of control charts) or for use in interlaboratory studies (see Section 3). CRMs are certified for specific parameters and are restricted to the verification of a measurement procedure or, in some cases, calibration purposes. The certification is based on specific procedures that are described in Section 8.

Basically, matrix RMs may represent all kinds of 'natural' matrices that are currently analysed by laboratories. For speciation studies, focus nowadays is towards the determination of chemical forms in biological and environmental matrices. Therefore, current available CRMs concern matrices such as biological tissues (mussel, fish, lobster, etc.), sediments (freshwater, coastal and marine) and soils of various compositions, which are certified for a range of chemical forms of elements (e.g. As, Hg, Pb and Sn). Details on the available matrix CRMs are given in the following sections.

2.4 Operationally defined reference materials

Among the matrix RMs, a particular category concerns reference materials related to operationally defined parameters. In this case, the assigned or certified values are directly linked to a specific method, following a strict analytical protocol. With respect to speciation of trace elements, as discussed elsewhere in this book (see IUPAC definition), this type of measurement is referred to as 'fractionation'. It does not correspond to a determination of chemical forms of elements *sensu stricto* but rather to the measurements of broad forms of elements defined according to specific extraction procedures [13], of which the results are used for the definition of forms such as *mobile, bioavailable, carbonate-bound*, etc.; one should note that these 'forms' correspond to interpretation of results rather than actual measurements. In other words, a laboratory will measure e.g. EDTA-extractable contents of trace metals or element contents determined following a sequential extraction scheme, and these measurement results will be interpreted to correspond to *forms*. Whatever the interpretation given, the only possibility of achieving comparability of measurements based on extraction methods is to use harmonised (method adopted by consensus among a group of laboratories) or standardised (standard methods adopted by an official standardisation organisation) methods [14]. CRMs based on harmonised extraction procedures have been produced for soil and sediment analyses and are the only examples to date for ensuring comparability of data in this specific 'speciation' field [14].

3 USE OF REFERENCE MATERIALS

3.1 Method validation

It is not the purpose here to describe in detail method validation aspects that are described in the literature [15] but to stress that reference materials play a crucial role at all stages of method validation (development, optimisation and quality control) for all types of chemical measurements, including speciation. Pure calibrating solutions will be used for testing performance criteria such as sensitivity, selectivity, linearity, detection and quantification limits. Extracts or matrix-matched solutions may be used for the same purpose, but also to detect possible interferences and check yields of chemical reactions (e.g. derivatisation). Finally, matrix RMs will be used for checking performance criteria such as accuracy (trueness, repeatability and reproducibility) and the application range of the method.

Examples of use (and misuse) of reference materials for validation purposes are discussed elsewhere [16], with emphasis on traceability aspects that are discussed in the following sections. Use of RMs for routine quality control purposes is also discussed below. At this stage, it should be stressed that CRMs are products of high added value, with very costly production and certification (typically several hundred thousand Euros). For this reason, they should in principle be reserved for the verification of the accuracy of analytical procedures and not for daily use (e.g. routine internal control of a laboratory). This aspect, however, is debatable since the price of CRMs may be marginal in comparison to the costs of running expensive analytical techniques; the final price also depends on the overall production and use of the 'products' by the customers. It is increasingly considered that the better use of CRMs is to sell the stock as rapidly as possible and to re-certify a new batch instead of getting old CRMs that have remained on the shelves of the producer. This strategy obviously depends on the real market for the CRMs.

3.2 Interlaboratory testing

The organisation of interlaboratory studies is described in the literature [17] and details and examples of studies related to speciation are also fully documented [2, 4]. Three types are generally considered, namely (1) proficiency testing schemes, (2) method performance studies, and (3) reference material certification studies. These different studies all have in common that they are centrally coordinated and that they involve several laboratories (from five to hundreds!). The coordination implies a planning of the work, discussions with the participants, preparation and distribution of clear participating instructions (see examples in ref. 4 with respect to speciation analysis), chairing of technical meetings and result evaluation. Interlaboratory studies are based on the distribution of homogeneous and stable reference materials. Examples of materials for speciation analyses as used in the three above-defined types of tests are given below.

Proficiency testing schemes aim to evaluate the performance of one or more methods applied by a group of laboratories. They are generally organised by official organisations, often in the framework of accreditation. With respect to speciation studies, very few tests are performed at the present stage. To the author's knowledge, the only scheme where reference materials are analysed (for MeHg and organotins) is the QUASIMEME (Quality Assurance of Information of Marine Environment Measurements) proficiency testing scheme, focusing on marine monitoring [18]. In this framework, reference materials are produced, characterised and distributed to participating laboratories for checking their performance at the time of routine analyses carried out for monitoring purposes. The particularity of these *ad hoc* studies is that it may involve reference materials with a limited shelf life, e.g. 'fresh' materials stable over a period of 3 months, which is sufficient for the purpose of the method evaluation. This aspect of stability requirement is further discussed below.

Method performance studies focus on the systematic evaluation of one or several methods. These studies are generally carried out on a limited scale, e.g. for standardisation purposes or prior to the certification of a reference material. Examples of systematic scrutiny of methods for speciation studies have been published elsewhere [2, 4]. They involved the distribution of various categories of reference materials of increasing complexity to sequentially test the various steps of speciation methods (extraction, derivatisation, separation and detection), i.e. pure solutions, raw and spiked extracts, spiked materials, and real samples. An example of a scheme followed for the improvement of analytical methods for MeHg is shown in Figure 7.2.1.

The certification of reference materials may be carried out through an interlaboratory study, the objective being to assign a reference value to a given material, along with a determined uncertainty. This last type of study, which represents one of the different possibilities for certifying a RM, is further discussed in Section 8.

3.3 Control charting

The establishment of control charts by laboratories for monitoring the reproducibility of methods is now generalised. The principle is to report

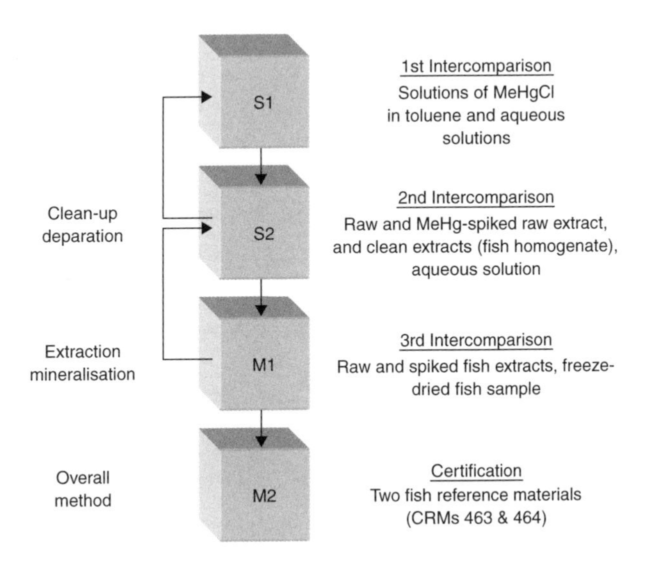

Figure 7.2.1. Example of improvement scheme for a method performance study: MeHg in fish tissue.

results of analyses of one or several control samples (LRM) on a chart as a function of time, and to monitor the variations of these results in comparison with reference values [19]. The user may define warning and action lines, which will serve as 'alarm bell' when the system is out of control. Several types of charts exist, e.g. Shewhart and Cusum charts, which are described in detail in the literature [19, 20]. Similarly to other reference materials, LRMs used in control charting must be homogeneous, stable, representative of the analytical problem(s) and available in sufficient amounts for long-term monitoring of methods. To date, most of the LRMs for daily quality control purpose are prepared on a limited scale by laboratories, i.e. there are few structures that make these (not certified) reference materials available for routine quality checks. With respect to speciation, as mentioned above, the only organisation that prepares materials on a regular basis for routine quality control purposes (including control charting) is the QUASIMEME network [18]. Other materials are prepared by some specialist laboratories to fulfil their own needs.

3.4 Evaluation of analytical results using a matrix CRM

This section examines how an analytical result may be evaluated in comparison with the certified value of a matrix CRM. This approach of course applies to speciation analysis; it is adapted from the procedure proposed by Walker and Lumley [16]. The general use of RMs in a validation process of a method is summarised in Figure 7.2.2.

The use of a matrix CRM will be based on the evaluation of an analytical result (x) as compared with a certified value (μ) of the CRM. The error in the analytical result (Δ) is calculated from the formula $\Delta = x - \mu$.

Considering the random errors of the method, the value of Δ will probably not be equal to zero, even if the result is not affected by any systematic error. The greater the random errors

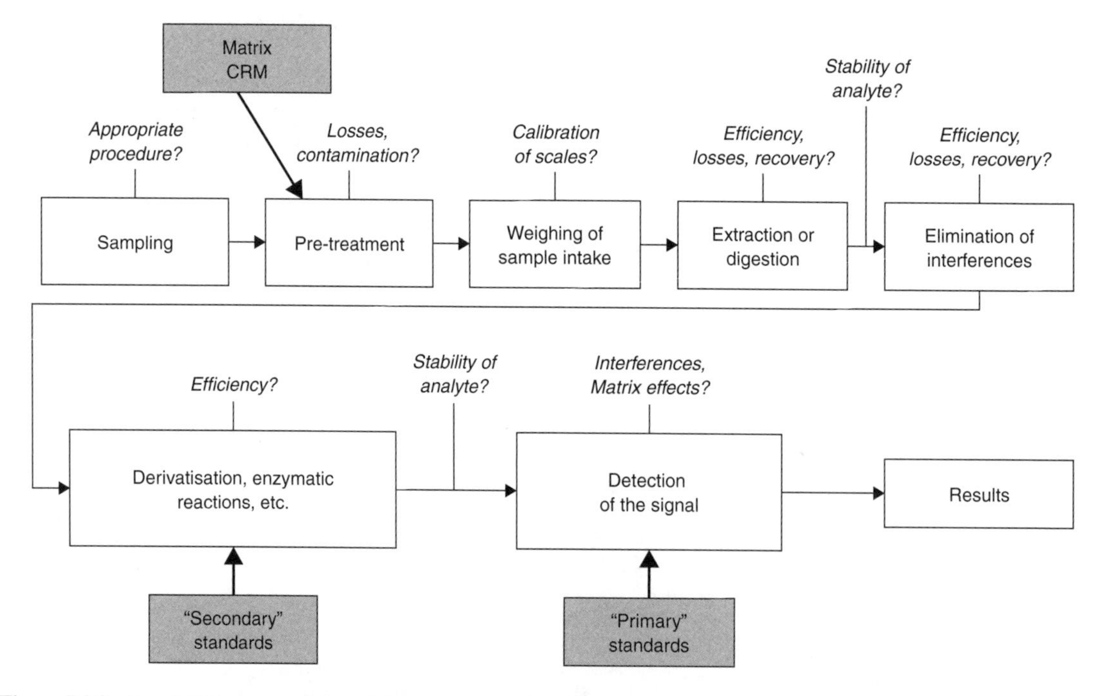

Figure 7.2.2. Use of RMs in a method validation process, with focus on speciation analysis (adapted from ref. 16). Reprinted from *Trends in Analytical Chemistry*, vol. 18, Walker and Lumley, Pitfalls in terminology and use of reference materials ..., pp. 594–616, Copyright 1999, with permission from Elsevier Science.

(i.e. the poorer the precision), the greater the value of Δ and hence the more difficult it is to detect the occurrence of a systematic error. The precision, therefore, is a critical parameter that should not be underestimated when evaluated the trueness of a method. Walker and Lumley [16] distinguish the intralaboratory standard deviation, s_i, characterised by the measurement repeatability of which the estimate should be calculated on the basis of at least seven repetitions of CRM analyses, and the between-laboratory standard deviation, s_e, which is more difficult to estimate. The authors propose several approaches to calculate the latter parameter:

(1) The reproducibility, s_R, may be estimated by replicate analyses (at least seven, preferably up to 20) carried out over a given period of time (if possible over 3 months). The between-laboratory standard deviation, s_e, may also be estimated in the framework of a method validation interlaboratory study in which the laboratory will know the repeatability values, s_r, and the reproducibility values, s_R, of the method according to the document summarising the results of the study. The value of s_e will hence be equal to $\sqrt{(s_R{}^2 - s_r{}^2)}$.
(2) When the CRM has been characterised in the framework of an interlaboratory study, information on the between-laboratory standard deviation is generally given in the certification report of the material. If the method to be tested is similar to one of those used for the certification of the RM, the value of s_e given in the report may be used.
(3) Predicted values found in the literature may also enable an estimate of s_e. This type of information is available in the agro-food sector [21] but comparatively few values exist in the sector of speciation analysis, with the exception of values obtained in the framework of certification campaigns [2, 4].
(4) In the absence of any information, an estimate of s_e may be obtained from the value of s_i according to the formula: $s_e \approx 2s_i$.

The precision σ of an analytical result of a matrix CRM will be calculated by combination of two components:

$$\sigma = \sqrt{s_e{}^2 + \frac{s_i{}^2}{n}}$$

where n is the number of replicates of CRM analyses. In general, the value s_i is smaller than the value of s_e (typically by a factor of 2 as indicated above). The fact that n is at least equal to 7 means that s_e will represent the main contribution of σ. At first sight, it may appear sufficient to base the estimate of the precision σ of a method used by an individual laboratory on the sole value of s_i. However, s_i reflects the random dispersion of results of a series around their mean, which is itself randomly distributed around the CRM certified value with a dispersion that is characterised by the value s_e. Therefore, the combination of s_i and s_e (as indicated above) is used to describe the overall dispersion of the results around the certified value, which is taken as true value [16].

The parameter s_e measures the sources of random errors that cannot be evaluated by replicate analyses in a single laboratory, but which contribute to the result dispersion around the certified value (true value or admitted as such). An example of random error is the possible variation of the final volume of a sample extract before its introduction in a measurement instrument, without taking care of the variations of ambient temperature. Such volume variations would not be significant for the estimate of the repeatability and would therefore not be considered in the calculation of s_i. However, the same measurements carried out by different laboratories (or by a single laboratory over a given period of time) would be subject to random errors due to variations of the ambient temperature. The effects of such variations would be included in the term s_e.

It is also useful to remember that, when a laboratory analyses a matrix CRM, it actually takes an effective part in an 'interlaboratory study' (if the certified values have indeed been measured on the basis of such study, see Section 8). Under these circumstances, it is clearly appropriate that the component s_e of the precision be considered when a laboratory compares its results with CRM values.

If the information on the value s_i is available (e.g. the repeatability value s_r of the method as validated through an interlaboratory study), a χ^2 test may then be carried out that will establish whether s_i (measured by the laboratory) is acceptable, i.e. whether the laboratory performs its method with a sufficient precision. However, even if s_i is significantly greater than s_r, if the measured value ${s_i}^2/\sqrt{n}$ is small in comparison with ${s_e}^2$, there will be little or no benefit in repeating a series of measurements of a CRM with the aim of obtaining a smaller value of s_i [16].

The estimate of the possible occurrence of systematic errors will be based on a statistical test aiming to evaluate whether the value Δ is significantly different from zero. If it is not the case, it is possible to conclude that no systematic error has been demonstrated. A test that is currently used is based on bracketing the value Δ in an interval with limits of $\pm 2\sigma$ in which it is estimated that no systematic error has occurred: $-2\sigma < \Delta < 2\sigma$

The affirmation that no systematic error has occurred has to be considered with some care. It is indeed possible that errors are left undetected, e.g. in the case of positive and negative errors which compensate each other.

As previously mentioned, the choice of the $\pm 2\sigma$ interval means that the confidence level of this conclusion is about 95 %. The adoption of limits $\pm 3\sigma$ would allow a confidence level of 99.7 %. This is equivalent to the calculation of z scores used in proficiency testing schemes (as a reminder, $z = (x - X)/\sigma$), where x is the measured content of the analyte, X is the reference value, and σ is based, in this case, on the standard deviation resulting from the test).

It is important that the value of σ be a reliable estimate of the measurement precision. Among the five approaches described above, the procedure (1) implies that at least seven replicate analyses be carried out (which is generally considered sufficient). However, if the method has previously been studied (enabling one to obtain a good estimate of the standard deviation of the measurement for the considered matrix) the number of CRM analyses may be less than seven, although the minimum is to duplicate the analysis. A single analysis may be envisaged where the laboratory is confident of its statistical control. The value of n used for the calculation of σ should obviously reflect the number of replicate analyses effectively carried out on the CRM.

To illustrate this purpose, consider the results of MeHg determinations in a fish CRM (CRM 463) as repeated five times. The certified value of MeHg in this CRM is equal to (3.04 ± 0.16) mg kg^{-1} [22] and the values obtained by the laboratory are, respectively, 3.02, 3.07, 3.04, 3.26 and 3.34, corresponding to a mean content of 3.15 mg kg^{-1} and a standard deviation of 0.15 mg kg^{-1}. The adopted value for s_e is 0.30 mg kg^{-1}, based on the measurement of the measurement reproducibility by the laboratory. The value of σ is therefore equal to: $\sigma = [(0.30)^2 + (0.15)^2/5]^{1/2} = 0.31$ mg kg^{-1}.

The calculated value of Δ is obtained as: $3.15 - 3.04 = 0.11$ mg kg^{-1}

It is hence verified that this value corresponds to the conditions of acceptability of the method, i.e. $-0.62 < 0.11 < 0.62$.

Let us note once more that the validity of the test described above depends upon the validity of the adopted values for s_i and s_e. If these values are erroneous, the value of σ will be also erroneous, and the test will lead to wrong conclusions.

In some cases, it appears necessary to take into account the uncertainty of the certified value of the CRM (if this uncertainty is significantly different from σ) and to add a term corresponding to an enlarged uncertainty. Further details can be found in the literature [16, 23].

As a conclusion, let us recall that the error may be expressed in two different ways in the framework of a method validation: (1) as an absolute value $|x - x_0|$ where a positive error indicates a higher value, or (more often in the case of method validation), (2) as a recovery factor, i.e. a fraction or a percentage, x/x_0 or $100x/x_0$, where x is the measured value and x_0 the certified value. This type of approach is particularly useful when several tests or materials are subject to similar and proportional errors.

4 CALIBRATING REFERENCE MATERIALS

4.1 Pure substances

Pure compounds of chemical forms of elements are rarely available commercially. Examples mainly concern 'classical' speciation analyses such as methylmercury chloride and organotin compounds. In specific cases, individual laboratories or consortia decided to produce pure compounds for calibration purposes or as a mean to verify home-made calibrants. These substances are generally produced in small amounts and are unfortunately not available commercially.

An example, already quoted above, was the synthesis of derivatised organotin compounds for the purpose of a certification campaign of organotins in a freshwater sediment reference material [11]. Commercial organotins were used as starting materials and purification was achieved using a number of recrystallisation and/or distillation steps. Alkylated derivatives were prepared from the purified organotin salts. In total, 18 compounds were prepared each in 5 g quantities for alkylated derivatives (ethylated and pentylated mono-, di- and tributyl- and phenyltin compounds) and 10 g for purified chloride salts (also for six butyl- and phenyltin compounds, respectively).

Another example concerns arsenobetaine and arsenocholine, two compounds that are not commercially available. Their synthesis and characterisation were undertaken in the framework of a certification project [24], and the two compounds were made available in the form of crystals with a 99 % purity to the participating laboratories. This is another example of synthesis of pure compounds that were originally prepared for a limited number of laboratories. In the case of arsenobetaine, as mentioned below, the work did not stop with the pure crystal preparation.

4.2 Calibrating solutions

Considering the need to verify commercially available calibrants, calibrating solutions of verified stoichiometry and purity have been prepared in limited quantities in some speciation-certification projects, e.g. solutions of methylmercury chloride, solutions of trimethyllead chloride, etc. Details on the preparation of these solutions are given elsewhere [4]. These solutions were prepared on an *ad hoc* basis, without aiming to provide long lasting stocks.

Conversely, the preparation of pure arsenobetaine and arsenocholine (see above) led to the preparation of a batch of calibrating solutions of arsenobetaine in water, of which the value was certified on a gravimetric basis with a verification by inductively coupled plasma atomic emission spectrometry (ICP-AES) and energy-dispersive X-ray fluorescence spectroscopy (EDXRF). This is the only example to date of a calibrating CRM solution (CRM 626 produced by BCR) that is available commercially for As speciation studies [24].

4.3 Matrix-matched solutions

Reference materials of solutions may be prepared with the purpose of simulating natural waters. In this case, the materials are aimed to be used for checking the method performance rather than for calibrating purposes. With respect to speciation analyses, attempts were made by BCR to prepare and certify matrix-matched solutions.

The first example is the preparation of an artificial rainwater reference material to be certified for its trimethyllead content [25]. Solutions (stored in ampoules) were prepared, matching the composition of rainwater, and were first tested in the framework of interlaboratory studies. The excellent agreement found among various participating laboratories and methods justified that a large stock of ampoules be prepared for certification. However, the monitoring of the material (stored in the dark at 4 °C) over 3 years indicated an instability after storage for 1 year, which hampered certification.

Similarly, simulated freshwater solutions were prepared for tentative certification of inorganic Se(IV) and Se(VI). The same situation as above was encountered, i.e. the analytical state of the art was found to be excellent and suitable for certification and the solutions were found to be stable over 1 year, which justified the organisation of a certification campaign [26]. Unfortunately, the stability monitoring over 2 years showed an instability of the two prepared materials, which hampered the final certification.

The lesson of these above projects was not negative, however. It shows that matrix-matched solutions containing e.g. inorganic Se species may be prepared and kept stable over at least 1 year, which is sufficient for internal quality control purposes. The instability in the long term hampers the production of long-lasting stocks of reference materials.

4.4 Other types of RM for speciation studies

A procedure for immobilizing chemical species on solid supports has been proposed by Mena Fernández *et al.* [27] to be possibly used for the preparation of new types of reference material for speciation studies. The principle is related to the determination of mercury species by flow injection systems incorporating microcolumns of sulphydryl cotton fibre (SCF) to effect on-line sample preconcentration prior to quantification by cold vapour atomic fluorescence spectrometry (CV-AFS) or gas chromatography microwave-induced plasma atomic emission spectrometry (GC-MIP-AES). In the case of natural waters, the same microcolumn may be used in the field to collect and immobilize Hg species for final treatment and analysis. Besides using SCF microcolumns as an aid to sample processing and sampling, microcolumns charged with analyte could serve as external calibrants and, if integrity would be preserved, a new class of reference material could be developed. The stability of SCF microcolumns charged with methylmercury chloride and inorganic Hg was studied at 4 °C over 4 months, determining Hg species at regular intervals and showing a good stability of the species. This demonstrated the feasibility of preparation of calibrants and new types of RM based on this immobilization technology, but so far these findings have not been used for the production of such new possible quality control tools.

5 MATRIX REFERENCE MATERIALS

5.1 Representativeness

Several requirements have to be fulfilled for the preparation of matrix reference materials, which have to be representative of currently analysed samples, homogeneous and stable over long-term storage. The following sections describe general rules to be followed for the preparation of matrix CRMs with special focus on speciation analysis. Selected preparation procedures have to be adapted to each type of materials (liquids or solids) and they have to be fit for the purpose of the analytical work.

Sound conclusions on the performance of an analytical method or a laboratory call for the use of one or several reference materials with a composition as close as possible to those of the samples routinely analysed by the laboratory. This means that a matrix reference material should, in principle, pose similar difficulties, i.e. induce the same sources of error, to those encountered when analysing real samples. Requirements for the representativeness of a reference material imply in most cases a similarity of matrix composition, concentration range of substances of interest, binding states of the analytes, occurrence of interfering compounds, and physical status of the material.

When a matrix reference material is prepared, all these requirements should be considered with great care. For practical reasons, in many cases the similarity cannot be entirely achieved. The material should be homogeneous and stable to guarantee that the samples provided to the laboratories are similar, and compromises often have to be made at the stage of preparation to comply with this requirement, which is particularly acute for speciation measurements. Some interesting parameters, characteristic of real samples, may disappear. Unstable compounds or matrices can hardly be stabilised or their stabilisation may severely affect their representativeness. The producer will have to carefully evaluate the limitations of the preparation procedures to be followed and their impact on representativeness. The 'production chains' have to be

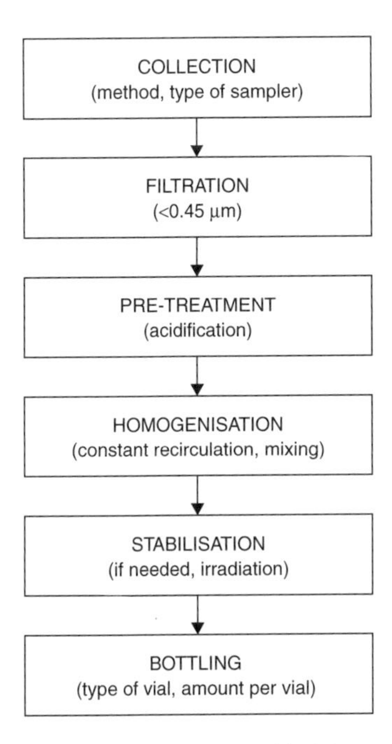

Figure 7.2.3. Production chain for liquid RMs.

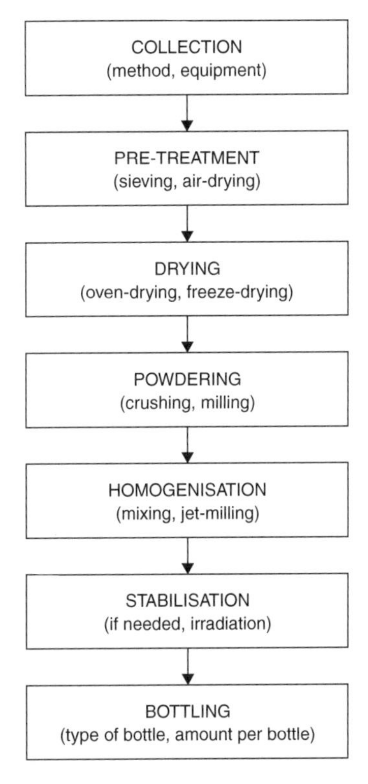

Figure 7.2.4. Production chain for solid RMs.

examined case by case as depicted in Figures 7.2.3 (liquid RMs) and 7.2.4 (solid RMs), the details of which are discussed in the following sections.

The degree of acceptance of compromises to be considered with respect to the material representativeness will depend upon the producer and the user's needs. The users should, in all cases, be informed about the real status of the sample, its treatment and possibly the treatment that has to be applied to bring the sample to a state that is more representative of a natural sample. An example of compromise was the stabilisation of Cr species in solutions through a freeze-drying process, requiring a reconstitution in a buffer solution [28]; this was the only way to achieve a reconstituted 'matrix-matched solution' recovering the certified values of the Cr species.

Other examples show that the compromises often correspond to the preparation of dried powdered materials. Of course, the best representativeness would be the availability of 'fresh' materials, but these are hardly stabilised over long-term periods or require expensive storage/handling procedures to preserve their integrity. Recent attempts to prepare such materials have been made for certifying trace organic compounds (PCBs) in fresh mussel material [29] and one may think that a similar approach could be used for speciation studies.

5.2 Collection

The amount of collected material has to be adapted to the aim of the analysis, and to various parameters such as the size of the current sample intakes, the stability, the frequency of use and the potential market (for CRMs). It is sometimes better to prepare a limited batch of samples to respond to the needs for a given period (e.g. 5 years) and to prepare a new batch of material when new requests are made to respond to needs of modern analytical techniques or to changes in regulations. This is particularly true for speciation analysis, for which it is unlikely that the stability may be ensured on a long-term basis (i.e. over 10 years). The amount collected may vary from a few litres or kilograms for the preparation of LRM (used for internal quality control) to cubic metres or hundreds of kilograms for materials to be used in interlaboratory studies or for the production of CRMs. The producer should be equipped to treat the appropriate amount of material without substantially changing its representativeness. Table 7.2.1 gives examples of the amounts of matrix RMs that have been collected for the preparation of some speciation-related CRMs.

5.3 Stabilisation

A most critical step for the preparation of matrix reference materials is the stabilisation. As stressed above, stabilisation may affect the material representativeness. This step is mandatory, however, to ensure the long-term stability of the material. Stabilisation has to be adapted to each particular case (matrix, type of substance) and should in principle be studied systematically before proceeding to the treatment of the bulk sample. Synthetic solutions containing mixtures of pure substances are generally assumed to be stable, not requiring particular treatment; experience has shown that it is not as straightforward when dealing with speciation measurements (see example on Se speciation in Section 4.3). Conversely, natural samples are often very unstable, in particular for compounds that are sensitive to long-term temperature variations or prone to chemical changes. Table 7.2.2 lists some

Table 7.2.1. Examples of amounts of material collected and collection methods for the preparation of matrix-RMs aimed at the certification of chemical forms of elements (adapted from refs 4 and 6).

Matrix	Chemical species	Collection, pretreatment and collected amount
Coastal sediment	Butyltins	Collection of 10 cm sediment layer, followed by air drying at ambient temperature. About 180 kg (dry mass) collected, transported in ice boxes, further dried at 55 °C in air for 100 h and sieved at 75 μm.
Mussel tissue	Organotins	1200 kg *Mytilus edulis* purchased from mussel farm, washed with cleaned freshwater, frozen in liquid nitrogen to facilitate shelling, resulting in about 325 kg of frozen tissues that were ground in a PTFE-coated mill.
Estuarine sediment	Methyl-mercury	Collection of 30–40 cm top sediment layer (about 250 kg) with a grab sampler, air drying at 25 °C, manual crushing, sieving at 2 mm, hammermill grinding and sieving at 90 μm.
Fish tissue	Methyl-mercury	Fish obtained from the market (around 300 kg), slicing, freezing at −25 °C and mincing with tungsten carbide blades.
Fish tissue	As species	Fish obtained from the market (around 300 kg), slicing, freezing at −25 °C and mincing with tungsten carbide blades.
Urban dust	Trimethyl-lead	Around 15 kg of dust collected by sweeping lay-by in road tunnel, sieving at 500 μm, air drying at ambient temperature for 5 days, ballmill grinding, sieving at 125 μm.
Welding dust on filter	Cr(VI)	Collection of welding dust on borosilicate microfibre glass filters mounted on air-sampling units, loading with welding fume from manual metal arc (by batches of 100 filters), flushing filter monitors with nitrogen and sealing. Total of 1100 filters produced.

Table 7.2.2. Examples of homogeneity tests and recommended sample intakes for matrix RMs aimed at the certification of chemical forms of elements (adapted from refs 4 and 6).

Matrix	Chemical species	Stabilisation
Coastal and freshwater sediment	Organotins	Pasteurisation (heating at 100 °C or 120 °C for 2 h after drying at 55–60 °C), resulting in a moisture of less than 3 %.
Mussel tissue	Organotins	Freeze-drying to a moisture of less than 4 %.
Estuarine sediment	Methyl-mercury	γ-irradiation with a ^{60}Co source (moisture content of about 3.5 %).
Fish tissue	Methyl-mercury	Freeze-drying to a moisture of less than 2.5 %.
Fish tissue	As species	γ-irradiation with a ^{60}Co source.
Urban dust	Trimethyl-lead	Freeze-drying.
Welding dust on filter	Cr(VI)	Flushing of filter monitors with nitrogen to ensure inert atmosphere.

stabilisation procedures that have been tested and used for the purpose of certification work.

5.4 Homogenisation

A material may be used as a reference only if on each occasion of analysis an identical portion of sample is available. Therefore, when a material is stabilised, it has to be homogenised to guarantee a homogeneity that is sufficient within and between each vials with respect to the certified properties [2]. Homogenisation is not the most difficult problem for water samples (it is usually achieved by constant recirculation of the solution by pumping or by continuous mixing over a defined period of time). For solid materials, various treatments have to be adapted case by case, e.g. crushing, sieving and grinding for soil and sediment analyses, cryogrinding for biological samples followed by milling, etc. [30]. Homogenisation may be achieved by mixing the (dry) powdered material in a drum (avoiding all contamination risks). A more sophisticated technique (used e.g. at the Institute for Reference Materials and Measurements) is jet-milling with a particle classifier, which guarantees to produce a powder with a defined range of particles, hence ensuring a good material homogeneity; this may, however, be at the expense of representativeness (in particular for soil samples which contain a wide range of particle sizes). Table 7.2.3 gives examples of homogenisation procedures used for matrix RMs certified for chemical forms of elements.

5.5 Bottling

A proper selection of vials for the storage of reference materials is vital for ensuring their long term stability, in particular avoiding contamination from containers or losses (e.g. through adsorption on container walls or degradation) and protecting the samples from UV light, shocks etc. Vials used for the storage of liquid RMs can be sealed ampoules or bottles (generally polyethylene, polycarbonate,

Table 7.2.3. Examples of homogenisation procedures used for matrix-RMs aimed at the certification of chemical forms of elements (adapted from refs 4 and 6).

Matrix	Chemical species	Stabilisation
Sediment	Organotins	Jet-milling, mixing in cone mixer.
Mussel tissue	Organotins	Mixing under Ar atmosphere (15 days) in polyethylene-lined drum (around 35 kg dry material processed).
Sediment	Methyl-mercury	Mixing under Ar atmosphere (15 days) in polyethylene-lined drum (around 80 kg dry material processed).
Fish tissue	Methyl-mercury	Mixing under Ar atmosphere (15 days) in polyethylene-lined drum (around 35 kg dry material processed).
Fish tissue	As species	Mixing under Ar atmosphere (15 days) in PVC drum.
Urban dust	Trimethyl-lead	Mechanical shaking for few hours of different subsamples, combination of subsamples (around 10 kg dry material processed).
Welding dust on filter	Cr(VI)	No homogenisation necessary.

Table 7.2.4. Examples of types of vials used for matrix RMs aimed at the certification of chemical forms of elements (adapted from refs 4 and 6).

Matrix	Chemical species	Stabilisation
Sediment	Organotins	Brown glass bottles with polyethylene inserts, each containing 25–40 g.
Mussel tissue	Organotins	Brown glass bottles with polyethylene inserts, each containing about 15 g.
Sediment	Methyl-mercury	Brown glass bottles with polyethylene inserts, each containing about 40 g.
Fish tissue	Methyl-mercury	Brown glass bottles with polyethylene inserts, each containing about 15 g.
Fish tissue	As species	Brown glass bottles with polyethylene inserts, each containing about 15 g.
Urban dust	Trimethyl-lead	Brown glass bottles with polyethylene inserts, each containing about 15 g.
Welding dust on filter	Cr(VI)	Filters stored in monitor cases.

more rarely glass). It is generally recommended to protect the materials from light, and amber glass or high-density polymers are generally used. In case risks of lixiviation from the walls of the flasks are suspected (e.g. from glass), quartz may be recommended. In the case of dry powdered material, brown glass bottles with polyethylene inserts and screwcaps are most generally used. In some cases, small polytetrafluoroethylene (PTFE) balls are inserted in the bottle to facilitate rehomogenisation before taking intakes for analysis. Examples of types of bottles used for CRMs related to speciation analysis are given in Table 7.2.4.

6 CHARACTERISATION

6.1 Analysis of main composition

Interfering substances present in the RM matrix may have an effect on the determination of chemical forms of elements, e.g. high organic matter content affecting the yield of derivatisation reactions. Therefore it is often useful to determine major matrix components in the RM to warn the user about possible analytical difficulties posed by this particular matrix. Also users are often interested in being informed about the total content of the element(s) of which species have been determined. In some cases indeed it is possible to calculate balances for a given element (e.g. sum of methylmercury and inorganic mercury, versus the total Hg in a fish matrix), which may serve as additional quality control verification. However, this information is not relevant for species that are at the ultratrace level in the matrix when this is compared with high amounts of the inorganic form (e.g. trace of butyltins versus the content of inorganic Sn in sediment).

6.2 Homogeneity testing

During a chemical analysis, the sample intake of a given material can only be used once since it is generally destroyed during the analysis. The amount of material in a bottle or an ampoule therefore has to be sufficient to carry out several determinations. Moreover, the producer has to guarantee that the material is similar from the first vial prepared to the last one. Therefore, the homogeneity of the material should be verified between vials (in the case of water samples) and within vials (for solid samples) of the same batch to guarantee that no significant difference may occur between sample intakes taken from different vials. The (in)homogeneity may be estimated by comparing the coefficients of variations of repeated measurements on samples from different vials with those of repeated measurements of samples taken from a single vial (which, in the case of water analysis, are considered as the uncertainty of the analytical method). Homogeneity assessment of (solid) matrix CRMs produced by BCR for speciation analyses has generally been based on ten replicate measurements in a single vial (within-vial homogeneity), as compared to one measurement in each of 20 vials that were randomly taken during the bottling procedure (between-vial homogeneity) and five replicate analyses of an extract (method uncertainty). It should be noted that the CV resulting from five extract replicate analyses is always underestimated in comparison to the within-vial and between-vial CVs since it does not include the uncertainty of the extraction step.

The analytical method used for a homogeneity study should be sufficiently precise (suitable repeatability and reproducibility). A high level of trueness is usually not required since the interesting parameter is, in this case, the existing difference between the samples. Homogeneity may be tested at several levels of sample intakes. A properly conducted testing enables the producer to evaluate and recommend minimum sample intakes for the analysis of the CRM. Examples of homogeneity tests carried out for various types of CRMs and resulting minimum sample intakes are shown in Table 7.2.5.

Microhomogeneity studies may be undertaken by solid sampling Zeeman electrothermal atomic absorption spectrometry on a range of elements, enabling fine-tuning of the homogeneity study to sample intakes of less than 50 mg for some

Table 7.2.5. Examples of homogeneity tests and recommended sample intakes for matrix-RMs aimed at the certification of chemical forms of elements (adapted after (4) and (6)). These amounts correspond to the actual sample intakes taken for verifying the reference material's homogeneity.

Matrix	Chemical species	Homogeneity tests
Sediment	Organotins	Ten replicates (within-bottle) as compared to 20 replicates (one analysis in each of 20 bottles) at 1 or 0.5 g sample intake level.
Mussel tissue	Organotins	Ten replicates (within-bottle) as compared to 20 replicates (one analysis in each of 20 bottles) at 0.5 g sample intake level.
Estuarine sediment	Methyl-mercury	Ten replicates (within-bottle) as compared to 20 replicates (one analysis in each of 20 bottles) at levels of 50, 100 and 250 mg. Recommended sample intake of 50 mg.
Fish tissue	Methyl-mercury	Ten replicates (within-bottle) as compared to 20 replicates (one analysis in each of 20 bottles) at 0.2 g sample intake level.
Fish tissue	As species	Ten replicates in each of two bottles (within-bottle) as compared to 20 replicates (two analyses in each of 20 bottles) at 1 g sample intake level.
Urban dust	Trimethyl-lead	Ten replicates (within-bottle) as compared to 15 replicates (one analysis in each of 15 bottles) at 1 g sample intake level.
Welding dust on filter	Cr(VI)	Full sample to be used (filter leaching).

elements as shown for a certification of trace elements in a fish CRM [31]. This evaluation concerns total contents of element, but it may give useful indications for the homogeneity of their chemical forms.

6.3 Stability testing

The composition of a reference material and the parameters studied should remain stable over the entire period of utilisation of the material. The extent of the study of the temporal stability will depend upon the use of the material. If a material is to be used in a short-term interlaboratory trial (e.g. 6 months), its stability should only be verified for the duration of the exercise. Additional studies may be needed, e.g. to simulate conditions that may be encountered during the transport of the material (e.g. severe climatic conditions with temperature rise). In the case of a CRM, the stability study has to be planned over some years. The stability (or instability) has to be studied or known before producing the reference material on a large scale, and it has to be verified on the entire batch of material (taking a given number of samples randomly over the whole batch). Analyses for studying the stability of a CRM may start at the beginning of the storage period and after various intervals, e.g. 1, 3, 6, 12 months or more, if necessary.

A currently used procedure (e.g. by BCR) for testing the stability of CRMs for speciation parameters is to use samples stored at low temperature (e.g. at −20 °C) as reference for studying samples stored at e.g. +4 °C or +20 °C [2, 4]. A study at higher temperature (e.g. +40 °C) over a short-term period (e.g. 15 days) may be foreseen to evaluate possible instability risks during worst-case transport conditions (in principle, this would also allow a researcher to extrapolate the stability of a RM, taking into account that a temperature increase usually brings an increase in reaction rate, e.g. decomposition reactions). By calculating ratios of results obtained, over a given period, at different temperatures versus results obtained at a reference temperature (at which it is supposed that no major changes will occur), it is possible to get rid of possible variations that are due only to the analytical method (reproducibility). Indeed, these variations are in principle similar, at a given period, for the analysis of CRMs stored at the reference temperature and those stored at higher temperatures (+4 °C and/or +20 °C). In the ideal case, the ratios R_T of the measurements should be equal to 1. In practice, an uncertainty on these ratios is calculated, taking into account random errors on measurements, which enables us to estimate that the CRM is stable if the value 1 is between the values of $(R_T - U_T)$ and $(R_T + U_T)$. An example of a stability study is given in Figure 7.2.5, which illustrates a case (TBT in sediment) where the

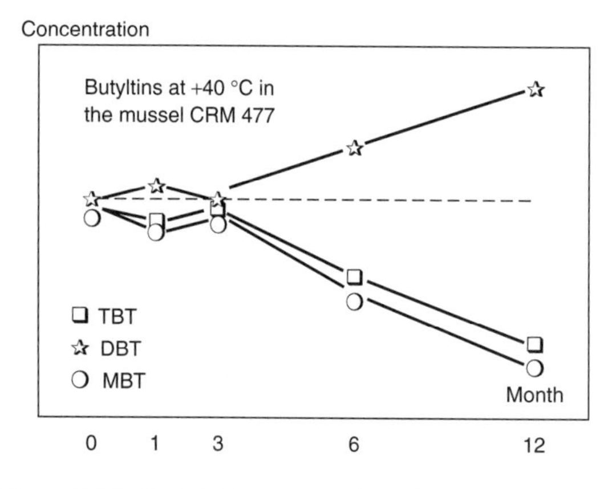

Figure 7.2.5. Graphical representation of a stability study using the ratio methods.

compound was stable at +20 °C over a period of 12 months but strongly degraded after storage at +40 °C for 1 month.

The reference to samples stored at low temperature has, however, some limitations. This approach is not applicable to the study of water samples in which some compounds may precipitate at low temperature without the possibility of redissolving them in a reproducible way upon warming of the sample. In such situations, other better adapted approaches have to be developed, e.g. to detect the formation of products of chemical reactions or metabolites. In addition, freezing soil or sediment samples may result in extractability changes upon thawing, which has an effect on stability experiments on extractable trace element contents; such an effect was not noticed for organometallic compounds [4].

An alternative method for studying stability (which is also based on storing 'reference samples' at low temperature, with the limitations expressed above), consists of storing all samples at −20 °C (or less) and placing them at higher temperatures for a given period of time (e.g. 12 or 24 months) before the date of the analysis. The same storage time at higher temperatures as in the above-described approach is considered but all the measurements are planned at the end of the study [32]. Measurements are then performed over a shorter period of time, limiting the risks for long-term analytical variations. This approach also allows the laboratory to concentrate on this analytical work over a short period of time. It does not completely eliminate the reproducibility problem since

Table 7.2.6. Examples of stability tests and results for matrix RMs and matrix-matched solutions aimed at the certification of chemical forms of elements (adapted from refs 4 and 6).

Matrix	Chemical species	Stability tests and results
Coastal sediment	DBT and TBT	Stability verified for DBT and TBT from −20 °C to +20 °C over 12 months. Unstable at +40 °C after 3 months and at +20 °C after 24 months.
Freshwater sediment	Phenyltins	Stability verified at −20 °C over 12 months. Unstable at +20 °C and +40 °C (also tested over 12 months).
Mussel tissue	Organotins	Stability verified for MBT, DBT and TBT at −20 °C over 44 months. Unstable at +20 °C and +40 °C, and to a lesser extent at +4 °C. For phenyltins, unstable at −20 °C and above.
Sediment	Methyl-mercury	Stability verified from −80 °C to +40 °C over 15 months.
Fish tissue	Methyl-mercury	Stability verified from −20 °C to +40 °C over 12 months.
Fish tissue	As species	Stability verified from −20 °C to +40 °C over 9 months.
Aqueous solution	Asbetaine	Stability verified from −20 °C to +40 °C over 9 months.
Urban dust	Trimethyl-lead	Stability verified from +4 °C to +20 °C over 37 months. Instability at +40 °C.
Artificial rainwater	Trimethyl-lead	Stability verified from +4 °C to +20 °C over 24 months. Unstable at these temperatures after 37 months, and at +40 °C after 1 month.
Welding dust on filter	Cr(VI)	Stability verified from +5 °C to +20 °C over 12 months.
Lyophilised solution	Cr(III) and Cr(VI)	Stability verified from +5 °C to +20 °C over 12 months.
Aqueous solution	Se(IV) and Se(VI)	Stability verified at +20 °C over 12 months but instability shown after 24 months.
Soil and sediment	Extractable trace element contents	Stability verified from −20 °C to +40 °C over 12 months. Instability suspected at +40 °C. Extractability changes suspected at −20 °C, stressing the need to carry out stability studies at +4 °C instead of −20 °C.

up to 50 or more samples have to be analysed, which may take several days or weeks for some complex speciation measurements, thus rendering the argument of better precision a bit marginal. Nevertheless, this approach certainly represents an advantage for managing repeated challenge tests in laboratories in charge of maintaining large stocks of CRMs [2]. This approach has been proposed and used for controlling the stability of BCR CRMs [32].

In addition to the stability study carried out at the stage of preparation of the material, the stability should in principle be verified throughout the entire utilisation period of the material. The above section has already quoted an example of a material that had been stable over 12 months but was then detected to be unstable, i.e. CRM solutions containing inorganic Se(IV) and Se(VI) species [23]. Other compounds, e.g. phenyltins, have been shown to be unstable at temperatures above −20 °C. Table 7.2.6 summarises some aspects of experience gained by BCR with respect to stability of chemical forms of elements. Experience on stability experiments has also been reviewed for speciation analysis in calibrating solutions [33] and environmental RMs [34].

7 STORAGE AND TRANSPORT

As stressed in Section 5.5, the parameters related to the homogeneity and stability of the reference material are implicitly linked to the vial used for the long-term storage. Storage and transport imply precautions for avoiding an exposure of the samples to elevated temperatures, UV light and shocks.

Table 7.2.7. Examples of storage/transport conditions for matrix RMs and matrix-matched solutions for chemical forms of elements (adapted from refs 4 and 6).

Matrix	Chemical species	Storage and transport conditions
Coastal sediment	DBT and TBT	Recommended storage at −20 °C in the dark. Transport possible at +20 °C (in the dark) if less than 15 days.
Freshwater sediment	Phenyltins	Recommended storage at −20 °C in the dark. Transport possible at +20 °C (in the dark) if less than 15 days.
Mussel tissue	TBT, DBT and MBT	Recommended storage at −20 °C in the dark. Transport in cool conditions required.
Mussel tissue	Phenyltins	Not stable at −20 °C. Storage studies at temperatures below −20 °C are necessary.
Sediment	Methyl-mercury	Recommended storage at +20 °C or below in the dark. Transport possible at temperature up to +40 °C (in the dark).
Fish tissue	Methyl-mercury	Recommended storage at +20 °C or below in the dark. Transport possible at temperature up to +40 °C (in the dark).
Fish tissue	As species	Recommended storage at +20 °C or below in the dark. Transport possible at temperature up to +40 °C (in the dark).
Aqueous solution	Asbetaine	Recommended storage at +20 °C or below in the dark. Transport possible at temperature up to +40 °C (in the dark).
Urban dust	Trimethyl-lead	Recommended storage at +4 °C or below in the dark. Transport possible at temperature up to +20 °C (in the dark).
Artificial rainwater	Trimethyl-lead	Not stable at +4 °C over 37 months but storage possible at +4 °C in the dark over 12 months. Transport possible at +20 °C in the dark if less than 15 days.
Welding dust on filter	Cr(VI)	Recommended storage at +20 °C or below in the dark. Transport possible at temperature up to +40 °C (in the dark).
Lyophilised solution	Cr(III) and Cr(VI)	Recommended storage at +20 °C or below in the dark. Transport possible at temperature up to +40 °C (in the dark).
Aqueous solution	Se(IV) and Se(VI)	Not stable at +20 °C over 24 months but storage possible at +20 °C in the dark over 12 months. Transport possible at +20 °C in the dark if less than 15 days. Storage studies are necessary at +4 °C to optimise long-term storage conditions.
Soil and sediment	Extractable trace element contents	Recommended storage at +20 °C or below in the dark. Transport possible at temperature up to +40 °C (in the dark), if less than 15 days. Freezing not recommended owing to risks of extractability changes.

The storage temperature should be appropriate for ensuring sufficient stability of the reference material. As shown in Table 7.2.6, low temperatures are often recommended but are not always necessary, e.g. organometallic compounds such as MeHg in matrix CRMs are fairly stable at ambient temperature whereas other compounds (e.g. phenyltins) may rapidly degrade when exposed to temperatures above −20 °C. As previously highlighted, cooling of liquid materials may sometimes affect some parameters, e.g. precipitation of dissolved compounds, and some parameters (e.g. extractability of trace elements) may be changed upon freezing/thawing of sediment and soil samples. Storage conditions, as well as the selected transport means, should be deduced from a well-designed stability study that has been adapted to each type of matrix and parameter. A preliminary study on various storage conditions (different temperatures and flask types) is often recommended, in particular for the preparation of CRMs.

The transport has to be performed in the shortest possible time window. Express distribution systems are unfortunately expensive and can only be used in particular cases (e.g. samples that would only be stable for some hours or days). The material should in principle be accompanied by a form to be sent back to the organiser of the interlaboratory tests or the producer (for a CRM), indicating the status of receipt of the material. Temperature indicators may be added to the sample in order to detect high temperatures that possibly occurred during transport. Examples of storage/transport conditions used for CRMs related to speciation parameters are given in Table 7.2.7.

8 ASSIGNING/CERTIFYING VALUES

The certification of reference materials follows strict rules that are described in the ISO Guide 35 [35] and other guidelines made available by some producers [36, 37]. Various approaches may be followed in relation to the types of parameters and matrices to be certified, which are summarised below.

8.1 Pure substances and calibrating solutions

The certification of chemical forms of elements in calibrating solutions will rely on the identification of the compounds, the evaluation of their purity and stoichiometry, and gravimetric measurements with analytical checks. Examples of materials for speciation studies belonging to this category have been described in the above sections, e.g. arsenobetaine solution [24], simulated freshwater for Se species [27] and artificial rainwater for trimethyllead [38]. In the first case, interlaboratory studies had permitted evaluation of the state of the art of the analytical methods currently used for determining the species in question [4], but the actual certification was based on gravimetric measurements. In the two other cases, a similar approach had been followed, with verification of the certified values by results obtained by a range of techniques; as indicated above, the certification could not be achieved owing to long-term instability of the species.

The certification of Cr-species in freeze-dried samples [26] was based on the analysis of reconstituted solution (HCO_3^-/H_2CO_3 buffer solution) by a range of independent techniques, i.e. in this case the certified value was not established on a gravimetric basis but rather as the means of laboratory means. Among the techniques used, isotope dilution mass spectrometry (IDMS) as the primary method guaranteed the traceability of this material to SI units (see discussion below).

Matrix CRMs cannot be certified on the basis of gravimetric methods since the samples are generally analysed after partial or total transformation of the matrix. In this case, other approaches exist, which are described in the following sections.

8.2 Matrix reference materials – single laboratory approach

The certification of chemical parameters in matrix reference materials may be carried out in a single laboratory, either using a so-called 'definitive method' applied by one or more independent analysts (e.g. IDMS) or using one or more reference methods applied by one or more independent analysts. One should stress that the certification carried out in one single laboratory, even using a 'definitive method', does not eliminate risks of systematic errors related to the human factor (manipulation error). A supplementary confirmation by interlaboratory testing – even limited – is therefore recommended.

For some chemical forms of elements (mainly inorganic), so-called 'definitive' methods are available, e.g. isotope dilution mass spectrometry [39]. The certification of matrix reference materials using a single 'definitive' method (e.g. as applied to Cr speciation) would not give the user, who does not apply this technique in his routine work, a good estimate of the uncertainty obtained with more classical techniques. A telling example is the certification of Cr(III) and Cr(VI) in lyophilised solution (CRM 544) where the uncertainty (95 % confidence interval of the means of laboratory means) of the certified values (as CV) were, respectively, 4.4 % and 3.7 %, whereas the uncertainty of the IDMS (based on five replicate measurements) was, respectively, 3.1 % and 2.3 % [28].

However, it may be questioned whether a technique such as IDMS may be considered as a 'definitive method' as applied to speciation analysis. Indeed, as long as the sample preparation is the critical step (since the matrix cannot be decomposed in the same way as for inorganic analyses owing to risks of affecting the speciation of the element) and the variance of the measurements is much smaller than the variance of sample preparation, there is no point in using e.g. ID-ICPMS as a 'definitive technique' to produce high-level results. Moreover, the application field of 'definitive methods' is limited with respect to the types of matrices and parameters that may be certified. These techniques do not yet exist for the certification of organometallic compounds (mainly because isotopically labelled organometallic species are not yet commercially available). A recent attempt has been made to develop an HPLC-ID-ICPMS technique for the purpose of certification of trimethyllead and butyltins in matrix-matched RMs [40]. However, the required extraction and its associated uncertainty on recovery do not qualify this technique as 'definitive' and the certification through interlaboratory studies remains the most adapted method.

8.3 Matrix reference materials – collaborative approach

A more general approach used for the certification of chemical forms of elements is through interlaboratory studies, using one or more independent methods, if possible including 'definitive methods' (for inorganic species). In this case, the certifications are based on interlaboratory studies that are organised following the same basic principles as classical interlaboratory studies [2, 17] but only involving specialised laboratories. According to the level of analytical complexity, it may be necessary that the participating laboratories demonstrate their capabilities in preliminary exercises. The organiser should also work according to well-defined rules and his ability to organise such exercises should be recognised. The best way to check the reliability of participating laboratories is to request them to demonstrate their performance in interlaboratory improvement schemes [2, 4]. This approach has been followed by the BCR programme for all new reference materials (related to speciation analysis and other analytical fields) that had to be certified for the first time, in particular the matrix CRMs [2].

In each interlaboratory study, detailed instructions and forms to submit results should be prepared, requesting each participant to demonstrate the quality of the analyses performed. Important information should be given, in particular the validity of calibration (including the calibration of weighing scales, volumetric flasks etc., the use of calibrants of suitable purity and known stoichiometry, sufficiently pure solvents and reagents etc.). Absence of contamination should also be demonstrated by blank measurements, and yields of chemical reactions (e.g. derivatisation) should in principle be accurately known and demonstrated. All precautions should be taken to avoid losses (e.g. formation of insoluble or volatile compounds). If results of totally independent methods such as gas chromatography (GC) or high performance liquid chromatography (HPLC) with different extraction procedures and detectors (between-method variations) applied by laboratories working independently (between-laboratory variations) are in good agreement, it can be concluded that the risk of systematic error related to each technique is negligible and that the mean value of the results obtained is the closest approximation of the true value. In the case of solid matrix CRMs, there is always a risk, however, that a remaining bias is left undetected since, at present, it is not possible to firmly demonstrate the completeness of extraction methods used in speciation analysis. Therefore, even if an excellent agreement is obtained between different techniques, the certified values based on consensus may not necessarily correspond to the 'true values', at least in the case of organometallic determinations involving an extraction step. Adversely, with respect to inorganic species in water, the certified values may indeed correspond to true values when they are based either on gravimetric measurements or on measurements involving a 'definitive' method. The aspects of traceability of CRMs are discussed below.

One should stress that, to date, it is hardly possible to speak about 'reference methods' when dealing with speciation analysis. It is more appropriate to refer to 'validated' methods that have been tested in collaborative exercises (hence the interest of interlaboratory study organisations). Examples of methods successfully used in certification campaigns are given in Table 7.2.8.

A particularity is to be noted with respect to speciation measurements, that is the determination/

Table 7.2.8. Examples of methods used for the certification of matrix RMs for their contents of chemical forms of elements (adapted from refs 4 and 6).

Matrix	Chemical species	Methods
Sediment	Organotins	HG-GC-QFAAS; HG-CGC-FPD; Et-CGC-FPD; SFE-CGC-FPD; Pe-CGC-FPD; Pe-CGC-QFAAS; CGC-AED; Pe-CGC-MIP-AES; Pe-CGC-MS; HPLC-ICPMS; HPLC-FLUO; POL
Mussel tissue	Organotins	HG-GC-QFAAS; Et-CGC-AAS; HG-GC-ICPMS; HG-CGC-FPD; Et-CGC-FPD; SFE-CGC-FPD; Pe-CGC-FPD; Pe-CGC-MIP-AES; Et-CGC-MS; Pe-CGC-MS; HPLC-ICPAES; HPLC-ICPMS; HPLC-FLUO
Estuarine sediment	Methyl-mercury	CGC-ECD; CGC-CVAAS; GC-QFAAS; CGC-CVAFS; HPLC-CVAAS; HPLC-ICPMS; HPLC-CVAFS; SFE-CGC-MIP-AES
Fish tissue	Methyl-mercury	IE-AAS; CGC-ECD; CGC-CVAFS; CGC-FTIR; CGC-MIP-AES; GLC-CVAAS
Fish tissue	As species	UV-HG-ICPAES; UV-HG-QFAAS; HG-GC-QFAAS; HG-LC-ICPAES; LC-QFAAS; LC-ICPMS
Urban dust	Trimethyl-lead	Et-CG-QFAAS; Pr-GC-QFAAS; Et-CGC-MIP-AES; SFE-CGC-MS; Pe-CGC-MS; Bu-CGC-MS; HPLC-ID-ICPMS
Welding dust on filter	Cr(VI)	IC-SPEC; AE-ETAAS; FAAS; ICPMS; AE-IDMS
Lyophilised solution	Cr(III) and Cr(VI)	IC-SPEC; IC-CHEM; UV-DPCSV; AE-ETAAS; AE-IDMS; Microcolumn-ICPMS
Aqueous solution	Se(IV) and Se(VI)	HG-AAS; HG-AFS; HG-ICPMS; Et-GC-MIP-AES; Microcolumn-ETAAS; microcolumn-ICPMS; HPLC-ICPMS; HG-HPLC-AAS
Soil and sediment	Extractable trace element contents	Single or sequential extraction protocols followed by FAAS, ETAAS, ICPAES or ICPMS detection

Abbreviations: AAS, atomic absorption spectrometry; AE, anion exchange chromatography; AES, atomic emission spectrometry; AFS, atomic fluorescence spectrometry; Bu, butylation; CGC, capillary gas chromatography; CHEM, chemiluminescence; CVAAS, cold vapour AAS; CVAFS, cold vapour AFS; DPCSV, differential pulse cathodic stripping voltammetry; ECD, electron capture detection; Et, ethylation; ETAAS, electrothermal AAS; FAAS, flame AAS; FLUO, fluorimetry; FPD, flame photometric detection; FTIR, fourier transform infrared spectroscopy; GC, packed gas chromatography; GLC, gas liquid chromatography, HG, hydride generation; HPLC, high performance liquid chromatography; IC, ion chromatography; ICPAES, inductively coupled plasma atomic emission spectrometry; ICPMS, inductively coupled plasma mass spectrometry; ID, isotope dilution; IDMS, isotope dilution mass spectrometry; IE, ion exchange chromatography; LC, liquid chromatography; MIP-AES, microwave induced plasma AES; MS, mass spectrometry; QFAAS, quartz furnace AAS; Pe, pentylation; Pr, propylation; POL, polarography; SFE, supercritical fluid extraction; SPEC, spectrometric detection; UV, ultraviolet.

certification of extractable forms of elements [13, 14]. In this case, the parameters determined will be dependent on the methods used, which then are real 'reference methods'. Examples are EDTA, DTPA and acetic acid single extraction protocols applied to the certification of extractable trace elements in soils, and a harmonised three-step sequential extraction procedure used for the certification of extractable trace elements in sediment [4, 14].

With respect to not-certified materials, there is an interest in obtaining good reference values (assigned values). The same approach and rules as those used for certification are in principle also needed to obtain good assigned values. A high degree of accuracy for these values is rarely mandatory for a Laboratory RM used for routine quality control checks (control charts) but it should be attempted for each RM that is used in method performance studies. Assigned values may be established through measurements carried out in the framework of interlaboratory studies involving experienced laboratories (they hence correspond to 'consensus' values), which is very similar indeed to the approach followed for certification. The main difference between a good assigned value and a certified value is actually linked to the (legally binding) guarantee given by the producer (certificate of analysis) and the procedure used to obtain this guarantee.

8.4 Uncertainty

The certification of a given parameter in a reference material leads to a certified value that is typically the mean of several determinations or the result of a metrologically valid procedure, e.g. weighing. In many instances, the uncertainty of the certified values is taken as the halfwidth of the 95 % confidence interval of the mean of laboratory means. Increasingly, however, uncertainty calculations of certified values are based on the GUM (Guide to the Expression of Uncertainty in Measurement) developed by ISO, which combines different sources of uncertainty (e.g. related to possible inhomogeneity of the material). An example of such an approach as applied to the certified value of butyltin compounds in a sediment CRM has been described elsewhere [41]. Calculations were carried out from the results of the homogeneity study of the material to see whether the uncertainty of the certified values should take into account a possible (quantified) material inhomogeneity. Two extreme cases were considered: (1) the inhomogeneity is very small (<1/3) compared with the standard error of the mean of laboratory means, in the case of which inclusion will not change the uncertainty; (2) the uncertainty is large compared with the mean of laboratory means and, in this case, it should be included in the uncertainty of the certified values which, otherwise, would be unrealistically low. The calculations (described in detail in ref. 41) showed that the differences between the original uncertainty and the combined uncertainty values were in most cases insignificant, with the exception of TBT. However, for consistency the combined uncertainty U_{CRM} was considered for all compounds. The certified values and their combined uncertainties are given in Table 7.2.9 [41].

Table 7.2.9. Uncertainties of organotin certified values in BCR-646 sediment CRM (95 % CI and combined uncertainties), adapted from ref. [41].

Certified compounds	Certified values ($\mu g\,kg^{-1}$ as cations)	Uncertainty ($\mu g\,kg^{-1}$ as cations)	
		Original	Combined
TBT	480	50	80
DBT	770	80	90
MBT	610	110	120
TPhT	29	9	11
DPhT	36	7	8
MPhT	69	17	18

Examples of uncertainty values of certified values for various chemical forms of elements have been published in the literature, e.g. for MeHg [42], organotins [43], trimethyllead [44] and inorganic forms of Cr and Se [45]; relevant figures are examplified in Table 7.2.10.

8.5 Certificate of analysis

Supplementary information to be provided to the user is described in the ISO Guide 31 [46] and covers, in particular:

Table 7.2.10. Uncertainties of certified values for chemical forms of elements on CRMs produced by the BCR (adapted from refs [4] and [6]).

Matrix/CRM	Chemical species	Certified value ± uncertainty (95 % CI)	CV (%)
Sediment/CRM 462	TBT	(54 ± 15) $\mu g\,kg^{-1}$ as TBT	27.8
	DBT	(68 ± 12) $\mu g\ kg^{-1}$ as DBT	17.6
Sediment/ BCR-646	TBT	(491 ± 65) $\mu g\,kg^{-1}$ as TBT	13.2
	DBT	(770 ± 117) $\mu g\,kg^{-1}$ as DBT	15.2
	MBT	(674 ± 102) $\mu g\,kg^{-1}$ as MBT	15.1
	TPhT	(35 ± 6) $\mu g\,kg^{-1}$ as TPhT	17.1
	DPhT	(38 ± 10) $\mu g\,kg^{-1}$ as DPhT	26.3
	MPhT	(74 ± 22) $\mu g\,kg^{-1}$ as MPhT	29.7
Mussel tissue/ CRM 477	TBT	(2.20 ± 0.19) $mg\,kg^{-1}$ as TBT	8.6
	DBT	(1.54 ± 0.12) $mg\,kg^{-1}$ as DBT	7.8
	MBT	(1.50 ± 0.27) $mg\,kg^{-1}$ as MBT	18.0
Fish tissue/CRM 463 CRM 464	Methyl-mercury	(3.04 ± 0.16) $mg\,kg^{-1}$ as MeHg	5.3
		(5.50 ± 0.17) $mg\,kg^{-1}$ as MeHg	3.1
Sediment/ CRM 580	Methyl-mercury	(75.5 ± 3.7) $\mu g\,kg^{-1}$ as MeHg	4.9
Urban dust/ CRM 605	Trimethyl-lead	(7.9 ± 1.2) $\mu g\,kg^{-1}$ as TML	15.2
Welding dust/CRM 545	Cr(VI)	(40.16 ± 0.56) $mg\,kg^{-1}$ welding dust (as Cr)	1.4
Lyophilised solution/ CRM 544	Cr(III)	(22.8 ± 1.0) $\mu g\ L^{-1}$ as Cr	4.4
	Cr(VI)	(26.8 ± 1.0) $\mu g\,L^{-1}$ as Cr	3.7
Fish tissue/ CRM 627	Asbetaine	(51.5 ± 2.1) $mmol\,kg^{-1}$ as Asbet	4.1
	DMA	(2.04 ± 0.27) $mmol\,kg^{-1}$ as DMA	13.2
Aqueous solution/ CRM 628	Asbetaine	(5.77 ± 0.03) $mmol\,kg^{-1}$ as Asbet	0.5

- administrative information on the producer and the material;
- a brief description of the material, including the characterisation of its main properties and its preparation;
- the expected use of the material;
- information on correct use and storage of the CRM;
- certified values and confidence intervals;
- other not-certified values (optional);
- analytical methods used for certification;
- identification of laboratories participating in the certification;
- legal notice and signature of the certification body.

Other information, potentially useful to the user of the CRM, cannot be given in a simple certificate. Therefore, some producers (e.g. BCR, European Commission) provide the materials with a certification report including details on the information given in the certificate. This report underlines, in particular, the difficulties encountered during certification and the typical errors that may occur when analysing the material with current analytical techniques. The overall work described in the certification report should, in principle, be examined by an independent group of experts so that all the possibly unacceptable practices are detected and removed. The experts should have in-depth knowledge of metrology as well as a good background expertise in analytical chemistry; they have to decide whether or not the CRM can be certified. In the framework of BCR, the certification committee is composed of representatives from EU countries and associated states, covering a wide field of expertise in chemical, biological and physical measurement sectors [2].

9 TRACEABILITY OF REFERENCE MATERIALS

CRMs and traceability are closely interconnected since certified values and their uncertainty should, in principle, be linked to established references. In theory, the certified value of a CRM should be traceable to the amount of substance of the element or compound of concern, i.e. the SI unit that underpins chemical measurements, the mole [47]. However, the traceability is actually possible only in relation to the kilogram since there is no available '^{12}C mole' standard [48].

It is difficult, if not impossible, to trace all the matrix CRMs to primary RMs, because of matrix effects, the variety of sample composition and substances etc. In addition, factors influencing the analytical process (e.g. homogeneity of the CRM) have an effect on the certified values. A classification is proposed by Pan [49] in Table 7.2.11, giving the main criteria for establishing a hierarchy in the traceability chain for CRMs:

- metrological quality of methods used for certifying values of the CRM;
- homogeneity and stability;
- calculation of uncertainty;
- metrological competence and recognition of the producer at the national and/or international level;
- demonstration of traceability.

Examining this scheme as applied to speciation measurements, one may distinguish:

- Primary RMs may correspond to the materials described in the Section 2.2, i.e. pure substances and calibrating materials, with respect to the certification and traceability criteria. No national metrological laboratory, however, is presently involved in this kind of work, which is instead undertaken by specialised laboratories.
- Certified RMs represent the vast majority of the materials that have been described in the present chapter. The mention of 'traceability' refers to a requested demonstration of traceability of the measurements to primary RMs.
- Working or Laboratory RMs correspond, for our purposes, to the materials produced in the framework of interlaboratory studies (i.e. for method performance studies). The mention of an 'accredited' organisation in this table, however, is not applicable to laboratories working in this particular field.

Numerous chemical measurements exist, for which certified reference materials can hardly

Table 7.2.11. Classification of chemical reference materials (adapted from ref. [49]). Note that the requirement of accreditation is, in many cases, questionable (e.g. many laboratories producing Laboratory RM are not accredited).

Level	Category	Criteria
I	Primary RM	• Materials with the highest metrological qualities of which the values are determined (certified) by a primary method. • Developed by a national metrological laboratory. • Recognised by national decision. • Traceable to SI units and verified through international intercomparisons.
II	Certified RM	• Fulfills all the conditions of the definition given by ISO Guide 30. • Generally developed by a national reference laboratory or a specialised organisation. • Certified by reference methods, a comparison of different methods, or a combination of the two approaches. • Recognised by national or specialised organisations. • Accompanied by a certificate indicating the uncertainty of the certified values and describing their traceability.
III	Working RM (or Laboratory RM)	• Fulfills all the conditions of the definition given by ISO Guide 30. • Produced by an accredited organisation. • Reference values established by one or several validated methods. • Accompanied by a description of the material traceability and giving an uncertainty estimate.

be prepared owing to their instability [50]. As discussed above, this is the case for species such as inorganic Se species, phenyltin compounds etc. (One should note, however, that an RM may be prepared for short-term use, e.g. in an interlaboratory study or as laboratory control material. This may tolerate the lack of long-term stability, providing that stability is demonstrated over the intended period of use.) In other cases, RMs may be available but their matrix is too different from that of the analysed sample, and the reference used to demonstrate the traceability of the results is then questionable. Trivial errors are still made even by experienced laboratories (often in good faith!), e.g. using a fish CRM to check a method for the determination of MeHg in a sediment RM because no sediment material is available. The wrong conclusion on the method validation actually only adds the confusion!

The comments in Table 7.2.11 indicate that some CRMs are directly traceable to SI units and opens the possibility of traceability of measurements to these units, e.g. high purity substances, playing the role of primary RMs for related speciation measurements. Two examples of traceability chains for relatively 'simple' speciation analysis, i.e. solution not requiring extensive chemical pretreatment, and complex determinations, i.e. solid matrix requiring an extraction step, are shown in Figures 7.2.6 and 7.2.7. The first case applies to the certification of Cr species in lyophilised solution (CRM 544): the discussion above already stressed that IDMS was used as the primary method along with other independent methods [28]. The traceability chain is maintained in this measurement situation, the only risk of traceability loss being at the level of sample reconstitution. One may assume, therefore, that the certified Cr species in this material are traceable to SI units, i.e. the true values of amount of substances of Cr(III) and Cr(VI). With respect to the second case (Figure 7.2.7), the complexity of measurements may be exemplified by

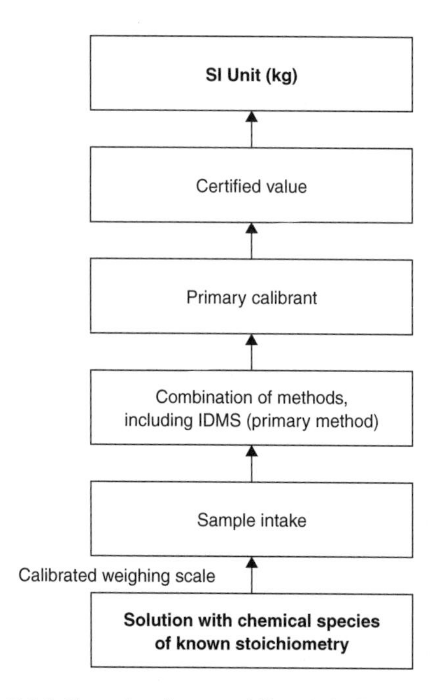

Figure 7.2.6. Example of a traceability chain for a relatively 'simple' speciation analysis of a matrix CRM, Cr species in solution.

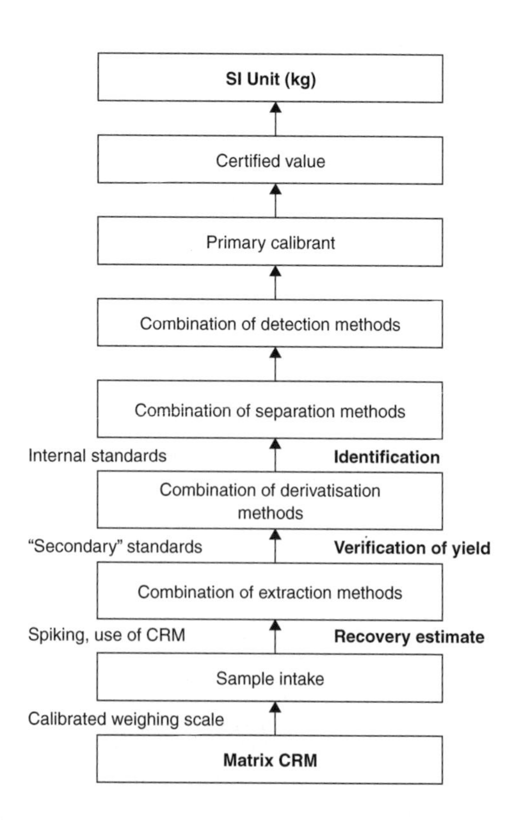

Figure 7.2.7. Example of a traceability chain for a complex speciation analysis of a matrix CRM, organotins in mussel tissue.

the certification of butyltin compounds in the mussel tissue CRM 477. In this case, HPLC-ID-ICPMS was used along with a range of independent methods (see Table 7.2.8) but, as stressed above, the uncertainty on extraction recovery does not qualify this method *sensu stricto* as a primary method. As shown in the figure, the traceability link is broken at several stages (e.g. extraction, derivatisation) and the attempts to control this loss of traceability (evaluation of extraction recovery and derivatisation yields), albeit of high value, are not sufficient to bring a firm demonstration of full traceability. The reason for this is mainly that, at present, it is virtually impossible to demonstrate that the extraction methods (and derivatisation reactions) enable a 100 % recovery to be obtained. The best of the state of the art is presently achieved, but it is based on spiking procedures and not on materials containing incurred compounds. One way to demonstrate full extraction recovery would be to prepare materials in such a way that isotopically labelled compounds would be 'naturally' bound to the matrix, and to test various extraction methodologies using isotope dilution HPLC-ICPMS as the measurement method.

The user of a CRM and of certified values should be informed about all the aspects of traceability that have directed the preparation and certification of the RM, the technical explanations on the rejection of outlying results, the sources of error, the procedures of recovery evaluation (based on a spiking procedure or the analysis of another CRM), the available documentation on the CRMs used to validate the certification methods etc.

10 REFERENCE MATERIAL PRODUCERS FOR SPECIATION STUDIES

More than 150 reference material producers exist world wide, but very few of them are dedicated to speciation analysis. To the author's knowledge, only five major organisations have produced CRMs related to speciation analysis to date and ensure a continuity of the stocks. They are listed below in their alphabetical order:

- European Commission, BCR: the production of speciation-related CRMs has been so far managed by a Research program in Brussels (Measurements and Testing, which is a continuation of the BCR program) through collaborative studies [4], in close cooperation with the Institute for Reference Materials and Measurements (IRMM), which is now responsible for the stock maintenance, the stability monitoring, the re-production and re-certification of exhausted CRMs and CRM sales (IRMM, European Commission Joint Research Centre, Retieseweg, B-2440 Geel, Belgium).
- International Atomic Energy Agency (IAEA), which organises wide-scale interlaboratory studies and certifies speciation-related CRMs on the basis of collaborative testing (IAEA, P.O. Box 100, A-1400 Vienna, Austria).
- National Institute for Environmental Studies (NIES), also producing speciation-related CRMs following an interlaboratory approach (NIES, 16-2 Onogawa, Tsukuba, Ibaraki 305, Japan).
- National Institute for Standards and Technology (NIST), which generally produces materials following a single laboratory approach (NIST, Standard Reference Materials Program, Building 202, Gaithersburg, MD 20 899, USA).
- National Research Council Canada (NRCC), which produces marine CRMs following a collaborative approach (NRCC, Institute for Environmental Chemistry, Montreal Road, K1A 0R6 Ottawa, Canada).

11 AVAILABILITY OF SPECIATION-RELATED CRMs

Table 7.2.12 gives a (nonexhaustive) list of available CRMs for the quality control of speciation analyses. Many of these CRMs are the result of improvement schemes and/or have been produced in the context of interlaboratory studies involving expert laboratories. They hence represent the best of the state of the art of speciation analysis and internationally recognised references to establish traceability of measurements of the chemical forms of elements in various matrices.

Table 7.2.12. Examples of reference materials certified for their contents in chemical forms of elements. The table lists CRMs available at BCR (European Commission), IAEA (international organisation), NIES (Japan), NIST (USA) and NRCC (Canada). This list is not exhaustive.

CRM	Certified parameters and matrices	Producer
SRM 2108	Cr(III) and Cr(VI) in solution	NIST
SRM 2109	Cr(III) and Cr(VI) in solution	NIST
CRM 544	Cr(III) and Cr(VI) in lyophilised solution	BCR
CRM 545	Cr(VI) in welding dust (loaded on a filter)	BCR
SRM 1974a	Total mercury and methylmercury in mussel tissue	NIST
SRM 2974	Total mercury and methylmercury in mussel tissue	NIST
SRM 2976	Total mercury and methylmercury in mussel tissue	NIST
IAEA-142	Total mercury and methylmercury in mussel tissue	IAEA
DORM-1	Methylmercury in fish muscle (dogfish)	NRCC
DORM-2	Total mercury and methylmercury in fish muscle (dogfish)	NRCC
DOLT-1	Total mercury and methylmercury in dogfish liver	NRCC
DOLT-2	Total mercury and methylmercury in dogfish liver	NRCC
CRM 463	Total mercury and methylmercury in fish muscle (tuna)	BCR
CRM 464	Total mercury and methylmercury in fish muscle (tuna)	BCR
IAEA-350	Total mercury and methylmercury in fish muscle (tuna)	IAEA
TORT-1	Trace elements and methylmercury in lobster tissue	NRCC
LUTS-1	Trace elements and methylmercury in lobster tissue	NRCC
IAEA-140	Total mercury and methylmercury in sea plant	IAEA
CRM 580	Total mercury and methylmercury in sediment	BCR
IAEA 356	Total mercury and methylmercury in sediment	IAEA
IAEA-085	Total mercury and methylmercury in human hair (spiked)	IAEA
IAEA-086	Total mercury and methylmercury in human hair	IAEA
NIES 13	Total mercury and methylmercury in human hair	NIES
PACS-1	Butyltin compounds in marine sediment	NRCC
CRM 462	Butyltin compounds in coastal sediment	BCR
CRM 646	Butyl- and phenyltins in freshwater sediment	BCR
CRM 477	Butyltin compounds in mussel tissue	BCR
NIES 11	Total tin and tributyltin compounds in fish tissue	NIES
CRM 627	Organoarsenic compounds in fish muscle	BCR
CRM 605	Trimethyllead in urban dust	BCR

Efforts remain to be made to produce CRMs for other chemical species [7]. Also, materials certified for a range of chemical forms of elements are strongly needed [51]. A project has been carried out in the framework of the Measurements and Testing Programme (European Commission), focusing on the certification of butyltins, MeHg and As species in an oyster reference material (date of completion: end of 2002). Further efforts should obviously be made in this direction.

12 ACKNOWLEDGEMENTS

The experience summarised in this chapter represents a sum of collaborations with hundreds of expert laboratories and I gratefully acknowledge the fruitful exchanges that I had with many experts over the last few years. Special thanks go to the project coordinators of the 'BCR speciation' projects and their teams, namely Iver Drabaek (methylmercury project), Michel Astruc, Wim Cofino, Freek Ariese and Roberto Morabito (organotin project), Freddy Adams and Roy Harrison (trimethyllead project), Les Ebdon (organotin and trimethyllead projects), Maurice Leroy and Alain Lamotte (arsenic speciation project), Carmen Cámara (selenium speciation project), Rita Cornelis (chromium speciation project), Allan Ure and Gemma Rauret (project on single and sequential extractions). Last but not least, I would like to warmly acknowledge 10 years of close cooperation with my EC colleagues, Ben Griepink, Herbert Muntau and Eddie Maier; this chapter would never have been possible without their guidance and exchanges over the years.

13 REFERENCES

1. Stoeppler, M. and Bowen, H. J. M., Introduction, in *Reference Materials for Chemical Analysis – Certification, Availability and Proper Usage*, Stoeppler, M., Wolf, W. R. and Jenks, P. J. (Eds), Wiley-VCH, Weinheim, 2001, Chapter 1, pp. 1–19.
2. Quevauviller, Ph. and Maier, E. A., *Interlaboratory Studies and Certified Reference Materials for Environmental Analysis – The BCR Approach*, Elsevier, Amsterdam, 1999.
3. Ebdon, L., Pitts, L., Cornelis, R., Crews, H., Donard, O. F. X. and Quevauviller, Ph. (Eds), *Trace Element Speciation for Environment, Food and Health*, The Royal Society of Chemistry, Cambridge, 2001.
4. Quevauviller, Ph., *Method Performance Studies for Speciation Analysis*, The Royal Society of Chemistry, Cambridge, 1998.
5. Quevauviller, Ph., *Spectrochim. Acta*, **53**, 1261 (1998).
6. Quevauviller, Ph., Certified reference materials: a tool for quality control of elemental speciation analysis, in *Elemental Speciation – New Approaches for Trace Element Analysis*, Caruso, J. A., Sutton, K. L. and Ackley, K. L. (Eds), Elsevier, Amsterdam, 2000, Chapter 15, pp. 531–569.
7. Cornelis, R., Horvat, M. and Quevauviller, Ph., Certification of organometallic and other species, in *Reference Materials for Chemical Analysis – Certification, Availability and Proper Usage*, Stoeppler, M., Wolf, W. R. and Jenks, P. J. (Eds), Wiley-VCH, Weinheim, 2001, Chapter 3.3, pp. 75–83.
8. ISO Guide 30, *Terms and Definitions Used in Connection with Reference Materials*, ISO/REMCO Resolution 1/91, 2nd Edition, International Standardisation Organisation, Geneva, 1991.
9. de Guillebon, B., Pannier, F., Seby, F., Bennink, D. and Quevauviller, Ph., *Trends Anal. Chem.*, **20**, 160 (2001).
10. Quevauviller, Ph., Wang, Y., Turnbull, A. B., Dirkx, W. M. R., Harrison, R. M. and Adams, F. C., *Appl. Organomet. Chem.*, **9**, 89 (1995).
11. Ariese, F., Cofino, W. P., Gómez-Ariza, J. L., Kramer, G. N. and Quevauviller, Ph., *J. Environ. Monitor.*, **1**, 191 (1999).
12. Lagarde, F., Asfari, Z., Leroy, M. J. F., Demesmay, C., Ollé, M., Lamotte, A., Leperchec, P. and Maier, E. A., *Fresenius J. Anal. Chem.*, **363**, 12 (1999).
13. Ure, A. M. and Davidson, C. M. (Eds), *Chemical Speciation in the Environment*, Blackie, London, 1995.
14. Quevauviller, Ph., *Trends Anal. Chem.*, **17**, 289 (1998).
15. Wood, R., *Trends Anal. Chem.*, **18**, 624 (1999).
16. Walker, R. and Lumley, I., *Trends Anal. Chem.*, **18**, 594 (1999).
17. Maier, E. A., Quevauviller, Ph. and Griepink, B., *Anal. Chim. Acta*, **283**, 590 (1993).
18. Wells, D. E. and Cofino, W. P., An holistic structure for quality management: A model for marine environmental monitoring, in *Quality Assurance in Environmental Monitoring – Sampling and Sample Pretreatment*, Quevauviller, Ph. (Ed.), VCH, Weinheim, 1995, Chapter 10, p. 255.
19. Hartley, T. H., *Computerized Quality Control: Programs for the Analytical Laboratory*, 2nd edn, Ellis Horwood, Chichester, 1990.
20. Prichard, E., *Quality in the Analytical Chemistry Laboratory*, John Wiley & Sons, Ltd, Chichester, 1995, p. 307.
21. Horwitz, W., Kamps, L. R. and Boyer, K. W., *J. Assoc. Off. Anal. Chem.*, **1344**, 63 (1980).
22. Quevauviller, Ph., Drabaek, I., Bianchi, M., Muntau, H. and Griepink, B., *Trends Anal. Chem.*, **15**, 390 (1996).
23. ISO Guide 33, *Uses of Certified Reference Materials*, International Standardisation Organisation, Geneva, 1989.
24. Lagarde, F., Amran, M. B., Leroy, M. J. F., Demesmay, C., Ollé, M., Lamotte, A., Muntau, H., Michel, P., Thomas, P., Caroli, S., Larsen, E., Boner, P., Rauret, G., Foulkes, M., Howard, A., Griepink, B. and Maier, E. A., *Fresenius' J. Anal. Chem.*, **363**, 18 (1999).
25. Quevauviller, Ph., Ebdon, L., Harrison, R. and Wang, Y., *The Analyst*, **123**, 971 (1998).
26. Cámara, C., Quevauviller, Ph., Palacios, M. A. and Cobo, G., *The Analyst*, **123**, 947 (1998).
27. Mena-Fernández, M. L., Morales-Rubio, A., Cox, A. G., McLeod, C. W. and Quevauviller, Ph., *Quím. Anal.*, **14**, 164 (1995).
28. Vercoutere, K., Cornelis, R., Mees, L. and Quevauviller, Ph., *The Analyst*, **123**, 965 (1998).
29. de Boer, J. and McGovern, E., *Trends Anal. Chem.*, **20** (2001).
30. Kramer, G. N. and Pauwels, J., *Mikrochim. Acta*, **123**, 87 (1996).
31. Pauwels, J., Kurfürst, U., Grobecker, K. H. and Ph. Quevauviller, *Fresenius' J. Anal. Chem.*, **345**, 478 (1993).
32. Lamberty, A., Schimmel, H. and Pauwels, J., *Fresenius' J. Anal. Chem.*, **360**, 359 (1998).
33. Quevauviller, Ph. de la Calle-Guntiñas, M. B., Maier, E. A. and Cámara, C., *Mikrochim. Acta*, **118**, 131 (1995).
34. Gómez Ariza, J. L., Morales, E., Sánchez-Rodas, D. and Giráldez, I., *Trends Anal. Chem.*, **19**, 200 (2000).
35. ISO Guide 35, *Certification of Reference Materials – General and Statistical Principles*, International Standardisation Organisation, Geneva, 1985.
36. Taylor, J. K., *Handbook for SRM Users*, National Institute for Standards and Technology, Gaithersburg, Special Publication 260-100, 1993.
37. *Guidelines for the Production of BCR Reference Materials*, Doc. BCR/48/93, European Commission, Brussels, 1994.
38. Wang, Y., Harrison, R. M. and Quevauviller, Ph., *Appl. Organomet. Chem.*, **10**, 69 (1996).
39. Nusko, R. and Heumann, K. G., *Anal. Chim. Acta*, **286**, 283 (1994).
40. Brown, A. A., Ebdon, L. and Hill, S. J., *Anal. Chim. Acta*, **286**, 391 (1994).

41. Quevauviller, Ph. and Ariese, F., *Trends Anal. Chem.*, **20**, 207 (2001).
42. Quevauviller, Ph., Filippelli, M. and Horvat, M., *Trends Anal. Chem.*, **19**, 157 (2000).
43. Ph. Quevauviller, Astruc, M., Morabito, R., Ariese, F. and Ebdon, L., *Trends Anal. Chem.*, **19**, 180 (2000).
44. Quevauviller, Ph., Ebdon, L., Harrison, R. and Adams, F. C., *Trends Anal. Chem.*, **19**, 195 (2000).
45. Cámara, C., Cornelis, R. and Quevauviller, Ph., *Trends Anal. Chem.*, **19**, 189 (2000).
46. ISO Guide 31, *Contents of Certificates of Reference Materials*, International Standardisation Organisation, Geneva, 1981.
47. Valcárcel, M., Ríos, A., Maier, E., Grasserbauer, M., Nieto de Castro, C., Walsh, M. C., Rius, F. X., Niemelä, R., Voulgaropoulos, A., Vialle, J., Kaarls, R., Adams, F. and Albus, H., *Metrology in Chemistry and Biology: A Practical Approach, EUR Report 18405 EN*, European Commission, Brussels, 1998.
48. De Bièvre, P. and Taylor, P. D. P., *Metrologia*, **34**, 67 (1997).
49. Pan, X. R., *Metrologia*, **34**, 35 (1997).
50. Richter, W. and Dube, G., *Metrologia*, **34**, 13 (1997).
51. Cornelis, R., Crews, H., Donard, O., Ebdon, L. and Quevauviller, Ph., *Fresenius' J. Anal. Chem.*, **370**, 120 (2001).

CHAPTER 8

Screening Methods for Semi-quantitative Speciation Analysis

M. E. Foulkes

University of Plymouth, UK

1 INTRODUCTION

1.1 'Divining' the definitions

The advent of any new and important technique in analytical chemistry brings with it a need to publicly define it. This has been the case for the phrase 'chemical speciation'. To meet this challenge, a plethora of attempts have appeared in numerous books and journal articles on this topic. Recent IUPAC commissions have now given rise to recommended definitions for terms such as 'chemical species' and 'speciation analysis' [1]. These focus on the clearly identified and specific form (or forms) of an element and its quantification. This obviates the inclusion of any 'operationally' defined methodology which, by its very nature,

Handbook of Elemental Speciation: Techniques and Methodology R. Cornelis, H. Crews, J. Caruso and K. Heumann
 ISBN: 0-471-49214-0

may be the basis of an important screening method used in speciation analysis.

A screening method may be considered to be 'a process' that extracts, isolates or identifies a component or set of components that possess a defined set of characteristics. These can be collectively quantified if required. At its most basic level, a screening method could take the form of a simple yes/no qualitative measurement [2]. This can be extended to include various levels of quantitative sophistication. However, the process ceases to be a 'simple' screening method if a full multicomponent speciation/mass balance procedure is performed (no matter how rapidly it is made). For example in chemical terms one may need to consider only the organometallic species of an element in a sample, or only those in the same oxidation state, or all those species leachable in a specific solvent. With these points in mind, the broader discussion and definition of 'speciation' offered by Ure [3] is of more practical benefit; considering it to be 'a process' of identifying and quantifying the different defined species, forms or phases present in a material. This also allows the species, forms and phases to be defined functionally (e.g. forms which are considered to be plant available [4, 5], biologically active [6], mobile [4], nutritional [5] etc.), operationally (e.g. soluble in specified solvent or solutions [7]) or as specific compounds [8] or oxidation states [9]. It is important to note that, within the IUPAC recommendations previously cited [1], the term 'fractionation' has been defined in order to consider the above restrictions upon the use of the term 'speciation'. This definition states fractionation as 'a process of classification of an analyte or group of analytes from a certain sample according to physical or chemical properties'. This clearly includes the functional approach previously stated and encompasses the focus of this chapter.

1.1 Why do we perform speciation analyses?

There are various reasons for performing speciation analysis. These can include the need to understand the environment and how it operates (biogeochemical systems and their links [10]), to discover particular sources for commerce as in geochemical exploration [11], to conduct nutritional [5, 12], food [13, 14], clinical [15]

Table 8.1. A range of target levels for some metals and species in different samples.

Element	Water	Soil/Sediment [21]	Foodstuffs
As	Drinking Water 50 µg L^{-1} (EU and USA) 10 µg L^{-1} (WHO) [22]	50 mg kg^{-1} (UK) 40 mg kg^{-1} (parks, open etc.) 10 mg kg^{-1} (gardens etc.) (ICRCL)	1 mg kg^{-1} (total) [23] (under review as Inorganic)
Se	10 µg L^{-1} (DW UK) 50 µg L^{-1} (DW USA)	3 mg kg^{-1}	70 µg RDA (USA) 40 µg RDA (UK)
Sn	25 µg L^{-1} Inorg (fresh)		200 mg kg^{-1} (UK)
Sn(Bu)$_3$ (TBT)	2 ng L^{-1} DW (UK) 20 ng L^{-1} fresh (UK) 30 ng L^{-1} salt (UK) (~8 x in USA for fresh and salt)	250 µg kg^{-1} (USA)	
Hg	1–2 µg L^{-1} (total/inorg)	1.0 to 1.5 mg kg^{-1} (EU) 1.0 mg kg^{-1} (UK) sludge amended	0.5 mg kg^{-1} Total (EU) 0.3 mg kg^{-1} 1.0 mg kg^{-1} (some fish)
MeHg	under revision (USA)		0.3 µg kg^{-1} BW/day (MRL) 0.3 mg kg^{-1} (wet wt, USA)
Pb	50 µg L^{-1} DW 15 µg L^{-1} DW (action) 10 µg L^{-1} DW (WHO) (~10 × for blood) 100–400 µg L^{-1} child/adult	400 mg kg^{-1} (child) 1200 mg kg^{-1} (adult) 40–250 µg ft^{-2} dust (indoors)(USA)	1 mg kg^{-1} (fish, UK) up to 10 mg kg^{-1} (shellfish) limit depends on source

and health [16, 17], quality assurance and quality control [18] and toxicological studies [19] as well as for legal requirements for monitoring and policing [20]. One of the complications encountered by the analyst is the broad range of definitions that can be associated with a particular 'target' level of an element or its species in a sample. Examples include recommended, allowable, enforceable, actionable, minimum risk, daily intake, based on body weight, based on fresh or dry weight, national or international directives, etc. Table 8.1 illustrates a small sample of this range for a number of target elements in various matrices. It may be possible to encompass more than one of these definitions within a sample measurement using a suitable screening method.

1.3 Why use screening analysis?

It is important to emphasise that the analytical approach taken to make a measurement should be 'fit for purpose' and this should be reflected in the protocol adopted. Hence the need to identify those 'requirements' for solving the defined analytical problem. For example from a legal standpoint it may not be necessary to perform a full speciation breakdown of an element in a sample (at some considerable cost in time and resources) if the 'total' content of, say, a targeted metal is less than the concentration permitted for a particular species of that metal [24]. From a toxicological viewpoint, if the inorganic forms of an element are considered the more important [23], an extraction process that selectively acquires inorganic and rejects organoforms would allow the 'total' inorganic species to be quantified.

Some fully quantitative methods are highly dependent upon the instrumentation employed to make them rapid. While such systems may be relatively complex (e.g. two- or three-stage hyphenated techniques), in terms of time of analysis, a screening process may appear not to be necessary. However, it is also important to consider a more 'cost-effective' approach. For example, the case where a screening/semi-quantitative method can utilise a US$7000–15 000 instrument system rather than resort to a complex instrument that is expensive to run and maintain, and costs 10–20 times more. If there is no loss in overall quality of values obtained, i.e. usefulness, or information gained in order to meet the requirements of the analysis, then the screening method could be implemented.

Screening analysis may also be used for field-based measurements [25]. This could include preliminary investigations to identify suitable sampling points prior to a larger detailed study [26], or for samples whose stability is in question when removed from their source [27] and also for samples where single, rapid threshold or coarse resolution field values are sufficient to assess risk and facilitate rapid decision making [28]. These cases then bring in the question of 'level' of quantification.

1.4 What is meant by semi-quantitative?

The term 'semi-quantitative', when used in an analytical screening method, may be considered to include a broad range of applications. Some of these are defined in terms of the instrument used. Examples can range from the rapid acquisition of data using semi-quantitative ICP-MS, which can give a nominal value within a factor of 2 of the correct or accepted value [29], on through laser ablation-ICP-MS [30, 31], where matrix matching of standards with samples may prove problematic, and even to include those techniques that might employ standardless measurement [32] (e.g. XRF, XRD and GD-MS). This is not to imply that the figure given in all semi-quantitative measurements is inaccurate or imprecise. Indeed, the term semi-quantitative can be applied where the information it gives in terms of its 'value' is only part of a speciation 'mass balance' for a sample. It may be unpopular to point this out, but the supposedly full speciation of an element in an extract that is only 80 % efficient from a solid is, in effect, still only semi-quantitative in 'value'; no matter how reproducible it is! And if it is reproducible at only 80 % efficiency for a given solvent or range of extractants, then these values may be considered as 'operationally defined'. However, from a practical point, the semi-quantitative screening method may include, for example, a 'less than' value for an element or a species in a sample (threshold values), or it may include a measure of a group of related elemental species rather than their individual quantification. An example of the latter case could be represented by the accurate and precise values determined for the total organic versus total inorganic species content of an element in a material.

1.5 How do we perform a speciation screening analysis?

In most chemical speciation cases the analyst is looking for particular characteristics of a species in order to make the overall method specific and 'fit for purpose'. The method chosen for screening analysis should therefore also target certain characteristics of the species under investigation. Examples include a species' or range of species' solubility in organic [20] or inorganic solvents. This solubility may be modified by the addition of complexing [33] or coupling agents which may also alter chemical activity. Hence, species may be retained or released during any back extraction procedure. In addition, solid-phase extraction (SPE) can be used to target particular character, e.g. reversed phase, cationic/anionic, complexation/chelation etc. [34].

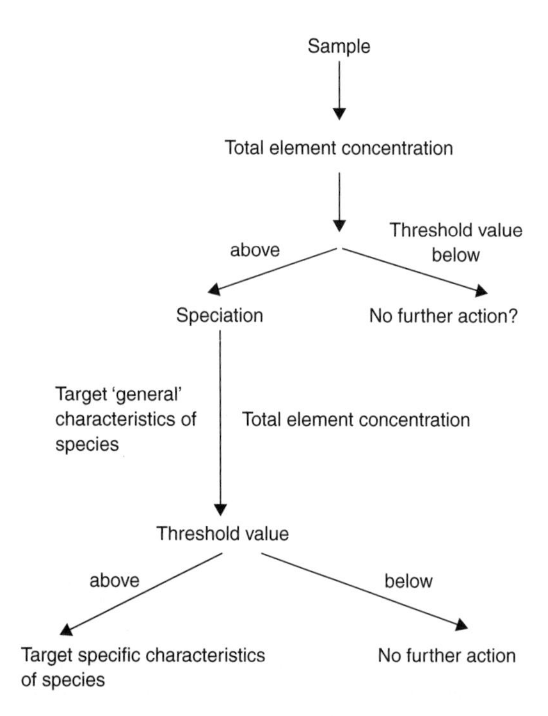

Figure 8.1. Protocol showing the progression through a speciation analysis which demonstrates inherent screening steps.

A logical protocol that takes account of the progression from 'total' element concentration in a sample through to its full speciation, may be presented in the form shown in Figure 8.1. The screening method, whether its result is considered semi-quantitative or not, should often form part of this protocol. A set of simplified criteria would guide this screening method to ensure the 'value' of the chemical information obtained still meets the overall requirements of the analysis.

1.6 Criteria of merit for screening methods

Any screening method should meet some, if not all, of the ideal criteria of merit. These being that it is simple, reliable and portable, rapid, efficient (100 % extraction and target on a particular characteristic), cost effective compared with the fully quantitative parent method, relatively free from interferences and with suitable sensitivity and limits of detection.

Again, the requirement of being 'fit for purpose' means that some form of overall performance evaluation should be made using a set of performance criteria [25]. Simplification of the screening/speciation process may therefore be brought about by use of a digestion/extraction technique that targets the general characteristic of the species directly; for example the solubility of all organotin species from water or sediment samples in a suitable organic solvent. While direct measurement of the total levels of tin in this extract may include a small contribution from inorganic tin, suitably rapid threshold values may be obtained. Recourse to further work would be judged accordingly.

2 INSTRUMENTAL REQUIREMENTS

2.1 What is available and what are the criteria for choice?

In general terms it is noted that the more specific a detector is, the less demanding the extraction process needs to be for the screening process. Instrumentally, the detection systems are based upon being element (atomic), molecular (e.g. functional group) or structurally specific, or at least selective. Of equal importance are the requirements of acceptable limits of detection and the presence of relatively few interferences. If recourse is made to 'selective' rather than 'specific' detectors, then selectivity factors need to be taken into account (as considered when using ion selective electrodes). Some instrumental examples with references are shown in Table 8.2. The possibility of hyphenated or coupled techniques is noted, resulting in improved selectivity and/or screening.

In all cases involving a semi-quantitative screening measurement, the final detection technique used and, if needed, the wet chemical extraction method employed require careful consideration and matching. For example, it is unhelpful to employ toluene to target the important organic mercury species in an extraction process, if the finishing technique required, to meet the criteria of sensitivity and specificity, is cold vapour atomic fluorescence spectrometry. Here, the reduction to mercury vapour in an organic solvent is hindered severely and the presence of even small quantities of toluene vapour in the AFS system can result in substantial interference. Use of a preconcentrating stage in the method will, of course, allow a normally less sensitive detection system

Table 8.2. Examples of instrumentation used in speciation screening analysis.

Mode of speciation operation	Example of application		Limit of detection[a]
	Ref.	Species	
Molecular			
FT-IR	35	As(V)	$mg\,kg^{-1}$
FT-NMR	36	Butyltin	$g\,kg^{-1}$
UV-Vis	27	Cr(VI)	$\mu g\,dm^{-3}$
HPLC/GC/CE–hyphenated	37	Metals/organic pollutants	Variable[b]
Molecular fluorescence	38	Al in waters	$\mu g\,dm^{-3}$ (down to 10^{-18} g)
Atomic			
FAAS/ETAAS	39	General review	$\mu g\,dm^{-3}/<\mu g\,dm^{-3}$
AFS	8	Hg	$ng\,dm^{-3}$
XRF	9	Cr(III) and Cr(VI)	$\mu g\,dm^{-3}$
ICP-AES	40	d block elements and Pb (soils)	$\mu g\,dm^{-3}$
ICP-MS	40	d block elements and Pb (soils)	$ng\,dm^{-3}$
Electron microscopy	41	Chemical speciation	$g\,kg^{-1}$
Electroanalytical			
ISE	42	Cyanide	$mg\,dm^{-3}$
ASV	43	Organotin	$ng\,dm^{-3}$
CSV	44	Inorganic As	$ng\,dm^{-3}$
Polarography	43	Organotin	$mg\,dm^{-3}$
CZE–hyphenated	12	Organocobalt	Variable[b]
Structural			
XRD	32	Elemental, oxide (lime)	$g\,kg^{-1}$
NMR	36	Organotin	$g\,kg^{-1}$
Radiochemical			
Simple α and β detectors	28	^{134}Cs, ^{137}Cs, ^{99}Tc, and ^{90}Sr	Variable, circa $\mu g\,kg^{-1}$

[a]These limits of detection are nominal and are from a direct measurement. They refer to the technique and not necessarily to the example reference.
[b]Dependent upon the coupled end-detection technique.

to be introduced and will subsequently lower the limit of detection for the method.

3 ISOLATION AND EXTRACTION TECHNIQUES

The extraction technique used must not change the species present in a sample or their equilibrium concentrations if in admixture. This is a baseline requirement for speciation analysis. For most screening procedures, it is preferable that one adheres to this requirement despite the fact that the analytical measurement may be considered as semi-quantitative. However, the requirements of the analysis will dictate what is 'fit for purpose' and may allow the above edict to be 'broken' under particular conditions. This assumes a sound knowledge of the chemical system under investigation and judgement at this stage may be considered critical. For example, the 'total' organoelement concentration, associated with a procedure that selectively extracts the organic species, can still be measured even if those species present change form (and hence concentration) after the extraction procedure has been performed. However, this type of 'general' targeting does not allow further species measurement to be made.

3.1 Validation

It is also important that the extraction efficiency for the target species (or characteristic of such) is high, is reproducible and is independent of the matrix. While highly desirable, all these requirements are rarely met in practise. These must be validated in some way, or even the semi-quantitative screening method may be in error. The use of certified reference materials (CRMs), or at least well-characterised in-house reference materials, during development of and routine work involving screening steps is assumed to be an inherent part of the analysis. The range of CRMs available that contain some speciation data is increasing. Examples include alkyl lead in urban dust (CRM 605), chromium species (Cr(VI) and total leachable) in welding dust (CRM 545), arsenic species in tuna (arsenobetaine, dimethylarsenate and total, CRM 627), mercury species in tuna (CRM 464) and sediment (CRM 580) and of course the sequential extraction materials (CRMs 483, 484, 600, 601).

3.2 Isolation and extraction techniques

As stated previously, a digestion or extraction technique that targets 'general characteristics' of a species directly for a screening process but can, if required, be used later for a full speciation protocol, may help to simplify the overall process. Examples include the use of methanolic KOH for the digestion of fish tissues [45], where all the major mercury species are directly extracted and solubilised unchanged, or the use of enzymolysis digestion such as with trypsin at pH 8 for meats, fish and proteinaceous samples [46], and the use of cellulase [47] at pH 5 for vegetables, plants etc. for the direct extraction of arsenic and tin species.

While sample types can be many and varied it is also noted that more than one phase may be present. The latter is often manifest in the so-called 'simple' determination of 'total' elements and their species, where a correction has to be made due to the presence of associated water in a sample. Even a simple screening for the 'water-soluble species' in a two-phase system may depend on a definition, i.e. those particulates that pass through a 0.45 or 0.22 μm filter. However, in this case, a centrifugation-based approach may be a suitable alternative.

To summarise, the methodology adopted in a screening process should be 'fit for purpose' and its envelope of applicability (or boundary range) should be suitably validated using a similar sample type. A method which also allows for any further, as necessary, more targeted speciation should be considered, rather than having to repeat the extraction/digestion process by an alternative and comparatively time consuming methodology.

One may consider the following approaches in terms of their ease of application:

(1) Direct or nondestructive techniques for screening/semi-quantitative speciation [31], e.g. liquid samples introduced directly to element specific detectors or *via* a suitable separation technique, such as HPLC or GC [37], or for various phases using sample introduction such as laser ablation coupled with ICP-MS [31], or instrumentation such as XRF and XRD [32].

(2) Extraction/digestion techniques for screening/semi-quantitative speciation. This includes the sequential extraction from solids [40], through solvent extraction using organic liquids [20], aqueous solutions and their mixtures, acidic/alkaline (mild or strong) solutions, chelation (aqueous and organic) systems [7] and even enzymolysis digestion [46]. The sorbent-based solid-phase extraction (SPE) systems, e.g. ion exchange, chelation, reversed phase etc. [34], have become important additions in the analyst's armoury for speciation determination. They may also be used effectively in screening processes either directly with liquid samples [28], or indirectly following an extraction into suitable liquid media which may be aqueous or organic based [15]. Use of SPE techniques allows both targeting of certain characteristics of species and, as previously stated, the possibility of preconcentration; thereby improving limits of detection for the overall process and allowing a broader range of instrumentation to be employed, for example, the direct measurement of chromium species adsorbed from waters onto ion exchange resins using XRF [9]. Use of such in the field would broaden the scope of portable XRF systems [26].

Screening processes should generally be as rapid as possible. The use of microwave energy or ultrasonication in multiple extractions can improve efficiency in addition to, or instead of, simple shaking [48]. Commercially available instrumentation-based accelerated extraction techniques can also be employed. However, great care must be taken when considering their use as it may be possible to alter the species present and their equilibrium concentrations due to the energies involved. Their use would, of course, be tailored within the frame of the analytical requirements.

4 SELECTED EXAMPLES

While the underlying analytical philosophy discussed in previous sections is important, it is often more instructive to look at a selected range of general speciation techniques or those for targeted elements that employ a screening methodology. Many methods may also include this possibility, i.e. the measurement range of fully quantitative down to qualitative, as part of their requirement to be 'fit for purpose'. Hence, a semi-quantitative/screening method may then be considered an inherent part of simplifying the specific approach taken. The examples presented have only been chosen to illustrate some of the principles discussed. The literature describes many fully quantitative methods which can be adapted to serve the analyst's needs. A suitable search of the relevant sources will reveal the large range of possible methods available.

4.1 Arsenic and selenium: the inorganic problem and a suitable case for reduction

4.1.1 Arsenic

A review of arsenic speciation in various sample types can often give guidance on suitable screening techniques to meet those requirements for its measurement [10]. Toxicological studies (Table 8.3) have resulted in specified levels for total inorganic arsenic in food, $1\,mg\,kg^{-1}$ [23], and total arsenic in water, $50\,\mu g\,dm^{-3}$ (under consideration to be lowered to the WHO level, $10\,\mu g\,dm^{-3}$, in some countries, see Table 8.1). The measurement of total arsenic in aqueous samples may be achieved rapidly and directly using suitable instrumentation. Possible interferences include (i) the presence of chloride when using ICP-MS, and (ii) the variable response when using a hydride generation technique due to the presence of transition metals, or the different inorganic oxidation states, organoarsenic species and their nonreducible forms.

While the efficient extraction and measurement of inorganic arsenic in foodstuffs may prove problematic, one possible screening method takes

Table 8.3. Toxicity of selected arsenic species [49, 50].

LD_{50} (rat)	Arsenite: 1.5 mg kg^{-1}
	Arsenate: 5 mg kg^{-1}
	MMAA: 50 mg kg^{-1}
	DMAA: 600 mg kg^{-1}
	TeMA: 890 mg kg^{-1}
LD_{50} (mice)	AsB: >10 g kg^{-1}
	AsRbF: nontoxic?
LDL_0 (child)	Arsenite: 2 mg kg^{-1}
Lethal dose	Arsenite: 100 mg

MMAA = monomethylarsonic acid; DMAA = dimethylarsinic acid; TeMA = tetramethylarsonium; AsB = arsenobetaine; AsRbF = arsenoribofuranosides.

Table 8.4. Toxicity versus reducible character of arsenic species.

Toxic	Nontoxic?
Reducible	Nonreducible
As(III)	AsB
As(V)	AsCh
DMAA	AsRbF
MMAA	
TMAO	

TMAO = trimethylarsine oxide; AsCh = arsenocholine.

advantage of the presence of the nonreducible (or involatile or less toxic) organoarsenic forms such as arsenobetaine and the arsenoribofuranosides. These can often constitute the major species found in samples such as fish, poultry and certain macroalgae [19]. From Table 8.4, it is seen that a hydride generation technique coupled with a suitable element-specific detector, will give threshold values for the inorganic arsenic content of extracts from foodstuffs, following a pre-reduction of the As(V) to As(III) using iodide or L-cysteine. Theoretically, no false negative values are obtained. This is because any contribution from the often lower concentration reducible organoarsenic species will result in artificially high values for the inorganic arsenic that is assumed only to be present.

Use of enzymes such as trypsin [46] or cellulase [47] both of which solubilise the major arsenic species found in solid biological samples unchanged, also allow further screening speciation studies to be made based upon the reducible arsenic values (assumed inorganic). For example, low pressure mini-column chromatography with a resin based strong anion exchanger at pH 10 can be used to separate organoarsenic from inorganic arsenic species as two simple peaks [51]. In this case, advantage is taken of the species' pK_a values (Table 8.5). The character which is targeted is the charge on any arsenic anions formed at this high pH and their ease of removal from the column using a competitive eluent such as an aqueous sulfate solution.

A full speciation separating all the generally encountered inorganic and organic arsenic species may be performed on the same extract using high performance liquid chromatography [53]. In all the above cases, suitable element-specific detection methods include AAS, AFS, ICP-AES and ICP–MS; these are dependent upon the sensitivity or LOD required and the possible interferences that may be encountered, e.g. effect of chloride, during the 'direct' measurement of extracts from biological samples for total arsenic species using ICP-MS. The addition of N_2 to the carrier gas can remove this effect.

4.1.2 Selenium

There has been considerable interest in this element and its species in recent years. This is because it is both an 'essential' element in our diet and shows toxic properties at certain levels. This relatively narrow band of efficacy (RDA 70 μg (USA), 40 μg (UK) but advisory level not greater than 1000 μg) shows species dependency. Clinical and nutritional studies have focussed on a number of organic and inorganic forms of this element [53, 54]. It is noted that for a range of biological samples a similar regime to that used for the extraction of arsenic species can be adopted [51] to yield unchanged selenium species for measurement. Again a hydride generation technique coupled with a suitable element-specific detection technique (AAS, AFS, ICP systems etc.) can be used to estimate the 'total' inorganic selenium content of a sample in the presence of organoselenium species which have limited reduction properties [55, 56]. Also in common with inorganic As(V), the higher oxidation state of inorganic selenium, Se(VI), requires a reduction step if total inorganic selenium is to be determined [56]. If no reduction

Table 8.5. pK_a values for inorganic and organic arsenic and selenium species [52, 75].

Arsenous acid	$HAsO_2 \rightarrow AsO_2^-$	pK_a 9.23	
Arsenic acid	$H_3AsO_4 \rightarrow AsO_4^-$	pK_{a1} 2.20 pK_{a2} 6.97 pK_{a3} 11.53	
Monomethylarsonic acid	$CH_3As(=O)(OH)_2$ (OH above and below As)	pK_{a1} 3.6 pK_{a2} 8.2	
Dimethylarsinic acid	$(CH_3)_2As(=O)-OH$	pK_{a1} 1.28 pK_{a2} 6.2	
Arsenobetaine	$(CH_3)_3As^+ CH_2COOH$	pK_a 2.18	
Selenous acid	H_2SeO_3	pK_{a1} 2.46	pK_{a2} 7.31
Selenic acid	H_2SeO_4	pK_a 1.92	
Selenomethionine	$CH_3SeCH_2CH_2CH(COOH)NH_2$	pK_{a1} 2.6	pK_{a2} 8.9
Selenocystine	$(H_2N–CH(COOH)–CH_2–Se–)_2$	pK_{a1} 2.4	pK_{a2} 8.9

step is employed, then an estimate of the inorganic Se(IV) can be obtained directly and hence the Se(VI) obtained indirectly by difference. Simple resin based anion mini-column systems can be used to target the anionic/reversed-phase character of the selenium species at pH 5–10 (pK_a values, Table 8.5) and rapidly separate inorganic and organic species. Selenium detection may be brought about by using the same range of suitable element-specific instrumentation.

The use of SPE to target a range of selenium and arsenic species is a popular approach and has been reviewed [53, 54]. Much of the SPE targeting chemistry follows the principles underlying the fully quantitative chromatographic methods used to separate species, i.e. cationic, anionic, chelation, reversed-phase character.

4.2 Tin, mercury and lead: the organic problem

These metals each have a number of species which are considered to be environmentally and biologically important. Of higher priority in their lists are the organometallic species because of their toxicological activity at such low levels. For example it is noted that 'imposex' effects (endocrinal disruption) can be seen in shellfish in the presence of tributyltin as low as 2–3 ng L^{-1} and that there is evidence for immunotoxicity in humans due to exposure to butyltin compounds [57]; that methylmercury in foods shall not exceed the risk level which is limited to providing not more than 0.3 μg per kg body weight per day, and that lead alkyls should not exceed 50 μg m^{-3} air (TLV) in the workplace. In most cases the methods employed to look at these species require low limits of detection. The use of sensitive element-specific detectors and preconcentration steps are common. Hence, the use of solid-phase extraction (SPE) [58] with subsequent elution followed by HG/CV techniques to instruments such as GFAAS [59] or to GC-MS [60], GC-ICP-AES [33] and ICP-MS [61].

Screening methods may be based upon 'total metal' and 'total organometal' using targeting of species character, i.e. solubility in certain solvents, e.g. toluene for butyltin and inorganic tin using sodium diethydithiocarbamate (NaDEDC) complexing agent [33], or *n*-hexane to target mainly organotin species either before or after SPE has been used.

Sediments can yield full recoveries for 'available' organic and inorganic tin species if microwave energy is used for 3 min [61] with acetic acid/tropolone solvents or acetic acid/sodium pentanesulfonate mixtures; thereby allowing threshold values to be determined.

The signal from inorganic lead in an extract can be suppressed using EDTA/HCl [59] (or citric acid) when a HG technique is employed. This allows the total volatile organolead species to be measured. The use of DEDC to complex lead species in water prior to SPE allows low levels of inorganic and organic lead to be immobilised [60]. Preconcentration onto a C60 fullerene SPE medium and subsequent elution with isobutylmethyl ketone (IBMK) or n-hexane allows screening for total/inorganic/organic lead species dependent upon the solvent system and finishing technique used. Mercury species may be released from biological samples using mild digestants such as methanolic KOH [45], enzymes or even strong HCl/toluene [62]. In the latter case, microwave-assisted extraction can improve the speed and efficiency in acquiring methylmercury species from reference fish material. Direct measurement of these extracts using suitable element-specific detectors will allow threshold values to again be determined rapidly. The need for further speciation can therefore be appraised. Finishing techniques for the above include GC-MS and GC-AFS.

4.3 Colorimetric analysis: the 'Cinderella' technique

Visible and UV spectrophotometric methods have been employed for screening and/or semi-quantitative analysis for many years. The advantages of acquiring rapid, field based as well as laboratory based measurements using simplified instrumentation have been utilised for a variety of sample types and in a variety of locations. Use of specific complexing agents and SPE has improved the technique's selectivity and lowered its limits of detection. Hence, a sound understanding of the underlying chemical principles involved, together with regard to interferences and matrix effects, has helped maintain this technique's popularity. Examples include simple quantitative yes/no results [63] through to semi-quantitative threshold values [27, 63] for Pb in paint samples [63] and for Cr(VI) species in waters, soil extracts and airborne particulates [27]. The use of sorptive preconcentration extended the latter's measurement to the sub-$\mu g\,L^{-1}$ level. Both these examples are field based screening techniques. The screening of inorganic Cu, Pb and Cd in river and ground waters using two SPE cartridges, one to screen out interferences and the other to retain target species has also been employed [64]. Preconcentration also allowed measurement down to the low $\mu g\,L^{-1}$ level and employed FIA methodology.

4.3.1 Rapid screening tests for element selective ions

Simple, single-step (in the case of waters) or two-step (if extracted from solids) rapid screening tests for certain element based ions have been employed in the laboratory and field for many years. The early chemical based tests for nitrogen, phosphorus, sulfur, chlorine and iron are among the environmental field based analyst's armoury. These tests usually centre upon the formation of particular complexes or compounds which can be identified, and if need be quantified, using visible or ultraviolet spectrophotometry. Hence, the technically undemanding colorimetric measurement affords, in many cases, an ultrarapid screening method for the above elements (and many others) in their particular chemical form. This has been made possible by the introduction of commercially available 'spot-test' kits, which require little more than adding a preformulated sachet (or 'pillow') of mixed compounds directly to a known volume of liquid sample (or extract from a solid) prior to measurement. As stated previously, a sound knowledge of the chemistry involved and possible interferences is required in order to interpret the results accordingly. It is noted that some formulations have only a limited shelf life which could obviously effect results.

4.3.1.1 Nitrogen

Common nitrogen speciation in aqueous environmental samples includes NO_2^-, NO_3^- and NH_4^+/NH_3. Other, more exotic, forms include N_2H_4 and some amino acids. Tests for N–NO_3^- and N-NO_2^- are often based upon the formation of first, diazo compounds and then on coupled azo dyes, or on nitro complexes etc., all buffered to a specific pH range and oxidation state [65]. By pretreating the sample to change the oxidation state (e.g. reduction of NO_3^- to NO_2^-), it is possible to differentiate between nitrite and nitrate either directly or by difference. Certain organic pollutants, however, can modify the response. NH_4^+ can be measured when a sample reacts with phenol–alkaline hypochlorite to form indophenol blue. The colour can be intensified with sodium nitroferricyanide. An alternative method is to use a solution of mercury(II) iodide in alkaline potassium iodide (Nessler's reagent), which in the presence of ammonia/ammonium, forms a brown colouration or precipitate.

4.3.1.2 *Phosphorus*

This is mainly detected as the orthophosphate (PO_4). For rapid testing of this element, the direct measurement of the PO_4 species (or production of it from another phosphorus species) is *via* a coloured complex involving molybdenum. The two most common approaches involve the creation of either the molybdenum blue complex [65, 66] or the yellow vanadyl phosphomolybdate [65] system under particular pH and reducing conditions. It is important to note that arsenate and silicate can interfere with the molybdenum blue reaction, giving rise to false positive results.

4.3.1.3 *Silicon*

Under acidic conditions it is possible to determine molybdate-reactive silica. Ammonium molybdate and silica, together with a suitable reducing agent, will form a blue colour which may be measured colorimetrically. Phosphate interference can be reduced by the addition of certain organic acids, such as citric [65].

4.3.1.4 *Sulfur*

SO_4^{2-} may be determined colorimetrically in fresh waters using barium chromate reagent. The determination relies upon the measurement of the chromate spectrum under particular pH conditions [67].

SO_3^{2-} may be determined colorimetrically in aqueous samples using potassium hexacyanoferrate and sodium nitroprusside to form a red complex in the presence of a suitable metal ion, e.g. zinc [65].

S^{2-} the colorimetric method commonly employed for soluble sulfide measurement in aqueous samples is the methylene blue method. Sulfides react with dimethyl *p*-phenylenediamine in the presence of a suitable oxidising agent, such as ferric chloride, to produce the required methylene blue [65].

4.3.1.5 *Chloride*

A colorimetric method for the determination of the chloride ion in biological samples (blood, serum, plasma, urine) is that based upon the quantitative reduction of free mercuric ions in solution by chloride ions. Free mercuric ions form a purple complex with diphenylcarbazone. The introduction of chloride ions, which form a nondissociable complex with mercury ions, will reduce the intensity of the coloured complex; which is inversely proportional to the chloride concentration [68].

4.3.1.6 *Metal ions [65]*

Suitable screening tests exist for the identification and determination of metal ions in relatively simple liquid matrices. For example, iron can be measured colorimetrically in the presence of the chelating reagent 1,10-phenanthroline (or 2,2-bipyridyl) after reduction to the +2 state, producing the red coloured complex. Suitable use of buffering and reducing agents will allow both Fe(II) and Fe(III) to be determined. Copper, used in some fungicides, can be determined in solution using the complexing reagents cuprazone or bathocuproine. The former reagent, as the hydrazone, forms a blue complex, whereas the latter reagent, as the disulfonate, forms an orange coloured complex. Chromium can be determined colorimetrically in its hexavalent state as the red/violet compound formed with diphenylcarbazide under acid conditions. The importance of chromium in the (VI) state as a possible carcinogen is well documented.

4.4 Transition and main group metals by electroanalytical methods

4.4.1 Copper, cadmium, lead, zinc, iron and manganese

No discussion of screening techniques for speciation in its broader sense would be complete without some further examples involving electroanalytical techniques. Polarography, including anodic and cathodic stripping voltammetry and related adsorption techniques, can offer rapid field based analyses. In the right hands and with a good understanding of its capabilities for a given sample system, it is a powerful technique that offers low limits of detection. This is demonstrated in those examples involving copper, cadmium, lead and zinc measurements such as in lake waters [69] (labile versus nonlabile Cu and Zn species) and in extracts from hazardous waste sites, soils and sediments [70] (available Cd, Zn, Cu and Pb). Electroanalytical techniques have been extended to include a wide variety of metals, metalloids and their species (organo-As and Sn, and available Fe, Mn, Al etc.) [43, 44, 71] and have been employed in a wide range of field environments [70, 71].

4.4.2 ISE methods

It has been noted in previous sections, that ion selective electrodes (ISEs), an electroanalytical technique, can be employed as part of a rapid screening method for nonmetallic ions. It is commonplace to find in the modern laboratory and field kit, ion selective electrodes for the measurement of NO_3^- [72], F^-, Cl^-, SO_4^{2-}, H^+ etc. The technique has developed to now include selective electrodes for metal ions, albeit under rather strict conditions. Examples include Ca, Zn and K and Na [73]. Control of their environment and sample pretreatment can be used to extend information about their 'speciation'.

5 CONCLUSIONS

Speciation screening methods, whether they are qualitative or semi-quantitative, should follow a given set of criteria that make them 'fit for purpose'. In common with their more complex, and time consuming, fully quantitative parent methods they may often share part of the same mass balance route. This allows further detailed speciation to be performed, when necessary, on the same sample/extract etc. Also in common, they share the need for a validation step which will inform the analyst of the 'envelope of applicability' or boundary conditions of the screening method. The advantages of speedy, threshold information, cost effectiveness and in many cases the ability to be taken into the field [74] to measure at or near the sample source (and limit any effect from sample instability) make this analytical approach worth pursuing.

6 REFERENCES

1. Templeton, D. M., Ariese, F., Cornelis, R., Danielsson, L. G., Muntau, H., Van Leeuwen, H. P. and Lobinski, R., *Pure Appl. Chem.*, **72**, 1453 (2000).
2. Cutrufelli, M. E., Mageau, R. P., Schwab, B. and Johnston, R. W., *J. AOAC Int.*, **76**, 1022 (1993).
3. Ure, A. M., Methods of analysis for heavy metals in soils, in *Heavy Metals in Soils*, Alloway, B. J., (Ed.), Blackie, Glasgow, 1990, pp. 40–80.
4. Shann, J. R. and Bertsch, P. M., *Soil Sci. Soc. Am. J.*, **57**, 116 (1993).
5. De Rijck, G. and Schrevens, E., *J. Plant Nutr.*, **21**, 849 (1998).
6. Blum, U. and Schwedt, G., *Acta Hydrochim. Hydrobiol.*, **26**, 235 (1998).
7. Galvez-Cloutier, R. and Dube, J. S., *J. Water Air Soil Pollut.*, **102**, 281 (1998).
8. Cano-Pavon, J. M., De Torres, A. G., Sanchez-Rojas, F. and Canada-Rudner, P., *Int. J. Environ. Anal. Chem.*, **75**, 93 (1999).
9. Menendez-Alonso, E., Hill, S. J., Foulkes, M. E. and Crighton, J. S., *J. Anal. At. Spectrom.*, **14**, 187 (1999).
10. Szpunar, J. and Lobinski, R. S., *Fresenius' J. Anal. Chem.*, **363**, 550 (1999).
11. Hirner, A. V., Krupp, E., Schulz, F., Koziol, M. and Hofmeister, W., *J. Geochem. Explor.*, **64**, 133 (1998).
12. Baker, S. A. and Miller-Ihli, N. J., *Spectrochim. Acta, Part B*, **55**, 1823 (2000).

13. Gunther, K. and VonBohlen, A., *Spectrochim. Acta, Part B*, **46**, 1413 (1991).
14. Shiraishi, K., *J. Radioanal. Nucl. Chem.*, **238**, 67 (1998).
15. Drummer, O. H., *J. Chromatogr. B*, **733**, 27 (1999).
16. Werner, M. A., Thomassen, Y., Hetland, S., Norseth, T., Berge, S. R. and Vincent, J. H., *J. Environ. Monit.*, **1**, 557 (1999).
17. Vaughan, M. A., Baines, A. D. and Templeton, D. M., *Clin. Chem.*, **37**, 210 (1991).
18. Quevauviller, P. and Ariese, F., *Trends Anal. Chem.*, **20**, 207 (2001).
19. Benramdane, L., Bressolle, F. and Vallon, J. J., *J. Chromatogr. Sci.*, **37**, 330 (1999).
20. Colume, A., Cardenas, S., Gallego, M. and Valcarcel, M., *Anal. Chim. Acta*, **436**, 153 (2001).
21. Royal Commission on Environmental Pollution, Sustainable Use of Soil, 19th Report, HMSO, London, 1996.
22. US EPA drinking water priority, www.epa.gov/safewater/ars/arsenic.html, June 2002.
23. DEFRA, Regulations for chemical contaminants and naturally occurring toxicants in food. Summary 16, Section III. Web:http://www.defra.gov.uk/research/econeval/chemcont/sec_iii pdf.
24. Binato, G., Biancotto, G. J., Piro, R. and Angeletti, R., *Fresenius' J. Anal. Chem*, **361**, 333 (1998).
25. Song, R., Schlecht, P. C. and Ashley, K., *J. Hazard. Mater.*, **83**, 29 (2001).
26. Kirtay, V. J., Kellum, J. H. and Apitz, S. E., *Water Sci. Technol.*, **37**, 141 (1998).
27. Frenzel, W., *Fresenius' J. Anal. Chem.*, **361**, 774 (1998).
28. Beals, D. M., Hofstetter, K. J., Johnson, V. G., Patton, G. W. and Seely, D. C., *J. Radioanal. Nucl. Chem.*, **248**, 315 (2001).
29. Hu, Y., Vanhaecke, F., Moens, L. and Dams, R., *Anal. Chim. Acta*, **355**, 105 (1997).
30. Alteyrac, J., Augagneur, S., Medina, B., Vivas, N. and Glories, Y., *Analusis*, **23**, 523 (1995).
31. Lochner, F., Appleton, J., Keenan, F. and Cooke, M., *Anal. Chim. Acta*, **401**, 299 (1999).
32. Yellepeddi, R. and Kohler, A., *ZKG Int.*, **49**, 522 (1996).
33. Allan, L. M., Verma, D. K., Yang, F., Chau, Y. K. and Maguire, R. J., *AIHAJ.*, **61**, 820 (2000).
34. van Doornmalen, J., Kroon, K. J., van Elteren, J. T. and de Goeij, J. J. M., *J. Radioanal. Nucl. Chem.*, **249**, 349 (2001).
35. Mollah, M. Y. A., Lu, F. and Cocke, D. L., *Sci. Tot. Environ.*, **224**, 57 (1998).
36. Pettinari, C., Marchetti, F., Pettinari, R., Gindulyte, A., Massa, L., Rossi, M. and Caruso, F., *Eur. J. Inorg. Chem.*, **6**, 1447 (2002).
37. de la Guardia, M. and Garrigues, S., *Trends Anal. Chem.*, **17**, 263 (1998).
38. Fairman, B., Sanz-Medel, A., Jones, P. and Evans, E. H., *Analyst*, **123**, 699 (1998).
39. Welz, B., *J. Anal. At. Spectrom.*, **13**, 413 (1998).
40. Payaperez, A., Sala, J. and Mousty, F., *Int. J. Environ. Anal. Chem.*, **51**, 223 (1993).
41. Newbury, D., Wollman, D., Irwin, K., Hilton, G. and Martinis, J., *Ultramicroscopy*, **78**, 73 (1999).
42. Gattrell, M., Cheng, S. C., Guena, T. and MacDougall, B., *J. Electroanal. Chem.*, **508**, 97 (2001).
43. Ochsenkuhn, K. M., Ochsenkuhn-Petropoulou, M., Tsopelas, F. and Mendrinos, L., *Mikrochim. Acta*, **136**, 129 (2001).
44. Barra, C. M. and dos Santos, M. M. C., *Electroanal.*, **13**, 1098 (2001).
45. Horvat, M., Liang, L., Azemard, S., Mandie, V., Villeneuve, J. P. and Coquery, M., *Fresenius' J. Anal. Chem.*, **358**, 411 (1997).
46. Dean, J. R., Ebdon, L., Foulkes, M. E., Crews, H. M. and Massey, R. C., *J. Anal. At. Spectrom.*, **9**, 615 (1994).
47. Menendez Alonso, E., Investigacion de Nuevos Metodos Enzimaticos de Extraccion para el Analisis de Especies de Arsenico en Alimentos, Tesis de Licenciatura (Dissertation), Universidad de Oviedo, Oviedo, Spain, 1995.
48. Zheng, J., Iijima, A. and Furuta, N., *J. Anal. At. Spectrom.*, **16**, 812 (2001).
49. Vela, N. P. and Caruso, J. A., *J. Anal. At. Spectrom.*, **8**, 787 (1993).
50. Kaise, T., Watanabe, S. and Itoh, K., *Chemosphere*, **14**, 1327 (1985).
51. Ebdon, L., Fitzpatrick, S. and Foulkes, M. E., *Chemia Analityczna*, **47**, 179 (2002).
52. Hansen, S. H., Larsen, E. H., Pritzl, G. and Cornett, C., *J. Anal. At. Spectrom.*, **7**, 629 (1992).
53. Guerin, T., Astruc, A. and Astruc, M., *Talanta*, **50**, 1 (1999).
54. Pyrzynska, K., *Solvent Extr. Ion. Exch.*, **13**, 369 (1995).
55. Zhang, D. Q., Sun, H. W. and Yang, L. L., *Fresenius' J. Anal. Chem.*, **359**, 492 (1997).
56. Brunori, C., de la Calle-Guntinas, M. B. and Morabito, R., *Fresenius' J. Anal. Chem.*, **360**, 26 (1998).
57. Nielsen, J. B. and Strand, J., *Environ. Res.*, **88**, 129 (2002).
58. Mester, Z., Sturgeon, R. and Pawliszyn, J., *Spectrochim. Acta, Part B*, **56**, 233 (2001).
59. Bettmer, J. and Cammann, K., *Appl. Organomet. Chem.*, **8**, 615 (1994).
60. Baena, J. R., Cardenas, S., Gallego, M. and Valcarcel, M., *Anal. Chem.*, **72**, 1510 (2000).
61. Chao, W. S. and Jiang, S. J., *J. Anal. At. Spectrom.*, **13**, 1337 (1998).
62. Vazquez, M. J., Abuin, M., Carro, A. M., Lorenzo, R. A. and Cela, R., *Chemosphere*, **39**, 1211 (1999).
63. Hutter, G. and Moshman, D., *J. Hazard. Mater.*, **40**, 1 (1995).
64. Castillo, E., Cortina, J. L., Beltran, J. L., Prat, M. D. and Granados, M., *Analyst*, **126**, 1149 (2001).
65. *APHA Standard Methods*, 19th edn, Methods section 3500 and 4500, 1995.
66. Frenzel, W. and Dantan, N., *Chemia Analityczna*, **44**, 539 (1999).

67. Golterman, H. L. and Bierbrauwerwurtz, I. M. D., *Hydrobiologica*, **228**, 111 (1992).
68. Yokoi, K., *Biol. Trace Elem. Res.*, **85**, 87 (2002).
69. Xie, H. P., Kistler, D. and Sigg, L., *Limnol. Oceanogr.*, **40**, 1142 (1995).
70. Olsen, K. B., Wang, J., Setiadji, R. and Lu, J. M., *Environ. Sci. Technol.*, **28**, 2074 (1994).
71. de Jong, J. T. M., Boye, M., Schoemann, V. F., Nolting, R. F. and de Baar, H. J. W., *J. Environ. Monit.*, **2**, 496 (2000).
72. LeGoff, T., Braven, J., Ebdon, L., Chilcott, N. P., Scholefield, D. and Wood, J. W., *Analyst*, **127**, 507 (2002).
73. Burnett, R. W., Covington, A. K., Fough-Andersen, N., Kulpmann, W. R., Lewenstam, A., Maas, A. H. J., Muller-Plathe, O., Sachs, C., Siggaard-Andersen, O., Van Kessel, A. L. and Zijlstra, W. G., *Clin. Chem. Lab. Med.*, **38**, 1065 (2000).
74. Hou, X. and Jones, B. T., *Microchem. J.*, **66**, 115 (2000).
75. Larsen, E. H., Hansen, M., Fan, T. and Vahl, M., *J. Anal. At. Spectrom.*, **16**, 1403 (2001).

CHAPTER 9

Risk Assessments/Regulations

9.1 Environmental Risk Assessment and the Bioavailability of Elemental Species

John H. Duffus

The Edinburgh Centre for Toxicology, Edinburgh, UK

1 GENERAL INTRODUCTION

Assessment of the environmental risks involved in the production, use and disposal of chemicals is essential if they are to be used safely. Environmental risk assessment is the essential basis for 'risk management' [1]. Risk management for effects on the environment has been defined

Handbook of Elemental Speciation: Techniques and Methodology R. Cornelis, H. Crews, J. Caruso and K. Heumann
 ISBN: 0-471-49214-0

as being management to protect the environment, preferably by preventing pollution but also by providing suitable protective measures to prevent damage from pollution, with appropriate remediation measures to repair damage once caused [2]. Effective risk management must deal with public concerns over potential causes of environmental pollution and must aim to reduce risk to the practical minimum. In practice, risk management must take into account socioeconomic considerations, as well as balancing risk against benefit, but these considerations are outside the scope of the present review which will concentrate on the assessment process as it relates to the bioavailability of chemical species of the elements other than carbon which are essential to life. Bioavailable chemical species are those which can be taken up by living cells and organisms from their environment. Chemical species which are not bioavailable may cause physical damage or may alter the availability of other substances but cannot contribute to nutrition or have toxic effects. Thus, bioavailability is the prime consideration in environmental risk assessment for toxicity. The present author has recently reviewed the relevant fundamental scientific considerations relating to chemical speciation of trace elements in relation to their bioavailability and risk assessment [3]. Another useful review on persistence and availability of metals in aquatic environments has been published in 2001 by Di Toro *et al.* [4]. This chapter will therefore build on these two previous reviews and will concentrate on reviewing the application of our present knowledge to current and proposed approaches to environmental risk assessment with a view to identifying future developments.

The generally accepted model for risk assessment of potentially toxic chemicals relates primarily to exposure of human beings and is shown in Figure 9.1.1 (based on ref. 5). The key steps are hazard identification, dose–response assessment, and exposure assessment. These steps are followed by evaluation of the information and risk characterization, an estimate of the probability of

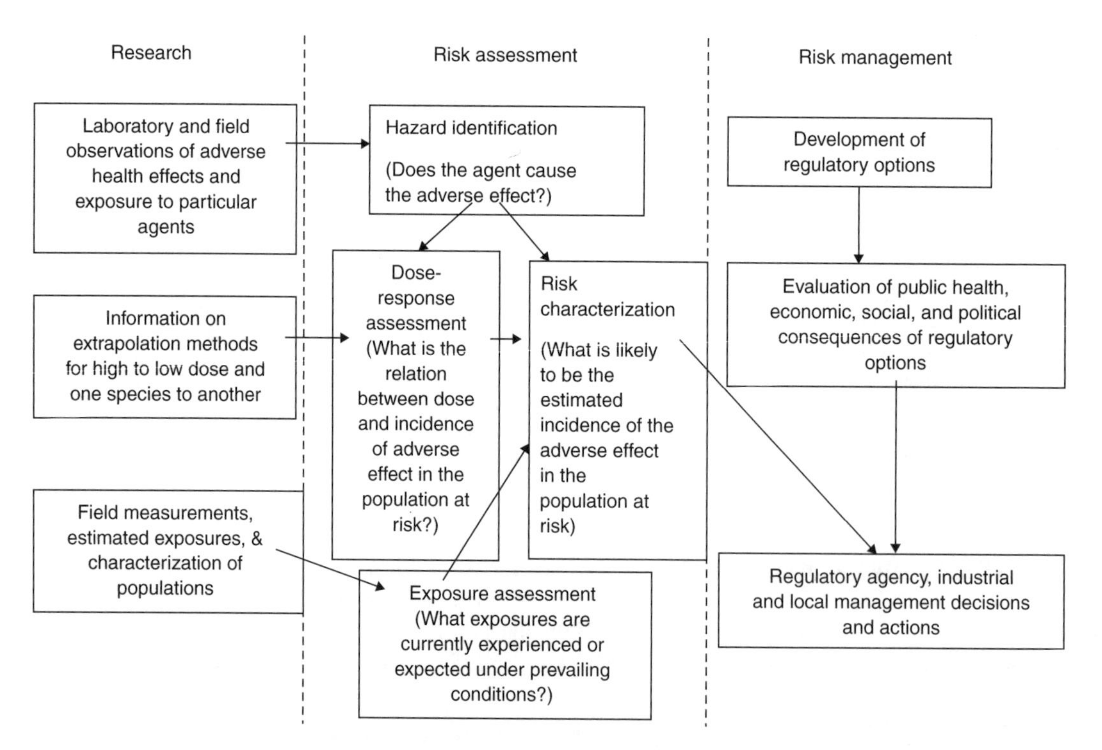

Figure 9.1.1. Risk assessment in relation to research and risk management, based on ref. [5].

harm to any individual in the population at risk. This is a developing process and is dependent on continuing research.

Here we shall consider risk assessment for populations of species in the natural environment. Essentially the same approach has been adopted as for human populations. However, there is a fundamental difference between consideration of populations of a single species, such as the human species, and consideration of populations of various species belonging to different Genera that may have become adapted to a wide range of environmental conditions, including physicochemical extremes. In this review, I shall consider how the complexity of the problems involved in risk assessment for the natural environment have been tackled in the past and, from this, how approaches may develop in the future to allow for the effects of both chemical and biological speciation.

2 NATURAL BACKGROUND CONCENTRATIONS OF BIOAVAILABLE ELEMENTAL SPECIES

It is important to note that in the lithosphere many elements are present in large amounts but are not bioavailable until disturbed by nature or by human activities. Naturally occurring substances in the lithosphere are brought into the biological environment by geological processes such as volcanic action, gradual chemical change (aging) and weathering [6]. The great variability in the origins of surface material is reflected in the variability of natural background levels from place to place (on both a macro and a micro scale). From their points of origin, natural substances may be dispersed by weathering followed by transport in water (including soil water) or through the atmosphere as particulates, vapours and gases. In areas without recent volcanic action, terrestrial soils are formed from local superficial sources that have been exposed. Weathering and aging may break down rocks to produce soil in various ways, including freezing and thawing, oxidation, carbonic acid solution, and aqueous erosion. Water-soluble substances in soil may be washed away in rain water, or by rivers, and enter lakes and their sediments, becoming bioavailable immediately or subsequently following further physicochemical changes. Ultimately, some dissolved substances may enter marine sediments. Throughout these geochemical processes changes occur in chemical speciation and hence in bioavailability.

Human activities, such as mining, smelting and metal refining, the use of natural products, the combustion of fossil fuels, and the release of sewage and agricultural wastes, may lead to increased bioavailability of naturally occurring substances, for example by acidification of water. Bioavailability of substances in the environment depends on the intrinsic properties of the substances, the type of medium in which they occur, the natural or man-made variations in the properties of that medium, and the biochemistry of the organisms at risk. Changes in the bioavailability of any substance will alter the relationship between exposure and effect for that substance. Thus, application of uniform environmental exposure standards over a large and variable environmental area with differing ecosystems, though common and administratively convenient, cannot be scientifically justified.

All elements may exist in a wide variety of physical and chemical forms [7]. Some of these forms will be ionic, but many will be organic and inorganic complexes, and many will be particulate. There is a common belief that, for metallic elements, free metal ions in aqueous solution are the elemental species that form the bioavailable fraction. However, a brief consideration of our knowledge of bioavailability indicates that this may be true only for the few metallic elements which dissolve very readily in water to provide free ions under normal environmental conditions (for a detailed consideration of this, see ref. 3). Copper, discussed below, is one of these elements. Physicochemical conditions in environmental media, such as hardness, acidity (pH), salinity, redox potential and the presence of other substances, determine the chemical species [8] and hence the bioavailability of naturally occurring substances. Figure 9.1.2 illustrates how the oxidation state of various metals varies with pH as a result of hydrolysis [8].

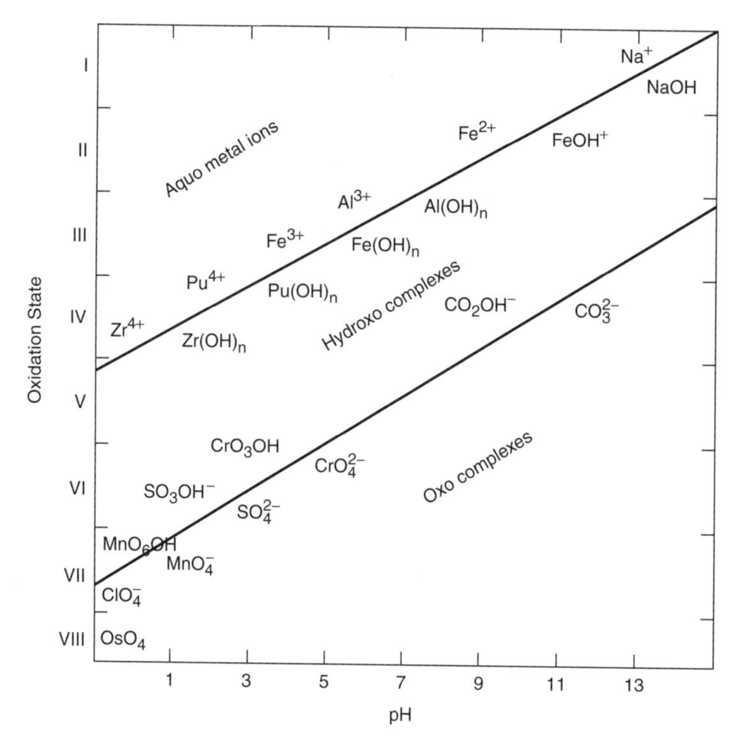

Figure 9.1.2. Chemical speciation of some metals in water showing how the distribution of the elements between aquo metal ions, hydroxo, hydroxo-oxo, and oxo complexes varies with pH.

The importance of such considerations has been recognized by the United States Environmental Protection Agency in its regulations [9] and this will be discussed here further in relation to the biotic ligand model for risk assessment.

In addition to the effects of environmental chemistry, metabolic processes may also influence the bioavailability of substances in the natural environment by causing changes in chemical speciation. For example, the pH in aquatic systems may rise as a result of primary production (growth of algae and plants). Plants can lower the pH in the adjacent soil solution by exudation of protons. Plants may also release organic acids and excrete metal ion complexing agents [10]. If the complexed metal ions are taken up by plant roots and transported in this form to the leaves, repeated leaf fall will result in enhanced accumulation of the complexed ions in the top layer of the soil. Since the complexed ions in this circumstance are bioavailable to the plants, the possibility of such an accumulation in the soil surface layers should be given careful consideration in any risk assessment. It may lead eventually to toxicity for any animals that eat the leaves of such plants.

3 ENVIRONMENTAL RISK ASSESSMENT AND ECOLOGY

The aim of an environmental risk assessment is to estimate an acceptable environmental concentration at and below which there will be no harm to any threatened ecosystem and to people who may be exposed through their environment. It is normally assumed that protection of the most sensitive tested species in the environment at risk will ensure protection of most (at least 95 %) of the species present. In the commonly used methods of risk assessment, an ecosystem is considered to be simply a collection of species covering several taxonomic groups and different trophic levels.

Functional aspects of the ecosystem, such as ecological processes and species interactions are usually ignored. Therefore, the effects of substances on an ecosystem are estimated from the sensitivity of selected species rather than from that of the ecosystem as a whole. Although this approach may take into account biodiversity at the species level, it does not include consideration of ecological diversity and the functional consequences of species and community interactions. Each of the communities in a given environment will be adapted to naturally occurring background concentrations of the bioavailable chemical species of essential elements. Community adaptation may involve biological species interactions which cannot be observed in the single species toxicity tests usually applied.

Because an environment at risk maintains an ecosystem characterized by a specific group of species, toxicity data used in environmental risk assessment should reflect the variation in chemical sensitivity within this particular group. An approximate estimate of the variation in sensitivity is normally obtained by using toxicity data for a selection of species, tested independently, covering a range of taxonomic groups and trophic levels. However, within any species there is a wide range of sensitivity to a potential toxicant and test systems themselves have intrinsic variability. Variation in dose and exposure time, variation in examined effect parameters, and variation in culture and test media contribute to the overall variation in toxicity data. Following the OECD Guidelines for the Testing of Chemicals and for Good Laboratory Practice [11] helps to make the data more reliable but they may still lack consistency from one laboratory to another.

Variation in toxicity data due to within-species differences may result from different life-stages being used in tests, and from clonal variation. Variation related to exposure may be caused by the way exposure concentration is defined, the exposure time, and the way in which a potentially toxic substance is applied. The exposure concentration may be given as a nominal concentration or as an actual concentration. If the actual measured concentration is given, it may simply be the initial concentration or it may be the arithmetic or geometric mean concentration over the period of the test. Further, it may refer to dissolved or to total concentration. The type of salt used, or the use of solvents, may affect the exposure level and also the bioavailable fraction. Environmental factors, such as pH, temperature, redox potential, salinity, water hardness, and the nature of the organic fraction of a soil or sediment, may influence the bioavailability of compounds in soils, sediments, water and air. These environmental factors may also affect the sensitivity of test organisms. In this respect, it must be mentioned that the Organization for Economic Cooperation and Development (OECD) recently recommended that metals and metal compounds should be tested in natural waters or in artificial waters that approximate to natural waters [12]. However, the current OECD Test Guidelines do not make it clear how this is to be achieved or what the key factors in the approximation should be.

Environmental factors during laboratory tests are usually held within predefined ranges. Unfortunately, the range may be quite large. The pH in aquatic tests, even those following OECD Guidelines, can vary by 1 pH unit (tenfold in terms of hydrogen ion concentration) or more. Similar variation is permitted in redox potential. Further, these variations have usually been unmonitored and so we know little of their influence on most of the data currently reported in the literature. Indeed, it is likely that most data in the literature are seriously compromised because of this.

In natural ecosystems, environmental factors vary from place to place and with time, through the day, through the seasons and through the years. There is a need for more sophisticated laboratory studies to elucidate the effects of environmental variations and the derivation of valid models for valid extrapolations from laboratory data to field predictions. As it is, to allow for the weaknesses in laboratory data and the differences between laboratory and field conditions, in deriving predicted no effect concentrations (PNECs – see later) we are forced to apply uncertainty factors which result in values that, since they deliberately err on the side of safety, are probably very far below the true no effect concentrations.

In testing substances essential to life, the concentrations used in the maintenance culture medium may alter the sensitivity of the organisms to be used. Organisms can become conditioned (adapted) to prevailing concentrations of the bioavailable chemical species of the essential element to be tested and thus may be more or less sensitive in the laboratory than in the field. The test organisms may also be influenced by the concentration of bioavailable chemical species of other elements that interact with uptake and elimination mechanisms for the test substance [3]. To allow for environmental factors that influence the toxicity of elements, empirically derived relationships describing this influence in a quantitative way may be applied to correct estimated toxicity levels. For example, the USEPA has applied a correction for water hardness in those cases where water hardness in an area to be assessed differs from that of the water used in toxicity tests [13].

4 DETERMINATION OF THE PREDICTED NO EFFECT CONCENTRATION (PNEC) FROM DOSE–RESPONSE DATA

4.1 The uncertainty factor approach

Once a substance has been identified as a potential hazard, the next step traditionally has been to establish a PNEC from toxicological dose–response data by the application of uncertainty (assessment, modifying) factors appropriate to the uncertainty of the data (see next paragraph). An early example of this approach is to be found in the original USEPA assessment scheme [14] and a similar scheme can be found in the EC Technical Guidance Document on Risk Assessment [15]. In the USEPA approach, an uncertainty factor is applied to an EC50, the concentration producing a defined effect in 50 % of a test population, or the no observed effect concentration (NOEC) value from short-term (acute) laboratory tests. Ideally long-term (chronic) tests would be better but in practice they are difficult to carry out in a repeatable way. The uncertainty factor is chosen to take account of the differences between laboratory and field populations, the differences between acute and chronic exposures, statistical uncertainties, and inter- and intraspecies variation. The uncertainty factor approach was given international approval in 1992 by the OECD workshop on the extrapolation of laboratory aquatic toxicity data to the natural environment [16]. Table 9.1.1 shows the OECD scheme for the calculation of a PNEC for aquatic exposure calculated from data in the form of acute or chronic EC50s.

The participants in the OECD meeting suggested that the following qualifications should be applied to the table:

(1) Most weight should be given to endpoints with direct ecological relevance (survival, growth, reproduction).
(2) The lowest chronic NOEC should be divided by a factor of 10 to extrapolate to the field situation.
(3) An acute effect concentration should be divided by a factor of 10 to obtain a chronic effect concentration.
(4) An L(E)C50 for a species from one of the three taxa used may be divided by a factor of 10 to extrapolate effects to the all of them.

Table 9.1.1. The OECD uncertainty factor scheme, used for determining an extrapolation factor in calculating a PNEC for aquatic exposure [14].

Available information	Assessment factor
Lowest acute LC50, EC50, or QSAR estimate within a data set on one or two aquatic species	1000
Lowest acute LC50, EC50, or QSAR estimate within a data set comprising at minimum algae, crustaceans and fish	100
Lowest chronic LC50, EC50, or QSAR estimate within a data set comprising at minimum algae, crustaceans and fish	10

(5) If multiple effect concentrations for a given species and a given endpoint are available, it was recommended that the geometric mean of the available effect concentrations be used in any extrapolation.
(6) It is permissible to use freshwater data in those cases where marine data are not available, subject to expert consideration.
(7) A list of substances should be drawn up for which different uncertainty factors may be required.

It must again be emphasized that the precautionary nature of the uncertainty factor approach inevitably leads to the calculation of a PNEC value which is much lower than the true value. Partly, this is because the bioavailability of the substance tested is usually much higher in the laboratory than in the field [17]. In fact, it is generally assumed that there is 100 % bioavailability in the test medium. Anything like 100 % bioavailability is most unlikely in natural media. This problem will be discussed in more detail below in relation to cupric ions. There are other sources of error in laboratory methods for testing aquatic toxicity that may contribute to error in the PNEC value. In particular, the exposure concentration is rarely defined properly. In a static tank system the concentration falls steadily and the 'true' exposure concentration must always be lower than the initial concentration. This error will tend to give a PNEC which is higher than the true value. In a water replacement (renewal) system, where the test solutions and control water are renewed periodically, the water concentration will fall and rise periodically. It is difficult to say what the 'true' exposure concentration is in these circumstances and the stress factor imposed by the fluctuating exposure may enhance the toxicity. Only in a flowthrough system can the exposure concentration remain constant, but these systems are difficult to maintain and lead to considerable problems of effluent disposal.

While it is probable that the current aquatic toxicity tests tend to overestimate toxicity of naturally occurring elements other than carbon because of the unnaturally high bioavailability of the chemical species used in the tests, there are also factors that may lead to an underestimate of toxicity, for example, the use of static test systems (see the preceding paragraph). Also, since only a few species are used to produce laboratory toxicity data, they are unlikely to include the most susceptible species in the field. On the other hand, the strains tested may be particularly susceptible to certain naturally occurring elemental species from which they have been removed by many generations of laboratory culture.

Most importantly, laboratory toxicity tests relate poorly to the natural environment because they are usually carried out with a single substance. In the natural environment multisubstance exposure is the norm. Thus, synergies and additive or antagonistic effects cannot be observed or deduced from most laboratory test data. From theoretical considerations, such interactions should occur in the field but it is unclear to what extent they occur in practice.

4.2 Statistical extrapolation methods

Probabilistic approaches to the extrapolation of laboratory toxicity data have been developed as an addition or alternative to the uncertainty factor approach. Kooijman (1987) [18] introduced a method that calculates the hazardous concentration (HC) as the lower bound of concentrations that may be harmful to a given community. It is based upon the calculation of an assessment factor (T) that is applied to the geometric mean of the available L(E)C50 values (the LC50 is the median concentration lethal to 50 % of a test population). The assessment factor is derived by fitting a log-logistic distribution to the available toxicity data (as LC50s) [19] and is chosen to protect a predefined fraction of the community. The derivation is based on the assumption that the LC50 values for both the test species and for the community species can be treated as independent random trials from the log-logistic distribution. Kooijman introduced a factor, based on the number of species for which toxicity data is available, that allows for the uncertainty in the mean and the variance of the LC50. This assessment factor decreases (down to zero) as the number of species for which toxicity data is available increases.

Van Straalen and Denneman (1989) [20] adapted the model of Kooijman (1987). They questioned the assumption that, in order to protect an ecosystem, it is necessary to protect the most sensitive species. They argued that small effects on numbers of sensitive species may be tolerable because of the resilience and the regulatory capacity of ecosystems. They proposed that a NOEC should be the base for an ecological risk assessment, and that test organisms should be selected on the basis of how well they represented the community to be protected. They suggested three criteria for selection of organisms:

(1) Ecological function: the set chosen should include primary producers (the base of all food webs), consumers (the mediators of the food web) and saprotrophs (which feed on dead organisms and recycle their components).
(2) Anatomical design: the set chosen should include species from different taxonomic groups.
(3) Exposure routes: the set should include species exposed to substances in different ways.

Van Straalen and Denneman also suggested that NOEC toxicity data should be normalized when test media that are qualitatively different are used. They further suggested that a laboratory-to-field extrapolation factor should be estimated to allow for modification of toxicity under field conditions. The model of Van Straalen and Denneman estimates the HCp, the hazardous concentration for p % of the species in a community, by applying an assessment factor (T) to the geometric mean NOEC. The uncertainty factor in this model is comparable with the uncertainty factor estimated by Kooijman and is also derived by fitting a log-logistic distribution to the available toxicity data. However, it must be emphasized that the data used by van Straalen and Denneman are NOECs and not the L(E)C50s used by Kooijman. The use of the NOEC raises the key question of whether the effect studied is truly an adverse effect or not and may tend to define an HCp which is much lower than the true threshold concentration for harm to the ecosystem at risk.

Aldenberg and Slob [21] also used the method of Kooijman, as adapted by Van Straalen and Denneman as a basis for the calculation of a toxicological protection level. Their protection level is called L, the left confidence limit (a statistical term corresponding with a concentration at which a significant part of the species, more than 95 %, will be protected). Aldenberg and Slob introduced e_{m} a confidence factor that can be derived using Monte Carlo analysis (Bayesian statistics).

When the number of species is large, there are no differences between the method of van Straalen and Denneman (1989) and the modification of Aldenberg and Slob (1991). When fewer toxicity data are available, the extrapolation factor used by Aldenberg and Slob is much higher than the extrapolation factor calculated by Kooijman or by van Straalen and Denneman.

The statistical methodologies described do not take into account the natural background levels of potential toxicants. In fact, the use of the log-logistic distribution carries with it the assumption of the possible occurrence of infinitely low concentrations. Recently, increasing emphasis has been given to consideration of the naturally occurring levels of substances such as cupric and zinc ions, especially because the statistical extrapolation methods described above can produce hazardous concentration values that are lower than the background environmental levels in which the species tested normally live.

4.3 Homeostasis (adaptation) approach

The methodologies described above were originally developed for, and applied to, the assessment of the environmental risk associated with xenobiotic organic substances, for which living organisms are unlikely to have homeostatic regulatory mechanisms. Most organisms well established in a given habitat have been selected for adaptation to the naturally occurring concentration range of substances in the environmental media. They can usually regulate their uptake of substances from natural background levels in such a way that their internal concentration is kept relatively stable (homeostasis). This implies that organisms tested for risk assessment of naturally occurring

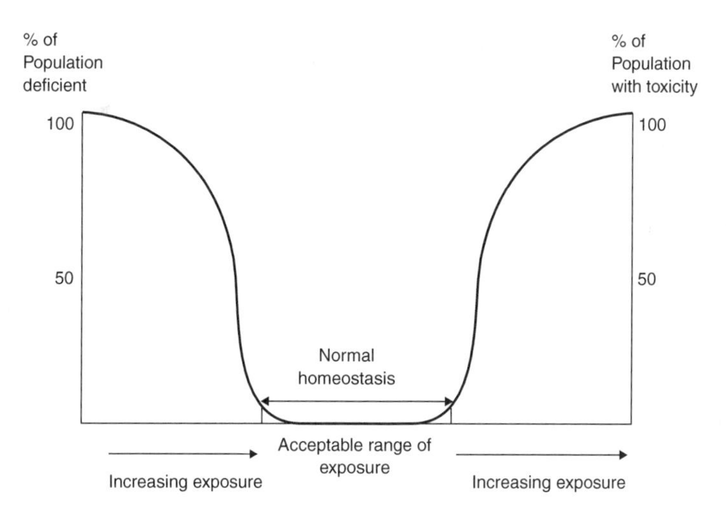

Figure 9.1.3. Exposure (dose)–response curve for an essential nutrient, showing a homeostatic optimal range of exposure which the exposed population can tolerate.

substances should originate from an environment similar to that to be assessed and that they should be cultivated and tested under the conditions of chemical exposure appropriate to that environment. These points were taken up by Van Assche *et al.* (1996) [22] in reconsidering the above approaches so that they could be applied to naturally occurring chemical species of essential elements. Their aim was to derive an environmental threshold level for toxicity, without going below the deficiency threshold, taking into account the optimal concentration for these substances (Figure 9.1.3). The area of overlap of the optimum concentration ranges for various species was defined as the no risk area (NRA), assuming that, within its boundaries, none of the biological species will be subject to toxicity or deficiency stress. The NRA has two boundaries (deficiency and toxicity) that are both determined by the species with the most limited homeostasis regulation capacity with respect to either toxicity or deficiency. The authors argue that the NRA is linked to a naturally occurring concentration range of bioavailable species of essential elements, and should therefore be derived from toxicity data relevant to the area under study. For this reason they introduced the consideration of habitat type to define geochemical (or metallo-) regions with similar concentrations of the bioavailable species of essential elements, especially metallic elements. Within such geochemical (or metallo-) regions, all biota may be assumed to be conditioned to comparable bioavailable background concentration levels.

The method of Van Assche *et al.* takes the lowest available NOEC of a set of more than ten species from different trophic levels (including at least one autotrophic organism) as the PNEC value that should be used in environmental risk assessment. No uncertainty factor is used as it is assumed that major uncertainties have been allowed for in the NRA approach, since:

(1) Intraspecies variability is covered by the species homeostatic control of essential elements (adaptation).
(2) Interspecies variability is covered by the NRA – the homeostatic (adaptation) overlap between species.
(3) Unlike the PNEC, the NRA is based on chronic toxicity data, eliminating any need for acute to chronic extrapolation.
(4) If the following criteria for relevance of data are applied, the natural essential bioavailable chemical species concentration range will be covered by the same homeostatic mechanisms in the laboratory as in the field, making extrapolation unnecessary.

(a) The organisms used should originate from the area (or a comparable area with respect to the natural background concentration) for which the assessment is being made.
(b) The organisms used should be cultured and tested in the laboratory in physicochemical conditions comparable to their natural environment.
(c) Testing should cover the entire life cycles of the organisms chosen.

Van Assche *et al.* in taking the NOEC as the basis of determination of the PNEC are assuming that any effect observed which is not clearly beneficial must be adverse. This implies that changes resulting from the process of homeostatic regulation itself may be considered to be adverse effects. Clearly the induction of regulatory mechanisms is a dose-related effect of a potential toxicant and it may result in a reduction of energy available for other processes such as growth and reproduction but, if we are looking at true homeostasis, the total effect should not alter species survival significantly.

4.4 Comparison of the applicability of different risk assessment procedures to naturally occurring chemical species of trace elements

4.4.1 Incorporation of background environmental concentration and species adaptation

It is clear that there is no one optimal environmental concentration of any elemental species that suits all biological species. Under any given set of environmental conditions, some biological species or populations of those species will have to cope with a deficiency of essential substances whilst others may suffer toxicity from an excess. All species must cope with natural fluctuations in the bioavailability of substances on which they depend. In other words, the natural background chemistry of any given region determines which species are present and when.

The natural background chemistry may vary seasonally. For example, increasing temperatures in summer may simultaneously increase concentrations of potentially toxic substances in water while reducing the ambient concentration of oxygen. Such natural variations may lead to periods of stress for particular species. This must be taken into account in any risk assessment. If changes in concentrations are within a consistent limited range, the organism(s) present may experience stress for only a short time. However, care must be taken not to impose additional stress by effluent discharge at these times. Advice to this effect should be part of any risk assessment. Unfortunately, current test regimes do not give any useful information to assist in giving such advice.

Ideally, risk assessment methods should incorporate the natural background concentration range as an explicit factor in the model applied and use test data for appropriate species cultured at naturally relevant background concentrations of the substances of concern. Only the homeostasis and statistical extrapolation approaches comply with this criterion. However, even then, the statistical extrapolation approach may still give a PNEC that is lower than the background concentration.

4.4.2 Incorporation of bioavailability considerations

The bioavailable amount of any substance is almost always less than the total measured amount in the environment. Oxidation state, hydrolysis, binding to dissolved compounds, adsorption onto sediment surfaces, and inclusion in organic films surrounding particles or at a water surface, amongst other things, may make substances unavailable. Site-specific variations in bioavailability occur as a result of physicochemical and geochemical differences in natural environments. Thus, risk assessment procedures must allow not only for differing bioavailability of different chemical species of elements, but must also take into account how bioavailability varies from place to place and time to time. Bioavailability can be accounted for by the selection of appropriate toxicity data in which the natural bioavailability is sufficiently represented, or by the application of uncertainty factors to adjust the experimental test conditions to compare with those of the natural situation.

The concentration limitation method of Aldenberg and Slob and the NRA approach of Van Assche *et al.* acknowledge the importance and influence of bioavailability on the calculation of an environmental protection level. In the NRA approach, the importance of acknowledging bioavailability in toxicity tests is emphasized. In the concentration limitation method, the calculated maximum permissible concentration can be adjusted according to the bioavailable fraction of the substance in the area under study.

Another aspect of bioavailability, which is dependent on chemical speciation and which is often overlooked in risk assessment for substances based on elements other than carbon, is partition between media. For example, substances tightly bound to soil particles may be not be bioavailable and become bioavailable only when they enter the soil water or dissolve in gastric juices or inside phagosomes. In each case, the degree of bioavailability will be different although the original particles may be identical. If there is vaporization from solution, the substance becomes more available through the air and less so from the aqueous solution. There are distribution models which can describe such environmental movement quite well for carbon compounds but they cannot be applied to most inorganic chemical species.

4.4.3 *Incorporation of interspecies variation*

All current risk assessment methods make some allowance for interspecies variation. In the assessment factor approach, an additional uncertainty factor is used when the toxicity data are not thought to represent adequately the ecosystem (without a minimum set of data from at least one species of algae, one crustacean species, and one fish species). Interspecies variation is further covered by using the lowest available toxicity value as the baseline for deriving a PNEC since protecting the most sensitive species should guarantee protection of the least sensitive species. With methods based on statistical extrapolation (probabilistic approaches), the interspecies variation is represented by an estimated log-logistic frequency distribution. On the basis of available toxicity data (the geometric mean value of the most sensitive effect parameter) a frequency distribution is fitted which is assumed to represent the complete ecosystem.

The NRA approach is based on the optimum concentration curves for at least ten species. The interspecies variation determines the size and location of the NRA. The environmental protection level is set to the lowest available NOEC value. This follows the principle of aiming to protect all species by protection of the most sensitive species and selecting an effect which is sensitive but not necessarily adverse.

4.4.4 *Incorporation of laboratory–field extrapolation*

The toxicity of substances and the sensitivity of biota in laboratory toxicity tests may differ from those in natural environments owing to differences in the species tested, different bioavailability of the substances tested, and differences in environmental factors. The uncertainty in the size of effects in the field compared with those found in the laboratory is usually allowed for by using uncertainty factors. The laboratory–field extrapolation factors are intended to prevent underestimation of effects in the field. Toxicity data derived from field or semi-field (mesocosm) data may be used in risk assessment procedures to improve the realism of test data for natural environments. It should also be possible to use data from laboratory tests that take into account and simulate natural environmental conditions.

The need for extrapolation from laboratory data to field data in environmental risk assessment is acknowledged in all the current methods of risk assessment apart from that of Kooijman and the NRA method. In most cases, an uncertainty factor is used to make allowance for the differences in environmental conditions between the laboratory and the field. The use of field data instead of laboratory data would be preferable to the use of an uncertainty factor. In the NRA method, it is suggested that there is no need to use a laboratory–field uncertainty factor when the toxicity data are selected with care and give a good representation of the natural situation. Although this will decrease the uncertainty in the estimation, one cannot be certain that it makes sufficient allowance for differences in the conditions in the laboratory (which are kept as constant as possible) and in the field where environmental conditions fluctuate.

4.4.5 Incorporation of species representativeness

The term 'species representativeness' may have two different meanings. The first, the most commonly used, relates to the consideration of an adequate range of different taxonomic groups and trophic levels within the food webs at risk. Most current risk assessment methods use this definition. In the assessment factor approach, an additional safety factor is used in those cases where the set of available species is not representative of the ecosystem according to this concept. Although the method of Kooijman includes an uncertainty factor that will decrease when data on more species are used, Kooijman does not mention the need for toxicity data from a representative sample of species. Van Straalen *et al.* explicitly define species representativeness, according to the above definition, in their data requirements. Since the method of Aldenberg and Slob is based on the method of van Straalen *et al.* and the concentration limitation method is based upon the method of Aldenberg and Slob, both account for species representativeness in the same way. In the NRA approach, species representativeness according to the same definition is part of the data requirements.

The second definition of species representativeness relates to the appropriateness of the organisms tested to the area under consideration. This consideration is particularly relevant to substances, such as the chemical species of trace elements, for which variable natural background concentrations occur. Organisms from the environment being assessed may be selectively adapted to the background concentrations of naturally occurring substances and may be more or less sensitive than the arbitrarily chosen species commonly used in toxicity tests.

Apart from the NRA approach and the EU Technical Guidance Document, where it is explicitly stated in the data requirements, none of the methods require toxicity data derived from species endemic to the areas to which the risk assessment applies.

4.4.6 Incorporation of possible deficiency of essential substances

Enforcement of environmental standards based solely on toxicity may result in ambient concentrations which are too low for ecosystem health if the potential toxicants are also essential nutrients at low concentrations. Thus, appropriate risk assessment models must take into account the possible effects of deficiency of these substances. The models must also allow for the natural bioavailable background concentration to which organisms at risk have become adapted. Realistic models for risk assessment for essential substances in the environment should never result in the derivation of environmental standards that are below the natural background levels.

Of the methods of risk assessment described above, only the NRA approach complies with the deficiency toxicity criterion. This method was specifically designed to calculate maximum permissible concentrations for essential substances. The NRA approach is based on the dose–response curves for essential substances that, at low concentrations, include harmful effects caused by deficiency. The method of calculation uses the lowest available NOEC value to calculate the PNEC. No assessment factor is applied to this value since it is assumed that the lowest available NOEC defines the nonstress threshold of the most sensitive endemic species and thus protects the whole endemic community.

5 DETERMINATION OF A PREDICTED ENVIRONMENTAL CONCENTRATION (PEC)

The predicted or estimated environmental concentration (PEC or EEC) of the substance of concern is determined from its known chemical properties and the relevant characteristics of the receiving environment (discharge patterns, physicochemical properties, possible metabolism, particularly by microorganisms, and other potential environmental transformations) using worst-case assumptions. This often involves modelling the partitioning of the substance in the environment. Such modelling works fairly well for organic compounds with a limited potential for ionization but has so far not been successful for highly ionized organic compounds or for ionic inorganic compounds.

Once the PEC is known, the risk is expressed as the ratio of PEC or EEC to PNEC. If the resultant risk index is well below 1, the risk may be regarded as negligible. If the risk index is above 1, the situation must be further evaluated to determine whether the worst-case assumptions included in the calculations of both PNEC and PEC are realistic in the particular situation for which the risk assessment has been required. In principle, to err on the side of safety, the threshold concentration for the appearance of an adverse effect in the most sensitive stage of the most sensitive species should be determined and divided by an uncertainty factor of 10 to establish the PNEC. In practice, the information available is always deficient in various ways and greater uncertainty factors, or other precautionary approaches, must be applied to data that may only be available in terms of EC50 or LC50 values. In the majority of cases, because of the use of worst case assumptions in deriving the PNEC, a risk index of less than 10 probably does not indicate any significant level of risk.

It must be emphasized here that the PNEC calculation is based on the results of toxicity tests that were originally devised for organic pollutants. Factors affecting chemical speciation of inorganic substances are not always adequately controlled. Being based on assumptions appropriate to organic chemicals, aquatic toxicity tests usually fail to take into account the changes in chemical speciation and hence in bioavailability of trace elements in waters that are chemically different from those used in the test systems. Most concentrations of metallic elements in published tests have been calculated and not measured, chemical speciation has been ignored, and it is often assumed that all of the element in whatever form is bioavailable as a simple ion. For example, the pH in many test systems, including those recommended by the OECD [11], may change by 1 pH unit or more during a test and so chemical speciation may also change considerably in the course of a single test [8]. Thus, any extrapolation of values obtained in such tests to the natural aquatic environment for risk assessment must be very cautious. Other important factors affecting speciation, such as redox potential, are also often poorly controlled in most current test systems, adding further doubts about their validity as a basis for risk assessment. Such tests give us little clue as to the exact nature of the chemical species producing the effects observed.

In practice, nearly all environmental risk assessments have been related to single substances, assuming no interactions with other substances that may be present. The reality is that there are multiple complex exposures in the natural environment, and this is particularly the case for the bioavailable chemical species of trace elements.

It must also be remembered that the bioavailable fraction of any substance is determined not only by the physicochemical characteristics of the medium in which it occurs, but also varies from species to species of organism depending on the relevant uptake route and available binding sites on or in the organisms at risk [3]. The bioavailable fraction of the substance of concern is always what matters biologically and all studies should be related to this. This is one of the strongest arguments in favour of the biotic ligand model, presented below.

6 PROBLEMS WITH CURRENT RISK ASSESSMENT PROCEDURES

Some of the weaknesses in current risk assessment procedures have already been discussed above. In particular, they are absolutely dependent upon toxicity data from a very limited range of organisms, with little relevance to the environment to be assessed, tested in laboratory media that may be very different physicochemically from the natural media in the environment at risk.

The most commonly applied methods for assessment of risk (Section 2.1 above) involve the assumption, often unstated, that zero risk of harm applies to situations of zero exposure. It is further assumed that exposure and bioavailability are the same thing. In fact, exposure and bioavailability are not the same thing. For example, ionic aluminium in water at pH 7.0 is largely not bioavailable because it is in the form of unionized hydroxides. Exposure may be high under these conditions but bioavailability and toxicity are very low. On the other hand, at pHs below 6, the aluminium is bioavailable as free trivalent ions and an exposure concentration that is harmless at pH 7.0 can poison fish at pH 5.5 [8]. Thus, the same exposure in terms of total aluminium at different pHs is associated with different levels of bioavailability and potential for harm. Schemes of risk assessment must take this into account.

Another difficulty arises when considering substances for which there is a range of natural background levels. Are there background levels that pose a risk? It has been argued that since life has evolved and adapted to the background levels of naturally occurring substances, zero risk of harm must apply to natural background levels. This may well be true for the organisms that have evolved over millions of years in areas of high background level, but is unlikely to be true for organisms adapted to lower background levels which have migrated to these areas fairly recently. For an extreme example, consider the physiological problems faced by a freshwater fish suddenly placed in the sea. Thus, for naturally occurring substances, risk assessment should be related to the populations of organisms likely to be at risk and not simply to individual species which may well have both sensitive and resistant strains. Risk assessment should also take into account possible physicochemical and biological changes in the environment to be protected which might affect chemical speciation over time and thus alter the bioavailability of potentially harmful substances.

Many naturally occurring substances are essential for life. There will be a measurable risk of environmental harm to be associated with levels of exposure to these substances that are too low to preserve the health of key organisms. Amongst these essential substances are the so-called *essential elements*, a list of which might include such elements as Cu, Co, Mn, Ni, Se and Zn. The process of risk assessment for essential elements is complicated further by another assumption, never clearly stated, that has confounded the interpretation of most previous research into their biological function. This assumption is that elements other than carbon can be considered as giving generic properties to their derivatives. With this assumption, derivatives of trace elements have been classified identically for toxicological (and hence risk assessment) purposes. The faults in this assumption are most obvious in relation to the metallic elements. For metallic elements, the assumption has developed into the idea that the bioavailable and hence the biologically active form (chemical species) of any metal is the ionic form. In general, it is further assumed that all of any metal in an environmental sample is either in the ionic form or can readily be transformed entirely to this form. Only with this assumption can you justify the current regulatory regimes, which generally rely on monitoring and control of total elemental content and rarely refer to defined chemical species. In any natural environment, there is always a range of chemical species of any element and not all of these are readily bioavailable. The range of bioavailable species is a function of the ecosystem in which they occur and reflects both its physicochemical and biological properties. As with the biological species in the ecosystem, the range of bioavailable chemical species is not constant but changes with changing environmental conditions.

The ability of living organisms to control the internal concentrations of naturally occurring substances in a homeostatic manner is obviously a key factor in determining the relationship between toxic effects and the level of exposure. In many cases, mechanisms have evolved to prevent the accumulation of a harmful internal dose, for example by regulated excretion or by sequestration in vacuoles. In other cases, where naturally occurring levels of an essential substance are low in the natural environment, living organisms may have developed effective mechanisms of bioaccumulation and even a small increase in concentration of such substances in the environment of an organism may be potentially harmful [23]. Thus, for any threatened environment, it is essential to determine, at least for the key organisms, bioavailable exposure – response curves for both beneficial and toxic effects in order to define the exposure range and the corresponding range of tolerance which will permit the organisms to survive with optimal health. This, of course, is the objective of the NRA approach described above (Figure 9.1.3).

7 COPPER – THE WAY AHEAD FOR RISK ASSESSMENT OF METALLIC ELEMENTS?

7.1 The USEPA water quality criteria for copper

In the USA, the appropriate PNECs for water, designed to protect aquatic life, are the Water Quality Criteria (WQC). The USEPA has established WQC for copper in 1985 [24], 1995 [25], 1996 [26] and 1998 [27] based on the uncertainty factor approach (see Section 4.1 above [28]) applied to laboratory toxicity test results obtained with a variety of organism types. The test duration is typically 48 or 96 h for acute toxicity tests, and longer periods of time for chronic toxicity tests. The effects monitored include survival (or equivalently, lethality/mortality), reproduction, and growth. The lethality results of the toxicity test are used to determine the LC50. The LC50 is often within about a factor of 2 of the maximum concentration associated with the same mortality as in the untreated control system and this observation is used in calculating the WQC (see below).

The LC50 results are used to evaluate the WQC using standardized data analysis and computational procedures as described in the '*Guidelines for Deriving Numerical Water Quality Criteria for Protection of Aquatic Organisms and Their Uses*' (the 'Guidelines' [28]). In general terms, the approach is as follows. First, a literature review is performed to identify all the relevant, valid toxicity test data that are available. Then the LC50 results are averaged by test species to determine the species mean acute values (SMAVs). From the SMAVs, genus mean acute values (GMAVs) are evaluated as the geometric means of the SMAVs within each genus. The results are plotted from the lowest (most sensitive) to highest (least sensitive) GMAVs. The horizontal axis (the x-axis) indicates the percentile associated with the corresponding GMAV, the y-axis value. The WQC is intended to protect 95 % of the genera, so the 5 percentile GMAV is used to set what is referred to as the final acute value (FAV). According to the Guidelines, the fifth percentile is evaluated from the four GMAVs that are closest to the 5 percentile level of the probability distribution of GMAVs. Since the FAV corresponds to an LC50, it is divided by a factor of 2 to estimate the highest concentration along the dose–response curve that corresponds to the absence of an effect, i.e., control level mortality, for the 5 percentile genus. The resulting concentration, referred to as the criterion maximum concentration, or CMC, is the acute water quality criterion. A similar procedure is followed to evaluate a chronic criterion if sufficient data exist. More typically, as in the case of copper, the chronic criterion is determined by dividing the FAV by the final acute-to-chronic ratio (FACR), a factor that is determined on the basis of available test data.

The WQCs for some metals, including copper, cadmium, nickel, lead and zinc, are formulated in a way that accounts for the observation that the toxicity of these metals is inversely related to water hardness. Thus, as hardness (calcium + magnesium) increases, it takes a higher metal concentration to achieve the same effect and as a result, the LC50 increases [24]. The reason that this effect occurs will be discussed in greater detail later. It is allowed for in the current WQC for copper by expressing the WQC [26] as an equation that is a function of water hardness. The current freshwater WQC for copper is commonly expressed as:

$$\text{Acute WQC} = \text{CMC} = \tfrac{1}{2}\text{FAV} = e^{0.9422(\ln H)-1.700}$$

where H = hardness.

The preceding equation is used to evaluate the magnitude of the WQC for copper. The WQCs for copper and for other constituents also specify an allowable duration and frequency of exceedance of the criteria. The acute WQC is specified as a 1 h average concentration that may be exceeded once every 3 years, on average. However, for practical purposes, it is sometimes applied as a daily average concentration. The allowable frequency of exceedances of once every 3 years, on average, is based on the judgement that most aquatic ecosystems can probably recover from most exceedances in about 3 years [28].

The chronic WQC is referred to as the criterion continuous concentration (CCC). The available data are insufficient to calculate a final chronic value (FCV) for copper in the same way that was used to evaluate the FAV. It is therefore evaluated as FCV = FAV/FACR, where the FACR = 2.823 [26]. When expressed as a function of hardness, the freshwater WQC is:

$$\text{Chronic WQC} = \text{CCC} = e^{0.8545(\ln H)-1.702}$$

The chronic WQC is taken to be a 4 day average concentration that is not to be exceeded more than once every 3 years, on average.

Freshwater WQCs are currently expressed in terms of total recoverable copper. Procedures have been published to convert the WQC to a dissolved basis (conversion factor = 0.96), and for converting total recoverable limits for copper to a dissolved concentration basis in the receiving water, such that they can be compared on a common basis [26]. The dissolved metal concentration measurement is considered to be more representative of the bioavailable fraction than the total recoverable measurement [29].

For salt water, the acute WQC is not a function of hardness. The final acute value is slightly less than the 5 percentile GMAV ($10.39\,\mu g\,L^{-1}$). This is because, in accordance with the procedures in the Guidelines, the FAV is reduced to the GMAV of the blue mussel, *Mytilus edulis*, in order to protect this commercially important species. The CMC is then 1/2 of the FAV expressed as dissolved copper. The saltwater addendum to the copper criteria document indicates that this is a 24 h average concentration that is not to be exceeded more than once every 3 years, on average [25]. Dividing the FAV by an FACR of 3.127, the CCC is obtained in terms of dissolved copper. As with the freshwater CCC, this is a 4 day average, not to be exceeded more than once every 3 years as well.

7.2 Limitations of the current water quality criteria for copper

USEPA freshwater WQCs are site specific to the extent that the measured hardness in site water is used in their evaluation. However, it was recognized in the 1984 WQC document for copper [24] that water quality characteristics other than hardness, including both pH and dissolved organic carbon (DOC), affect the toxicity of copper. Methods to allow for the effects of these aspects of water quality have been developed, including the water effect ratio (WER) procedure [9, 30]. The WER is a factor that is evaluated on the basis of side-by-side toxicity tests that are performed in laboratory water and in site water. It is applied to the national water quality criterion to make allowance for site-specific variability of water quality and hence of bioavailability and toxicity. Evaluation of a WER requires that toxicity test measurements (e.g. LC50s) be made at three different times during the year, using two types of organisms. The paired toxicity endpoints are determined using a sample of the receiving water and a laboratory water approved by EPA. The WER is evaluated from the ratio of the LC50 in the receiving water at the site (LC50 SITE) to the LC50 in the laboratory water (LC50 LAB):

$$\text{WER} = \text{LC50 SITE}/\text{LC50 LAB}$$

The site-specific WQC, WQCSITE, is evaluated by multiplying the national WQC by the WER to adjust the WQC to reflect local conditions:

$$\text{WQC SITE} = \text{WER} \times \text{WQC}$$

Although the WER methodology provides a rational basis for adjusting the WQC to site-specific conditions, the methods may be costly and time consuming to perform and can give results that are erratic and difficult to interpret and apply. Further, basing the WQC on bioassays performed with only three samples provides little information about how site water affects bioavailability over time, and it does not provide any clue as to why site-specific conditions are affecting bioavailability and toxicity.

It is clear that to set accurate environmental standards for elements other than carbon, such as copper, we need to understand how water chemistry affects their chemical speciation and bioavailability. Nearly 30 years ago, Pagenkopf and co-workers used a chemical equilibrium model in an attempt to explain how water chemistry affects the chemical speciation of copper and how the speciation is related to the toxicity of copper to fish [31]. They concluded that divalent ionic copper (Cu^{2+}) and copper hydroxide ($CuOH^{-}$) were both possible toxic species.

Sunda and Hansen [32], showed that the level of dissolved organic carbon (DOC) affects copper speciation. They estimated that, at pH = 7, the Cu:DOC complex accounts for most of the copper in solution, amounting to more than 98 % when the DOC exceeds $1\,mg\,L^{-1}$, a very low level for most natural waters. Ionic copper, the chemical species that Pagenkopf and coworkers [31] identified as being most directly related to toxicity, accounts for only about 1 % of the total copper at $1\,mg\,L^{-1}$ of DOC, and about 10 times less (0.1 % of total copper) at $5\,mg\,L^{-1}$ of DOC, a typical level of DOC in many natural waters. The rest of the copper present in solution exists in the form of the copper hydroxide and carbonato complexes, primarily $CuOH^{-}$ and $CuCO_3$, respectively. Bearing these facts in mind, attention is concentrated on free divalent copper ion activity since it appears to be the main toxic chemical species. The general distribution of copper among the various metal species is qualitatively much the same over a wide range of pH.

Pagenkopf *et al.* [31] proposed that the bioavailability and aquatic toxicity of metals, and of copper in particular, are related to ionic activity. Since then, many examples of the significance of ionic activity have been described. USEPA [33] and Campbell [34] describe observations, not only for copper [35–38] but also for cadmium [39], zinc [40]) and other metals. Figure 9.1.4 illustrates a well-known example, using data from Anderson and Morel [37] to illustrate how this applies to copper. Dose–response curves are shown for the acute toxicity of copper to a dinoflagellate (*Gonyaulax tamarensis*), with the absence of motility used as the indicator of death. Four sets of tests were run, two with ethylenediaminetetraacetic acid (EDTA) added (△), shown furthest to the right, and two without EDTA added (○), the

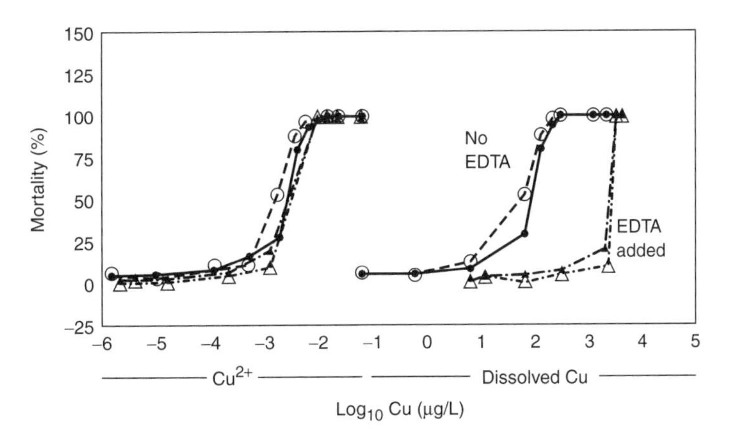

Figure 9.1.4. Acute toxicity of copper to the dinoflagellate *Gonyaulax tamarensis* in the presence and absence of EDTA, after ref. [37].

second pair of curves from the right. EDTA is a complexing agent that is analogous to DOC, in that it forms a strong complex with copper. This leads to there being a relatively low level of copper in the ionic form when EDTA is present. As shown, without EDTA added, the LC50 is somewhat less than 100 μg L^{-1}. A concentration of dissolved copper that is acutely toxic in the absence of EDTA (100 % mortality at 1000 μg L^{-1}) produces no effect in the presence of EDTA.

The set of four curves shown on the left in the figure represents these same data, but plotted versus free cupric ions. It can be seen that the free cupric ion levels are more than four orders of magnitude lower than the dissolved copper levels (i.e., free copper activity is less than 0.01 % of dissolved copper). More importantly, the four sets of data converge to a relatively narrow band of curves, indicating that copper toxicity is a unique function of free cupric ion activity. Thus, the experimental observations show clearly that total dissolved copper is not directly related to toxicity and is not the correct parameter to regulate in order to avoid aquatic toxicity.

The finding that cupric ion activity is directly related to so-called 'copper' toxicity was a significant advance in understanding how to assess the bioavailability and toxicity of copper. It seems to follow that, if the copper activity in a water sample can be predicted by use of a suitable chemical equilibrium computer program, the likelihood of toxic effects can be assessed simply on this basis. However, for some metals, the USA national WQC is expressed as a function of hardness, and as hardness increases, the water quality criterion increases proportionately. In other words, as hardness increases, toxicity is reduced and higher concentrations of dissolved metal are required to produce the same toxicity that is observed at lower hardness levels. This is *not* because the divalent ions Ca^{2+} and Mg^{2+} react with Cu^{2+} and affect the speciation of copper or its activity in solution.

Consider the dose–response results of Figure 9.1.5. The two curves on the right show test results for *C. dubia* exposed to copper in laboratory control water at hardness levels of 75 and 200 mg L^{-1} as $CaCO_3$ [41]. The low hardness results (○) yield an LC50 of 4.2 μg L^{-1} of dissolved copper, while the high hardness results (∗) yield an LC50 of 45 μg L^{-1}. When the results are converted to cupric ion activity calculated by a chemical equilibrium model, the dose response curves, shown on the left, do not converge to a single curve, as was the case in the preceding example with EDTA (Figure 9.1.4). Thus, cupric ion activity does not appear to be enough to define toxicity in the presence of varying levels of hardness.

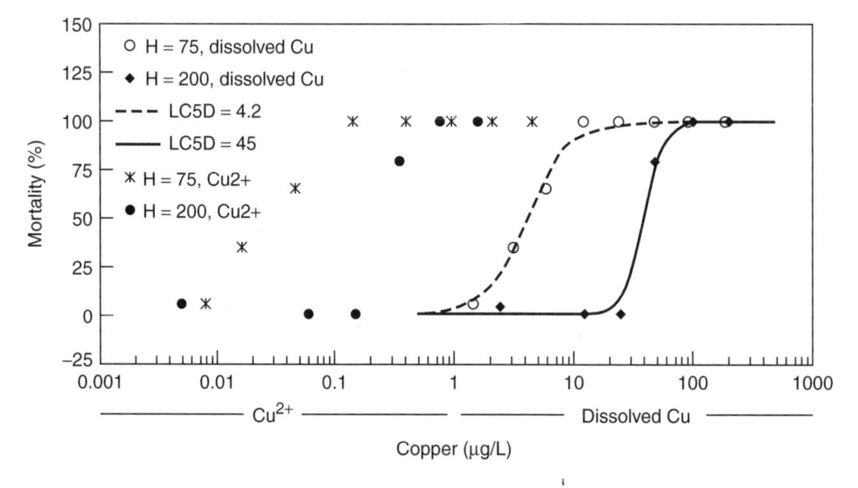

Figure 9.1.5. Acute toxicity of copper to the microcrustacean cladoceran *Ceriodaphnia dubia* in laboratory water at hardness = 75 and 200 mg L^{-1} as calcium carbonate, after ref. [41].

It has been shown that in addition to calcium and magnesium cations, other cations, such as sodium and potassium, also affect toxicity by a competitive effect [42], without affecting cupric ion activity in solution. Even protons can exert a competitive effect as well as having a marked influence on copper speciation. These effects, in combination with complexation by organic matter and by other constituents present in water, all affect copper bioavailability and thus toxicity. They must, therefore, be taken into account in predicting the possibility of toxicity for any particular given set of water quality conditions.

7.3 The biotic ligand model of acute toxicity

Allen and Hansen in 1996 [43] proposed the use of a chemical equilibrium model to predict the water effect ratio (WER) and to replace the use of bioassays. They considered the effect of complexation of copper by organic matter on bioavailability and toxicity, but they did not take into account the competitive effects of hardness and other cations on the same properties.

Pagenkopf and coworkers [31] had noted the effect of hardness on toxicity but they suggested that it was due to increasing alkalinity, with increasing hardness and increasing total inorganic carbon (TIC) in the form of carbonate. Thus, an increase in inorganic carbonato complexes of copper ($CuCO_3$, $Cu(CO_3)_2^{2-}$ etc.) reduces free cupric ion activity. Pagenkopf and coworkers thought that increased water hardness might also have the effect of increasing cupric ion activity and hence toxicity, because calcium and magnesium cations would prevent cupric ions binding to the carbonate ion. However, they concluded that this effect must be negligibly small in comparison with other interactions that take place. Later, Pagenkopf proposed the gill surface interaction model (GSIM) for trace metal toxicity to fish [44]. The conceptual representation of metal–gill interactions in the GSIM is consistent with our current understanding of how hardness affects toxicity and with the biotic ligand model (BLM). It is now clear that the site of action of copper and other metals producing acute toxicity in fish is the gill [45]. It has been proved experimentally that toxicity decreases as hardness increases because calcium and magnesium cations compete with copper for binding to physiologically active, negatively charged sites on the gill. These sites control the fish's ability to regulate the transport of essential ions across the gill membrane [45, 46].

Playle and coworkers showed that cations in solution compete for a limited number of binding sites on the gill, and that this competition affects the degree of accumulation of metal at the gill surface [47]. Based on these and other related experiments, showing similar effects by Na^+ and H^+ on the level of metal accumulation, Playle and coworkers [48, 49] evaluated site densities and other conditional stability constants that could be used in a chemical equilibrium model to predict the interactive effects of copper exposure concentration, competition, and complexation on accumulation of copper at the fish gill. The level of metal accumulation at the gill reflects the integration of effects resulting from both complexation of the metal in the water and competition of the metal with other cations for binding sites at the gill surface.

In order to carry out risk assessment, the gill accumulation levels must be related quantitatively to toxicity. This was achieved by experiments where juvenile rainbow trout were exposed to a constant total dissolved copper concentration, while adjusting free copper by varying the amount of dissolved organic carbon (DOC) in the water [50, 51]. Using MacRae's measurements, a dose–response curve was defined relating mortality to gill copper accumulation, rather than to total dissolved copper or copper activity in the water. Based on these results, the lethal accumulation (LA) of copper on the gill that results in 50 % mortality, the LA50, was estimated to be about $10\,nmol\,g^{-1}$, in excess of background gill Cu levels (about $12\,nmol\,g^{-1}$ w). With this information, it was possible to produce the BLM that can predict the accumulation of copper on the gill over a range of water quality conditions (hardness, pH, DOC, alkalinity, Na^+, etc.). From the predicted level of accumulation and the corresponding dose response curve, it is possible to predict toxic effects. For a given set of water quality characteristics, the predicted LC50 will correspond to the copper concentration in the water that results in a gill Cu accumulation equal to the LA50.

The BLM provides the necessary framework relating LC50 to LA50 [52]. The general term 'biotic ligand' is employed rather than 'gill' because the site of action producing toxicity may not necessarily be the gill but may vary with substance, chemical species, organism, life stage etc. The overall BLM framework is illustrated in Figure 9.1.6 [52]. The BLM incorporates many of the factors that are included in the GSIM proposed by Pagenkopf. The chemical equilibrium computations in the BLM are based on the chemical equilibrium in soils and solutions model, CHESS [52], modified to include the calculation of metal–DOC interactions that is included in model V of the Windermere humic aqueous model, WHAM [53]. The approach used by Playle *et al.* [48] for prediction of metal accumulation at the site of toxic action is fundamental to the BLM. The BLM takes into account metal–organic matter and other known interactions (as appropriate) to predict cupric ion activity in solution. The competitive effects of other cations in solution are used in determining cupric ion accumulation at the gill that, in turn, is used to predict the end point, lethality.

A key difference between the BLM and the GSIM is that the BLM actually computes the level of accumulation of metal at the site of action, the biotic ligand, while the GSIM does not. Rather, the GSIM utilizes trace metal speciation, gill surface interaction and the effects of competition to predict an 'effective toxicant concentration' in water, the ETC, that is constant across water types for a given test organism, exposure duration and biological effect. Also, although the GSIM conceptual framework could incorporate the effects of metal complexation by dissolved organic carbon (DOC), the example applications of the model that have been published have been for tests performed in water of low DOC content.

The BLM has been developed for copper [54, 55] and for silver [52]. The BLM for copper has been developed for fathead minnows and for the crustacea, *Daphnia pulex* and *Ceriodaphnia dubia*. It seems to be reliable for use in predicting the LC50 of the organisms to within about a factor of 2 over a range of water quality conditions and over a range of LC50 concentrations spanning several orders of magnitude. Initial steps in developing a copper model for use in marine waters, based on the toxicity of copper to *Mytilus edulis*, have been completed. Data showing that the same approach

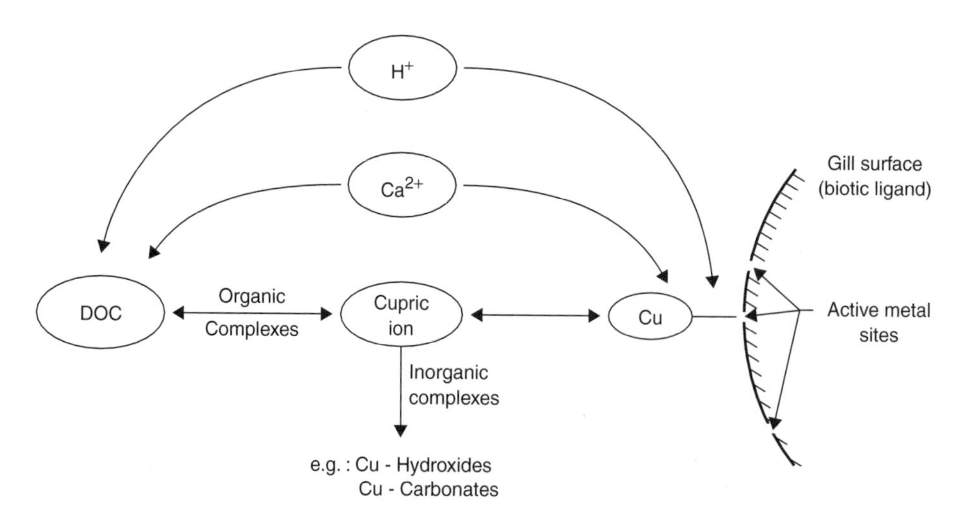

Figure 9.1.6. Diagram to show how the BLM incorporates inorganic and organic complexation and the interaction of cations at the biotic ligand binding site, after ref. 52 (reproduced by permission of The International Copper Association).

should be applicable to cadmium [48, 49] and to nickel [55] have been reported. More recently, the BLM approach has been applied to assess the bioavailability of metals (copper and cadmium) in soils [56]. The BLM is currently under review by the USEPA Science Advisory Board [54] and is being considered for use in the development of refined water quality criteria in the United States, Europe, South America and elsewhere.

8 CONCLUSIONS

It is clear that risk assessment methodologies originally designed for application to xenobiotic organic substances cannot be applied without modification to naturally occurring chemical species of essential elements. It is also clear that risk assessment for any increased exposure to naturally occurring substances should be based on PNECs that are appropriate to the environment at risk. In order to achieve such realism, we must start from more realistic fundamental data. Thus, we must look carefully at our test systems to ensure that they give us a sound basis for our calculations. We must use organisms and genetic strains that are endemic to the environment at risk, representing different taxonomic groups and trophic levels (at least primary producers, primary and secondary consumers, and saprophytes), and different exposure routes. Culture and test conditions for these organisms should relate as closely as possible to those in the natural environment at risk to minimize stress and to ensure that chemical speciation and hence bioavailability of the substances being assessed is nearly the same. In this respect, it must be mentioned that the OECD recently recommended that metals and metal compounds should be tested in natural waters or in artificial waters that approximate to natural waters [12]. Perhaps the revised Guidelines to be published soon will make it clear how this is to be done. Testing should cover the entire life cycles of the selected organisms. Test data should be normalized with respect to the most important abiotic and biotic factors affecting bioavailability and toxicity. Where available, mesocosm and field data should be taken into consideration, since each of the communities in a given environment will be adapted to the naturally occurring background concentrations of the bioavailable chemical species of essential elements. Further, there may be alterations to biological species interactions, for example through semiotic chemical interactions, which cannot occur in the normal single-species toxicity tests. The closer we can come in our test systems to simulating the real environment, the less we shall have to rely on extrapolation with uncertainty factors and the better our risk assessment and risk management will be.

Having obtained dose–response data which are trustworthy and appropriate, there remains the question of the toxicological reference point to be used in calculating the PNEC. Ideally, one should like to use the threshold dose or concentration of the appropriate chemical species of the element of interest for the most sensitive adverse effect produced by that species. In practice, a threshold for toxicity is virtually impossible to determine accurately, even if there is agreement on the effect to use. To get round this problem, Van Assche *et al.* based their risk assessment method on NOEC values. Inevitably, these must be below the threshold values and may refer to effects that are not adverse. Thus, use of NOEC values will always result in a PNEC or PNEC equivalent that is much lower than it should be. It seems better to follow the trend in human toxicology [57–61] and to use, instead of a NOEC, a benchmark dose or concentration, defined as the upper 95 % confidence limit dose or concentration giving a certain percentage response (usually 5 or 10 %) for the most sensitive adverse effect. Calculation of the benchmark dose uses all the available dose–response data and incorporates statistical confidence considerations.

Having obtained reliable, relevant and realistic toxicity test data, the question of bioavailability of elemental species in the test medium must be faced. Here the BLM of elemental species uptake becomes the key to subsequent use of the data. No matter how close the conditions in the test media are to the average conditions for the environment at risk, they cannot replicate the whole range of variation possible. However, the BLM can be applied to permit the extrapolation that will still be necessary. As argued above, I believe that application of the BLM for fish should be based on determining a gill benchmark dose (or concentration, preferably in moles per/unit dry weight). This must be related to the concentration of the toxic chemical species in the test medium. Although the most toxic elemental species may often be a free ionic form [3], it should not be assumed that this is necessarily the case in all circumstances. More research may be needed to determine the bioavailable species of any given element under differing physicochemical and biological conditions (especially microbiological conditions – the conversion of mercuric ions to methyl mercury should not be forgotten). Once the benchmark dose of the toxic chemical species is known, it can be related to the total elemental concentration or the dissolved elemental concentration in the test medium or in any natural sample of the same medium under consideration. Differences in physical chemistry can be allowed for by application of a suitable computerized chemical equilibrium model. This approach should enable regulatory toxicologists to apply different environmental exposure standards to environmental areas with differing geochemistry and ecosystems, making possible a more flexible and realistic system of environmental quality standards.

Of the risk assessment methods reviewed, the NRA approach seems to be the best option for essential elements and their bioavailable chemical species. Even applying the most stringent set of data requirements, the assessment factor approach may still yield an environmental threshold level within the deficiency range. This is also true for the statistical extrapolation methods. Since a log-logistic distribution curve has no lower boundary, the left tail always stretches into the deficiency area regardless of the data requirements that are used. The assessment factor and the statistical extrapolation methods may all result in the derivation of environmental standards that are below natural background levels. Realistic models for risk assessment should never do this.

The NRA approach has been developed specifically for the essential elements and incorporates relevance to the ecology of the area under consideration. However, as initially proposed, it is based on NOEC values. In view of the problems in defining these data accurately, it seems clear that the use of benchmark concentrations would be more realistic and would also include the statistical background that van Straalen and Denneman suggest is necessary but did not include in their original method.

9 ACKNOWLEDGEMENTS

I am very grateful to Ilse Schoeters of Eurometaux for her help in providing me with copies of relevant Eurometaux publications and also those of ICME, The International Council of Metals and the Environment.

I am also grateful to Dr Christopher M. Lee, the Environmental Program Director of the International Copper Association, Ltd., for providing me with a copy of ref. 52. 'I am also grateful to the International Copper Association, Ltd. for permission to copy figures from this publication.

10 REFERENCES

1. Duffus, J. H., *Chem. Int.*, **23**, 34 (2001).
2. Illing, H. P. A., *Toxicity and Risk – Context, Principles and Practice*, Taylor & Francis, Basingstoke, 2001.
3. Duffus, J. H., in *Trace Element Speciation for Environment, Food and Health*, Ebdon, L., Pitts, L., Cornelis, R., Crews, H., Donard, O. F. X. and Quevauvillier, P., (Eds), Royal Society of Chemistry, Cambridge, 2001, pp. 354–372.
4. Di Toro, D. M., Kavvadas, C. D., Mathew, R., Paquin, P. R. and Winfield, R. P., *The Persistence and Availability of Metals in Aquatic Environments*, International Council on Metals and the Environment, Ottawa, 2001.
5. US National Research Council, *Risk Assessment in the Federal Government: Managing the Process*, National Academy Press, Washington, DC, 1983.
6. VanLoon, G. W. and Duffy, S. J., *Environmental Chemistry: a Global Perspective*, Oxford University Press, Oxford, 2000.

7. Greenwood, N. N. and Earnshaw, A., *Chemistry of the Elements*, 2nd edn, Butterworth-Heinemann, Oxford, 1997.
8. Stumm, W. and Morgan, J. J., *Aquatic Chemistry: Chemical Equilibria and Rates in Natural Waters*, 3rd edn, Wiley-Interscience, New York, 1996.
9. USEPA, *Water Quality Standards Handbook*, USEPA, Washington, DC, 1994.
10. Martin, M. H. and Coughtrey, P. J., *Biological Monitoring of Heavy Metal Pollution*, Applied Science, London, 1982.
11. OECD, *OECD Guidelines for Testing of Chemicals*, OECD, Paris, 1981–date.
12. OECD *Report of the OECD Workshop on Environmental Hazard Assessment*, OECD, Paris, 1995.
13. Larson, L. and Hyland, J., *Ambient Aquatic Life Water Quality Criteria for Zinc*, USEPA, Washington, DC, 1987.
14. USEPA, *Estimating Concern Levels for Concentrations of Chemical Substances in the Environment*, USEPA, Washington, DC, 1984.
15. European Commission, *Technical Guidance Document in Support of Commission Directive 93/67/EEC on Risk Assessment for New Notified Substances and Commission Regulation (EC) No 1488/94 on Risk Assessment for Existing Substances*, Office for Official Publications of the European Communities, Luxembourg, 1996.
16. OECD, *Report of the OECD Workshop on the Extrapolation of Laboratory Aquatic Data to the Real Environment*, OECD, Paris, 1992.
17. ECETOC, *Environmental Hazard Assessment of Substances*, ECETOC, Brussels, 1993.
18. Kooijman, S. A. L. M., *Water Res.*, **21**, 269 (1987).
19. ACGIH Worldwide, *TLVs® and BEIs® – Threshold Limit Values for Chemical Substances and Physical Agents and Biological Exposure Indices*, ACGIH Worldwide, Cincinnati, OH, 2001.
20. Straalen, N. M. v. and Denneman, C. A. J., *Ecotoxicol. Environ. Safety*, **18**, 241 (1989).
21. Aldenberg, T. and Slob, W., *Confidence Limits for Hazardous Concentrations Based on Logistically Distributed NOEC Toxicity Data*, RIVM, Bilthoven, 1991.
22. Assche, F. v., Tilborg, W. J. M. v. and Waeterschoot, H., Environmental risk assessment for essential elements case study, zinc, In *International Workshop on Risk Assessment of Metals and their Inorganic Compounds*, Angers, November 13–15, 1996, The International Council on Metals and the Environment, Ottawa, 1996.
23. Simkiss, K. and Taylor, M. G., *Rev. Aquat. Sci.*, **1**, 173 (1989).
24. USEPA, *Ambient Water Quality Criteria for Copper – 1984*, USEPA, Washington, DC, 1985.
25. USEPA, *Ambient Water Quality Criteria – Saltwater Copper Addendum*, USEPA, Washington, DC, 1995.
26. USEPA, *1995 Updates: Water Quality Criteria Documents for the Protection of Aquatic Life in Ambient Water*, USEPA, Washington, DC, 1996.
27. USEPA, *Fed. Regist.*, **63**, 68354 (1998).
28. Stephan, C. E., Mount, D. I., Hansen, D. J., Gentile, J. H., Chapman, G. A. and Brungs, W. A., *Guidelines for Deriving Numerical National Water Quality Criteria for the Protection of Aquatic Organisms and their Uses*, USEPA Office of Research and Development, Environmental Research Laboratories, Duluth, MN; Narragansett, RI; Corvallis, OR, 1985.
29. Prothro, M. G., *Office of Water Policy and Technical Guidance on Interpretation and Implementation of Aquatic Life Metals Criteria, a Memorandum from USEPA Acting Assistant Administrator for Water to Water Management Division Directors and Environmental Services Division Directors, Regions I–X*, US Office of Water, Washington, DC, 1993.
30. Renner, R., *Environ. Sci. Technol.*, **31**, 466 (1997).
31. Pagenkopf, G. K., Russo, R. C. and Thurston, R. V., *Fish. Res. Board Can.*, **31**, 462 (1974).
32. Sunda, W. G. and Hansen, P. J., in *Chemical Modeling in Aqueous Systems*, Jenne, E. A. (Ed.), American Chemical Society, Washington, DC, 1979, pp. 147–180.
33. USEPA, *Briefing Report to the EPA Science Advisory Board on the Equilibrium Partitioning Approach to Predicting Metal Bio-availability in Sediments and the Derivation of Sediment Quality Criteria for Metals*, USEPA, Washington, DC, 1994.
34. Campbell, P. G. C., in *Metal Speciation and Bio-availability in Aquatic Systems*, Tessier, A. and Turner, D. R. (Eds), John Wiley & Sons Inc., New York, 1995, pp. 45–102.
35. Sunda, W. G. and Guillard, R. R. L., *J. Mar. Res.*, **34**, 511 (1976).
36. Sunda, W. G. and Lewis, J. A. M., *Limnol. Oceanogr.*, **23**, 870 (1978).
37. Anderson, D. M. and Morel, F. M. M., *Limnol. Oceanogr.*, **23**, 283 (1978).
38. Sunda, W. G. and Gillespie, P. A., *J. Mar. Res.*, **37**, 761 (1979).
39. Sunda, W. G., Engel, D. W. and Thoutte, R. M., *Environ. Sci. Technol.*, **12**, 409 (1978).
40. Allen, H. E., Hall, R. H. and Brisbin, T. D., *Environ. Sci. Technol.*, **14**, 441 (1980).
41. Hall and Associates and EEMA, *Evaluation of Copper Toxicity and Water Effect Ratio for Treated Municipal Wastewater*, Pennsylvania Copper Group, Pittsburg, PA, 1998.
42. Erickson, R. J., Benoit, D. A., Mattson, V. R., Nelson, H. P. Jr. and Leonard, E. N., *Environ. Toxicol. Chem.*, **15**, 181 (1996).
43. Allen, H. E. and Hansen, D. J., *Water Environ. Res.*, **68**, 42 (1996).
44. Pagenkopf, G. K., *Environ. Sci. Technol.*, **17**, 342 (1983).
45. McDonald, D. G., Reader, J. P. and Dalziel, T. R. K., in *Acid Toxicity and Aquatic Animals*, Morris, R., Taylor, E. W., Brown, D. J. A. and Brown, J. A., (Eds), Cambridge University Press, Cambridge, 1989, pp. 221–242.

46. Wood, C. M., Adams, W. J., Ankley, G. T., DiBona, D. R., Luoma, S. N., Playle, R. C., Stubblefield, W. A., Bergman, H. L., Erickson, R. J., Mattice, J. S. and Schlekat, C. E., in *Reassessment of Metals Criteria for Aquatic Life Protection: Priorities for Research and Implementation. Proceedings of the Pellston Workshop on Reassessment of Metals Criteria for Aquatic Life Protection*, February 10–14, 1996, Bergman, H. L. and Dorward-King, E. J. (Eds.), SETAC Press, Pensacola, 1997.
47. Playle, R. C., Gensemer, R. W. and Dixon, D. G., *Environ. Toxicol. Chem.*, **11**, 381 (1992).
48. Playle, R. C., Dixon, D. G. and Burnison, K., *Can. J. Fish. Aquat. Sci.*, **50**, 2667 (1993).
49. Playle, R. C., Dixon, D. G. and Burnison, K., *Can. J. Fish. Aquat. Sci.*, **50**, 2678 (1993).
50. MacRae, R. K., The copper binding affinity of rainbow trout (*Onchorhyncus mykiss*) and brook trout (*Salvelinus fontinalis*) gills, University of Wyoming, 1994.
51. MacRae, R. K., Smith, D. E., Swoboda-Colberg, N., Meyer, J. S. and Bergman, H. L., *Environ. Toxicol. Chem.*, **18**, 1180 (1999).
52. Di Toro, D. M., Allen, H. A., Bergman, H. L., Meyer, J. S. and Santore, R. S., *The Biotic Ligand Model – a Computational Approach for Assessing the Ecological Effects of Copper and Other Metals in Aquatic Systems*, International Copper Association, New York, 2000.
53. Tipping, E., *Comput. Geosci.*, **20**, 973 (1994).
54. USEPA, *Integrated Approach to Assessing the Bioavailability and Toxicity of Metals in Surface Waters and Sediments, a Report* – USEPA-822-E-99-001, USEPA, Washington, DC, 1999.
55. Meyer, J. S., Santore, R. S., Bobbitt, J. P., DeBrey, L. D., Boese, C. J., Paquin, P. R., Allen, H. E., Bergman, H. L. and Di Toro, D. M., *Environ. Sci. Technol.*, **33**, 913 (1998).
56. Plette, A. C. C., Nederlof, M. M., Temminghoff, E. J. M. and van Riemsdijk, W. H., *Environ. Sci. Technol.*, **18**, 1882 (1999).
57. Crump, K. S., *Fundam. Appl. Toxicol.*, **4**, 854 (1984).
58. Allen, B. C., Kavlock, R. J., Kimmel, C. A. and Faustman, E. M., *Fundam. Appl. Toxicol.*, **23**, 487 (1994).
59. Faustman, E. M., Allen, B. C., Kavlock, R. J. and Kimmel, C. A., *Fundam. Appl. Toxicol.*, **23**, 478 (1994).
60. Allen, B. C., Kavlock, R. J., Kimmel, C. A. and Faustman, E. M., *Fundam. Appl. Toxicol.*, **23**, 496 (1994).
61. Kavlock, R. J., Allen, B. C., Faustman, E. M. and Kimmel, C. A., *Fundam. Appl. Toxicol.*, **26**, 211 (1995).

9.2 Speciation and Legislation

Torsten Berg

Danish Veterinary and Food Administration, Søborg, Denmark

1 INTRODUCTION

Most international legislation concerning trace elements in food, in the environment or in occupational health regulations is based on total element contents, and is frequently given as maximum limits or guideline levels. Only a few regulations pay attention to the molecular species in which the elements are bound, let alone to their oxidation state or complex or molecular structure.

The scarcity of species-specific regulations does not necessarily mean that the issue of speciation was not considered in the scientific reports containing the basis for the legislation. Background papers such as those presented by the International Agency of Research on Cancer and the FAO/WHO Joint Expert Committee of Food Additives and Contaminants (JECFA) contain information concerning the chemical nature as well as the uptake and fate of the various species ingested. Lack of relevant speciation data, however, has directed legislators to cover, as a first approximation, the total content of an element only. Thus, data generated will not support future development of speciation-oriented legislative norms [1, 2].

The international legislation concerning contaminants in food is presently being established in the Codex Alimentarius, which is an independent United Nations organisation under the joint FAO/WHO Food Standards Programme. Development of the Codex General Standard for Contaminants and Toxins in Food provides the framework for future international legislation on metals as well as other contaminants in food [3, 4]. For certain food additives, which include some essential minerals, speciation is an integral part of the specification criteria, because only certain defined chemical compounds are permitted as sources of the essential element.

2 SPECIATION IN INTERNATIONAL FOOD LEGISLATION IN CODEX ALIMENTARIUS AND IN THE EU

International food legislation is laid down under the auspices of the Codex Alimentarius Commission (CAC), a United Nations organisation

Handbook of Elemental Speciation: Techniques and Methodology R. Cornelis, H. Crews, J. Caruso and K. Heumann
 ISBN: 0-471-49214-0

established by the FAO and the WHO. The statutes of the CAC describe the purpose of the Food Standards Programme to be *inter alia*: 'protecting the health of the consumers and ensuring fair practices in the food trade'. Since 1962 the CAC has developed specific standards for individual foods or groups of foods, also called vertical or commodity standards, but the trend over the last decade has been to develop horizontal or general standards, such as the General Standard for Contaminants and Toxins in Food and the General Standard for Food Additives, both developed by the Codex Committee for Food Additives and Contaminants (CCFAC) [3].

Procedures for adopting food legislation in the European Union are complex, and different procedures may be applied for contaminants in food and for food additives. Legislative instruments are mainly regulations and directives. Regulations apply directly in all Member States, whereas directives are transposed into national legislation in the Member States and their implementation may therefore vary slightly from one Member State to another.

2.1 Contaminants

The Codex Alimentarius commodity standards frequently contain provisions concerning contaminants, such as a Codex maximum level for total lead in orange juice or for total lead in rapeseed oil. For fish, however, Codex in 1991 laid down a guideline level of $0.5\,mg\,kg^{-1}$ for methylmercury for all fish, except for predatory fish where the guideline level is $1.0\,mg\,kg^{-1}$. Most mercury in fish will be present as methylmercury.

The Codex Alimentarius General Standard for Contaminants and Toxins in Food is at present a legislative framework to be filled with maximum limits for relevant contaminants in specific foods and food groups, covering all foods relevant for that contaminant, i.e. foods covering at least 80 % of the diet and at the same time 80 % of the intake of the contaminant. The basis for setting maximum limits and for other provisions for each contaminant is a review called a Position Paper with information about the contaminant, including relevant general chemistry, toxicology, levels in different foods, including speciation considerations, intake in different parts of the world and fair trade considerations. Finally, there may be concluding remarks including recommendations for source-related measures to reduce potential pollution problems as well as proposals for maximum limits. There are Position Papers on lead, cadmium, arsenic and tin, as well as for certain mycotoxins and organic contaminants.

In the European Union there is a general (so-called framework) regulation on contaminants in food [5]. Specific regulations setting maximum levels and laying down sampling methods and methods of analysis for official control of the levels of lead, cadmium and mercury in specific foods have recently been adopted and will come into force in 2002 [6]. Whereas the maximum levels are laid down for mercury and apply a maximum limit of $0.5\,mg\,kg^{-1}$ for mercury in fish and fishery products in general (except for some fish species, where the limit is $1.0\,mg\,kg^{-1}$), it is noteworthy that the preamble of the Directive makes reference to methylmercury.

2.2 Food additives

The EU Directives on the identity and purity of food additives specify the requirements for the additives intended to be used for foods. They include provisions for the total content of trace elements in the food additives. These specifications for identity and purity are symbolized by an E preceding the number of the food additive, e.g. for the food colour E140i chlorophyll, the specifications concerning trace elements are As $< 3\,mg\,kg^{-1}$, Pb $< 10\,mg\,kg^{-1}$ and Hg and Cd $< 1\,mg\,kg^{-1}$. Furthermore, the directives on foods for special dietary uses, such as the directive on infant formulae and follow-up formulae, contain a list of substances permitted as sources for the essential minerals added to these foods. The provisions of this and similar directives, however, were agreed upon without much attention being paid to the bioavailability and biological efficacy of the trace elements in the mineral sources, and thus chemical species were not included.

2.3 Arsenic

Developing Codex Alimentarius Maximum Levels for a contaminant such as arsenic in food within the framework of the Codex General Standard for Contaminants and Toxins in Food starts with the compilation of the Position Paper [7]. For a trace element the Position Paper will contain general coordination chemistry considerations, taking into account the chemical properties of the particular element and the stepwise formation of coordination compounds with various ligands available in foods, the gastrointestinal tract, or in the human body fluids. Dissociation constants for aquo complexes at relevant pH, and stability and kinetics for robust transition metals, form the basis of classical abiotic chemical considerations and possibly also for prediction of the chemistry in real but complex media, such as food. Knowledge of the pathway for biosynthesis of for example arsenic-containing ribosides and the non-toxic quaternary arsonium compounds in the marine environment predicts a number of arsenic species that may be present in seafood products such as fish, shellfish and seaweed. The proposed biochemical pathway also suggests that the toxic inorganic arsenic will largely be absent in such commodities.

There is a Codex Position Paper on arsenic outlining present food toxicology knowledge regarding the content of total arsenic and arsenic species [7, 8]. The toxicity of arsenic is expressed by the FAO/WHO Joint Expert Committee on Food Additives and Contaminants (JECFA) as a provisional tolerable weekly intake (PTWI) at $15\,\mu g\,kg^{-1}$ bodyweight for inorganic arsenic, corresponding to a tolerable daily intake of $150\,\mu g$ for a 70 kg person. The data used for this risk assessment refer to inorganic arsenic. However, WHO could not establish a similar recommendation for organic arsenic species due to lack of appropriate toxicological data. The recommended WHO guideline level for inorganic arsenic in drinking water is $10\,\mu g\,L^{-1}$.

The toxic species are inorganic arsenite and arsenate, whereas other species found in fish and shellfish are the less toxic monomethylarsonic acid, dimethylarsinic acid and arsenosugars as well as largely nontoxic arsenicals such as arsenobetaine, arsenocholine and the tetramethylarsonium ion. The literature data available on speciation and toxicology, however, do not allow a detailed toxicological evaluation for the organoarsenicals. This emphasises the need for speciation data to be available to facilitate meaningful decisions based on the differences between the various arsenic species that may be present in food.

In the marine environment the total arsenic concentrations in animals and plants typically range between 0.5 and $50\,mg\,kg^{-1}$, with freshwater fish having a much lower content than oceanic fish, normally less than $0.01\,mg\,kg^{-1}$. In plants used for food arsenic is generally found in low concentrations, below $0.02\,mg\,kg^{-1}$, except in rice – around $0.2\,mg\,kg^{-1}$ – and in certain mushrooms, where the content can be several $mg\,kg^{-1}$. Levels and speciation of arsenic in drinking water are a matter of concern, as arsenic levels above $0.2\,mg\,L^{-1}$ have been reported from several countries [7, 8].

Humans are mainly exposed to arsenic via the diet and drinking water. A small group of the population may additionally be occupationally exposed. The average adult intake from food has been found or estimated to be $30–120\,\mu g\,day^{-1}$. The intake may possibly be considerably higher in countries with large consumption of seafood and seaweed products, as the data reflect that the total arsenic intake in a given population is largely determined by the amount of seafood consumed [7, 8].

Drinking water, however, including natural mineral water, is not included in the intake figures cited above and may well remain a problem, because significant concentrations of inorganic arsenic may be found in water. There may also be foods produced under circumstances where industrial contamination by inorganic arsenic prevails, such as mining areas, or where conversions of non-toxic to more toxic species may occur, for example during processing (canning or freezing) and storage, and thus give rise to toxicological concern.

When the daily intake of total arsenic is well below the tolerable intake of inorganic arsenic, there should be no potential health risks associated with the consumption of the foods included in the surveys.

In order to arrive at a meaningful estimate of the intake of the different relevant arsenic species, which in turn can be evaluated toxicologically, speciation information is strongly needed. However, such data for arsenic species in market foods are rarely available in the literature.

2.4 Selenium

Selenium is essential to man, and recent research has indicated that a high selenium intake may reduce the incidence of some cancers. When an essential trace element such as selenium is permitted as a food additive/micronutrient in e.g. food supplements, several compounds will be given as sources for this trace element. The concentration of selenium in soil is low in many parts of Europe, resulting in low levels in foods such as cereals. Cereals are a most important selenium source in many national diets.

The Danish Positive List for Food Additives includes both sodium selenite and sodium selenate as well as L-selenomethionine as permitted selenium sources, and the use of selenium yeast may also be approved on a case by case basis. This has led to the study of the organoselenium compounds in selenised yeast supplements in order to monitor the claim that selenium is organically bound. Using anion exchange or cation exchange HPLC separation with selenium-selective detection by ICP-MS, it can be shown whether or not a specific selenium source actually contains organically bound selenium such as selenomethionine [9]. Speciation analysis may thus be used as a tool to determine whether selenium-containing food supplements really contain those organic selenium species, which are claimed by the manufacturer to be particularly valuable [10].

2.5 Iron

There are also speciation considerations behind the recommendations concerning iron in the diet. The bioavailability of iron in various foods varies substantially. The haem-iron found in meat is one of the more bioavailable forms, as approximately 25 % is absorbed. The average absorption of iron in general in the diet is only approximately 10 %. As women need more absorbed iron (1–2 mg per day) than men (1 mg per day), the situation for vegetarians, and in particular female vegetarians, could give rise to concern, because the bioavailability of the iron species found in plant based foods is fairly low. However, the absorption of iron(II) species is in general higher than the absorption of the corresponding iron(III) species. Since the reducing ability of the vegetarian diet, which can be rich in ascorbic acid, should favour the formation of the iron(II) compounds, different dietary recommendations for men and woman concerning iron have not yet been established, but the problem will be monitored.

2.6 Chromium

As food additives – chromium sources for food supplements – some Cr(III) salts only are permitted, but no chromium(VI) compounds.

3 SPECIATION IN INTERNATIONAL ENVIRONMENTAL, OCCUPATIONAL HEALTH AND OTHER LEGISLATION

International legislation concerning trace elements in the environment and in the workplace is under development, and only to a very limited extent are speciation considerations included when details are laid down. So far, reference is most frequently made to specific contaminants 'and their compounds', without distinction.

The problems concerning organotin compounds as biocides in antifouling paints for have led to inclusion of tributyltin and triphenyltin in the US EPA Environmental Endocrine Disruptors List and in the European Priority Pollutant Lists (the annexes to EU Council Directive 76/464, with subsequent amendments).

Reference is made to chemical substances and to compounds in the European Commission Directive 93/67/EEC on risk assessment for new notified substances, in the European Commission Regulation 1488/94 on risk assessment for existing substances and in the related legislation, including the supporting Technical Guidance documents.

When trace element compounds are assessed, this is done on a compound by compound basis, i. e. the six different zinc compounds evaluated under the regulation are treated individually. However, sections of the evaluation reports will be identical. This type of approach may be considered as a very first attempt to take speciation into account.

Chromium is an essential trace element for man and animal, and in view of the toxic properties of Cr(VI), as compared to the less toxic Cr(III), chromium speciation has attracted scientific and regulatory interest. There is evidence for carcinogenicity for several chromium(VI) compounds. Hexavalent chromium is regulated in Europe through the Dangerous Substances Directive, 67/458/EEC, with subsequent amendments. Occupational exposure in the USA is regulated by the Occupational Safety and Health Administration, who set Permissible Exposure Limits, which are different for chromic acid and chromates, chromium(III) compounds and chromium metal [11].

As a rare example in legislation concerning speciation and occupational health, there are different Occupational Exposure Limits for different nickel compounds, including nickel subsulfide and nickel carbonyl, as well as for classes of compounds, such as 'insoluble nickel' in the UK and USA legislation [12].

4 TRENDS AND FUTURE DEVELOPMENT

Nobody would regulate and control residues of organochlorine or organophosphorus pesticides on the basis of the content of total chlorine or phosphorus in the sample analysed. Similarly, it is anticipated that, when adequate methods of analysis are or become available, regulation and control of trace elements and trace element compounds in food, in the environment, and in occupational and other health related areas, should be based on the chemical species actually present rather than on total element as is the case, in general, today. Only when the trace element coordination compounds are labile and are easily transformed will the total element content remain the relevant parameter.

The present situation has been described as a vicious circle, where the progress of research in speciation fails to influence legislation [1, 13]. Consequently, legislators continue to build, almost exclusively, on expressing the contents of trace elements in a matrix by their total content, disregarding any chemical species that may be present in the sample. Hence, analytical control of the legislation will provide data for the total content of the trace elements only. Consequently, there will be no challenge to produce simple and fast analytical methods, which could in turn provide the more relevant speciation data. The challenge is to break the vicious circle, which represents an obstacle to the necessary development.

First and foremost it remains necessary to continue to provide more and better species-specific analytical and toxicological/nutritional research data for contaminants and mineral nutrients. The data must be relevant to recognised problems in society. Particularly, problems with relation to food, to the environment and to occupational health are of great concern today. The second step – and possibly the most difficult one – is to communicate these findings to legislators, so that species-specific provisions based upon research results are laid down when relevant.

Such species-specific regulations must be enforceable. For the control of the provisions there is a need for validated routine control methods, including species-specific certified reference materials. Instrumentation that provides simple and reliable ways to monitor and enforce legislation should be introduced, so that more reliable species-specific data can be generated. This would facilitate the necessary and relevant progress in determining chemical species and developing appropriate legislation.

5 REFERENCES

1. Berg, T. and Larsen, E. H., Speciation and legislation – where are we today and what do we need for tomorrow?, *Fresenius' J. Anal. Chem.*, **363**, 431–434 (1999) and references cited therein.

2. Ebdon, L., Pitts, L., Cornelis, R., Crews, H., Donard, O. F. X. and Quevauviller, P. (Eds.), *Trace Element Speciation for Environment, Food and Health*, Royal Society of Chemistry, Cambridge, 2001.
3. Codex Alimentarius (1998), *General Standard for Contaminants and Toxins in Food*, CODEX-STAN 193-1995, (Rev. 1-1997), Joint FAO/WHO Food Standards Programme, FAO, Rome, 2000.
4. Berg, T., Development of the Codex Standard for Contaminants and Toxins in Food, in Rees, N. and Watson, D. (Eds), *International Standards for Food Safety*, Aspen Publishers Inc., Gaithersburg, 2000.
5. European Council Regulation EC 315/93 of 8 February 1993 on Community Procedures for Contaminants in Food (1993) L 37, pp. 1–3.
6. European Commission Regulation No 466/2001 of 8 March 2001 setting maximum levels for certain contaminants in foodstuffs (2001), L 77, pp. 1–13, with subsequent amendments and European Commission Directive 2001/22/EC of 8 March 2001 laying down sampling methods and the methods of analysis for the official control of the levels of head, cadmium mercury and 3-MCPD in foodstuffs (2001), L77, pp. 14–21.
7. Larsen, E. H. and Berg, T., Trace element speciation and international food legislation – A Codex Alimentarius Position Paper on arsenic as a contaminant, in Ebdon, L., Pitts, L., Cornelis, R., Crews, H., Donard, O. F. X. and Quevauviller, P. (Eds.), *Trace Element Speciation for Environment, Food and Health*, Royal Society of Chemistry, Cambridge, 2001, pp. 251–260 and references cited therein.
8. Codex Committee on Food Additives and Contaminants (1998), *Position Paper on Arsenic*, CX/FAC 98/23.
9. Larsen, E. H., Hansen, M., Fan, T. and Vahl, M., *J. Anal. At. Spectrom.*, **16**, 1403 (2001).
10. B'Hymer, C. and Caruso, J. A., Evaluation of yeast-based selenium food supplements using HPLC and inductively coupled plasma mass spectrometry, *J. Anal. At. Spectrom.*, **15**, 1531–1539 (2000).
11. Darrie, G., The importance of chromium in occupational health, in Ebdon, L., Pitts, L., Cornelis, R., Crews, H., Donard, O. F. X. and Quevauviller, P. (Eds.), *Trace Element Speciation for Environment, Food and Health*, Royal Society of Chemistry, Cambridge, 2001, pp. 315–330.
12. Williams, S. P., Occupational health and speciation using nickel and nickel compounds, in Ebdon, L., Pitts, L., Cornelis, R., Crews, H., Donard, O. F. X. and Quevauviller, P. (Eds), *Trace Element Speciation for Environment, Food and Health*, Royal Society of Chemistry, Cambridge, 2001, pp. 297–307.
13. Welz, B., Personal communication, 1998.

Index

Note: Figures and Tables are indicated by *italic page numbers.*

With kind thanks to Paul Nash for compilation of this index.